Paleoclimate and Evolution, with Emphasis on Human Origins

Paleoclimate and Evolution, with Emphasis on Human Origins

Elisabeth S. Vrba, George H. Denton,
Timothy C. Partridge, Lloyd H. Burckle, *editors*

Yale University Press / New Haven and London

Set in Ehrhardt Roman and Futura type by The Composing Room of Michigan, Inc., Grand Rapids, Michigan. Printed in the United States of America by BookCrafters, Inc., Chelsea, Michigan.

A catalogue record for this book is available from the British Library.

The paper in this book meets the guidelines for permanence and durability of the Committee on Production Guidelines for Book Longevity of the Council on Library Resources.

Library of Congress Cataloging-in-Publication Data

Paleoclimate and evolution, with emphasis on human origins / Elisabeth S. Vrba . . . [et al.], editors.
p. cm.
Includes bibliographical references and index.
ISBN 0-300-06348-2

1. Human evolution. 2. Paleoclimatology. 3. Evolution (Biology) I. Vrba, E. S. (Elisabeth S.)
GN281.4.P35 1996
573.2—dc20 95-24462
CIP

10 9 8 7 6 5 4 3 2 1

Contents

Preface

The more distant in time the subject matter of earth scientists, the more explicitly they seem to accept that climate and organic evolution are so interlinked by reciprocal causes and effects that they must be studied together.[1] The profound influence of climate on organisms is already obvious in the most basal branches of life's tree, the Archaebacteria studied by Carl Woese: the fact that many of their living representatives grow at the boiling point of water and are fastidious anaerobes provides the earliest paleoclimatic record associated with life, of high ambient temperatures and highly reducing environments, probably before 4.0 billion years (byr) ago. This biological record of paleoclimate is preserved in the structure, metabolism, ecology, and genealogy of organisms, and it predates the physical paleoclimatic record of ca. 3.9 byr in the oldest rocks now recognized. The earlier parts of the earth's history also show spectacular instances of the reciprocal influence of organisms on climate, such as the undoubted initiation of the oxygen-rich atmosphere by the proliferation of photosynthetic bacteria before 2 byr. In Woese's words, the "intimacy between the evolution of the planet and the life forms thereon" underlines the need for "a close relationship between the geologist, not merely the paleontologist, and the evolutionist."[2]

As the subject matter of physical geologists and evolutionists moves forward in time, these subdisciplines become increasingly divergent and intellectually isolated. The need to bridge this isolation was our major motivation in starting a series of workshops entitled "Paleoclimate and Evolution" in the mid-1980s; these workshops eventually led to the conference held in May 1993 in Airlie, Virginia, on which this volume is based. We chose the Middle Miocene to Pleistocene as a focus for an interdisciplinary onslaught on outstanding paleoclimatic and evolutionary questions. This record allows synthetic research that combines abundant and well-dated data on paleoclimate and fossil species with information on the same or closely related living ecosystems and species. It is also the interval of hominid evolution. The contributors are leading authorities on subjects ranging from geology and paleoceanography, through palynology and mammalian paleontology, to paleoanthropology. We came together from many parts of the world, and our data extend across all continents and oceans.

It is best to admit in the preface that we found many problems of interpretation daunting in spite of the comparatively rich data base of the Late Neogene. At the conference we discovered, paradoxically, that although it is comparatively easy to forge agreement on vague generalizations, as the data increase in volume and precision, it becomes increasingly difficult to agree on what they mean. Thus, in this book readers will find some basic disagreements on which parts of the data can be trusted, on the proximal processes indicated by the data, and especially on how paleoclimate and evolution might be linked. This aspect of the book is valuable, because it is precisely the discrepancies between data sets and the fierce arguments about patterns and processes that should provide encouragement and direction to future

1. The word *paleoclimate* in the title refers to all the local and global conditions and changes in the atmosphere and in the surface layers of the earth that impinge on the habitats of organisms and species, including climatic changes from astronomical sources and from geological changes such as uplift, volcanism, and continental suturing and breakup.

2. C. W. Woese, 1987, Bacterial evolution, *Microbiological Reviews* 51: 223.

research. We confronted our disagreements by addressing the data.

Readers will find a great deal of basic information in this volume that can be compared within subdisciplines through the common denominator of the best chronological estimates now available for Late Neogene climatology and paleontology. The book should be of interest to specialists who seek a current overview of Late Neogene records and hypotheses, many new and unpublished, on paleoclimatic changes and on floral and mammalian evolution. Paleoanthropologists will appreciate the syntheses of hominid evolution that include reference to recent new fossil finds and are set in state-of-the-art evaluations of systematic, stratigraphic, and chronological contexts that encompass the entire hominid record. But we suspect that this volume will be most useful to those who, like us, are dissatisfied with studying these components in isolation. The unique opportunity to see advances in these diverse subdisciplines summarized in a single volume allows such readers to pose and evaluate their own integrative hypotheses and to compare them with similar proposals in the present chapters. In sum, we hope that our specialist colleagues will find this book useful for research as well as teaching and that graduate and undergraduate students will use it as a reference for course work, projects, and theses.

A truly synthetic understanding of the processes involved would have resulted in a layout of chapters under such general headings as "Stochastic Processes in the Evolution of Climate and Life." But our field has not yet approached that level of synthesis. The division into sections on the Middle Miocene, Pliocene, and Pleistocene, with particular emphasis on the Pliocene, reflects a more modest level of synthesis that allows the worldwide evidence for particular time intervals to be assessed together, thereby enabling one to study the underlying processes from a broader perspective than was possible previously.

Finally, after a conference such as Airlie and a volume such as this are completed, there are many aspects that one wishes one had covered better. We mention several that we resolve to address more fully should we have the opportunity to continue this series. The impact of vulcanism and various forms of tectonism on climatic changes, and ultimately on biotic (including hominid) evolutionary changes, deserves further attention. The effects of biotic macroevolution on global climate are of deep importance (as evidenced by the devastating effects of the proliferation and behavioral evolution of humans on the earth's atmosphere and environments). Yet this topic receives far less attention within our subdisciplines as a whole than does the potential effect of climate on evolution.

The history of this project goes back to 1983, when at a conference on paleoclimate held in Swaziland three of us (George Denton, Timothy Partridge, and Elisabeth Vrba) got together—in spite of very different backgrounds in glaciology, geomorphology, and mammal paleontology—to hatch plans for a series of workshops on paleoclimate and evolution. The fourth member of our team (Lloyd Burckle), who joined us soon after that, added yet another research area, marine micropaleontology. Between 1984 and 1985 we held three workshops on aspects of paleoclimate and evolution: at the Lamont-Doherty Earth Observatory, Palisades, New York, in September 1984; at Sun City, Bophutatswana, South Africa, in February 1985; and again at the Lamont-Doherty Earth Observatory in May 1985. The proceedings were published as extended abstracts in three dedicated issues of the *South African Journal of Science* (vols. 81 [5], 1985; 82 [2], 1986; and 82 [9], 1986).

The launching and execution of broadly interdisciplinary projects has always been more difficult than is the case with more narrowly focused efforts. People in control of funding as well as reviewers in diverse disciplines tend to be skeptical about whether scientists from such different research traditions can really talk to one another. We were fortunate in receiving early and sustained understanding of the importance of the connections between paleoclimate and evolution from several key individuals and institutions. Our earliest support and funding for this project came from Reinhardt Arndt, President of the Foundation for Research Development of the Council for Scientific and Industrial Research in South Africa, who continued his generous sponsorship right up to funding the travel costs of the South African participants at the Airlie conference. Another early supporter was John Yellen, Director of the Anthropology Program of the National Science Foundation (NSF), who encouraged us to apply for NSF funding. We gratefully acknowledge that the NSF grant made up the bulk of all funds available for the conference. Sydel Silverman, President of the Wenner-Gren Foundation, and other officials of that organization generously granted us the full financial contribution that we requested. Without that amount the conference and this volume in its present form would have been impossible. The officials of the L.S.B. Leakey Foundation for Anthropological Research are thanked also for their crucial support. Without the generosity and interest of Graham Baker, Editor of the *South African Journal of Science,* the volumes of extended abstracts of the earlier three workshops could not have been published. We are aware that these volumes have been much used in our various subdisciplines, and we thank Graham for making that possible. We acknowledge the important input of the scientists who contributed to the early workshops and to the resulting proceedings in the *South African Journal of Science* but who were not present at the Airlie conference: N. Abrams, R. P. Ackert, D. A. Adamson, J. B. Anderson, N. T. Boaz, J. J. Carroll, C. M. Clapperton, J. A. Coetzee, B. A. Cohn, J. E. Cronin, H. J. Deacon, J. Deacon, R. K. Dell, E. Delson, R. G. Fairbanks, R. Gersonde, F. E. Grine, J. M. Harris, D. M. Harwood, D. A. Hodell, W. R. Howard, T. J. Hughes, J. Kappelman, L. D. Keigwin, D. E. Kellogg, T. B. Kellog, G. Kukla, J. M. Lowenstein, M. C. McKenna, I. K. McMillan, M. C. G. Mabin, R. K. Matthews, R. R. Maud, J. H. Mercer, G. H. Miller, K. G. Miller, H.E.H. Paterson, J. Pickard, M. L. Prentice, P. G. Quilty, J. H. Schwartz, S. M. Stanley, D. E. Sugden, R. H. Tedford, P. V. Tobias, A. Turner, E. M. Van Zinderen Bakker, J. C. Vogel, R. Weed, J. E. Yellen, and A. L. Zihlman. We thank these contributors for their valuable participation in the workshops. We are grateful that the Lamont-Doherty Earth Observatory was able to host two of the workshops. Additional generous support was received from the Yale University Press, the Peabody Museum, and the Department of Geology and Geophysics at Yale University. The volume is among the first major projects associated with the ECOSAVE Center affiliated with the Biospherics Institute and the Peabody Museum at Yale University. We also thank Elizabeth Lofquist, Business Manager of the Department of Geology and Geophysics at Yale University, for her wise and efficient handling of fund disbursements and accounting, and Joan Borrelli and David Hodgins of the Peabody Museum at Yale University for their valuable assistance.

Part One

Introduction

Chapter 1

Current Issues in Pliocene Paleoclimatology

Lloyd H. Burckle

As early as the nineteenth century, the Pliocene was recognized as the epoch during which the modern world began to emerge. Sir Archibald Geikie (1893) noted that the Pliocene covers the transition between the warm temperate climate of the Miocene and the cold Pleistocene. The Pliocene is still considered a period of transition. Consider, for example, such diverse and seemingly unrelated events as the final linkage of North and South America, the accelerated uplift of mountains and plateaus, the growth of ice sheets on Greenland and West Antarctica, the emergence of humans and, in the Pleistocene, the periodic growth of continental ice sheets that extended into midlatitudes of the Northern Hemisphere. There is little doubt that events such as mountain and plateau uplift along with the closing of low-latitude oceanic "choke points" (e.g., the formation of the Isthmus of Panama) must have had an impact upon the climate of the Pliocene epoch, which, in turn, must have influenced various evolutionary pathways leading to modern fauna and flora. In spite of the fact that the general long-term trends of both climatic and biotic evolution during the Tertiary are reasonably well known, many details must still be worked out. The global climate of the past 50 million years (myr) has generally shown a long-term cooling trend, and only during a few intervals has this trend accelerated in a steplike fashion. One such step occurred in the late Paleogene, when, presumably, the East Antarctic ice sheet enlarged. Another period of accelerated growth of this ice sheet is seen at the Early to Middle Miocene boundary. The response to this step was global, and the fundamental mechanics of present-day oceanic surface circulation were established. In contrast to previous epochs that experienced ice-sheet growth in high latitudes, the Late Pliocene was marked by growth of a terrestrial ice sheet (the Greenland ice sheet) that extended south, as well as one (the West Antarctic ice sheet) that was largely marine-based.

In spite of this, the Pliocene also apparently underwent unusually warm periods—possibly the warmest of the last 5 myr. Overall, the Pliocene can be viewed as an epoch of climatic contrasts; warm and wet during the first half and cooler and drier during the second. The Early Pliocene is characterized by a relatively low diversity of life, climate, and landforms; during the Late Pliocene, however, diversity increased as life was fragmented into a complex mosaic, owing both to climatic change (cooling and drying in Africa, for example) and to vertical uplift and rifting. Below, we discuss the factors that need to be studied more fully if we are to understand the role that climate has played in the emergence of the genus *Homo* and the associated flora and fauna.

The Role of Tectonics

To explain the Tertiary deterioration of climate, Berner et al. (1983) invoked a progressive reduction in atmospheric CO_2. In their model (referred to as the BLAG model, after Berner, Lasaga, and Garrels) they related the flux of CO_2 in the atmosphere to rates of sea-floor spreading, subduction, and volcanic outgassing. The resulting concentration of CO_2 in the atmosphere was further tempered by increased chemical erosion caused by globally warmer temperatures. Thus, in their model, higher rates of sea-floor spreading should have resulted in higher atmospheric CO_2 and in higher average global

ISBN 0–300–06348–2.

temperatures. Although this model is compelling, geological data do not entirely support it. As Raymo and Ruddiman (1992) pointed out, if rates of sea-floor spreading drive average global climate, the timing of the Tertiary-climate deterioration does not coincide with periods of lowered rates. The mechanism described in the BLAG model, or a modification of it, may indeed have played a role in climate change of the past 50 myr, but it is still not clear what that role was.

Although not without problems, the suggestion of Ruddiman and Kutzbach (1989) that mountain and plateau uplift played a role in the climatic change of the past 50 myr seems to be more widely supported in the geologic literature. It has long been recognized that high plateaus and mountains play a significant role in dictating global climate (Manabe and Terpstra, 1974), but only recently have modelers realized that the most important uplift has occurred largely during the Tertiary. Uplift in key regions of the world (i.e., in the Northern Hemisphere and in the path of the prevailing westerlies) appears to be important in the evolution of Tertiary climate. Two regions of the world are identified as being instrumental in forcing climatic cooling: the Tibetan Plateau and the mountainous region of western North America centered on the Colorado Plateau. That these features are important to modern-day climate is not contested; uplift in critical areas of the Northern Hemisphere, besides influencing local climate, will divert the jet stream winds and low-level winds northward around them.

Ruddiman and Kutzbach (1989) pointed out that when they applied General Circulation Models (GCMs) to the climate system before and after uplift, the post-uplift circulation accurately reflected Tertiary climatic trends recorded in geologic data. What is lacking is a detailed history of both the timings and rates of uplift. Although it is generally agreed that uplift in both the Tibetan Plateau and the American West began about 30 to 40 myr ago (Ruddiman and Kutzbach, 1989), there may have been periods when the rate of uplift increased or when rates of erosion exceeded uplift. Ruddiman and Kutzbach (1991) also noted that the predictions of their model do not always agree with the geologic evidence; nevertheless, the overall agreement is quite compelling. Molnar and England (1990) developed a modification of the uplift model in which they argued that uplift and high rates of both chemical and mechanical erosion could be the result of global cooling and not the cause of this cooling. Raymo (1991) agreed with this view to the extent that mountain-glacier erosion could have further enhanced Tertiary global cooling associated with regional uplift of, for example, the Tibetan Plateau.

The scenario of plateau and mountain uplift would not be complete, however, without also considering some of the consequences of uplift (i.e., erosion, particularly chemical). Although Chamberlin (1899) was the first to suggest a link among enhanced chemical erosion, drawdown of atmospheric CO_2, and glacial periods, it was Raymo et al. (1988) and Raymo (1991) who put this idea into a modern context. Basically, Chamberlin (1899) suggested that periods of increased tectonism and globally higher continental elevations were associated with increased rates of chemical weathering, particularly of crystalline rocks. As shown below, the reaction driven primarily by chemical erosion would lead to increased burial of calcium carbonate, resulting in drawdown of atmospheric CO_2.

$$CaSiO_3 + CO_2 \rightarrow CaCO_3 + SiO_2$$

Raymo (1991) also pointed out that the relationship among mountain and plateau uplift, increased chemical weathering, and global climatic cooling not only holds for the Tertiary but also seems to be true for the Paleozoic.

Early Pliocene Warming and the Stability of the East Antarctic Ice Sheet

With respect to high southern latitudes, there are three current views on global climate during the Pliocene, two of which are in sharp contrast. Webb et al. (1984) reported the recovery of Pliocene diatoms in glaciogenic sediment (Sirius Group) from the Transantarctic Mountains, which essentially separate the East and West Antarctic ice sheets. To accommodate the presence of diatoms of this age in the till, they postulated that during the Pliocene the East Antarctic ice sheet largely collapsed, raising global sea level by as much as 60 m and allowing marine waters to flood the interior seaways of East Antarctica. When the ice sheet was reestablished, in effect it scoured these interior seaways of their sediments and redeposited them on the slopes of the Transantarctic Mountains in the Sirius Group deposits. This theme of major ice-sheet collapse has been explored further by Harwood (1983, 1986a, 1986b), LeMasurier et al. (1994), McKelvey et al. (1991), Webb (1990), Webb et al. (1984, 1986, 1987) and Webb and Harwood (1991).

The stability hypothesis (Denton et al., 1993) suggests that meltdown did not occur during the Early to Middle Pliocene and that any Pliocene global climatic change had little or no influence on a stable East Antarctic ice sheet. A third view suggests that although the Early Pliocene was a warm interval, there was only limited ice sheet drawdown (Burckle and Pokras, 1991; Kennett and Hodell, 1993). This latter view appears weak, however, in that it is suggestive of a compromise, one that is not supported by modeling data (Huybrechts, 1993), which indicates that for a modest temperature rise (less than 5°C) the East Antarctic ice sheet actually increased in volume during increased snowfall. The pressing question is one of resolving the extent of warming and the degree, if any, of ice-sheet meltdown in East Antarctica. A major collapse of the East Antarctic ice sheet would not only greatly raise sea levels (and thereby destroy or constrict existing landbridges); it would also imply a considerable warming (or even an intense warming) during most of Early Pliocene time. Because of the physical and biogeographical implications, the issue of Pliocene stability of the East Antarctic ice sheet should be explored more fully. Resolution of these questions (i.e., what was the duration and intensity of Early Pliocene warming?) will help investigators address what one reviewer termed the "vexed question of Pliocene Antarctic deglaciation."

The Pliocene history of the East Antarctic ice sheet also needs to be explored in other contexts. Those who propose that this feature suffered a major collapse during the Pliocene draw most of their support from the study of glacial deposits in East Antarctica; on the other hand, those who propose that no major ice-sheet collapse took place during the Early Pliocene base their belief on the study of deep-sea sediments and glacial deposits on Antarctica. Proponents of the collapse hypothesis have suggested that the problem is with the calibration of $\delta^{18}O$ and that recalibration brings the deep-sea record for the Early Pliocene in harmony with the record of Antarctic ice-sheet collapse (Harwood and Webb, 1991). To my knowledge, geochemists do not question the rationale behind the isotopic method; however, the apparent discrepancy between the deep-sea record and the continental record needs further investigation. There are also discrepancies in the interpretation of Pliocene microfossil spectra in the marine record, particularly in high latitudes, that need to be addressed. For example, Abelmann et al. (1988) suggested that Early Pliocene seas around Antarctica were some 5°–10°C warmer than at present, whereas Burckle (this vol.), interpreting much the same data set, suggested a rise in summer sea-surface temperatures of less than 2°C over today's values during the same interval.

Climatic Development of the West Antarctic and Greenland Ice Sheets

Although it is generally agreed that the West Antarctic ice sheet formed during the Early Pliocene and the Greenland ice sheet during the early part of the Late Pliocene, little is known of their subsequent histories. Specifically, the East Antarctic ice sheet is believed to have expanded during glacial intervals, yet no one is certain of its history during interglacials or during brief, very warm intervals that may have occurred during the Pliocene. Scherer (1991), for example, has argued that the West Antarctic ice sheet may have collapsed at least once during warmer-than-present Pleistocene interglacials. Although the collapse of one of these two ice sheets (West Antarctic or Greenland) would raise sea level some 6 m, collapse of both would add 13 m to global sea level. There is another point to consider, however. We know from study of the Medieval Optimum, the last warm interval on earth, that humans as well as other biota greatly extended their geographic range. This was true not only for midlatitude biota, which extended their ranges poleward, but also for subtropical forms, which extended their range into midlatitudes.

The Time When Low-Latitude Climate First Responded to High-Latitude Cycles

The role of low latitudes in any scenarios of climatic change has generally been considered a passive one, in the sense that no major temperature change was considered to have taken place between glacial and interglacial intervals. Studies conducted in the 1970s and early 1980s (CLIMAP, 1976, 1981), for example, concluded that, compared to today, the low latitudes did not cool significantly during the Last Glacial Maximum. In some instances, no change was seen in sea-surface temperatures between the Holocene and the Last Glacial Maximum, whereas in others no more than a 2°C difference was indicated (CLIMAP, 1976, 1981). Recently, however, it has been demonstrated that, at least in the Atlantic, sea-surface temperatures cooled by as much as 5°C during the Last Glacial Maximum (Guilderson et al., 1994). The recognition that low latitudes responded to global climatic change prompts us to pose a number of questions about the Pliocene. Did the low latitudes warm up significantly during the presumed Early Pliocene warming? What was the low-latitude response during the Late Pliocene, when the Greenland ice sheet developed? What was the role of the Indian Ocean and West African monsoons in modulating midlatitude rainfall patterns (e.g., in varying the degree to which precessional changes drove the Indian Ocean and West African monsoons)? What role did albedo changes and atmospheric turbidity (as documented by charcoal and dust fluxes) play? Finally, we need to know when and how high-latitude climatic change influenced low-latitude climate during the Pliocene-Pleistocene.

Global Change and the Evolution of Human Cultural Complexity

The theme of this volume is climate and evolution. Unfortunately, climate is generally found in one context and evolution in another. A time series of proxy climate records is best recovered from deep-sea sediments, where, in some cases, more than a million years of climatic history is present. Such records are invaluable because they permit one to follow and analyze the tempo and mode of climatic change—not only for Milankovitch time scales but also for higher frequencies. An important component of these records is their continuity, which usually allows them to be set within a relatively high-resolution time frame. Evidence of evolution (including evolution of cultural complexity), on the other hand, is generally found not in a time series but rather in a series of discrete "snapshots" that frequently cover a broad geographic area.

The task is to interrelate these two records so as to assess the influence that one (climate) has upon the other (evolution). From my viewpoint, the key to bringing about such a union is a high-resolution age model. Fortunately, new and, in some cases, old tools have been applied to this problem—including magnetostratigraphy, tephrachronology, and radiometric dating techniques. One drawback is that lithologic sections are frequently encountered that are not amenable to any of these techniques. Because a key concern is to identify the external forces that drive evolution, it is encumbent upon the community to develop dating tools that will tie continental records more accurately to the deep-sea record; one will then be able to address this issue properly.

References

Abelmann, A., Gersonde, R., and Spiess, V. 1988. Pliocene-Pleistocene paleoceanography in the Weddell Sea: Siliceous microfossil evidence. In *Geological history of the polar oceans: Arctic versus Antarctica,* pp. 729–759 (Ed. U. Bleil and J. Thiede). Kluwer Academic, Dordrecht.

Barrett, P. J., Adams, C. J., McIntosh, W. C., Swisher, C. C., III, and Wilson, G. S. 1992. Geochronological evidence supporting Antarctic deglaciation three million years ago. *Nature* 359:816–818.

Berner, R. A., Lasaga, A. C., and Garrels, R. M. 1983. The carbonate-silicate cycle and its effect on atmospheric carbon dioxide over the past 100 million years. *American Journal of Science* 283:641–683.

Burckle, L. H., and Pokras, E. M. (1991). Implications of a Pliocene stand of Nothofagus (southern beech) within 500 kilometres of the South Pole. *Antarctic Science* 3:389–403.

Chamberlin, T. C. 1899. An attempt to frame a working hypothesis of the cause of glacial periods on an atmospheric basis. *Journal of Geology* 7:545–584, 667–685, 751–787.

CLIMAP Project Members. 1976. The surface of the ice age earth. *Science* 191: 1131–1137.

CLIMAP Project Members. 1981. Maps of Northern and Southern Hemisphere continental ice, sea ice, and sea surface temperatures in August for the modern and the last glacial maximum. *Geological Society of America Map and Chart Series,* MC-36. Geological Society of America, Boulder.

Denton, G. H., Sugden, D. E., Marchant, D. R., Hall, B. L., and Wilch, T. I. 1993. East Antarctic ice sheet sensitivity to Pliocene climatic change from a Dry Valleys perspective. *Geografiska Annaler* 75:155–204.

Geikie, A., 1893. *Text-book of geology.* Macmillan, London.

Gordon, A. L. 1988. The south Atlantic: An overview of results from 1983–1988 research. *Oceanography,* 1:12–17.

Guilderson, T. P., Fairbanks, R. G., and Rubenstone, J. L. (1994). Tropical temperature variations since 20,000 years ago: Modulating interhemispheric climate change. *Science* 263:663–665.

Harwood, D. M. 1983. Diatoms from the Sirius Formation, Transantarctic Mountains. *Antarctic Journal of the United States* 18:98–100.

———. 1986a. Diatom biostratigraphy and paleoecology and a Cenozoic history of Antarctic ice sheets. Ph.D. diss., Ohio State University.

———. 1986b. Recycled siliceous microfossils from the Sirius Formation. *Antarctic Journal of the United States* 21:101–103.

Harwood, D. M., and Webb, P.-N. 1991. Early interpretations of Arctic and Antarctic glacial history: Uniformitarian biases mask the dynamic history of Cenozoic ice volume variations. *Abstracts of the Geological Society of America* 23:A106.

Huybrechts, P. 1993. Glaciological modelling of the late Cenozoic East Antarctic ice sheet: Stability or dynamism? *Geografiska Annaler* 75:221–238.

Kennett, J. P., and Hodell, D. A. 1993. Evidence for relative climatic stability of Antarctica during the early Pliocene: A marine perspective. *Geografiska Annaler* 75:205–220.

LeMasurier, W. E., Harwood, D. M., and Rex, D. C. 1994. Geology of Mount Murphy volcano: An 8-m.y. history of interaction between a rift volcano and the West Antarctic ice sheet. *Geological Society of America Bulletin* 106:265–280.

McKelvey, B. C., Webb, P.-N., Harwood, D. M., and Mabin, M. C. G. 1991. The Dominion Range Sirius Group: A record of the Late Pliocene–Early Pleistocene Beardmore glacier. In *Geological evolution of Antarctica,* pp. 675–682 (ed. M. R. A. Thomson, J. A. Crame, and J. W. Thomson). Cambridge University Press, Cambridge.

Manabe, S., and Terpstra, T. B. 1974. The effects of mountains on the general circulation of the atmosphere as evidenced by numerical experiments. *Journal of Atmospheric Science* 31:3–24.

Molnar, P., and England, P. 1990. Late Cenozoic uplift of mountain ranges and global climate change: Chicken or egg? *Nature* 346:29–34.

Raymo, M. E. 1991. Geochemical evidence supporting T. C. Chamberlin's theory of glaciation. *Geology* 19:344–347.

Raymo, M. E., and Ruddiman, W. F. 1992. Tectonic forcing of Late Cenozoic climate. *Nature* 359:117–122.

Raymo, M. E., Ruddiman, W. F., and Froelich, P. N. 1988. Influence of Late Cenozoic mountain building on ocean geochemical cycles. *Geology* 16:649–653.

Ruddiman, W. F., and Kutzbach, J. E. 1989. Forcing of Late Cenozoic Northern Hemisphere climate by plateau uplift in southern Asia and the American West. *Journal of Geophysical Research* 94:18409–18427.

Ruddiman, W. F., and Kutzbach, J. E. 1991. Plateau uplift and climate change. *Scientific American* 265:66–76.

Webb, P.-N. 1990. The Cenozoic history of Antarctica and its global impact. *Antarctic Science* 2:3–21.

Webb, P.-N., and Harwood, D. M. 1991. Late Cenozoic glacial history of the Ross embayment, Antarctica. *Quarternary Science Reviews* 10:225–237.

Webb, P.-N., Harwood, D. M., McKelvey, B. C., Mabin, M. C. G., and Mercer, J. H. 1986. Late Cenozoic tectonic and glacial history of the Transantarctic Mountains. *Antarctic Journal of the United States* 21:99–100.

Webb, P.-N., Harwood, D. M., McKelvey, B. C., and Stott, L. D. 1984. Cenozoic marine sedimentation and ice-volume variation on the East Antarctic craton. *Geology* 12:287–291.

Webb, P.-N., McKelvey, B. C., Harwood, D. M., Mabin, M. C. G., and Mercer, J. H. 1987. Sirius formation of the Beardmore glacier region. *Antarctic Journal of the United States* 22:8–13.

Chapter 2

Climatic Effects of Late Neogene Tectonism and Volcanism

Timothy C. Partridge, Gerard C. Bond, Christopher J. H. Hartnady, Peter B. deMenocal, and William F. Ruddiman

The belief that the onset of continental glaciation, toward the end of the Neogene, was linked to the rise of major mountain chains has long been embedded in the geological literature. The reconstruction of paleotemperatures for the whole of the Cenozoic (i.e., the last 65 million years [myr]) from isotope ratios in marine microorganisms has revealed that this period was characterized by a progressive shift toward cooler conditions, particularly over the last 50 myr. In detail, the Cenozoic temperature curve displays a number of steplike episodes during which temperatures declined markedly in comparison with intervening periods (fig. 2.1). It is now clear that the abrupt decreases that occurred during the Neogene, at ca. 14 myr in the Miocene, and again at ca. 2.8 myr near the end of the Pliocene, were responsible for major increases in global ice volume (see Burckle, chap. 1, this vol.): the Miocene event brought about a significant extension of the previously impermanent Antarctic ice sheet, whereas the Late Pliocene cooling marked the onset of continental glaciation in the Northern Hemisphere. On shorter time scales, of 100 thousand years (kyr) or less, the extent of glacial ice has been modulated by changes in receipts of solar radiation induced by variations in the earth's orbit.

A number of researchers have linked the creation of the earth's major areas of high relief, most of which came into being during the past 50 myr, to global cooling during the Cenozoic. Several have also suggested that the significant temperature declines evident during the Neogene were the result of pulses of tectonic activity that caused the rise of large mountain massifs such as those of Tibet and western North America, with concomitant cooling and disruption of global patterns of atmospheric circulation (e.g., Ruddiman and Raymo, 1988; Ruddiman and Kutzbach, 1989). These suggestions were supported by inferences, based on a variety of evidence, that, particularly in Tibet, rates of uplift accelerated during the Neogene (e.g., Hsu, 1978; Liu and Menglin, 1984; West, 1984; Mercier et al., 1987). This notion has been challenged by Molnar and England (1990), who pointed out that much of the evidence used to infer uplift could be explained by climatic changes of a global nature acting on already elevated, or more or less continuously rising, areas. Hence, temperature declines inferred from paleobotanical data, increased coarse sedimentation, and the youthful dissection of uplands, which had been regarded conventionally as indicators of recent elevation within mountain and plateau areas, could equally well result from a global shift to cooler and stormier climates. In support of their alternative interpretation, Molnar and England pointed out that "late Cenozoic uplift has been inferred for mountain ranges throughout the world, yet globally synchronous changes in plate motions, if they have occurred, have been small. Climate change, on the other hand, has been a global phenomenon" (p. 33).

Although Molnar and England's caveats must be applauded, mountain uplift and the construction of volcanoes did not cease during the Late Neogene, and the influence of global orogenesis on the major climatic shifts that characterized this period remains to be elucidated. Partridge believes that available evidence, particularly from the Southern Hemisphere, points to a number of tectonic and volcanogenic interludes during the Neogene that are likely to have exerted a significant influence on regional, if not global, patterns of climate.

ISBN 0-300-06348-2.

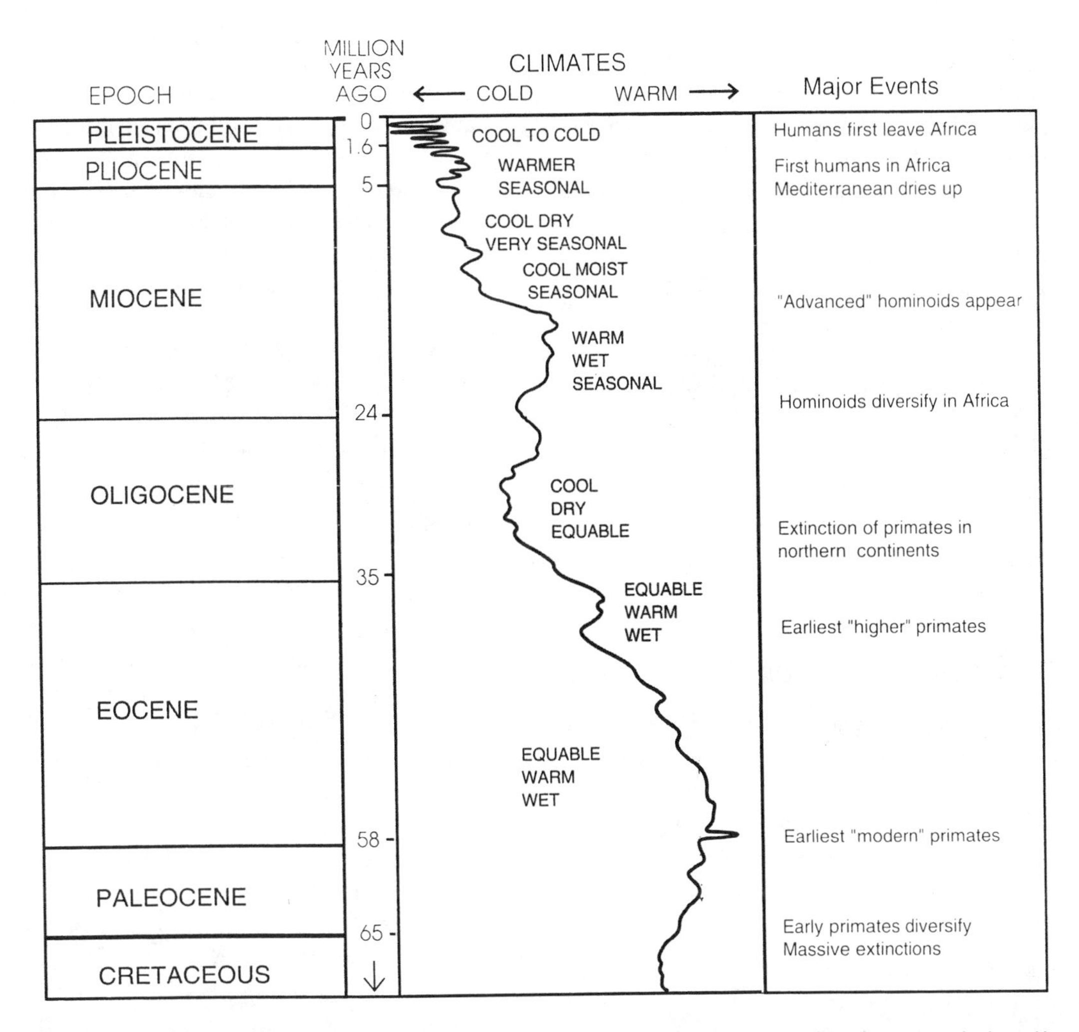

Fig. 2.1. The climatic trend within the Cenozoic, showing the general cooling that eventually culminated in the ice ages of the Pleistocene epoch. The world temperature curve is extrapolated from the ratio of heavy and light oxygen isotope increases as measured in the shells of foraminifera. Shifts to heavy values in the curve indicate changes to cooler temperatures. Major events in primate evolution are indicated along the right side of the diagram (Tattersall, 1993).

In their recent analysis of uplift of the Tibetan Plateau, Molnar et al. (1993) in fact present evidence, from a variety of sources, for rapid uplift of the plateau by between 1,000 and 2,500 m a little before 8 myr; they go on to argue that this event may have played a major role in the intensification of the Indian Ocean monsoon at about this time, which is well documented by evidence from oceanic cores, and in a reduction in the CO_2 content of the atmosphere, which is inferred from the more or less simultaneous global spread of plants using the C_4 photosynthetic pathway. In chapter 24 of this volume further evidence is presented for large-scale uplifts in the eastern hinterland of Africa and in western South America during the Neogene; in the former, and probably also in the latter, a significant component of this uplift can be referred to the latter part of the Neogene.

The rise of most major mountain chains during the last 50 myr can be linked to global patterns of tectonism arising from the interaction of crustal plates (these patterns are examined further below). Many such active plate margins have been characterized by the building of chains of volcanic edifices on a large scale (e.g., in the Andes of South America). Regionally important uplifts have also occurred within individual plates in response to thermal and isostatic influences.

Cenozoic uplift has been shown to be both a potential cause and regional modifier of global climatic change (Ruddiman and Kutzbach, 1989), and the global climatic impacts of episodes of volcanism, although fairly transitory, are well documented. To the extent that evolutionary events are driven by climatic change, the influences of tectonism and volcanism must clearly be taken into account in any analysis of hominid and faunal evolution during the Neogene; a major problem has existed, however, in defining periods of uplift at temporal scales that are compatible with relatively rapid pulses of evolutionary change. Although the timing of major tectonic movements in some areas may be too uncertain to permit the establishment of causal links with regionally or globally important events in the evolutionary record, the geological history of other major foci of Neogene orogenesis is sufficiently well documented to distinguish the effects of regional tectonics from global climatic influences. One such area is the eastern hinterland of Africa (see Partridge et al., chap. 24, this vol.).

Plate Motions during the Neogene

Although latitudinal shifts of continental masses through the Neogene have been generally too small to be of regional climatic significance, plate-boundary reorganizations involving collisions, rifting, and changes in patterns of intraplate stress provide several possible tectonic mechanisms on a scale sufficient to influence regional and global climates:

(1) Depending on where such reorganizations occur along particular interplate boundaries, they may open or close seaways and thus crucially affect major patterns of ocean current circulation. Late Neogene examples are the isolation of the Mediterranean at the end of the Miocene and the closure of the Panama isthmus, which affected the direct exchange of surface waters between the Pacific and Atlantic Oceans.

(2) Large-scale tectonic uplift within or marginal to critically located orogenic belts may influence patterns of wind and rainfall on a regional or hemispheric scale (Ruddiman and Kutzbach, 1989). The Himalayas and the Tibetan Plateau are the most comprehensively analyzed cases.

(3) More generally, the tectonically increased exposure of larger volumes of crustal rock to chemical weathering may lower levels of the main atmospheric "greenhouse" gas, CO_2, and thus promote a global cooling trend (Raymo et al., 1988).

(4) Large-scale epeirogenic uplift of continental interiors may occur in an entirely intraplate setting owing to the buoyancy and viscous-tractional forces associated with mantle plumes (Westaway, 1993). Climatic changes may result from adiabatic lowering of temperature and topographic influences on regional patterns of atmospheric circulation, and associated volcanism may be sufficient to influence the composition of trace gases or aerosols in the atmosphere and thus perturb climate, at least in the short term.

(5) Differential epeirogenic warping along intraplate passive margins, resulting from substantial changes in the general level of lithospheric stress or from isostatic adjustments, may also have climatic consequences.

One should bear in mind that the global nature of lithosphere-plate tectonics implies that any major change to the balance of driving forces on any one plate may have worldwide ramifications on the motions and the boundary kinematics of all or most other plates (fig. 2.2). In this connection Cathles and Hallam (1991) and Cloetingh and Kooi (1992) have drawn attention to the notion that the state of stress within the lithosphere is continually changing in response to changes in plate dynamics. The latter authors ascribe rapid vertical motions and accelerations in tectonic subsidence in the Late Cenozoic record to an increase in the level of compressional stress in the plates: "It could well be that the sediments of the rifted basins around the Atlantic record a phase of intensive global compressional tectonics, associated with an important late Cenozoic plate reorganization of possibly global nature" (Cloetingh and Kooi, 1992, p. 346). Their evidence is drawn from basins on old "passive" margins of the North American, Eurasian, and African plates, in intraplate settings not currently subject to active tectonics. Cloetingh et al. (1985, p. 164) refer specifically to a significant change in the stress state of the Indo-Australian plate at about 5–6 myr. This alteration may have been related to the culmination, at about 7 myr, of Himalayan-Tibetan uplift that began along the Eurasian boundary of this plate at about 20 myr (Harrison et al., 1992, 1993; Amano and Taira, 1992). Molnar et al. (1993) have proposed an alternative model of isostatic uplift of the Tibetan Plateau at about 8 myr, as a result of convective removal of deeper lithosphere. Amano and Taira (1992) relate two pulses of sediment input in the Bengal Fan to the direct influence of uplift in the higher Himalayas. They date the first, Late Miocene pulse between 10.9 and 7.5 myr and identify a second, Pleistocene pulse beginning at 0.9 myr. When calculated

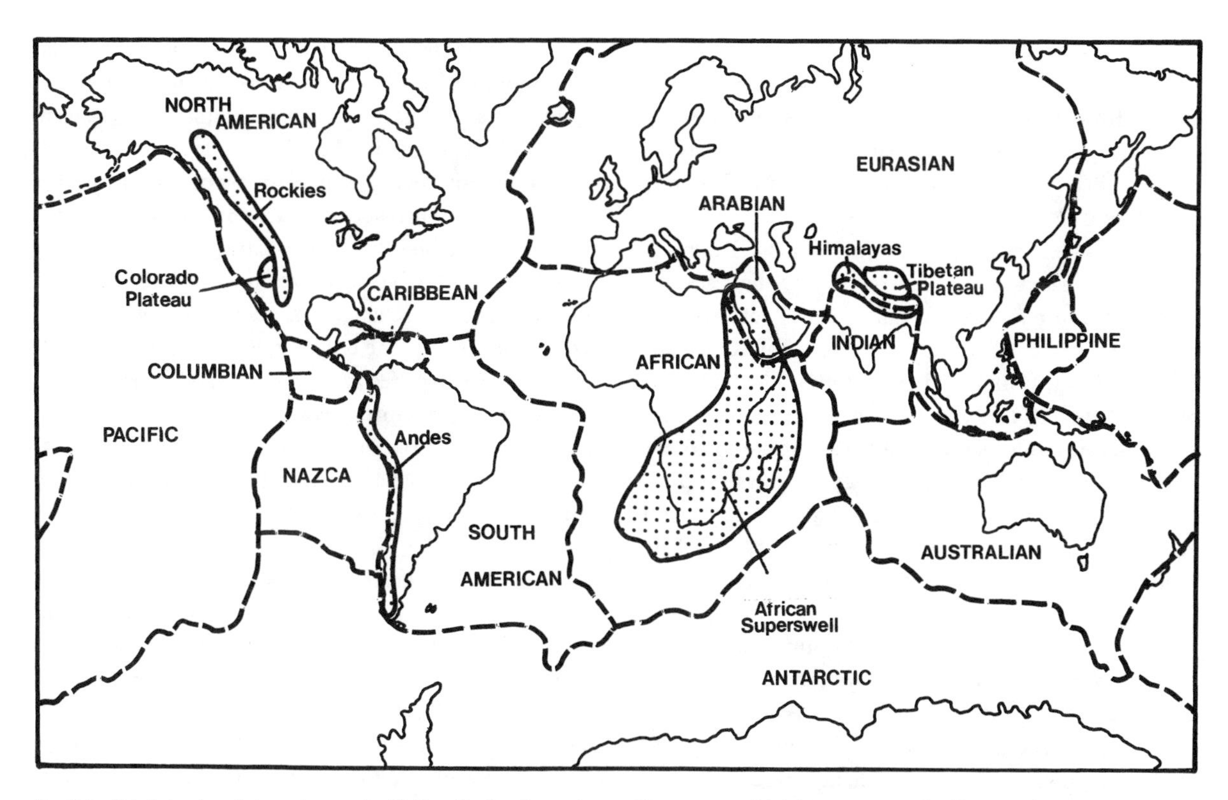

Fig. 2.2. Global plate boundaries and areas of uplift (dotted) referred to in the text. Plate names are labeled in uppercase, and uplift areas in lowercase.

on magnetostratigraphy relative to the new geomagnetic polarity time scale (Cande and Kent 1992), the rates of sediment accumulation in the Siwilak foreland basin show the most prominent maxima (>10 mm/yr) at 7.4 and 6.8 myr and the most pronounced minima (<2 mm/yr) between 6.6 and 6.0 myr (Harrison et al., 1993). A steplike increase in the $\delta^{13}C$ values of pedogenic carbonates in the sequence occurred at about 7 myr (Harrison et al., 1993, fig. 9; Cerling, 1991). Cerling et al. (1993) have also presented evidence for a global decrease in atmospheric CO_2 between 7 and 5 myr; this reduction may be linked to exhumation of fresh rock surfaces during rapid mountain building in conjunction with accelerated weathering following upon intensification of the monsoon as a result of plateau uplift (Molnar et al., 1993). Changes in the composition of terrestrial sediments in the Bay of Bengal support the occurrence of enhanced chemical weathering after ca. 7 myr (Bouquillon et al., 1990; France-Lanord et al., 1993).

A major zone of compressional folding and faulting developed within the oceanic lithosphere of the central Indian Ocean and postdates a major deep-sea unconformity dated at 7–8 myr (Bull and Scrutton, 1992). This wide zone is currently associated with a diffuse boundary between the now-separate Indian and Australian plates (DeMets et al., 1990). A possible explanation for the deformation along this belt is that after Tibet achieved its maximum thickness and elevation, other mechanisms were needed to accommodate the continuing convergence between the western part of the former Indo-Australian plate and the Eurasian plate (Harrison et al., 1992).

The onset of oblique convergence across the Pacific boundary of the former Indo-Australian plate is dated by sea-floor magnetic evidence to anomaly 5 time (ca. 9 myr), and fission-track analyses indicate that rapid tectonic uplift commenced shortly thereafter along the southern part of the Alpine Fault in New Zealand (Tippett and Kamp, 1993). If a first-order trend surface is used to model the data, the onset of uplift apparently migrated northward along the fault from about 8 to 5 myr (Tippett and Kamp, 1993, fig. 12), the period that encompasses both the apparent acme of Himalayan-Tibetan uplift and the onset at about 7 myr of intraplate deformation in the central Indian Ocean. Individual point estimates of uplift age show maxima of 7.4 myr near the southern end and 6.8 myr near the northern end of the Alpine Fault. It is therefore possible that the change from pure strike-slip to transpression along the Australian-Pacific plate boundary is causally linked to the coeval changes along and near the Indian-Eurasian boundary.

In the wider Pacific realm, synchronous changes in sea-floor spreading rates occurred during the Late Miocene, around anomaly 3A time (ca. 5.6 myr) on the Pacific-Antarctic Ridge, the southeast Indian Ridge, and the Chile Ridge (Cande and Kent, 1992). These changes may all have been related to a change, originally estimated to date around 4 myr, in the absolute motion of the Pacific plate (Pollitz, 1986), the largest and fastest moving of all present-day plates. Interactions across the Pacific–North America plate boundary may also have produced an episodic change in the system of plates around the Atlantic (Pollitz, 1988, 1991).

Pollitz (1988) has modeled the absolute motion of the North America plate between 20 myr and the present day and concluded that a significant change occurred during the Late Miocene at about 9 myr. Among the regional geological consequences of this change were: (1) rifting and accelerated uplift in the western United States around what is sometimes called the Sierra microplate; (2) intense compressional deformation along the northern Caribbean boundary, which may have produced the conditions leading to later closure of the Panama isthmus; (3) a hiatus in Aleutian-arc volcanic activity; and (4) a change in the orientation of the Lesser Antilles arc. Closure of the Panama isthmus around 4–3 myr has been proposed as a factor that would influence North Atlantic sea-surface temperatures and deep-water formation, and there is little doubt that it played a role in climatic change in the Northern Hemisphere during the Pliocene. Major climatic trends in North Africa (de Menocal and Bloemendal; Partridge et al. [chap. 24], both in this vol.) do not, however, display any demonstrable response to this event.

The additional force that caused the 9 myr change in North American plate motion is surprisingly similar to the force that apparently caused the later extensive change in the absolute motion of the Pacific plate (Pollitz, 1986). Both forces are thought to have been applied in the Kurile-Kamchatka region, near the triple junction of the North American, Pacific, and Eurasian plates. Although Pollitz (1988) has speculated about possible deep alterations to the slab-pull force along the subducting North American–Pacific boundary, it is possible to postulate, as an alternative, that the perturbing force arose from additions to the collision resistance along the nearby Eurasian–North American boundary in eastern Siberia. This may, in turn, be related to the coeval change in collision resistance elsewhere on the boundaries of the Eurasian plate, but especially around the Himalayan-Tibetan uplift along the Indian-Eurasian segment.

Pollitz (1991, table 2, p. 92) links a Late to end-Miocene (ca. 6 myr) change in the absolute motion of the African plate to: (1) the opening of the modern Gulf of Aden and the Red Sea (5–4 myr); (2) a change in the orientation of principal stresses on the European platform (8–4 myr); (3) a slowdown in spreading rate along the southern Mid-Atlantic ridge (8–4 myr); and (4) the initiation of the extensional phase of the East African Rift (7–4 myr).

The first three correlations associate the change in the absolute motion of the plate with relative-motion changes along its boundaries with the Indian, Eurasian, and (North-South?) American plates, respectively. The last correlation, based primarily on chronology of the Kenya (Gregory) Rift, relates it to the ongoing breakup of the African plate into separate Nubian and Somalian plates, which the current model of global plate motions (DeMets et al., 1990) fails to resolve.

In connection with changes in stresses on the European platform and the possible relaying of Pacific–North American kinematic changes to the African plate via coupling along the North American–African plate boundary, Sloan and Patriat (1992) have documented a change in axial orientation, spreading direction, and spreading rate that occurred along the Mid-Atlantic ridge at approximately anomaly 4 time (7.01 myr on a polarity time scale calibrated by Patriat). The African plate is, of course, directly coupled to the Eurasian plate in the Alpine-Mediterranean belt. If, as speculated above, the changes in Pacific and North American plate motion were induced by forces acting around the Eurasian–Pacific–North American triple junction, then alterations to the balance of forces along the Eurasian-African plate boundary may have been more directly responsible for the change at 6 myr in African absolute motion. These events along the Eurasian–African boundary include the transformation of the larger Indo-Australian plate to the Indian plate and the latter's interaction with the Arabian plate.

During the terminal Miocene (6.0–5.0 myr) the link between the deep Mediterranean Basin and the Atlantic Ocean was repeatedly closed by a combination of glacioeustatic lowering of sea level and tectonic movements at the western end of the Mediterranean connected to its gradual narrowing during the late Cenozoic, as Africa drifted northward (Stanley and Wezel, 1985; Hodell et al., 1986). The resulting catastrophic drying of the Mediterranean in the so-called Messinian Salinity Crisis is not apparent in nearby oceanic records of aeolian dust fluxes: there is no "eventlike" feature that begins around 5.6 myr and ends near 4.8 myr. The oceanic signal that heralded the aridification of the Sahara occurs somewhat later, in the Early Pliocene (see Partridge et al., chap. 24, this vol.); this casts serious doubt on the notion that Late Neogene tectonic events in the Mediterranean Basin had any significant effect on North African climate. The penecontemporaneous establishment of a permanent connection between the Red Sea and the Gulf of Aden seems likewise to have had little impact on the climate of adjacent areas to the south.

Pollitz (1991) indicates that the 6 myr change in African motion corresponds to a southwesterly directed force acting over the whole plate, which may be consistent with enhanced ridge-push along the Indian-Arabian plate boundary and the (accelerated?) opening of the Red Sea and Gulf of Aden. However, Jestin and Huchon (1992) model Arabia-Africa (Nubia) motion as occurring at a relatively constant rate of about 0.4°/myr about a pole near 32°N 23°E since the Middle Miocene (13 myr). The initial opening of the Red Sea apparently occurred at the much slower rate of 0.1°/myr, assuming that it was distributed uniformly over the whole interval from 30 to 13 myr (Jestin and Huchon, 1992). The maximum estimate of 13 myr for this proposed kinematic change appears to be constrained mainly by the age of important rhyolitic and phonolitic eruptions in the Omo River area of Ethiopia, which preceded the main rifting event in that area. WoldeGabriel et al. (1990) maintain that the main Ethiopian rift underwent an "ancestral" stage during the Late Oligocene to Middle Miocene period (ca. 30–15 myr), in which a series of alternating half-grabens developed, and that by 8.3 myr and 9.7 myr, respectively, the eastern and western faulted margins of the present symmetrical rift had come into being. In the light of these possibly lower age limits, the timing of accelerated rifting around the Afar triple junction may well be approximately synchronous with the North American plate-motion change at 9 myr (Pollitz, 1988), and the estimated 6 myr age of the African kinematic event (Pollitz, 1991) may be in need of some revision.

Hartnady and Partridge interpret the evidence reviewed above as documenting a worldwide sequence of episodic tectonic changes during Late Miocene to earliest Pliocene times, between about 10 and 5 myr. These changes appear mostly to follow temporally upon, and may therefore be a geodynamic consequence of, an important tectonic episode involving the culmination of major uplift along the Indian-Eurasian plate boundary around 8 myr. Perhaps the analogy of a tectonic "domino effect" can be drawn, but in the absence of precise age constraints, it may not be appropriate to speculate on which of these tectonic episodes has causal priority. Because the system of lithospheric plates is a rigidly interconnected global entity on a closed surface, any local tectonic mechanism that allows all related episodic changes to proceed worldwide is, in a sense, a cause of them.

Late Neogene Events in the Context of Longer-Term Changes during the Cenozoic

In addition to the Late Neogene plate reorganizations and mountain-plateau uplifts discussed above, a number of changes that might have affected continental climates occurred on somewhat longer time scales, spanning most of the Cenozoic. These alterations include an absolute fall in sea level of several tens of meters, a decrease in rates of sea-floor spreading, and net increases in continental hypsometry caused by uplifts of some continental interiors. What is intriguing about these changes is that they appear to have occurred on about the same time scale as the puzzling decrease in global temperature during the Cenozoic (fig. 2.1). This temperature drop began about 50 myr ago and has continued, with a number of brief warmings, through the dramatic deterioration of climate of the Late Neogene. As the Late Neogene climate may be, at least in part, a continuation of the preceding cooling, we briefly explore possible connections between climate and the long-term trends in Cenozoic tectonics.

The first two trends, a fall in sea level and a decrease in spreading rates, are thought to be closely linked, because a drop in spreading rates leads to a decrease in the volume of spreading ridges, causing the volume of the ocean basins to increase and, hence, sea level to fall. An appreciation of the magnitudes of change involved is given by Engebretson et al. (1992), who estimated that between 50 myr and the present, subduction rates, which are directly related to spreading rates, fell from close to 3.5 km^2/yr to slightly less than 2.5 km^2/yr. They calculated that a change of this magnitude would cause sea level to fall about 200 m (fig. 2.3).

Both the fall in sea level and the decrease in spreading rates could have indirectly affected Cenozoic climates. A falling sea level increases the ratio of land to ocean and, therefore, the surface albedo. Although Barron et al. (1985) concluded from a series of atmospheric circulation model experiments that the temperature effect of change in albedo since the Eocene might be very small, they emphasized that their models were primitive and did not involve a coupled ocean-atmosphere. No attempt has been made since to reexamine the albedo effect with more sophisticated models. A drop in spreading rates will be accompanied by a drop in the rate of mantle upwelling and, hence, a decrease in mantle outgassing at ridge crests. Some geochemists have argued that this process might be large enough to cause a decrease in atmospheric CO_2 (Berner et al., 1983). Using the recent plate-motion data of Engebretson et al. (1992), Berner (1994) found that reduced CO_2 emission along the oceanic ridges could have contributed to the Cenozoic cooling from 50 myr to the present.

A net increase in hypsometry, which is the distribution of elevation with respect to surface area, is one of the most striking changes that took place within some continents during the Cenozoic. This change is the consequence of true uplift—that is, uplift relative to the geoid as distinct from that caused by a fall in sea level. Parts of two continents in particular appear to have undergone significant increases in hypsometry since about 50 myr; the western interior of North America and nearly all of eastern Africa and Arabia (Bond, 1978, 1979, 1985).

Direct evidence for changes in hypsometry is based on the distribution of shallow marine Cretaceous and

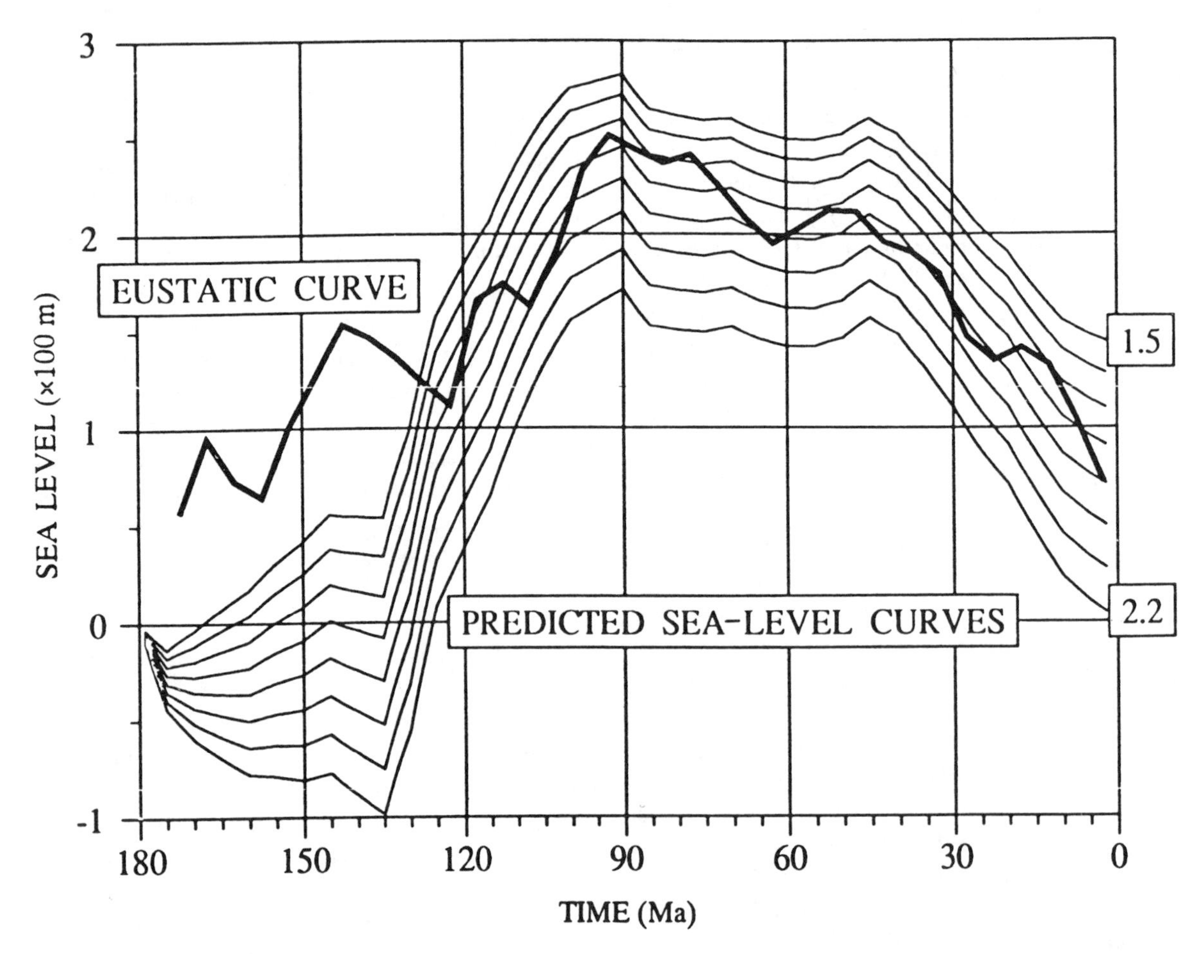

Fig. 2.3. Sea-level change (light lines) predicted from the subduction-rate calculations of Engebretson et al. (1992), compared with the eustatic sea-level curve from Haq et al. (1987).

younger sediments relative to elevation. In both North America and Africa, these sediments, which were deposited near what was then sea level, now lie at elevations exceeding any reasonable estimate of sea level highstands at the time they were laid down. The magnitude of uplift of these sediments was not large, probably of the order of several hundred meters to about a kilometer, but the surface areas involved appear to have been enormous, far exceeding the areas of mountain-plateau uplift in Asia. When the areas of uplifted Late Cretaceous to Miocene sediments in Africa are included with the uplift inferred for the "African Superswell," it appears that a major part of the African continent was uplifted by at least several hundred meters to as much as a kilometer during the Cenozoic (Fig. 2.4; see also Partridge et al., chap. 24, this vol., for a more detailed account). This uplift appears to have been the main cause of the unusual hypsometry of Africa relative to the other continents, which gives the eastern part of the continent a topography resembling that of a relatively high plateau (Bond, 1978).

These uplifts may have been too small in area and amplitude to have had a significant effect on global cli-

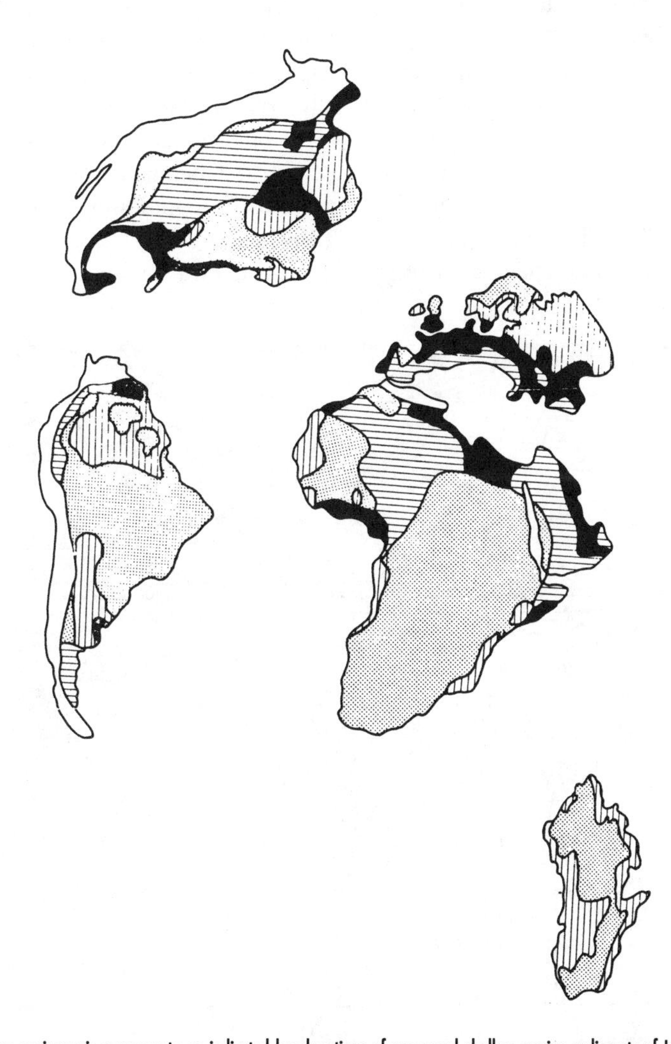

Fig. 2.4. Areas of post-Cretaceous epeirogenic movements, as indicated by elevations of preserved shallow marine sediments of Late Cretaceous and Early Tertiary age (Bond 1979). Assuming that these shallow marine sediments were deposited essentially at sea level, they provide a datum that can be used to infer true uplift or subsidence. That is, if the present elevation of the deposits is higher than any reasonable maximum elevation of the Late Cretaceous sea, then the area covered by those deposits must have been uplifted since the late Cretaceous. The maximum elevation of Late Cretaceous sea level was probably about 200 m (see also fig. 2.3). Uplift can therefore be inferred where the deposits lie above 200 m (indicated by horizontal lines). On the other hand, if areas that are not covered by those deposits presently lie below 200 m, those areas probably subsided after the late Cretaceous (indicated by vertical lines). Two other situations are indeterminate: 1) those covered by the deposits and now lying below 200 m (dotted areas); and 2) those lacking the deposits and now lying above 200 m (solid areas). An assumption inherent in this argument is that erosive loss of the Late Cretaceous deposits has been relatively small (see Bond, 1979, for a discussion of error in the assumption).

mate, but they could have had important local effects. Assuming a lapse rate of 6°C/km, the uplift alone could cause temperatures over these areas to drop several degrees, a change that would be accompanied by a decrease in continental humidity as well.

Unfortunately, the age distributions of the elevated shallow marine sediments do not always tightly constrain the timing of the uplifts. In North America, the uplift occurred sometime after the Late Cretaceous or earliest Tertiary. Uplift within the coastal hinterland of eastern Africa and most of the Arabian peninsula, however, must have occurred after the Early to Middle Miocene (see Partridge et al., chap. 24, this vol.)

The causes of large epeirogenic uplifts within continents have recently been investigated using techniques from the emerging field of seismic tomography (Dziewonski and Anderson, 1984). Some of the results are leading to exciting new insights on the uplift mechanisms, especially in Africa. Seismic tomography is the study of the seismic structure of the earth's mantle using velocities of different earthquake waves, especially shear waves and surface waves. The technique, which is analogous to the medical technique called CAT scanning, uses information from a large number of crisscrossing waves to construct three-dimensional images of the mantle.

What is distinctive about the seismic tomography of Africa is that beneath its northern part, where a large amount of uplift occurred, the shear-wave mantle velocities are slow relative to average mantle (fig. 2.5). Because shear-wave velocities slow with decreasing viscosity, the slow-velocity anomalies beneath northern Africa are taken as evidence that the mantle there is warmer and hence less viscous than average mantle. Two other related features of Africa are noteworthy. First, an unusually large number of the world's hot spots (rising plumes of mantle material) are present within the continent (Zhang and Tanimoto, 1993). Second, the continent appears to be significantly elevated with respect to the geoid (Richards and Hager, 1988). This geoid anomaly is thought to be a consequence of a dynamic uplift of the

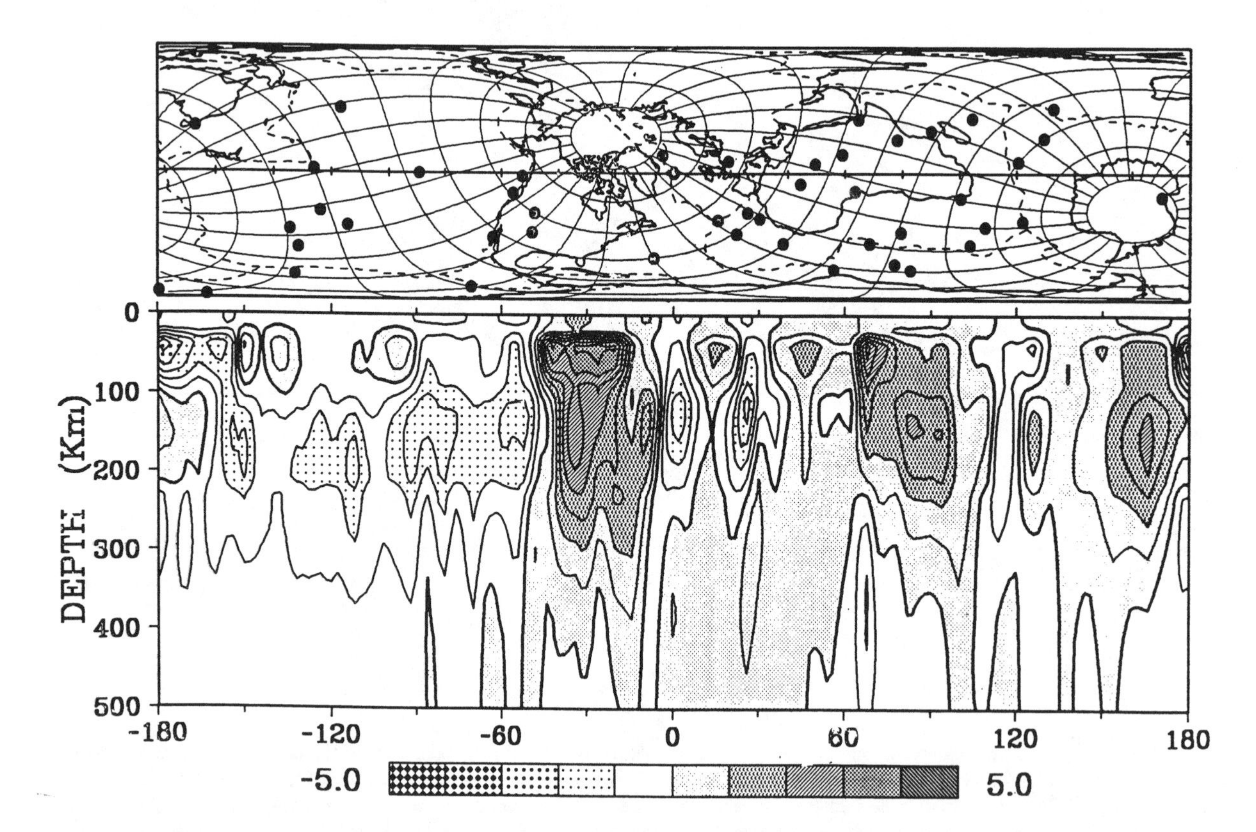

Fig. 2.5. Vertical section of mantle seismic velocities based on a seismic tomographic model. Section is along the solid line in the map in the upper half of the diagram. Negative values in the bar indicate slow velocities, and positive values indicate fast velocities, relative to the average. Values in the bar are in kilometers per second. Note the very slow velocities throughout the mantle beneath northern Africa relative to that beneath northern North America. Velocities are somewhat slower, relative to northern North America, beneath southern Africa and, notably, beneath Antarctica (Zhang and Tanimoto, 1993).

lithosphere resulting from vertical forces associated with rising mantle material beneath the continent. The explanation of these observations is that the mantle beneath Africa has been heated enough to become buoyant and upwell, causing widespread uplift and volcanism to break out over a large part of the continent's surface (e.g., Anderson, 1990). The cause of this heating is thought to be a consequence of the slow motion of the African continental lithosphere relative to the mantle, combined with the general absence of plate subduction beneath the continent (Anderson, 1990). Continental lithosphere acts as an insulator, and if it is moving slowly relative to mantle, as in the case for Africa, it would cause the mantle to warm. Subduction of cold lithosphere appears to be a significant factor in cooling the mantle; hence, where subduction is absent or slow, as around most of Africa, the mantle tends to become hotter than usual.

It appears that no other continent experienced such an unusual combination of epeirogenic uplift, heating of subcontinental mantle, and surface volcanism during the Cenozoic. This observation is intriguing when combined with the evidence of evolution of hominids during the Late Cenozoic in Africa, as shown in figure 2.1. The implications of such uplift and volcanism for hominid evolution are investigated more fully in Partridge et al., chap. 24, this vol.

Late Neogene Volcanism

Just as the coupling between crustal plates has ensured that some major tectonic events are manifested at a continental or even subglobal scale, so, during certain periods in the past, volcanism appears to have occurred in discrete pulses over large areas. The synchronous emission of large volumes of volcanic aerosols has been shown to be capable of exerting an influence on global climate, although with the exception of a few cases, its duration would probably have been too short to have had a major evolutionary impact.

Although analyses indicate that the volumes of volcanic material extruded during the evolution of the East African rifts were relatively modest (Williams, 1978; Ebinger, 1989), the volume generated during the Neogene in the circum-Pacific region was large (Kennett et al., 1977). Analyses of dated terrestrial-volcanic sequences and records of volcanic-ash horizons preserved in deep-sea cores provide a useful record of both extrusive and explosive volcanism, from which Kennett et al. were able to demonstrate that activity in the southwestern Pacific, Central America, and western North America occurred in discrete pulses. The most notable of these date to the Mid-Miocene (16–14 myr), when a major decline in global temperature is also evident (fig. 2.1), and the Quaternary (<2 myr), with less important episodes in the Late Miocene (11–8 myr) and between 6 and 3 myr. It has been suggested that such synchronous episodes of volcanism are related to changes in rates of sea-floor spreading and subduction (Kennett and Thunell, 1977).

The Impact of Major Uplifts on Global and Regional Climate

The tectonic elevation of major plateau areas and mountain chains has had a major impact on past climates, not only in the immediate vicinity of the uplifted areas but as a result of the redirection of tropospheric planetary waves into a meandering configuration (Ruddiman and Raymo, 1988; Ruddiman and Kutzbach, 1989). Hence, major tectonic events have played an important role as modifiers of hemispheric or even global climate and have sometimes acted as triggers of modal changes in climatic regime.

In the context of the evolution of climatic patterns in Africa during the Late Neogene, three major foci of uplift may have been important: 1) the Himalayas and Tibetan Plateau, 2) the Andes of South America, and 3) the African Superswell (fig. 2.2). The origin and timing of Tibetan uplift in terms of plate motions has been considered in the preceding section; of importance is the fact that although this uplift has occurred episodically throughout the last 40 myr, the main pulse of Neogene movement appears to date to the Late Miocene, ca. 8 myr (Molnar et al., 1993). The influence of the rise of the Andes may have been equally profound in the context of the development of climates in the Southern Hemisphere. This event, the most spectacular example of a continental-margin orogen, has a latitudinal extent of some 9,000 km, with elevations in excess of 5 km. During the evolution of this mountain chain, some 2×10^6 km^3 of granite were intruded, and the presence of more than 900 Plio-Pleistocene volcanoes attests also to the importance of exogenetic processes in the generation of its relief. Within the central part of the range, subduction of the Nazca Plate beneath its South American counterpart has probably occurred since the Mesozoic; the modern phase of orogeny, however, appears to have begun in the

Miocene at about 11 myr and was associated with a pulse of volcanism and granitic intrusion that peaked near the end of the Early Pliocene (Vicente, 1970, 1972; Strecker et al., 1989; see also Partridge et al., chap. 24, this vol.). This latter orogenic episode involved all of the Andean mountain belts and resulted in the development of a series of high, narrow ranges as the Eastern Cordillera was pressed against the Brazilian and Patagonian shields.

Late Neogene tectonism in East and southern Africa is considered in some detail in chapter 24. Of overriding importance is the existence, over most of this area, of anomalously elevated topography that spans the continent-ocean boundary and that has been referred to as the African Superswell (Nyblade and Robinson, 1994). As is indicated above, over large parts of eastern Africa the amplitude of this anomaly approaches or exceeds 1 km, and its area surpasses that of the Tibetan Plateau by almost an order of magnitude (see fig. 24.12, this vol.). A wealth of recent evidence suggests that the rise of the Afar and East African Plateaus, which has been directly linked with the evolution of the East African Rift, began in the Miocene but had its major component in the latter part of the Pliocene. In southern Africa, although some of the anomalously elevated topography inland of the Great Escarpment can be linked to uplift of rift shoulders prior to the fragmentation of Gondwanaland, the high plateaus of the eastern hinterland are chiefly the product of Neogene uplifts, the largest of which occurred between 5 and 3 myr. There is, therefore, a temporally coherent record of Neogene plateau uplift throughout the eastern part of Africa, with a considerable body of evidence pointing to the restriction of the largest movements to the Pliocene. As discussed below and in chapter 24, the local climatic effects of these movements were important and served to augment or modify contemporaneous climatic changes of a global nature.

Of particular significance from the evolutionary viewpoint were the local effects of rift generation in East Africa, which resulted in a high degree of environmental fragmentation as a result of both local topographic and resulting rain-shadow effects. Parallels can be seen in southern Africa in proximity to the Bushveld Basin of the Transvaal, whose subsidence in Pliocene times gave rise to a diverse environmental mosaic in an area that documents events important in the late Neogene evolution of hominids in this part of Africa (Partridge et al., chap. 24, this vol.). Analogies with the climatic deterioration and fragmentation of South American ecosystems following the rise of the Andes and the associated evolution of specialized faunal taxa are noteworthy.

No comprehensive studies of the likely influence of Southern Hemispheric uplifts on global and regional patterns of atmospheric circulation have yet been carried out using currently available numerical modeling techniques. The examples presented below illustrate how important such studies can be for unraveling the relative contributions of uplift and other forcing mechanisms to patterns of climatic change.

Effects of Uplift of the Tibetan Plateau

Creation of the Tibetan Plateau has been occurring since about 50 myr ago, when plate-tectonic motions caused the Indian subcontinent to make initial collisional contact with southern Asia (Molnar and Tapponier, 1975). There is still considerable disagreement about the specific timing of net uplift of the surface of Tibet within the Late Cenozoic, ranging from one extreme view that almost all of the uplift of the plateau has occurred within the last 3 myr (Li, 1991) to a contrasting proposition that no uplift of the main part of the plateau has occurred within the past 5 myr (Molnar and England, 1990). The most recent evidence suggests that the main pulse of Neogene uplift, amounting to between 1,000 and 2,500 m, occurred slightly before 8 myr (Molnar et al., 1993).

The Tibetan Plateau, although located far from the continent of Africa, is sufficiently massive to be able to alter climate over at least the portions of Africa in the Northern Hemisphere. The nature of these effects has been demonstrated by sensitivity tests using General Circulation Models (GCMs), in which separate experiments are run with and without high topography. Ruddiman and Kutzbach (1989) first related the results from these experiments to the geologic evolution of the Tibetan Plateau and climatic evolution of the Northern Hemisphere, although many of these findings are implicit in earlier experiments reported by Manabe and Terpstra (1974) and Hahn and Manabe (1975).

The GCM experiments show that in these parts of Africa the strongest climatic effects are in summer (Ruddiman and Kutzbach, 1989). These climatic alterations occur because strong summer-insolation heating of the Tibetan Plateau forms an elevated heat source that emits sensible and latent heat to the upper atmosphere and causes a powerful low-pressure cell to become established over southern Asia. This low-pressure system totally rearranges the circulation over that area, with associated effects reaching well into Africa. These effects have particular relevance to the long-range evolution of climate in Africa, because summer is the only wet season over a broad region centered on the Sahel.

Two potentially major changes in atmospheric circulation over Africa owing to Tibetan uplift were summarized by Ruddiman et al. (1989). First, the counterclockwise circulation around the Asian low-pressure cell drives a strong flow of warm, dry air from east-central Asia southeastward into northeastern and east-central Africa. These summer winds from the continental interior replace two kinds of moisture-bearing oceanic winds that the GCM experiments indicate would flow in the absence of the plateau: westerlies that would otherwise flow from the Atlantic eastward across northern Africa, and southern trade winds that would otherwise flow from the western Indian Ocean northward into eastern North Africa. Replacement of these moist, low-level oceanic winds by drier winds from the continental interior of Asia suggest that aridification of northeastern Africa should occur because of the uplift of Tibet.

Second, the GCM simulation indicates that Tibetan uplift would enhance the strength of the summer Hadley cell over western North Africa, causing increased rising and precipitation over western equatorial Africa and increased subsidence and drying over the northern and northwestern portion of the continent (now the region of the central and northern Sahara). This Hadley-cell intensification occurs in direct association with the creation of an upper-tropospheric easterly jet driven by the increasing high-altitude pressure differential between the northern Indian Ocean and southern Asia.

These GCM results in large measure support a conceptual model first put forward by Flohn and Nicholson (1980), who also inferred that this intensification of the Hadley cell would be restricted to the western part of North Africa because it is dynamically associated with the "exit region" from the local jet-stream velocity maximum created over the African continent. The combined effects of these two tendencies, drying in the northern Sahara and increased moisture along the Ivory Coast and southern Sahel, should be to increase the summer moisture gradient in the Sahel.

Tibetan uplift may also have had an impact on both global temperatures and climate over Africa by increasing global-mean chemical weathering of silicate rocks and thereby drawing CO_2 out of the atmosphere (Raymo et al., 1988). The increase in chemical weathering was ascribed to the combined effects of several factors: increased exposure of relatively fresh, unweathered sedimentary rocks by folding and faulting; intensification of monosoonal rainfall on these exposed rock surfaces; fragmentation of erosion products from the rocks at high altitudes by cryogenic and other mechanical erosion processes; and increasingly rapid flushing of the chemical erosion products by the enhanced monsoonal rains and steepened slopes that promote quick runoff of chemically dissolved products (Stallard and Edmond, 1983; Raymo and Ruddiman, 1992). A variety of chemical indexes from oceanic sediments are consistent with a substantial Late Cenozoic increase in global mean rates of chemical weathering (Delaney and Boyle, 1988).

GCM experiments indicate that decreasing CO_2 levels in the atmosphere should affect both global and regional climate (Schlesinger and Mitchell, 1987). The largest cooling owing to falling CO_2 levels should occur in middle and higher latitudes as a result of albedo-temperature feedback in regions of growing snow and sea-ice cover. There should also be some cooling over Africa, however, and this would probably lead to drying because of the diminished amounts of water vapor carried by cooler air. These tendencies toward a cooling-induced aridity appear to have been magnified during glacial intervals of the Plio-Pliestocene climatic cycles (Sarnthein et al., 1982; deMenocal et al., 1993).

In summary, experiments using General Circulation Models indicate that the most likely climatic effects of Tibetan uplift on North African climate would be (1) a modest cooling of climate; and (2) enhanced aridity, particularly in the summer over both the northern and northwestern Sahara and the northeastern part of the continent. There may have been a circulation-induced tendency toward greater precipitation over western tropical Africa near the Ivory Coast, although this may have been counteracted by the overall trend toward greater aridity owing to an uplift-driven, CO_2-induced global cooling.

Comparative Roles of Ice-Sheet Size, North Atlantic Sea-Surface Temperatures, and Northern Hemispheric Plateau Uplift in Forcing Climate Change in Northeastern Africa. Records of aeolian dust supply from the African continent to adjacent marine basins demonstrate that subtropical Africa experienced profound shifts in climate variability centered near 2.8 myr and again near 1.0 myr (deMenocal and Bloemendal, this vol.). Specifically, the marine-dust records document the onset of low-latitude episodes of coolness and dryness tied to high-latitude glaciations, beginning about 2.8 myr ago. Detailed records of Plio-Pleistocene aeolian dust supply from both West African and Arabian source regions have been reconstructed from several deep-sea-sediment drill sites in the eastern equatorial Atlantic (Bloemendal and deMenocal, 1989; deMenocal et al., 1993; Tiedemann et al., 1994) and the Arabian Sea (Bloemendal and deMenocal, 1989; deMenocal et al., 1991; Clemens and Prell, 1990). Chronostratigraphic control in these records has been established through a combination of oxygen isotopic stratigraphy and bio- and magnetostratigraphy.

These detailed records indicate that variability in the subtropical African climate prior to ca. 3 myr was tied to earth orbital precession, which regulates the seasonal receipt of low-latitude solar insolation and, hence, the intensities of the African and Asian monsoons. Orbital precession varies the season when the earth passes closest to the sun and thus regulates the heat supply driving the monsoon circulation. Climate models have demonstrated the extreme sensitivity of monosoonal circulation to precessional insolation forcing (e.g., Prell and Kutzbach, 1987). Aeolian records imply that terrestrial African climate prior to 3 myr was characterized by moderate-amplitude wet and dry cycles at 23–19 kyr precessional periodicities.

Aeolian records from both western and eastern margins of subtropical Africa demonstrate that after 2.8 myr African climate became dependent, for the first time, upon the onset and subsequent amplification of high-latitude glacial-interglacial cycles. The growth of large high-latitude glaciers and ice sheets is indicated by first occurrences of ice-rafted material in polar sea sediments near 2.7 myr. Glacial-interglacial climatic cycles were sustained at a regular 41 kyr periodicity until ca. 1 myr, when climatic cycles shifted to a dominant 100 kyr periodicity. Aeolian dust records from both margins of Africa demonstrate marked increases in variance at the characteristic 41 kyr periodicity after 2.8 myr and increases in 100 kyr variance after 1 myr. These records point to a connection between high- and low-latitude climate that which was established only after ca. 2.8 myr, when high-latitude ice sheets became sufficiently large to affect low-latitude climate. Experiments using climate models have demonstrated that subtropical Africa is sensitive to specific components of high-latitude climatic variability (deMenocal and Rind, 1993).

Numerical models have been used to assess the sensitivity of low-latitude climatic systems to changes in high-latitude glacial-boundary conditions, such as the size and extent of glacial ice sheets and sea-surface temperatures (SSTs) of the North Atlantic (e.g., deMenocal and Rind, 1993; Rind et al., 1986; Manabe and Hahn, 1977) and to changes in Tibetan-Himalayan elevation (Ruddiman and Kutzbach, 1989; Manabe and Terpstra, 1977). These experiments demonstrate that models of both West and East African climates are sensitive to remote changes in climate and orography, which are likely to have combined with local topographic influences in forcing the changes at ca. 2.8 and 1.0 myr documented above.

The model results suggest that uplift in southern Asia would enhance seasonality and significantly cool and dry East Africa as well as the Mediterranean and Middle Eastern regions, as discussed in the preceding section (Ruddiman and Kutzbach, 1989). A number of studies have examined the sensitivity of low-latitude climate to changes in high-latitude ice cover and SSTs. African climate is relatively insensitive to the isolated effects of the large ice sheets that covered much of North America and northern Europe. Arabia and northeastern Africa were both subject to modest downstream cooling and drying effects (deMenocal and Rind, 1993) related to the European ice sheet. North Atlantic SSTs were cooler by as much as 15°–20°C during glacial maxima and these cool SST anomalies did have considerable impact on West African climate. The cooler SSTs enhanced winter season Hadley circulation over northwestern Africa, resulting in increased advection of cool, dry European air over West Africa. The cool North Atlantic SSTs also reduced the intensity of the summer monsoon, thereby depriving the region of its main source of annual precipitation. Model simulations therefore confirm that Late Neogene climatic changes in East Africa were driven by a complex interplay between tectonic influences and other forcing mechanisms of a global nature. This interplay, and the resulting history of climatic changes in areas of Africa where humankind's earliest ancestors first emerged, are considered more fully in chapter 24.

Conclusions

There is little doubt that climatic change during the Cenozoic has been influenced, in some cases dramatically, by large-scale tectonic movements, which affected both the distribution of terrestrial relief and the geometry of the ocean basins. The mechanisms linking atmospheric responses to such movements may, however, involve complex feedbacks. Current views on the effects of uplift of the Tibetan Plateau illustrate this point. Major elevation of the plateau surface around 8 myr, as inferred by Molnar et al. (1993), appears to have caused marked strengthening of the monsoon and also seems to be linked in some way with a global change in ^{13}C in paleosols, which, in turn, suggests a worldwide expansion of C_4 plants. From this, Cerling et al. (1993) have inferred that the CO_2 content of the atmosphere may have decreased through a critical threshold of around 400–500 ppm at about this time. Raymo et al. (1988) and Raymo and Ruddiman (1992) have proposed that increased weathering associated with uplift may have withdrawn CO_2 from the atmosphere, both through an increase in rainfall, brought about by strengthening of the monsoon, and through accelerated erosion and exhumation of susceptible silicate rocks. Evidence for such increased exhumation and weathering is forthcoming from studies of sediments within the Bengal Fan and other oceanic data, as discussed previously. Theoretical calculations suggest that drawdown of CO_2 from the atmosphere on a scale sufficient to cause pronounced global cooling could have occurred as a result of the accelerated release of silicate minerals that followed Tibetan uplift around 8 myr ago (Molnar et al., 1993).

The Late Neogene rise of the African Superswell, although less dramatic, is likely to have influenced climate in a similar, if less profound, manner. And although less satisfactorily constrained in time, the building of the Andean mountain chain may well have been associated with a comparable climatic response. These topics deserve the kind of research that has yielded such valuable insights into the paleoclimatic significance of high Tibet. Such studies are vital for a better understanding of the factors that forced changes in global climate throughout the Cenozoic.

References

Amano, K., and Taira, A. 1992. Two-phase uplift of Higher Himalayas since 17 Ma. *Geology,* 20:391–394.

Anderson, D. L. 1990. Geophysics of the continental mantle: An historical perspective. In *Continental mantle,* Monographs on Geology and Geophysics, vol. 16, pp. 1–30 (ed. M. A. Menzies). Oxford University Press, Oxford.

Barron, E. J. 1985. Explanations of the Tertiary global cooling trend. *Palaeogeography, Palaeoclimatology, Palaeoecology,* 15:45–61.

Berner, R. A. 1994. 3GEOCARB II: A revised model of the atmosphere CO_2 over Phanerozoic time. *Journal of Geology,* 294:56–91.

Berner, R. A., Lasaga, A. C., and Garrels, R. M. 1983. The carbonate-silicate geochemical cycle and its effect on atmospheric carbon dioxide over the past 100 million years. *American Journal of Science,* 283:641–683.

Bond, G. C. 1978. Evidence for late Tertiary uplift of Africa relative to North America, South America, Australia and Europe. *Journal of Geology,* 86:47–65.

———. 1979. Evidence for some uplifts of large magnitude in continental platforms. *Tectonophysics,* 61:285–305.

———. 1985. Comment on "Continental Hypsography," by C. G. A. Harrison et al. *Tectonics,* 4:251–255.

Bloemendal, J., and deMenocal, P. B. 1989. Evidence for a change in the periodicity of tropical climate cycles at 2.4 myr from whole-core magnetic susceptibility measurements. *Nature,* 342:897–900.

Bouquillon, A., France-Lanord, C., Michard, A., and Tiercelin, J.-J. 1990. Sedimentology and isotopic chemistry of the Bengal Fan sediments: The denudation of the Himalaya. In *Proceedings of the Ocean Drilling Program: Scientific results,* vol. 116, pp. 43–58 (ed. J. R. Cochran, D. A. V. Stow et al.). Ocean Drilling Program, College Station, Tex.

Bull, J. M., and Scrutton, R. A. 1992. Seismic reflection images of intraplate deformation, central Indian Ocean, and their tectonic significance. *Journal of Geological Society of London,* 149:955–966.

Cande, S. C., and Kent, D. V. 1992. A new geomagnetic polarity time scale for the Late Cretaceous and Cenozoic. *Journal of Geophysical Research,* 97:13917–13951.

Cathles, L. M., and Hallam, A. 1991. Stress-induced changes in plate density, epeirogeny, Vail sequences, and short-lived global sea-level fluctuations. *Tectonics,* 10:659–671.

Cerling, T. E. (1991). Carbon dioxide in the atmosphere: Evidence from Cenozoic and Mesozoic paleosols. *American Journal of Science,* 291:377–400.

Cerling, T. E., Wang, Y., and Quade, J. 1993. Expansion of C_4 ecosystems as an indicator of global ecological change in the late Miocene. *Nature,* 361:344–345.

Clemens, S., and Prell, W. L. 1990. Late Pleistocene variability of Arabian Sea summer monsoon winds and continental aridity: Eolian records from the lithogenic component of deep-sea sediments. *Paleoceanography,* 5:109–146.

Cloetingh, S., and Kooi, H. 1992. Tectonics and global change: Inferences from Late Cenozoic subsidence and uplift patterns in the Atlantic/Mediterranean region. *Terra Nova,* 4:340–350.

Cloetingh, S., McQueen, H., and Lambeck, K. 1985. On a tectonic mechanism for regional sea-level variations. *Earth and Planetary Science Letters,* 75:157–166.

Delaney, M. L., and Boyle, E. A. 1988. Tertiary paleoceanic chemical variability: Unintended consequences of simple geochemical models. *Paleoceanography,* 3:137–156.

deMenocal, P. B., Bloemendal, J., and King. J. W. 1991. A rock-magnetic record of monosoonal dust deposition to the Arabian Sea: Evidence of a shift in the mode of deposition at 2.4 Ma. In *Proceedings of the Ocean Drilling Program, Scientific results,* vol. 117, pp. 389–407 (ed. W. L. Prell and N. Niitsuma et al.). Ocean Drilling Program, College Station, Tex.

deMenocal, P. B., and Rind, D. 1993. Sensitivity of Asian and

African climate to variations in seasonal insolation, glacial ice cover, sea-surface temperature, and Asian orography. *Journal of Geophysical Research,* 98:7265–7287.

deMenocal, P. B., Ruddiman, W. F., and Pokras, E. M. 1993. Influences of high- and low-altitude processes on African terrestrial climate: Pleistocene eolian records from equatorial Atlantic Ocean Drilling Program Site 663. *Paleoceanography,* 8:209–242.

DeMets, C., Gordon, R. G., Argus, D. F., and Stein, S. 1990. Current plate motions. *Geophysical Journal International,* 101:425–478.

Dziewonski, A. M., and Anderson, D. L. 1984. Seismic tomography of the earth's interior. *American Scientist,* 72:483–494.

Ebinger, C. J. 1989. Tectonic development of the western branch of the East African rift system. *Geological Society of America Bulletin,* 101:885–903.

Engebretson, D. C., Kelley, K. P., Cahsman, H. J., and Richards, M. A. 1992. 180 million years of subduction. *Geological Society of America Today,* 2:93–95, 100.

Flohn, H., and Nicholson, S. E. (1980). Climatic fluctuation in the arid belt of the "Old World" since the last glacial maximum: Possible causes and implications. *Palaeoecology of Africa,* 12:13–22.

France-Lanord, C., Derry, L., and Michard, A. 1993. Evolution of the Himalayas since Miocene time: Isotopic and sedimentological evidence from the Bengal Fan. In *Himalayan tectonics,* 74:605–623 (ed. P. J. Treloar and M. P. Searle). Geological Society Special Publication, London.

Hahn, D. G., and Manabe, S. 1975. The role of mountains in the South Asian monsoon circulation. *Journal of Atmospheric Science,* 32:1515–1641.

Haq, B.U., Hardenbol, J., and Vail, P. R. 1987. Chronology of fluctuating sea levels since the Triassic. *Science,* 235:1136–1165.

Harrison, T. M., Copeland, P., Hall, S. A., Quade, J., Burner, S., Ojha, T. P., and Kidd, W. S. F. 1993. Isotopic preservation of Himalayan/Tibetan uplift, denudation, and climatic histories of two molasse deposits. *Journal of Geology,* 101:157–175.

Harrison, T. M., Copeland, P., Kidd, W. S. F., and Yin, A. 1992. Raising Tibet. *Science,* 255:1663–1670.

Hodell, D. A., Elmstrom, K. M., and Kennett, J. P. 1986. Latest Miocene benthic delta O^{18} changes, global ice volume, sea level and the "Messinian Salinity Crisis." *Nature,* 320:411–414.

Hsu, J. (Xu Ren). 1978. On the paleobotanical evidence for continental drift and Himalayan uplift. *Palaeobotanist,* 25:131–145.

Jestin, F., and Huchon, P. 1992. Cinématique et déformation de la jonction triple mer Rouge–golfe d'Aden–rift éthiopien depuis l'Oligocène. *Geological Society of France Bulletin,* 163:125–133.

Kennett, J. P., McBirney, A. R., and Thunell, R. C. 1977. Episodes of Cenozoic volcanism in the circum-Pacific region. *Journal of Vulcanology and Geothermal Research,* 2:145–163.

Kennett, J. P., and Thunell, R. C. 1977. Comments on Cenozoic explosive volcanism related to east and southeast Asian arcs. In *Island arcs, deep sea trenches and back-arc basins,* Maurice Ewing Series. American Geophysical Union, 1:348–352.

Li, J. 1991. The environmental effects of the uplift of the Qinghai-Xizang Plateau. *Quarternary Science Reviews,* 10:479–484.

Liu, D-S., and Menglin, D. 1984. The characteristics and evolution of the palaeoenvironment of China since the late Tertiary. In *The evolution of the East Asian environment,* pp. 11–40 (eds. R. O. Whyte, T-N. Chiu, C-K. Leung, and C-L. So). University of Hong Kong, Center of Asian Studies, Hong Kong.

Manabe, S., and Terpstra, T. B. 1974. The effects of mountains on the general circulation of the atmosphere as identified by numerical experiments. *Journal of Atmospheric Science,* 31:3–42.

Mercier, J-L., Armijo, R., Tapponier, P., Covey-Gailhardis, E., and Tonglin, H. 1987. Change from Tertiary compression to Quaternary extension in southern Tibet during the India-Asia collision. *Tectonics,* 6:275–304.

Molnar, P., and England, P. 1990. Late Cenozoic uplift of mountain ranges and global climate change: Chicken or egg? *Nature,* 346:29–34.

Molnar, P., England, P., and Martinod, J. 1993. Mantle dynamics, uplift of the Tibetan Plateau, and the Indian monsoon. *Reviews of Geophysics,* 31:357–396.

Molnar, P., and Tapponier, P. 1975. Cenozoic tectonics of Asia: Effects of a continental collision. *Science,* 189:419–426.

Nyblade, A. A., and Robinson, S. W. 1994. The African Superswell. *Geophysical Research Letters,* 21:765–768.

Pollitz, F. F. 1986. Pliocene change in Pacific plate motion. *Nature,* 320:738–741.

———. 1988. Episodic North America and Pacific plate motions. *Tectonics,* 7:711–726.

———. 1991. Two-stage model of African absolute motion during the last 30 million years. *Tectonophysics,* 194:91–106.

Prell, W. L., and Kutzbach, J. E. 1987. Monsoon variability over the past 150,000 years. *Journal of Geophysical Research,* 92.

Raymo, M. E., and Ruddiman, W. F. 1992. Tectonic forcing of late Cenozoic climate. *Nature,* 359:117–122.

Raymo, M. E., Ruddiman, W. F., and Froelich, P. N. 1988. Influence of Late Cenozoic mountain building on ocean geochemical cycles. *Geology,* 16:649–653.

Richards, M. A., and Hager, B. H. 1988. Dynamically supported geoid highs over hotspots: Observations and theory. *Journal of Geophysical Research,* 93:7690–7708.

Rind, D. 1987. Components of the Ice Age circulation. *Journal of Geophysical Research,* 92:4241–4281.

Rind, D., Peteet, D., Broecker, W., McIntyre, A., and Ruddiman, W. 1986. The impact of cold North Atlantic sea surface temperatures on climate: Implications for the Younger Dryas cooling (11–10 kyr). *Climate Dynamics,* 1:3–34.

Ruddiman, W. F., and Kutzbach, J. E. 1989. Forcing of Late Cenozoic northern hemisphere climate by plateau uplift in southern Asia and the American West. *Journal of Geophysical Research,* 94:18409–18427.

Ruddiman, W. F., Prell, W. L., and Raymo, M. E. 1989. Late Cenozoic uplift of southern Asia and the American West: Rationale for general circulation modeling experiments. *Journal of Geophysical Research,* 94:18379–18391.

Ruddiman, W. F., and Raymo, M. E. 1988. Northern hemisphere climate regimes during the past 3 Ma: Possible tectonic connections. *Philosophical Transactions of the Royal Society of London,* ser. B, 318:411–430.

Ruddiman, W. F., Sarnthein, M., Backman, J., Baldauf, J. G., Curry, W., Dupont, L. M., Janecek, T., Pokras, E. M., Raymo, M. E., Stabell, B., Stein, R., and Tiedemann, R. 1989. Late Miocene to Pleistocene evolution of climate in Africa and the low-latitude Atlantic: Overview of Leg 108 results. *Ocean Drilling Program: Scientific results,* vol. 108, pp. 463–484 (ed. W. Ruddiman, M. Sarnthein et al.). College Station, Tex.

Sarnthein, M., Thiede, J., Pflaumann, U., Erlendeuser, H., Futterer, D., Koopman, B., Lange, H., and Seibold, E. 1982. Atmospheric and oceanic circulation patterns off northwest Africa during the past 25 million years. In *Geology of the northwest African continental margin,* pp. 545–603 (ed. U. von Rad, K. Hinz, M. Sarnthein, and E. Seibold). Springer-Verlag, Berlin.

Schlesinger, M. E., and Mitchell, J. F. B. 1987. Climate model simulations of the equilibrium response to increased carbon dioxide. *Reviews of Geophysics,* 25:760–798.

Sloan, H., and Patriat, P. 1992. Kinematics of the North American–African plate boundary between 28° and 29°N during the last 10 Ma: Evolution of the axial geometry and spreading rate and direction. *Earth and Planetary Science Letters,* 113:323–341.

Stallard, R. F., and Edmond, J. M. 1983. Geochemistry of the Amazon 2: The influence of geology and weathering environment on the dissolved load. *Journal of Geophysical Research,* 88:9671–9688.

Stanley, D. J., and Wezel, F.-C. (eds.). 1985. *Geological evolution of the Mediterranean basin.* Springer-Verlag, Berlin.

Strecker, M. R., Cervery, P., Bloom, A. L., and Malizia, D. 1989. Late Cenozoic tectonism and landscape development in the foreland of the Andes: Northern Sierras Pampeanas (26°–28°S), Argentina. *Tectonics,* 8:517–534.

Tattersall, I. 1993. *The human odyssey: Four million years of human evolution.* Prentice Hall, New York.

Tiedemann, R., Sarnthein, M., and Shackleton, N. J. 1994. Astronomic timescales for the Pliocene Atlantic $\delta^{18}O$ and dusk flux records of ODP Site 659. *Palaeoceanography,* 9:619–638.

Tippett, J. M., and Kamp, P. J. J. 1993. Fission track analysis of the Late Cenozoic vertical tectonics of continental Pacific crust, South Island, New Zealand. *Journal of Geophysical Research,* 98:16119–16148.

Vicente, J. C. 1970. Reflexiones sobre la porción meriodional del Sistema Peropacifico Oriental. Conferencia sobre problemas de la Tierra Sólida (Buenos Aires, October 26–31, 1970). *Projecto Internacional del Manto Superior,* 35:158–184.

———. 1972. Aperçu sur l'organisation des Andes argentino–chiliennes centrales au parallèle de l'Aconcagua. *Proceedings of the Twenty-fourth International Geological Congress (Montreal, Canada, 1972),* 3:424–436.

West, R. M. 1984. Siwalik faunas from Nepal: Paleoecological and paleoclimatic interpretations. In *The evolution of the East Asian environment,* pp. 724–744 (ed. R. O. Whyte, T-N. Chiu, C-K. Leung, and C-L. So). University of Hong Kong, Center of Asian Studies, Hong Kong.

Westaway, R. 1993. Forces associated with mantle plumes. *Earth and Planetary Science Letters,* 19:331–348.

Williams, L. A. J. 1978. The volcanological development of the Kenya Rift. In *Petrology and geochemistry of continental rifts: Proceedings of the NATO Advanced Study Institute on Paleorift Systems,* no. 36 (ed. H. J. Neumann and I. B. Ramberg). NATO Advanced Study Institute, Reidel, Dordrecht.

WoldeGabriel, G., Aronson, J. L., and Walter, R. C. 1990. Geology, geochronology, and rift basin development in the central sector of the main Ethiopian Rift. *Geological Society of America Bulletin,* 102:439–458.

Zhang, Y., and Tanimoto, T. 1993. High-resolution global upper mantle structure and plate tectonics. *Journal of Geophysical Research,* 98:9793–9823.

Chapter 3

On the Connections between Paleoclimate and Evolution

Elisabeth S. Vrba

A fundamental question on how the physical world relates to evolution remains unsolved. Is physical change the necessary pacemaker of speciations and extinctions, or do living entities drive themselves to evolve and disappear even in its absence? Views on each side accept that biotic systems are constantly interacting and changing. The first holds that most of the time such organismal and population processes confer on species and communities a long-term dynamic equilibrium that is reset to a radically different state—by extinction or speciation—only upon extrinsic forcing. The focus is on species-specific habitats and resources and on how habitat changes in space and time bring about species turnover, that is, speciation, extinction, and migration (see definitions in app. 3.1). The second view sees biotic interactions as the sufficient engines of evolution, although climatic changes are recognized by some to accelerate it. One can describe it as competition-centered, because all biotic interactions—competition, parasitism, predation, mutualism—are aspects of competition *sensu lato.*

Subsidiary questions, raised by one or other of the two theories, include: How does speciation relate to the Milankovitch climatic cycles and to severe shifts in the mean, amplitude, and power spectrum of these cycles? Are there general rules governing the similarities and differences among lineages in the incidence of speciation and its timing and sequence of steps? What kinds of causal connections exist between long-distance migrations of species' distributions, climatic shifts, and speciation? Have intervals of strongly increased turnover been a pervasive and orderly feature of the history of life, brought on again and again by common causes? Or is there no consistent order in macroevolution, with the major turnover events rare, random clusters of independent microevents? Can we study these problems fruitfully using the fossil record? Or are the effects of taphonomic distortion—particularly changes in preservation potential that result in gaps in the fossil record—too severe to distinguish between alternative hypotheses? So far there is little consensus in our field on answers to these questions or even on how best to set about answering them, as is apparent in comparing the paleontological and paleoanthropological contributions in this volume. I prefer to let readers discover for themselves the diversity of evolutionary and taphonomic arguments in these chapters rather than to summarize them (and possibly misrepresent them) in an introduction. After a brief review of the beginnings of the competition- and habitat-based dichotomy, I shall concentrate on new hypotheses and tests that might be used to distinguish between those paradigms. A particular challenge is the understanding of species' originations, and it is this manifestation of turnover that I shall emphasize. The hypotheses I propose are compatible with previous hypotheses (Vrba, 1992) that also center on the habitat concept. Together they are part of an internally consistent larger theory, "habitat theory," for brief reference (Vrba, 1992).

In choosing a title for this volume, the editors had in mind climatic changes both from astronomical sources (Milankovitch and other quasi-periodic variations and occasionally impacts of extraterrestrial bodies on earth; Raup and Sepkoski, 1984; Hut et al., 1987) and from structural changes in the earth's crust (such as continental suturing and breakup, uplift, and volcanism; see Partridge et al., chap. 2, this vol.). The scales of climatic

ISBN 0-300-06348-2.

changes vary from local fluctuations in weather to massive and long-lasting global alterations. In discussing whether climatic change might have influenced evolution, all of these sources and scales are potentially included. In fact, at the outset it is appropriate to recall that evolution occurs at different temporal and geographic scales and at different organizational levels of the biological hierarchy—including the levels of genes, cells, organisms, populations, and species. I shall argue that a cardinal difference between the competition- and habitat-based theories, and therefore between the ways my opening question has been answered in the past, lies in whether the causal focus is restricted to the level of organisms or whether it includes the larger temporal, geographic, and structural scales of population and species.

Speciation: From Early Concepts to Limited Modern Consensus

Darwin's Views on Turnover and Climate. The evolutionary process that Darwin (1859) emphasized almost exclusively is natural selection acting through competition among organisms: "The theory of natural selection is grounded on the belief that . . . each new species is produced by having some advantage over those with which it comes into competition; and the consequent extinction of less-favored forms almost inevitably follows" (p. 320). He was not aware of continental drift. But his concept of climatic changes already included an exceptionally warm period during the "newer Pliocene period" (reminiscent of the last "golden age" of some recent authors; see PRISM, this vol.) that preceded "the Glacial epoch . . . of extreme cold [when] I believe that the climate under the equator at the level of the sea was about the same with that now felt there at the height of six or seven thousand feet . . . [and] large spaces of the tropical lowlands were clothed with a mingled tropical and temperate vegetation" (p. 378). He recognized climate-induced changes in geographic distributions and vicariance (pp. 351, 353). (Vicariance refers to the fragmentation, into spatially and genetically isolated—or allopatric—populations, of a formerly continuous species' distribution by the appearance of tectonic and climate barriers within that distribution.) Yet Darwin did not consider allopatry and population structure to be important to speciation, perhaps because he saw climate as subservient to and merely as an accelerator of competition: "As climate chiefly acts in reducing food, it brings about the most severe struggle between the individuals" (p. 68). Thus, in Darwin's view, the ubiquitous and chief evolutionary process is natural selection through competition, which is not only necessary but also sufficient for speciation.

Modern adherents of the competition paradigm accept that physical changes in temperature, substrate, and the like do result in selection such that population divergence and—more rarely—speciation may result. But even in the absence of physical change, biotic interactions are held to constitute selective forces that promote increasingly competitive characteristics. According to some models (e.g., Van Valen, 1973; Stenseth and Maynard Smith, 1984), this biotic force at the level of organisms constantly works toward speciation and extinction, irrespective of population structure and even in the absence of physical change. That is, the biota is akin to a perpetual motion machine that inexorably drives itself to evolution.

Gulick's and Wright's Emphasis on Allopatry and Population Structure. Carson (1987) pointed out the early insight of John Thomas Gulick, a missionary and naturalist who grew up in the Hawaiian islands. Inspired by Darwin's (1839) account of the voyage on the *Beagle,* Gulick studied the local land snails. In a lecture delivered in 1853 and in later papers (Gulick 1872, 1873, 1905) he argued strongly that speciation is promoted by physical separation of populations and by chance apportionment among such small allopatric populations of genetic variation on which natural selection can act. Thus, in Gulick's view, both natural selection and population allopatry are necessary for speciation. Neither is sufficient on its own.

Gulick anticipated the "shifting balance theory of evolution" that Sewall Wright developed mathematically during and after the 1930s (e.g., Wright, 1932). Carson (1987) summarized the theory: "Although a small population may occupy (i.e., be balanced on) a peak of adaptation, chance events caused by size bottlenecks will cause it to wander genetically off its peak, temporarily operating against the effects of selection that operate to keep it on the peak. The population may end up in a less adaptive valley. When this happens, the genetic changes that have occurred by chance may provide natural selection with new combinations of genetic materials, allowing a new start. Thus, the population may become adapted differently, that is, it climbs by natural selection onto a different genetically determined peak" (p. 719). Gulick's and Wright's arguments imply structure and rules at hierarchical levels above that of the organism, as explicit in Wright's (1967) analogy: just as mutations are random with respect to the direction of selection in ecological time, so might population differentiation and speciation be random with respect to longer-term evolutionary directions and forces that sort among species.

Speciation Is Predominantly Allopatric, a Point of Modern Consensus, and the Implications of that for Macrorevolutionary Theory. Earlier theoretical treatments of sympatric speciation (e.g., Maynard Smith, 1981), as well as claims by Bush (1975) and White (1978), aroused discussion on the possibility that two populations might diverge to speciation while in contact. But subsequently many authors agreed that the various non-allopatric models are neither supported by empirical evidence nor theoretically likely. The stress on strongly predominant allopatric speciation originated with Mayr (1942, 1963) and was later supported by Paterson (1978, 1982), Carson (1982), Futuyma and Mayer (1980), and Templeton (1981). Cases that suggest sympatric speciation continue to be reported occasionally, for instance, the monophyletic cichlid species inhabiting Lake Barombi Mbo in Cameroon (Schliewen et al., 1994, who also cite some recent models of sympatric speciation). Nevertheless, it is fair to say that the notion of strongly predominant allopatric speciation is consistent with the evidence as a whole and continues to enjoy widespread consensus.

Three kinds of processes can in principle result in allopatric populations:

1. A species may be dispersed over preexisting barriers, the onset of which, I suggest, implies the causal influence of physical change. For instance, chance dispersals of *Drosophila* flies over the ocean always occur, whether there are islands within reach or not; but it took the production of the precursor islands of the Hawaiian Archipelago for the founding of those first allopatric populations of Hawaiian drosophilids (see Carson et al., 1970).
2. The chief cause of allopatric populations in the history of life has probably been fragmentation of formerly continuous species' distributions by the appearance of tectonic and climatic barriers within those distribution, namely, vicariance. (This is how vicariance was originally defined; see Croizat et al., 1974). Vicariance produces allopatry in situ, that is, the location of the resulting allopatric population was not reached by dispersal.
3. One should also consider the possibility that allopatric populations might result in situ from biotic interactions—such as competition for resources among variant phenotypes, or disease—that arise within a continuous distribution of a species. (This would be biotic initiation of allopatry and thereby possibly of allopatric speciation. It should not be confused with biotic initiation of sympatric speciation, for instance, "competitive speciation" as modeled by Rosenzweig [1978].) However, it is difficult to take this third model seriously, because in Paterson's terms (1982, 1985), it is a Class II model, the term he used for any model that is not supported by any good evidence, in distinction to a Class I model, for which at least some reliable evidence exists. The notions of allopatry by dispersal and by vicariance are Class I models. In fact, each is strongly supported by a large body of data. In contrast, I am aware of no evidence that allopatry was ever produced in situ by biotic interactions in the absence of physical change.

I conclude that, at least in the vast majority of cases, allopatry and consequent population restructuring are not brought about by biotic interactions alone but are quintessentially the results of physical change. If allopatric speciation is strongly predominant, then so must physical initiation of speciation be. These inferences are consistent with the comparatively recent evidence that astronomical climatic oscillations have affected most species' habitats strongly, regularly, and frequently. Information on the limits of species' habitat specificities, together with the new paleoclimatic data, implies that species must have been in a constantly dynamic state as they responded to habitat alterations by regular and frequent changes in geographic distributions. From this joint new perspective, macroevolutionary theory should not be restricted to processes operating among organisms, such as biotic interactions and selection. The evolutionary models that are most relevant are those that include the population context, which focus on how organismal interactions, selection, and genetic drift interact to produce different results in shrinking and separating versus expanding and coalescing populations and on the long-term consequences for lineages of regular alternation and repetition of such population dynamics. Within the modern evolutionary synthesis, Wright's (1932, 1967, 1988) work was seminal in this respect. He broadened the theoretical emphasis to include modes of differentiation and sorting among populations long before the evidence for highly dynamic paleoclimates appeared.

In sum, the consensus on predominant allopatric speciation signifies that Gulick's (1872) early views have triumphed. However, it is a qualified triumph. Allopatry may be necessary for most speciation, but it is clearly not sufficient, as shown by the following evidence.

Changes in Species' Distributions Are Prevalent Responses to Climatic Cycles, a Second Point of Modern Consensus. Many studies have traced fossil morphologies through parts of the Milankovitch cycles, particularly for the Late Pleistocene. These results show that a common response by many different kinds of species has been shifting and fragmentation, alternating with reconstitution, of their geographic distributions. (For plants, see Hooghiemstra, this vol.; Dupont and Leroy, this vol.; Huntley and Webb, 1989; for beetles, see Coope, 1979; for marine plankton, Howard, 1985.) The migratory shifts involved up to thousands of kilometers in many organisms (e.g., see Sutcliffe, 1985, for mammals). The physical initiation of species migrations to new places can be visualized to occur in various ways. It may occur by a Milankovitch change that shifts and spreads a species' habitat to new areas in the ocean or on land or that links similar, formerly discontinuous environments between which the migrants then can move. It may occur by sea-level lowering between previously isolated landmasses, by sea-level rise uniting hitherto separate marine basins, or by mountain uplift that obliterates the low-lying area between previously disjunct high plateaus. The geographic distributions of species may respond to climatic changes by expansion, by contraction, by fragmentation, or simply by tracking habitats latitudinally or altitudinally without much change in net size; some combinations of these responses are possible and supported by evidence. There is also strong evidence that successive expansion and retreat of species' distributions provided opportunities for allopatry in new places and in many cases for speciation, as populations were "left behind" in habitat islands during the retreating phase. (See Haffer, 1982, for Andean birds; Coope and Brophy, 1979, for Eurasian beetles; Grubb, 1978, for African mammals.)

It is clear that most species during the Late Neogene were frequently "attacked" and fragmented into isolated populations by climatic changes. Yet, many such species survived numerous climatic cycles and bouts of vicariance without net change and without noticeably giving rise to new species. In other words, although allopatry is usually necessary for speciation, it is apparently not sufficient. Thus, we need models of speciation that invoke particular kinds of allopatry promoted by particular kinds of climatic change. In a later section I propose that long-term, continuous vicariance, and long-term increases and decreases in the minimum- and maximum-value envelopes of the astronomical climatic cycles, may be the kinds of allopatry and climatic change that are particularly important for speciation. But first I shall propose a model of interland migrations in relation to climatic cycles.

Climate and Migration: The "Traffic Light" Model

One outstanding question is why more migrants should move in one direction than in another between two landmasses. Darwin (1859) was intrigued by the "remarkable fact . . . that many more . . . forms have apparently migrated from the north to the south, than in the reversed direction" (p. 379). He speculated that this was largely due to "the northern forms having . . . been advanced through natural selection and competition to a higher stage of perfection or dominating power, than the southern forms. And thus, . . . the northern forms were enabled to beat the less powerful southern forms." Precisely the same explanation was suggested more recently for the "Great American Interchange" of mammals after the Isthmus of Panama emerged (Marshall et al., 1982); but a reanalysis suggests that an explanation based on habitat theory that does not draw on competitive advantages of northern forms may fit the data better (Vrba, 1992). Habitat theory may also apply to interchanges between Eurasia and Africa, one particular focus in this volume, which are also marked by far more mammalian migrants from north to south than vice versa. For instance, during the Middle-Late Miocene eight bovid taxa immigrated to Africa from Eurasia, and only one migrated in the reverse direction (Thomas, 1984); during the Plio-Pleistocene the corresponding imbalance was thirteen to six (Vrba, this vol., fig. 27.8).

I propose what might be called the "traffic light" model of interland migration (fig. 3.1). Land migration can occur only when both (1) a landbridge is present and (2) suitable habitat extends across it from the ancestral habitat. The model predicts unequal lengths of such times of possible migration, akin to a biased traffic light. It draws on observations (e.g., Denton, this vol.) that in

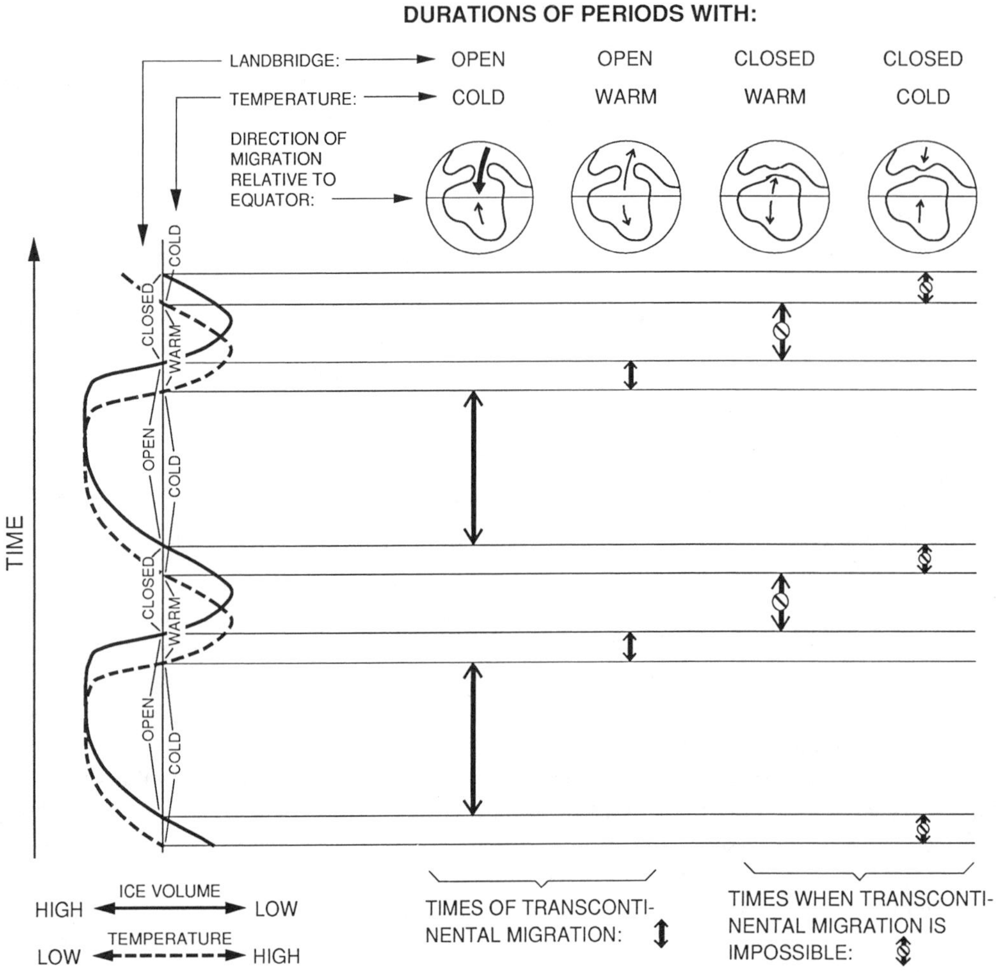

Fig. 3.1. The "traffic light" model of transcontinental land migrations. Migrations require both a landbridge and suitable habitat across that landbridge. When the earth is cold, the landbridge is open (the traffic light is green) for a long time, because cool habitats extend to the equator and sea level is low. In contrast, when the earth is warm, the landbridge is reduced or absent for most of the time. Only during the short lag between onset of global temperature change and ice-volume response can migration from the equator toward higher latitudes of warmer-adapted forms occur. The cold periods are shown as relatively longer in this diagram in accord with current climatic data for the Plio-Pleistocene. But even if interglacials and glacials were equivalent in length, the interplay of eustasy and temperature would ensure that more species should move toward lower latitudes than vice versa.

each case of cooling or warming on an earth with polar ice, the ice-volume response (and with it, eustatic sea-level change) is delayed relative to land cooling or warming and biome response. For the African-Eurasian case, I accept that during global warming the landbridge was significantly reduced from its maximal glacial extent and at times flooded. During glacial maxima, the land connection was much more extensive than it is today, especially southward.

During global cooling, changes to new, seasonally cooler habitats on either side of the persistent water barrier precede polar ice buildup, sea-level lowering, and landbridge enlargement. Thus, the first arrivals from north to south are limited by landbridge enlargement, not by habitat in the immigrant area. The traffic light turns green for migrants only when both the habitat and landbridge condition are met and then stays green for a long time while cooler habitat conditions and lowered sea level coincide throughout most of a glacial period. On a cooling earth, migrants that traverse toward the equator, here Eurasia to Africa, are strongly favored because the vector of moving habitats carries along more higher latitude migrants to the landmass at a lower latitude than vice versa. Fewer lower-latitude species, here African species (fig. 3.1), that were confined to high altitudes during the previous warm phase would also gain habitat passage to Asia as their habitats link up through lowlands with similar ones moving in from the north.

During global warming, polar melting lags behind land warming and biome change: the vegetation and other habitat changes should spread to higher latitudes and altitudes relatively fast (e.g., Dupont and Leroy, and Hooghiemstra, both in this vol.). Melting of ice caps is likely to be delayed for a short while (Denton, this vol.), thereby allowing a brief interval during which landbridges remain open and are already covered by suitable habitat for some species. Thus, the first arrival in Eurasia of African immigrants should record the spread to higher latitudes of mesic habitats before landbridge reduction. The abrupt ending of such a migration episode would signify the time when ice melting has progressed sufficiently to sever the landbridge or at least those parts of it that were formerly habitable and traversable by migrants. This lag period, between land warming and sea-level rise, is the only time during the entire warmer phase of the cycle when both conditions for passage of warm-adapted species are met simultaneously (fig. 3.1). This brief interval favors moving away from the equator, the direction of moving habitats. But such immigrants to Eurasia are likely to be few, for example, the lone species of *Hippotragus* among antelopes that appeared in Eurasia before 3 myr ago (Vrba, chap. 27, this vol.), because this traffic light turns red again quickly as sea-level rise reduces landbridges.

Given a suitable habitat, and a means to traverse to a new area, mammals can move very quickly in geological terms (Kurten, 1957; Savage and Russell, 1983). Thus, the earliest arrivals of immigration episodes can date landbridge emergence during cooling and land-biome response during warming (see Bernor and Lipscomb, this vol.).

In sum, there has been a strong bias for an equatorward migration vector for terrestrial species, dictated by the linked climatic effects of global temperature and sea level. An added factor during the Cainozoic is that there have been more frequent and, on average, longer global cooling events than warming excursions. On an earth with most landmasses situated at high northern latitudes, through long time a preponderance of north to south movements results, as Darwin (1859) noted. But this has little to do with the northern forms having "been advanced . . . to a higher stage of perfection or dominating power, than the southern forms" (Darwin, 1859, p. 379) but rather with the relative lengths of cooler and warmer time during which thoroughfares of suitable habitat were available.

One can similarly explore land migration between east and west, and marine migration between north and south and between east and west. Essentially, land migration overall favors cooler-adapted species, and warmer-adapted forms generally have more restricted opportunities for land passage, namely, only during the first onset of global warming, before polar ice masses melt sufficiently to obliterate landbridges. For instance, land passage across the Bering landbridge by temperate forms is expected to have been restricted to the earlier parts of interglacials. In contrast, marine migration overall favors warmer-adapted species, and cooler-adapted forms have briefer opportunities for passage between marine basins, during the first onsets of global cooling before polar ice masses build up sufficiently to sever sea connections.

Climate, Allopatric Speciation, and Extinction: The "Relay Model"

The model I propose here predicts that lineages in contrasting habitat categories have new species starting up (by speciation) at displaced times and old species ending (by extinction) at differing times, rather like runners in a relay race. I shall therefore, for brief reference, term this the model of "relay turnover" or simply the "relay model."

Species Differ in Habitats and Therefore in Timing of Vicariance and Speciation during the Climatic Oscillations. The geographic distributions of some kinds of species are largest and most continuous around glacial extremes, and smallest and most prone to vicariance during interglacial extremes (e.g., muskoxen and collared lemmings; Sutcliffe, 1985). The converse is true for other kinds of species (e.g., *Cercopithecus* monkeys of the African rain forest; Grubb, 1978). A similar contrast between cooler- and warmer-adapted species exists in cases that differ less extremely than do Arctic and rainforest forms. Within the African-savanna biomes, for instance, hartebeests, springbuck, and other antelopes in the contingent adapted to cool, arid, open grasslands had more extensive geographic distributions during glacials, whereas taxa such as lechwes and duikers within the category loving warmth, moisture, and wood cover were more widespread during interglacial extremes in areas where they do not occur today (e.g., Klein, 1984; Avery, this vol.). One can express such information, on habitat continuity versus fragmentation for individual species or groups of species, directly in terms of the climatic signals in long records that reflect the astronomical cycles (e.g., oscillations in $\delta^{18}O$ values in deep-sea cores; oscillations between loess and soil as in China [Kukla, 1987]; or oscillations between forest, shrubland, and grassland pollen in Bolivia [Hooghiemstra, this vol.] and in western Africa [Dupont and Leroy, this vol.]). One can in principle specify the predicted upper and lower limits along a given axis of climatic variation over which each species' habitat and geographic distribution is maximally continuous and beyond which increased vicariance occurs. I refer to these upper and lower limits as the "vicariance thresholds" and to the range between them as the "optimal range." The model assumes that a minimum interval of continuous vicariance is necessary for most speciations and also precedes most extinctions. That is, within that interval, speciation cannot occur, although extinction can occur with a given probability, and at the end of that interval, new species with evolved (changed) vicariance thresholds will originate with a given probability. One can run the model using various combinations of groups of species and climatic data, as well as specifications of vicariance thresholds, minimum intervals of continuous vicariance, and probabilities of speciation and extinction, and predict the differences among groups of species in timing of vicariance episodes and in timing of turnover events. The model can be used to explore the internal consistency of combinations of particular postulated conditions and to compare the predictions with real patterns.[1]

Illustration of a Modeling Experiment. Figure 3.2 illustrates the model only with respect to the timing of speciation events. Figure 3.3 offers a more general overview of many lineages that differ in speciation and extinction rhythms across successive warming and cooling trends. I will discuss the model first relative to the cooling curve in figure 3.2a. I explore a simple dichotomy between two categories: warmer-adapted and colder-adapted lineages of species. The experiment starts at the right of figure 3.2 during interval t_0 to t_1, when the ancestral species in both categories are the only species present. The warmer-adapted ancestors have vicariance thresholds at x°C and y°C, that is, they have continuous geographic distributions only over the optimal range x°C to y°C and undergo increased vicariance outside those values, as shown in figure 3.2 by the geographic distributions at the right. The colder-adapted ancestral species are most continuously distributed between w°C and x°C. The minimum time of continuous vicariance needed for speciation in both groups of species is indicated by the length of the horizontal bars marked I through IV (ca. 80–100 thousand years [kyr] in the present example). That is, new species can originate only at the end of such a vicariant phase. As a new species originates, so its vicariance thresholds evolve in the direction of the climatic trend: In figure 3.2a, warmer-adapted lineages evolve from an ancestral optimal range of x°C and y°C to a descendant one of w°C–x°C, and colder-adapted lineages evolve from an ancestral optimal range of w°C and x°C to a descendant one of v°C–w°C.

During the earliest segment, t_0 to t_1, on the right side, both kinds of ancestral lineages are never in the vicariant state for long, because the warm cyclic extremes keep recurring quickly to roughly similar values, and so do the cold cyclic extremes. The vicariance events are too brief to culminate in the genetic and phenotypic divergence necessary for speciation.

During a major cooling trend (fig. 3.2a), the warmer-adapted species are the first to enter long-term vicariance (vicariant phase I, time t_1 to t_2). During this interval, the allopatric populations in each warmer-adapted species shrink or disappear with cooling and expand with relative warming yet never coalesce again, because the maxima remain below x°C, the threshold temperature for distributional continuity of the warmer-adapted ancestral species. Long-term vicariance of many small populations is hypothesized to promote two kinds of processes important to speciation, in accordance with parts of the models proposed by Gulick (1872), Wright (1932), Mayr (1942), Carson (1987), and others: 1) size bottlenecks associated with the decreasing populations result in random genetic sampling and increased genetic diversity among the vicariant populations; and 2) the environmental changes exert strong selection pressure for a shift in habitat specificity toward the conditions surrounding each vicariant population. Such a small allopatric population represents only to a limited extent a refugial island that maintains the ancestral habitat specificity: Around its periphery and even within it, the new environmental conditions intrude in the form of new selection pressures, and the unusual population constitution resulting from chance genetic sampling is further expected to alter the selective environment. The new selection regime has an unusually high genetic and phe-

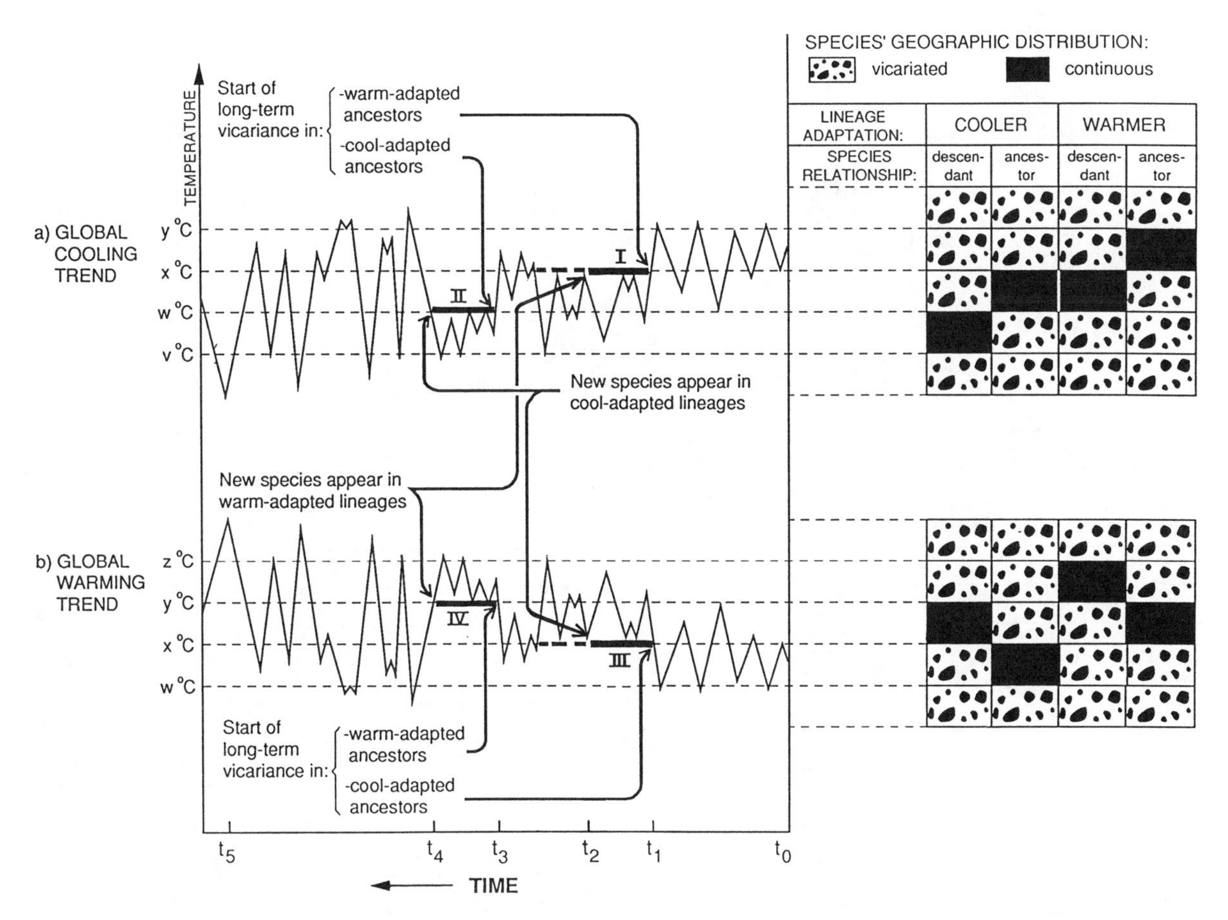

Fig. 3.2. An illustration of the model of "relay speciation," which assumes that speciation is initiated by continuous, long-term vicariance. The cooling curve in a) is adapted from the ca. 3.1–2.4 myr data of Shackleton (this vol.). The warming curve in b) is hypothetical, having been obtained by rotation of the curve in a). We can start by looking at the earliest segment, t_0 to t_1, on the right side, over which interval the vicariance events are too brief to culminate in the genetic and phenotypic divergence necessary for speciation. I chose two categories of species with temperature tolerance ranges at the start (t_0 to t_1) as follows: continuous geographic distributions in warmer-adapted species occur only over x°C–y°C and in colder-adapted species over w°C–x°C (compare the key for geographic distributions, at the right, which is linked by dashed lines to the temperature scale). The minimum time of continuous vicariance needed for speciation in both categories of species is here modeled as the length of horizontal bars I through IV (ca. 80–100 kyr in the present example; the dashed lines produced from bars I and III indicate the total continuous vicariance phase relative to the ancestral species). Only at the end of such intervals of long-term continuous vicariance do new species appear, with new tolerance ranges having evolved toward the next temperature division (e.g., cooler in the case of the cooling trend). See text for further discussion.

notypic variation to act upon, apportioned among the numerous vicariant populations, and is expected to act in the direction of the new surrounding environment. As a result, a few such previously warmer-adapted populations (adapted to spread during the climatic phases between x°C and y°C) are likely by the end of the vicariant phase t_1 to t_2 to have adapted sufficiently to be able to achieve widespread and continuous distributions during the cooler times with temperatures between w°C and x°C. If such adaptation involved divergence of the fertilization system in a population (Paterson, 1985), speciation will have resulted.

During the course of the same major cooling episode in figure 3.2a, the cooler-adapted species enter long-term vicariance only at a later stage, vicariance phase II, starting at time t_3, once the Milankovitch maxima remain for a longer interval below w°C, the lower threshold temperature for distributional continuity of the cooler-adapted ancestors. Previous to time t_3, those species still resumed geographic continuity regularly, over short time intervals. Processes similar to those outlined in the previous paragraph are then hypothesized to affect the persistently vicariated populations of the cooler-adapted ancestors, culminating in new, colder-adapted species by the end of vicariant phase II.

Vicariance is always the precursor not only of specia-

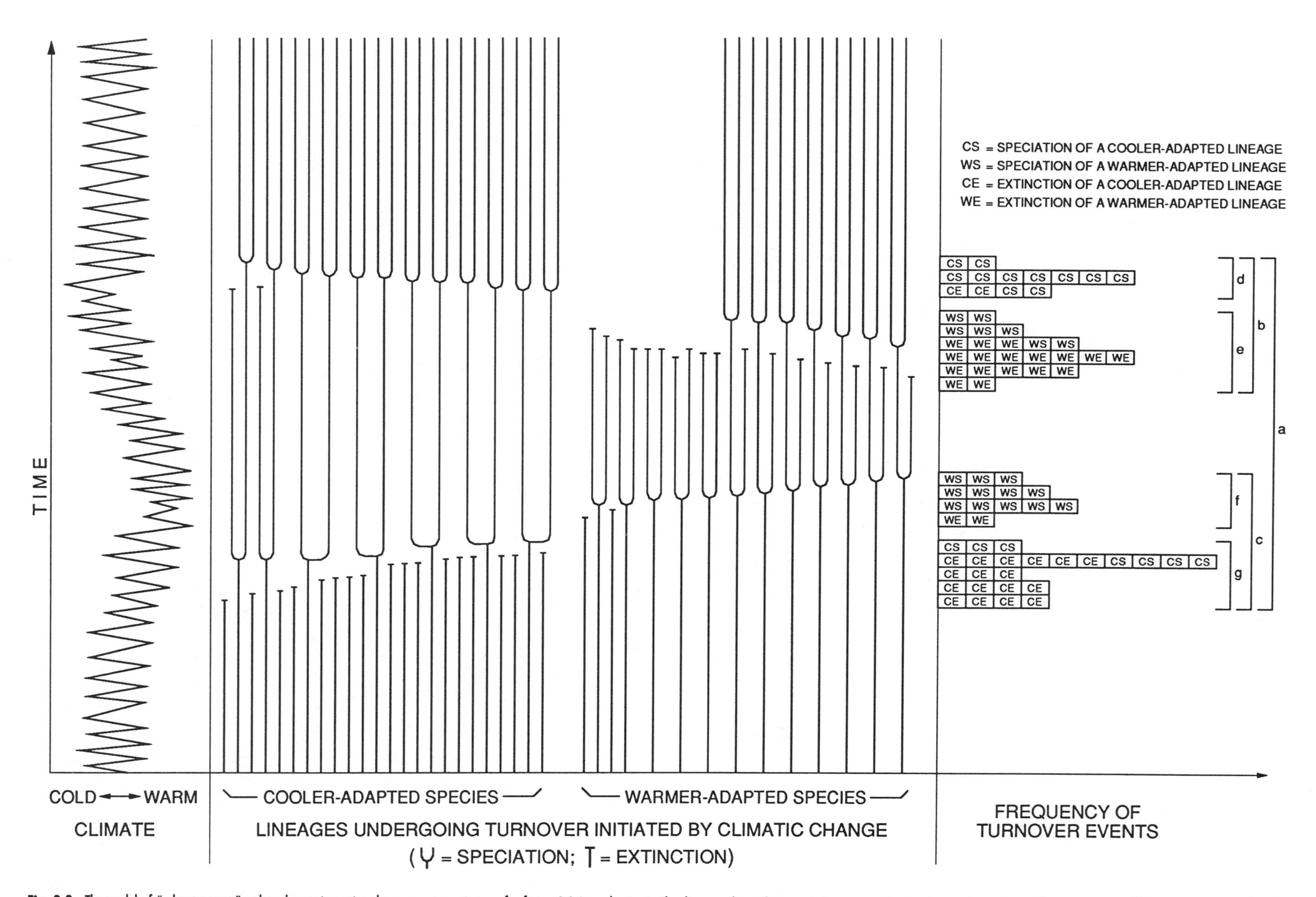

Fig. 3.3. The model of "relay turnover": a broad overview using the same assumptions as for figure 3.2 (see also text). The diagram shows the expected sequence of extinctions and speciations of two categories of lineages, warmer- and cooler-adapted species, during a major warming followed by a major cooling shift of the astronomical climatic cycles. Note that there are smaller turnover pulses nested inside larger ones.

tion but also of extinction. Extinction is expected to precede speciation in any one turnover event. Extinction can in principle result rapidly, by simple accumulation of organismal deaths as their tolerance limits are exceeded. The lineages that are least tolerant of the direction of the climatic trend are those expected to succumb rather than evolve through persistent vicariance. In contrast, speciation results by divergence processes that are expected to take longer. Thus, the first extinctions predicted during a cooling trend such as in figure 3.2a are those of warmer-adapted species, within long-term vicariance phase I, and before new warmer-adapted species appear at the end of that phase (figure 3.3). Similarly, the least cold-tolerant among the cooler-adapted species are expected to become extinct during phase II, preceding the speciations in other cooler-adapted lineages near the end of that phase. Overall, during a major cooling trend many more extinctions and fewer speciations are expected in warmer-adapted than in cooler-adapted lineages.

During a major warming trend (fig. 3.2b), similar but inverse arguments imply that cooler-adapted species should enter long-term vicariance (at time t_1) before warmer-adapted species do (t_3) and that colder-adapted species should also speciate (at t_2) before warmer-adapted species do (t_4). Also, during the course of the same major warming episode, extinctions within each habitat category should precede speciations, and cooler-adapted species should not only enter vicariance earlier but also suffer earlier extinction events than the warmer-adapted species. Overall, many more extinctions and fewer speciations are expected in cooler-adapted than in warmer-adapted lineages during a major warming trend (fig. 3.3).

Predictions and implications include the following:

A. Major trends of cooling or warming are particularly associated with clusters of speciations and extinctions. That is, when the climatic maxima show a decreasing trend (or minima an increasing trend), the probability rises that a series of successive cyclic maxima remain below (or minima above) the vicariance threshold of numerous species (see horizontal bars I–IV in fig. 3.2).

B. Turnover is characterized by particular sequential and frequency patterns. Long-term warming is accompanied by successive events roughly in the following order: 1) extinction of cooler-adapted species; 2) speciation of cooler-adapted species; 3) extinction of warmer-adapted species; and 4) speciation of warmer-adapted species. A higher proportion of warmer-adapted species should appear, and more cooler-adapted species should be last recorded. By contrast, long-term cooling elicits a turnover pattern in which warmer-adapted species turn over before cooler-adapted species, higher proportions of cooler-adapted species appear, and higher proportions of warmer-adapted species become extinct.

C. If prolonged vicariance is important for turnover and especially for speciation, then major shifts in the mode, or periodicity pattern, of the astronomical cycles should be examined for associated turnover (Vrba, 1992). DeMenocal et al. (this vol.) document a shift from dominant climatic influence occurring at 23–19 kyr periodicity prior to about 2.8 million years (myr) to one at 41 kyr variance thereafter, with further increases in 100 kyr variance after 0.9 myr (see also Ruddiman and Raymo, 1988). They suggest that any strong climatic influence on African faunal and hominid evolution starting near 2.8 myr should be sought in a change in mode of subtropical climatic variability rather than in a long-term trend in the cyclic mean. From the point of view of the relay model, changes in dominance from a cyclic pattern of shorter period to one of longer period, together with progressive severity of the cyclic extremes, are especially significant for prolongation of vicariance across many lineages. Such global changes appear to have coincided around 2.8–2.7 myr and again 0.9 myr ago. The model based on prolonged vicariance promoting speciation predicts a high rate of speciation at these times.

In fact, the African bovid pattern 2.7–2.5 myr ago does seem to support the relay sequence expected during a net cooling trend: at 2.7 myr, a significantly higher proportion of warmer- and more closed-adapted taxa have FADs, whereas the interval of 2.6–2.5 myr includes more FADs of cooler- and more open-adapted forms (Vrba, table 27.3, this vol.; FAD or LAD = first or last appearance datum). Ancestral species of climatic generalists (for instance, with a broad tolerance of most of the temperature range from v°C to z°C in fig. 3.2) are expected to have the lowest incidence of turnover and often to last right through climatic changes (see Vrba, 1987, for some support from the African mammal record).

There are also relays *among* successive major cooling and warming episodes over much longer periods (fig. 3.3). Major warming is accompanied by the appearance of relatively more species from warmer-adapted ancestors and more extinctions of species from cooler-adapted ancestors, whereas major cooling has the converse associations. Thus, the smaller-scale relays occurring at higher frequencies are embedded within larger-scale relays with lower frequencies. I suggest (Vrba, chap. 27, this vol.) that such relay turnover may explain why lineages of pigs, monkeys, and giraffids tend to have first appeared before and up to ca. 3.0 myr ago, during major warming, whereas in open-adapted forms, such as most bovids and rodents, waves of first appearance occurred later, during cooling, between 2.7 and 2.4 myr ago. The relay model predicts that these two categories of taxa should be climatically out of phase with each other: the generally warmer-adapted pigs, monkeys, and giraffids should have evolutionary heydays during exceptionally warm, wet, wooded intervals such as the one before and near 3.0 myr, while the cooler- and more open-adapted forms, like most bovids and rodents, should diversify during major cooling trends such as that in the Late Pliocene.

Over the Cenozoic and particularly over the last few millions of years, there has been a strong emphasis on episodes of global cooling (e.g., Kennett, this vol.). Thus, the relay succession has been very unequal, heavily favoring diversification in increasingly cooler-adapted lineages.

Climate and Turnover Pulses

The essential point of the turnover pulse hypothesis is that "evolution is normally conservative at least in relation to speciation and extinction. Speciation does not occur unless forced by changes in the physical environment. Similarly, forcing by the physical environment is required to produce extinctions and most migration events" (Vrba, 1993, p. 428). The prediction, stripped to its bare essentials, is that most lineage turnover has occurred in pulses, varying from minute to massive in scale, across disparate groups of organisms and in predictable temporal association with changes in the physical environment. Previous discussions included other basic elements and corollaries: global climatic and tectonic changes are causal agents of turnover (Vrba, 1985, table 2); pulses occur at different scales of time, geography, and numbers of lineages involved; major shifts in the mean, amplitude, and mode or periodicity pattern of the astronomical cycles accounted for a high proportion of all turnover events (Vrba, 1992); biotic interactions are important in influencing the nature of turnover, although the initiation is reserved for physical change; and lineages respond differently, depending on intrinsic properties. I shall reexamine these earlier statements.[2]

The underlying theory of turnover pulses is based on ecological, behavioral, and genetic processes that are held to promote net equilibrium of species as long as their habitats persist (in spite of geographic movements). Disruption of habitats and species, by fragmentation, and qualitative changes within habitats are needed for speciation or extinction. There is general agreement that equilibrium in constant habitats and speciation in changing habitats are perfectly in accord with established genetic and ecological theory (Paterson, 1978, 1982, 1985; Lande, 1980, 1985, 1986; Wright, 1982, 1988; Charlesworth et al. 1982; Wake et al., 1983; Carson and Templeton, 1984; Newman et al., 1985; Turner, 1986; Maynard Smith, 1989; review in Larson, 1989). But these theories vary in how they combine the following three factors: 1) stabilizing selection acts in constant habitats, and directional selection in changing habitats; 2) the effects of genetic drift differ between large, demographically stable populations and those that are undergoing vicariance and shrinking recurrently in response to environmental changes (following Wright, 1932); and 3) the resilient epigenetic and / or polygenic bases of many characters normally confer stasis yet may evolve rapidly given the strong directional selection and the population structures resulting from habitat change. The evidence that has accumulated, since Eldredge and Gould (1972), that punctuated equilibria obtains in at least a substantial proportion of lineages (reviewed in Gould and Eldredge, 1993) strongly suggests that these three processes in some combination are indeed important. My hypothesis is that all three classes of factors play a part in evolution, particularly during speciation.

Much of our biological data base confirms that species are, so to speak, habitat addicts that are pushed into an evolutionary response only when habitat withdrawal by climatic change exceeds their thresholds of tolerance. These thresholds vary among species and clades in two basic ways that may be illustrated with reference to the temperature scale. Taxa differ with respect to the breadth they can tolerate along the scale of variation (specialist clades tolerate a narrower band and therefore reach threshold values more often, and generalists tolerate a broader one); and they differ in being situated across different ranges along that scale.

Comments on the Structure of Turnover Pulses: A Nested Hierarchy of Pulses? Similar phenomena at different scales are relevant to this hypothesis. Scientists have become accustomed to recognizing hierarchical nesting of similar phenomena or similarity among phenomena at different scales, especially recently, with the growing exploration of fractal structures and processes—from those involved in the sutures of ammonites, through tree growth, to coastlines (Mandelbrot et al., 1979; Gleick, 1987). There is an hierarchical geographical layering relevant to turnover: habitat plates shifting rapidly over the continental plates that drift more sedately beneath them. There is also hierarchical nesting of climatic change, which contains several tiers of higher-frequency cycles that are nested within lower-frequency cycles. The variation among taxa in breadth of habitat tolerance dictates that the more ecologically specialized should respond with turnover to smaller climatic changes more often than would more generalized taxa. In fact, both the hierarchically nested pattern of the climate signal itself, and the variation in breadth of habitat tolerance among taxa, suggest that biotic responses to climate should present an hierarchically nested pattern at different scales as well.

Comments on the Structure of Turnover Pulses in Relation to the Climatic Signal. Organismal phenotypes are not distributed evenly throughout phenotype space but show marked clustering within limited parts of that space (e.g., Raup, 1966). We should expect this to also be true of those phenotypes that determine habitat thresholds. Thus, one cannot expect a simple one-to-one correspondence between the magnitude of climatic variation and turnover response: some parts of the climatic scale are likely to e exceptionally thinly populated by any turnover events or by turnover events of a particular kind (e.g., change to very extreme climate may elicit mainly extinctions and few speciations). At the same time, the fact that the habitat specificities of taxa differ in being situated at different parts of the temperature scale (or of another habitat-determining scale) predicts that the turnover response has a particular structure in relation to decreasing and increasing climatic variables. This is exemplified by earlier speciation of warm- than of cooler-adapted taxa during a cooling trend (figs. 3.2, 3.3). And as mentioned above, in spite of the lack of a simple one-to-one correspondence, turnover pulses, like climatic events, are predicted to be hierarchically nested phenomena. If one views the turnover frequency pattern in figure 3.3 from a broad perspective, one can refer to the bimodal aggregation marked *a* as a turnover pulse. With increasing resolution, the component turnover pulses *b* and *c* (with their different contents in terms of kinds of lineages) within *a* become evident, as do the subpulses *d* and *e* within *b*, and *f* and *g* within *c*, and so on. In sum, in discussions of whether or not an observed pattern in the fossil record is a turnover pulse, the scale of resolution in the fossil data relative to that in the climatic data needs to be considered.

Comments on Testing of Predictions. The turnover pulse hypothesis is testable in principle both by modeling and by comparative tests. The example in Vrba (1985) of what the signature of a turnover pulse might look like referred to FADs and LADs concentrated within 100 kyr or less (app. 3.2, no. 2). This description is inadequate, as pointed out by the preceding discussion and by the example explored in figures 3.2 and 3.3. Of course, one can model the predictions of physical initiation of turnover in ways different from the particular model conditions adopted in figure 3.2. But I suggest that all such models will lead to the conclusion that turnover pulses are not identical phenomena that can be categorized by a rigid requirement of 100 kyr or less. They are predicted to contain subevents at different scales that bear particular and potentially complex relations to one another. An example is the partly ecological prediction that within a larger pulse speciations of warmer-adapted taxa should precede those of cooler-adapted taxa during long-term cooling (fig. 3.3). In fact, it is this complexity that adds to the prognosis of successful testing. To confirm only a general correlation of increased turnover with time is a weaker corroboration of the hypothesis than the observation of a predicted ecological pattern.

Let us say that, in a given case, there really were turnover pulses nested hierarchically at different scales of magnitude, for instance, in response to climatic cycles of periodicities 20 kyr, 40 kyr, 100 kyr, 800 kyr, and 2 myr. Very few data sets would be sufficiently resolved to reflect pulses at the highest frequencies, whereas many should detect the larger-scale patterns. It is valid to study turnover pulses only at the larger scales in a given record, whether or not the subevents are visible. The escape from the danger of seeing turnover pulses everywhere as the permissible time scales are expanded lies in the test of whether the frequency of such a set of turnover events across more than one kind of organismal lineage (across whatever time scale is in question) stands out as statistically significant against surrounding intervals and whether the hypothesized, physical, initiating cause bears the predicted temporal relation to the turnover events. Of course, many records are currently still too incomplete and badly dated to allow a test even at the highest temporal scales of turnover. But one needs to separate the issue of whether or not an hypothesis is the appropriate one from the difficulty of testing it on given data available at a given time. After all, to recall a famous example, Wegener's (1924) hypothesis was no less significant because the technology for decisive testing was not developed until the 1960s, as reviewed in McKenna (1983). The point is that the hypothesis was always testable in principle.

What the Turnover Pulse Hypothesis Does Not Imply. The hypothesis does not imply that only global climatic change initiates turnover. Tectonic and climatic changes from various sources and at various scales may do so. It does not predict that a turnover pulse must occupy strictly one instant in time, nor that all kinds of lineages must participate in each pulse, nor that speciations and extinctions must be strictly coeval. Also, in evaluating whether or not a given data set rejects the hypothesis, it is necessary to examine both the paleontological and climatic signals critically. For instance, based on less resolved data than are now available, I had suggested that a turnover pulse in Africa and elsewhere might have occurred near 2.5 myr (Vrba, 1985, 1988). Recent revisions indicate that both the climatic trend and the fossil patterns show a more complex and protracted pattern. Whether my conclusion (see chap. 27, this vol.) that these patterns are consistent with the turnover pulse predictions is correct or not is not at issue here. The point is that any particular dates focused on in isolation are irrelevant to the hypothesis. A turnover pulse needs to bear a predictable and logical relationship to the climatic record. The essence of the hypothesis itself, and of testing it, is a significant relationship between speciation and extinction frequencies and the documented data on physical change. (It is not trivial to decide what is or is not a "significant relationship." But this is a problem common to all scientific tests that employ statistics.) Finally, an attempt to test the hypothesis does not imply a priori acceptance that the FADs and LADs in the fossil record need to represent times of speciations and extinctions, as discussed in the next section.

Climate and Taphonomic Bias

Three kinds of biases that alone or in combination can result in gaps and therefore mimic a FAD or LAD pulse even if there was no change in species turnover are illustrated in figure 3.4. Under "taphonomic bias," I include biases with respect to FAD and LAD times due to differential rarity of those species and biases due to distortion by preservational factors of the underlying relative abundances in the living community, as well as other kinds of biases commonly called "preservational biases." Even if turnover in the real world were constant, a gap in the record could create artifactual LAD and FAD pulses (fig. 3.5a, b). The converse may also occur (fig. 3.5c, d): the FADs and LADs of a true turnover pulse can be "smeared out" by "Signor-Lipps effects" (Signor and Lipps, 1982) to appear gradual. Taphonomic biases are always operating. To the extent that their effects have remained averagely constant, any observed FAD pulse must be accounted for by true species turnover. Thus, the danger for interpretation resides not in biases per se but in significant large-scale changes in biases (app. 3.3). In sum, the aim is to distinguish two kinds of potential causes of an apparent increase in FADs (or LADs): changes in biases and true species turnover. I suggest a testing protocol in three steps to make this distinction in a record of FADs. (The basic arguments on the meaning of a pattern of FADs apply with slight changes to a pattern of LADs.) If the cumulative number of FADs in a given fossil record is plotted against time (as when the FADs are plotted in rank temporal order; see Vrba, fig. 27.3, this vol.), the resulting line through the FADs can be represented by the equation $N = N_0e^{At}$ (a logarithmic representation of $\ln N = At$ is given in fig. 3.6, together with details of the variables). This equation represents the null hypothesis, H_0, that the exponential rate of accumulation, A, which is potentially influenced for each time interval by turnover and/or taphonomic factors, is on average constant through long time (see Vrba, chap. 27, this vol., for why this model represents a reasonable null hypothesis). The statistical tests relate to three questions: 1) Is there a significant FAD pulse in the FAD accumulation curve? 2) Given that a FAD pulse is identified, could it be an artifact of one or more taphonomic gaps (preceding or succeeding it, or both), or was it caused by turnover, including speciation and immigration? 3) What proportion of a FAD pulse represents speciation rather than immigration, and what proportion of a FAD pulse represents speciation close before the pulse? I shall present only brief outlines of the reasoning behind the methods. Additional discussion and applications to the African bovid data can be found in Vrba (chap. 27, this vol.).

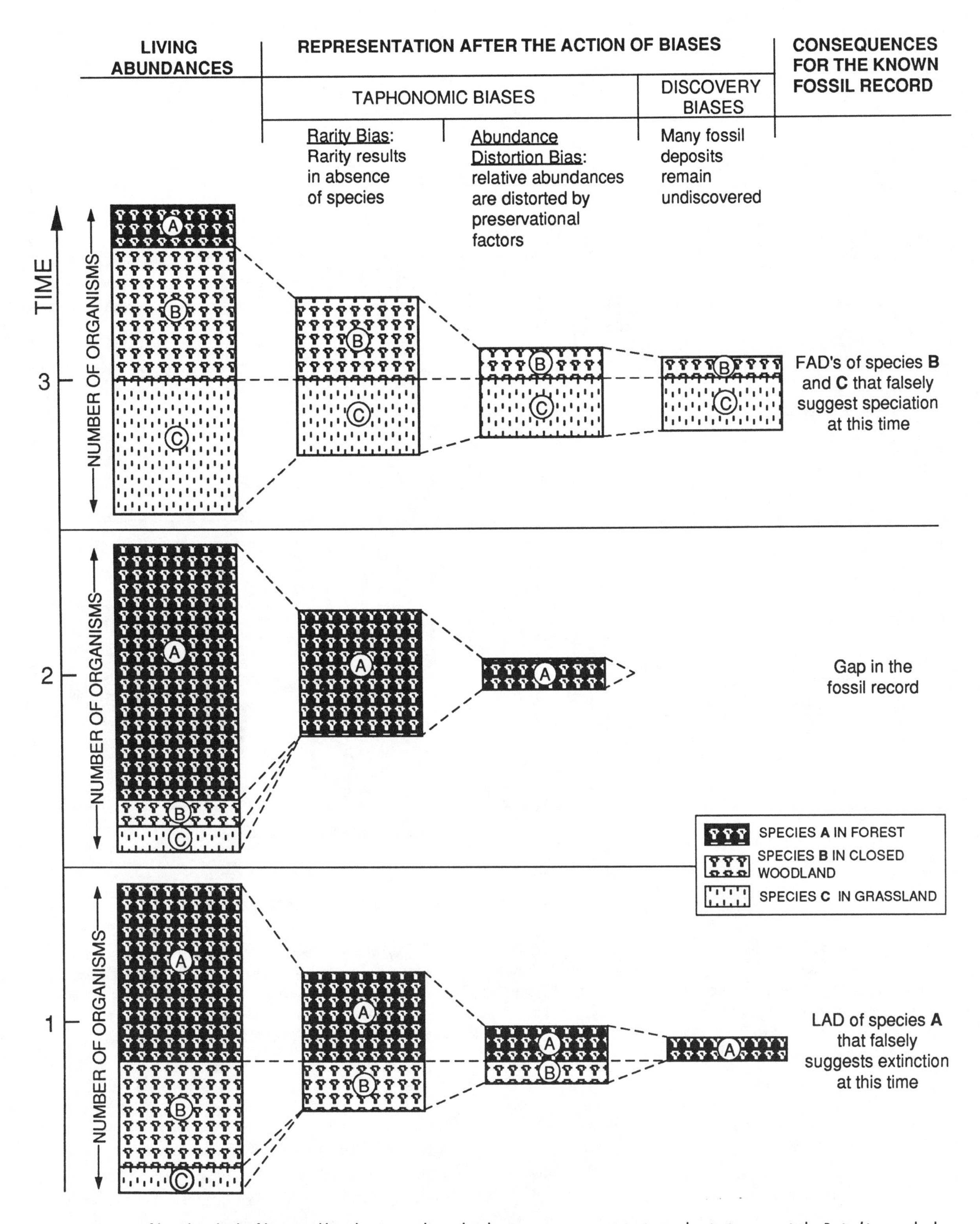

Fig. 3.4. Diagram of how three kinds of biases could result in FADS and LADS that do not represent true originations and extinctions respectively. *Rarity bias:* grassland species C provides an example. This species is rare during intervals 1 and 2 and greatly increases by interval 3, because climatic change has increased grasslands. Even assuming equivalent sampling probability for all taxa, species C remains unrepresented in the record of intervals 1 and 2 because of rarity. It has a FAD in interval 3 that falsely mimics turnover. *Abundance distortion bias:* a deviation in the fossil record from the relative abundances of taxa in the living biota at the time of deposition. For instance, open, arid areas tend to preserve vertebrate fossils better than do wetter, more forested ones (Hare, 1980). This is reflected by the lower preservation rate of forest species A (indicated in the column headed Abundance Distortion Bias) that contributes to a LAD of species A at time 1 that falsely mimics extinction. *Discovery bias:* fossils from strata of a particular age, although present in reality, have for geological (or political) reasons not yet found their way into fossil collections, whereas earlier and later fossils are well represented.

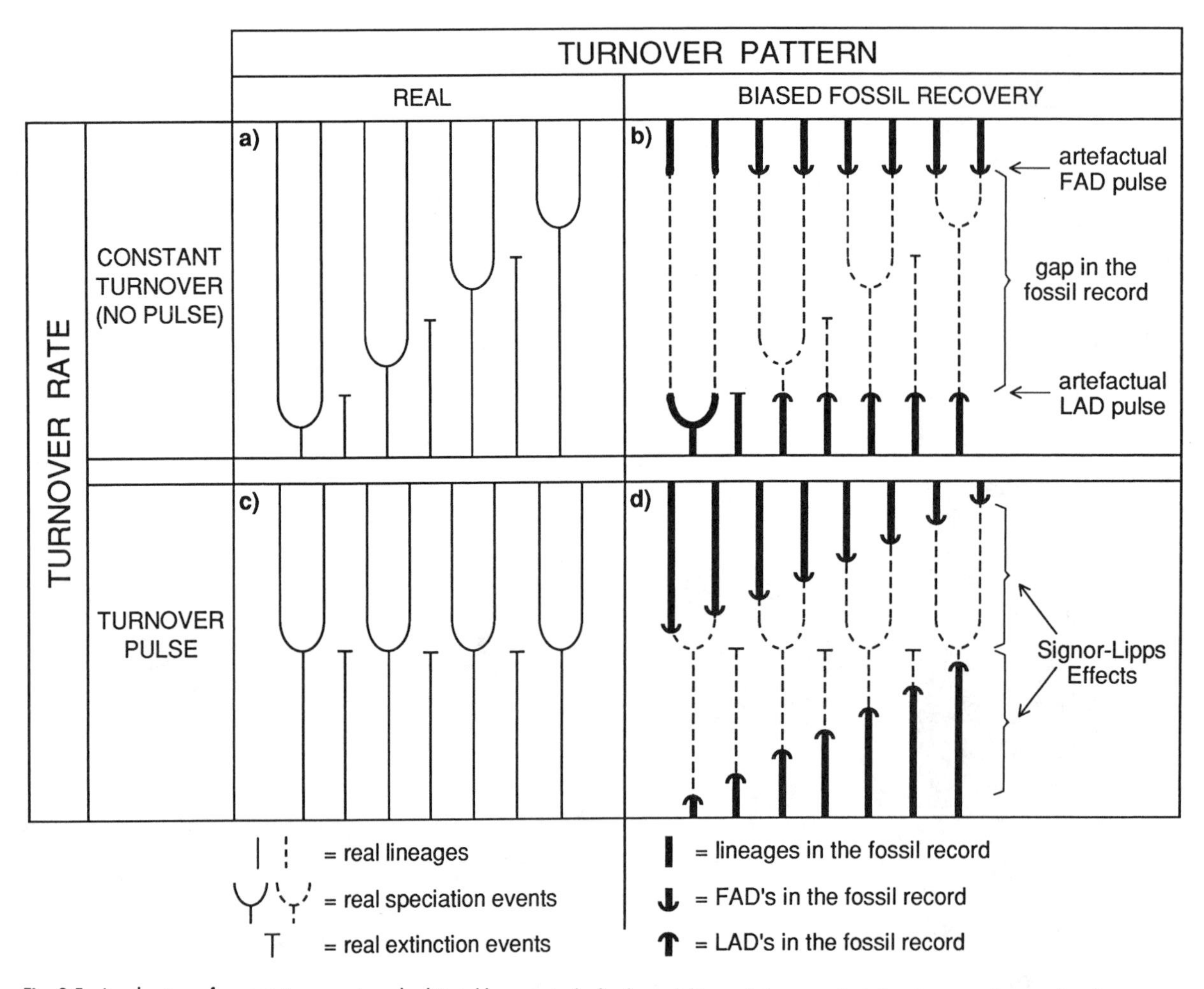

Fig. 3.5. A real pattern of constant turnover, a), can be distorted by a gap in the fossil record, b), to mimic a LAD pulse before the gap and a FAD pulse of origins after the gap. Independent evidence that the *speciation events* fall close to the same time would falsify an hypothesis of constant turnover. The hypothesis would also be falsified by evidence that a pulse of FADS *and* a pulse of LADS date to the same time. A real turnover pulse, c), can be distorted by a random pattern of taphonomic loss of taxa, as in d), to mimic constant turnover. Signor and Lipps (1982) argued that a true mass extinction might give the appearance of gradual extinction, and this "Signor-Lipps effect" applies equally to a true mass episode of splitting events. Other combinations of real patterns and recovery biases, such as a turnover pulse distorted by a gap in the record and the like, are possible.

1. *Is there a significant FAD pulse?* One can use an iterative χ^2 method to test H_0 and to identify which, if any, time segments of the accumulation curve show significant positive or negative deviations (see Vrba, chap. 27, this vol., for an example). A significant positive or negative deviation is a significant FAD pulse or FAD gap. A FAD and/or LAD pulse is only a pattern observed in the fossil record. It is not synonymous with a turnover pulse; nor does it necessarily reflect a turnover pulse, although it may do so. The challenge is to distinguish the taphonomic and turnover contributions to that pulse. Given that a significant FAD pulse is identified, the range of potential causes includes combinations of climatic change, biotic interaction, taphonomic bias, and turnover. I argue that if any significant pulses of FADs are found in the fossil record, then climate-driven widespread change in taphonomic conditions and/or climate-driven change in turnover rates are the candidate causes that we need to unravel. A more expanded and formal statement of my reasons is given in appendix 3.3. (Recall that climatic change may have more than one cause, including local or widespread tectonism. Thus, the concept of climatic forcing of taphonomic bias or turnover includes the possibility that tectonic changes forced the climatic changes. Also, taphonomic bias here includes both biases with respect to FAD and LAD times due to differential rarity of those species and biases

due to distortion by preservational factors of the underlying relative abundances in the living community, as well as other kinds of biases commonly called "preservational biases.")

2. *Is an observed FAD pulse an artifact of one or more taphonomic gaps, or does it reflect increased turnover, or both?* Figure 3.6 shows some of the main forms that the logarithmic graphs, $\ln N = At$, of FAD accumulation can theoretically resemble. It is useful that the slope of the line over any time interval (the coefficient of t) equals the accumulation rate of FADs, or A. This allows quick visual assessment of any deviations from the straight line expected under H_0. (This technique is similar to the logarithmic conversion used in allometry.) There are distinct differences between the pure signatures in figure 3.6 of b), a FAD pulse produced only by a prior gap in the record; of c), a FAD pulse produced only by a subsequent gap; and of e), a true pulse of new species. Note that a gap in the record within a time interval, t_{i-1} to t_i, which might mimic a turnover pulse at time t_i, would be pointed out by this representation. One can, then, apply various significance tests to the observed FAD accumulation curve plotted in the form $\ln N = At$ to determine which of the theoretical patterns in figure 3.6, or additional ones, most closely resemble the fossil record. Recall that the null hypothesis, H_0, is that the exponential rate of accumulation, A, which is potentially influenced for each time interval by turnover and/or taphonomic factors, is on average constant through long time. If the tests of H_0 identify time segments of the accumulation curve that show significant positive or negative deviations (see Vrba, app. 27.4, this vol., for an example), one can next apply detailed tests of whether slopes of the different line segments differ significantly and whether there are significant displacements along the ln N axis as required, for instance, by the hypothesis of a true speciation pulse. Tests for slope changes can be done by using standard tests for significant differences between regression coefficients (e.g., Hoel, 1964). Significant displacements along the ln N axis can be assessed by inspection of standard errors of estimate of the relevant line segments (see Vrba, fig. 27.6, this vol., for an example).

3. *What proportion of a FAD pulse represents speciation closely preceding the FADs?* Take the hypothesis that a FAD pulse seen in the record is an artifact of a gap and that the turnover underlying those FADs in the real

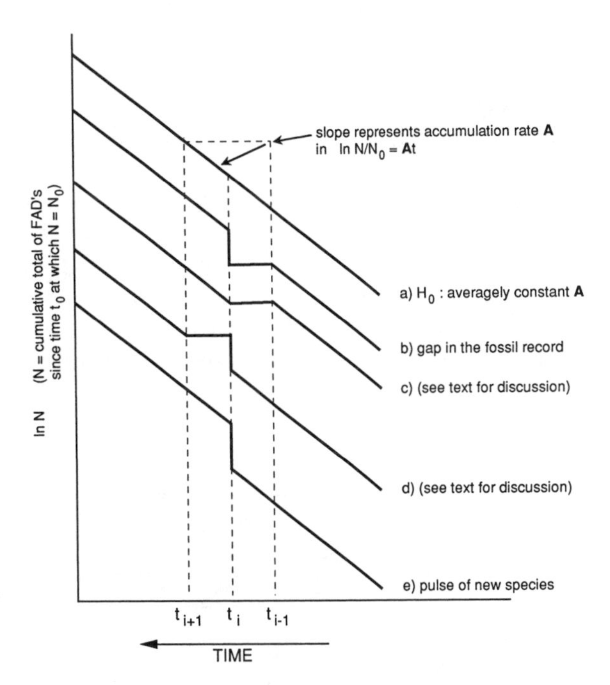

Fig. 3.6. Diagrammatic examples of the curve of the cumulative total of FADS (N, here converted to ln N) through time. The slope $A = (\ln N_i - \ln N_0)/(t_i - t_0)$ of the line over any time interval t_i to t_0 is the FAD accumulation rate. (Based on $N = N_0e^{At}$, where $t = t_1 - t_0$, or time elapsed since the start of the interval, N_0 = total number of taxa at the start.) These curves are meant to represent sampling of FADS across a large area and many localities. The interpretations of a) to e) are given as examples. One can think of other examples and combinations of the a) to e) versions. Turnover here refers to speciation and/or immigration.

a) H_0: no significant changes in background rate from either taphonomic or turnover sources;
b) A FAD pulse at time t_i that is an artifact of a gap in the fossil record between t_{i-1} and t_i, with recovery only from t_i onward of species that were added by turnover in the real world during t_{i-1} to t_i. An alternative, that turnover decreased only during t_{i-1} to t_i and showed an equal increase at t_i, is regarded as much less likely.
c) A decrease in rates of fossil preservation, a gap, without subsequent recovery of the taxa missing during the gap and/or a decrease in turnover rates during t_{i-1} to t_i, with resumption of previous net rates from t_i onward. This can resemble a FAD pulse at time t_{i-1} in data that are not transformed into logarithms.
d) An increase in rates of fossil preservation and/or turnover at t_i, followed by an equal decrease in one or both factors during t_i to t_{i+1}, with resumption of previous net rates from t_i onward.
e) A pulse of speciation and/or immigration at time t_i, with the accumulation pattern before and after that agreeing with H_0. An alternative, that only once in a long time did fossil preservation increase greatly, is regarded as much less likely given sampling of FADS across a large area and many localities.

world was constant (fig. 3.5a, b). Using cladistic analyses, one can falsify this, provided one makes three assumptions (illustrated in fig. 3.7): 1) The cladistic hypothesis is correct. To assume this implies that taxon A in a) cannot be ancestral to taxa B-C because it has autapomorphies and that the taxa labeled A in cases b) and c) are hypothetical or potential direct ancestors because each is wholly plesiomorphic with respect to its sister taxon. Such a hypothesis of ancestry is falsifiable by such additional characters. 2) The hypotheses of direct ancestry in cases b) and c) are correct. That is, it is assumed that additional characters will not be found to be autapomorphic for the taxon labeled A. 3.) The inferred ancestor A gave rise by dichotomous branching to taxa B and C, and A became extinct in that process (fig. 3.7b, c). Given these assumptions, the time window in which speciation occurred is small in case c), larger in case b), while in case a) only the minimum age limit is known. Note that the application of this method to an observed FAD pulse in the record circumvents the taphonomic bias of a previous gap in the fossil record in those cases of FADs for which cladistics and fossil chronology indicate a small time window during which speciation occurred.

One can apply this kind of analysis to all the taxa in an observed FAD pulse and calculate the following ratio:

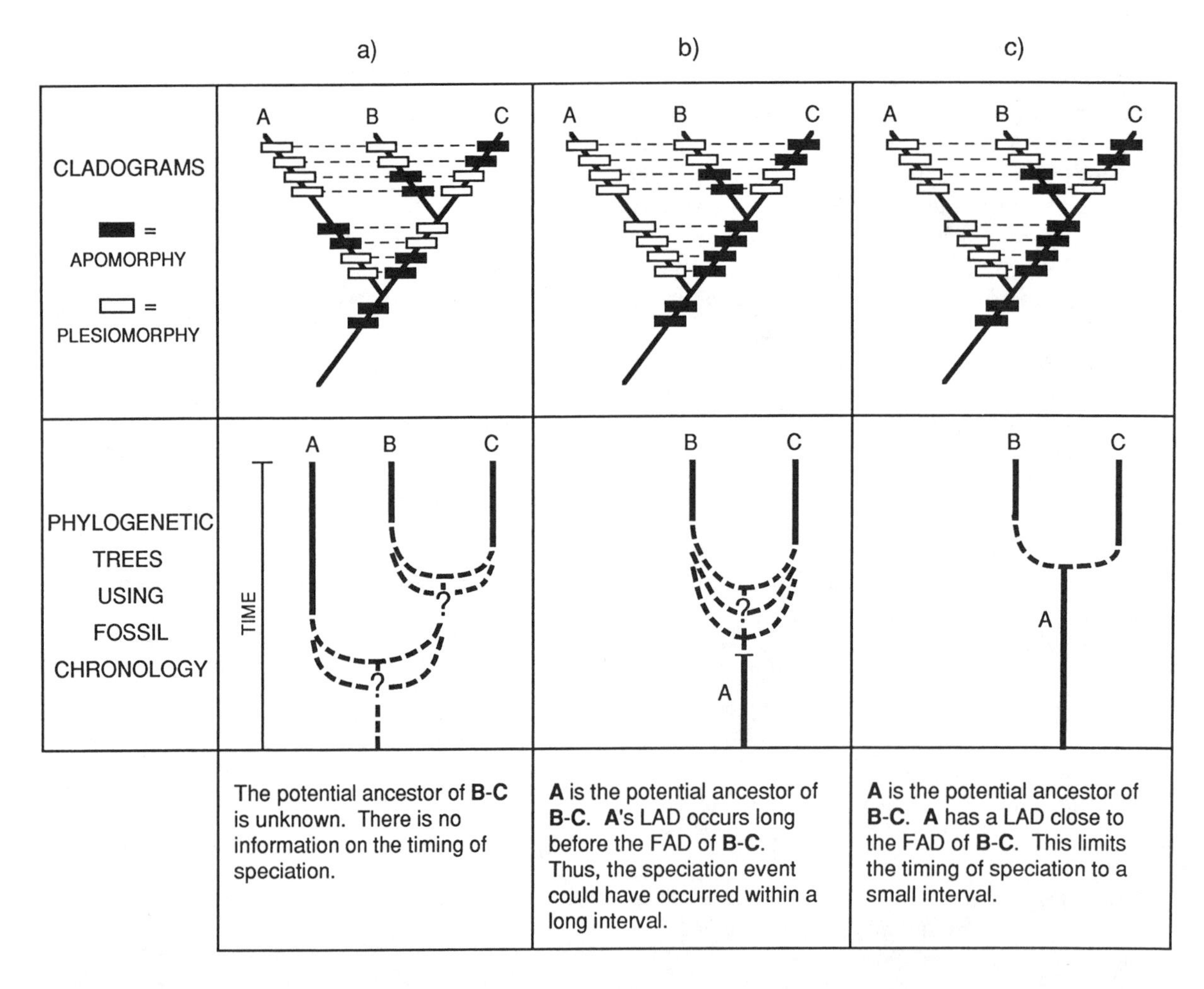

Fig. 3.7. An illustration of the use of character distributions in cladograms to deduce which taxa included in the analysis are possible candidate taxa for true ancestry of perceived FADs (in each of b and c, but not in a, the taxon labeled A is such a potential ancestor). The duration in the fossil record of all taxa in the three cases is shown in the phylogenetic trees at the base. In cases b and c, the fossil chronology does not falsify an hypothesis that the potential ancestors A gave rise by dichotomous branching to taxa B and C and that A became extinct in that process. One can use all those cases of potential ancestors for which there is a well-dated fossil record to determine what proportion of those ancestors have LADs close before a FAD pulse (only A in c fulfills this condition).

(number of FADs for which potential ancestors indicate speciation events timed close to the time interval of the FAD pulse) : (number of FADs in the pulse). This is amenable to further testing. Take the case of an observed pulse in a time interval t_1 to t_2 that includes twenty FADs among which are taxa B and C in figure 3.7. If a change in taphonomic regimes accounts for all of it (by hiding from our view species long since present until climatic change resulted in dramatically improved preservation), the splitting events that gave rise to those taxa should not be nearly coeval closely before the FAD pulse but, instead, strung out in time before that. Let us say that a statistical test has suggested that at least ten of the twenty FADs need to have splitting events between t_1 and t_2 for rejection of the null hypothesis of constant speciation rate distorted by a taphonomic gap (fig. 3.5a, b). If the minimum estimate of the number of taxa with inferred origination between t_1 and t_2 is greater than ten, H_0 is contradicted and the alternative hypothesis of a speciation pulse is supported. If for most of the FADs in a pulse no potential ancestors are known, neither H_0 of a taphonomic gap nor the alternative hypothesis of a turnover pulse can be rejected.

Summary

The modern consensus that most speciation occurs in allopatry implies a strong role for physical initiation of speciation. I have explored ways in which physical changes might influence turnover, particularly speciation, and also migration.

The traffic light model of interland migration provides a possible explanation for why more Neogene species migrated from north to south than in the reverse direction. It invokes the interplay between climate-induced habitat movements and eustasy to predict unequal lengths of possible passage between landmasses at different latitudes, akin to a biased traffic light.

A second model, the relay model, postulates that long-term vicariance needs to precede most speciation and often also precedes extinction. The term "relay" conveys the notion that different categories of lineages are predicted to speciate and become extinct during different phases of the climatic oscillations. For instance, during long-term global warming, extinction and speciation of cooler-adapted species are expected to precede extinction and speciation of warmer-adapted species, with a converse sequence holding true during long-term global cooling.

The turnover pulse hypothesis is reexamined and expanded. The essence of the hypothesis itself, and of tests of it, is a set of turnover events across more than one kind of organismal lineage (across whatever time scale is in question) that stands out as statistically significant against surrounding intervals and a significant relationship between those turnover events and the documented data on physical change. Both the hierarchically nested pattern of the climate signal itself, and the variation in habitat tolerances among taxa, suggest that biotic responses to climate should present an hierarchically nested pattern at different scales as well. Thus, turnover pulses are expected to be nested inside each other and to exhibit certain other internal patterns of scale and sequence. Together these patterns allow testing that goes beyond checking a crude correlation of gross turnover frequency with time.

On taphonomy, I argue that if any significant pulses of first appearances of taxa are found in the fossil record, then climate-driven change in general taphonomic conditions and/or climate-driven change in turnover rates are the only candidate causes that we need to unravel. (The concept of climatic forcing of taphonomic bias or turnover includes the possibility that tectonic changes forced the climatic changes.)

Acknowledgments

I thank Nick Shackleton for allowing me to use a version of his oxygen isotope data (Shackleton, this vol.) in figure 3.2. I received very useful comments from John Barry, Lloyd Burckle, George Denton, Jelle Reumer, Tim White, and Bernard Wood. Susan Hochgraf prepared the illustrations.

Notes

1. A quantitative exploration of this model is currently being carried out by E. Vrba with K. Lindgren and M. Nordahl at the University of Göteborg in Sweden.
2. Some quotations on parts of the hypothesis that I expand and revise are given in appendix 3.2.

References

Behrensmeyer, A. K., and Hill, A. P. (eds.). 1980. *Fossils in the making.* University of Chicago Press, Chicago.

Bush, G. L. 1975. Modes of animal speciation. *Ann. Rev. Ecol. Syst.* 6:339–364.

Carson, H. L. 1982. Speciation as a major reorganization of polygenic balances. In *Mechanisms of speciation,* pp. 411–433 (ed. C. Barigozzi). Alan R. Liss, New York.

———. 1987. The process whereby species originate. *BioScience* 37:715–720.

Carson, H. L., Hardy, D. E., Spieth, H. T., and Stone, W. S.

(1970). The evolutionary biology of the Hawaiian Drosophilidae. In *Essays in evolution and genetics in honour of Theodosius Dobzhansky,* pp. 437–543 (ed. M. K. Hecht and W. C. Steere). Appleton-Century-Crofts, New York.

Carson, H. L., and Templeton, A. R. 1984. Genetic revolutions in relation to speciation phenomena: The founding of new populations. *Ann. Rev. Ecol. Syst.* 15:97–131.

Charlesworth, B. R., Lande, B. R., and Slatkin, M. 1982. A neo-Darwinian commentary on macroevolution. *Evolution* 36:474–498.

Coope, G. R. 1979. Late Cenozoic fossil Coleoptera: Evolution, biogeography, ecology. *Ann. Rev. Ecol. Syst.* 10:247–267.

Coope, G. R., and Brophy, J. A. 1972. Late glacial environmental changes indicated by a coleopteran succession from North Wales. *Boreas* 1:97–142.

Croizat, L., Nelson, G., and Rosen, D. E. 1974. Centers of origin and related concepts. *Syst. Zool.* 23:265–287.

Darwin, C. 1839. *Journal of researches into the natural history and geology of the countries visited by* HMS Beagle, *under the command of Captain FitzRoy, R. N. from 1832 to 1836.* Henry Colburn, London.

———. 1859. *On the origin of species by means of natural selection, or the preservation of favoured races in the struggle for life.* John Murray, London.

Eldredge, N., and Gould, S. J. 1972. Punctuated equilibria: An alternative to phyletic gradualism. In *Models in paleobiology,* pp. 82–115 (ed. T. J. M. Schopt). W. H. Freeman, San Francisco.

Futuyma, D. J., and Mayer, G. C. 1980. Non-allopatric speciation in animals. *Syst. Zool.* 29:254–271.

Gleick, J. 1987. *Chaos: Making a new science.* Viking, New York.

Gould, S. J., and Eldredge, N. 1993. Punctuated equilibrium comes of age. *Nature* 366:223–227.

Grubb, P. 1978. Patterns of speciation in African mammals. *Bull. Carn. Mus. Nat. Hist.* 6:152–167.

Gulick, J. T. 1872. On the variation of species as related to their geographical distribution, illustrated by the *Achatinellidae. Nature* 6:222–224.

———. 1905. *Evolution, racial and habitudinal,* Publ. no. 25. Carnegie Institution of Washington, Washington, D.C.

Haffer, J. 1981. Aspects of neotropical bird speciation during the Cenozoic. In *Vicariance biogeography,* pp. 371–394 (ed. G. Nelson and D. E. Rosen). Columbia University Press, New York.

Hare, P. E. 1980. Organic geochemistry of bone and its relation to the survival of bone in the natural environment. In *Fossils in the making: Vertebrate taphonomy and paleoecology,* pp. 208–219 (ed. A. K. Behrensmeyer and A. P. Hill). University of Chicago Press, Chicago.

Howard, W. R. 1985. Late Quaternary southern Indian Ocean circulation. *S. Afr. J. Sci.* 81:253–254.

Huntley, B., and Webb, T., III. 1989. Migration: Species' response to climatic variations caused by changes in the earth's orbit. *J. Biogeog.* 16:5–19.

Hut, P., Alvarez, W., Elder, W. P., Hansen, T., Kauffman, E. G., Keller, G., Shoemaker, E. M., and Weissman, P. R. 1987. Comet showers as a cause of mass extinctions. *Nature* 329:118–126.

Klein, R. G. 1984. The large mammals of southern Africa: Late Pliocene to Recent. In *Southern African prehistory and paleoenvironments,* pp. 107–146 (ed. R. G. Klein). A. A. Balkema, Rotterdam.

Kukla, G. 1987. Loess stratigraphy in central China. *Quat. Sci. Rev.* 6:191–219.

Kurten, B. 1957. Mammal migrations: Cenozoic stratigraphy, and the age of Peking man and the australopithecines. *J. Paleontol.* 31:215–227.

Lande, R. 1980. Genetic variation and phenotypic evolution during allopatric speciation. *Am. Nat.* 116:463–479.

———. 1985. Expected time for random genetic drift of a population between stable phenotypic rates. *Proc. Natl. Acad. Sci.* 82:7641–7645.

———. 1986. The dynamics of peak shifts and the pattern of morphological evolution. *Paleobiology* 12:343–354.

McKenna, M. C. 1983. Holarctic landmass rearrangement, cosmic events, and Cenozoic terrestrial organisms. *Ann. Missouri Bot. Gard.* 70:459–489.

Mandelbrot, B., Laff, M., and Hubbard, J. H. 1979. Fractals and the rebirth of iteration theory. In *The beauty of fractals,* pp. 151–160 (ed. H. Peitgen and P. H. Richter). Springer-Verlag, Berlin.

Marshall, L. G., Webb, S. D., Sepkoski, J. J., and Raup, D. M. 1982. Mammalian evolution and the great American interchange. *Science* 215:1351–1357.

Maynard Smith, J. 1981. Sympatric speciation. *American Naturalist* 100:386–392.

———. 1989. *Evolutionary genetics.* Oxford University Press, Oxford.

Mayr, E. 1942. *Systematics and the origin of species.* Columbia University Press, New York.

———. 1963. *Animal species and evolution.* Harvard University Press, Cambridge.

Newman, C. M., Cohen, J. E. and Kipnis, C. 1985. Neo-darwinian evolution implies punctuated equilibria. *Nature* 315:400–401.

Paterson, H. E. H. 1978. More evidence against speciation by reinforcement. *S. Afr. J. Sci.* 74:369–371.

———. 1982. Perspective on speciation by reinforcement. *S. Afr. J. Sci.* 78:53–57.

———. 1985. The recognition concept of species. In *Species and speciation,* Transvaal Museum Monograph 4, pp. 21–34 (ed. E. S. Vrba). Transvaal Museum, Pretoria.

Raup, D. M. 1966. Geometric analysis of shell coiling: General problems. *J. Paleont.* 40:1178–1190.

Raup, D. M., and Sepkoski, J. J. 1984. Periodicity of extinctions in the geologic past. *Proc. Natl. Acad. Sci. U.S.A.* 81:801–805.

Rosenzweig, M. L. 1978. Competitive speciation. *Biol. J. Linn. Soc.* 10:275–289.

Ruddiman, W. F., and Raymo, M. 1988. Northern Hemisphere climate regimes during the last 3 million years: Possible tectonic connections. *Phil. Trans. Roy. Soc. Lond.*, ser. B, 318:411–430.

Savage, D. E., and Russell, D. E. 1983. *Mammalian paleofaunas of the world.* Addison-Wesley, London.

Schliewen, U. K., Tautz, D., and Paabo, S. 1994. Sympatric spe-

ciation suggested by monophyly of crater lake cichlids. *Nature* 368:629–632.
Signor, P. W., and Lipps, J. H. 1982. Sampling bias, gradual extinction patterns and catastrophes in the fossil record. *Geol. Soc. Am. Special Paper* 190:291–296.
Stenseth, N. C., and Maynard Smith, J. 1984. Coevolution in ecosystems: Red Queen evolution or stasis. *Evolution* 38:870–880.
Sutcliffe, A. J. 1985. *On the track of Ice Age mammals.* Harvard University Press, Cambridge.
Templeton, A. R. 1981. Mechanisms of speciation: A population genetic approach. *Ann. Rev. Ecol. Syst.* 12:23–48.
Thomas, H. 1984. Les Bovidae (Artiodactyla: Mammalia) du Miocène du sous-continent indien, de la peninsule arabique et de l'Afrique, biostratigraphie, biogéographie et écologie. *Palaeo., Palaeo., Palaeo.* 45:251–299.
Turner, J. R. G. 1986. The genetics of adaptive radiation: A neo-Darwinian theory of punctuated equilibrium. In *Patterns and processes in the history of life,* pp. 183–207 (eds. D. M. Raup and D. Jablonski). Springer-Verlag, Heidelberg.
Van Valen, L. 1973. A new evolutionary law. *Evol. Theory* 1:1–30.
Vrba, E. S. 1985. Environment and evolution: Alternative causes of the temporal distribution of evolutionary events. *S. Afr. J. Sci.* 81:229–236.
———. 1987. Ecology in relation to speciation rates: Some case histories of Miocene-recent mammal clades. *Evol. Ecology* 1:283–300.
———. 1992. Mammals as a key to evolutionary theory. *J. Mamm.* 73:1–28.
———. 1993. Turnover-pulses, the Red Queen, and related topics. *Am. J. Sci.* 293-a:418–452.
Wake, D. B., Roth, G., and Wake, M. H. 1983. On the problem of stasis in organismal evolution. *J. Theoret. Biol.* 101:211–224.
Wegener, A. 1924. *The origin of continents and oceans.* Trans. J. G. A. Skerl. Methuen, London.
White, M. J. D. 1978. *Modes of speciation.* W. H. Freeman, San Francisco.
Wright, S. 1932. The roles of mutation, inbreeding, crossbreeding, and a selection in evolution. *Proceedings of the Sixth International Congress on Genetics* 1:356–366.
———. 1967. Comments on the preliminary working papers of Eden and Waddington. In *Mathematical challenges to the neo-Darwinian theory of evolution,* pp. 117–120 (eds. P. S. Moorehead and M. M. Kaplan). Wistar Inst. Symp., 5.
———. 1982. Character change, speciation, and the higher taxa. *Evolution* 36:427–443.
———. 1988. Surfaces of selective value revisited. *Am. Nat.* 131:115–123.

Appendix 3.1. Glossary

habitat: A habitat of an organism or species includes the places and the resources that are necessary for life of that organism or species. Habitat specificity of an organism or species refers to the resource requirements of that organism or species.

resources: Any components of the environment that can be used by an organism in its metabolism and activities are considered resources, including temperature, light, inorganic ions and molecules, pH, salinity, relative humidity, stream flow velocity, substrate characteristics, all kinds of organic foods (such as prey), mates, and other mutualist organisms in the same or different species; and places for living, nesting and sheltering.

turnover: Used without qualification, this term refers to the actual appearances and disappearances of taxa from particular areas in the real world through a) evolutionary changes and b) migration. In my discussion of distorted first and last taxic appearances in the fossil record, that is, temporal patterns that do not correspond to the actual times of evolutionary and migratory turnover, I use such terms as *artifactual turnover.* Turnover can theoretically be initiated by a large variety of causes derived from physical change, from the biotic interactions (such as competition and predation), or from both. *Evolutionary turnover* of species includes i) *speciation* by lineage splitting and final disappearance of species after terminal *extinction;* and ii) *anagenetic species turnover* as new morphologies appear and old ones disappear as a result of evolutionary change in unbranching lineages. *Migratory turnover* includes the appearance of a species in a given area by *immigration* after a previous duration elsewhere, and the local disappearance of a species by *emigration* to persist elsewhere.

Appendix 3.2. Turnover Pulses: Previous Statements (Vrba, 1993) That Have Been Partly Revised and Expanded in This Chapter

1. *Basic prediction.* If nearly all speciation and extinction requires initiation by physical changes, "most lineage turnover has occurred in pulses, nearly synchronous across diverse groups of organisms, and in predictable synchrony with changes in the physical environment" (p. 428).

2. *A previous definition* does not do justice to the concept: "turnover-pulse—concentration of turnover events against time. For example, if a high number of first and last records of species in different lineages occur together within a time interval of 100,000 years or less, preceded and postdated by a million years of predominant stasis in the same monophyletic groups, I would regard this as evidence of a turnover-pulse" (p. 449).

3. *On turnover pulses at different scales of magnitude and time.* "Most turnover pulses are small peaks involving few lineages and/or restricted geographic areas. Some of them are massive and of global extent" (p. 428). The possibility exists "that episodic physical changes did indeed produce turnover-pulses but with very short intervals in between. For instance, turnover-pulses might coincide with the Milankovitch cyclic extremes, the longest of which has a period of only about 100,000 yrs. Turnover pulses with such short time spacing may well be discernible in some data sets, such as the Plio-Pleistocene record of foraminifers. But in most records their signatures would be indistinguishable from a random blur" (p. 432).

4. *Major physical changes and turnover pulses.* "Major global climatic changes, that in the climatic record occurred from one to several million years apart, accounted for the vast majority of speciation [and extinction] events" (p. 432).

5. *Biotic interactions participate in turnover pulses.* The hypothesis "allows the possibilities of gradual microevolution, coevolution, competition, and other proximal biotic interactions but reserves the *initiation* of turnover for physical changes. Thus, the disappearance of grass species and the appearance of bush with a

new set of parasites and diseases might be crucial proximal causes of turnover of grassland-adapted herbivores. But under the turnover-pulse hypothesis the question would be: what physical changes caused the biotic changes in the first place?" (p. 428).

6. *Lineages respond differently based on their intrinsic properties and on where they occur.* "Environmental change has a deterministic role in that turnover would not occur without it. But other, internal factors contribute to the *probability* of speciations and together with local biotic interactions must provide strong proximal causes of the *nature* of speciation change" (p. 432). "Differences between lineages [in turnover response] may be causally influenced by intrinsic biological factors" (p. 431). Not "all geographic areas and similar biotas in different areas are affected equally by the same global climatic change" (p. 431). "Geographic factors determine the spatial distribution of turnover events during a pulse" (p. 432).

Appendix 3.3. Possible Causes of a Pulse of FADs in the Fossil Record

In tables 3.1 and 3.2 below, the following symbols for possible causal factors of an observed significant FAD pulse are used:

A = A significant increase in the *Accumulation Rate* of first appearances (FADs) of distinct morphologies in the fossil record under study. A is assumed in each case considered in tables 1 and 2. It is the result that needs explanation.
S = A significant increase in *Speciation Rate and/or Immigration Rate* has contributed crucially to observation A. Speciation here includes both lineage splitting and unbranching origin or new morphologies and recognizes that a new species may have originated in the area under study or speciated just before or during immigration to that area. ("Crucial" here means that without S, A would not have occurred.)
T = A significant change in *Taphonomic Biases* has contributed crucially to observation A. For instance, species that were long previously present in the area under study may appear in the record only by the time of observation A because preservational potential increased by then.
B = A significant change in biotic interactions, including competition, predation, and disease, has contributed crucially to observation A.
C = Climatic change is a crucial initiating cause of A.

Assumptions in setting up different potential causal permutations in tables 3.1 and 3.2 include the following:

1. T and S are always intermediate causes that must each be initiated by either C or B. One could plausibly think of factors other than C or B that might initiate particular taphonomic biases or speciation and migration events. But it is difficult to come up with causes other than C or B for an observed significant FAD *pulse across lineages.*
2. B cannot cause A directly. That is, $B \rightarrow A$ is impossible. But B can cause either T or S, and thereby result in C.

The result of the assumptions is that some categories of combinations of cause and effect in the tables remain empty. Some of the entries under (a) through (l) in table 3.1 are highly unlikely to occur. The entry $B \rightarrow S \rightarrow A$ implies the onset of a new set of biotic interactions that were not caused by any climatic changes that might have occurred and that on their own sufficed to precipitate an unusually high rate of speciation as the proximal cause of the FAD pulse. I am not aware of any biological theory and observations that support such a proposition. Something similar is true for the entry $B \rightarrow T \rightarrow A$ in table 3.1. Given the same species composition and climate, for instance, an owl species might discover a rodent species as a new item of prey and therefore influence a taphonomic

Table 3.1. Possible crucial causes of an observed significant increase in FADs: the broader version. ("Crucial" here means that without that cause, the significant increase in FADs would not have occurred.) See text for details of letter codes: A = FAD pulse; S = increase in speciation rate; T = change in taphonomic bias; B = change in biotic interactions; C = climate change; arrows denote causal directions; no = the factor is not a significant cause of A; yes = the factor is a significant cause of A. The causal chains under cases (a)-(l) have to be combined to conform with the stated conditions. For instance, under (a), $C \rightarrow A$ could not be the sole cause but would have to be accompanied by $C \rightarrow S \rightarrow A$ and $B \rightarrow T \rightarrow A$ or by another combination that satisfies the requirements of category (a): S yes, T yes, C yes, B yes.

	Intermediate causes of A			
Initiating causes of A	S yes T yes	S yes T no	S no T yes	S no T no
C yes B yes	(a) $C \rightarrow S \rightarrow A$ $C \rightarrow T \rightarrow A$ $B \rightarrow S \rightarrow A$ $B \rightarrow T \rightarrow A$ $C \rightarrow B \rightarrow S \rightarrow A$ $C \rightarrow B \rightarrow T \rightarrow A$	(b) $C \rightarrow S \rightarrow A$ $B \rightarrow S \rightarrow A$ $C \rightarrow B \rightarrow S \rightarrow A$	(c) $C \rightarrow T \rightarrow A$ $B \rightarrow T \rightarrow A$ $C \rightarrow B \rightarrow T \rightarrow A$	(d) empty
C yes B no	(e) $C \rightarrow S \rightarrow A$ $C \rightarrow T \rightarrow A$	(f) $C \rightarrow S \rightarrow A$	(g) $C \rightarrow T \rightarrow A$	(h) empty
C no B yes	(i) $B \rightarrow S \rightarrow A$ $B \rightarrow T \rightarrow A$	(j) $B \rightarrow S \rightarrow A$	(k) $B \rightarrow T \rightarrow A$	(l) empty

bias affecting that rodent species. But the leap from that notion to one of a geologically sudden increase in biotic interactions of *many lineages,* culminating in a taphonomically enhanced FAD *pulse,* is very farfetched. If this notion is accepted, then the initiation of a significant FAD pulse through a change in taphonomy and/or in speciation rate can come only from climatic change. That is, climatic change can result in biotic and/or taphonomic bias, which in turn results in a FAD pulse. But neither biotic bias nor taphonomic bias, nor a combination of these two, is sufficient in the absence of climatic change. Let us also allow the statement $C \rightarrow S \rightarrow A$ to include $C \rightarrow B \rightarrow S \rightarrow A$, and $C \rightarrow T \rightarrow A$ to include $C \rightarrow B \rightarrow T \rightarrow A$. Then the problem of distinguishing the potential causes of an observed episode of increase in FADs reduces to the matrix in table 3.2.

Table 3.2. A restricted range of possible crucial causes of an observed significant increase in FADs. ("Crucial" here means that without that cause, the significant increase in FADs would not have occurred.) The cases in table 3.1 that are highly unlikely or redundant (according to my arguments given in this appendix) have been omitted.

	Taphonomic bias (T)	
Speciation rate (S)	T yes	T no
S yes	(a) $C \rightarrow S \rightarrow A$ turnover pulse and $C \rightarrow T \rightarrow A$ taphonomic pulse	(b) $C \rightarrow S \rightarrow A$ turnover pulse
S no	(c) $C \rightarrow T \rightarrow A$ taphonomic pulse	(d) empty

Part Two

The Middle to Late Miocene

Chapter 4

A Review of Polar Climatic Evolution during the Neogene, Based on the Marine Sediment Record

James P. Kennett

Knowledge of the climatic, paleoceanographic, and cryospheric evolution of the polar regions during the Cenozoic is crucial to a broader understanding of global environmental history and modern oceanographic and climatic processes. The changing character of the Cenozoic world during the last 50 million years (myr) is marked by global climatic and associated biotic changes driven primarily by progressive and sequential development of the polar cryosphere. The evolution of the terrestrial and marine biota over broad areas of the globe has been strongly influenced by these climatic changes.

This chapter is a brief overview of the Neogene paleoenvironmental evolution of the polar regions. Information was compiled largely from the results and interpretations of many workers who have studied deep-sea sediment sequences from high to middle southern latitudes. A significant body of sedimentary, fossil, and isotopic evidence (fig. 4.1) indicates a sequential cooling and cryospheric development of Antarctica and the Southern Ocean during much of the Cenozoic, followed, in the latest Neogene, by the development of major ice sheets in the Northern Hemisphere. These events profoundly affected the course of global oceanic and atmospheric circulation, sediment deposition in the oceans, and biotic evolution during the Cenozoic. Cooling and cryospheric development did not proceed uniformly but were punctuated by periods that represent more rapid transitions from one climatic state to another. These transitions represent threshold events that incorporated strong, positive feedback mechanisms that advanced the climate system to a new steady state. Important cooling steps occurred during the Middle to Late Eocene, the earliest Oligocene, the Middle Oligocene, the Middle Miocene (ca. 14.5–14 myr), the Late Miocene (ca. 6.5–5 myr), and the Late Pliocene (3.2–2.7 myr). Further important steps that occurred during the Quaternary are not summarized here.

Major accumulation of ice on East Antarctica began during the earliest Oligocene. This early ice sheet exhibited much variability throughout the Oligocene and Early Miocene. During the Neogene there were three major episodes of polar cryospheric expansion: the development of the East Antarctic ice sheet to a near-permanent state in the Middle Miocene, the development of the West Antarctic ice sheet during the Late Miocene, and the development of Northern Hemispheric ice sheets during the Late Pliocene and Quaternary (fig. 4.1). Reversals were uncommon in the cooling trend associated with this Cenozoic evolution. Distinct warming episodes are known to have occurred during the latest Oligocene and the Early Miocene. Distinct Early Pliocene warmth occurred in the high northern latitudes and the temperate regions, but its expression appears to have been mild in the Southern Ocean and the Antarctic continent. The development of the polar cryosphere during the Neogene caused fundamental changes in global oceanic and atmospheric circulation that, in turn, caused a succession of increasing levels of aridification in the lower latitudes. This drying trend strongly affected the evolution of the terrestrial biota.

Numerous criteria have been employed to help reconstruct the history of climatic development in the Antarctic and Arctic regions. They include: the analysis of oxygen and carbon isotopes and other inorganic and organic geochemical tracers; the distribution of ice-rafted

ISBN 0-300-06348-2.

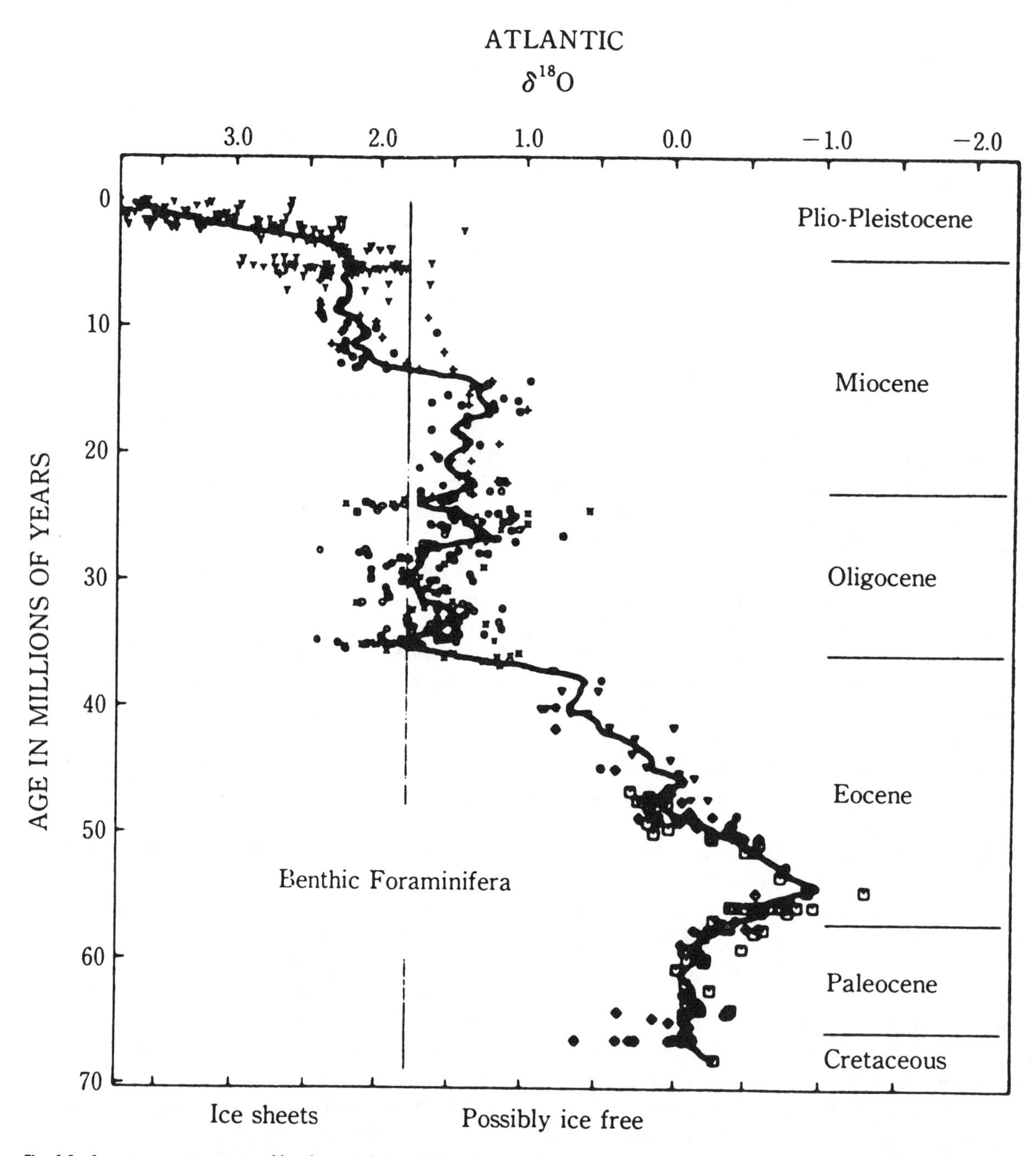

Fig. 4.1. Cenozoic oxygen isotopic record based on compilation of $\delta^{18}O$ values from deep-sea benthic foraminifera. The long-term increase in $\delta^{18}O$ since the Early Eocene reflects cooling of the deep ocean and growth of ice sheets at high latitudes. Note sharp increases in $\delta^{18}O$ during the earliest Oligocene, the Middle Miocene, and the Late Pliocene. Oxygen isotopic values are relative to Pee Dee Belemnite (PDB) (after Miller et al., 1987).

sediments and terrigenous sediments transported into the ocean by other processes; the character of biogenic sediments; and data from a wide range of marine microfossils and pollen and spores incorporated into marine sediments. The use of any one criterion for making paleoenvironmental interpretations is not an optimal approach because of the limitations imposed by each method. Paleoenvironmental interpretations are more reliable when a range of independent criteria have been collected, an approach used in the following summary.

The Development of Polar Paleoclimate during the Neogene

Early Miocene

During most of the Early Miocene (ca. 23–16 myr) $\delta^{18}O$ values were relatively low, indicating relative warmth and relatively low global ice volume. The lowest values for the entire Neogene were reached during the late Early to earliest Middle Miocene (ca. 19.5–15 myr), representing the climax of Neogene warmth. A range of paleontological evidence also indicates maximum warmth at the end of the Early Miocene (Kennett and von der Borch, 1985). In modern temperate southern latitudes, warm subtropical mollusca and foraminiferal assemblages spread southward (Hornibrook, 1992). Planktonic foraminiferal faunas then reached their peak of Miocene diversity (Jenkins, 1973), and calcareous nannofossil assemblages in the South Pacific (Edwards, 1968) and the Atlantic (Haq, 1980) reflect maximum Neogene warmth.

During the Early Miocene, Southern Ocean biosiliceous sediments continued to increase in importance, a trend that had begun during the Oligocene (Kennett and Barker, 1990). This trend reflects a continuing increase in intensity of upwelling in the Southern Ocean, resulting from increased wind strength in this region. One might expect that Early Miocene global warmth would lead to decreased upwelling in the Antarctic. It seems likely, therefore, that the development of the Antarctic Circumpolar Current at the beginning of the Neogene (Kennett, 1977) and the associated Polar Front Zone (Kennett, 1977; Lazarus and Caulet, 1993) had sufficiently isolated the climatic system of the Antarctic region from major climatic change to the north.

Antarctic waters continued to cool, as shown by a decrease in calcareous planktonic diversity. Further cooling is also indicated by an increase in the deposition of ice-rafted detritus in the ocean off Dronning Maud Land (East Antarctica), which is suggestive of additional ice accumulation near the margins of East Antarctica at this time. This increase in ice accumulation perhaps resulted from higher precipitation on Antarctica during the warmer global conditions (Grobe et al., 1990). A temporary return to slightly higher planktonic microfossil diversity during the later part of the Early Miocene reflects temporary warming, although this warming was more damped than that in lower latitudes, perhaps because of the developing thermal isolation of the Antarctic region.

The Middle Miocene

The early Middle Miocene (15.6–12.5 myr) was a critical threshold in Cenozoic climatic evolution, a time of renewed cooling at high latitudes and in the deep ocean, of major expansion of the East Antarctic ice sheet, and of important changes in deep-water circulation. This was one of the most significant intervals of cryospheric and climatic development of the Cenozoic and reflects the transition from the relative warmth of the Early Miocene to the colder climates of the Middle Miocene and later periods. In many respects the Middle Miocene marks the onset of the Late Neogene mode of climatic and oceanic circulation, marked by strong meridional and vertical thermal gradients and dominated by high-latitude, cold, deep-water sources (Kennett, 1977; Miller et al. 1987; Woodruff and Savin, 1991; Flower and Kennett, 1993).

The Middle Miocene is marked by a major increase in $\delta^{18}O$ values in deep-sea sequences (fig. 4.1), generally interpreted to represent major Antarctic ice growth and cooling of bottom waters (Shackleton and Kennett, 1975a; Savin et al., 1975; Kennett, 1985a). A combination of cooling at high latitudes and Antarctic ice growth caused a series of increases in oxygen isotopic values in deep-sea sequences between ca. 15 and 12.5 myr (fig. 4.2; Flower and Kennett, 1993). The total increase of 1.2 ‰ in the benthic oxygen isotopes during the Middle Miocene was largely incorporated in two steps, an initial increase of 0.8‰ from 14.5 to 14.0 myr and a second of 0.7‰ from 13.5 to 12.45 myr. A strong covariance between planktonic and benthic foraminiferal isotopic records during the first of these steps suggests major, permanent growth of the East Antarctic ice sheet (Flower and Kennett, 1993). This interval (ca. 14 myr)

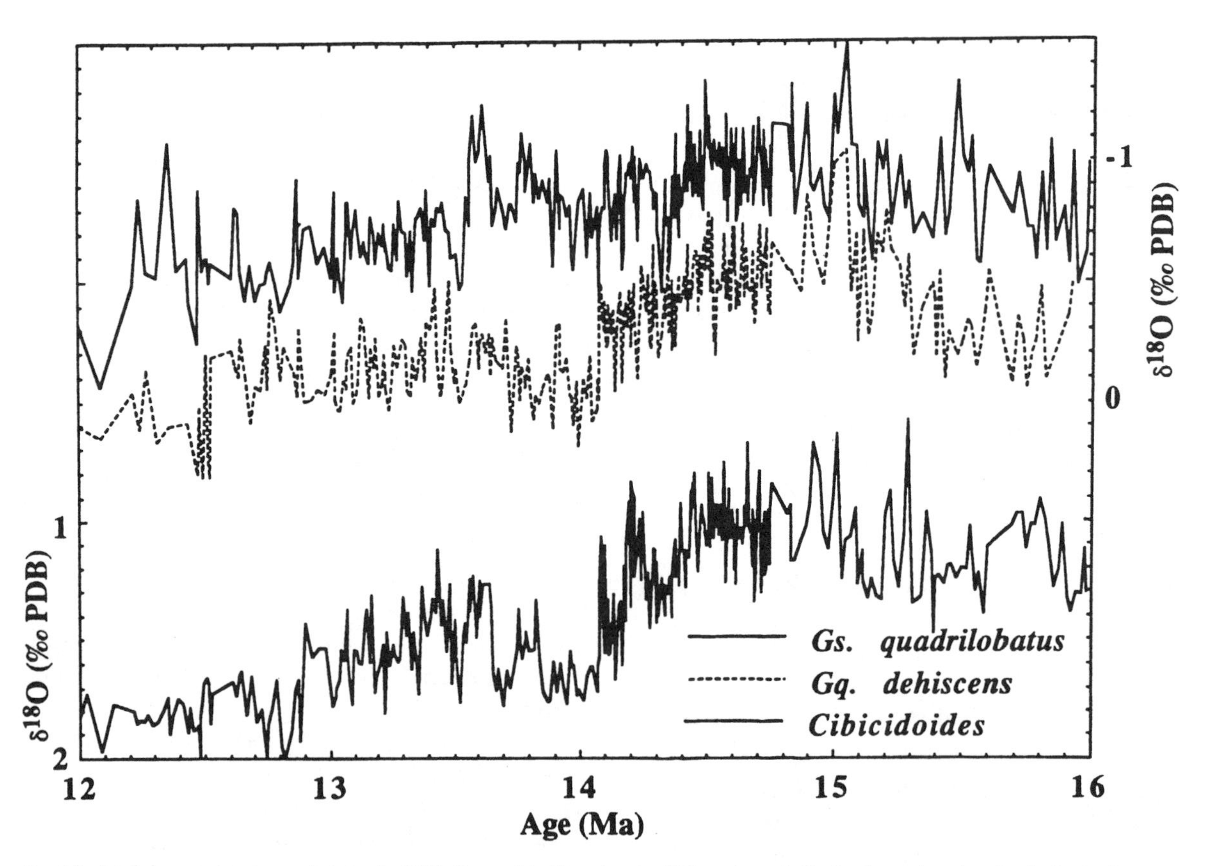

Fig. 4.2. Detailed oxygen isotopic records during the Middle Miocene (16–12 myr) at site 588A, southwest Pacific. Trends are shown for a benthic foraminifer, *Cibicidoides;* a shallow-dwelling planktonic foraminifer, *Globigerinoides quadrilobatus;* and a deeper-dwelling planktonic foraminifer, *Globoquadrina dehiscens.* Oxygen isotopic values are relative to Pee Dee Belemnite (PDB). Steps representing increased values after 14.5 myr were caused by cooling at high latitudes and major growth and increased stability of the East Antarctic ice sheet (after Flower and Kennett, 1993).

also marks the beginning, during the Neogene, of a trend toward lower sea level (Haq et al., 1987, 1988).

An increase in ice-rafted detritus and other sedimentological changes near East Antarctica during the Middle Miocene represents additional evidence for increased ice accumulation on East Antarctica (Kennett and Barker, 1990). Similar changes are not evident in the ocean adjacent to West Antarctica, where there is a lack of ice-rafted detritus in the Neogene sequence until the latest Miocene (Barker, Kennett et al., 1988). Siliceous biogenic sedimentation continued to expand around Antarctica, reflecting further intensification in the upwelling system. This trend was almost certainly in response to a strengthening of atmospheric circulation (Brewster, 1980). A Middle Miocene increase in upwelling has also been inferred from increased deposition of biosiliceous sediments at the Subtropical Divergence, well to the north of Antarctica (Stein and Robert, 1985). Global deep-ocean circulation also underwent important changes during the Middle Miocene, including a reduction of deep-water production in the middle latitudes and subsequent dominance of deep waters produced at high latitudes (Schnitker, 1980; Woodruff and Savin, 1989; Wright et al., 1992; Flower and Kennett, 1993, 1994).

During the Middle to Late Miocene, calcareous nannofossil and planktonic foraminiferal assemblages, when present, became essentially monospecific, reflecting further cooling of Antarctic surface waters. This cooling was also evident in changes in planktonic assemblages at middle latitudes (Jenkins, 1973; Haq, 1980; Thunell and Belyea, 1982) and in the disappearance of subtropical benthic foraminifera in New Zealand (Hornibrook, 1992). The Middle Miocene expansion of the East Antarctic ice sheet was associated with a number of other changes, including a marked steepening of the planetary temperature gradient (Savin et al., 1975, 1985; Loutit et al., 1983), a cooling of oceanic deep waters as indicated by the development of the characteristic Late Neogene deep-sea benthic foraminiferal assemblages (Woodruff et al., 1981; Thomas, 1985; Woodruff, 1985; Kurihara and Kennett, 1985, 1992; Boersma, 1985; Miller and Katz, 1987), an increase in hiatus abundance (van Andel et al., 1975), and a shoaling of the calcium carbonate compensation depth (CCD) in the equatorial Pacific (Berger and Winterer, 1974).

Marine sediments in the South Pacific also record significant changes in the terrestrial climate of Australia during the early Middle Miocene. Changes in clay mineral assemblages in wind-blown sediments transported to the deep sea east of Australia record increased aridity in northern and central Australia (Stein and Robert, 1985). The first appearance, during the Middle Miocene, of opal phytoliths in deep-sea sediments to the east of Australia suggests the initial development of open grasslands in Australia, almost certainly related to increasing continental dryness (Locker and Martini, 1985). Cooling and increased aridity during the early Middle Miocene seem to have affected terrestrial biotas on other continents. Initial development of grasslands and grazing-adapted faunas seem to have occurred in South America (Pascual and Jaureguizar, 1990) and Africa (Retallack, 1992).

The Late Miocene

Although deep-sea benthic $\delta^{18}O$ values fluctuated during much of the Late Miocene (ca. 12–6.5 myr), average values remained consistently higher than in the Early Miocene, reflecting a prolonged interval of cool climate. The oxygen isotopic record, nevertheless, indicates two distinct cooling episodes close to the Middle-Late Miocene boundary. The earliest of these events occurred between 12.5 and 11.5 myr in the latest Middle Miocene. The second cooling event occurred from 11 to 9 myr and is marked by some of the highest $\delta^{18}O$ values of the entire Miocene. In New Zealand sequences, this interval is marked by distinct lithostratigraphic changes attributed by Wright and Vella (1988) to a glacioeustatic drop in sea level. The oxygen isotopic record indicates that this interval was followed by relative warmth during the middle Late Miocene (9–7 myr; Kennett, 1985a, 1985b), a trend noted also by Haq (1980) based on changes in Atlantic calcareous nannofossil distribution.

After the middle Late Miocene (ca. 8 myr), biosiliceous (diatomaceous) sediments dominated the Maud Rise sequence, near East Antarctica, reflecting high siliceous biogenic productivity and deposition resulting from vigorous oceanic upwelling (Kennett and Barker, 1990). Further increases also occurred in the abundance of ice-rafted detritus in sediment sequences adjacent to East Antarctica, indicating an expansion of iceberg activity (Grobe et al., 1990). There is no evidence to suggest major deglaciation of the East Antarctic ice sheet during the Late Miocene. Brief intervals of climatic amelioration occurred at about 10.5 and 8.5 myr in the Weddell Sea region (Maud Rise), as indicated by the deposition of calcareous nannofossils and planktonic foraminifera. The fossil assemblages in these calcareous deposits indicate that surface-water temperatures were warmer than 3°C (Burckle and Pokras, 1991) and perhaps as warm as about 5°–6°C (S. Wise, pers. comm.). Absence of calcareous nannofossils during the remaining Late Neogene suggests that surface-water temperatures near Antarctica were less than 3°C.

The distribution of planktonic foraminiferal faunas in the North Pacific indicates that gyral circulation and the Kuroshio Current had strongly intensified by the late Miocene, as compared with the Early Miocene, when the system was relatively weak (Kennett et al., 1985). This intensification was in part related to the steepening of the meridional temperature gradient during the Middle and Late Miocene, which increased climatic zonality and led to greater latitudinal provincialism in planktonic communities. Antarctic and Subantarctic surface waters continued to advance northward with the expansion of the Antarctic Circumpolar Current and further cooling of the Antarctic region.

The Antarctic cryosphere continued to expand during the Late Miocene with the development of the West Antarctic ice sheet that also then grounded below sea level (Ciesielski et al., 1982; Kennett and Barker, 1990). This ice sheet began to form at ca. 10–9 myr, causing increased deposition of hemipelagic sediments and ice-rafted detritus on the adjacent continental margins, rapid deposition of turbidite sequences in the Weddell and Bellingshausen abyssal plains, and a northward expansion of ice-rafted detritus. The sedimentological evidence reflects major instability of the West Antarctic ice sheet during the Late Miocene.

Further changes in clay mineral assemblages and increased rates of accumulation of wind-blown terrigenous sediments in deep-sea sediments east of Australia indicate continued southward expansion of arid areas in Australia during the Late Miocene (Stein and Robert, 1985).

Paleobotanical data also indicate increasing continental aridification (Kemp, 1978; Locker and Martini, 1985).

Terminal Miocene Events

The latest Miocene (ca. 6.5–5 myr) was marked by a general increase in $\delta^{18}O$ values, indicating further cooling at high to middle latitudes and additional ice accumulation on Antarctica (Shackleton and Kennett, 1975b; Hodell et al., 1986; Hodell and Kennett, 1986). This cooling episode is associated with global marine regression (Kennett, 1967; Haq et al., 1987, 1988) that almost certainly led to the isolation and desiccation of the Mediterranean Basin (Ryan et al., 1974; Adams et al., 1977; Hsü et al., 1977; Hodell et al., 1986, 1994). High-frequency variations (0.5‰) in $\delta^{18}O$ values during the latest Miocene suggest pulses of ice-sheet expansion and contraction inferred to be associated to periodic grounding and ungrounding of ice on West Antarctica (Hodell and Kennett, 1986). The extent of ice buildup on East Antarctica is unclear, but the terrestrial record suggests that no additional East Antarctic ice sheet expansion occurred during the latest Miocene (Denton et al., 1993). High-latitude cooling during the latest Miocene also led to a distinct and rapid northward movement (300 km) of the biosiliceous sediment province surrounding Antarctica and associated northward movement of the Polar Front Zone (Tucholke et al., 1976).

Many other changes occurred in association with the latest Miocene West Antarctic ice sheet expansion, including widespread cooling at high and middle latitudes (Kennett, 1967; Ingle, 1967; Barron, 1973; Kennett and Vella, 1975; Kennett et al., 1979; Keller, 1980) and major intensification of bottom-water circulation (Kaneps, 1979; Ciesielski et al., 1982). An increase in meridional temperature gradients intensified atmospheric circulation that led to increased wind-driven oceanic upwelling and biogenic fertility of the surface ocean (Vincent et al., 1980). In the polar regions these changes are reflected by increased rates of biosiliceous sediment accumulation (Kennett et al., 1975; Brewster, 1980; Sancetta, 1982). At 6 myr the northern limit of ice-rafted sediment detritus expanded northward into Subantarctic waters of the southwest Atlantic (Warnke et al., 1992), including the Falkland Plateau at 50°S (Ciesielski et al., 1982; Wise et al., 1985). This event probably reflects expansion of the Antarctic ice shelves and floating ice tongues, because tabular icebergs transport much more sediment than do outlet glaciers (Mercer, 1973). In southern Argentina and in Chile, Patagonia glaciers expanded beyond the Andean Mountain front between 7 and 5.2 myr, apparently for the first time (Mercer and Sutter, 1982). This continued expansion of the Antarctic cryosphere created further intensification of oceanic circulation (Ciesielski et al., 1982).

Near the beginning of latest Miocene, a permanent decrease also occurred in the $\delta^{13}C$ of oceanic dissolved inorganic carbon (DIC) values (Loutit and Kennett, 1979; Keigwin, 1979; Keigwin and Shackleton, 1980; Vincent et al., 1980; Haq et al., 1980) and in pedogenic carbonate values (Cerling, 1993). The origin of this Late Miocene carbon isotopic shift is still debated (Hodell et al., 1994), although several workers have proposed that it resulted from increased erosion and transport into the oceans of organic carbon from terrigenous regions during the latest Miocene glacioeustatic regression (Vincent et al., 1980; Loutit and Keigwin, 1982).

The trend toward increased continental aridity continued in New Zealand and Australia during the latest Miocene to Early Pliocene. Pollen and pore sequences in DSDP Site 594, to the east of southern New Zealand, document the replacement of warm, moist conditions of the early Late Miocene with cooler, drier conditions of the latest Miocene (Heusser, 1985). Further aridification of Australia is suggested by increased accumulation of wind-blown terrigenous sediments and opal phytoliths in deep-sea sediments to the east of the continent (Stein and Robert, 1985; Locker and Martini, 1985).

The Early Pliocene

Much evidence exists from regions outside of the Antarctic for significant warming during the Early Pliocene (ca. 5–3 myr) relative to the Late Miocene and the present day (Ingle, 1967; Kennett and Watkins, 1974; Kennett and Vella, 1975; Elmstrom and Kennett, 1985; Kennett, 1967, 1985a, 1985b; McKenzie et al., 1984; Hodell and Kennett, 1986; Sarnthein and Fenner, 1987; Hornibrook, 1992; Cronin, 1991; Dowsett and Poore, 1991; Dowsett et al., 1992). Global marine transgression (Kennett, 1967; Vail and Hardenbol, 1979; Haq et al., 1988) and relatively low $\delta^{18}O$ values in the deep-sea benthic foraminiferal record (fig. 4.3) reflect a decrease in polar-ice volume relative to the present day. Detailed oxygen isotopic studies in northwest Morocco indicate that the warming trend commenced near the end of the

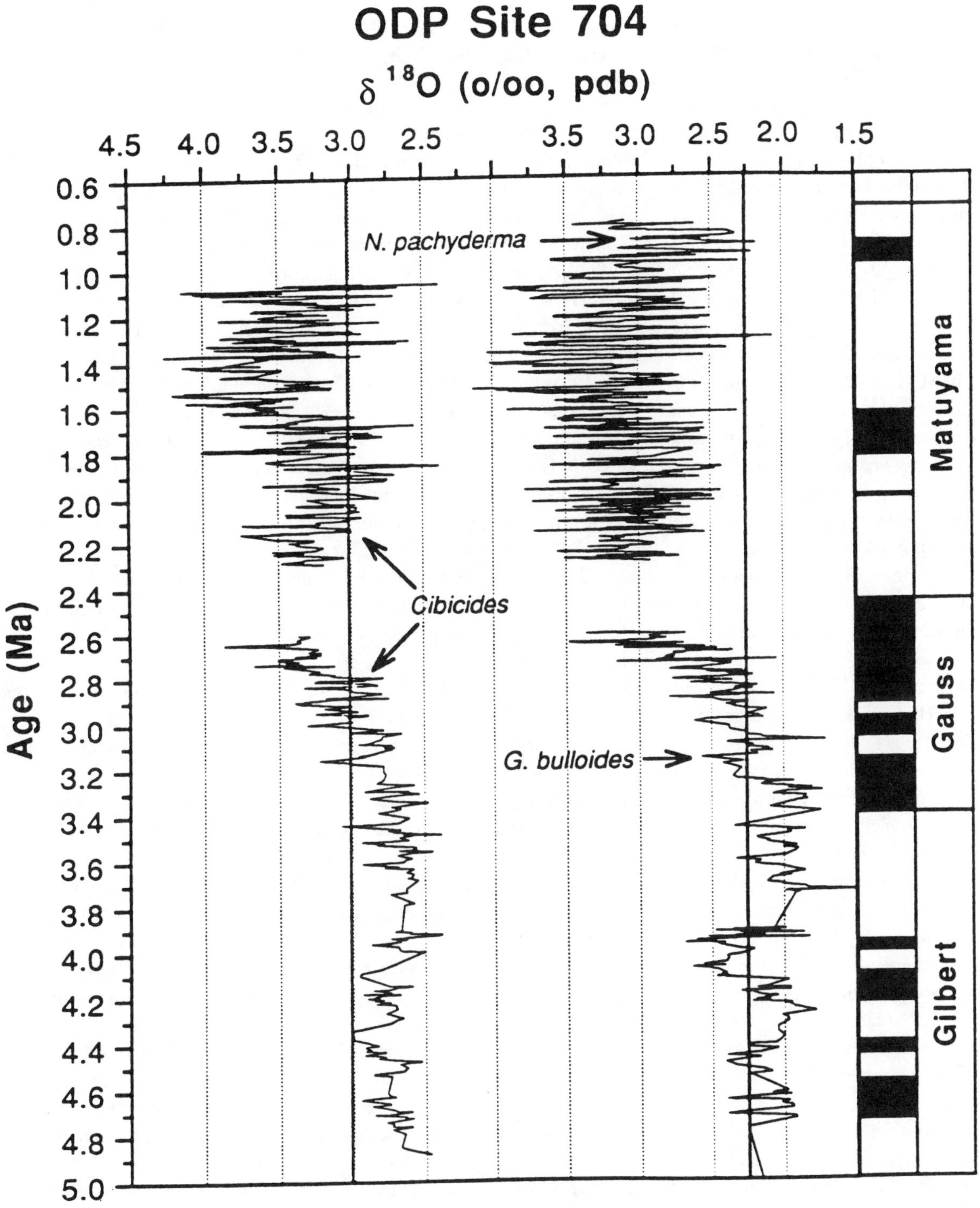

Fig. 4.3. Plio-Pleistocene (5–0.8 myr) oxygen isotopic record of ODP Site 704 (Subantarctic sector of southeast Atlantic Ocean) of the benthic foraminifer (*Cibicidoides*) and planktonic foraminifers (*Globigerina bulloides* and *Neogloboquadrina pachyderma*). Oxygen isotopic values are relative to Pee Dee Belemnite (PDB). The bold vertical lines represent the measured Holocene $\delta^{18}O$ values for the respective taxa at the top of Hole 704. Epoch boundaries and magnetostratigraphic chrons are shown at right (from Hodell and Venz, 1992).

Messinian Stage, ca. 0.3 myr before the Miocene-Pliocene boundary, which is dated at 5.3 myr (Hodell et al., 1994). Glaciers in Patagonia experienced Early Pliocene deglaciation until ca. 3.6 Ma, when renewed glaciation occurred (Mercer and Sutter, 1982).

Unlike interpretations of lower-latitude sequences, differences of opinion exist as to whether, according to the stability hypothesis, the Antarctic ice sheets and cold polar climate withstood the Early Pliocene warmth with relatively little change (Shackleton and Kennett, 1975b; Kennett, 1977; Clapperton and Sugden, 1990) or, according to the deglacial hypothesis (Webb and Harwood, 1991; Barrett et al., 1992), major reductions in ice volume occurred as the result of an episode of remarkable Antarctic warmth. The deglacial hypothesis advocates almost complete removal of the Antarctic ice sheets and the presence of marine epeiric seas deep into the East Antarctic craton, as far south as about 86°S. This hypothesis is rejected for several reasons. First, reconstructions by Drewry (1983), incorporating isostatic rebound upon removal of the ice sheet, show no East Antarctic basins at or below sea level south of 75°S. Indeed, most of the area inferred by Webb and Harwood (1991) to be represented by intracontinental marine basins during the Late Neogene were hundreds of meters above sea level (Drewry, 1983). Second, a wide range of excellent paleoclimatic data summarized by Kennett and Hodell (1993, 1994) from high-latitude marine sediment sequences provides strong evidence for the relative stability of the Antarctic cryospheric-climatic system during the Pliocene. Oxygen isotopic data and an absence of sediment-forming calcareous nannofossils, supported by many other data, indicate that average Antarctic sea-surface temperatures during the warmest Pliocene intervals could not have increased by more than about 3°C (Hodell and Venz, 1992; Burckle and Pokras, 1991; Kennett and Hodell, 1993, 1994). These data also indicate that the Antarctic climate system, as reflected in ocean-water temperatures and ice volume, operated within relatively narrow limits (Hodell and Venz, 1992; Shackleton et al., in press). Third, geomorphic, stratigraphic, and geochronologic data from Dry Valleys regions of the Transantarctic Mountains indicate the persistence of hyperarid, cold-desert conditions during the last 13.5 myr, reflecting the relatively stable climate of Antarctica during the Late Neogene (Clapperton and Sugden, 1990; Denton et al., 1993; Marchant et al., 1993; Sugden et al., 1993). Fourth, experiments performed using a 3-D model of the Antarctic cryosphere show that major deglaciation of East Antarctica requires average temperature increases of 17°–20°C above present levels. The rather modest temperature increases suggested by the Early Pliocene data from marine sediment sequences were insufficient to cause ice-sheet instability (Huybrechts, 1993).

Much evidence from the marine sedimentary record indicates the relative stability of the Antarctic climatic-cryospheric system during the Early Pliocene. Oxygen isotopic evidence from Subantarctic deep-sea sequences (fig. 4.3) suggests only relatively minor Antarctic deglaciation and warming of Subantarctic surface waters from 4.8 to 3.1 myr (Hodell and Warnke, 1991; Hodell and Venz, 1992). Evidence exists for a marine transgression during the Early Pliocene, but the magnitude of the sea-level rise is debated (Haq et al., 1988; Wardlaw and Quinn, 1991; Krantz, 1991; Dowsett and Cronin, 1990). A small rise in Antarctic sea-surface temperatures clearly caused limited ice-sheet melting and associated marine transgression, but the oxygen isotopic data constrains maximum sea-level rise to less than 25 m above the present for brief intervals (Kennett and Hodell, 1993, 1994). It is suggested that the marine transgression during the Early Pliocene resulted from limited thinning and marginal melting of the ice sheets.

A mild amelioration of climate in the Weddell Sea during the Early Pliocene reduced sea-ice extent and promoted diatom productivity and preservation (Barker, Kennett et al., 1988; Burckle et al., 1990). The construction of Early Pliocene Antarctic microfossil assemblages indicates that Antarctic surface-water temperatures increased no more than about 3°C, of insufficient magnitude to cause major deglaciation of the cryosphere. Ice rafting of sediments continued far to the north of Antarctica in the Subantarctic southwest Atlantic (Warnke et al., 1992), the Maud Rise (Kennett and Barker, 1990), the Kerguelen Plateau (Breza, 1992), and south of Australia (Blank and Margolis, 1975). There is no suggestion of significant reduction in northward distribution of iceberg sediment transport, indicating the relative stability of the Antarctic climatic system and the position of the water masses during the Pliocene. Oxygen isotopic analysis of Pacific Deep Sea Drilling Project (DSDP) sequences also indicates that the vertical temperature gradient of the Pacific Ocean during the Early Pliocene was similar to that of the modern ocean (Keigwin et al., 1979).

The relative stability of the Antarctic cryosphere during the Early Pliocene is also supported by changes in the

Late Neogene sediment record in the Weddell abyssal plain and possibly the Bellingshausen Basin. This sequence is marked by a near-complete cessation of turbidite deposition at ca. 4.8 myr. Because the turbidity currents are considered to have been derived from West Antarctica, it has been suggested that sediment starvation of the abyssal plains resulted from the development of a relatively stable ice sheet on West Antarctica. The development of ice-sheet stability reduced erosion of West Antarctic continental rocks (Kennett and Barker, 1990). Persistent Late Neogene iceberg transportation of sediments to Subantarctic latitudes of the South Atlantic also supports the continued presence of the West Antarctic ice sheet during the last 6 myr. Thus, the West Antarctic ice sheet may have been relatively stable during most of the Pliocene and Quaternary; this hypothesis requires further testing, however.

Although there is much evidence from marine sedimentary sequences for increased climatic warmth outside of the Antarctic during the Early Pliocene, evidence exists for the maintenance of relative aridity in Australia. In deep-sea sequences east of Australia, the interval between 4 and 3 myr was marked by maximum accumulation rates of wind-blown terrigenous sediment, suggesting increased wind strength and/or aridity (Stein and Robert, 1985). On the other hand, average precipitation in continental areas adjacent to the North Atlantic during the Early Pliocene was higher than the present (Suc, 1984; Dowsett et al., 1992).

The Late Pliocene

The last major cooling trend in the Antarctic occurred during the Late Pliocene between 3.2 and 2.4 myr. The earliest part of this trend (ca. 3.2 and 2.7 myr; Middle of Gauss Chron; fig. 4.3) is marked by a distinct increase in $\delta^{18}O$ in deep-sea sediments (Shackleton and Opdyke, 1977; Kennett, 1985a; Elmstrom and Kennett, 1986; Hodell and Venz, 1992; Raymo, 1992a), representing further cooling and ice accumulation. This period is a transitional interval linking generally warmer, relatively stable climate before ca. 3.2 myr with colder, more variable climate after 2.7 myr (Hodell and Venz, 1992; Kennett and Hodell, 1993). In the Weddell Sea, another regional decrease in sedimentation rates occurred at ca. 3 myr, resulting from further reduction in supply of terrigenous sediments to the Antarctic continental margins (Kennett and Barker, 1990). This decrease may have been caused by further increase in stability of the West Antarctic ice sheet and the ice shelves. Rates of accumulation of ice-rafted sediments increased again in the Subantarctic (Warnke et al., 1992). The Late Pliocene is also marked by the beginning of major expansion in sea ice, initially to the south and later to the north in the Weddell Sea. Expansion of sea-ice cover limited light in surface waters, suppressed diatom productivity, and drastically reduced the rate of deposition of biosiliceous sediment and the quality of diatom preservation (Burckle et al., 1990). Deep-water temperatures in the South Atlantic also decreased (Hodell et al., 1985).

The cooling trend in Antarctic and Subantarctic waters continued from 2.7 to 2.4 myr (latest Gauss Chron; Ciesielski and Grinstead, 1986; Hodell and Venz, 1992), with further northward advance in the Polar Front Zone (Hodell and Venz, 1992), increased sediment ice rafting in the Subantarctic (Warnke et al., 1992) and cooling of Subantarctic surface waters (Hodell and Venz, 1992). These events represent the final major changes in Antarctica leading to environmental conditions like those of the present day.

The Late Pliocene also was marked by progressive cooling in the temperate latitudes in the Southern Hemisphere (Kennett and Vella, 1975; Elmstrom and Kennett, 1985; Hornibrook, 1992), accompanied by the replacement of warm, shallow-water benthic faunas with cool Southern Ocean faunas in New Zealand (Hornibrook, 1992). The earliest significant cooling during the Late Pliocene in New Zealand was at 2.4 myr (Beu et al., 1987). Patagonian glaciers began to expand again after ca. 3.5 myr (Mercer and Sutter, 1982).

The distinct increase in $\delta^{18}O$ during the late Pliocene in deep-sea sediments (figs. 4.1 and 4.3) coincided with general cooling at high latitudes of the North Atlantic (Backman, 1979; Backman and Pestiaux, 1987; Raymo et al., 1987; Einarsson and Albertsson, 1988) and the Mediterranean Sea (Thunell, 1979; Keigwin and Thunell, 1979), the development of extensive glaciation in Iceland (McDougall and Wensink, 1966), and the first loess deposition in Alaska (Westgate et al., 1990). Such interpretations seem to be in conflict with those of Dowsett et al. (1992), who suggest that high-latitude North Atlantic surface waters and surrounding landmasses remained distinctly warm until 3 myr. Significant evidence exists at high latitudes of the Northern Hemisphere for diachronous development of the cryosphere during the Late Neogene. Warnke (1982) described earliest Pliocene (5 myr) ice-rafted sediments at high Subarctic latitudes, which is close to a 4.5 myr age for the earliest observed ice-rafted sediments in central Arctic Ocean sediment cores (Margolis and Herman, 1980). By then glaciers had reached the ocean and produced some ice rafting. This early evidence for glacial development preceded major ice accumulation in the Northern Hemisphere.

Although a clear record exists for major expansion of the cryosphere in both hemispheres during the Late Pliocene, oxygen isotopic data from marine sequences indicates that the majority of ice accumulation must have occurred in the Northern Hemisphere (Raymo, 1992b). The increase in $\delta^{18}O$ in deep-sea sequences after ca. 3 myr is usually interpreted to represent the sequential development of major Northern Hemispheric ice sheets (Berggren, 1972; Poore and Berggren, 1975; Shackleton and Opdyke, 1976; Backman, 1979; Keigwin and Thunell, 1979; Poore, 1981; Prell, 1985; Keigwin, 1986; Joyce et al., 1990; Hodell and Venz, 1992; Kennett and Hodell, 1993). At this time the global cryosphere became truly bipolar. The drop in sea level that resulted from the development of ice sheets in the Northern Hemisphere, in turn, would have caused the lateral expansion of the Antarctic ice sheets and ice shelves owing to grounding on the continental shelf. Accumulation of large ice sheets in the Northern Hemisphere is usually considered to have occurred between 2.7 and 2.4 myr. During this time ice sheets were about one-fourth to one-half as large as those of the Late Quaternary and delivered abundant ice-rafted sediments to the Norwegian Sea (Jansen and Sjoholm, 1991). At ca. 2.4 myr ice-rafted detritus began to be delivered to the open North Atlantic (Shackleton et al., 1984). Evidence for sequential cooling and cryospheric development in the Northern Hemisphere is described by Thiede et al. (1990).

From ca. 3 myr to the present, an abundance of aeolian terrigenous sediments and particular clay mineral assemblages on the Lord Howe Rise suggest that Australian continental climate became dominantly arid (Stein and Robert, 1985). Opal phytoliths also significantly increased during the Late Pliocene, peaking at 2.5 myr, indicating further expansion of Australian grassland areas and intensified transport of dust to the Lord Howe Rise (Locker and Martini, 1985).

Discussion

The material presented above shows that the climatic evolution of the earth during the Cenozoic largely reflects a progressive and sequential trend toward lower temperatures and cryospheric development in the polar regions, initially in Antarctica and later in the Northern Hemisphere (Kennett, 1982). The evolution of the polar cryosphere profoundly influenced global climate, sea-level history, the earth's heat budget, atmospheric composition and circulation, and thermohaline circulation. These factors in turn directly and indirectly affected the direction of marine and terrestrial biotic evolution of the planet. An increase in the meridional thermal gradient, which resulted from major cooling at high latitudes, caused increased wind strength, intensification of oceanic circulation, and related near-shore cold-water upwelling in certain regions during the Neogene. A major consequence of these changes was an extensive increase in aridification in the middle- to low-latitude regions, leading to major evolutionary changes in the terrestrial biota. The developing detailed climatic history from the deep-sea record provides a useful framework for more thorough understanding of global Cenozoic biotic evolution, including that of the terrestrial realm.

The climatic development of Antarctica and the Southern Ocean resulted, in part, from the rearrangement of landmasses in the Southern Hemisphere. Antarctica became increasingly isolated as fragments of Gondwanaland moved northward and Antarctic circumpolar circulation developed, allowing unrestricted latitudinal flow. Development of the Antarctic Circumpolar Current during the Middle Cenozoic effectively isolated Antarctica thermally by decoupling warmer waters of the subtropical gyres from the Antarctic continent (Kennett, 1977, 1982; Lawver et al., 1992). This isolation led to major cooling of the Antarctic region during the Middle Cenozoic. Continued northward expansion of the Antarctic Circumpolar Current, coupled with positive environmental feedbacks owing to the development of the cryosphere itself, led to the development of the major Neogene Antarctic ice sheets. Associated climatic changes included the cooling of surface waters surrounding the continent, production of increasingly cold deep waters flowing to the ocean basins, extensive sea-ice production, and wind-driven upwelling of nutrient-rich intermediate waters, which increased the biological productivity of surface water in the Southern Ocean.

The expansion of the ocean basins around Antarctica, and especially the development of seaways for Antarctic circumpolar circulation, have created a unique set of ocean-continent boundary conditions that have increased isolation and provided much stability to the

Antarctic environmental system (Kennett and Hodell, 1995). The northern limits of the Antarctic Circumpolar Current became relatively stable once vigorous circumpolar flow was established after the opening of the Drake Passage (ca. 20 myr) and a major East Antarctic ice sheet developed (14 myr). With the establishment of both East and West Antarctic ice sheets by the latest Miocene, the basic modern climatic oceanic system became well established in the Southern Ocean. This situation created widespread thermal inertia in the Antarctic climatic system, which contributed to its observed stability during Early Pliocene warmth (Kennett and Hodell, 1993). The relative robustness of the Late Neogene Antarctic climatic system is supported by the modeling experiments of Robin (1988), who suggested that the Antarctic cryosphere expands in steps as a result of feedback loops that drive the system toward greater stability (Kennett and Hodell, 1995). This model suggests that each step in ice accumulation becomes permanent unless later conditions become much warmer.

Global climatic change remained largely asymmetric between the polar regions until the Late Pliocene. Major climatic steps in the Antarctic, including the development of the cryosphere, were not matched by equivalent change in the Arctic. Conversely, significant warming in high northern latitudes, such as occurred in the Early Pliocene, was not matched by equivalent warming in Antarctica. Early Pliocene climate was distinctly different in high latitudes of each hemisphere. With the development of major ice sheets in the Northern Hemisphere during about the last 3 myr, the interpolar climate differences were reduced. During interglacial maxima such as the Holocene, however, major assymetry in polar climates is evident, as is the relative robustness of the Antarctic climatic-oceanic system. No convincing evidence yet exists for deglaciation of west Antarctica during warmer-than-present interglacials of the Quaternary (Burckle, 1993).

Summary

The climate of Antarctica and the Southern Ocean since the Middle Eocene (ca. 50 myr) has experienced generally progressive and sequential decreases in temperature and increases in ice accumulation. Reversals of this cooling trend were unusual. The most distinct warming occurred during the Late Paleogene through the Early Neogene. No major climatic reversals in Antarctic cooling occurred after the early Middle Miocene (ca. 15 myr). The development of the polar cryosphere caused fundamental changes in global oceanic and atmospheric circulation, leading in stepwise fashion to widespread aridity in middle- to low-latitude regions.

From Middle Miocene time, East Antarctica has been covered by a major and relatively permanent ice sheet. Evidence from marine and terrestrial records indicates the considerable stability of this ice sheet and of a cold Antarctic climate during the Middle and Late Neogene, even during times of considerable warmth north of the Antarctic region. The stability of the climatic-cryospheric system resulted largely from the thermal isolation of the continent imparted by the Antarctic Circumpolar Current, which became well established by the beginning of the Neogene. The development of the Antarctic cryosphere also produced environmental feedbacks that reinforced the development of colder conditions marked by additional ice accumulation. These effects include increased albedo owing to ice expansion and transfer of heat from Antarctica to the deep ocean basins during formation of bottom and intermediate waters.

Warming in the Antarctic during certain intervals after the Middle Miocene was insufficient to cause significant melting of the East Antarctic ice sheet, even during the Early Pliocene (ca. 5–3 myr), a time of warmth to the north of the Antarctic region, including high North Atlantic latitudes. The thermal isolation of Antarctica strengthened during the Neogene as a result of the continued expansion northward of the Southern Ocean and the strengthening of the Antarctic Circumpolar Current and the Antarctic Convergence (Polar Front Zone). Further associated cooling led to the development, during the Late Miocene (ca. 9–5 myr), of the marine-based West Antarctic ice sheet. Initially this sheet exhibited high glacial-interglacial instability, but it appears to have evolved into a relatively stable structure near the beginning of the Pliocene (ca. 5 myr), although this hypothesis requires rigorous testing. The sequential development of the polar cryosphere continued with the accumulation of major, though unstable, ice sheets in the Northern Hemisphere beginning in the Late Pliocene (ca. 3 myr). The sequence of these changes suggests that earlier development of a large Antarctic cryosphere was an essential basis for the later formation of ice sheets in the Northern Hemisphere. The establishment of a bipolar cryosphere during the Late Pliocene to Quaternary led to a significant increase in aridity in middle- to low-latitude regions. During the last 3 myr, global climate has exhibited the largest variation of the Neogene, owing to the large-scale glacial-interglacial variations displayed by the Northern Hemispheric ice sheets.

Global climatic evolution during the Cenozoic has been profoundly affected by the development of the polar cryosphere. The cryosphere has, in turn, fundamentally affected the course of terrestrial and marine biotic evolution.

Acknowledgments

This research was supported by National Science Foundation grant DPP92-18720. The work has also strongly benefited from my long and fruitful association with the Ocean Drilling Program and the Deep Sea Drilling Project.

References

Adams, C. G., Benson, R. H., Kidd, R. B., Ryan, W. B. F., and Wright, R. C. 1977. The Messinian Salinity Crisis and evidence of Late Miocene eustatic changes in the world ocean. *Nature,* 269:383–386.

Backman, J. 1979. Pliocene biostratigraphy of DSDP Sites 111 and 116 from the North Atlantic Ocean and the age of Northern Hemisphere glaciation. *Stockholm Contributions to Geology,* 32(3):115–137.

Backman, J., and Pestiaux, P. 1987. Pliocene *Discoaster* abundance variations, Deep Sea Drilling Project Site 606: Biochronology and palaeoenvironmental implications. *In* Ruddiman, W. F., Kidd, R. B., et al. (eds.), *Init. Repts. Deep Sea Drill. Proj.,* Washington, D.C.: U.S. Government Printing Office, 94:903–909.

Barker, P. F., Kennett, J. P., et al. (eds.). 1988. *Proc. Ocean Drill. Prog., Init. Repts.,* College Station, Tex., 113.

Barrett, P. J., Adams, C. J., McIntosh, W. C., Swisher, C. C., and Wilson, G. S. 1992. Geochronological evidence supporting Antarctic deglaciation three million years ago. *Nature,* 359:816–818.

Barron, J. A. 1973. Late Miocene–Early Pliocene paleotemperatures for California from marine diatom evidence. *Palaeogeogr., Palaeoclimatol., Palaeoecol.,* 14:277–291.

Berger, W. H., and Winterer, E. L. 1974. Plate stratigraphy and the fluctuating carbonate line. *In* Hsü, K. J., and Jenkyns, H. C. (eds.), *Pelagic sediments: On land and under the sea,* Oxford: Blackwell, 11–98.

Berggren, W. A. 1972. Late Pliocene–Pleistocene glaciation. *In* Laughton, A. S., Berggren, W. A., et al. (eds.), *Init. Repts. Deep Sea Drill. Proj.,* Washington, D.C.: U.S. Government Printing Office, 12:953–963.

Beu, A. G., Edwards, A. R., and Pillans, B. J. 1987. A review of New Zealand Pleistocene stratigraphy with emphasis on marine rocks. *In* Itihara, M., and Kamei, T. (eds.), *Proceedings of the First International Colloquium on Quaternary Stratigraphy of Asia and Pacific Area, Osaka, 1986,* Osaka: INQUA Commission on Quaternary Stratigraphy, Subcommission on Quaternary Stratigraphy of Asia and Pacific Area and National Working Group of Quaternary Stratigraphy, Science Council of Japan.

Blank, R. G., and Margolis, S. V. 1975. Pliocene climatic and glacial history of Antarctica as revealed by southeast Indian Ocean deep-sea cores. *Geol. Soc. Am. Bull.,* 86:1058–1066.

Boersma, A. 1985. Biostratigraphy and biogeography of Tertiary bathyal benthic foraminifers: Tasman Sea, Coral Sea, and on the Chatham Rise (Deep Sea Drilling Project, Leg 90). *In* Kennett, J. P., and von der Borch, C. C., et al. (eds.), *Init. Repts. Deep Sea Drill. Proj.,* Washington, D.C.: U.S. Government Printing Office, 90:961–991.

Brewster, N. A. 1980. Cenozoic biogenic silica sedimentation in the Antarctic Ocean. *Geol. Soc. Am. Bull.,* 91:337–347.

Breza, J. R. 1992. High resolution study of ice-rafted debris, ODP Leg 120, Site 751 southern Kerguelen Plateau. *In* Schlich, R., Wise, S. W., et al. (eds.), *Proc. Ocean Drill. Prog., scientific results,* College Station, Tex., 120:207–221.

Burckle, L. H. 1993. Is there direct evidence for late Quaternary collapse of the west Antarctic ice sheet? *J. Glaciology* 39:491–494.

Burckle, L. H., Gersonde, R., and Abrams, N. 1990. Late Pliocene-Pleistocene palaeoclimate in the Jane Basin region: ODP Site 697. In Barker, P. F., Kennett, J. P., et al. (eds.), *Proc. Ocean Drill. Prog., scientific results,* College Station, Tex., 113:803–812.

Burckle, L. H., and Pokras, E. M. 1991. Implications of a Pliocene stand of *Nothofagus* (southern beech) within 500 kilometres of the South Pole. *Antarctic Science,* 3(4):389–403.

Cerling, T. E. 1993. Expansion of C_4 ecosystems as an indication of global ecological change in the Late Miocene. *Nature,* 361:344–345.

Ciesielski, P. F., and Grinstead, G. P. 1986. Pliocene variations in the position of the Antarctic Convergence in the southwest Atlantic. *Paleoceanography,* 1:197–232.

Ciesielski, P. F., Ledbetter, M. T., and Ellwood, B. B. 1982. The development of Antarctic glaciation and the Neogene palaeoenvironment of the Maurice Ewing Bank. *Mar. Geol.,* 46:1–51.

Clapperton, C. M., and Sugden, D. E. 1990. Late Cenozoic glacial history of the Ross Embayment, Antarctica. *Quat. Sci. Rev.,* 9:252–272.

Cronin, T. M. 1991. Pliocene shallow water paleoceanography of the North Atlantic Ocean based on marine ostracodes. *Quat. Sci. Rev.,* 10:175–188.

Denton, G. H., Sugden, D. E., Marchant, D. R., Hall, B. L., and Wilch, T. I. 1993. East Antarctic ice sheet sensitivity to Pliocene climatic change from a Dry Valleys perspective. *In* Sugden, D. E., Marchant, D. R., and Denton, G. H. (eds.), *The case for a stable East Antarctic ice sheet, Geografiska Annaler,* Special Vol., 75A(4):155–204.

Dowsett, H. J., and Cronin, T. M. 1990. High eustatic sea level during the Middle Pliocene: Evidence from the southeastern U.S. Atlantic Coastal Plain. *Geology,* 18:435–438.

Dowsett, H. J., Cronin, T. M., Poore, R. Z., Thompson, R. S., Whatley, R. C., and Wood, A. M. 1992. Micropaleontological evidence for increased meridional heat transport in the North Atlantic Ocean during the Pliocene. *Science,* 258:1133–1135.

Dowsett, H. J., and Poore, R. Z. 1991. Pliocene Sea surface temperatures of the North Atlantic Ocean at 3.0 Ma. *In* Cronin, T. M., and Dowsett, H. J. (eds.), *Quat. Sci. Rev.*, 10(2/3): 189–204.

Drewry, D. J., ed. 1983. *Antarctica: Glaciological and geophysical folio.* Cambridge: Cambridge University Press.

Edwards, A. R. 1968. The calcareous nannoplankton evidence for New Zealand Tertiary climate. *Tuatara,* 16(1):26–31.

Einarsson, T., and Albertsson, K. 1988. The glacial history of Iceland during the past three million years. *In* Shackleton, N. J., West, R. G., and Bowen, D. Q. (eds.), *The past three million years: Evolution of climatic variability in the North Atlantic region. Philosophical Transactions of the Royal Society* (London), ser. B, 318(1191):637–644.

Elmstrom, K., and Kennett, J. P. 1985. Late Neogene palaeoceanographic evolution of Site 590: Southwest Pacific. *In* Kennett, J. P., von der Borch, C. C., et al. (eds.), *Init. Repts. Deep Sea Drill. Proj.,* Washington, D.C.: U.S. Government Printing Office, 90:1361–1382.

Flower, B. P., and Kennett, J. P. 1993. Middle Miocene ocean/climate transition: High-resolution oxygen and carbon isotopic records from DSDP Site 588A, southwest Pacific. *Paleoceanography,* 8(6):811–843.

———. 1994. The Middle Miocene climatic transition: East Antarctica ice sheet development, deep ocean circulation and global carbon cycling. *Paleogeogr. Paleoclimatol., Palaeoecol.,* 108:537–555.

Grobe, H. Futterer, D. K., and Spiess, V. 1990. Oligocene to Quaternary processes on the Antarctic continental margin, ODP Leg 113, Site 693. *Proc. Ocean Drill. Program., scientific results,* College Station, Tex., 113:121–131.

Haq, B. U. 1980. Biogeographic history of Miocene calcareous nannoplankton and paleoceanography of the Atlantic Ocean. *Micropaleontology,* 26(4):414–443.

Haq, B. U., Hardenbol, J., and Vail, P. R. 1987. Chronology of fluctuating sea levels since the Triassic. *Science,* 235:1156–1167.

———. 1988. Mesozoic and Cenozoic chronostratigraphy and cycles of sea-level change. *In* Wilgus, D. K., Hastings, B S., et al. (eds.), *Sea-level changes: An integrated approach.* Society of Economic Paleontologists and Mineralogists, Special Publ., 42:71–108.

Haq, B. U. Worsley, T. R., Burckle, L. H., Douglas, R. G., Keigwin, L. D., Jr., Opdyke, N. D., Savin, S. M., Sommer, M. A., II, Vincent, E., and Woodruff, F. 1980. Late Miocene marine carbon-isotope shift and synchroneity of some phytoplanktonic biostratigraphic events. *Geology,* 8:427–431.

Heusser, L. E. 1985. Palynology of selected Neogene samples from Holes 594 and 594A, Chatham Rise. *In* Kennett, J. P., von der Borch, C. C., et al. (eds.), *Init. Repts. Deep Sea Drill. Proj.,* Washington, D.C.: U.S. Government Printing Office, 90:1085–1092.

Hodell, D. A., Benson, R. H., Kent, D. V., Boersma, A., and Rakic-el Bied, K. 1994. Magnetostratigraphic, biostratigraphic and stable isotope stratigraphy of an Upper Miocene drill core from the Salé Briqueterie (northwestern Morocco): A high resolution chronology for the Messinian Stage. *Paleoceanography,* 9:835–855.

Hodell, D. A., Elmstrom, K. M., and Kennett, J. P. 1986. Latest Miocene benthic $\delta^{18}O$ changes, global ice volume, sea level and the Messinian Salinity Crisis. *Nature,* 320(6061):411–414.

Hodell, D. A., and Kennett, J. P. 1986. Late Miocene–Early Pliocene stratigraphy and palaeoceanography of the south Atlantic and southwest Pacific oceans: A synthesis. *Paleoceanography,* 1:285–311.

Hodell, D. A., and Venz, K. 1992. Toward a high-resolution stable isotopic record of the Southern Ocean during the Pliocene-Pleistocene (4.8 to 0.8 Ma). *In* Kennett, J. P., and Warnke, D. A. (eds.), *The Antarctic paleoenvironment: A perspective on global change,* pt. 1, Antarctic Research Series, 56:265–310.

Hodell, D. A., and Warnke, D. A. 1991. Climatic evolution of the Southern Ocean during the Pliocene epoch from 4.8 to 2.6 million years ago. *Quat. Sci. Rev.,* 10:205–214.

Hodell, D. A., Williams, D. F., and Kennett, J. P. 1985. Late Pliocene reorganization of deep vertical water-mass structure in the western South Atlantic: Faunal and isotopic evidence. *Geol. Soc. Am. Bull.,* 96:495–503.

Hornibrook, N. de B. 1992. New Zealand Cenozoic marine paleoclimates: A review based on the distribution of some shallow water and terrestrial biota. *In* Tsuchi, R., and Ingle, J. C., Jr. (eds.), *Pacific Neogene,* Tokyo: University of Tokyo Press, 83–106.

Hsü, K. J., et al. 1977. History of the Mediterranean Salinity Crisis. *Nature,* 267:399–403.

Huybrechts, P. 1993. Glaciological modelling of the Late Cenozoic East Antarctic ice sheet: Stability or dynamism? *In* Sugden, D. E., Marchant, D. R., and Denton, G. H. (eds.), *The case for a stable East Antarctic ice sheet, Geografiska Annaler,* Special Vol., 75A(4):221–238.

Ingle, J. C., Jr. 1967. Foraminiferal biofacies variation and the Miocene-Pliocene boundary in southern California. *Am. Paleont. Bull.,* 52:217–394.

Jansen, E., and Sjoholm, J. 1991. Reconstruction of glaciation over the past 6 myr from ice-borne deposits in the Norwegian Sea. *Nature,* 349:600–603.

Jenkins, D. G. 1973. Diversity changes in New Zealand Cenozoic foraminifera. *J. Foraminif. Res.,* 3(2):78–88.

Joyce, J. E., Tjalsma, L. R. C., and Prutzman, J. J. 1990. High-resolution planktic stable isotope record and spectral analysis for the last 5.35 m.y.: Ocean Drilling Program Site 625, northeast Gulf of Mexico. *Paleoceanography,* 5(4):507–529.

Kaneps, A. G. 1979. Gulf Stream: Velocity fluctuations during the latest Cenozoic. *Science,* 204:297–301.

Keigwin, L. D., Jr. 1979. Late Cenozoic stable isotope stratigraphy and paleoceanography of Deep Sea Drilling Project Sites from the east equatorial and central North Pacific Oceans. *Earth Planet. Sci. Lett.,* 45:361–382.

———. 1986. Pliocene stable isotope record of Deep Sea Drilling Project Site 606: Sequential events of ^{18}O enrichment beginning at 3.1 Ma. *In* Ruddiman, W. F., Kidd, R. B., et al. (eds.), *Init. Repts. Deep Sea Drill. Proj.,* Washington, D.C.: U.S. Government Printing Office, 94:911–920.

Keigwin, L. D., Jr., Bender, M. L., and Kennett, J. P. 1979.

Thermal structure of the deep Pacific Ocean in the Early Pliocene. *Science,* 205:1386–1388.

Keigwin, L. D., Jr., and Shackleton, N. J. 1980. Uppermost Miocene carbon isotope stratigraphy and paleoceanography of a piston core in the equatorial Pacific. *Nature,* 284:313–314.

Keigwin, L. D., Jr., and Thunell, R. 1979. Middle Pliocene climatic change in the western Mediterranean from faunal and oxygen isotopic trends. *Nature,* 282:294–296.

Keller, G. 1980. Middle to Late Miocene planktonic foraminiferal datum levels and paleoceanography of the north and southwestern Pacific Ocean. *Mar. Micropaleont.,* 5:249–281.

Kemp, E. M. 1978. Tertiary climatic evolution and vegetation history in the southeast Indian Ocean region. *Palaeogeogr., Palaeoclimatol., Palaeoecol.,* 24:169–208.

Kennett, J. P. 1967. Recognition and correlation of the Kapitean Stage (Upper Miocene, New Zealand). *New Zealand Journal of Geology and Geophysics,* 10(4):1051–1063.

———. 1977. Cenozoic evolution of Antarctic glaciation, the Circum-Antarctic Ocean, and their impact on global paleoceanography. *J. Geophys. Res.,* 82(27):3843–3860.

———. 1982. *Marine geology,* Englewood Cliffs, N.J.: Prentice-Hall.

———. 1985a. Miocene and Early Pliocene oxygen and carbon isotopes stratigraphy of the southwest Pacific, of the Deep Sea Drilling Project, Leg 90. *In* Kennett, J. P., von der Borch, C. C., et al. (eds.), *Init. Repts. Deep Sea Drill. Proj.,* Washington, D.C.: U.S. Government Printing Office, 29:1383–1411.

———. 1985b. Neogene palaeoceanography and plankton evolution. *Suid-Afrikaanse Tydskrif vir Wetenskap,* 81:251–253.

Kennett, J. P., and Barker, P. F. 1990. Latest Cretaceous to Cenozoic climate and oceanographic developments in the Weddell Sea, Antarctica: An ocean-drilling perspective. *In* Barker, R. F., Kennett, J. P., et al. (eds.), *Proc. Ocean Drill. Program, scientific results,* College Station, Tex., 113:937–960.

Kennett, J. P., and Hodell, D. A. 1993. Evidence for relative climatic stability of Antarctica during the Early Pliocene: A marine perspective. *In* Sugden, D. E. Marchant, D. R., and Denton, G. H. (eds.), *The case for a stable East Antarctic ice sheet, Geografiska Annaler,* Special Vol., 75A(4):205–220.

———. 1995. Stability or instability of Antarctic ice sheets during warm climates of the Pliocene? *GSA Today,* 5(1):1, 10–13, 22.

Kennett, J. P., Houtz, R. E., et al., 1975. Cenozoic paleoceanography in the Southwest Pacific Ocean, Antarctic glaciation and the development of the Circum-Antarctic Current. *In* Kennett, J. P., Houtz, R. E., et al. (eds.), *Init. Repts. Deep Sea Drill. Proj.,* Washington, D.C.: U.S. Government Printing Office, 29:1155–1169.

Kennett, J. P., Keller, G., and Srinivasan, M. S. 1985. Miocene planktonic foraminiferal biogeography and paleoceanographic development of the Indo-Pacific region. *Mem. Geol. Soc. Am.,* 163:197–235.

Kennett, J. P., Shackleton, N. J., Margolis, S. V., Goodney, D. E., Dudley, W. C., and Kroopnick, P. M. 1979. Late Cenozoic oxygen and carbon isotopic history and volcanic ash stratigraphy: DSDP Site 284, South Pacific. *Amer. J. Sci.,* 279:52–69.

Kennett, J. P., and Vella, P. 1975. Late Cenozoic planktonic foraminifera and palaeoceanography at DSDP Site 284 in the cool subtropical South Pacific. *In* Kennett, J. P., Houtz, R. E., et al. (eds.), *Init. Repts. Deep Sea Drill. Proj.,* Washington, D.C.: U.S. Government Printing Office, 29:869–782.

Kennett, J. P., and von der Borch, C. C. 1985. Southwest Pacific Cenozoic paleoceanography. *In* Kennett, J. P., von der Borch, C. C., et al. (eds.), *Init. Repts. Deep Sea Drill. Proj.,* Washington, D.C.: U.S. Government Printing Office, 90:1493–1517.

Kennett, J. P., and Watkins, N. D. 1974. Late Miocene–Early Pliocene paleomagnetic stratigraphy, paleoclimatology, and biostratigraphy in New Zealand. *Geol. Soc. Am. Bull.,* 85:1385–1398.

Krantz, D. E. 1991. A chronology of Pliocene sea-level fluctuations: The U.S. Middle Atlantic Coastal Plain record. *Quat. Sci. Rev.,* 10:163–174.

Kurihara, K., and Kennett, J. P. 1985. Neogene benthic foraminifers: Distribution in depth traverse. *In* Kennett, J. P., von der Borch, C. C., et al. (eds.), *Init. Repts. Deep Sea Drill. Proj.,* Washington, D.C.: U.S. Government Printing Office, 90:1037–1078.

———. 1992. Paleoceanographic significance of Neogene benthic foraminiferal changes in a southwest Pacific bathyal depth transect. *Mar. Micopaleontol.,* 19:181–199.

Lawver, L. A., Gahagan, L. M., and Coffin, M. F. 1992. The development of paleoseaways around Antarctica. *In* Kennett, J. P., and Warnke, D. A. (eds.), *The Antarctic paleoenvironment: A perspective on global change,* pt. 1, Antarctic Research Series, 56:7–30.

Larzarus, D., and Caulet, J. P. 1993. Cenozoic Southern Ocean reconstructions from sedimentologic, radiolarian, and other microfossil data. *In* Kennett, J. P., and Warnke, D. A. (eds.), *The Antarctic paleoenvironment: A perspective on global change,* pt. 2, Antarctic Research Series, 60:145–174.

Locker, S., and Martini, E. 1985. Phytoliths from the southwest Pacific, Site 591. *In* Kennett, J. P., von der Borch, C. C., et al. (eds.), *Init. Repts. Deep Sea Drill. Proj.,* Washington, D.C.: U.S. Government Printing Office, 90:1079–1084.

Loutit, T., and Keigwin, L. D., Jr. 1982. Stable isotopic evidence for latest Miocene sea-level fall in the Mediterranean region. *Nature,* 300(5888):163–166.

Loutit, T., and Kennett, J. P. 1979. Application of carbon isotope stratigraphy and correlation to Late Miocene shallow marine sediments, New Zealand. *Science,* 204:1196–1199.

Loutit, T. S., Kennett, J. P., and Savin, S. M. 1983. Miocene equatorial and southwest Pacific paleoceanography from stable isotope evidence. *Mar. Micropaleontol.,* 8:215–233.

McDougall, I., and Wensink, H. 1966. Paleomagnetism and geochronology of the Pliocene-Pleistocene lavas in Iceland. *Earth and Planetary Sci. Lett.,* 1:232–236.

McKenzie, J. A., Weissert, H., Poore, R. Z., Wright, R. C., Percival, S. F., Jr., Oberhansli, H., and Casey, M. 1984. Paleoceanographic implications of stable-isotope data from upper Miocene–Lower Pliocene sediments from the southeast Atlantic (Deep Sea Drilling Project Site 519). *In* Hsü, K. J., La Breque, J. L., et al. (eds.), *Init. Repts. Deep Sea Drill. Proj.,* Washington, D.C.: U.S. Government Printing Office, 79:717–724.

Marchant, D. R., Denton, G. H., and Swisher, C. C. 1993. Miocene-Pliocene-Pleistocene glacial history of Arena Valley, Quatermain Mountains, Antarctica. *In* Sugden, D. E., Marchant, D. R., and Denton, G. H. (eds.), *The case for a stable east Antarctic ice sheet, Geografiska Annaler,* Special Vol., 75A(4):269–302.

Margolis, S. V., and Herman, Y. 1980. Northern Hemisphere sea-ice and glacial development in the Late Cenozoic. *Nature,* 286:145–149.

Mercer, J. H. 1973. Cainozoic temperature trends in the Southern Hemisphere: Antarctic and Andean glacial evidence. *In* van Zinderen Bakker, E. M. (ed.), *Palaeoecology of Africa,* Rotterdam: A. A. Balkema, 85–114.

Mercer, J. H., and Sutter, J. F. 1982. Late Miocene-earliest Pliocene glaciation in southern Argentina: Implications for global ice-sheet history. *Palaeogeogr., Palaeoclimatol., Palaeoecol.,* 38:185–206.

Miller, K. G., Fairbanks, R. G., and Mountain, G. S. 1987. Tertiary oxygen isotope synthesis, seal level history, and continental margin erosion. *Paleoceanography,* 2(1):1–19.

Miller, K. G., and Katz, M. E. 1987. Oligocene to Miocene benthic foraminiferal and abyssal circulation in the North Atlantic. *Micropaleontology* 33:97–149.

Pascual, R., and Juareguizar, E. O. 1990. Evolving climates and mammal faunas in Cenozoic South America. *J. Hum. Evol.,* 19:23–60.

Poore, R. Z. 1981. Temporal and spatial distribution of ice-rafted mineral grains in Pliocene sediments of the North Atlantic: Implications for Late Cenozoic climatic history. *SEPM Special Publ.,* 32:505–515.

Poore, R., and Berggren, W. 1975. Late Cenozoic planktonic foraminiferal biostratigraphy and paleoclimatology of Hatton-Rockall Basin: DSDP Site 116. *J. Foraminif. Res.,* 5:270–293.

Prell, W. L. 1985. Pliocene stable isotope and carbonate stratigraphy (Holes 572C and 573A): Paleoceanographic data bearing on the question of Pliocene glaciation. *In* Mayer, L., Theyer, F., et al. (eds.), *Init. Repts. Deep Sea Drill. Proj.,* Washington, D.C.: U.S. Government Printing Office, 85:723–734.

Raymo, M. E. 1992a. Late Cenozoic evolution of global change. *In* Tsuchi, R., and Ingle, J. C. Jr. (eds.), *Pacific Neogene: Environment, evolution, and events,* Tokyo: University of Tokyo Press, 107–116.

———. 1992b. Global climate change: A three million year perspective. *In* Kukla, G. J., and Went, E. (eds.), *Start of a glacial,* Berlin: Springer-Verlag, 207–223.

Raymo, M. E., Ruddiman, W. F., and Clement, B. M. 1987. Pliocene-Pleistocene palaeoceanography of the North Atlantic at Deep Sea Drilling Project Site 609. *In* Ruddiman, W. F., Kidd, R. B., et al. (eds.), *Init. Repts. Deep Sea Drill. Proj.,* Washington, D.C.: U.S. Government Printing Office, 94:895–901.

Retallack, G. J. 1992. Middle Miocene fossil plants from Fort Ternan (Kenya) and evolution of African grasslands. *Paleobiology,* 18:383–400.

Robin, G. de Q. 1988. The Antarctic ice sheet, its history and response to sea level and climatic changes over the past 100 million years. *Palaeogeogr., Palaeoclimatol., Palaeoecol.,* 67:31–50.

Ryan, W. B. F., Cita, M. B., Rawson, M. D., Burckle, L. H., and Saito, T. 1974. A paleomagnetic assignment of Neogene stage boundaries and the development of isochronous datum planes between the Mediterranean, the Pacific and the Indian Oceans in order to investigate the response of world ocean to Mediterranean "salinity crisis." *Riv. Ital. Paleont.,* 80:631–688.

Sancetta, C. A. 1982. Distribution of diatom species in surface sediments of the Bering and Okhotsk Seas. *Micropaleontology,* 28(3):221–257.

Sarnthein, M., and Fenner, J. 1987. Global wind-induced change of deep-sea sediment budgets, new ocean production and CO_2 reservoirs ca. 3.3–2.35 Ma. *In* Shackleton, N. J., West, R. G., and Bowen, D. Q. (eds.), *The past three million years: Evolution of climatic variability in the North Atlantic region. Philosophical Transactions of the Royal Society* (London), 318:487–504.

Savin, S. M., Abel, L., Barrera, E., Hodell, D., Keller, G., Kennett, J. P., Killingley, J., Murphy, M., and Vincent, E. 1985. The evolution of Miocene surface and near-surface marine temperatures: Oxygen isotopic evidence. *In* Kennett, J. P. (ed.), *The Miocene ocean: Paleoceanography and biogeography, Mem. Geol. Soc. Am.,* 163:49–82.

Savin, S. M., Douglas, R. G., and Stehli, F. G. 1975. Tertiary marine paleotemperatures. *Geol. Soc. Am. Bull.,* 86:1499–1510.

Schnitker, D. 1980. North Atlantic oceanography as possible cause of Antarctic glaciation and eutrophication. *Nature,* 284:615–616.

Shackleton, N. J., Backman, J., Zimmerman, H., Kent, D. V., Hall, M. A., Roberts, D. G., and Baldauf, J. 1984. Oxygen isotope calibration of the onset of ice-rafting and history of glaciation in the North Atlantic region. *Nature,* 307:620–623.

Shackleton, N. J., Hall, M. A., and Pate, D. (in press). Pliocene stable isotope stratigraphy of ODP Site 846. *Proc. Ocean Drill. Prog., scientific results,* 138.

Shackleton, N. J., and Kennett, J. P. 1975a. Paleotemperature history of the Cenozoic and the initiation of Antarctic glaciation: Oxygen and carbon isotope analyses in DSDP Sites 277, 279 and 281. *In* Kennett, J. P. Houtz, R. E., et al. (eds.), *Init. Repts. Deep Sea Drill. Proj.,* Washington, D.C.: U.S. Government Printing Office, 29:743–756.

———. 1975b. Late Cenozoic oxygen and carbon isotopic changes at DSDP Site 284: Implications for glacial history of the Northern Hemisphere and Antarctica. *In* Kennett, J. P., Houtz, R. E., et al. (eds.), *Init. Repts. Deep Sea Drill. Proj.,* Washington, D.C.: U.S. Government Printing Office, 29:801–807.

Shackleton, N. J., and Opdyke, N. D. 1976. Oxygen isotope and paleomagnetic stratigraphy of equatorial Pacific core V28-239, Late Pliocene to latest Pleistocene. *In* Cline, R. M., and Hays, J. D. (eds.), *Mem. Geol. Soc. Am.,* 145:449–464.

———. 1977. Oxygen isotope and palaeomagnetic evidence for early Northern Hemisphere glaciation. *Nature,* 270:216–219.

Stein, R., and Robert, C. 1985. Siliciclastic sediments at Sites 588, 590, and 591: Neogene and Paleogene evolution in the southwest Pacific and Australian climate. *In* Kennett, J. P., von der Borch, C. C., et al. (eds.), *Init. Repts. Deep Sea Drill. Proj.*, Washington, D.C.: U.S. Government Printing Office, 90:1437–1454.

Suc, J.-P. 1984. Origin and evolution of the Mediterranean vegetation and climate in Europe. *Nature,* 307:429–432.

Sugden, D. E., Marchant, D. R., and Denton, G. H. (eds.), 1993. *The case for a stable East Antarctic ice sheet, Geografiska Annaler,* Special Vol., 75A(4).

Thiede, J., Clark, D. L., and Herman, Y. 1990. Late Mesozoic and Cenozoic paleoceanography of the northern polar oceans. *In* Grantz, A., Johnson, L., and Sweeney, J. F. (eds.), *The Arctic ocean region, the Geology of North America,* 50:427–458, Geological Society of America, Boulder, Colo.

Thomas, E. 1985. Late Eocene to Recent deep-sea benthic foraminifers from the central equatorial Pacific Ocean. *In* Mayer, L., and Theyer, F. (eds.), *Init. Repts. Deep Sea Drill. Proj.*, Washington, D.C.: U.S. Government Printing Office, 85:655–694.

Thunell, R. C. 1979. Pliocene-Pleistocene paleotemperatures and paleosalinity history of the Mediterranean Sea, results from Deep Sea Drilling Project Sites 125 and 132. *Mar. Micropaleontol.,* 4:173.

Thunell, R., and Belyea, P. 1982. Neogene planktonic foraminiferal biogeography of the Atlantic Ocean. *Micropaleontology,* 28:381–398.

Tucholke, B. E., Hollister, C. D., Weaver, F. M., and Vennum, W. R. 1976. Continental rise and abyssal plain sedimentation in the southeast Pacific basin, Leg 35, Deep Sea Drilling Project. *In* Hollister, C. D., Craddock, C., et al. (eds.), *Init. Repts. Deep Sea Drill. Proj.*, Washington, D.C.: U.S. Government Printing Office, 35:359–400.

Vail, P. R., and Hardenbol, J. 1979. Sea-level changes during the Tertiary. *Oceanus,* 22:71–79.

van Andel, T. H., Heath, G. R., and Moore, T. C., Jr. 1975. *Cenozoic tectonics, sedimentation, and paleoceanography of the central equatorial Pacific, Mem. Geol. Soc. Am.,* 143.

Vincent, E., Killingly, J. S., and Berger, W. H. 1980. The magnetic epoch-6 carbon shift, a change in the ocean's $^{13}C/^{12}C$ ratio 6.2 million years ago. *Mar. Micropaleontol.,* 5:185–203.

Wardlaw, B. R., and Quinn, T. M. 1991. The record of Pliocene sea-level change at Enewetak Atoll. *Quat. Sci. Rev.,* 10:247–258.

Warnke, D. A. 1982. Pre-Middle Pliocene sediments of glacial and preglacial origin in the Norwegian-Greenland Seas: Results of the DSDP Leg 38. *Earth Evol. Scis.,* 2:69–78.

Warnke, D. A., Allen, C. P., Muller, D. W., Hodell, D. A., and Brunner, C. A. 1992. Miocene-Pliocene Antarctic glacial evolution: A synthesis of ice-rafted debris, stable isotope, and planktonic foraminiferal indicators, ODP Leg 114. *In* Kennett, J. P., and Warnke, D. A. (eds.), *The Antarctic paleoenvironment: A perspective on global change,* pt. 1, Antarctic Research Series, 56:311–325.

Webb, P-N., and Harwood, D. M. 1991. Late Cenozoic glacial history of the Ross Embayment, Antarctica. *Quat. Sci. Rev.,* 10:215–224.

Westgate, J. A., Stemper, B. A., and Pewe, T. L. 1990. A 3 m.y. record of Pliocene-Pleistocene loess in interior Alaska. *Geology,* 18:858–861.

Wise, S. W., Jr., Gombos, A. M., and Muza, J. P. 1985. Cenozoic evolution of polar water masses, southwest Atlantic Ocean. *In* Hsü, K. J., and Weissert, H. J. (eds.), *South Atlantic paleoceanography,* Cambridge: Cambridge University Press, 283–324.

Woodruff, F. 1985. Changes in Miocene deep-sea benthic foraminiferal distribution in the Pacific Ocean: Relationship to paleoceanography. *In* Kennett, J. P. (ed.), *The Miocene ocean: Paleoceanography and biogeography, Mem. Geol. Soc. Am.,* 163:131–176.

Woodruff, F., and Savin, S. M. 1989. Miocene deepwater oceanography. *Paleoceanography,* 4:87–140.

———. 1991. Mid-Miocene isotope stratigraphy in the deep sea: High-resolution correlations, paleoclimatic cycles, and sediment preservation. *Paleoceanography,* 6:755–806.

Woodruff, F., Savin, S., and Douglas, R. G. 1981. Miocene stable isotope record: A detailed deep Pacific Ocean study and its paleoclimatic implications. *Science,* 212:665–668.

Wright, I. C., and Vella, P. 1988. A New Zealand Late Miocene magnetic stratigraphy: Glacioeustatic and biostratigraphic correlations. *Earth Planet. Sci. Lett.,* 87:193–204.

Wright, J. D., Miller, K. G., and Fairbanks, R. G. 1992. Early and Middle Miocene stable isotopes: Implications for deepwater circulation and climate. *Paleoceanography,* 7:357–389.

Chapter 5

Pollen Evidence for Vegetational and Climatic Change in Southern Africa during the Neogene and Quaternary

Louis Scott

In southern Africa, plant remains such as seeds or pollen grains are very scarce at fossil sites associated with human evolution. Therefore the interpretation of past vegetational change as a background for evolution relies less on direct evidence than on interpretations of the fauna (Vrba, 1985). Knowledge of the development of vegetation in the central interior of the subcontinent during the Neogene and Pleistocene is limited, because the erosional nature and climate of the interior region were generally not favorable for accumulation and long-term preservation of organic material. Unlike the grassland and savanna plateau biomes of the interior, early hominid fossils are not available in Plio-Pleistocene deposits from fynbos (macchia) and karoo shrubland biomes of the southern coastal areas (Hendey, 1984). Neogene pollen data, however, are readily available from the coastal and offshore areas where anaerobic conditions prevailed in depositional basins.

Dating of polleniferous Neogene deposits, especially from the coastal areas, is difficult, because datable fossils are often scarce and materials for chronometric dating, such as volcanic remains, are not always available. Offshore sediment cores provide a better potential because they contain microfossils and possibilities of palaeomagnetic dating (Sancetta et al., 1992).

To summarize Neogene and Quaternary vegetational changes in southern Africa, I discuss published and new palynological data from four areas in the subcontinent: the southern and southwestern Cape region, the relatively arid western coastal area of Namaqualand, the offshore area near the northern Namibian coast, and the interior plateau (fig. 5.1).

ISBN 0-300-06348-2.

The Southern and Southwestern Cape

Knysna Lignites. Thiergart et al. (1962), Helgren and Butzer (1977), and Coetzee et al. (1983) reported on pollen in different layers of lignite deposits from Knysna, in the southern Cape Province. Lowland swamp vegetation with Restionaceae, surrounded by Proteaceae woodland, is indicated by the pollen assemblages. The pollen composition in the lignites differs markedly from that of modern vegetation in the area consisting of forests. Some lignite layers show high numbers of palm-pollen grains similar to *Jubaeopsis* and *Nypa*, suggesting tropical conditions, whereas others contain Casuarinaceae (Helgren and Butzer, 1977; Coetzee et al., 1983).

Age indications for the Knysna lignite deposits are contradictory. Initial palynological interpretations of grass and Restionaceae pollen and the absence of Asteraceae (Compositae) suggested that they date to the interval between the Eocene and the Miocene (Thiergart et al., 1964; Helgren and Butzer, 1977; Coetzee et al., 1983). Studies of tectonics indicate an age no older than Middle Miocene, because these deposits rest on a surface that was probably formed by erosion after uplift in the Early Miocene (Maud and Partridge, 1987; Partridge and Maud, 1987).

Noordhoek. The pollen sequence from Noordhoek in the winter rainfall region of the southwestern Cape Province, which supports sclerophyllous fynbos (macchia), is informative, although chronometric dating of the deposits is not available (Coetzee, 1978a, 1978b; Coetzee and Rogers, 1982; Coetzee et al., 1983; Coetzee and Muller, 1984). The fossil-pollen data show that a markedly different vegetation, with subtropical forests

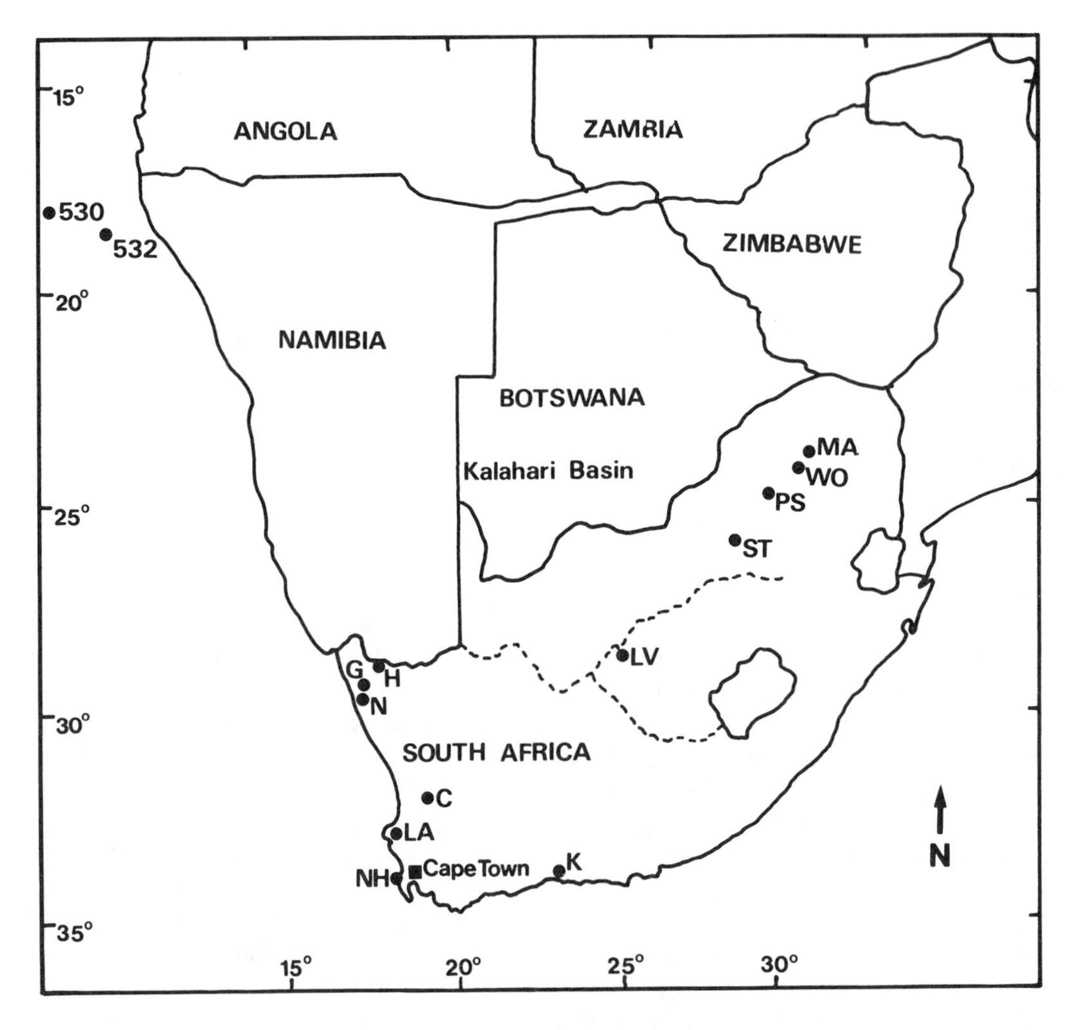

Fig. 5.1. Neogene and Quaternary palynological sites in southern Africa. K = Knysna; NH = Noordhoek; LA = Langebaanweg (Varswater Formation); C = Cederberg; N = Noup; G = Graafwater; H = Henkries; LV = Lower Vaal River; ST = Sterkfontein; PS = Pretoria Saltpan; WO = Wonderkrater; MA = Makapansgat; 530 and 532 = Deep Sea Drilling Project Sites.

that included palms, was present during the Early Neogene in the southwestern Cape region. The change from subtropical forest to modern fynbos vegetation has been linked to the growth of the Antarctic ice sheet and to the related development of the circum-Antarctic current in the southern ocean during the Miocene and Pliocene. Age estimates are based on correlation of pollen contents with the $\delta^{18}O$ chronology from the southeastern Indian Ocean and global sea-level fluctuations (Shackleton and Kennet, 1975; van Zinderen Bakker, 1975; Coetzee, 1978a, 1978b; Siesser, 1978; Vail and Hardenbol, 1979). The similarity of the older section of southwestern Cape pollen sequences with assemblages from Knysna suggests that they are possibly of a similar age, namely, Miocene.

Langebaanweg. A single pollen spectrum, 92 percent of which is made up of an unidentified pollen type, from the fossil-bearing Varswater Formation (Tankard and Rogers, 1978) represents the only published botanical data associated directly with the well-known fauna from the Early Pliocene (Hendey, 1984). Evidence for the vegetation associated with a peaty layer in the underlying Miocene Elandsfontyn Formation has, however, been published (Coetzee and Rogers, 1982). The Varswater deposits are not rich in pollen. To understand the nature of the vegetation, the original microscopic slide preparations of the spectrum described by Tankard and Rogers (1987) have been reexamined, together with two additional samples from the Varswater Formation (one taken from the quarry floor, the other from about 6.4 m below the floor). The dominant pollen type belongs to the Ranunculaceae (buttercup family; fig. 5.2), which probably occurred locally as aquatic or semiaquatic plants in the swampy basin that also trapped the faunal assemblage. The sample contains very few other diagnostic elements of an open vegetation and includes *Myrica*, Poaceae, Restionaceae and *Cliffortia* pollen. The quarry-floor sample indicates swampy conditions with Cyperaceae, and the presence of fynbos vegetation is confirmed by small percentages of Proteaceae, *Cliffortia*, Ericaceae, and Restionaceae pollen. The high proportions of Asteraceae (Compositae, long-spine type) and Cheno/Ams (Chenopodiaceae and Amaranthaceae type), however, indicate relative dryness. *Cliffortia*, Restionaceae, and palm-pollen types are found in the 6.4-m sample. Tree pollen types (*Podocarpus*, *Olea*, and Proteaceae) were recorded in small numbers and indicate the past existence of some woodland in the general surroundings. This level seems to be transitional between the pollen types cited above and the Miocene pollen spectra of the underlying Elandsfontyn Formation, which suggests subtropical swampy conditions comparable to those indicated in the Noordhoek sequence (Coetzee and Rogers, 1982).

Changes in the pollen sequence confirm the wider regional interpretation, based on pollen profiles and faunal indications, that subtropical forest in the Miocene was replaced by more open vegetation in the Early Pliocene (Coetzee and Rogers, 1982; Hendey, 1984). Coastal-plains elements like Ranunculaceae, Cyperaceae, Umbelliferae, and Asteraceae are recorded, and fynbos probably occurred in the wider surroundings. Restionaceae and *Cliffortia* in the Pliocene samples are probably of local origin. The palm pollen in the deepest level (fig. 5.2) is transitional to the Elandsfontyn material and gives a suggestion of more tropical conditions than exist at present.

Cederberg. The contrast between glacial and interglacial cycles in the coastal region during the Quaternary can be assessed in a new pollen sequence of about 20 thousand years (kyr) from the Cape Province. This sequence was obtained from fossil hyrax middens preserved in a rock shelter in the Cederberg (Scott, 1994). Essentially, the sequence shows altitudinal lowering of vegetation zones, with more prominent *Elytropappus* (*Stoebe*-type pollen) in the Cape fynbos area during the cold phase than at present.

Namaqualand

Only isolated pollen spectra from different alluvial sequences are available from the Namaqualand-Namibia region (fig. 5.1). The data presented here are preliminary, because these deposits have not been dated. These pollen spectra (fig. 5.3) are compared with average modern values in the region derived from the Richtersveld (Scott and Cooremans, 1992).

First, pollen in channel clays from a borehole drilled at Noup suggests a moist climate with woodland and forest elements in the area, differing entirely from the vegetation in the present hyperarid situation (fig. 5.3). The assemblage consists of pollen of *Podocarpus*, *Olea*, Proteaceae, and Myrtaceae and many fern spores (table 5.1). Some Asteraceae types (of both the Mutisiae type and typical long-spine groups) occur in small numbers. The similarities with the southern and southwestern Cape pollen spectra suggest a probable Miocene age.

Laminated clays from an alluvial sequence drilled at Graafwater, also in Namaqualand, contain mainly Asteraceae pollen (long-spine, *Pentzia*, and *Pacourina* types), as well as that of the Cheno/Ams, Cyperaceae, and *Typha* (fig. 5.3). Fynbos elements are represented by *Passerina* and *Cliffortia*, and arboreal pollen by *Podocarpus*, *Olea*, *Rhus*, *Celtis*, and Capparaceae (table 5.1). Fern spores of Gleicheniaceae and Schizaeaceae are also present. The total combination suggests both dry and wet conditions, either in seasonal or in local and nonlocal contexts.

Diatomaceous clays from Henkries (fig. 5.1), in the same region, contain mainly pollen of grasses, Cyperaceae, long-spine Asteraceae, Cheno/Ams, Restionaceae, *Anthospermum*, Ericaceae, Aizoaceae, *Boscia*, and some fern spores (*Ophioglossum;* fig. 5.3). This assemblage can apparently be correlated with the period after the main change recognized in the southwestern Cape sequence by Coetzee (1978a, 1978b) and is therefore possibly no older than Late Miocene. It seems to indicate, however, that slightly moister conditions occurred in this semi-arid area than in the present higher-lying karoo region of the interior.

Offshore Areas

Neogene vegetation along the southwestern African coast can be reconstructed in part on the basis of palynological data from offshore regions. Neogene pollen spectra from Deep Sea Drilling Project cores in marine

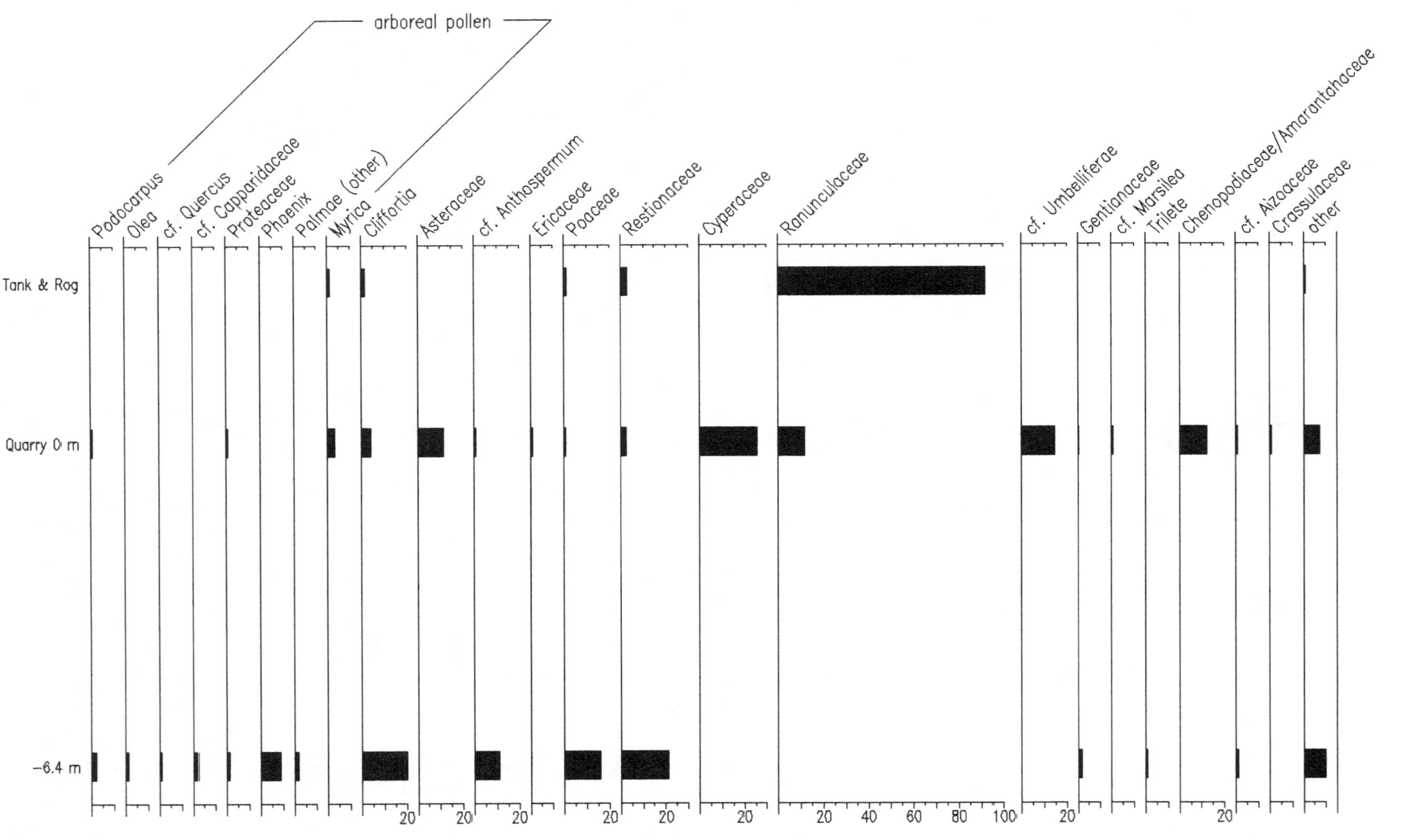

Fig. 5.2. Pollen percentages in samples from the Pliocene Varswater Formation at Langebaanweg. Tank & Rog = Tankard and Rogers (1987); Quarry 0 m = sample from quarry surface; −6.4 m = depth below quarry surface.

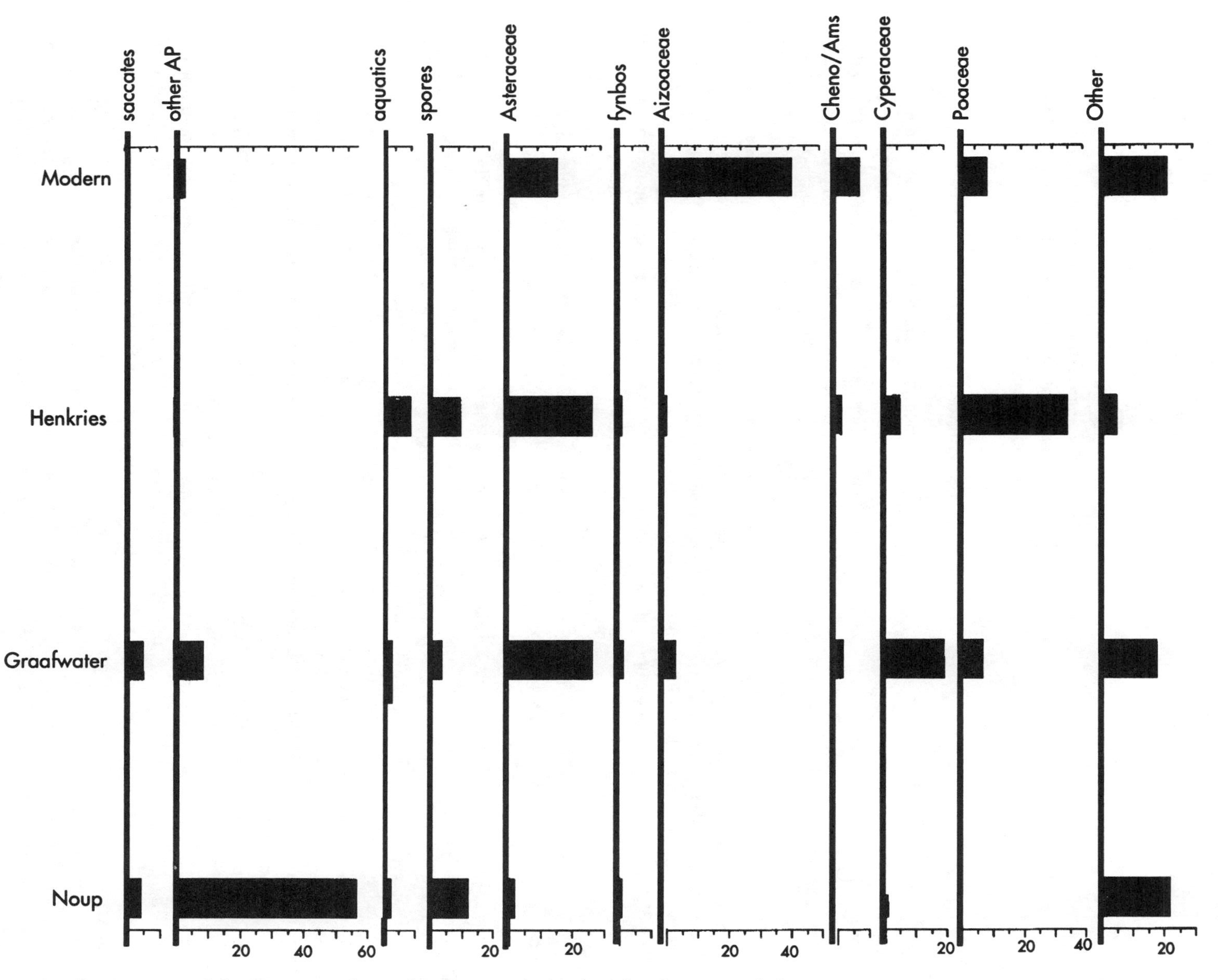

Fig. 5.3. Pollen percentages in sediments from Noup, Graafwater, and Henkries compared with modern pollen values in Namaqualand.

Table 5.1. Some Neogene Arboreal Pollen Types (% of Total Spectra), Namaqualand

	Graafwater	Noup
Microcachryidites*		1
Podocarpus	5	3
Syzygium		2
Eugenia		1
Sapotaceae		1
Trema		1
Olea	3	15
Proteaceae	1	5
Palmae		5
Alchornea	1	
cf. Isoberlinia		1
Celastraceae		7
Combretaceae		1
Capparaceae	1	12

*Extinct genus, present until ca. Late Miocene (Coetzee et al., 1983).

deposits at Sites 530 and 532, off the northern Namibian and southern Angolan coast (fig. 5.1), have been described by van Zinderen Bakker (1984) and Sancetta et al. (1992). The former gave a description of pollen in twelve core samples from the two sites ranging in age from Late Miocene to Quaternary (fig. 5.4), while Sancetta et al. (1992) provided fluctuations of three pollen types in a relatively closely sampled section of the 532 core. Improved dating by Sancetta et al. (1992) suggests that this interval ranges in age between 2.6 and 2.2 million years (myr).

The results of the first study show that pollen types of grass (Poaceae), Chenopodiaceae, and Asteraceae were dominant beginning with the Late Miocene, suggesting that relatively open desertic conditions had already developed by then. Chenopodiaceae pollen has its lowest values in van Zinderen Bakker's (1984) Late Miocene sample, possibly indicating relatively wet conditions.

Four samples from Early to Late Pliocene levels, which should be older than 2.6 myr, contain relatively high proportions of Chenopodiaceae pollen (almost 40 percent; fig. 5.4). They have been interpreted by van Zinderen Bakker to represent the vegetation of salt flats on an exposed ocean floor during a period of low sea levels. There are indications, however, that at least some parts of the Early to Mid-Pleistocene sea levels were relatively high (Hendey, 1984). The presence of palm and other tree pollen such as Combretaceae may argue for a tropical climate, possibly associated with intense seasonal evaporation and dryness. Sancetta et al. (1992) show relatively high Chenopodiaceae pollen (up to 30 percent) between 2.35 and 2.5 myr and attribute it to a decrease in precipitation. The pollen samples suggest wet, grassy conditions with fewer Asteraceae (Compositae) between 2.25 and 2.35 myr and also ca. 2.5 myr. High Asteraceae values between 2.32 and 2.4 myr are attributed to increased winter rainfall during glacial events. Oscillations of grass pollen in the profile seem to follow the obliquity cycle.

The Interior Plateau

Large-scale accumulation of Neogene sediments occurred in the Kalahari Basin (fig. 5.1), but they were not suitable for pollen preservation. Sands in the Older Gravels of the lower Vaal River, which produced a grassland pollen spectrum (Scott, 1986), have been resampled, but the occurrence of open vegetation could not be confirmed. The spectrum is probably not in situ, because the pollen occurs in the form of clumps, which suggests postdepositional accumulation in bee tunnels (Bottema, 1975).

A problem with modern contamination has been identified in the Plio-Pleistocene, calcareous, hominid cave deposits from the interior plateau (Scott, 1982a; Scott and Bonnefille, 1986). The presence of recrystallized carbonate suggests that ground water moved through the deposits in the past. If this is true, oxygen in the water probably destroyed most of the pollen in the sediments. In recent times the porous breccia surfaces have been exposed to the atmosphere by excavations and have become penetrated by modern pollen rain, as is suggested by historically introduced exotic elements like *Pinus.* Fresh samples from deep below the excavation surfaces yield neither in situ nor modern pollen but only rare contaminants contracted during sampling or laboratory processing.

Stalagmites or travertines that occur in breccias are not porous, and their contents are therefore more effectively sealed off from ground water and the atmosphere. As a result, pollen that was trapped in this material has a better chance of being preserved. Scott and Bonnefille (1986) have recently focused attention on travertines and speleothems as potential sources of in situ pollen. Those that have been formed in closed chambers and not in proximity to past cave openings and atmospheric dust, however, cannot be expected to be rich in pollen.

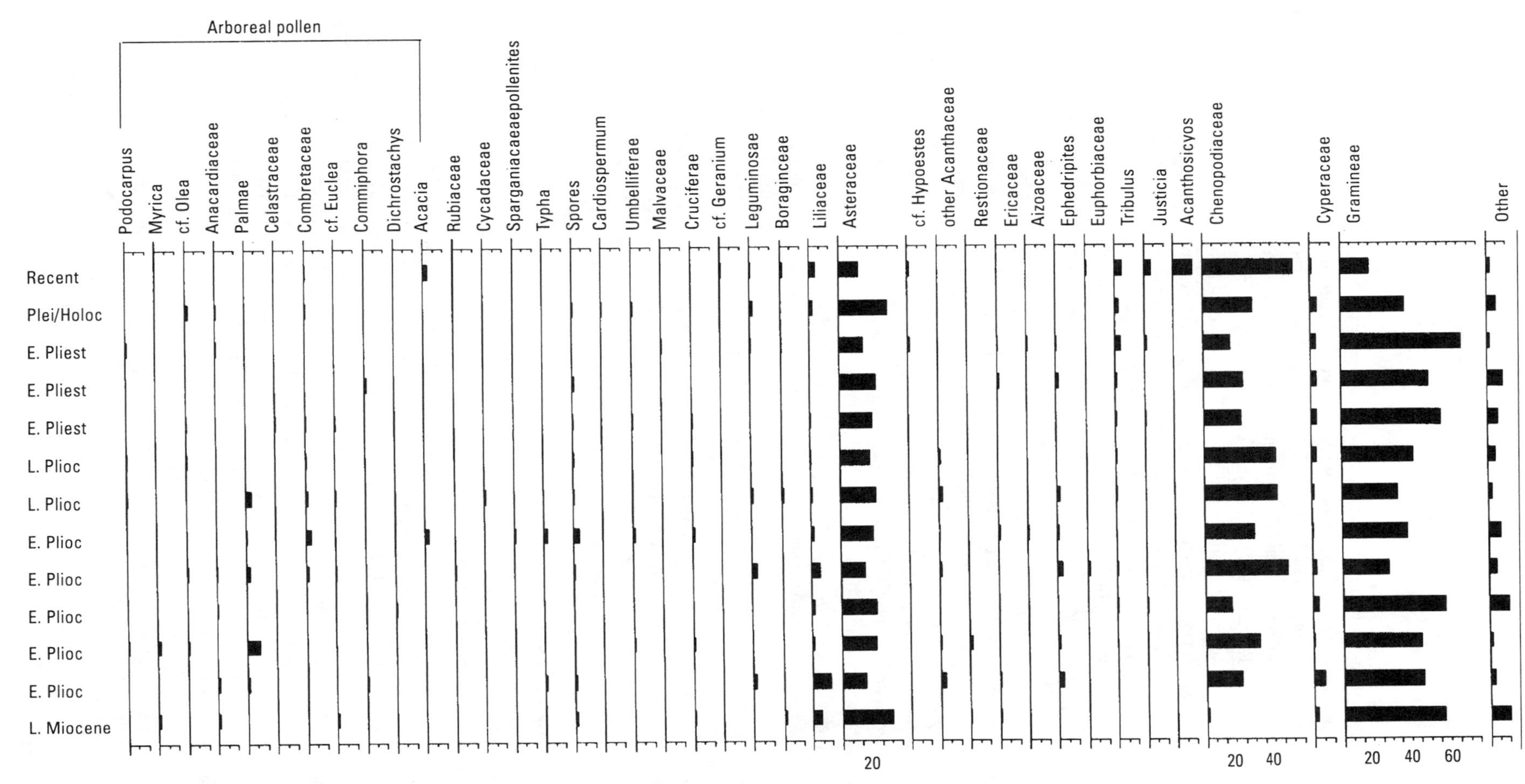

Fig. 5.4. Pollen percentages in marine sediments from DSDP Sites 530 and 532, based on data published by van Zinderen Bakker (1984).

Makapansgat. The 3 myr-old hominid-bearing breccias from the dolomite cave Makapansgat, in the interior summer-rainfall plateau of northern Transvaal (fig. 5.1), produced bushveld and exotic *Pinus* pollen (Cadman and Rayner, 1989). Cadman and Rayner (1989) and Rayner et al. (1993) claim that these spectra provide evidence of wet conditions and forest vegetation around 3 myr ago. This explanation does not account, however, for the relatively high proportion of Cheno / Am-type pollen (indicative of dryness and strong evaporation) and the lack of *Podocarpus* pollen. The pollen types (*Celtis* and *Myrica*) considered by Rayner et al. (1993) as forest indicators can be found in different regions of the Transvaal and the Orange Free State and in areas other than forest (Scott, 1982b). Further, the pollen spectra from the Makapansgat sediments are strongly suggestive of the modern environment, and the presence of *Pinus* indicates that they were probably contaminated recently by the atmosphere. Because Makapansgat is surrounded by natural bush and is far from *Pinus* plantations, the presence of its pollen can be considered as a sign of relatively severe contamination.

Furthermore, I found no convincing fossil pollen in 23 hyena coprolites from the Member 3 breccia in Makapansgat (supplied by Alun Hughes in 1981). One produced a spectrum with 36 percent Asteraceae, 23 percent Poaceae, and 25 percent arboreal pollen. The latter includes *Kirkia* (4 percent), *Burkea* (5 percent), Combretaceae (7 percent), *Euclea* (2 percent), Mimosoideae (including *Acacia*, 2 percent), *Boscia* (2 percent), *Tarchonathus* (1 percent), and *Pinus* (1 percent). The spectrum can be interpreted as a modern dust contamination trapped by the archaeological fixative on the coprolite. The pollen combination is probably of the modern assemblage at Makapansgat and does not differ too much from Cadman and Rainer's (1989) spectra.

Kromdraai and Sterkfontein. The presence of exotic *Pinus* pollen, in samples from the southern Transvaal hominid breccias at Kromdraai and Sterkfontein, is indicative of contamination derived from many exotic trees in the vicinity (Scott, 1982a; Scott and Bonnefille, 1986). Updated pollen data from Sterkfontein and Kromdraai are presented in figure 5.5 and include previously published results (Scott and Bonnefille, 1986), new results of a Sterkfontein Member 5 travertine, and modern spectra representing different seasons in the area. A pollen spectrum indicating open Proteaceae savanna in travertines from the basal part of a Member 2/3 stalagmite in Kromdraai (Scott and Bonnefille, 1986) contains a relatively low *Pinus* pollen percentage (fig. 5.5). A similar assemblage is found in the Member 5 travertines from Sterkfontein. The lack of Proteaceae pollen at Sterkfontein and Kromdraai today shows that this pollen is an unlikely contaminant from the modern environment in the travertines and is probably partly in situ, representing the past vegetation at the site during the Plio-Pleistocene.

Proteaceae pollen of Plio-Pleistocene age at Sterkfontein and Kromdraai is assigned to *Protea* rather than to *Faurea*, because it is accompanied by grassland rather than woodland elements. Today open *Protea* savanna occurs in the vicinity, on quartzite outcrops, but in the past it might have been denser or more widely distributed, covering the dolomite area near the caves. The environment must have differed only slightly from the modern one, but with the evidence available it is difficult to interpret the change in terms of palaeoclimate. An open grassland savanna, however, conforms with the faunal indications of a change from dense woodland to more open vegetation around Plio-Pleistocene times (Vrba, 1985).

Wonderkrater and the Pretoria Saltpan. Long Quaternary pollen records have been obtained from the organic warm spring at Wonderkrater (Scott, 1982c) and the crater lake of the Pretoria Saltpan (Partridge et al., 1993). Both sites show strong cyclic patterns in their pollen sequences during the Late and Middle Pleistocene, suggesting alternations between dry woodland savanna, cool, open upland grassland (including fynbos elements), and mesic woodland with extensive forests.

Preliminary counts of microscopic charcoal in the Pretoria Saltpan indicate that phases of intense fires occurred in the bushveld environment in the past (fig. 5.6). Periods with high percentages of tree pollen, suggesting denser bush cover, do not always coincide with high charcoal. Increased fires can possibly be related to the human activities and may indicate pulses of occupation of the area alternating with long phases with no occupation.

Discussion

Salient aspects of the palynological evidence for southern Africa are summarized in table 5.2. It is almost certain that during the Miocene, when subtropical elements flourished in the southwestern coastal and western de-a.

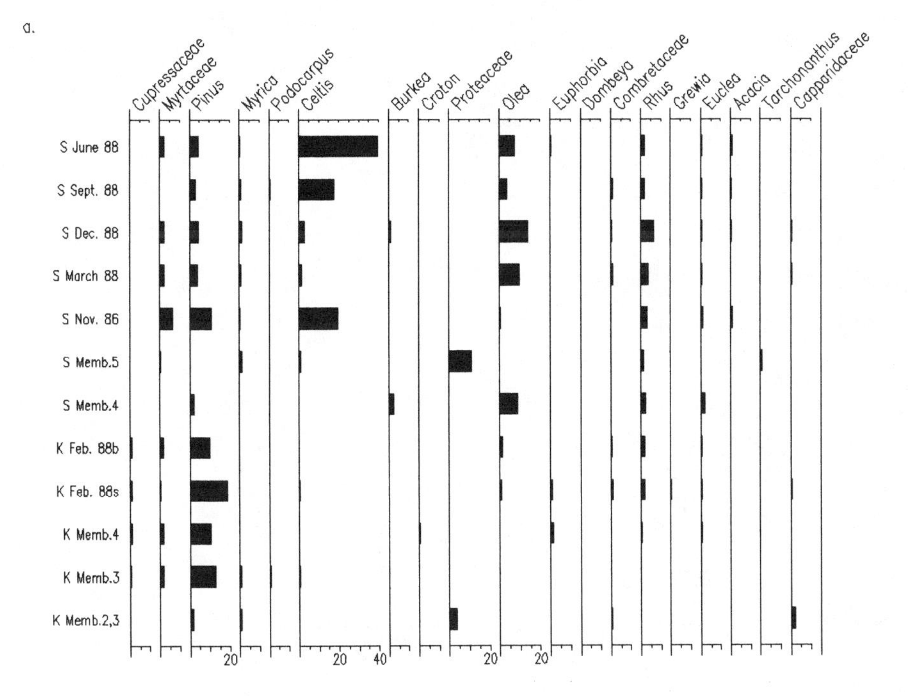

b.

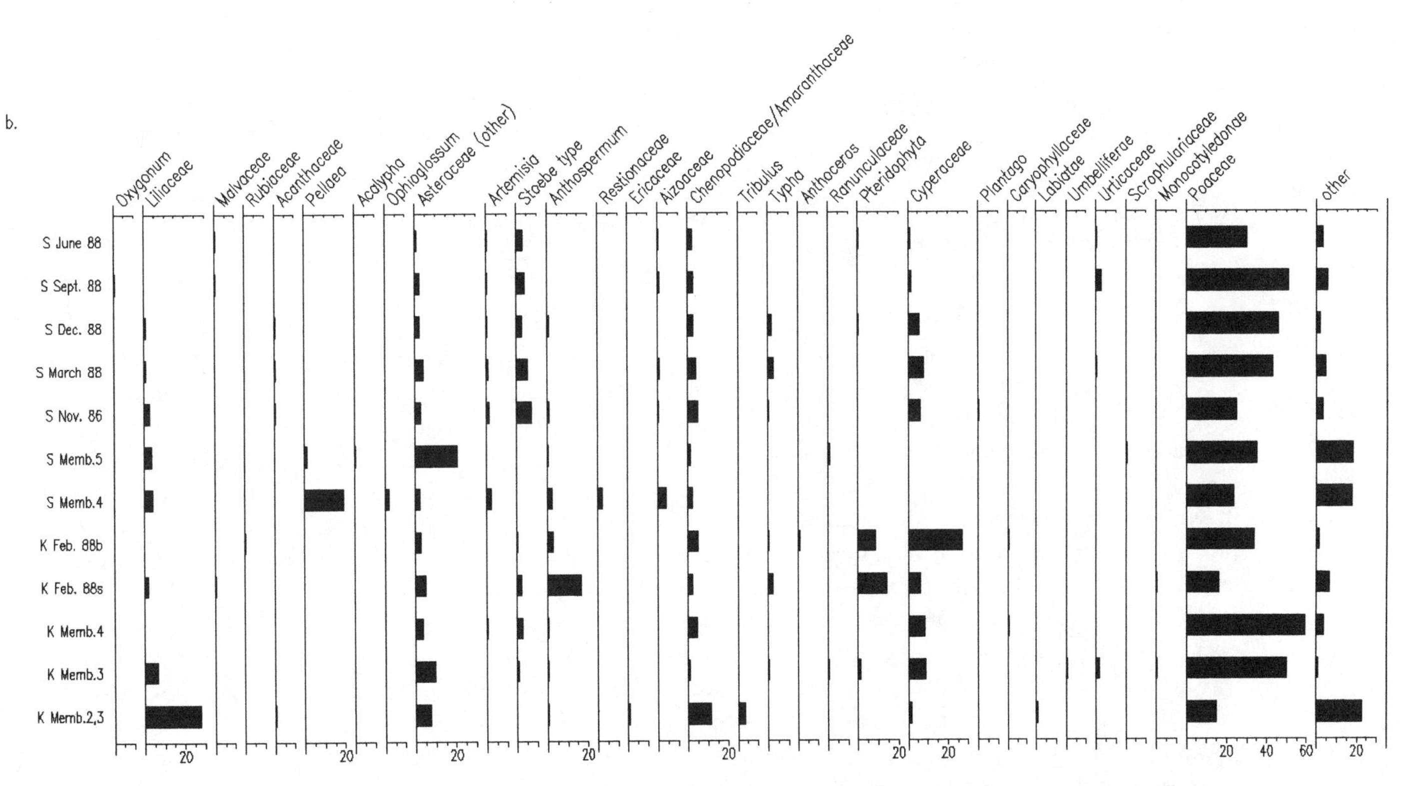

Fig. 5.5. Pollen percentages (part a = arboreal pollen; part b = nonarboreal pollen) in samples from Kromdraai (K) and Sterkfontein (S), representing different modern seasons, breccias (S, Members 4; K, Members 3 and 4), and travertines (S, Member 5; K, Member 2, 3).

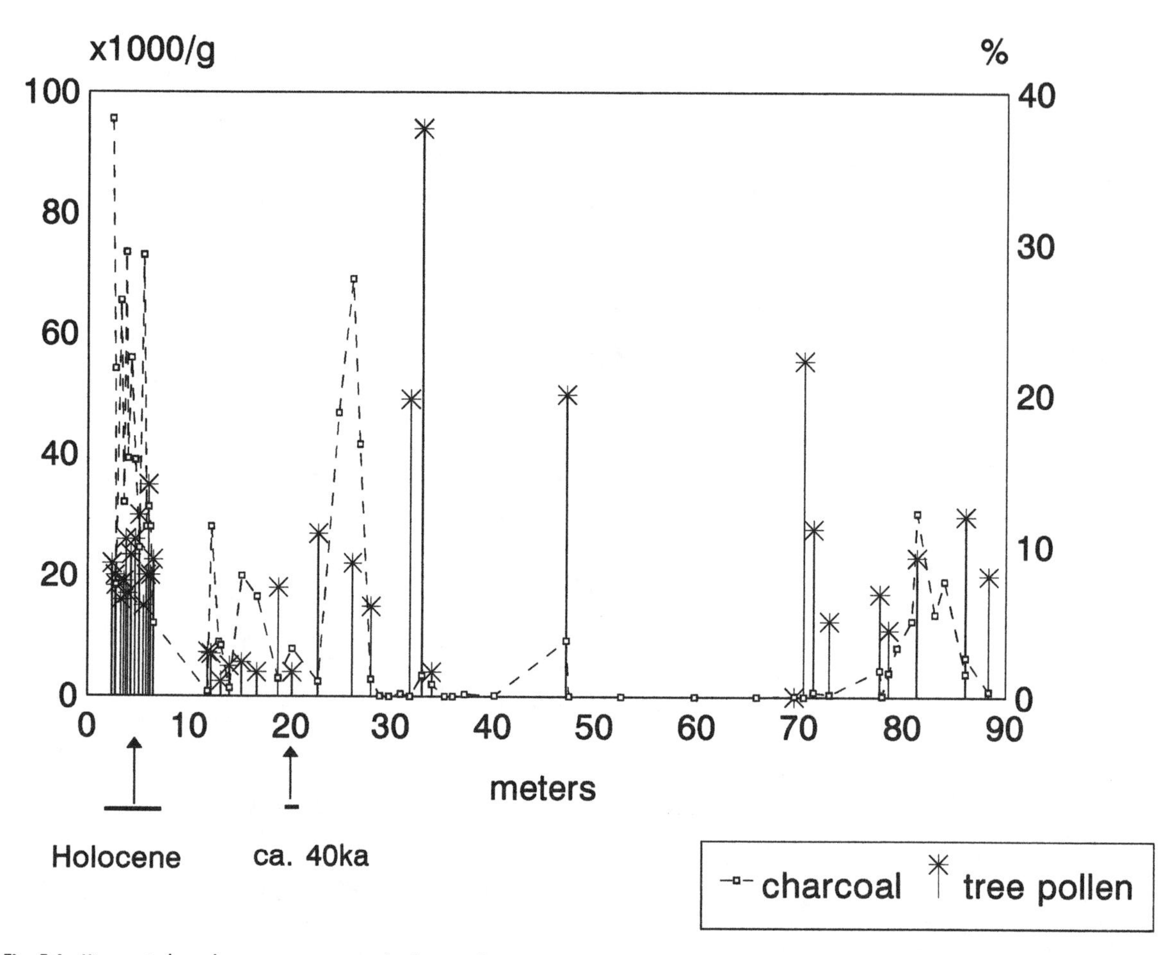

Fig. 5.6. Microscopic charcoal concentrations compared with tree-pollen percentages in the Pretoria Saltpan. The positions of the Holocene section and the 40,000-year level, based on radiocarbon dates, are indicated by arrows.

sert regions of the subcontinent, conditions in the interior must also have been relatively warm and wet and probably supported subtropical or tropical woodlands and forests. The vegetation must have consisted of an entirely different assemblage of species than that found in the modern environment. This past woodland was probably replaced by open bushveld savanna and grassland by the end of the Miocene; according to faunal evidence, the vegetation continued to become even more open during Plio-Pleistocene (Vrba, 1985).

Environments during the Pliocene seem to have been relatively close to the modern ones, as is suggested by pollen from the southwestern Cape region and the offshore pollen sequences from northern Namibia. Glacial and interglacial cycles, which are related to orbital forcing, were manifested in vegetational changes (Sancetta et al., 1992). The scarce pollen data from the calcareous cave sites in the interior regions suggest little deviation from present environmental conditions. However, cyclic changes comparable to those in Quaternary pollen sequences like those at Wonderkrater and the Pretoria Saltpan must also have been manifested as far back as the Pliocene period in the interior. An example of such a cyclic pattern in the Neogene period can be seen in the pollen record from Bogotá (Hooghiemstra, this vol.).

It is doubtful whether long pollen records of the Neogene period will be found in the interior of southern Africa, although deposits in the great lakes further to the north in Africa have the potential of producing some information. In the productive coastal area of the southwestern Cape and in the offshore areas, new detailed numerical analyses of pollen and descriptions of Neogene assemblages may in future help to refine interpretations of environmental change in southern Africa.

Table 5.2. Indications of Vegetation in Parts of Southern Africa during the Neogene, Based on Pollen Data

Period	Southern and Southerwestern Cape	Namaqualand	Interior Plateau	Marine Area of Namibia, off West Coast
Quaternary	Fynbos (macchia)	Succulent rich dwarf shrubland or grassland	Woodland savanna or upland grassland or moist mesic woodland	Desert vegetation or dry grassland
Pliocene	Fynbos		Nearly similar to Quaternary vegetation	Open desert or dry woodland or shrubland vegetation or dry grassland
Late Miocene/Pliocene	Transition from subtropical woodland to fynbos	Karoid shrubland with fynbos and woodland elements		Development of desert elements like Chenopodiaceae
Miocene	Subtropical woodland with swamps	Subhumid subtropical woodland		

Summary

In southern Africa the palynological record from the Cape coastal area and the southern desert show that major changes in vegetation, from tropical forms to open fynbos or desert plants, occurred before the Pliocene period. Regular cyclic vegetational change apparently related to orbital parameters occurred throughout the Pliocene and Pleistocene and have been reported in Pliocene marine-pollen sequences off the west coast and in Quaternary lake and spring deposits in the interior. Palynological interpretations from Pliocene and Pleistocene hominid deposits are hampered by limited pollen preservation and contamination by modern dust.

Acknowledgments

Langebaanweg deposits were supplied by the Geological Survey of South Africa, and those from Namaqualand by De Beers Ltd. Research on the palynology of hominid deposits from the Transvaal were conducted under the auspices of the Palaeoanthropological Research Unit of the University of the Witwatersrand, which is directed by P. V. Tobias.

References

Bottema, S. 1975. The interpretation of pollen spectra from prehistoric settlements (with special attention to Liguliflorae). *Palaeohistoria* 17:18–35.

Cadman, A., and Rayner, R. J. 1989. Climate change and the appearance of *Australopithecus africanus* in the Makapansgat sediments. *Journal of Human Evolution* 18:107–113.

Coetzee, J. A. 1978a. Late Cainozoic palaeoenvironments of southern Africa. In *Antarctic glacial history and world palaeoenvironments*, pp. 115–127 (ed. E. M. van Zinderen Bakker). Balkema, Rotterdam.

———. 1978b. Climate and biological changes in south-western Africa during the Late Cainozoic. *Palaeoecology of Africa* 10:13–29.

Coetzee, J. A., and Muller, J. 1984. The phytogeographic significance of some extinct Gondwana pollen types from the Tertiary of the southwestern Cape (South Africa). *Annals of the Missouri Botanical Gardens* 71:1088–1099.

Coetzee, J. A., and Rogers, J. 1982. Palynological and lithological evidence for the Miocene palaeoenvironment in the Saldanha region (South Africa). *Palaeogeography, Palaeoclimatology, Palaeoecology* 39:71–85.

Coetzee, J. A., Scholtz, A., and Deacon, H. J. 1983. Palynological studies and the vegetation history of the fynbos. In *Fynbos palaeoecology: A preliminary synthesis*, South African Scientific Programmes Report 75, pp. 156–173 (ed. H. J. Deacon), C.S.I.R., Pretoria.

Helgren, D. M., and Butzer, K. W. 1977. Palaeosols of the southern Cape coast, South Africa: Implications for laterite definition, genesis, and age. *Geographical Review* 67:430–445.

Hendey, Q. B. 1984. Southern African late Tertiary vertebrates. In *Southern African prehistory and palaeoenvironments*, pp. 81–106 (ed. R. G. Klein). Balkema, Rotterdam.

Maud, R. R., and Partridge, T. C. 1987. Regional evidence for climate change in southern Africa since the Mesozoic. *Palaeoecology of Africa* 18:337–348.

Partridge, T. C., Kerr, S. J., Metcalfe, S. E., Scott, L., Talma, A. S., and Vogel, J. C. 1993. The Pretoria Saltpan: A 200,000 year southern African lacustrine sequence. *Palaeogeography, Palaeoclimatology, Palaeoecology* 101:317–337.

Partridge, T. C., and Maud, R. R. 1987. Geomorphic evolution of southern African since the Mesozoic. *South African Journal of Geology* 902:179–208.

Rayner, R. J., Moon, B. P., and Masters, J. C. 1993. The Makapansgat australopithecine environment. *Journal of human evolution* 24:219–231.

Sancetta, C., Heusser, L., and Hall, M. A. 1992. Late Pliocene climate in the southeast Atlantic: Preliminary results from a multi-disciplinary study of DSDP Site 532. *Marine Micropaleontology* 20:59–75.

Scott, L. 1982a. Pollen analyses of Late Cainozoic deposits in the Transvaal, South Africa, and their bearing on palaeoclimates. *Palaeoecology of Africa* 15:101–107.

———. 1982b. Late Quaternary fossil pollen grains from the Transvaal, South Africa. *Review of Palaeobotany and Palynology* 36:241–278.

———. 1982c. A Late Quarternary pollen record from the Transvaal bushveld, South Africa. *Quaternary Research* 17:339–370.

———. 1986. The Late Tertiary and Quaternary pollen record in the interior of South Africa. *South African Journal of Science* 82(2):73.

———. 1994. Palynology of late Pleistocene hyrax middens, south-western Cape Provence, South Africa: A preliminary report. *Historical Biology* 9:71–81.

Scott, L., and Bonnefille, R. 1986. A search for pollen from the hominid deposits of Kromdraai, Sterkfontein and Swartkrans: Some problems and preliminary results. *South African Journal of Science* 82:380–382.

Scott, L., and Cooremans, B. 1992. Pollen in recent Procavia (hyrax), Petromus (dassie rat) and bird dung in South Africa. *Journal of Biogeography* 19:205–215.

Shackleton, N. J., and Kennet, J. P. 1975. Palaeotemperature history of the Cenozoic and the initiation of Antarctic glaciation: Oxygen and carbon isotope analysis in D.S.D.P. sites 277, 279 and 281. *Initial Reports of the Deep Sea Drilling Project* 29:743–756.

Siesser, W. G. 1978. Aridification of the Namib Desert: Evidence from oceanic cores. In *Antarctic glacial history and world palaeoenvironments,* pp. 105–114 (ed. E. M. van Zinderen Bakker). Balkema, Rotterdam.

Tankard, A. J., and Rogers, J. 1978. Late Cenozoic palaeoenvironments on the west coast of Southern Africa. *Journal of Biogeography* 5:319–337.

Thiergart, F., Frantz, U., and Raukopf, K. 1962. Palynologishe Untersuchungen von Teriärkohlen und einer Oberflächenprobe nahe Knysna, Südafrika. *Advancing Frontiers of Plant Sciences* 4:151–178.

Vail, P. R., and Hardenbol, J. 1979. Sea-level changes during the Tertiary. *Oceanus* 22:71–79.

van Zinderen Bakker, E. M. 1975. The origin and palaeoenvironment of the Namib Desert biome. *Journal of Biogeography* 2:65–73.

———. 1984. Palynological evidence for Late Cenozoic arid conditions along the Namibia coast from holes 532 and 530A, leg 75, Deep Sea Drilling Project. *Initial Reports of the Deep Sea Drilling Project* 75:763–768.

Vrba, E. S. 1985. Early hominids in southern Africa: Updated observations on chronological and ecological background. In *Hominid evolution,* pp. 195–200 (ed. P. H. Tobias). Alan R. Liss, New York.

Chapter 6

Tertiary Environmental and Biotic Change in Australia

Michael Archer, Suzanne J. Hand, and Henk Godthelp

Australia has one of the most distinctive biotas in the world, partly because of its Gondwanan origins and partly because of its long period of isolation following separation from Antarctica between 35 and 45 million years (myr) ago. During this Indian Ocean voyage, its northward movement into lower latitudes counteracted the net cooling of world climates and enabled it to maintain, for much of the Tertiary, reasonably equable climates—at least until the northern edge of the continent began to buckle (as the highlands of New Guinea did) following collision with the continental fragments of southeastern Asia about 15 myr ago.

Yet, in spite of this long and generally pleasant climatic sojourn, Tertiary Australian mammal-bearing ecosystems have left a frustratingly uneven record and one largely restricted to the eastern half of the continent. Following the Mesozoic (with only one Early Cretaceous local fauna, from Lightning Ridge, New South Wales), there is one Early Eocene local fauna (Murgon, Queensland). The record improves in the Middle Tertiary, with many Late Oligocene to Middle Miocene local faunas (South Australia, Queensland, Tasmania), only to degenerate again in the Late Miocene, with only three significant local faunas (Northern Territory, Victoria, New South Wales). The Pliocene record, particularly the earliest part, improves (Queensland, Victoria, South Australia, New South Wales, New Guinea), leading up to a plethora of Pleistocene assemblages from all parts of the continent, including Western Australia and New Guinea. Major gaps in the Australian mammal record are the pre-Early Cretaceous; Middle Cretaceous to Late Paleocene; Middle Eocene to Middle Oligocene; and the pre-Late Pliocene record for New Guinea. Uncertainties in dating pertain to most local faunas and in particular to those interpreted to be Late Miocene to Middle Pleistocene. Because of the paucity of well-dated Late Miocene and post-Early Pliocene to pre-Late Pleistocene sites, it is difficult to date precisely three major episodes of faunal turnover, which appear to have occurred sometime between 15 and 8 myr, between 8 and 5 myr, and between 5 myr and the Late Pleistocene. The major biotic turnover in the Late Pleistocene, approximately 30 to 17 thousand years (kyr), is better understood.

We summarize here current understanding of changes in Australian paleoenvironments and paleoclimates, with a focus on the Late Oligocene–Early Miocene to the Late Pliocene record of mammals. This chapter highlights the need for more information about events between 15 and 2 myr, a critical period during which major changes occurred in Australian climates, ecosystems, and biodiversity, and the need for a better-dated and more substantial correlative framework for the whole of Australia's Middle and Late Tertiary record. We also consider congruence with the broad features of the climatic oscillation model of McGowran (e.g., Frakes et al., 1987) and with relevant parts of the Australian paleobotanical and paleoclimatic records. Finally, we attempt, within the limitations of the record, to relate faunal turnover events in Australia to those of the rest of the world.

We use the term Meganesia (*sensu* Filewood, 1984) for the area comprising Australia, New Guinea, and surrounding islands on the continental shelf. Higher-level systematic nomenclature of marsupials follows Aplin and Archer (1987). Biostratigraphic nomenclature follows Woodburne et al. (1985), Archer et al. (1989, 1994a,

ISBN 0-300-06348-2.

1994b), Rich et al. (1991), and Tedford et al. (1992). Unless otherwise indicated, epoch boundaries and their ages in millions of years follow the scheme promoted by Harland et al. (1990). The concept of local fauna used in this chapter follows Tedford (1970).

The Australian Mammal Record: Evidence for Biotic Turnover and Paleoecological Interpretation

Because of the patchy nature of the Australian record, evolution of this continent's mammals as a function of time has seven discrete stages separated by gaps of either little or no knowledge (fig. 6.1). They are: 1) the Early Cretaceous; 2) the Early Eocene; 3) the Late Oligocene to Middle Miocene; 4) the Late Miocene; 5) the Early Pliocene; 6) the Late Pleistocene; and 7) the Holocene. The precise timing and manner of transition from one stage to the next are poorly understood.

Early Cretaceous. Australia's pre-Cenozoic mammal record can be readily summarized: it consists of one undoubted mammal, the ornithorhynchid monotreme *Steropodon galmani,* based on a dentary fragment from Lightning Ridge, New South Wales, which is approximately 120 myr old (Archer et al., 1985); plus a second, more distinctive Lightning Ridge mammal presently under study (Flannery, Archer, Rich, and Jones, in prep.). Cretaceous biotas of South America also include only non-therian mammal groups (Bonaparte, 1990). In southern Victoria, there is evidence (Rich et al., 1988) to suggest that regional climates of this time were cold, with mean annual temperatures close to 0°C. The next youngest and only Early Tertiary monotreme known is from the Early Paleocene of southern Argentina (the Banco Negro Inferior faunal assemblage; Pascual et al., 1992); all younger monotremes known are from Australia (Archer et al., 1992).

Early Eocene. The Tingamarra local fauna from southeastern Queensland is the only Early Tertiary mammal fauna known from Australia (Godthelp et al., 1992). Radiometric dating of illites in the clay containing the fossils provides the age estimate of 54.6 ± 0.05 myr. This fauna contains the following mammals: paleochiropterygoid-like bats (Hand et al., 1994); *Tingamarra porterorum,* a condylarth-like placental (Godthelp et al., 1992); and marsupials, including *Thylacotinga bartholomaii,* a bunodont, frugivore-omnivore (Archer et al., 1993), a *Pachybiotherium*-like, probably omnivorous microbiotheriid, a didelphoid-like, probable insectivore, a dasyurid-like insectivore, a perameloid-like insectivore, and a caroloameghiniid-like omnivore. All of these mammals are small to tiny. The paleocommunity was a shallow lake or swamp with abundant frogs, turtles, crocodiles, and birds. There is as yet no information about temperature, rainfall, or vegetation of the immediate region, but McGowran (1991) indicates that at this time Australia was in a "greenhouse" state, presumably with warm, wet conditions.

Late Oligocene to Early Miocene. For present purposes and following Harland et al. (1990), Late Oligocene is accepted as the interval between 29.3 and 23.3 myr; Early Miocene, between 23.3 and 16.3 myr; Middle Miocene, between about 16.3 and 10.4 myr; and Late Miocene, between about 10.4 and 5.2 myr. Late Oligocene and Early Miocene mammal assemblages are known from South Australia (Lake Eyre Basin and Lake Frome Embayment; Woodburne et al., 1985, 1994), Queensland (Riversleigh; Archer et al., 1989, 1994a, 1994b), New South Wales (Canadian Lead; Rich et al., 1991), Tasmania (Wynyard and Geilston Bay; Tedford et al., 1975) and the Northern Territory (Kangaroo Well; Woodburne et al., 1985). The Late Oligocene age of the oldest assemblages in South Australia and Queensland (e.g., Archer et al., 1989; Rich et al., 1991; Archer et al., 1993, 1994a, 1994b; Woodburne et al., 1994) is based on biocorrelative studies of foraminferans (Etadunna Formation; Lindsay, 1987), K/Ar dating of illite clay (Etadunna Formation), magnetostratigraphy, biocorrelation of marsupials, and international correlation of hipposiderid bats (Riversleigh; Sigé et al., 1982). In the interests of making specific points, we have been selective in our coverage of sites and faunal assemblages. Faunal lists can be found in Rich et al. (1991), Archer et al. (1994a), and Woodburne et al. (1994) and in the accumulating data base for the assemblages of Riversleigh.

The central Australian Late Oligocene to Early Miocene assemblages (Woodburne et al., 1985, 1994) occur as allopatric suites in South Australia, the Lake Eyre Basin, and the Lake Frome Embayment, and as an isolated spot in the southern Northern Territory (Kangaroo Well). Very little is known about the Northern Territory assemblage (Rich et al., 1991), which comprises a few fragments of perhaps four species. The richer South Australian assemblages may represent relatively open,

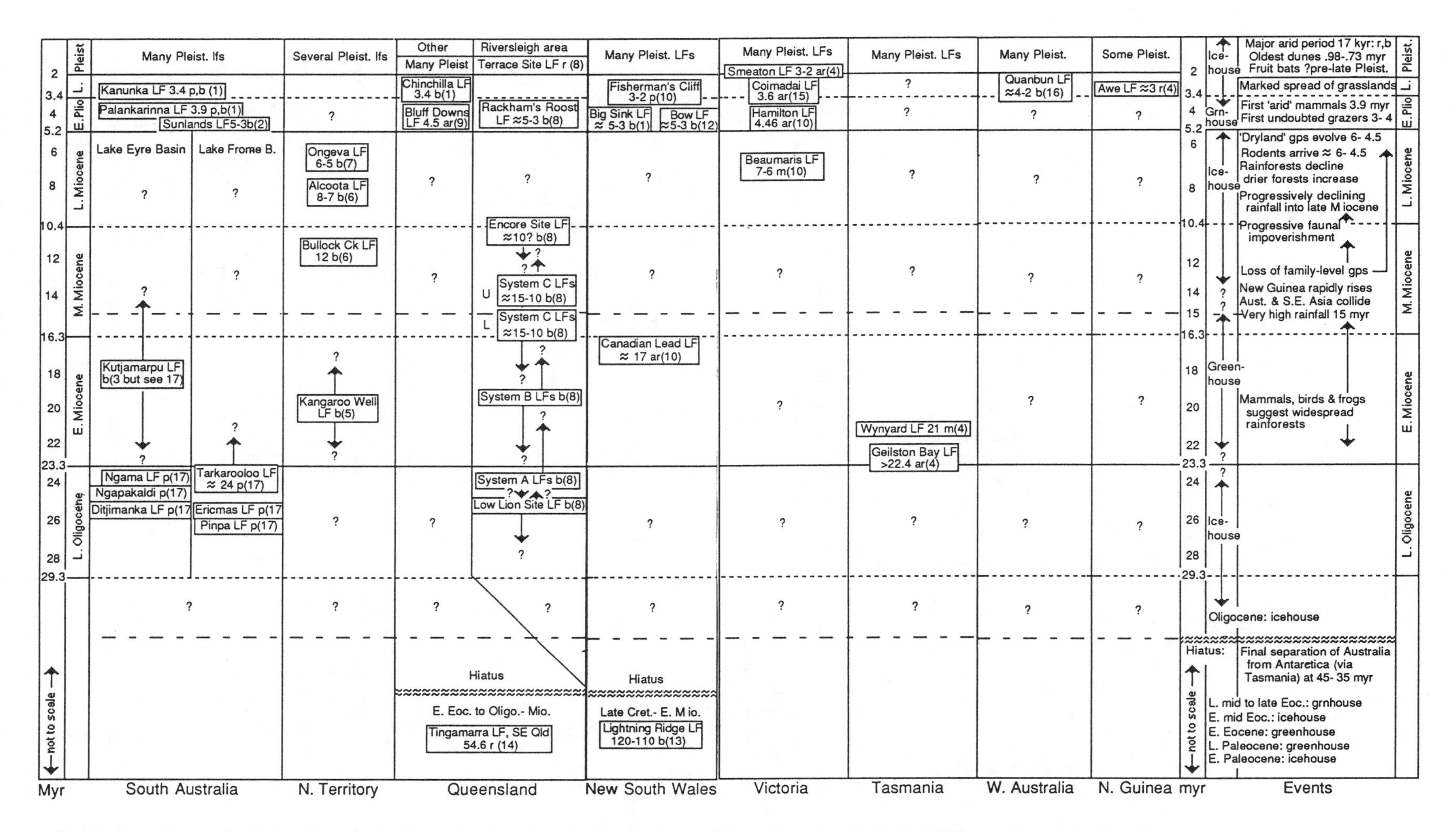

Fig. 6.1. Mammal-bearing deposits from Meganesia (Australia plus New Guinea) noted in the text, with estimated ages. The entries in boxes give the following information: local faunal name; approximate age, in myr; basis for age assessment; reference. Abbreviations: b = dated by biocorrelation with terrestrial taxa; B. = Basin; E. = early; gps = groups; L. = late; Grnhouse = Greenhouse; LF(s) = local fauna(s); m = dated by marine correlations; M. = middle; p = dated by paleomagnetic association; Qld = Queensland; r = directly associated with radiometric date; ar = in sediment below unit with radiometric date; SE = southeastern. Sources are: 1) Tedford, Wells, and Barghoorn (1992); 2) Pledge (1987); 3) Megirian (1982); 4) Woodburne et al. (1985); 5) Woodburne et al. (1985), as modified by Megirian (1992); 6) Murray and Megirian (1992); 7) Murray and Megirian (1994); 8) Archer et al. (1989, 1994a, 1994b); 9) Archer and Wade (1976); 10) Rich et al. (1991); 11) Hand, Dawson, and Augee (1988); 12) Flannery and Archer (1984); 13) Archer et al. (1985); 14) Godthelp et al. (1991); 15) Turnbull, Lundelius, and Tedford (1992); 16) Flannery (1984); 17) Woodburne et al. (1994).

wet forest communities, as opposed to the rain-forest communities of Riversleigh (see below). Mammalian species-level diversity in all central Australian assemblages is low relative to those in Riversleigh, although it is not clear that this difference reflects open versus closed forest communities. It is probable that most South Australian depositional sites (particularly those in the Etadunna Formation, which produced the Ditjimanka, Ngapakaldi, and Ngama local faunas) were more distant from the terrestrial biotas they sampled than were those at Riversleigh. Sediments and vertebrate assemblages suggest the presence of extensive freshwater lakes, swamps, and broad sandy channels. Freshwater lakes in the Frome Basin contained rhabdosteid dolphins indicating former or intermittent connections to the southern ocean via a large river system. Some South Australian Oligo-Miocene sites (e.g., Ngapakaldi local fauna; Stirton et al., 1961, Woodburne et al., 1994) are dominated by dog-sized palorchestid browsers; others (e.g., Ditjimanka local fauna; Woodburne et al., 1985, 1994) contain a more even balance of arboreal and terrestrial taxa, but relatively low numbers of each. Although a few South Australian deposits (e.g., some at Lakes Ngapakaldi and Pinpa) contain articulated skeletons, most mammals are found as rare fragments among masses of fish, turtle, and crocodile remains.

The age of the Kutjamarpu local fauna (Stirton et al., 1967) is uncertain. Woodburne et al. (1994 and pers. comm.) suggest it may be Late Oligocene or Early Miocene. It shares species such as *Neohelos tirarensis, Wakiewakie lawsoni,* and *Wakaleo oldfieldi* with Riversleigh's system B and low system C local faunas. The local fauna comes from the Wipijiri channel deposit cut into the top of the Etadunna Formation. Mammal diversity in this assemblage is comparable to that of some of the Riversleigh Early and Middle Miocene assemblages, although because of its demonstrably transported nature, some of its taxa may have been members of more distal communities. The vegetation of the proximal community was probably either a riparian rain forest or a wet sclerophyll forest. The absence of grazers argues against the presence of any significant herbaceous or grassland habitats in either the proximal or distal communities.

Two Early Miocene assemblages from Tasmania, the Geilston Bay local fauna (Tedford et al., 1975) of southeastern Tasmania and the Wynyard local fauna (Woodburne et al., 1985) of northwestern Tasmania, are too poorly known to provide useful paleoclimatic or paleobotanic information, although the former contains two taxa that may have been arboreal. Similarly, the Canadian Lead local fauna of New South Wales contains fragments of a single long-beaked echidna (*Megalibgwilia robusta*).

Riversleigh's geographically and / or stratigraphically discrete assemblages, interpreted to be Late Oligocene to Early Miocene (Archer et al., 1989, 1994a), have been allocated to two lithostratigraphic / biostratigraphic systems that are superpositionally related; the older System A and younger System B, and to a few assemblages that appear to underlie sediments of System A such as Low Lion Site. The majority of these were accumulated in shallow lakes, lime-rich pools, or caves cut into preexisting Miocene or Cambrian limestones. As presently understood, most appear to represent rain-forest communities biotically more diverse than any other Tertiary or modern Australian local faunal assemblages. Their diversity is approximated, however, by *regional* species-level diversity in modern rain-forest communities of the wet tropics and mid-montane New Guinea. The Early Miocene Upper Site local fauna, recovered from a piece of limestone 2 $m^2 \times 60$ cm deep, contains sixty-four species of mammals as well as comparably high numbers of other vertebrate and invertebrate taxa. It exhibits high species diversity, low species abundances (number of individuals per species) and indications of feeding guilds in which congeneric and sympatric species (e.g., perameloid bandicoots; J. Muirhead, 1994) are separated from each other by standard intervals of size.

In terms of niche diversity, in some of the larger Early Miocene fossil assemblages at Riversleigh (those of system B), marsupials make up about 83 percent of the mammal species (versus approximately 50 percent of modern mammal assemblages), placentals (bats) 16 percent (versus about 25 percent of modern), and monotremes 1 percent (versus less than 1 percent of modern). Rodents that currently make up approximately 25 percent of modern species were absent from Australia until Early Pliocene time. All Early Miocene herbivores at Riversleigh appear to have been either arboreal folivores (e.g., pseudocheirids), terrestrial browsers (e.g., macropodids and diprotodontids), frugivores (e.g., pilkipildrids), or omnivores (e.g., phalangerids). Grazers, as judged by crown height and development of accessory enamel ridges, were absent. Mammalian carnivores in these assemblages were small to medium in size (none being larger than a German shepherd). Reptilian carnivores, in contrast, included large as well as tiny crocodiles (commonly three species per local fauna) and large as well as tiny pythonid and matdsoiid snakes. Flannery

(1991) has suggested that the relative paucity of large mammalian carnivores in Australia in general relates to this continent's nutrient-deficient soils and unpredictable resources, conditions that might favor reptilian over mammalian carnivores. Large mammalian carnivores, however, are rare in all of the world's rain forests. If, as we have suggested, the dominant vegetation of most of pre-Late Miocene Australia was rain forest, the paucity of large carnivores may better reflect the short time that elapsed following the collapse of these forests and their replacement in the latest Miocene to Early Pliocene by widespread woodlands and later woodland savannas.

Megirian (1992) has proposed that high mammal diversity in Riversleigh's Miocene assemblages (Systems B and C) is the result of species from drier distal habitats being preserved with those from the local, closed forest environments. This hypothesis is difficult to support for several reasons. First, there is no sedimentological or osteological evidence for postmortem transport of organic remains from more distal communities, such as occurs in the Middle Miocene Wipajiri Formation of South Australia, which contains the diverse Kutjamarpu local fauna (see below). Second, it is implausible that non-rain-forest arboreal (e.g., ringtail possums) or fossorial (e.g., marsupial moles) taxa dispersed from drier distal habitats to the Riversleigh rain forests. Third, the Riversleigh assemblages are seemingly balanced local faunas with evidence of feeding guilds, structures that facilitate species packing into resource-diverse communities. We have also pointed out (Archer et al., 1989) that overall mammal diversity in some modern rain forests (e.g., mid-montane New Guinea and lowland Sarawak) is comparable to that of the Early to Middle Miocene assemblages of Riversleigh; the latter therefore do not present an anomaly that requires an extraordinary explanation. It is, we suggest, a mistake to presume that marsupial diversity in Australia's modern tropical rain forests is the "normal" condition for Australia and that departures from this level require explanation. Many authors have pointed out (e.g., Winter, 1988) that these modern rain-forest refugia underwent severe climatic crises during at least the Late Pleistocene, almost to the point of being entirely lost during the last glacial maximum. These crises would have had a severe impact on the mammalian communities of the rain forest, contributing significantly to what is now atypically low diversity at all taxonomic levels. If there is any unusual aspect of the Early and Middle Miocene mammalian communities of Riversleigh, it is the high marsupial diversity (relative to modern Australia) at the family and generic levels. This feature, however, may be explained by the absence of murid rodents and perhaps megachiropteran bats, which today make up more than 30 percent of Australia's mammalian communities. Although it is possible that one or two species in the Miocene Riversleigh assemblages may have originated from other, more distant, non-rain-forest habitats, the balance of evidence suggests that these assemblages fairly represent species-level diversity within the immediate area of the fossil deposits and that this diversity is not extraordinary for unstressed, low-latitude rain forests in this region of the world.

The Middle Miocene. There are two principle areas in Australia with Middle Miocene assemblages: Queensland (Riversleigh local faunas of System C) and the Northern Territory (Bullock Creek local fauna). Riversleigh's System C local faunas (e.g., Dwornamor, Ringtail, and Henk's Hollow), like those of System B, contain high species diversity and indications that the fossils accumulated in pools and caves within rain forests. Diverse frog assemblages from Riversleigh's Systems B and C local faunas provide evidence that during this period the Riversleigh region was uniformly wet throughout the year. The diversity of frog species whose modern relatives create bubble nests for egg laying suggests that temperatures in this region did not exceed 20°C for extended periods, the temperature above which modern bubble nests cannot be built (M. Tyler, pers. comm.). Arboreal herbivorous mammals were abundant, with up to twelve sympatric species of arboreal folivores (pseudocheirids and phascolarctids) in some cases (e.g., Dwornamor local fauna, System C). This abundance has been interpreted (Archer et al., 1989, 1994a, 1994b) to indicate high floral diversity within small regions, a feature characteristic of rain forests but not of open forests. Our rough estimate of the age of the Middle Miocene System C local faunas, based on biocorrelation, is an interval of time somewhere between 15 and 10 myr ago.

Between Systems B and C, there is a net loss of family- and generic-level diversity in mammals. For example, wynyardiids, ilariids, ngapakaldine palorchestids, and many genera characteristic of Systems A and B are absent from System C. Conversely, some groups unrepresented in Systems A or B (e.g., species of *Bettongia*) make their first appearance in System C. There is increasing evidence for a climatic shift within System C, near its base. Generic diversity (e.g., in arboreal pseudocheirid possums) undergoes a marked decline following deposition of a flowstone travertine (Archer, 1994). This widespread feature defines a boundary between lower and upper System C local faunas. The flowstone may flag the last and marked interval of high rainfall in the early Middle Miocene identified in coastal foraminiferal assemblages by McGowran and Li (1994) at 15 myr. The onset of the Middle and Late Miocene icehouse phase followed this interval of high rainfall and may be the explanation for the decline in diversity within upper System C assemblages. By the time of the uppermost assemblages of System C, rain-forest frogs were a rarity. Vombatoids (wombats), a group that may more directly reflect change in vegetation, demonstrate a similar pattern of change within the Riversleigh region. In System B, they were very low-crowned (an undescribed family) or slightly hypsodont (small rhizophascolonids). In most of the lower and upper System C assemblages, only hypsodont forms (large rhizophascolonids) are present. In the uppermost levels of System C (e.g., the ?early Late Miocene Encore Site; see below), only hypselodont (rootless) vombatids are present. This sequence could be interpreted to indicate a response to an increase in dryness, seasonality, and the xeromorphic (e.g., more heavily sclerotinized and hence abrasive) nature of some of the plants.

The Bullock Creek local fauna from the Northern Territory, interpreted by Murray and Megirian (1992) to be approximately 12 myr old (i.e., late Middle Miocene), differs from the Middle Miocene System C assemblages of Riversleigh primarily in being much less diverse, having proportionately fewer arboreal taxa and a distinct imbalance in species abundances, with the sheep- to cow-sized *Neohelos* sp. (a new species that is also present in the upper System C assemblages of Riversleigh) dominant. This finding presumes that Bullock Creek samples processed so far fairly represent the contemporary biota of the region. The paleoenvironment has been interpreted as fluvio-lacustrine, with some permanent and other seasonally abundant warm riverine waters. Some of the evidence suggests seasonal drying. The primary vegetation surrounding the water bodies may have been sclerophyll forest, judging from the abundance of large terrestrial browsers. But there is no unequivocal evidence for grazing adaptations—with the possible exception of a single, relatively high-crowned macropodid kangaroo molar from Bullock Creek—and no independent evidence for grasslands at this time (see below). Frog (M. Tyler, pers. comm.) and to some extent bird assemblages from Bullock Creek resemble (at the generic level) assemblages from the Oligo-Miocene of South Australia, and both contrast with the Oligo-Miocene assemblages of Riversleigh. This situation suggests that Early to Middle Miocene forest structure may not have been uniform across the continent and that open forests may have been present in central and north-central Australia, and closed forests in northern Queensland and eastern Australia.

Late Miocene. Australia's Late Miocene mammal assemblages are uncommon. The best known are from the Northern Territory (e.g., Alcoota). Much smaller assemblages are known from Victoria (Beaumaris) and possibly Queensland (Encore Site).

Of this period, the 8–7 myr-old (Mitchellian Stage; Murray and Megirian, 1992) Alcoota local fauna is similar in some structural aspects to the Bullock Creek local fauna but differs in being dominated by even larger diprotodontids and enormous dromornithid birds (one of which, *Dromornis stirtoni,* is a contender for the title of largest bird in the world). Arboreal mammals are almost entirely absent. Only one dentary of a ringtail possum is known. Based on sedimentology, Murray and Megirian (1992) speculate that the region supported an ephemeral fluvio-lacustrine environment with a large lake that was perhaps several kilometers in diameter when full. They further suggest that the region was subjected periodically to severe aridity and that the surrounding area was a subtropical savanna with patches of forest only in protected gullies. This conclusion, however, seems at odds with the low-crowned dentitions of the Alcoota macropodids and diprotodontids, which suggest browsing adaptations more compatible with an open sclerophyll forest habitat than with savanna. Although the paucity of arboreal mammals in the deposit appears to challenge any notion that the area was forested, fluviatile sorting before deposition may have resulted in loss of smaller skeletal elements of arboreal forms. Martin (1990a, 1990b) challenged earlier paleobotanical reviews that Murray and Megirian (1992) cited as corroborative evidence for the conclusion that "aridification" was in progress in northern Australia at least as early as the Middle Miocene and probably much earlier.

A younger Late Miocene assemblage from the Northern Territory, the Ongeva local fauna from sediments about 7 m above the Alcoota local fauna level, is possibly 6 to 5 myr old (Cheltenhamian; Megirian et al., 1994). This assemblage suggests that conditions in this inland region of the southern Northern Territory may have been similar to those present during Alcoota time.

The Beaumaris local fauna, from marine sediments in coastal Victoria, appears to be approximately of the same age as the Alcoota and Ongeva local faunas (Murray and Megirian, 1992). Too little is known about this assemblage, however, to provide an indication of Late Miocene terrestrial paleoenvironments or paleoclimates of southern Victoria (Woodburne et al., 1985).

Encore Site from the Gag Plateau of Riversleigh *may* be Late Middle to early Late Miocene in age—that is, perhaps 10 myr old. Although material from this site is only now being processed, it includes mammals more derived than those that characterize typical Middle Miocene System C assemblages. For example, it contains Riversleigh's only hypselodont vombatid (an undescribed species related to the Late Cenozoic *Warendja wakefieldi*); a giant species of the carnivorous kangaroo genus *Ekaltadeta* that appears to be annectent between the Early to Middle Miocene *E. ima* and the Plio-Pleistocene species of *Propleopus;* a phascolarctid that appears to be congeneric with the modern Koala (*Phascolarctos*); and a unique dasyurid that is larger than any previously known from the region. Other Encore taxa appear to represent lineages similar to those known from upper System C assemblages. Hence, our first estimate is that the Encore local fauna falls somewhere in time between those of upper System C and the Alcoota local fauna.

Early Pliocene. The Early Pliocene is here taken as approximately 5.2–3.4 myr and the Late Pliocene as approximately 3.4–2.0 myr. Assemblages interpreted to be Early Pliocene are known from South Australia (Sunlands, Curramulka, and Palankarinna), Queensland (Bluff Downs and possibly Rackham's Roost), New South Wales (Big Sink and Bow), and Victoria (Hamilton and Coimadai).

The Bluff Downs local fauna from northeastern Queensland has been obtained from sediments capped by a basalt interpreted to be Early Pliocene in age (Archer and Wade, 1976). The age of this basalt is being reassessed (B. Mackness, pers. comm.) and may be younger, although it is still Early Pliocene. Little is known about the microfauna of this assemblage. Bird fossils suggest (W. Boles and B. Mackness, pers. comm.) a fluvio-lacustrine environment similar to the modern Kakadu wetlands of the northern part of the Northern Territory, with open sclerophyll forest and woodlands. There is no clear indication of remnant rain forest. The presence of saltwater crocodiles (*C. porosus*) and file snakes (acrochordids; J. Scanlon, pers. comm.) suggests at least seasonal connections to the northern oceans. The mammal assemblage (Archer and Wade, 1976; B. Mackness, pers. comm.) includes mostly browsers among the herbivores but also some taxa that were probably grazers, such as hypselodont vombatids, and some high-crowned macropodids. The Early Pliocene Curramulka local fauna from southern South Australia (Pledge, 1992) also appears to represent a sclerophyll forest assemblage. Contemporaneous continental diversity is indicated by the Hamilton local fauna (sealed in by a 4.5 myr-old basalt; Rich et al., 1991) from Victoria and the Sunlands local fauna (dated by microfossils; Pledge, 1987) of South Australia, which appear to represent rain-forest assemblages. The late Early Pliocene Palankarinna local fauna from northern South Australia (3.9 myr old; Tedford et al., 1992) appears to have accumulated in a broad, shallow basin with ephemeral saline lakes and marginal mud flats. This is the earliest indication in central Australia of relatively arid environments. The age of the Bow local fauna (in east-central New South Wales) is in doubt, though it probably represents the late Early (or Middle) Pliocene (Rich et al., 1991). Its taxa suggest open forest habitats with a large, meandering stream. Kangaroos that were probably grazers or grazer-browsers (*Macropus* spp.) were present but not dominant in this assemblage.

On balance, the Early Pliocene mammal assemblages are far more modern in composition than those of the Middle to Late Miocene, in part because of loss of nearly one-third of the families present in Middle Miocene assemblages. Further, Early Pliocene mammal assemblages contain, on average, herbivores that were larger and in most cases more terrestrial than arboreal. The Hamilton rain-forest assemblage may be an exception. These differences suggest that mammals of the Early to Middle Pliocene in at least central and northern Australia were adapting to more open and probably more herbaceous vegetation than that which characterized the Miocene. There is evidence, however, that wet-forest assemblages persisted in areas of southern and southeastern Australia. The late Early Pliocene assemblages, at 3.9 myr old, are the first to exhibit reasonably clear evidence of the onset of aridity.

Late Pliocene. Local faunas interpreted to be Late Pliocene in age include: Kanunka (Victoria); Chichilla

(Queensland); Fisherman's Cliff (New South Wales); Smeaton and Dog Rocks (Victoria); Quanbun (Western Australia); and Awe (New Guinea).

The Kanunka local fauna (3.4 myr ago; Tedford et al., 1992; Rich et al., 1991) from northern South Australia suggests a transition to a more mesic environment, even though some taxa (*Thylogale* sp., cf. *Dendrolagus* sp.) suggest persistent pockets of wetter forest. The more diverse Chinchilla local fauna of southeastern Queensland (3.4 myr old; Tedford et al., 1992) suggests an open forest environment. The Fisherman's Cliff local fauna of southwestern New South Wales (Marshall, 1973) contains some taxa indicative of drier habitats (e.g., *Dasyuroides achilpatna, Lagostrophus* sp. cf. *L. fasciatus*), as well as others of uncertain habitat requirements. Smeaton contains a single dasyurid (*Glaucodon ballaratensis*) of uncertain paleoclimatic significance (Rich et al., 1991). The Dog Rocks local fauna of Victoria also contains a blend of taxa (Rich et al., 1991) that suggest forest on the one hand (*Phalanger* sp., *Vombatus ursinus*, and two pseudocheirids) and herbaceous plants or grasslands (*Macropus* (*Macropus*) spp.) on the other. The Awe local fauna of Papua, uncertainly dated as Late Pliocene (Woodburne et al., 1985), lacks any evidence of grazing mammals (Rich et al., 1991). This island assemblage is dominated instead by browsing kangaroos and diprotodontids and suggests late persistence of a more equable, forested environment.

On balance, the Late Pliocene assemblages suggest that the central regions of the continent were becoming arid while east-coastal and some northern regions were forested, though with open rather than closed forests. Rain-forest refugia must have persisted in at least northeastern Queensland, however, because of the modern occurrence there of relictual, endemic rain-forest groups such as hypsiprymnodontine kangaroos and woolly ringtails of the genus *Pseudochirops*.

Australia's Cenozoic Paleobotanical Record

In broad terms, the Cenozoic record for Australia's terrestrial plants corresponds with that for mammals, except that there is a better record for the Early Tertiary. Rather than attempt to review all of the significant deposits here, the following discussion is drawn primarily from the reviews of Martin (1990a, 1990b, 1994) and Hill (1992), which update earlier reviews (e.g., Kershaw, 1988). Martin's studies tend to focus on the Oligocene to Pliocene and hence are of particular importance in the present context.

Paleocene vegetation is southeastern Australia (e.g., in southeastern New South Wales; Taylor et al., 1990) was cool, temperate rain forest with a mean annual temperature of 14°–20°C and 1,200–2,400 mm of precipitation. There was a distinct cold season, precipitation was uniform, and there were no droughts. During the Early Eocene, *Nothofagus* pollens became abundant. During the Oligocene, *Nothofagus* pollens were still abundant, but the floras were not as diverse as those of the Eocene. Oligocene climates in the southeast were very wet, although a slight decrease in precipitation occurred during the Late Oligocene to Early Miocene, from about 1,800 to 1,500 mm per year, still above the lower limit for widespread rain forest in New South Wales. During the Early to Middle Miocene, *Nothofagus* pollens decreased slightly in inland regions but stayed high in coastal and southeastern areas of the continent. At this time a maximum diversity of forest types existed. By the Late Miocene, drastic changes had taken place, with the disappearance of *Nothofagus* pollens and an increase in Myrtaceae. Kershaw (1988) concluded that rain forest declined from the Early Miocene in central Queensland but not until the Late Miocene in northern New South Wales and Victoria. Wet sclerophyll forest, with a tall open eucalypt canopy and rain-forest species in the understory, gradually replaced rain forest. Herbaceous and Poaceae taxa were present in open forests, though in limited quantities. This change coincided with an increase in charcoal dust, an indicator that the flora was becoming more pyrophilic. Precipitation in southeastern Australia was probably down to 1,500–1,000 mm per year, below the lower limit for rain forests. There was a dry season, and burning was probably a regular event. A brief resurgence of rain forest occurred in the Early Pliocene but only to a fraction of the extent of rain forests of the Early to Middle Miocene. Precipitation in southeastern Australia increased to just over 1,500 mm per year. By the mid-Late Pliocene, vegetation had returned to wet sclerophyll forest, and annual precipitation in southeastern Australia to 1,500–1,000 mm. By latest Pliocene time, rain-forest taxa including gymnosperms had disappeared from most of the continent. Myrtaceous forests became more like dry sclerophyll forests. Herbaceous taxa increased. By the latest Pliocene to Pleistocene, a dramatic change had taken place in the vegetation, with open, arid woodlands and grasslands and herb fields becoming dominant. Precipitation in southeastern Australia had decreased to 500–800 yearly, the lower limit for eucalypts in wet sclerophyll forest.

Martin (1990b) challenged the widely noted report (W. Harris, in Callen and Tedford, 1976) that pollens from the Late Oligocene–Early Miocene base of the Namba Formation in central Australia indicate grasslands as well as rain-forest taxa. Her reanalysis of the same material indicates that most of the "grass" pollens actually belong to Restionaceae (swamp plants) and that the only true grasses may have been swamp grasses. The whole assemblage therefore suggests a swamp environment rather than the "extensive areas of grassland" reported in Callen and Tedford (1976). Other pollens (mostly rain-forest gymnosperms, Casuarinaceae, and a little *Nothofagus brassii*) from the same sample indicate the presence of rain forest. Eucalypt-type pollen is present in trace amounts. Pollens from near the top (Late Miocene to ?Pliocene) of the Namba Formation also represent a swamp flora, but other pollens from this upper section clearly indicate the presence of dry sclerophyll forest rather than rain forest. From a study of a borehole in central southeastern Australia, Truswell et al. (1985) reported similarly low quantities (2 percent) of grass pollens throughout the Miocene. At Lake George in southeastern Australia, grasses began to increase between 3.0 and 2.5 myr (McEwen Mason, 1990; cited in Martin, 1994). Two mid-Late Miocene pollen assemblages from northwestern New South Wales do not contain any grasses. Similarly, pollen samples from a number of Early Miocene boreholes in western New South Wales contained only minor quantities of grass pollens, insufficient to represent grasslands (Martin, 1988). Well-dated sea cores from northwestern Australia show an initial small increase in grass pollen in the Late Miocene to Early Pliocene continuing into the Late Pliocene. On balance, there is no evidence for Early to Middle Miocene grasslands in Australia. No Tertiary assemblages of plants in Australia can be classified as arid prior to at least the Pliocene, although some from the Late Miocene to Pliocene appear to represent dry sclerophyll forests.

In summary, Martin (1994) concluded that Australia was widely forested throughout the Tertiary and that most of these forests were rain forests prior to the Middle Miocene. The nature of these forests, however, were not the same in all parts of the continent, with far more sclerophyllous taxa represented earlier in the West. In some regions during the Late Miocene, drier forests became dominant. In the Late Plio-Pleistocene, herbaceous elements increased, and many forests opened up into woodlands, grasslands and herb fields. Climatic gradients parallel to those evident today (wetter on the coast and drier inland) probably existed throughout the Tertiary such that the decline of rain forests occurred first in central (and probably western) Australia.

Marine Evidence for Climatic Change in Australia

Feary et al. (1991) review oxygen isotope data for northeastern Australia. Subsequent reviews (McGowran and Li, 1994) are used here to modify some of the conclusions of Feary et al. (1991). During the early Late Oligocene (28 myr ago), there was a marked drop in sea temperatures. During the Late Oligocene (26–24 myr), although temperatures remained low, they fluctuated slightly. By the earliest Early Miocene (24–22 myr), a warming period developed. In the Early Miocene (24–16 myr), temperatures fluctuated slightly, although a gradual warming in all latitudes occurred with a climatic optimum at 16–15 myr ago (Chron 5B, zones N8–N9) coincident with a massive influx of fresh water (from high precipitation) into coastal marine environments (B. McGowran, pers. comm.). By the Middle Miocene (16–12 myr), high-latitude temperatures fell rapidly (possibly coincident with the establishment of the East Antarctic ice cap), but a slight warming occurred in low and middle latitudes. Into the late Middle Miocene and Late Miocene (12–6.2 myr), temperatures continued to drop in the western Pacific and northern Atlantic, with minor fluctuations periodically. By the latest Late Miocene to Early Pliocene (6.2–4.3 myr), there was continued cooling in high latitudes and slight cooling or constant temperature in low and middle latitudes. By the Early Pliocene (4.3–3.4 myr), brief episodes of warmer temperatures occurred that were most pronounced in low and middle latitudes. By the Late Pliocene (3.4–1.6 myr), temperatures again dropped, a trend that persisted into the Early Pleistocene.

Late Cenozoic Australian Paleoclimates and the Appearance of Arid Land Forms

Bowler (1982) posited that as Australia drifted northward, it was overtaken from the south by anticyclonic high-pressure cells, with subsequent development of aridity from south and north. He suggested that there was a significant decline in humidity between 5 and 6 myr ago, that the threshold to aridity was crossed near 4 myr ago, and that fluctuations thereafter increased in magnitude through the Late Pliocene and Pleistocene. Kershaw (1988) supported Bowler's hypothesis and concluded that this climatic trend led to strongly seasonal climates and, in the south, to patterns of summer rainfall of the kind that today characterize northern Australia. By 2.5 myr ago, this belt would have been in northern Australia; and for the first time a low-pressure system would have dominated southern Australia, bringing cool, wet winters and westerly winds. Although Bowler's model has received wide support, there is doubt about when this sequence of events began to affect Australia's climates. The original estimates of age were based on cores from near the edge of Lake George, New South Wales. Cores from more centrally situated positions currently under study suggest that the change to arid conditions took place much later. Chen and Barton (1991) reported that the oldest dated dunes in Australia are 0.98–0.73 myr old and that the onset of arid-zone facies in central Australia may have predated the onset in the southeastern part of the continent by 0.4 myr but may still date only to about 1.1 myr ago.

The McGowran "Icehouse-Greenhouse" Model of Climatic Change

McGowran (e.g., as discussed in Frakes et al., 1987) suggested that Australia has undergone at least three cycles of alternating "greenhouse" and "icehouse" conditions since the Paleocene. His model is based on a series of well-dated excursions of large foraminiferans into high latitudes. These excursions and alternating periods of retreat correlate with other marine and terrestrial events. Periods of expansion were characterized by (among other things) warm, humid, and equable (i.e., greenhouse) climatic conditions, marine transgressions, increased biotic diversity, increased evolutionary turnover, and increased chances of fossilization. Periods of contraction are characterized by cool, relatively arid, and less equable (icehouse) climatic conditions, marine regressions, decreased biotic diversity, decreased evolutionary turnover, and decreased chances of fossilization.

Broad aspects of Australia's terrestrial fossil record support McGowran's model, although his hypothetical "arid" conditions for the early Middle Eocene and Oligocene are perhaps better described as relatively drier conditions, there being no unambiguous evidence for aridity in Australia's terrestrial ecosystems prior to the Plio-Pleistocene. The hypothesized episodes of icehouse conditions during the Oligocene, Late Miocene, and Late Pliocene–Pleistocene were accompanied by low biotic diversity and a paucity of terrestrial vertebrate and plant assemblages. The greenhouse episodes of warming and high rainfall during the Early Eocene, Early and Middle Miocene, and Early Pliocene were matched by high diversity in the vertebrate and plant records and an abundance of fossil sites. Because pulses of foraminiferan species ought to reflect widespread changes in at least the Southern Hemisphere, his model may be of value in helping to correlate climatic and biotic events on the other southern continents.

Australian Mammal Events: Wider Correlations?

The events that tugged bipedal humans out of the forests and into the African savanna, and bipedal (but perhaps not quite so bright) grazing kangaroos into the Australian savanna, may well have been driven by the same global engine of environmental change. Although precise timing of events in Australia is often difficult, it is worth considering events that might be intercontinentally correlated.

Too little is known about the timing of events in Australia's Early Tertiary, apart from the generalization that climatic conditions were, on balance, equable, with most of the continent covered by either rain forest or wet sclerophyll forest until at least the Middle Miocene. It is also apparent that although Australia is relatively flat topographically, with few physical barriers to partition regional biotas, the western half and center may have differed from the eastern half in forest structure, in the timing of the loss of inland forests, and in the structure of contemporaneous mammal assemblages. Unfortunately, because no mammal assemblages that definitely predate the Pliocene are known from western Australia, it is difficult to compare contemporaneous Tertiary communities across the breadth of the continent (Archer, 1994b).

On the strength of the foraminiferan record from southern Australia (McGowran and Li, 1994), 15 myr ago (Middle Miocene), a climatic divide seems to have occurred separating a period of high rainfall in the Early Miocene from a period of decreasing rainfall in the Late Miocene. Vast tracts of closed forest became fragmented or disappeared from the interior, northern, and eastern areas of the continent primarily because of declining precipitation. In at least central and southern Australia, however, assemblages 8–6 myr in age were still dominated by browsing rather than grazing herbivores. This finding suggests that widespread increase in grasslands, and therefore in coarse plant food (which might be expected to have influenced the evolution of grazing mammals), did not occur before the middle of the late Miocene (i.e., no earlier than 6 myr ago), despite an earlier onset in the decline of rainfall. Other coincident Late Miocene biological changes include gigantism in many lineages of mammals. The number of arboreal marsupials also appears to have declined markedly between the Middle and Late Miocene. Many groups that disappeared first in the center of the continent may have persisted in forest refuges on the periphery. Rodents initially entered Australia possibly in the latest Miocene but certainly by the earliest Pliocene, presumably via a dry sclerophyll forest corridor in north-central Australia. They subsequently diversified in dry and wet forest habitats. They do not appear to have invaded remnant rain forests until the Pleistocene. Although no Late Miocene bats are known, there is a marked transition in the community structure of cave bats between the early Middle Miocene and the Pliocene. Assemblages in which hipposiderids were common and vespertilionids were rare gave way to assemblages in which vespertilionids were as common as hipposiderids. This change presumably reflects either a decline in humidity or an increase in seasonality, or both.

Correlated, physical changes of the Late Miocene record in Australia include the rapid rise of New Guinea to high elevations (the highest rocks being Middle Miocene marine limestones), with an attendant rain shadow developing across at least part of northern Australia. Inland lakes (such as in northeastern South Australia) either vanished or underwent seasonal decline. Southern marine transgressions (e.g., across the region now comprising the Nullarbor Plain) retreated. The effects and timing of anticyclonic high-pressure cells on the northward-drifting Australia are controversial, but they may have contributed to a decrease in continental precipitation and to a Late Miocene shift of summer rainfall from the southern to the northern half of the continent. Widespread development of silcretes appears to have occurred as evaporation exceeded precipitation. In summary, greenhouse conditions of the Early and Middle Miocene appear to have given way to icehouse conditions during the Middle and Late Miocene.

The Late Miocene drying syndrome and scarcity of Late Miocene terrestrial fossil localities in Australia probably correspond with the drying out of the Mediterranean, with the buildup of ice in West Antarctica (about 6–5 myr), with the general drop in sea level, and with other symptoms of an icehouse phase in the alternating cycle model of Frakes et al. (1986). In Africa, the Middle Miocene marked the onset of aridity and (*fide* Dugas and Retallack, 1993) a major expansion of grasslands (at 15 myr), a change related to the closing of the Tethys Sea and orogenic events that left Africa with a more continental climate. In Pakistan, grasslands appear to have undergone a major expansion in the Late Miocene (6 myr ago).

The Early Pliocene record (5–4 myr ago) in Australia suggests a brief period of climatic amelioration with expansion of wet forests. Although grazing wombats and a few kangaroos (the first undoubted grazers to appear in the record) were present, most herbivorous mammals were browsers. Undoubted areas of rain forest were present in at least peripheral areas of the southeastern corner of the continent. There was an expansion of rain forest into what are now semiarid to arid regions of New South Wales (Martin, 1987). A brief Early Pliocene expansion of large foraminiferans into high latitudes (McGowran, 1986) suggests warming of oceanic waters and a brief greenhouse phase. This phase may correspond with an interval of ice melting (and *Nothofagus*) in Antarctica.

By late Early Pliocene, about 3.9 myr ago, mammal assemblages from central Australia were exhibiting the first signs of the onset of aridity, suggesting that Australia had returned to icehouse conditions. However, younger assemblages (such as at 3.4 myr), some of which are more mesic in character, suggest that, at least in the central areas of the continent, there were periods of major climatic fluctuation. Yet the first evidence of arid land forms (dunes) in Australia does not appear until the Pleistocene, approximately 1.1 myr ago. This supports other indications that the widespread aridity today (44 percent of the continent) is a product of events less than 2 myr old.

Summary: Climatic Change and Evolutionary Response

On balance, although there appear to have been regional differences in biodiversity between the central and northern parts of the continent, little change in regional diversity (e.g., within the Riversleigh region) is apparent during the climatically equable Early Miocene. The beginnings of a change to icehouse conditions in the ?late

Middle to Late Miocene, sometime between 15 and 10 myr ago (the precise timing and rate of change being unclear), appear to coincide with an overall drop in biodiversity (e.g., from Riversleigh's System C assemblages to those of Bullock Creek) and a marked decline (at least in northern and central Australia) in most lineages of arboreal mammals. These declines resulted in transition to biotically simpler communities with fewer higher-level taxa. During the period from 12 to 8 myr ago (e.g., Bullock Creek to Alcoota time), decline in higher-level diversity continued. Throughout these periods of decline, it is not evident that significant higher-level expansions into new niches were taking place, although many lineages were exhibiting trends to gigantism.

By Early Pliocene time (4.5 myr ago), however, a great deal of innovation had taken place, and many modern genera were appearing for the first time. If the Ongeva local faunas of the Northern Territory is approximately 6 myr old (Murray and Megirian, 1994), its basic similarity to the Alcoota local fauna and marked distinction from the Early Pliocene faunas suggest that a major innovative phase of evolution of more xeric-adapted lineages started or accelerated sometime between 6 and 4.5 myr ago. Rodents which today make up about 25 percent of Australia's native mammal species, also invaded from the north sometime in this interval, probably during a Late Miocene low stand of sea level. Form 4.5 myr ago until the Late Pleistocene, the mammal faunas of Australia took on a progressively more modern, semiarid aspect with evolution of grazers phasing out browsers, macropodid kangaroos undergoing a massive radiation as browsing diprotodontids declined, burrowing mammals replacing forest-floor occupants, and vespertilionid bats diversifying as rhinolophoid bats declined. Gigantism, which has been interpreted (e.g., Main, 1978; Archer, 1984) as an ecophysiological response to declining nutrient levels in food, became a conspicuous feature of almost all lineages of terrestrial mammals.

A subset of the higher forest taxa that characterized the Miocene persisted through this period in rain-forest refuges in southeastern and northeastern Australia and montane New Guinea. It is perhaps significant that in these wet forest refuges, very few fundamentally new groups evolved as replacements for those that were lost, possible exceptions being tree kangaroos (*Dendrolagus*), forest bandicoots (peroryctids), a group of ringtail possums (*Pseudocheirulus*), and striped possums (dactylopsiline petaurids). There were, however, Late Tertiary –Quaternary adaptive radiations within Miocene rain-forest groups such as cuscuses (*Phalanger sensu lato*) and woolly ringtail possums (*Pseudochirops*) as well as invasions of groups new to Australia's rain forests, including hydromyine rodents in the Pleistocene and pteropodid bats in the Plio-Pleistocene. On balance, however, Australia's modern rain-forest mammal communities more closely resemble those of the Miocene than do the newer mammal communities of the rest of the continent.

Acknowledgments

We wish to acknowledge the vital support of the Australian Research Grant Scheme; the National Estate Grants Scheme (Queensland); the Department of Environment, Sports and Territories; the Queensland National Parks and Wildlife Service; the University of New South Wales; ICI Australia Pty. Ltd.; the Australian Geographic society; Century Zinc Ltd. CRA; Wang Australia Pty. Ltd.; the Queensland Museum; the Australian Museum; Mount Isa Mines Pty. Ltd.; and the Riversleigh Society Inc. In addition, we are indebted to Elaine Clark, Martin Dickson, Sue and Jim Lavarack, Sue and Don Scott-Orr, and Margaret Beavis; and to our research colleagues, including Alan Bartholomai, Alex Baynes, Karen Black, Walter Boles, Jenni Brammall, Bernie Cooke, Phil Creaser, Tim Flannery, Anna Gillespie, Miranda Gott, Mark Hutchinson, Anne Kemp, Alan Krikmann, Brian Mackness, Helene Martin, Dirk Megirian, Jeanette Muirhead, Peter Murray, Cathy Nock, Neville Pledge, Tom Rich, David Ride, John Scanlon, Richard Tedford, Michael Tyler, Arthur White, Stephan Williams, Paul Willis, Michael Woodburne, and Steve Wroe. We also gratefully acknowledge support from Elisabeth Vrba for travel and accommodation funds that enabled Michael Archer to attend the 1993 conference on which this book is based and to benefit from discussions with many international colleagues.

References

Aplin, K., and Archer, M. 1987. Recent advances in marsupial systematics with a new syncretic classification. In *Possums and opossums: Studies in evolution,* pp. xv–lxxii (ed. M. Archer). Royal Zoological Society of New South Wales and Surrey Beatty and Sons Pty. Ltd., Sydney.

Archer, M. 1984. The Australian marsupial radiation. In *Vertebrate zoogeography and evolution in Australia,* pp. 633–808 (ed. M. Archer and G. Clayton). Hesperian Press, Perth.

Archer, M. (in press). Refugial vertebrates from the Miocene rainforests of Western Australia: Where are they? (ed. S. D. Hopper). Kings Park Gardens, Perth.

Archer, M., Flannery, T., Ritchie, A., and Molnar, R. 1985. First Mesozoic mammal from Australia—an early Cretaceous monotreme. *Nature* 318:363–366.

Archer, M., Godthelp, H., and Hand, S. J. Early Eocene marsupial from Australia: Kaupia. *Darmstadter Beitrage zur Naturgeschichte* 2:193–200.

Archer, M., Godthelp, H., Hand, S. J., and Megirian, D. 1989. Fossil mammals of Riversleigh, northwestern Queensland:

Preliminary overview of biostratigraphy, correlation and environmental change. *Australian Zoologist* 25:29–64.

Archer, M., Hand, S. J., and Godthelp, H. 1994a. *Riversleigh.* 2d ed. rev. Reed Books, Sydney.

———. 1994b. Patterns in the history of Australia's mammals and inferences about palaeohabitats. In *History of the Australian vegetation,* pp. 80–103 (ed. R. Hill). Cambridge University Press, Cambridge.

Archer, M., Murray, P., Hand, S. J., and Godthelp, H. 1993. Reconsideration of monotreme relationships based on the skull and dentition of the Miocene *Obdurodon dicksoni.* In *Mammal phylogeny,* vol. 1, *Mesozoic differentiation, monotremes, early therians, and marsupials,* pp. 75–94 (ed. F. S. Szalay, M. J. Novacek, and M. C. McKenna). Springer-Verlag, New York.

Archer, M., and Wade, M. 1976. Results of the Ray E. Lemley Expeditions. Pt. 1, The Allingham Formation and a new Pliocene vertebrate fauna from northern Queensland. *Memoirs of the Queensland Museum* 17:379–397.

Bonaparté, J. 1990. New late Cretaceous mammals from the Los Alamitos Formation, northern Patagonia, and their significance. *National Geographic Research* 6:63–93.

Bowler, J. M. 1982. Aridity in the late Tertiary and Quaternary of Australia. In *Evolution of the flora and fauna of arid Australia,* pp. 35–45 (ed. W. R. Barker and P. J. M. Greenslade). Peacock Publications, Adelaide.

Callen, R. A., and Tedford, R. H. 1976. New late Cainozoic rock unit and depositional environments, Lake Frome area, South Australia. *Transactions of the Royal Society of South Australia* 100:125–167.

Chen, X. Y., and Barton, C. E. 1991. Onset of aridity and dune-building in central Australia: Sedimentological and magnetostratigraphic evidence from Lake Amadeus. *Palaeogeography Palaeoclimatology Palaeoecology* 84:55–73.

Dugas, D. P., and Retallack, G. J. 1993. Middle Miocene fossil grasses from Fort Ternan, Kenya. *Journal of Paleontology* 67:113–128.

Feary, D. A., Davies, P. J., Pigram, C. J., and Symonds, P.A. 1991. Climatic evolution and control on carbonate deposition in northeast Australia. *Palaeogeography Palaeoclimatology Palaeoecology* 89:341–361.

Filewood, W. 1984. The Torres connection: Zoogeography of New Guinea. In *Vertebrate zoogeography and evolution in Australia,* pp. 1121–1131 (ed. M. Archer and G. Clayton). Hesperian Press, Perth.

Flannery, T. F., 1984. Re-examination of the Quanbun Local Fauna, a Late Cenozoic vertebrate fauna from Western Australia. *Records of the Western Australian Museum* 11:119–128.

———. 1991. The mystery of the Meganesian meat-eaters. *Australian Natural History* 23:722–729.

Flannery, T. F., and Archer, M. 1984. The macropodoids (Marsupialia) of the Early Pliocene Bow Local Fauna, central eastern New South Wales. *Australian Zoologist* 21:357–383.

Frakes, L. A., McGowran, B., and Bowler, J. M. 1987. Evolution of Australian environments. In *Fauna of Australia,* vol. 1A, General Articles, pp. 1–16 (ed. G. R. Dyne and D. W. Walton). Australian Government Publishing Service, Canberra.

Godthelp, H., Archer, M., Cifelli, R., Hand, S. J., and Gilkeson, C. F. 1992. Earliest known Australian mammal fauna. *Nature* 356:514–516.

Hand, S. J., Dawson, L., and Augee, M. 1988. *Macroderma koppa,* a new Pliocene species of false vampire bat (Microchiroptera: Megadermatidae) from Wellington Caves, New South Wales. *Records of the Australian Museum* 40:343–351.

Hand, S. J., Novacek, M., Godthelp, H., and Archer, M. 1994. First Eocene bat from Australia. *Journal of Vertebrate Paleontology* 14:375–381.

Harland, W. B., Armstrong, R. L., Cox, A. V., Craig, L. E., Smith, A. G., and Smith, D. G. 1990. *A geologic time scale.* Cambridge University Press, Cambridge.

Hill, R. S. 1992. Australian vegetation during the Tertiary: Macrofossil evidence. *The Beagle, Records of the Northern Territory Museum of Arts and Sciences* 9:1–10.

Kershaw, A. P. 1988. Australasia. In *Vegetation history,* pp. 237–306 (ed. B. Huntley and T. Webb III). Kluwer, Dordrecht.

Lindsay, J. M. 1987. Age and habitat of a monospecific foraminiferal fauna from near-type Etadunna Formation, Lake Palankarinna, Lake Eyre Basin. Report no. 87/93. Department of Mines and Energy, South Australia.

McGowran, B. 1986. Cainozoic oceanic events: The Indo-Pacific biostratigraphic record. *Palaeogeography Palaeoclimatology Palaeoecology* 55:247–265.

———. 1991. Evolution and environment in the early Palaeogene. *Geological Society of India Memoir* 20:21–53.

McGowran, B., and Li. Q. 1994. The Miocene oscillation in southern Australia. *Records of the South Australian Museum* 27:197–212.

Main, A. R. 1978. Ecophysiology: Towards an understanding of late Pleistocene marsupial extinction. In *Vertebrate zoogeography and evolution in Australasia,* pp. 169–184 (ed. M. Archer and G. Clayton). Hesperian Press, Perth.

Marshall, L. G. 1973. Fossil vertebrate faunas from the Lake Victoria region, S.W. New South Wales, Australia. *Memoirs of the National Museum of Victoria* 34:151–171.

Martin, H. A. 1987. The Cainozoic history of the vegetation and climate of the Lachlan River Region, New South Wales. *Proceedings of the Linnean Society of New South Wales* 109:313–357.

———. 1990a. The palynology of the Namba Formation in the Wooltana-1 bore, Callabona Basin (Lake Frome), South Australia, and its relevance to Miocene grasslands in central Australia. *Alcheringa* 14:247–255.

———. 1990b. Tertiary climatic phytogeography of southeastern Australia. *Review of Palaeobotany and Palynology* 65:47–55.

———. 1994. Australian Tertiary phytogeography: Evidence from palynology. In *History of the Australian vegetation,* pp. 104–142 (ed. R. Hill). Cambridge University Press, Cambridge.

Megirian, D. 1992. Interpretation of the Miocene Carl Creek Limestone, northwestern Queensland. *The Beagle, Records of the Northern Territory Museum of Arts and Sciences* 9:219–248.

Megirian, D., Murray, P. F., and Wells, R. T. 1994. The Late Miocene Ongeva Local Fauna from the Waite Formation of central Australia. *Records of the South Australian Museum* 27:225.

Muirhead, J. 1994. Riversleigh bandicoots (Marsupialia, Perameloidea). *Abstracts Riversleigh Symposium 1994,* pp. 18–19. University of New South Wales, Sydney.

Murray, P., and Megirian, D. 1992. Continuity and contrast in Middle and Late Miocene vertebrate communities from the Northern Territory. *The Beagle, Records of the Northern Territory Museum of Arts and Sciences* 9:195–218.

Pascual, R., Archer, M., Ortiz Jaureguizar, E., Prado, J. L., Godthelp, H., and Hand, S. J. 1992. First discovery of monotremes in South America. *Nature* 356:704–706.

Pledge, N. S. 1987. *Phascolarctos maris,* a new species of koala (Marsupialia: Phascolarctidae) from the Early Pliocene of South Australia. In *Possums and opossums: Studies in evolution,* pp. 327–330 (ed. M. Archer). Royal Zoological Society of New South Wales and Surrey Beatty and Sons Pty. Ltd., Sydney.

———. 1992. The Curramulka local fauna: A new late Tertiary fossil assemblage from Yorke Peninsula, South Australia. *The Beagle, Records of the Northern Territory Museum of Arts and Sciences* 9:115–142.

Rich, P. V., Rich, T. H., Wagstaff, B., McEwan-Mason, J., Douthitt, C. G., Gregory, R. T., and Felton, E. A. 1988. Biotic and geochemical evidence for low temperatures and biologic diversity in Cretaceous high latitudes of Australia. *Science* 242:1403–1406.

Rich, T. H., Archer, M., Hand, S. J., Godthelp, H., Muirhead, J., Pledge, N. S., Flannery, T. F., Woodburne, M. O., Case, J. A., Tedford, R. H., Turnbull, W. D., Lundelius, E. L., Jr. Rich, L. S. V., Whitelaw, M. J., Kemp, A., and Rich, P. V. 1991. Appendix 1 of *Vertebrate paleontology of Australia,* pp. 1005–1058 (ed. P. Vickers-Rich, J. M. Monaghan, R. F. Baird, and T. H. Rich). Pioneer Design Studies, Melbourne.

Sigé, B., Hand, S. J., and Archer, M. 1982. An Australian Miocene *Brachipposideros* (Mammalia, Chiroptera) related to Miocene representatives from France. *Palaeovertebrata* 12:149–172.

Stirton, R. A., Tedford, R. H., and Miller, A. H. 1961. Cenozoic stratigraphy and vertebrate paleontology of the Tirari Desert. *Records of the South Australian Museum* 14:19–61.

Stirton, R. A., Tedford, R. H., and Woodburne, M. O. 1967. A new Tertiary formation and fauna from the Tirari Desert, South Australia. *Records of the South Australian Museum* 15:427–462.

Taylor, G., Truswell, E. M., McQueen, K. G., and Brown, M. C. 1990. Early Tertiary palaeogeography, landform evolution and palaeoclimates of the southern Monaro, N.S.W., Australia. *Palaeogeography Palaeoclimatology Palaeoecology* 78:109–134.

Tedford, R. H. 1970. Principles and practices of mammalian geochronology in North America. In *Proceedings of the North American Paleontological Convention, Field Museum of Natural History, Chicago, Sept. 5–7, 1969,* pp. 666–703 (ed. E. L. Yochelson). Allen Press, Lawrence.

Tedford, R. H., Banks, M. R., Kemp, N. R., McDougall, I., and Sutherland, F. L. 1975. Recognition of the oldest known fossil marsupials from Australia. *Nature* (London) 255:141–142.

Tedford, R. H., Wells, R. T., and Barghoorn, S. F. 1992. Tirari Formation and contained faunas, Pliocene of the Lake Eyre Basin, South Australia. *The Beagle, Records of the Northern Territory Museum of Arts and Sciences* 9:173–194.

Truswell, E. M., Sluiter, I. R., and Harris, W. K. 1985. Palynology of the Oligo-Miocene sequence in Oakvale: 1 corehole, western Murray Basin, South Australia. *Journal of Australian Geology and Geophysics* 9:267–295.

Turnbull, W. D., Lundelius, E. L., Jr., and Tedford, R. H. 1992. A Pleistocene marsupial fauna from Limeburner's Point, Victoria, Australia. *The Beagle, Records of the Northern Territory Museum of Arts and Sciences* 9:143–172.

Winter, J. W. 1988. Ecological specialization of mammals in Australian tropical and sub-tropical rainforest: Refugial or ecological determinism? *Proceedings of the Ecological Society of Australia* 15:127–138.

Woodburne, M. O. 1967. The Alcoota fauna, central Australia: An integrated palaeontological and geological study. *Journal of Australian Geology and Geophysics* 87:1–187.

Woodburne, M. O., Tedford, R. H., Archer, M., Turnbull, W. D., Plane, M. D., and Lundelius, E. L., Jr. 1985. Biochronology of the continental mammal record of Australia and New Guinea. *Special Publications of the South Australian Department of Mines and Energy* 5:347–363.

Woodburne, M. O., McFadden, B. J., Case, J. A., Springer, M. S., Pledge, N. S., Power, J. D., Woodburne, J. M., and Springer, K. B. 1994. Land mammal biostratigraphy and magnetostratigraphy of the Etadunna Formation (late Oligocene) of South Australia. *Journal of Vertebrate Paleontology* 13:483–515.

Chapter 7

Climatic Implications of Large-Herbivore Distributions in the Miocene of North America

S. David Webb, Richard C. Hulbert, Jr., and W. David Lambert

The rise and fall of a great ungulate fauna took place in North America largely within the Miocene. Large-herbivore diversity reached its apogee in the Late Barstovian, about 15 million years (myr) ago, stayed high through the Clarendonian, and then entered a series of devastating extinctions during the Middle and Late Hemphillian (about 5 myr ago). The broad continuity of this fauna, from 15 to 5 myr ago, warrants its designation as the "Clarendonian Chronofauna." It is characterized especially by its diversity of native North American ungulates in the families Equidae, Camelidae, and Antilocapridae.

Any student of large mammals must admire the rich array of ungulates that dominated the North American landscape in the Middle Miocene; camels as tall as giraffes; horses as diverse as African antelopes; rhinos proportioned like hippos and living a similar semiaquatic existence. The success of the Clarendonian Chronofauna calls for an explanation, and its demise carries a dire warning. Exploration of the analogy with living African savanna faunas may provide valuable insights (Webb, 1983a). On the other hand that is only an analogy: the North American Miocene and the African modern faunas share no land-mammal taxa below the family level, and no ecosystem exactly replicates another (Janis, 1993). Even so, because both continents similarly supported a diverse ungulate fauna on a savanna-woodland mosaic, one may expect a comparable interplay of browsers and grazers, with seasonal migrations to diversify resources. Insights drawn from modern Serengeti studies must be interpolated with the 10 myr perspectives derived from the rise and fall of this marvelous Miocene fauna.

ISBN 0-300-06348-2.

In this chapter we explore the hypothesis that cooling and drying climatic trends governed first the rise, and later the fall, of the great North American ungulate fauna. Six ungulate families are the major players in this drama, namely, horses, camels, protoceratids, pronghorn antelopes, dromomerycids, and gomphotheres. After outlining their chronologic and biogeographic ranges, we discuss the climatic implications of their evolving patterns of adaptations, diversity, and distribution. The regional distributions that emerged in the Late Miocene show that several taxa surviving only in the Gulf Coastal Plain reflect a persistently humid climate and refugia consisting of woodland-savanna habitats. Close analysis of such regional differences helps illuminate and elaborate the climatic hypothesis of mammal extinctions.

A Brief History of Miocene Horses

North American horses represent an autochthonous radiation into a diversity of dental and muzzle morphologies, body sizes, and limb proportions. Presumably these features reflect ecologic specializations that account for the high levels of sympatric species during the Miocene. Seven taxonomic and morphologic groups of taxa are used in this study; their genera are listed in table 7.1. Two of the seven, the parahippine and merychippine horses, are not true clades, as each is demonstrably paraphyletic, but both form relative speciose and morphologically cohesive groups. The seven groups are united by only one known derived character state, in which the metaloph on their upper cheek teeth is connected to the ectoloph (Evander, 1989). This synapomorphy distinguishes

Table 7.1. Miocene Horse Taxa in Seven Taxonomic-Morphologic Groups

Group	Taxa
Anchithere Group	*Anchitherium*, *Kalobatippus*, *Hypohippus*, *Megahippus*
Archaeohippine Group	*Archaeohippus*
Hipparionine Group	*Neohipparion*, *Pseudhipparion*, *Nannippus*, *Cormohipparion*, *"Hipparion"*, *"Merychippus" coloradense*
Equine Group	*Pliohippus*, *Dinohippus*, *Astrohippus*, *Onohippidium*, *Hippidion*, *Equus*, *"Merychippus" stylodontus*
Parahippine Group	*Parahippus*, *Anchippus*, *Desmatippus*
Merychippine Group	*Merychippus* (sensu stricto), *"Merychippus" gunteri*, *"Merychippus" primus*, *"Merychippus" carrizoensis*, *"Merychippus" sejunctus*, *"Merychippus" isonesus*, *"Merychippus" gorrisi*, *"Merychippus" tertius*
Protohippine Group	*Protohippus*, *Calippus*, *"Merychippus" intermontanus*

them from the typical Oligocene horses of North America, *Mesohippus* and *Miohippus* (Prothero and Shubin, 1989).

Three of the seven groups of North American Miocene horses appeared prior to the end of the Arikareean (very Early Miocene, ca. 20 myr): the anchitheres, parahippines, and archaeohippines (fig. 7.1). In terms of their Late Arikareean biogeography, anchitheres are known from the Far West (e.g., *Anchitherium praestans* from Oregon) and the Great Plains (*Anchitherium* and *Kalobatippus* from Nebraska); parahippines from the Great Plains (e.g., *Parahippus pristenus, P. tyleri,* and others from Nebraska and South Dakota) and the Gulf Coastal Plain (*Anchippus* from Texas and Florida); and archaeohippines from Florida only. The restricted range of Arikareean archaeohippines assumes that the referral of *Miohippus equinanus* from South Dakota to *Archaeohippus* by Skinner et al. (1968) was not correct, for the reasons stated by Prothero and Shubin (1989). The Florida specimens, isolated teeth from the Middle Arikareean Cowhouse Slough local fauna, are of *Archaeohippus* size and, unlike *M. equinanus,* have metalophs with connected ectolophs. In the Late Arikareean, differences in dental and skull morphology among the three groups were relatively minor, with variation in body size being the key separating factor. Archaeohippines were very small, anchitheres large, and parahippines intermediate in size, though with more variation than the other two groups. Equids were not typically numerous or dominant among Late Arikareean ungulate faunas (especially in contrast to later in the Miocene), whereas oreodonts, camelids, and rhinoceroses evidently occurred in much greater numbers.

All three Late Arikareean equid groups continued into the Hemingfordian (20–16.3 myr), expanded their geographic ranges, and diversified. In particular, two species groups emerged within the parahippine group: one characterized by increased size and brachyodont, simple cheek teeth without a covering of cement; the other by a decrease in size and more complex, cement-covered cheek teeth with greater crown heights. A species of the latter group gave rise to the merychippine group about 18 myr ago, probably from a form close to *Parahippus leonensis* (Hulbert and McFadden, 1992). Merychippine species rapidly diversified in the Late Hemingfordian, became widespread, ranging from Florida to California, and often dominated over the older groups in most faunas. The overall relative representation of Hemingfordian equids among ungulates is notably much greater than in the Late Arikareean, as exemplified by such equid-dominated faunas as Thomas Farm (Florida), Garvin Gully (Texas), Box Butte (Nebraska), and Sheep Creek (Nebraska).

During the Early Barstovian (16.3–14.5 myr) the three original Miocene horse groups remained a consistent but relatively rare component in ungulate communities across North America. If present, archaeohippines were represented by a single species in a fauna (such as *Archaeohippus mourningi* or *A. ultimus* in the West or *A. penultimus* in the Great Plains). Archaeohippines are not recorded in the Gulf Coastal Plain after the Hemingfordian. Anchitheres were more diverse in the southern Great Basin, with many faunas containing both *Hypohippus* and *Megahippus* and some having additional *Anchitherium*-like holdovers (Tedford and Barghoorn, 1993), but were also present in lesser numbers in faunas of the Far West, Great Plains, and Gulf Coastal Plain. Parahippines continued to decline in numbers in the Early Barstovian but remained widespread. Only the Mascall Fauna of Oregon contains more than one species of parahippine, producing both *Parahippus brevidens* and *Desmatippus avus.* The merychippine group is the dominant Early Barstovian horse group, their adaptive radiation that began in the Late Hemingfordian having culminated in the Early Barstovian (Hulbert, 1993).

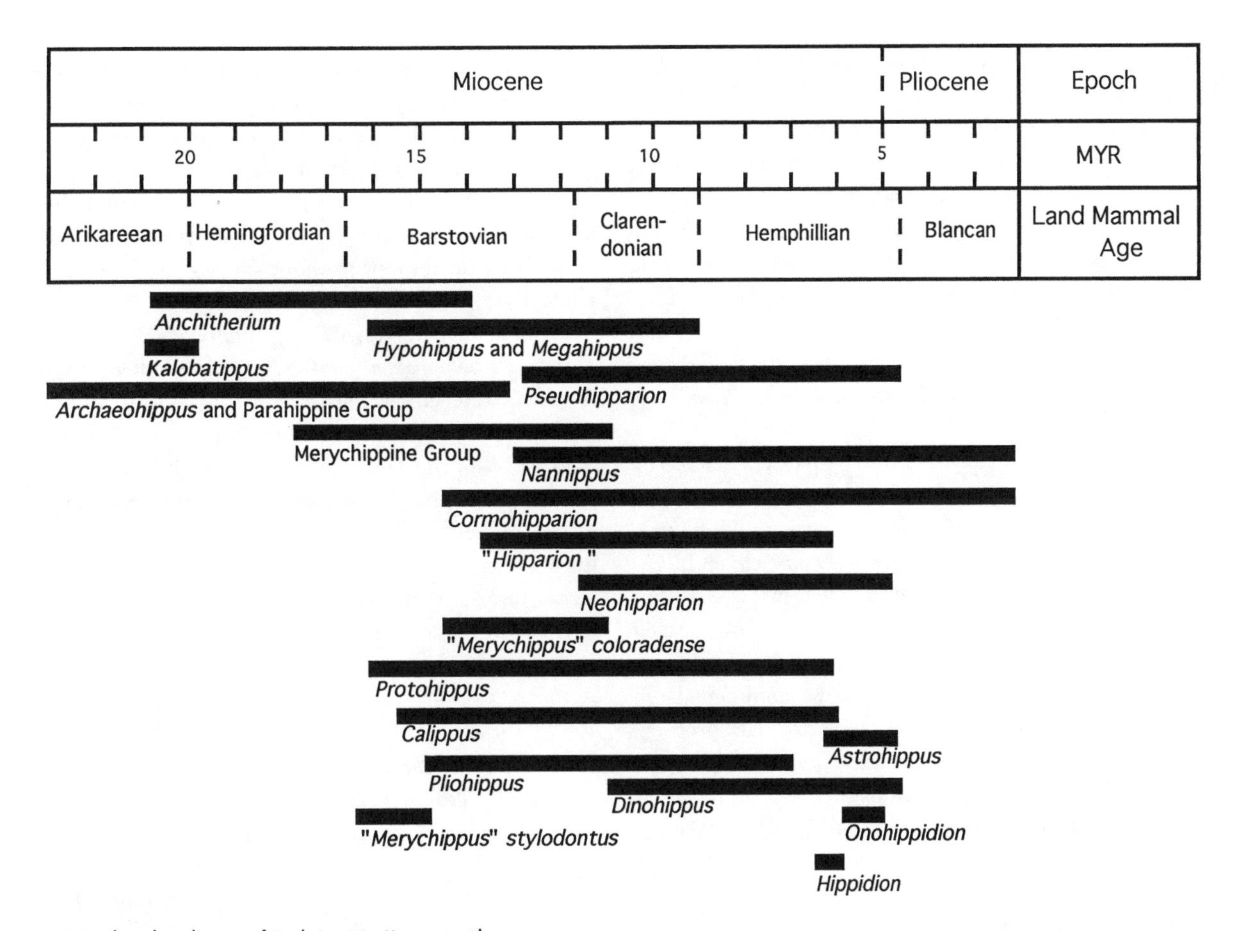

Fig. 7.1. Chronological ranges of North American Miocene equids.

Also, the three remaining horse groups, the hipparionines, protohippines, and equines, evolved out of the merychippine complex during the Early Barstovian. Species with truly hypsodont cheek teeth first appeared during this interval as well.

The Late Barstovian (14.5–11.8 myr) marked the acme of equid taxonomic richness, in numbers of contemporaneous species and genera (Hulbert, 1993). The ratio of hypsodont to mesodont and brachyodont species changed dramatically through this interval. At its start, species with low-crowned teeth (brachyodont or mesodont) made up about 70 percent of all species, but this had declined to about 25 percent by the end of the Late Barstovian. This dramatic change resulted from both greater extinction rates among low-crowned lineages and evolution of hypsodont teeth in other genera. The parahippine and archaeohippine groups made their last appearance in the earlier half of the Late Barstovian, at the Norden Bridge Fauna in Nebraska (Voorhies, 1990). Two anchithere genera, *Hypohippus* and *Megahippus*, persisted in rare numbers through this interval. Although mesodont merychippines also persisted in all regions of North America, their diversity was far below that of the Early Barstovian. Replacing them were more hypsodont members of the hipparionine, protohippine, and equine groups. Especially abundant genera were *Cormohipparion, Pseudhipparion, Protohippus, Calippus,* and *Pliohippus.* Species of these five genera dominated the equid portion of the Clarendonian Chronofauna (Late Barstovian through Early Hemingfordian).

The Clarendonian (11.8–9.0 myr) was for equids essentially a continuation of the trends begun in the Late Barstovian. Mesodont merychippines became increasingly rare and finally became extinct in the middle of the age, while anchitheres persisted a little longer, until the end of the Clarendonian. Although overall equid species richness declined from what it had been in the Late Barstovian, individual Clarendonian faunas continued to have exceptional numbers of sympatric species. Values of eight to twelve hypsodont equid species in single faunas or sites are the rule, not the exception (Webb, 1969). The hipparionine group was the most diverse in the Great Plains and Gulf Coastal Plain: *Cormohipparion* and *Pseudhipparion* remained the most common genera overall, but *Neohipparion, Nannipus,* and "*Hipparion*" all increased in numbers and dominated a few sites. Second in abundance were the protohippines, with *Calippus* being especially speciose. Western faunas (e.g., Black Hawk Ranch, Tejon Hills, Ricardo) differed in that equines were codominant with hipparionines, *Pseudhipparion* and protohippines were absent, and "*Hipparion*" was common.

The Early Hemphillian (9.0–6.0 myr) was marked by a second peak in overall equid species richness, owing in part to diversification in the equine group and in some lineages of the hipparionine group (*Nannippus* and *Neohipparion*) and in part to increasing provincialism between the West Coast, Great Plains, and Gulf Coastal Plain biogeographic realms. In the Great Plains, *Neohipparion* replaced *Cormohipparion* as the dominant large hipparionine, although the latter persisted with progressively fewer numbers into the early Late Hemphillian. *Pseudhipparion,* the other dominant hipparionine genus of the Clarendonian, is absent and probably became temporarily extinct in the Great Plains. Also absent was another clade of small body size, the subgenus *Calippus* (*Calippus*). Monodactyl equines, both the regionally autochthonous *Pliohippus* and the western allochthonous *Dinohippus,* increased in importance through the Early Hemphillian, but hipparionines and protohippines continued to dominate most faunas. Numbers of equid species found in single faunas were about one-third to one-half less than Clarendonian values.

The boundary between the Early and Late Hemphillian (ca. 6.0 myr) records a mass extinction event for equids, when about ten of the existing eighteen species lineages vanished (Hulbert, 1993). The surviving species, with only two exceptions, were those with very hypsodont cheek teeth, that is, unworn crown heights four or more times greater than basal crown length. No member of the protohippine group survived into the Late Hemphillian. This extinction event occurred across North America, apparently over a very short interval. In the ensuing Late Hemphillian interval (6.0–4.5 myr), Great Plains and Far West equid faunas were dominated by monodactyl equines (*Dinohippus, Astrohippus,* and *Onohippidium*), and usually one species of the three accounted for 80 percent or more of the equid individuals in the fauna. Thus, equid diversity in these sites had declined dramatically. Very hypsodont hipparionine species such as *Nannippus lenticularis* and *Neohipparion eurystyle* added to the species richness but made up only a small percentage of the individuals.

Late Hemphillian faunas of the Gulf Coastal Plain stand in marked contrast to those of the rest of the continent, because they were more species-rich and much more diverse. Hipparionines accounted for well over 95 percent of the individuals, with *Nannippus, Cormohipparion, Pseudhipparion,* and *Neohipparion* all common. Equines were either uncommon (*Dinohippus*) or very rare (*Astrohippus*). In spite of the superficial Clarendonian aspect of this fauna owing to the high diversity and dominance of the hipparionines, all of the species possessed numerous derived character states in terms of hypsodonty, enamel pattern, reduction or loss of facial fossae, and / or reduction of lateral digits when compared to their Clarendonian ancestors.

Miocene Camels

Camels became the prevalent group of even-toed ungulates in the North American Miocene, just as horses became the major group of odd-toed ungulates. There are several intriguing resemblances in the phylogenetic and distribution patterns of these two families. Both camels and horses experienced autochthonous radiations into diverse adaptive roles mainly in the Miocene; both reached their acme in the Middle Miocene; both experienced similar stepwise declines in the Mio-Pliocene; both dispersed key genera to other continents in the Late Pliocene, where they still survive; and both became wholly extinct in North America in the Pleistocene. Presumably these parallels between horse and camel evolution reflect similar adaptive responses to the same environmental history.

Miocene camels are divided here into six subfamilies, the principle genera of which are listed in table 7.2. The following five subfamilies appeared prior to the end of the Arikareean (very Early Miocene, ca. 20 myr): stenomylines, floridatragulines, miolabidines, protolabidines, and aepycamelines (fig. 7.2). Camelids are more abundant than equids in most Early Miocene faunas. The little gazelle-like stenomylines were precociously hypsodont and ranged throughout the Miocene in the Great Plains and the Far West. When they occur, they occur abundantly and tend to be associated with aeolian sands and semidesertic environments (Hunt, 1990). Protolabidine adaptations include reduced premolar rows,

Table 7.2. Miocene Camel Genera in Six Subfamilies

Stenomylinae	Protolabidinae	Camelinae
Dyseotylopus	*Protolabis*	*Alforjas*
Stenomylus	*Michenia*	*Procamelus*
Blickomylus	*Tanymykter*	*Pliauchenia*
Rakomylus		*Hemiauchenia*
	Aepycamelinae	*Megatylopus*
Floridatragulinae	*Aepycamelus*	*Megacamelus*
Floridatragulus	*Oxydactylus*	
Aguascalientia	*Priscocamelus*	
Gentilicamelus	*Australocamelus*	
Miolabidinae		
Miolabis		
Nothokemas		
Nothotylopus		

and deep, relatively hypsodont molars; they too are restricted to the Great Plains and Far West. Floridatragulines, with brachyodont teeth and extremely elongate jaws and limbs, are restricted to the Gulf Coastal Plain. The other two groups are more widespread. These five subfamilies continued into the Hemingfordian (20–16.3 myr), with strong emphasis in most groups on increasing size and greater hypsodonty.

Camel evolution is marked in the Barstovian (16.3–11.8 myr) by a maximum diversity of lineages (Webb, 1983a). During that interval there were at least twelve contemporaneous genera and perhaps twice that number of species. While the five earlier subfamilies persisted, the first of the large, essentially modern camelines appeared and began to diversify. In the Gulf Coastal Plain the last *Floridatragulus* species occurred in the Cold Spring Fauna. The giraffe-camels reached truly giraffe-like size and proportions in the genus *Aepycamelus*, which ranged from the eastern slopes of the Rocky

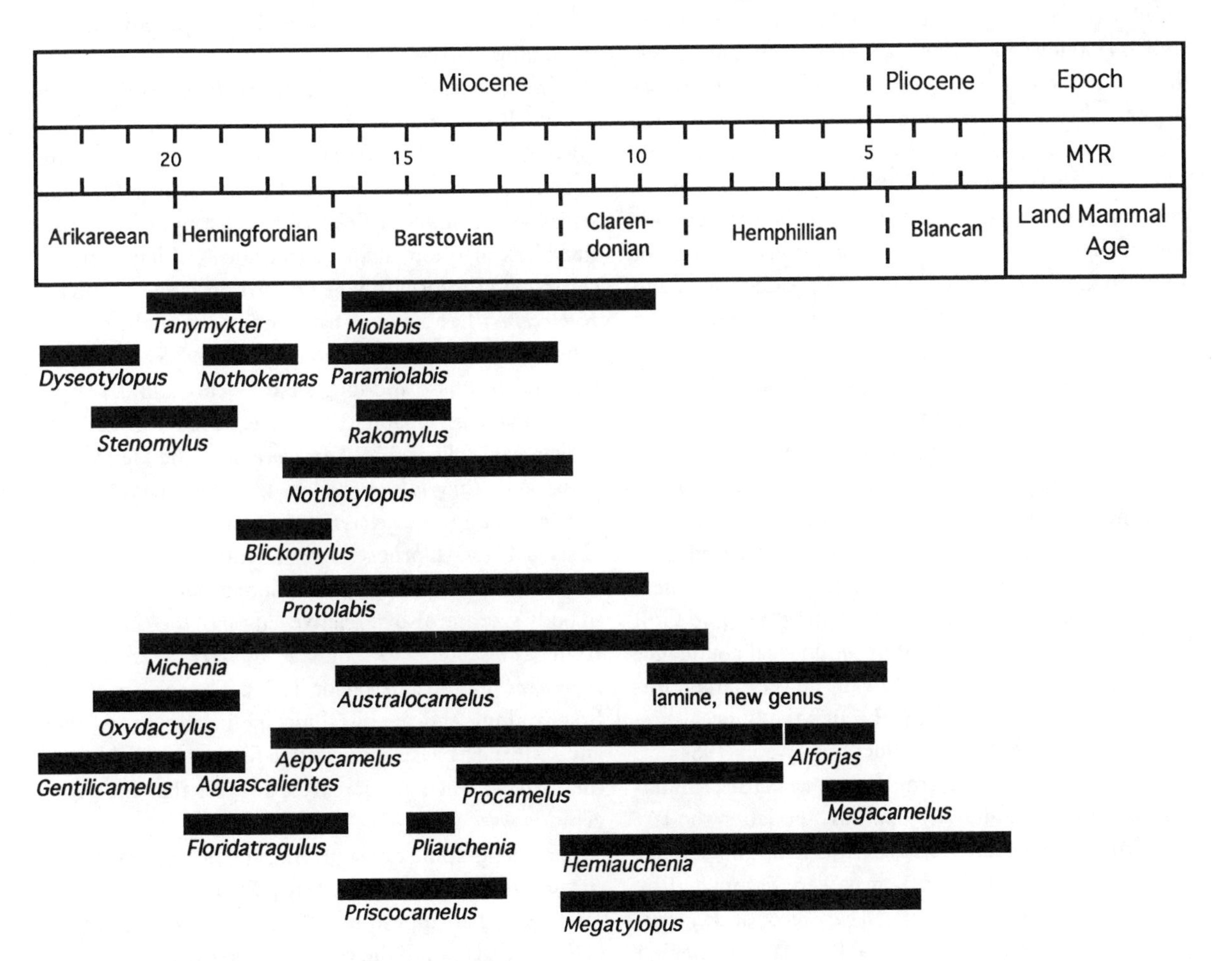

Fig. 7.2. Chronological ranges of North American Miocene camelids.

Mountains through the Great Plains to the Gulf Coastal Plain. The smaller *Hesperocamelus* occurred in the Far West. The large, hypsodont, essentially modern camelines were represented throughout the continent by *Procamelus,* which appeared in the Late Barstovian. In the Santa Fe beds of New Mexico, thanks to the extraordinary collecting campaigns of Childs Frick's teams, the rich record of Mid-Miocene camelids is known in exquisite detail; there eight genera occur together in the Late Barstovian and Early Clarendonian and far outnumber horses in the same beds (Tedford and Barghoorn, 1993).

By the Clarendonian (11.8–9.0 myr) half of the older, mesodont subfamilies had gone extinct. The aepycamelines were gone from the Far West but persisted in the Great Plains and Gulf Coastal Plain. In the Far West the last of the very hypsodont stenomylines persisted. Protolabidines reached their peak of abundance and diversity in the Clarendonian, showing considerable advances in hypsodonty, in their distinctive crania with flared nares and pinched rostra, and in their fused metapodials (Honey and Taylor, 1978). The camelines diversified and showed recurrent tendencies to attain very large size. Two widespread genera that appeared in the Clarendonian were *Hemiauchenia* and *Megatylopus.* They were characterized by loss of upper incisors, disengagement of large, procumbent lower incisors, elongation of diastemata with corresponding loss of anterior premolars, very elongate limbs, large padded feet, and regular use of a pacing gait, all adaptations for occupying open-country habitats (Webb, 1972). The first smaller, llama-like forms appear in the Gulf Coastal Plain at the Love Bone Bed (Webb et al. 1980).

The Early Hemphillian (9.0–6.0 myr) was marked by the final loss of all of the earlier subfamilies except aepycamelines, which held on mainly in Florida. The larger camelines persisted and were generally quite abundant in the Great Plains and Far West; and some new genera such as *Alforjas* appeared (Harrison, 1985). The Gulf Coastal Plain is characterized by an unusual abundance of llama-related genera and none of the very large camelines, a pattern that persisted through the Pliocene and up to the Late Pleistocene extinctions (Webb, 1984a).

The end of the Early Hemphillian marks the termination of *Procamelus* and *Aepycamelus,* the latter the last high browser. Thereafter, in the Late Hemphillian (6.0–4.5 myr), North American faunas were dominated by two large cameline genera, *Megatylopus* and *Hemiauchenia.* In addition, Gulf Coastal Plain faunas retained a small, undescribed llama-like form (Webb et al., 1980). Late Hemphillian camel-like and llama-like forms have living relatives characterized by mixed feeding in semiarid to arid habitats. Such environmental preferences are suggested not only by their modern relatives but also by such adaptations as long limbs, pacing gaits, procumbent incisors, and hypsodont dentitions.

Horned Cameloids

The autochthonous family Protoceratidae may be described succinctly as horned cameloids. Along with camelids, oreodonts, agriochoerids, and several other families, protoceratids appeared in the late Eocene on this continent as part of the first wave of selenodont (crescent-toothed artiodactyl) immigrations from Asia. (We discuss below two families from the second wave of selenodont immigrations.) Although the protoceratids were never diverse and often quite rare, we treat them here because they have distinctive adaptations and an intriguing distribution.

The Miocene Synthetoceratinae differ from the Oligocene Protoceratinae in possessing only two pairs of horns (rostral and frontal) instead of three (rostral, frontal, and occipital) or even five. The Miocene genera consist of two distinct clades (tribes), distinguished by the thickness of the median rostral horn (which remained separate in Protoceratinae). One group (*Syndyoceras* and *Kyptoceras*) has a broad base and short shaft, and the other (*Lambdoceras, Prosynthetoceras,* and *Synthetoceras*) has a narrow base and long shaft (Webb, 1981).

Syndyoceras inhabited the Great Plains in the Late Arikareean, followed by *Lambdoceras* in the Hemingfordian, and *Prosynthetoceras* in the Barstovian. *Lambdoceras* and *Prosynthetoceras* also ranged into the Gulf Coastal Plain. A progression of synthetoceratines from Hemingfordian through Clarendonian shows an increase in body size and also in relative length of the rostral horn from *Lambdoceras* through *Prosynthetoceras* to *Synthetoceras* (Patton and Taylor, 1971). The teeth remained brachyodont, and the distal metapodials relatively short and unfused. These conservative features, together with the rarity of their occurrences, suggest that the protoceratids were forest-dwelling browsers.

The geographic range of Miocene protoceratids contracted southward from the Great Plains after the medial Hemingfordian appearance of *Lambdoceras* at Flint Hill in South Dakota and the very rare Barstovian appearance of *Prosynthetoceras* at Norden Bridge, Nebraska (Voorhies, 1990). Thereafter protoceratids were known in the Texas Panhandle around the Gulf Coastal Plain, to the north in New Jersey, and to the south in Panama. Three genera coexisted during the Barstovian in the Trinity River Fauna of Texas (Patton and Taylor, 1973). The last records of the group are *Synthetoceras,* from the Early Hemphillian at McGehee Farm in Florida, and *Kyptoceras,* from the Late Hemphillian Bone Valley fauna in Florida and coastal North Carolina (Webb, 1981). The last genus was larger and somewhat more hypsodont (mesodont) than its predecessors.

Pronghorn Antelopes

In the second wave of selenodont immigration, also from Asia, antilocaprids, palaeomerycids, and moschids reached North America in the Arikareean (20 myr), near the beginning of the Miocene. We consider first the antilocaprids and then the palaeomerycids. In phylogenetic perspective the Antilocapridae represent the New World branch of Cervidae and these Palaeomerycidae an early New World clade of Giraffoidea.

Antilocaprids form a prominent group in many parts of the North American Miocene record. The group consists of two successive subfamilies, with genera as indicated in table 7.3. The merycodontines are small, mesodont to moderately hypsodont, with velvet-covered horns (the covering of which was frequently replaced, as indicated by worn tips) and multiple superficial burrs. The merycodonts ranged from Arikareean through Barstovian in the Great Plains and the Far West (figure 7.3). In the Late Arikareean and Hemingfordian the small mesodont genus *Paracosoryx* occurred abundantly in most faunas in the Great Plains. By Late Hemingfordian the moderately hypsodont *Merycodus* had appeared, accompanied by the multitined *Merriamoceras,* in the Great Plains and Far West. It is remarkable that, as far as any records show, no merycodonts reached the Gulf Coastal Plain.

Table 7.3. Miocene Antilocaprid Genera in Two Subfamilies

Merycodontinae	Antilocaprinae
Paracosoryx	*Plioceros*
Cosoryx	*Proantilocapra*
Ramoceros	*Texoceros*
Meryceros	*Ottoceros*
Merycodus	*Ilingoceros*
Merriamoceros	*Subantilocapra*
	Hexameryx
	Osborneceros

In Early Barstovian time *Meryceros* also appeared in western faunas. Merycodont diversity reached its acme, with as many as five species occurring together in diverse faunas in the Great Plains, New Mexico, and the Far West.

By Clarendonian time there had been a bottleneck and almost complete turnover in antilocaprid subfamilies, as the advanced antilocaprines eclipsed the merycodontines. Only *Merycodus* itself occurs with the earliest antilocaprines in transitional Clarendonian faunas such as the Minnechaduza in Nebraska (Webb, 1969). Antilocaprines, including *Antilocapra,* the living western pronghorn, are characterized by larger size, very hypsodont dentitions, and horn-sheathed horns. Male pronghorns with multiple-branched horns, such as *Hexameryx,* had separate sheaths above their common base (Webb, 1973). The key transitional antilocaprine is *Plioceros* of the Clarendonian. The rare *Proantilocapra* appears in the Late Clarendonian of Nebraska. One of these early antilocaprines also occurs in Florida as a rare (marginal) member of the Love Bone Bed (Webb et al., 1980), where it provides the first evidence of antilocaprids in the Gulf Coastal Plain.

Hemphillian antilocaprines are represented by at least eight genera (table 7.3). It is common to have two genera per fauna, for example, *Ilingoceros* and *Sphenophalos* at Thousand Creek in Nevada, *Osbornoceros* and *Plioceros* at Leyden in New Mexico, and *Subantilocapra* and *Hexameryx* in the Palmetto Fauna (upper Bone Valley Formation) of Florida. Most of these genera are notable for their narrow geographic ranges. Evidently the antilocaprines did not suffer from the Late Clarendonian and Mid-Hemphillian extinctions in the same manner as did other ungulate groups considered here.

American Giraffoids

Another group of large ruminants that immigrated from Asia to North America near the beginning of the Miocene (at about the same time as pronghorn antelopes) were the Palaeomerycids (formerly given their own family name, Dromomerycidae), the largest ruminants of the North American Tertiary. They are divided into three subfamilies, as indicated in table 7.4. Palaeomerycids tend to be rare members of most faunas in North America, although some Barstovian faunas yield exceptionally large samples of *Cranioceras.* In the Great Plains, where

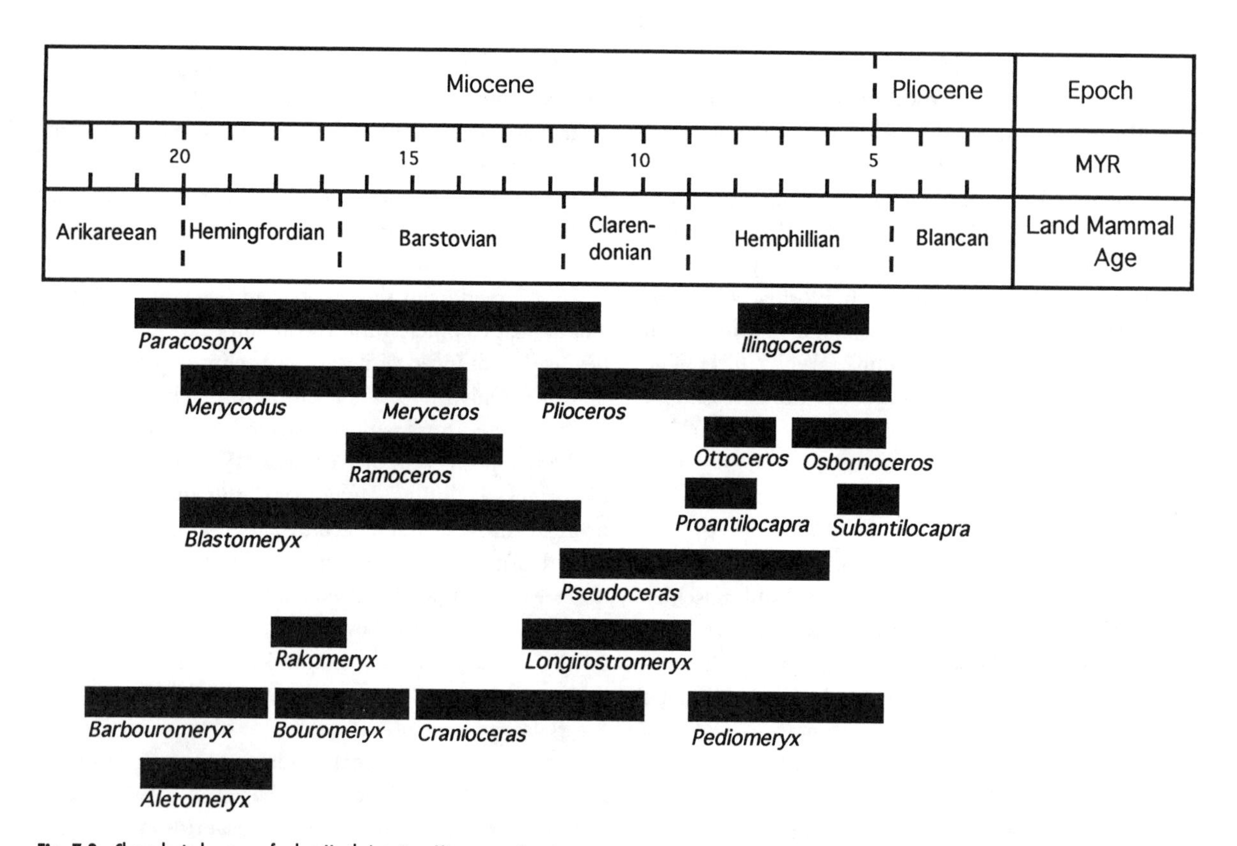

Fig. 7.3. Chronological ranges of select North American Miocene ruminants.

palaeomerycids are best and most consistently known, as many as three genera (one from each subfamily) may occur together during the Hemingfordian and Barstovian. After the Barstovian only the cranioceratines survive, ranging widely throughout the continent. From Clarendonian into Late Hemphillian faunas the last surviving genus, *Pediomeryx,* occurs as a rare form in just a handful of faunas. Even so, its geographic range extends from Mt. Eden in California through the Coffee Ranch Fauna in Texas and the ZX Bar Fauna in Nebraska to the Moss Acres Racetrack Fauna in Florida (Webb, 1983).

Diagnostic features of palaeomerycids include large size, multiple giraffe-like horns (with a long median occipital horn in cranioceratines), and brachydont to mesodont dentitions. Notable trends in *Pediomeryx* after the Mid-Hemphillian extinction event include rapid size reduction (counter to the trend in all earlier cranioceratines), premolar reduction, and a modest increase in hypsodonty (Webb, 1983b).

Table 7.4. Miocene Paleomerycid Genera in Three Subfamilies

Dromomerycinae	Aletomerycinae	Cranioceratinae
Dromomeryx	*Aletomeryx*	*Barbouromeryx*
Drepanomeryx	*Sinclairomeryx*	*Bouromeryx*
Rakomeryx		*Procranioceras*
		Cranioceras
		Pediomeryx

A Brief History of Miocene Gomphotheres

A single gomphothere genus, *Gomphotherium,* immigrated from Asia to North America across the Bering Landbridge during the Barstovian (Middle Miocene) and soon became widespread across the continent. Subsequent diversification of genera in four subfamilies is listed in table 7.5. Barstovian-age faunas containing gomphotheres range geographically from Oregon to Maryland and at least as far south as Florida, with a particular abundance in the Great Plains (Gazin and Collins, 1950; Downs, 1952; Tedford, et al., 1987; Webb, 1969; Lambert, in press). Although gomphotheres became faunally abundant and widespread during the Barstovian, their generic diversity was relatively low, with

Table 7.5. Miocene Gomphothere Genera in Four Groups

Gomphotheriinae	Amebelodontinae	"Rhynchotheriinae"
Gomphotherium	*Amebelodon*	*Rhynchotherium*
Gnathabelodon	*Serbelodon*	*Stegomastodon*
Eubelodon	*Torynobelodon*	**Tetralophodontidae**
Megabelodon	*Platybelodon*	*Tetralophodon*

only *Eubelodon* and *Megabelodon* (rare autochthons from the Great Plains) known in addition to *Gomphotherium* (fig. 7.4; Lambert and Shoshani, in press).

During the Clarendonian the generic diversity of gomphotheres in North America increased. New immigrants from Asia include *Platybelodon* (a Late Clarendonian immigrant), "*Tetralophodon*" (of unclear affinities to Old World *Tetralophodon*), and *Serbelodon* (a presumed immigrant). *Gnathabelodon*, an autochthonous form characterized by loss of its lower tusks and its broadened lower symphsis, ranged through most of the Clarendonian. The combination of these new taxa with the two Barstovian survivors *Megabelodon* (which extended only into the Early Clarendonian) and *Gomphotherium* yielded a Clarendonian fauna with at least six genera. The Late Clarendonian Black Butte Fauna of Oregon (Shotwell, 1963), includes a gomphothere erroneously referred to *Platybelodon*, which may represent a new, undescribed genus. The geographical center of Clarendonian gomphothere diversity lies in the Great Plains, where six of these genera occur. Clarendonian faunas in both the Far West and Gulf Coastal Plain were distinctly poor in gomphothere genera as compared to the Great Plains, with only two genera known from each region: *Gomphotherium* and *Serbelodon* in the Far West, and *Gomphotherium* and *Gnathabelodon* in the Gulf Coastal Plain. Of these genera, only *Gomphotherium* was fairly common (Osborn, 1933; Frick, 1933; Osborn,

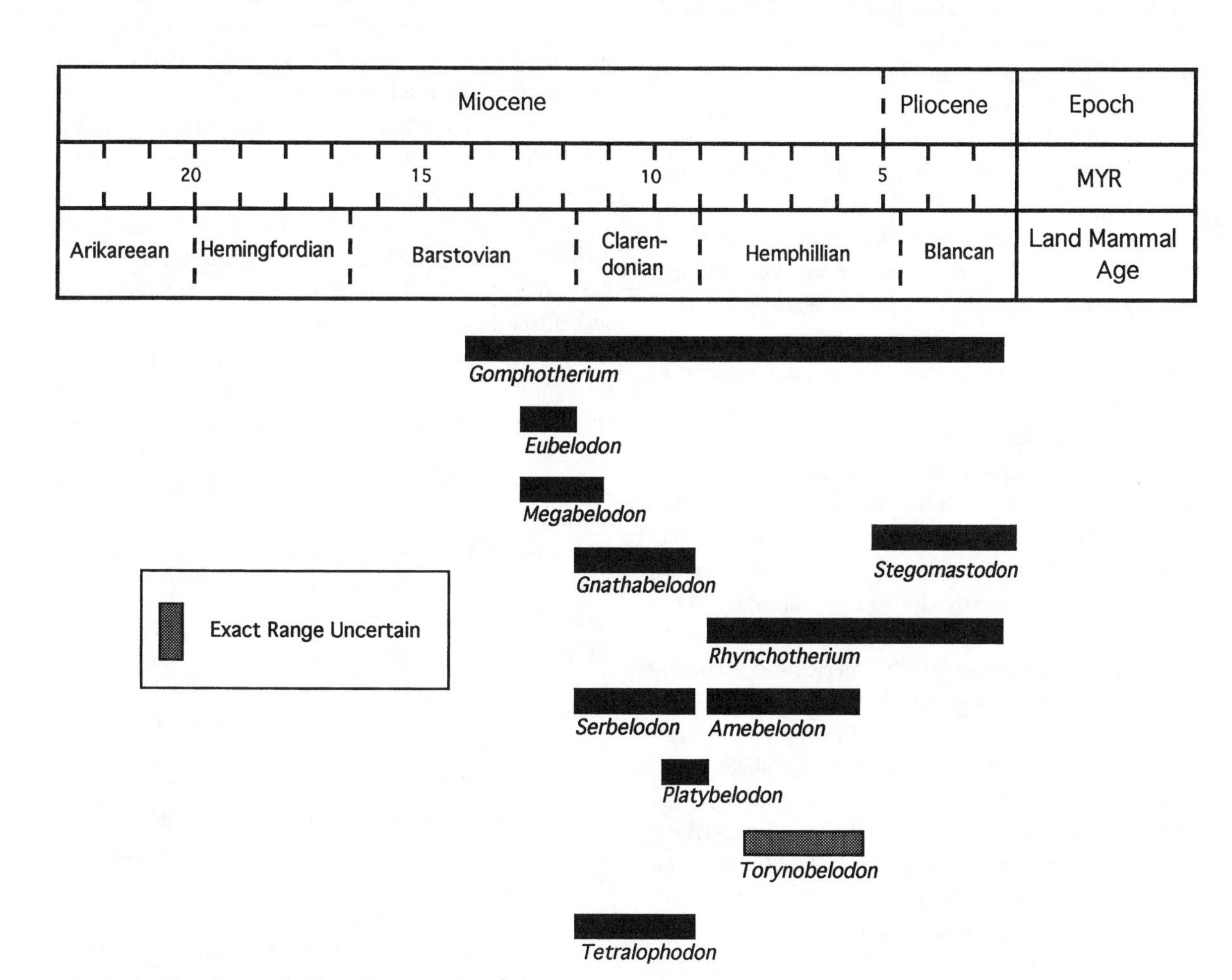

Fig. 7.4. Chronological ranges of North American Miocene gomphotheres.

1936; Sellards, 1940; Tobien, 1973; Whistler and Burbank, 1992; Lambert, in press).

Generic diversity in this family reached its peak in North America during the latest Clarendonian and Early Hemphillian, with at least seven genera represented (Frick, 1933; Osborn, 1936; Tobien, 1972, 1973; Mebrate, 1987; Tedford et al., 1987 Whistler and Burbank, 1992; Lambert, in press; Lambert and Shoshani, in press). Among survivors from earlier in the Clarendonian were *Gomphotherium, Gnathabelodon, Serbelodon,* and *Platybelodon. Torynobelodon* appears as a probable immigrant, and *Amebelodon* as a possible rapidly evolving autochthon (although Tedford et al. [1987] called it an immigrant). Faunas of this interval supported three shovel-tusked genera, *Amebelodon, Platybelodon* and *Torynobelodon,* but only one kind occurs in any one sample. Gomphothere diversity was still high in the Great Plains during this time, with all of the aforementioned genera represented. Diversity had increased moderately in the Far West, consisting of *Gomphotherium, Serbelodon,* and *Amebelodon,* and in the Gulf Coastal Plain, consisting of *Gomphotherium, Amebelodon,* and either *Platybelodon* or *Torynobelodon* (Frick, 1933; Shotwell, 1963; Mebrate, 1987; Lambert, 1990; Whistler and Burbank, 1992). Tedford et al. (1987) cite *Rhynchotherium* as diagnostic for the Late Hemphillian in temperate North America. Webb and Perrigo (1984), however, document the earlier (Early Hemphillian) presence of *Rhynchotherium* in a subtropical fauna from the Gracias Formation of Honduras, suggesting tropical origin and northward range extension.

Late Hemphillian gomphothere diversity in North America underwent both a dramatic decline and a geographical redistribution. In the Great Plains, only *Amebelodon* of genera present during the Early Hemphillian survived into the Late Hemphillian, where it is known only from the Rhinoceros Hill Fauna (Lambert, 1990). In the Late Hemphillian *Rhynchotherium* spread northward, reaching temperate North America and becoming locally abundant in the southern half of the continent, from California across the southern Great Plains to Florida (Schultz, 1977; Tedford et al., 1987; Lambert, 1990; Hulbert, 1992; Lambert, in press). *Gomphotherium* survived as a rare holdover in the Palmetto Fauna (Bone Valley Formation) of Florida and possibly the Santa Fe marls of New Mexico. The massive-toothed, hypsodont genus *Stegomastodon* appeared as a possible autochthon in the Great Plains (like *Amebelodon,* its origins are not completely resolved) but not in the Far West nor in the Gulf Coastal Plain. Thus, in a relatively brief span gomphotheres were transformed from an abundant, diverse, and widespread component of the North American fauna to one of relative unimportance, restricted to the southern half of the continent with only *Rhynchotherium* even moderately common. In the Pliocene *Rhynchotherium* gave rise to a major radiation of elephant-like, short-jawed gomphotheres that shifted southward and dominated the American tropics until the end of the Pleistocene.

Climatic Implications of Ungulate Patterns

Early Miocene Rise in Diversity. The evolution of ungulate faunas depends heavily on the distribution and productivity of vegetational formations, and these in turn are governed by regional climatic patterns. In the Early Miocene the prevailing climate across temperate North America supported a mosaic of riparian forest and woodland savanna (Axelrod, 1985; Wolfe, 1985). Global cooling trends favored more xeric habitats at the expense of more mesic habitats; this effect was intensified in the Great Plains and Great Basin as Cordilleran uplift increased rain-shadow effects.

Under such circumstances it is not surprising to find diversification of many native ungulate groups in the Early Miocene, including a few groups that may have specialized behaviorally in open-country existence. The Early Miocene horses and camels show an array of new subfamilies, yet all remain brachyodont and none attain very large size. Only *Stenomylus,* the little gazelle-camel, reveals direct morphological evidence of scrub-adaptation, notably in its precociously hypsodont third molars.

In the late Arikareean (about 20 myr) came the second wave of selenodont artiodactyl immigrations, suggesting continuous corridors of deciduous forest from Eurasia across Beringia into temperate North America. The important contributions of pronghorn antelopes, palaeomerycids, and other immigrants have been noted by Tedford et al. (1987) and by Webb and Opdyke (1995). The high-temperate distribution of cold-intolerant vertebrates, such as *Alligator* and large tortoises, in North America indicates that winter temperatures were still quite moderate.

By Hemingfordian time, as climatic deterioration intensified, the first mesodont horses, camels, and antilocaprids appeared. Horses in this radiation were the small-sized parahippines and the first merychippines. The more precocious camels of this time were stenomylines and the first protolabidines. Similarly, there is a distinct increase in hypsodonty among the Hemingfordian merycodont antelopes. Presumably these progressive ungulates moved into ecotonal environments along the forest edge, feeding in a mixed mode, opportunistically eating grasses, forbs, and leaves. Such diets are consistent with tooth-wear rates demonstrated in *Parahippus leonensis* (Hulbert, 1984) and dental microwear features in *Merychippus insignis* (Hayek et al., 1992).

Middle Miocene Savanna Optimum. Late Barstovian conditions (about 15 myr) produced the savanna optimum, when a rich mosaic of riparian forest and woodland savanna sustained the greatest diversity of grazers, mixed feeders, and browsers in North America. Webb (1983a) showed that the number of North American ungulate genera had doubled from the Early Miocene to this acme at ca. 15 myr. In horses and camels one finds a substantial number of brachyodont browsing forms still persisting alongside the more progressive hypsodont groups. The medial Barstovian also marks the appearance of *Gomphotherium,* which spread widely throughout North America, enriching still further the ungulate fauna of the savanna optimum.

The best direct evidence of the conditions that fostered this savanna optimum come from the Late Barstovian Kilgore pollen and leaf flora from the Valentine Formation in Nebraska. There MacGinitie (1962) showed that dense riparian forests near the sites of deposition gave way to "open grassy forests of pine-oak woodland on the interstream divides."

Late Miocene Decline in Diversity. Clarendonian ungulate faunas differ little from Late Barstovian faunas and are only slightly less diverse. For example, most differences between the Burge (latest Barstovian) and Minnechaduza (Clarendonian) faunas in the sandhills of northern Nebraska consist of subtle chronoclinal evolution in shared genera (Webb, 1969). Nevertheless, when the entire record is carefully perused, it is evident that the process of winnowing out the brachyodont groups had begun by the end of the Barstovian. Two horse genera (*Anchitherium* and *Desmatippus*) disappeared at that time, and three more (*Merychippus, Megahippus,* and *Hypohippus*) vanished by the end of the Clarendonian. Two camel genera became extinct (*Miolabis* and *Protolabis*); all merycodont antilocaprids vanished; and only one of the three lines of palaeomerycids continued. Both major groups of protoceratids retreated to the Gulf of Mexico (and probably Central America). Thus the secular trend toward cooler, drier conditions produced inexorable results, eliminating a majority of the browsers with low-crowned teeth.

By the Early Hemphillian the dominant vegetative pattern in the Great Plains was open savanna, with trees confined mainly to riparian corridors (Axelrod, 1985; Thomasson et al., 1990). In the Far West the Madro-Tertiary floristic trends had produced semidesertic scrub in regions of rain shadow. The effect on ungulates was to limit browsers severely but also to reduce the diversity of mixed feeders and grazers. The more progressive groups of horses, tending toward pure grazing, showed the following features: very hypsodont cheek teeth, enamel patterns dominated by long, thick plates of enamel aligned perpendicularly to the transverse shearing direction (Rensberger et al., 1984), shallow or absent facial fossae (allowing the deep, bulging maxilla to house tall teeth), and long, medial metapodials, tending to lose side toes. The key lineages exemplifying these trends were *Dinohippus, Astrohippus, Neohipparion,* and *Nannippus.* Among camels, four of the five early subfamilies completely vanished, leaving only *Aepycamelus,* the largest of the giraffe camels, to represent the Early Miocene radiation. The dominant forms were essentially modern camelines: *Hemiauchenia* and *Megatylopus,* characterized by large size, procumbent incisors, very hypsodont molars, long limbs, and padded digitigrade feet, became the dominant mixed feeders. Gomphotheres reached a peak of diversity in the Early Hemphillian. *Amebelodon* became a dominant form in the Great Plains and Gulf Coastal Plain. With its expanded lower tusks, heavy enamel-banded upper tusks, and massive hypsodont cheek teeth with as many as seven lophids, *Amebelodon* stripped bark, dug tubers, and ate all manner of vegetation (Lambert, 1992).

Many of the browsers and generalized feeders that did survive into the Early Hemphillian were restricted to the Great Plains and the Gulf Coastal Plain. Among horses, only the Gulf Coastal Plain maintained a high level of species richness. There *Cormohipparion, Pseudhipparion,* and little *Calippus* remained abundant, accompanied by *Nannippus, Neohipparion,* and *Dinohippus. Dinohippus* made its first appearance in the Early Hemphillian at Moss Acres in Florida (Hulbert, 1988c). *Aepycamelus* and *Amebelodon* pulled back from the west. The Miocene Protoceratidae had never reached the Far West. Among gomphotheres the center of diversity was clearly the Great Plains, where as many as six genera coexisted. Presumably the continued success of these diverse ungulate faunas depended on the continued existence of sufficient browse along floodplains and mountain slopes to maintain mixed feeders during the dry season. More arid conditions in the Far West foreclosed their existence there.

At the end of the Early Hemphillian (about 6.0 myr) North America experienced both an abrupt increase in aridity and a sharp increase in seasonal temperature extremes (Axelrod, 1985; Wolfe, 1985). This is also the time when arid-adapted tropical grasses that use the four-carbon photosynthesis pathway assume a dominant role in temperate North America (Thomasson et al., 1990). By then these events evidently triggered the most extensive land-mammal extinction episode in the entire Neogene record, eliminating most of the prominent elements of the Clarendonian Chronofauna (Webb, 1984a; Tedford et al., 1987, p. 191). Grazing equids that were merely hypsodont (species of *Protohippus, Calippus,* "*Hipparion,*" and *Cormohipparion*) suffered a mass extinction (Hulbert, 1993). On the other hand, the horses that were very hypsodont and had the other characteristics noted above were affected positively by these changes, increasing in numbers of individuals and species. Among camels, *Aepycamelus,* the last large browser, and the ubiquitous and abundant *Procamelus* vanished at this time. Also missing were the moderate-sized pronghorns *Texoceros* and *Plioceros.* The diversity of gomphotheres also dropped dramatically from six to three genera at the end of the Early Hemphillian. *Rhynchotherium* became the only widespread genus of the Late Hemphillian. The mass land-mammal extinctions of the Mid-Hemphillian were even larger than the mass extinctions of the latest Pleistocene. The major differences are that during the Miocene mammals of all sizes were eliminated and that the hand of human hunting could not be blamed (Webb, 1984a).

The Gulf Coastal Plain Refuge of the Latest Miocene

On a global scale the latest Miocene is noted for its dramatically cooler and drier climates, presaging glacial episodes of the Quaternary. During the Messinian, the European Late Miocene stage most precisely correlative with the Late Hemphillian, the Mediterranean nearly dried up, producing severe climatic effects in that region. Expanded glaciers in East Antarctica and new ones in West Antarctica evidently correlate with the strong isotopic signals approaching modern glacial levels (Kennett, this vol.; Shackleton, this vol.). Similar indications of increased aridity across much of North America were presumably driven by similar causes (Webb, 1984b). In the Late Hemphillian (ca. 5.5 myr) steppe conditions supplanted woodland savanna over much of the Great Plains. Aeolian sands, loess, caliche deposits, and an abundance of grass hulls characterize the Ogallala deposits. In the Far West, the Mt. Eden flora exemplifies the desert chaparral that supported the depauperate large-mammal fauna of the Late Miocene (Axelrod, 1985).

In the Late Hemphillian (about 5.5 myr) North American mammals became more extremely differentiated into regional faunas than at any previous time. Most clearly, the Gulf Coastal Plain became a refuge for several groups that became extinct in the Great Plains and the Far West (table 7.6). Undoubtedly their persistence reflects their access to more productive woodland habitats in the moister, more maritime conditions of the southeastern region. Several of these refugial species are illustrated in figure 7.5.

The only three widespread equid genera of the Late Hemphillian were *Dinohippus, Astrohippus* (both large monodactyl forms), and *Neohipparion. Nannippus* occurred in the southern Great Plains, Mexico, and the Gulf Coastal Plain but evidently had become extinct in the Far West. In the Gulf Coastal Plain two additional hipparionine species were regularly present, namely,

Table 7.6. Ungulate Taxa of the Late Hemphilian Gulf Coastal Refuge

Equidae *Cormohipparion emsliei* *Pseudhipparion simpsoni* *Nannippus minor*	**Cervidae** *Odocoileus* (undescribed)
Camelidae small llama (undescribed; also in Chihuahua, Mexico)	**Moschidae** *Pseudoceras*
Protoceratidae *Kyptoceras amatorum*	**Tayassuidae** *Mylohyus elmorei*
Antilocapridae *Hexameryx simpsoni* *Subantilocapra garciae*	**Gomphotheriidae** *Gomphotherium*, small sp. (may occur in New Mexico)

Fig. 7.5. Reconstruction of early Hemphillian animals from the Great Plains or the Gulf Coastal Plain, showing the camelid *Aepycamelus* (right), the gomphothere *Amebelodon* (below), and a pair of the hipparionine equid *Nannippus* (upper left).

Cormohipparion emsliei and *Pseudhipparion simpsoni*. (*Pseudhipparion simpsoni* also occurred in Kansas as an extremely rare form [Webb and Hulbert, 1986].) We note that the extremely small size of this *Pseudhipparion*, its incipiently hypselodont (ever-growing hypsodont) cheek teeth, and its narrow hypselodont incisors indicate that it was a selective grazer-browser (Webb and Hulbert, 1986; fig. 7.6). Other, more general differences in the equid faunas of the Gulf Coastal Plain are the dominance and diversity of hipparionines, whereas one species of monodactyl equine invariably dominates the latest Miocene faunas in the Great Plains and the Far West.

The Late Hemphillian was a bottleneck for camel diversity, as for horse diversity, in North America. The two essentially modern forms that persisted widely were *Megatylopus* and *Hemiauchenia*. In addition, a small, undescribed llama-like camelid survived in the Late Hemphillian of the Gulf Coastal Plain, evidently the continuation of a lineage first seen in the late Clarendonian Love Bone Bed (Webb et al., 1980).

Ever since the Oligocene the geographic range of protoceratids had shrunk southward and eastward. Evidently they had become endemic to the Gulf Coastal Plain and Central America after about 14 myr, the last record in the Great Plains being a specimen of *Prosynthetoceras* from the Mid-Barstovian Norden Bridge Fauna (Voorhies, 1990). *Synthetoceras* occurred only as far north as the Texas Panhandle in the Clarendonian and became restricted to the Gulf Coastal Plain in the Early Hemphillian (Patton and Taylor, 1971). *Kyptoceras*, representing a different subfamily, survived into the late Hemphillian in Florida and as far north as the coastal plain of North Carolina (Webb, 1981; fig. 7.6).

Antilocaprines are represented by very few records in the Late Hemphillian, as if the group had nearly vanished. *Texoceras* occurs in the Great Plains and *Ilingoceras* in the Great Basin. The Gulf Coastal Plain supported two endemic genera during the Late Hemphillian: *Hexameryx*, the large six-horned antelope; and *Subantilocapra*, a somewhat smaller form close to the ancestry of the modern American pronghorn. Another Late Hemphillian genus closely related to *Hexameryx* is *Hexobelomeryx* from the Yepomera Fauna in Chihuahua, Mexico (Webb, 1973).

The last palaeomerycid, *Pediomeryx hemphillensis*, survived the Early Hemphillian, becoming smaller in size and developing longer diastemata and more hypsodont molars. Its geographic distribution consists of a spotty southern record, including but not confined to the Gulf Coastal Plain. The last records are from Mt. Eden in Southern California, Coffee Ranch in the Texas Panhandle, and Moss Acres in Florida, where the only known *Pediomeryx* cranium with horns was recently discovered. Webb (1983b) speculated that in the Bone Valley Fauna of Florida the first appearance and local abundance of immigrant deer from Asia had displaced *Pediomeryx* in the very latest Hemphillian.

Among Late Hemphillian gomphotheres, the Gulf Coastal Plain is notable for the survival of a small species of true *Gomphotherium* (fig. 7.6). The only other possible Late Hemphillian record of this genus is a *Gomphotherium* from the Chamita Formation in New Mexico.

Other groups of large herbivores not specifically treated in this chapter also fall into the pattern of a Late Hemphillian refuge in more mesic (or subtropical) habitats on the Gulf Coastal Plain. We note the presence of abundant tapirs; the rare, little, browsing moschid, *Pseudoceras;* and *Mylohyus elmorei*, a forest-adapted peccary, which is larger in size and has more widely flared (more primitive) zygomatic arches than the familiar Quaternary species (Wright and Webb, 1984).

Another notable inhabitant of the Gulf Coastal Plain Refuge is the giant flying squirrel, *Cryptopterus webbi* (Robertson, 1976). This very large petauristine squirrel, with heavily crenulated teeth, represents an Old World group that immigrated to eastern North America. Such species are absolutely dependent on rich mesic forests and thus bespeak Miocene continuity of such habitats. This Florida species was originally recovered from Late Blancan spring deposits at Haile 15A but is now also known from the Late Hemphillian Bone Valley Fauna.

Conclusions

Climatic deterioration during the Miocene governed the progression of ungulate faunas in North America. During the Early Miocene most ungulate taxa diversified, producing browsing, mixed feeding, and grazing branches. The acme of horses, camels, and pronghorn antelope diversity occurred in the Late Barstovian, about 15 myr. At that time the prevailing mosaic of woodland savanna and riparian forest supported an optimally rich ungulate fauna composed of diverse browsers as well as more progressive mixed feeders and grazers. Conditions during the Clarendonian (about 10 myr) were not greatly different, but a few browsers were lost at the end of that mammal age. Most ungulate groups were still broadly distributed across the continent. By Late Miocene time a

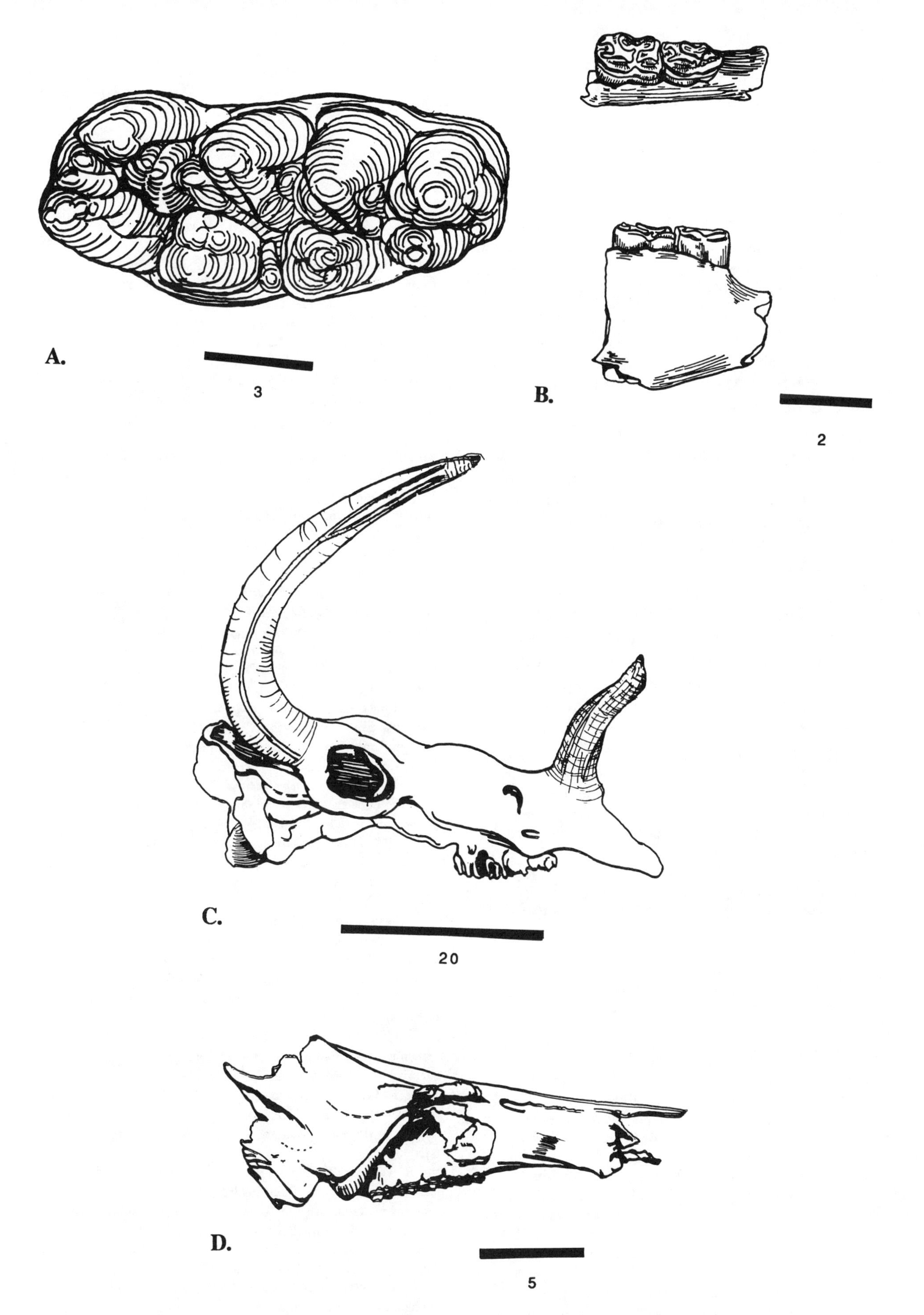

Fig. 7.6. Illustrations of specimens representing relictual taxa present in the Gulf Coastal Plain during the Late Miocene and Early Pliocene (numbers on bars represent scale in cm): A) *Gomphotherium,* small sp., lower third molar; B) *Pseudhipparion simpsoni,* partial mandible with the second and third premolars, side and occlusal views; (C) *Kyptoceras amatorum,* skull (after Webb, 1981); and D) *Mylohyus elmorei,* skull (after Wright and Webb, 1984).

decline in ungulate diversity became apparent. The Mid-Hemphillian mass extinction was the most severe in the record of North American land-mammal genera (Webb, 1984a). Most browsing taxa went extinct, presumably because climatic conditions became cooler and drier. The diversity of grazing and mixed feeding was also decimated.

In the latest Miocene (Late Hemphillian) North American climatic effects became even more devastating, evidently part of a global cascade of climatic deterioration. In the Far West semidesertic habitats became widespread, and in the Great Plains steppe habitats predominated. The Gulf Coastal Plain was the only region where mesic conditions sustained large expanses of woodland. It therefore became a Late Hemphillian refuge for many ungulate taxa. Such groups as hipparion horses, tapirs, llamas, the last protoceratids, two kinds of large antilocaprines, a small gomphothere, a browsing peccary, and the earliest relatives of white-tailed deer endowed the Gulf Coastal Plain with a uniquely rich ungulate fauna in the latest Miocene.

Acknowledgments

We thank John Eisenberg, Richard Tedford, Bruce MacFadden, and Gary Morgan for helpful discussion of stratigraphic and phylogenetic problems. Laurie Walz provided some of the illustrations used in this chapter. We are especially grateful to the many amateurs and professionals who, by working together, have added so much to the Florida Miocene record. *Kyptoceras amatorum*, as the species name indicates, was named in honor of the amateur paleontologists whose efforts brought the type skull of this animal and many others to a safe haven in the Florida Museum of Natural History.
This chapter is contribution 446 in paleobiology from the Florida Museum of Natural History. The research was supported in part by grants from the National Science Foundation (BSR 8918065) and the McKenna Foundation.

References

Axelrod, D. I. 1985. Rise of the grassland biome, central North America. *Botanical Review* 51:163–201.

Downs, T. 1952. A new mastodont from the Miocene of Oregon. *University of California Publications in Geological Sciences* 29:1–20.

Dowsett, H. J., Cronin, T. M., Poore, R. Z., Thompson, R. S., Whatley, R. C., and Wood, A. M. 1992. Micropaleontological evidence for increased meridional heat transport in the North Atlantic Ocean during the Pliocene. *Science* 258:1133–1135.

Evander, R. L. 1989. Phylogeny of the family Equidae. *In* Prothero, D. R., and R. M. Schoch (eds.), *The evolution of perissodactyls.* Oxford University Press, New York. Pp. 109–127.

Frick, C. 1933. New remains of trilophodont-tetrabelodont mastodons. *Bulletin of the American Museum of Natural History* 59:505–552.

Gazin, C. L., and Collins, R. E. 1950. Remains of land mammals from the Miocene of the Chesapeake Bay region. *Smithsonian Miscellaneous Collections* 116:1–21.

Graham, A. 1975. Late Cenozoic evolution of tropical lowland vegetation in Veracruz, Mexico. *Evolution* 29:723–735.

Gregory, J. T. 1971. Speculations on the significance of fossil vertebrates for the antiquity of the Great Plains of North America. *Abhandlungen Hessisches Landesamt für Bodenforschung* 60:64–72.

Harrison, J. A. 1985. Giant camels from the Cenozoic of North America. *Smithsonian Contributions to Paleobiology* 57:1–29.

Hayek, L. A., Bernor, R. L., Solounias, N., and Steirgerwald, P. 1992. Preliminary studies of hipparione horse diet as measured by tooth microwear. *Annales Zoologici Fennici* 28:187–200.

Honey, J. G., and Taylor, B. E. 1978. A generic revision of the Protolabidini (Mammalia, Camelidae), with a description of two new Protolabidines. *Bulletin of the American Museum of Natural History* 161:367–426.

Hulbert, R. C. 1984. Paleoecology and population dynamics of the Early Miocene (Hemingfordian) horse *Parahippus leonensis* from the Thomas Farm Site, Florida. *Journal of Vertebrate Paleontology* 4:547–558.

———. 1988a. *Calippus* and *Protohippus* (Mammalia, Perissodactyla, Equidae) from the Miocene (Barstovian–Early Hemphillian) of the Gulf Coastal Plain. *Bulletin of the Florida State Museum,* Biological Sciences, 32:221–340.

———. 1988b. A new *Cormohipparion* (Mammalia, Equidae) from the Pliocene (latest Hemphillian and Blancan) of Florida. *Journal of Vertebrate Paleontology* 7:451–468.

———. 1988c. *Cormohipparion* and *Hipparion* (Mammalia, Perissodactyla, Equidae) from the Late Neogene of Florida. *Bulletin of the Florida State Museum,* Biological Sciences, 33:229–338.

———. 1989. Phylogenetic interrelationships and evolution of North America Late Neogene Equinae. *In* Prothero, D. R., and R. M. Schoch (eds.), *The evolution of perissodactyls.* Oxford University Press, New York. Pp. 176–196.

———. 1992. A checklist of fossil vertebrates in Florida. *Papers in Florida Paleontology* 6:1–35.

———. 1993. Taxonomic evolution in North American Neogene horses (subfamily Equinae): The rise and fall of an adaptive radiation. *Paleobiology* 19:216–234.

Hulbert, R. C., and MacFadden, B. J. 1991. Morphological transformation and cladogenesis at the base of the adaptive radiation of Miocene hypsodont horses. *American Museum Novitates* 3000:1–61.

Hunt, R. M., Jr. 1990. Taphonomy and sedimentology of Arikaree (lower Miocene) fluvial, eolian, and lacustrine paleoenvironments, Nebraska and Wyoming: A paleobiota entombed in fine-grained volcaniclastic rocks. *In* Lockley, M. G., and A. Rice (eds.), *Volcanism and fossil biotas.* Geological Society of America Special Paper 244, pp. 69–111.

Janis, C. M. 1993. Tertiary mammal evolution in the context of changing climates, vegetation and tectonic events. *Annual Review of Ecology and Systematics* 24:467–500.

Lambert, W. D. 1990. Rediagnosis of the genus *Amebelodon* (Mammalia, Proboscidea, Gomphotheridae), with a new subgenus and species *Amebelodon* (*Konobelodon*) britti. *Journal of Paleontology* 64:1032–1040.

———. 1992. The feeding habits of the shovel-tusked gomphotheres: Evidence from tusk wear patterns. *Paleobiology* 18:132–147.

———. (in press). The biogeography of the gomphotheriid proboscideans of North America. *In* Shoshani J., and P. Tassy (eds.), *The Proboscidea: The paleoecology and evolution of elephants and their relatives.* Oxford University Press: Oxford.

Lambert, W. D., and Shoshani, J. (in press). Proboscidea. *In* Janis, C., and K. Scott (eds.), *Tertiary mammals of North America,* vol. 1. Cambridge University Press, Cambridge.

Leopold, E. B., and Denton, M. F. 1987. Comparative age of grassland and steppe east and west of the northern Rocky Mountains. *Annals of the Missouri Botanical Garden* 74:841–867.

MacFadden, B. J. 1992. *Fossil horses: Systematics, paleobiology, and evolution of the family Equidae.* Cambridge University Press, New York.

Mebrate, A. 1987. The long-jawed gomphotheres of North America. Ph.D. diss., University of Kansas.

Osborn, H. F. 1933. *Serbelodon burnhami,* a new shovel-tusker from California. *American Museum Novitates* 639:1–5.

———. 1936. *Proboscidea,* vol. 1. American Museum Press, New York.

Patton, T. H., and Taylor, B. E. 1971. The Synthetoceratinae (Mammalia, Tylopoda, Protoceratidae). *Bulletin of the American Museum of Natural History* 145:119–218.

Prothero, D. R., and Shubin, N. 1989. The evolution of Oligocene horses. *In* Prothero, D. R., and R. M. Schoch (eds.), *The evolution of perissodactyls.* Oxford University Press, New York. Pp. 142–175.

Rensberger, J. M., Forsten, A., and Fortelius, M. 1984. Functional evolution of the cheek tooth pattern and chewing direction in Tertiary horses. *Paleobiology* 10:439–452.

Retallack, G. J. 1982. Paleopedological perspectives on the development of grasslands during the Tertiary. *Third North American Paleontological Convention Proceedings* 2:417–421.

Robertson, J. R. 1976. Latest Pliocene mammals from Haile XVA, Alachua County, Florida. *Bulletin of the Florida State Museum, Biological Sciences,* 20:111–186.

Sarmiento, G. 1984. *The ecology of neotropical savannas.* Harvard University Press, Cambridge.

Schultz, G. E. 1977. Guidebook: Field conference on Late Cenozoic biostratigraphy of the Texas Panhandle and adjacent Oklahoma, August 4–6, 1977. *Kilgore Research Center Special Publication* 1:1–160.

Sellards, E. H. 1940. New Pliocene mastodon. *Bulletin of the Geological Society of America* 51:1659–1664.

Shotwell, J. A. 1963. The Juntura Basin: Studies in earth history and paleoecology. *Transactions of the American Philosophical Society* 53:1–77.

Skinner, M. F., Skinner, S. M., and Gooris, R. J. 1968. Cenozoic rocks and faunas of Turtle Butte, south-central South Dakota. *Bulletin of the American Museum of Natural History* 138:379–436.

Stebbins, G. L. 1981. Coevolution of grasses and herbivores. *Annals of the Missouri Botanical Garden* 68:75–86.

Stirton, R. A. 1940. Phylogeny of North American Equidae. *Bulletin of the Department of Geological Sciences, University of California* 25:165–198.

Tedford, R. H., and Barghoorn, S. F. 1993. Neogene stratigraphy and mammalian biochronology of the Española Basin, New Mexico. *Bulletin of the New Mexico Museum of Natural History and Science* 2:159–168.

Tedford, R. H., Skinner, M. F., Fields, R. W., Rensberger, J. M., Whistler, D. P., Galusha, T., Taylor, B. E., MacDonald, J. R., and Webb, S. D. 1987. Faunal succession and biochronology of the Arikarrean through Hemphillian interval (Late Oligocene–earliest Pliocene epochs) in North America. *In* Woodburne, M. O. (ed.), *Cenozoic mammals of North America.* University of California Press, Berkeley. Pp. 153–210.

Thomasson, J. R., Zakrzewski, R. J., Lagarry, H. W., and Mergern, D. E. 1990. A Late Miocene (late Early Hemphillian) biota from northwestern Kansas. *National Geographic Research* 6:231–244.

Tobien, H. 1972. The status of the genus *Serridentinus* Osborn, 1923 (Proboscidea, Mammalia) and related forms. *Mainzer Geowissenschaftliche* 1:143–191.

———. 1973. On the evolution of the mastodonts (Proboscidea, Mammalia). Pt. 1, The bunodont trilophodont groups. *Hessisches Landesamt für Bodenforschung Notizblatt* 101:202–276.

Voorhies, M. R. 1990. Vertebrate paleontology of the proposed Norden Reservoir area, Brown, Cherry, and Keya Paha counties, Nebraska. Technical Report 82-09, Division of Archeological Research, University of Nebraska, Lincoln.

Webb, S. D. 1969. The Burge and Minnechaduza Clarendonian mammalian faunas of north-central Nebraska. *University of California Publication in Geological Science* 78:1–191.

———. 1972. Locomotor evolution in camels. *Forma et Functio* 5:99–112.

———. 1973. Pliocene pronghorns of Florida. *Journal of Mammalogy* 54:203–221.

———. 1977. A history of savanna vertebrates in the New World. Pt. 1, North America. *Annual Review of Ecology and Systematics* 8:355–380.

———. 1981. *Kyptoceras amatorum,* new genus and species from the Pliocene of Florida, the last Protoceratid Artiodactyl. *Journal of Vertebrate Paleontology* 1:357–365.

———. 1983a. The rise and fall of the Late Miocene ungulate fauna in North America. *In* Nitecki, M. H. (ed.), *Coevolution.* University of Chicago Press, Chicago. Pp. 267–306.

———. 1983b. A new species of *Pediomeryx* from the Late Miocene of Florida, and its relationships within the subfamily Cranioceratinae (Ruminantia: Dromomerycidae). *Journal of Mammalogy* 64:261–276.

———. 1984a. Ten million years of mammal extinctions in North America. *In* Martin, P. S., and R. G. Klein (eds.), *Quaternary extinctions: A prehistoric revolution.* University of Arizona Press, Tucson. Pp. 189–210.

———. 1984b. On two kinds of rapid faunal turnover. *In* Berggren, W. A., and J. A. Van Couvering (eds.), *Catastrophes and earth history.* Princeton University Press, Princeton. Pp. 417–436.

———. 1992. A brief history of New World Proboscidea with emphasis on their adaptations and interactions with man. *In* Fox, J. W., C. B. Smith and K. T. Wilkins (eds.), *Proboscidean and Paleoindian interactions.* Baylor University Press, Waco, Tex. Pp. 15–34.

Webb, S. D., and Hulbert, R. C. 1986. Systematics and evolution of *Pseudhipparion* (Mammalia, Equidae) from the Late Neogene of the Gulf Coastal Plain and the Great Plains. *In* Flanagan, K. M., and J. W. Lillegraven (eds.), *Vertebrates, phylogeny, and philosophy.* University of Wyoming, Laramie. Pp. 237–285.

Webb, S. D., and Opdyke, N. D. 1995. Global climatic influence on Cenozoic land mammal faunas. *In* Kennett, J. P., and S. M. Stanley (eds.), *Effects of past global change on life: Studies in geophysics.* National Academy of Sciences, Washington, D.C. Pp. 184–208.

Webb, S. D., and Perrigo, S. C. 1984. Late Cenozoic vertebrates from Honduras and El Salvador. *Journal of Vertebrate Paleontology* 4:237–254.

Wolf, J. A. 1985. Distribution of major vegetational types during the Tertiary. *Geophysical Monograph* 32:357–375.

Whistler, D. P., and Burbank, D. W. 1992. Miocene biostratigraphy and biochronology of the Dove Spring Formation, Mojave Desert, California, and characterization of the Clarendonian mammal age (Late Miocene) in California. *Bulletin of the Geological Society of America* 104:644–658.

Wright, D. B., and Webb, S. D. 1984. Primitive *Mylohyus* (Artiodactyla: Tayassuidae) from the Late Hemphillian Bone Valley of Florida. *Journal of Vertebrate Paleontology* 3:152–159.

Chapter 8

Mammalian Migration and Climate over the Last Seven Million Years

Neil D. Opdyke

It has long been supposed that a relationship must exist between climatic change and mammalian migration and evolution (Barry et al., 1984; Pickford, 1989; Bernor et al., 1987; Opdyke, 1990; McKenna, 1975; Flynn et al., 1991; and many others). Mammalian migrations among Eurasia and North America and Africa and among Eurasia and North and South America were all sensitive to sea-level change, as they continue to be, because potential corridors between these continents were at or near the present sea level, and each has a particular climatic barrier that mammal groups must overcome to make the journey: tropical rain forest in Panama, cold Arctic steppe conditions in Beringia, and aridity in the Suez corridor. The severity of these climatic filters will vary with climatic change.

The Climatic Record

The climatic record for the last 7 million years (myr) is now known in great detail, thanks to the efforts of Shackleton and his many collaborators (Shackleton and Opdyke, 1973; Shackleton et al., 1990; Shackleton et al., 1995), who have produced a proxy record for climatic change using $\delta^{18}O$ records going back to the Late Miocene. This high-quality record has been extended further into the Miocene by Hodell and Kennett (1986) and by Hodell et al. (1994), who have analyzed pre-Messinian sediments in Morocco. The value of this record has been enhanced by the increasing precision of the magnetic polarity time scale through the application of astronomical dating (Johnson, 1982; Shackleton et al., 1990; Shackleton et al., 1995). The precision of the time scale has been confirmed by $^{40}Ar/^{39}Ar$ dating (Baksi, 1995). The $\delta^{18}O$ record is presented in figure 8.1 to the base of the Gilbert. It is against this climatic record that mammalian evolution and migration must be determined. The very complexity and completeness of this proxy record makes it much more thorough than the faunal record. Nevertheless, through high-precision $^{40}Ar/^{39}Ar$ dates and magnetic stratigraphy, faunal events can be directly compared with this climatic record.

It is natural to characterize the paleoclimatic record of the past 7 myr with respect to the high-frequency signals of 105 kyr and less, which are so prominent in figure 8.2. Longer-term trends are also apparent in this data set and show clear changes in amplitude and base level.

At about 6.9 myr within C3AR a shift occurs to heavier $\delta^{18}O$ values (Hodell et al., 1994). This change is coincident with the $\delta^{13}C$ shift in the world's oceans previously noted by Haq et al. (1980) and Vincent et al. (1989). It is also coincident with a striking shift $\delta^{13}O$ in paleosol carbonate in the Siwalik sediments of Pakistan, which Quade et al. (1989) attribute to the onset of the Asian monsoon and to the dominance of plants with a C_4 photosynthetic pathway (grasses) in the ecosystem (Cerling and Quade, 1993).

The next youngest climatic event is a sharp double event in the oxygen isotope record between 5.7 and 5.8 myr, given the designation TG20 and TG22 by Shackleton et al. (1995), which seems to represent a significant glacial event. The isotopic record then indicates a significant warming with $\delta^{18}O$ values becoming lighter and peaking at TG9 at 5.45 myr, just prior to the Miocene-Pliocene boundary. If this spike represents glacial retreat, then rising sea level at this time may have initiated the

ISBN 0-300-06348-2.

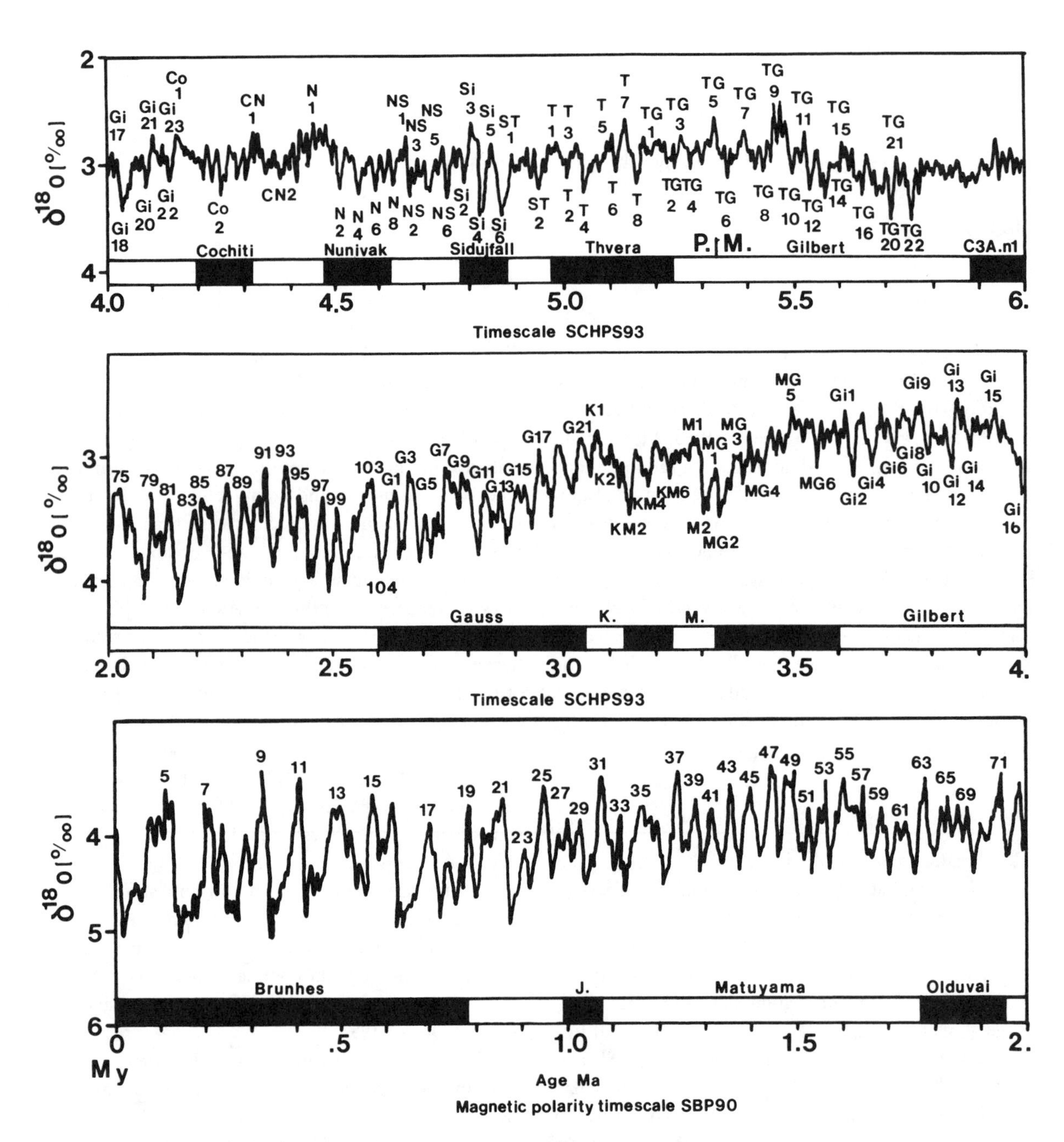

Fig. 8.1. The δ¹⁸O isotopic record for the last 6 myr, modified from data from Shackleton et al. (1990, 1994).

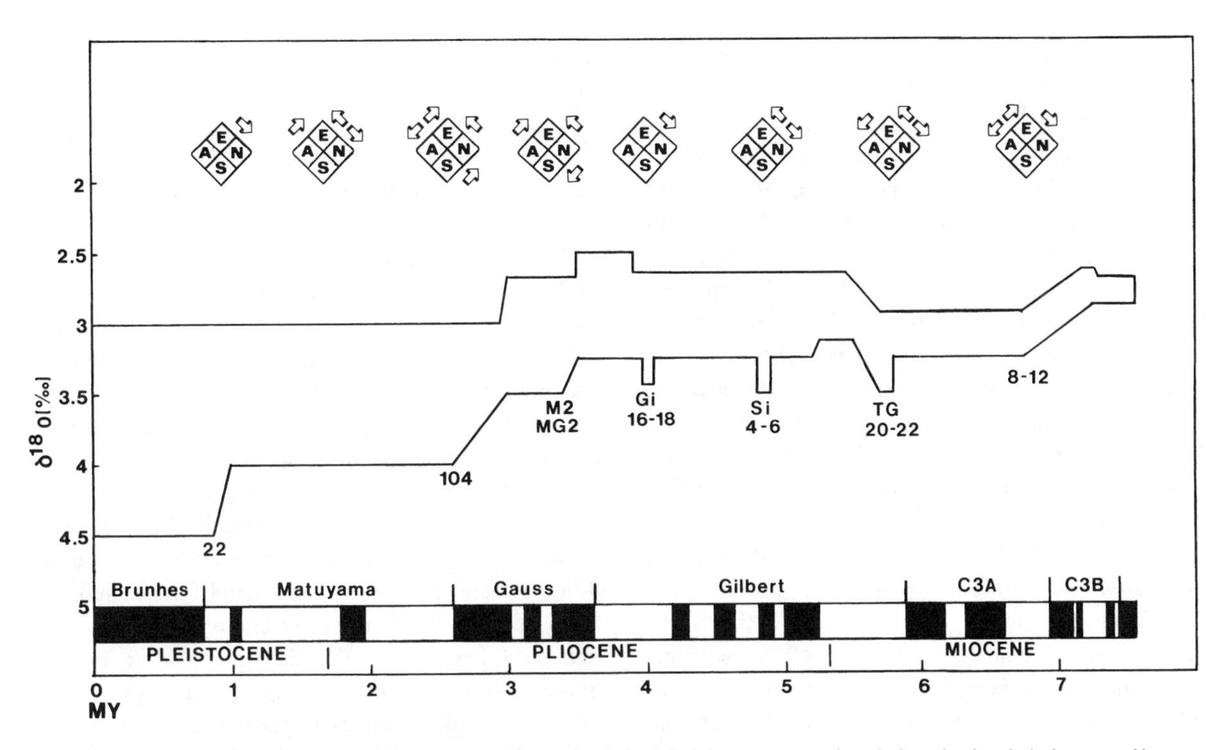

Fig. 8.2. Simplification of the isotopic record for the last 7 myr. The upper line encloses the lightest (warmest) values; the lower line bounds the heaviest (colder) values of the isotopic record given in fig. 8.1. The record from 6 to 7 myr is derived from the isotopic record of Hodell et al. (1994). The isotopic stages that give significantly heavier (colder) values above background are indicated by the isotopic stage number (see fig. 8.1). The squares above the upper line indicate times of significant intercontinental mammalian migration. The continents are abbreviated as follows: A = Africa, E = Eurasia, N = North America, S = South America; arrows indicate the direction of migration.

refilling of the Mediterranean Sea (Hodell et al., 1994). The next significant isotopic change occurs within the Sidufjall subchron, where two sharp, positive excursions occur—designated Si4 and Si6, which date between 4.8 and 4.9 myr. The next significant positive excursions take place at about 4 myr, designated Gi16 and Gi18. These changes at 4 myr do not appear to be as strong as those that preceded them. The variability seems to range about .6‰, and following these events at 4 myr, the $\delta^{18}O$ curve ramps up to lighter values, where the record stays until about the Mammoth subchron, where the beginning of heavier (colder) values occur at M2 and MG2. The half million years between 4 and 3.4 myr would seem to represent the warmest Pliocene temperatures. The heavy values begin to increase after the Mammoth subchron in a steady cooling trend to isotope stage 104, which may mark the beginning of glaciation in the Northern Hemisphere. The glacial stages within Matuyama have been shown by Ruddiman et al. (1986) to have a frequency of 40 thousand years (kyr), in phase with the changes in tilt of the earth's axis. The final change in the isotope record occurs between the Jaramillo and the top of the Brunhes, where the 100 kyr frequency of major glaciation begins and the amplitude, which presumably indicates an increase in glacial ice cover, increases. It is interesting to note that the glacial pulses increase in amplitude in three steps, beginning in the Mammoth subchron, while the warm interglacial stages remain more or less constant. If climate were forcing mammalian migration (and faunal change), one might expect the major interchanges to occur at 6.5–7 myr, 5.7–5.8 myr, 4.8–4.9 myr, about 4 myr, 3.25–3.35 myr, 2.6–2.7 myr, and .85–.9 myr.

The Mammalian Record

Our understanding of faunal record is, unfortunately, not nearly as complete as our understanding of the climatic record and lacks the precision of the isotope record. Nevertheless, the first appearance in Eurasia of *Equus,* which traveled to Eurasia from North America, required it to cross the Bering straits prior to the time of its first appearance datum in Eurasia. Probably the highest level of precision to be expected in the mammalian record is about 200 kyr. In some areas, such as the Siwaliks of Pakistan, higher precision may be attained.

The earliest migratory event that can be associated with the climatic events discussed above is the Late Miocene carbon shift (Mi9). Lindsay et al. (1984) have placed the Late Hemphillian–Early Hemphillian boundary at this position in the North American record. Tedford et al. (1987) record eight taxa that arrived in North America at about 7 myr, and Webb and Opdyke (in press) have termed this a first-order migratory episode. In Pakistan a very important faunal change occurs at this horizon (Barry et al., 1985); it is marked by the first appearance of *Hexaprotodont* and perhaps of *Stegodon*. In China the magnetostratigraphy and sediments of the Yushi Basin do not reach this time level. Nevertheless, Murids and *Stegodon* are present at 6 myr at the base of the succession (Flynn et al., 1991). In Africa the faunal record shows a significant turnover at this time (Hill, this vol.), with Bovini migrating from Eurasia via Spain and with *Hippopotamus, Macaca,* and Ruduncini going from Africa to Spain (Moyà-Solà and Agustí, 1989). This exchange occurs after the beginning of the Messinian (Hodell et al., 1994), and faunal turnovers at this time are clearly related to the beginning of the drying out of the Mediterranean.

The final isotope events of the Miocene, TG20–22, occur between 5.7 and 5.8 myr in the lower Gilbert. In Asia the appearance of Camelidae and *Sinomastodon* from North America occurs within the lower Gilbert in the Yushi Basin (Flynn et al., 1991). The Pakistan record does not record significant immigration events at this level. However, Ovibovini migrate into Africa from Eurasia (Vrba, chap. 27, this vol.). North African and southern European climates are dominated by the Messinian Salinity Crisis. The big faunal change in the Mediterranean Basin occurs at the end of the Miocene and is associated with the refilling of the Mediterranean Sea at the beginning of the Pliocene. Tedford et al. (1987) have recorded the final wave of Miocene immigrants into North America, which they place at 6 myr; this dating is not significantly different from the age about the TG20, TG22 isotope event. This first-order migratory event includes *Castor,* among others (Lindsay et al., 1984).

The earliest isotopic event in the Pliocene occurs in the Sidufjall subchron. In North America this isotopic event may well correspond to the position of the Blancan-Hemphillian boundary, as suggested by Lindsay et al. (1984). Tedford et al. (1987) record over seven immigrant taxa at this level, including *Ursus* (bear) and *Odocoileus* (deer). In China and East Asia *Microtoscoptes* and *Canidae* may have appeared at this time, although the time of appearance of these animals in the geomagnetic polarity time scale is imprecise (Flynn et al., 1991).

The last cooling event of the Gilbert occurs at 4 myr, (Gi16–18 of Shackleton et al., 1995). In North America a dispersal event, termed the *Trigonicthys* by Lindsay et al. (1984), occurs at this time; on most other continents evidence for mammalian migration at this time is lacking. In the Yushe Basin, for example, the Late Gilbert is mostly missing because of an unconformity.

The climate becomes very warm during the latest Gilbert and Early Gauss time. This period probably represents the warmest interval in the Pliocene. Migration of faunas between North America and Eurasia and between Africa and Eurasia is probably curtailed because of high sea levels. It is at this time that shallow marine faunas with North Pacific origins appear in the North Atlantic. This development implies a major marine barrier resulting from high sea level in the Bering straits. It is probable that the same is true for the Gulf of Suez.

This warm climatic optimum terminates at 3.3 myr, during the Mammoth subchron at isotope stages MG2 and M2. After this point isotopic values never return to the lighter values recorded in the Early Gauss. It is at this time that the elephant *Archidiskodon* disperses from Africa and is recorded throughout Eurasia and that its first occurrence is securely dated by magnetostratigraphy in the Siwaliks (Opdyke et al., 1979), China (Flynn et al., 1991), and the Crimea (Kochegura and Zubakov, 1978). This datum is clearly correlative with the cooling within the Mammoth subchron. In the Yushi Basin *Vulpis* and *Canis* appear, with *Archidiskodon* having migrated from North America.

Evidence is now available that the first important mammalian interchange across the Isthmus of Panama takes place in the Middle Gauss, during the Kaena and Mammoth subchrons. Magnetostratigraphic studies from sediments of the type Uguian in Argentina indicate that the first wave of immigrants from North America to South America, such as *Equus,* appear at this time (Orgeira, 1990). South American forms appear later in North America at the Gauss-Matuyama boundary. The inter-American interchange seems to have been asymmetric. Interestingly, Hooghiemstra (this vol.) has noted that the first appearance of North American plants in Columbia occurs in the Middle Gauss. The reason for the delayed arrival of South American forms in North America is not clear.

The migratory events at and around the Gauss-Matuyama boundary are very significant and are clearly coincident with the onset of Northern Hemispheric glaciation. Lindsay et al. (1980) have called this dispersal event the *Equus* datum, based on the spread of *Equus* throughout Eurasia. In the Siwaliks this datum is associated with the appearance of *Cervids*. In western Europe two tribes of bovids appear at 2.5 myr, the Alcelaphini and Hippotragini, both of which have an African origin. One group of antelope (the Caprini) and *Canis* appear in Africa from Eurasia (Vrba, chap. 27, this vol.). In North America the wave of South American migrants appears coincident with the Gauss-Matuyama boundary (Galusha et al., 1984). It is clear that mammalian faunas were moving freely between continents near to the Gauss-Matuyama boundary, undoubtedly because of a sea-level drop and climatic forcing owing to the onset of Northern Hemispheric glaciation at isotope stage 104.

The frequent reoccurrence of glaciation over the next 1.5 myr would lead one to believe that continuous movement of fauna between continents would be the norm. There appear to be times, however, when migratory pulses are concentrated, the most conspicuous being before and after the Olduvai subchron. Isotope stages 78 and 82 at 2.05 and 2.15 myr, respectively, appear to have heavier or cooler values than isotope stages 100 and 104 at the Gauss-Matuyama boundary. Important migratory events took place at this time, both into and out of Africa. *Equus* appears in the faunal record of East Africa, and the earliest migration of *Homo erectus-meganthropus* out of Africa to Southeast Asia may have taken place at this level, as suggested by Ninkovich et al. (1982) and confirmed by Swisher et al. (1994). In both North America and Europe migratory events precede and follow the Olduvai subchron. This change delineates the transition between the Blancan and Irvingtonian faunas in North America and the change from the Middle to late Villafrachian in Europe. Azzaroli et al. (1988) and Massine and Torre (1990) note the spread of *Pantera Toscana, Pliohyaena brevirostris,* and the wolf *Canis Etuscus* to Europe; in North America important Eurasian immigrants such as *Mammutus, Enceraterium,* and jaguars appear in the stratigraphic record. This faunal change in Eurasia and North America is coincident with the Plio-Pleistocene boundary, at isotope stage 60.

The final major climatic change and migratory exchange occur following the Jaramillo subchron and coincide closely with the intensification of the Northern Hemispheric glaciation at isotope stage 23. This represents the end of the Villafranchian in Europe, the transition to the Galerian fauna of western Europe and the Triaspolian faunas of Eurasia, and the beginning of the Irvingtonian II event of Repenning (1987), which is marked by the appearance of *Microtus paroperarius* and *Soergelia*.

Conclusions

The thesis that climatic change from the Miocene to the Recent dramatically influences the faunas of all continents is beyond doubt. Because of the complexity of the system, however, the problem of exactly how this takes place must be resolved. The improved understanding of the climatic record and the dramatic increase in its precision, combined with the increasing precision of dates in the faunal record provided by magnetic stratigraphy and $^{40}Ar/^{39}Ar$ dating will allow a more precise correlation of faunal dynamics to the climatic record.

Faunal turnover is characterized by evolution, extinction, and immigration, and this chapter has attempted to address only intercontinental migration. As I have shown above, migratory events are in fact not randomly distributed in time but are closely associated with climatic events related to a rise and fall in sea level associated with increasing and decreasing glaciation. Intercontinental interchange adds to turnover occurring at these times. There is clearly a first-order correlation between these events. Future progress in understanding these changes will require a better understanding of the temporal and spatial aspects of the faunal record on all continents.

References

Azzaroli, A., De Giuli, C., Ficcarelli, G., and Torre, D. 1988. Late Pliocene to early Mid-Pleistocene mammals in Eurasia: Faunal succession and dispersal events. *Palaeogeo. Palaeoclim. Palaeoecol.*, 66:77–100.

Baksi, A. K., 1995. Fine tuning the radiometrically derived geomagnetic polarity time scale. *Geophys. Res. Lett.*, 22:457–460.

Barry, J. C., Johnson, N. M., Raza, S. M., and Jacobs, L. L. 1985. Neogene faunal change in southern Asia: Correlations with climatic tectonic and instatic events. *Geology,* 13:637–640.

Bernor, R. L., Brunet, M., Ginsburg, L., Mein, P., Pickford, M., Rogl, F., Sen, S., Steininger, F., and Thomas, H. 1987. A consideration of some major topics concerning Old World Miocene mammalian chronology, migrations and paleogeography. *Geobios,* 20:431–439.

Cerling, T. E., and Quade, J. 1993. Stable carbon and oxygen isotopes in soil carbonates. In *Climate change in continental isotopic records* (ed. P. K. Swart, K. C. Lohman, J. McKensie, and S. Savin) pp. 217–231. Geophysical Monograph 78, American Geophysical Union, Washington, D.C.

Flynn, L. J., Tedford, R. H., and Zhanxiang, Q. 1991. Enrichment and stability in the Pliocene mammalian fauna of North China. *Paleobiology,* 17:246–265.

Galusha, T., Johnson, N. M., Lindsay, E. H., Opdyke, N. D., and Tedford, R. H. 1984. Biostratigraphy and magnetostratigraphy, Late Pliocene rocks, 111 Ranch Arizona. *Geol. Soc. Am. Bull.,* 95:714–722.

Haq, B. U., Worsley, T. R., Burckle, L. H., Douglas, R. G., Keigwin, L. D., Opdyke, N. D., Savin, S. M., Sommer, M. A., Vincent, E., and Woodruff, F. 1980. Late Miocene marine carbon-isotopic shift and synchroneity of some phytoplanktonic biostratigraphic events. *Geology,* 8:427–431.

Hodell, D. A., Benson, R. H., Kent, D. V., Boersma, A., and Rakic-El Bied, K. 1994. Magnetostratigraphic, biostratigraphic and stable isotope stratigraphy of an Upper Miocene drill core from the Salé Briqueterie (northwestern Morocco): A high-resolution chronology for the Messinian stage. *Paleoceanography,* 9:835–855.

Hodell, D. A., and Kennett, J. P. 1986. Latest Miocene benthic $\delta^{18}O$ changes in global ice volume, sea level, and the Messinian Salinity Crises. *Nature,* 32:411–414.

Johnson, R. G. 1982. Brunhes-Matuyama magnetic reversal dated at 790,000 yr B.P. by marine-astronomical correlations. *J. Quat. Res.,* 17:135–147.

Kochegura, V. V., and Zubakov, V. A. 1978. Palaeomagnetic time scale of the Ponto-Caspian Plio-Pleistocene deposits. *Palaeogeo. Palaeoclim. Palaeoecol.,* 23:151–160.

Lindsay, E. H., Opdyke, N. D., and Johnson, N. M. 1980. Pliocene dispersal of the horse *Equus* and Late Cenozoic mammalian dispersal events. *Nature,* 287:135–138.

———. 1984. Blancan-Hemphillian land mammal ages and Late Cenozoic mammal dispersal events. *Ann. Rev. Earth and Planet Sci.,* 12:445–488.

McKenna, M. C. 1975. Fossil mammals and Early Eocene Atlantic land continuity. *Ann. Missouri Botanical Gardens,* 62:335–353.

Massini, F., and Torre, D. 1990. Large mammal dispersal events at the beginning of the Late Villafranchian. In *European Neogene Mammal Chronology* (ed. E. H. Lindsay, V. Fahlbush, and P. Mein), pp. 131–138. Plenum Press, New York.

Moyà-Solà, S., and Agustí, J. 1989. Movements and mammal successions in the Spanish Miocene. In *European Neogene mammal chronology* (ed. E. H. Lindsay, V. Fahlbush and P. Mein), pp. 357–373. Plenum Press, New York.

Ninkovich, D., Burckle, L. H., and Opdyke, N. D. 1982. Paleographic and geologic setting for early man in Java. In *The Ocean Floor* (ed. R. A. Scrutton and M. Talwani), pp. 211–228. John Wiley, New York.

Opdyke, N. D. 1990. Magnetic stratigraphy of Cenozoic terrestrial sediments and mammalian dispersal. *Jour. of Geol.,* 98:621–637.

Opdyke, N. D., Lindsay, E., Johnson, G. D., Johnson, N., Tahirkheli, R. A. K., and Mirza, M. A. 1979. Magnetic polarity stratigraphy and vertebrate paleontology of the Upper Siwalik subgroup of northern Pakistan. *Palaeogeo. Palaeoclim. Palaeoecol.,* 27:1–34.

Opdyke, N. D., Mein, P., Moissenet, E., Perez-Gonzalez, A., Lindsay, E., and Petko, M. 1989. The magnetic stratigraphy of the Late Miocene sediments of the Cabriel Basin, Spain. In *European Neogene mammal chronology* (ed. E. H. Lindsay, V. Fahlbush, and P. Mein), pp. 507–514. Plenum Press, New York.

Orgeira, M. J., 1990. Paleomagnetism of Late Cenozoic fossiliferous sediments from Barranca de los Lobos (Buenos Aires Province, Argentina): The magnetic ages of the South American land-mammal ages. *Phys. Earth Planet. Int.,* 64:121–132.

Pickford, M. 1989. Dynamics of Old World biogeographic realms during the Neogene; Implications for biostratigraphy. In *European Neogene mammal chronology* (ed. E. H. Lindsay, V. Fahlbush, and P. Mein), pp. 413–442. Plenum Press, New York.

Quade, J., Cerling, T. E., and Bowman, J. R. 1989. Development of Asian monsoon revealed by marked ecological shift during the latest Miocene in northern Pakistan. *Nature,* 342:163–166.

Repenning, C. A., 1987. Biochronology of the microtine rodents of the United States. In *Cenozoic mammals of North America* (ed. M. D. Woodburne), pp. 236–268. University of California Press, Berkeley.

Ruddiman, W. F., McIntyre, A., and Raymo, M. 1986. Matuyama 41,000 year cycles North Atlantic Ocean and Northern Hemisphere ice sheets. *Earth and Planet. Sci. Lett.,* 80:117–129.

Shackleton, N. J., Berger, A., and Peltier, W. R. 1990. An alternative astronomical calibration of the lower Pleistocene timescale based on ODP Site 677. *Trans. R. Soc. Edinburgh,* 81:251–261.

Shackleton, N. J., Hall, M. A., and Pate, D. 1995. Pliocene stable isotope stratigraphy of ODP Site 846. *Proc. Ocean Drill. Prog., sci. res.,* 138.

Shackleton, N. J., and Opdyke, N. D. 1973. Oxygen isotope and paleomagnetic stratigraphy of equatorial pacific core V28-238: Oxygen isotope temperatures and ice volumes of a 10^5 and 10^6 year scale. *J. Quat. Res.,* 3(1):39–55.

Swisher, C. C., III, Curtis, G. H., Jacob, T., Getty, A. G., A. Suprijo, Widiasmoro, 1994. Age of the earliest known hominids in Java, Indonesia. *Science,* 263:1118–1121.

Tedford, R. H., Skinner, M. F., Fields, R. W., Rensberger, J. M., Whistler, D. P., Galusha, T., Taylor, B. E., Macdonald, J. R., and Webb, S. D. 1987. Faunal succession and biochronology of the Arikareean through Hemphillian interval (Late Oligocene through earliest Pliocene epochs) in North America. In *Cenozoic mammals of North America* (ed. M. D. Woodburne), pp. 153–210. University of California Press, Berkeley.

Vincent, E., Killingly, J. S., and Berger, W. H. 1985. Miocene oxygen and carbon isotope stratigraphy of the tropic Indian Ocean. *Geol. Soc. Am. Mem.,* 163:103–130.

Webb, S. D., and Opdyke, N. D. (in press). Global climatic influence on Cenozoic land mammal faunas. Nat. Aca. Sci.

Chapter 9

Faunal Turnover and Diversity in the Terrestrial Neogene of Pakistan

John C. Barry

Fossil vertebrates from the Neogene Siwalik formations of northern India and Pakistan were first reported in the 1830s. These formations have since been shown to comprise one of the longest, richest sequences of terrestrial vertebrate faunas known. The past two decades in particular have been marked by considerable activity in Pakistan, where fossiliferous Siwalik sediments are especially well exposed in the north on the Potwar Plateau. In this region the Neogene sediments are typically 2,000–5,000 m thick and are exposed as broad bands of outcrop that extend laterally for tens of kilometers. Relative stratigraphic positions of most fossil sites are easily determined, and the ages of the fossil sites estimated using the established magnetostratigraphic framework of Opdyke et al. (1979), G. D. Johnson et al. (1982), Tauxe and Opdyke (1982), N. M. Johnson et al. (1982, 1985), and Kappelman (1986).

Typical Siwalik fossil localities are small concentrations that formed as attritional assemblages of disarticulated bones (Badgley, 1986; Behrensmeyer, 1987), although some may contain hundreds or even thousands of bones. Plants are extremely rare, but snails, clams, and other invertebrates are relatively common. The vertebrates include bony fish, turtles, and crocodiles, as well as mammals and, more rarely, amphibians and birds. Remains of both aquatic and terrestrial taxa are most abundant in the fills of small channels and in crevasse splays in the interval between 18 and 7 million years (myr), which is therefore the most sampled paleontologically. Both the channel and crevasse-splay facies become rare after 7 myr (Behrensmeyer, 1987), resulting in the preservation of fewer fossils between 7 and 3 myr.

Among other objectives, recent paleontological research in the Potwar Siwaliks has focused on documenting the patterns and timing of faunal change among the mammals. In particular, attention has been given to the patterns of species turnover, changes in species richness, and their relationship to environmental events. This work has been of special interest, because the temporal resolution of the stratigraphic sequence allows some limited tests of hypotheses relating faunal and environmental changes (Vrba, 1985, 1988). Terrestrial deposits such as the Siwaliks, however, are usually incomplete at finer scales of resolution, and, in addition, the abundance of fossil remains varies considerably throughout any sequence. Estimates of record quality must therefore be used to weigh the significance of any observed faunal patterns; this approach has become a major part of the research agenda in the Siwaliks.

A second focus of research has been analysis of Siwalik paleoecology, especially the changing ecological character of the faunal assemblages through time. Points of interest include determining ecomorphic profiles, relative abundances, and body-size distributions for coexisting species, as well as examining the relationship of fossil occurrences to depositional environments. Our ability to determine these attributes and relationships is significantly influenced by the quality of preservation and the completeness of the fossil material and taphonomic biases that may have been operating. Thus, in this context as well, consideration of the reliability of the preserved record has also become an important research focus.

In this chapter I review the stratigraphy of the Siwalik

ISBN 0-300-06348-2.

deposits as well as recent work on the patterns of faunal change, with emphasis on the critical Late Miocene–Early Pliocene transition, which is a time of considerable environmental change. I also consider the completeness and quality of the Siwalik record as it relates to our understanding of the significance of the observed faunal patterns.

Overview of Siwalik Stratigraphy

Terrestrial Neogene sediments, all of which can be referred to loosely as "the Siwaliks," are present throughout India, Nepal, and Pakistan, where they are associated with young, active orogenic belts resulting from the collision of India and Asia. The important sedimentologic and taphonomic features of the rocks have been extensively discussed by others (Behrensmeyer, 1987; Behrensmeyer et al., 1995; Willis and Behrensmeyer, 1995; Willis, 1993a, 1993b), and need only a brief review here. The formations are fluvial in origin and comprise alternating sandstones and fine-grained sediments, with occasional conglomerates, especially in the upper parts of the section. Typical depositional environments include channels, crevasse splays, fills, and floodplain soils. Individual formations are somewhat arbitrarily distinguished on the basis of ratios of sand to clay and silt, and because local depositional conditions vary greatly, the lithostratigraphy of the formations is complex. Although some lithostratigraphic units (e.g., the "Chinji" or "Nagri" Formations) are recognized over a broad region, most units are very restricted in their geographic extent. This aspect, which has only recently been recognized, means that the lithostratigraphic correlations used in earlier syntheses of Siwalik stratigraphy (e.g., Pilgrim, 1910) are of limited use, if not actually misleading.

Understanding of Siwalik geology, stratigraphy, sedimentation, and chronology has greatly improved in the past two decades. The classic formulation of Siwalik stratigraphy was developed by G. E. Pilgrim in a series of papers on the occurrence of fossils and sediments throughout the Indian subcontinent (Pilgrim, 1908, 1910, 1913, 1917, 1926, 1934). He recognized seven successive "faunal zones" for the Early Miocene through Early Pleistocene: Gaj, Kamlial, Chinji, Nagri, Dhok Pathan, Tatrot, and Pinjor. These terms were at first used only as faunal units, but they subsequently became confused with lithostratigraphic units and are now used mainly with reference to some of the lithological formations recognized throughout Pakistan and India (Colbert, 1935; Shah, 1977).

Age estimates by Pilgrim and others for the formations and their contained fauna were based on imprecise faunal correlations to European and even North American Neogene sequences (Pilgrim, 1926, 1934; Matthew, 1929; Colbert, 1935). Consequently, since 1973 much effort has been directed at refining Pilgrim's biostratigraphic sequence and placing the faunas within a secure chronostatigraphic framework. Because the sediments are well suited for magnetostratigraphic studies, it has been possible to establish a regional chronostratigraphic framework (fig. 9.1) that now spans the Early Miocene through the Pleistocene (Opdyke et al., 1979; G. D. Johnson et al., 1982; Tauxe and Opdyke, 1982; N. M. Johnson et al., 1982, 1985; Kappelman, 1986; Friedman et al., 1992; Downing et al., 1993). This framework also allows Siwalik faunal assemblages to be more precisely correlated to well-dated sequences in Africa, Europe, and elsewhere (Flynn et al., 1990).

The Potwar Plateau sequence is probably the most complete sequence depositionally and is currently the best documented and most intensely collected of the widespread Siwalik deposits. Noye Johnson and others (1982, 1985) have shown that the Potwar Siwalik sediments span the interval between 18.3 and 0.6 myr, during which only one major depositional hiatus has been identified (Opdyke et al., 1979). (To be consistent with the earlier work I shall be discussing, all ages in this chapter are stated with reference to the Geomagnetic Reversal Time Scale of Berggren et al. [1985]. Alternative time scales make some of these ages as much as 400,000 years older.) As a result, the ages of the Miocene Potwar formations (the Kamlial, Chinji, Nagri, and Dhok Pathan Formations) and the Late Pliocene through Pleistocene formations are now fairly well understood (fig. 9.1).

Older, fossiliferous formations are also known in Pakistan outside the Potwar Plateau but in most cases are not well dated. These include parts of the Murree Formation near Banda Daud Shah (de Bruijn et al., 1981); the Chitarwata and lower Vihowa Formations in the Zinda Pir Dome near Dera Ghazi Khan (Friedman et al., 1992; Downing et al., 1993) and correlative rocks at Dera Bugti (Flynn et al., 1986); and the upper Gaj and Manchar Formations in Sindh (Raza et al., 1984; de Bruijn and Hussain, 1984). The Zinda Pir Dome sequence is now the best dated, with a series of superposed large- and small-mammal localities spanning the interval between

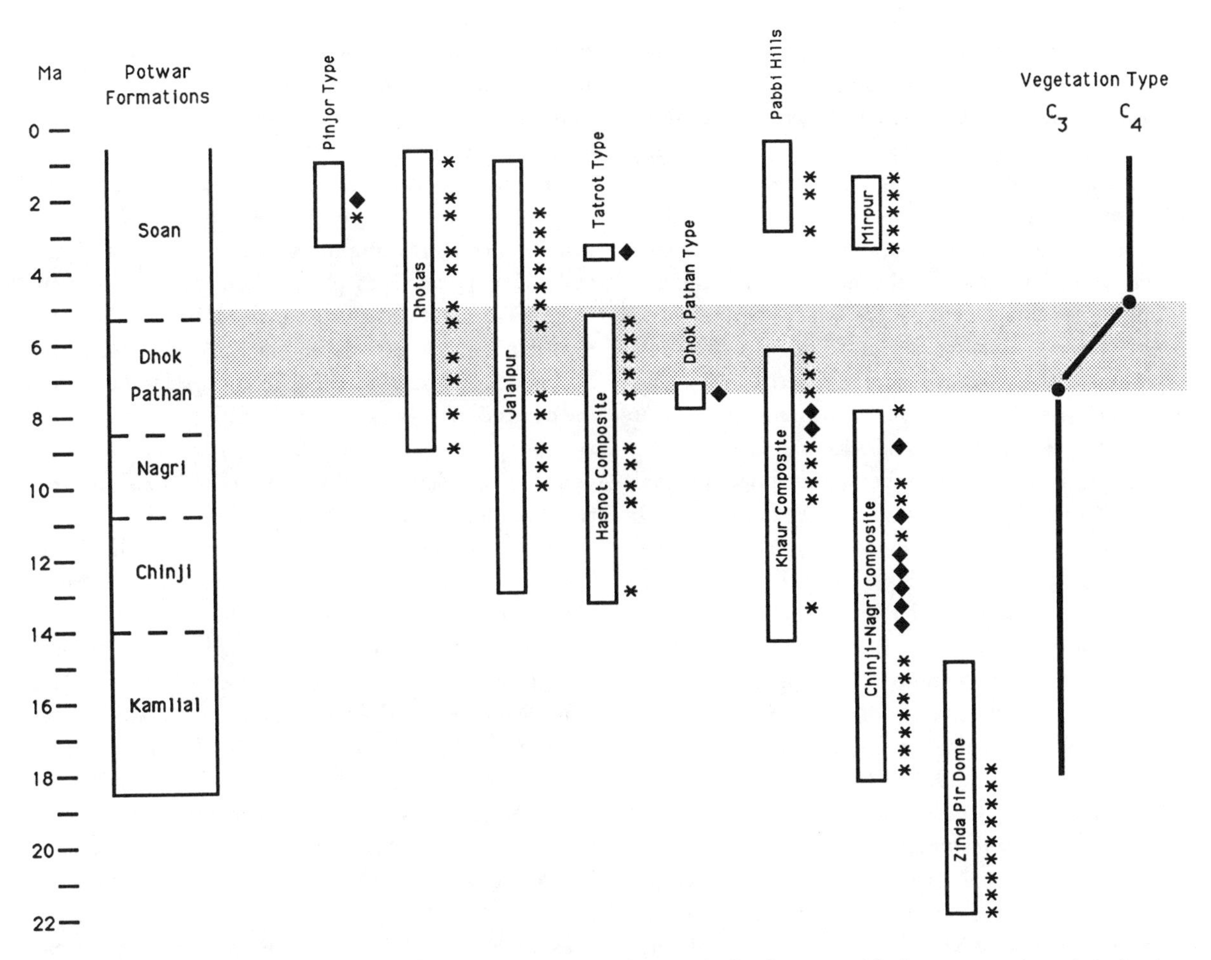

Fig. 9.1. Stratigraphic sections and occurrences of fossils in Siwalik sediments. Asterisks indicate levels with poor-to-good fossil representation; diamonds, levels with very good representation. The horizontal bar indicates transition period between vegetation types as inferred from carbon isotope ratios.

22 and ca. 15 myr and extending the mammal sequence of southern Asian another 4 million years.

Changes in the rates of sediment accumulation and an apparent widespread depositional hiatus have made the stratigraphic relations and ages of the Early Pliocene formations of the Potwar less certain. These include the Tatrot Formation and its equivalents, which contain an important fauna recording occurrences of *Elephas, Hippohyus, Sivachoerus,* and *Sus* (Barry et al., 1982; Hussain et al., 1992) and somewhat older sediments near Rhotas and Jalalpur. Previously the Tatrot was interpreted as being in the upper part of the Gauss Chron (Barry et al., 1982), while critical parts of the Rhotas and Jalalpur sequences have been interpreted as being in the Gilbert Chron (Opdyke et al., 1979; Johnson et al., 1982). The upper part of the Gauss is between 2.5 and 2.9 myr on the Berggren et al. (1985) time scale, whereas the Gilbert spans the considerably longer interval between 3.4 and 5.35 myr. Recent work by Hussain et al. (1992) suggests that the Tatrot Formation might be older than previously thought and could be in the lower part of the Gauss, between 3.2 and 3.4 myr. Support for this older age also comes from new, unpublished collections of the Harvard–Geological Survey of Pakistan group. In the accompanying tables I assign fossils from the Tatrot type area an age of 3.3 myr. Similarly, unpublished work by Kappelman and Stubblefield suggests problems in the dating of the Rothas section, and in the following I have used only some of the data available for it. This leaves only the Jalalpur section as spanning 5.4 to 3.4 myr; in this chapter I have used a new correlation to the geomagnetic time scale that makes the Pliocene fossiliferous horizons between 0.5 and 1.5 million years older than previously thought (Barry et al., 1982).

Also unresolved are the stratigraphic relationships between the younger Potwar formations (referred to under various formational names, including the Samwal, Kakra, and Mirpur Formations of Hussain et al., 1992) and the Pliocene and Pleistocene formations of northwestern India and Kasmir. The latter includes the important Pinjor and younger faunas, which are thought to span the interval between 2.5 and 0.9 myr (Azzaroli and Napoleone, 1982; Tandon et al., 1984; Barry, 1987). The well-dated fossiliferous Potwar formations exposed at Mirpur and in the Pabbi Hills, as well as elsewhere throughout the Potwar, span the slightly longer interval of 3.4–0.6 myr (Opdyke et al., 1979; West, 1981; G. D. Johnson et al., 1982; N. M. Johnson et al., 1982), but it is not yet clear where within this interval the Indian sediments and fossils lie. Because Pinjor and younger formations contain both cervids and *Equus,* which are known to appear in the Potwar sections in the upper Gauss and near the Gauss-Matuyama boundary, respectively (Opdyke et al., 1979; Hussain et al., 1992), a likely age for the oldest appearance of the Pinjor fauna is 2.4–2.5 myr.

Some progress has been made in developing a comprehensive Siwalik biostratigraphy that is separate from current lithostratigraphic usage. In 1982, Barry et al., proposed a series of biostratigraphic interval zones that begin with the first appearance of equids and end with the local extinction of three-toed hipparionine equids. These zones were defined in two designated reference sections on the Potwar Plateau, where they span approximately the interval between 10.0 and 1.5 myr. More recently, Hussain et al. (1992) have modified the 1982 scheme by dividing the youngest zone into two range zones and showing its lower boundary to be older than previously thought. Aside from unpublished work by Raza (1983), no attempt has been made to subdivide biostratigraphically the strata that are older than the first appearance of equids, probably because the widespread application of magnetostratigraphy has made it less imperative.

Quality of the Siwalik Fossil Record

Although the abundance and quality of fossils vary, throughout the Middle and Late Miocene and much of the Pliocene all stratigraphic levels have some fossils and the record of many subintervals is good to excellent (figs. 9.2 and 9.3). As a consequence, patterns of faunal turnover and changes in diversity can be documented and analyzed using relatively short chronological subdivisions (currently of 0.5 million years duration). Data quality, however, critically affects inferences that can be made about the patterns, especially with regard to measures of first and last occurrences and species richness, which depend on assumptions about the fraction of original living assemblages preserved in the fossil assemblages. Factors that affect the reliability of the fossil record include the original abundance of the species, the likelihood of preservation of different species' habitats in the depositional system, and the relative ease with which various species can be identified, which is influenced by both the completeness of the fossils and their intrinsic recognizability. Taphonomic factors influence how abundant and well preserved fossils are at localities and, in turn, which taxa are most likely to be fossilized, recognized, and collected. Localities or horizons with abundant and well-preserved fossils usually have more taxa than those with fewer or poorly preserved fossils and are more likely to record first or last occurrences (Badgley and Gingerich, 1988; Badgley, 1990). Taphonomic processes therefore determine how well fossil assemblages record original diversity and how accurately the magnitude and timing of changes in diversity can be estimated. It is important for that reason to assess the quality of preserved records. Furthermore, because there is no way to estimate original diversity from the number of species found in a fossil assemblage, we must remember that even in the best of circumstances we cannot confidently know how closely a fossil assemblage will record the original life assemblage. Nevertheless, modern faunas from geographic regions of comparable size have similar numbers of small- and large-mammal species as the Siwalik fossil assemblages.

In previous papers, two methods have been used to express how reliably the Siwalik fossil assemblages might represent the original living assemblages and to determine whether data from different intervals can be compared. One method involved the interval-quality scores used by Barry et al. (1990), whereby each interval (0.5 myr subdivision) was graded on criteria that included the number of specimens, number of localities, and preservation of the fossils occurring within it. The rationale and limitations of this method have been discussed by Barry et al. (1990). Because of differences in taphonomy and collecting techniques, small and large mammals have always been evaluated separately. We now recognize seven quality classes, ranging from Class 0 (no data) to

Siwalik Rodents

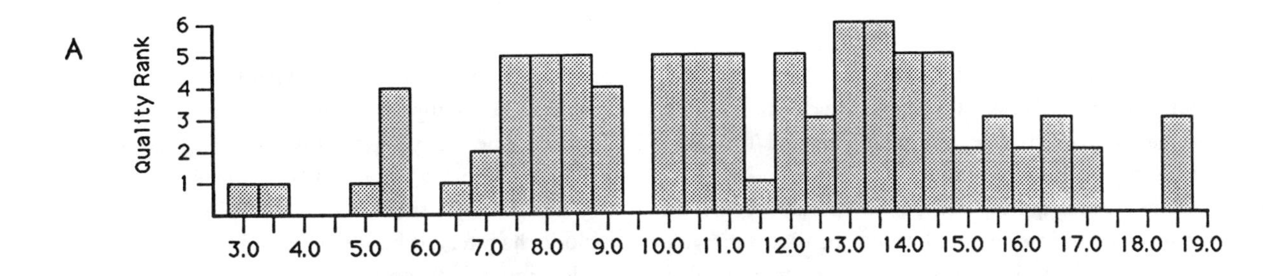

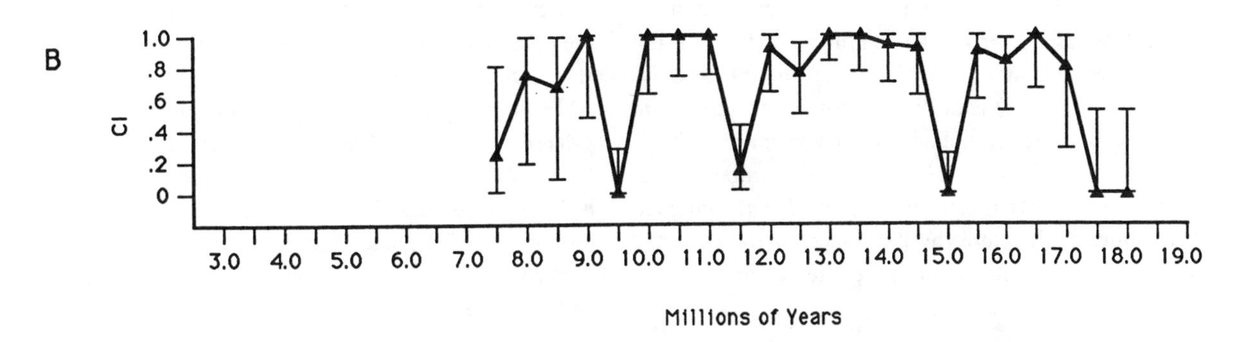

Fig 9.2. Record quality for rodents in the Potwar Siwalik sequence: A) Rankings of interval quality; B) Completeness Index with 95 percent confidence limits.

Siwalik Artiodactyls

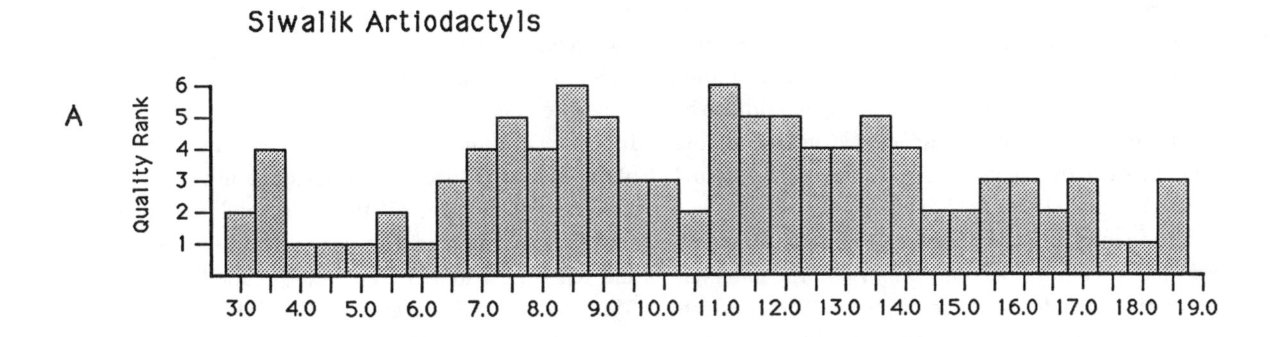

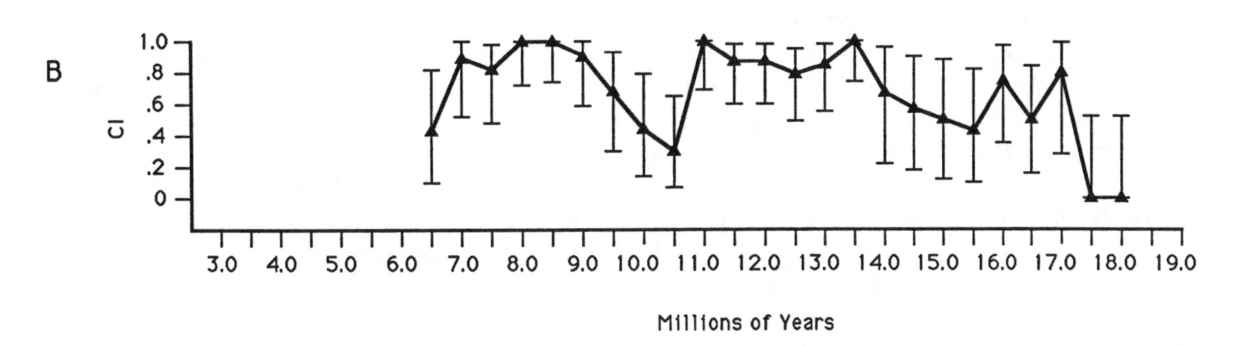

Fig. 9.3. Record quality for artiodactyls in the Potwar Siwalik sequence: A) Rankings of interval quality; B) Completeness Index with 95 percent confidence limits.

Class 6 (Barry et al., 1995). For both large and small mammals only Classes 4 through 6 are considered to be reliable and comparable, although Classes 1 through 3 have some useful information.

The interval-quality method incorporates several different kinds of information but has the disadvantage of being sequence-specific, because scores for intervals in one sequence cannot be compared to scores from other sequences (Barry et al., 1995). An index using a measure of expected diversity and a measure of observed or recorded diversity overcomes this limitation (figs. 9.2 and 9.3). One such index (CI) developed for the Siwaliks (Barry et al., 1995) uses the number of taxa known in intervals both before and after the target interval (N_{ba}), with observed diversity (N_{bda}) being the number of species that are known before, during, and after the interval. The index is therefore the percentage of the expected taxa that are actually found in the interval, or

$$CI = N_{bda}/N_{ba}.$$

Because the expected values are independent of the observed values, confidence limits can be placed on the estimates of completeness, allowing the effects of sample size to be taken into consideration. The confidence limits are calculated using the binomial distribution, assuming (1) that taxa in older and younger intervals should also be present in the interval of interest; and (2) that the discovery of a taxon in the interval is independent of the discovery of other taxa (Barry et al., 1995).

Data on interval quality between 18.5 and 3.0 myr are presented in figures 9.2 and 9.3. Both the interval-quality scores and completeness indexes demonstrate considerable variation in data quality for both large and small mammals. The small-mammal record is the most incomplete, having 6 intervals with no data and another 9 with poor Class 1 or 2 records. However, both the small- and large-mammal records have 13 Class 4 or better intervals that may be well enough sampled to give reliable estimates of the original diversity of the assemblages. CI values range from 0 to 1, and the 95 percent confidence limits on the CI values range from .16 to .90, indicating considerable variation in the quality of the estimates of completeness. That is, some of the CI values are good estimates of completeness for that interval, whereas others are very poor.

The indexes of figures 9.2 and 9.3 are based on a data set of occurrences that is older than data sets now available; these indexes were calculated only for the Miocene, because after 6.5 myr very few species are found in successive intervals. Updated and extended data are unlikely to alter radically any conclusions based on the older data, however. The indexes generally track the quality ranks, although there are important differences between the two measures for small mammals. Points of difference include the intervals between 7.5 and 9.5 myr, where the quality ranks indicate a better quality record, and between 15.5 and 17.0 myr, where the quality ranks indicate a more mediocre record. In both cases, the 95 percent confidence limits indicate that the values for CI are poor estimators. This is especially true for the intervals younger than 9.5 myr.

In summary, thirteen of the Siwalik intervals are considered to be of good quality for both small and large mammals. In both cases, all but one of the better intervals lie between 14.5 and 7.0 myr, which is therefore the best-known part of the sequence. It is noteworthy that after 7 myr the intervals are particularly poor for both large and small mammals.

Review of Siwalik Faunal Change

There is a long history of study of fossil vertebrates from the Neogene of southern Asia, beginning with discoveries in the 1830s by British colonial officers and continuing into this century with the work of Pakistani and Indian institutions and foreign scientific missions. Most of the early work focused on taxonomic and phylogenetic aspects, but Pilgrim (1910, 1926) and later Colbert (1935) also commented on some of the biogeographic and temporal patterns displayed by Siwalik mammals. More recently, (Barry et al., 1990, 1995; Jacobs et al., 1990; Flynn et al., 1995; Morgan et al., 1995), species diversity and the dynamics of faunal turnover have become central issues, as more refined data on stratigraphy, relative abundance, and body size have become available.

Faunal turnovers and changes in diversity are consequences of local extinction, immigration, anagenetic evolution, and cladogenesis. Immigrant species and lineages are those that have dispersed from outlying geographic areas, but although they have a prior history elsewhere, their appearance in a stratigraphic sequence cannot always be distinguished from appearances owing to in situ originations. Our studies of Siwalik faunal change, therefore, have treated appearances resúlting from immigration, anagenetic evolution, and cladogenesis as "first occurrences." Preliminary analyses indicate, however, that as many as half of all Siwalik first appearances could be immigration events, and thus immigration and extinction appear to be the dominant processes of faunal change in southern Asia.

During the Neogene, the Siwaliks belonged to an isolated zoogeographic province that may have been a precursor of the modern Oriental Province (Bernor, 1983, 1984; Thomas, 1985; Barry et al., 1985). Throughout its history the degree of isolation of this province varied, with exchanges with African and other Eurasian provinces occurring occasionally. Isolation was most complete in the earliest Miocene, whereas appearances of cricetid rodents and ruminant artiodactyls accompanied substantial faunal exchange with Eurasia and Africa at ca. 20 myr. Episodic faunal exchange continued between 18 and 7.5 myr, ending with a long-lasting faunal turnover that radically altered the taxonomic composition of the Siwalik mammal fauna. This last event is possibly part of a restriction of the Oriental biogeographic province to more southeastern regions and may have been linked to the beginning or intensification of the Indian monsoon (Quade et al., 1989).

Thirteen orders of mammals have been identified in the Siwalik fossil assemblages, although most species are rodents, artiodactyls, or perissodactyls. Among large mammals, the ruminant artiodactyls and equids are the most common and speciose, while murid and cricetid rodents dominate the small-mammal assemblages. Complete lists of the mammals are not yet available for all intervals, but well-sampled intervals ordinarily have 50 or more species. Table 9.1 presents a comprehensive list for three of the best-known intervals, at 13.5, 11.0, and 8.5 myr, with 72, 64, and 61 species, respectively. In the following discussion I refer primarily to data on the occurrences of rodents and artiodactyls (summarized in table 9.2). The data are derived from fossil collections made since 1973 as part of research in Pakistan and have been slightly updated from previous studies (Barry et al., 1990, 1991, 1995). The data include only rodents and artiodactyls because at present only they provide a reliable species-level data set. These two orders, however, typically comprise about 60 percent of the species in any stratigraphic interval, and I assume they reflect patterns prevalent within the complete mammal fauna. Although perissodactyls, and especially equids, also have many species, they are not included because of the difficulties of identifying their species from fragmentary material.

The stratigraphic-range data for artiodactyls (given in table 9.2) are displayed in figure 9.4; Flynn et al. (1995) have prepared a similar figure for rodents. These are the ranges as we currently understand them, and they

Table 9.1. Siwalik Mammal Species Present in the 13.5, 11.0, and 8.5 myr Intervals

	Stratigraphic Interval		
	13.5 myr	11 myr	8.5 myr
Chiroptera			
Chiroptera, indet.	x	x	x
Pipistrellus sp.	x		
Insectivora			
Erinaceidae, indet.			x
Galerix rutlandae	x	x	
Galerix sp. B		x	
Amphechinus sp.	x		
Soricidae, indet.	x		
Crocidura sp. A		x	x
Crocidura sp. B		x	x
Scandentia			
Tupaiidae, indet.	x		x
Dendrogale sp.		x	
Tupaia sp.		x	
Primates			
Sivaladapis sp.	x	x	
Lorisidae, indet. A			x
Lorisidae, indet. B	x		
Sivapithecus indicus		x	
Sivapithecus sivalensis			x
Creodonta			
Dissopsalis carnifex	x	x	
Hyainailourous sulzeri	x	x	
Metapterodon sp.	x	x	
Carnivora			
Amphicyonidae, indet.	x	x	x
Agnotherium antiguum		x	
cf. *Ischyrictis* sp.			x
Plesiogulo sp.			x
Martes lydekkeri	x		
Vishnuonyx chinjiensis	x	x	
Sivaonyx bathygnathus			x
Nimravidae, indet.		x	
Sansanosmilus sp.	x		
Barbourofelis sp.			x
Hyaenidae, sm. sp. A		x	
Hyaenidae, sm. sp. B			x
Percrocuta carnifex		x	
Percrocuta grandis			x
Felidae, indet.		x	x
Viverra, sm. sp.			x
Viverra lge. sp.			x
Herpestes med. sp.			x
Tubulidentata			
Orycteropus small sp.	x	x	
Orycteropus sp.			x
Proboscidea			
Deinotherium sp.	x	x	x

(*continued*)

Table 9.1. (*Continued*)

	Stratigraphic Interval		
	13.5 myr	11 myr	8.5 myr
Zygolophodon metachinjiensis	x		
Protanancus chinjiensis	x		
"Gomphotherium" browni	x	x	
"Gomphotherium" sp.			x
cf. *Choerolophodon* sp.	x	x	
Choerolophodon corrugatus			x
cf. *Stegolophodon* sp.		x	x
Perissodactyla			
Aprotodon fatehjangense	x		
Brachypotherium perimense	x	x	x
Chilotherium intermedium	x	x	x
Gaindatherium browni	x	x	x
Caementodon oettingenae	x	x	
Chalicotherium salinum	x	x	x
Sivalhippus sp.			x
Sivalhippus perimense			x
?Hipparion antelopinum			x
Artiodactyla			
"Anthracotherium" punjabiense	x	x	x
Hemimeryx spp.	x	x	x
Schizochoerus gandakasensis			x
Listriodon pentapotamiae	x	x	
Conohyus sindiensis	x	x	
Tetraconodon magnus			x
Hyotherium pilgrimi	x	x	
Hippopotamodon sivalense		x	x
Propotamochoerus hysudricus			x
Dorcabune small sp.	x		
Dorcabune anthracotherioides	x	x	
Dorcabune nagrii			x
Dorcatherium minimus	x		
Dorcatherium very small sp. B		x	
Dorcatherium nagrii			x
Dorcatherium small sp. complex	x	x	
Dorcatherium sp. D			x
Dorcatherium minus	x		
Dorcatherium cf. *minus*		x	
Dorcatherium majus			x
Giraffokeryx punjabiensis	x	x	
Bramatherium megacephalum			x
cf. *Oioceros* sp.	x		
Sivoreas eremita	x	x	
?Gazella spp.	x	x	
Gazella lydekkeri			x
cf. *Elachistoceras khauristanensis*	x	x	x
Miotragocerus unnamed sp.		x	
Sivaceros gradiens		x	
Protragocerus gluten	x	x	
Tragoceridus pilgrimi			x
Helicoportax praecox		x	
Selenoportax vexillarius			x
Tragoportax salmontanus			x
Rodentia			
Sciurid, sp. B	x		
Petauristine, sp. B	x		
cf. *Ratufa* sp.	x		x
Ratufa sp.			x
cf. *Hylopetes* sp.		x	x
Eutamias urialis	x	x	
cf. *Eutamias* sp.	x		
Eutamias sp.			x
Gliridae, indet.	x	x	x
Sayimys spp.	x	x	
Kochalia geespei	x		
Prokanisamys "benjavuni"	x		
Kanisamys indicus	x	x	
Kanisamys sivalensis			x
Kanisamys nagrii		x	
Kanisamys potwarensis	x		
"Brachyrhizomys" nagrii			x
"Brachyrhizomys" micrus			x
"Brachyrhizomys" blacki			x
"Brachyrhizomys" tetracharax			x
"Brachyrhizomys", cf. *"B." pilgrimi*			x
Democricetodon kohatensis	x	x	
Democricetodon sp. A	x		
Democricetodon sp. B+C	x	x	
Democricetodon sp. E	x	x	
Democricetodon sp. F	x	x	
Democricetodon sp. G		x	
Democricetodon sp. H	x	x	
Punjabemys mikros	x		
Punjabemys downsi	x		
Megacricetodon aguilari	x		
Megacricetodon mythikos	x		
Megacricetodon sivalensis	x		
Megacricetodon daamsi	x		
Myocricetodon sivalensis	x		
Myocricetodon sp.	x		
Dakkamyoides lavocati	x		
Dakkamyoides perplexus	x	x	
Dakkamys barryi	x		
Dakkamys asiaticus		x	
Paradakkamys chinjiensis		x	
Muridae, indet.		x	
Antemus chinjiensis	x		
Progonomys debruijni			x
Progonomys sp.		x	
Karnimata darwini			x
Karnimata large sp.			x
Parapodemus sp.			x

Table 9.2. Ages of First and Last Occurrences of Artiodactyls and Rodents in the Siwalik Formations of Pakistan

Artiodactyls	First Occurrences (myr)	Last Occurrences (myr)
Sanitherium schlagintweiti	>18.3	15.3
"Anthracotherium" punjabiense	>18.3	8.3
Hemimeryx spp.	>18.3	5.7
Merycopotamus dissimilis	5.2	2.5
Hexaprotodon sivalensis	5.7	2.1
Schizochoerus gandakasensis	9.4	7.9
cf. *Listriodon* sp.	17.3	17.3
Listriodon pentapotamiae	16.9	9.6
Conohyus sindiensis	16.3	9.6
Tetraconodon magnus	9.2	8.5
Sivachoerus prior	3.3	2.5
Sivahyus punjabiensis	6.6	3.3
Hyotherium pilgrimi	13.7	10.5
Hippopotamodon sivalense	10.8	6.6
Propotamochoerus hysudricus	9.1	5.7
Potamochoerus sp.	5.2	3.3
cf. *Kolpochoerus* sp.	3.3	3.3
?Sus sp.	5.7	3.3
Hippohyus spp.	>4.0	2.7
Dorcabune small sp.	17.0	13.5
Dorcabune anthracotherioides	13.9	10.0
Dorcabune nagrii	9.8	7.6
Dorcatherium very sm. sp. A	16.3	13.8
Dorcatherium minimus	13.7	12.9
Dorcatherium very sm. sp. B	12.6	11.1
Dorcatherium nagrii	9.2	7.1
Dorcatherium sm. sp. complex	18.3	10.8
Dorcatherium sp. D	9.8	6.1
Dorcatherium minus	13.9	11.5
Dorcatherium cf. *minus*	11.2	10.8
Dorcatherium majus	10.6	6.6
"Dorcatherium" sp. E	5.4	3.3
Cervid, indet.	2.8	—
Giraffoid, indet.	>18.3	14.6
Giraffokeryx punjabiensis	14.1	10.0
Bramatherium megacephalum	9.6	6.9
Giraffa punjabiensis	7.1	5.8
Kubanotragus sokolovi	13.8	13.8
cf. *Oioceros* sp.	13.5	13.5
Pachytragus sp. A	12.8	12.0
Pachytragus sp. C	12.5	12.5
Sivoreas eremita	13.9	10.0
Caprotragoides potwaricus	9.2	9.2
Prostrepsiceros vinayaki	7.5	7.1
?Gazella spp.	15.2	6.0
Antilope subtorta	5.2	3.3
Eotragus unnamed sp.	18.3	18.3
cf. *Boselaphus* unnamed sp.	7.9	7.9
cf. *Elachistoceras khauristanensis*	13.8	6.7
Miotragocerus unnamed sp.	10.8	10.8
Sivaceros gradiens	12.0	10.8
Protragocerus gluten	13.9	10.8
Tragoceridus spp.	10.7	5.7
Helicoportax praecox	13.1	10.8
Selenoportax vexillarius	9.5	ca. 8.5
Selenoportax lydekkeri	7.8	6.4
Tragoportax salmontanus	8.5	8.1
cf. *Leptobos* spp.	3.3	2.9?
Hemibos spp.	2.8	—
cf. *Dorcadoxa porrecticornis*	7.4	3.3
Reduncini, indet.	ca. 4	—
Vishnucobus patulicornis	3.3	3.3
Hippotragini, indet.	5.7	ca. 4
cf. *Oryx sivalensis*	3.3	3.3
Hippotragus brevicornis	3.3	3.3

Rodents	First Occurrences (myr)	Last Occurrences (myr)
Rodentia, unnamed sp.	17.0	17.0
Sciuridae, large sp.	14.3	14.0
Sciuridae, indet. A	18.3	18.3
Sciuridae, sp. A	17.0	15.3
Sciuridae, indet. B	7.2	7.2
Sciuridae, sp. B	13.8	12.5
Petauristine, sp. A	15.3	15.3
Petauristine, sp. B	13.8	11.9
Petauristine, tiny sp.	14.3	14.3
Hylopetes sp.	7.9	7.3
cf. *Hylopetes* sp.	11.1	7.9
Ratufa sp.	9.8	8.5
cf. *Ratufa* sp.	13.5	13.5
Eutamias urialis	13.6	10.6
Eutamias sp.	9.8	7.8
cf. *Eutamias* sp.	14.3	13.8
Heteroxerus sp.	13.0	13.0
Gliridae, indet.	13.8	7.2
Sayimys spp.	18.3	9.2
cf. *Diatomys* sp.	10.6	10.6
Hystrix sp.	7.2	7.0
Kochalia geespei	16.3	12.1
Paraulacodus indicus	12.9	12.5
Prokanisamys sp.	>18.3	18.3
Prokanisamys "benjavuni"	17.0	11.9
Kanisamys indicus	17.0	10.8
Kanisamys potwarensis	14.3	13.2
Kanisamys sivalensis	8.5	7.2
Kanisamys nagrii	11.1	8.8
Eicooryctes kaulialensis	7.0	6.5
Protachyoryctes tatroti	7.0	6.7
Rhizomyides punjabiensis	9.8	9.5
Rhizomyides, unnamed sp.	7.4	7.4
Rhizomyides sivalensis	6.5	5.7
"Brachyrhizomys" nagrii	8.7	8.3

(*continued*)

Table 9.2. (*Continued*)

Rodents	First Occurrences (myr)	Last Occurrences (myr)
"Brachyrhizomys" micrus	8.5	8.5
"Brachyrhizomys" blacki	8.4	8.4
"Brachyrhizomys" choristos	7.4	7.4
"Brachyrhizomys" tetracharax	8.4	7.5
"Brachyrhizomys," cf. "B." pilgrimi	8.5	7.2
Democricetodon kohatensis	16.3	9.5
Democricetodon A	18.3	13.6
Democricetodon B + C	17.0	9.1
Democricetodon E	14.1	9.8
Democricetodon F	13.6 (7.9)	9.1 (7.8)
Democricetodon G	13.8 (7.9)	9.1 (7.8)
Democricetodon H	13.8	9.8
Punjabemys mikros	16.3	13.0
Punjabemys downsi	14.3	12.6
Megacricetodon aguilari	18.3	13.5
Megacricetodon mythikos	15.3	12.5
Megacricetodon sivalensis	13.8	12.5
Megacricetodon daamsi	13.8	12.5
cf. *Protatera* sp.	7.8	7.8
Myocricetodon sivalensis	18.3	12.9
Myocricetodon sp.	14.3	13.8
Dakkamyoides lavocati	15.3	12.5
Dakkamyoides perplexus	13.6	10.6
Dakkamys barryi	13.6	12.1
Dakkamys asiaticus	13.0	9.8
Paradakkamys chinjiensis	12.6	9.8
Potwarmus primitivus	18.3	14.3
New Muroid	14.1	14.0
Muridae, indet.	10.8	10.8
Muridae, unnamed genus	9.5	9.1
Antemus chinjiensis	13.8	12.5
cf. *Progonomys* sp.	12.1	10.6
Progonomys sp. A	9.8	9.1
Progonomys debruijni	8.5	7.8
Progonomys sp. B	7.4	7.1
Mus auctor	5.7	5.7
Karnimata sp. A	10.6	9.1
Karnimata darwini	8.5	7.8
Karnimata sp. B	7.4	7.1
Karnimata huxleyi	5.7	5.7
Karnimata large sp.	8.4	7.8
cf. *Parapelomys* sp.	7.4	7.1
Parapelomys robertsi	5.7	5.7
Parapodemus sp.	8.4	8.4

are liable to change as we collect more material, especially in the less fossiliferous parts of the section. The ranges may also change as the ages of localities are made more precise and as taxa are revised and delimited differently. Comments on the artiodactyls relevant to the latter two issues are contained in appendix 9.1.

Diversity of the Fossil Faunas. Diversity, or species richness, can be expressed several ways. One commonly used measure is the maximum number of species that have been found in the interval, plus the number of species known from both older and younger horizons, but not within the interval. (These are the so-called "range-through" taxa. They are assumed to have been present but not found owing to sampling or taphonomic reasons.) Alternatively, diversity can be calculated as an estimate of the number of taxa that might have coexisted in the middle of an interval, which will be referred to as "standing diversity." This estimate is found by counting the number of taxa known before and after the interval and adding it to the sum of first, last, and only occurrences divided by two. (Note that in this case the taxa known before and after the interval may also include taxa actually found within the interval.) This method assumes that first and last occurrences are symmetrically distributed about the midpoint of the interval, which may not be true, but it seems less problematic than assuming that all the first occurrences are older than all the last occurrences (an assumption of the first method).

I have used the data on the occurrences of rodents and artiodactyls to derive maximum and standing-diversity curves for the best-known intervals between 14.5 and 7 myr (fig. 9.5). (Note that the data used to derive these curves and those of fig. 9.6 differ slightly from the more recently updated data in table 9.2.) The two curves are very similar, although the standing-diversity curve is smoother and peaks an interval later. When compared to figures 9.2 and 9.3, both diversity curves clearly show a record-quality effect in that poor intervals tend to have fewer taxa. Nevertheless, between 14.5 and 13.0 myr there is an increase in the number of species that is at least partly independent of record quality; subsequently there is a steep decline that continues to 9 myr, followed by a shallower decline lasting until at least 7 myr. These changes in species numbers are due primarily to changes in the number of rodents, because the numbers of artiodactyls and other mammals remain approximately constant despite changes in their species composition (Barry et al., 1995). The trends in artiodactyl and rodent diversity are concordant with the changes in the total number of recorded species in the three best-known intervals (table 9.1).

Patterns of Faunal Turnover. The stratigraphic range data for artiodactyls and rodents (table 9.2) also docu-

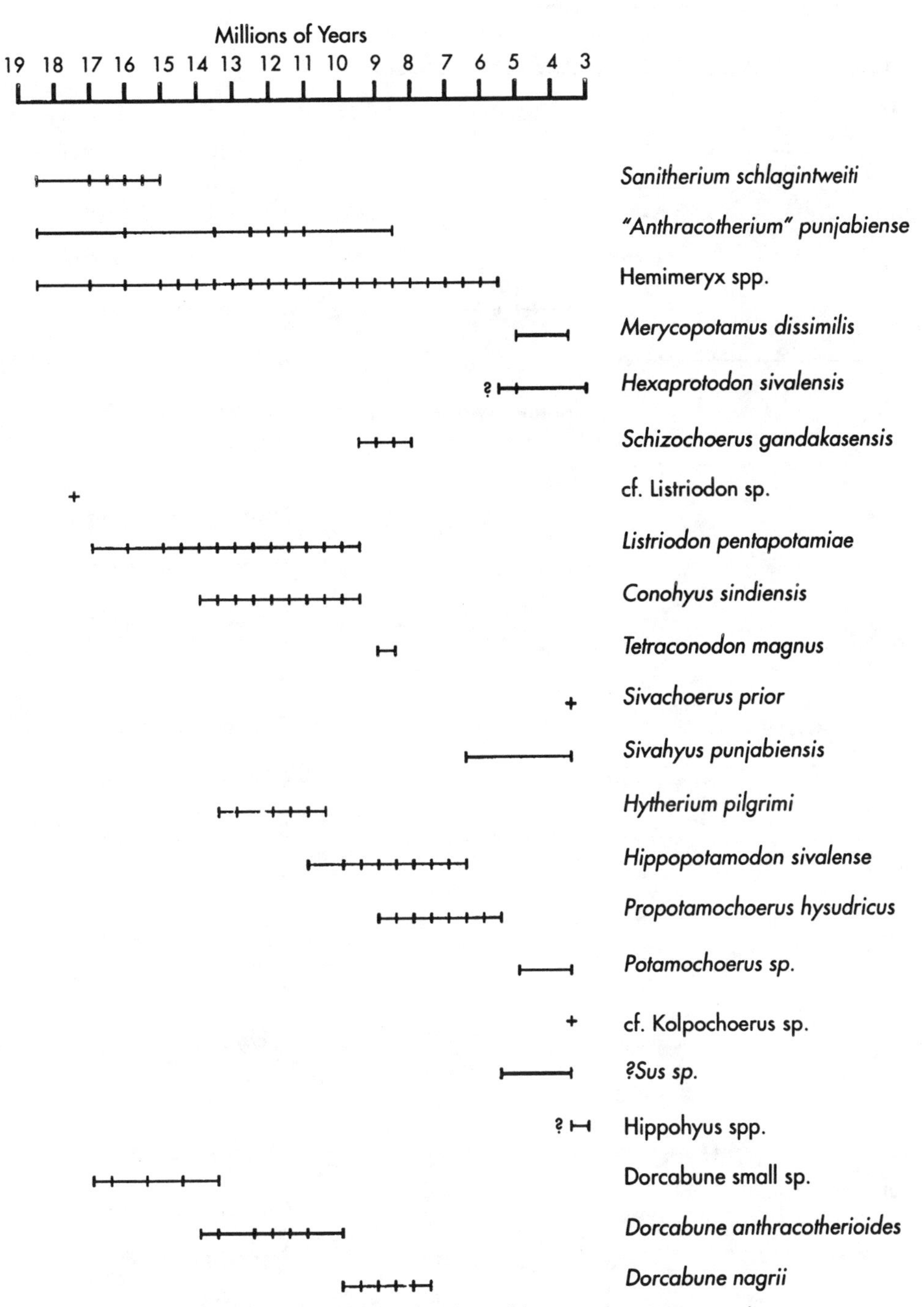

Fig. 9.4. Stratigraphic ranges of artiodactyl species. Cross ticks indicate the presence of the taxon within a 0.5 myr interval.

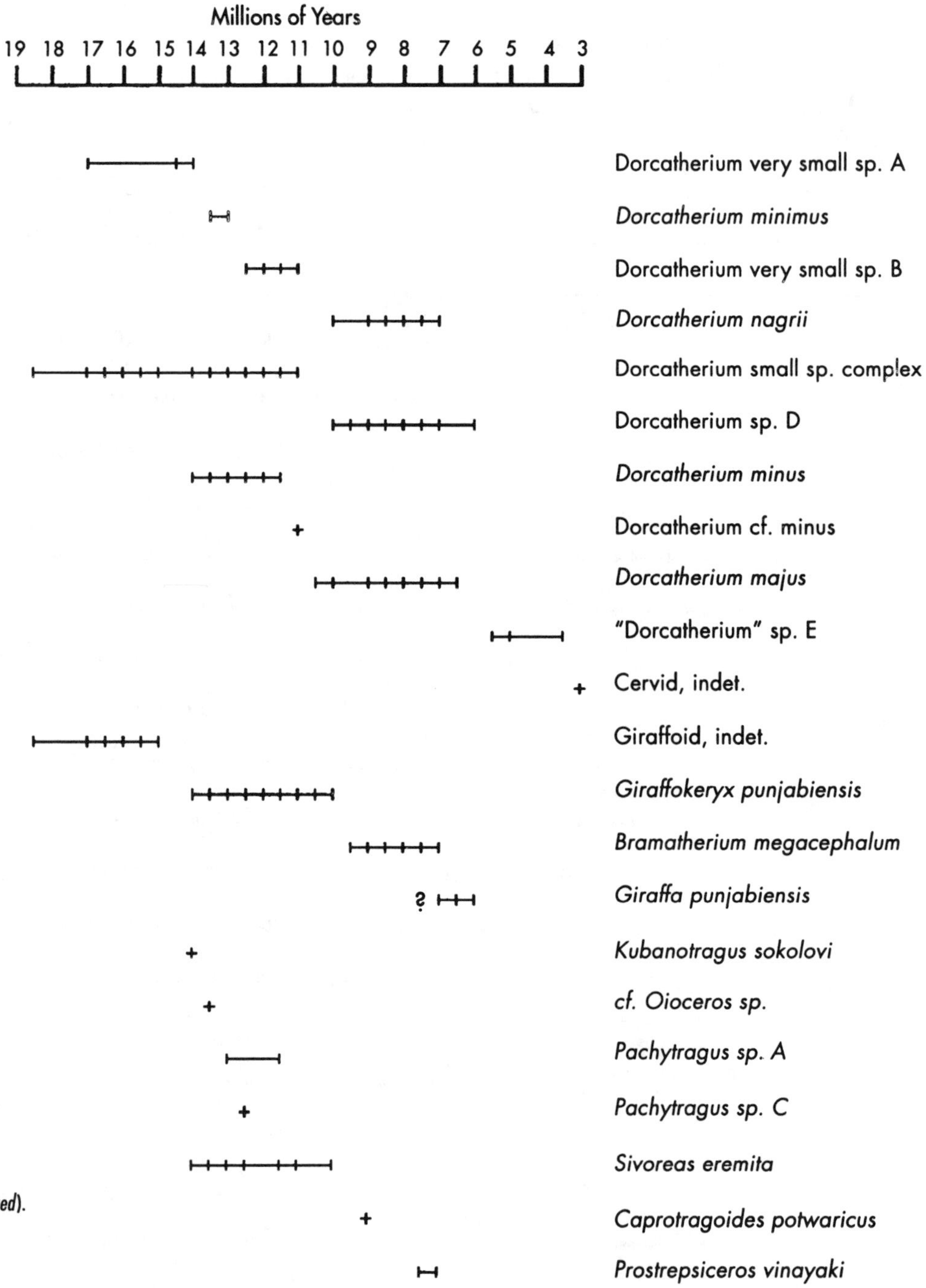

Millions of Years

19 18 17 16 15 14 13 12 11 10 9 8 7 6 5 4 3

Dorcatherium very small sp. A

Dorcatherium minimus

Dorcatherium very small sp. B

Dorcatherium nagrii

Dorcatherium small sp. complex

Dorcatherium sp. D

Dorcatherium minus

Dorcatherium cf. minus

Dorcatherium majus

"Dorcatherium" sp. E

Cervid, indet.

Giraffoid, indet.

Giraffokeryx punjabiensis

Bramatherium megacephalum

? *Giraffa punjabiensis*

Kubanotragus sokolovi

cf. Oioceros sp.

Pachytragus sp. A

Pachytragus sp. C

Sivoreas eremita

Caprotragoides potwaricus

Prostrepsiceros vinayaki

Fig. 9.4. (*Continued*).

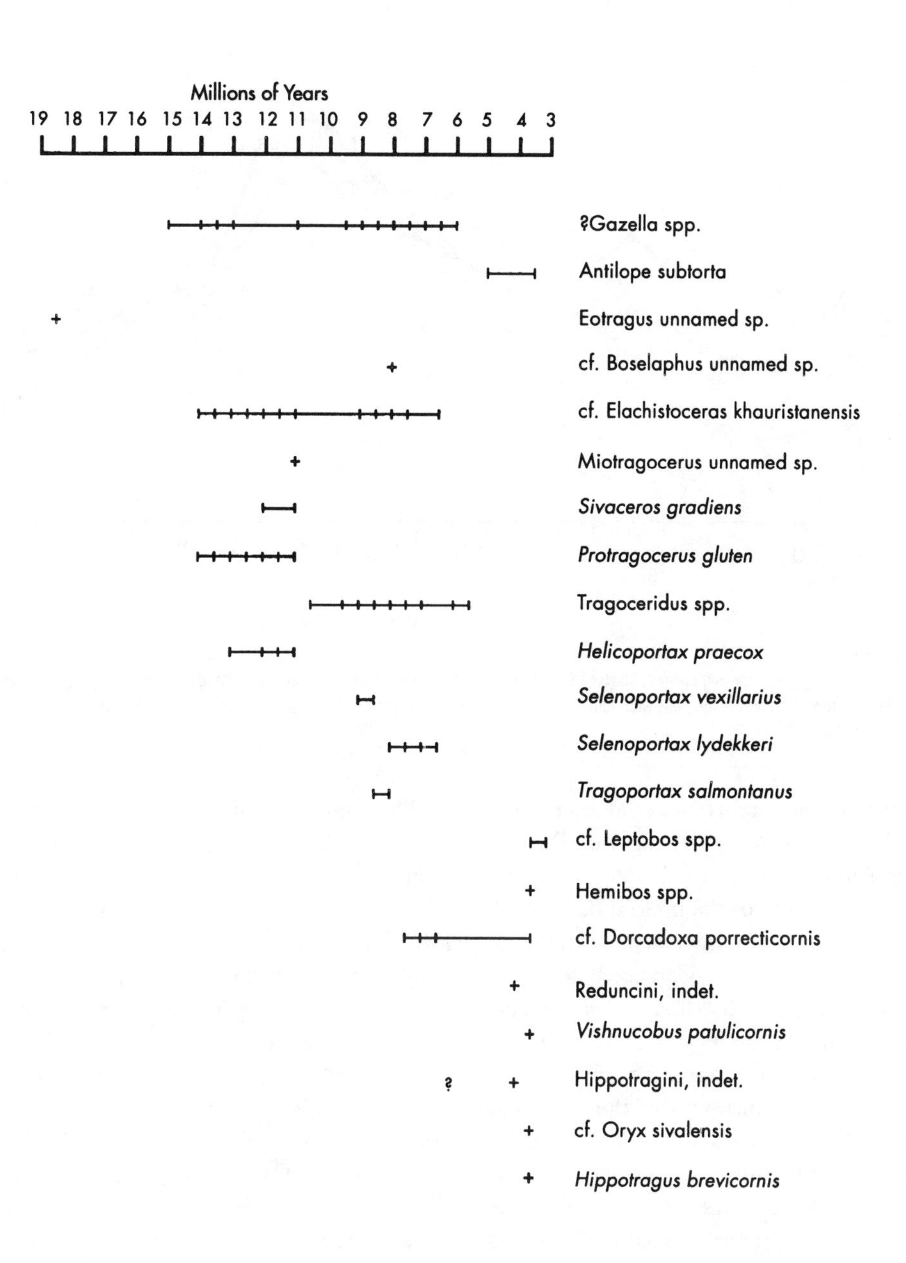

Fig. 9.4. (*Continued*).

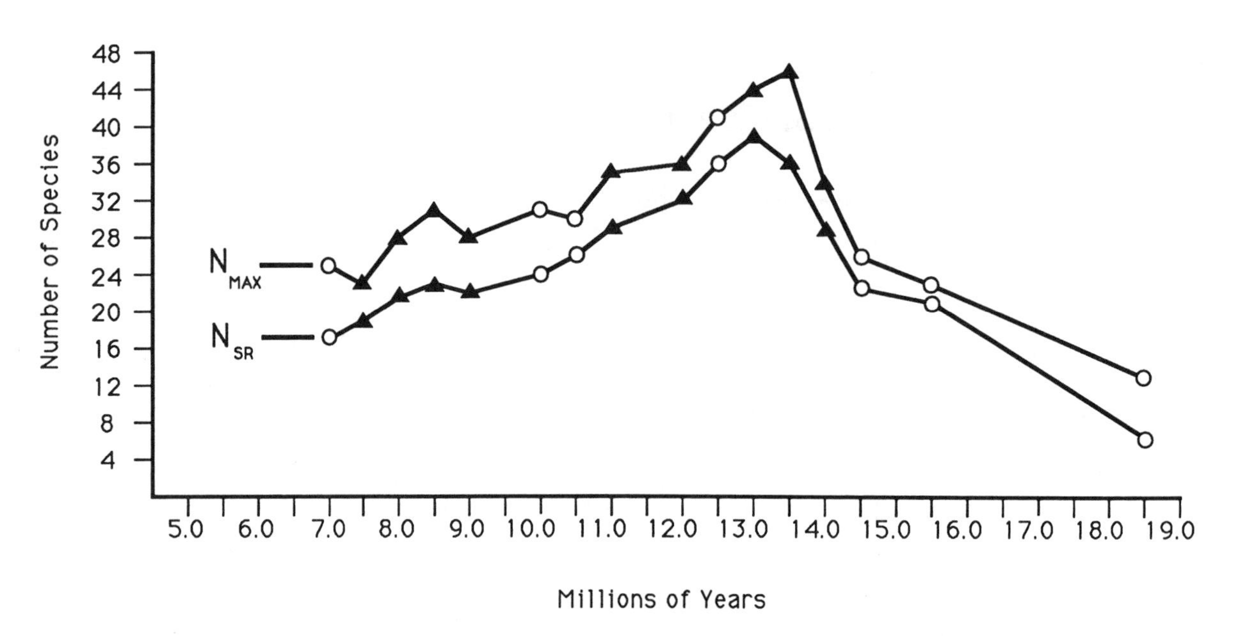

Fig. 9.5. Maximum number (N_{max}) and standing species richness (N_{sr}) of rodent and artiodactyl species found or inferred to be present using the range-through method for 0.5 myr intervals. Triangles indicate intervals with both rodent and artiodactyl records of Class 4 or better; circles indicate intervals with one record of Class 3 and the other of Class 4 or better.

ment apparent first and last occurrences and can be used to examine patterns of faunal turnover (fig. 9.6). Because of the severe limitations of the record after 5 myr, this analysis has been done only for the interval between 19 and 5 myr, and even within this interval my analyses do not reflect very recent changes documented for some taxa. Such analyses assumed that first and last occurrences fall close to times of species origins/arrivals and extinctions. Nevertheless, the apparent patterns are strongly biased by the completeness of the particular fossil record (Badgley and Gingerich, 1988; Badgley, 1990), with poorly fossiliferous intervals making some last occurrences seem too old and some first occurrences seem too young. As a result, the actual record of observed first and last occurrences will in some degree be misleading. In an effort to correct for the bias, I have extended the observed stratigraphic range of each taxon by using the estimates of interval quality to include some period of variable length during which the taxon may have been present in the region. Details are discussed in Barry et al. (1990), but the approach depends on two assumptions; first, that the potential range of a taxon can be determined from the estimates of data quality; and second, that the probability of finding the true first and last occurrence is equal for each additional interval within the potential range. These assumptions are conservative in respect to my conclusions.

There was substantial variation in the number of first and last occurrences between intervals (fig. 9.6), with some having very high turnover and others essentially none. When expressed as rate quotients (Gingerich, 1987; Barry et al., 1995), the values range from one-half to three and one-half times what would be expected if the turnover events were independent and equally likely in all intervals. They also produce a highly significant χ^2 value in a goodness-of-fit test (Barry et al., 1995). Thus, both first and last occurrences appear to be clustered in brief pulses separated by longer periods of low turnover, as predicted by Vrba's (1985) turnover pulse hypothesis. The statistically significant maxima of first appearances, however, occur at 13.5 and 8.5 myr, whereas the maxima of last appearances come at 12.5 and 8.0 myr. It is therefore apparent that in this record increased extinction did not accompany or closely follow maxima of first appearances. That is, pulses of increased extinction and first appearances were not closely synchronous. There also appears to be a subtle alternation of the pattern after 11 myr, as distinctly pulsed turnover gives way to more constant turnover.

Other Aspects of Faunal Change. Significant changes in relative abundance among families of artiodactyls and genera of muroid rodents have also been documented in the Potwar sequence (Barry et al., 1991). The patterns

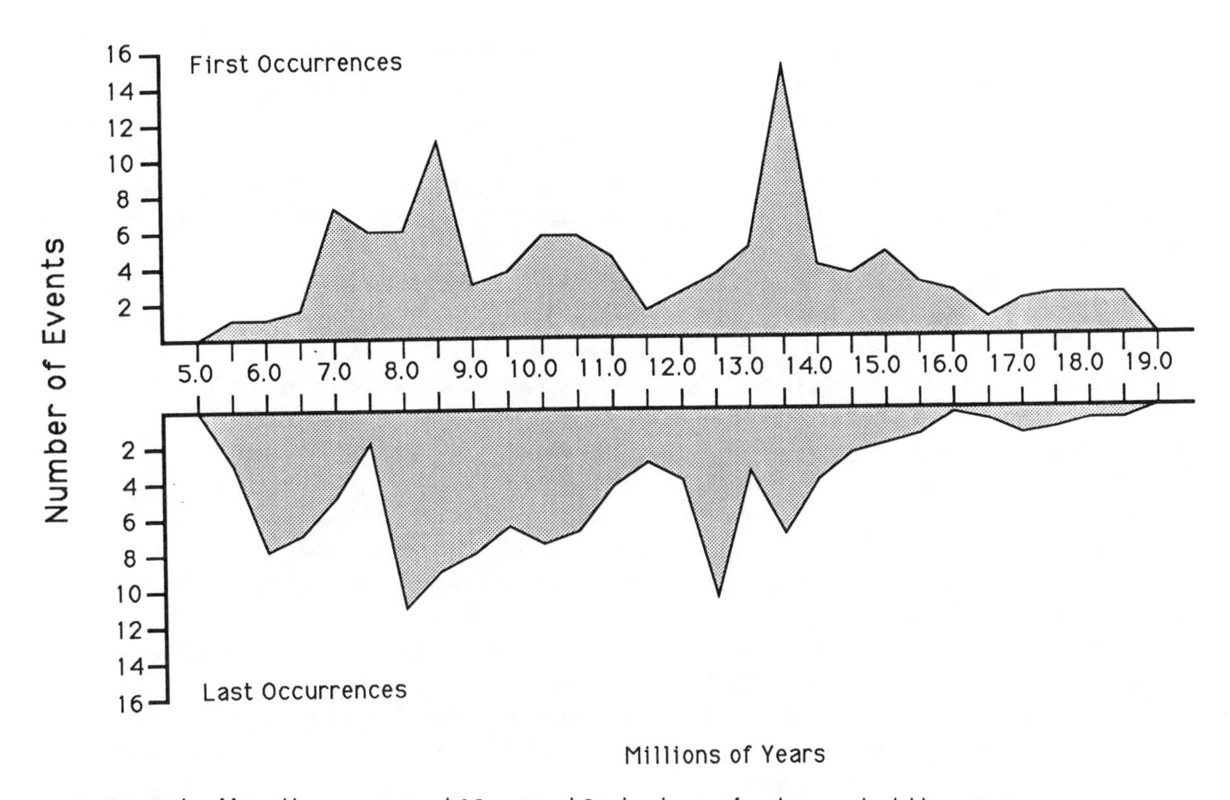

Fig. 9.6. Number of first and last occurrences in each 0.5 myr interval. Data have been transformed using equal-probability partitioning.

are complex and need further analysis, which could now be done at the species level. Noteworthy features include rapid increases in the abundance of bovids as compared to tragulids between 16.5 and 15 myr and again between 9.5 and 7.5 myr, as well as a very abrupt increase in the murids at 12 myr (at which time they are represented by at most 2 species), followed by a slower increase ending at ca. 7.5 myr with the local extinction of all cricetids. Barry et al. (1991) also documented trends in body-size increase among ruminant artiodactyls, and Morgan (1994) has now expanded their data base and extended it to all Siwalik artiodactyls. Again, the patterns are complex and need further analysis. Both studies show a pattern of progressive increase in size, as ever-larger species appear over time. The trend is present in each of the families included and, with the persistence of small species, sometimes also involves an increase in the range of body sizes. Significantly, there are indications that size increase in some taxa may occur in rapid steps that accompany the maxima of first appearances at 13.5 and 8.5 myr.

Attempts at reconstructing aspects of community structure and documenting community change over time have also been made on the same half-million-year-interval time scale (Badgley, 1986a, 1986b; Morgan et al., 1995). Although these studies are preliminary, they show that some periods, such as those after 8 myr, seem to have been times of substantial community change, whereas others, such as between 13 and 11 myr, seem to have been periods of stability. Similarly, Flynn et al. (1995) have shown that patterns of species longevity change after 10 myr, with the older period characterized by long-lived species and the younger by species with much shorter durations.

The Late Miocene–Early Pliocene Faunal Transition

Correlations of the faunal events discussed above to global climatic records are generally weak owing to problems in dating (Barry et al., 1990, 1995), but it is clear that the Late Miocene events documented in the Potwar Siwaliks accompanied a global episode of cooling and increasing aridity. This period was also a time of considerable environmental change in the Siwaliks. Quade and others (Quade et al., 1989; Quade and Cerling, 1995), for instance, have documented a trend in carbon isotope ratios that shows significant changes beginning between

7.4 and 7.0 myr and ending at 5 myr (fig. 9.1). They argue that this trend resulted from a shift from vegetation dominated by C_3 plants to vegetation dominated by C_4 plants, which they interpret as a shift from closed or mixed forest habitats before 7.4 myr to more open grasslands with minor bush or trees after 5.0 myr. Vegetation in the interval between was probably a mixture of both C_3- and C_4-dominated communities, although after 6.4 myr consistently positive isotopic values suggest that pure C_3 communities no longer existed in the area (Quade et al., 1989; Quade and Cerling, 1995). Other aspects of the sediments also change during the same stratigraphic interval, such as the depth of soil leaching, soil colors, and the size and frequencies of stream channels (Quade et al., 1989; Quade and Cerling, 1995; Behrensmeyer, 1987).

On the Potwar there are several sections that span all or part of the interval between 7.4 and 5 myr, including, most critically, sections near Khaur, Hasnot, Jalalpur, and Rhotas. (fig. 9.1). Unfortunately, except for the first half million years, the vertebrate faunas between 7.4 and 5.0 myr are poorly known and continue to receive little attention. This time interval has been neglected partly because the sediments younger than 7 myr are poorly fossiliferous (figs. 9.2, 9.3), owing to the changes in depositional environments noted by Behrensmeyer (1987), and partly because this is a difficult part of the magnetic time scale, without distinctive magnetic landmarks such as the Gauss or C5N Chron. There also appears to have been considerable local variation in rates of sediment accumulation, which acts to distort the preserved stratigraphic record. The ages of fossil localities are therefore less secure. To compound the dating problem, the lack of paleontological focus has meant that these sections have not been resampled to the extent necessary to fully resolve conflicts in age interpretation.

In spite of these problems, recent collections and work on the fossil material allow us to add details to earlier discussions of faunal change in this interval. In terms of the general patterns, there is a continuing decline in the diversity of species, turnover seems to have become more continuous when compared to the Middle Miocene, and a greater predominance of first occurrences come before the bulk of the extinctions (figs. 9.5, 9.6). All three of these attributes are influenced to a great extent by the declining fossil abundance, but they also are parts of an older pattern and are not entirely artificial. Recent studies of rodents and large mammals also show significant patterns as to which taxa turn over. Among the rodents, last records of likely closed-habitat species occur just before or soon after 7.5 myr (sciurids, glirids, and cricetids), to be replaced by a greater diversity of murids (Flynn and Jacobs, 1982; Flynn et al., 1995). Significantly, a large species of *Hystrix* also appears shortly before 7.0 myr, and hypsodont, grazing tachyoryctines replace lower-crowned rhizomines (Flynn, 1982). Among the large mammals the picture is also mainly one of loss of species, as would be expected from the declining number of fossils. Most notably the suids *Hippopotamodon sivalense* and *Propotamochoerus hysudricus*, all but one tragulid, the giraffid *Bramatherium megacephalum*, and the bovid lineage of *Tragoceridus* disappear, all of which had been the most abundant elements in the pre-7.5 myr faunas. Subsequent appearances include the hippopotamus *Hexaprotodon sivalensis*, true giraffes, reduncines, and very large bovids (fig. 9.4); rabbits, colobine monkeys, ursids, and species of *Lycyaena*, *Plesiogulo*, and *Machairodus* also appear. At least four of the artiodactyl lineages after 7.5 myr are hypsodont (*Dorcatherium* sp. *D. Tragoceridus punjabicus*, *Selenoportax lydekkeri*, and cf. *Dorcadoxa porrecticornis*).

The above summary suggests the appearance of a faunal assemblage that inhabited more open country and was most similar to slightly older assemblages in southwestern Asia and Europe. The nature of the transition, with appearances and extinctions occurring over a considerable interval, suggests a temporal gradient of environmental change, with species persisting or appearing at different times because of differences in their ecological tolerances. This temporal gradient may also parallel a geographic gradient within a transition zone between zoogeographic provinces and reflect a time-dependent shift of a major zoogeographic boundary toward the east. Causally, this Siwalik turnover event may be linked to changes in the Indian monsoon and, somewhat more speculatively, to changes in atmospheric composition. The isotopic excursions documented by Quade et al. (1989) were first explained as resulting from vegetational changes brought about by the origin or strengthening of the monsoon system of southern Asia, which caused greater seasonality of precipitation. More recently, however, Cerling and others (Cerling et al., 1993; Quade and Cerling, 1995) have suggested that the isotopic changes reflect more global climatic events and/or a decrease in atmospheric CO_2 concentration that altered the competitive balance between C_3 and C_4 grasses.

Summary

The fluvial Neogene Siwalik formations of northern Pakistan comprise a thick sedimentary sequence with abundant terrestrial vertebrate faunas. Patterns of faunal turnover and changes in diversity can be documented with a resolution of 0.5 million years. In northern Pakistan the complete sequence extends from slightly older than 18 to less than 1 myr, the best-sampled interval being that between 14 and 7 myr. Deposits elsewhere on the subcontinent allow this faunal sequence to be extended to about 22 myr and greatly improve the overall quality of the record, especially in the youngest horizons.

Thirteen orders of Siwalik mammals have been identified in the fossil collections, but most are either rodents or artiodactyls. Well-sampled subintervals of 0.5 million years duration typically have 50 or more species. Bovids are the most common and most speciose of the larger mammals, while murid and "cricetid" rodents dominate the small-mammal assemblages.

The quality of the fossil record, as determined by the number of fossils and the nature of their preservation, varies considerably throughout the course of the more than 20 myr represented by the Siwalik deposits. Most of the 0.5 myr subintervals used to analyze the data, and especially those between 14 and 7 myr, are well represented, and all have some fossils. However, the critical intervals between 7 and 3.5 myr, which is a time of important environmental change between the Miocene and the Mid-Pliocene, are poorly known for the most part.

Focusing on the best-known part of the sequence, the stratigraphic-range data on rodents and artiodactyls show that between 18 and 7 myr species diversity varied considerably. In these two orders the number of species first increased between 15 and 13 myr and then steadily decreased to reach an overall low at 7.5 myr. Faunal change in the Siwaliks is also shown to have been episodic, with short intervals of high turnover as well as longer periods of very little change. Significant maxima of first appearances occurred at approximately 13.5 and 8.5 myr; significant maxima of last occurrences were at 12.5 and 8.0 myr. The data therefore indicate that intervals of high extinction did not coincide with intervals having many first appearances. Some of the observed faunal events can be correlated to climatic and environmental changes. The latest Miocene decline in species diversity and increased turnover accompanied oxygen and carbon isotopic changes that correlate to global climatic events. These isotopic changes reflect changes in Siwalik vegetation and may be related to changes in atmospheric composition and the southern Asian monsoon. Both the observed changes in the taxonomic composition and ecological structure of the mammal communities and the inferred changes in Siwalik vegetation are consistent with increasing seasonality and aridity. Other correlations between Siwalik faunal and global climatic events are ambiguous. The Middle Miocene diversification occurred during a period of global cooling, but the marked decrease in diversity and the major turnover events between 13 and 8 myr do not match known local or global events.

Acknowledgments

I wish to thank the organizers of the Airlie conference, especially Elisabeth Vrba, for the opportunity to participate. Close collaboration with the Geological Survey of Pakistan, in particular S. Mahmood Raza, made this research possible. Financial support was provided by grant BNS-8812306 from the National Science Foundation and grant 7087120000 from the Smithsonian Foreign Currency Program.

References

Azzaroli, A., and Napoleone, G. 1982. Magnetostratigraphic investigation of the upper Sivaliks near Pinjor, India. *Rivista Italiana di Paleontologia e Stratigrafia* 87:739–762.

Badgley, C. 1986a. Taphonomy of mammalian fossil remains from Siwalik rocks of Pakistan. *Paleobiology* 12:119–142.

———. 1986b. Counting individuals in mammalian fossil assemblages from fluvial environments. *Palaios* 1:328–338.

———. 1990. A statistical assessment of last appearances in the Eocene record of mammals. Geological Society of America, Special Paper 243:153–167.

Badgley, C., and Gingerich, P. D. 1988. Sampling and faunal turnover in early Eocene mammals. *Palaeogeography, Palaeoclimatology, Palaeoecology* 63:141–157.

Barry, J. C. 1987. The history and chronology of Siwalik cercopithecoids. *Human Evolution* 2:47–58.

Barry, J. C., Flynn, L. J., and Pilbeam, D. R. 1990. Faunal diversity and turnover in a Miocene terrestrial sequence. In *Causes of evolution: A paleontological perspective,* pp. 381–421 (ed. R. Ross and W. Allmon). University of Chicago Press, Chicago.

Barry, J. C., Johnson, N. M., Raza, S. M., and Jacobs, L. L. 1985. Neogene mammalian faunal change in southern Asia: Correlations with climatic, tectonic, and eustatic events. *Geology* 13:637–640.

Barry, J. C., Lindsay, E. H., and Jacobs, L. L. 1982. A biostratigraphic zonation of the middle and upper Siwaliks of the Potwar Plateau of northern Pakistan. *Palaeogeography, Palaeoclimatology, Palaeoecology* 37:95–130.

Barry, J. C., Morgan, M. E., Flynn, L. J., Pilbeam, D., Jacobs, L. L., Lindsay, E. H., Raza, S. M., and Solounias, N. 1995. Patterns of faunal turnover and diversity in the Neogene

Siwaliks of northern Pakistan. *Palaeogeography, Palaeoclimatology, Palaeoecology* 114:209–226.

Barry, J. C., Morgan, M., Winkler, A. J., Flynn, L. J., Lindsay, E. H., Jacobs, L. L., and Pilbeam, D. 1991. Faunal interchange and Miocene terrestrial vertebrates of southern Asia. *Paleobiology* 17:231–245.

Behrensmeyer, A. K. 1987. Miocene fluvial facies and vertebrate taphonomy in northern Pakistan. In *Recent developments in fluvial sedimentology* (ed. F. G. Ethridge, R. M. Flores, and M. D. Harvey). Society of Economic Paleontologists and Mineralogists, Special Publication 39:169–176.

Behrensmeyer, A. K., Willis, B., and Quade, J. 1995. Floodplains and paleosols in the Siwalik Neogene and Wyoming Paleogene deposits: A comparative study. *Palaeogeography, Palaeoclimatology, Palaeoecology* 114:37–60.

Berggren, W. A., Kent, D. V., Flynn, J. J., and Van Couvering, J. A. 1985. Cenozoic geochronology. *Geological Society of America Bulletin* 96:1407–1418.

Bernor, R. L. 1983. Geochronology and zoogeographic relationships of Miocene Hominoidea. In *New interpretations of ape and human ancestry*, pp. 21–64 (ed. R. L. Ciochon and R. S. Corruccini). Plenum Press, New York.

———. 1984. A zoogeographic theater and biochronologic play: The time/biofaces phenomena of Eurasian and African Miocene mammal provinces. *Paléobiologie Continentale* 14:121–142.

de Bruijn, H., and Hussain, S. T. 1984. The succession of rodent faunas from the Lower Manchar Formation southern Pakistan and its relevance for the biostratigraphy of the Mediterranean Miocene. *Paléobiologie Continentale* 14:191–204.

de Bruijn, H., Hussain, S. T., and Leinders, J. J. M. 1981. Fossil rodents from the Murree Formation near Banda Daud Shah, Kohat, Pakistan. *Proceedings of the Koninklijke Nederlandse Akademie van Wetenschappen*, ser. B, 84:71–99.

Cerling, T. E., Yang Wang, and Quade, J. 1993. Expansion of C_4 ecosystems as an indicator of global ecological change in the late Miocene. *Nature* 361:344–345.

Colbert, E. H. 1935. Siwalik mammals in the American Museum of Natural History. *Transactions of the American Philosophical Society* 26:1–401.

Downing, K. F., Lindsay, E. H., Downs, W. R., and Speyer, S. E. 1993. Lithostratigraphy and vertebrate biostratigraphy of the early Miocene Himalayan Foreland, Zinda Pir Dome, Pakistan. *Sedimentary Geology* 87:25–37.

Flynn, L. J. 1982. Systematic revision of Siwalik *Rhizomyidae* (*Rodentia*). *Geobios* 15:327–389.

Flynn, L. J., Barry, J. C., Morgan, M. E., Pilbeam, D., Jacobs, L. L., and Lindsay, E. H. 1995. Neogene Siwalik mammalian lineages: Species longevities, rates of change, and modes of speciation. *Palaeogeography, Palaeoclimatology, Palaeoecology* 114:249–264.

Flynn, L. J., and Jacobs, L. L. 1982. Effects of changing environments on Siwalik rodent faunas of northern Pakistan. *Palaeogeography, Palaeoclimatology, Palaeoecology* 38:129–138.

Flynn, L. J., Jacobs, L. L., and Cheema, I. U. 1986. Baluchimyinae: A new ctenodactyloid rodent subfamily from the Miocene of Baluchistan. *American Museum of Natural History Novitates* 2841:1–58.

Flynn, L. J., Pilbeam, D., Jacobs, L. L., Barry, J. C., Behrensmeyer, A. K., and Kappelman, J. W. 1990. The Siwaliks of Pakistan: Time and faunas in a Miocene terrestrial setting. *Journal of Geology* 98:589–604.

Friedman, R., Gee, J., Tauxe, L., Downing, K., and Lindsay, E. 1992. The magnetostratigraphy of the Chitarwata and lower Vihowa formations of the Dera Ghazi Khan area, Pakistan. *Sedimentary Geology* 81:253–268.

Gingerich, P. D. 1987. Extinction of Phanerozoic marine families. Abstracts with Programs, Geological Society of America 1987 Annual Meeting 19:677.

Hussain, S. T., van den Bergh, G. D., Steensma, K. J., de Visser, J. A., de Vos, J., Arif, M., van Dam, J., Sondaar, P. Y., and Malik, S. M. 1992. Biostratigraphy of the Plio-Pleistocene continental sediments (Upper Siwaliks) of the Mangla-Samwal anticline, Azad Kashmir, Pakistan. *Proceedings of the Koninklijke Nederlandse Akademie van Wetenschappen*, ser. B, 95:65–80.

Jacobs, L. L., Flynn, L. J., Downs, W. R., and Barry, J. C. 1990. *Quo vadis, Antemus*? The Siwalik muroid record. In *European Neogene mammal chronology*, pp. 573–586 (ed. E. H. Lindsay, V. Fahlbusch, and P. Mein). Plenum Press, New York.

Johnson, G. D., Zeitler, P., Naeser, C. W., Johnson, N. M., Summers, D. M., Frost, C. D., Opdyke, N. D., and Tahirkheli, R. A. K. 1982. The occurrence and fission-track ages of Late Neogene and Quaternary volcanic sediments, Siwalik Group, northern Pakistan. *Palaeogeography, Palaeoclimatology, Palaeoecology* 37:63–93.

Johnson, N. M., Opdyke, N. D., Johnson, G. D., Lindsay, E. H., and Tahirkheli, R. A. K. 1982. Magnetic polarity stratigraphy and ages of Siwalik Group rocks of the Potwar Plateau, Pakistan. *Palaeogeography, Palaeoclimatology, Palaeoecology* 37:17–42.

Johnson, N. M., Stix, J., Tauxe, L., Cerveny, P. F., and Tahirkheli, R. A. K. 1985. Paleomagnetic chronology, fluvial processes and tectonic implications of the Siwalik deposits near Chinji Village, Pakistan. *Journal of Geology* 93:27–40.

Kappelman, J. W. 1986. The paleoecology and chronology of the Middle Miocene hominoids from the Chinji Formation of Pakistan. Ph.D. diss., Harvard University.

Made, J. van der, and Hussain, S. T. 1989. *"Microstonyx" major* (Suidae, Artiodactyla) from the type area of the Nagri Formation, Siwalik Group, Pakistan. *Estudios Geológicos* 45:409–416.

Matthew, W. D. 1929. Critical observations upon Siwalik mammals. *Bulletin of the American Museum of Natural History* 56:437–560.

Morgan, M. E. 1994. Paleoecology of Siwalik Miocene hominoid communities: Stable carbon isotope, dental microwear, and body size analyses. Ph.D. diss., Harvard University.

Morgan, M. E., Badgley, C. E., Gunnell, G. F., Gingerich, P. D., Kappelman, J., and Maas, M. C. 1995. Comparative paleoecology of Paleogene and Neogene mammalian faunas: Body-size structure. *Palaeogeography, Palaeoclimatology, Palaeoecology* 114:287–317.

Opdyke, N. D., Lindsay, E., Johnson, G. D., Johnson, N., Tahirkheli, R. A. K., and Mirza, M. A. 1979. Magnetic polarity stratigraphy and vertebrate paleontology of the Upper

Siwalik Subgroup of northern Pakistan. *Palaeogeography, Palaeoclimatology, Palaeoecology* 27:1–34.
Pilgrim, G. E. 1908. The Tertiary and post-Tertiary freshwater deposits of Baluchistan and Sind, with notices of new vertebrates. *Records of the Geological Survey of India* 37:139–166.
———. 1910. Preliminary note on a revised classification of the Tertiary freshwater deposits of India. *Records of the Geological Survey of India* 40:185–205.
———. 1913. The correlation of the Siwaliks with mammal horizons of Europe. *Records of the Geological Survey of India* 43:264–326.
———. 1917. Preliminary note on some recent mammal collections from the basal beds of the Siwaliks. *Records of the Geological Survey of India* 48:98–101.
———. 1926. The Tertiary formations of India and the interrelation of marine and terrestrial deposits. *Proceedings of the Pan-Pacific Science Congress, Australia, 1923* 2:896–931.
———. 1934. Correlation of the ossiferous sections in the Upper Cenozoic of India. *American Museum of Natural History Novitates* 704:1–5.
Quade, J., and Cerling, T. E. 1995. Expansion of C_4 grasses in the Late Miocene of northern Pakistan: Evidence from stable isotopes in paleosols. *Palaeogeography, Palaeoclimatology, Palaeoecology* 114:91–116.
Quade, J., Cerling, T. E., and Bowman, J. R. 1989. Development of Asian monsoon revealed by marked ecological shift during the latest Miocene in northern Pakistan. *Nature* 342:163–166.
Raza, S. M. 1983. Taphonomy and paleoecology of Middle Miocene vertebrate assemblages, southern Potwar Plateau, Pakistan. Ph.D. diss., Yale University.
Raza, S. M., Barry, J. C., Meyer, G. E., and Martin, L. 1984. Preliminary report on the geology and vertebrate fauna of the Miocene Manchar Formation, Sind, Pakistan. *Journal of Vertebrate Paleontology* 4:584–599.
Shah, S. M. I. (ed.) 1977. Stratigraphy of Pakistan. *Memoirs of the Geological Survey of Pakistan* 12:1–138.
Tandon, S. K., Kumar, R., Koyama, M., and Niitsuma, N. 1984. Magnetic polarity stratigraphy of the Upper Siwalik Subgroup, east of Chandigarh, Punjab sub-Himalaya, India. *Journal of the Geological Society of India* 25:45–55.
Tauxe, L., and Opdyke, N. D. 1982. A time framework based on magnetostratigraphy for the Siwalik sediments of the Khaur area, northern Pakistan. *Palaeogeography, Palaeoclimatology, Palaeoecology* 37:43–61.
Thomas, H. 1977. Un nouveau Bovidé dans les couches à Hominoidea du Nagri (Siwaliks moyens), Plateau de Potwar, Pakistan: *Elachistoceras khauristanensis* gen. et sp. nov. (Bovidae, Artiodactyla, Mammalia). *Bulletin de la Société Géologique de France* 19:375–383.
———. 1984. Les Bovidés anté-hipparions des Siwaliks inférieures (Plateau du Potwar, Pakistan). *Mémoires de la Société Géologique de France,* n.s., 145:1–67.
———. 1985. The Early and Middle Miocene land connection of the Afro-Arabian plate and Asia: A major event for hominoid dispersal? In *Ancestors: The hard evidence,* pp. 42–50 (ed. E. Delson). Alan R. Liss, New York.
Vrba, E. S. 1985. Environment and evolution: Alternative causes of the temporal distribution of evolutionary events. *South African Journal of Science* 81:229–236.
———. 1988. Late Pliocene climatic events and hominid evolution. In *Evolutionary history of the "robust" australopithecines,* pp. 405–426 (ed. F. E. Grine). Aldine de Gruyter, New York.
West, R. M. 1981. Plio-Pleistocene fossil vertebrates and biostratigraphy, Bhittanni and Marwat Ranges, north-west Pakistan. In *Proceedings of the Field Conference on Neogene-Quaternary Boundary, India, 1979,* pp. 211–215 (ed. M. V. A. Sastry, T. K. Kurien, A. K. Dutta, and S. Biswas). Geological Survey of India, Calcutta.
Willis, B. 1993a. Ancient river systems in the Himalayan foredeep, Chinji Village area, northern Pakistan. *Sedimentary Geology* 88:1–76.
———. 1993b. Evolution of Miocene fluvial systems in the Himalayan foredeep through a two kilometer-thick succession in northern Pakistan. *Sedimentary Geology* 88:77–121.
Willis, B. J., and Behrensmeyer, A. K. 1995. Fluvial systems in the Siwalik Neogene and Wyoming Paleogene. *Palaeogeography, Palaeoclimatology, Palaeoecology* 114:13–35.

Appendix 9.1. Comments on Taxa Listed in Figure 9.4 and Table 9.2

ANTHRACOTHERIIDAE

"Anthracotherium" punjabiense probably includes two or even three species, with specimens in the 8.5 myr interval differing the most from older specimens. These putative taxa may represent a single evolving lineage.

Similarly, *Merycopotamus dissimilis* and the species of *Hemimeryx* may represent a succession of morphologies within a single lineage.

HIPPOPOTAMIDAE

The oldest certain record of *Hexaprotodon sivalensis* is in the 5.5 myr interval, but there are records in the Rhotas section that could be 6 or even 7 myr. The uncertainty results from problems in the dating of localities in the Rhotas and equivalent sections.

SUIDAE

The two taxa cf. *Listriodon* sp. and *Listriodon pentapotamiae* may form a lineage, while *Listriodon pentapotamiae* itself may be composed of a succession of closely related species. I have attributed material as old as 17 myr to *Listriodon pentapotamiae,* which is considerably older than the date estimated for the first appearance of fully lophodont *Listriodon* given by Bernor and Tobien (1990). Although fragmentary, this material is from a lophodont suid.

The oldest specimens attributed to *Sivahyus punjabiensis* may belong to a different taxon than the younger ones.

Specimens attributed to *Hyotherium pilgrimi* possibly belong to two species, both with uncertain relationships to younger taxa.

Specimens identified as *Hippopotamodon sivalense* show a trend toward increasing size throughout its stratigraphic range. Specimens from the Nagri Formation identified as *"Microstyonx" major* by Made and Hussain (1989) fit within this evolving taxon.

It is difficult to separate the available material from the 5.5 myr interval of *Propotamochoerus hysudricus* and *Sus;* these may represent a lineage with the transition between the species occurring at 5.5 myr.

The age of the first appearance of *Hippohyus* spp. is uncertain owing to problems dating the Rhotas section. The species is certainly older than 4.0 myr and may be as old as 5.4 myr.

TRAGULIDAE

Dorcabune anthracotherioides and *Dorcabune nagrii* occur in the same interval, though at slightly different levels, and do not appear to overlap in time.

The species labeled *Dorcatherium* very small sp. A, *Dorcatherium minimus, Dorcatherium* very small sp. B, and *Dorcatherium nagrii* are distinct from each other but might represent a single lineage. *Dorcatherium* very small sp. A may be a composite of several species.

The *Dorcatherium* small sp. complex is larger than the preceding complex of very small species, but it too may be a composite of more than one species.

The relationship between *Dorcatherium* sp. D and "*Dorcatherium*" sp. E is not clear. Both are about the same size. "*Dorcatherium*" sp. E. is characterized by high crowned molars, which are also seen in late members of *Dorcatherium* sp. D, but not in earlier ones. *Dorcatherium* sp. D may therefore be two distinct species, with the higher-crowned forms first appearing in the 7.5 myr interval; it may give rise to "*Dorcatherium*" sp. E.

Dorcatherium minus, Dorcatherium cf. *minus,* and *Dorcatherium majus* may be a single lineage that becomes progressively larger between 11.5 and 10.5 myr.

GIRAFFIDAE

Giraffa punjabiensis is definitely known from an ossicone at 7 myr. The species is possibly also present in the 7.5 myr interval. It is not clear where the top of the range of *Bramatherium megacephalum* lies, because teeth and postcranials of this species are difficult to separate from those of *Giraffa*. There is an ossicone of *Bramatherium megacephalum* in the 7.0 myr interval at a level slightly above the occurrence of the *Giraffa* ossicone.

BOVIDAE

All specimens referred to *Kubanotragus sokolovi* collected by the Harvard–Geological Survey of Pakistan project come from a single locality in the base of the Chinji Formation, which suggests that this taxon has a very short time span in the Siwaliks.

Some of fossil material now identified as cf. *Dorcadoxa porrecticornis* might belong to *Prostrepsiceros vinayaki*. Reassignment of the material would extend the stratigraphic range of *Prostrepsiceros vinayaki* into the 6.5-myr interval.

The type specimen of *Caprotragoides potwaricus* is assumed to have come from the Y311 locality in the type section of the Nagri Formation.

Early members of what I refer to as *?Gazella* spp. may not belong to *Gazella* and may not form either a single species or a lineage with later members. Some of the horns could belong to *Caprotragoides potwaricus*. Specimens that can more confidently be identified as *Gazella* appear at 9.5 myr, and the taxon is certainly present by 7.5 myr. The 9.5 and 7.5 myr forms are probably different species.

Thomas's (1977) hypodigm for *Elachistoceras khauristanensis* included horncores, plus dental and postcranial material. Most of the material came from horizons younger than 9.0 myr. Material of very small bovids is also present throughout the interval between 14 and 11 myr, some of which he referred to this species. It is likely that these represent a different taxon, or even several taxa. No horncores older than 8.5 myr are known for the taxon.

Protragocerus gluten is the most abundant species between 14 and 11 myr. Our collections include a confusing variety of horn sizes and shapes. The material may belong to two or more distinct species.

Species of *Tragoceridus,* which may be the sister group of *Protragocerus* (Thomas, 1984), are the most abundant bovids after 10.5 myr. Their dentitions are more hypsodont after 7.5 myr, which may be when the transition occurred between forms assigned to *T. pilgrimi* and *T. punjabicus* (see Thomas, 1984).

The teeth assigned to *Selenoportax lydekkeri* are relatively hypsodont and indicate that it was a very large species. It marks the first occurrence in the Siwalik sequence of a bovid approaching the largest living bovids in body size.

Species of cf. *Leptobos* and *Hemibos* are reported by Hussain et al. (1992) from upper-Gauss-age sediments near New Mirpur. *Leptobos* is also known from the type area of Tatrot, which is now placed in the lower normal zone of the Gauss at ca. 3.3 myr.

Material identified as belonging to cf. *Dorcadoxa porrecticornis* includes both horncores and lower dentitions. The teeth and horncores are not, however, closely associated and might belong to different taxa. The teeth have some but not all the characteristics of reduncine teeth. They are exceptionally high-crowned, even though the species is small.

A fragment of a frontlet and horn base from about 4 myr belongs to a reduncine different from cf. *Dorcadoxa porrecticornis.* This may be the same taxon as *Vishnucobus patulicornis,* otherwise first reported from the Tatrot at 3.3 myr.

Teeth from the 5.5 myr interval could belong either to a reduncine very much larger than cf. *Dorcadoxa porrecticornis* or to an otherwise indeterminate hippotragine. Because the teeth are very large relative to the size of the mandible, I have identified this taxon as a hippotragine and grouped them with a frontlet and horn from the 4 myr levels at Rhotas. The teeth and frontlet are the correct size to go with either cf. *Oryx sivalensis* or *Hippotragus brevicornis.*

CRICETIDAE

Democricetodon sp. F and *Democricetodon* sp. G become locally extinct at 9.1 myr and subsequently reappear at 7.9 myr.

Chapter 10

The Effect of Paleoclimate on the Evolution of the Soricidae (Mammalia, Insectivora)

Jelle W. F. Reumer

As a result of continued research, the evolutionary history of the Soricidae is becoming better known. Yet many hiatuses exist in our knowledge, not because the fossil record is insufficient but because so few paleontologists have studied shrews. A general picture does emerge, however, allowing us to look beyond the field of pure systematical paleontology and to investigate where, when, and why shrews evolved.

Shrews are among the smallest living homoiothermic animals; as a result, their physiology must cope with an extremely unfavorable surface-to-volume ratio. Shrews maintain a very high metabolic rate and have a relatively high oxygen consumption and a nearly constant need for food. They require environments that have temperate climates and provide an uninterrupted food source. Because their diet consists mainly of small invertebrates, the environment must not be too arid to support such life. Ecological studies have shown that environmental moisture may be the ultimate determinant of within-habitat diversity and numerical abundance of soricids (Feldhamer et al., 1993; see also Vogel, 1980; Reumer, 1985, 1989; Churchfield, 1990, for details).

A survey of the recent biogeography of the Soricidae (which found the greatest diversity in tropical Africa, where conditions are both warm and humid) suggests that shrew evolution would have been favored by relatively warm and humid paleoclimates.

In earlier studies (Reumer, 1984, 1985, 1989) I investigated the contrary effect: shrew migrations and extinctions during periods of paleoclimatic cooling. Cooling took place, for example, at the Pliocene-Pleistocene boundary, ca. 2.4 million years (myr) ago, and during periods of advancing glaciations in the Pleistocene.

In the present chapter, I examine the effect of paleoclimate on shrew evolution. These case studies of the paleoecology and of the possible climatic causes of evolution, extinction, and migration of a single mammalian family are intended to contribute to the overall understanding of the influence of paleoclimate on mammals.

Methods

Taxonomical Framework. Soricidae are defined by the following characters: they are generally small-sized Insectivora, possessing a deeply pocketed internal temporal fossa in the mandible, lacking a zygomatic arch, and having a dorsoventrally separated mandibular condyle. The anterior dentition is reduced, leaving a single large incisor and several (2–5) small antemolars in both the upper and lower jaw. The upper fourth premolar is molarized, whereas the lower one is not (Reumer, 1987). This definition implies that the Heterosoricidae are not included in the Soricidae (Reumer, 1987) and that they are omitted from the present line of reasoning. For this study, the following systematic subdivision of the family is used:

Family: Soricidae Gray, 1821
- Subfamily: Crocidosoricinae Reumer, 1987
- Subfamily: Allosoricinae Fejfar, 1966
- Subfamily: Limnoecinae Repenning, 1967
- Subfamily: Soricinae Fischer von Waldheim, 1817
  - Tribus: Soricini Fischer von Waldheim, 1817
  - Tribus: Blarinini Kretzoi, 1965
  - Tribus: Notiosoricini Reumer, 1984

ISBN 0-300-06348-2.

Tribus: Amblycoptini Kormos, 1926
Tribus: Beremendini Reumer, 1984
Tribus: Soriculini Kretzoi, 1965
Subfamily: Crocidurinae Milne-Edwards, 1868–74

In addition to this subdivision, I recognize a separate group of shrews within the tribus Soricini. This group comprises the genera *Petenyia* Kormos, 1934; *Blarinella* Thomas, 1911; *Hemisorex* Baudelot, 1967; *Tregosorex* Hibbard and Jammot, 1971; *Anchiblarinella* Hibbard and Jammot, 1971; and *Parydrosorex* Wilson, 1968. Researchers have yet to determine whether the genus *Alluvisorex* Hutchison, 1966, also belongs to this group. A detailed study of this group is forthcoming; it may warrant separation as a distinct soricine tribe. Use of the term Soricini in this chapter is meant to exclude this *Petenyia-Blarinella* group and is hence used *sensu stricto.* A total of eleven distinct groups of shrews can thus be used in this study. Taxa grouped here under Crocidosoricinae gave rise to the other subfamilies of shrews (Limnoecinae, Allosoricinae, Soricinae, Crocidurinae) during the Miocene (Reumer, 1987, 1989, 1992). So far, we have not attempted to match the origins of these subfamilies to a biochronological framework, let alone to the absolute time scale. Because Eurasian as well as American taxa and localities are involved, a correlated European and American biostratigraphy is needed as the framework for this study.

Biostratigraphical Framework. In figure 10.1, the chart that is the nucleus of this study, I have used the absolute time scale (in myr) as the starting point (column A). It is plotted against several stratigraphical columns, of which the usual Tertiary subdivision is shown in column B. Two biozonations are added: the North American zonation (column C) is after Tedford et al. (1987). The European biozonation (column D) is compiled after Mein (1990) and De Bruijn et al. (1992). The European MN zonation (MN = Mammalian Neogene, column E) is added, using Mein (1990) and also Steininger et al. (1990) and Van der Meulen and Daams (1992). A few European and American fossil-vertebrate localities are shown (column F). The stratigraphical attribution of the localities is mostly after Woodburne (1987) for American localities after and De Bruijn et al. (1992) for European localities.

The right-hand part of the chart shows two paleoclimatic curves: a temperature curve (column J) and a relative humidity curve (column K). Both are composed of an uninterrupted line and a broken line. The uninterrupted lines are taken from Van der Meulen and Daams (1992), who published detailed curves for the Ramblian, Aragonian, and part of the Vallesian based on their research on Spanish rodent faunas. These curves are chosen 1) because they are based on land vertebrates and 2) because they are relatively fine in scale. The broken lines, after Daams et al. (1988), are less precise; they are used only for the time span not covered by the curves of Van der Meulen and Daams (1992). Finally, the dotted line in the humidity curve is the result of a study by Calvo et al. (1993). It represents a refinement of the other curve in that the Ruscinian humidity peak is divided in two.

In column G the chart shows First Appearance Dates (FADs) and Last Appearance Dates (LADs) of the eleven supergeneric taxa mentioned above. The data for this compilation come from various literature, among which Repenning (1967), Reumer (1984), Tedford et al. (1987), and Ziegler (1989) are the most important, and from my own work on the shrews from Maramena, in Greece (Doukas et al., 1995). Each of these biostratigraphic ranges plays a crucial role in this study.

Ranges

In this section, the ranges for the eleven taxa will be discussed in the order of their first appearances.

Crocidosoricinae. The evolutionary history of the subfamily Crocidosoricinae is reviewed by Reumer (1994). This group of plesiomorphic shrews originated in Asia, where representatives are found in strata of Early Oligocene age. They reached Europe sometime after the "Grande Coupure" event, which corresponds roughly to the MP20-MP21 boundary (MP = Mammalian Paleogene), about 33 myr ago. The European FAD is determined by the finds of *Srinitium marteli* Hugueney, 1976, from Saint-Martin-de-Castillon, in southern France, dated to MP23 (Schmidt-Kittler, 1987). Note that this biozone is beyond the lower margin of figure 10.1.

The Crocidosoricinae witnessed their maximum abundance both in terms of biodiversity and geographical distribution in the Early Miocene (Ramblian-Aragonian; biozones MN3 and MN4). Then they began to decrease, to disappear during the early Late Miocene (Early Vallesian, MN9, ca. 11 myr ago). By that time they lived only in Spain, after having been spread all over Europe during Early and Middle Miocene times. Only an island relic survived in Italy, possibly into Pliocene

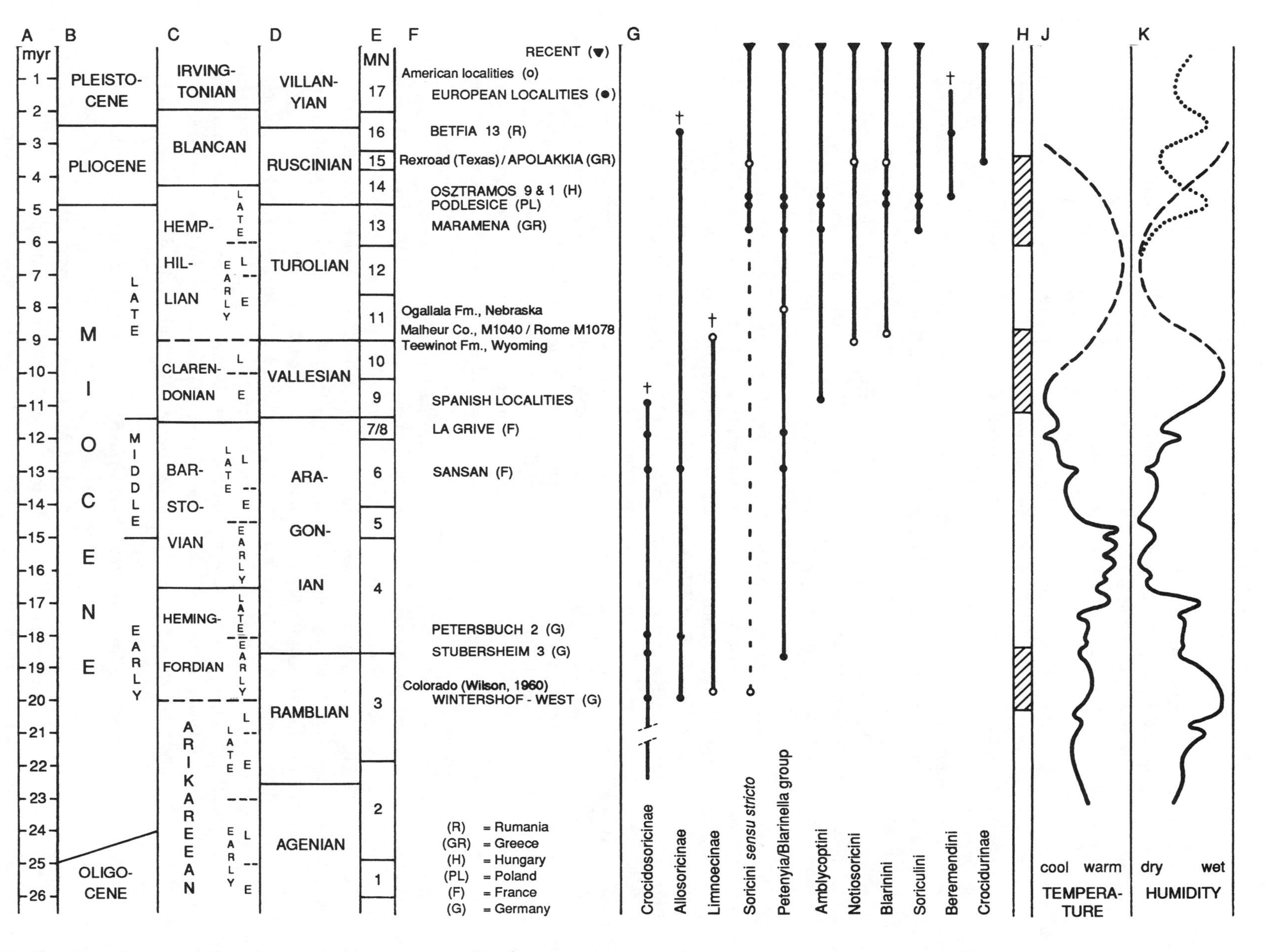

Fig. 10.1. Ranges of major groups of Soricidae in relation to paleoclimatic parameters. Columns denote the following: A = age in million years (myr); B = ages; C = North American continental zones; D = European continental zones; E = European MN zonation; F = major localities (American localities in lowercase; European localities in capitals); G = ranges; H = periods of enhanced speciation; J = temperature; K = humidity. See text for further explanation.

times, but this record has been omitted from the chart because the exact stratigraphic correlation of this Gargano island fauna to the continental biozonation is unknown. (See Reumer, 1994, and references therein.)

Allosoricinae. The taxonomy of the subfamily Allosoricinae was revised recently (Reumer, 1992). Two genera are attributed: *Paenelimnoecus* Baudelot, 1972, and *Allosorex* Fejfar, 1966. *Allosorex* is known only from the Middle Pliocene (Ruscinian, MN15) locality of Ivanovce A in Czechoslovakia. *Paenelimnoecus,* on the other hand, witnessed a range from MN3 (Wintershof-West: *P. micromorphus*) up till MN16 (*P. pannonicus*). This range implies an origin for this subfamily of some 20 myr ago (Ramblian or Orleanian). *P. micromorphus* is also known from Petersbuch 2 and Erkertshofen 2, both MN4 (Ziegler, 1989), which means that by the Early Aragonian the genus *Paenelimnoecus* was well established, at least in central Europe. The last occurrences of Allosoricinae in the fossil record are finds of *P. pannonicus* from Early Villányian localities in central Europe (e.g., Mala Cave, Poland: Sulimski et al., 1979; and Betfia 13, Romania: Terzea and Jurcsák, 1976). This extinction came after the soricid extinction wave around the Pliocene-Pleistocene boundary (Reumer, 1985, and below). Allosoricinae are found only in Europe.

Angustidens and Antesorex. Two genera of soricids appear in the North American fossil record at the onset of the Hemingfordian, ca. 20 myr ago. Tedford et al. (1987, p. 186) noted that "the early Hemingfordian (and hence the Hemingfordian as a whole) can be defined by the earliest appearance of the following immigrants: the soricine (*Antesorex*) and limnoecine (*Angustidens*) shrews." Both shrews immigrated from Eurasia over the temporary Bering landbridge that connected the two continents and that brought, among other mammals, the horse *Anchitherium* into Eurasia (Ramblian, see Van der Meulen and Daams, 1992). As far as is known, no true Soricidae older than these existed in North America, so it may be safely assumed that this Early Hemingfordian immigration wave brought the first shrews from Asia into America.

Limnoecinae. Wilson (1960) described *Angustidens vireti,* which stands at the base of the subfamily Limnoecinae Repenning, 1967. *Angustidens* was followed in the fossil record by the genus *Limnoecus* Stirton, 1930, of Barstovian to Hemphillian age. The latest record known to me is from United States Geological Survey (USGS) Vertebrate Locality M1078, Rome, Malheur County, Oregon, which is of Early Hemphillian age (ca. 9 myr ago; Repenning, 1967).

Records of Limnoecinae from Europe (e.g., Doben-Florin, 1964; Baudelot, 1972) are based on erroneous interpretations; the present subfamily is here considered to be an exclusively American taxon.

Soricini. The genus *Antesorex* Repenning, 1967, was classified as a Soricini. If we exclude the above-mentioned *Petenyia-Blarinella* group from the Soricini *sensu* Repenning (1967), then *Antesorex* still belongs to the Soricini *sensu stricto.* This attribution may need revision. As Repenning (1967) stated, the only known species (*A. compressus*) closely resembles *Crocidosorex antiquus* from Europe. This latter taxon is known from Early Miocene localities (primarily belonging to zones MN1 and MN2) and thus antedates *Antesorex.* A direct descent of *Antesorex* from a *Crocidosorex*-like shrew therefore seems highly plausible. If this is indeed the case, and if at the same time *Antesorex* does rightly belong to the Soricinae, tribe Soricini, then we may conclude that this tribe originated around the Arikareean-Hemingfordian boundary, that is, ca. 20 myr ago. The problem that arises is that from that time onward, the record of the Soricini is obscure, both in Eurasia and in North America. True and clearly identifiable Soricini do not occur until the latest Miocene or Early Pliocene (Late Turolian in Europe, Early Blancan in North America).

The genus *Sorex* Linnaeus, 1758, is found in America from the Early Blancan onward (Repenning, 1967). In Europe, the Late Turolian witnesses the first occurrence of the genera *Sorex* and *Deinsdorfia* Heller, 1963 (Maramena, Greece: Doukas et al., 1995).

From the Early Hemingfordian to the Early Blancan in America, and up till the Late Turolian in Europe, no unambiguous Soricini are known. For that reason the line in the Soricini *sensu stricto* in column G of figure 10.1 is dashed between *Antesorex* and the other Soricini.

***Petenyia-Blarinella* Group.** Ziegler (1989) described *Hemisorex* sp. from the German locality of Stubersheim 3 (MN3-MN4). If the attribution of the scanty material (only one ramus with two molars) to the genus *Hemisorex* is correct, it may represent the earliest occurrence of this group of shrews. It would be taxonomically and phylogenetically interesting to investigate whether the crocidosoricine *Florinia stehlini* (Doben-Florin, 1964) belongs to this group or whether *F. stehlini* is ancestral to the *Petenyia-Blarinella* group. *F. stehlini*, however, is known from localities dated to MN3 and MN4 (e.g., Wintershof-West, MN3, and Petersbuch 2 and Erkertshofen 2, both MN4) and therefore partly postdates the *Hemisorex* find from Stubersheim 3 (see Doben-Florin, 1964; Ziegler, 1989). Nevertheless, we may conclude that the *Petenyia-Blarinella* group originated around the MN3-MN4 boundary, namely, the Ramblian-Aragonian boundary, roughly 18–19 myr ago, and probably in MN3, some 20 myr ago.

Amblycoptini. This tribe, as well as the next two, came into existence during Late Miocene times. The soricine tribe Amblycoptini, of which one representative is extant (*Anourosorex squamipes* Milne-Edwards, 1872), is first encountered in localities in Spain in the form of *Crusafontina endemica* Gibert, 1975. The type locality for this species is Can Llobateres (Gibert, 1975), which is also the stratotype for mammal zone MN9 (Early Vallesian; see De Bruijn et al., 1992). This fact means that this lineage started ca. 11 myr ago. Several genera lived in Eurasia during Late Miocene and Pliocene times, the most successful of which was *Amblycoptus* Kormos, 1926 (Reumer, 1984). This tribe never reached America.

Notiosoricini. The Notiosoricini, on the other hand, are exclusively American. There are still two living members of this lineage: *Notiosorex crawfordi* (Coures, 1877) and *Megasorex gigas* (Merriam, 1897). A fossil *Notiosorex* (*N. jacksoni* Hibbard, 1950) is reported from the Rexroad fauna (Blancan) of Kansas (Hibbard, 1950). Most probably, as stated before (Reumer, 1984), also *Hesperosorex lovei* Hibbard, 1957, belonged to this tribe. *H. lovei* was found in the Teewinot Formation, Wyoming. This formation has yielded a local fauna that has been dated around the Clarendonian-Hemphillian boundary, ca. 9.0 myr ago, by being deposited upon a 9.2 myr tuff (Repenning, 1987), making it the oldest date for a Notiosoricini.

Blarinini. The two species—*Adeloblarina berklandi* Repenning, 1967, and *Paracryptotis rex* Hibbard, 1950—are the oldest known representatives of the tribe Blarinini. *A. berklandi* is described from USGS Vertebrate Locality M1040, Malheur County, Oregon, and was attributed a "Late Miocene (Barstovian, possibly late Barstovian)" age (Repenning, 1967 p. 38). Repenning (1987) attributed to the USGS M localities of Malheur County an Early Hemphillian age. If we consider this latter and later correlation to be correct, the Blarinini came into existence ca. 9 myr ago. *P. rex* comes from the same region, USGS Vertebrate Locality M1078, which has approximately the same age. This species is also found in the Blancan Rexroad fauna (Kansas), thereby suggesting a persistence of the species of at least 5.5 myr (the entire Hemphillian and the earliest half of the Blancan).

During the Early Ruscinian of Europe, Blarinini of the genera *Blarinoides* Sulimski, 1959, and *Mafia* Reumer, 1984, appeared, followed in the Late Ruscinian by *Sulimskia* Reumer, 1984. It is not known how and when the Blarinini crossed the Beringian barrier between America and Europe. Given the early presence of Blarinini in America ca. 9 myr ago and the first appearance in Europe ca. 4.5 myr ago, it can be assumed that the migration took place from America into Eurasia. The only fossil records of Asian Blarinini that have come to my attention are the Japanese *Shikamainosorex densicingulata* Hasagawa, 1957, and the Chinese *Peisorex pohaiensis* Kowalski and Li, 1963 (Repenning, 1967). Both finds are from the Middle Pleistocene and do not help in answering this question.

Presently, Blarinini live only in America: specifically, the genera *Blarina* Gray, 1838, and *Cryptotis* Pomel, 1848. Blarinini are the only shrews to have reached South America.

Soriculini. The tribe Soriculini is exclusively Eurasian. Many forms are extant, especially in Asia, which could suggest an Asiatic origin for this tribe. The earliest record of a Soriculini is from the Late Turolian of Greece: *Episoriculus gibberodon* (Petenyi, 1864) from Maramena (Doukas et al., 1995), making this lineage one of a series of developments that appeared around the Miocene-Pliocene boundary. Other developments are the European first records of the Soricini *sensu stricto* and of the Blarinini, as well as the tribe Beremendini (see below).

Within the European realm, the Soriculini were never very diverse; the genus *Episoriculus* developed one major species (*E. gibberodon* [Petenyi, 1864]) and gave rise to several island endemics of the genus *Nesiotites* Bate, 1944 (see below). They were replaced in the continental Pleistocene by the genus *Neomys* Kaup, 1829, with one or two extinct species and two extant ones. As mentioned, many more Soriculini still exist in Asia, but the Asiatic fossil record is extremely scanty.

Beremendini. The last distinct lineage within the Soricinae is formed by the tribe Beremendini, which is monogeneric. Its only genus is *Beremendia* Kormos, 1934, with four described species of Eurasiatic origin. The Early Ruscinian fauna from Osztramos 1 (Hungary) can be considered the earliest record of Beremendini: both species *B. minor* Rzebik, 1976, and *B. fissidens* (Petenyi, 1864) are found in this fauna (Reumer, 1984). *B. fissidens* in particular proved to be an extremely successful species: it is to be found in many Ruscinian and Villányian localities and continued well into the Pleistocene. The tribe is now extinct.

Crocidurinae. Finally, the subfamily Crocidurinae comes into the fossil record. This late appearance comes as a great surprise, because at present the crocidurines are by far the most diverse group of shrews, in terms of both biogeographic distribution and morphology. The surprise becomes even greater if one considers the rather plesiomorphic nature of several characters of the dentition and mandibular osteology. The Crocidurinae must form an old group, but the fossil record does not confirm this idea. The oldest find that came to my attention is a *Crocidura* from the Aegean island of Rhodes, in the Apolakkia Local Fauna, dated to MN15 (Middle Ruscinian, Van de Weerd et al., 1982).

Results of Range Analysis

All data discussed above (FADs, LADs, ranges) are plotted in figure 10.1. The result is a range chart for the eleven shrew groups discussed. (The onset of the Crocidosoricinae is omitted, because it would merely extend the bottom of the chart without adding much information.) In studying the ranges, it becomes clear that we can discern three periods in which groups first appeared and which are interpreted here as periods of enhanced speciation. I have indicated these periods with hatched bands in column H.

The first band is found in the Early Miocene, some 19–20 myr ago. We notice the first appearance of the Allosoricinae in Wintershof-West and the first appearances (both coinciding with the arrival in America) of the Soricinae (tribe Soricini) and of the Limnoecinae. Three of the four subfamilies that arose from crocidosoricine stock therefore appeared in this time span (the fourth subfamily, Crocidurinae, has a rather more enigmatic origin; see above).

The second period of speciation occurred in the early Late Miocene, more or less coinciding with the Vallesian stage in Europe and the Clarendonian in America, roughly between 9 and 11 myr ago. During this period, we record the disappearance of the Crocidosoricinae in continental Europe, the first appearances of the tribes Amblycoptini (in Spain), Notiosoricini (North America), and Blarinini (North America), and the last occurrence of the Limnoecinae in America.

The third band coincides more or less with the Pliocene: Ruscinian in Europe, Blancan in America (somewhere between 6.0 and 2.4 myr ago). The tribes Soriculini and Beremendini appear, and the first Crocidurinae is noted in Europe. Around or just beyond the end of the Ruscinian the last Allosoricinae disappear, perhaps somewhat later than the 2.4 myr extinction wave but certainly as a part of the shrew crisis that occurred near the Pliocene-Pleistocene boundary (discussed below; see also Reumer, 1985).

It is interesting to see what these three periods have in common and to identify prevailing circumstances that might have stimulated shrew evolution. Comparison of the hatched bands of column H with the climatic curves in columns J and K shows that periods of high relative humidity coincide with the periods during which the Soricidae proliferate. It is therefore concluded that the proliferation of shrews is stimulated by humid paleoclimates, which partly confirms the hypothesis put forward in the introduction (namely, that the evolution of shrews was favored by warm and humid paleoclimates).

Quaternary Developments

Previous studies (Reumer, 1984, 1985) have yielded results that warrant reexamination in the framework of the present chapter. They concern the influence of paleotemperature (rather than humidity, although both parameters are to some degree linked) on the biodiversity and biogeography of the Soricidae from the latest Pliocene onwards.

Biodiversity around the Pliocene-Pleistocene Boundary. Figure 10.2 gives information about the Soricidae from (mostly central) Europe and their biostratigraphic context in the time span between MN zones 14 and 17 (and even somewhat later: zone Q1 = Quaternary 1). This is roughly between 5 and 1.5 myr. The figure is taken from Reumer (1985) and compiled after Reumer (1984) and correlation charts published by Suc and Zagwijn (1983).

The Ruscinian shrew faunules in Europe are often

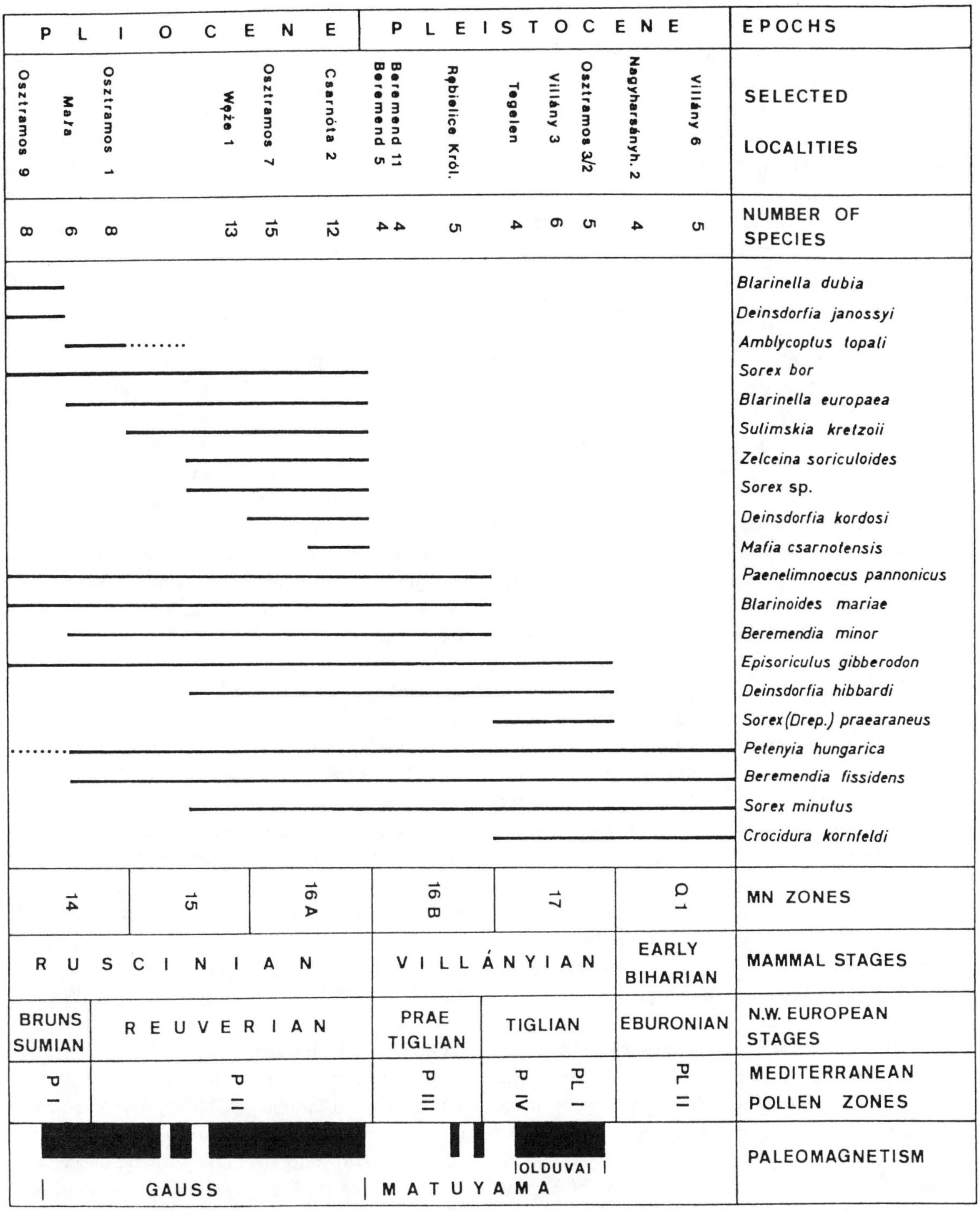

EPOCHS	PLIOCENE							PLEISTOCENE						
SELECTED LOCALITIES	Osztramos 9	Mała	Osztramos 1	Węże 1	Osztramos 7	Csarnóta 2	Beremend 11	Beremend 5	Rębielice Król.	Tegelen	Villány 3	Osztramos 3/2	Nagyharsányh. 2	Villány 6
NUMBER OF SPECIES	8	6	8	13	15	12	4	4	5	4	6	5	4	5

Blarinella dubia
Deinsdorfia janossyi
Amblycoptus topali
Sorex bor
Blarinella europaea
Sulimskia kretzoii
Zelceina soriculoides
Sorex sp.
Deinsdorfia kordosi
Mafia csarnotensis
Paenelimnoecus pannonicus
Blarinoides mariae
Beremendia minor
Episoriculus gibberodon
Deinsdorfia hibbardi
Sorex (Drep.) praearaneus
Petenyia hungarica
Beremendia fissidens
Sorex minutus
Crocidura kornfeldi

MN ZONES: 14 | 15 | 16 A | 16 B | 17 | Q 1

MAMMAL STAGES: RUSCINIAN | VILLÁNYIAN | EARLY BIHARIAN

N.W. EUROPEAN STAGES: BRUNSSUMIAN | REUVERIAN | PRAE TIGLIAN | TIGLIAN | EBURONIAN

MEDITERRANEAN POLLEN ZONES: P I | P II | P III | P IV | PL I | PL II

PALEOMAGNETISM: GAUSS | MATUYAMA | OLDUVAI

Fig. 10.2. Diversity of several European Pliocene and Pleistocene shrew faunules. Indicated are the number of species per locality, the ranges of individual species as derived from Reumer (1984), and several stratigraphic correlations (partly after Suc and Zagwijn, 1983). Owing to unrepresented or indeterminate species, the number of species per locality does not always correspond to the number derived by counting the species' ranges. (Reumer, 1985)

very rich in taxa. In the Early Ruscinian localities, we find six to eight different species. The Late Ruscinian localities Węże 1 (in Poland) and Osztramos 7 and Csarnóta 2 (both in Hungary) have all yielded at least a dozen species (thirteen, fifteen, and twelve, respectively). The Villányian faunules, however, are much poorer. Beremend 5 and 11 (Hungary) both contain four species, as does Tegelen (The Netherlands, stratotype of the Tiglian). Rębielice Królewskie (Poland) contains five species: Villány 3 (Hungary), six species; and Osztramos 3/2 (also Hungary), five species (Jánossy, 1979; Kowalski, 1960; Reumer, 1984; Sulimski, 1962; Sulimski, 1962; Sulimski et al., 1979).

Apparently, a considerable impoverishment of the Soricidae occurred at the Ruscinian-Villányian boundary in central Europe some 2.5–2.4 myr ago. The reduction in biodiversity coincides with a number of extinctions. Figure 10.2 shows the ranges of the shrew taxa studied in the framework of a previous paper (Reumer, 1984), arranged according to their LADs. At the Ruscinian-Villányian boundary, three genera disappeared: *Sulimskia, Zelceina,* and *Mafia.* A fourth genus, *Blarinella,* disappeared from Europe. Two more genera, *Blarinoides* and *Paenelimnoecus,* disappeared during the Villányian. These disappearances were hardly balanced by new developments or immigrations. The void was filled partly by the evolution of *Sorex* (*Drepanosorex*) spp. and by the immigration of *Crocidura,* but the species richness of the Ruscinian was not restored. The impoverishment of the Soricidae around 2.4 myr ago is interpreted (Reumer, 1985) as having been caused by the worldwide climatic deterioration (decreasing temperatures and increasing aridity) that took place between 2.7 and 2.3 myr ago (Ruddiman et al., 1989; Bloemendal and deMenocal, 1989; references in Reumer, 1985; see also Bonnefille, Dupont and Leroy, Hooghiemstra, Shackleton, Vrba, and Wesselman, this vol.).

Biogeography of Crocidura. The spread of the crocidurine genus *Crocidura* through Europe shows northward movement throughout the warmer parts of the Pleistocene and a southward retreat during the last glacial period. As noted above, the oldest record of *Crocidura* in Europe is from the Greek island of Rhodes, close to the Turkish mainland (Apolakkia; see Van de Weerd et al., 1982).

By the Late Villányian, *Crocidura* had reached the Carpathian Basin. It is known from Villány 3 and Osztramos 3 (which yielded the species *C. kornfeldi* Kormos, 1934; see Reumer, 1984). In the Middle Biharian, the genus is present in northern Italy (*C. zorzii,* from the Cava Sud, near Verona; Pasa, 1947). A long time elapses before further movements are recorded, apparently because of the less favorable climatic situation in northwestern and central Europe. We then have two records from England, both from sediments that are correlated to the Ipswichian-Eemian interglacial. The first is from the Vivian Vault of Tornewton Cave, Devon (*Crocidura* sp.; Rzebik, 1968). The second is from Aveley, Essex, a smaller species described as *C.* cf. *suaveolens* (Stuart, 1976).

The onset of the Weichselian (last) glacial period caused the apparent retreat of the genus from England; it is still absent from the British Isles. On the European continent, the genus *Crocidura* does not live at latitudes above ca. 53°N (Reumer, 1984). This restriction in its biogeography has a physiological explanation (see Vogel, 1980): this genus avoids low temperatures. The present-day climate of southern England certainly favors the presence of *Crocidura,* but the climatic improvement of the Holocene also flooded the North Sea Basin, thereby blocking a new invasion into England. Figure 10.3 summarizes the European biogeography of the genus.

Biogeography of Episoriculus and Nesiotites. A comparable situation is recorded for some Soriculini. In the Ruscinian, *Episoriculus gibberodon* was widespread in Europe. It is known from many localities—among them Podlesice, in Poland (Kowalski, 1956; Rzebik-Kowalska, 1981); Osztramos 1 and 7 and Csarnóta 2, in Hungary (Reumer, 1984); and Maritsa and Apolakkia, in Rhodes, Greece (De Bruijn et al., 1970; Van de Weerd et al., 1982). It is, however, absent from Late Ruscinian Polish faunas. Węże 1 contains an abundance of Soricidae, but *Episoriculus* is lacking (Sulimski, 1959, 1962; Rzebik-Kowalska, 1981). The youngest Polish locality from which the genus is known is the early Ruscinian Podlesice. In the Hungarian-Czech-Slovakian area, *Episoriculus* persists until the beginning of the Biharian. The youngest localities from which the genus is recorded in this area are Plešivec (Fejfar, 1961) and Osztramos 14 (Jánossy, 1979). After the Early Biharian, the genus disappeared from the Carpathian Basin, but it persists in Italy and in some western Mediterranean islands. Two of the Middle Biharian localities in Italy that contain *Episoriculus* (the species *E. castellarini*) are the Cava Sud, near Verona (Pasa, 1947), and Monte Peglia (Van der Meulen, 1973).

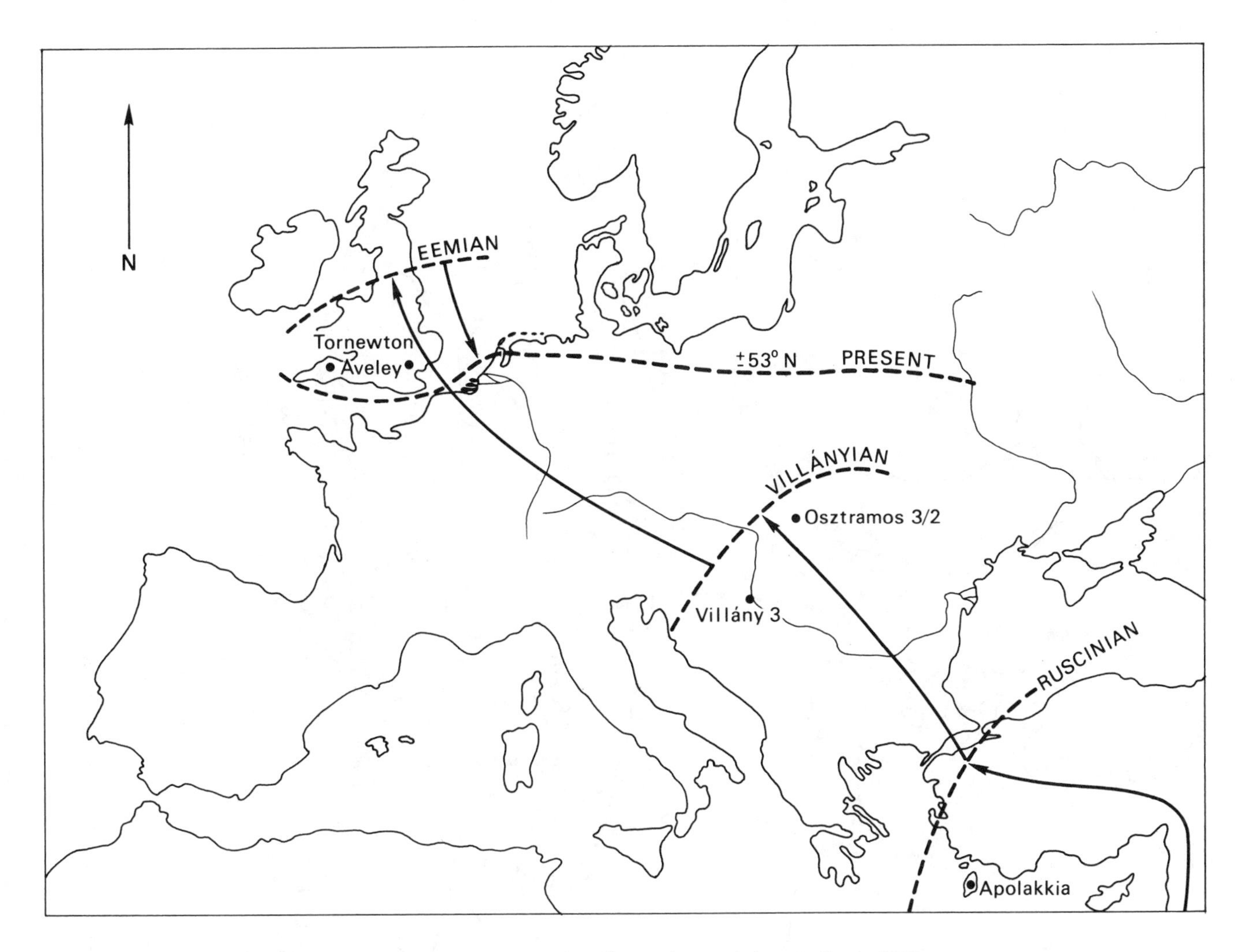

Fig. 10.3. Reconstruction of the migration of *Crocidura* through Europe during the Late Pliocene and Pleistocene. (Reumer, 1984)

After the Middle Biharian, the genus disappeared from the European continent. It had, however, a direct descendant: the genus *Nesiotites* Bate, 1944, which lived in four Mediterranean islands. Known species are *N. corsicanus* Bate, 1944, from Corsica; *N. similis* (Hensel, 1855) from Sardinia; *N. ponsi* Reumer, 1979; *N. meloussae* Pons and Moyà, 1980; and *N. hidalgo* Bate, 1944, from the Baleares (Majorca and Menorca). Although biostratigraphic correlations of island faunules are normally rather difficult to execute, it is assumed that these species were present in the islands from the Late Pliocene onward. They lived on into the Holocene, to become extinct after the introduction by humans of predators and competitors (Reumer, 1984).

The biogeography of *Episoriculus* is summarized in figure 10.4, after Reumer (1984). It is assumed that the southward retreat of this taxon is caused by the ongoing climatic deterioration of the latest Tertiary and the Quaternary.

Discussion

The use of paleoclimatic curves derived from research on Spanish rodent faunas in a study on worldwide evolutionary phenomena within a different group of mammals (*in casu* Soricidae) may seem somewhat risky. The general climatic inferences that were derived by Van der Meulen and Daams (1992), however, are found in various other regions and by using paleontological data sets other than rodent faunas. The overwhelming amount of paleoclimatic data accumulated in this volume testifies to this statement. Certainly, many more data apply to paleotem-

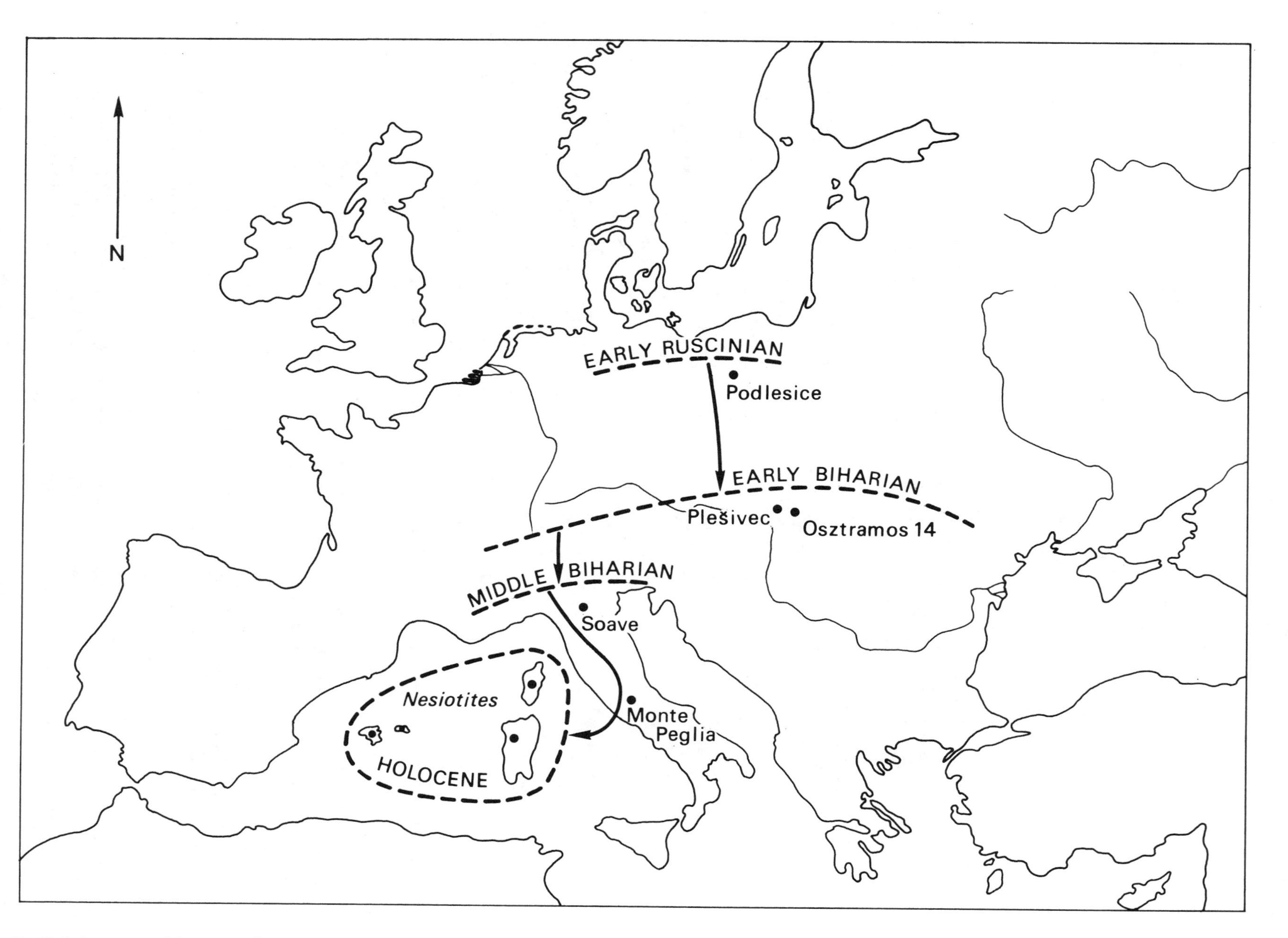

Fig. 10.4. Reconstruction of the migration of *Episoriculus* through Europe during the Pliocene and Pleistocene. (Reumer, 1984)

perature than to paleohumidity, although both parameters are to some degree linked.

Both the Vallesian and the Ruscinian show humidity peaks that are well documented. A period of increasing humidity, starting around 12 myr and lasting into the Vallesian, was reported by Chamley et al. (1986), using clay minerals, stable isotopes, and planktonic foraminifera from Sicily. Van der Meulen and Van Kolfschoten (1986), who studied entire mammal faunas, reported a wet period at the Early Ruscinian in Greece. These are only two examples of a considerable literature that confirms the general trends observed by Van der Meulen and Daams (1992).

The conclusion that humid paleoclimates per se stimulated shrew proliferation is interesting but needs to be considered with some caution. It cannot be excluded that although such humid conditions coincide with general abundance and geographic spreading of shrews, the actual speciation events took place before that. Cooler and more arid periods preceding the warmer and humid conditions could have forced the shrews into smaller, allopatric populations that were under climatic pressure. This situation would generally be considered to be conducive to speciation.

Conclusion

The present study, based partly on previous studies (Reumer 1984, 1985, 1989), shows that the composition of soricid faunas can be taken as a measure of ecological adequacy of the environment for shrews. This conclusion can be used in both directions. In the first place, an abundance of Soricidae in an association of fossil mammals can be taken as evidence of a relatively warm and humid paleoclimate. In the second place, if a certain area is known to have had a relatively warm and humid paleoclimate during a given period, one should not be surprised to find a high biodiversity of Soricidae in the fossil mammal association.

Changes in the composition of an association of fossil Soricidae can be interpreted in a paleoclimatic sense. Observed increasing biodiversity should have been caused by a paleoclimatic change toward warmer and/or more humid situations. A drop in soricid diversity should have been caused by increasing aridity and/or decreasing temperatures.

This general conclusion implies that Soricidae can be used as one of the biotic parameters in paleoclimatic interpretations. They warrant more attention than they usually get.

Summary

Shrews (Soricidae) are very small mammals that are maladapted to extremely cold or extremely arid environments. The hypothesis is tested that shrew evolution is favored by warm and/or humid paleoclimates. When the ranges of eleven distinct groups of shrews are plotted against a paleoecological background, it is shown that most events (First Appearances and Last Appearances) took place when humid paleoclimates prevailed. The hypothesis is confirmed as far as humidity is concerned. Enhanced proliferation is shown to have occurred ca. 19–20 myr ago, between 11 and 9 myr ago, and between 6 and 2.5 myr ago. Climatic deterioration around the Pliocene-Pleistocene boundary (ca. 2.7–2.3 myr ago) caused the European shrew fauna to become severely impoverished, thus demonstrating the effect of temperature as well. Finally, during the Pleistocene, several migrations can be observed that are related to the movement of climatic belts as a result of temperature fluctuations.

Acknowledgments

The author wishes to thank Dr. John de Vos (Leiden), Dr. Albert van der Meulen, and Mr. Jan van Dam (Utrecht) for their stimulating discussions. Professor Remmert Daams (Madrid) allowed me to use unpublished material on paleohumidity from Spain. John de Vos read an early version of this paper and provided helpful suggestions. The discussions during the conference at Airlie, Virginia, with Elisabeth Vrba and the other participants proved very rewarding; I wish to thank them all. Finally, Elisabeth Vrba is to be thanked for her linguistic, editorial, and scientific suggestions.

References

Baudelot, S. 1972. Etude des chiroptères, insectivores et rongeurs du Miocène de Sansan (Gers). Ph.D. diss., University Paul Sabatier, Toulouse.

Bloemendal, J., and deMenocal, P. 1989. Evidence for a change in the periodicity of tropical climate cycles at 2.4 Ma from whole-core magnetic susceptibility measurements. *Nature* 342:897–900.

Calvo, J. P. et al. 1993. Up-to-date Spanish continental Neogene synthesis and a paleoclimatic interpretation. *Revista de la Sociedad Geológica de España* 6(3–4):29–40.

Chamley, H., Meulenkamp, J. E., Zachariasse, W. J., and van der Zwaan, G. J. 1986. Middle to Late Miocene marine ecostratigraphy: Clay minerals, planktonic foraminifera and stable isotopes from Sicily. *Oceanologica Acta* 9(3):227–238.

Churchfield, S. 1990. *The natural history of shrews.* Cornell University Press, Ithaca.

Daams, R., Freudenthal, M., and Van der Meulen, A. J. 1988. Ecostratigraphy of micromammal faunas from the Neogene of Spain. In *Biostratigraphy and paleoecology of the Neogene micromammalian faunas from the Calatayud-Teruel Basin (Spain)* (ed. M. Freudenthal). Scripta Geologica, special issue, 1:287–302, Leiden.

De Bruijn, H., Daams, R., Daxner-Höck, G., Fahlbusch, V., Ginsburg, L., Mein, P., and Morales, J. 1992. Report of the RCMNS working group on fossil mammals, Reisensburg 1990. *Newsletter on Stratigraphy* 26(2/3):65–118.

De Bruijn, H., Dawson, M. R., and Mein, P. 1970. Upper Pliocene Rodentia, Lagomorpha and Insectivora (Mammalia) from the Isle of Rhodes (Greece). *Proceedings Koninklijke Nederlandse Academie van Wetenschappen,* ser. B, 73:535–584.

Doben-Florin, U. 1964. Die spitzmäuse aus dem Alt-Burdigalium von Wintershof-West bei Eichstätt in Bayern. *Abhandlungen Bayerische Akademie der Wissenschaften, Math.-naturw. Klasse,* n.s., 117:1–82.

Doukas, C. S., van den Hoek Ostende, L. H., Theocharopoulos, C., and Reumer, J. W. F. 1995. The vertebrate locality Maramena (Macedonia, Greece) at the Turolian-Ruscinian boundary (Neogene). 5. Insectivora. *Münchner geowissenschaftliche Abhandlungen,* in press.

Fejfar, O. 1961. Review of Quaternary Vertebrata in Czechoslovakia. *Czwartorzęd Europy Środkowej i Wschodniej, Część* 1(34):109–118.

Feldhamer, G. A., Klann, R. S., Gerard, A. S., Driskell, A. C. 1993. Habitat partitioning, body size, and timing of parturition in pygmy shrews and associated soricids. *Journal of Mammalogy* 74(2):403–411.

Gibert, J. 1975. New insectivores from the Miocene of Spain. *Proceedings Koninklijke Nederlandse Academie van Wetenschappen,* ser. B, 78:108–133.

Hibbard, C. 1950. Mammals of the Rexroad formation from Fox Canyon, Meade County, Kansas. *Michigan University Museum Paleontological Contributions* 8(6):113–192.

Jánossy, D. 1979. *A Magyarországi Pleisztocén tagolása gerinces faunák alapján.* Akadémiai Kiadó, Budapest.

Kowalski, K. 1956. Insectivores, bats and rodents from the Early Pleistocene bone breccia of Podlesice near Kroczyce (Poland). *Acta Palaeontologica Polonica* 1(4):331–398.

———. 1960. Pliocene insectivores and rodents from Rębielice Królewskie (Poland). *Acta Zoologica Cracoviense* 5(5):155–201.

Mein, P. 1990. Updating of MN zones. In *European Neogene mammal chronology,* pp. 73–90 (ed. E. H. Lindsay, V. Fahlbusch, and P. Mein). Plenum Press, New York.

Pasa, A. 1947. I mammiferi di alcune antiche brecce veronesi. *Memoria Museo Civico di Storia Naturale Verona* 1:1–111.

Repenning, C. A. 1967. Subfamilies and genera of the Soricidae. *U.S. Geological Survey Professional Paper* 565:1–74.

———. 1987. Biochronology of the Microtine rodents of the United States. In *Cenozoic mammals of North America: Geochronology and biostratigraphy,* pp. 236–268 (ed. M. O. Woodburne). University of California Press, Berkeley.

Reumer, J. W. F. 1984. Ruscinian and Early Pleistocene Soricidae (Insectivora, Mammalia) from Tegelen (the Netherlands) and Hungary. *Scripta Geologica* 73:1–173.

———. 1985. The paleoecology of Soricidae (Insectivora, Mammalia) and its application to the debate on the Plio-Pleistocene boundary. *Revue de Paléobiologie* 4(2):211–214.

———. 1987. Redefinition of the Soricidae and the Heterosoricidae (Insectivora, Mammalia), with the description of the Crocidosoricinae, a new subfamily of Soricidae. *Revue de Paléobiologie* 6(2):189–192.

———. 1989. Speciation and evolution in the Soricidae (Mammalia: Insectivora) in relation with the paleoclimate. *Revue Suisse de Zoologie* 96(1):81–90.

———. 1992. The taxonomical position of the genus *Paenelimnoecus* Baudelot, 1972 (Mammalia: Soricidae): A resurrection of the subfamily Allosoricinae. *Journal of Vertebrate Paleontology* 12(1):103–106.

———. 1994. Phylogeny and distribution of the Crocidosoricinae. In *Advances in the biology of shrews,* pp. 345–356 (ed. J. F. Merritt, G. L. Kirkland, and R. K. Rose). Carnegie Museum of Natural History, special publ. 18, Pittsburgh.

Ruddiman, W. F., Sarnthein, M., Backman, J., Baldauf, J. G., Curry, W., Dupont, L. M., Janecek, T., Pokras, E. M., Raymo, M. E., Stabell, B., Stein, R., and Tiedemann, R. 1989. Late Miocene to Pleistocene evolution of climate in Africa and the low-altitude Atlantic: Overview of Leg 108 results. In *Proceedings of the Ocean Drilling Program, scientific results,* vol. 108, pp. 463–484.

Rzebik, B. 1968. *Crocidura* Wagler and other Insectivora (Mammalia) from the Quaternary deposits of Tornewton Cave in England. *Acta Zoologica Cracoviense* 13(10):251–263.

Rzebik-Kowalska, B. 1981. The Pliocene and Pleistocene Insectivora (Mammalia) of Poland. IV. Soricidae: *Neomysorex* n. gen. and *Episoriculus* Ellerman et Morrison-Scott, 1951. *Acta Zoologica Cracoviense* 25(8):227–250.

Schmidt-Kittler, N. (ed.). 1987. International symposium on mammalian biostratigraphy and paleoecology of the European Paleogene. *Münchner Geowissenschaftliche Abhandlungen* A 10:1–312.

Steininger, F. F., Bernor, R. L., and Fahlbusch, V. 1990. European Neogene marine / continental chronological correlations. In *European Neogene mammal chronology,* pp. 15–46. (ed. E. H. Lindsay, V. Fahlbusch, and P. Mein). Plenum Press, New York.

Stuart, A. J. 1976. The history of the mammal fauna during the Ipswichian / last interglacial in England. *Philosophical Transactions of the Royal Society,* London, ser. B, 276(945):221–250.

Suc, J.-P., and Zagwijn, W. H. 1983. Plio-Pleistocene correlations between the northwestern Mediterranean region and northwestern Europe according to recent biostratigraphic and paleoclimatic data. *Boreas* 12:153–166.

Sulimski, A. 1959. Pliocene insectivores from Węże. *Acta Paleontologica Polonica* 4(2):119–177.

———. 1962. Supplementary studies on the insectivores from Węże 1 (Poland). *Acta Paleontologica Polonica* 7(3–4):441–502.

Sulimski, A., Szynkiewicz, A., and Woloszyn, B. 1979. The Middle Pliocene micromammals from Central Poland. *Acta Paleontologica Polonica* 24(3):377–403.

Tedford, R. H., Skinner, M. F., Fields, R. W., Rensberger, J. M.,

Whistler, D. P., Galusha, T., Taylor, B. E., Macdonald, J. R., and Webb, S. D. 1987. Faunal succession and biochronology of the Arikareean through Hemphillian interval (Late Oligocene through earliest Pliocene epochs) in North America. In *Cenozoic mammals of North America: Geochronology and biostratigraphy,* pp. 153–210 (ed. M. O. Woodburne). University of California Press, Berkeley.

Terzea, E., and Jurcsák, T. 1976. Fauna de Mammifères de Betfia-XIII (Bihor, Roumanie) et son âge géologique. *Travaux de l'Institut de Spéléologie "Emile Racovitza"* 15:175–185.

Van de Weerd, A., Reumer, J. W. F., and de Vos, J. 1982. Pliocene mammals from the Apolakkia Formation (Rhodes, Greece). *Proceedings Koninklijke Nederlandse Academie van Wetenschappen,* ser. B, 85(1):89–112.

Van der Meulen, A. J. 1973. Middle Pleistocene smaller mammals from the Monte Peglia (Orvieto, Italy), with special reference to the phylogeny of *Microtus* (Arvicolidae, Rodentia). *Quaternaria* 17:1–144.

Van der Meulen, A. J., and Daams, R. 1992. Evolution of Early-Middle Miocene rodent faunas in relation to long term palaeoenvironmental changes. *Palaeogeography, Palaeoclimatology, Palaeoecology* 93:227–253.

Van der Meulen, A. J., and van Kolfschoten, T. 1986. Review of the Late Turolian to Early Biharian mammal faunas from Greece and Turkey. *Memoria Societa Geologica Italiana* 31:201–211.

Vogel, P. 1980. Metabolic levels and biological strategies in shrews. In *Comparative physiology: Primitive mammals,* pp. 170–180 (ed. K. Schmidt-Nielsen, L. Bolis, and C. R. Taylor). Cambridge University Press, Cambridge.

Wilson, R. W. 1960. Early Miocene rodents and insectivores from northeastern Colorado. Kansas University Publications, Paleontological Contributions, *Vertebrata* 7:1–92.

Woodburne, M. O. (ed.). 1987. *Cenozoic mammals of North America: Geochronology and biostratigraphy.* University of California Press, Berkeley.

Ziegler, R. 1989. Heterosoricidae and Soricidae (Insectivora, Mammalia) aus dem Oberoligozän und Untermiozän Süddeutschlands. *Stuttgarter Beiträge zur Naturkunde,* ser. B, 154:1–73.

Chapter 11

What Metapodial Morphometry Has to Say about Some Miocene Hipparions

Véra Eisenmann

Several myths have crystallized concerning the origins, migrations, diversifications, and adaptations of the Old World tridactyl horses named hipparions. Although new data have produced evidence to counter some of these myths, a few are still broadly accepted.

Since the late 1970s (Skinner and MacFadden, 1977), an enormous amount of work has been devoted to the morphology, systematics, and phylogeny of hipparions, most of which has been based on skull characters, particularly the pit situated in front of the orbit known as the preorbital fossa. Hipparions without any fossa, as well as hipparions not represented by skulls, have been more or less bypassed by the main flow of studies.

This chapter discusses some myths and looks at some rarely mentioned hipparions, taking the other end of the animal—not the skull but the foot—as the leading element of comparison. Appendixes 11.1–11.5 give metrical data for the forms discussed, either specimen by specimen or after statistical elaboration when the number of specimens is large. The system of measurements is illustrated in figures 11.1–11.5. Comparisons are made using ratio diagrams (Simpson, 1941) in which the reference line represents *H. mediterraneum* from Koufos (1987a). A more detailed study of the Pikermi material, however, convinced me that the fossils of two species were grouped under that name: *H. mediterraneum sensu stricto* and *H. dietrichi*. For this reason, the ratio diagram for what I believe are *H. mediterraneum* metacarpals (fig. 11.7) is not identical to the Koufos's *H. mediterraneum* reference line.

Naming morphotypes of metapodials that are not associated with skulls is rather like attempting a taxonomy of footprints. In the present chapter, in some cases I am fairly certain that the metapodials belong to a particular species, whereas in other cases my evaluation is more of a guess. Some of the resemblances may be due to parallelism. A study restricted to metapodials also precludes the application of formal taxonomy. Indeed, it would be premature to put species in synonymy just because the metapodials belong to the same morphotype. The specific or subspecific names used should therefore be considered as identifying terms for morphotypes, not as taxonomically valid names. Each subspecific name in the appendixes is included to remind the reader of the name previously used for the hipparion of the relevant locality. The biochronology is based on Mein (1990).

Monotypy of Vallesian Hipparions

Following the studies of Forsten (1968), there has been a general tendency to contrast the Vallesian large monotypic *H. primigenium*, with rather robust limb bones, with the diverse Turolian hipparions, among which figure small (*H. matthewi*), slender (*H. mediterraneum*), or very slender (*H. dietrichi*) forms.

It appears, however, that even in the Lower Vallesian (MN 9 zone of Mein) there are at least three morphotypes: 1) the very small *H. minus* (Pavlow, 1890) from Sebastopol; 2) the usually robust *H. primigenium* (Eppelsheim, Höwenegg, Germany; Can Llobateres, Spain; Yassiören, Turkey) possibly including very large variants (Nombrevilla, Spain), more slender variants (Rudabanya, Hungary; Gritzev, Russia), and variants with a large anteroposterior proximal diameter (Bou Hanifia, Algeria; Yassiören, Turkey); and 3) the slender

ISBN 0–300–06348–2.

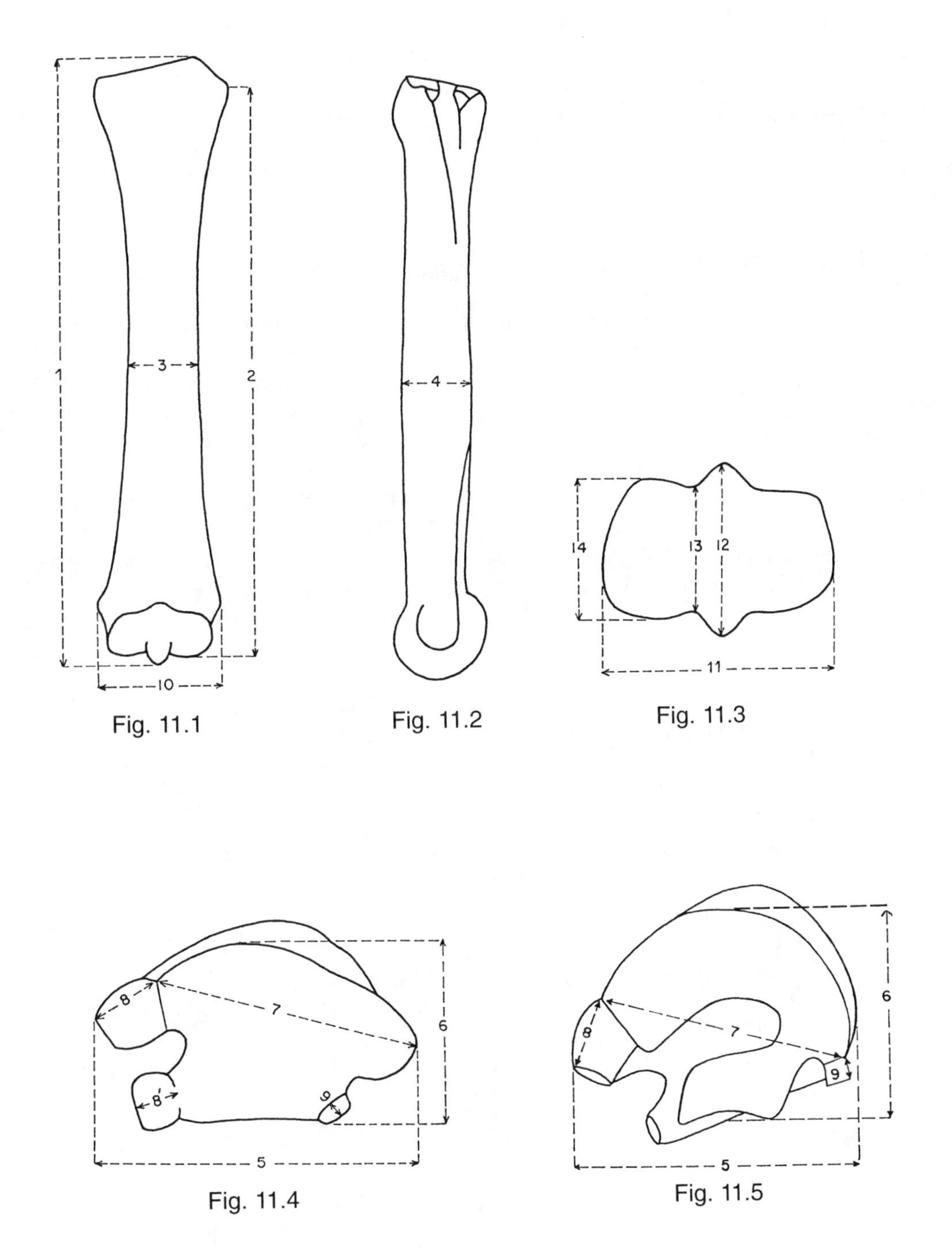

Figs. 11.1 to 11.5. System of measurements of equid metapodials. The names of the measurements are given in appendixes 11.1 through 11.5.

H. sebastopolitanum, from Russia (Borisiak, 1914), also present at Yassiören (Turkey), Can Llobateres, Los Valles de Fuentidueña, and El Lugarejo (Spain). *H. melendezi* (Alberdi, 1974) described at Los Valles de Fuentidueña and El Lugarejo is probably a younger synonym of *H. sebastopolitanum. H. melendezi* was not mentioned at Can Llobateres, but at least one metatarsal probably belongs to this form.

During the MN 10 zone, *H. primigenium* is probably still present at Ravin de la Pluie, Greece, and Montredon, France (*H. depereti* of Sondaar, 1974, and Eisenmann, 1988), one variant or morphologically close species (*H. garedzicum,* Gabunia, 1959) is present at Udabno, Georgia, and two morphologically close species are present at Samburu, Kenya (Nakaya and Watabe, 1990). *H. sebastopolitanum* is still present at Masia del Barbo, Spain (Sondaar, 1974), and a new small species, *H. macedonicum* (Koufos, 1984), appears at Ravin de la Pluie and possibly at Montredon (Eisenmann, 1988).

Consequently, the presence during the Vallesian of at least four, and possibly seven, metapodial morphs of hipparion (fig. 11.6, table 11.1), some of which may be associated in the same site, does not support the classical concept of Vallesian monotypy. Neither does it support a contrast between "large-robust Vallesian" forms and "small-slender Turolian" forms. It is true that the most slender forms do not appear before the Turolian and that there are more morphotypes in the Turolian than in the Vallesian, but the diversity pattern based on metapodial morphology tends to minimize the usually accepted differences between Vallesian and Turolian hipparions and the classical notion of explosive diversification during the Turolian.

Turolian Diversity

I distinguish at least nine metacarpal morphotypes (figs. 11.7, 11.8, 11.9, table 11.2); if two, probably endemic Spanish hipparions, *H. periafricanum* and *H. gromovae* from El Arquillo (MN 13), are included, the number of morphotypes becomes eleven.

The *H. brachypus* morphotype is similar to, and sometimes difficult to distinguish from, *H. primigenium* (figs. 11.6 and 11.7). It is the only Vallesian morphotype that can be traced clearly into the Turolian. The metacarpal of *H. minus* is larger than that of *H. periafricanum* and flatter than that of *H. matthewi* (figs. 11.6, 11.8, 11.9). The metacarpal of *H. macedonicum* is about the same size

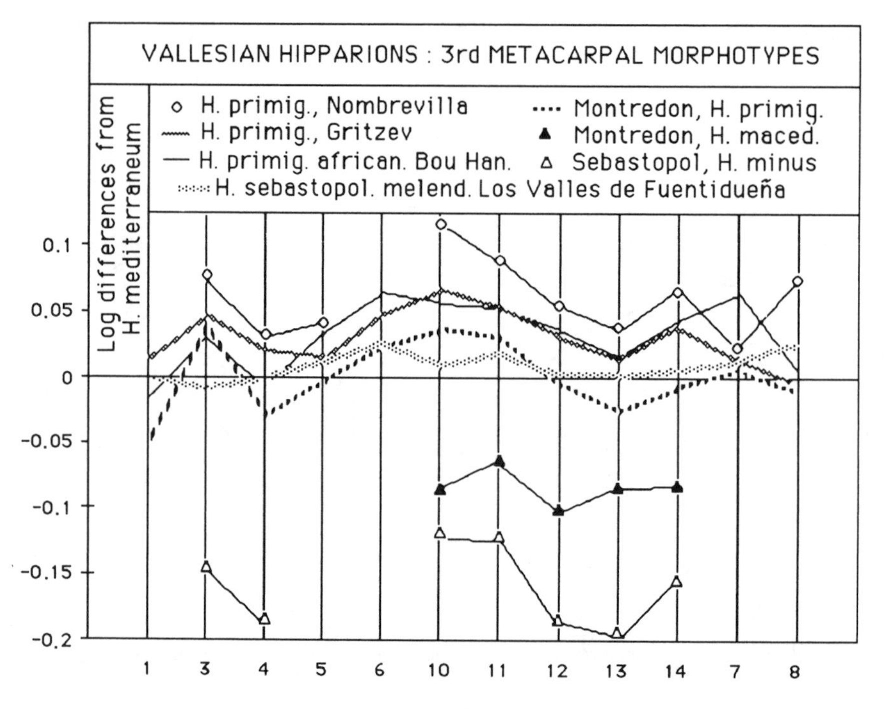

Fig. 11.6. Ratio diagrams of Vallesian hipparions metacarpals: H. = *Hipparion;* primig. = *primigenium;* african. = *africanum;* maced. = *macedonicum;* sebastopol. = *sebastopolitanum;* melend. = *melendezi;* Bou Han. = Bou Hanifia = Oued el Hammam. The names of measurements 1 through 14 are given in appendixes 11.1 through 11.5.

Table 11.1. Occurrences of Different Hipparion Metapodial Morphotypes during the Vallesian

	H. primigenium						
	1 Big	2 Normal	3 Slender	4 Deep	*H. sebast.*	*H. maced.*	*H. minus*
	Samburu		Samburu				
		Montredon				Montredon	
MN 10					Masia Barbo		
	Udabno	RPl				RPl	
		Can Llobat.					
				Bou Hanifia	Sebastopol		Sebastopol
MN 9		Yassiören	Rudabanya	Yassiören	Yassiören		
	Nombrevilla	Höwenegg	Gritzev		El Lugarejo		
		Eppelsheim			Los Valles F.		

Note: The biochronology is taken form Mein (1990). H. = *Hipparion*; sebast. = *sebastopolitanum*; maced. = *macedonicum*; RPl = Ravin de la Pluie; Can Llobat. = Can Llobateres; Los Valles F. = Los Valles de Fuentidueña.

as that of *H. matthewi* but flatter (figs. 11.6 and 11.8). Contrary to Forsten (1982), I do not think that the hipparions from Los Valles de Fuentidueña (MN 9) and Masia del Barbo (MN 10) are conspecific with the species from Concud (MN 12). In my opinion, the first two do belong to the same morphotype (*H. sebastopolitanum*), but *H. concudense* is more robust and lacks the proximal depth of the Spanish *H. sebastopolitanum*.

According to data published by Forsten (1968), *H. plocodus* is characterized by relatively small epiphyses (measurements 5, 6, 10 to 14; the numbered measurements referred to here and subsequently in the text are given in the appendixes) when compared to the diaphysis length (measurement 1). The diaphysis breadth is unknown. I tentatively refer to this morphotype (fig. 11.8) two specimens (FAM 23054 and FAM 23054e) from Samos 5 (Bernor and Tobien, 1989, table 5), although they would be much younger (MN 13 instead of MN 11, according to Qiu et al., 1987). The same figure shows how the *H. plocodus* morphotype differs from the other small hipparion metacarpals from Samos (Bernor and Tobien, 1989, table 5), which may be referred to *H. matthewi*.

Figures 11.8 and 11.9 show the differences between the small hipparions of Greece and Spain. Although *H. gromovae* (fig. 11.9) has about the same breadths as *H. matthewi*, it is much shorter and therefore more robust. *H. periafricanum* (fig. 11.9) is notably smaller.

On the whole, the Turolian hipparions are more slender than the Vallesian forms, sometimes very much so, as in the case of *H. dietrichi* and *H. matthewi* (fig. 11.8). They are often deeper, either at the diaphysis level (measurement 4), at the proximal end (measurement 6), or at the distal end (measurements 12, 13, and 14). Usually, the anterior articular facet for the second metacarpal is more developed (measurement 8).

Functional, Evolutionary, and Ecological Interpretations of Metapodial Morphology

The gracility of the diaphysis may be just a matter of diminished breadth (measurement 3) relative to the length (measurement 1), or it may be partly related to the deepening of the diaphysis (measurement 4) or to the lengthening of the bone (measurement 1) relative to the other segments of the limb. In the first case, the character would indicate a drier climate (Gromova, 1952); in the second, a posterior shifting of the lateral metapodials (see below); and in the third, an adaptation for running (Gregory, 1912; Osborn, 1929).

The shifting of the lateral metapodials from a lateral to a posterolateral position relative to the third metapodial can be considered as an evolutionary change toward functional monodactyly and, thus, as a better adaptation for running (Gromova, 1952). This shifting is usually accompanied by a deepening of the whole bone (measurements 4, 6, 12, 13, and 14) and an effacement of the distal supra-articular tuberosities. As a result, the preeminence of the supra-articular breadth (measurement 10) appears decreased relative to the articular breadth (measurement 11).

The development of the sagittal keel (measurement 12) is a character that enhances pendular movements of the limbs and, again, is a sign of better adaptation for

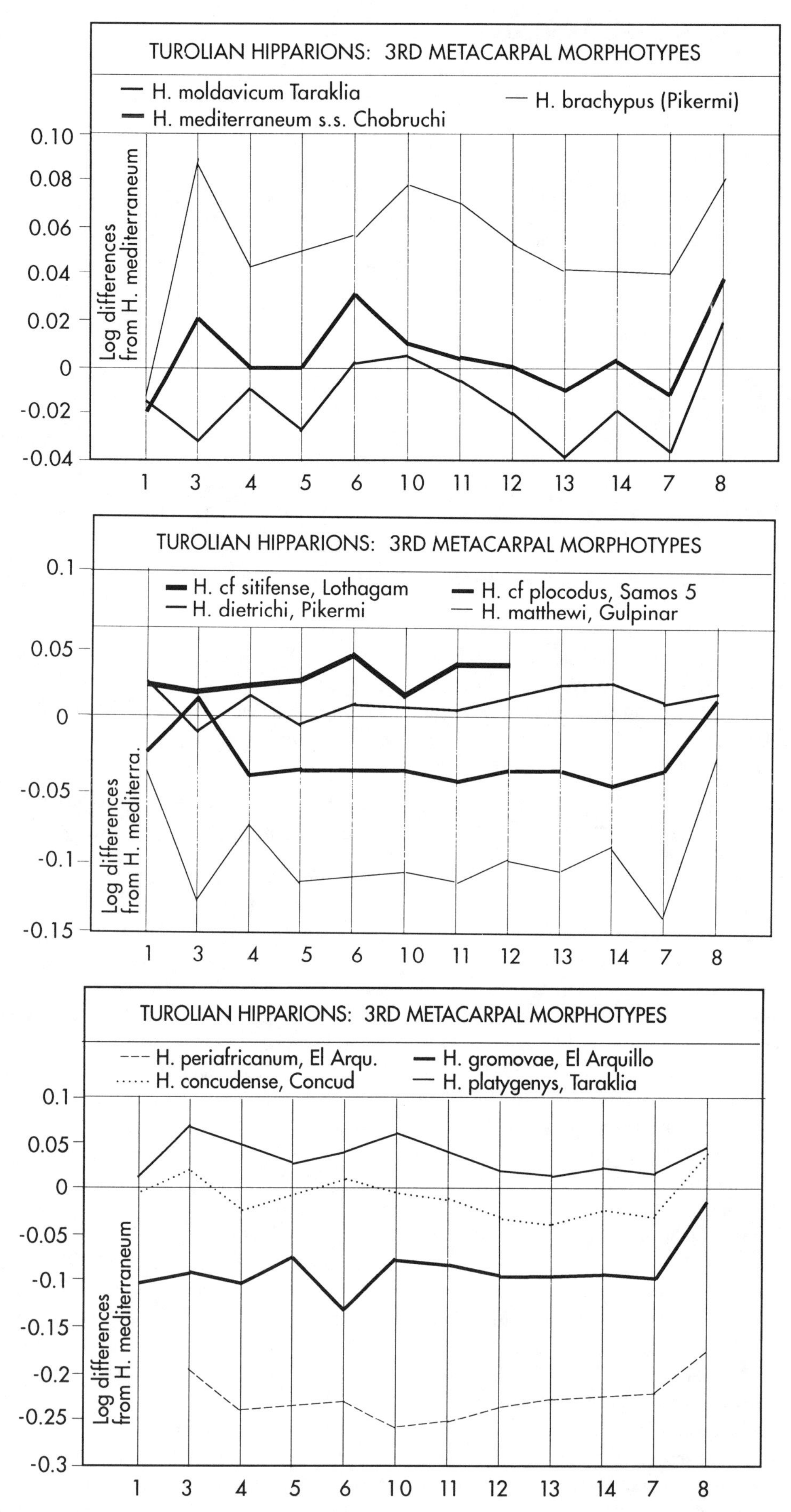

Figs. 11.7 to 11.9. Ratio diagrams of third metacarpals of Turolian hipparions.

Table 11.2. Occurrences of Different Hipparion Metapodial Morphotypes during the Turolian

	H. primig. *H. brachyp.*	*H. platyg.*	*H. medit.* ss	*H. concud.*	*H.* cf. *plocod.*	*H. mold*	*H.* cf. *sitif.*	*H. dietrichi*	*H. matthewi*	*H. gromovae*	*H. periaf.*
	Sahabi						Sahabi				
							Lothagam	Pavlodar			
MN 13	Baltavar					Dytiko	Lukeino	Venta Moro	Dytiko	Arquillo	Arquillo
							Baccinello				
	Polgardi				Samos 5		Kujal'nik		Samos 5		
	Pikermi		Pikermi			Tudorovo		Pikermi	Gulpinar		
		Saloniki	Lubéron					Saloniki			
MN 12			Chobruchi	Concud					Kinik		
		Taraklia				Taraklia					
								Samos Andr			
MN 11	Sumeg										
	Piera							Vathylakkos	Vathylakkos		
	Grebeniki			Grebeniki				Ravin Zo.	Ravin Zo.		

Note: H. = *Hipparion*; primig. = *primigenium*; brachyp. = *brachypus*; platyg. = *platygenys*; medit. = *mediterraneum*; concud. = *concudense*; plocod. = *plocodus*; mold. = *moldavicum*; sitif. = sitifense; periaf. = *periafricanum*; Andr = Andrianos Quarry; Ravin Zo. = Ravin des Zouaves.

running (Gromova, 1952; Staesche and Sondaar, 1979; Eisenmann and Sondaar, 1989). This development may be accompanied by reduction of the minimal distal depth (measurement 13), together with enlargement of the maximal condyle depth (measurement 14). Altogether, these changes diminish lateral mobility still more and create better conditions for anteroposterior movements.

Unfortunately, relative lengths of the limb bones, which can be compared easily in modern equids and even in some fossil species (Eisenmann, 1984), are difficult to compare in hipparions, because the long bones are usually broken or distorted. Other proportions can be compared, however, using ratio diagrams.

Figure 11.10 illustrates the morphological and functional differences between a Vallesian *H. primigenium* (Yassiören, Turkey) and the Turolian *H. brachypus* (Pikermi, Greece). Although the Turolian form is more robust (measurement 3 is large relative to measurement 1), in the Vallesian hipparion the diaphysis is flatter (measurement 4), the preeminence of the distal supra-

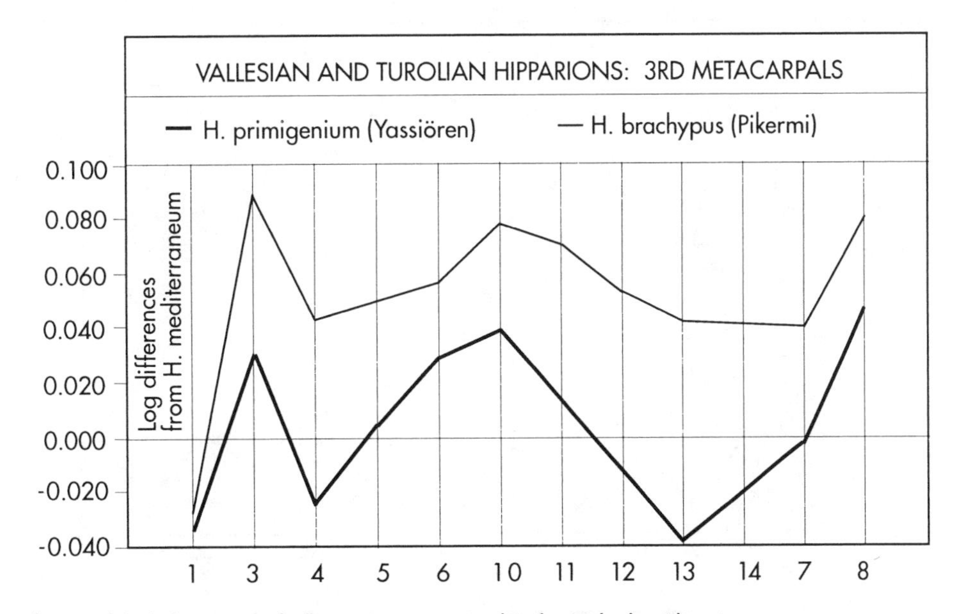

Fig. 11.10. Ratio diagrams of the third metacarpals of Vallesian (*H. primigenium*) and Turolian (*H. brachypus*) hipparions.

articular breadth (measurement 10) on the distal articular breadth (measure 11) is more pronounced, and the distal depths (12, 13, 14) are less developed. In the Vallesian form, these differences are probably linked to a lateral position of the lateral metapodials, whereas in the Turolian hipparion they have shifted to a more posterior position.

Figure 11.11 illustrates the differences between two Turolian forms, *H. brachypus* from Pikermi and *H. dietrichi* from Saloniki. The Saloniki hipparion is a good example of a slender (3), deep (4) diaphysis, with a deep proximal end (6) and two deep anteroposterior distal diameters (12 and 14) on each side of the condylar constriction (13). The distal breadths (10 and 11) are subequal and not very developed.

The anatomical characters allow only tenuous inferences of climate. Gracility may well be the only character that directly indicates climatic conditions (Gromova, 1952). The inference that deep metapodials are associated with savanna conditions is based on several probable, but not quite certain, suppositions: that the depth is the result of the shifting of the lateral metapodials; that this shift is beneficial to rapid locomotion; that rapid locomotion is beneficial only in open landscapes; and that open landscapes are dry. The assumption linking robust, flat metapodials to forest environments is valid only if 1) flatness is linked with lateral metapodials in a lateral position such that lateral movements are not altogether suppressed; 2) lateral mobility is more beneficial on uneven ground on which there are obstacles such as trees; and 3) trees grow mostly in humid conditions.

If all these suppositions are true, the study of the metapodials of some Miocene hipparions suggests the following conclusions (fig. 11.12):

1. The *H. primigenium* populations from the Vallesian (Yassiören, Nombrevilla, Höwenegg, Eppelsheim, Ravin de la Pluie, Montredon) and from the Turolian (Piera, Sumeg, Baltavar), noted as *H. primigenium* 1 and 2 in figure 11.12, were in general forest dwellers. Because of the flatness of the diaphysis, this is also probably true of the very small Vallesian *H. minus* and the small Vallesian *H. macedonicum.* But because we do not know the metapodial length, these species are not included in figure 11.12, where morphotypes are listed according to their robustness or gracility.
2. The slenderness of some Vallesian variants of the *H. primigenium* morphotype indicate drier conditions: Samburu, Gritzev, Rudabanya (*H. primigenium* 3 in fig. 11.12).
3. Some Vallesian variants of the same morphotype (Can Llobateres, Bou Hanifia, Yassiören) were better adapted to open landscapes (*H. primigenium* 4 in fig. 11.12).
4. This adaptation is even more pronounced in the Vallesian *H. sebastopolitanum* morphotype (Los Valles de Fuentidueña, El Lugarejo, Masia del Barbo,

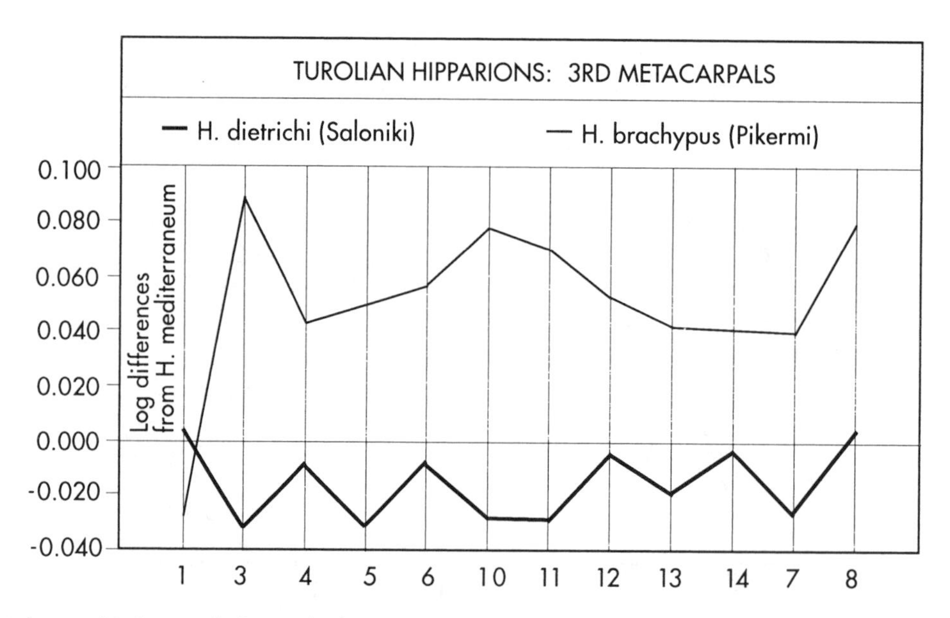

Fig. 11.11. Ratio diagrams of third metacarpals of two Turolian hipparions.

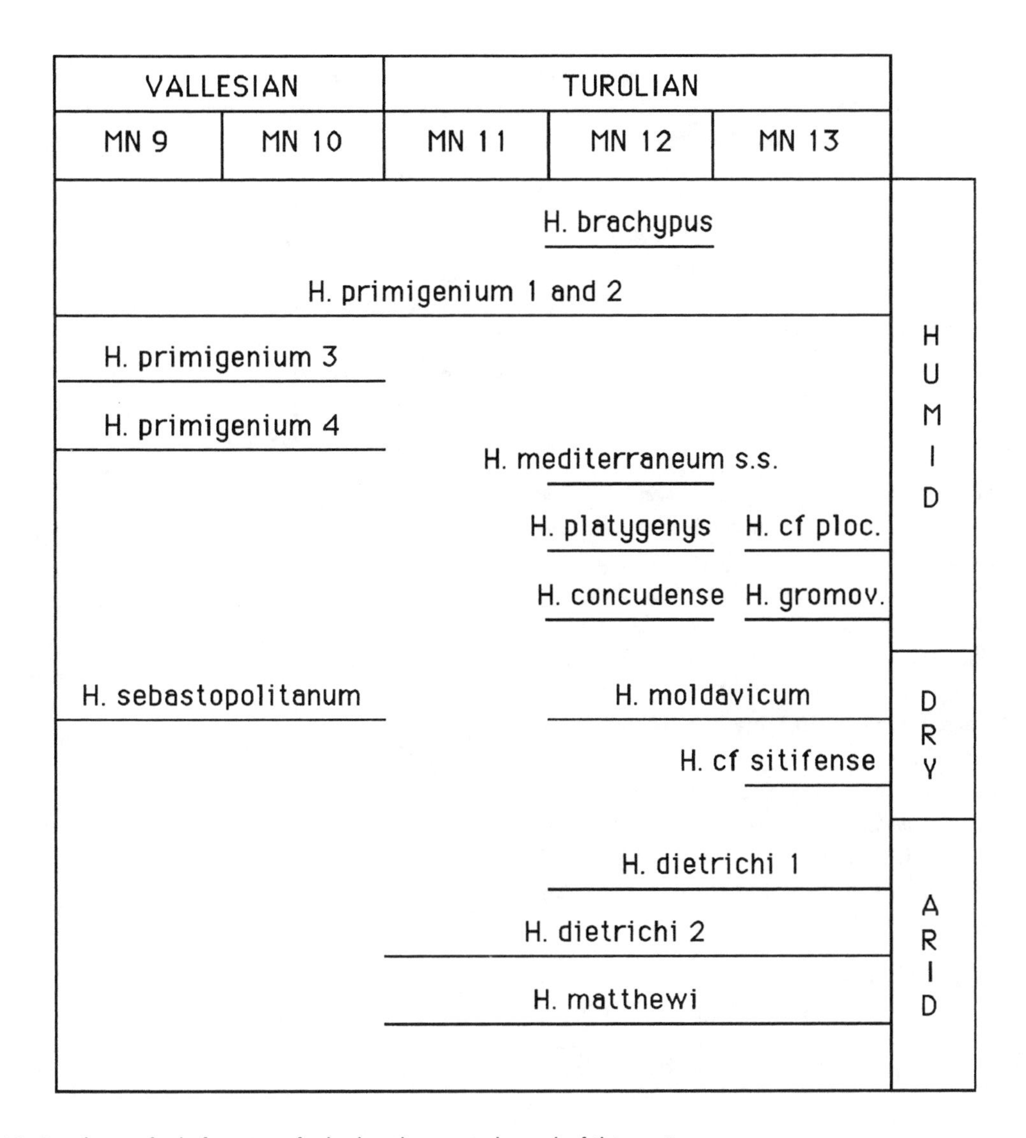

Fig. 11.12. Ratio diagrams for the four species of tridactyl equids present in the sample of Christmas Quarry.

Sebastopol, Yassiören), which also indicates dry conditions.

5. During the Turolian, the morphotype of *H. matthewi* indicates very dry conditions and open landscapes (Vathylakkos, Ravin des Zouaves, Kinik, Gulpinar, Samos 5, and Dytiko). The Turolian *H. dietrichi* (Vathylakkos, Ravin des Zouaves, Samos Andrianos, Saloniki, Pikermi, and Venta del Moro) suggests dry conditions and variable adaptations to open landscapes, more so in the Thessaloniki area (Vathylakkos, Ravin des Zouaves, Saloniki = *H. dietrichi* 2 in fig. 11.12) than elsewhere (Samos Andrianos, Pikermi, Venta del Moro = *H. dietrichi* 1).
6. Continuing dry conditions, though not so pronounced, and imperfect adaptation to open landscapes are suggested by the *H.* cf. *sitifense* morphotype (Lothagam, Sahabi, Baccinello, Kujal'nik). It is interesting to note that the same morphotype seems to be represented by the *Neohipparion* from Chihuahua, Mexico (Hemphillian-Blancan according to Lundelius et al., 1987).
7. The *H. moldavicum* morphotype (Taraklia, Tudorovo, Dytiko) is close to that of *H.* cf. *sitifense.*
8. The remaining morphotypes (table 11.2) indicate more humid conditions—very humid in the case of *H. brachypus* (Pikermi)—but with better adaptation to open landscapes than in the typical *H. primigenium.*
9. On the whole, there is certainly evidence for drier

conditions and more open landscapes in the Turolian than in the Vallesian and a progression of the number of dry morphotypes during the Turolian: two in the Early (*H. dietrichi* and *H. matthewi*); three in the Middle (*H. dietrichi, H. matthewi,* and *H. moldavicum*); and four in the Late Turolian (*H. dietrichi, H. matthewi, H. moldavicum,* and *H.* cf. *sitifense*).

10. There is no evidence for abrupt climatic changes, possibly because the dating of the fossils is not precise enough.

Origins

The North American *Cormohipparion occidentale* is richly represented at Christmas Quarry, Nebraska (Skinner and Johnson, 1984). Christmas Quarry is situated in the higher parts of the Ash Hollow Formation, above the Cap Rock Member, and is probably younger than 10 myr (Tedford et al., 1987). *H. primigenium,* the most ancient Old World Hipparion, belongs in the MN 9 zone (Mein, 1990) and is dated at around 10.5 myr (Sen, 1990; Bernor and Lipscomb, this vol.).

The questions of how many species are included under the name *Cormohipparion occidentale* at Christmas Quarry, and of which one might be the sister group of Old World hipparions, have been discussed previously (Eisenmann et al., 1987). More recently, Bernor et al. (1990, p. 294), using scatter diagrams of distal widths to maximal lengths of third metapodials from the same quarry, concluded that a detailed morphological study "is needed to determine . . . which of these morphs is . . . more similar to *H. primigenium*."

Such a study has been done. The late C. De Giuli and I measured all the relevant metapodials from Christmas Quarry, and I have completed individual ratio diagrams with twelve measurements for each bone, as well as ratio diagrams for the average of each group. These diagrams (fig. 11.13) clearly show four different kinds of metacarpals. One has the *H. primigenium* morphology. A second one is close in size to the smallest European hipparion: *H. periafricanum* from El Arquillo, Spain (MN 13). A third resembles, in size and slenderness, *H. matthewi* from Samos Quarry 5 and Dytiko, both in Greece, and Gulpinar, in Turkey (all MN 13 zone). The fourth morphotype is very close in size and proportions to *H. mediterraneum* of Upper Chobruchi, Moldova, and to the slender hipparion from Lubéron, France (MN 12).

If the respective dates are confirmed, the existence of an *H. primigenium* morphotype earlier in Europe than in Christmas Quarry suggests that the immediate ancestor of *H. primigenium* will be found in older American sites.

If Christmas Quarry tridactyl horses belonged to

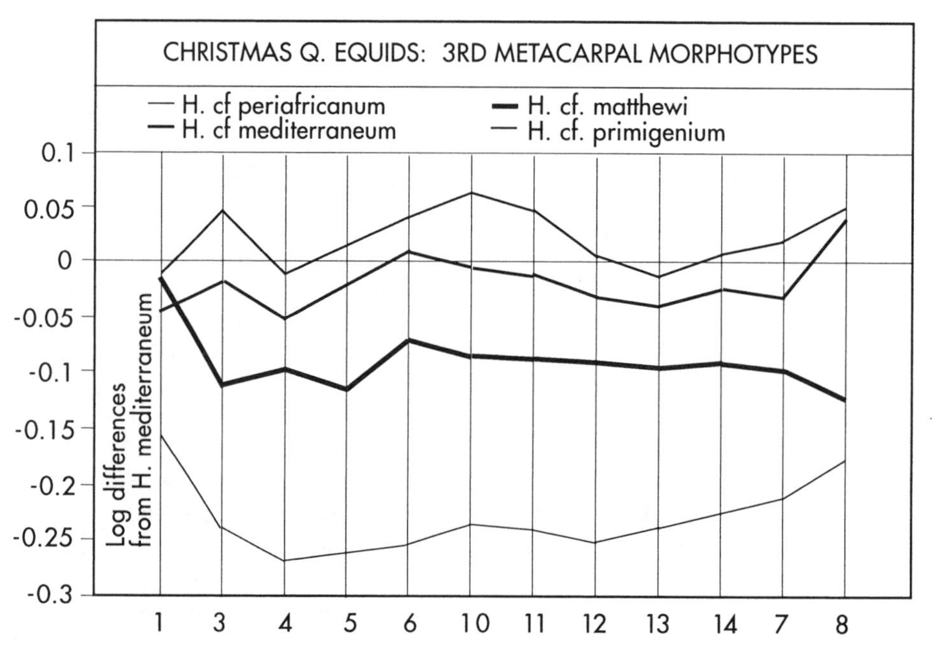

Fig. 11.13. Chronostratigraphical distribution of Vallesian and Turolian hipparions according to the robustness or slenderness of their metacarpals and the inferred humid or dry climatic conditions.

European instead of North American sites, they would indicate a Middle, or even Late, Turolian rather than a Vallesian age. Their age being fixed at somewhere around 9.5 myr, it seems that the conditions existing at this time in Nebraska had led to adaptations to dry conditions, although marked adaptation to open landscapes is not present in these forms. With the exception of the *H. primigenium* morphotype, already present in the European Vallesian, and the *H. periafricanum* morphotype probably representing a *Calippus* (Sondaar, pers. comm.), the other forms found at Christmas Quarry may have migrated later in Europe, when conditions there became dry enough.

Alternatively, parallel evolution may have resulted in European morphotypes similar to North American ones. Such a parallel evolution seems possible in the case of the Mexican (Chihuahua) and Old World hipparions belonging to the *H.* cf. *sitifense* morphotype. As is often the case, the answers to these questions depend on better data, particularly with regard to Asiatic hipparions.

References

Alberdi, M.-T. 1974. El género *Hipparion* en España. Nuevas formas de Castilla y Andalucia, revisión e historia evolutiva. *Trabajos sobre Neogeno-Cuaternario,* 1:1–146.

Bernor, R. L., and Tobien, H. 1989. Two small species of *Cremohipparion* (Equidae, Mamm.) from Samos, Greece. *Mitteilungen der bayerischen Staatssammlung für Paläontologie und historische Geologie,* 29:207–226.

Bernor, R. L., Tobien, H., and Woodburne, M. O. 1990. Patterns of Old World hipparionine evolutionary diversification and biogeographic extension. In *European Neogene mammal chronology,* ed. E. H. Lindsay et al., Plenum Press, New York. Pp. 263–319.

Borisiak, A. A. 1914. Sevastopol'skaja fauna mlekopitajuschchikh, vol. 1. *Trudy Geologicheskovo Komiteta,* n.s., 87:1–154.

Eisenmann, V. 1984. Sur quelques caractères adaptatifs du squelette d'*Equus* et leurs implications paléoécologiques. *Bulletin du Muséum National d'Histoire Naturelle,* 4th ser., 6, sec. C:185–195.

———. 1988. Les Périssodactyles Equidae. In *Contribution à l'étude du gisement miocène supérieur de Montredon (Hérault): Les grands mammifères. Palaeovertebrata,* special issue, 65–96.

Eisenmann, V., Sondaar, P., Alberdi, M.-T., and De Giuli, C. 1987. Is horse phylogeny becoming a playfield in the game of theoretical evolution? *Journal of Vertebrate Paleontology,* 7:224–229.

Forsten, A.-M. 1968. Revision of the Palearctic Hipparion. *Acta Zoologica Fennica,* 119:1–134.

———. 1982. *Hipparion primigenium melendezi* Alberdi reconsidered. *Annales Zoologici Fennici,* 19:109–113.

Gabunia, L. K. 1959. *K istorii gipparionov (po materialam Neogena SSSR).* Izdatel'stvo Akademii Nauk SSSR, Moscow.

Gregory, W. K. 1912. Notes on the principles of quadrupedal locomotion and on the mechanism of the limb bones in hoofed animals. *Annals of the New York Academy of Sciences,* 22:267–294.

Gromova, V. I. 1952. Gippariony (rod *Hipparion*) po materialam Taraklii, Pavlodara i drugim. *Trudy Paleontologischeskogo Instituta Akademii Nauk SSSR,* 36:1–475.

Koufos, G. D. 1984. A new *Hipparion* (Mammalia, Perissodactyla) from the Vallesian (Late Miocene) of Greece. *Paläontologische Zeitschrift,* 58:307–317.

———. 1987a. Study of the Pikermi Hipparions. Pt. 1, Generalities and taxonomy. *Bulletin du Muséum National d'Histoire Naturelle,* 4th ser., 9, sec. C:197–252.

———. 1987b. Study of the Pikermi Hipparions. Pt. 2, Comparisons and odontograms. *Bulletin du Muséum National d'Histoire Naturelle,* 4th ser., 9, sec. C:327–363.

Lundelius, E. L., Churcher, C. S., Downs, T., Harington, C. R., Lindsay, E. H., Schultz, G. E., Semken, H. A., Webb, S. D., Zakrzewski, R. J. 1987. The North American Quaternary sequence. In *Cenozoic mammals of North America,* ed. M. O. Woodburne, University of California Press, Berkeley. Pp. 211–235.

Mein, P. 1990. Updating of MN zones. In *European Neogene mammal chronology,* ed. E. H. Lindsay et al., Plenum Press, New York. Pp. 73–90.

Nakaya, H., and Watabe, M. 1990. Hipparion from the Upper Miocene Namurungule Formation, Samburu Hills, Kenya: Phylogenetic significance of newly discovered skull. *Géobios,* 23:195–219.

Osborn, H. F. 1929. *The Titanotheres of ancient Wyoming, Dakota and Nebraska.* 2 vols. Monographs of the United States Geological Survey, no. 55, vol. 2. Washington, D.C.

Pavlow, M. 1890. Etudes sur l'histoire paléontologique des Ongulés. Pt. 4, Hipparions de la Russie. Pt. 5, Chevaux pléistocènes de la Russie. *Bulletin de la Société Impériale des Naturalistes de Moscou,* n.s., 3:653–716.

Qiu Zhansiang, Huang Weilong, and Guo Zhihui. 1987. The Chinese hipparionine fossils. *Palaeontologica Sinica,* 175, ser. C, 25:1–250.

Sen, S. 1990. *Hipparion* datum and its chronologic evidence in the Mediterranean area. In *European Neogene mammal chronology,* ed. E. H. Lindsay et al., Plenum Press, New York. Pp. 495–505.

Simpson, G. G. 1941. Large Pleistocene felines of North America. *American Museum Novitates,* 1136:1–27.

Skinner, M. F., and Johnson, F. W. 1984. Tertiary stratigraphy and the Frick collection of fossil vertebrates from north-central Nebraska. *Bulletin of the American Museum of Natural History,* 178:215–368.

Skinner, M. F., and MacFadden, B. J. 1977. *Cormohipparion* n. gen. (Mammalia, Equidae) from the North American Miocene (Barstovian-Clarendonian). *Journal of Paleontology,* 51:912–926.

Sondaar, P. Y. 1974. The Hipparion of the Rhone Valley. *Geobios,* 7:289–306.

Sondaar, P. Y., and Eisenmann, V. 1989. *L'évolution de la famille du Cheval.* Instituut voor Aardwetenschappen, Universiteit Utrecht. Pp. 1–43.

Staesche, U., and Sondaar, P. Y. 1979. *Hipparion* aus dem Vallesium und Turolium (Jüngtertiär) der Türkei. *Geologisches Jahrbuch*, 33:35–79.

Tedford, R. H., Skinner, M. S., Fields, R. W., Rensberger, J. M., Whistler, D. P., Galusha, T., Taylor, B. E., Macdonald, J. R., and Webb, S. D. 1987. Faunal succession and biochronology through Hemphillian interval (Late Oligocene through earliest Pliocene epochs) in North America. In *Cenozoic mammals of North America*, ed. M. O. Woodburne, University of California Press, Berkeley. Pp. 153–210.

Appendix 11.1. In appendixes 11.1–11.4: Measurements (in mm) of third metacarpals (MC III) and third metatarsals (MT III) of tridactyl equids. Abbreviations: n = number of measures; x = mean; min = minimal observed value; max = maximal observed values; s = standard deviation; v = coefficient of variation (v = 100x/s); sebastopol. = *sebastopolitanum.*

		H. primigenium africanum, Bou Hanifia						*H. primigenium catalaunicum*, Can Llobateres				
MC III	n	x	min	max	s	v	n	x	min	max	s	v
1: Maximal length	4	207.5	204.0	211.0	3.11	1.50	8	207.5	199.0	215.0	5.53	2.67
3: Minimal breadth	5	27.0	25.5	28.0	1.06	3.93	11	28.0	26.0	29.0	1.07	2.82
4: Depth at level of 3	5	21.1	20.0	22.0	0.74	3.51	10	21.6	19.0	23.0	1.36	6.30
5: Proximal articular breadth	3	39.3	38.0	41.0	1.53	3.88	12	38.6	37.0	41.0	1.51	3.91
6: Prox. art. depth	3	28.7	26.0	30.0	2.31	8.06	11	27.0	25.0	29.0	1.24	4.59
10: Distal max. supra-art. breadth	4	37.3	37.0	38.0	0.50	1.34	10	38.0	36.0	42.0	1.87	4.92
11: Dist. max. art. breadth	4	35.9	35.5	36.0	0.25	0.70	11	36.7	35.0	40.1	1.44	3.92
12: Dist. max. depth of keel	4	28.8	28.0	30.0	0.87	3.01	11	28.1	26.2	29.5	1.03	3.67
13: Dist. min. depth of medial condyle	4	24.0	23.5	24.5	0.41	1.70	11	24.4	23.0	25.6	0.95	3.89
14: Dist. max. depth of med. condyle	4	26.8	26.0	27.5	0.65	2.41	11	26.2	23.6	28.5	1.23	4.69
7: Max. diameter facet 3rd carpal	3	36.2	34.0	38.0	2.02	5.59	12	33.1	31.0	35.0	1.58	4.77
8: Diam. anterior facet 2nd carpal	2	9.8	9.0	10.5	1.06	10.88	13	10.6	9.5	12.0	0.81	7.64
MT III												
1: Maximal length	5	240.7	234.0	247.0	4.66	1.94	4	234.5	228.0	245.0	7.33	3.13
3: Minimal breadth	7	26.1	25.0	27.0	0.61	2.33	11	29.4	27.0	32.0	1.43	4.86
4: Depth at level of 3	4	28.2	26.0	30.0	1.72	6.10	10	27.7	26.0	30.0	1.29	4.66
5: Proximal articular breadth	7	38.1	36.5	40.0	1.17	3.07	6	43.1	41.0	46.0	2.01	4.66
10: Distal max. supra-art. breadth	4	36.5	35.5	37.5	0.91	2.50	9	39.4	37.0	43.0	2.28	5.79
11: Dist. max. art. breadth	5	35.6	34.5	37.0	0.96	2.70	9	37.0	34.0	40.1	2.13	5.76
12: Dist. max. depth of keel	6	29.8	29.0	31.0	0.75	2.52	10	30.4	27.8	32.0	1.26	4.14
13: Dist. min. depth of medial condyle	6	23.9	23.0	26.0	1.11	4.66	8	24.8	23.5	27.2	1.09	4.40
14: Dist. max. depth of med. condyle	6	27.3	27.0	28.0	0.52	1.89	8	28.1	26.0	31.5	1.77	6.30
7: Max. diameter facet 3rd tarsal	7	36.0	34.0	38.0	1.41	3.93	6	39.6	38.0	41.0	1.02	2.58
8: Diam. facet 2nd tarsal	5	8.0	7.0	9.5	1.06	13.26	8	9.1	7.0	10.0	1.25	13.74

		H. sebastopol. melendezi, Los Valles de Fuentidueña						*H. primigenium*, Piera				
MC III	n	x	min	max	s	v	n	x	min	max	s	v
1: Maximal length	7	215.7	207.0	222.0	5.15	2.39	10	199.0	187.0	203.5	5.59	2.81
3: Minimal breadth	17	24.7	22.7	26.0	1.10	4.43	16	26.6	24.7	29.0	1.21	4.55
4: Depth at level of 3	20	21.4	19.0	23.7	1.27	5.93	15	19.9	17.5	22.5	1.60	8.04
5: Proximal articular breadth	13	37.4	36.0	41.0	1.41	3.77	15	37.1	35.5	38.5	0.90	2.43
6: Prox. art. depth	14	26.3	24.2	29.0	1.34	5.11	13	25.5	24.0	26.5	0.84	3.29
10: Distal max. supra-art. breadth	18	33.6	31.0	36.0	1.49	4.44	18	35.7	32.0	38.0	1.64	4.59
11: Dist. max. art. breadth	18	33.2	31.0	36.0	1.35	4.08	18	33.8	32.0	35.5	1.20	3.55
12: Dist. max. depth of keel	15	26.6	24.0	29.5	1.23	4.62	15	27.0	25.0	29.0	1.32	4.89
13: Dist. min. depth of medial condyle	18	23.2	22.0	25.0	0.81	3.49	19	22.9	20.0	25.5	1.52	6.64
14: Dist. max. depth of med. condyle	15	24.6	23.0	26.0	0.78	3.16	16	24.7	22.8	27.0	1.37	5.55
7: Max. diameter facet 3rd carpal	14	32.3	31.0	34.5	1.11	3.43	13	31.4	28.0	33.0	1.27	4.04
8: Diam. anterior facet 2nd carpal	13	10.3	8.5	12.0	1.09	10.63	12	10.9	10.0	12.5	0.76	6.97
MT III												
1: Maximal length	10	249.0	242.0	255.0	4.88	1.96	15	232.1	222.0	242.0	6.76	2.91
3: Minimal breadth	31	24.9	22.0	27.0	1.25	5.02	37	27.2	23.5	31.0	1.61	5.92

	H. sebastopol. melendezi, Los Valles de Fuentidueña						*H. primigenium,* Piera					
MT III	n	x	min	max	s	v	n	x	min	max	s	v
4: Depth at level of 3	28	26.3	24.0	28.0	1.19	4.52	36	26.9	24.0	31.0	1.42	5.28
5: Proximal articular breadth	25	38.8	36.0	42.0	1.54	3.97	27	38.8	35.0	44.0	2.04	5.26
10: Distal max. supra-art. breadth	29	34.8	32.0	38.0	1.73	4.97	28	37.7	32.7	41.0	1.94	5.15
11: Dist. max. art. breadth	28	34.0	30.0	37.0	1.81	5.32	26	35.3	33.0	38.0	1.18	3.34
12: Dist. max. depth of keel	15	29.0	27.0	31.0	1.41	4.86	26	29.0	27.0	33.0	1.51	5.21
13: Dist. min. depth of medial condyle	27	24.1	21.0	26.0	1.36	5.64	27	23.9	22.0	27.0	1.27	5.31
14: Dist. max. depth of med. condyle	26	26.7	23.0	29.0	1.54	5.77	27	26.8	25.0	29.0	1.30	4.85
7: Max. diameter facet 3rd tarsal	23	36.2	34.0	42.0	1.74	4.81	22	36.7	34.0	40.5	1.69	4.60
8: Diam. facet 2nd tarsal	24	9.3	7.0	11.0	1.12	12.04	21	8.6	6.7	11.2	1.41	16.40

Appendix 11.2. In appendixes 11.1–11.4: Measurements (in mm) of third metacarpals (MC III) and third metatarsals (MT III) of tridactyl equids. Abbreviations: n = number of measures; x = mean; min = minimal observed value; max = maximal observed values; s = standard deviation; v = coefficient of variation (v = 100x/s); sebastopol. = *sebastopolitanum.*

	H. brachypus giganteum, Grebeniki						*H. concudense verae,* Grebeniki					
MC III	n	x	min	max	s	v	n	x	min	max	s	v
1: Maximal length	9	213.8	201.0	230.0	8.87	4.15	16	211.7	199.0	222.0	6.55	3.09
3: Minimal breadth	11	28.0	27.0	29.0	0.80	2.86	18	25.5	23.5	27.1	1.14	4.46
4: Depth at level of 3	9	22.0	20.0	24.0	1.23	5.60	17	20.8	19.0	22.0	0.88	4.24
5: Proximal articular breadth	10	38.8	36.0	40.5	1.23	3.18	14	37.2	35.0	40.0	1.74	4.68
6: Prox. art. depth	6	27.0	25.0	28.1	1.28	4.74	15	25.7	24.0	28.0	1.14	4.46
10: Distal max. supra-art. breadth	11	37.6	36.7	39.0	0.80	2.12	18	33.9	31.5	35.1	1.13	3.34
11: Dist. max. art. breadth	10	36.1	34.0	39.0	1.65	4.56	18	33.4	31.7	35.0	1.05	3.15
12: Dist. max. depth of keel	8	28.8	28.0	30.0	0.76	2.63	16	26.9	26.0	28.0	0.84	3.11
13: Dist. min. depth of medial condyle	8	24.7	23.7	26.0	0.75	3.05	18	22.7	21.0	24.0	0.80	3.54
14: Dist. max. depth of med. condyle	9	26.4	25.0	27.0	0.73	2.76	17	24.7	23.0	26.0	0.77	3.12
7: Max. diameter facet 3rd carpal	8	32.8	32.0	35.0	0.98	2.97	14	31.8	28.5	34.0	1.79	5.65
8: Diam. anterior facet 2nd carpal	9	11.4	10.0	12.5	0.77	6.82	15	10.9	10.0	12.1	0.77	7.02
MT III												
1: Maximal length	9	248.6	235.0	259.0	8.13	3.27	16	247.4	236.0	254.0	6.33	2.56
3: Minimal breadth	11	31.2	28.0	33.5	1.85	5.94	19	26.6	24.0	29.0	1.25	4.71
4: Depth at level of 3	8	30.2	26.7	32.0	1.82	6.00	18	26.4	24.5	30.0	1.44	5.43
5: Proximal articular breadth	10	44.4	43.0	47.0	1.15	2.59	16	40.0	37.0	43.0	1.59	3.96
10: Distal max. supra-art. breadth	11	42.6	38.1	45.0	1.87	4.39	19	36.0	33.0	38.0	1.64	4.56
11: Dist. max. art. breadth	11	39.9	37.3	42.0	1.62	4.07	19	35.0	33.0	37.0	1.16	3.32
12: Dist. max. depth of keel	9	33.3	31.5	36.0	1.36	4.08	19	29.2	27.0	31.0	1.01	3.47
13: Dist. min. depth of medial condyle	11	26.5	24.5	28.5	1.10	4.15	19	23.8	22.0	25.5	1.07	4.51
14: Dist. max. depth of med. condyle	10	30.1	28.0	32.0	1.29	4.27	19	26.7	24.5	28.5	0.97	3.64
7: Max. diameter facet 3rd tarsal	8	41.4	39.0	43.3	1.32	3.18	14	36.1	32.5	38.0	1.50	4.14
8: Diam. facet 2nd tarsal	7	11.3	7.5	13.0	1.78	15.78	14	10.0	9.0	12.0	1.07	10.69

	H. moldavicum moldavicum, Taraklia						*H. moldavicum tudorovense,* Tudorovo					
MC III	n	x	min	max	s	v	n	x	min	max	s	v
1: Maximal length	13	210.4	193.0	219.0	6.32	3.00	11	206.5	199.0	216.0	5.95	2.88
3: Minimal breadth	14	24.1	22.5	25.0	0.78	3.23	12	24.2	22.0	26.0	1.33	5.50
4: Depth at level of 3	14	21.0	20.0	22.0	0.66	3.12	12	20.2	18.0	21.5	0.99	4.90
5: Proximal articular breadth	13	34.2	31.5	36.0	1.30	3.80	11	34.6	32.0	39.0	2.17	6.27
6: Prox. art. depth	10	24.9	23.0	27.0	1.27	5.11	9	24.2	23.5	25.0	0.53	2.20
10: Distal max. supra-art. breadth	13	33.2	31.7	35.0	1.02	3.07	11	32.7	31.2	34.7	1.06	3.25

(*Continued*)

Appendix 11.2. (*Continued*)

MC III	*H. moldavicum moldavicum*, Taraklia						*H. moldavicum tudorovense*, Tudorovo					
	n	x	min	max	s	v	n	x	min	max	s	v
11: Dist. max. art. breadth	13	31.4	29.8	33.0	0.87	2.78	10	31.4	29.5	33.0	1.34	4.28
12; Dist. max. depth of keel	12	25.2	24.0	27.0	0.73	2.89	11	25.6	24.0	28.0	1.15	4.47
13: Dist. min. depth of medial condyle	14	21.1	20.0	22.0	0.49	2.30	11	21.8	21.0	24.0	0.96	4.40
14: Dist. max. depth of med. condyle	12	23.4	23.0	25.0	0.63	2.71	10	23.6	22.0	25.2	1.02	4.32
7: Max. diameter facet 3rd carpal	12	28.7	26.5	30.0	1.03	3.60	11	24.8	26.0	30.7	1.37	4.81
8: Diam. anterior facet 2nd carpal	13	10.1	7.5	12.0	1.24	12.30	10	9.9	9.0	13.0	1.19	12.06
MT III												
1: Maximal length	14	231.1	216.0	241.0	7.45	3.22	9	240.3	228.0	257.0	8.03	3.34
3: Minimal breadth	14	23.6	22.0	25.0	0.75	3.18	11	25.0	23.0	27.0	1.41	5.65
4: Depth at level of 3	14	24.7	23.0	26.5	0.98	3.98	10	25.6	23.0	27.1	1.44	5.61
5: Proximal articular breadth	13	36.1	33.5	39.0	1.81	5.02	8	36.0	32.0	38.0	1.98	5.51
10: Distal max. supra-art. breadth	14	33.4	31.0	35.0	1.37	4.11	13	34.2	33.0	37.0	1.11	3.24
11: Dist. max. art. breadth	14	31.4	30.0	33.0	0.88	2.82	13	32.3	31.0	34.2	1.00	3.09
12: Dist. max. depth of keel	14	26.9	24.0	29.0	1.31	4.88	13	27.2	24.5	29.0	1.03	3.79
13: Dist. min. depth of medial condyle	14	21.6	19.0	23.0	1.14	5.29	13	22.5	20.2	24.0	0.89	3.97
14: Dist. max. depth of med. condyle	13	24.3	22.5	26.0	1.16	4.77	12	24.9	22.7	26.0	0.94	3.78
7: Max. diameter facet 3rd tarsal	13	32.7	30.0	35.0	1.40	4.26	8	33.2	30.0	35.5	1.82	5.48
8: Diam. facet 2nd tarsal	13	7.7	6.0	9.5	1.07	13.88	7	9.4	8.0	10.5	1.02	10.79

Appendix 11.3. In appendixes 11.1–11.4: Measurements (in mm) of third metacarpals (MC III) and third metatarsals (MT III) of tridactyl equids. Abbreviations: n = number of measures; x = mean; min = minimal observed value; max = maximal observed values; s = standard deviation; v = coefficient of variation (v = 100x/s); sebastopol. = *sebastopolitanum.*

MC III	*H. mediterraneum* s.s., Chobruchi						*H. concudense concudense*, Concud					
	n	x	min	max	s	v	n	x	min	max	s	v
1: Maximal legnth	4	206.8	203.0	210.0	2.99	1.44	5	211.9	187.5	230.0	18.02	8.50
3: Minimal breadth	10	26.0	23.0	27.7	1.49	5.72	9	26.0	24.0	29.0	1.71	6.57
4: Depth at level of 3	10	21.4	20.0	22.2	0.72	3.37	10	20.3	19.0	21.6	0.76	3.76
5: Proximal articular breadth	7	36.4	36.0	37.5	0.73	2.01	11	35.8	32.8	38.0	1.75	4.90
6: Prox. art. depth	5	26.6	25.0	28.0	1.14	4.29	9	25.3	23.0	27.5	1.60	6.32
10: Distal max. supra-art. breadth	8	33.7	32.0	36.0	1.35	3.99	19	34.2	31.0	36.5	1.44	4.22
11: Dist. max. art. breadth	8	32.3	30.0	34.0	1.33	4.12	18	33.7	31.7	35.3	1.11	3.29
12: Dist. max. depth of keel	8	26.5	24.0	28.7	1.37	5.16	19	26.8	25.1	28.0	0.83	3.10
13: Dist. min. depth of medial condyle	8	22.5	21.0	24.5	1.02	4.54	19	22.7	21.8	24.0	0.67	2.95
14: Dist. max. depth of med. condyle	8	24.5	22.7	26.5	1.10	4.49	17	24.5	23.5	26.0	0.67	2.74
7: Max. diameter facet 3rd carpal	5	30.5	29.0	32.0	1.11	3.63	11	31.1	29.0	33.0	1.23	3.94
8: Diam. anterior facet 2nd carpal	6	10.5	9.0	11.5	0.99	9.45	13	10.1	9.0	11.7	0.97	9.67
MT III												
1: Maximal length	4	250.0	247.0	255.0	3.56	1.42	6	237.8	230.0	248.0	8.68	3.65
3: Minimal breadth	10	24.9	22.0	27.0	1.47	5.90	21	27.1	24.0	31.0	1.93	7.12
4: Depth at level of 3	10	25.5	24.0	27.0	1.22	4.78	17	26.4	22.0	29.0	1.82	6.91
5: Proximal articular breadth	11	38.0	33.0	42.0	2.72	7.16	11	38.8	36.3	42.0	1.67	4.31
10: Distal max. supra-art. breadth	7	34.9	34.0	37.0	1.02	2.91	29	35.4	30.0	39.0	2.14	6.04
11: Dist. max. art. breadth	7	33.3	32.0	36.7	1.57	4.72	29	34.2	30.0	37.0	1.89	5.53
12: Dist. max. depth of keel	6	28.8	28.0	30.0	0.71	2.46	25	29.3	25.0	32.5	1.86	6.35
13: Dist. min. depth of medial condyle	6	23.3	23.0	24.0	0.52	2.21	30	23.5	21.0	27.0	1.46	6.19
14: Dist. max. detp of med. condyle	6	25.8	25.2	26.0	0.32	1.24	28	25.8	22.5	29.0	1.31	5.08
7: Max. diameter facet 3rd tarsal	9	34.4	30.0	38.0	2.50	7.26	11	36.2	34.0	39.0	1.60	4.42
8: Diam. facet 2nd tarsal	9	9.6	9.0	10.0	0.49	5.06	11	9.2	7.5	11.8	1.27	13.78

	H. gromovae, El Arquillo						*H. dietrichi*, Venta del Moro					
MC III	n	x	min	max	s	v	n	x	min	max	s	v
1: Maximal length	1	169.0					3	206.7	200.0	215.0	7.64	3.70
3: Minimal breadth	8	20.7	19.0	22.5	1.06	5.11	6	21.8	20.1	23.5	1.15	5.27
4: Depth at level of 3	7	17.0	15.0	18.0	1.04	6.12	5	19.4	18.2	20.0	0.82	4.20
5: Proximal articular breadth	4	31.3	29.0	32.0	1.50	4.80	7	32.6	31.0	34.7	1.47	4.50
6: Prox. art. depth	1	18.0					7	22.8	22.0	23.0	0.39	1.73
10: Distal max. supra-art. breadth	7	27.9	27.0	30.0	1.03	3.70	6	30.2	29.0	31.0	0.71	2.36
11: Dist. max. art. breadth	7	26.8	25.0	28.0	1.22	4.53	6	29.5	28.7	30.0	0.61	2.08
12: Dist. max. depth of keel	7	21.5	20.0	23.0	0.96	4.45	6	24.5	23.5	26.0	1.04	4.25
13: Dist. min. depth of medial condyle	7	18.8	17.7	20.0	0.71	3.75	6	20.8	20.0	22.0	0.91	4.35
14: Dist. max. depth of med. condyle	6	19.9	18.3	21.5	1.11	5.57	6	21.8	21.0	22.7	0.60	2.76
7: Max. diameter facet 3rd carpal	3	25.5	23.0	27.5	2.29	8.99	4	28.4	27.0	29.5	1.11	3.91
8: Diam. anterior facet 2nd carpal	4	9.3	8.0	10.0	0.96	10.35	7	8.4	7.0	9.0	0.81	9.62
MT III												
1: Maximal length	1	203.0					2	235.0	233.0	237.0	2.83	1.20
3: Minimal breadth	7	21.7	21.0	22.5	0.70	3.22	14	22.1	20.5	23.0	0.68	3.06
4: Depth at level of 3	6	22.7	20.0	25.0	1.66	7.34	13	24.2	22.5	25.2	0.87	3.61
5: Proximal articular breadth	8	32.5	30.0	36.7	2.39	7.36	7	34.4	32.5	37.0	1.97	5.71
10: Distal max. supra-art. breadth	3	28.5	27.5	29.0	0.87	3.04	6	31.2	29.0	33.7	1.82	5.85
11: Dist. max. art. breadth	3	27.9	27.7	28.0	0.17	0.62	8	30.3	28.3	33.0	1.78	5.88
12: Dist. max. depth of keel	3	23.6	23.0	24.5	0.79	3.36	7	26.6	24.5	29.0	1.68	6.30
13: Dist. min. depth of medial condyle	3	20.0	19.0	21.0	1.00	5.00	8	21.1	19.0	23.0	1.25	5.90
14: Dist. max. depth of med. condyle	3	21.8	21.7	22.0	0.15	0.70	7	23.4	22.0	25.5	1.28	5.47
7: Max. diameter facet 3rd tarsal	5	29.9	27.0	33.5	2.46	8.23	6	32.5	30.0	35.0	1.89	5.83
8: Diam. facet 2nd tarsal	5	9.0	8.0	10.5	1.17	13.03	5	8.3	6.5	9.0	1.10	13.20

Appendix 11.4. In appendixes 11.1–11.4: Measurements (in mm) of third metacarpals (MC III) and third metatarsals (MT III) of tridactyl equids. Abbreviations: n = number of measures; x = mean; min = minimal observed value; max = maximal observed values; s = standard deviation; v = coefficient of variation (v = 100x/s); sebastopol. = *sebastopolitanum.*

	H. cf. *primigenium*, Christmas Quarry						*H.* cf. *mediterraneum*, Christmas Quarry					
MC III	n	x	min	max	s	v	n	x	min	max	s	v
1: Maximal length	23	209.2	200.8	216.9	4.87	2.33	7	194.6	188.5	201.5	4.76	2.44
3: Minimal breadth	22	29.2	26.7	32.9	1.82	6.25	7	24.5	23.8	25.4	0.55	2.24
4: Depth at level of 3	22	20.9	18.9	23.2	1.19	5.69	7	19.1	17.5	20.3	1.02	5.37
5: Proximal articular breadth	19	37.6	35.4	40.1	1.27	3.39	6	34.1	32.8	36.9	1.62	4.77
6: Prox. art. depth	20	27.5	25.0	30.2	1.56	5.68	7	25.4	24.2	28.2	1.45	5.68
10: Distal max. supra-art. breadth	23	37.4	34.7	40.8	1.76	4.72	7	32.3	31.2	34.4	1.02	3.16
11: Dist. max. art. breadth	22	34.9	31.8	37.4	1.62	4.64	7	30.9	29.9	32.2	0.93	3.02
12: Dist. max. depth of keel	21	26.8	24.4	29.0	1.13	4.23	7	24.6	22.9	27.2	1.50	6.10
13: Dist. min. depth of medial condyle	22	22.3	20.9	23.8	0.81	3.65	7	21.1	20.1	22.4	0.81	3.83
14: Dist. max. depth of med. condyle	20	24.7	22.2	26.6	1.22	4.95	7	22.9	21.8	23.5	0.55	2.40
7: Max. diameter facet 3rd carpal	17	32.4	29.0	34.5	1.60	4.93	6	29.2	27.1	31.8	1.73	5.94
8: Diam. anterior facet 2nd carpal	18	10.3	8.6	12.4	0.94	9.11	6	10.2	8.9	12.5	1.46	14.40

	H. cf. *matthewi*, Christmas Quarry						*H.* cf. *periafricanum*, Christmas Quarry					
MC III	n	x	min	max	s	v	n	x	min	max	s	v
1: Maximal length	6	208.2	200.4	218.2	5.82	2.80		147.5				
3: Minimal breadth	6	19.8	17.7	22.5	1.71	8.66		14.8				
4: Depth at level of 3	5	17.7	15.9	18.5	1.06	5.97		12.0				

(*Continued*)

Appendix 11.4. (*Continued*)

MC III	n	x	min	max	s	v	n	x	min	max	s	v
	H. cf. *matthewi*, Christmas Quarry						*H.* cf. *periafricanum*, Christmas Quarry					
5: Proximal articular breadth	6	29.2	27.0	32.5	1.92	6.59		20.6				
6: Prox. art. depth	6	22.2	20.1	25.5	1.81	8.13		14.8				
10: Distal max. supra-art. breadth	6	27.2	24.9	31.6	2.28	8.40		20.1				
11: Dist. max. art. breadth	6	26.3	24.3	31.0	2.46	9.35		19.2				
12: Dist. max. depth of keel	6	22.7	21.1	25.8	1.79	7.91		15.2				
13: Dist. min. depth of medial condyle	6	19.6	18.2	22.1	1.38	7.06		13.4				
14: Dist. max. depth of med. condyle	6	21.0	19.5	23.8	1.67	7.93		14.5				
7: Max. diameter facet 3rd carpal	6	26.1	25.0	28.4	1.29	4.95		19.2				
8: Diam. anterior facet 2nd carpal	6	7.4	6.4	8.1	0.73	9.99		6.5				

MC III	n	x	min	max	s	v	n	x	min	max	s	v
	H. dietrichi, Saloniki						*H. mediterraneum* s.s., Pikermi					
1: Maximal length	5	220.0	218.0	222.0	1.41	0.64	21	214.9	202.0	221.7	5.55	2.58
3: Minimal breadth	6	23.5	22.5	24.0	0.77	3.30	21	25.5	23.5	27.0	1.10	4.31
4: Depth at level of 3	6	21.2	20.0	23.0	1.17	5.52	20	21.3	20.1	23.5	0.84	3.93
5: Proximal articular breadth	5	34.1	32.0	36.5	1.75	5.12	22	36.7	33.5	40.0	1.94	5.28
6: Prox. art. depth	6	24.5	23.0	27.0	1.52	6.19	19	25.0	22.0	29.0	1.57	6.28
10: Distal max. supra-art. breadth	5	30.9	29.0	32.0	1.34	4.34	20	32.5	30.5	34.5	1.30	3.99
11: Dist. max. art. breadth	5	29.9	29.0	31.0	0.89	2.99	21	31.6	30.0	33.5	1.05	3.31
12: Dist. max. depth of keel	4	26.0	25.0	26.8	0.81	3.12	22	26.3	23.0	29.5	1.46	5.54
13: Dist. min. depth of medial condyle	4	21.9	21.7	22.0	0.15	0.68	22	22.9	20.1	26.5	1.26	5.51
14: Dist. max. depth of med. condyle	3	24.0	23.0	25.0	1.00	4.17	20	24.1	21.0	27.0	1.31	5.46
7: Max. diameter facet 3rd carpal	5	29.4	27.0	31.0	1.52	5.16	20	31.2	29.0	34.5	1.58	5.08
8: Diam. anterior facet 2nd carpal	5	9.8	8.3	11.0	0.98	10.05	20	9.9	8.0	11.5	0.99	10.00

MC III	n	x	min	max	s	v	n	x	min	max	s	v
	H. primigenium, Sumeg						*H. mediterraneum*, Lubéron					
1: Maximal length	1	191.0	191.0	191.0			4	200.8	191.5	207.0	6.58	3.28
3: Minimal breadth	14	29.3	26.5	33.0	1.92	6.54	13	24.1	22.0	25.0	0.96	3.97
4: Depth at level of 3	14	20.9	19.0	22.1	1.10	5.25	13	20.1	18.0	21.0	1.02	5.09
5: Proximal articular breadth	14	37.5	35.3	41.0	1.75	4.66	10	34.1	33.0	36.0	1.06	3.10
6: Prox. art. depth	12	26.5	24.0	28.0	1.27	4.79	11	24.5	23.0	26.0	0.94	3.82
10: Distal max. supra-art. breadth	11	36.6	33.0	39.0	1.88	5.14	11	31.1	29.7	34.1	1.55	4.99
11: Dist. max. art. breadth	11	35.0	33.0	37.6	1.66	4.76	10	29.9	28.0	32.5	1.45	4.86
12: Dist. max. depth of keel	9	27.1	25.0	29.0	1.20	4.43	11	25.2	24.0	27.0	0.74	2.94
13: Dist. min. depth of medial condyle	10	23.1	21.0	24.5	1.10	4.75	11	22.0	20.7	23.8	0.99	4.51
14: Dist. max. depth of med. condyle	10	24.9	23.0	26.0	0.95	3.80	8	23.1	22.0	24.0	0.75	3.26
7: Max. diameter facet 3rd carpal	12	31.8	30.0	34.0	1.21	3.80	11	29.2	28.0	30.5	0.91	3.13
8: Diam. anterior facet 2nd carpal	12	10.6	10.0	12.0	0.65	6.15	8	9.6	9.0	12.0	1.06	11.02

Appendix 11.5. Measurements (in mm) of third metacarpals (MC III) of hipparions. Abbreviations: H. = *Hipparion*; primig. = *primigenium*; sebast. = *sebastopolitanum*; periaf. = *periafricanum*; platyg. = *platygenys*; sitif. = *sitifense*; concud. = *concudense*; Höwen. = Höwenegg; Nomb. = Nombrevilla; Yass. = Yassiören; Gritz. = Gritzev; El Lug. = El Lugarejo; Arqu. = Arquillo; Piker. = Pikermi; Tarak. = Taraklia; Salon. = Saloniki; Baccin. = Baccinello; Lubér. = Lubéron; n = number of measures.

MC III	*H. primig.* Eppelsheim	*H. primig.* Höwen. n = 3	*H. primig.* Nomb. n = 1–3	*H. primig.* Yass. 13	*H. primig* Rudabanya	*H. primig.* Gritz. n = 2–3	*H. primig.* Yass. n = 3–4	*H. sebast.* El Lug. n = 2–4	*H. sebast.* Yass. 12
1: Maximal length	215.0	214.0		198.0	226.0	224.0	205.9	210.5	222.0
3: Minimal breadth	28.6	31.5	30.0	27.0	27.1	28.0	27.5	25.3	24.2
4: Depth at level of 3	23.0	22.1	23.0	20.0	22.0	22.4	21.3	21.6	21.1
5: Proximal articular breadth	40.1	40.6	40.0	37.0	33.5	37.7	37.2	35.0	36.2
6: Prox. art. depth	28.9	28.6		26.5		27.7	26.6	25.1	23.3
10: Distal max. supra-art. breadth	40.3	39.3	43.0	36.0	38.0	38.3	34.9	33.4	34.2
11: Dist. max. art. breadth	37.1	37.3	39.0	33.0	36.0	35.9	33.5	32.6	33.0
12: Dist. max. depth of keel	28.0	28.0	30.0	25.5	28.0	28.4	27.2	25.3	27.8
13: Dist. min. depth of medial condyle	24.1	23.3	25.2	21.0	22.5	23.8	24.0	22.4	23.1
14: Dist. max. depth of med. condyle	26.0	26.8	28.1	23.0	24.2	26.4	26.1	24.0	25.0
7: Max. diameter facet 3rd carpal	36.0	33.5	33.0	31.0	29.3	32.3	31.8	30.6	31.0
8: Diam. anterior facet 2nd carpal	10.0	10.5	11.5	11.0	10.0	9.6	9.7	10.8	9.5

MC III	*H. minus* Sbastopol	*H. periaf.* Arqu. n = 1–2	*H. matthewi* Gulpinar n = 1	*H. dietrichi* Piker. n = 3–6	*H. platyg.* Tarak. n = 2–3	*H. platyg.* Salon. n = 1–2	*H.* cf. *sitif.* Baccin. n = 2	*H.* cf. *concud.* Lubér. n = 1–5	*H. brachypus* Polgardi
1: Maximal length			198.0	229.8	226.0	223.5		210.4	210.0
3: Minimal breadth	18.0	16.1	19.0	24.3	29.5	29.5		27.8	28.2
4: Depth at level of 3	14.0	12.5	17.9	22.3	24.0	24.0		20.5	22.3
5: Proximal articular breadth		22.0	28.0	35.8	39.5	38.0		36.1	40.0
6: Prox. art. depth		15.2	19.5	25.6	27.2	28.0			
10: Distal max. supra-art. breadth	25.0	18.0	26.1	33.2	38.0	38.0	33.7	33.1	38.7
11: Dist. max. art. breadth	24.0	17.6	25.0	31.9	35.3	33.5	34.5	31.6	37.0
12: Dist. max. depth of keel	17.3	15.3	22.0	27.6	27.8	26.5	27.4	26.3	28.2
13: Dist. min. depth of medial condyle	14.8	13.5	19.0	24.7	24.0	23.1	22.6	23.3	25.0
14: Dist. max. depth of med. condyle	17.0	14.3	20.7	25.7	25.8	25.1	24.4	24.6	26.0
7: Max. diameter facet 3rd carpal		18.3	23.0	31.8	32.5	33.0		29.4	33.0
8: Diam. anterior facet 2nd carpal		6.3	9.0	10.0	10.8	9.0		10.5	11.0

Chapter 12

A Consideration of Old World Hipparionine Horse Phylogeny and Global Abiotic Processes

Raymond L. Bernor and Diana Lipscomb

Hipparionine horses are among the most abundant ungulates in Old World later Neogene faunas. Until recently, their systematic relationships have been poorly understood. After nearly 150 years of isolated taxonomic studies, the group was critically reviewed by Gromova (1952), Gabunja (1959), and Forsten (1968), but without any phylogenetic analysis. In their initial analysis of Eurasian and North African hipparionine evolutionary relationships, Woodburne and Bernor (1980) discovered that the genus "*Hipparion*" is a paraphyletic grade composed of several single-to-multiple species lineages. Further detailed revisions of the relationships between local faunas of different provinces and continents included hipparions from Iran (Bernor, 1985a), Indo-Pakistan (MacFadden and Woodburne, 1982; Bernor and Hussain, 1985), Greece (Koufos, 1984, 1986, 1987a, 1987b, 1987c), North Africa (Bernor et al., 1987b), China (Qiu et al., 1987; Bernor et al., 1990b), and Europe (Bernor et al., 1980, 1988b, 1990, in press; also see reviews by Alberdi, 1989; Bernor et al., 1989; Woodburne, 1989). As the species-diverse lineages of Old World hipparionines are more highly resolved systematically, they become more useful for chronologic and biogeographic interpretations.

Continental mammalian fossils have been widely used for stratigraphic and chronologic correlations. However, they are never distributed continuously through the stratigraphic record: there are always stratigraphic, and hence temporal, gaps. As a result, mammalian paleontologists correlate discrete segments of the stratigraphic record by inferring the evolutionary stage of occurring taxa that are distributed broadly through time and space. In fact, the taxa used are often a paraphyletic grade: they represent incomplete lineages. This circumstance raises the need for paleontologists to examine carefully the bases for determining evolutionary relationships before inferring biochronologic ranking.

There have been some recent proposals to use congruence between phylogenetic (cladistic) trees and temporal duration in the stratigraphic record to test for preservational bias (Novacek and Norell, 1982; Eldredge and Novacek, 1985; Norell, 1992; Norell and Novacek, 1992). Here, we use cladistic trees of the "*Hippotherium*" and "*Sivalhippus*" Complexes to reconstruct the sequence of evolutionary events. These evolutionary events are then compared to independent geochronologic ages of the taxa's first known occurrences; from this framework we seek to infer the relationship between biogeographic distribution and global abiotic processes. Our uses of the name *Hipparion* and other terms are defined in appendix 12.1.

Chronological Framework

European, west Asian, and North African faunas rarely occur in sections that are dated radioisotopically and are usually of insufficient length to allow magnetostratigraphic correlations. Rather, they have been age-ranked using a biochronologic correlation system: Mammal Neogene (MN) units (*sensu* Fahlbusch, 1991). Thaler (1966) was the first to rank several western European mammal faunas using a "stage-of-evolution" methodology. His system was subsequently refined by Sudre (1969, 1972), Hartenberger (1969), and Thaler (1972).

Mein (1975, 1979, 1989) revolutionized the emerging

ISBN 0–300–06348–2.

mammalian biochronologic system by characterizing zones using three criteria: 1) characteristic taxa belonging to broadly distributed mammalian groups; 2) well-known generic first appearances; and 3) characteristic associations of genera. Lindsay and Tedford (1989) and Fahlbusch (1991) have subsequently argued correctly that MN "zones" do not have a physical stratigraphic basis and are therefore purely biochronological. Fahlbusch (1991) has suggested that MN "zones" be termed MN units. Steininger et al. (1989) identified radioisotopic, magnetostratigraphic, and marine and continental interdigitations to give the MN system some independent chronologic tie points through several intervals in the system.

Daams and Freudenthal (1981) in particular have criticized the use of MN units because of the lack of a strictly biostratigraphic framework. In Europe, such a biostratigraphic basis exists only in Spain (Van der Weerd, 1976; Daams et al., 1977; Daams and Freudenthal, 1988, 1989; Daams and van der Meulen, 1984). Correlations between Spain and other European districts are confounded by the striking provinciality and diachroneity of mammalian events (Bernor, 1978, 1983, 1984), which are believed to be due to tectonically based regional paleogeographic and paleoenvironmental biases (Rögl and Steininger, 1983). South Asia has a well-developed magnetostratigraphic framework, but biochronologic correlations are also made difficult by marked provinciality. This provinciality is episodically punctuated by intercontinental migration, which allows a limited basis for extraprovincial correlations (Bernor, 1983, 1984; Barry et al., 1985). In contrast, East Asia had broader biogeographic connections with Europe during the Neogene, which have permitted Qiu (1989) to correlate a series of diverse Chinese mammalian faunas with European MN units. Recent magneto- and biostratigraphic research in China has begun to achieve increased precision in Chinese correlations (Tedford et al., 1991). An extensive radioisotopic chronology exists for East Africa, within which the mammalian record has been integrated (see Harris, 1983; Pickford, 1983; Brown and Feibel, 1991, for a recent review). South African strata are thus far only biochronologically correlated to distant strata (Vrba, 1976). Of the various Old World correlation methods used, mammalian biochronology has clearly been the one most broadly applied. Bernor and Tobien (1990) have argued that the ability to correlate localities separated by intercontinental distances is due to the relationship between episodic short-interval global eustatic events (Haq et al., 1987) and mammalian biogeographic extensions. Such correlations, however, usually lack resolution in the absence of a well-resolved systematic basis such as provided by cladistic analyses. Only a few such analyses for Old World mammals have appeared to date, including the Bovidae (Vrba, 1976); Proboscidea (Tassy, 1989), hipparionine horses (Bernor et al., 1989; Bernor and Lipscomb, 1991), and eucatarrhine primates (Begun, 1992). We compare cladograms of the "*Hippotherium*" and "*Sivalhippus*" Complexes to independent chronologic estimates for first occurrences of subclades. Thereafter, we examine the potential relationships between continental migratory extensions and short-interval global eustatic events.

Cladistic Analysis

Forty-nine characters of the skull, maxillary, and mandibular cheek teeth (tables 12.1 and 12.2) were used to analyze the taxa under consideration. Character polarity was determined by the outgroup comparison method, using taxa that were progressively further removed from Old World hipparions: *Cormohipparion occidentale, C. sphenodus, C. goorisi, Merychippus insignis,* and *Parahippus leonensis* (fig. 12.1; tables 12.1 and 12.2). Character state order of multistate characters was determined using the method described by Lipscomb (1992).

The analysis produced a single tree with a consistency index equal to 70 and a residual index equal to 86 (fig. 12.1). In figures 12.2 and 12.3 the advanced characters (apomorphies) that define each monophyletic group are indicated.

Old World Hipparionine Phylogenies and Their Chronology

The "Hippotherium" Complex. Berggren and Van Couvering (1974) proposed that "*Hipparion*" first appeared across Eurasia and Africa during a geochronologic "instant." These authors reported a number of Old World localities where this event was recorded by directly associated radioisotopic calibrations and marine and continental correlations (see also Sen, 1989; Bernor et al., in press b; Swisher, in press; Woodburne et al., in press). Currently, it is believed that the "Hipparion Datum" represents a rapid, widespread migration of these horses across Eurasia and Africa.

The "Hipparion Datum" includes a group of Mio-

Table 12.1. Character State Distribution of Hipparinine Horses

<table>
<tr><td colspan="2">North American Outgroup Taxa</td></tr>
<tr><td></td><td>. . . . | . . . . 1 . . . . | . . . . 2 . . . . | . . . . 3 . . . . | . . . . 4 . . . . | . . . . 5</td></tr>
<tr><td>Parahippus leonensis</td><td>A A A A D C A C B A A A A A A A A A D A D A A A A A A B A A A A A A B A B B A C F A B A A A A A A</td></tr>
<tr><td>Merychippus insignis</td><td>A B A B C B A B A A A A A A A A A A C B B A B B A A A A A A A A A A B A B B B C D A B A B A A A A</td></tr>
<tr><td>Cormohipparion goorisi</td><td>B C A C A A A A A B B A A A A A A B B B B B B B B A A A A A A A A A B A B A B B D B B A B A A A A</td></tr>
<tr><td>Cormohipparion sphenodus</td><td>B C A C A A A A A B B A A A B A B B B B B B C B C B A A A A A A A A B A B A B B A C B A B A A A A</td></tr>
<tr><td>Cormohipparion occidentale</td><td>C C A D A A A A A B B A A A C A C C B B B B D B C B B A A A A A A A B A B A B B A C B A C A A A A</td></tr>
<tr><td colspan="2">Hippotherium and Related "Group 1" Taxa</td></tr>
<tr><td></td><td>. . . . | . . . . 1 . . . . | . . . . 2 . . . . | . . . . 3 . . . . | . . . . 4 . . . . | . . . . 5</td></tr>
<tr><td>Hippotherium primigenium</td><td>C C B D A A A A A B B A A A C A C C A B A B E B C B B A A A A A A A A A B A B A A C B A C B B A A</td></tr>
<tr><td>"Hipparion" weihoense</td><td>C C B E A A A A A B B A A A C B C C A B A B D B C B B A -</td></tr>
<tr><td>"Hipparion" catalaunicum 1</td><td>C C B J A A A A A B B A A A C B C C A B A B E B C B B A - - - - - - - - - - - - - C - - - - - - -</td></tr>
<tr><td>"Hipparion" africanum</td><td>C C B J A A A A A B B A A A D B C C A B A B F B C B B A A A A A A A B A B A B B A C B B C B B A A</td></tr>
<tr><td>"Hipparion" dermatorhinum</td><td>D C B D B B A A A B B A A A F B C C A B B B J B C B B A A A A B B B B B B A B B B C B B C B B A A</td></tr>
<tr><td>"Hipparion" giganteum</td><td>C C B D B B A A B B B A A A C B C C A B A C E B C B B A - - - A A A - A - - - - - - - - - B B A A</td></tr>
<tr><td>Hipparion gettyi</td><td>C C B E A B A B B B B A A A C B C C B B A C C B C B B A -</td></tr>
<tr><td>Hipparion melendezi</td><td>C C B G A B A B B B B A A A C B C C A B A B E B C B B A -</td></tr>
<tr><td>"Hipparion" catalaunicum 2</td><td>C C B J D C A C B B B A A A C B C C B B A B K B C B B A -</td></tr>
<tr><td colspan="2">"Sivalhippus" Complex</td></tr>
<tr><td></td><td>. . . . | . . . . 1 . . . . | . . . . 2 . . . . | . . . . 3 . . . . | . . . . 4 . . . . | . . . . 5</td></tr>
<tr><td>"Sivalhippus" platyodus</td><td>C C B D B B A A A B B A A A C B D C A B A B E B C B B A A A A C C C A C B A B B A C B A D B B A A</td></tr>
<tr><td>"Sivalhippus" perimense</td><td>G C B F C B A A A B B A A A C B D D A B A B H B C B B A A A A C C C - C B A B A C - B D D B B A B</td></tr>
<tr><td>"Sivalhippus" turkanense</td><td>H C B H C C A D B B B A A A C B D D A B A B H B C B B A - - - C - C B - B A - C - - B D - B B A B</td></tr>
<tr><td>"Plesiohipparion" houfenense</td><td>H C B H C C A D B B B A A A C B D D A B A B H B C B B A B A A C C C B C B A B C A C B D D A A A B</td></tr>
<tr><td>Eurygnathohippus afarense</td><td>H C B H C C A D B B B A A A C B D D A B A B K B C B B A B B B C C C B C B A B A B C A D D A A A B</td></tr>
<tr><td>"Plesiohipparion" crusafonti</td><td>I C B H D D A E B B B A A A - B D D A B A B H B C B B A B A A C C C - C B A B - - - B D D - - - B</td></tr>
<tr><td>Proboscidipparion pater</td><td>I C B I D D A E B B B A A A G B D D A B A B E B C B B A B A A C C C B C B A B C B C B D D B B A B</td></tr>
<tr><td>Proboscidipparion sinense</td><td>I C B I D D A E B B B A A A H B D D E B C B I B C B B A B A A C C C B C B A B C B C B D D B B A B</td></tr>
<tr><td>"Plesiohipparion" aff. houfenense</td><td>I C B I D D A E B B B A A A C B D D B B A C H B C B B A B A A C C F B F B A B C B C B D D A A B B</td></tr>
<tr><td>"Plesiohipparion" huangheense</td><td>- A B A A F F F B F B A C B A E B D D A A B B</td></tr>
<tr><td>"Plesiohipparion" aff. huangheense</td><td>- - - - - - - - - - - - - - - - - - A B A C H B C B B - - - - F F F B F B A C B A E B E E A A B B</td></tr>
</table>

Definitions of Hipparionine Character States

The transformation series for each multistate character is included in parentheses.

1) Relationship of lacrimal to the preorbital fossa: A = lacrimal large, rectangularly shaped, invades medial wall and posterior aspect of preorbital fossa; B = lacrimal reduced in size, slightly invades or touches posterior border of preorbital fossa; C = preorbital bar long with the anterior edge of the lacrimal placed more than half the distance from the anterior orbital rim to the posterior rim of the fossa; D = preorbital bar reduced slightly in length but with the anterior edge of the lacrimal placed still more than half the distance from the anterior orbital rim to the posterior rim of the fossa; E = preorbital bar vestigal, but lacrimal as in D; F = preorbital bar absent; G = preorbital bar very long, with anterior edge of lacrimal placed less than half the distance from the anterior orbital rim to the posterior rim of the fossa; H = preorbital bar vestigal, but lacrimal as in G; I = preorbital bar absent, but lacrimal as in G; (A-B-C-[D-E-F],G-H-I).
2) Nasolacrimal fossa: A = preorbital fossa large, ovoid shape and separated by a distinct medially placed, dorsoventrally oriented ridge, dividing preorbital fossa into equal anterior (nasomaxillary) and posterior (nasolacrimal) fossae; B = nasomaxillary fossa sharply reduced compared to nasolacrimal fossa; C = nasomaxillary fossa absent (lost), leaving only nasolacrimal portion (when a preorbital fossa is present); (A-B-C).
3) Orbital surface of lacrimal bone: A = with foramen; B = reduced or lacking foramen.
4) Preorbital fossa morphology: A = large, ovoid shape, anteroposteriorly oriented; B = preorbital fossa truncated anteriorly; C = preorbital fossa further truncated, dorsoventrally restricted at anterior limit; D = subtriangularly shaped and anteroventrally oriented; E = subtriangularly shaped and anteroposteriorly oriented; F = egg-shaped and anteroposteriorly oriented; G = C-shaped and anteroposteriorly oriented; H = vestigial but with a C-shaped or egg-shaped outline; I = vestigial without C-shape outline, or absent; J = elongate, anteroposteriorly oriented; K = small, rounded structure; L = posterior rim straight, with non-oriented medial depression; (A-B-C-D-[E-G],[F-H-I],J-K).
5) Fossa posterior pocketing: A = deeply pocketed, greater than 15 mm in deepest place; B = pocketing reduced, moderate to slight depth, less than 15 mm; C = not pocketed but with a posterior rim; D = absent; (A-B-C-D).
6) Fossa medial depth: A = deep, greater than 15 mm. in deepest place; B = moderate depth, 10–15 mm in deepest place; C = shallow depth, less than 10 mm in deepest place; D = absent; (D-C-B-A).
7) Preorbital fossa medial wall morphology: A = without internal pits; B = with internal pits (C-B-A-D-E).

8) Fossa peripheral border outline: A = strong, strongly delineated around entire periphery; B = moderately delineated around periphery; C = weakly defined around periphery; D = absent with a remnant depression; E = absent, no remnant depression (C-B-A-D-E).
9) Anterior rim morphology: A = present; B = absent.
10) Placement of infraorbital foramen: A = placed distinctly ventral to approximately half the distance between the preorbital fossa's anteriormost and posteriormost extent; B = inferior to, or encroaching upon, anteroventral border or the preorbital fossa.
11) Confluence of buccinator and canine fossae: A = present; B = absent, buccinator fossa is distinctly delimited.
12) Buccinator fossa: A = unpocketed posteriorly; B = pocketed posteriorly.
13) Caninus (intermediate) fossa: A = absent; B = present.
14) Malar fossa: A = absent; B = present.
15) Nasal notch position: A = at posterior border of canine or slightly posterior to canine border; B = approximately half the distance between canine and P2; C = at or near the anterior border of P2; D = above P2; E = above P3; F = above P4; G = above M1; H = posterior to M1; (A-B-C-D-E-F-G-H).
16) Presence of P1: A = persistent and functional; B = lost early
17) Curvature of maxillary cheek teeth: A = very curved; B = moderately curved; C = slightly curved; D = straight; (A-B-C-D).
18) Maximum cheek tooth crown height: A = < 30 mm; B = 30–40 mm; C = 40–60 mm; D = > 60 mm maximum crown height; (A-B-C-D).
19) Maxillary cheek tooth fossette ornamentation: A = complex, with several deeply amplified plications; B = moderately complex with fewer, more shortly amplified, thinly banded plications; C = simple complexity with few, shortly amplified plications; D = generally no plis; E = very complex; (D-C-B-A-E).
20) Posterior wall of postfossette: A = may not be distinct; B = always distinct.
21) Pli caballin morphology: A = double; B = single or occasionally poorly defined double; C = complex; D = plis not well formed; (D-B-A-C).
22) Hypoglyph: A = hypocone frequently encircled by hypoglyph; B = deeply incised, infrequently encircled hypocone; C = moderately deeply incised; D = shallowly incised (A-B-,C-D).
23) Protocone shape: A = round q-shape; B = oval q-shape; C = oval; D = elongate-oval; E = lingually flattened–labially rounded; F = compressed or ovate; G = rounded; H = triangular; I = triangular-elongate; J = lenticular; K = triangular with rounded corners; (A-B-C-D-E-F,G,J,H-I,K).
24) Isolationof protocone: A = connected to protoloph; B = isolated from protoloph.
25) Protoconal spur: A = elongate, strongly present; B = reduced, but usually present; C = very rare to absent; (A-B-C).
26) Premolar protocone/hypocone alignment: A = anteroposteriorly aligned; B = protocone more lingually placed.
27) Molar protocone/hypocone alignment: A = anteroposteriorly aligned; B = protocone more lingually placed.
28) P2 anterostyle (26U)/paraconid(26L): A = elongate; B = short and rounded.
29) Mandibular incisor morphology: A = not grooved; B = grooved.
30) Mandibular incisor curvature: A = curved; B = straight.
31) I3 lateral aspect: A = elongate, not transversely constricted; B = very elongate, transversely constricted; C = atrophied; (A-B-C).
32) Premolar metaconid: A = rounded; B = elongated; C = angular on distal surface; D = irregularly shaped; E = square-shaped; F = pointed (A-B,[C-F][E-D]).
33) Molar metaconid: A = rounded; B = elongated; C = angular on distal surface; D = irregularly shaped; E = square-shaped; F = pointed; (A-B,[C-F][E-D]).
34) Premolar metastylid: A = rounded; B = elongated; C = angular on proximal surface; D = irregularly shaped; E = square-shaped; F = pointed; (A-B,[C-F][E-D]).
35) Premolar metastylid spur: A = present; B = absent
36) Molar metastylid: A = rounded; B = elongate; C = angular on proximal surface; D = irregularly shaped; E = square-shaped; F = pointed; (A-B,[C-F][E-D]).
37) Molar metastylid spur: A = present; B = absent.
38) Premolar ectoflexid: A = does not separate metaconid and metastylid; B = separates metaconid and metastylid.
39) Molar ectoflexid: A = does not separate metaconid and metastylid; B = separates metaconid and metastylid; C = converges with preflexid and postflexid to abut against metaconid and metastylid (A-B-C).
40) Pli caballinid: A = complex; B = rudimentary or single; C = absent; (A-B-C).
41) Protostylid: A = present on occlusal surface; B = absent on occlusal surface, but may be on side of crown buried in cement; C = strong, columnar; D = a loop; E = a small, poorly developed loop; F = a small, pointed projection continuous with the buccal cingulum; (F-D-B-A-C).
42) Protostylid orientation: A = courses obliquely to anterior surface of tooth; B = less obliquely coursing, placed on anterior surface of tooth; C = vertically placed, lies flush with protoconid enamel band; D = vertically placed, lying lateral to protoconid band; E = open loop extending posterolabially; (A-B-C-D,E).
43) Ectostylids: A = present; B = absent.
44) Premolar linguaflexid: A = shallow; B = deeper, V-shaped; C = shallow U shape; D = deep, broad U shape; E = very broad and deep U shape; (A-B-C-D-E).
45) Molar linguaflexid: A = shallow; B = V-shaped; C = shallow U shape; D = deep, broad U shape; E = very broad and deep U shape; (A-B-C-D-E).
46) Preflexid morphology: A = simple margins; B = complex margins; C = very complex; (A-B-C).
47) Postflexid morphology: A = simple margins; B = complex margins; C = very complex; (A-B-C).
48) Postflexid invades metaconid-metastylid by anteriormost portion, bending sharply ligually: A = no; B = yes.
49) Protoconid enamel band morphology: A = rounded; B = flattened.

Table 12.2. First and Last Appearance Estimates for Hipparionine Horses Used in This Study

	First Appearance Datum	Last Appearance Datum
North American Taxa		
Parahippus leonensis	19.0	18.0
Merychippus insignis	16.0	15.0
Cormohipparion goorisi	15.0	14.0
Cormohipparion sphenodus	14.0	12.5
Cormohipparion occidentale	12.5	8.0
Eurasian "Group 1" Taxa		
Hippotherium primigenium	10.5	9.0
"Hipparion" weihoense	10.0	?
"Hipparion" catalaunicum 1	10.0	?
"Hipparion" catalaunicum 2	8.6	?
Hipparion gettyi	9.0	8.2
Hipparion melendezi	9.0	?
"Hipparion" giganteum	8.0	?
"Hipparion" africanum	9.6	?5.0
"Hipparion" dermatorhinum	8.0	?
"Sivalhippus" Complex		
"Sivalhippus" platyodus	9.5	?5.0
"Sivalhippus" perimense	8.0	?
"Sivalhippus" turkanense	6.0	?5.0
"Plesiohipparion" houfenense	6.0	?5.0
Eurygnathohippus afarense	3.4	?2.6
"Plesiohipparion" aff. houfenense	2.0	?0.7
"Plesiohipparion" crusafonti	5.0	?3.5
Proboscidipparion pater	5.0	3.0
Proboscidipparion sinense	3.0	?0.7
"Plesiohipparion" huangheense	2.6	?
"Plesiohipparion" aff. huangheense	2.6	?

cene Old World horses, "Group 1" of Woodburne and Bernor (1980), which Bernor et al. (1988, p. 440, fig. 10) identified as a paraphyletic group. Our cladistic analysis revises several aspects of previous cladistic hypotheses (MacFadden, 1984; Bernor and Lipscomb, 1991) using an expanded data set of forty-nine (tables 12.1 and 12.2; app. 12.2).

The order of divergence given for the "*Hippotherium*" Complex taxa is congruent with their respective independent ages (Bernor et al., 1989; Bernor et al., in press b). These age criteria are various, including: radioisotopic (r), magnetostratigraphic (m), biochronologic (b), and biostratigraphic in a strictly superpositional sense (s). The cladistic-chronologic ranking (first known occurrence) is: *Hippotherium primigenium*, 10.5 myr (r, m, b); "*Hipparion*" *weihoense*, 10 myr (b); "*Hipparion*" *catalaunicum* 1, 10 myr (b); "*Hipparion*" *africanum*, ≤ 9.6 myr (r, m, s, b); "*Hipparion*" *dermatorhinum*, ≤ 8 myr (b); "*Hipparion*" *giganteum*, 8 myr (r, m, b); *Hipparion gettyi*, ≤ 9 myr (r, b); "*Hipparion*" *melendezi*, ≥ 9 myr (b); "*Hipparion*" *catalaunicum* 2, ≤ 8.6 myr (s, b).

The polytomy "*Hipparion*" *dermatorhinum*—"*Hipparion*" *giganteum*–*Hipparion gettyi*–"*Hipparion*" *melendezi*–"*Hipparion*" *catalaunicum* is less secure than other portions of the cladogram. "*Hipparion*" *dermatorhinum* and "*Hipparion*" aff. *catalaunicum* both exhibit a number of autopomorphies that suggest a significant gap in our knowledge about their evolutionary history. This gap may or may not represent a significant chronologic hiatus, depending on the evolutionary processes that shaped their history. Nevertheless, the presence of several autopomorphies diminishes these taxa's biochronologic utility and is an aspect of their record that would be unknown, or underappreciated, without a cladistic analysis.

The "Sivalhippus" Complex. The "Sivalhippus" Complex has become recognized through a series of independent investigations (Bernor and Hussain, 1985; Qiu et al., 1987; Bernor et al., 1987c; Flynn and Bernor, 1987; Alberdi, 1989; Bernor et al., 1989, 1990; Bernor and Lipscomb, 1991). Alberdi has suggested that all taxa belonging to the "*Sivalhippus*" Complex should be maintained within the genus *Hipparion*, despite the fact that she posits their origin from a hipparionine clade separate from other Old World hipparions, which emigrated from North America ca. 5.5 myr (Alberdi, 1989, p. 252, fig. 13.5, p. 254, fig. 13.6). We reject this paraphyletic grouping.

Figure 12.1 presents the relationship of the "*Sivalhippus*" Complex to other "Group 1" horses. Figure 12.3 shows the sister-group relationships of the horses in the "*Sivalhippus*" Complex based on the cladistic analysis of forty-nine characters (see app. 12.3 for details of the character state distribution).

The order of cladistic divergence for the "*Sivalhippus*" Complex is for the most part congruent with independent age criteria given elsewhere (Bernor et al., 1989; Bernor and Lipscomb, 1991). These ages are: "*Sivalhippus*" *platyodus*, 9.5 myr (b); "*Sivalhippus*" *perimense*, 8 myr (s, m); "*Sivalhippus*" *turkanense*, 6 myr (r); *Plesiohipparion houfenense*, 6 myr (b); *Eurygnathohippus afarense*, ≥ 3.4 myr (s, r, m); *Plesiohipparion crusafonti*, < 5 myr (b); *Proboscidipparion pater*, 5 myr (b); *Proboscidipparion*

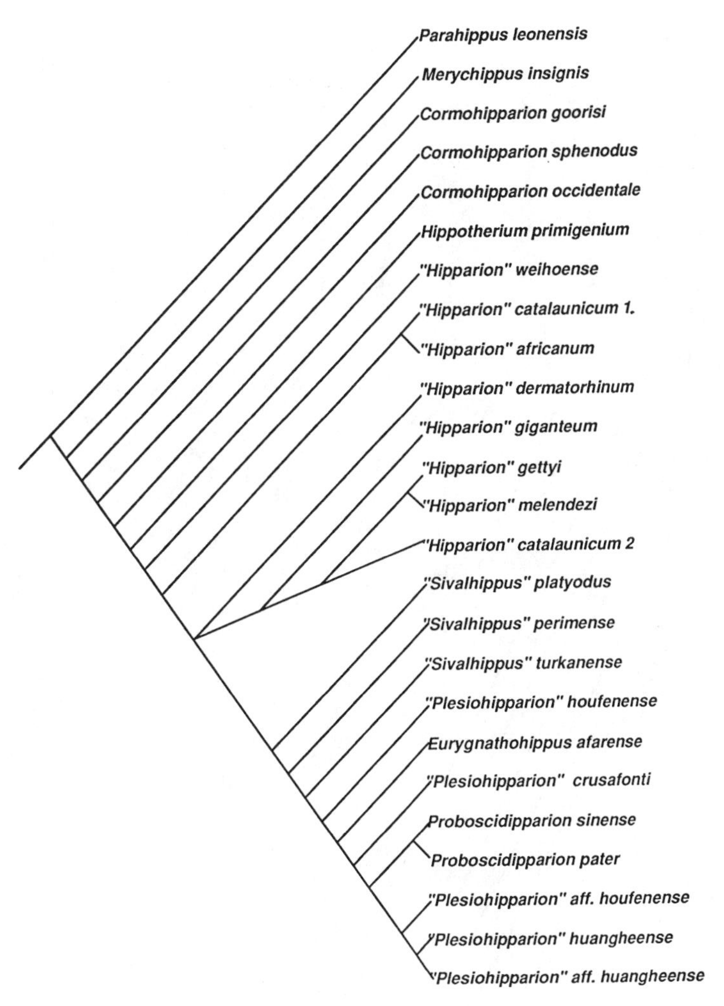

Fig. 12.1. Cladogram of 25 taxa using characters 1–49 (see legend to table 12.1). The cladogram has a length of 154 steps, a consistency index of 0.70, and a retention index of 0.86.

sinense, 3 myr (b); *Plesiohipparion* aff. *houfenense,* > 2 myr (b); *Plesiohipparion huangheense,* 2.6 myr (b, m); *Plesiohipparion* aff. *huangheense,* 2.6 myr (b, m). Of these taxa, the divergence of *Eurygnathohippus afarense* from the cladogram is incongruent with its age of first known occurrence. This incongruency can be due to: 1) unrecognized earlier occurrence of this lineage; or 2) incorrect placement in the cladogram, which will be discovered as more information is gathered and analyzed. For example, there may be a closer relationship with "*Sivalhippus*" *turkanense* than is indicated by known characters.

Present evidence suggests that the "*Sivalhippus*" Complex first evolved in Asia, with the earliest occurrence being "*Sivalhippus*" *platyodus,* ca. 9.5 myr. Between 8 and 5.5 myr, an advanced member of the clade not dissimilar in morphology to "*Sivalhippus*" *perimense* would appear to have extended its southern Asian range to East Africa ("*Sivalhippus*" *turkanense*). Bernor and Lipscomb (1991, p. 118) have noted that an 8-myr biogeographic extension would correlate with a period of global sea lowering (Haq et al., 1987). This hypothesis should be testable with other Eurasian and African mammalian lineages of medial Turolian (MN 12) age. A second "*Sivalhippus*" Complex biogeographic extension is the "*Plesiohipparion*" group, which currently is first known to have occurred in China (6 myr; Qiu et al., 1987) and to have extended into Europe, including Spain (Alberdi, 1989; Bernor et al., 1989; Bernor and Lipscomb, 1991), Hungary (Bernor, personal observation). The Plio-Pleistocene *Eurygnathohippus* clade may have origi-

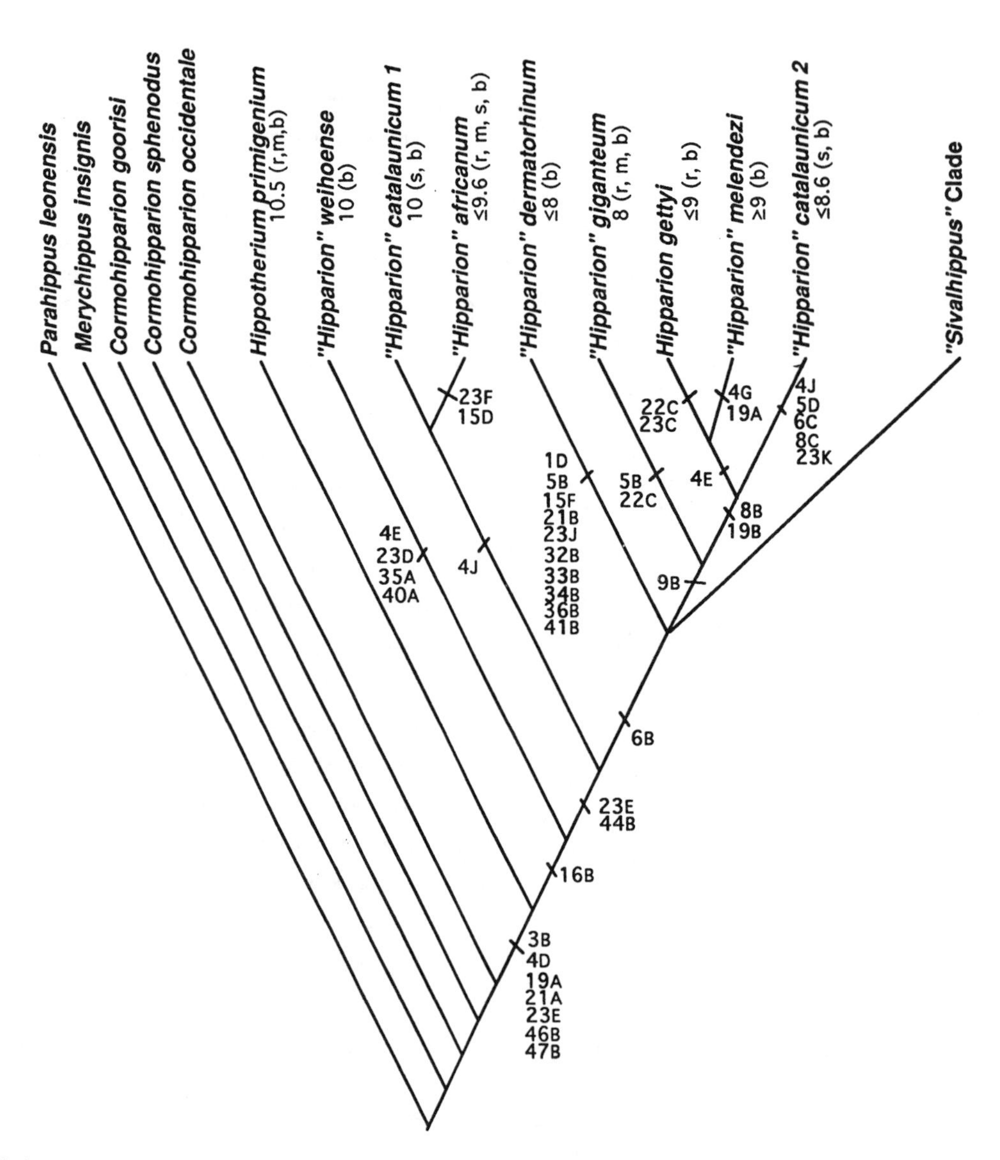

Fig. 12.2. The lower portion of the cladogram (fig. 12.1) is expanded and redrawn here to show the character states that define each clade.

nated from the "*Plesiohipparion*" clade's latest Miocene extension. Or, if *Eurygnathohippus afarense* was derived directly from an ancestor like "*Sivalhippus*" *turkanense,* there was a single extension of the "*Sivalhippus*" Complex into Africa between 8 and 5.5 myr. A Late Miocene extension of the "*Plesiohipparion*" clade is also closely correlated with a global regression event, the Messinian Salinity Crisis (Haq et al., 1987; Bernor et al., 1989; Bernor and Lipscomb, 1991).

The *Proboscidipparion* lineage was apparently endemic to East Asia, ca. 5–2 myr. The lineage of "*Plesiohipparion*" aff. *houfenense,* "*Plesiohipparion*" *huangheense,* and "*Plesiohipparion*" aff. *huangheense* may have evolved from an ancestor similar to "*Plesiohipparion*" *houfenense* and remained endemic to East Asia except for a medial Pliocene extension into western Asia (Gülyazi, Turkey).

Bernor and Lipscomb (1991), following Meulen and Kolfschoten (1986), have referred the Gülyazi horse, "*Plesiohipparion*" aff. *huangheense,* to the medial Pliocene (MN 16b) based on rodent biochronology and on the occurrence of a primitive *Equus* (first appearance datum [FAD] = MN 16b). Lindsay et al. (1980) have magnetostratigraphically calibrated the Old World *Equus* FAD as

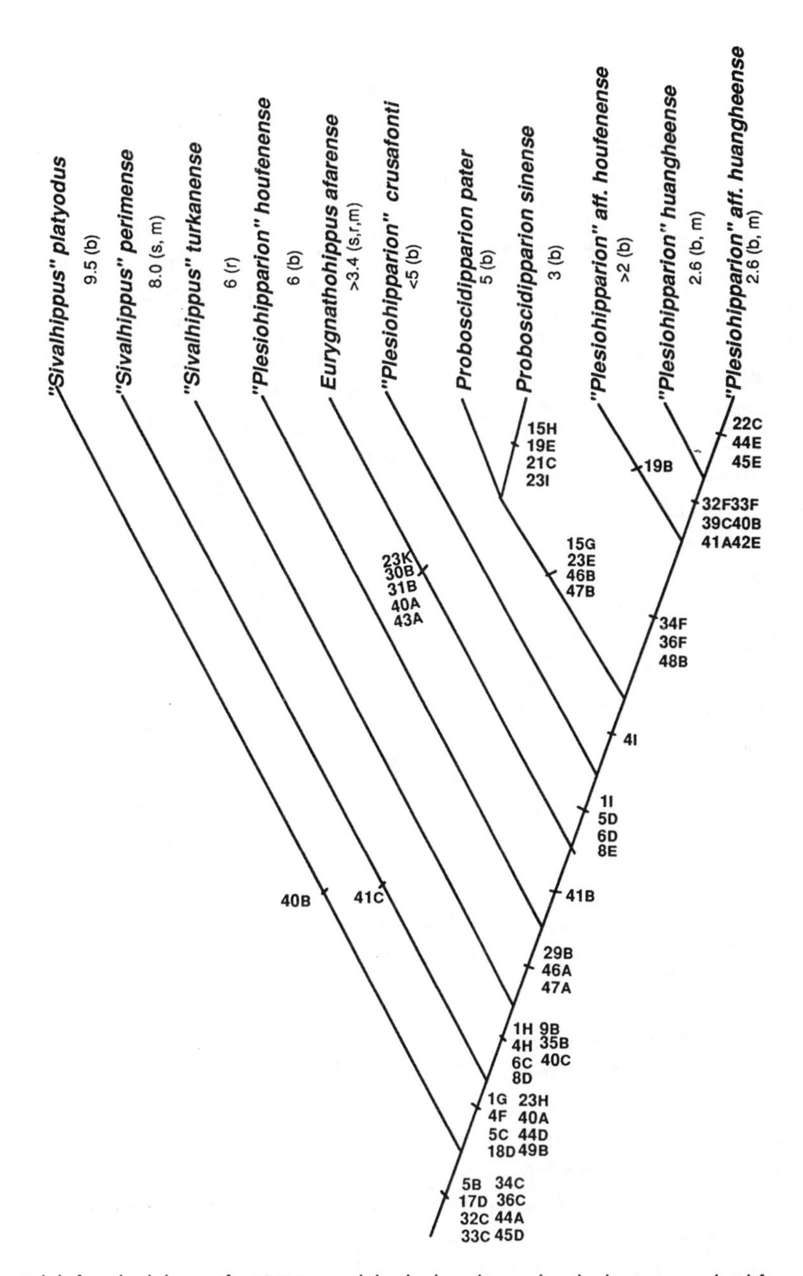

Fig. 12.3. The "*Sivalhippus*" clade from the cladogram (fig. 12.1) is expanded and redrawn here to show the character states that define each clade.

being closely correlative with the Gauss-Matuyama boundary, ca. 2.6 myr (magnetic recalibration after Cande and Kent, 1992).

The 2.6 myr datum is also correlated with a global sea-lowering event in which Arctic and Antarctic ice-volume expansion is coupled for the first time (Prentice and Denton, 1988, p. 398). Suc and Zagwijn (1983) and Meulen and Kolfschoten (1986) have suggested that the 2.6 myr interval correlates with a major European continental-wide environmental shift whereby forests were substantially reduced and a characteristic seasonal Mediterranean climate, with marked summer drought, appeared. Vrba (1988) has suggested that the 2.6 myr global sea-lowering event precipitated dramatic environmental shifts and accompanying mammalian speciation and extinction events in Africa. Thus, the climatic event correlated with the biogeographic extension of "*Plesiohipparion*" *huangheense* into Turkey has a global correlation.

Discussion

We find that those Old World hipparionine lineages that were restricted to smaller provinces, such as Plio-Pleistocene Asian *Proboscidipparion* or the later Vallesian and Turolian species of the *Hippotherium primigenium* evolutionary lineage, may reveal adaptations to restricted biotopes. In contrast, geographically extensive lineages such as the "*Sivalhippus*" Complex reveal adaptability to intercontinental-scale environments and are useful for long-range correlations. Both restricted and broadly ranging lineages have their use for correlation. For example, because *Proboscidipparion* species appear to be restricted to East Asia, the evolutionary history of this lineage could become increasingly useful for provincial correlations. On the other hand, the "*Sivalhippus*" Complex includes a number of closely related taxa (that is, species of "*Sivalhippus*" and "*Plesiohipparion*") that span Eurasia and Africa, making them particularly useful for correlating between Eurasia and Africa during the Late Miocene to Middle Pliocene. During this time, species belonging to this complex appear to have made major intercontinental extensions ca. 8 myr, 5 myr, and 2.6 myr. Each of these extensions correlates with a moment of global eustatic sea lowering. These eustatic events, and the continental species extensions that accompany them, were apparently of short geochronologic duration and were followed by endemic evolution (the African Late Miocene–Pleistocene *Eurygnathohippus* lineage being the best example).

The examples given here have general applicability for mammalian geochronology and biogeography. We have shown that cladistic analyses provide the basis for evaluating biochronologic correlations and biogeographic reconstructions. Furthermore, hypotheses proposed here can be tested and accepted using multiple mammalian lineages as a rigorous basis for deeper insights into the climatic and tectonic phenomena that may regulate evolution and biogeographic extensions.

Summary

This chapter has demonstrated that mammalian biochronologic correlations can gain precision by employing a cladistic methodology on densely sampled species derived from faunas that have a preponderance of independent chronologic information. These analyses have enabled us to hypothesize moments of intercontinental migration and to strike correlations between European MN biochronologic units and faunas in Asia and Africa. The intercontinental extensions discussed here for hipparionine horses (Bernor et al., 1989, in press b), and those previously discussed by other authors for Bovidae (Vrba, 1976), Proboscidea (Tassy, 1989), eucatarrhine primates (Bernor et al., 1988a; Begun, 1992), the Pasalar fauna (Turkey; Bernor and Tobien, 1990), and Siwalik faunas (Barry et al., 1985), would all appear to be related to events involving the global lowering of sea level. On the other hand, intraprovincial restriction of lineages such as *Proboscidipparion* may have been regulated by tectonic or climatic factors that maintained strikingly different biotopes in adjacent provinces. Whether a lineage is provincially restricted or ranges at the intercontinental scale, the information gained provides insights into the biology of the species under study, and provides further opportunities for a broad range of paleobiological studies.

Acknowledgments

We wish to thank Dr. Elisabeth Vrba for asking R. Bernor to participate in the symposium on paleoclimate and for inviting us to present our analysis on Old World hipparionine evolution and biogeography in this volume. We would like to thank Drs. John Barry, Daryl Domning, Robert Emry, Volker Fahlbusch, Larry Flynn, Everett H. Lindsay, Carl Swisher, and Michael O. Woodburne for providing useful criticisms as well as background discussion for this work. Hennig86 is under separate licenses to both R. Bernor and D. Lipscomb. This work was supported by grants to Bernor from the Alexander Von Humboldt Stiftung; NSF (BSR88-06645); NATO (CG85/0045); the National Geographic Society; and the Institute of Human Origins. Howard University provided research leave for Bernor to pursue field and museum research in central Europe and East Africa. The Staatliches Museum für Naturkunde, Karlsruhe; the Paleoanthropology Laboratories of the CRCCH, Addis Ababa, Ethiopia; and the National Museums of Kenya provided research facilities and equipment critical for developing the data presented here.

References

Alberdi, M. T. 1989. A review of Old World hipparionine horses. In *The evolution of perissodactyls*, pp. 234–261 (ed. D. R. Prothero and R. M. Schoch). Oxford, New York.

Barry, J. C., Johnson, N. M., Raza, S. M., and Jacobs, L. L. 1985. Neogene mammalian faunal change in southern Asia: Correlations with climatic, tectonic, and eustatic events. *Geology* 13:637–640.

Begun, D. 1992. Phyletic diversity and locomotion in primitive European hominids. *American Journal of Physical Anthropology* 87:311–340.

Berggren, W. A., and Van Couvering, J. A. 1974. The Late Neogene: Biostratigraphy, geochronology and paleoclimatology of the last 15 million years in marine and continental sequences. *Palaeogeography, Palaeoclimatology, Palaeoecology* 16:1–216.

Bernor, R. L. 1978. The mammalian systematics, biostratigraphy and biochronology of Maragheh and its importance for understanding Late Miocene hominoid zoogeography and evolution. Ph.D. diss., University of California, Los Angeles.

———. 1983. Geochronology and zoogeographic relationships of Miocene Hominoidea. In *New interpretations of ape and human ancestry,* pp. 21–64 (ed. R. L. Ciochon and R. S. Corruccini). Plenum, New York.

———. 1984. A zoogeographic theater and biochronologic play: The time/biofacies phenomena of Eurasian and African Miocene mammalian provinces. *Paléobiologie Continentale* 14: 121–142.

———. 1985. Systematic and evolutionary relationships of the hipparionine horses from Maragheh, Iran (Late Miocene, Turolian age). *Palaeovertebrata* 15:173–269.

Bernor, R. L., Fahlbusch, V., and Mittmann, H.-W. (in press a). *The evolution of western Eurasian later Neogene faunas.* Columbia University Press, New York.

Bernor, R. L., Flynn, L. J., Harrison, T., Hussain, S. T., and Kelley, J. 1988. *Dionysopithecus shuangouensis* (Catarrhini, Primates): A new element in the Kamlial fauna of southern Sind, Pakistan. *Journal of Human Evolution* 17:339–358.

Bernor, R. L., and Hussain, S. T. 1985. An assessment of the systematic, phylogenetic and biogeographic relationships of Siwalik hipparionine horses. *Journal of Vertebrate Paleontology* 5:32–87.

Bernor, R. L., Koufos, G., Woodburne, M. O., and Fortelius, M. (in press b). The evolutionary history and biochronology of European and southwestern Asian Late Miocene and Pliocene hipparionine horses. In *The evolution of western Eurasian later Neogene faunas.* (ed. R. L. Bernor, V. Fahlbusch, and H.-W. Mittmann). Columbia University Press, New York.

Bernor, R. L., Kovar-Eder, J., Lipscomb, D., Rögl, F., Sen, S., and Tobien, H. 1988b. Systematics, stratigraphic and paleoenvironmental contexts of first-appearing Hipparion in the Vienna Basin, Austria. *Journal of Vertebrate Paleontology* 8:427–452.

Bernor, R. L., Kovar-Eder, J., Suc, J.-P., and Tobien, H. 1990. A contribution to the evolutionary history of European Late Miocene age hipparionines (Mammalia, Equidae). *Paléobiologie Continentale* 17:291–309.

Bernor, R. L., and Lipscomb, D. 1991. The systematic position of "*Plesiohipparion*" aff. *huangheense* (Equidae, Hipparionini) from Gülyazi, Turkey. *Mitteilung der Bayerischen Staatsslammlung für Paläontologie und historische Geologie* 31:107–123.

Bernor, R. L., Qiu, Z., and Hayek, L.-A. 1990b. Systematic revision of Chinese hipparion species described by Sefve, 1927. *American Museum of Natural History Novitates* 2984:1–60.

Bernor, R. L., Qiu, Z., and Tobien, H. 1987c. Phylogenetic and biogeographic bases for an Old World hipparionine horse geochronology. *Annales Instituti Geologici Publici Hungarici* 70:43–53.

Bernor, R. L., and Tobien, H. 1990. The mammalian geochronology and biogeography of Pasalar (Middle Miocene, Turkey). *Journal of Human Evolution* 19:551–568.

Bernor, R. L., Tobien, H., Hayek, L., and Mittmann, H.-W. (in press c). *Hippotherium primigenium* (Equidae, Mammalia) from the Late Miocene of Höwenegg (Hegau), Germany. *Andrias.*

Bernor, R. L., Tobien, H., and Woodburne, M. O. 1989. Patterns of Old World hipparionine evolutionary diversification and biogeographic extension. In *Topics on European mammalian chronology,* pp. 263–319 (ed. E. H. Lindsay, V. Fahlbusch, and P. Mein). Plenum, New York.

Bernor, R. L., Woodburne, M. O., and Van Couvering, J. A. 1980. A contribution to the chronology of some Old World Miocene faunas based on hipparionine horses. *Géobios* 13:25–59.

Brown, F. H., and Feibel, C. S. 1991. Stratigraphy, depositional environments and palaeogeography of the Koobi Fora Formation. In *Koobi Fora Research Project.* Vol. 3, *Fossil ungulates: Geology, fossil artiodactyls and palaeoenvironments,* pp. 1–30 (ed. J. M. Harris). Clarendon, Oxford.

Cande, S. C., and Kent, D. V. 1992. A new geomagnetic polarity time scale for the Late Cretaceous and Cenozoic. *Journal of Geophysical Research* 97 (B10):13917–13951.

Daams, R., and Freudenthal, M. 1981. Aragonian: The stage concept versus Neogene mammal zones. *Scripta Geologica* 62:152–182.

———. 1988. Synopsis of the Dutch-Spanish collaboration program in the Aragonian type area. *Scripta Geologica,* special issue, 1:3–18.

———. 1989. The Ramblian and Aragonian: Limits, subdivision, geographical and temporal extension. In *Topics on European mammalian chronology,* pp. 263–319 (ed. E. H. Lindsay, V. Fahlbusch, and P. Mein). Plenum, New York.

Daams, R., Freudenthal, M., and Sierra, M. A. 1987. Ramblian: A new stage for continental deposits of Early Miocene age. *Geologie Mijnbouw* 65:297–308.

Daams, R., Freudenthal, M., and van der Weerd, A. 1977. Aragonian: A new stage for continental deposits of Miocene age. *Newsletters in Stratigraphy* 6:42–55.

Daams, R., and van der Meulen, A. 1984. Paleoenvironmental and paleoclimatic interpretation of micromammal faunal successions in the Upper Oligocene and Miocene of north central Spain. *Paléobiologie Continentale* 14:241–257.

Eisenmann, V., Alberdi, M.-T., de Giuli, C., and Staesche, U. 1988. *Studying fossil horses.* Vol. 1, *Methodology* (ed. M. O. Woodburne and P. Y. Sondaar). Brill, Leiden.

Eldredge, N., and Novacek, M. 1985. Systematics and paleontology. *Paleobiology* 11:65–74.

Fahlbusch, V. 1991. The meaning of MN-zonation: Considerations for a subdivision of the European continental Tertiary using mammals. *Newsletters in Stratigraphy* 24:159–173.

Flynn, L. J., and Bernor, R. L. 1987. Late Tertiary mammals from the Mongolian People's Republic. *American Museum of Natural History Novitates* 2872:1–16.

Forsten, A. M. 1968. Revision of the Palearctic *Hipparion. Acta Zoologicia Fennica* 119:1–134.

———. 1972. *Hipparion primigenium* from southern Tunisia. *Service Geologie Notes* 35:7–28.

Gabunja, L. K. 1959. *History of the genus Hipparion, on materials from the USSR* (in Russian). Academy of Science, Moscow.

Gromova, V. 1952. Les *Hipparion* d'après les matériaux de Tarkalia, Pavlodar et autres. Bureau de Recherches Géologiques et Minières. French translation of Traveaux de l'Institute Paléontologique Academie des Sciences de l'URAS, vol. 26, Géologie et Minéralogie.

Haq, B. U., Hardenbol, J., and Vail, P. R. 1987. Chronology of fluctuating sea levels since the Triassic. *Science* 235:1156–1167.

Harris, J. M. 1983. Correlation of the Koobi Fora succession. In *Koobi Fora research project.* Vol. 2, *The fossil ungulates: Proboscidea, Perissodactyla and Suidae,* pp. 303–318 (ed. J. M. Harris). Clarendon, Oxford.

Hartenberger, J.-L. 1969. Les Pseudosciuridae (Mammalia, Rodentia) de L'Eocène moyen de Bouxwiller, Egerkinger et Lissieu. *Palaeovertébrata* 3:27–91.

Hulbert, R. 1987. A new *Cormohipparion* (Mammalia, Equidae) from the Pliocene (Latest Hemphillian and Blancan) of Florida. *Journal of Vertebrate Paleontology* 7:451–468.

Hulbert, R., and MacFadden, B. J. 1991. Morphological transformation and cladogenesis at the base of the adaptive radiation of Miocene hypsodont horses. *American Museum of Natural History Novitates* 3000:1–61.

Koufos, G. 1984. A new hipparion (Mammalia, Perissodactyla) from the Vallesian (Late Miocene) of Greece. *Paläontologische Zeitschrift* 58:307–317.

———. 1986. Study of the Vallesian hipparions of the lower Axios Valley (Macedonia, Greece). *Géobios* 19:61–79.

———. 1987a. Study of Turolian hipparions of the lower Axios Valley (Macedonia, Greece). Pt. 2, Locality "Prochoma-1" (PXM). *Paläontologische Zeitschrift* 61:339–358.

———. 1987b. Study of the Pikermi hipparions. Pt. 1, Generalities and taxonomy. *Bulletin du Muséum National d'Histoire Naturelle,* 4th ser., 9:197–252.

Lindsay, E. H., Opdyke, N. D., and Johnson, N. M. 1980. Pliocene dispersal of the horse *Equus* and Late Cenozoic mammalian dispersal events. *Nature* 287:135–138.

Lindsay, E. H., and Tedford, R. 1989. Development and application of land mammal ages in North America and Europe: A comparison. In *Topics on European mammalian chronology,* pp. 601–624 (ed. E. H. Lindsay, V. Fahlbusch, and P. Mein). Plenum, New York.

Lipscomb, D. 1992. Parsimony, homology and the analysis of multistate characters. *Cladistics* 8:45–65.

MacFadden, B. J. 1980. The Miocene horse *Hipparion* from North America and from the type locality in southern France. *Palaeontology* 23:617–635.

———. 1984. Systematics and phylogeny of *Hipparion, Neohipparion, Nannippus,* and *Cormohipparion* (Mammalia, Equidae) from the Miocene and Pliocene of the New World. *Bulletin of the American Museum of Natural History* 179:1–196.

MacFadden, B. J., and Woodburne, M. O. 1982. Systematics of the Neogene Siwalik hipparions (Mammalia, Equidae) based on cranial and dental morphology. *Journal of Vertebrate Paleontology* 2:185–218.

Mein, P. 1975. Résultats du Groupe de Travail des Vertébrés. In *Report on Activity of the RCMNS Working Groups (1971–1975), Bratislava: Vertebrata,* pp. 78–81 (ed. J. Senes). I.U.G.S. Commission on Stratigraphy, Subcommission on Neogene Stratigraphy, Bratislava.

———. 1979. Rapport d'activité du Groupe de Travail des Vertébrés: Mise à jour de la biostratigraphie du Néogène basée sur les mammifères. *Annales Géologie Pays Hellénica* 1973(3):1367–1372.

———. 1989. European mammal correlations. In *Topics on European mammalian geochronology,* pp. 73–90 (ed. E. H. Lindsay, V. Fahlbusch, and P. Mein). Plenum, New York.

Meulen, A. J. van der, and van Kolfschoten, T. 1986. Review of the Late Turolian to Early Biharian mammalian faunas from Greece and Turkey. *Memoire Società Geologica Italiana* 31:201–211.

Norell, M. 1992. Taxic origin and temporal diversity: The effect of phylogeny. In *Extinction and phylogeny,* pp. 89–118 (ed. M. J. Novacek and Q. E. Wheeler). Columbia University Press, New York.

Norell, M., and Novacek, M. J. 1992. Congruence between superpositional and phylogenetic patterns: Comparing cladistic patterns with fossil records. *Cladistics* 8:319–338.

Novacek, M., and Norell, M. 1982. Fossils, phylogeny and taxonomic rates of evolution. *Systematic Zoology* 31:366–375.

Pickford, M. 1983. Sequence and environments of the Lower and Middle Miocene hominoids of western Kenya. In *New interpretations of ape and human ancestry,* pp. 421–439 (ed. R. L. Ciochon and R. S. Corruccini). Plenum, New York.

Prentice, M. L., and Denton, G. H. 1988. The deep-sea oxygen isotope record, the global ice sheet system and hominid evolution. In *Evolutionary history of the "robust" australopithecines,* pp. 383–404 (ed. F. E. Grine). Aldine, New York.

Qiu, Z. 1989. The Chinese Neogene mammalian biochronology: Its correlation with the European Neogene mammalian zonation. In *Topics on European mammalian chronology,* pp. 527–556 (ed. E. H. Lindsay, V. Fahlbusch, and P. Mein). Plenum, New York.

Qui, Z., Weilong, H., and Zhihui, G. 1987. The Chinese hipparionine fossils. *Palaeontologica Sinica,* ser. C, 175:1–250.

Rögl, F., and Steininger, F. F. 1983. Vom Zerfall der Tethys zu Mediterran und Paratethys. Die Neogene Paläogeographie und Palinspastik des zirkum-mediterranen Raum. Annalen des Naturhistorischen Museums in Wien, ser. A, 85:135–163.

Sen, S. 1989. *Hipparion* Datum and its chronologic evidence in the Mediterranean area. In *Topics on European mammalian chronology,* pp. 495–505 (ed. E. H. Lindsay, V. Fahlbusch, and P. Mein). Plenum, New York.

Steininger, F. F., Bernor, R. L., and Fahlbusch, V. 1989. European Neogene marine/continental chronologic correlations. In *Topics on European mammalian geochronology,* pp. 15–46 (ed. E. H. Lindsay, V. Fahlbusch, and P. Mein). Plenum, New York.

Suc, J. P., and Zagwijn, W. H. 1983. Plio-Pleistocene correlations between the northwestern Mediterranean region and northwestern Europe according to the recent biostratigraphic and paleoclimatic data. *Boreas* 12:153–166.

Sudre, J. 1969. Les gisements de Robiac (Eocène supérieur) et leurs faunes de mammifères. *Palaeovertébrata* 2:95–165.

———. 1972. Révision des artiodactyles de l'Eocène moyen de Lissieu (Rhône). *Palaeovertébrata* 5:111–156.

Swisher, C. C., III (in press). New $^{40}Ar/^{39}Ar$ dates and their contribution toward a revised chronology for the Late Miocene of Europe and West Asia. In *The evolution of western Eurasian later Neogene faunas.* (ed. R. L. Bernor, V. Fahlbusch, and H.-W. Mittmann). Columbia University Press, New York.

Tassy, P. 1989. The "Proboscidean Datum Event": How many proboscideans and how many events? In *Topics on European mammalian chronology,* pp. 237–252 (ed. E. H. Lindsay, V. Fahlbusch, and P. Mein). Plenum, New York.

Tedford, R. H., Flynn, L. J., Zhanxiang, Q., Opdyke, N. D., and Downs, W. R. 1991. Yushe Basin, China: Paleomagnetically calibrated mammalian standard from the Late Neogene of eastern Asia. *Journal of Vertebrate Paleontology* 11:519–525.

Thaler, L. 1966. Les rongeurs fossiles du Bas-Languedoc dans leurs rapports avec l'histoire des faunes et la stratigraphie du Tertiaire d'Europe. *Mémoire Muséum National d'Histoire Naturelle,* ser. C, 17:1–295.

———. 1972. Datation, zonation et mammifères. *Mémoire Bureau Recherche Géologie et Minérologie* 77:711–724.

Vrba, E. 1976. The fossil Bovidae of Sterkfontein, Swartkrans and Kromdraai. *Transvaal Museum Memoires* 27:1–166.

———. 1988. Late Pliocene climatic events and hominid evolution. In *Evolutionary history of the "robust" australopithecines,* pp. 405–426 (ed. F. D. Grine). Aldine, New York.

Webb, S. D., and Hulbert, R. C. 1986. Systematics and evolution of *Pseudhipparion* (Mammalia, Equidae) from the Late Neogene of the Gulf Coastal Plain and the Great Plains. *Contributions in Geology,* University of Wyoming Special Paper, 3:237–272.

Weerd, A. van der. 1976. Rodent faunas of the Mio-Pliocene continental sediments of the Teruel-Alfambra region, Spain. *Utrecht Micropaleontology Bulletin,* special issue, 2:1–217.

Woodburne, M. O. 1989. Hipparion horses: A pattern of endemic evolution and intercontinental dispersal. In *The Evolution of perissodactyls,* pp. 197–233 (ed. D. R. Prothero and R. M. Schoch). Oxford, New York.

Woodburne, M. O., and Bernor, R. L. 1980. On superspecific groups of some Old World hipparionine horses. *Journal of Paleontology* 8:315–327.

Woodburne, M. O., Bernor, R. L., and Swisher, C. C. III (in press). An appraisal of the stratigraphic and phylogenetic bases for the "Hipparion Datum" in the Old World. In *The evolution of western Eurasian later Neogene faunas* (ed. R. L. Bernor, V. Fahlbusch, and H.-M. Mittmann). Columbia University Press, New York.

Woodburne, M. O., MacFadden, B. J., and Skinner, M. F. 1981. The North American "*Hipparion*" Datum and implications for the Neogene of the Old World. *Geobios* 14:493–524.

Woodburne, M. O., Theobald, G., Bernor, R. L., Swisher, C. C. III, König, H., and Tobien, H. (in press). Advances in the geology and stratigraphy at Höwenegg, southwestern Germany. In *The evolution of western Eurasian later Neogene faunas* (ed. R. L. Bernor, V. Fahlbusch, and H.-W. Mittmann). Columbia University Press, New York.

Appendix 12.1. Definitions of Hipparionine Horses

Hipparionine or hipparion — refers to horses with an isolated protocone on maxillary premolar and molar teeth and, as far as known, tridactyl feet, including species of the following genera: *Hipparion, Neohipparion, Nannippus, Cormohipparion, Hippotherium,* "*Proboscidipparion,*" "*Plesiohipparion,*" "*Sivalhippus,*" *Pseudhipparion,* "*Eurygnathohippus*" (senior synonym of "*Stylohipparion*"), and *Cremohipparion.* Characterizations of these taxa can be found in MacFadden (1984), Bernor and Hussain (1985), Webb and Hulbert (1986), Hulbert (1987), Qiu et al. (1987), Bernor et al. (1988b, 1989), Woodburne (1989), and Hulbert and MacFadden (1991).

Hipparion sensu strictu — restricted to a specific lineage of horses with the facial fossa positioned high on the face (MacFadden, 1980, 1984, p. 53; Woodburne and Bernor, 1980, p. 1329; Woodburne et al., 1981, p. 496; MacFadden and Woodburne, 1982, p. 187; Bernor and Hussain, 1985, p. 134;.Bernor, 1985a, p. 180; Bernor et al., 1987c, p. 46 and fig. 4; Bernor et al., 1989, p. 298; Woodburne, 1989, p. 205). The posterior pocket becomes reduced (and is eventually lost) and confluent with the adjacent facial surface (includes Group 3 of Woodburne and Bernor, 1980, p. 1329). We differ from some of the previous authors in that we do not recognize North American species of *Hipparion s.s.:* Bernor (1985) and Bernor (in Bernor et al., 1989) do not recognize any North American taxon as belonging to *Hipparion s.s.;* any morphologic similarity is argued to be homoplasious.

"*Hipparion*" — used for several distinct and separate lineages of Old World hipparionine horses once considered to be referrable to the genus *Hipparion* (Woodburne and Bernor, 1980, p. 1328; MacFadden and Woodburne, 1982, p. 187; Bernor and Hussain, 1985, p. 34; Bernor, 1985a, p. 1980; Bernor et al., 1988, p. 428; Bernor et al., 1989).

Appendix 12.2. Character State Distribution of "*Hippotherium*" *primigenium* Complex Species

In this new analysis the Old World hipparions are united by six characters. *Hippotherium primigenium* is defined as follows: the orbital surface of lacrimal bone is reduced or lacking a foramen (3B); preorbital fossa is subtriangular in shape and anteroventrally oriented (4D); maxillary cheek tooth fossette ornamentation is complex, with several deeply amplified plications (19A); pli caballin morphology is double (21A); preflexid morphology is complex (46B; this character reverses to state A in some taxa); postflexid enamel margins are complex (47B; this character reverses to state A in some taxa).

"*Hipparion*" *weihoense* is united with the rest of the group by the loss of P1 (16B). "*Hipparion*" *weihoense* itself exhibits the presence of an elongate-oval protocone (23D), a metastylid spur (35A), and a complex pli caballinid (40A) as unique features. Also, it exhibits a subtriangular-shaped preorbital fossa (4E), which occurs as a homoplasy in *Hipparion gettyi* further up the tree.

"*Hipparion*" *catalaunicum* and "*Hipparion*" *africanum* are united with the rest of the group by a lingually flattened and labially rounded protocone (23E) and a deeper premolar linguaflexid (44B). "*Hipparion*" *catalaunicum* and "*Hipparion*" *africanum* form a clade defined by an elongate, anteroposteriorly oriented preorbital fossa (4J). "*Hipparion*" *africanum* exhibits two autopomorphies: a nasal notch that is placed above P2 (15D) and protocones that are compressed or ovate (23F).

The remaining Old World hipparions and the "*Sivalhippus*" clade are united by a moderately deep medial fossa aspect (6B). "*Hipparion*" *dermatorhinum* has ten autapomorphies, including presence of a long preorbital bar with the anterior edge of the lacrimal placed more than half the distance from the anterior orbital rim to the posterior rim of the preorbital fossa (1D); preorbital fossa with reduced pocketing and moderate to slight depth, less than 15 mm (5B); nasal notch incised to a position above P4 (15F); pli caballin single or occasionally poorly defined double (21B); protocone lenticular shaped (23J); premolar and molar metaconids and metastylids elongate (32B, 33B, 34B, 36B); protostylid absent on the occlusal surface (41B). The clade including "*Hipparion*" *giganteum, Hipparion gettyi,* "*Hipparion*" *melendezi,* and "*Hipparion*" *catalaunicum* 2 is united by the absence of a preorbital fossa anterior rim (9B). "*Hipparion*" *giganteum* exhibits two autapomorphies, a preorbital fossa that exhibits reduced posterior pocketing (5B) and a hypoglyph that exhibits reduced, moderate incision (22C). *Hipparion gettyi,* "*Hipparion*" *melendezi,* and "*Hipparion*" *catalaunicum* 2 are united by a moderately delineated fossa peripheral border outline (8B) and the presence of moderately complex, shortly amplified, thinly banded plications (19B). *Hipparion gettyi* and "*Hipparion*" *melendezi* are further united by a subtriangularly shaped, anteroposteriorly oriented preorbital fossa (4E) and preorbital fossa peripheral border outline that is only moderately delineated (8B). *Hipparion gettyi* exhibits two autapomorphies: the hypoglyph is incised moderately deeply (22C) and the protocone is oval in shape (23C). "*Hipparion*" *melendezi* also has two autapomorphies: preorbital fossa is C-shaped and anteroposteriorly oriented (4G), and the maxillary cheek teeth are complexly ornamented (19A). "*Hipparion*" *catalaunicum* 2 has five autapomorphies: the preorbital fossa is elongate and anteroposteriorly oriented (4J; a homoplasy shared with the "*H.*" *catalaunicum*–"*H.*" *africanum* clade), and four autapomorphies: the preorbital fossa lacks posterior pocketing (5D); the preorbital fossa has a shallow medial depth, less than 10 mm in the deepest place (6C); the preorbital fossa has a weakly defined peripheral border (8C); the protocone is triangular-shaped with rounded corners (23K).

Appendix 12.3. Character State Distribution of "*Sivalhippus*" Complex Species

The synapomorphies that unite all the horses of the "*Sivalhippus*" Complex include: reduced preorbital fossa pocketing (5B); straight maxillary cheek teeth (17D); premolar (32C) and molar (33C) metaconids become angular on their distal surfaces; premolar (34C) and molar (36C) metastylids become angular on their proximal surfaces; premolar linguaflexids are shallow (44A); molar linguaflexids are deep and broadly U-shaped (45D). "*Sivalhippus*" *platyodus* exhibits a single autapomorphy: the pli caballinid is single (40B).

The next step unites all the members of the "*Sivalhippus*" Complex except "*Sivalhippus*" *platyodus* by a large suite of synapomorphies that include the following: the preorbital bar lengthens so that the lacrimal is placed further posterior to preorbital fossa (1G); the preorbital fossa reduces its dorsoventral dimension, transforming it to an egg shape (4F); preorbital fossa's posterior pocketing is further reduced (5C); there is an increased crown height (< 60 mm; 18D); protocones increase in length giving a triangular shape (23H); pli caballinid morphology reverses to the state found in "*Hippotherium*" *primigenium* to become complex (40A); premolar linguaflexid morphology becomes a broad, deep U shape (44D); the protoconid enamel band becomes flattened (49B).

"*Sivalhippus*" *perimense* exhibits one autapomorphy: protostylids are very strongly developed and columnar in morphology (41C).

All of the "*Sivalhippus*" taxa except "*S.*" *platyodus* and "*S.*" *perimense* form a group united by a number of characters, including: the preorbital bar becomes vestigial (1H); the preorbital fossa exhibits a C-shaped or egg-shaped outline (4H); preorbital fossa depth becomes shallow (6C); preorbital fossa peripheral border outline is absent, leaving a remnant depression (8D); preorbital fossa anterior rim is lost (9B); premolar metastylid spur is lost (35B); pli caballinid is lost (40C).

When "*S.*" *turkanense* is excluded, the remaining taxa are united by three characters: the incisors become grooved (29B); preflexids (46A) and postflexids (47A) acquire simple margins.

The clade that remains when "*Plesiohipparion*" *houfenense* is excluded is united by a single character: the protostylid is reduced, usually being absent on the crown's surface, but buried in the protoconid wall (41B). At this step, *Eurygnathohippus afarense* exhibits five autapomorphies: protocones become triangular-shaped with rounded corners (23K); incisors lose their curvature, becoming straight (30B); mandibular I3 becomes reduced by the posterior aspect becoming transversely restricted (31B); pli caballinid again reverses to become more complex (40A); ectostylids become present in adult cheek teeth (43A).

The next step unites remaining taxa by four characters: the preorbital bar is lost (1I); preorbital fossa pocketing is lost (5D); preorbital fossa medial depth is absent (6D); preorbital fossa peripheral border outline is absent (8E).

When "*P.*" *crusafonti* is excluded, remaining taxa are united by the absence of the preorbital fossa (4I). Within this suite of taxa there are two major groups: the *Proboscidipparion* clade and the "*Plesiohipparion*" aff. *houfenense*–"*Plesiohipparion*" *huangheense* clade.

The *Proboscidipparion* clade is well supported, being united by strong retraction of nasals, initially to a level of M1 (15G); protocone reacquires a lingually flattened and labially rounded morphology by reducing its length (23E; this character becomes further modified in *P. sinense*); cheek tooth pre- (46B) and postflexids (47B) undergo a reversal reacquiring complex enamel plications (possibly a dietary adaptation to eating more browse). *Proboscidipparion sinense* acquires four autapomorphies including: hyperretraction of nasals posterior to M1 (to orbital level; 15H); fossettes become very complex (19E); pli caballins become complex (21C); protocone becomes triangular-elongate (23I).

The "*Plesiohipparion*" aff. *houfenense*–"*Plesiohipparion*" *huangheense* and "*Plesiohipparion*" aff. *huangheense* group is united by three synapomorphies: premolar (34F) and molar (36F) metastylids are lingually pointed; postflexid invades metaconid/metastylid interface by the anterior most portion, bending sharply lingually (48B; but note, no skulls known of "*P.*" *huangheense* so that the presence of 4I is inferred because of congruence of other characters). "*Plesiohipparion*" aff. *houfenense* exhibits a single autapomorphy: fossette enamel is moderately complex (19B).

The "*Plesiohipparion*" *huangheense*–"*Plesiohipparion*" aff. *huangheense* group exhibits a series of synapomorphies: premolar (32F) and molar (33F) metaconids become lingually pointed; molar ectoflexid converges with pre- and postflexid (39C); pli caballinid reverses to become single or rudimentary (40B); protostylid reverses to become present on the occlusal surface (41A) and is expressed as an open loop extending posterolabially (42E). "*Plesiohipparion*" aff. *huangheense* has three potential autapomorphies, which if verified by further collection and analysis could lead to its taxonomic distinction at the species level: hypoglyph incision is moderate (22C); premolar (44E) and molar (45E) linguaflexids become a very broad, deep U shape (the greatest of all "*Sivalhippus*" Complex members). However, because no skulls are known for "*P.*" *huangheense*, further collection and analysis are needed before these characters can be used for taxonomic distinction at the species level.

Chapter 13

Faunal and Environmental Change in the Neogene of East Africa: Evidence from the Tugen Hills Sequence, Baringo District, Kenya

Andrew Hill

In considering the patterns of change in terrestrial faunas through time, it is necessary know what those faunas are, precisely how old they are, and which species are present. Accurate dating is essential for documenting the patterns of change in lineages and faunas in order to estimate reliable first and last appearance data (FADs and LADs). Precise dating is also important in explaining these patterns in terms of extrinsic factors such as climate. Signals of extensive climatic change come from sources that are often remote from the faunas being studied, such as deep-sea cores, Antarctic ice, or climate models. Before addressing hypotheses of causality, we must be able to correlate events in a number of disparate regions so as to establish synchronicity.

A major problem with this endeavor in Africa is the lack of detailed documentation of terrestrial faunas and faunal change for most of the Neogene. There are clusters of sites, mainly in western and northern Kenya and in northeastern Uganda, that provide information about animal communities from around 25 to 14 million years (myr). Starting with Laetoli, in Tanzania, at 3.7 myr, Pliocene faunas have been documented to about 3 myr at the Hadar, Ethiopia. Information concerning the interval from 3 myr or so into the Pleistocene is provided by the Omo sequence in south Ethiopia, by sites on the eastern and western margins of Lake Turkana, Kenya, and from Olduvia in Tanzania. South Africa has relatively poorly dated sites that may extend back to 3 myr.

Consequently, until recently there has been a large gap in the record between 14 and 4 myr; the time between 3 and 2 myr is also poorly known. The great majority of the sites falling into the 10 myr gap between 14 and 4 myr are concentrated at the recent end of this range, between about 5.5 and 4 myr. Most of these have been investigated only in the past decade. They include Allia Bay, Aterir, Kanam, Kanapoi, and Lothagam, in Kenya; the western Rift of Uganda; the Middle Awash, in Ethiopia; the Manonga Valley, in Tanzania; and Langebaanweg, in South Africa. Only a few isolated sites fall into the period between 14 and 5.5 myr, among them Berg Aukas, in Namibia, at possibly 13 myr; Samburu Hills, Kenya, at around 9 myr; and Nakali, Kenya, at perhaps about 8 myr.

Only one region preserves a more or less continuous succession through this whole period, and that is the Tugen Hills in the Kenya Rift Valley.[1] This sequence is a complicated fault block west of Lake Baringo, Kenya, extending about 60 km north and south along the northern part of the Rift Valley (fig. 13.1). The block is tilted to the west; sediments and lavas over 3,000 m thick are exposed in its fault scarps and in the foothills on the eastern side. Fossiliferous sediments are extensive, ranging in age from about 16 myr into the Pleistocene. The Tugen Hills are the exclusive source of data for certain quite extensive time intervals. For these periods, if something is not known from the Tugen Hills succession, we do not know of it from anywhere else in sub-Saharan Africa.

The succession consists of six major sedimentary formations and a number of lesser ones. They are, from bottom to top: Muruyur, Ngorora, Mpesida, Lukeino, Chemeron, and Kapthurin (fig. 13.2). The Kapthurin Formation is an important Mid-Pleistocene unit containing a good record of fauna (including hominids) and artifacts, but it will not be discussed here further, be-

ISBN 0–300–06348–2.

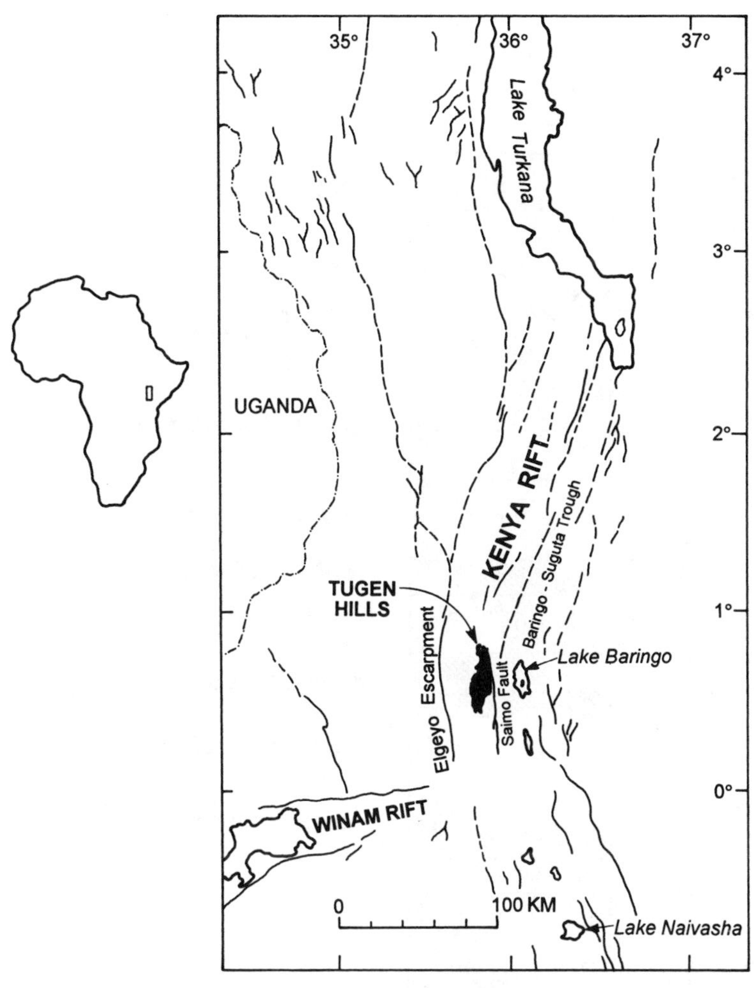

Fig. 13.1. Map of northern Kenya, showing the Tugen Hills, the Rift Valley, and other major structural features. (From Kingston et al., 1994)

cause it is the subject of an independent expedition (see McBrearty et al., in press). It is important to stress that these units are not single localities but extensive sets of exposures, consisting mainly of lacustrine and riverine sediments. The formations often represent a considerable amount of time and contain many individual fossil localities. They are usually separated from one another by volcanic lavas, the dates on which form the overall temporal framework; within the sedimentary units themselves are tuffaceous horizons, which facilitate more refined calibration. Our original assumption was that relatively large lacunae existed in the sequence, which stratigraphic work would reveal and document. On the contrary, we have found that gaps in the succession are fewer and of much briefer duration than expected. In fact, except for one or two significant time gaps, we have very good representation from about 15.5 myr onward.

During the collection of fossils in the Tugen Hills, detailed attention has always been given to location and stratigraphic level. Before our project began, however, taxonomic faunal lists were usually assembled for formations as a whole. A formational framework is basic to any investigation, but the practice has in some cases resulted in an inability to state which taxa are definitely diagnosed as present at particular sites; such knowledge has become more critical as our temporal control has improved. Consequently, we have been confirming identifications for each specimen in the faunal collection so as to establish faunal lists in which we can have taxonomic confidence for each site and horizon. We have already compiled lists for many of the sites that we have discovered ourselves or

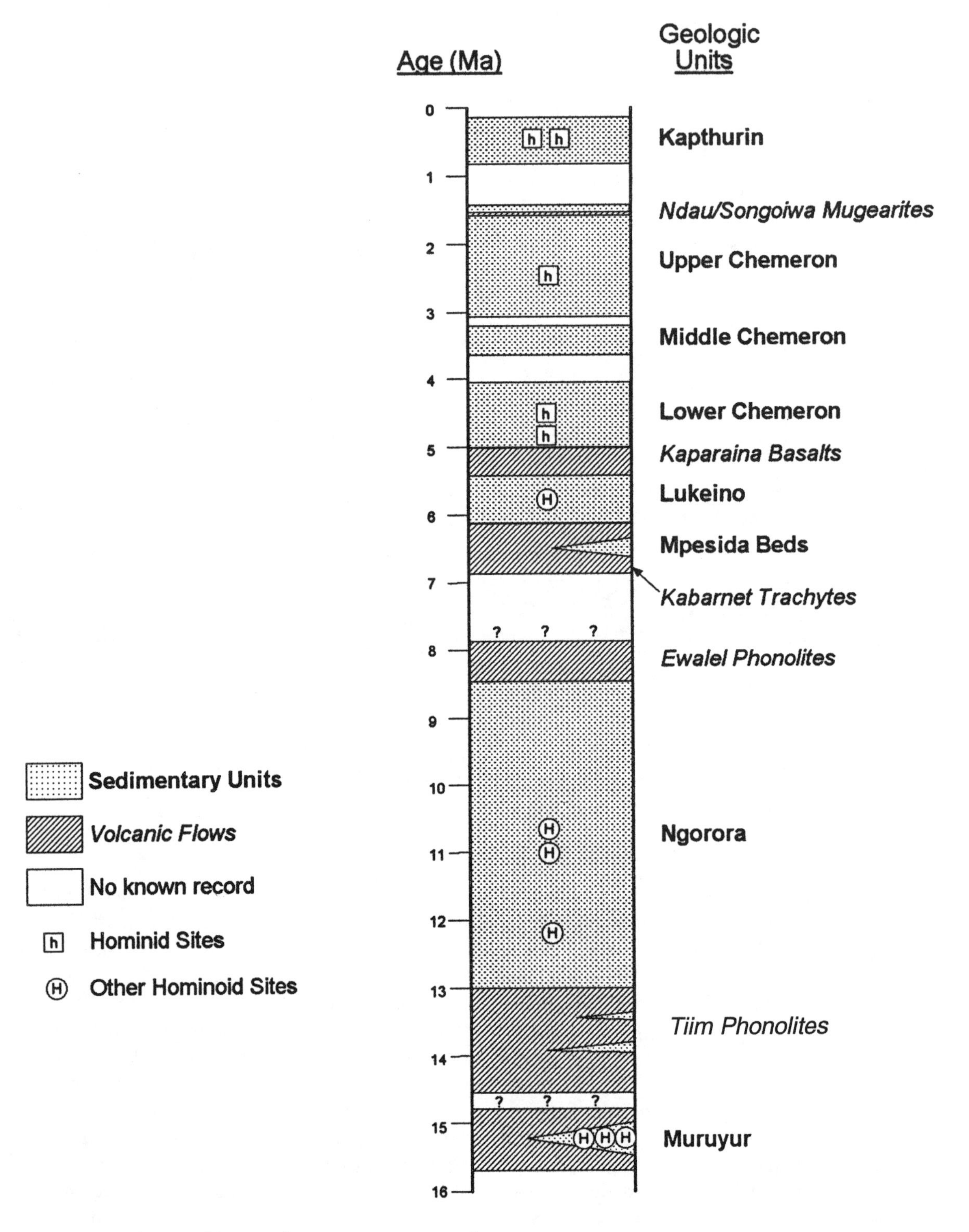

Fig. 13.2. Stratigraphic section of the Tugen Hills succession. (From Kingston et al., 1994)

worked on intensively, particularly those associated with hominoids. Because a number of the Tugen Hills formations span a considerable period, faunal lists for whole formations may give a false impression of faunal homogeneity. I do not provide such faunal lists here but instead discuss particular taxa that arise or disappear during the course of each formation. Eventually we will have taxonomically reliable faunal lists and accurate dating for every site in the succession.

The use of faunal lists for entire formations has had unfortunate effects. For example, those who refer to, and use, such concepts as the "Ngororan" should remember that they are discussing a fauna as if it were from a single time horizon, when in fact it dates from 13 to ca. 8 myr, a period of over 4 myr. Use of the term "Ngororan" is equivalent to lumping together the modern African fauna with all the fossil fauna from Lake Turkana, Omo, Olduvai, Hadar, Middle Awash, and Laetoli and treating it as if it were a single animal community. I think the value of such stage concepts is limited in a region with excellent time calibration, especially when such "faunas" are then compared to faunas from much shorter intervals, such as the "present" or the "Lothagamian."

In the Tugen Hills extensive faulting is an advantageous structural feature, because it results in the exposure of rocks covering a long period. But it also has a practical disadvantage, because it fragments strata and renders lateral correlation problematic. Often sites cannot be correlated directly, and radiometric dating has to be carried out in several separate sections of the same formation. Although the region can now be reached by a major road, local access within the Hills is very difficult. There is great and varied topographic relief, and unlike more northern areas, such as Lake Turkana, extensive plant cover, often thick thorn bush, obscures sites and sections.

Despite these logistical difficulties, the long Tugen Hills succession of fossiliferous sediments, now becoming more finely calibrated, provides us with the opportunity to investigate the pattern of African Neogene faunal change through time in some detail. Significant faunal change can be detected among all five major formations, within the duration of the 4 myr-long Ngorora Formation, and within the Chemeron Formation (Hill, 1985a; Hill et al., 1985, 1986; Barry et al., 1985a).

The Muruyur Beds

The Muruyur Beds were originally described by Chapman (1971) during his work with the East African Geological Research Unit (EAGRU; see also Bishop et al., 1971). Pickford (1975a, 1988) carried out additional work on the unit, discovering more exposures and adding considerably to the faunal list. So far, however, these beds remain a relatively little known formation at the base of the sequence toward the northern end of the Tugen Hills. The unit is paleontologically important principally because of localities in the Kipsaramon site complex. One horizon in the complex is a bone bed, which crops out at a number of localities (BPRP#89A to BPRP#89L). The first exposure of the bone bed was originally discovered by Pickford and Cheboi, and although some surface fossils were collected, it was not further investigated. Our more recent work (Hill, 1989; Hill et al., 1991) has shown that it is very extensive indeed, being up to 20 cm thick and extending over at least 2,500 m^2. Above the bone bed is a silty layer containing compressed leaves and fruits. The bone horizon itself is solid with interlocked fossils; there is very little sedimentary matrix. The fossils range in size from proboscidean maxillae to isolated rodent teeth and include a wide range of taxa. Proboscideans include *Choerolophodon*, *Protanancus*, and *Deinotherium*. There are at least two genera of rhinoceros, *Aceratherium* and *Paradiceros*. Artiodactyls include a possible hippopotamid and possibly the suid *Listriodon*. There is the tragulid *Dorcatherium*, a species of giraffid, and at least two bovids, including *Protragoceros*. Among the at least seven families of rodents present, a scaly-tailed flying squirrel suggests widespread tropical forest, although the presence of a springhare implies some open habitat.

There are at least two species of hominoid from the bone bed, *Kalepithecus* and a species similar to *Proconsul major*. The monkey *Victoriapithecus* also occurs. Pickford (1988) has reported a hominoid talus from another site (BPRP#91) at about this same level (see also Hill and Ward, 1988; Hill et al., 1991), which is also similar to *Proconsul major*. BPRP#122, a site about 10 m higher in the succession and about 1 km away, has produced over forty hominoid teeth comparable to those of *Kenyapithecus africanus* (Brown et al., 1991; Hill et al., 1992a). These were supplemented in 1993 and 1994 with a mandible and postcranials of another individual. Up to now, at least seven hominoid individuals are represented from this younger horizon. Other taxa are sparse at BPRP#122.

We have now produced a detailed stratigraphy for the Kipsaramon region and have begun a reevaluation of the type section (Behrensmeyer et al., in prep.). The unit was known to be below the Ngorora Formation, the base of which is dated at 13 myr (Deino et al., 1990). It was also thought to overlie lavas dated at 14.3 myr. Consequently, the fauna was originally believed to date between ca. 14 and 13 myr (Bishop et al., 1971; Pickford, 1988). Our recent work has shown it to be about 15.5 myr (Hill et al., 1991). BPRP#122, with cf. *Kenyapithecus,* is also likely to be older than 15 myr, and determinations are in progress for other sites and the type section (Deino et al., in prep.).

The Muruyur sites are important in providing links with the better-known sites of similar age in western Kenya and with Buluk and Kalodirr, in the Turkana Basin (Leakey and Walker, 1985; Leakey and Leakey, 1986a, 1986b). The fauna collected so far seems to ally Kipsaramon provisionally with the western Kenya occurrences rather than with the more northern sites. Nevertheless, some of the fauna does resemble material from Kalodirr, in west Turkana (Hill et al., 1991). The Muruyur Beds are also particularly relevant to clarifying the early history of Kenyan rifting and volcanicity, both of which are likely to have had a considerable effect on East African Miocene environments and ecology.

The Ngorora Formation

The Muruyur Beds probably represent a span of less than 1 myr and are relatively limited in outcrop. In contrast, the Ngorora Formation is extensive spatially and temporally. It documents over 4 myr of continuous sedimentation, a remarkably long period for African continental deposits and one that is otherwise totally unknown in Africa. Currently there is a gap of about 2 myr in our sequence, between the youngest Muruyur sediments and the oldest Ngorora sediments.

The Ngorora Formation was first investigated by Chapman (Bishop and Chapman, 1970; Chapman, 1971; Bishop et al., 1971) working with EAGRU. Subsequent work was carried out by Pickford (Bishop and Pickford, 1975; Pickford, 1975b, 1978a). Since then, BPRP has carried out further studies. Most outcrops of the formation are at the northern end of the Tugen Hills, but they also extend far to the south. Outcrops occur in distinct fault-bounded basins, separated by up to 50 km, and the faults appear to have been active at the time of deposition. This situation makes correlation between fault blocks difficult in that there can be abrupt changes in facies across fault boundaries. Such changes are an important factor in the dating of sites, because the successions in each fault block must be dated separately. The Ngorora type section is at Kabasero, in the north, where about 450 m of sediments are exposed, representing at least 4 myr of more or less continuous sedimentation. Near Bartabwa, a little to the west, the formation is only about 200 m thick.

The Ngorora Formation is bounded and defined by the Tiim Phonolite Formation, below, and the Ewalel Phonolites, above. Recently, Deino et al. (1990) redescribed the paleomagnetic stratigraphy and lithostratigraphy of much of the formation and presented a recalibration of the sequence based upon $^{40}Ar/^{39}Ar$ single-crystal laser-fusion methods. This recalibration improves upon the earlier results achieved by conventional means (Hill et al., 1985; Tauxe et al., 1985). Tuff horizons have also been dated in two new long sections. Age determinations on the Tiim Phonolite flow underlying the base of the formation in the type section gives a date of 13.15 myr. Ngorora sedimentation began shortly after this. Concordant dates on tuffs through the section, in about 100–200 thousand years (kyr) increments, document sedimentation up to 10.51 myr in the type section. About 100–130 m of sediment continue above this dated horizon, which on paleomagnetic grounds extend to younger than 10 myr. At Ngeringerowa, an outcrop of the formation to the south, we have obtained dates that indicate an age range of between 10 and 8 myr. The new set of dates on tuffs throughout the formation gives us much greater control over the ages of sites and of significant fossils through this long time span. The relation of the radiometric information and the paleomagnetic stratigraphy correlated to the Geomagnetic Reversal Time Scale (GRTS) has implications for deep-sea spreading rates, and the nature and quality of the data suggest revisions to the GRTS for this portion of time.

The many fossil sites throughout the Ngorora succession are not distributed evenly through time. There are relatively few near the base, a part of the section characterized by coarse volcanic agglomerates and high sedimentation rates. They are more frequent from about 12.7 myr onward, until perhaps about 10.5 myr, after which we know of fewer sites, at least in the type area. The Ngeringerowa sites, however, date to as recently as about 8 myr.

The fauna from many Ngorora localities essentially resembles those from earlier Miocene sites, although it includes many taxa that are unique to the formation. One example is site BPRP#38 in the Ngorora section, dated at 12.5 myr. It is a small locality with a particularly diverse assemblage of fossils, mainly teeth. The site includes at least one species of hominoid and a new but unnamed species of monkey, similar to *Victoriapithecus* (Hill et al., in MS). There are at least four species of carnivore represented: the aardvark *Orycteropus chemeldoi*, the hyrax *Parapliohyrax ngororaensis*, a proboscidean, and a rhinoceros. Among the artiodactyls present is an anthracothere, the suid *Lopholistriodon kidogosana*, the palaeomerycid *Climacoceros gentryi*, the giraffid *Paleotragus*, the tragulid *Dorcatherium*, and at least four species of bovid; these include *Protragocerus*, *Homoiodorcas tugenium*, and *Pseudotragus gentryi*. There is a minimum of three species of cricetid rodent.

Other taxa are known from other Ngorora sites that are fairly low in the section. There are insectivores, other carnivores and rodents, creodonts (including a giant hyaenodontine), proboscideans such as *Prodeinotherium* and *Choerolophodon ngororae*, and three species of rhinoceros. Among the artiodactyls is another species of suid, the hippopotamid *Kenyapotamus coryndoni*, two species of the tragulid genus *Dorcatherium*, giraffids, and a variety of bovids.

Later Ngorora sites include more advanced or modern elements. The sites at Ngeringerowa probably overlap in time with the top of the Ngorora type section at Kabasero but also include sediments from a period more recent than any represented there. These later sites include what may be the first equids in sub-Saharan Africa. Equids are not known from the type section, in which there are sites to about 10 myr, and probably all Ngorora sites that include equids postdate 10 myr. These later faunas include other, more modern groups, such as possible reduncine bovids. Ngeringerowa has also produced the oldest known colobine monkeys, *Microcolobus tugenensis* (Benefit and Pickford, 1986). So we have an indication of important faunal changes within the Ngorora Formation itself.

The formation contains several specimens of small-bodied hominoids and four specimens from larger species (Hill and Ward, 1988; Hill, 1994). All are isolated teeth. One left upper molar (KNM-BN 1378; Bishop and Chapman, 1970), not yet well dated, is unlike that of any other known hominoid, though it most resembles the second molar of a large chimpanzee. A lower left fourth premolar (KNM-BN 10489), from a site just younger than 12.42 myr (Hill et al., 1985; Deino et al., 1990), appears to resemble *Proconsul*. If it does belong to that taxon, it would be the most recent representative of that genus in the record. An incisor tooth (KNM-BN 1461) similar to *Proconsul* comes from site BPRP#38 (see above), dated at 12.5 myr (Hill et al., in MS). From the same site is a canine tooth (KNM-TH 23144), which, if hominoid, is probably not *Proconsul* (Hill et al., in MS).

Another important feature of the Ngorora Formation is the existence of what may well be the best fossil macroflora in Africa, now dated at 12.59 myr. Direct evidence of fossil plants is extremely rare in Africa beyond about 30 kyr, so this occurrence is particularly interesting. It consists of an autochthonous assemblage of leaves, sometimes still attached to stems, that accumulated in a subaerial airfall tuff (Jacobs and Winkler, 1992). This tuff can be traced laterally for over 1 km. The fossil leaves are extremely well preserved and permit the observation of leaf margins and venation and the study of epidermal micromorphology and other fine structure through scanning electron microscopy. Jacobs and Kabuye (Kabuye and Jacobs, 1986; Jacobs and Kabuye, 1987) have identified over fifty-five tree species that suggest a lowland rain-forest habitat at that time.

The Mpesida Beds

A further significant time gap in the Tugen Hills sequence appears to exist at the top of the Ngorora Formation. The overlying Mpesida Beds are not as extensive as the Ngorora Formation, but recently we have shown them to be much more extensive than was previously thought. Not only have we discovered new areas of exposure, but some sediments originally mapped as part of the Lukeino Formation (Chapman, 1971) in fact belong to the Mpesida Beds. The unit is of relatively short duration, being bracketed by flows of the Kabarnet Trachytes. Lower flows, which lie with an angular unconformity on the Ewalel Phonolites, have produced dates of between 7.5 and 7.0 myr; flows above the Mpesida Beds give dates of 6.2–6.3 myr (Chapman and Brook, 1978; Hill et al., 1985, 1986).

Most Mpesida faunal sites appear to have formed in the deeper portions of a lake or lakes. Crocodiles and Hippopotamidae are common, and exclusively terrestrial mammals fairly rare. But the unit is significant, because the Mpesida Beds show the earliest evidence of a major change from the bulk of the Ngorora assemblages. The latter have a typical Late Miocene character, even those that contain equids. In the Mpesida Beds whole new families appear, such as the Elephantidae (*Stegotetrabelodon orbus*) and the first leporids in sub-Saharan Africa. Among new genera are the "gomphothere" *An-*

ancus kenyensis and a modern genus of rhinoceros, *Ceratotherium praecox*. Also some new species arise within already established genera, such as *Brachypotherium* cf. *lewisi* and within *Hipparion*. There is also the chalicothere *Chemositia tugenensis*, a hippopotamid, a giraffid, and bovids, which include *Tragelaphus*, *Kobus*, *Madoqua*, and *Gazella* or *Raphiceros*. We have one colobine, but as yet no hominoids are known from this unit, which is unfortunate because a hominoid from this period would be especially intriguing.

Another particularly interesting feature of this unit is the presence of many examples of fossil wood; some of the trees are even in place, having been buried in an airfall tuff. Through these fossils, we hope to discover details of tree distribution and density at the time. The wood preserves its internal structure and will be the subject of further work. A preliminary analysis shows one specimen to belong to what is presently a monospecific genus found in West African forests.

The Lukeino Formation

The Lukeino Formation was originally named as a member of the Kabarnet Trachyte Formation by Chapman (1971), and other relevant outcrops were described by Martyn (1969) and McClenaghan (1971). Pickford carried out further work on the unit, gave it formational status, and published a type section (Pickford, 1975a, 1978b). The Lukeino Formation overlies the Kabarnet Trachytes, dated in its upper flows at 6.2 and 6.3 myr (Chapman and Brook, 1978; Hill et al., 1985, 1986). For most of its outcrop it is overlain by the Kaparaina Basalt, which is dated at around 5.6 myr (Hill et al., 1985, 1986). The sediments are diverse and widespread, even though exposure is often poor, and there are many fossil sites.

The suggestion of significant faunal change seen in the Mpesida Beds is emphasized by the better-sampled Lukeino Formation. The listriodont pigs found in the Ngorora Formation have been replaced by *Nyanzachoerus syrticus*. There is a new elephant, *Primelephas gomphotheroides;* the oldest *Ancylotherium*, a chalicothere; the oldest porcupine; and a number of new carnivores. There are leporids, cercopithecoids, the aardvark *Orycteropus*, and deinotheres; among elephants *Stegotetrabelodon orbus* continues. Perissodactyls include rhinocerotids and hipparionine equids. Among artiodactyls are hippopotamids; giraffids, namely, *Giraffa* sp.; and the bovid genera *Tragelaphus*, *Ugandax*, *Cephalophus*, and *Aepyceros*.

One hominoid is known from the formation—a lower left first molar (KNM-LU 335). It is unlike any known taxon but has some resemblance to that of a chimpanzee (McHenry and Corruccini, 1980; Hill and Ward, 1988; Hill, 1994; Ungar et al., 1994). Ungar et al. (1994) also indicate similarities to specimens of *Australopithecus* cf. *afarensis* from Lake Turkana, Kenya.

The Chemeron Formation

The Lukeino Formation is superseded by the Chemeron Formation, lying for most of its exposure on flows of the Kaparaina Basalt, dated at 5.6 myr (Hill et al., 1985, 1986). There are a number of separate regions of outcrop, and in the past the relationship between them has not always been clear. Rocks now placed in this formation were first discussed by Gregory (1896, 1921) and formed part of his Kamasia sediments. Fuchs (1934, 1950) also worked on this unit, but McCall et al. (1967) were the first to recognize two distinct formations separated by an angular unconformity. The lower unit became the Chemeron beds; the upper, the Kapthurin. Martyn (1967, 1969), working with EAGRU, undertook more detailed research and formalized these terms. The type area is exposed by a set of rivers, such as the Barsemoi, Ndau, Chemeron, and Kapthurin, which flow east from the Tugen Hills to Lake Baringo. Martyn also added to the formation sediments in a separate set of outcrops to the west, known as the Kipcherere Basin. Later, Pickford (1975a; Pickford et al., 1983) added to the unit other sediments that are exposed farther north. These are broadly equivalent to Martyn's Kipcherere sediments in the south. In the past the relationship between the Kipcherere region and the type area, both of which lie on the Kaparaina Basalt, was confused (e.g., Bishop, 1972). We have now shown that they are of different ages. Some sediments in the Kipcherere succession begin at around 5 myr, whereas sediments to the east, such as those in the Kapthurin River valley, begin at about 3 myr. This finding led us to think that there would be two formations in rocks labelled Chemeron, in distinct mappable outcrops and of distinct, nonoverlapping age. However, our recent work on extensive exposures of the unit suggests that sediments also extend from the top of the previously known "lower Chemeron" to the base of the "upper Chemeron." Therefore, it appears that we have another formation extending for a long period. The unit is unique in Africa in that it encompasses the whole of the Pliocene.

In many parts of the type area the Chemeron Formation strata are faulted and exposure is intermittent. Recently, however, we located a continuous section of outcrop through at least the whole upper part of the unit, which we anticipate will extend from before 4 myr to 1.6 myr. Soon we will be dating this section in detail and examining the faunas and repeated diatomite units within it. We are also beginning a program of tuff fingerprinting to establish correlations with areas distal to the Tugen Hills, such as the Lake Turkana Basin. Namwamba (1993), working with and assisted by our project, carried out some preliminary work in this field and identified four tuff horizons in the Chemeron Formation that also occur elsewhere (see also Brown, 1994). At Kipcherere there are outcrops of the Lokochot Tuff, defined in the Turkana Basin and also occurring as the Kyampanga Tuff in the Lake Albert Basin, in western Uganda (Pickford et al., 1991), and in DSDP cores from the Gulf of Aden. Slightly higher in the succession we have the Sidi Hakoma Tuff of the Hadar and Middle Awash, Ethiopia, also known in the Turkana Basin as the Tulu Bor Tuff and from Gulf of Aden cores, dated at 3.32 myr (Haileab and Brown, 1992). Also in the Tugen Hills succession is the Lokalalei Tuff from Lake Turkana (Tuff D in the Shungura Formation), dated at 2.52 myr, which also occurs near the base of the succession at Gadeb, Ethiopia (Haileab and Brown, 1994). We are investigating these and other tephrostratigraphic links that will enable us to correlate events in different regions over wide areas of eastern Africa. In addition, we have good radiometric control over the relevant sections, which may permit us to date other important sites and regions that have so far proved intractable, such as Kanapoi, Gadeb, and some horizons in the Middle Awash.

Species from sites in the lower part of the Chemeron Formation are broadly similar to those in the Lukeino Formation, though some differences in species and genera do occur. Among the suids, *Nyanzachoerus syrticus* has given way to its probable descendants *N. jaegeri* and *N. kanamensis.* Bovids include *Tragelaphus, Kobus,* and *Aepyceros.* The first species of the modern genus of African elephant, *Loxodonta adaurora,* is represented, but one of the earlier elephants, *Stegotetrabelodon,* is absent. Other proboscideans include *Anancus kenyensis* and *Deinotherium bozasi.* The chalicothere *Ancylotherium hennegi* continues, as do the rhinoceros *Ceratotherium praecox* and hipparionine equids. There are rodents and cercopithecoids.

Further changes are apparent by the times of the upper part of the unit, shown principally from sites between just over 3 myr and 2.4 myr (Hill et al., 1992). There is the modern species of rhinoceros, *Ceratotherium simum.* Among suids are *Kolpochoerus afarensis, K. limnetes, Notochoerus scotti,* and *N. euilus.* Proboscideans are represented by *Loxodonta exoptata* and *Elephas recki.* Among nonhominoid primates are the types and other examples of *Paracolobus chemeroni* and *Theropithecus baringensis,* as well as other papionines.

Three hominoids are known from the Chemeron, two of them from sites near the base of the formation. One (KNM-TH 13150) comes from the site of Tabarin (BPRP#77; Hill, 1985b; Ward and Hill, 1987; Hill and Ward, 1988; Hill, 1994). It is a fragmentary right mandibular corpus with first and second molars, which was described as indistinguishable from those in the small size range of *Australopithecus afarensis.* Although many diagnostic parts are obviously missing, this specimen is a good candidate for Hominidae. If it is Hominidae, it is one of the earliest specimens of the family known, dated between 5 and 4.15 myr. Only the fragmentary mandible from Lothagam, also a plausible hominid, might be older (Hill et al., 1992b). At the time of discovery I remarked (Hill, 1985b) that because the Tabarin specimen was up to 1 myr older than the oldest specimens in the *Au. afarensis* hypodigm and because it preserved only a small number of relatively undiagnostic morphological traits, its attribution might change when further material became available. Such material has become available from the Middle Awash, and the specimen has recently been named *Ardipithecus ramidus* (White et al., 1994, 1995). Although the Tabarin mandible cannot be definitely diagnosed as this species either, in that it is probably slightly older than the Ethiopian *Ar. ramidus* material, it may best be identified provisionally with that taxon, as White et al. (1994) suggest. Another good early hominid candidate is a proximal humerus (KNM-BC 1745) discovered by Pickford at another Chemeron site (BPRP#37; Pickford et al., 1983). This is around the same age and shows features, such as a shallow intertubercular groove, that may be derived characters for Hominidae (Hill and Ward, 1988; Hill, 1994). Again, it resembles specimens of *Au. afarensis,* although it more probably belongs to *Ar. ramidus.*

A third hominid specimen (KNM-BC 1), also important for the timing of the origin of clades, comes from higher in the sequence. This specimen was found in 1965 (Martyn, 1967) and described as Hominidae *indet.* (Tobias, 1967). At that time the site (BPRP#2) was not well dated, nor was there much early hominid material with which to compare the specimen. Recently we reexamined the fossil, a right temporal fragment, identified it as *Homo*, and dated the relevant horizon at 2.43 myr (Hill et al., 1992). The age has been further substantiated by additional dates within the local succession (Deino and Hill, in press). KNM-BC 1 is, therefore, the oldest known representative of the genus *Homo*.

Schrenk et al. (1993) described the hominid mandible (UR 501) from Uraha, Malawi, as "oldest *Homo*" and gave it a date of 2.4 myr, so a digression here is perhaps necessary. Their date is based on associated fauna from Unit 3A of the Chiwondo Formation, although it is not clear how closely associated the fauna is with the actual level of the hominid, which appears in a condensed section. Schrenk et al. suggest that this fauna is similar to the fauna that Hill et al. (1992) cite as being associated with the Chemeron temporal and implicitly use this similarity to reinforce an age estimate of 2.4 myr. As is clearly stated in that paper, however, the list we provided covers the whole of the upper part of the Chemeron Formation, not just the time horizon of the *Homo* site. The fauna listed came from a sequence of sites that are precisely dated by radiometric means and that fall into the range of 3.2–1.6 myr. This is an example of the misapprehensions that can result from publishing composite lists of fauna based on several sites at different levels in formations with considerable age range. The age ranges of individual taxa from Chiwondo 3A are not much more informative. Several bovids identified only to genera are identified with genera that are extant. Most of those identified to species are only tentatively identified as having an affinity with particular fossil or modern species, although they have an archaic character (Kaufulu et al., 1981). The most reliable taxa from the point of view of dating are the suids, which include the following species: *Notochoerus capensis, Notochoerus scotti, Notochoerus euilus, Kolpochoerus limnetes, Metridiochoerus compactus, Metridiochoerus andrewsi*. The age ranges for these taxa provided by Schrenk et al. can be refined (Bishop, 1994; White, this vol.). Bishop's and White's estimates of ranges differ slightly, but both of them show that ranges of the suids associated with the Uraha mandible all extend to as recent as 2 myr or younger (*Notochoerus capensis* is an exception, but its known range is poorly constrained). And *Metridiochoerus compactus* is known only at sites younger than 1.92 myr, according to Bishop (1994), or 1.6 myr, according to White (this vol.). These facts suggest that the assemblage, including the hominid, could in fact be relatively young in age. Excluding *N. capensis*, the only time during which all these species could plausibly overlap is from about 2.1 to 2.0 myr. More probably the Chiwondo 3A fauna is a mixed assemblage representing a considerable period. In this case the hominid could be 2.4 myr, younger than that, or older than 2.4 myr. Until additional information becomes available, however, it is best to regard the Uraha mandible as imprecisely dated. At present the Chemeron temporal remains the oldest securely dated member of the genus *Homo*, being about half a million years older than other well-dated examples.

Carbon Isotopes

Apart from paleontology, we are also pursuing other approaches relevant to the study of paleoclimate and environment. Aspects of paleovegetation can be reconstructed by an analysis of paleosol carbonates and associated organic material. Such analyses indicate the relative proportions of flora that use the Hatch-Slack (C_4) or the Calvin cycle (C_3) photosynthetic pathways. In equatorial regions, C_3 plants include most trees, shrubs, and montane grasses, whereas C_4 plants almost exclusively comprise grasses favoring warm growing seasons. Quade et al. (1989) demonstrated a dramatic shift from C_3- to C_4-dominated floodplain biomass beginning about 7.4–7 myr ago in the Siwaliks of Pakistan. They interpreted this shift as reflecting a change from trees and shrubs to grasslands and hypothesized that this striking ecological transformation might be correlated with the development of the monsoon system. Later Cerling et al., 1993 suggested that it might be triggered by a global lowering of atmospheric CO_2. The event also correlates with the local disappearance of large hominoids, the arrival of cercopithecoids, and other faunal changes. Cerling, one of the authors of the Siwalik work, visited the Tugen Hills project briefly to gather similar data on the African succession. His work suggested that this sudden shift is not present in East Africa, and although he believed the proportion of C_4 vegetation increased in the Late Pliocene and Pleistocene, there is no evidence for the development of a pure C_4-dominated biomass, such as savannas, until quite late in the Pleistocene (Cerling, 1992). This finding agreed with a conclusion I had been advocating for some time (e.g., Hill, 1986, 1987), based on macrofloral evidence and other indications.

Kingston has been investigating carbon isotope variation through the Tugen Hills sequence in more detail

(Kingston, 1992; Kingston et al., 1992, 1994; Morgan et al., 1994). Pedogenic carbonate was sampled at ninety-five sites and revealed extensive lateral variation throughout the succession (fig. 13.3), suggesting a persistent mosaic of C_3 and C_4 plants. There appears to be no statistically significant difference in variation through time, except for the modern carbonate sample, which indicates a slightly greater proportion of C_4 vegetation. Also, there is no detectable reflection in the sequence of any postulated global shifts, such as the Messinian Salinity Crises at 6.5–5.3 myr; in this context the apparent association of some of the earliest East African hominids with wooded habitats and colobine monkeys is interesting (WoldeGabriel et al., 1994). Nor is the cooling trend that others in this volume have noted during the interval between 2.8 and 2.3 myr reflected in this Tugen Hills data. This lack of correlation need not be seen as contradictory evidence; it may be that African terrestrial vertebrate habitats were to some extent buffered from climatic changes seen elsewhere. There is no sign of the savannas at 8 myr required in the Rift itself by Coppens's theory (Coppens, 1994), nor, of course, is the dramatic shift at 7.4 myr in the Siwaliks documented in the Rift Valley sequence. We also have independent macrobotanical evidence of forest in our sequence at 6.5 myr.

Signs of C_4 vegetation, however, are present throughout the succession. The Tugen Hills signal at 15.3 myr represents the oldest known record of C_4 vegetation in the world. But Kingston's work on $^{13}C/^{12}C$ ratios in tooth enamel through the sequence suggests that although C_4 grasses were present by 15.3 myr, they were not utilized exclusively by herbivores until 7 myr (Morgan et al., 1994). This evidence need not signify a vegetational change at that time but could reflect faunal immigration, in situ speciation and changes in diet, and so on.

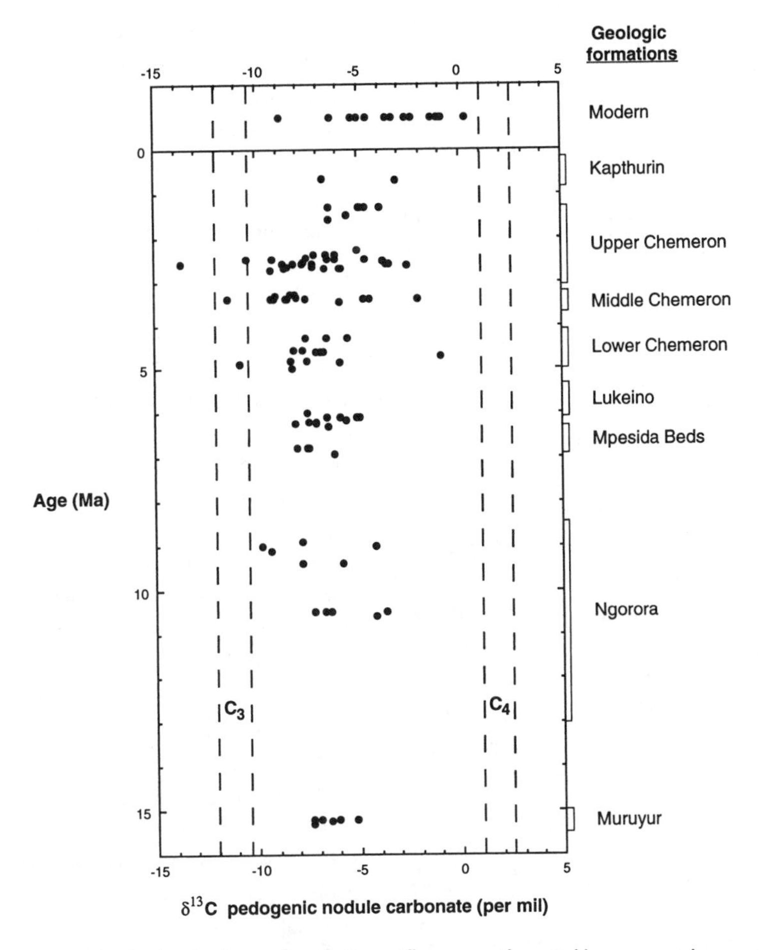

Fig. 13.3. Stable carbon isotope composition of paleosol carbonates from the Tugen Hills sequence. The vertical bars represent the average δ^{13} values of pedogenic carbonates forming in C_3-dominated (left) and C_4-dominated environments (right). (From Kingston et al., 1994)

Patterns and Problems in Faunal Change

A major issue in vertebrate evolution concerns the degree to which extrinsic environmental factors influence the pattern of change. After Croll (1864 and other works) related ice ages and climatic change to changes in orbital eccentricity and calculated the related periodicities, Wallace (1870) went on to link the mechanism to rates of species change more generally. Wallace postulated that these periodic and radical changes in world climate would result in periodic bursts of general faunal change. Climatic change induced by high eccentricity would "lead to a rapid change of species, low eccentricity to a persistence of the same forms" (Wallace, 1870, p. 454). Incidentally, this view also appears to be an early injection of the idea of punctuated equilibria into the evolutionary debate.

In the African situation much interest centers on the possible influence of climatic and other environmental stimuli on the evolution of hominoids. Darwin (1871, p. 433) commented that bipedalism might have arisen in an ape owing to "some change in the surrounding conditions." Since then a number of people have discussed Hominoidea as possible examples of evolutionary innovation stimulated by climatic changes, both local or global. Brain (1981) was one of the first to link in any detail changes in Hominoidea with climatic shifts inferred from deep-sea cores. Did Hominidae arise as a result of the Messinian Salinity Crises? Were the origins of *Homo* and *Paranthropus* responses to Late Pliocene cooling? These ideas, which incorporate a strongly environmental approach to faunal change, with respect not only to hominoids but to terrestrial faunas generally, have been made explicit and refined by Vrba (e.g., 1985a, 1985b, 1985c, 1988, 1993, 1994, and this vol.; Vrba et al., 1989) in her turnover pulse hypothesis and its variants. In its simplest and most original form it suggests that "speciations and extinctions across diverse lineages should occur as concerted pulses in predictable synchrony with changes in physical environment, chiefly in global temperature" (Vrba, 1985a, pp. 70–71). Although hers is a general theory, one important and prominent example focused on the apparent radiation of hominids into the robust lineage and *Homo*. Vrba placed this event at about 2.5 myr, when "a particularly marked and widespread environmental change occurred in Africa" (Vrba, 1985a, p. 70; see also Vrba, 1993).

There are several problems with testing hypotheses of evolutionary pattern in terrestrial mammalian faunas, particularly in testing whether faunal shifts are triggered by global climatic events. No one denies that global climate has changed and fluctuated radically in the course of the Neogene. But how suddenly and at what times? The first question is, therefore, to what extent do sudden climatic "events" exist? For example, a number of authors in this volume indicate that there is no particular event at 2.5 myr and suggest that the Pliocene cooling was spread out and protracted over a much greater period, between perhaps 2.8 and 2 myr. No one brief interval within that span can be selected as more generally significant than any other. This somewhat erodes the force of the turnover pulse hypothesis as originally framed.

Other problems with testing relate mainly to the imperfections and resolution of the fossil record. The quality of the record in time and space influences our ability to date FADs and LADs reliably and to make sufficiently accurate correlations with other events in the world (Hill, 1987). These issues still have not been resolved, partly because some of them cannot be addressed given the nature of the relevant fossil record (White, this vol.). Unless fossils occur at a high density in a section spanning a significant amount of time, little confidence can be placed in the accuracy of FADs and LADs. It is no coincidence that in the Tugen Hills succession the greatest apparent faunal shifts coincide with the greatest time gaps in the section.

How well can we estimate FADs and LADs? Unfortunately, in the case of hominoids, which are rare, FADs may be nowhere near their true times of origin. Our recognition of new taxa depends very much upon unpredictable factors of preservation and discovery. A silly, yet pertinent, example is the gorilla, the earliest scientific record of which is about 1847. What major climatic shift in the early 1800s caused this sudden speciation? Similarly, *Sivapithecus* disappears from the fossil record in Asia around 7.4 myr. Were it not for its probable descendant (broadly speaking) in the form of contemporary orangutan the clade would appear to have become extinct in the Late Miocene. Our own expedition in the Tugen Hills provides examples. In the past ten years our work there has pushed back the origin of Hominidae (Hill, 1985b) and of the genus *Homo* (Hill et al., 1992) by half a million years each. This is a large percentage shift in their previous temporal ranges and demonstrates how unreliable our estimated FADs are for some important hominoid clades.

Hominid lineages appear to behave "appropriately" around 2.5 myr, with *Homo* appearing in the record at 2.4 myr (Hill et al., 1992) and *Paranthropus* at 2.6 myr (Walker et al., 1985). Given the uncertainty of their FADs, however, I think their appearances are coincidental. Suids form a better test. Although they may in some ways be like hominoids ecologically, fossil pigs, unlike hominoids, are abundant, diverse, easy to identify to species, and present at most sites. Therefore, we can have confidence that their FADs and LADs reflect true appearance and disappearance times more accurately than do those for rarer taxa. Bishop (1993, 1994) has just completed a study of the time distribution of East African Suidae, including those from the later part of the Tugen Hills succession. Her data form a good test of the turnover pulse hypothesis at 2.5 myr. Like the independent work of White (this vol.), her results show clearly that in this one major and well-sampled East African family the turnover pulse hypothesis in its classical formulation is not supported. Less specifically, looking at the whole Turkana Basin fauna, Feibel et al. (1991, p. 345) also suggest that a change at 2.5 myr "cannot be substantiated from the fossil record of the Turkana Basin," though they do report the possibility of faunal turnover at about 2 myr.

A further problem relates to geographical sampling. All our samples for significant sections of the African Neogene come from a small area of the Rift Valley on the eastern margin of the continent. What is the likelihood that all taxa originated there or moved there instantaneously from their point of origin? No doubt one or two important evolutionary events occurred in the over 99 percent of Africa not represented by Neogene fossil sites. Furthermore, the area represented by sites is highly unusual compared to the rest of the continent. It has been very active geologically through the Neogene, and events in that region are likely to have had a profound effect upon local faunas. Elevation has risen from near sea level to 3,000 m. Widespread mobile basalts have at times flooded the entire area and once filled up the developing rift. Changing topography due to volcanism and tectonics will have repeatedly modified local climate. The African Neogene record essentially monitors very local events, though it may also reflect more widespread, or even global, ones. But the two categories are difficult to distinguish.

This problem is better documented elsewhere. In the Siwaliks of Pakistan, for example (Barry et al., 1985b, 1990, 1991; Barry and Flynn, 1990; Flynn and Jacobs, 1982; Flynn et al., 1990), pulses of faunal change can be detected, but probably not all are directly related to global climatic change. They are seen as being modified by local conditions and neighboring geography. One fruitful source of testing will be to compare long sections on different continents, such as the Tugen Hills and the Siwaliks. Through such comparisons we may detect synchronous changes in remote areas that will lead to more confidence in assessing global influences.

I am not opposed to the idea that climatic change on a major scale influences the evolution of terrestrial faunas. The turnover pulse hypothesis in its original form is a very elegant and attractive idea. However, testing it and related ideas about evolutionary pattern in Africa remains difficult at present. I have always agreed that significant climatic change may have occurred at the Messinian and in the Middle to Late Pliocene and that these changes may have had an effect on terrestrial vertebrate faunas. The idea that the origin of Hominidae and its later radiation are in some way connected with environmental change accompanying global climatic shifts has been an attractive one since Brain (1981) proposed it. My objections have been concerned more with the suddenness of the change originally proposed at 2.5 myr and with the idea that taxa should be affected synchronously. I also believe that many apparent correlations could be the artifacts of poor temporal resolution, given the poor record of Neogene vertebrates in Africa. It has always seemed likely that the situation was more complicated than the turnover pulse originally suggested (Hill, 1987). Various modifications can be made to the turnover pulse hypothesis, but then the idea loses much of its originality and force, becoming essentially similar to other, earlier hypotheses that incorporate environmental forcing.

Summary

If we are to assess reasons for the pattern of terrestrial vertebrate evolution through time, we must know what that pattern is. For this we need high-density data on taxonomic representation through long sequences and accurate dating of first and last appearances. It is also important to collect other kinds of data relating to environment and paleoecology. For much of the Neogene in Africa, the only succession that has the potential to provide such information is the Tugen Hills sequence, which extends from about 16 myr into the Pleistocene and, with very few gaps, includes periods not documented elsewhere in sub-Saharan Africa. Many parts of this succession are now well calibrated; the density of faunal sampling is improving. We document, with varying degrees of precision, the first and last appearances of numerous major and minor taxa in the Ethiopian fauna: the first equids, the first elephantids, the first colobines, for example. The Tugen Hills have produced all the African monkeys known between 14 and 5.5 myr. In Hominoidea, we have a fine sample of cf. *Kenyapithecus,* and we may have the last *Proconsul.* At the base of the Chemeron Formation are two of the earliest known hominids, probably most closely identifiable with *Ardipithecus ramidus.* Higher in the Chemeron Formation is the earliest known member of the genus *Homo.* At Chesowanja, another site in the area, is what may be the latest *Paranthropus boisei.* There are excellent collections of plants at 12.59 myr, and more are being found at other levels in the succession. We also have data on vegetation derived from carbon isotope analyses.

We are not directing our research in the Tugen Hills explicitly toward falsifying the turnover pulse hypothesis or any other model of faunal change. Rather, our work is informed by such theories and remains open to different alternatives. The effects of global climatic change on vegetation and faunas can be mediated by shifts in geography; tectonics and biotic factors also play a part. I feel that in the Neogene of East Africa we have not yet excluded the influence on faunas of local geographical or other changes (Hill, 1987). All these ideas have different outcomes that are important in their effects upon the course of evolution in our own and other mammalian lineages. What is needed is continued work in the relevant geological sections to refine dating, to increase the density of our faunal samples, and to establish more reliable FADs and LADs. Hypotheses must be well constrained by rigorous falsifiable criteria. We have to discover and understand more thoroughly and objectively the pattern of evolution we are trying to explain. We need to collect more critical information than our present fragmentary data provides for testing the pattern suggested by any hypothesis of evolutionary mode.

Acknowledgments

I wish to thank Liz Vrba for organizing what was a truly excellent and productive conference at Airlie. The work summarized here forms part of the Baringo Paleontological Research Project, based at Yale University and jointly affiliated with the National Museums of Kenya. We thank the Office of the President, Republic of Kenya, for permission to conduct research in the Tugen Hills and the National Museum for much logistic support. The work is funded by NSF SBR-9208903 and grants from the Louise Brown Foundation and J. Clayton Stephenson. The project is a collaborative endeavor involving diverse participants. I thank them all, in particular the two postdoctoral associates at Yale, Barbara Brown and John Kingston. Al Deino is responsible for our radiometric dating program, and I am grateful for his comments. I thank Laura Bishop for her advice, particularly about the Tugen Hills pigs. John Reader reminded me of Kelvin's relevance. Sally McBrearty has not yet found artifacts at 3 myr in the Chemeron Formation but has always offered valuable counsel on the project, especially on this manuscript. Tom Gundling also provided valuable comments on the draft.

Notes

1. The Tugen Hills were one of the first regions in Kenya to be commented on geologically, first by Thomson (1884), then by Gregory (1896, 1921). Fuchs (1934, 1950) later visited the area, but the region remained essentially unknown until the late 1960s, when the East African Geological Research Unit, based at London University under the direction of Basil King, began a program of geological mapping. Members of the unit began to discover fossil sites, and starting in 1967 a subsidiary project directed by Bill Bishop, including myself and, later, Martin Pickford, was formed to investigate these. The London University work provided the basic stratigraphic framework for the region (King and Chapman, 1972; Chapman et al., 1978; Williams and Chapman, 1986), radiometric dating (Chapman and Brook, 1978), and collections of fossil material (Bishop et al., 1971). At the beginning of the 1980s work was resumed at the Baringo Paleontological Research Project (BPRP). The project is now based at Yale University and affiliated with the National Museums of Kenya.

References

Barry, J., and Flynn, L. 1990. Key biostratigraphic events in the Siwalik sequence. In *European Neogene mammal chronology,* pp. 557–571 (ed. E. Lindsay). Plenum Press, New York.

Barry, J., Flynn, L., and Pilbeam, D. 1990. Faunal diversity and turnover in a Miocene terrestrial sequence. In *Causes of evolution: A paleontological perspective,* pp. 381–421 (ed. R. Ross and W. Allmon). University of Chicago Press, Chicago.

Barry, J., Hill, A., and Flynn, L. 1985a. Variation de la faune au Miocène inférieur et moyen de l'Afrique de l'est. *L'Anthropologie* (Paris) 89:271–273.

Barry, J., Johnson, N. M., Raza, S. M., and Jacobs, L. L. 1985b. Neogene mammalian faunal change in southern Asia: Correlations with climatic, tectonic, and eustatic events. *Geology* 13:637–640.

Barry, J., Morgan, M., Winkler, A., Flynn, L., Lindsay, E., Jacobs, L., and Pilbeam D. 1991. Faunal interchange and Miocene terrestrial vertebrates of southern Asia. *Paleobiology* 17:231–245.

Benefit, B., and Pickford, M. 1986. Miocene fossil cercopithecoids from Kenya. *American Journal of Physical Anthropology* 69:441–464.

Bishop, L. C. 1993. Hominids of the East African Rift Valley in a macroevolutionary context. *American Journal of Physical Anthropology*, suppl., 16:57.

———. 1994. Pigs and the ancestors: Hominids, suids and environments during the Plio-Pleistocene of East Africa. Ph.D. diss., Yale University.

Bishop, W. W. 1972. Stratigraphic succession "versus" calibration in East Africa. In *Calibration of hominid evolution*, pp. 219–246 (ed. W. W. Bishop and J. A. Miller). Scottish Academic Press, Edinburgh.

Bishop, W. W., and Chapman, G. R. 1970. Early Pliocene sediments and fossils from the northern Kenya Rift Valley. *Nature* 226:914–918.

Bishop, W. W., Chapman, G. R., Hill, A., and Miller, J. A. 1971. Succession of Cainozoic vertebrate assemblages from the northern Kenya Rift Valley. *Nature* 233:389–394.

Bishop, W. W., and Pickford, M. 1975. Geology, fauna and palaeoenvironments of the Ngorora Formation, Kenya Rift Valley. *Nature* 254:185–192.

Brain, C. K. 1981. The evolution of man in Africa: Was it a consequence of Cainozoic cooling? *Annals of the Geological Society of South Africa* 84:1–19.

Brown, B., Hill, A., and Ward, S. 1991. New Miocene large hominoids from the Tugen Hills, Baringo District, Kenya. *American Journal of Physical Anthropology*, suppl., 12:55 (abstract).

Brown, F. 1994. Development of Pliocene and Pleistocene chronology of the Turkana Basin, East Africa, and its relation to other sites. In *Integrative paths to the past: Paleoanthropological advances in honor of F. Clark Howell*, pp. 285–312 (ed. R. S. Corruccini and R. L. Ciochon). Prentice-Hall, Englewood Cliffs, N. J.

Cerling, T. 1992. Development of grasslands and savannas in East Africa during the Neogene. *Palaeogeography, Palaeoclimatology, Palaeoecology* (Global and Planetary Change Section) 97:241–247.

Cerling, T., Wang, Y., and Quade, J. 1993. Expansion of C_4 ecosystems as an indicator of global ecological change in the Late Miocene. *Nature* 361:344–345.

Chapman, G. R. 1971. The geological evolution of the northern Kamasia Hills, Baringo District, Kenya. Ph.D. diss., University of London.

Chapman, G. R., and Brook, M. 1978. Chronostratigraphy of the Baringo Basin, Kenya Rift Valley. In *Geological background to fossil man*, pp. 207–222 (ed. W. W. Bishop). Geological Society of London, London.

Chapman, G. R., Lippard, S. J., and Martyn, J. E. 1978. The stratigraphy and structure of the Kamasia Range, Kenya Rift Valley. *Journal of the Geological Society of London* 135:265–281.

Coppens, Y. 1994. East side story: The origin of mankind. *Scientific American* 88–95.

Croll, 1864. On the physical cause of the change of climate during geological epochs. *Philosophical Magazine* 31:26–28.

Darwin, C. 1871. *The descent of man, and selection in relation to sex*. John Murray, London.

Deino, A., and Hill, A. (in press). $^{40}Ar/^{39}Ar$ dating of the Chemeron Formation strata encompassing the site of hominid KNM-BC 1, Tugen Hills, Kenya. *Journal of Human Evolution.*

Deino, A., Tauxe, L., Monaghan, M., and Drake, R. 1990. $^{40}Ar/^{39}Ar$ age calibration of the litho- and paleomagnetic stratigraphies of the Ngorora Formation, Kenya. *Journal of Geology* 98:567–587.

Feibel, C. S., Harris, J. M., and Brown, F. H. 1991. Palaeoenvironmental context for the Late Neogene of the Turkana Basin. In *Koobi Fora Research Project*. Vol. 3, *The fossil ungulates: Geology, fossil artiodactyls, and palaeoenvironments*, pp. 321–346. (ed. J. M. Harris). Clarendon Press, Oxford.

Flynn, L., and Jacobs, L. 1982. Effects of changing environments on Siwalik rodent faunas of northern Pakistan. *Palaeogeography, Palaeoclimatology, Palaeoecology* 38:129–138.

Flynn, L., Pilbeam, D., Jacobs, L., Barry, J., Behrensmeyer, A. K., and Kappelman, J. 1990. The Siwaliks of Pakistan: Time and faunas in a Miocene terrestrial setting. *Journal of Geology* 98:589–604.

Fuchs, V. E. 1934. The geological work of the Cambridge Expedition to the East African Lakes, 1930–1931. *Geological Magazine* 71:97–112.

———. 1950. Pleistocene events in the Baringo Basin. *Geological Magazine* 87:149–174.

Gregory, J. W. 1896. *The Great Rift Valley*. John Murray, London.

———. 1921. The rift valleys and geology of East Africa. Seeley Service and Co., London.

Haileab, B., and Brown, F. H. 1992. Turkana Basin: Middle Awash Valley correlations and the age of the Sagantole and Hadar Formations. *Journal of Human Evolution* 22:453–468.

———. 1994. Tephra correlations between the Gadeb prehistoric site and the Turkana Basin. *Journal of Human Evolution* 26:167–173.

Hill, A. 1985a. Les variations de la faune du Miocène récent et du Pliocène d'Afrique de l'est. *L'Anthropologie* (Paris) 89:275–279.

———. 1985b. Early hominid from Baringo, Kenya. *Nature* 315:222–224.

———. 1986. The Tugen Hills sequence: Environmental change in the Late Miocene and Pliocene. In *The longest record: The human career in Africa*. A Conference in Honour of J. Desmond Clark, Berkeley. Abstracts, pp. 45–46.

———. 1987. Causes of perceived faunal change in the later Neogene of East Africa. *Journal of Human Evolution* 16:583–596.

———. 1989. Kipsaramon: A Miocene hominoid site in Kenya. *American Journal of Physical Anthropology* 78:241.

———. 1994. Late Miocene and Early Pliocene Hominoids from Africa. In *Integrative paths to the past: Paleoanthropological advances in honor of F. Clark Howell*, pp. 123–145 (ed. R. S. Corruccini and R. L. Ciochon). Prentice-Hall, Englewood Cliffs, N.J.

Hill, A., Behrensmeyer, K., Brown, B., Deino, A., Rose, M., Saunders, J., Ward, S., and Winkler, A. 1991. Kipsaramon: A lower Miocene hominoid site in the Tugen Hills, Baringo District, Kenya. *Journal of Human Evolution* 20:67–75.

Hill, A., Brown, B., and Ward, S. 1992a. Miocene large hominoids from Kipsaramon, Tugen Hills, Kenya. In *Apes or ances-*

tors? pp. 19–20 (ed. J. Van Couvering). American Museum of Natural History, New York.

Hill, A., Curtis, G., and Drake, R. 1986. Sedimentary stratigraphy of the Tugen Hills, Baringo District, Kenya. In *Sedimentation in the African rifts,* pp. 285–295 (ed. L. Frostick, R. W. Renaut, I. Reid, and J.-J. Tiercelin). Geological Society of London Special Publication, no. 25. Blackwell, Oxford.

Hill, A., Drake, R., Tauxe, L., Monaghan, M., Barry, J. C., Behrensmeyer, A. K., Curtis, G., Fine Jacobs, B., Jacobs, L., Johnson, N., and Pilbeam, D. 1985. Neogene palaeontology and geochronology of the Baringo Basin, Kenya. *Journal of Human Evolution* 14:749–773.

Hill, A., Leakey, M., Kingston, J., and Ward, S. (in MS). New cercopithecoids and a hominoid from 12.5 Ma in the Tugen Hills succession, Kenya. *Journal of Human Evolution* (submitted).

Hill, A., and Ward, S. 1988. Origin of the Hominidae: The record of African large hominoid evolution between 14 My and 4 My. *Yearbook of Physical Anthropology* 31:49–83.

Hill, A., Ward, S., and Brown, B. 1992b. Anatomy and age of the Lothagam mandible. *Journal of Human Evolution* 22:439–451.

Hill, A., Ward, S., Deino, A., Curtis, G., and Drake, R. 1992. Earliest *Homo. Nature* 335:719–722.

Jacobs, B. F., and Kabuye, C. H. S. 1987. Environments of early hominoids: Evidence for Middle Miocene forest in East Africa. *Journal of Human Evolution* 16:147–155.

Jacobs, B. F., and Winkler, D. A. 1992. Taphonomy of a Middle Miocene autochthonous forest assemblage, Ngorora Formation, central Kenya. *Palaeogeography, Palaeoclimatology, Palaeoecology* 99:31–40.

Kabuye, C. H. S., and Jacobs, B. F. 1986. An interesting record of the genus *Leptaspis:* Bambusoideae from Middle Miocene flora deposits in Kenya, East Africa. *Abstracts: International Symposium on Grass Systematics and Evolution,* p. 32. Smithsonian Institution, Washington, D.C.

Kaufulu, Z., Vrba, E., and White, T. D. 1981. Age of the Chiwondo Beds, northern Malawi. *Annals of the Transvaal Museum* 33:1–8.

King, B. C., and Chapman, G. R. 1972. Volcanism of the Kenya Rift Valley. *Philosophical Transactions of the Royal Society,* ser. A, 271:185–208.

Kingston, J. D. 1992. Stable isotopic evidence for hominid paleoenvironments in East Africa. PhD. diss., Harvard University.

Kingston, J. D., Hill, A., and Marino, B. 1992. Isotopic evidence of Late Miocene/Pliocene vegetation in the East African Rift Valley. *American Journal of Physical Anthropology,* suppl., 13:100–101.

Kingston, J. D., Marino, B., and Hill, A. 1994. Isotopic evidence for Neogene hominid paleoenvironments in the Kenya Rift Valley. *Science* 264:955–959.

Leakey, R. E., and Leakey, M. G. 1986a. A new Miocene hominoid from Kenya. *Nature* 324:143–146.

———. 1986b. A second new hominoid from Kenya. *Nature* 324:146–148.

Leakey, R. E., and Walker, A. 1985. New higher primates from the Early Miocene of Buluk, Kenya. *Nature* 318:173–175.

McBrearty, S., Bishop, L. C., and Kingston, J. D. (in press). Variability in traces of Middle Pleistocene hominid behavior in the Kapthurin Formation, Baringo, Kenya. *Journal of Human Evolution.*

McCall, G. J. H., Baker, B. H., and Walsh, J. 1967. Late Tertiary and Quaternary sediments of the Kenya Rift Valley. In *Background to evolution in Africa,* pp. 191–220 (ed. W. W. Bishop and J. D. Clark). University of Chicago Press, Chicago.

McClenaghan, M. P. 1971. Geology of the Ribkwo area, Baringo District, Kenya. Ph.D. diss., University of London.

McHenry, H. M., and Corruccini, R. S. 1980. Late Tertiary hominoids and human origins. *Nature* 285:397–398.

Martyn, J. E. 1967. Pleistocene deposits and new fossil localities in Kenya. *Nature* 215:476–477.

———. 1969. The geological history of the country between Lake Baringo and the Kerio River, Baringo District, Kenya. Ph.D. diss., University of London.

Morgan, M. E., Kingston, J. D., and Marino, B. D. 1994. Carbon isotope evidence for the emergence of C_4 plants in the Neogene from Pakistan and Kenya. *Nature* 367:162–165.

Namwamba, F. L. 1993. Tephrostratigraphy of the Chemeron Formation, Baringo Basin, Kenya. Master's thesis. University of Utah.

Pickford, M. 1975a. Stratigraphy and palaeoecology of five Late Cainozoic formations in the Kenya Rift Valley. Ph.D. diss., University of London.

———. 1975b. Late Miocene sediments and fossils from the northern Kenya Rift Valley. *Nature* 256:279–284.

———. 1978a. Geology, palaeoenvironments and vertebrate faunas of the Mid-Miocene Ngorora Formation, Kenya. In *Geological background to fossil man,* pp. 237–262 (ed. W. W. Bishop). Geological Society of London, London.

———. 1978b. Stratigraphy and mammalian palaeontology of the Late-Miocene Lukeino Formation, Kenya. In *Geological background to fossil man,* pp. 263–278 (ed. W. W. Bishop). Geological Society of London, London.

———. 1988. Geology and fauna of the Middle Miocene hominoid site at Muruyur, Baringo District, Kenya. *Human Evolution* 3:381–390.

Pickford, M., Johanson, D. C., Lovejoy, C. O., White, T. D., and Aronson, J. L. 1983. A hominid humeral fragment from the Pliocene of Kenya. *American Journal of Physical Anthropology* 60:337–346.

Pickford, M., Senut, B., Poupeau, G., Brown, F. H., and Haileab, B. 1991. Correlation of tephra layers from the Western Rift Valley (Uganda) to the Turkana Basin (Ethiopia/Kenya) and the Gulf of Aden. *C. R. Acad. Sci.* (Paris) 313:223–229.

Quade, J., Cerling, T. E., and Bowman, J. R. 1989. Development of Asian monsoon revealed by marked ecological shift during the latest Miocene in northern Pakistan. *Nature* 342:163–166.

Schrenk, F., Bromage, T. G., Betzler, C. G., Ring, U., and Juwayeyi, Y. M. (1993) Oldest *Homo* and Pliocene biogeography of the Malawi Rift. *Nature* 365:833–836.

Tauxe, L., Monaghan, M., Drake, R., Curtis, G., and Staudigel, H. 1985. Paleomagnetism of Miocene East African Rift sediments and the calibration of the Geomagnetic Reversal Timescale. *Journal of Geophysical Research* 90:4639–4646.

Thomson, J. 1884. *Through Masailand.* Samson Low, London.

Tobias, P. V. 1967. Pleistocene deposits and new fossil localities in Kenya. *Nature* 215:478–480.

Ungar, P. S., Walker, A., and Coffing, K. 1994. Reanalysis of the Lukeino Molar (KNM-LU 335). *American Journal of Physical Anthropology* 94:165–173.

Vrba, E. 1985a. Ecological and adaptive changes associated with early hominid evolution. In *Ancestors: The hard evidence,* pp. 63–71 (ed. E. Delson). Liss, New York.

———. 1985b. Environment and evolution: Alternative causes of the temporal distribution of evolutionary events. *South African Journal of Science* 81:229–236.

———. 1985c. Early hominids in southern Africa: Updated observations on chronological and ecological background. In *Hominid evolution: Past, present and future,* pp. 195–200 (ed. P. V. Tobias). Liss, New York.

———. 1988. Late Pliocene climatic events and hominid evolution. In *Evolutionary history of the "robust" australopithecines,* pp. 405–426 (ed. F. Grine) Aldine de Gruyter, New York.

———. 1993. The pulse that produced us. *Natural History* 102:47–51.

———. 1994. An hypothesis of heterochrony in response to climatic cooling and its relevance to early hominid evolution. In *Integrative paths to the past: Paleoanthropological advances in honor of F. Clark Howell,* pp. 345–376 (ed. R. S. Corruccini and R. L. Ciochon). Prentice-Hall, Englewood Cliffs, N. J.

Vrba, E., Denton, G., and Prentice, M. 1989. Climatic influences on early hominid behavior. *Ossa* 14:127–156.

Walker, A., Leakey, R., Harris, J., and Brown, F. 1986. 2.5 Myr *Australopithecus boisei* from west of Lake Turkana, Kenya. *Nature* 322:517–522.

Wallace, A. R. 1870. The measurement of geological time. *Nature* 1:399–401, 452–455.

Ward, S., and Hill, A. 1987. Pliocene hominid partial mandible from Tabarin, Baringo, Kenya. *American Journal of Physical Anthropology* 72:21–37.

White, T. D., Suwa, G., and Asfaw, B. 1994. *Australopithecus ramidus:* A new species of early hominid from Aramis, Ethiopia. *Nature* 371:306–312.

———. 1995. *Australopithecus ramidus,* a new species of early hominid from Aramis, Ethiopia (corrigendum). *Nature* 375:88.

Williams, L. A. J., and Chapman, G. R. 1986. Relationships between major structures, salic volcanism and sedimentation in the Kenya Rift from the equator northwards to Lake Turkana. In *Sedimentation in the African rifts,* pp. 59–74 (ed. L. Frostick, R. W. Renaut, I. Reid, and J.-J. Tiercelin). Geological Society of London Special Publication, no. 25. Blackwell, Oxford.

WoldeGabriel, G., White, T. D., Suwa, G., Renne, P., de Heinzelin, J., Hart, W. K., and Helken, G. 1994. Ecological and temporal placement of Early Pliocene hominids at Aramis, Ethiopia. *Nature* 371:330–333.

Part Three

The Pliocene

Chapter 14

Middle Pliocene Paleoenvironments of the Northern Hemisphere

PRISM Project Members

The Pliocene spans an important transition from an interval of relatively warm global climates, when glaciers and sea ice were greatly reduced in the Northern Hemisphere, to the generally cooler climates of the Pleistocene, with prominent glacial and interglacial cycles involving the periodic buildup and wasting of massive ice sheets in the Northern Hemisphere. The PRISM (Pliocene Research, Interpretation, and Synoptic Mapping) project has been investigating the magnitude and variability of Middle-to-Late Pliocene climatic changes from marine and terrestrial records in order to document and understand the environmental conditions and changes during this interval.

This chapter summarizes an initial attempt to create an integrated synoptic reconstruction of environmental conditions in the Northern Hemisphere during a warm interval centered on 3.0 million years (myr; see Dowsett et al., 1994). A primary reason for the reconstruction is the establishment of boundary conditions and benchmark data for model simulations of this past global warming. The methodology and parameters selected were therefore tailored to meet the needs of modeling experiments. The reconstruction also provides a general frame of reference for individual local records of environments and changing conditions during the Middle-to-Late Pliocene.

This synthesis represents a compilation of our own and other published data. As we point out in the following discussion, the quality of environmental estimates, age control, and geographic coverage varies greatly among sources and types of data. Because we do not have enough data at this time to provide estimates of Pliocene conditions in the Southern Hemisphere, the reconstruction is limited to the Northern Hemisphere.

We have delineated with confidence some major departures from modern conditions, but there are gaps in data in all geographic areas. This work should be viewed as a progress report in our continuing effort to develop a global Pliocene paleoenvironmental reconstruction.

Selection of a "Time Slab"

The interval between 3.15 and 2.85 myr (using the geomagnetic polarity time scale of Berggren et al., 1985) was selected as the basis for a Pliocene paleoclimatic reconstruction for several reasons. Down-core studies of marine microfossils (Dowsett and Poore, 1991; Dowsett et al., 1992; Dowsett and Loubère, 1992; Cronin, 1991a; Barron, 1992a) and studies of Middle Pliocene high-latitude vegetation (Matthews and Ovenden, 1990; Webb and Harwood, 1991) suggest a substantially warmer climate than we have today. This warm interval occurs prior to the 2.5–2.4 myr oxygen isotope excursion and Northern Hemispheric glaciation event, which represents a major step toward modern conditions, when polar fronts were strengthened and glacial-interglacial variation intensified (Sancetta and Silvestri, 1986; Raymo et al., 1989; Hodell and Ciesielski, 1991).

Although the interval between 3.15 and 2.85 myr (hereafter referred to as the PRISM time slab) is distinct in that mean conditions differed from those of the intervals

The PRISM group is made up of John Barron, Thomas Cronin, Harry Dowsett, Farley Fleming, Thomas Holtz, Jr., Scott Ishman, Richard Poore, Robert Thompson, and Debra Willard, all of whom are with the U.S. Geological Survey.

ISBN 0–300–06348–2.

immediately surrounding it, there is a high degree of variability within it (Dowsett and Poore, 1991; Barron, 1992a, Hodell and Venz, 1992; Shackleton et al., in press). For most of our marine records, we have adopted a strategy whereby we develop an estimate of mean "interglacial" conditions within the time slab. This approach minimizes the problems associated with attempting point-to-point correlations between data sites separated by large geographic distances. The Late Pleistocene analog would be to provide a single sea-surface temperature (SST) value (for winter and summer) representing average interglacial conditions at each site (for example, average SST of isotope stages 5, 7, and 9).

Because of its proximity to a number of biostratigraphic and magnetostratigraphic events, the PRISM time slab is long enough to be identified and reliably correlated between marine sequences, independent of local climatic characteristics (Berggren et al., 1985; Dowsett, 1989a, 1989b). Deep-sea records and, to a lesser extent, ocean margin records can be correlated with some confidence to this interval. Many of our terrestrial records come from short sequences that rely on limited radiometric dates and magnetostratigraphy for chronology. The sparseness of long terrestrial time series with multiple age-control points makes identification of high-frequency variability and integration of our terrestrial paleoclimatic estimates into the PRISM time slab less certain than that of our marine estimates (see below).

Because the selected time period is geologically recent, many fossil taxa encountered are extant. Thus, environmental interpretations derived from 3.0 myr fossil assemblages are more likely to be reliable estimates of physical conditions than are interpretations derived from older assemblages containing greater numbers of extinct taxa.

In the remainder of this chapter, we use the terms "3.0 myr" and "Middle Pliocene" to indicate our reconstruction interval.

The PRISM Middle Pliocene Data Set

Parameters included in the reconstruction include sea-surface temperature, vegetation (land cover), sea level (shoreline), sea ice, and land ice. The data discussed below have been used to develop gridded data sets at 8° × 10° resolution for use in model experiments. Dowsett et al. (1994) provide an explanation of the data sets and a more in-depth discussion than is given here. Chandler et al. (1994) present results of an initial model experiment using these data sets.

The locations of key control points used in this reconstruction are shown in figure 14.1. Data come from deep-sea core material, discrete sedimentary sections, and regional syntheses. Table 14.1 lists environmental estimates (marine in the form of temperature; terrestrial in the form of temperature and effective moisture) for these key sites as deviations from modern conditions. Details on data sets and interpretations that were summarized and integrated into the reconstruction from these sites are contained in publications listed in Table 14.1.

Sea-Surface Temperature

SSTs for this reconstruction are based on quantitative and semiquantitative analyses of planktic foraminifers, shallow-water ostracods, and diatoms. Methodology for

Data Localities

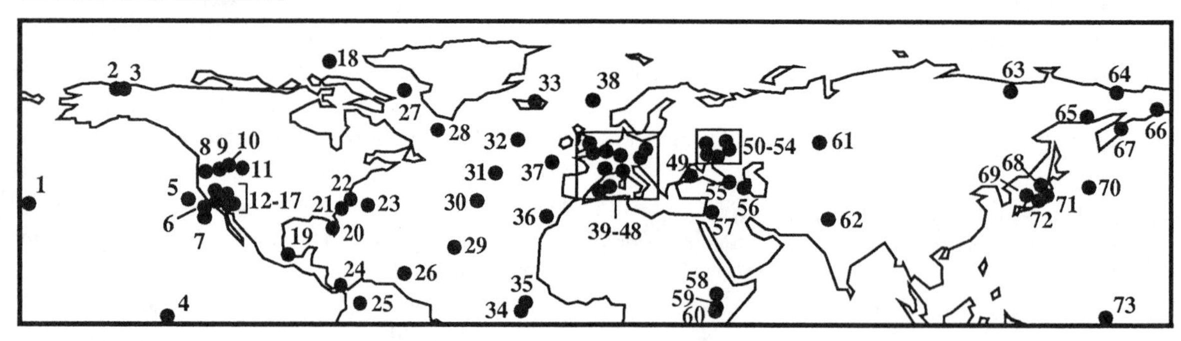

Fig. 14.1. Sites used in climatic reconstruction of Middle Pliocene. Locality numbers refer to table 14.1. Figure from Dowsett et al. (1994).

Table 14.1. Localities and Data Used in the Mid-Pliocene Northern Hemispheric Reconstruction

			Marine		Terrestrial			
Locality	Latitude	Longitude	ΔFeb.	ΔAug	ΔMoisture	ΔTemp	Vegetation Data	References
1 DSDP 310	36.52	−176.54	3.0	2.0	·	·	·	Reinterpretation of Keller (1978)
2 Ocean Point	70.83	−151.40	·	·	>	>	√	Nelson and Carter (1985)
3 Colvillian	70.29	−150.42	2.0	4.0	·	·	·	Cronin et al. (1993)
4 DSDP 573	0.49	−133.30	−1.0	−0.2	·	·	·	Hays et al. (1989)
5 DSDP 32	37.13	−127.56	·	·	·	·	√	Fleming (1992)
6 DSDP 467	33.86	−120.76	3.0	3.0	·	·	√	Barron (1992b)
7 DSDP 469	32.62	−120.56	3.0	3.0	·	·	√	Barron (1992b)
8 Tule Lake	41.58	−121.50	·	·	<	>	√	Adam et al. (1989, 1990)
9 Bruneau	42.92	−115.82	·	·	>	<	√	Thompson (1992)
10 Fossil Gulch	42.75	−114.95	·	·	>	<	√	Leopold and Wright (1985)
11 INEL	43.64	−113.25	·	·	>	<	√	Thompson (1991)
12 Searles Lake	35.68	−117.34	·	·	>	·	·	Smith (1984)
13 Furnace Creek	36.47	−116.89	·	·	>	·	·	Winograd et al. (1985)
14 Anza Borrego	32.97	−116.17	·	·	>	<	√	Remeika et al. (1988)
15 Amargosa Desert	36.00	−116.00	·	·	>	·	·	Hay et al. (1986)
16 Verde Valley	34.80	−112.00	·	·	>	·	·	Nations et al. (1981)
17 San Pedro Valley	31.90	−110.80	·	·	>	·	·	Smith et al. (1993)
18 Meighen Island	79.50	−100.00	·	·	>	>	√	Matthews (1987, 1990); Matthews and Ovenden (1990)
19 Paraje Solo	19.00	−96.00	·	·	·	<	√	Graham (1989)
20 Pinecrest Beds	27.35	−82.43	0.0	0.0	=	=	√	Willard et al. (1993)
21 Lee Creek	76.75	−35.38	2.3	3.1	·	·	·	Cronin (1991a)
22 Yorktown	37.00	−76.50	5.2	1.8	>	>	√	Dowsett & Wiggs (1992); Cronin (1991a); Groot (1991); Litwin and Andrele (1992a, 1992b); Willard (1992, 1994)
23 DSDP 603	35.49	−70.03	5.4	0.5	·	·	·	Dowsett & Poore (1991)
24 DSDP 502	11.49	−79.38	0.2	−0.1	·	·	·	Dowsett & Poore (1991)
25 Plain of Bogotá	4.83	−74.20	·	·	>	·	√	Hooghiemstra (1989); Hooghiemstra and Sarmiento (1991)
26 ODP 672	15.50	−58.50	0.7	0.7	·	·	·	Dowsett & Poore (1991)
27 ODP 645	70.45	−64.66	·	·	>	>	√	de Vernal and Mudie (1989a, 1989b)
28 ODP 646	58.25	−48.33	0.0	0.0	>	>	√	de Vernal and Mudie (1989b); Dowsett & Poore (1991); Willard (1992, 1994)
29 DSDP 396	22.90	−43.50	0.2	0.7	·	·	·	Dowsett & Poore (1991)
30 DSDP 606	37.34	−35.50	2.0	2.6	·	·	·	Dowsett & Poore (1991)
31 DSDP 410	45.51	−29.48	1.0	3.4	·	·	·	Dowsett & Poore (1991)
32 DSDP 552	56.04	−23.23	3.6	7.9	·	·	·	Dowsett & Poore (1990, 1991)
33 Tjornes	66.16	−17.25	4.0	6.2	>	>	√	Cronin (1991a, 1991b); Schwarzbach and Pflug (1957); Akhmetiev et al. (1978); Akhmetiev (1991); Willard (1992, 1994)
34 ODP 667	4.55	−21.90	0.5	0.0	·	·	·	Dowsett & Poore (1991)
35 DSDP 366	5.68	−19.85	·	·	·	·	√	Sarnthein et al. (1982)
36 DSDP 397	26.84	−15.18	·	·	·	·	√	Sarnthein et al. (1982)
37 DSDP 548	48.85	−12.00	6.7	9.7	·	·	·	Dowsett & Poore (1991); Dowsett & Loubere (1992)
38 ODP 642	67.23	2.93	·	·	>	>	√	Willard (1992, 1994)
39 North Sea	52.50	1.50	4.7	0.4	·	·	·	Wood et al. (1994)
40 Red and Walton Crags	55.18	1.25	·	·	·	·	√	Zalasiewicz et al. (1988); Hunt (1989)
41 Reuver	44.25	6.10	·	·	>	·	√	Suc and Zagwijn (1983); Zagwijn (1992)
42 Northwest Germany	51.00	6.50	·	·	·	·	√	Mohr (1986)
43 Bresse Basin	47.00	5.00	·	·	>	·	√	Rousseau et al. (1992)
44 Stirone River Section	44.75	10.00	·	·	·	·	√	Bertolani Marchetti (1975); Bertolani Marchetti et al. (1979); Gregor (1990)

(*continued*)

Table 14.1. (*Continued*)

Locality	Latitude	Longitude	Marine		Terrestrial		Vegetation Data	References
			ΔFeb.	ΔAug	ΔMoisture	ΔTemp		
45 Garraf 1	40.50	−0.50	·	·	>	<	√	Suc (1984)
46 Autan 1	43.00	4.00	·	·	>	·	√	Cravatte and Suc (1981); Suc and Zagwijn (1983)
47 Southern Poland	?	?	·	·	·	·	√	Stuchlik and Shatilova (1987)
48 Slovakia	?	?	·	·	·	·	√	Planderová (1974)
49 DSDP 380	42.09	29.60	·	·	>	·	√	Traverse (1982)
50 Russian Plain #1	54.00	40.00	·	·	>	·	√	Grichuk (1991)
51 Russian Plain #2	52.50	40.00	·	·	>	·	√	Grichuk (1991)
52 Russian Plain #3	54.50	54.00	·	·	>	·	√	Grichuk (1991)
53 Russian Plain #4	52.50	51.00	·	·	>	·	√	Grichuk (1991)
54 Russian Plain #5	49.00	45.00	·	·	>	·	√	Grichuk (1991)
55 Western Georgia	42.00	42.00	·	·	·	·	√	Shatilova (1980, 1986); Shatilova et al. (1991)
56 Azerbaijan	40.00	49.00	·	·	·	·	√	Mamedov (1991)
57 Hula Basin	33.05	35.60	·	·	>	<	√	Horowitz (1989); Horowitz and Horowitz (1985)
58 Hadar	11.0	40.50	·	·	>	<	√	Bonnefille et al. (1987)
59 Turkana Basin	4.00	36.00	·	·	·	·	√	Williamson (1985)
60 East Africa	4.00	37.00	·	·	·	·	√	Cerling et al. (1988); Cerling (1992)
61 West Siberia	63.00	70.00	·	·	>	>	√	Volkova (1991)
62 Northern Pakistan	33.50	73.00	·	·	·	·	√	Quade et al. (1989)
63 Lena River	73.00	129.00	·	·	>	>	√	Fradkina (1991)
64 Kolyma Basin	68.00	159.00	·	·	>	>	√	Giterman et al. (1982)
65 Magadan District	60.00	150.00	·	·	>	>	√	Fradkina (1991)
66 Anadyr Basin	65.00	176.00	·	·	>	>	√	Fradkina (1991)
67 Karaginsky	58.85	164.04	4.0	2.0	·	·	·	Gladenkov et al. (1991)
68 ODP 794A	40.19	138.23	·	·	·	=	√	Heusser (1992)
69 ODP 797B	38.61	134.53	·	·	·	=	√	Heusser (1992)
70 DSDP 580	41.63	153.98	5.0	3.0	·	·	·	Barron (1992a)
71 Sasaoka	39.50	140.50	1.5	2.0	·	·	·	Ikeya & Cronin (1993)
72 Lake Biwa	35.25	136.00	·	·	·	·	√	Tanai and Huzioka (1967)
73 DSDP 586	−0.50	158.50	0.0	0.0	·	·	·	Jenkins (1992a, 1992b)

Note: All latitudes are north. West longitude is indicated by a minus sign; east longitude is positive. Estimates for SST changes at marine localities are in degrees centigrade. Estimates for terrestrial localities are given in terms of moisture and temperature being the same (=), greater than (>), or less than (<) modern conditions. A check mark under the vegetation column indicates that data are available. When available vegetation data do not allow environmental estimates for moisture or temperature, a period is placed under the appropriate column(s).
Source: Dowsett et al. (1994).

estimating Pliocene temperatures using these fossil groups is documented in Dowsett and Poore (1990, 1991), Dowsett (1991), Cronin and Dowsett (1990), and Barron (1992a). Estimates from individual sites in each ocean basin were then used to construct temperature amplification fields to translate modern SST values into Middle Pliocene SST estimates.

North Atlantic. The basic pattern and magnitude of SST variability in the North Atlantic region through the Pliocene were established from several high-resolution time series throughout the basin (Dowsett and Poore, 1990, 1991; Dowsett and Loubère, 1992; Poore and Gosnell, 1990). Quantitative analyses of planktic foraminifer assemblages in these cores reveal two types of SST variability (fig. 14.2). SST estimates from all cores exhibit low-amplitude, high-frequency variability on times scales of tens of thousands of years. SST estimates from low-latitude areas show little or no change from modern values. In contrast, SST estimates from cores from mid- to high-latitude areas also show longer-term changes in mean values upon which high-frequency variability is superimposed. The interval chosen for this reconstruction is one of these longer-term "warm" intervals observed at mid-to-high latitudes (fig. 14.2). Results from the North Atlantic sites show a consistent pattern of little or no SST change (relative to today) in low latitudes, with marked increases in SST in mid- and high-latitude areas.

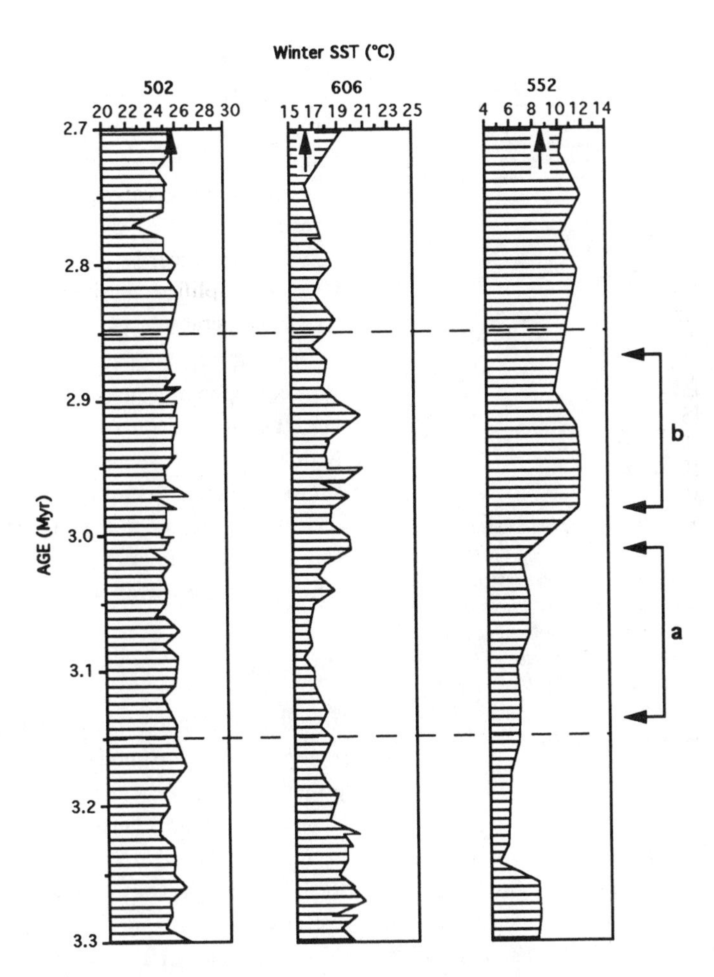

Fig. 14.2. Comparison of marine time series for the period surrounding 3 myr. These records from the North Atlantic illustrate the magnitude and variability of winter SSTs at different latitudes. Dashed horizontal lines mark the limits of our time-slab reconstruction. Arrow at top of each column indicates the modern winter (February) SST for site. At Sites 606 and 552, the winter SST varies about a lower mean during the earlier part of the time slice (a) than it does during the later part (b). Figure from Dowsett et al. (1994).

We compared North Atlantic temperature gradients from the modern, last glacial, last interglacial, and Mid-Pliocene (3.0 myr) warm interval (fig. 14.3). These gradients and amplifications of warming over modern conditions, along with supplemental site information from areas off the transect (Cronin and Dowsett, 1990; Cronin, 1991b; Dowsett and Wiggs, 1992; Wood et al., 1994), were used to develop seasonal amplification fields for the North Atlantic region. Modern SST values were then increased by the amplification field to obtain Pliocene SST (see Dowsett et al., 1994).

North Pacific. Middle Pliocene SST estimates for the North Pacific Ocean are sparse, but the basic pattern and magnitude of SST variability in the North pacific are similar to those of the North Atlantic: low-latitude areas show little or no change, and mid- to high-latitude sites show warming.

Data on diatom abundance at Deep Sea Drilling Project (DSDP) Site 580 have been interpreted to indicate a warming of 3.0°–5.5°C over Holocene conditions (Barron, 1992a). Quantitative faunal analysis of radiolarian assemblages at Site 580 (Morley and Dworetzky, 1991) indicates that climate conditions between 3.0 and 2.47 myr were relatively mild in the North Pacific compared to today's. This finding is further corroborated by our unpublished quantitative reinterpretation of planktic foraminiferal census data of Keller (1978) from DSDP Site 310, which suggests a 3°C warming over Holocene conditions.

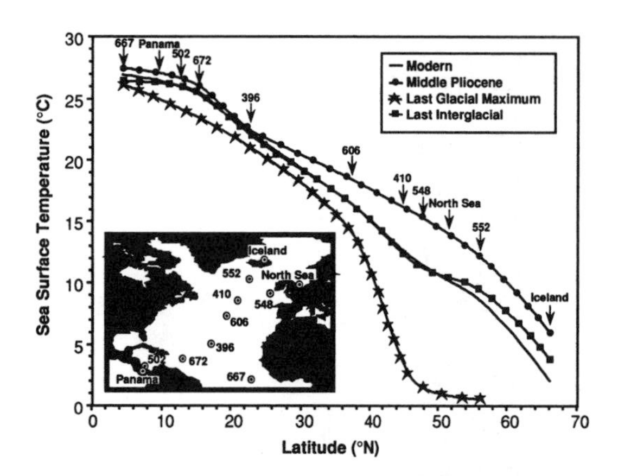

Fig. 14.3. Comparison of latitudinal transects showing North Atlantic SSTs during four times and climatic states: present-day climate, last glacial maximum (CLIMAP, 1981), last interglacial (CLIMAP, 1984), and Middle Pliocene (Dowsett et al., 1992). Circles on inset map show sites used to construct Middle Pliocene transect. The last glacial maximum shows temperatures that were consistently cooler than modern ones from the equator to the mid-latitudes and a sharply steeper gradient than today's between 35° and 45° N. The Middle Pliocene transect shows that low-latitude marine temperatures were essentially similar to modern conditions. There is a general increase in temperature at middle and high latitudes. Figure from Dowsett et al. (1992).

Along the western Pacific margin, quantitative ostracod and molluscan data from Karaginsky Island (Gladenkov et al., 1991) indicate a +4°C deviation from modern winter conditions near 55° N. Farther south, on the Japanese margin of the Japan Sea, quantitative estimates of shallow, mixed-zone bottom temperature (Cronin et al., 1994) indicate a warming of up to 2°C. Together, these data suggest that warming of the Kuroshio Current system was coincident with the warming of the Atlantic Gulf Stream system. Along the eastern margin of the North Pacific, intervals of poor diatom preservation (DSDP Sites 467, 469, and 32) suggest an August SST up to 3°C warmer than that of today (Barron, 1992b).

Hays et al. (1989) used radiolarian faunal data to show that Middle Pliocene equatorial Pacific SST at DSDP Site 573 was essentially identical to that of modern conditions. Our own unpublished quantitative analysis of planktic foraminiferal faunas at DSDP Site 586 and paleobiogeographic distribution of planktic foraminifers in the southwestern Pacific suggests no evidence that tropical surface-water temperatures were different from modern ones throughout the Late Cenozoic (Jenkins, 1992a, 1992b).

Arctic. Pliocene SST estimates for the open ocean are not yet available, but ample evidence for warming in the Arctic can be found from marginal deposits containing thermophilic ostracod immigrants from the Pacific and Atlantic Oceans (Brouwers et al., 1991; Brouwers, 1994; Cronin et al., 1993). Based on this information and on palynological records from the Arctic margin (see below), we amplified marginal Arctic SST but left central Arctic temperatures at near-modern values.

In summary, compared to modern conditions, Middle Pliocene open and marginal marine records show a coherent pattern of warmer SST in mid- and high-latitudes, with little or no change in low latitudes. Figure 14.4 shows temperature anomalies between modern and Middle Pliocene winter and summer SSTs derived from PRISM estimates. The anomalies and warming are most pronounced in the northeastern North Atlantic.

Vegetation

To adapt our Pliocene vegetational reconstruction to fit general-circulation-model parameters, we used a simplified version of the global vegetation scheme of Matthews (1985) and Hansen et al. (1983), which divides vegetation into eight categories: desert, tundra, grassland, shrub-grassland, tree-grassland, deciduous forest, evergreen forest, and rain forest. These categories are used to specify surface hydrology and surface albedo for model runs. Because our vegetational reconstruction relies primarily on palynological data and because of difficulties in distinguishing among the three types of grassland from the pollen record, we merged them into a single grassland category for our purposes. In addition, the vegetational summary had to take into account short-term variations that may be related to orbitally induced climatic changes (fig. 14.5).

We started with a representation of modern vegetation in the eight-fold classification of Hansen et al. (1983) at 8° × 10° scale and adjusted the modern gridded data set according to the Pliocene data from sites listed in table 14.1. Below we review our current knowledge of the distribution of Pliocene vegetation of the Northern Hemisphere in the context of the six general categories of vegetation. Several significant differences from present-day vegetational distribution are evident in the Pliocene reconstruction (fig. 14.6). In general, vegetation shows a climatic pattern similar to that of marine microfossils. At high altitudes the vegetation is charac-

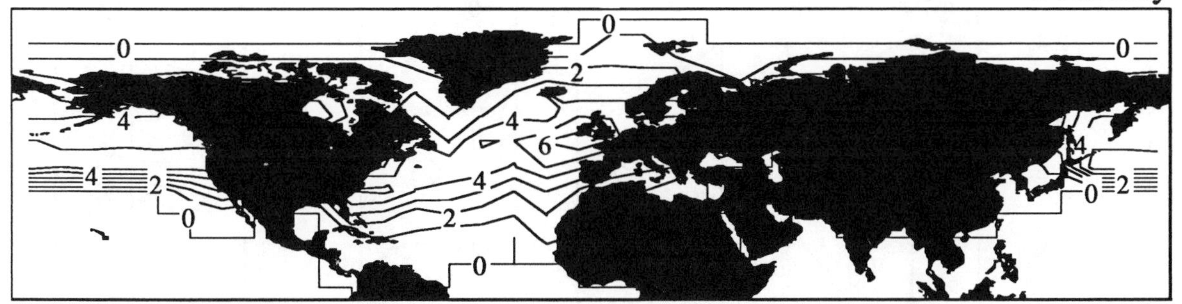

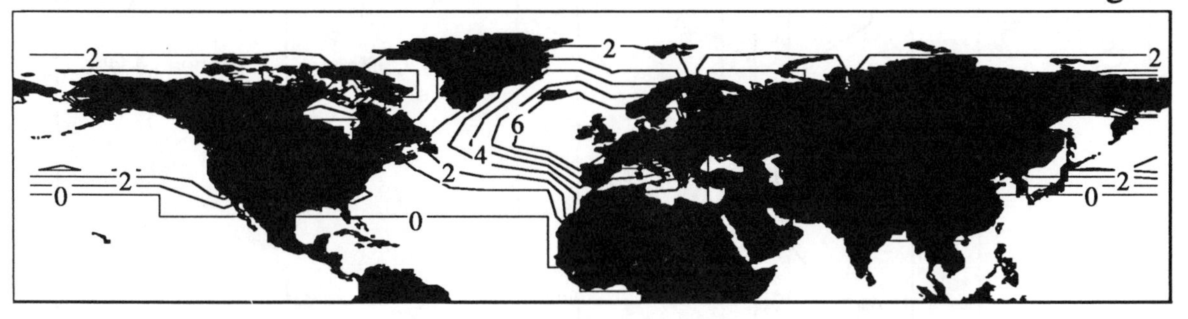

Fig. 14.4. Differences in SST estimates between Middle Pliocene and modern Northern Hemispheric oceans derived from PRISM reconstruction for the winter season (February, above) and the summer season (August, below). Figure from Dowsett et al. (1994).

teristic of much warmer temperatures than today's; at low latitudes there is little change.

Tundra. This cold-adapted treeless vegetational assemblage covers large areas of the circum-Arctic region (fig. 14.6), whereas during the Middle Pliocene this complex was a minor component of even the northernmost fossil floras. At 80° N in northern Canada a Middle Pliocene assemblage contains tundra associated with boreal and temperate conifers (Matthews, 1987, 1990; Matthews and Ovenden, 1990; Fyles et al., 1991). Elsewhere in the modern tundra zone, conifer forest was present during the period surrounding 3 myr (Volkova, 1991; Giterman et al., 1982; Fradkina, 1991).

Evergreen Forest. As noted above, coniferous evergreen forest extended as far as 80° N during the Middle Pliocene. Marine palynological assemblages indicate that boreal forest with temperate elements was also present in the region surrounding Baffin Bay and on Labrador Sea (de Vernal and Mudie, 1989a, 1989b), although the southern limit of the boreal forest may have been farther north than it is today (Willard, 1992).

Northern Iceland supported forests with *Pinus* and *Picea* prior to 3 myr (Akhmetiev et al., 1978; Akhmetiev, 1991; Schwarzbach and Pflug, 1957; Willard, 1992), although tundralike conditions may have developed soon after this time (Akhmetiev, 1991). In Siberia and in the far northeast of Arctic Russia, coniferous forests were present through the Pliocene until ca. 2.5 myr (Volkova, 1991; Giterman et al., 1982; Fradkina, 1991).

Coniferous forests alternated with steppe-forest mosaics through the Late Gilbert and Gauss chrons in western North America (Thompson, 1991; Leopold and Wright, 1985). In eastern North America, conifers were present in mixed hardwood-conifer forests (Groot, 1991), and *Pinus* was important in the vegetation of the southern coastal plain (Litwin and Andrle, 1992a,

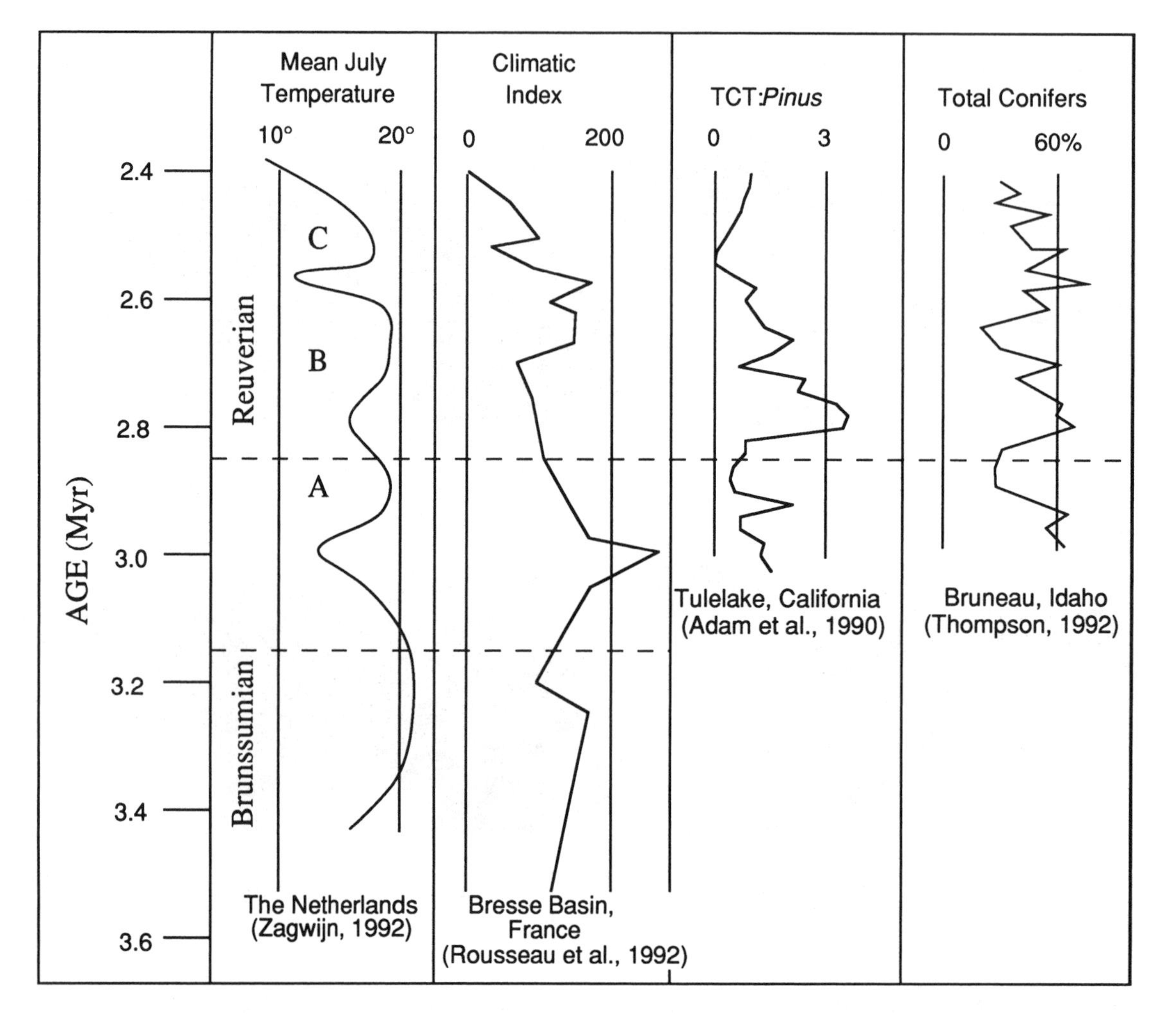

Fig. 14.5. Comparison of terrestrial climate time series for the period surrounding 3.0 myr. Dashed horizontal lines mark the limits of our time-slab reconstruction. These records from different regions illustrate the known climatic and vegetational variability for the 1.2 myr period centered on 3.0 myr. The panel to the left illustrates the interpreted paleoclimatic variations in the Netherlands (the Brunssumian-Reuverian boundary shown as in Zagwijn, 1992; earlier papers show this boundary closer to 3.25 myr). The second panel from the left demonstrates the climatic variability inferred from the pollen record from the Bresse Basin, France (here the Climatic Index is calculated from the changing percentages of herbs, *Pinus, Quercus,* and *Engelhardtia*). The second panel from the right illustrates the changing forest composition near Tulelake, in northern California; this ratio of TCT (the grouping of *Taxodiaceae, Cupressaceae,* and *Taxaceae*) to *Pinus* may reflect temperature variations but has no simple translation into paleoclimatic estimates. The panel to the right shows generalized changing forest composition inferred from the total conifer component in pollen records from central Idaho. Figure from Dowsett et al. (1994).

1992b). In Vera Cruz, Mexico, palynological data indicate expanded coverage of coniferous elements (relative to today's) during the Pliocene (Graham, 1989).

In northwestern Europe temperate conifers were important in a mixed conifer-hardwood assemblage during part of the Reuverian period (ca. 3.2 to 2.4 myr; Suc and Zagwijn, 1983; Zagwijn, 1974, 1975, 1992). Portions of this period were more boreal than coniferous (Zalasiewicz et al., 1988; Hunt, 1989). In central Europe boreal and temperate coniferous trees grew with mixed hardwoods (Planderová, 1974; Stuchlik and Shatilova, 1987), and conifers were present on the Russian Plain, with *Pinus* dominant in the west and *Picea* in the east (Borisova, 1991; Grichuk, 1991).

Conifers were important forest components in the Middle Pliocene of Italy (Bertolani Marchetti, 1975; Bertolani Marchetti et al., 1979; Gregor 1990) and the western Mediterranean (Suc 1984). Palynological data from the Black Sea (Traverse, 1982) demonstrate the dominance of Taxodiaceae and other conifers through the Pliocene. Middle Pliocene records from Georgia and Azerbaijan indicate that the higher reaches of the Cau-

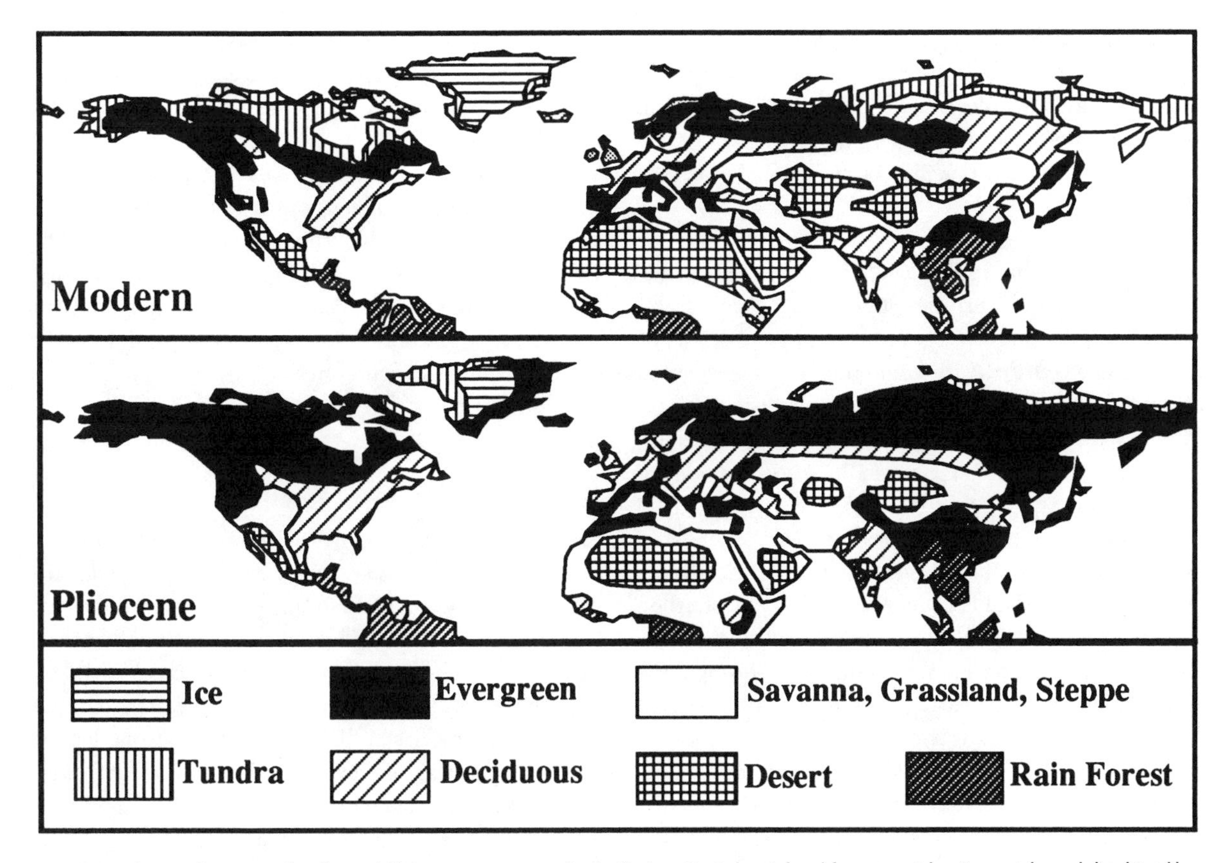

Fig. 14.6. Schematic illustrations of modern and Pliocene vegetation cover for the Northern Hemisphere inferred from terrestrial environmental records listed in table 14.1. Figure from Dowsett et al. (1994).

casus Mountains supported conifer forests (Stuchlik and Shatilova, 1987; Mamedov, 1991). In the modern desert of northeastern Israel, coniferous taxa were present during the Middle Pliocene (Horowitz and Horowitz, 1985; Horowitz, 1989).

The Pliocene record of Japan indicates a mix of deciduous trees and conifers (Tanai and Huzioka, 1967), and the basal spectra (ca. 3 myr) of the Lake Biwa pollen record indicate a climate somewhat warmer than that of the modern day, with temperate, warm-temperate, and boreal plants growing near the site (Fuji, 1988).

In addition to evergreen coniferous vegetation, evergreen sclerophyllous broad-leaved elements are recorded in the Pliocene flora of southern France (Suc, 1984) and of East Africa (Bonnefille et al., 1987; Bonnefille, this vol.) and were probably also present in portions of southwestern North America, North Africa, and southern Asia.

Deciduous Forest. During the Middle Pliocene deciduous forests extended farther north in eastern North America than they do today, but their southern borders were about the same as today's (Willard, 1994). In the central part of North America deciduous forests extended farther south into Mexico (Graham, 1989). The vegetation of the Plain of Bogotá was Andean forest similar to that of today (without North American immigrants that are now important [*Quercus, Alnus;* Hooghiemstra, 1989]).

In northwestern Europe during the Reuverian, mixed conifer-hardwood and hardwood forests included a diverse group of deciduous forest plants (Zagwijn, 1974, 1992; Mohr, 1986; Rousseau et al., 1992; Hunt, 1989). Mixed deciduous forests were present at least as far east as southern Poland throughout most of the Pliocene (Stuchlik and Shatilova, 1987).

In the western Mediterranean region, palynological data indicate that 3 myr was a time of transition, when temperate deciduous trees declined rapidly in abundance and *Pinus* increased (Suc and Zagwijn, 1983). Deciduous trees were abundant in the middle elevational zones of the Caucasus region during the Middle Pliocene

(Stuchlik and Shatilova, 1987; Mamedov, 1991), although conifers were dominant during some intervals around 3 myr (Shatilova et al., 1991). In East Africa, montane forest was present in modern, subdesertic steppe near 3 myr (Bonnefille et al., 1987; Bonnefille, this vol.).

Grassland, Steppe, and Savanna. As used here grassland combines the categories of grassland, shrub-grassland (steppe), and tree-grassland (savanna). Steppe vegetation alternated with coniferous forest vegetation through the Pliocene in the interior of western North America (Adam et al., 1989, 1990; Leopold and Wright, 1985; Thompson, 1991). Similar fluctuations occurred in the Black Sea region (Traverse, 1982), although in both regions, steppe coverage was less than that of today. Savanna was present in the modern deserts of Azerbaijan through the Middle Pliocene (Mamedov, 1991). Carbon isotopes from soil carbonates in northern Pakistan suggest that C_4 grasslands were dominant through the Pliocene and into the Pleistocene (Quade et al., 1989). In East Africa, palynological data suggest that grassland was present at 3.7 myr and after 2.5 myr (Bonnefille, 1985; Bonnefille, this vol.), and other data suggest that savanna may have persisted in this region for most of the last 4 myr (Williamson, 1985).

Desert. The category of desert has two components—polar desert and middle- or low-latitude desert. Because coniferous trees advanced as far north as the northern Canadian Arctic archipelago around 3 myr, we eliminated polar desert from our Pliocene reconstruction. In western North America, paleohydrological and paleontological data suggest that Middle Pliocene conditions were more humid than they are today (Thompson, 1991). Similarly, sites surrounding the Sahara Desert indicate levels of effective moisture that were higher than present levels (Bonnefille et al., 1987; Cerling et al., 1988; Suc, 1984; Cravatte and Suc, 1981; Suc and Zagwijn, 1983; Sarnthein et al., 1982; van Zinderen Bakker and Mercer, 1986). Although a trend toward aridification began prior to 3 myr, it seems likely that at that time the Sahara was smaller than it is today. However, palynological data of Pliocene age from Algeria and Egypt suggest near-modern vegetation in some portions of North Africa (Marley, 1980).

Farther east, palynological data from the modern desert in Israel (Horowitz and Horowitz, 1985; Horowitz, 1989) reveal that dominance through the period near 3 myr alternated between broad-leaved and coniferous trees and steppe. Thus, we reduced the areal extents around the peripheries of the modern deserts (fig. 14.6).

Rain Forest. Palynological data from southeastern Mexico (Graham, 1989) suggest that tropical rain-forest coverage was reduced during portions of the Middle Pliocene. In East Africa, rain forest may have expanded eastward from ca. 3.4 to 3.3 myr but was apparently in near-modern position before 3 myr (Williamson, 1985). Thus, our reconstruction makes only minor changes to the distribution of rain forest.

Global Sea Level

Middle Pliocene high-stands of sea level have been documented by several lines of evidence. Based on sequence stratigraphy, Haq et al. (1987a, 1987b) reported that sea levels around 3.8–2.9 myr were up to 60 m higher than those of the present. The Orangeburg scarp, a Middle Pliocene paleoshoreline along the southeastern United States Atlantic Coastal Plain, and micropaleontological data form the basis of Dowsett and Cronin's (1990) estimate of sea-level stands between 3.5 and 3.0 myr that were about 35 ± 18 m higher than those of the present. The stratigraphic distribution of disconformities at Enewetak Atoll indicate a Pliocene sea-level maximum of 36 m in the equatorial Pacific Ocean (Wardlaw and Quinn, 1991). Krantz (1991) correlated Atlantic Coastal Plain sediments with $\delta^{18}O$ isotopic records from deep-sea cores and estimated sea-level to have been 30–35 m higher at 4.0–3.2 myr and about 25 m higher at 3.0 myr. Pliocene sea-level changes were controlled primarily by glacioeustatic events, but their precise correlation to changes in the polar regions is still unclear. Webb and Harwood (1991) and Webb et al. (1984) have argued for a reduction in Antarctic ice volume between 5 and 3 myr to about one-third of present volume, which would have raised sea level approximately 44 m (Oerlemans and van der Veen, 1984). Shackleton et al. (in press) found evidence for cyclic changes in the size of the Antarctic ice sheet during the Mid-Pliocene. Based on the studies and data cited above, we chose +25 m as a conservative estimate of sea-level rise for our reconstruction.

The greatest changes in land-sea distribution occur on low-lying passive margins, peninsulas, and isthmuses and cannot be readily seen at the scale of the figures presented in this chapter.

Northern Hemispheric Sea Ice and Land Ice

Shallow marine ostracods in marine sediments deposited during several Middle- to-Late Pliocene transgressions along the northern slope of Alaska and on the northern tip of Greenland (Brouwers et al., 1991; Brouwers, in press; Cronin et al., 1993; Brigham-Grette and Carter, 1992) indicate that sea ice was greatly reduced during these warm Pliocene intervals. Similarly, pollen and plant macroflora (discussed above) show that evergreen forests extended to the edge of the Arctic Ocean at this time. Our North Atlantic and North Pacific SST estimates indicate that relatively warm waters were entering the Arctic during warm intervals of the Pliocene. All these observations indicate a great reduction in sea-ice extent and suggest that the Arctic was at least seasonally ice-free during the Middle Pliocene warm interval (3.0 myr). For Pliocene winter, we infer that sea ice was similar to modern summer sea-ice conditions (fig. 14.7; see Dowsett et al., 1994, for additional discussion).

The Middle Pliocene presents a controversial period with respect to land-ice volume. The multiple lines of evidence indicating that Middle Pliocene sea level was at least 25 m higher than it is today were the main factor used to determine Pliocene ice volume. At present, almost all major continental ice is located in Greenland and Antarctica. The estimated volume of sea-level rise equivalent (SLE) from all other small ice caps and glaciers is only 0.3–0.6 m (Meier, 1985). Therefore, the observed sea-level increases in the Pliocene must be related to mass wastage of the Greenland and Antarctic ice sheets.

The estimated sea-level rise with total removal of the Greenland ice sheet is 7.4 m (Reeh, 1985). There is little data directly documenting the extent of Pliocene land ice on Greenland during the PRISM time slab. SSTs from some high-latitude North Atlantic deep-sea cores (Dowsett and Poore, 1991) document significantly warmer conditions and evidence of boreal forests extending to the Arctic Ocean and suggest lower ice volume in Greenland. However, Middle Pliocene SST estimates from Ocean Drilling Program (ODP) Site 646 in the Labrador Sea are not significantly warmer than today's temperatures and ice-rafted detritus (IRD) occurs as far back as 6 myr at that site. Jansen et al. (1990) also documented IRD pulses associated with glacial cycles at ODP Site 642 in the Norwegian Sea during the Early Pliocene. Thus, some

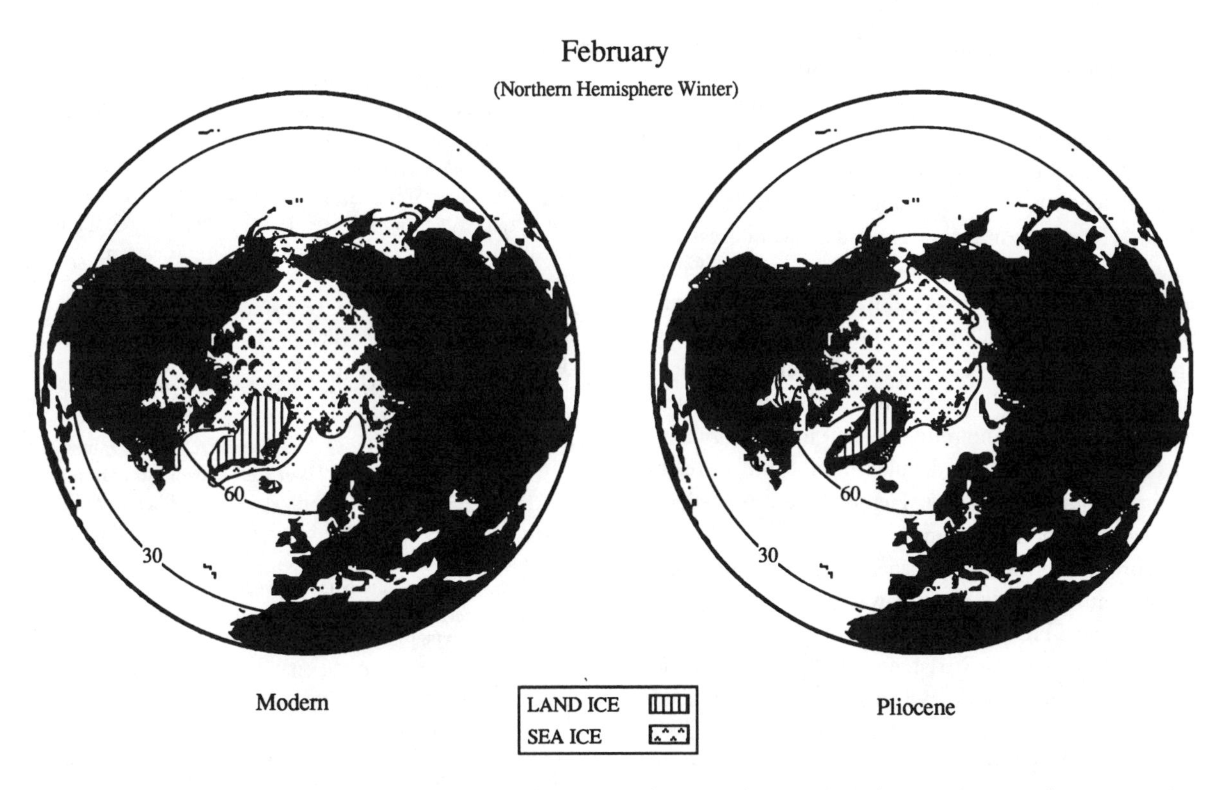

Fig. 14.7. Generalized winter distribution of modern and Pliocene land ice and sea ice for winter (February) in the Northern Hemisphere. Figure from Dowsett et al. (1994).

ice must have been present on Greenland during the Middle Pliocene. Marine and freshwater diatoms recovered from sediments at the base of the Camp Century core indicate that this site was free of ice during the stage 5e warm event (125 myr; Harwood, 1986), which indicates a reduction of about 50 percent in the Greenland ice sheet. In our reconstruction, we make the conservative assumption that the Middle Pliocene Greenland ice sheet was also reduced by 50 percent (fig. 14.7), which would result in a sea-level rise of about 4 m.

Our interpretation of Middle Pliocene sea-level and Northern Hemispheric land-ice reconstruction infers that about 21 m of sea-level rise was caused by reduction in the Antarctic ice sheet. Discussion of the Antarctic record is far beyond the scope of this chapter. We note, however, that the glacial history of Antarctica during the Pliocene is a topic of considerable debate (e.g., Denton et al., 1989; Denton, this vol.; Clapperton and Sugden, 1990; Webb and Harwood, 1991).

Summary and Conclusions

The data and interpretations summarized here represent a first step toward a global reconstruction of a warm interval of the Middle Pliocene. The PRISM reconstruction of the Northern Hemisphere during the Middle Pliocene, though preliminary, reveals a number of significant contrasts with the modern world. Multiple marine and terrestrial records depict a coherent pattern of increased warmth in middle- and high-latitude regions. Records from the Arctic margin and adjacent areas indicate that the Arctic was seasonally ice-free and that the ice sheet on Greenland was reduced to 50 percent or less of its current extent. Our data from the Northern Hemisphere indicate that sea level was about 25 m higher during the Middle Pliocene. Our interpretation requires a reduction of the Antarctic ice sheet to obtain much of that sea-level rise. More detailed marine and terrestrial records from the Southern Hemisphere are needed to confirm that interpretation.

During the Middle Pliocene, evergreen (and in particular coniferous) vegetation was more widespread in the Northern Hemisphere than it is today. Boreal forest extended to the Arctic Coast, essentially occupying most areas of modern tundra. Boreal forest extended southward to the northern middle latitudes, where it graded into temperate mixed-conifer forest and conifer-hardwood forest. At middle latitudes, our reconstruction shows that desert areas of Africa and Asia were reduced during the Middle Pliocene.

PRISM studies of Middle Pliocene climates and environments continue to expand to include records from the Southern Hemisphere and to increase the geographic resolution of Northern Hemispheric records. Combining paleoenvironmental studies with modeling experiments will lead to better documentation of the causes and mechanisms by which the earth's climate evolved and of the effects of climatic change on regional and global ecosystems.

Acknowledgments

We thank P. Schweitzer and Bruce Wardlaw for constructive comments and suggestions on this chapter; Y. Gladenkov, R. Whatley, A. Wood, A. Coates, J. Jackson, N. Ikeya, T. Quinn, D. Rind, M. Chandler, D. Harwood, P. Schweitzer, and G. Wilson for assisting with the data sets; Kevin Foley, Andrew Shuckstes, Stephanie West, Emerson Polanco, Jennifer Buchner, Jean Self-Trail, Gary Belair, Shawn Freisen, Keith Goggin, and Frank Mossburg for help with various aspects of data reduction; and Stephanie West for much of the drafting. We appreciate the opportunity to participate in the Airlie conference and especially thank Dr. Elisabeth Vrba for her encouragement and helpful comments on this work.

References

Adam, D. P., Bradbury, Rieck, H. J., and Sarna-Wojcicki, A. M. 1990. Environmental changes in the Tulelake Basin, Siskiyou and Modoc Counties, California, from 3 to 2 myr before present. *U.S. Geological Survey Bulletin,* no. 1933.

Adam, D. P., Sarna-Wojcicki, A. M., Rieck, H. J., Bradbury, J. P., Dean, W. E., and Forester, R. M. 1989. Tulelake, California: The last 3 myr. *Palaeogeography, Palaeoclimatology, Palaeoecology* 72:89–103.

Akhmetiev, M. A. 1991. Flora, vegetation, and climate of Iceland during the Pliocene: Pliocene climates of the Northern Hemisphere. *Abstracts of the Joint US / USSR Workshop on Pliocene Paleoclimates, Moscow, USSR, April, 1990.* U.S. Geological Survey Open-File Report 91-447, pp. 8–9.

Akhmetiev, M. A., Bratoeva, G. M., Giterman, R. E., Golubeva, L. V., and Moiseeva, A. L. 1978. Late Cenozoic stratigraphy and flora of Iceland. *Transactions, Academy of Science of the USSR,* no. 316 (published in English by the National Research Council, Reykjavik, Iceland, 1981).

Barnosky, C. W. 1984. Late Miocene vegetational and climatic variations inferred from a pollen record in northwest Wyoming. *Science* 223:49–51.

———. 1987. Response of vegetation to climatic changes of different duration in the late Neogene. *Trends in Ecology and Evolution* 2(8):247–250.

Barron, J. A. 1992a. Pliocene paleoclimatic interpretation of DSDP Site 580 (NW Pacific) using diatoms. *Marine Micropaleontology* 20:23–44.

———. 1992b. Paleoceanographic and tectonic controls on the Pliocene diatom record of California. In *Proceedings of the Fifth International Congress on Pacific Neogene Stratigraphy and IGCP 246, Shizuoka, Japan (Oct. 6–10, 1991)*, pp. 25–41 (ed. R. Tsuchi and J. C. Ingle, Jr.). University of Tokyo Press, Tokyo.

Berggren, W. A., Kent, D. V., and Van Couvering, J. A. 1985. Neogene geochronology and chronostratigraphy. In *The Chronology of the geological record*, pp. 211–260 (ed. N. J. Snelling). Blackwell Scientific Publications, London.

Bertolani Marchetti D. 1975. Preliminary palynological data on the proposed Plio-Pleistocene boundary type-section of La Castella. *L'Atheneo Parmense Acta Naturalia* 11:467–485.

Bertolani Marchetti, D., Accorsi, C. A., Pelosio, G., and Raffi, S. 1979. Palynology and stratigraphy of the Plio-Pleistocene sequence of the Stirone River (northern Italy). *Pollen et Spores* 21:149–167.

Bonnefille, R. 1985. Evolution of the continental vegetation: The paleobotanical record from East Africa. *South African Journal of Science* 81(5):267–270.

Bonnefille, R., Vincens, A., and Buchet, G. 1987. Palynology, stratigraphy, and palaeoenvironment of a Pliocene hominid site (2.9–3.3 myr) at Hadar, Ethiopia. *Palaeogeography, Palaeoclimatology, Palaeocology* 60:249–281.

Borisova, O. K. 1991. Neogene temperature fluctuations on the southeastern Russian Plain: Pliocene climates of the Northern Hemisphere. *Abstracts of the Joint US / USSR Workshop on Pliocene Paleoclimates, Moscow, USSR, April, 1990.* U.S. Geological Survey Open-File Report 91-447, pp. 14–15.

Brigham-Grette, J., and Carter, L. D. 1992. Pliocene marine transgressions of Northern Alaska: Circumarctic correlations and paleoclimatic interpretations. *Arctic* 45:74–89.

Brouwers, E. M. 1994. Late Pliocene paleoecologic reconstructions based on ostracode assemblages from the Sagavanirktok and Gubik Formations, Alaskan North Slope. *Arctic* 47:16–33.

Brouwers, E. M., Jørgensen, N. O., and Cronin, T. M. 1991. Climatic significance of the ostracode fauna from the Pliocene Kap København Formation, North Greenland. *Micropaleontology* 37:245–267.

Cerling, T. E. 1992. Development of grasslands and savannas in East Africa during the Neogene. *Palaeogeography, Palaeoclimatology, Palaeoecology, Global and Planetary Change* 97:241–147.

Cerling, T. E., Bowman, J. R., and O'Neil, J. R. 1988. An isotopic study of a fluvial-lacustrine sequence: The Plio-Pleistocene Koobi Fora Sequence, East Africa. *Palaeogeography, Palaeoclimatology, Palaeoecology* 63:335–356.

Chandler, M., Rind, D., and Thompson, R. S. 1994. Joint investigations of the Middle Pliocene climate II: GISS GCM Northern Hemisphere results. *Global and Planetary Change* 9:197–219.

Clapperton, C. M., and Sugden, D. E. 1990. Late Cenozoic glacial history of the Ross Embayment, Antarctica. *Quaternary Science Reviews* 9:253–272.

CLIMAP 1981. Seasonal reconstruction of the earth's surface at the last glacial maximum. *Geological Society of America Map and Chart Series* MC-36:1–18.

———. 1984. The last interglacial ocean. *Quaternary Research* 21:123–224.

Cravatte, J., and Suc, J.-P. 1981. Climatic evolution of northwestern Mediterranean area during Pliocene and Early Pleistocene by pollen-analysis and forams of drill autan 1: Chronostratigraphic correlations. *Pollen et Spores* 23(2):247–258.

Cronin, T. M. 1991a. Late Neogene marine ostracoda from Tjörnes, Iceland. *Journal of Paleontology* 65:767–794.

———. 1991b. Pliocene shallow water paleoceanography of the North Atlantic Ocean based on marine ostracodes. *Quaternary Science Reviews* 10:175–188.

Cronin, T. M., and Dowsett, H. J. 1990. A quantitative micropaleontologic method for shallow marine paleoclimatology: Application to Pliocene deposits of the western North Atlantic Ocean. *Marine Micropaleontology* 16(1/2):117–148.

Cronin, T. M., Kitamura, A., Ikeya, N., Watanabe, M., and Kamiya, T. 1994. Late Pliocene paleoceanography, Sea of Japan: The Yabuta Formation. *Palaeogeography, Palaeoclimatology, Palaeoecology* 108:437–455.

Cronin, T. M., Whatley, R. C., Wood, A., Tsukagoshi, A., Ikeya, N., Brouwers, E. M., and Briggs, W. M., Jr. 1993. Microfaunal evidence for elevated Mid-Pliocene temperatures in the Arctic Ocean. *Paleoceanography* 8:161–173.

Denton, G. H., Bockheim, J. G., Wilson, S. C., Leide, J. E., and Andersen, B. G. 1989. Late Quaternary ice-surface fluctuations of Beardmore Glacier, Transantarctic Mountains. *Quaternary Research* 31:183–209.

de Vernal, A., and Mudie, P. J. 1989a. Late Pliocene to Holocene palynostratigraphy at ODP sites 645, Baffin Bay. *Proceedings of the Ocean Drilling Program: Scientific results* 105:387–399.

———. 1989b. Pliocene and Pleistocene palynostratigraphy at ODP Sites 646 and 647, eastern and southern Labrador Sea. *Proceedings of the Ocean Drilling Program: Scientific results* 105:401–422.

Dowsett, H. J. 1989a. Application of the graphic correlation method to Pliocene marine sequences. *Marine Micropaleontology* 14:3–32.

———. 1989b. Improved dating of the Pliocene of the eastern South Atlantic using graphic correlation: Implications for paleobiogeography and paleoceanography. *Micropaleontology* 35:279–292.

———. 1991. The development of a long-range foraminifer transfer function and application to Late Pleistocene North Atlantic climatic extremes. *Paleoceanography* 6:259–273.

Dowsett, H. J., and Cronin, T. M. 1990. High eustatic sea-level during the Middle Pliocene: Evidence from the southeastern U.S. Atlantic Coastal Plain. *Geology* 18:435–438.

Dowsett, H. J., Cronin, T. M., Poore, R. Z., Thompson, R. S., Whatley, R. C., and Wood, A. M. 1992. Micropaleontological evidence for increased meridional heat transport in the North Atlantic Ocean during the Pliocene. *Science* 258:1133–1135.

Dowsett, H. J., and Loubère, P. 1992. High resolution Late Pliocene sea-surface temperature record from the Northeast Atlantic Ocean. *Marine Micropaleontology* 20:91–105.

Dowsett, H. J., and Poore, R. Z. 1990. A new planktic foraminifer transfer function for estimating Pliocene through Holocene sea-surface temperatures. *Marine Micropaleontology* 16:1–23.

———. 1991. Pliocene sea-surface temperatures of the North

Atlantic Ocean at 3.0 myr. *Quaternary Science Reviews* 10:189–204.

Dowsett, H. J., Thompson, R. S., Barron, J. A., Cronin, T. M., Ishman, S. E., Poore, R. Z., Willard, D. A., and Holtz, T. R., Jr. 1994. Joint investigations of the Middle Pliocene climate, I: PRISM paleoenvironmental reconstructions. *Global and Planetary Change* 9:169–195.

Dowsett, H. J., and Wiggs, L. B. 1992. Planktonic foraminiferal assemblage of the Yorktown Formation, USA. *Micropaleontology* 38:75–86.

Fradkina, A. F. 1991. Pliocene climatic fluctuations in the far north-east of the USSR: Pliocene climates of the Northern Hemisphere. *Abstracts of the Joint US / USSR Workshop on Pliocene Paleoclimates, Moscow, USSR, April, 1990.* U.S. Geological Survey Open-File Report 91-447, p. 22.

Fuji, N. 1988. Palaeovegetation and palaeoclimate changes around Lake Biwa, Japan during the last ca. 3 myr. *Quaternary Science Reviews* 7:21–28.

Fyles, J. G., Marincovich, L., Jr., Matthews, J. V., Jr., and Barendregt, R. 1991. Unique mollusc find in the Beaufort Formation (Pliocene) on Meighen Island, Arctic Canada. *Geological Survey of Canada Current Research,* 91-B:105–112.

Giterman, R. E., Sher, A. V., and Matthews, J. V. 1982. Comparison of the development of tundra-steppe environment in west and east Beringia: Pollen and macrofossil evidence from key sections. In *Paleoecology of Beringia,* pp. 43–73 (ed. D. M. Hopkins, J. V. Matthews, Jr., C. E. Schweger, and S. B. Young). Academic Press, New York.

Gladenkov, Y. B., Barinov, K. B., Basilian, A. E., and Cronin, T. M. 1991. Stratigraphy and paleoceanography of Pliocene deposits of Karaginsky Island, eastern Kamchatka, USSR. *Quaternary Science Reviews* 10:239–246.

Graham, A. 1989. Late Tertiary paleoaltitudes and vegetational zonation in Mexico and Central America. *Acta Botanica Neerlandica* 38(4):417–424.

Gregor, H.-J. 1990. Contributions to the Late Neogene and Early Quaternary floral history of the Mediterranean. *Review of Palaeobotany and Palynology* 62:309–338.

Grichuk, V. P. 1991. Vegetation and climate of the Middle Akchaghylian (Late Pliocene) on the Russian Plain: Pliocene climates of the Northern Hemisphere. *Abstracts of the Joint US / USSR Workshop on Pliocene Paleoclimates, Moscow, USSR, April, 1990.* U.S. Geological Survey Open-File Report 91-447, pp. 26–27.

Groot, J. J. 1991. Palynological evidence for Late Miocene, Pliocene, and Early Pleistocene climate changes in the middle U.S. Atlantic Coastal Plain. *Quaternary Science Reviews* 10:147–162.

Hansen, J. E., Russell, G., Rind, D., Stone, P., Lacis, A., Lebedeff, S., Ruedy, R., and Travis, L. 1983. Efficient three-dimensional global models for climate studies: Models I and II. *Monthly Weather Review* 111:609–662.

Haq, B. U., Hardenbol, J., and Vail, P. R. 1987a. Chronology of fluctuating sea-levels since the Triassic. *Science* 235:1156–1167.

———. 1987b. *The new chronostratigraphic basis of Cenozoic and Mesozoic sea-level cycles.* Cushman Foundation Foraminiferal Research Special Publication 24:7–13.

Harwood, D. M. 1986. Diatom biostratigraphy and paleoecology with a Cenozoic history of Antarctic ice sheets. Ph.D. diss., Ohio State University.

Hay, R. L., Pexton, R. E., Teague, T. T., and Kyser, T. T. 1986. Spring-related carbonate rocks, Mg clays, and associated minerals in Pliocene deposits of the Amargosa Desert, Nevada and California. *Geological Society of America Bulletin* 97:1488–1503.

Hays, P. E., Pisias, N. G., and Roelofs, A. K. 1989. Paleoceanography of the eastern equatorial Pacific during the Pliocene: A high resolution radiolarian study. *Paleoceanography* 4:57–73.

Heusser, L. E. 1992. Neogene palynology of Holes 794A, 795A, and 797B in the Sea of Japan: Stratigraphic and paleoenvironmental implications of the preliminary results. *Proceedings of the Ocean Drilling Program: Scientific results* 127 / 128:325–339.

Hodell, D. A., and Ciesielski, P. F. 1991. Stable isotopic and carbonate stratigraphy of the Plio-Pleistocene of Ocean Drilling Program (ODP) Hole 704A: Eastern subantarctic South Atlantic. *Proceedings of the Ocean Drilling Program: Scientific results* 114:409–436.

Hodell, D. A., and Venz, K. 1991. Toward a high-resolution stable isotopic record of the Southern Ocean during the Pliocene-Pleistocene (4.8 to 0.8 myr). *Antarctic Research Series* 56:265–310.

Hooghiemstra, H. 1989. Quaternary and upper-Pliocene glaciations and forest development in the tropical Andes: Evidence from a long high-resolution pollen record from the sedimentary basin of Bogotá, Columbia. *Palaeogeography, Palaeoclimatology, Palaeoecology* 72:11–26.

Hooghiemstra, H., and Sarmiento, G. 1991. Long continental pollen record from a tropical intermontane basin: Late Pliocene and Pleistocene history from a 540-meter core. *Episodes* 14:107–115.

Horowitz, A. 1989. Continuous pollen diagrams for the last 3.5 myr from Israel: Vegetation, climate, and correlation with the oxygen isotope record. *Palaeogeography, Palaeoclimatology, Palaeoecology* 72:63–78.

Horowitz, A., and Horowitz, M. 1985. Subsurface late Cenozoic palynostratigraphy of the Hula Basin, Israel. *Pollen et Spores* 27(3–4):365–390.

Hunt, C. O. 1989. The palynology and correlation of the Walton Crag (Red Crag Formation, Pliocene). *Journal of the Geological Society* (London) 146:743–745.

Jansen, E., Sjoholm, J., Bleil, U., and Erichsen, J. A. 1990. Neogene and Pleistocene glaciations in the Northern Hemisphere and Late Miocene–Pliocene global ice volume fluctuations: Evidence from the Norwegian Sea. In *Geologic history of the polar oceans: Arctic vs Antarctic,* pp. 677–705 (ed. U. Bleil and J. Theide). Kluwer, Amsterdam.

Jenkins, D. G. 1992a. The paleogeography, evolution, and extinction of Late Miocene–Pleistocene planktonic foraminifera from the southwest Pacific. In *Centenary of Japanese micropaleontology,* pp. 27–35 (ed. K. Ishizaki and T. Saito). Terra Scientific Publishing Co., Tokyo.

———. 1992b. Predicting extinctions of some extant planktic foraminifera. *Marine Micropaleontology* 19:239–243.

Keller, G. 1978. Late Neogene biostratigraphy and paleoceanography of DSDP Site 310, central North Pacific, and correlation with the southwest Pacific. *Marine Micropaleontology* 3:97–119.

Krantz, D. E. 1991. A chronology of Pliocene sea-level fluctuations: The U.S. Middle Atlantic Coastal Plain record. *Quaternary Science Reviews* 10:163–174.

Leopold, E. B., and Denton, M. F. 1987. Comparative age of grassland and steppe east and west of the northern Rocky Mountains. *Annals of the Missouri Botanical Garden* 74:841–867.

Leopold, E. B., and Wright, V. C. 1985. Pollen profiles of the Plio-Pleistocene transition in the Snake River Plain, Idaho. In *Late Cenozoic history of the Pacific Northwest*, pp. 323–348 (ed. C. J. Smiley). American Association for the Advancement of Science, Pacific Division.

Litwin, R. J., and Andrle, V. A. S. 1992a. *Modern palynomorph and weather census data from the U.S. Atlantic Coast (Continental Margin Program samples and selected NOAA weather stations)*. U.S. Geological Survey Open-File report 92-263.

———. 1992b. *Palynomorph census data from Pliocene strata of the U.S. Atlantic Coastal Plain (Massachusetts to central Florida)*. U.S. Geological Survey Open-File Report 92-262.

Maley, J. 1980. Les changements climatiques de la fin du Tertiaire en Afrique: Leur conséquence sur l'apparition du Sahara et de sa végétation. In *The Sahara and the Nile: Quaternary environments and prehistoric occupation in northern Africa*, pp. 63–86 (ed. M. A. J. Williams and H. Faure). A. A. Balkema, Rotterdam.

Mamedov, A. V. 1991. The paleogeography of the Trans-Caucasus region during the Pliocene climatic optimum: Pliocene climates of the Northern Hemisphere. *Abstracts of the Joint US / USSR Workshop on Pliocene Paleoclimates, Moscow, USSR, April, 1990*. U.S. Geological Survey Open-File Report 91-447, pp. 28–31.

Matthews, E. 1985. Prescription of land-surface boundary conditions in GISS GCM II: A simple method based on high-resolution vegetation data bases. NASA Report No. TM 86096.

Matthews, J. V., Jr. 1987. Plant macrofossils from the Neogene Beaufort Formation on Banks and Meighen Islands, District of Franklin: Current research, Part A. *Geological Survey of Canada*, paper 87-1A, pp. 73–87.

———. 1990. New data on Pliocene floras / faunas from the Canadian Arctic and Greenland. Pliocene climates: Scenario for global warming. *Abstracts from USGS workshop, Denver, Colorado, October 23–25, 1989, Washington, D.C.* U.S. Geological Survey Open-File Report 90-64, pp. 29–33.

Matthews, J. V., Jr., and Ovenden, L. E. 1990. Late Tertiary plant macrofossils from localities in Arctic / Subarctic North America: A review of the data. *Arctic* 43:364–392.

Meier, M. F. 1985. Mass balance of the glaciers and small ice caps of the world. In *Glaciers, ice sheets, and sea level: Effect of a CO_2-induced climatic change*, pp. 139–144 (ed. M. F. Meier et al.). DOE / EV / 60235-1.

Mercer, J. H. 1978. West Antarctic ice sheet and CO_2 greenhouse effect: A threat of disaster. *Nature* 271:321–325.

Mohr, B. A. R. 1986. Die Mikroflora der Oberpliozänen Tone Von Willershausen (Kreis Northeim, Niedersachsen). *Palaeontographica*, pt. B, 198:133–156.

Morley, J., and Dworetzky, B. 1991. Evolving Pliocene-Pleistocene climate: A North Pacific perspective. *Quaternary Science Reviews* 10:225–237.

Nations, J. D., Hevley, R. H., Blinn, D. W., and Landye, J. J. 1981. Paleontology, paleoecology, and depositional history of the Miocene-Pliocene Verde Formation, Yavapai County, Arizona. *Arizona Geological Society Digest* 13:133–149.

Nelson, R. E., and Carter, L. D. 1985. Pollen analysis of a Late Pliocene and Early Pleistocene section from the Gubik Formation of Arctic Alaska. *Quaternary Research* 24:295–306.

Oerlemans, J., and van der Veen, C. J. 1984. *Ice sheets and climate*. D. Reidel, Dordrecht.

Pazzaglia, E. J. 1993. Stratigraphy, petrography, and correlation of Late Cenozoic Middle Atlantic Coastal Plain deposits: Implications for late-stage passive-margin geologic evolution. *Geological Society of America Bulletin* 105:1617–1634.

Planderová, E. 1974. The problem of the floristic boundary between Pliocene-Pleistocene in western Carpathians mounts on the basis of palynological examination. *Bureau des Recherches Géologiques et Minières, Mémoires* 78(2):547–551.

Poore, R. Z., and Gosnell, L. B. 1990. Quantitative planktic foraminifer record from Caribbean DSDP Site 502: 3 to 2 myr. *Eos, Transactions, American Geophysical Union* 71:1383.

Quade, J., Cerling, T. E., and Bowman, J. R. 1989. Development of Asian Monsoon revealed by marked ecological shift during the latest Miocene in northern Pakistan. *Nature* 342:163–166.

Raymo, M. E., Ruddiman, W. F., Backman, J., Clement, B. M., and Martinson, D. G. 1989. Late Pliocene variation in Northern Hemisphere ice sheets and North Atlantic deep water circulation. *Paleoceanography* 4:413–446.

Reeh, N. 1985. Greenland-ice-sheet mass balance and sea-level change. In *Glaciers, ice sheets, and sea level: Effect of a CO_2-induced climatic change*, pp. 1155–1171 (ed. M. F. Meier et al.). DOE / EV / 60235-1.

Remeika, P., Fischbein, I. W., and Fischbein, S. A. 1988. Lower Pliocene petrified wood from the Palm Springs Formation, Anza Borrego Desert State Park, California. *Review of Palaeobotany and Palynology* 56:183–198.

Rousseau, D.-D., Taoufiq, N. B., Petit, C., Farjanel, G., Meon, H., and Puissegun, J. 1992. Continental Late Pliocene paleoclimatic history recorded in the Bresse Basin (France). *Palaeogeography, Palaeoclimatology, Palaeocology* 95:253–261.

Sancetta, C., and Silvestri, S. 1986. Pliocene-Pleistocene evolution of the North Pacific ocean-atmosphere system, interpreted from fossil diatoms. *Palaeoceanography* 1:163–180.

Sarnthein, M., Thiede, J., Pflaumann, U., Erlenkeusen, H., Fütterer, D., Koopman, B., Lange, H. and Seibold, E. 1982. Atmospheric and oceanic circulation patterns off Northwest Africa during the past 25 myr: Geology of the northwest African Continental margin. In *Geology of the northwest African continental margin*, pp. 545–604 (ed. V. von Rad, K. Hinz, M. Sarnthein, and E. Seibold). Springer-Verlag, Berlin.

Schwarzbach, M., and H. D. Pflug. 1957. Das Klima des jüngeren Tertiärs in Island. *Neues Jahrbuch für Geologie und Paläontologie* 104:279–298.

Shackleton, N. J., Hall, M. A., and Pate, D. (in press). Pliocene stable isotope stratigraphy of Site 846. *Proceedings of the Ocean Drilling Program: Scientific results* 138.

Shatilova, I. I. 1980. Palynolgic study of Late Cenozoic and modern deposits in the eastern part of the Black Sea area. *Fourth International Palynological Conference, Lucknow, India.*

———. 1986. The palynological base of stratigraphical subdivision of Late Cenozoic deposits of the western Transcaucasus. *Review of Palaeobotany and Palynology* 48:409–414.

Shatilova, I. I., Macharadze, N. V., and Davitashvili, L. S. 1991. The Pliocene climate of western Georgia: Pliocene climates of the Northern Hemisphere. *Abstracts of the Joint US / USSR Workshop on Pliocene Paleoclimates, Moscow, USSR, April, 1990.* U.S. Geological Survey Open-File Report 91-447, pp. 35–37.

Smith, G. A., Wang, Y., Cerling, T. E., and Geissman, J. W. 1993. Comparison of a paleosol-carbonate isotope record to other records of Pliocene–Early Pleistocene climate in the western United States. *Geology* 21:691–694.

Smith, G. I. 1984. Paleohydrologic regimes in the southwestern Great Basin, 0–3.2 myr, compared with other long records of "global" climate. *Quaternary Research* 22:1–17.

Stuchlik, L., and Shatilova, I. I. 1987. Palynological study of Neogene Deposits of southern Poland and western Georgia. *Acta Palaeobotanica* 27(2):21–52.

Suc, J. P. 1984. Origin and evolution of the Mediterranean vegetation and climate in Europe. *Nature* 307:429–432.

Suc, J. P., and Zagwijn, W. H. 1983. Plio-Pleistocene correlations between the northwestern Mediterranean region and northwestern Europe according to recent biostratigraphic and palaeoclimatic data. *Boreas* 12:153–166.

Tanai, T., and Huzioka, K. 1967. Climatic implications of Tertiary floras in Japan. In *Tertiary correlations and climatic changes in the Pacific,* pp. 89–94 (ed. K. Hatai). Eleventh Pacific Science Congress, Symposium 25, Tokyo.

Thomasson, J. R. 1979. Late Cenozoic grasses and other angiosperms from Kansas, Nebraska, and Colorado: Biostratigraphy and relationships to living taxa. *Kansas Geological Survey Bulletin* 218:68.

Thompson, R. S. 1991. Pliocene environments and climates in the western United States. *Quaternary Science Reviews* 10(2/3):115–132.

———. 1992. *Palynological data from a 989-ft (301-m) core of Pliocene and Early Pleistocene sediments from Bruneau, Idaho.* U.S. Geological Survey Open-File Report 92-713.

Traverse, A. 1982. Response of world vegetation to Neogene tectonic and climatic events. *Alcheringa* 6:197–209.

van Zinderen Bakker, E. M., and Mercer, J. H. 1986. Major Late Cenozoic climatic events and palaeoenvironmental changes in Africa viewed in a world-wide context. *Palaeogeography, Palaeoclimatology, Palaeoecology* 56:217–235.

Volkova, V. S. 1991. Pliocene climates of west Siberia: Pliocene climates of the Northern Hemisphere. *Abstracts of the Joint US / USSR Workshop on Pliocene Paleoclimates, Moscow, USSR, April, 1990.* U.S. Geological Survey Open-File Report 91-447, pp. 44–45.

Wardlaw, B. R., and Quinn, T. M. 1991. The record of Pliocene sea-level change at Enewetak Atoll. *Quaternary Science Reviews* 10:247–258.

Webb, P.-N., and Harwood, D. M. 1991. Late Cenozoic history of the Ross Embayment, Antarctica. *Quaternary Science Reviews* 10:215–223.

Willard, D. A. 1992. *Late Pliocene pollen assemblages from Ocean Drilling Project Hole 646B: Census data and paleoclimatic estimates.* U.S. Geological Survey Open-File Report 92-405.

———. 1994. Palynological record from the North Atlantic region at 3 myr: Vegetational distribution during a period of global warmth. *Review of Palaeobotany and Palynology* 83:275–297.

Willard, D. A., Cronin, T. M., Ishman, S. E., and Litwin, R. J. 1993. Terrestrial and marine records of climate and environmental change during the Pliocene in subtropical Florida. *Geology* 21:679–682.

Williamson, P. G. 1985. Evidence for an Early Plio-Pleistocene rain-forest expansion in East Africa. *Nature* 315:487–489.

Winograd, I. J., Szabo, B. J., Coplen, T. B., Riggs, A. C., and Kolesar, P. T. 1985. Two-myr record of deuterium depletion in Great Basin ground waters. *Science* 227:519–522.

Wood, A. M., Whatley, R. C., Cronin, T. M., and Holtz, T. 1994. Pliocene palaeotemperature recombination for the southern North Sea based on Ostracoda. *Quaternary Science Reviews* 12:747–767.

Zagwijn, W. H. 1974. The Pliocene-Pleistocene boundary in western and southern Europe. *Boreas* 3:75–97.

———. 1975. Variations in climate as shown by pollen analysis, especially in the lower Pleistocene of Europe. In *Ice ages: Ancient and modern,* pp. 137–152 (ed. A. E. Wright and F. Moseley; ser. ed. G. Newall). Seel House Press, Liverpool.

———. 1992. The beginning of the Ice Age in Europe and its major subdivisions. *Quaternary Science Reviews* 11:583–591.

Zalasiewicz, J. A., S. J. Mathers, Hughes, M. J., Gibbard, P. L., Peglan, S. M., Harland, R., Nicholson, R. A., Boulton, G. S., Cambridge, P., and Wealthall, G. P. 1988. Stratigraphy and palaeoenvironments of the Red Crag and Norwich Crag Formations between Aldeburgh and Sizewell, Suffolk, England. *Philosophical Transactions of the Royal Society of London,* Ser. B., Biological Sciences, 322:221–272.

Zubakov, V. A., and Borzenkova, I. I. 1990. *Global paleoclimate of the Late Cenozoic.* Elsevier, Amsterdam.

Chapter 15

The Problem of Pliocene Paleoclimate and Ice-Sheet Evolution in Antarctica

George H. Denton

A fundamental question addressed in this volume is the degree to which environmental change drives evolution. Of particular interest is the role of paleoenvironmental changes in precipitating major Pliocene evolutionary events among mammal groups, including hominids, in Africa and elsewhere on the planet. Vrba (1985), for example, has argued that climatic cooling and environmental change centered around 2.5 million years (myr) ago triggered the evolution of early *Homo* on the African continent.

The overall problem of Pliocene environmental change and hominid evolution rests to an important degree on interpretations of Pliocene paleoclimate (a major control on paleoenvironment) and sea level (a major regulator of landbridges). Antarctica is important in this regard, because it is located in the center of the Southern Hemisphere and contained the only substantial ice sheet prior to 2.5 myr ago. As such, it should be a premier recorder of Pliocene climates in the Southern Hemisphere. Unfortunately, there is still no agreement as to whether Pliocene paleoclimate in Antarctica was temperate or polar. At the heart of the controversy lies the question of the stability of the Antarctic Ice Sheet and its role in global Pliocene sea-level changes.

The question of Pliocene ice-sheet stability focuses not only on the small marine-based ice sheet in West Antarctica (3.3×10^6 km^3; 6 m sea-level equivalent) but also on the possibility that the huge terrestrial East Antarctic Ice Sheet (14×10^6 km^2; 20×10^6 km^3; 60 m sea-level equivalent), which today dominates the geometry and environment of the continent, responded to Pliocene climatic warming by unstable meltdown (fig. 15.1; Drewry, 1982, 1983). Until very recently the prevailing view held that East Antarctica had been relatively stable since the inception of a polar ice sheet about 14 myr ago, in Middle Miocene time (Kennett, 1982). But a new hypothesis—that unstable ice-sheet behavior persisted in East Antarctica until Late Pliocene time—is now thought to represent a paradigm shift in our understanding of the behavior of polar ice sheets (Webb and Harwood, 1991).

The problem involves not only the evolution of polar ice sheets and Antarctic sensitivity to global warming but also the evolution of the Transantarctic Mountains, because much of the critical terrestrial evidence (geomorphic, rock, and fossil) of past ice sheets can be interpreted only in the context of landscape evolution. The questions of ice-sheet stability and mountain evolution are therefore intertwined. In turn, the evolution of the Transantarctic Mountains must be viewed in the context of the codevelopment of the West Antarctic Rift System and the Antarctic Ice Sheet during the Cenozoic era. The Transantarctic Mountains form the faulted, uplifted, and tilted shoulder of this highly asymmetric rift system. The problems of the timing and rates of uplift and subsidence of the rift system are complicated by the fact that the shoulder of the Transantarctic Mountains is comprised of numerous crustal blocks, with ten blocks in northern and southern Victoria Land alone (van der Wateren and Verbers, 1993). Conceivably, each could have had a discrete history and rate of uplift and subsidence. Thus it is possible that glacial features preserved in the various blocks of the Transantarctic Mountains reflect tectonism as well as paleoclimate.

One interpretation of available evidence in the Trans-

ISBN 0-300-06348-2.

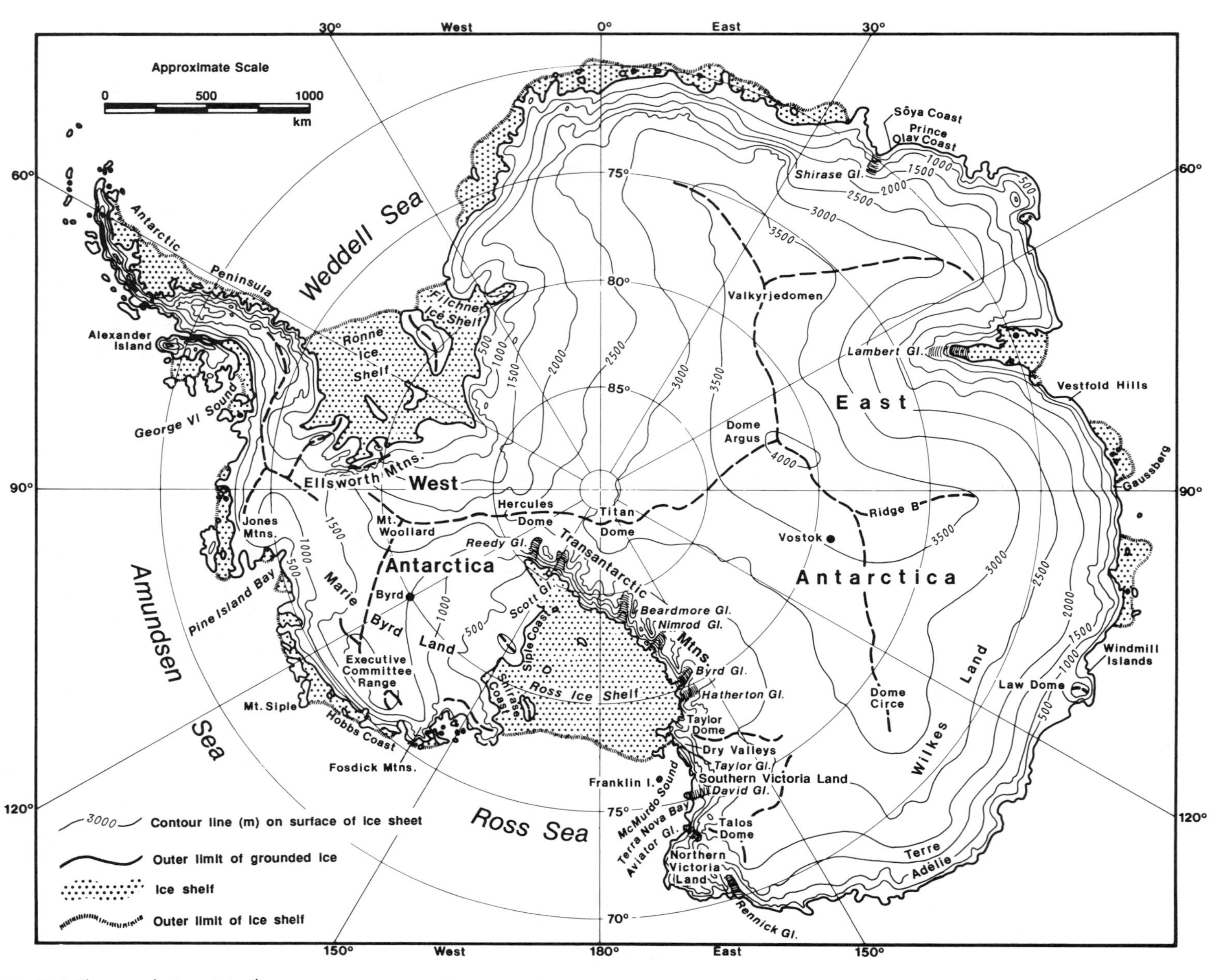

Fig. 15.1. The present-day Antarctic Ice Sheet.

antarctic Mountains is that Pliocene climate was significantly warmer than the present one (up to 20°–25°C above modern values, so that *Nothofagus* could have survived in Antarctica until Late Pliocene or Early Pleistocene time; Harwood, 1983, 1986; Webb and Harwood, 1987; Webb et al., 1984; Barrett, 1991; Barrett et al., 1992). By this scenario, a highly variable temperate Pliocene ice sheet existed in East Antarctic in a climate warm enough to allow the survival of *Nothofagus* until a stable polar ice sheet developed at or after 2.5 myr ago. Extensive Pliocene deglaciation was common (fig. 15.2), with the most recent meltdown occurring about 3.0 myr ago (Barrett et al., 1992), but also with extensive deglaciation during warm intervals of the Early Pliocene (Webb et al., 1984; Webb and Harwood, 1991). This interpretation postulates fluctuating temperate Pliocene climates in Antarctica at the center of the Southern Hemisphere until about 2.5 myr ago, when an enormous climatic deterioration induced polar conditions on the continent. Furthermore, periodic growth and collapse of a temperate East Antarctic Ice Sheet could explain Pliocene sea-level fluctuations and landbridge histories. Another, sharply contrasting interpretation of the paleoclimatic and glacial evidence of the Transantarctic Mountains is that a stable East Antarctic Ice Sheet existed under polar conditions throughout the Pliocene, without a dramatic Late Pliocene change in the character of East Antarctic paleoclimate or ice-sheet dynamics (Denton et al., 1993). The implication is that any Pliocene warmth above present-day values in the Southern Hemisphere was so moderate that it did not disturb the basic stability of the Antarctic climatic system and the Antarctic ice sheets. Moreover, any glacioeustatic component of Pliocene sea-level change must have come largely from the Greenland Ice Sheet and the marine-based West Antarctic Ice Sheet and would therefore have been relatively limited.

These two prevailing views of Antarctic paleoclimate and ice-sheet stability are mutually exclusive. Until this problem is resolved, important questions concerning the Pliocene climatic evolution of the Southern Hemisphere will remain unanswered. For example, interpretations of Pliocene oxygen isotope oscillations in benthic foraminifers from deep-sea cores in terms of polar sea-surface temperature, salinity, or ice volume depend critically on whether the Antarctic Ice Sheet was stable or variable. Likewise, the history of the Antarctic Ice Sheet would influence interpretations of the Pliocene record of sea-level variation and mammalian migrations across landbridges. Also the Pliocene paleoclimatic record of Antarctica can give important insights into the impact of environmental changes on a hemispheric scale. Today, Antarctica is thermally isolated by the Antarctic Circumpolar Current (fig. 15.3) and is characterized by a polar climate. Evidence for ice-sheet collapse and warmer-than-present Antarctic paleoclimatic in Early- and mid-Pliocene time would imply hemisphere-wide events strong enough to break down the Antarctic climatic system. Events such as these could have set the background for evolutionary change and would have curtailed mammalian migrations by causing landbridges to be flooded. The subsequent establishment of a stable polar climatic system and growth of polar ice sheets could have caused a pulse of mammalian migrations as landbridges were exposed by ice-sheet growth and sea-level drop; further, such changes could have signaled a major hemisphere-wide paleoclimatic event that in turn could have triggered a pulse of evolution. On the other hand, continuous cold Pliocene conditions and a stable ice sheet would indicate that any Pliocene climatic changes failed to perturb the Antarctic system in a major way. This stability would not mean that climatic changes in the Southern Hemisphere did not occur in the Pliocene but simply that the Antarctic climatic system was robust enough to withstand them, probably because of the strong thermal inertia of the Southern Ocean under the basic modern oceanographic processes that became established by Middle Miocene time (Kennett, 1982; Kennett and Hodell, 1993).

Antarctica Today

The separation of Antarctica from other continents in the Southern Hemisphere now allows the free passage of the Antarctic Circumpolar Current around the Southern Ocean (fig. 15.3). This current marks the polar front, which is an important factor in the thermal isolation of Antarctica. Sea ice in the Southern Ocean now covers 20 $\times 10^6$ km^2 in winter and 4 $\times 10^6$ km^2 in summer (Gordon, 1986). Today the Antarctic continent is the driest, windiest, and highest on earth. Mean annual surface temperatures range from -14°C near the coast to -55°C on the high interior plateau of East Antarctica. Precipitation in the form of snow ranges from about 40 cm (water equivalent) near the coast to about 3 cm on the high East Antarctic plateau.

The major feature of the continent is the huge Antarctic Ice Sheet, which consists of 30 $\times 10^6$ km^3 of ice

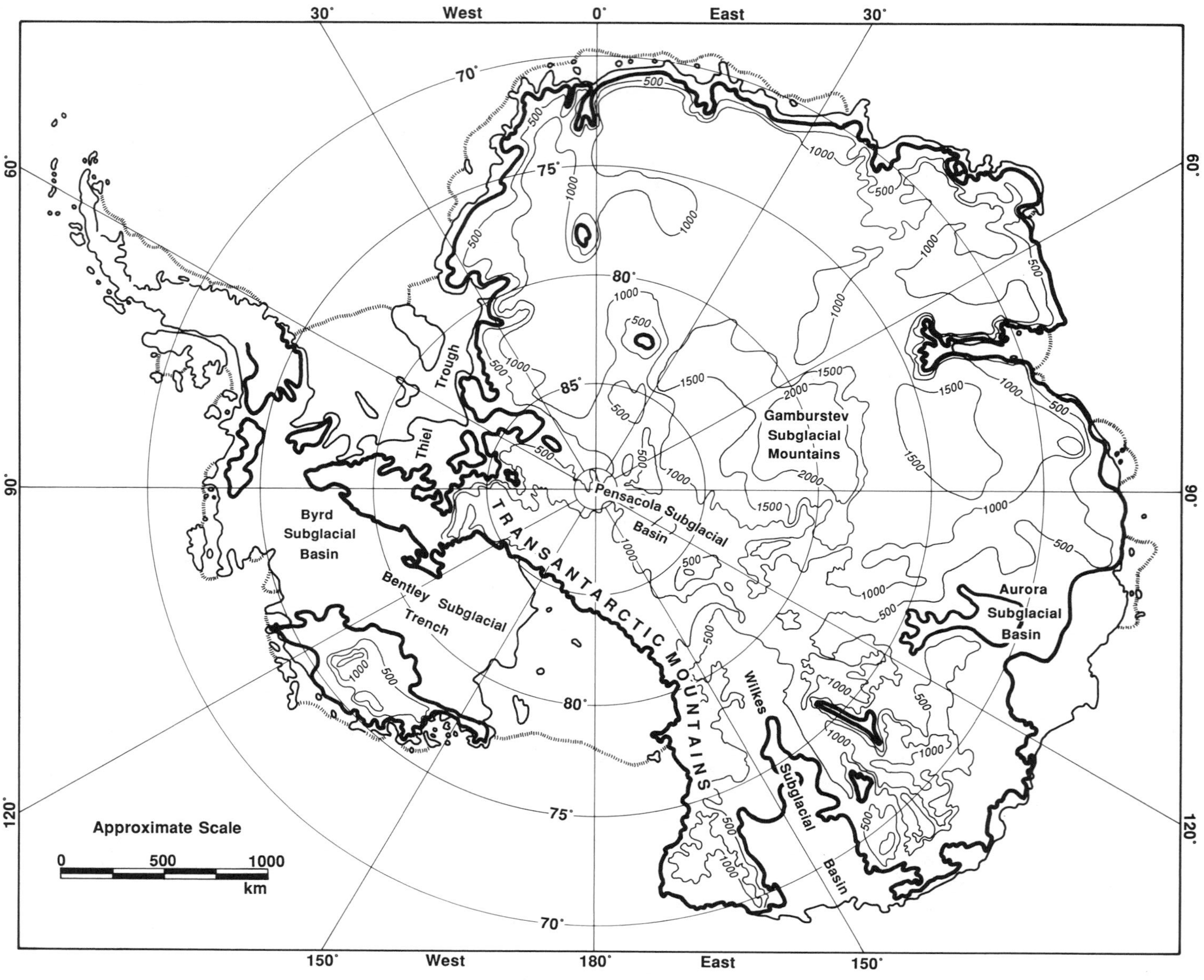

Fig. 15.2. (a) Map (adapted from Burckle and Pokras, 1991) showing topography of Antarctica with the ice sheet removed and the bedrock surface isostatically adjusted (after Drewry, 1983). Huybrechts (1993) shows that removal of ice from the Pensacola Basin requires an atmospheric temperature rise of 20°C. Such a rise yields an ice configuration smaller than that shown in (b), with the Pensacola Basin above sea level (Huybrechts, 1993). (b) Antarctic Ice Sheet during interval of extensive deglaciation in Pliocene time, as postulated by Webb et al. (1984; adapted from Denton et al., 1991). This reconstruction is based on the assumption that deglaciation of marine basins on the East Antarctic craton implies similar deglaciation in West Antarctica. The sketch, which is based on subglacial topography without isostatic adjustment (Drewry, 1983), is speculative, because the position of ice margins would have depended strongly on sea-surface temperature, sea level, mean annual atmospheric temperature, precipitation, and isostatic adjustment. This figure is adapted from Denton et al. (1993).

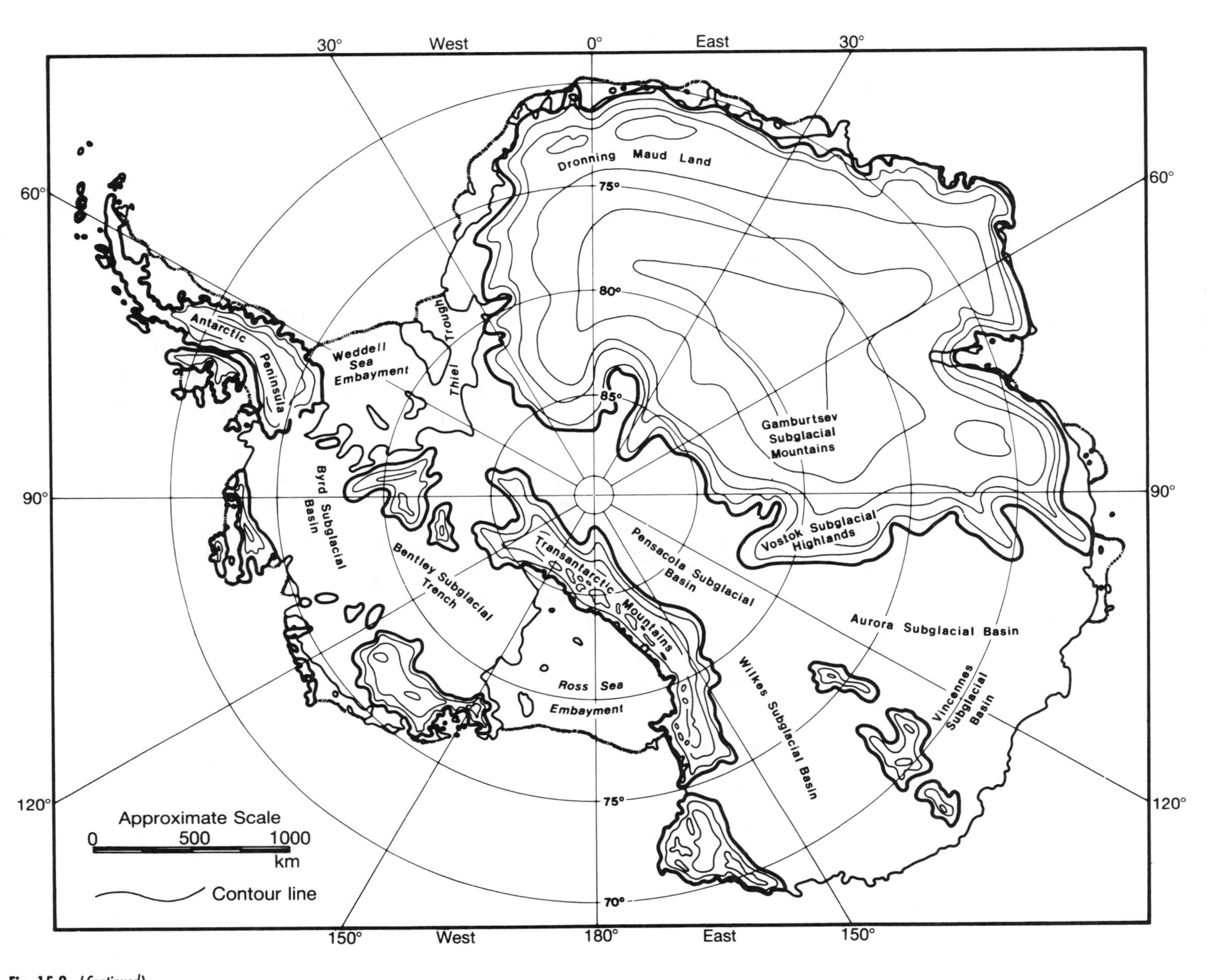

Fig. 15.2. (*Continued*)

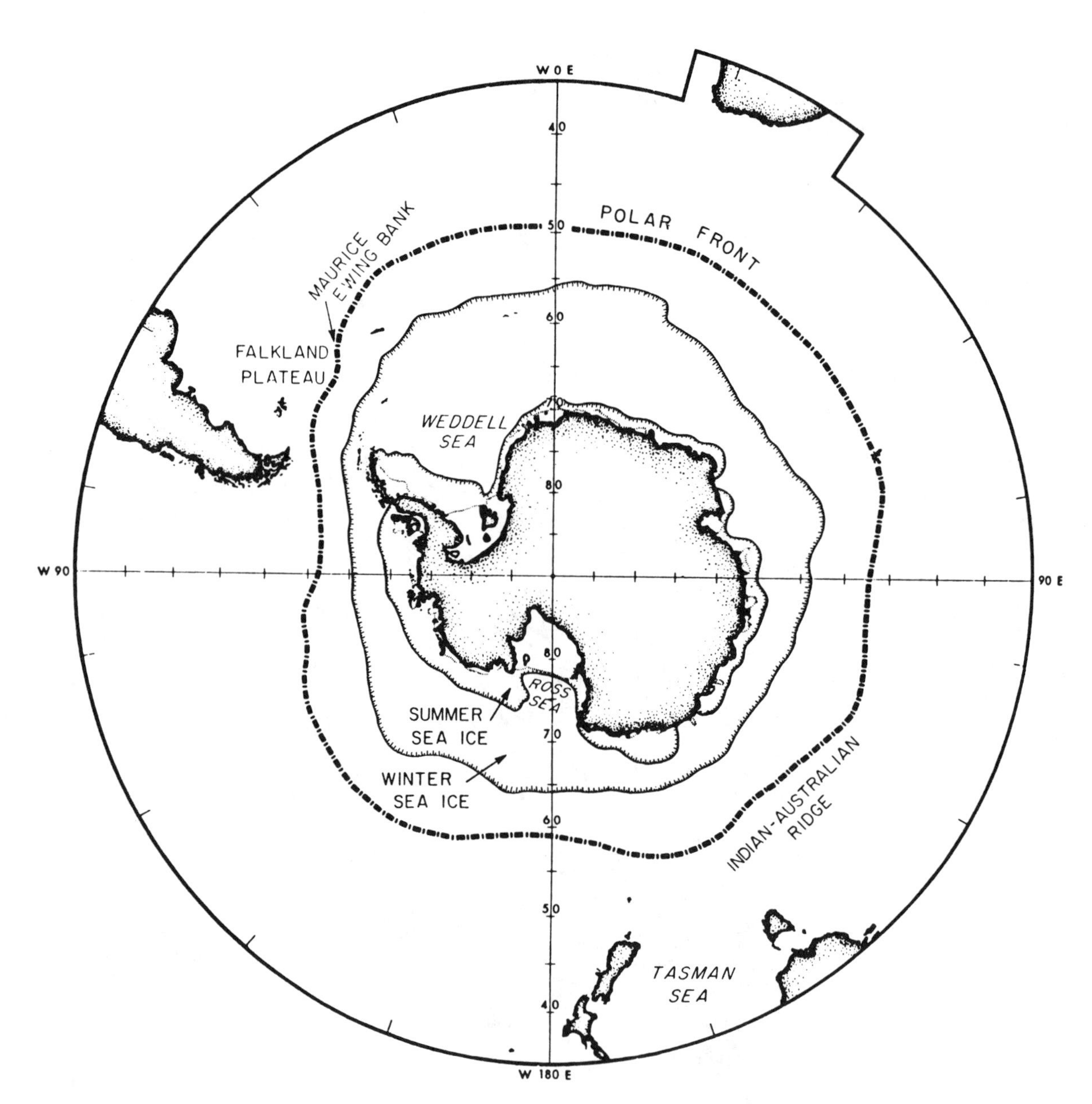

Fig. 15.3. The Southern Ocean. Sea-ice limits from Cooke and Hays (1982) and Zwally et al. (1983); adapted from Denton et al. (1991).

spread over 13.6 × 10^6 km^2 (Drewry, 1983; fig. 15.1). The Transantarctic Mountains divide the ice sheet into separate components in West and East Antarctica. The West Antarctic Ice Sheet (3.3 × 10^6 km^3; 6 m sea-level equivalent) rests on the rugged bedrock floor of the West Antarctic Rift System (Drewry, 1983). This is a marine ice sheet, because much of the bedrock surface on which it rests is well below sea level and would likely remain so if the ice sheet collapsed and isostatic compensation went to completion. Ice thickness is greatest in central West Antarctica, where subglacial rift basins reach −2,000 to −2,555 m. The domes of the West Antarctic Ice Sheet reach an elevation of 2,400–3,000 m and are anchored on bedrock highs. Most West Antarctic ice drains outward through ice streams. Many of the large ice streams drain into the Ross and Filchner-Ronne Ice Shelves, which are floating slabs of ice that occupy large embayments in the continent.

The East Antarctic Ice Sheet (26 × 10^6 km^3; 60 m sea level), the dominant topographic feature of the continent, is considered terrestrial because it rests on a base that is predominately above sea level (Drewry, 1983). The central domes reach an elevation of 3,200–4,000 m. On one flank the East Antarctic Ice Sheet is dammed behind the Transantarctic Mountains, but elsewhere it terminates either in the ocean or in floating ice shelves.

The Case for a Warm Pliocene Climate and a Variable East Antarctic Ice Sheet

A major shift in the climatic system of the earth occurred in the Pliocene epoch, between 2.9 and 2.5 myr ago (Shackleton, this vol.). Late Cenozoic ice ages began on a global scale during this decline, first with a 41 kyr and later, after 1 myr ago, with a 100 kyr pulse beat. Prior to 2.9 myr ago, repeated intervals of warmer-than-present climate marked Pliocene time, and large ice sheets were limited to Antarctica. The most recent of these warm intervals occurred about 3.0 myr ago, when high-latitude air and sea-surface temperatures in the Northern Hemisphere were elevated owing to increased meridional oceanic heat transfer (Dowsett et al., 1992). Other evidence for Pliocene warmth during this and earlier intervals comes from such diverse paleoclimatic indicators as European flora (van der Hammen et al., 1971; Grube et al., 1986); the lack of the significant ice-rafted detritus in North Atlantic Ocean cores (Ruddiman and Raymo, 1988); terrestrial plant fossils in the Arctic (Funder et al., 1985; Matthews, 1989); African pollen assemblages (Bonnefille, 1976, 1983), micromammals (Wesselman, 1985), and bovids (Vrba, 1988a, 1988b); molluscs along the western margin of the North Atlantic (Stanley, 1985); diatoms, radiolaria, and silicoflagellates from the Pacific and Southern Oceans (Hays and Opdyke, 1967; Ciesielski and Weaver, 1974); Andean flora (Hooghiemstra, this vol.); and some interpretations of the marine oxygen-isotope record (Prentice and Denton, 1988; Shackleton, 1993).

Warmer-than-present Antarctic conditions are inferred from Early-to-mid-Pliocene benthic foraminifers in cores from sites 10 and 11 of the Dry Valleys Drilling Project (DVDP), in eastern Taylor Valley (Ishman and Rieck, 1992); from Early Pliocene diatoms and silicoflagellates in the subantarctic Southern Ocean (Ciesielski and Weaver, 1974; Abelmann et al., 1990); and from microfossils in raised marine deposits in the Vestfold Hills (Pickard et al., 1988).

The hypothesis of Antarctic glaciation that calls for a highly variable temperate ice sheet in East Antarctica under warm Pliocene paleoclimatic conditions prior to 2.5–2.9 myr ago carries with it the implication that the East Antarctic Ice Sheet is susceptible to climatic warming of the magnitude last reached in the Pliocene. By analogy, Barrett et al. (1992) implied that the ice sheet could again nearly disappear in a greenhouse world a few degrees warmer than the present. This concept of a variable East Antarctic Ice Sheet existing under warm Pliocene paleoclimate is critically dependent on the inferred age and origin of reworked marine diatoms in Sirius Group outcrops along the Ross Sea sector of the Transantarctic Mountains (fig. 15.4; Webb et al., 1984; Harwood, 1986). The implication is that the diatoms originated in marine basins and that subsequently the East Antarctic Ice Sheet expanded across these basins, overrode the Transantarctic Mountains, and reworked marine microfossils into the Sirius Group. The only realistic location for such basins would be in the interior of East Antarctica—hence the argument that the ice sheet must have been sufficiently small to expose the Wilkes and Pensacola Subglacial Basins to the sea. Diatom floras that include elements indicative of warm (2°–5°C) interior East Antarctic seas are assigned ages varying from Early to Late Pliocene on the basis of stratigraphic ranges in Subantarctic deep-sea cores (Harwood, 1986). Isotopic dating of a volcanic ash layer in the CIROS-2 core in the Ferrar Glacier trough (fig. 15.5) in the Dry Valleys region confirms the biostratigraphic age control for cer-

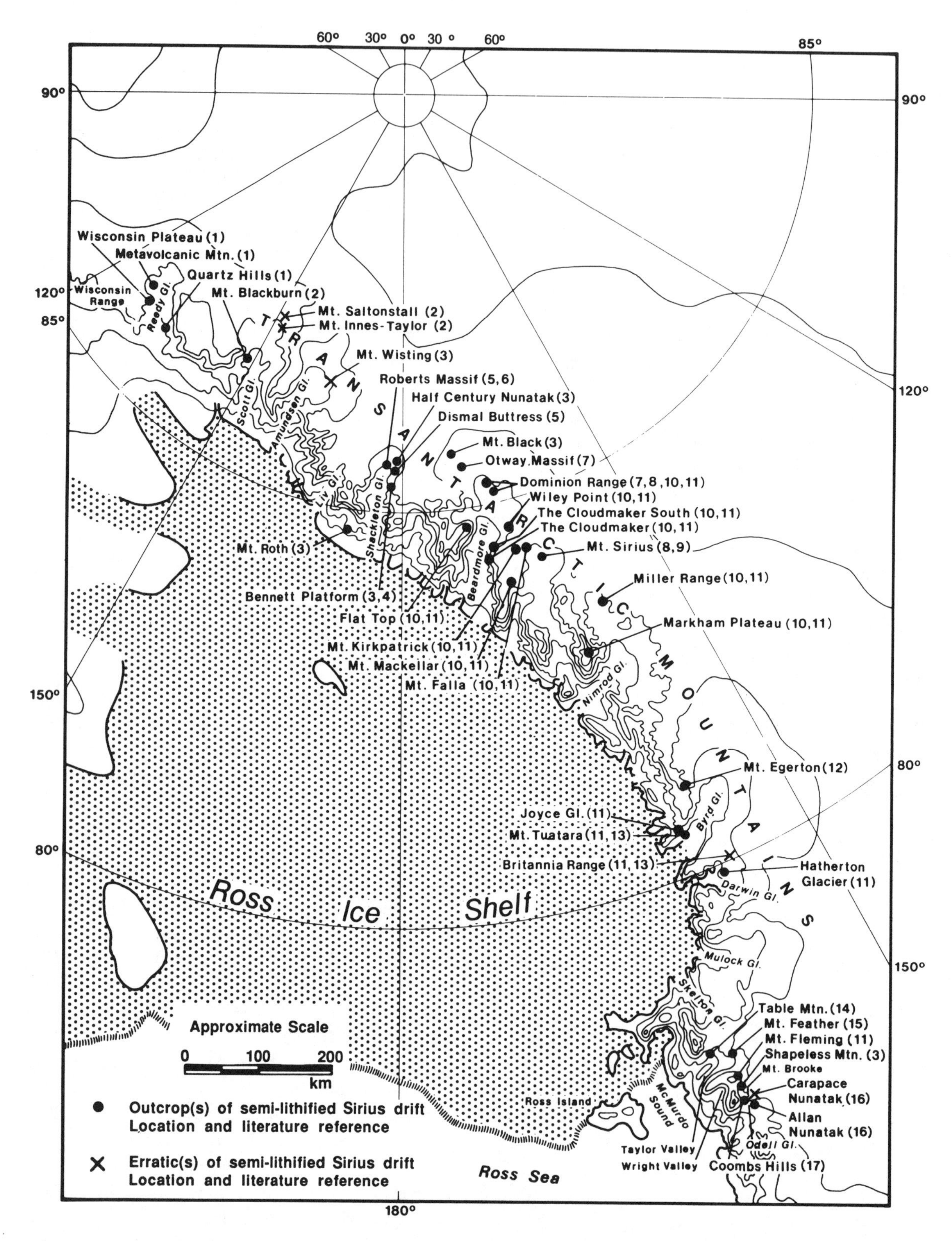

Fig. 15.4. Location of Sirius Group outcrops and erratics in the Transantarctic Mountains. References are as follows: 1. Mercer, 1968; 2. Doumani and Minshew, 1965; 3. Mayewski, 1975; 4. LaPrade, 1984; 5. McGregor, 1965; 6. Claridge and Campbell, 1968; 7. Elliot et al., 1974; 8. Mercer, 1972; 9. Barrett and Elliot, 1973; 10. Prentice et al., 1986; 11. Denton et al., 1991; 12. J. Anderson, unpublished; 13. Faure and Taylor, 1981; 14. Barrett and Powell, 1982; 15. Brady and McKelvey, 1979; 16. H. W. Borns, Jr., unpublished; 17. Brady and McKelvey, 1983. Adapted from Denton et al. (1991) and Denton et al. (1993).

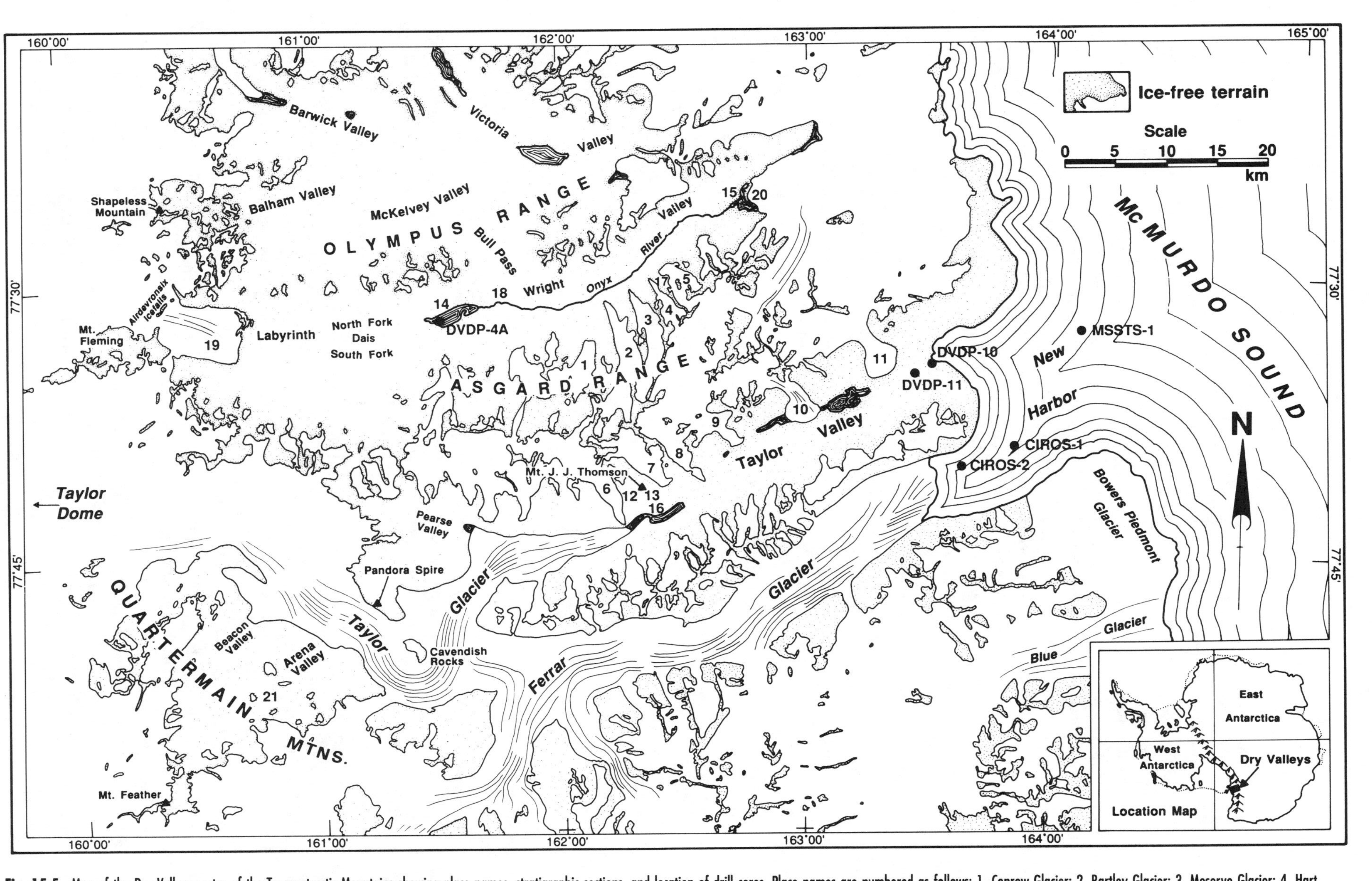

Fig. 15.5. Map of the Dry Valleys sector of the Transantarctic Mountains showing place names, stratigraphic sections, and location of drill cores. Place names are numbered as follows: 1. Conrow Glacier; 2. Bartley Glacier; 3. Meserve Glacier; 4. Hart Glacier; 5. Goodspeed Glacier; 6. Rhone Glacier; 7. Matterhorn Glacier; 8. Lacroix Glacier; 9. Suess Glacier; 10. Canada Glacier; 11. Commonwealth Glacier; 12. Rhone Platform; 13. Thomson moraine; 14. Lake Vanda; 15. Lake Brownworth; 16. Lake Bonney; 17. Hart Ash; 18. Prospect Mesa; 19. Wright Upper Glacier; 20. Wright Lower Glacier; 21. Arena Valley Ash.

tain critical diatoms in the age range of 2.5–3.0 myr ago (Barrett et al., 1992).

The concept of Pliocene deglaciation would be strengthened if the Pliocene-age Transantarctic Mountains were considerably lower than they are now. Such a configuration would make it much easier to explain overriding by an East Antarctic Ice Sheet postulated to have emplaced marine microfossils into the Sirius Group at or after 2.5 myr ago (Elliot et al., 1991). Overriding of low mountains by a relatively thin, temperate ice sheet would require neither polar ice shelves in the Ross Sea nor the existence of a West Antarctic Ice Sheet (Huybrechts, 1993). Because *Nothofagus* fossil wood occurs along with marine diatoms in Sirius Group deposits now at an elevation of 1,800 m in the Dominion Range near Beardmore Glacier (fig. 15.4), the overriding ice sheet is inferred to have advanced under temperate conditions into scrub vegetation containing *Nothofagus* (Webb et al., 1987; Webb and Harwood, 1987; McKelvey et al., 1987, 1991; Carlquist, 1987; Webb and Harwood, 1991; Hill et al., 1991). If it is assumed that Pliocene mountain elevations were indeed lower so that the *Nothofagus* grew near sea level, then the postulated growth of *Nothofagus* during the Late Pliocene and Early Pleistocene would imply atmospheric temperatures that were 20°–25°C warmer than at present (Barrett, 1991, p. 46), a value also suggested for Early Pliocene time (Webb and Harwood, 1991, p. 215). Without uplift, the Pliocene temperatures in Antarctica necessary to allow *Nothofagus* growth at the current high elevation of the fossil wood-bearing beds would have been at least 30°–35°C warmer than at present. In this regard, Pliocene-Pleistocene mountain uplift of 1,000–3,000 m has been postulated from faulting of Sirius Group deposits near Beardmore Glacier and from the current high elevation of the *Nothofagus* wood beds (Webb et al., 1986; Webb and Harwood, 1987), as well as from the "youthful-appearing" rift shoulder scarp along the mountain front (Behrendt and Cooper, 1991). Pliocene uplift, but of much less magnitude, is also inferred from benthic foraminifers in DVDP cores 10 and 11 (fig. 15.5; Ishman and Reick, 1992). A topographic barrier created by tectonic uplift of the Transantarctic Mountains is credited with forcing Late Pliocene climatic change and transforming the Antarctic Ice Sheet from temperate to polar by mountain uplift (Behrendt and Cooper, 1991).

The possible Pliocene collapse of the East Antarctic Ice Sheet has been taken up by other branches of earth science. High sea levels (35 ± 18 m) of Mid-Pliocene age (ca. 3.0–3.5 myr ago) on the Atlantic Coastal Plain (Orangeburg, Chippenham, and Thornburg Scarps) have been related to melting of the East Antarctic Ice Sheet (Dowsett and Cronin, 1990; Krantz, 1991). Wide oscillations of a dynamic East Antarctic Ice Sheet are used to explain fluctuations in the marine oxygen isotope record and in biogenic productivity noted in some deep-sea cores (Abelmann et al., 1990; Krants, 1991; Ishman and Rieck, 1992). PRISM (Pliocene Research, Interpretations, and Synoptic Mapping), a multidisciplinary study designed to assess the Pliocene warm peak at 3.0 myr ago (Cronin and Dowsett, 1991), has incorporated East Antarctic ice collapse into an emerging global reconstruction (Dowsett et al., 1992).

Tests of a Variable Pliocene Ice Sheet

To place the hypothesis of a variable East Antarctic Ice Sheet in perspective, it is first necessary to discuss the climatic conditions under which the East Antarctic Ice Sheet would become unstable and collapse. Numerical experiments with a three-dimensional model of the Antarctic Ice Sheet show persistent ice-sheet stability in the face of extensive warming (Huybrechts, 1993). In fact, the ice sheet expands for a temperature rise of up to 5°C above modern values owing to increased precipitation. With a rise of 8°–10°C in atmospheric temperatures, the marine-based West Antarctic Ice Sheet collapses. Only an atmospheric warming of 17°–20°C above modern values causes an extensive recession of the East Antarctic Ice Sheet. Even then, a distinct problem arises with regard to developing an open seaway across the interior Wilkes and Pensacola Basins in interior East Antarctica (fig. 15.2). Namely, late in deglaciation, by which time the Pensacola Basin becomes free of ice, it is isostatically uplifted above sea level, and only a small interior cap remains over the Gambertsev Mountains in the center of East Antarctica (Huybrechts, 1993).

It should be noted that the only feasible mechanism for collapse of the East Antarctic Ice Sheet (as opposed to the marine-based West Antarctic Ice Sheet) is by surface meltdown, not by sliding instability (Huybrechts, 1993). Therefore, ice-sheet collapse would require the imposition of surface-melting ablation zones on the East Antarctic Ice Sheet. This scenario requires an extensive rise of the snow line, now situated at equivalent elevations of −600 m in southern Antarctic latitudes (Robin, 1988). Such a change from present-day conditions should be easily discerned in Pliocene glacial sediments marginal to the ice sheet (outwash plains, kame terraces, ice-contact heads, moraine walls; Denton et al., 1993) and in Southern Ocean oxygen isotope records (Kennett and Hodell, 1993).

Within this framework, the hypothesis of a variable Pliocene ice sheet that underwent periodic collapse has several testable implications. Several such tests were carried out in the Dry Valleys sector of the Transantarctic Mountains (fig. 15.5; Denton et al., 1993). The first testable implication is of climatic warming so substantial that Pliocene environments as low as 85°S were at least subantarctic and perhaps even southern Patagonian in character (Mercer, 1986). Subantarctic diatoms in Sirius outcrops at Reedy Glacier imply sea-surface temperatures of 2°–5°C in the Pensacola Subglacial Basin near the South Pole, and *Nothofagus* fossil wood in Sirius deposits in the Dominion Range near Beardmore Glacier (fig. 15.4) imply air temperatures at least 20°–25°C above present-day values if the mountains were much lower than they are now. The second testable implication is that the East Antarctic Ice Sheet must have overtopped (or nearly overtopped) the Transantarctic Mountains in order to emplace Sirius Group outcrops on mountain surfaces now at high elevations. The amount of thickening required depends critically on the surface elevation of the mountains at the time of overriding. For reasons already described, the deglaciation hypothesis is more tenable if the Transantarctic Mountains underwent extensive Pliocene-Pleistocene surface uplift. Third, the variable ice-sheet hypothesis requires that mountain overriding occurred after the oldest part of the stratigraphic range of the youngest diatom included in Sirius deposits. This stratigraphic range means that overriding was younger than 3.0 myr ago (Barrett et al., 1992) and perhaps coincided with the Pliocene climatic deterioration and consequent growth of ice sheets in the Northern Hemisphere centered at 2.5 myr ago (Elliot et al., 1991). The final point is that Pliocene warming of the magnitude necessary to cause East Antarctic deglaciation should be easily discernible in the deep-sea sediment record of the Southern Ocean (Kennett and Hoddell, 1993).

The Case for a Cold Pliocene Climate in Antarctica and a Stable East Antarctic Ice Sheet

The case for continuous cold polar climatic conditions, and hence a relatively stable East Antarctic Ice Sheet, rests on paleoclimatic evidence from the Dry Valleys sector of the Transantarctic Mountains (figs. 15.1, 15.5). The pertinent data have been reviewed extensively in a dedicated issue of *Geografiska Annaler* (vol. 75A, 1993, which includes Denton et al., 1993; Marchant et al., 1993a, 1993b; Hall et al., 1993; and Wilch et al., 1993; see also Sugden et al., 1995). Only the main points are given here.

The Dry Valleys sector of the Transantarctic Mountain, which is adjacent to the East Antarctic Ice Sheet, contains several deposits critical for reconstructing Pliocene paleoclimate. First, it features high-elevation Sirius Group outcrops with enclosed marine Pliocene diatoms, including a Sirius deposit on Mount Feather, one of the original sites used to postulate the variable ice-sheet hypothesis (Webb et al., 1984). Second, it includes the CIROS-2 core, where the occurrence of critical diatoms that also occur in Sirius deposits is tied to an $^{40}Ar/^{39}Ar$ date of a volcanic ash layer (Barrett et al., 1992). Third, the Dry Valleys feature other cores (DVDP Sites 10 and 11; CIROS-1, at the mouth of Ferrar Valley) and stratigraphic sections (Prospect Mesa, in central Wright Valley) that contain microfossils used to infer warmer-than-present Pliocene marine paleoenvironments (Webb, 1972, 1974; Barrett, 1992; Ishman and Reick, 1992; Prentice et al., 1993). Fourth, widespread surficial Pliocene and Miocene drifts and volcanic ashes are still well exposed in the Dry Valleys, because Quaternary glacier expansions were so small in this area. Together, these factors make it possible to address issues of Pliocene paleoclimate and ice-sheet overriding of the mountains. A cautionary note is necessary, however, because the Transantarctic Mountains potentially have a complex history of denudation and surface uplift. Therefore, results from individual tectonic blocks, including the Dry Valleys, do not necessarily apply to the mountains as a whole.

The Dry Valleys feature the largest tract of ice-free terrain in the Transantarctic Mountains (fig. 15.5). Long trunk valleys now largely free of ice (Taylor Valley, Wright Valley, and the Victoria Valley system) cross the mountains from the inland ice sheet to the Ross Sea. These valleys are flanked by the Quartermain Mountains and by the Asgard and Olympus Ranges. The Wilson Piedmont Glacier blocks the mouths of Wright and Victoria Valleys. The Taylor Dome, on the periphery of the East Antarctic Ice Sheet, feeds local East Antarctic ice into the Taylor and Wright outlet glaciers, which flow into the western valleys. Inland ice flow drains north around the Dry Valleys by Mackay Glacier and south by Mulock Glacier.

Today the Dry Valleys are cold, windy, and hyperarid. An excess of sublimation over precipitation at all elevations yields a distinct precipitation deficit (Chinn, 1980). For example, beside Lake Vanda, at an elevation of 123 m on the floor of Wright Valley, mean annual temperature is −19.8°C (Schwerdtfeger, 1984), and mean annual precipitation is 10 mm (water equivalent; Keys, 1980). Snowfalls are light; they can occur at any time of year but are concentrated in the summer. Alpine glaciers occur where windblown snow is concentrated into preexisting theater-shaped embayments; such glaciers are largest near the coast, where snowfall (and therefore wind-concentrated accumulation) is greatest. Glaciers in alcoves close to the polar plateau, if they occur at all, are small. Because of their limited size and the cold environment in which they occur, these glaciers are frozen to underlying beds and are nearly free of debris (Meserve Glacier in Wright Valley has a basal temperature of −18°C; Bull and Carnein, 1968). In contrast, the larger outlet glaciers can attain basal melting conditions; for example, the deepest ice of Taylor Glacier is wet-based (Robinson, 1984). Today, 90 percent of glacier ablation in the Dry Valleys is due to sublimation and only 10 percent to melting (Chinn, 1980). Small meltwater streams, all at relatively low elevations, feed enclosed lakes in the Taylor and Victoria Valleys. In the Wright Valley, the Onyx River is a small stream that flows westward along the valley floor from Lake Brownworth (damned and fed by Wright Lower Glacier) to Lake Vanda (enclosed in the lowest part of the valley). Meltwater from alpine glaciers on the southern valley wall also feeds the Onyx River.

The Pliocene sediments in the Dry Valleys region are interpreted in two differing ways in terms of paleoclimate and ice-sheet history. The microfossil assemblages in marine cores, stratigraphic sections of uplifted marine sediments, and high-level Sirius Group deposits are interpreted to reflect a warm Pliocene paleoclimate consistent with a variable East Antarctic Ice Sheet. Yet much of the rest of the terrestrial paleoclimatic record is interpreted in terms of a cold Pliocene climate consistent with a stable East Antarctic Ice Sheet.

Pliocene marine water bodies occupied the eastern Ferrar Glacier Valley, eastern Taylor Valley, and Wright Valley. Foraminiferal and diatom biostratigraphy has been carried out for the following marine cores (fig. 15.5); CIROS-1 and CIROS-2 (Ferrar Glacier trough; Harwood, 1989; Webb, 1989; Barrett et al., 1992), DVDP-10 and DVDP-11 (eastern Taylor Valley; Ishman and Rieck, 1992), and DVDP-4A (Wright Valley; Brady, 1982). In addition, there is biostratigraphic control for the Prospect Mesa section that exposes Pliocene marine sediments (Prospect Mesa gravels) in central Wright Valley (Webb, 1972, 1974; Prentice et al., 1993). The biostratigraphic chronologies permit valley-to-valley correlation of marine sediments; equally important, paleowater temperatures have been inferred from the microfossil assemblages in the cores and the Prospect Mesa gravels.

One example comes from central Wright Valley. Webb (1972, 1974) concluded from the benthic foraminiferal assemblage in Prospect Mesa gravels (Pecten gravel) that bottom-water temperatures in a shallow Wright Valley fjord (ca. 100 m paleowater depth) were within the range of −2°C to +5°C, and perhaps as high as +10°C. The associated diatom assemblage, along with the absence of coccolithophores, is taken to indicate water temperatures of from 0°C to < 3°C (Burckle and Pokras, 1991; Prentice et al., 1993). Prospect Mesa gravels contain the mid-Pliocene foraminifera marker species *Ammoelphidiella antarctica* (*Trochoelphidiella onyxi* of Webb, 1972, 1974), which in Taylor Valley core DVDP-10 is taken to occur at 3.8–3.4 myr ago (Ishman and Rieck, 1992). Burckle et al. (1986) found Late Pliocene diatoms in Prospect Mesa gravels now taken to suggest an age of 2.5–3.0 myr from diatom biostratigraphy (Prentice et al., 1993). However, the $^{87}Sr/^{86}Sr$ ratios of two shells (*C. tuftsensis*) assumed to be unaltered from Prospect Mesa gravels suggest a date of 5.5 ± 0.4 myr (Prentice et al., 1993), which is considerably older than the age estimates from biostratigraphic ranges of the enclosed microfossils.

Another example comes from the CIROS-2 marine core from the Ferrar Glacier trough (Barrett et al., 1992). This core reveals the *Thalassiosira insigna / T. vulnifica* and *T. vulnifica* diatom zones. In Southern Ocean subantarctic cores, *T. insigna* and *T. vulnifica* occur together at 2.5–3.1 myr; *T. vulnifica* spans the range 2.2–3.1 myr. A volcanic ash in the CIROS-2 core interval containing *T. vulnifica* yielded single-crystal and glass laser-fusion $^{40}Ar/^{39}Ar$ dates of about 2.8 ± 0.3 myr, which is taken as the age of the *T. vulnifica* zone in the Ferrar Glacier trough. More important, however, the age can also be taken as a date of the inferred marine flooding of interior East Antarctic marine basins, because *T. insigna* and *T. vulnifica* occur together in Sirius Group outcrops in the Transantarctic Mountains. From other diatoms also enclosed in Sirius outcrops, these interior marine water bodies are inferred to have had surface temperatures of 2°–5°C (Harwood, 1986). The implication of such relatively warm interior water bodies is that similarly warm water must have also occurred in McMurdo Sound and the Dry Valleys fjords. By this scenario, the Dry Valleys region would have been nearly surrounded by relatively warm marine water in mid-Pliocene time. These water bodies would have included not only McMurdo Sound and its extension into the Ferrar, Taylor, and Wright Valleys but also a flooded interior Wilkes Basin on the western mountain flank, replacing part of the present-day East Antarctic Ice Sheet. Such a paleogeographic reconstruction is consistent with the growth of *Nothofagus* in the Transantarctic Mountains as low as 85°S alongside Beardmore Glacier.

The key to the hypothesis of warm Pliocene Antarctic climate is an unproven assumption—namely, that the critical marine diatoms in high-elevation Sirius Group deposits were derived from interior marine basins and deposited in the Sirius outcrops during a postulated Pliocene expansion of the East Antarctic Ice Sheet across the Transantarctic Mountains. Should this untested assumption prove incorrect in that the Sirius diatoms represent a contaminant introduced by wind or some other mechanism, then the underpinning would be removed for the hypothesis of warm Pliocene climate and a dynamic East Antarctic Ice Sheet. The evidence for warm interior seas and concurrent growth of *Nothofagus* in the adjacent mountains also would be rendered invalid for the Pliocene, because both concepts are based on the assumption that Sirius diatoms were emplaced by the East Antarctic Ice Sheet. The isotopic dating in the Ferrar Glacier trough of certain critical Pliocene diatoms could not then be applied to the Sirius Group deposits in the Transantarctic Mountains. The only remaining evidence for a warm Pliocene climate would be the interpretation of the paleoecological requirements of Pliocene marine microfossils in marine cores and stratigraphic sections in the Dry Valleys region. In this regard, the implications are ambiguous. An example is the wide range (−2° to +10°C) for water temperature estimates from the Pliocene foraminifer and diatom assemblages in the Prospect Mesa gravels in central Wright Valley. Overall, the Pliocene microfossil assemblages probably imply somewhat warmer-than-present marine waters but do not by themselves preclude polar desert conditions.

The Dry Valleys terrestrial record affords an independent perspective on Pliocene paleoclimate and ice-sheet variability (Denton et al., 1993; Marchant et al., 1993a, 1993b, 1993c; Marchant et al., 1994; Wilch et al., 1993; Hall et al., 1993). This record includes drift chronologies ($^{40}Ar/^{39}Ar$ dating of interbedded volcanic rocks; and ^{10}Be and ^{3}He exposure ages of surface boulders on moraines); remains of Pliocene cold deserts (thermal contraction cracks and tightly knit ventifact pavements); physical characteristics of glacial deposits; and evidence of water erosion of surface sediments. The major results derived from this terrestrial record are:

1. The Dry Valleys landscape is largely inherited from a previous climatic regime (Denton et al., 1993; Sugden et al., 1995). The upper-level escarpment landscape reflects rapid denudation associated with planation and escarpment retreat, all following a phase of rifting that began about 55 myr ago. The deep transverse valley systems graded to near sea level reflect downcutting by rivers, aided in places by glaciers and accompanied by faulting and tectonic backtilting. Although escarpment landscapes are widespread, there is also evidence that glaciers were important in landscape evolution. Semiconsolidated Sirius Group outcrops occur as remnants on the high-erosion surfaces of the western Dry Valleys. Unconsolidated glacial deposits of Middle Miocene to Pleistocene age occur from high-elevation theater-shaped embayments to the floors of the major transverse valleys. Fluvial spurs are truncated and some valley segments are straightened and overdeepened. Areal scouring beneath overriding ice is widespread at all elevations.
2. Paleoclimatic data indicate that a cold-desert climate was superimposed on the Dry Valleys by Middle Miocene time, prior to 13.6 myr ago, bringing large-scale landscape denudation to a halt (Marchant et al., 1993a, 1993b; Denton et al., 1993).
3. The areal distribution and chronology of drift sheets and moraines superimposed on the Dry Valleys landscape show that the last overriding of this sector of the Transantarctic Mountain was Middle Miocene in age (Marchant et al., 1993a, 1993b; Hall et al., 1993; Denton et al., 1993). Even more important, these data show that Pliocene glacier expansion was very limited in the Dry Valleys region (Wilch et al., 1993; Marchant et al., 1994). The implication is that the Sirius Group is Middle Miocene or older in age.
4. The areal distribution, isotopic age, and elevation of cold-desert indicators suggest that Pliocene mean annual atmospheric temperatures were at most only 3°–8°C above present-day values (Denton et al., 1993;

Marchant et al., 1993a, 1993b, 1993c; Hall et al., 1993). This hypothesis is consistent with the fact that Pliocene drift sheets lack morphologic features and sediments of temperate glaciers (outwash plains, ice-contact heads, lateral moraines with lodgment tills and steep interior walls, kame terraces, and glaciomarine sediments).

5. The current elevation of marine fjord deposits and paleo-shorelines, taken together with the age-and-elevation distribution of subaerially erupted volcanic deposits in Taylor Valley, show that surface uplift of the Dry Valleys block of the Transantarctic Mountains was only about 250–300 m (Hall et al., 1993).

These conclusions based on terrestrial data cast doubt on the concept of a highly variable East Antarctic Ice Sheet in Pliocene time on at least two counts. The first is that the terrestrial paleoclimatic indicators (Denton et al., 1993) give no hint of the marked climatic warming required for ice-sheet meltdown (Huybrechts, 1993). The second is that the East Antarctic Ice Sheet did not thicken nearly enough during the Pliocene to emplace Pliocene diatoms in the Sirius outcrop high on Mt. Feather. Hence, some other mechanism must be responsible for emplacement of the diatoms. If one can show by independent data that marine diatoms were not emplaced in the Mt. Feather Sirius outcrop by an expanded East Antarctic Ice Sheet, then it is not necessary to invoke East Antarctic ice to emplace diatoms in other Sirius outcrops. This argument removes the underlying rationale for warm interior Pliocene seas in East Antarctica and also casts doubt on the proposed Pliocene age for the *Nothofagus* fossil wood farther south in the Beardmore Glacier area.

Another test of the hypothesis of a variable East Antarctic Ice Sheet comes from deep-sea sediments in the Southern Ocean. If warming were of sufficient magnitude to induce East Antarctic deglaciation in the Pliocene prior to about 3 myr ago, then clear evidence of such warming should be evident in the marine stratigraphic data from high southern latitudes. Kennett and Hodell (1993) concluded that no such evidence exists on the basis of the following.

1. A high-resolution oxygen isotope record from near Antarctica shows that sea-surface temperatures could not have risen more than about 3°C above present-day values during the warmest intervals of the Pliocene, between 3.2 and 4.8 myr ago. Furthermore, this record shows no large-scale Pliocene fluctuations; instead, it is consistent with relative stability of the Pliocene Antarctic climate and ice-sheet system.
2. Biosiliceous faunas and floras dominated deposition in Antarctic deep-sea sediment cores during Pliocene time. Although Antarctic Pliocene sea-surface temperatures were somewhat warmer than at present, there was no return to calcareous nannofossil deposition, as would have occurred if surface waters had warmed by more than 5°C. This shows that Subantarctic did not replace Antarctic planktonic assemblages during the Pliocene.
3. The distinct changes in oceanic $\delta^{13}C$ records that should accompany large changes in sea level (because organic carbon is transferred between continental and oceanic reservoirs) are not present in Pliocene isotopic records from deep-sea cores. The implication is that the Antarctic ice sheets did not experience major deglaciation.

Discussion

There are two contrasting views concerning the Pliocene stability of the East Antarctic Ice Sheet and coeval Pliocene paleoclimate. Because both hypotheses stand at the present time, the Antarctic paleoclimatic record and ice-sheet history cannot yet contribute to the interlocked questions of paleoclimate and evolution.

Antarctic ice-sheet history is intertwined with the overall problems of rifting and landscape evolution in the Ross Embayment (van der Wateren and Verbers, 1993). Consequently, the Pliocene history of the Antarctic Ice Sheet cannot be understood in isolation from the problems of the development of the West Antarctic Rift System and the shoulder of this rift system formed by the Transantarctic Mountains. The timing and rates of uplift and basin development are not well known, but they are important in interpreting glacial deposits in terms of ice-sheet history. Different structural blocks could have had varying glacial and tectonic histories. Finally, it is not yet possible to relate terrestrial glacial sequences in the Transantarctic Mountains with offshore sedimentary sequences in the rift basins. A number of fundamental questions must, therefore, be answered before the Pliocene glacial history of the Antarctic can be understood.

Acknowledgments

The research for this work was supported by the Division of Polar Programs of the National Science Foundation. Richard

Kelly drafted the figures. Much of this chapter is based on a special issue of *Geografiska Annaler* (vol. 75A, 1993), entitled "The Case for a Stable East Antarctic Ice Sheet." I thank the editor of *Geografiska Annaler* and my coauthors for allowing me to borrow liberally from papers in that issue.

References

Abelmann, A., Gersonde, R., and Spieess, V. 1990. Pliocene-Pleistocene paleoceanography in the Weddell Sea–siliceous microfossil evidence. In *Geological history of the polar oceans: Arctic versus Antarctic,* pp. 729–759 (ed. U. Bleil and J. Theide). Kluwer, Dordrecht.

Barrett, P. J. 1991. Antarctica and global climatic change: a geological perspective. In *Antarctica and global climatic change,* pp. 35–50 (ed. C. Harris and B. Stonehouse), Belhaven Press, London.

Barrett, P. J., Adams, C. J., McIntosh, W. C., Swisher, C. C., III, and Wilson, G. S. 1992. Geochronological evidence supporting Antarctic deglaciation three million years ago. *Nature* 359:816–818.

Barrett, P. J., and Elliot, D. H. 1973. *Reconnaissance geological map of the Buckley Island quadrangle, Transantarctic Mountains, Antarctica.* Antarctic geologic map, no. A-3. United States Geological Survey, Washington, D.C.

Barrett, P. J., and Powell, R. D. 1982. Middle Cenozoic glacial beds at Table Mountain, southern Victoria Land. In *Antarctic geoscience,* pp. 1059–67 (ed. C. Craddock). University of Wisconsin Press, Madison.

Behrendt, J. C., and Cooper, A. K. 1991. Evidence of rapid Cenozoic uplift of the shoulder escarpment of the Cenozoic West Antarctic Rift System and a speculation on possible climate forcing. *Geology* 19:315–319.

Bonnefille, R. 1976. Palynological evidence for an important change in the vegetation of the Omo Basin between 2.5 and 2.0 million years ago. In *Earliest man and environments in the Lake Rudolph Basin,* pp. 421–431 (ed. Y. Coopens, F. C. Howell, G. L. Isaac, and R. E. F. Leakey). University of Chicago Press, Chicago.

———. 1983. Evidence for a cooler and drier climate in Ethiopia uplands towards 2.5 Ma ago. *Nature* 303:487–491.

Brady, H. 1982. Late Cenozoic history of Taylor and Wright Valleys and McMurdo Sound inferred from diatoms in Dry Valley Drilling Project cores. In *Antarctic geoscience,* pp. 1123–1131 (ed. C. Craddock). University of Wisconsin Press, Madison.

Brady, H. T., and McKelvey, B. C. 1979. The interpretation of a Tertiary tillite at Mount Feather, southern Victoria Land, Antarctica. *Journal of Glaciology* 22:189–193.

———. 1983. Some aspects of the Cenozoic glaciation of southern Victoria Land. *Journal of Glaciology* 29:343–349.

Bull, C., and Carnein, C. R. 1968. The mass balance of a cold glacier: Meserve Glacier, southern Victoria Land, Antarctica. *Journal of Glaciology* 13:415–429.

Burckle, L. H., and Pokras, E. M. 1991. Implications of a Pliocene stand of *Nothofagus* (southern beech) within 500 kilometers of the South Pole. *Antarctic Science* 3(4):389–403.

Burckle, L. H., Prentice, M. L., and Denton, G. H. 1986. Neogene Antarctic glacial history: New evidence from marine diatoms in continental deposits. *EOS, Transactions* 67:295. American Geophysical Union, Washington, D.C.

Carlquist, S. 1987. Upper Pliocene–lower Pleistocene *Nothofagus* wood from the Transantarctic Mountains. *Aliso* 11:571–583.

Chinn, T. J. 1980. Glacier balances in the Dry Valleys area, Victoria Land, Antarctica. In *Proceedings of the Riederalp Workshop,* pp. 237–247. IAHS-AISH Publication, no. 125.

Ciesielski, P. F., and Weaver, F. M. 1974. Early Pliocene temperature changes in the Antarctic seas. *Geology* 2:511–515.

Claridge, G. G. C., and Campbell, I. B. 1968. Soils of the Shackleton Glacier region, Queen Maud Range, Antarctica. *New Zealand Journal of Science* 11:11–15.

Cooke, D., and Hays, J. D. 1982. Estimates of Antarctic Ocean seasonal sea-ice cover during glacial intervals. In *Antarctic geoscience,* pp. 1017–1025 (ed. C. Craddock). University of Wisconsin Press, Madison.

Cronin, T. M., and Dowsett, H. J. (eds.) 1991. Pliocene climates. *Quaternary Science Reviews* 10(2/3):115–296.

Denton, G. H., Prentice, M. L. and Burckle, L. H. 1991. Cainozoic history of the Antarctic Ice Sheet. In *The geology of Antarctica,* pp. 365–433 (ed. R. J. Tingey). Clarendon Press, Oxford.

Denton, G. H., Sugden, D. M., Marchant, D. R., Hall, B. L., and Wilch, T. I. 1993. East Antarctic Ice Sheet sensitivity to Pliocene climatic change from a Dry Valleys perspective. *Geografiska Annaler* 75A:155–204.

Doumani, G. A., and Minshew, V. H. 1965. General geology of the Mt. Weaver area, Queen Maud Mountains, Antarctica. In *Geology and paleontology of the Antarctic,* pp. 127–139 (ed. J. B. Hadley). Antarctic Research Series, vol. 6. American Geophysical Union, Washington, D.C.

Dowsett, H. J., and Cronin, T. M. 1990. High eustatic sea level during the Middle Pliocene: Evidence from the southeastern U.S. Atlantic Coastal Plain. *Geology* 18:435–438.

Dowsett, H. J., Cronin, T. M., Poore, R. Z., Thompson, R. S., Whately, R. C., and Wood, A. M. 1992. Micropaleontological evidence for increased meridional heat transport in the North Atlantic Ocean during the Pliocene. *Science* 258:1133–1135.

Drewry, D. J. 1982. Ice flow, bedrock, and geothermal studies from radio-echo sounding inland of McMurdo Sound, Antarctica. In *Antarctic geoscience,* pp. 977–983 (ed. C. Craddock). University of Wisconsin Press, Madison.

Drewry, D. J. (ed.) 1983. *Antarctica: Glaciological and geophysical folio.* Scott Polar Research Institute, Cambridge, Eng.

Elliot, D., Barrett, P. J., and Mayewski, P. A. 1974. *Reconnaissance geologic map of the Plunket Point Quadrangle, Transantarctic Mountains, Antarctica.* Antarctic map, no. 4. United States Geological Survey, Washington, D.C.

Elliot, D. H., Bromwich, D. H., Tzeng Ren-Yow, Harwood, D. M., and Webb, P. N. 1991. The Sirius Group: A possible test of climate modeling results for Antarctica. In *International Conference on the Role of the Southern Ocean and Antarctica in Global Change: An ocean drilling perspective.* (Abstract).

Faure, G., and Taylor, K. S. 1981. Provenance of some glacial deposits in the Transantarctic Mountains based on Rb-Sr dating of feldspars. *Chemical Geology* 32:271–290.

Funder, S., Abrahamsen, N., Bennike, O., and Feyling-Hanssen, R. W. 1985. Forested Arctic: Evidence from North Greenland. *Geology* 13:542–546.
Gordon, A. L. 1986. Interocean exchange of thermocline water. *Journal of Geophysical Research* 91:5037–5046.
Grube, F., Christensen, S., and Vollmer, T. 1986. Glaciations in northwest Germany. *Quaternary Science Reviews* 5:347–358.
Hall, B. L., Denton, G. H., Lux, D. R., and Bockheim, J. G. 1993. Late Tertiary Antarctic paleoclimate and ice-sheet dynamics inferred from surficial deposits in Wright Valley. *Geografiska Annaler* 75A:239–267.
Harwood, D. M. 1983. Diatoms from the Sirius Formation, Transantarctic Mountains. *Antarctic Journal of the United States* 18(5):98–100.
———. 1986. Diatom biostratigraphy and paleoecology and a Cenozoic history of Antarctic ice sheets. Ph.D. diss., Ohio State University.
———. 1989. Siliceous microfossils. In *Antarctic Cenozoic history from the CIROS-1 drillhole, McMurdo Sound.* DSIR Bull. 245:67–97.
Hays, J. D., and Opdyke, N. D. 1967. Antarctic radiolaria, magnetic reversal and climate change. *Science* 158:1001–1011.
Hill, R. S., Harwood, D. M., and Webb, P. N. 1991. Last remnant of Antarctica's Cenozoic flora: Pliocene *Nothofagus* of the Sirius Group, Transantarctic Mountains. In *Eighth Gondwana Subcommission Symposium,* Hobart, Australia. (Abstract).
Huybrechts, P. 1993. Glaciological modelling of the Late Cenozoic glacial history of the East Antarctica Ice Sheet: Stability or dynamism? *Geografiska Annaler* 75A:221–238.
Ishman, S. E., and Rieck, H. J. 1992. A Late Neogene Antarctic glaco-eustatic record, Victoria Land Basin Margin, Antarctica. In *The Antarctic paleoenvironment: A perspective on global change,* pt. 1, pp. 327–347 (ed. J. P. Kennett and D. A. Warnke). Antarctic Research Series, no. 56, American Geophysical Union, Washington, D.C.
Kennett, J. P. 1982. *Marine geology.* Prentice-Hall, Englewood Cliffs, N.J.
Kennett, J. P., and Hodell, D. A. 1993. Evidence for relative climatic stability of Antarctica during the Early Pliocene: A marine perspective. *Geografiska Annaler* 75A:205–220.
Keys, J. R. 1980. *Air temperature, wind, precipitation and atmospheric humidity in the McMurdo region.* Antarctic Data Series, no. 9, Victoria University of Wellington.
Krantz, D. E. 1991. A chronology of Pliocene sea-level fluctuations: The U.S. Middle Atlantic Coastal Plain record. *Quaternary Science Reviews* 10:163–174.
LaPrade, K. E. 1984. Climate, geomorphology, and glaciology of the Shackleton Glacier Area, Queen Maud Mountains, Transantarctic Mountains, Antarctica. In *Geology of the Transantarctic Mountains,* p. 163–196 (ed. M. D. Turner and J. F. Splettstoesser). *Antarctic Research Series,* no. 36, American Geophysical Union, Washington, D.C.
McGregor, V. R. 1965. Notes on the geology of the area between the heads of the Beardmore and Shackleton Glaciers, Antarctica. *New Zealand Journal of Geology and Geophysics* 8:278–291.
McKelvy, B. C., Webb, P. N., Harwood, D. M., and Mabin, M. C. G. 1987. The Dominion Range Sirius Group: A late Pliocene–early Pleistocene record of the ancestral Beardmore Glacier. In *Fifth International Symposium on Antarctic Earth Sciences,* Cambridge. (Abstract 97).
———. 1991. The Dominion Range Sirius Group: A record of the Late Pliocene–Early Pleistocene Beardmore Glacier. In *Geological evolution of Antarctica,* pp. 675–682 (ed. M. R. A. Thomson, J. A. Crame, and J. W. Thomson. Cambridge University Press, Cambridge.
Marchant, D. R., Denton, G. H., Bockheim, J. G., Wilson, S. C., and Kerr, A. R. 1994. Quaternary ice-level changes of upper Taylor Glacier, Antarctica: Implications for paleoclimate and ice sheet dynamics. *Boreas* 23:29–42.
Marchant, D. R., Denton, G. H., Sugden, D. E., and Swisher, C. C., III. 1993a. Miocene glacial stratigraphy and landscape evolution of the western Asgard Range, Antarctica. *Geografiska Annaler* 75A:303–330.
Marchant, D. R., Denton, G. H., and Swisher, C. C., III. 1993b. Miocene-Pliocene-Pleistocene glacial history of Arena Valley, Quartermain Mountains, Antarctica. *Geografiska Annaler* 75A:269–302.
Marchant, D. R., Swisher, C. C., III, Lux, D. R., West, D. P., Jr., and Denton, G. H. 1993c. Pliocene paleoclimate and East Antarctic ice-sheet history from surficial ash deposits. *Science* 260:667–670.
Matthews, J. V. 1989. Late Tertiary Arctic environments: A vision of the future. *GEOS* 18:14–18.
Mayewski, P. A. 1975. *Glacial geology and Late Cenozoic history of the Transantarctic Mountains, Antarctica.* Institute of Polar Studies Report, no. 56, Ohio State University, Columbus.
Mercer, J. H. 1968. Glacial geology of the Reedy Glacier area, Antarctica. *Geological Society of America Bulletin* 79:471–486.
———. 1972. Some observations on the glacial geology of the Beardmore Glacier area. In *Antarctic geology and geophysics,* pp. 427–433 (ed. R. J. Adie). Universitetsforlaget, Oslo.
———. 1986. Southernmost Chile: A modern analog of the southern shores of the Ross Embayment during Pliocene warm intervals. *Antarctic Journal of the United States* 21:103–105.
Pickard, J., Adamson, D. A., Harwood, D. M., Miller, G. H., Quilty, P. G., and Dell, R. K. 1988. Early Pliocene marine sediments, coastline and climate of East Antarctica. *Geology* 16:158–161.
Prentice, M. L., Bockheim, J. C., Wilson, S. C., Burckle, L. H., Hodell, D. A., Schluchter, C., and Kellogg, D. E. 1993. Late Neogene Antarctic glacial history: Evidence from Central Wright Valley. In *The Antarctic paleoenvironment: A perspective on global change,* pt. 2, pp. 207–250 (ed. J. P. Kennett and D. A. Warnke). Antarctic Research Series, no. 60, American Geophysical Union, Washington, D.C.
Prentice, M. L., and Denton, G. H. 1988. The deep-sea oxygen isotope record, the global ice sheet system and hominid evolution. In *Evolutionary history of the "robust" australopithecines,* pp. 383–403 (ed. F. Grine). Aldine de Gruyter, New York.
Prentice, M. L., Denton, G. H., Lowell, T. V., Conway, H., and Heusser, L. E. 1986. Pre-Late Quaternary glaciation of the Beardmore Glacier region, Antarctica. *Antarctic Journal of the United States* 21:95–98.
Robin, G. de Q. 1988. The Antarctic Ice Sheet: Its history and

response to sea level and climatic changes over the past 100 million years. *Palaeogeography, Palaeoclimatology, Palaeoecology* 67:31–50.

Robinson, P. L. 1984. Ice dynamics and thermal regime of Taylor Glacier, South Victoria Land, Antarctica. *Journal of Glac.* 30:133–160.

Ruddiman, W. F., and Raymo, M. E. 1988. Northern Hemisphere climate regimes during the last 3 Ma: Possible tectonic connections. *Philosophical Transactions of the Royal Society of London,* ser. B, 318:411–430.

Schwerdtfeger, W. 1984. Weather and climate of the Antarctic. In *Developments in atmospheric science,* no. 15. Elsevier, Amsterdam.

Stanley, S. M. 1985. Climatic cooling and Plio-Pleistocene mass extinction of molluscs around the margins of the Atlantic. *South African Journal of Science* 81:266.

Sugden, D. E., Denton, G. H., and Marchant, D. R. 1995. Landscape evolution of the Dry Valleys, Transantarctic Mountains. *Journal of Geophysical Research* 100:9949–9967.

van der Hammen, T., Wijmstra, T. A., and Zagwijn, W. H. 1971. The floral record of the Late Cenozoic of Europe. In *Late Cenozoic glacial ages,* pp. 391–424 (ed. K. K. Turekian). Yale University Press, New Haven.

van der Wateren, F. M., and Verbers, A. L. L. M. 1993. Climate change, rifting, and landscape evolution in the Ross Embayment. *EOS, Transactions* 43:490–491.

Vrba, E. S. 1985. Early Hominidae in southern Africa: Updated observations on chronological and ecological background. In *Hominid evolution,* pp. 195–200 (ed. P. V. Tobias). Liss, New York.

———. 1988a. The environmental context of the evolution of early hominids and their culture. In *Bone modification,* pp. 27–42 (ed. R. Bonnichsen and M. H. Sorg). Center for the Study of Early Man, Orono, Maine.

———. 1988b. Late Pliocene climatic events and hominid evolution. In *Evolutionary history of the "robust" australopithecines,* pp. 405–426 (ed. F. Grine). Aldine de Gruyter, New York.

Webb, P. N. 1972. Pliocene marine invasion of an Antarctic dry valley. *Antarctic Journal of the United States* 7:215–224.

———. 1974. Micropaleontology, paleoecology, and correlation of the Pecten gravels, Wright Valley, Antarctica. *Journal of the Foraminiferal Research* 4:185–189.

———. 1989. Benthic foraminifera. In *Antarctic Cenozoic history from the CIROS-1 Drillhole, McMurdo Sound.* DSIR Bull. 245:99–118.

Webb, P. N., and Andreasen, J. E. 1986. Potassium/argon dating of volcanic material associated with the Pliocene Pecten Conglomerate (Cockburn Island) and Scallop Hill Formation (McMurdo Sound). *Antarctic Journal of the United States* 21:59.

Webb, P. N., and Harwood, D. M. 1987. Terrestrial flora of the Sirius Formation: Its significance for Late Cenozoic glacial history. *Antarctic Journal of the United States* 22:7–11.

———. 1991. Late Cenozoic glacial history of the Ross Embayment, Antarctica. *Quaternary Science Reviews* 10:215–223.

Webb, P. N., Harwood, D. M., McKelvey, B. C., Mercer, J. H., and Stott, L. D. 1984. Cenozoic marine sedimentation and ice-volume variation on the East Antarctic craton. *Geology* 12:287–291.

Webb, P. N., McKelvey, B. C., Harwood, D. M., Mabin, M. C. G., and Mercer, J. H. 1987. Sirius Formation of the Beardmore Glacier region. *Antarctic Journal of the United States* 22:8–13.

Wesselman, H. B. 1985. Fossil micromammals as indicators of climatic change about 2.4 myr ago in the Omo Valley, Ethiopia. *South African Journal of Science* 81:260–261.

Wilch, T. I., Denton, G. H., Lux, D. R., and McIntosh, W. C. 1993. Limited Pliocene glacial extent surface uplift in middle Taylor Valley, Antarctica. *Geogriska Annaler* 75A:331–351.

Zwally, H. J., Comiso, J. C., Parkinson, C. L., Campbell, J., Carsey, F. D., and Gloersen, P. 1983. *Antarctic sea ice, 1973–1976: Satellite passive microwave observations.* NASA-SP459. U.S. Government Printing Office, Washington, D.C.

Chapter 16

A Critical Review of the Micropaleontological Evidence Used to Infer a Major Drawdown of the East Antarctic Ice Sheet during the Early Pliocene

Lloyd H. Burckle

During the past ten years considerable debate has focused on the global climatic regime during the Pliocene. Although it is generally agreed that the Early Pliocene was warmer than the Late Pliocene, the debate has centered on the question of how warm the Early Pliocene really was. Major interest has focused on the behavior of the East Antarctic ice sheet during this time; certainly, a major drawdown would have global implications. Consider the following scenario: if both the East and West Antarctic ice sheets completely melted during the Early Pliocene, global sea level would have risen on the order of 70 m. The implications of such an event are not trivial. Taken within the context of evolution and/or migration in the biosphere, extreme warming during the Early Pliocene would have both regional and global implications. Regionally, warming would seriously deplete cold-adapted high-latitude fauna and flora; in global terms, the higher sea levels accompanying such an event would serve to limit migrations across narrow landbridges. Among such important "choke points" I include the Isthmus of Panama and the Suez region, areas which are recognized as important corridors dictating the migration of land plants and mammals. Similarly, the deepening of presently shallow-water seas north of Indonesia would influence surface-water exchange between major water bodies and have an impact upon climate, particularly in the Northern Hemisphere.

Investigators of early Pliocene climatic change generally fall into one of two categories. There are those who have concluded that Early Pliocene warming was significant and that all or a major part of the Antarctic ice sheet collapsed during this time, only to re-form during the Late Pliocene (see, e.g., Webb et al., 1984); in contrast, others believe that although Early Pliocene warming did take place, it was not as extreme as supposed and therefore had a less striking impact on high southern-latitude ice sheets and, by extension, on migration of lower-latitude continental faunas and floras (Burckle and Pokras, 1991; Kennett and Hodell, 1993). The original scenario postulating a major collapse of the East Antarctic ice sheet is based upon the occurrence of microfossils of reported Pliocene age in tills from Antarctica (Webb et al., 1984). Because biostratigraphy is my area of competence, I have chosen to do a critical review of this micropaleontological evidence and the associated literature used to support this scenario.

Webb et al. (1984) reported the presence of marine microfossils in sediments from the Sirius Formation (now called the Sirius Group [McKelvey et al., 1991]) recovered from the slopes of the Transantarctic Mountains (TAMS). Although many of the microfossils were recognized as coming from older Tertiary sediments, the presence of marine diatoms attributed to the Pliocene was rather unexpected. To account for these occurrences, Webb et al. (1984) postulated that a partial, but major, collapse of the East Antarctic ice sheet during the Pliocene allowed relatively warm marine waters (2°–6°C according to Harwood [1986a] and Webb and Harwood [1991]) to invade the presently subglacial Wilkes and Pensacola Basins (fig. 16.1). More recently, LeMasurier et al. (1994, p. 274) suggested that during the Early Pliocene "both East and West Antarctica were flooded by waters as warm as 6°–8°C." Marine microfossils (primarily diatoms) were deposited on the floor of these basins following this ice-sheet drawdown. During the

ISBN 0-300-06348-2.

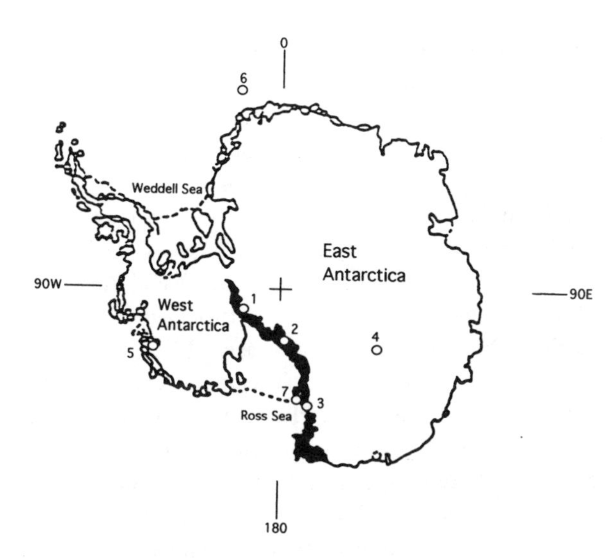

Fig. 16.1. Map of Antarctica showing the locations discussed in text: 1) Reedy Glacier area; 2) Beardmore Glacier; 3) Mt. Feather; 4) Dome C; 5) Mt. Murphy; 6) Core 1467 of Abelmann et al. (1988); 7) CIROS 2 of Barrett et al. (1992). The darkened strip just to the east and south of the Ross Sea is the approximate position of the Transantarctic Mountains (TAMs). The subglacial Wilkes and Pensacola Basins parallel the TAMs along the east and south sides.

late Pliocene–Early Quaternary reglaciation of East Antarctica these sediments were, in effect, bulldozed by advancing continental glaciers, which deposited them on the slopes of the TAMS. This scenario was further developed by Harwood (1986b), Webb et al. (1986, 1987), Webb and Harwood (1987), and Webb (1990). Current views are given in Webb and Harwood (1991). Basically, the former authors argue that the history of the East Antarctica ice sheet was dynamic and that it experienced several periods of major, but not total, deglaciation, the last of which occurred during the Early Pliocene; reglaciation of the continent occurred between 3 and 2.5 million years (myr) or later. In addition to multiple glaciations, and possibly multiple centers of glaciation, the deglacial hypothesis also included the possibility that Late Pliocene–Pleistocene uplift on the order of 1–3 km occurred along part of the then-coastal regions of East Antarctica (Webb et al., 1986; Webb and Harwood, 1991).

Burckle and Pokras (1991), Denton et al. (1993), Huybrechts (1993), and Kennett and Hodell (1993) take a different, but not necessarily unanimous, view of data they have collected from the Antarctic continent and the Southern Ocean. Burckle and Pokras (1991) argued that although the Southern Ocean was warmer during the Early Pliocene than at present, the deep-sea data do not support a sea-surface temperature as warm as that postulated by Harwood (1986a). Data discussed by Burckle and Pokras (1991) for the Early Pliocene suggest that surface-water temperatures during the austral summer were, at most, only a few degrees higher than they are presently. Kennett and Hodell (1993) summarized the sea-level evidence and suggested a rise of up to 30 m during the Early Pliocene, but for very short intervals of time. The sum of their evidence, however, suggests ice-sheet stability for this interval. Both studies view the East Antarctic ice sheet as dynamic but place more conservative limits on that dynamism. Denton et al. (1993) based their conclusions primarily on geomorphological considerations as well as on the physical setting of the Sirius Group in the Dry Valleys region (fig. 16.1). Their study indicates that there were hypoarid, cold-desert conditions from the Middle Miocene on, with little or no evidence for warmth. Huybrechts (1993) ran 3-D modeling experiments on the Antarctic ice sheet to determine its response to various climatic conditions and also to consider the physical mechanisms that might play a role in these responses. His results supported not only the view that the East Antarctic ice sheet was a stable feature but also the view that there were glaciological difficulties involved in trying to maintain an ice-free corridor over the Wilkes and Pensacola subglacial basins.

Discussion

Diatom Occurrences and Stratigraphy in the Sirius Group. Beginning with Harwood (1983), several researchers have addressed the evidence for, and implications of, a significant drawdown of the Antarctic ice sheets during the Pliocene. The bulk of the evidence on Pliocene deglaciation of Antarctica is found in Harwood (1986a; indeed, as recently as 1992, Barrett et al. referenced this work as a primary source of data to support Early Pliocene deglaciation of Antarctica; see, also LeMasurier et al., 1994). It is reasonable, therefore, to begin this chapter by examining Harwood in some detail. In particular, I shall examine the claim that "diatom assemblages from Mt. Feather, Reedy Glacier, and Beardmore Glacier are dominated by diverse Pliocene planktic marine taxa that were common in the circum-Antarctic diatom province" (Harwood, 1986a, p. 63; see my fig. 16.1 for locations). This claim is repeated by Harwood (1986a, p. 145), who in writing of the above-mentioned localities states that "the microfossil assemblages are dominated by Pliocene species"); and by Webb and Harwood (1991, p. 219), who state that "the majority of Sirius sites we have studied do contain Pliocene diatoms." Pliocene taxa refer to species found exclusively in sediments of Pliocene age (by the same token, species that are found only in Pleistocene sediments would be considered Pleistocene taxa; those found in both Pliocene and Pleistocene sediments would be considered Pliocene-Pleistocene taxa, and so on).

Harwood (1986a, table 9) indicated that diatoms from the Sirius Group at Mt. Fleming (Dry Valleys; fig. 16.1) represented a Late Miocene to Early Pliocene component based upon the joint occurrence of *Actinocyclus ingens* and *Denticulopsis hustedtii.* In Southern Ocean sediments, Baldauf and Barron (1991) date the first occurrence of *A. ingens* at about 16.4 myr and the first common occurrence of *D. hustedtii* at about 14 myr. The last common occurrence of the latter species is in the Late Miocene, at about 6.3 myr, whereas the former persisted into the Late Quaternary. Based upon these data, a minimum age of Middle to Late Miocene (not latest Miocene) is indicated for the Sirius Group at this locality. For the Mt. Feather locality (Dry Valleys; fig. 16.1) Harwood noted that thirteen samples were recovered over a period of some ten years. Yet his table 8 shows the results of an examination of six samples; one assumes, therefore, that the remaining seven samples did not contain microfossils. Harwood indicated that on the basis of his assemblage, the Sirius Group at this locality was Pliocene. However, no Pliocene taxa were recovered. None of the species that Harwood lists in table 8 to demonstrate a Pliocene age for this locality are exclusive to this time interval. They ranged either out of or into the Pliocene from sediments representing another interval. He points, for example, to the occurrence of *D. hustedtii Coscinodiscus marginatus, Eucampia antarctica, Hemidiscus karstenii, Nitzschia curta, N. kerguelensis, N. ritscheri, Rouxia antarctica, Stephanopyxis turris, Thalassiosira lentiginosa,* and *T. lentiginosa* var. *obovatus* as evidence that a Late Miocene through Pliocene component is present in his samples; Harwood mentions the last-named variety in the text (p. 135) but does not list it in table 8. Further, although Harwood and Maruyana (1992) indicate that this variety has a last occurrence datum of 2.4 myr, their assertion is not substantiated by an examination of their range tables. The other species have geologic ranges that exceed the time span of the Pliocene. For example, *E. antarctic, N. curta, N. kerguelensis,* and *N. ritscheri* are all important components of Pleistocene diatom assemblages in the Southern Ocean.

In the Beardmore Glacier area (fig. 16.1), Harwood (1986a) examined four Sirius Group samples from the Dominion Range. Although diatoms were few in number, he indicated that the reworked assemblage was from the Miocene and/or Pliocene, based upon the joint occurrence of *Thalassiosira* cf. *T. oestrupii* and *N. curta.* However, co-occurrence of these two forms indicates neither a Miocene nor a Pliocene age. Although *N. curta* does occur in the Pliocene (Gersonde and Burckle, 1990), it is abundant in the Pleistocene (Burckle, 1984), so that its occurrence cannot be used to indicate a Miocene or Pliocene age. Similarly, *Thalassiosira* cf. *T. oestrupii* cannot be used to indicate any age. Although *T. oestrupii* first appeared around the Miocene-Pliocene boundary and is still extant (Baldauf and Barron, 1991), the range of *Thalassiosira* cf. *T. oestrupii* is not known. Further, Harwood does not indicate why this form is referred to as *Thalassiosira* cf. *T. oestrupii* rather than as *T. oestrupii;* the former designation is used by paleontologists to indicate that specimens so named are different enough from the nominate taxon to be taxonomically in doubt. On the basis of such a meager recovery, one cannot place age constraints on this assemblage. Harwood examined sixteen samples from Mt. Sirius in the Beardmore Glacier region. Of these, three yielded rare diatom fragments, and one had abundant clasts containing many diatoms. According to Harwood, most of the clasts yielded diatoms representing a Late Oligocene to Early Miocene age, though at least one clast may contain Miocene or younger sediment. This latter age is based upon the occurrence of *Coscinodiscus margaritaceous?* and *Stephanopyxis* sp. However, one cannot identify an age younger than Miocene on the occurrence of a genus that was present throughout the entire Tertiary (*Stephanopyxis* sp.) and a species "question mark" (*Coscinodiscus margaritaceous?*).

Harwood (1986a) examined samples from a number of sections in the Reedy Glacier area (fig. 16.1), including four from the Metavolcanic Mt. section. Table 5 of Harwood gives results from an examination of three samples and, thus, it is assumed that one sample contained no diatoms. Although diatoms were rare and poorly preserved, Harwood identified *Coscinodiscus bullatus, Nitzschia angulata, N. curta, N. kerguelensis, Thalassiosira lentiginosa, T. torokina,* and *T. vulnifica* and, on the strength of these associations, stated that "a reworked upper Pliocene component is indicated in this deposit" (p. 115). The only species in this assemblage that is believed to be restricted to the Pliocene is *T. vulnifica*. Fenner (1991), however, found this species ranging into the Pleistocene in deep-sea cores recovered from the subantarctic part of the South Atlantic. Most of the other species, though present in the Late Pliocene, are most abundant in sediments of Pleistocene age. Again, one can easily argue for a Pleistocene age for these samples. The Quartz Hills section is also in the Reedy Glacier area. Harwood (1986a) examined three samples from this section, one of which contained microfossils. This sample was dated Late Pliocene on the basis of the occurrence of *Actinocyclus actinochilus, Thalassiosira lentiginosa, T. torokina*, and *T.* cf. *T oestrupii*. None of these species, however, is restricted to the Pliocene.

Nineteen samples were examined from Tillite Spur in the Reedy Glacier area, ten of which contained microfossils (Harwood, 1986a). Four of these ten contained what Harwood termed good microfossil assemblages. According to this author (p. 92), "Diverse assemblages from the lower Oligocene-uppermost Eocene?, possible upper Oligocene, middle Miocene, upper Miocene-lower Pliocene and mid-upper Pliocene were recovered" in sample 64-70, the sample with the largest number of diatoms. Indeed, some 90 percent of the diatoms recovered from Tillite Spur are from sample 64-70 (table 3 of Harwood, 1986a). The mid-Late Pliocene age is based upon the joint occurrence of a number of Pliocene species, notably *T. vulnifica, T. insigna*, and *N. praeinterfrigidaria*. But if one assumes that the Sirius Group was deposited over the interval suggested by Webb and Harwood (1991; i.e., between 3 and 2.5 myr), the co-occurrence, along with the aforementioned species, of *Coscinodiscus elliptipora* and *N. kerguelensis* in sample 64-70 presents a problem. According to Baldauf and Barron (1991, table 3), the former species first occurs in Southern Ocean sediments at about 2.2 myr, and the latter at about 2.7 myr. However, in Hole 745B (59.59°S), which was paleomagnetically dated, Baldauf and Barron (1991, table 12) show that the former species first occurred at 1.55–1.75 myr and the latter at 1.9–2.0 myr; both first occurrences were recorded in the Matuyama magnetic chron (Baldauf and Barron, 1991).

Fenner (1991) has also recorded a Pleistocene first occurrence for *Coscinodiscus elliptipora*. Her first abundant appearance datum is between 0.96 and 1.04 myr in three sites from the Atlantic sector of the Southern Ocean (Hole 699A, 51°32′S; Hole 701C, 51°59′S; and Hole 704A/B, 46°52′S). Similarly, Treppke (1989) found that this species first appears in the Early Pleistocene (between the Brunhes magnetochron and the Olduvai subchron) in a piston core recovered from the Atlantic sector (Core 1226-1; 54°31′S). Because there are also species in sample 64-70 that dominated during, but were not restricted to, the Pleistocene, one might argue for a Pleistocene age for this sample. In spite of these reservations about the most recent diatoms present in sample 64-70, the sample is unique in that it does contain at least four Pliocene diatoms. Indeed, it appears that the sole argument for an Early Pliocene collapse of the East Antarctic ice sheet and for reglaciation and emplacement of the Sirius Group sometime after 2.5 myr rests largely upon sample 64-70. However, if one examines the potential source areas proposed for Sirius Group sediments by Harwood (1986a, fig. 36), neither the Wilkes Basin nor the Pensacola Basin could have served as the possible source of sediments for the Tillite Spur outcrop.

Inferred Correlations with the Sirius Group. Harwood (1986a) also reported on diatoms found at Elephant Moraine, a concentration of boulders and sediments in the vicinity of the Allan Hills and near the boundary of the East Antarctic ice sheet. He assumed that this moraine was derived from a Sirius Group outcrop beneath the East Antarctic ice sheet; sixteen samples were examined, thirteen of which contained microfossils—including diatoms, some of which he identified as Pliocene (Harwood, 1986a, table 11). However, Harwood's evidence, the co-occurrence of *A. actinochilus* and *T. lentiginosa*, cannot be used to indicate a Pliocene age for emplacement of the source of Elephant Moraine sediments, because both are extant today and, in the case of the latter species, common to abundant in the Southern Ocean and the underlying Quaternary sediments. McKelvey and Stephenson (1990) studied the Pagodroma tillite in the Prince Charles Mountains, in East Antarctica. The presence of microfossils (attributed to the Miocene and Pliocene by Harwood [pers. comm. to McKelvey and Stephenson, 1990]) prompted these authors to postulate that the Pagodroma formed from reworking of "higher latitude Pliocene and Miocene strata by an expanding ancestral phase of Lambert Glacier" (McKelvey and Stephenson, 1990). Because Harwood (pers. comm. to McKelvey and Stephenson, 1990) determined that *T. insigna* and *N. kerguelensis* were present in this tillite, these authors suggested that its age of emplacement must be younger than about 3.5 myr. These findings are dismissed for the present, however, because no fossil data have been published.

LeMasurier et al. (1994) were the most recent to use displaced diatoms in sediments from the Antarctic continent to help document glacial history (in this case, the history of the West Antarctic ice sheet). Although part of their scenario is based upon K-Ar dates, it is also supported by displaced diatoms recovered from a 1 m-thick laminated mudstone recovered from the slopes of Mount Murphy, in West Antarctica (fig. 16.1). Using both K-Ar dates and dates based upon displaced diatoms and assuming that the occurrences of these diatoms in Mount Murphy sediments represented periods of open water where the West Antarctic ice sheet now stands, LeMasurier et al. (1994) postulated that this ice sheet was absent during the Early Miocene, the Early–Middle Miocene transition, the Early Pliocene, and possibly the Late Miocene. Final deposition of the laminated sediments was believed to have occurred during the Late Pliocene. The conclusion that ice was absent from West Antarctica during the Early Pliocene was based upon the presence of *N. praeinterfrigidaria,* a form that lived between 4.5 and 3.5 myr. However, the identification of this species is questionable. The specimen (LeMasurier et al., 1994, fig. 12, no. 18) referred to *N. praeinterfrigidaria* is a fragment, and the relevant features identifying the species cannot be observed; indeed, both apical ends are broken off, so it is not even possible to tell if the specimen is isopolar or heteropolar. Similarly, I question the identification of figure 12, number 22, as a *Denticulopsis maccollumii* (a species that ranges across the Early-Middle Miocene boundary). This specimen is a fragment and bears little resemblance to the nominate taxon. Figure 12, number 21 (LeMasurier et al., 1994), is a much smaller fragment and is even less convincing as a *D. maccollumii.* The specimens (fig. 12, nos. 3 and 6) referred to *Thalassionema nitzschioides* do not belong to that genus. *Thalassionema* is characterized by having isopolar valves (Round et al., 1990); clearly, these specimens have valves that are heteropolar. One cannot apply this criterion to the other specimens of *Thalassionema* (fig. 12, nos. 4 and 5), because these are fragments. The ages based upon diatoms (LeMasurier et al., 1994) rest upon a very fragile premise and, in the name of prudence, should be stricken.

Barrett et al. (1992) K-Ar dated a volcanic ash in CIROS 2, a core drilled in 211 m of water in the middle of Ferrar Fjord (77°41′S, 163°32′E; CIROS is an acronym for Cenozoic Investigations in the Western Ross Sea). The age of this ash was 3 myr, and because Harwood (1986a) had previously dated this core as Pliocene (based on the diatoms), Barrett et al. (1992) suggested that much of the CIROS 2 core was correlative with the Sirius Group and that it was the piece of evidence needed to prove that there had been a major drawdown of the East Antarctic ice sheet during the Pliocene (see also Sugden, 1992). As noted previously, I have no doubt about an Early Pliocene warming; there are doubts, however, that the study by Barrett et al. (1992) represents the "smoking gun." First, there is a problem with the identification of CIROS 2 diatoms. As noted previously, the key species used by Harwood (1986a) to limit part of the Sirius Group to the interval between approximately 3 and 2.5 myr are *T. insigna* and *T. vulnifica* (these are figured on plates 15 and 16 of Harwood, 1986a). The form shown in figure 10 of plate 15 is not, however, a *T. insigna;* rather, it is a *T. intersecta,* which ranges from the upper Miocene to the lower Pliocene (Gersonde and Burckle, 1990). Similarly, the form shown in figures 15–17 of plate 16 is very likely not a *T. vulnifica.* In my examination of CIROS 2 samples I found neither *T. insigna* nor *T. vulnifica.* Third, in table 12 (p. 228) *N. kerguelensis* is listed as being present in three CIROS 2 samples, at depths of 7.07 m (rare), 134.93 m (rare), and 151.56 m (very rare). In my examination of Sirius Group samples as well as samples from the Dry Valleys, *N. kerguelensis,* when present, was the most common diatom. In Sirius Group samples in my possession from Mt. Feather (fig. 16.1) 15 percent of the diatoms are *N. kerguelensis* and the average occurrence of this species in all of my samples (including samples from the Sirius Group and the Dry Valleys) is 23 percent. Yet, this species is very rare to rare in three samples from CIROS 2. Indeed, the dominant species present in CIROS 2 are not the dominant species found in Sirius Group sediments or in sediments from the Dry Valleys. It is very likely that Barrett et al. (1992) simply verified what is generally accepted, namely, that the Early Pliocene was a warm interval. The results of this study, however, cannot be used to support the deglacial thesis.

Emplacement Modes for Sirius Group Diatoms. Burckle et al. (1988) have been quoted widely by those who support a Pliocene age for deglaciation and reglaciation of East Antarctica (see, e.g., Webb and Harwood, 1991; Barrett et al., 1992; Le Masurier et al., 1994). Burckle et al. placed their imprimatur on the Pliocene age call for the Sirius and, using an analogy drawn from diatoms present in an ice core from the East Antarctic plateau (Dome C), they ruled out an aeolian source for Sirius Group diatoms. There are several problems with their findings, however. When Burckle et al. (1988) was prepared, I did not have access to the Harwood (1986a) thesis and did not realize how fragile the evidence was for a Pliocene age for emplacement of Sirius Group sediments. The main point of Burckle et al., however, dealt with the possibility that the occurrence of diatoms in the Sirius Group was due to aeolian transport. Using the aeolian origin of freshwater diatoms in the Dome C ice core (74°39′S, 124°10′E) as a model, these authors hypothesized that "if the Sirius diatoms are also eolian, then the two Antarctic sites (Dome C and the Sirius Group), equally remote from the Southern Ocean, will have the same exotic flora (i.e., windblown diatoms from either a fresh-water or marine source)" (p. 327). Because Dome C diatoms were largely freshwater and those from the Sirius largely marine, it was concluded that they must have had a different mode of emplacement and that aeolian transport of Sirius Group diatoms could be ruled out.

However, there is no reason for believing that the Sirius Group outcropping along the TAMS was always remote from the Southern Ocean. If the West Antarctic ice sheet had collapsed during the Pliocene or Pleistocene, then TAMS would be adjacent to open ocean. This would suggest several pathways for aeolian processes to introduce diatoms into Antarctic ice and sediments. Hadley circulation carries diatoms aloft in lower latitudes and toward the South Pole, where subsidence and weak inversion winds distribute them over the high plateau; such a circulation pattern would favor the entrainment of nonbiogenic dust and freshwater diatoms in rising air over continents of the Southern Hemisphere, which would then be carried aloft toward the south only to descend over the East Antarctic Plateau. The coastal regions of Antarctica, on the other hand, are frequently buffeted by storms that originate over the Southern Ocean and that could advect marine diatoms landward from the coast. This could account, along with other processes (see Kellogg and Kellogg, 1988), for the presence of marine diatoms in ice and sediments in coastal regions. Similarly, Tanimura (1992) has shown that open-ocean diatoms, including a large component of *N. kerguelensis*, can be carried under fast ice around the Antarctic continent.

As noted by Burckle and Pokras (1991), the identification of the Wilkes and Pensacola Basins (termed the Wilkes-Pensacola Basin by Webb et al., 1984) as the source of sediments for the Sirius Group presents problems. Webb et al. indicated that Early Pliocene deglaciation resulted in incursion of the Wilkes-Pensacola Basin by marine waters. Their argument on the source of the diatoms and the conclusion that East Antarctica was deglaciated during the Pliocene leans heavily, as it turns out, on sample 64-70 from Tillite Spur and upon their figure 1 (redrawn from Drewry, 1983), which shows an ice-free Antarctica after the continent is isostatically adjusted. Webb et al. (1984, p. 289) argued that "the Wilkes-Pensacola basin must have been marine basins until sea temperatures were low enough to generate ice shelves." Harwood (1986a, fig. 34, p. 153) further suggested that during maximum deglaciation "general flow from the Southwest Pacific to the Weddell sea may have been driven by the funneling of water from the Wilkes and Aurora seas into the narrow Pensacola sea." Webb et al. (1984, p. 289) proposed that during the "late Pliocene the ice sheet thickened over the Wilkes-Pensacola basin and grounded, onlapped against and overrode the Transantarctic Mountains." Burckle and Pokras (1991) previously noted, however, that the map shown by Webb et al. (1984), which purported to show the Antarctic continent with all ice removed and after isostatic adjustment, was not the map figured by Drewry (1983). This particular map by Drewry instead depicted bedrock topography with all ice removed but without isostatic adjustment. When this topography is isostatically uplifted (Drewry, 1983), much of the Wilkes-Pensacola Basin disappears (Burckle and Pokras, 1991). The deglacial hypothesis is then left without a source region for the diatoms and without a mechanism for implacing them into Sirius Group sediments.

Sea-Surface Temperature of the Early Pliocene Event. There is a second problem with identifying these interior basins as the source of diatoms in the Sirius Group. Harwood (1986a) used presumed temperature constraints on several extant diatom species (*Nitzschia kerguelensis* and *Thalassiosira lentiginosa*) to determine the probable sea-surface temperature for the interior basins of East Antarctica during the Early Pliocene (i.e., from about 5 to 3 myr). He noted that "the occurrence of these diatoms in the Sirius Formation suggests the presence of marine waters with temperatures similar to subantarctic values between 2° to 5°C in the Late Neogene Wilkes and Pensacola seas" (Harwood, 1986a, p. 151), whereas Webb and Harwood (1991, p. 219) noted that "confirmation of Harwood's (1986a, 1986b, 1987) Mid-Pliocene marine paleotemperature estimates for the interior basins is given by Burckle (1988) who writes: 'The Sirius Formation contains such open ocean diatoms as *Nitzschia kerguelensis* and *Thalassiosira lentiginosa.* Where the optimal temperature for the former species is around 5°C, the latter species is believed to have an optimal temperature between 5°C and 8°C. The sum total of our knowledge on reworked diatoms in the Sirius Group . . . indicates that the interior of East Antarctica was partially deglaciated, possibly during the Early Pliocene, with surface water temperatures similar to those in the vicinity of the present day Polar Front.'"

In their analysis, however, Webb and Harwood (1991) failed to take note of two points. First, neither *N. kerguelensis* nor *T. lentiginosa.* were prominent members of Early Pliocene diatom assemblages in the Southern Ocean. Although the first occurrence of the former species is estimated to be between 1.9 and 2.7 myr (Baldauf and Barron, 1991, tables 3 and 12), the latter species, though known from the Early Pliocene, does not become a major part of the Southern Ocean assemblage until the Late Pliocene–Early Pleistocene (Baldauf and Barron, 1991). Both species were prominent members of Southern Ocean diatom assemblages during the Quaternary (Donahue, 1967; Burckle, 1984; Treppke, 1989). Second, even if it can be shown that these species were common to abundant in the Early Pliocene, Webb and Harwood (1991) failed to note that Burckle (1988) was wrong in his estimate of the overlying surface-water temperature when these two species and *N. curta* occur together in bottom sediments. More recent data (summarized in Burckle and Pokras, 1991) indicate that *N. kerguelensis* and *T. lentiginosa* have a temperature range of 0°–8°C. Further, Burckle and Pokras (1991) pointed out that *N. curta,* which was also reported in Sirius Group samples (Harwood, 1986a), has a temperature range between −2° and 2°C. When one considers that coccoliths, which were not reported in Sirius Group samples (save for a few Paleogene forms), do not secrete shells below a temperature of 3°C, the surface-water temperature estimate that would result from the joint occurrence of these three diatom species together with the absence of coccoliths would have to be less than 3°C. Please note that this temperature estimate for surface waters of the Southern Ocean applies only to the growing season, when these forms live and reproduce.

Although there is evidence that silicoflagellates give a paleoproductivity rather than a paleotemperature signal (Takahashi, 1989), Harwood (1986a) used the work of Ciesielski and Weaver (1974) to demonstrate that Southern Ocean surface waters were warm enough during the Early Pliocene to support a significant Antarctic ice-sheet drawdown. The latter authors developed a paleotemperature curve based on the ratio of silicoflagellate genera *Dictyocha* and *Distephanus* (Mandra and Mandra, 1969; Jendrzejewski and Zarillo, 1971). Finding that a simple ratio gave too high a temperature estimate, Ciesielski (1974) determined the ratio in surface sediments of the southeast Indian sector of the Southern Ocean and, drawing a best-fit curve through these data, calibrated this line against annually averaged surface-water temperatures for this region. Using this procedure, Ciesielski and Weaver (1974) determined that annual sea-surface temperatures in the Southern Ocean during the Early Pliocene were as high as 10°C, even as far south as 68°S. Specifically, they found that the interval between 4.30 and 3.95 myr had very warm sea-surface temperatures, which, according to them, may have led to West Antarctic ice-sheet collapse during this time.

There are a number of problems with this conclusion, however. The first is taxonomic; the study by Weaver and Ciesielski (1974) is based upon the assumption that the extinct silicoflagellate *Distephanus speculum* forma *pseudofibula* belongs to the genus *Dictyocha.* This has been questioned by McCartney and Wise (1991). More substantive problems exist, however. In a study of both water-column and surface-sediment samples from the eastern equatorial Atlantic (and using the same methodology as Ciesielski, 1974—i.e. a best-fit curve), Schrader and Richert (1974) determined that the percentage of *Distephanus* rose exponentially over the temperature range of 16.5° to 8°C. They further found that water samples that contain 50 percent *Distephanus* (i.e., the ratio of *Dictyocha* to *Distephanus* is 1) have a surface-water temperature of 17°C, whereas in the underlying surface sediments, a *Dictyocha / Distephanus* ratio of 1 gives a temperature estimate of about 14.8°C; in the Southern Ocean, a ratio of 1 gives a temperature estimate of about 8°C (Ciesielski and Weaver, 1974).

We get yet a third type of response in the North Atlantic. The percentage of *Dictyocha* in twenty-nine surface-water samples was determined using data gathered over a period of 1.5 years and over temperatures ranging between 12.5° and 21°C (fig. 16.2). Water samples and surface-water temperatures were taken at approximately two-week intervals from the weather ship *Delta,* which remained in place (44°00′N, 41°00′W)

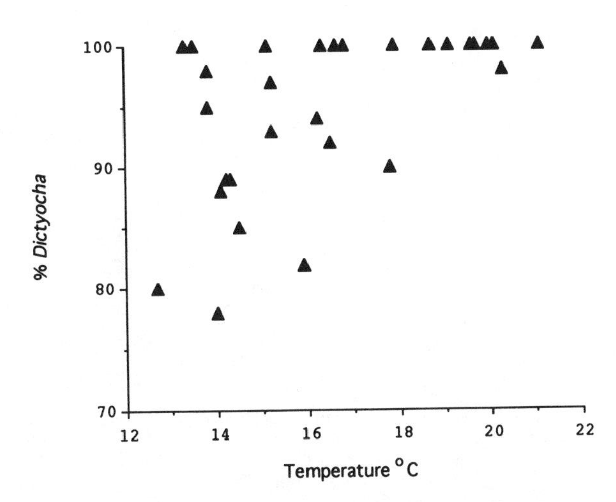

Fig. 16.2. Percentage of each sample containing *Dictyocha* in surface waters at weather ship *Delta* (44°00′N, 41°00′W). These figures can readily be converted into the *Dictyocha/Distephanus* ratio. Note that even at a temperature of less than 13°C, some 80 percent *Dictyocha* are present. Compare this with water-column data of Schrader and Richert (1974) on fig. 16.3.

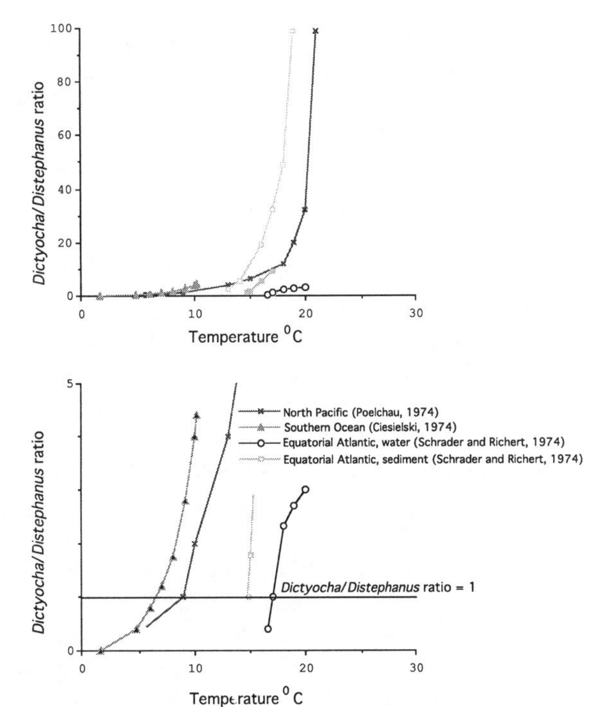

Fig. 16.3. *Dictyocha/Distephanus* ratio from water-column and surface-sediment studies. Upper diagram shows ratio values versus temperature from five studies; lower diagram shows ratio values between 0 and 5 versus temperature for four of those studies. The horizontal line in lower diagram indicates temperature estimates when the *Dictyocha/Distephanus* ratio is 1: this ratio gives estimates ranging from about 7°C to approximately 17°C.

year-round. A *Dictyocha / Distephanus* ratio of 1 was never reached during the period of observation in spite of the fact that surface-water temperatures reached below 13°C; in the eastern equatorial Atlantic a ratio of 1 occurs at a temperature of 14.8°C (Schrader and Richert, 1974). Similar comments can be made for the data of Poelchau (1974). Some of these data are summarized in figure 16.3. Takahashi (1989) monitored the flux of silicoflagellates to the seafloor at Station PAPA in the northeast Pacific (50°N, 145°W) over a four-year period; approximately twenty-four sediment trap samples were taken per year, at depths of 1,000 m and 3,500 m, and the surface-water temperature range was from less than 4° to almost 14°C. Takahashi (1989) concluded that *Dictyocha / Distephanus* ratios were more dependent on production rates than temperature.

Abelmann et al. (1988) carried out radiolarian, diatom, and magnetostratigraphic analysis of deep-sea cores taken along a north-south transect in the South Atlantic. The most southerly core was at about 68°S. On the basis of faunal and floral occurrences, these authors suggested that "during the early and middle early Pliocene, surface water temperatures were higher (5°–10°C)" (p. 751). They concluded, therefore, that they had found evidence from high southern-latitude deep-sea sediments for a warm Early Pliocene that lent credence to the deglacial hypothesis. Actually, their striking conclusion is based upon one core from the Weddell Sea (Core 1467, 64°06.51′S, 01°18.36′E). There are several problems with their conclusion, however. They noted that an extant radiolarian (*Spongotrochus glacialis*), which, according to them, was prominent in surface sediments beneath the subantarctic zone, was present at about 64°S in the Weddell Sea during the Early Pliocene. Working in the Southern Ocean, Lozano (1974, p. 303) has pointed out that this species is "an important constituent of the radiolarian fauna in all of our samples." This species is not restricted to the subantarctic north of the Polar Front Zone (PFZ). Rather, abundances of 10 percent or more appear to be characteristic of the subantarctic, but core-top abundances in excess of this value are reported well south of the present PFZ. Further, Morley and Stepien (1984, p. 367) found that in water samples recovered from beneath winter sea-ice cover from this region (i.e., in the Weddell Sea south of the PFZ) "*Spongotrochus glacialis* Popofsky is the most abundant species in samples from the upper 100 m interval and from 100 to 200 m." Similarly, this species is present over the same water depth beneath Arctic sea ice (Hülsemann, 1963;

Tibbs, 1967). It appears unlikely, therefore, that its presence in Southern Ocean sediments can be used as an indicator of surface-water temperatures during the Early Pliocene.

In addition, a number of diatom species cited by Abelmann et al. (1988) as being excluded from the present-day "Antarctic cold water belt" (*Actinocyclus octonarius, Azpeitia tabularis, Thalassionema nitzschioides, T. nitzschioides* var. *parva*, and *Thalassiosira oestrupii*) are all found south of the PFZ, although some occur in reduced numbers. Pichon et al. (1987) reported that of the twenty-four species used in their analysis of diatoms in surface sediments of the Southern Ocean, only two (*Hemidiscus cuneiformis* and *Roperia tesselata*) and the silicoflagellate genus *Dictyocha* occur exclusively north of the PFZ. Very likely the major reason why *T. nitzschioides* var. *parva* does not have a wider geographic distribution is that most workers lump this variety with *T. nitzschioides* var. *nitzschioides*. Although Abelmann et al. (1988) point to the presence of extinct, presumably warm-water radiolarian and diatom species in Early Pliocene sediments south of the PFZ, they present no independent evidence to indicate that these species are truly adapted to warm water rather than cosmopolitan. Further, a temperature rise of 5°–10°C would bring surface-water temperatures for much of the Southern Ocean well above 3°C. Abelmann et al. (1988) do not mention the presence of calcareous nannofossils in their lower Pliocene samples; however, because calcareous nannofossils secrete shells at temperatures above 3°C (A. McIntyre, pers. comm., in Burckle and Pokras, 1991), they should be present in lower Pliocene samples, and in considerable numbers. One might argue that during the Early Pliocene this core site was, as it is today, below the calcium carconate compensation depth. This situation seems unlikely, however, if, as Abelmann et al. (1988) contend, surface-water temperatures were some 5°–10°C higher than at present.

There are two additional problems with the claim by Abelmann et al. (1988) of a 5°–10°C rise in temperature during the Early Pliocene. It is not clear if they were referring to summer temperature or to average annual temperature. Because diatom growth in the higher-latitude Southern Ocean is seasonal (i.e., late spring–summer), one assumes that they were referring to a summer temperature rise; yet this is not clear from their text. One can see the problem that arises from imposing their 5°–10°C temperature rise onto the Weddell Sea (figs. 16.4 and 16.5). If one assumes that they were refer-

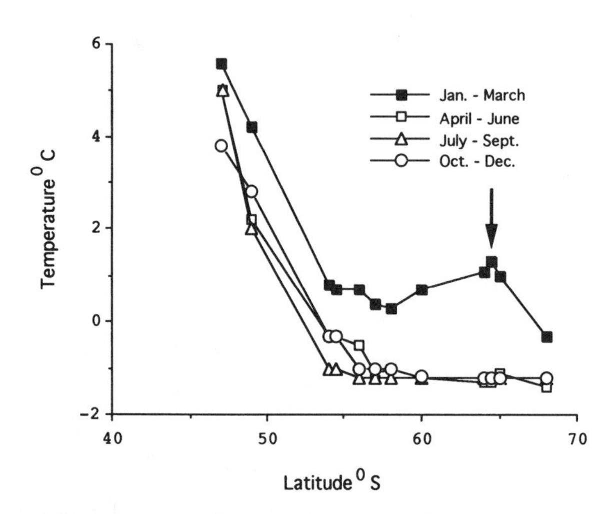

Fig. 16.4. Surface-water temperatures for each three-month period of the year, taken along the same transect as the piston cores studied by Abelmann et al. (1988). Temperature data are taken from Gordon and Molinelli (1982). Note that summer temperatures are high in the vicinity of Core 1467 (arrow), the core used by Abelmann et al. to demonstrate a surface-water warming of 5°–10°C during the Early Pliocene.

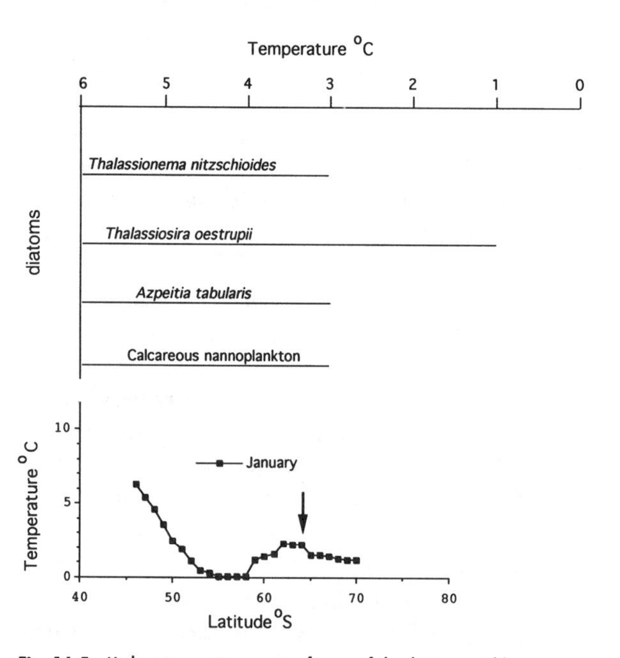

Fig. 16.5. Modern temperature range of some of the diatoms used by Abelmann et al. (1988) to postulate a temperature rise of 5°–10°C during the Early Pliocene in the Southern Ocean. The location of Core 1467 on this diagram is marked by an arrow. Note that a summer temperature rise of only a degree or two is needed to support these species at the location of Core 1467.

ring to a summer temperature rise of this magnitude, then temperatures during this season in the Weddell Sea at 64°S would be about 7.5°–12.5°C. Further, surface-water temperatures in the region of the PFZ would be about 10°–15°C. If one assumes that Abelmann et al. (1988) were referring to an annual average temperature rise, then the problem is even more acute. Such a scenario would imply that summer surface-water temperatures in the Weddell Sea and north of the PFZ (in the Atlantic sector) during the Early Pliocene were much more than 5°–10°C above present-day temperatures.

The second problem deals with the location of Core 1467 (64°06.51′S, 01°18.36′E), which is beneath the westward (i.e., return) flow of the Weddell Gyre. Plate 7 of the *Southern Ocean Atlas* (Gordon and Molinelli, 1982) shows average surface-water temperatures from January through March (i.e., during much of the austral summer; see fig. 16.4). In the region of Core 1467, summer surface-water temperatures reach above 2°C (Gordon and Molinelli, 1982), which is warmer than regions to the north or south of it. Summer surface-water temperatures of this region need only increase by a degree or two to approach those of the PFZ. This point is further illustrated in figure 16.5 where I have plotted the known temperature range of some of the diatoms used by Abelmann et al. (1988) to suggest a temperature rise of 5° to 10°C. If one acknowledges the constraint imposed by the absence of coccoliths, combined with the observed modern temperature range of those diatoms used by Abelmann et al. (1988), the late spring–summer temperature rise for the Early Pliocene recorded in Core 1467 need only be 1°C or 2°C. The presence of warmer-water diatoms and radiolarians in this region, therefore, cannot be used as evidence for an extreme general Southern Ocean warming.

Sea-Level Response to Pliocene Warming. Finally, there is the question of the sea-level response to Early Pliocene warming. Haq et al. (1988) indicated that sea level rose about 100 m (above the present datum) during the earliest Pliocene and almost 80 m at about 3 myr. Greenlee and Moore (1988) and Schroeder and Greenlee (1993) have questioned such an extreme sea-level rise. These authors, relying on data from a number of continental-shelf sections, but primarily from offshore New Jersey, suggested more modest changes in sea level (fig. 16.6). For example, the Early Pliocene sea-level rise of Greenlee and Moore (1988) is on the order of 30 m. Similarly, other studies suggest that maximum sea-level

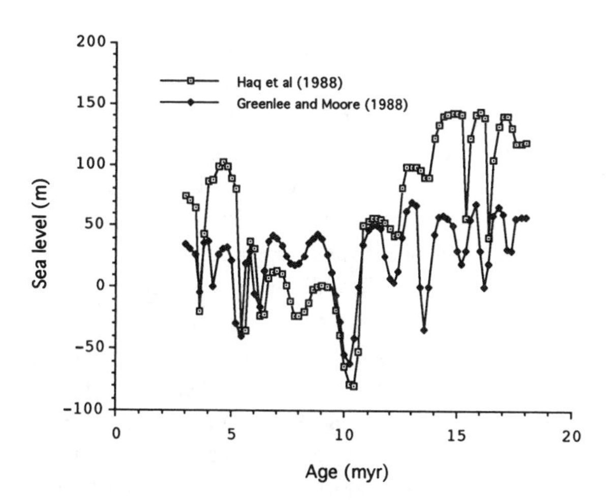

Fig. 16.6. Neogene sea-level history, after Haq et al. (1988) and Greenlee and Moore (1988). Note that these two data sets essentially parallel each other, although the latter shows less extreme swings.

estimates for the Pliocene of the East Coast of the United States are 35 ± 18 m (Dowsett and Cronin, 1991), 25–40 m (Krantz, 1990), and 30–45 m (Pazzaglia, 1993)—again, more in keeping with the view that there was not a major Early Pliocene drawdown of the East Antarctic ice sheet.

Conclusions

The claim of Harwood (1986, p. 145) that in Sirius Group samples "the microfossil assemblages are dominated by Pliocene species" is not supported by a critical examination of his data. Further, the claim that the Sirius Group represents sediments that were originally deposited in interior East Antarctic basins during the Early Pliocene is not supported by a careful examination of the same text. This claim, as it turns out, is based largely upon sample 64–70, recovered by John Mercer at Tillite Spur. Indeed, of the 350 occurrences of diatoms in those Sirius Group samples studied by Harwood (1986a), Pliocene diatoms account for 1.4 percent; Pliocene-Pleistocene diatoms (i.e., those diatoms that occur in both the Pliocene and Pleistocene) account for 6.8 percent; and Pleistocene diatoms represent .3 percent. Given these considerations, the often-cited view (see, e.g., LeMasurier et al., 1994) that the Early Pliocene marine (i.e., isotopic) record is at odds with the Antarctic continental record is not true. They are essentially in harmony. Excluding Tillite Spur, there seems little reason to embark on a major restudy of displaced diatoms in the Sirius. The history of the Antarctic cryosphere can be more fully and believably developed by study of deep-sea sediments recovered from the Southern Ocean and deep cores recovered from the Antarctic continent.

It is apparent that although there may have been drawdown of the East Antarctic ice sheet during the early Pliocene. it was not of the magnitude envisioned by the deglacial hypothesis. In addition, it is very likely that the West Antarctic ice sheet and alpine and tidewater glaciers from both hemispheres contributed to the accompanying sea-level rise. This has implications for Early Pliocene climate and paleogeography; a more limited sea-level rise during this time would be less restrictive of faunal and floral migrations across terrestrial choke points, such as the Suez region. Similarly, oceanic choke points such as the seas around the Indonesian archipelago would have been less affected by a more modest sea-level rise, and the Early Pliocene warming is seen as presenting less of a physical impediment to continental faunal and floral migration.

Acknowledgments

When I gave this manuscript to my internal readers, I asked that they be particularly harsh in their review. They were. I am exceedingly grateful for the comments of the following: G. Ashley, J. Barron, P. Bodén, G. Denton, D. Kellogg, T. Kellogg, J. Kennett, and D. Warnke. Colleagues at Lamont-Doherty Earth Observatory are also gratefully acknowledged. Nevertheless, responsibility for the content of this chapter rests solely with the author. Research support was provided by National Science Foundation Grant OPP91-18995. This is Lamont-Doherty Earth Observatory Contribution 5206.

References

Abelmann, A., Gersonde, R., and Spiess, V. 1988. Pliocene-Pleistocene paleoceanography in the Weddell Sea: Siliceous microfossil evidence. In *Geological history of the polar oceans: Arctic versus Antarctica,* pp. 729–759 (ed. U. Bleil and J. Thiede). Kluwer Academic, Dordrecht.

Baldauf, J. G., and Barron, J. A. 1991. Diatom biostratigraphy: Kerguelen Plateau and Prydz Bay regions of the Southern Ocean. *Initial Reports ODP 119,* B:547–598. Ocean Drilling Program, College Station, Tex.

Barrett, P. J., Adams, C. J., McIntosh, W. C., Swisher, C. C., III, and Wilson, G. S. 1992. Geochronological evidence supporting Antarctic deglaciation three million years ago. *Nature* 359:816–818.

Burckle, L. H. 1984. Diatom distribution and oceanographic reconstruction in the Southern Ocean: Present and last glacial maximum. *Marine Micropaleontology* 9:241–262.

———. 1988. Scientific rationale for sub-ice drilling in East Antarctica. In *Polar Drilling Workshop,* (ed. P.-N. Webb). Ohio State University, Columbus.

Burckle, L. H., Gayley, R. I., Ram, M., and Petit, J.-R. 1988. Diatoms in Antarctic ice cores: Some implications for the glacial history of Antarctica. *Geology* 16:326–329.

Burckle, L. H., and Pokras, E. M. 1991. Implications of a Pliocene stand of *Nothofagus* (southern beech) within 500 kilometres of the south pole. *Antarctic Science* 3:389–403.

Ciesielski, P. F. 1974. Silicoflagellate paleotemperature curve for the Southern Ocean. *Antarctic Journal of the United States* 9:269–270.

Ciesielski, P. F., and Weaver, F. M. 1974. Early Pliocene temperature changes in the Antarctic seas. *Geology* 23:511–515.

Denton, G. H., Sugden, D. E., Marchant, D. R., Hall, B. L., and Wilch, T. I. 1993. East Antarctic ice sheet sensitivity to Pliocene climatic change from a Dry Valleys perspective. *Geografiska Annaler* 75:155–204.

Donahue, J. G. 1967. Diatoms as indicators of Pleistocene climatic fluctuations in the Pacific sector of the Southern Ocean. In *Progress in oceanography: The Quaternary history of the ocean basins,* Vol. 4, pp. 133–140 (ed. B. C. Heezen). Pergamon, New York.

Dowsett, H. J., and Cronin, T. M. 1990. High eustatic sea level during the Middle Pliocene: Evidence from the southeastern U.S. Atlantic Coastal Plain. *Geology* 18:435–438.

Drewry, D. J. 1983. *Antarctica: Geological and geophysical folio.* Cambridge University Press, Cambridge.

Fenner, J. M. 1991. Late Pliocene–Quaternary quantitative diatom stratigraphy in the Atlantic sector of the Southern Ocean. *Initial Reports ODP 114,* B:97–121. Ocean Drilling Program, College Station, Tex.

Gersonde, R., and Burckle, L. H. 1990. Neogene diatom biostratigraphy of ODP Leg 113, Weddell Sea (Antarctic Ocean). *Initial Reports ODP 113,* B:761–783. Ocean Drilling Program, College Station, Tex.

Gordon, A. L., and Molinelli, E. J. 1982. *Southern ocean atlas.* Columbia University Press, New York.

Greenlee, S. M., and Moore, T. C. 1988. Recognition and interpretation of depositional sequences and calculation of sea level changes from stratigraphic data: Offshore New Jersey and Alabama Tertiary. In *Sea level changes: An integrated approach,* pp. 329–353 (ed. C. K. Wilgus, B. S. Hastings, C. G. Kendall, E. Posamentier, C. A. Ross, and J. C. Van Wagoner). Special Publication 42. Society of Economic Paleontologists and Mineralogists (SEPM), Tulsa.

Haq, B. U., Hardenbol, J., and Vail, P. R. 1988. Mesozoic and Cenozoic chronostratigraphy and cycles of sea level change. In *Sea level changes: An integrated approach,* pp. 71–108 (ed. C. K. Wilgus, B. S. Hastings, C. G. Kendall, E. Posamentier, C. A. Ross, and J. C. Van Wagoner). Special Publication 42. Society of Economic Paleontologists and Mineralogists (SEPM), Tulsa.

Harwood, D. M. 1983. Diatoms from the Sirius Formation, Transantarctic Mountains. *Antarctic Journal of the United States* 18:98–100.

———. 1986a. Diatom biostratigraphy and paleoecology and a Cenozoic history of Antarctic ice sheets. Ph.D. diss., Ohio State University.

———. 1986b. Recycled siliceous microfossils from the Sirius Formation. *Antarctic Journal of the United States* 21:101–103.

Harwood, D. M., and Maruyana, T. 1992. Middle Eocene to Pleistocene diatom biostratigraphy of Southern Ocean sediments from the Kerguelen Plateau, Leg 120. *Initial Reports ODP 120,* pp. 683–734. U.S. Government Printing Office, Washington, D.C.

Hülsemann, K. 1963. Radiolaria in plankton from the Arctic drifting station T-3, including the description of three new species. *Arctic Institute of North America,* Technical Paper, 13:1–52.

Huybrechts, P. 1993. Glaciological modeling of the Late Cenozoic East Antarctic ice sheet: Stability or dynamism? *Geografiska Annaler* 75:221–238.

Jendrzejewski, J. P., and Zarillo, G. A. 1971. Late Pleistocene paleotemperatures: Silicoflagellates and foraminiferal frequency changes in a subantarctic deep sea core. *Antarctic Journal of the United States* 6:178–179.

Kellogg, D. E., and Kellogg, T. B. 1988. Diatoms of the McMurdo ice shelf, Antarctica: Implications for sediment and biotic reworking. *Palaeogeography, Palaeoclimatology, Palaeoecology* 60:77–96.

Kennett, J. P., and Hodell, D. A. 1993. Evidence for relative climatic stability of Antarctica during the early Pliocene: A marine perspective. *Geografiska Annaler* 75:205–220.

Krantz, D. E. 1990. A chronology of Pliocene sea level fluctuations. *Quaternary Science Reviews* 10:163–174.

LeMasurier, W. E., Harwood, D. M., and Rex, D. C. 1994. Geology of Mount Murphy volcano: An 8-m.y. history of interaction between a rift volcano and the West Antarctic ice sheet. *Geological Society of America Bulletin* 106:265–280.

Lozano, J. 1974. Antarctic sedimentary, faunal, and sea surface temperature response during the last 230,000 years with emphasis on comparison between 18,000 years ago and today. Ph.D. diss., Columbia University.

McCartney, K., and Wise, S. W. 1991. Cenozoic silicoflagellates and Ebridians from ODP Leg 113: Biostratigraphy and notes on morphologic variability. *Initial Reports DSDP 113,* pp. 729–760. U.S. Government Printing Office, Washington, D.C.

McKelvey, B. C., and Stephenson, N. C. N. 1990. A geological reconnaissance of the Radok Lake area, Amery oasis, Prince Charles Mountains. *Antarctic Science* 2:53–66.

McKelvey, B. C., Webb, P.-N., Harwood, D. M., and Mabin, M. C. G. 1991. The Dominion Range Sirius Group: A record of the Late Pliocene–Early Pleistocene Beardmore Glacier. In *Geological evolution of Antarctica,* pp. 675–682 (ed. M. R. A. Thomson, J. A. Crame, and J. W. Thompson). Cambridge University Press. Cambridge.

Mandra, Y. T., and Mandra, H. 1969. Silicoflagellates, a new tool for the study of Antarctic Tertiary climates. *Antarctic Journal of the United States* 4:172–174.

Morley, J. J., and Stepien, J. C. 1984. Siliceous microfauna in waters beneath Antarctic sea ice. *Marine Ecology-Progress Series* 19:207–210.

Pazzaglia, F. J. 1993. Stratigraphy, petrography, and correlation of Late Cenozoic Middle Atlantic Coastal Plain deposits: Implications for late-stage passive-margin geologic evolution. *Geological Society of America Bulletin* 105:1617–1634.

Pichon, J.-J., Labracherie, M., Labeyrie, L. D., and Duprat, J. 1987. Transfer functions between diatom assemblages and surface hydrology in the Southern Ocean. *Palaeogeography, Palaeoclimatology, Palaeoecology* 61:79–95.

Poelchau, H. S. 1974. Distribution of Holocene silicoflagellates in North Pacific sediments. *Micropaleontology* 22:164–193.

Round, F. E., Crawford, R. M., and Mann, D. G. 1990. *The diatoms, biology and morphology of the genera.* Cambridge University Press, Cambridge.

Schrader, H.-J., and Richert, P. 1974. Paleotemperature interpretation by means of percent amount *Dictyocha / Distephanus* (Silicoflagellatae). *Third Planktonic Conference,* Kiel University. Abstracts, p. 65.

Schroeder, F. W., and Greenlee, S. M. 1993. Testing eustatic curves based on Baltimore canyon Neogene stratigraphy: An example application of basin-fill simulation. *American Association of Petroleum Geologists Bulletin* 77:638–656.

Sugden, D. 1992. Antarctic ice sheets at risk? *Nature* 359:775–776.

Takahashi, K. 1989. Silicoflagellates as productivity indicators: Evidence from long temporal and spatial flux variability responding to hydrography in the northeastern Pacific. *Global Biogeochemical Cycles* 3:43–61.

Tanimura, Y. 1992. Distribution of diatom species in the surface sediments of Lutzow-Holm Bay, Antarctica. In *Centenary of Japanese micropaleontology,* pp. 399–411 (ed. K. Ishizaki and T. Saito). Terra Scientific Publishing, Tokyo.

Tibbs, J. F. 1967. On some planktonic Protozoa taken from the track of drift station ARLIS 1, 1960–1961. *Arctic Institute of North America* 20:247–254.

Treppke, U. 1989. Biostratigraphische und palaookologische Untersuchungen van Diatomeenassoziatonen an Plio-bis Holozanen Sedimenten vom Atlantisch-Indischen Rucken, nordliches Weddelmeer. Master's thesis, Eberhard-Karl-Universität, Tübingen.

Weaver, F. M., and Ciesielski, P. F. 1974. Southern Ocean paleotemperatures based on silicoflagellates: Evidence for a Pliocene "interglacial" in Antarctica. *Third Planktonic Conference,* Kiel University. Abstracts, p. 78.

Webb, P.-N. 1990. The Cenozoic history of Antarctica and its global impact. *Antarctic Science* 2:3–21.

Webb, P.-N., and Harwood, D. M. 1987. Terrestrial flora of the Sirius Formation: Its significance for late Cenozoic glacial history. *Antarctic Journal of the United States* 22:7–11.

———. 1991. Late Cenozoic glacial history of the Ross embayment, Antarctica. *Quaternary Science Reviews* 10:225–237.

Webb, P.-N., Harwood, D. M., McKelvey, B. C., Mabin, M. C. G., and Mercer, J. H. 1986. Late Cenozoic tectonic and glacial history of the Transantarctic Mountains. *Antarctic Journal of the United States* 21:99–100.

Webb, P.-N., Harwood, D. M., McKelvey, B. C., and Stott, L. D. 1984. Cenozoic marine sedimentation and ice-volume variation on the East Antarctic craton. *Geology* 12:287–291.

Webb, P.-N., McKelvey, B. C., Harwood, D. M., Mabin, M. C. G., and Mercer, J. H. 1987. Sirius Formation of the Beardmore Glacier region. *Antarctic Journal of the United States* 22:8–13.

Chapter 17

New Data on the Evolution of Pliocene Climatic Variability

Nicholas J. Shackleton

The sequences recovered on Ocean Drilling Program (ODP) Leg 138 in summer 1991 provide wonderful opportunities for investigating Pliocene climatic variability with the resolution that is commonly devoted only to the Pleistocene. The cruise recovered over 5.5 km of sediment with unprecedented percent recovery and documented stratigraphic continuity. Mayer et al. (1992) have published the shipboard work, and the majority of the manuscripts for the Scientific Results volume (Pisias et al., 1994) are in press. In this chapter I highlight a few of the findings that relate to this book.

Studies that have been done on astronomical calibration of the time scale (Shackleton et al., 1994) provide further confirmation of the work of Hays et al. (1976), Imbrie et al. (1984), Shackleton et al. (1990), and Hilgen (1991a, 1991b), in addition to extending the astronomical calibration into the latest Miocene (see also Hilgen et al., 1993). These investigations have, of course, necessitated recalibrating the seafloor anomaly paleomagnetic chronology. This work is important not only for abstract interest in increasing the accuracy of our absolute chronology but also, in a very practical sense, for the correct simultaneous application of magnetostratigraphy and radiometric dating, as is evident in the elegant study by McDougall et al. (1992). Table 17.1 compares the new time scale with two earlier scales. Table 17.2 summarizes some recent radiometric calibrations, all of which tend to support the present version. The Reunion Event that Baksi et al. (1993) have redated is not included in table 17.2 because of uncertainties in relating the event they dated to the polarity scale of Cande and Kent (1992). As Baksi et al. (1993) point out, accurate $^{40}Ar/^{39}Ar$ dating has an important role to play in calibrating the polarity time scale. In addition, Wilson (1993) has shown that the pattern of seafloor magnetic anomalies is amenable to more precise investigation and that through the Pliocene a reevaluation of the spacing of magnetic anomalies tends to support the astronomically calibrated time scale rather than the earlier time scale of Cande and Kent (1992).

Another area in which our findings will be of interest is that of high-resolution oxygen isotope stratigraphy. The extension of high-resolution $\delta^{18}O$ records has proceeded very rapidly in recent years, and this has had an important impact on our understanding of climatic variability.

For the Middle and Upper Pleistocene (the Brunhes magnetochron), a large number of good records have become available since the pioneering work of Emiliani (1955, 1961) and its extension by Shackleton and Opdyke (1973); for example, Imbrie et al. (1992, 1993) make use of some twenty oxygen isotope records in their review of the nature and cause of climatic variability over this interval. Middle Pleistocene oxygen isotope records are characterized by a dominant cycle with a period of about 100 thousand years (kyr). It is interesting to note that even over this interval, the recalibration of the paleomagnetic chronology has had a significant impact, because although Imbrie et al. (1984) obtained a good orbitally tuned chronology for the past 600 kyr, they lacked records with adequate resolution spanning the Brunhes-Matuyama boundary. One consequence of the revised calibration of the interval between 0.6 and 1.0 myr is that the association between the 100 kyr cycles and variation in the eccentricity of the earth's orbit is

ISBN 0–300–06348–2.

Table 17.1. Ages for the Magnetic Polarity Changes Over the Past 15 myr (Million Years)

Name (BKF85 or Conventional Anomaly)	BKF85	CK92	SCHPS94	Name (Homogenized CK92)
Brunhes/Matuyama	0.73	0.780	0.780	C1n(o)
Jaramillo (t)	0.91	0.984	0.990	C1r.1n (t)
Jaramillo (o)	0.98	1.049	1.070	C1r.1n (o)
Olduvai (t)	1.66	1.757	1.770	C2n (t)
Olduvai (o)	1.88	1.983	1.950	C2n (o)
Matuyama/Gauss	2.47	2.600	2.600	C2An.1n (t)
Kaena (t)	2.92	3.054	3.046	C2An.1n (o)
Kaena (o)	2.99	3.127	3.131	C2An.2n (t)
Mammoth (t)	3.08	3.221	3.233	C2An.2n (o)
Mammoth (o)	3.18	3.325	3.331	C2An.3n (t)
Gauss/Gilbert	3.40	3.553	3.594	C2An.3n (o)
Cochiti (t)	3.88	4.033	4.199	C3n.1n (t)
Cochiti (o)	3.97	4.134	4.316	C3n.1n(o)
Nunivak (t)	4.10	4.265	4.479	C3n.2n (t)
Nunivak (o)	4.24	4.432	4.623	C3n.2n(o)
Sidufjall (t)	4.40	4.611	4.781	C3n.3n (t)
Sidufjall (o)	4.47	4.694	4.878	C3n.3n (o)
Thvera (t)	4.57	4.812	4.977	C3n.4n (t)
Thvera (o)	4.77	5.046	5.232	C3n.4n (o)
3A	5.35	5.705	5.875	C3An.1n (t)
3A	5.53	5.946	6.122	C3An.1n (o)
3A	5.68	6.078	6.256	C3An.2n (t)
3A	5.89	6.376	6.555	C3An.2n (o)
	6.37	6.744	6.919	C3Bn (t)
	6.50	6.901	7.072	C3Bn (o)
4	6.70	7.245	7.406	C4n.1n (t)
4	6.78	7.376	7.533	C4n.1n (o)
4	6.85	7.464	7.618	C4n.2n (t)
4	7.28	7.892	8.027	C4n.2n (o)
4	7.35	8.047	8.174	C4r.1n (t)
4	7.41	8.079	8.205	C4r.1n (o)
4A	7.90	8.529	8.631	C4An (t)
4A	8.21	8.861	8.945	C4An (o)
4A	8.41	9.069	9.142	C4Ar.1n (t)
4A	8.50	9.149	9.218	C4Ar.1n (o)
	8.71	9.428	9.482	C4Ar.2n (t)
	8.80	9.491	9.543	C4Ar.2n (o)
5	8.92	9.542	9.639	C5n.1n (t)
5	10.42	10.834	10.839	C5n.2n (o)
	10.54	10.940	10.943	C5r.1n (t)
	10.59	10.989	10.991	C5r.1n (o)
	11.03	11.378	11.373	C5r.2n (t)
	11.09	11.434	11.428	C5r.2n (o)
5A	11.55	11.852	11.841	C5An.1n (t)
5A	11.73	12.000	11.988	C5An.1n (o)
5A	11.86	12.108	12.096	C5An.2n (t)
5A	12.12	12.333	12.320	C5An.2n (o)
	12.46	12.618	12.605	C5Ar.1n (t)
	12.49	12.649	12.637	C5Ar.1n (o)
	12.58	12.718	12.705	C5Ar.2n (t)
	12.62	12.764	12.752	C5Ar.2n (o)
5AA	12.83	12.941	12.929	C5AAn (t)
5AA	13.01	13.094	13.083	C5AAn (o)
5AB	13.20	13.263	13.252	C5ABn (t)
5AB	13.46	13.476	13.466	C5ABn (o)
5AC	13.69	13.674	13.666	C5ACn (t)
5AC	14.08	14.059	14.053	C5ACn (o)
5AD	14.20	14.164	14.159	C5ADn (t)
5AD	14.66	14.608	14.607	C5ADn (o)
5B	14.87	14.800	14.800	C5Bn.1n (t)
5B	14.96	14.890	14.890	C5Bn.1n (o)

Note: (o) = onset; (t) = termination.
Source: BKF85: Berggren et al. (1985); CK92: Cande and Kent (1992); SCHPS94: Shackleton et al. (in press).

strengthened. The origin of this cycle is discussed by Imbrie et al. (1993).

The 100 kyr cycle appears to dominate the paleoclimatic record of the past million years in most regions of the globe, although this cycle can be established only in areas where some semblance of a complete record can be compiled. Areas from which a characteristic 100 kyr cycle is documented include the loess sections of central Europe (Kukla, 1970), China (Kukla et al., 1990), and central Asia (Dodonov, 1991; Shackleton et al., in prep). Long vegetational records dominated by a 100 kyr cycle are known from southern Europe (Wijmstra and Groenhart, 1983) and the Andes (Hooghiemstra, 1989). The fact that the marine oxygen isotope record is dominated by a 100 kyr cycle establishes this as the main rhythm of the great ice sheets. In addition, it is known that the New Zealand mountain glacial record is dominated by this cycle (Nelson et al., 1985). Marine oxygen isotope and other records from deep-sea sediments also contain evidence for variability associated with the 41 kyr obliquity cycle and the 23/19 kyr precession cycle. Imbrie et al. (1992) have explored the spatial distribution of global climate response to these cycles, and it is interesting to note that a few records have been discovered that are truly dominated by one or the other of these cycles rather than by the 100 kyr glacial cycle. In particular, those regions affected by the monsoon system appear to be especially affected by the precession cycle (Clemens and Prell, 1991; deMenocal et al., 1993; Pokras and Mix, 1987; Tiedemann et al. 1989).

For the Lower Pleistocene and latest Pliocene, the low-resolution records of piston cores V16-205 and

Table 17.2. New Radiometric Data (in myr) Relating to the Pliocene Magnetic Polarity Time Scale

Reversal	BKF85	SCHPS94	Radiometric
C1n (o)	0.73	0.78	0.79 ± 0.02[1]
C1n (o)	0.73	0.78	0.746[2]
C1n (o)	0.73	0.78	0.78 ± 0.01[3]
C1n (o)	0.73	0.78	0.783 ± 0.011[4]
C1n (o)	0.73	0.78	0.773 ± 0.009[5]
C1r.1n (t)	0.91	0.99	0.992 ± 0.039[2]
C1r.1n (t)	0.91	0.99	0.915[3]
C1r.1n (o)	0.97	1.07	1.05 ± 0.11[6]
C1r.1n (o)	0.97	1.07	1.010[3]
C2n (t)	1.66	1.77	1.78 ± 0.04[7]
C2n (o)	1.88	1.95	1.96 ± 0.04[7]
C2n (o)	1.88	1.95	>1.98[8]
C2An.1n (t)	2.47	2.60	2.60 ± 0.06[7]
C2An.1n (o)	2.92	3.046	3.02 ± 0.06[7]
C2An.2n (t)	2.99	3.131	3.09 ± 0.06[7]
C2An.2n (o)	3.08	3.23	>3.18[9]
C2An.2n (o)	3.08	3.23	3.21 ± 0.06[7]
C2An.3n (t)	3.18	3.33	>3.22[9]
C2An.3n (t)	3.18	3.33	3.29 ± 0.06[7]
C2An.3n (o)	3.40	3.594	3.57 ± 0.05[7]
C3n.2n (t)	4.10	4.478	4.51 ± 0.08[10]

Sources: 1) Izett and Obradovich, 1991; 2) Tauxe et al., 1992; 3) Spell and McDougall, 1992; 4) Baksi et al., 1992; 5) Hall and Farrell, 1993; 6) Glass et al., 1991; 7) McDougall et al., 1992; 8) Walter et al., 1991; 9) Walter et al., 1992; 10) Hall and Farrell, 1994.

V28-238 published by van Donk (1976) and by Shackleton and Opdyke (1976), respectively, are now far surpassed in terms of detail by those obtained in Deep Sea Drilling Project (DSDP) and ODP Site 607 (Ruddiman et al., 1987); Site 677 (Shackleton and Hall, 1989); Site 659 (Tiedemann et al., 1989); Site 714 (Hodell and Venz, 1992); Site 846 (Mix et al., 1994); and others (table 17.3). The overriding characteristic of this interval is the dominance of 41 kyr cycles (Ruddiman et al., 1986). The amplitude of the $\delta^{18}O$ cycles during this period is much less than for the recent glacial cycles, and it is clear that this reflects the accumulation of smaller ice sheets prior

Table 17.3. Locations of DSDP and ODP Sites Referred to in the Text

Site	Latitude	Longitude	Depth
DSDP552A	56°02.56′N	23°13.88′W	2,301 m
DSDP607	41°00.07′N	32°57.44′W	3,427 m
ODP659	18°04.63′N	21°01.57′W	3,070 m
ODP677	1°12.14′N	83°44.22′W	3,461 m
ODP714	5°03.6′N	73°47.2′E	2,038 m
ODP846	3°5.696′S	90°49.078′W	3,296 m

to oxygen isotope stage 24. Maximum ice volumes in the Northern Hemisphere during this period were probably between one-half and two-thirds of the last glacial maximum value. However, glacial stages back to stage 100 (about 2.5 myr) are associated with ice-rafted debris in the North Atlantic as far south as Rockall (Shackleton et al., 1984), as well as in the North Pacific (Rea et al., 1993). The earliest pure loess on the loess plateau in China was also probably deposited during stage 100.

Areas of the globe that experienced intense variability associated with the precession cycle during the last million years display the same characteristic during the Early Pleistocene, suggesting that this feature is indeed a direct response to forcing by the precession cycle, as was implied by the modeling experiments of Prell and Kutzbach (1987; see also deMenocal et al., 1991; Tiedemann et al., 1989). Until recently, this was the limit of high-resolution records, although the important transitional interval preceding stage 100 was well covered by the investigations of Raymo et al. (1989) in Site 607. Currently, Site 846 in the East Pacific is providing a detailed oxygen isotope record for the entire Pliocene, whereas DSDP Site 659 (Tiedemann et al., 1994) provides a comparable record for the equatorial Atlantic.

Figure 17.1 shows the $\delta^{18}O$ record of the Pleistocene and very latest Pliocene. The SPECMAP stack of oxygen isotope records (Imbrie et al., 1984) is used for the interval 0 to 0.62 myr. This record was published reduced to zero mean and unit variance. In figure 17.1 it is rescaled to have the same mean and variance as the benthic oxygen isotope data from ODP Site 677 for the same time interval; the SPECMAP record is extended to 2 myr with the ODP Site 677 record (from Shackleton et al., 1990). The reason for adopting this procedure is that the SPECMAP stack has become a standard for the Middle Pleistocene, while the record of Site 677 is probably the best available for the Lower Pleistocene. Figures 17.2 and 17.3 show the oxygen isotope record from ODP Site 846. Across the buildup of glacial cycles in the Northern Hemisphere, figure 17.2 confirms the picture of a gradual buildup in glacial intensity over several cycles (Raymo et al., 1980), in contrast to the picture of a sudden intensification suggested by the record of DSDP Site 552A (Shackleton and Hall, 1984), which is an artifact of incomplete recovery at that site.

Figure 17.2 also shows the extent of variability in the interval that was described by Sarnthein and Fenner (1988) as the Pliocene Golden Age (about 3.3 to 3.6 myr on the time scale used here). The stability that was im-

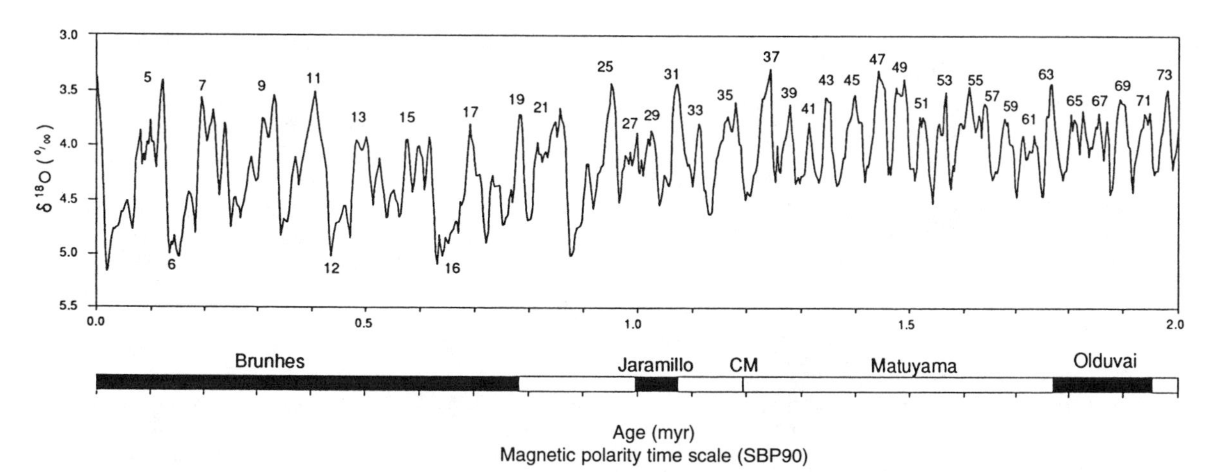

Fig. 17.1. Oxygen isotope record for the SPECMAP stack for the interval 0–0.62 myr and ODP Site 677 for the interval 0.62–2.0 myr. The SPECMAP record is scaled to have the same mean and variance as data from the same time interval in Site 677.

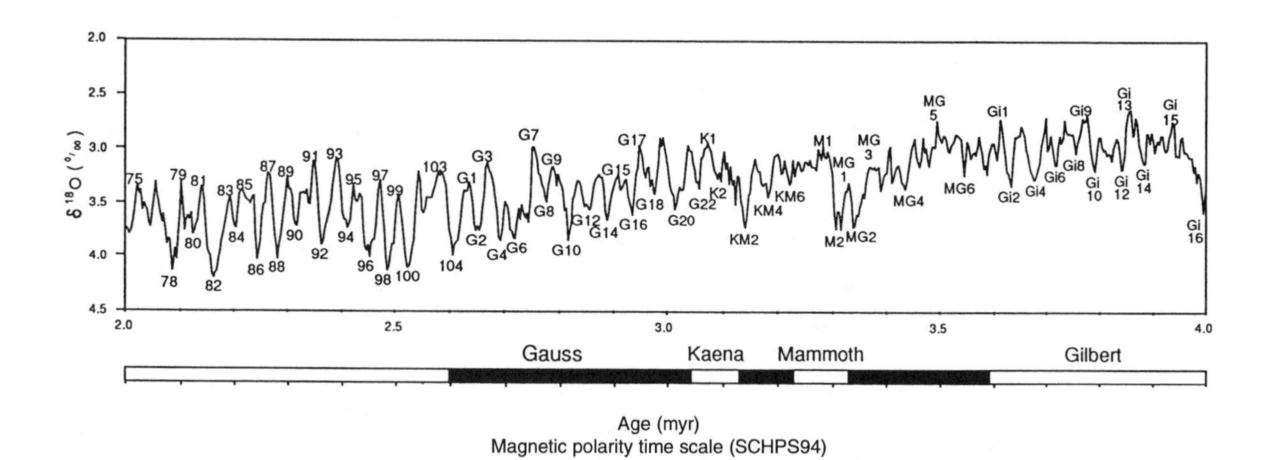

Fig. 17.2. Oxygen isotope record for benthic foraminifera from ODP Site 846 for the interval 2.0–4.0 myr.

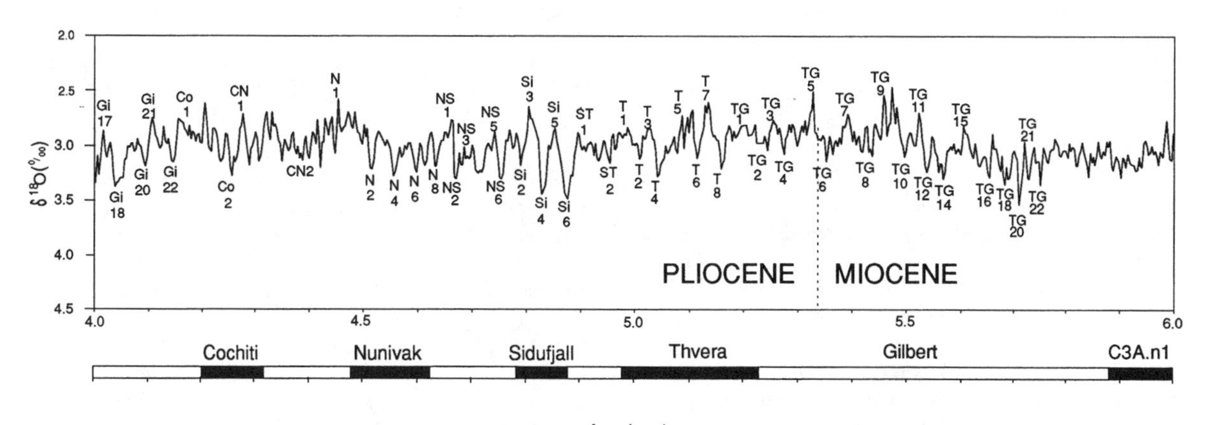

Fig. 17.3. Oxygen isotope record for benthic foraminifera from ODP Site 846 for the interval 4.0–6.0 myr.

plied by that epithet was an artifact of the low resolution that was available in records across that interval. The nature of climatic variability at this time certainly deserves further investigation. The major unsolved question is the extent to which variations in ice volume are reflected in the Pliocene oxygen isotope record. The excursions in the isotopically light direction are consistent with significant reductions in the volume of ice stored on Antarctica but can equally well be explained by warmer deep waters. It is, however, very difficult to envisage significant deglaciation of Antarctica without a warming of the deep water (Shackleton and Kennett, 1974; Hodell, 1992). On the other hand, excursions in the warm direction in deep-water temperature associated with a stable Antarctic ice mass are relatively easily explained in terms of varying heat input either in the North Atlantic or from the Mediterranean Sea.

Figure 17.3 shows the data for the interval to below the base of the Pliocene. This figure provides additional support for the idea that a glacio-eustatic sea-level fall might have initiated the final Messinian dessication, because there is evidence for a positive excursion implying an increasing ice volume that would have been associated with a sea-level fall that could have been as large as a few tens of meters. The studies of Leg 138 sequences will contribute much else to our understanding of Late Neogene climate evolution as well as refining our ability to place events in an increasingly accurate historical sequence.

References

Baksi, A. J., Hoffman, K. A., and McWilliams, M. 1993. Testing the accuracy of the geomagnetic polarity time-scale (GPTS) at 2–5 Ma, utilizing $^{40}Ar/^{39}Ar$ incremental heating data on whole-rock basalts. *Earth and Planetary Science Letters* 118:135–144.

Baksi, A. K., Hsu, V., McWilliams, M. O., and Farrar, E. 1992. $^{40}Ar/^{39}Ar$ dating of the Brunhes-Matuyama geomagnetic field reversal. *Science* 256:356–357.

Berggren, W. A., Kent, D. V., and Flynn, J. J. 1985. Jurassic to Paleogene. Pt. 2, Paleogene geochronology and chronostratigraphy. In *The chronology of the geological record: Geological Society of London Memoir* (ed. N. J. Snelling), 10:141–195.

Cande, S. C., and Kent, D. V. 1992. A new geomagnetic polarity time scale for the Late Cretaceous and Cenozoic. *Journal of Geophysical Research* 97:13917–13951.

Clemens, S. C., and Prell, W. L. 1991. One million year record of summer monsoon winds and continental aridity from the Owen Ridge (Site 722), northwest Arabian Sea. In *Proceedings ODP, scientific results 117* (ed. W. L. Prell, N. Niitsuma et al.). College Station, Tex. (Ocean Drilling Program), 365–388.

deMenocal, P. B., Bloemendal, J., and King, J. 1991. A rock-magnetic record of monsoonal dust deposition to the Arabian Sea: Evidence for a shift in the mode of deposition at 2.4 Ma. In *Proceedings ODP, scientific results 117* (ed. W. L. Prell, N. Niitsuma et al.). College Station, Tex. (Ocean Drilling Program), 389–407.

deMenocal, P. B., Ruddiman, W. F., and Pokras, E. M. 1993. Influences of high- and low-latitude processes on African terrestrial climate: Pleistocene eolian records from equatorial Atlantic Ocean Drilling Program Site 663. *Paleoceanography* 8:209–242.

Dodonov, A. E. 1991. Loess of central Asia. *GeoJournal* 24:185–194.

Emiliani, C. 1955. Pleistocene temperatures. *Journal of Geology* 63:538–578.

———. 1966. Paleotemperature analysis of Caribbean cores P6304-8 and P6304-9 and a generalized temperature curve of the past 425,000 years. *Journal of Geology* 74:109–126.

Glass, B. P., Kent, D. V., Schneider, D. A., and Tauxe, L. 1991. Ivory Coast microtektite strewn field: Description and relation to the Jaramillo geomagnetic event. *Earth and Planetary Science Letters* 107:182–196.

Hall, C. M., and Farrell, J. W. 1993. Laser $^{40}Ar/^{39}Ar$ age from Ash D of ODP Site 758: Dating the Brunhes-Matuyama reversal and oxygen isotope stage 19.1. Supplement *Eos, Transactions, American Geophysical Union,* April 20, 1993, 110. (Abstract)

Hall, C. M., and Farrell, J. W. 1994. Laser $^{40}/^{39}Ar$ dating of tephra in marine sediments: Calibrating the GPTS. International Conference on Geochemistry, 8, Berkeley, San Francisco. (Abstract)

Hays, J. D., Imbrie, J., and Shackleton, N. J. 1976. Variations in the earth's orbit: Pacemaker of the ice ages. *Science* 194:1121–1131.

Hilgen, F. J. 1991a. Astronomical calibration of Gauss to Matuyama sapropels in the Mediterranean and implication for the geomagnetic polarity time scale. *Earth and Planetary Science Letters* 104:226–244.

———. 1991b. Extension of the astronomically calibrated (polarity) time scale to the Miocene-Pliocene boundary. *Earth and Planetary Science Letters* 107:349–368.

Hilgen, F. J., Lourens, L. J., Berger, A., and Loutre, M. F. 1993. Evaluation of the astronomically calibrated time scale for the Late Pliocene and earliest Pleistocene. *Paleoceanography* 8:549–565.

Hodell, D. A., and Venz, K. 1992. Toward a high-resolution stable isotopic record of the southern ocean during the Plio-Pleistocene (4.8 to 0.8 Ma). *Antarctic Research Series* 56:265–310.

Hooghiemstra, H. 1989. Quaternary and Upper-Pliocene glaciations and forest development in the tropical Andes: Evidence from a long high-resolution pollen record from the sedimentary basin of Bogotá, Columbia. *Palaeogeography, Palaeoclimatology, Palaeoecology* 72:11–26.

Imbrie, J., Berger, A., Boyle, E. A., Clemens, S. C., Duffy, A., Howard, W. R., Kukla, G., Kutzbach, J., Martinson, D. G., McIntyre, A., Mix, A. C., Molfino, B., Morley, J. J., Peterson, L. C., Pisias, N. G., Prell, W. L., Raymo, M. E., Shackleton,

N. J., and Toggweiler, J. R. 1993. On the structure and origin of major glaciation cycles. Pt. 2, The 100,000-year cycle. *Paleoceanography* 8:699–735.

Imbrie, J., Boyle, E. A., Clemens, S. C., Duffy, A., Howard, W. R., Kukla, G., Kutzbach, J., Martinson, D. G., McIntyre, A., Mix, A. C., Molfino, B., Morley, J. J., Peterson, L. C., Pisias, N. G., Prell, W. L., Raymo, M. E., Shackleton, N. J., and Toggweiler, J. R. 1992. On the structure and origin of major glaciation cycles. Pt. 1, Linear responses to Milankovitch forcing. *Paleoceanography* 7:701–738.

Imbrie, J., Hays, J. D., Martinson, D. G., McIntyre, A., Mix, A., Morley, J. J., Pisias, N. G., Prell, W., and Shackleton, N. J. 1984. The orbital theory of Pleistocene climate: support from a revised chronology of the marine $\delta^{18}O$ record. In *Milankovitch and climate,* pt. 1, NATO ASI Series, 269–305 (ed. A. Berger, J. Imbrie, J. Hays, G. Kukla, and B. Saltzman). Reidel, Dordrecht.

Izett, G. A., and Obradovich, J. D. 1991. Dating of the Matuyama-Brunhes Boundary based on ^{40}Ar-^{39}Ar ages of the Bishop Tuff and Cerro San Luis rhyolite. *Geological Society of America Abstract Programs* 23:A106. (Abstract)

Kukla, G., An, Z. S., Melice, J. L., Gavin, J., and Xiao, J. L. 1990. Chronostratigraphy of Chinese loess. *Transactions of the Royal Society of Edinburgh: Earth Sciences* 81:263–288.

Kukla, J. 1970. Correlations between loesses and deep-sea sediments. *Geologiska Föreningen i Stockholm Förhandlingar* 92:148–180.

McDougall, I., Brown, F. H., Cerling, T. E., and Hillhouse, J. W. 1992. A reappraisal of the geomagnetic polarity time scale to 4 Ma using data from the Turkana Basin, East Africa. *Geophysical Research Letters* 19:2349–2352.

Mayer, L., Pisias, N., Janecek, T., et al. 1992. *Proceedings ODP, initial reports 138.* College Station, Tex. (Ocean Drilling Program).

Mix, A. C., Le, J., and Shackleton, N. J. (in press). Benthic foraminiferal stable isotope stratigraphy of Site 846: 0–1.8 Ma. In *Proceedings of the ODP, scientific results 138* (ed. N. G. Pisias, L. A. Mayer, and T. R. Janecek et al.). College Station, Tex. (Ocean Drilling Program).

Nelson, C. S., Hendy, C. H., Jarrett, G. R., and Cuthbertson, A. M. 1985. Near-synchroneity of New Zealand alpine glaciations and Northern Hemisphere continental glaciations during the past 750 kyr. *Nature* 318:361–363.

Obradovich, J. D., and Izett, G. A. 1992. The geomagnetic polarity time scale (GPTS) and the astronomical time scale (ATS) now in near accord. Supplement *EOS, Transactions, American Geophysical Union,* October 27, 1992, 630. (Abstract)

Pisias, N. G., Mayer, L. A., Janecek, T. R., et al. (in prep.). *Proceedings ODP, scientific results 138.* College Station, Tex. (Ocean Drilling Program).

Pokras, E. M., and Mix, A. C. 1987. Earth's precession cycle and Quaternary climatic change in tropical Africa. *Nature* 326:486–487.

Prell, W. L., and Kutzbach, J. E. 1987. Monsoon variability over the past 150,000 years. *Journal of Geophysical Research* 92:8411–8425.

Raymo, M. E., Hodell, D., and Jansen, E. 1992. Response of deep ocean circulation to initiation of Northern Hemisphere glaciation (3-2 Ma). *Paleoceanography* 7:645–672.

Raymo, M., Ruddiman, W. F., Backman, J., Clement, B., and Martinson, D. G. 1989. Late Pliocene variation in Northern Hemisphere ice sheets and North Atlantic deep circulation. *Paleoceanography* 4:413–446.

Raymo, M. E., Ruddiman, W. F., Shackleton, N. J., and Oppo, D. W. 1990. Evolution of global ice volume and Atlantic-Pacific $\delta^{13}C$ gradients over the last 2.5 M.Y. *Earth and Planetary Science Letters* 97:353–368.

Rea, D. K., Basov, I. A., Janecek, T. R., and Leg 145 Scientific Party. 1993. Paleoceanographic record of North Pacific quantified. *EOS, Transactions, American Geophysical Union* 74:406–411.

Ruddiman, W. F., McIntyre, A., and Raymo, M. 1987. Paleoenvironmental results from North Atlantic Sites 607 and 609. In *Initial reports DSDP, 94* (ed. W. F. Ruddiman and R. B. Kidd). U.S. Government Printing Office, Washington, D.C., 855–879.

Ruddiman, W. F., Raymo, M., and McIntyre, A. 1986. Matuyama 41,000-year cycles: North Atlantic Ocean and Northern Hemisphere ice sheets. *Earth and Planetary Science Letters* 80:117–129.

Sarnthein, M., and Fenner, J. 1988. Global wind-induced change of deep-sea sediment budgets, new ocean production and CO_2 reservoirs ca. 3.3–2.35 Ma BP. *Philosophical Transactions of the Royal Society London* 318:487–504.

Shackleton, N. J., Backman, J., Zimmerman, H., Kent, D. V., Hall, M. A., Roberts, D. G., Schnitker, D., Baldauf, J. G., Desprairies, A., Homrighausen, R., Huddlestun, P., Keene, J. B., Kaltenback, A. J., Krumsiek, K. A. O., Morton, A. C., Murray, J. W., and Westberg-Smith, J. 1984. Oxygen isotope calibration of the onset of ice-rafting and history of glaciation in the North Atlantic region. *Nature* 307:620–623.

Shackleton, N. J., Berger, A., and Peltier, W. R. 1990. An alternative astronomical calibration of the Lower Pleistocene timescale based on ODP Site 677. *Transactions of the Royal Society of Edinburgh: Earth Sciences* 81:251–261.

Shackleton, N. J., Crowhurst, S., Hagelberg, T., Pisias, N., and Schneider, D. A. (in press). A new Late Neogene time scale: Application to ODP Leg 138 sites. In *Proceedings ODP, scientific results 138* (ed. N. G. Pisias, L. A. Mayer, T. R. Janecek et al.). College Station, Tex. (Ocean Drilling Program).

Shackleton, N. J., and Hall, M. A. 1984. Oxygen and carbon isotope stratigraphy of Deep Sea Drilling Project Hole 552A: Plio-Pleistocene glacial history. In *Initial reports DSDP, 81* (ed. D. G. Roberts, D. Schnitker et al.). U.S. Government Printing Office, Washington, D.C., 599–609.

Shackleton, N. J., and Hall, M. A. 1989. Stable isotope history of the Pleistocene at ODP Site 677. In *Proceedings of the ODP, scientific results 111* (ed. K. Becker, H. Sakai et al.). College Station, Tex. (Ocean Drilling Program), 295–316.

Shackleton, N. J., and Opdyke, N. D. 1973. Oxygen isotope and palaeomagnetic stratigraphy of equatorial Pacific core V28-238: Oxygen isotope temperatures and ice volumes on a 10^5 and 10^6 year scale. *Quaternary Research* 3:39–55.

Shackleton, N. J., and Opdyke, N. D. 1976. Oxygen isotope and paleomagnetic stratigraphy of Pacific core V28-239, Late

Pliocene to latest Pleistocene. In *Geological Society of America Memoir* (ed. R. M. Cline and J. D. Hays), 145:449–464.

Spell, T. L., and McDougall, I. 1992. Revisions to the age of the Brunhes-Matuyama Boundary and the Pleistocene geomagnetic polarity timescale. *Geophysical Research Letters* 19:1181–1184.

Tauxe, L., Deino, A. D., Behrensmeyer, A. K., and Potts, R. 1992. Pinning down the Brunhes/Matuyama and upper Jaramillo boundaries: A reconciliation of orbital and isotopic time scales. *Earth and Planetary Science Letters* 109:561–572.

Tiedemann, R., Sarnthein, M., and Shackleton, N. J. 1994. Astronomic timescale for the Pliocene Atlantic $\delta^{18}O$ and dust flux records of Ocean Drilling Program Site 659. *Paleoceanography* 9:619–638.

Tiedemann, R., Sarnthein, M., and Stein, R. 1989. Climatic changes in the western Sahara: Aeolo-marine sediment record of the last 8 million years (Sites 657–661). In *Proceedings ODP, scientific results 108* (ed. W. Ruddiman, M. Sarnthein et al.). College Station, Tex. (Ocean Drilling Program), 241–277.

van Donk, J. 1976. An ^{18}O record of the Atlantic Ocean for the entire Pleistocene. In *Geological Society of America Memoir* (ed. R. M. Cline and J. D. Hays), 145:147–164.

Walter, R. C., Brown, F. H., and Hillhouse, J. W. 1992. Refining the Plio-Pleistocene GPTS using laser-fusion $^{40}Ar/^{39}Ar$ tephrochronology: Case studies from the East African Rift. Supplement *EOS, Transactions, American Geophysical Union,* October 27, 1992, 629. (Abstract)

Walter, R. C., Manega, P. C., Hay, R. L., Drake, R. E., and Curtis, G. H. 1991. Laser-fusion $^{40}Ar/^{39}Ar$ dating of Bed 1, Olduvai Gorge, Tanzania. *Nature* 354:145–149.

Wijmstra, T. A., and Groenhart, M. C. 1983. Record of 700,000 years vegetational history in eastern Macedonia (Greece). *Revista de la Academia Colombiana de Ciencias Exactas, Físicas y Naturales* 15:87–98.

Wilson, D. S. 1993. Confirmation of the astronomical calibration of the polarity time scale from rates of seafloor spreading. *Nature* 364:788–790.

Chapter 18

Environmental and Paleoclimatic Evolution in Late Pliocene–Quaternary Colombia

Henry Hooghiemstra

Long continental pollen cores with records of climatic, environmental, and biotic change are of great importance in understanding present-day ecosystems and in facilitating comparisons of land-based and ocean-based climatic histories. In the Eastern Cordillera of Colombia, the high plain of Bogotá (measuring ca. 25 × 40 km) represents the bottom of a former lake that occupied a subsiding intermontane basin at an elevation of 2,550 m. Here, pollen records have been retrieved from two boreholes known as Funza I (357 m deep) and Funza II (586 m, which reached the bedrock), representing the period from the Late Pliocene to the latest Pleistocene (Hooghiemstra and Ran, 1994b). About 28 thousand years (kyr) ago the paleolake of Bogotá drained, possibly as a result of erosion of the Tequendama Falls, which forms the only outlet of this high plain.

As a result of changes in the elevation of the main vegetation belts, the high plain of Bogotá was situated in the Andean forest belt during interglacials and in the subparamo or grassparamo belt during glacials. These changes in the regional vegetation cover around the paleolake reflect mainly a temperature record, which is documented by the pollen that is conserved in slowly accumulating lake sediments in the sedimentary basin of Bogotá.

Tropical mountains especially seem to be in a favorable position for the recording of terrestrial climatic changes, because a change in temperature results mainly in a vertical shift of vegetation belts over the mountain slopes. The different vegetation belts stay in the vicinity and are registered continuously by their intercepted pollen. The sediments of the Bogotá basin accumulated at an elevation that lies halfway between the highest position of the upper forest line during interglacial conditions (ca. 3,600 m) and the lowest position of the upper forest line during glacial conditions (ca. 1,800 m), rendering the Bogotá sediments a sensitive recorder of past climatic change. The objective of this chapter is to provide an overview of Late Pliocene–Pleistocene climatic, environmental, and biotic change in the Eastern Cordillera of Colombia and to show relationships with the adjacent Amazonian lowlands.

ISBN 0-300-06348-2.

Present Vegetation and Climate

The present-day altitudinal zonation of the vegetation in the Eastern Cordillera of Colombia is summarized in order to understand the changes documented by the pollen record (fig. 18.1). The following vegetation zones can be recognized (after Van der Hammen, 1974):

—tropical lowland rain forest from 0 to 1,000 m altitude (main taxa: *Byrsonima, Iriartea, Mauritia*)
—subandean forest belt (lower montane forest) from 1,000 to 2,300 m altitude (main taxa: *Acalypha, Alchornea, Cecropia*)
—Andean forest belt (upper montane forest) from 2,300 to 3,200–3,500 m altitude (main taxa: *Podocarpus, Hedyosmum, Weinmannia, Quercus, Alnus, Vallea, Myrsine* (= *Rapanea*), *Symplocos, Ilex, Juglans, Miconia, Eugenia, Myrica*)
—subparamo belt from 3,200–3,500 to 3,400–3,600 m altitude (main elements: Ericaceae, *Hypericum*, Compositae, *Polylepis-Acaena*)
—grassparamo belt from 3,400–3,600 to 4,000–4,200 m altitude (main elements: Gramineae, *Valeriana*, Car-

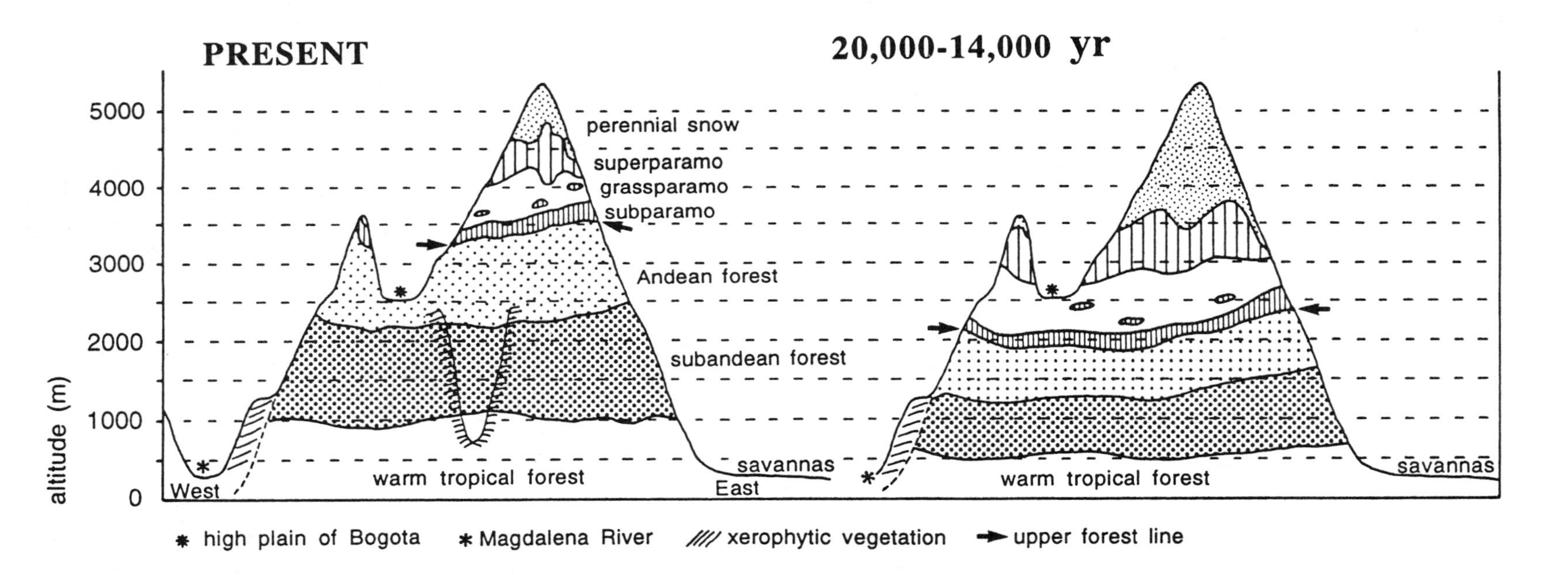

Fig. 18.1. Altitudinal distribution of vegetation belts in the Eastern Cordillera of Colombia at present and during the last glacial maximum. Vertical shifts of the vegetation belts are mainly related to changes in temperature (after Van der Hammen, 1974).

yophyllaceae, Gentianaceae, *Plantago, Aragoa, Geranium, Ranunculus, Lycopodium* taxa with foveolate spores)
—and the superparamo belt from 4,000 to 4,200 m altitude (main taxa: *Draba,* mosses, and blue algae).

The nival zone proper, practically devoid of vegetation, extends from 4,500 to 4,800 m upward. The highest areas of the Eastern Cordillera in the Sierra Nevada de Cocuy, about 200 km north of Bogotá, extending up to 5,500 m, may be permanently covered by snow.

The modern upper forest line in the area of Bogotá, at an altitude of 3,200 m, corresponds with the 9.5°C annual isotherm. Thus temperature changes at the level of the high plain of Bogotá (2,550 m; modern average annual temperature 13°–14°C) can be calculated when changes of the altitudinal position of the upper forest line are estimated on the basis of the pollen record, using a lapse rate of 0.66°C per 100-m displacement of the upper forest line.

Time Control of the Bogotá Sediments and Land-Sea Correlation

The 1984 time frame (Hooghiemstra, 1984, 1989) of the sediments of Bogotá was recently improved, based on new fission track dates (Andriessen et al., 1993). The revised time frame is based on 11 zircon fission track dates that were obtained both from exposed ash layers in the outer valleys of the high plain and from a series of ashes from the Funza II core. The dating results of 5.33 ± 1.02 million years (myr), 3.67 ± 0.50 myr, and 2.77 ± 0.50 myr, for sediments that are considered to have been deposited before, at the beginning, and shortly after the final major upheaval of the Eastern Cordillera, respectively, provide absolute chronological control for the older part of the sequence (6.0–2.5 myr). The following dates from the Funza II core give absolute dating control for the younger part of the sequence (3.0–0 myr), and these are coherent with the fission track dates of the older part of the sediment sequence: 67.7 m: 0.20 ± 0.12 myr; 298–307 m: 1.02 ± 0.23 myr; 317 m: 1.44 ± 0.33 myr; 322 m: 1.01 ± 0.21 myr; 506 m: 2.74 ± 0.63 myr. Nevertheless, three fission track dates on zircon, at 329 m (0.26 ± 0.18 myr), 250 m (0.27 ± 0.11 myr), and 270–277 m (0.53 ± 0.15 myr) are considered as too young (fig. 18.2; see the discussion in Andriessen et al., 1933). The Neogene-Quaternary sediments of the Bogotá area span a period of at least the last 6 myr. The fluvial-lacustrine sediment record registers a major tectonic uplift of the Eastern Cordillera for the period between 5 and 3 myr and the development of the large sedimentary basin of Bogotá after 3.5 myr (Helmens, 1990; Andriessen et al., 1993). The pollen record of the last 3 myr may be interpreted in terms of climatic and biotic change.

A challenging similarity between the climate records registered in Andean Colombia and those in the deep sea is apparent. Graphical correlation of the Funza I arboreal pollen record with the oxygen isotope record of ODP Site 677, by Shackleton and Hooghiemstra, is shown in figure 18.3 (Hooghiemstra et al., 1993). The intervals that represent the oxygen isotope stages 3 through 25 show the best correlation between climatic oscillations. Correlation of the lower part of the pollen record, where the climatic oscillations are of a higher frequency, needs a higher temporal resolution, which is in preparation.

Development of Flora and Vegetation of Montane Forests and Paramo: Implications of the Closure of the Panamanian Isthmus

Around 4–3 myr the principal upheaval of the area had ceased. The high plain of Bogotá was by then an extensive lake at approximately 2,500 m altitude. Because of tectonic subsidence of the bottom of the lake, which process was more or less in equilibrium with sediment accumulation, the basin accumulated almost 600 m of lake sediments during the past 3 myr, providing a long and continuous palynological record of the vegetational and climatic changes. In the next section, a concise interpretation of the Funza I pollen record (recovered in 1976) and the Funza II pollen record (recovered in 1988; fig. 18.4) is given. The depth interval given refer to the Funza II core. The estimated ages are based on the fission track dates of the Funza II core, in combination with the tentative land-sea correlation of the upper 1.2 myr (fig. 18.3). A preliminary correlation (also based on ODP Site 677) is used for the lower part of the Funza II record; but because we presently have available only data at 1-m distances throughout the lower part of the core, these ages have to be regarded as provisional. The following reconstruction is largely based on Hooghiemstra and Cleef (submitted) and Hooghiemstra and Ran (1994a).

The Interval 540–465 m Core Depth (3.2–2.7 myr) shows warm climatic conditions. The basin had just started to accumulate lacustrine and river sediments, af-

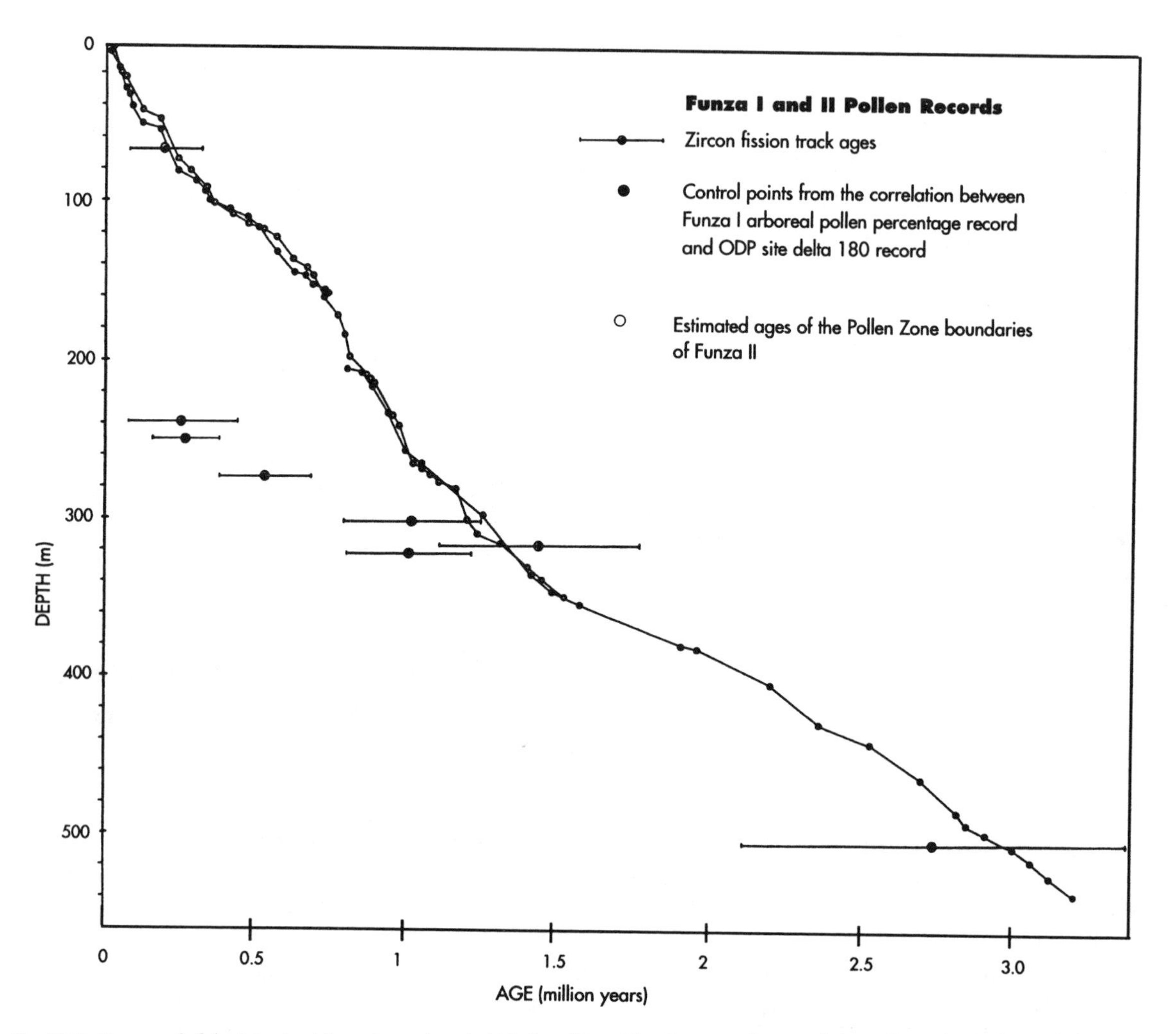

Fig. 18.2. Time control of the Funza I and Funza II cores from the high plain of Bogotá based on zircon fission track dates of intercalated volcanic ash horizons of the Funza II core (after Andriessen et al., 1993). Open circles indicate the control points of the graphical correlation between Funza I arboreal pollen percentage record and ODP Site 677 $\delta^{18}O$ record (fig. 18.3). Solid circles indicate the estimated ages of the boundaries of the Funza II pollen zones (see fig. 18.5). For the upper part of the Funza II core, age estimate are based on a provisional correlation with the SPECMAP time scale (Imbrie et al., 1984). For the lower part of the Funza II core, age estimations are based on a provisional visual correlation with ODP Site 677 $\delta^{18}O$ record (Shackleton et al., 1990). (Figure adapted from Hooghiemstra and Ran, 1994b.)

ter a period in which sediment only accumulated in the outer valleys. The upper limit of the subandean forest belt was situated at an elevation 500 m lower than it is today. In the Andean forest belt *Podocarpus*-rich forest, *Hedyosmum-Weinmannia* forest (a precursor of the modern *Weinmannietum*), and *Vallea-Miconia* forest were in this sequence the main constituents with increasing elevation. *Hypericum* and *Myrica* played an important part in the timberline dwarf forests, which possibly constituted a substantial transitional zone from the early Andean forest belt (upper-montane forest belt) to the open grassparamo belt. The contribution of herbs to the paramo vegetation, dominated by Gramineae and Compositae, seems less diverse than during the Upper Quaternary. The Late Pliocene Andean forests were more open than during the Middle and Upper Quaternary, because heliophytic elements, such as *Borreria*, were abundant. The composition of forests on the high plain was dynamic: at irregular intervals arboreal taxa with pioneer qualities (*Dodonaea*, *Eugenia*) and other taxa (*Symplocos*, *Ilex*) apparently constituted azonal forests in the basin. The upper forest line oscillated most of the time from 2,800 to 3,600 m in elevation. The average annual temperature on the high plain was ca. 11.5°–16.5°C.

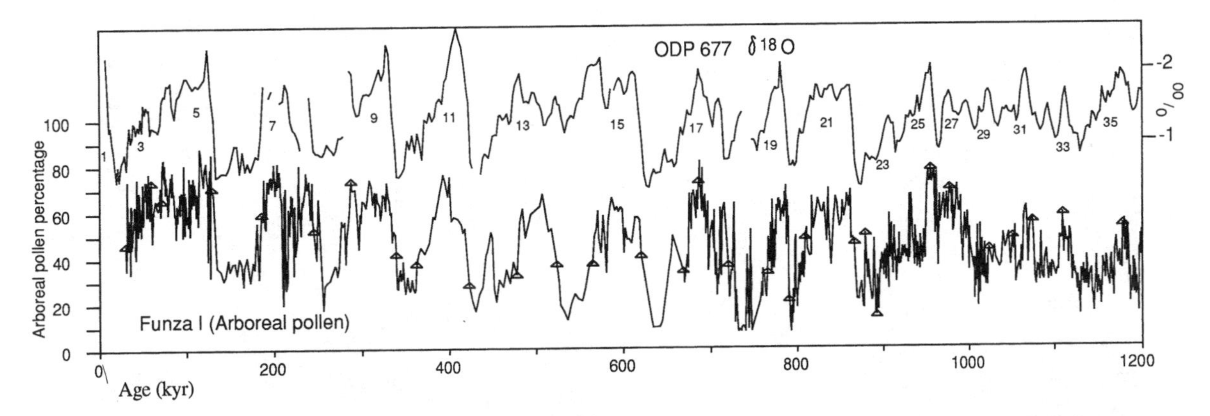

Fig. 18.3. Graphical correlation of the Funza I arboreal pollen record (pollen data based on Hooghiemstra, 1984, 1989; time control based on Andriessen et al., 1993) and $\delta^{18}O$ record ODP Site 677 (Shackleton et al., 1990). Using the absolute time control of the Funza sediments and the ODP Site 677 $\delta^{18}O$ record, the intervals representing stage 22 in both records and the core tops were correlated. Subsequently, the pollen record was minimally stretched and squeezed between 36 control points (indicated as triangles). The parallel records show the provisional correlation for $\delta^{18}O$ stages 3 through 25. (Figure adapted from Hooghiemstra and Ran, 1994b.)

The Interval 465–415 m Core Depth (2.7–2.2 myr) shows colder climatic conditions. The upper forest line oscillated most of the time from 2,600 to 2,800 m in the first half of this period, and from 2,400 to 2,800 m in the second half. Average annual temperatures were ca. 10.5°–11.5°C and ca. 8.8°–11.5°C, respectively. The *Podocarpus*-rich forest type occurred until the end of this period. *Weinmannia* was almost absent, and for the first time *Hedyosmum* completely dominated the *Hedyosmum-Weinmannia* forest type. *Miconia* dominated in the *Vallea-Miconia* forest type, in which *Ilex*, *Myrsine* (= *Rapanea*) and *Daphnopsis* were probably associated elements. For the first time in the Pleistocene, paramo vegetation became widespread in the Eastern Cordillera near Bogotá. Caryophyllaceae and *Valeriana* were the most dominant paramo herbs at that time, whereas the contribution of *Plantago* and *Aragoa* increased. During intervals with a low water level, marsh vegetation, including Cyperaceae, *Polygonum*, *Hydrocotyle*, *Ludwigia*, *Myriophyllum*, *Sphagnum* and *Azolla*, was abundant on the high plain.

The Interval 415–337 m Core Depth (2.2–1.42 myr) shows for the first time in the record a rather persistently cold climate. The upper forest line oscillated most of the time from 1,900 to 2,500 m, corresponding to an average annual temperature of 5.5°–9.5°C on the high plain. *Podocarpus* had lost its dominance in the Andean forest belt. *Hedyosmum-Weinmannia* forest and *Miconia-Vallea* forest were most important. *Daphnopsis* had almost disappeared from the Andean forest belt and *Borreria* became less common during this period, which would suggest that forests became denser in structure. *Hypericum* was for the first time in the record the most important element in the dwarf forest, but at the end of this period *Polylepis* dwarf forest started to increase near the upper forest line and in the lower paramo. *Juglans* appeared regularly for the first time, with low frequency, and *Styloceras* became a more regular component of the Andean forest belt. On a local scale, *Plantago* became very abundant in the basin and probably replaced a great part of the local grassparamo.

The Interval 337–257 m Core Depth (1.42–1.0 myr) shows a long period with mainly cold climatic conditions. The upper forest line oscillated most of the time from 2,200 to 2,600 m, slightly increasing at the end of this period to 2,400–2,800 m most of the time. This corresponds to an average annual temperature on the high plain of ca. 7.5°–10°C and ca. 9°–11.5°C, respectively. *Weinmannia* was almost absent in the Andean forest belt, and a *Hedyosmum* forest, possibly with important contribution of *Eugenia*, *Myrsine* (= *Rapanea*), and Ericaceae constituted a precursor of the present-day *Weinmannia* forest. For the first time in the record *Vallea* contributed in the same proportion as *Miconia* to the *Vallea-Miconia* forest type. *Borreria* occurred at low frequency and disappeared almost at the end of this period, indicating that the forest structure was more dense and unsuitable for heliophytic elements. *Polylepis* dwarf forest was important at the upper forest line and, in the lower paramo, *Myrica* (*M. pubescens*) probably contributed substantially to the upper part of the Andean forest

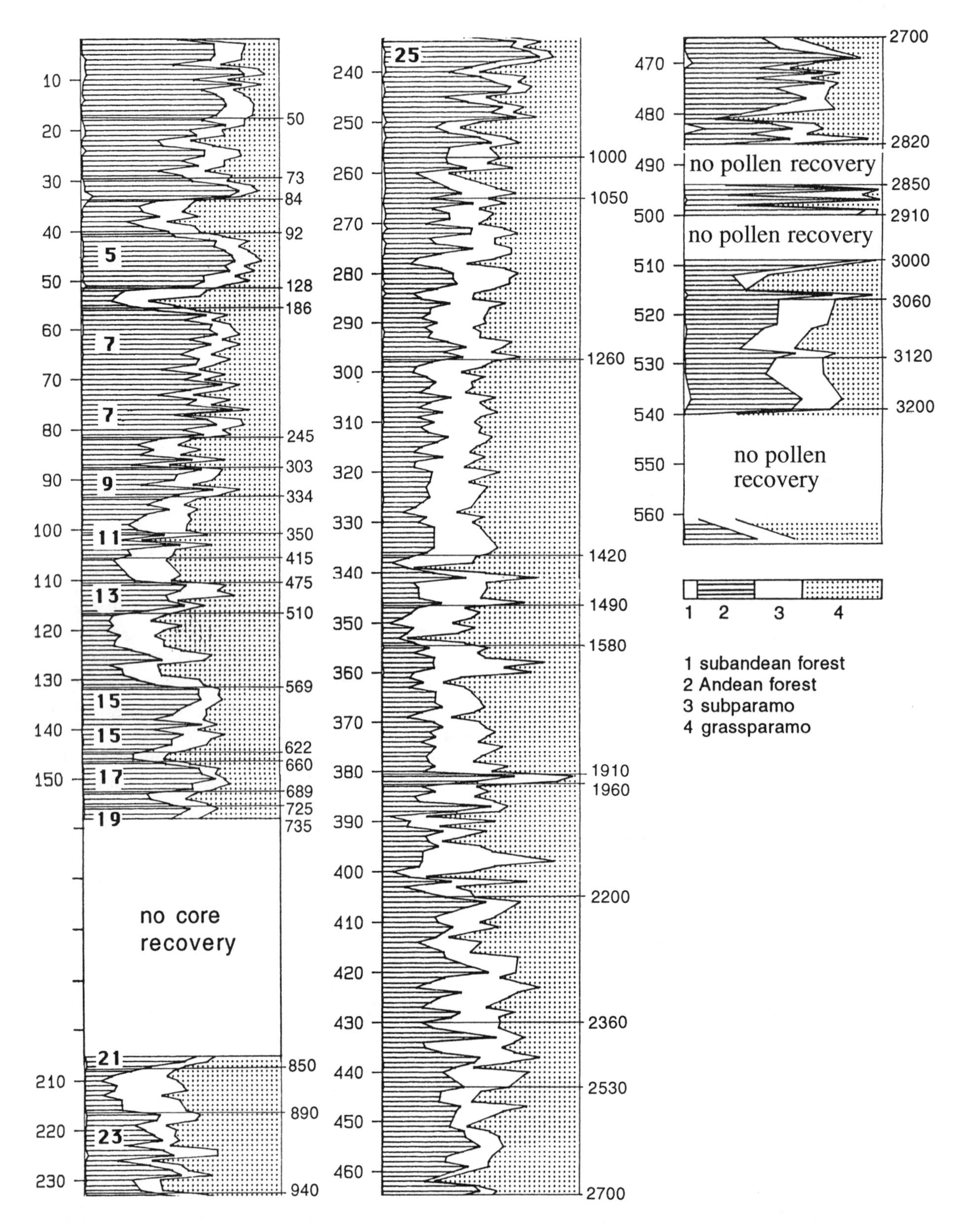

Fig. 18.4. Summary diagram of pollen record Funza II (565–2 m core interval) for the period of ca. 3.2 myr to ca. 27 kyr, with a time resolution of ca. 5–6 kyr (sample distance 100 cm). Estimated ages (kyr) of pollen zone boundaries are based on absolute time control and land-sea correlation (Andriessen et al., 1993). For convenience, $\delta^{18}O$ stage numbers 5 through 25 are indicated in the diagram based on a provisional correlation. Downcore variation in the representation of vegetation belts is shown for subandean forest (modern range 1,000–2,300 m), Andean forest (modern range 2,300–3,200/3,500 m), subparamo (modern range 3,200/3,500–3,400/3,600 m), and grassparamo (modern range 3,400/3,600–4,000/4,200 m). The graph of downcore changes of the percentage of total arboral pollen (calculations based on a pollen sum including *Alnus*) shows oscillations that, in fact, represent vertical shifts of the upper forest line as a response to mainly temperature change. Data based on Hooghiemstra and Cleef (submitted) and Hooghiemstra and Ran (1994a). (Figure adapted from Hooghiemstra and Ran, 1994b.)

belt. The upper limit of subandean forest reached higher elevations (modern conditions were reached after the immigration of *Quercus*, later in the record). The lake on the high plain was shallow, and extensive marsh prevailed during the first half of this period. Possibly owing to tectonic adjustments of the basin, the lake was deeper during the past 1.26 myr, and *Isoetes* vegetation and algae became abundant, both changes apparently occurring over a short time interval.

The Interval 257–205 m Core Depth (1.0–0.85 myr) shows the first major glacial-interglacial cycles, characteristic of the Middle and Upper Quaternary (Hooghiemstra et al., 1993). Comparison with the marine oxygen isotope record indicates that climatic change in the interval 240–205 m correlates with stages 25 to 21. The interglacial (stage 25) to glacial (stage 22) lowering of the upper forest line is ca. 800 m (from ca. 3,000 to ca. 1,800 m, maximally), corresponding to a temperature decrease of ca. 4.8°C. After the immigration of *Alnus*, a characteristic genus of the Northern Hemisphere, large areas of swamp forest developed on the wet flats around the lake. *Alnus* probably also occurred at lower frequency as an element of the zonal forests. The contribution of *Myrica* became considerably reduced, indicating that *Myrica* contributed earlier to the azonal vegetation (*M. parvifolia*) as well as to the zonal forests (*M. pubescens*). The large altitudinal shifts of all montane vegetation belts, in response to the main glacial-interglacial climatic cycles, in the remaining part of the record place the high plain alternately in the Andean forest belt and in the grassparamo belt.

The Interval 158–131 m Core Depth (Estimated Age 735–569 kyr) shows warm climatic conditions most of the time and is tentatively correlated with the oxygen isotope stages 19.1 to 15.1. The pollen spectra have no direct modern analogues because of the absence of *Quercus* and related conditions. The upper forest line oscillated from 2,100 to 2,700 m during most of that time. The corresponding average annual temperature on the high plain is 6.5°–11°C. The high plain was situated in the Andean forest belt most of that time. The upper limit of the subandean forest belt (*Acalypha, Alchornea*) was situated hundreds of meters below the modern elevation. *Podocarpus* was most important in the lower part of the Andean forest belt. *Weinmannia* forest, the precursor of the modern *Weinmannietum*, included a substantial contribution of *Hedyosmum* and, with lower frequency, *Myrsine* (= *Rapanea*) and *Eugenia*. A type of *Vallea-Miconia* forest, including low presence of *Ilex* and *Myrsine* (= *Rapanea*), could have occurred on the drier parts of the high plain. The lake was mostly shallow, with local marsh vegetation of cyperaceous reed swamp and *Hydrocotyle*. *Myrica* thickets (*M. parvifolia*) and *Alnus* carr covered the wet flats around the lake. *Myrica* (*M. pubescens*) and *Alnus* possibly contributed also, with low frequency, to the zonal Andean forest belt. Dwarf forest of *Polylepis*, *Myrica*, and Compositae scrub occurred at the upper forest line.

The Interval 131–100 m Core Depth (Estimated Age 569–350 kyr) shows cold climatic conditions most of that time and is tentatively correlated with oxygen isotope stages 14.4 to 11.1. The upper forest line oscillated mostly from 1,800 to 2,500 m. The corresponding average temperature on the high plain is ca. 5°–9.5°C. The high plain was situated in the grassparamo belt most of that time. Apart from Gramineae (e.g., *Calamagrostis*, *Chusquea*) and woody stem rosettes of *Espeletia* (Compositae), a variety of paramo herbs (*Valeriana*, Caryophyllaceae, *Geranium*, *Aragoa*, *Lycopodium* fov.) were present with substantial frequencies, as were abundant cushion bogs of *Plantago* (*P. rigida*). The water level in the lake was high, and marsh vegetation limited. *Polylepis* dwarf forest occurred in the subparamo belt, along with shrub of Compositae, *Hypericum* and Ericaceae. In the Andean forest belt, *Vallea-Miconia* forest and *Weinmannia-Hedyosmum* forest were most important.

The Interval 100–57 m Core Depth (Estimated Age 350–186 kyr) shows warm climatic conditions most of that time and is tentatively correlated with oxygen isotope stages 10.2 to 7.1. The upper forest line oscillated from 2,000 to 2,600 m in the first part and from 2,600 to 2,900 m during the last part of this interval. The corresponding average annual temperatures are 6°–10° and 10°–12°C, respectively. During the first part of this interval, the high plain was situated mainly in the paramo, and in the last part of this interval, in the Andean forest belt. During this interval, *Quercus* immigrated into the area of the high plain. *Quercus* forest occurred in a wide altitudinal range from 1,000 to 2,800 m in elevation and constituted at first local patches of forest. At the end of this interval, zonal *Quercus* forest was a major part of the Andean forest belt (Hooghiemstra and Ran, 1994a; Van't Veer et al., 1994). *Acalypha* and *Alchornea* reached higher elevations in the *Quercus* forests, and the upper limit of subandean forest rose to modern elevations. *Weinmannia* dominated in the *Weinmannia-Hedyosmum* forest type. At the end of this interval, the contribution of *Vallea-Miconia* forest increased markedly and replaced *Weinmannia* forest. *Podocarpus*-rich forest occurred in the lower part of the Andean forest belt. *Alnus* swamp forest and *Myrica* thickets were abundant around the lake, which was shallow. Algae (*Botryococcus*) became very abundant from the beginning of this interval to the top of the record.

The Interval 57–2 m Core Depth (Estimated Age 186–24 kyr shows, for the first time in the record, abundant presence of zonal *Quercus* forests. The composition of the upper part of the Andean forest belt in particular had changed dramatically by this time. Based on arboreal percentages, climatic conditions seem to have been warm most of that time. But it should be noted that the frequency of *Quercus,* a wind pollinator that produces large amounts of pollen, is probably unrealistically high in the core. This interval is tentatively correlated with oxygen isotope stages 6 to 3. The upper forest line oscillated mostly from 2,000 to 3,000 m. This corresponds to average annual temperatures of 6°–12.5°C. *Quercus* forests, resembling the modern *Saurauia-Quercus humboldtii* forest, and *Weinmannia-Hedyosmum* forest, resembling the modern *Weinmannietum,* dominated in the Andean forest belt. *Vallea-Miconia* forest probably resembled the modern *Xylosma-Duranta-Vallea* forest, but the latter is palynologically difficult to recognize (Cleef and Hooghiemstra, 1984). *Eugenia, Ilex,* and *Myrsine* (= *Rapanea*) contributed substantially to this rather dry forest type of low stature. *Polylepis* dwarf forest was frequent at the forest line and possibly also in the paramo belt up to 4,000 m. *Alnus* swamp forest dominated completely the flat parts of the high plain. *Myrica* thickets and marsh vegetation became reduced during the last part of this period. Sediment accumulation was very high (up to 60 cm per 1,000 years). Supposedly, erosion of the Tequendama Falls, in the Rio Bogotá, led to the final drainage of the lake between ca. 28 and 22 kyr.

Evolution of Biomes and Biostratigraphy

Graham (1992) evaluated for the Cenozoic the selective influence of altitudes and climatic change on the utilization by migrants of the Panamanian isthmus. Van der Hammen et al. (1973) established for the last 5 myr a biozonation for the Eastern Cordillera based on the stratigraphic distribution of important floral elements (see also Helmens, 1990). The two most recent biozones were based on the periodic immigration of the two characteristic Northern Hemispheric trees *Alnus* (alder) and *Quercus* (oak), which arrived in Colombia after passage of the Panamanian isthmus. The Andean biozones IV to VII are represented in the Funza records (fig. 18.5). Biozone IV (540–415 m core interval; estimated age 3.2–2.2 myr) is characterized mainly by high percentages of *Borreria*. *Alnus* and *Quercus* are absent. In biozone V (415–257 m core interval; estimated age 2.2–1.0 myr), *Polylepis* replaced *Hypericum* as a major element in the dwarf forest zone, *Weinmannia* replaced *Hedyosmum* as the most important element in the *Weinmannia-Hedyosmum* forests (precursor of the modern *Weinmannietum*), and the upper limit of the subandean forest belt had increased by several hundreds of meters. *Alnus* and *Quercus* are still absent. Biozone VI (257–94 m core interval; estimated age 1.0–0.34 myr) is characterized by the immigration of *Alnus* (first appearance date for the study area 1.0 myr). Biozone VII (100–0 m core interval; estimated age 0.34 myr to Recent) is characterized by the immigration of *Quercus.* The first appearance date for the study area is 0.34 myr; *Quercus* formed zonal forest in the upper part of the upper montane belt only during the last 0.2 myr. It seems that by this time subandean forest elements, such as *Alchornea* and *Acalypha,* could reach higher elevation in the oak forest, and the upper limit of the subandean forest belt reached modern elevations.

Figure 18.5 shows the evolution of the composition of montane forest. During biozone IV, the arboreal genera *Vallea, Miconia, Ilex, Eugenia,* and *Daphnopsis* were the major elements making up the upper montane forest. From the Middle Pleistocene up to modern time, these elements became of minor importance and the arboreal

Fig. 18.5. Summary pollen diagram of core Funza II, pollen zones, estimated ages (provisionally; after Hooghiemstra and Ran, 1994a, and Hooghiemstra and Cleef, submitted), and biozones (after Van der Hammen et al., 1973). Changes of the contribution of 21 selected taxa to the zonal (regional) and azonal (local) montane vegetation throughout the Late Pliocene–Pleistocene is indicated. (Figure adapted from Hooghiemstra and Cleef, 1995). Solid lines: presence with relatively high percentages in the vegetation (note the first appearance dates of *Alnus* and *Quercus*). Densely dotted lines: continuous presence with low percentages in the vegetation. Spaciously dotted lines: discontinuous presence with low percentages in the vegetation. →

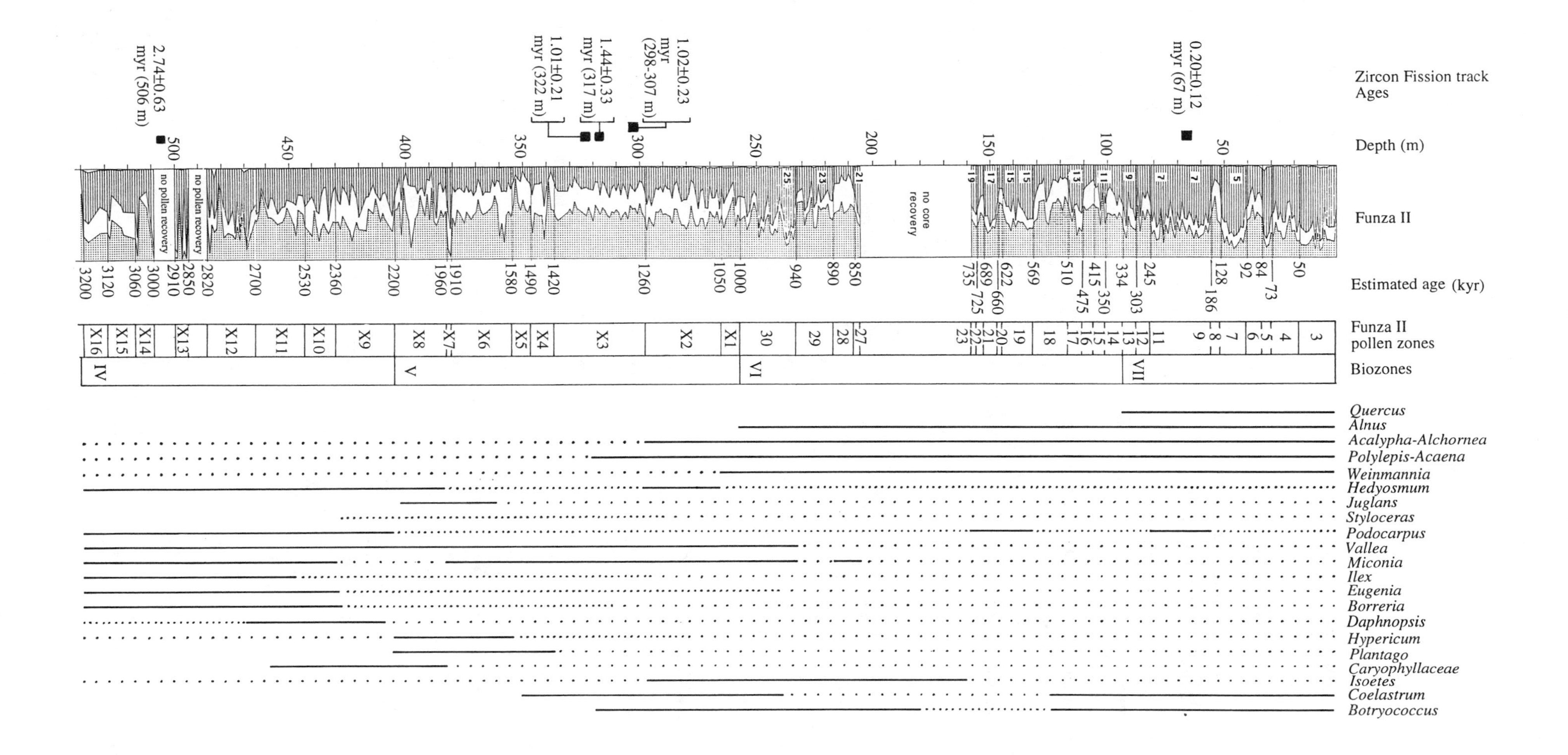

Zircon Fission track Ages
Depth (m)
Funza II
Estimated age (kyr)
Funza II pollen zones
Biozones
Quercus
Alnus
Acalypha-Alchornea
Polylepis-Acaena
Weinmannia
Hedyosmum
Juglans
Styloceras
Podocarpus
Vallea
Miconia
Ilex
Eugenia
Borreria
Daphnopsis
Hypericum
Plantago
Caryophyllaceae
Isoetes
Coelastrum
Botryococcus
0.20±0.12 myr (67 m)
1.02±0.23 myr (298-307 m)
1.44±0.33 myr (317 m)
1.01±0.21 myr (322 m)
2.74±0.63 myr (506 m)
no core recovery
no pollen recovery
VII
VI
V
IV

genera *Weinmannia* and *Polylepis,* in combination with the immigrant *Alnus* and, later on, *Quercus,* dominated the northern Andean forests.

Dynamics of Montane Vegetation and Forest-Savanna Transitions in Amazonian Lowlands

The greatest extension of the glaciers was registered (Van der Hammen et al., 1980/81) during the Middle Pleniglacial (ca. 45–25 kyr). The climate was humid and cold, mountain lakes had relatively high water levels, and the upper forest line was about 800–1,000 m lower than at present. The paramo belt was narrow and wet, with abundant *Polylepis* dwarf forest in the lower parts. These vegetational conditions are well documented in the pollen record of Lake Fuquene, at an elevation of 2,580 m in the Eastern Cordillera (Van Geel and Van der Hammen, 1973; fig. 18.6). The Fuquene diagram also shows a clear transition to a period of very cold and dry conditions during the Upper Pleniglacial (ca. 21–14 kyr). There was a moderate extension of the glaciers, and a broad and relatively dry paramo belt was present. The upper forest line was about 1,200 to 1,500 m lower than it is at present. Lower water levels prevailed in the mountain lakes. The oscillating width of the paramo zone was previously discussed by Van der Hammen (1981). The 60 kyr pollen record from Carajas in the northeastern part of the Brazilian tropical rain forest (longitude 50°W, latitude 6°S; Absy et al., 1991) shows a sequence of forested and open savanna phases, correlating with humid and dry periods documented in the Fuquene record. Records of both sites show a clear transition from humid to dry climatic conditions from ca. 30 to 20 kyr.

From the frequency analysis of the Funza I core, it is shown that in northern South America climatic change is markedly influenced by 23 kyr and 19 kyr periods belonging to the precession band of orbital forcing (Hooghiemstra et al. 1993). Berger (1989) discussed Pleistocene climatic variability at astronomical frequencies and Berger et al. (1993) showed that the effect of precession variations is indeed expected to be distinctly present at equatorial latitudes. Temporal variation in precipitation at equatorial latitudes, as documented in the Fuquene and Carajas pollen records, is most possibly related to changes in the latitudinal position of the intertropical convergence zone (ITCZ). The area influenced by the ITCZ, with its accompanying rain belt, oscillates through time and was minimally extended around 18 kyr and maximally extended during the Early Holocene, separated in time by half a precession cycle. This mechanism may explain dry conditions around 18 kyr in the Colombian Andes and in parts of the Amazon Basin, when both sites are outside the ITCZ-influenced zone. This mechanism also predicts humid conditions at half a precession cycle earlier and later than 18 kyr, namely, around 28 kyr and 6 kyr, when both sites are inside the ITCZ-influenced zone. The pollen records of Carajas and Fuquene indeed show humid phases around 28 kyr, characterized by a forest phase in Carajas and a wide subparamo belt in Fuquene. The Carajas record shows also a forest phase around 6 kyr, but the Fuquene pollen record is unable to register the width of the subparamo belt around 6 kyr because Fuquene became surrounded by forest at the beginning of the Holocene, ca. 10 kyr ago. Therefore, pollen records from an elevation of about 1,000 m higher are necessary to confirm the expected wide subparamo belt in the Colombian Andes around 6 kyr. Several radiocarbon-dated pollen records, such as La Primavera (3,525 m; Melief, 1985; see fig. 18.6, this vol.), TPN 36C (3,770 m; Salomons, 1986) and TPN 21B (3,950 m; Salomons, 1986; see fig. 18.6, this vol.) provide evidence of a wide subparamo belt around 6 kyr and fit the hypothesized relationship among observed changes in vegetation and precipitation, supposed atmospheric circulation, and the forcing mechanism of the astronomical precession cycle.

The ITCZ influence on precipitation is expected to fade out to higher latitudes. In this respect, it is interesting to note that the El Valle pollen record from lowland Panama (Bush and Colinvaux, 1990), at the northernmost rim of the zone with potential ITCZ influence, shows only little evidence for marked changes in climatic humidity between 30 and 20 kyr. The Costa Rican pollen record La Chonta, in the Cordillera de Talamanca (Hooghiemstra et al., 1992), shows no evidence for marked changes in climatic humidity. This fits our expectation in that La Chonta lies north of the oscillating zone with ITCZ influence. The ITCZ influence in the southern direction is more complicated, because two convergence zones have to be considered (C. Nobre [Brazil], oral communication): the latitudinally oscillating belt of the equatorial ITCZ and an Amazonian convergence zone. The latter has a stable geographical position over a glacial-interglacial cycle, owing to the concave shape of the Andean mountain range at Amazonian latitudes, which force traveling air masses into a fixed geographical trajectory. As a consequence, this model suggests an area

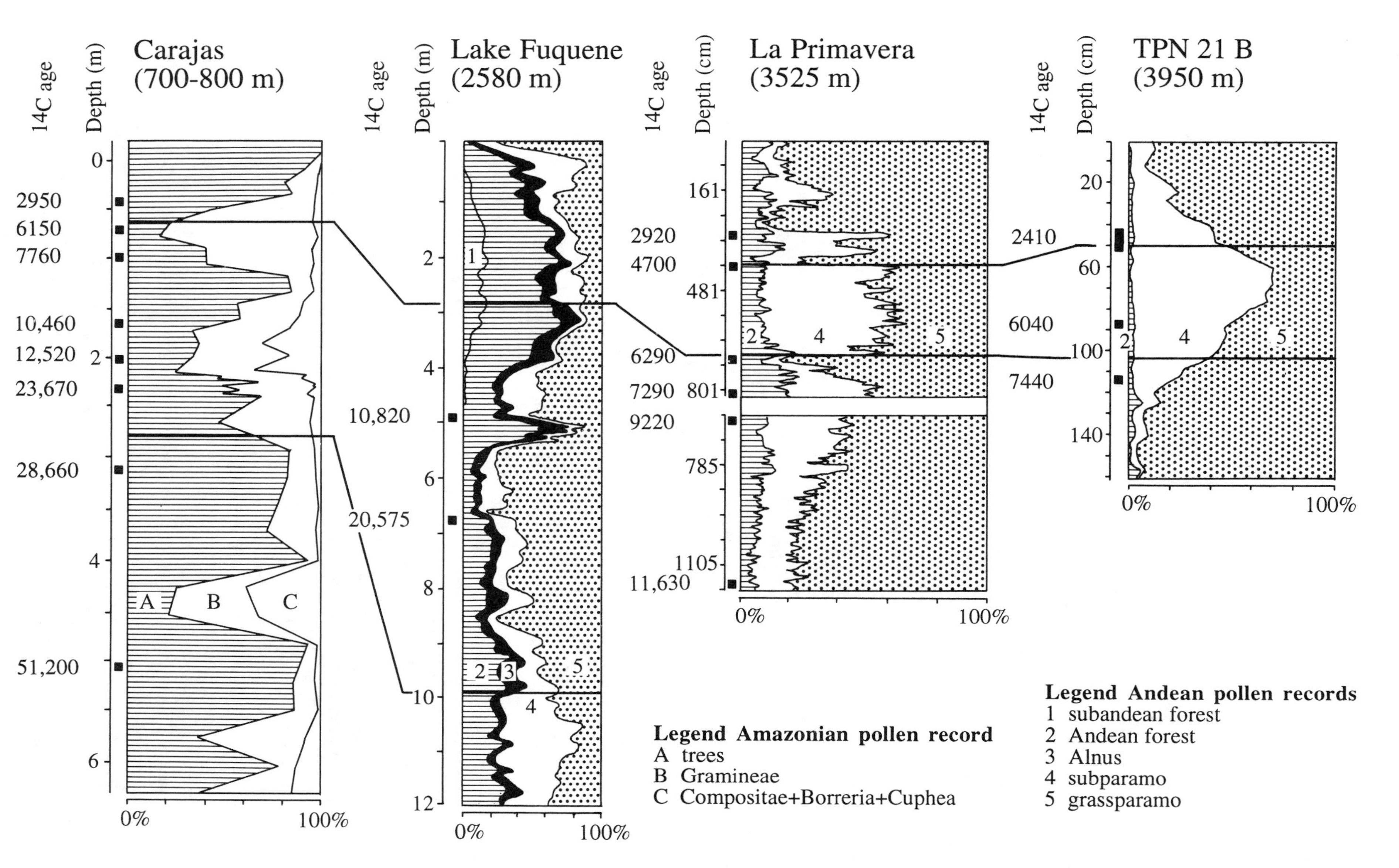

Fig. 18.6 Summary pollen diagrams of the lowland site at Carajas (after Absy et al., 1991) and high-elevation sites at Lake Fuquene (after Van Geel and Van der Hammen, 1973), La Primavera (after Melief, 1985), and TPN 21B (after Salomons, 1986). Vegetation indicative of high precipitation is present around 27 kyr (forest in Carajas and a wide subparamo belt in Fuquene). About 21 kyr later, at ca. 6 kyr, vegetation in Carajas (forest) and in the Andes (a wide subparamo belt in the pollen records La Primavera and TPN 21B) again indicates high precipitation. Vegetation indicative of low precipitation (a savanna phase in Carajas and an average width of the subparamo belt in the Andes) is present around 18 kyr.

in the west-central part of the Amazon Basin with a stable high precipitation regime over a glacial-interglacial cycle. This model gives support to the recently postulated hypothesis of Van der Hammen and Absy (1994). They explain the evidence of last glacial maximum (LGM) distribution of rain forest and savanna by accepting the present-day distribution pattern of precipitation also valid for 18 kyr. They do assume an overall reduction of 25 to 40 percent, compared to modern precipitation values, to explain the available LGM pollen evidence that shows the presence of rain forest in Rondonia (Absy and Van der Hammen, 1976) and Colombian Amazonas (see references in Van der Hammen and Absy, 1994) and the presence of savanna in Carajas (Absy et al., 1991).

Vegetation belts serve as the corridors for migrating paleoflora and paleofauna. It can be inferred that different kinds of organisms might have been presented with ecological opportunities for migratory passage (for instance, across the Isthmus of Panama) at particular times of full glacial cycles, depending on their habitat preferences.

Summary

An overview is given of the climatic history and evolution of Andean ecosystems of Colombia during the last 3 myr. The reconstruction is based on the long pollen records from the high plain of Bogotá, which have a high temporal resolution, and on a time control based on fission track ages of volcanic ashes and land-sea correlation of climatic change. Glacial-interglacial climatic change caused altitudinal shifts of the vegetation belts: the upper forest line shifted from ca. 1,800 m during glacial conditions to 3,500 m during interglacial conditions, reflecting a temperature change of ca. 8°C at an altitude of 2,550 m. At the Pliocene-Pleistocene transition, from ca. 2.7 to 2.3 myr, an estimated temperature decrease of ca. 4°C is inferred. Around 0.8–0.9 myr, high-frequency climatic oscillations, with 23 kyr and 19 kyr periods of the precession band, became overruled by ca. 100 kyr periods of the eccentricity band, leading to the rhythm of the classical Middle and Late Pleistocene ice ages. Periods of the precession band relate to changes in precipitation and seem to cause the Andean subparamo vegetation belt to oscillate in width and a selected part of Amazonian lowland vegetation to alternate between forest phases and savanna phases. Dynamics in the altitudinal (Andes) and latitudinal (Amazon Basin) range of vegetation and climatic zones, constituting the corridors for migrating flora and fauna, must have had important bearing on the modern biogeography and biodiversity of northern South American biomes. Emergence of the Panamanian isthmus and Pleistocene climatic change led to important changes in the composition of the northern Andean forest belts and paramo vegetation, which are documented in the Colombian pollen record.

Acknowledgments

I thank the Netherlands Foundation for Scientific Research (NWO, The Hague, grant H75-284), INGEOMINAS (Bogotá), and COLCIENCIAS (Bogotá) for supporting the Funza II coring program. I am indebted to A. M. Cleef (Amsterdam), J. L. Melice (Louvain-la-Neuvre/Rio de Janeiro), E. T. H. Ran (Amsterdam), N. J. Shackleton (Cambridge), T. Van der Hammen (Amsterdam/Bogotá) for fruitful cooperation. I also wish to thank N. J. Shackelton for preparing an improved version of the land-sea correlation graph and E. S. Vrba for the opportunity to participate in the conference "Paleoclimate and Evolution, with Emphasis on Human Origins."

References

Absy, M. L., Cleef, A. M., Fournier, M., Martin, L., Servant, M., Sifeddine, A., Ferreira da Silva, M., Soubies, F., Suguio, K., Turcq, B., and Van der Hammen, T. 1991. Mise en évidence de quatre phases d'ouverture de la forêt dense dans le sud-est de l'Amazonie au cours des 60,000 dernières années: Première comparaison avec d'autres régions tropicales (abridged English version: Occurrence of four episodes of rain-forest regression in southeastern Amazonia during the last 60,000 yrs.: First comparison with other tropical regions). *Comptes Rendus de l'Académie des Sciences* (Paris), ser. 2, 312: 673–678.

Absy, M. L., and Van der Hammen, T. 1976. Some paleoecological data from Rondonia, southern part of the Amazon Basin. *Acta Amazonica* 6:293–299.

Andriessen, P. A. M., Helmens, K. F., Hooghiemstra, H., Riezebos, P. A., and Van der Hammen, T. 1993. Absolute chronology of the Pliocene-Quaternary sediment sequence of the Bogotá area, Colombia. *Quaternary Science Reviews* 12:483–501.

Berger, A. 1989. Pleistocene climatic variability at astronomical frequencies. *Quaternary Science Reviews* 2:1–14.

Berger, A., Loutre, M. F., and Tricot, C. 1993. Insolation and earth's orbital periods. *Journal of Geophysical Research* 96:10341–10362.

Bush, M. B., and Colinvaux, P. A. 1990. A pollen record of a complete glacial cycle from lowland Panama. *Journal of Vegetation Science* 1:105–118.

Cleef, A. M., and Hooghiemstra, H. 1984. Present vegetation of the area of the high plain of Bogotá. In H. Hooghiemstra, *Vegetational and climatic history of the high plain of Bogotá, Colombia: A continuous record of the last 3.5 million years.* Dissertaciones Botanicae, no. 79. J. Cramer, Vaduz. Pp. 42–66.

Graham, A. 1992. Utilization of the isthmian land bridge during the Cenozoic-paleobotanical evidence for timing, and the selective influence of altitudes and climate. *Review of Palaeobotany and Palynology* 72:119–128.

Helmens, K. F. 1990. Neogene-Quaternary geology of the high plain of Bogotá, Eastern Cordillera, Colombia (stratigraphy, paleoenvironments and landscape evolution). Dissertaciones Botanicae, 163. Berlin: J. Cramer.

Hooghiemstra, H. 1984. *Vegetational and climatic history of the high plain of Bogotá, Colombia: A continuous record of the last 3.5 million years.* Dissertaciones Botanicae, no. 79. J. Cramer, Vaduz.

———. 1989. Quaternary and Upper-Pliocene glaciations and forest development in the tropical Andes: Evidence from a long high-resolution pollen record from the sedimentary basin of Bogotá, Colombia. *Palaeogeography Palaeoclimatology Palaeoecology* 72:11–26.

Hooghiemstra, H., and Cleef, A. M. 1995. Pleistocene climatic change, environmental and generic dynamics in the north Andean montane forest and paramo. In S. P. Churchill et al., eds., *Biodiversity and conservation of neotropical montane forests.* The New York Botanical Garden, New York.

———. (submitted). Lower Pleistocene and Upper Pliocene climatic change and forest development in the Eastern Cordillera of Colombia: Pollen record Funza II (205–540 m core interval). *Palaeogeography Palaeoclimatology Palaeocology.*

Hooghiemstra, H., Cleef, A. M., Noldus, G. W., and Kappelle, M. 1992. Upper Quaternary vegetation dynamics and paleoclimatology of the La Chonta bog area (Cordillera de Talamanca, Costa Rica). *Journal of Quaternary Science* 7:205–225.

Hooghiemstra, H., Melice, J. L., Berger, A., and Shackleton, N. J. 1993. Frequency spectra and paleoclimatic variability of the high-resolution 30–1450 kyr Funza I pollen record (Eastern Cordillera, Colombia). *Quaternary Science Reviews* 12:141–156.

Hooghiemstra, H., and Ran, E. T. H. 1994a. Upper and Middle Pleistocene climatic change and forest development in the Eastern Cordillera of Colombia: Pollen record Funza II (2–158 m core interval). *Palaeogeography Palaeoclimatology Palaeoecology* 109:211–246.

———. 1994b. Late Pliocene–Pleistocene high resolution pollen sequence of Colombia: An overview of climatic change. *Quaternary International* 21:63–80.

Hooghiemstra, H., and Sarmiento, G. 1991. New long continental pollen record from a tropical intermontane basin: Late Pliocene and Pleistocene history from a 540 m-core. *Episodes* 14:107–115.

Imbrie, J. Hays, J. D., Martinson, D. G., McIntyre, A., Mix, A. C., Morley, J. J., Pisias, N. G., Prell, W. L., and Shackleton, N. J. 1984. The orbital theory of Pleistocene climate: Support from a revised chronology of the marine $\delta^{18}O$ record. In: A. Berger et al., ed., *Milankovitch and Climate,* pt. 1. Reidel, Dordrecht. Pp. 269–305.

Melief, A. B. M. 1985. Late Quaternary paleoecology of the Parque Nacional Natural los Nevados (Cordillera Central), and Sumapaz (Cordillera Oriental) areas, Colombia. Ph.D. diss., University of Amsterdam.

Salomons, J. B. 1986. *Paleoecology of volcanic soils in the Colombian Central Cordillera* (Parque Nacional Natural de los Nevados). Disertaciones Botanicae, no. 95. J. Cramer (Borntraeger), Berlin-Stuttgart.

Van der Hammen, T. 1974. The Pleistocene changes of vegetation and climate in tropical South America. *Journal of Biogeography,* 1:3–26.

———. 1981. Glaciales y glaciaciones en el Cuaternario de Colombia: Paleoecologia y estratigrafia. *Revista CIAF* (Bogotá) 6:635–638.

Van der Hammen, T., and Absy, M. L. 1994. Amazonia during the last glacial. *Palaeogeography Palaeoclimatology Palaeoecology,* 109:247–261.

Van der Hammen, T., Barelds, J., De Jong, H., and De Veer, A. A. 1980/1981. Glacial sequence and environmental history in the Sierra Nevada del Cocuy (Colombia). *Palaeogeography Palaeoclimatology Palaeoecology* 32:247–340.

Van der Hammen, T., Werner, J. H., and Van Dommelen, H. 1973. Palynological record of the upheaval of the northern Andes: A study of the Pliocene and Lower Quaternary of the Colombian Eastern Cordillera and the early evolution of its high-Andean biota. *Review of Palaeobotany and Palynology* 16:1–122.

Van Geel, B., and Van der Hammen, T. 1973. Upper Quaternary vegetational and climatic sequence of the Fuquene area (Eastern Cordillera, Colombia). *Review of Palaeobotany and Palynology* 14:9–92.

Van't Veer, R., Ran, E. T. H., Mommersteeg, H. J. P. M., and Hooghiemstra, H. 1994. Multivariate analysis of the Middle and Late Pleistocene Funza pollen records of Colombia. *Mededelingen Rijks Geologische Dienst* 52:1–19.

Chapter 19

Plio-Pleistocene Climatic Variability in Subtropical Africa and the Paleoenvironment of Hominid Evolution: A Combined Data-Model Approach

Peter B. deMenocal and Jan Bloemendal

Plio-Pleistocene climate is characterized by dramatic shifts in high-latitude climate near 2.8 and 0.9 million years (myr). Polar ice sheets in both hemispheres grew sufficiently large to develop calving margins and deposit ice-rafted debris in polar oceans just prior to the Matuyama-Gauss magnetic polarity reversal near 2.7 myr (e.g., Shackleton et al., 1984; Jansen et al., 1988; Rea and Schrader, 1985). Marine oxygen isotopic records of continental ice-volume variability demonstrate that ice growth was gradual, beginning near 3.0 myr and culminating near 2.6 myr. The primary mechanism responsible for the onset of bipolar glaciation remains elusive, and several hypotheses have been advanced (Keigwin, 1979; Ruddiman et al., 1989a; Raymo et al., 1988). Following this bipolar ice buildup, glacial-interglacial cycles of moderate amplitude were sustained at the orbital periodicity of 41 thousand years (kyr) until 0.9 myr. Other indexes of high-latitude climatic variability, including North Atlantic sea-surface temperature (SST), ice rafting, and Atlantic deep circulation (Ruddiman et al., 1989b; Raymo et al., 1989), are highly covariant with the isotopic record over this interval, indicating coherent high-latitude climatic responses to orbital insolation forcing.

After 0.9 myr, isotopic and other high-latitude paleoclimatic records demonstrate that glacial maxima became more extreme and that the dominant periodicity of variation shifted from 41 kyr to 100 kyr. The net result was a dramatic increase in glacial-interglacial amplitude and the development of strongly nonlinear ice-sheet dynamics. The nonlinear component comprises fully 50 percent of the total ice volume variance over the last 600 kyr (Imbrie et al., 1993). The strong 100 kyr variance is difficult to explain in terms of a linear orbital response, because there is almost no insolation forcing at this periodicity. A recent analysis of the "100 kyr problem" suggests that the strong 100 kyr response may be attributed to massive Northern Hemispheric ice sheets, which act as nonlinear amplifiers of otherwise linear climatic responses to orbital variations at periodicities of 41 kyr and 23–19 kyr (Imbrie et al., 1993).

In contrast to high latitudes, where glacial-interglacial climatic change has been reasonably well defined for the last several million years, a similar understanding of low-latitude climatic sensitivity during this same interval has not been forthcoming. Whereas temperature is the dominant climatic variable at high latitudes, low-latitude (subtropical) climatic variability is best expressed in terms of variations in precipitation. A number of investigators have examined the latest Pleistocene (ca. the last 300 kyr) succession of arid-humid cycles in Africa and its environs; these studies have provided a basis for understanding African climatic responses to orbital insolation changes and the effects of greatly expanded high-latitude ice sheets. Available Late Pleistocene low-latitude paleoclimatic studies can be grouped generally into at least two classes of paleoclimatic response. In the first, subtropical climate is effectively dependent upon high-latitude glacial-interglacial change (e.g., Hays and Perruzza, 1972; Kolla et al., 1979; Sarnthein et al., 1981; Clemens et al., 1990; Gasse et al., 1990; Street-Perrott and Perrott, 1990; Lezine, 1991; Anderson and Prell,

ISBN 0-300-06348-2.

1993; deMenocal et al., 1993); in the second, subtropical climate responds primarily to direct low-latitude orbital insolation variations (e.g., Rossignol-Strick, 1983; Street-Perrot and Harrison, 1984; Pokras and Mix, 1985; Prell and Kutzbach, 1987; Molfino and McIntyre, 1990; Clemens et al., 1992; deMenocal et al., 1993). There is strong evidence that both of these mechanisms have affected West African climate over the last 0.9 myr, although the high-latitude mechanism apparently exerted a dominant influence over African aridity and vegetation (deMenocal et al., 1993).

A number of long (covering several myr) but relatively low-resolution records of African aeolian dust transport have been reconstructed from deep-sea sediment records off West Africa (e.g., Sarnthein et al., 1982; Stein, 1985; Stein and Sarnthein, 1986; Teidemann et al., 1989). These studies are derived primarily from drill sites off northwestern Africa (15°–20°N; downwind of Saharan dust sources) and sites further south (0°–5°N; downwind of Sahelian dust sources). Many of these lower-resolution records suffer from aliasing effects, which commonly mask or misrepresent higher-frequency climatic oscillations present in the geologic record (e.g., Pisias and Mix, 1988). This problem is further compounded when flux records are calculated from data with sparse age control. Nonetheless, these records do provide valuable constraints on the long-term evolution of African climate. An overview of the drill sites off northwest Africa suggests that the Quaternary pattern of increased Saharan aridity and stronger trade winds began near 3.4 myr and further intensified near 2.6 myr and again near 0.9 myr (e.g., Sarnthein et al., 1982; Stein, 1985). Although the long-term evolution of West African aridity has responded to Northern Hemispheric glaciation, dust supply from the southern Saharan region has been significant since at least the Late Miocene (Tiedemann et al., 1989; Ruddiman et al., 1989c; Tiedemann et al., 1994). Increased aeolian dust concentrations were also observed at these same intervals at several sites located further south within the Sahelian dust plume (Ruddiman and Janecek, 1989). Tiedemann et al. (1989, 1994) presented additional dust-flux data from this region (Sites 658 and 659), which largely confirmed these results, and identified other discrete arid episodes between 4.5 and 3.0 myr. Although broad similarities between the different records are evident, there are also significant differences, which may be real or, in part, artifacts of chronology, sample aliasing, and differing environments of deposition (e.g., coastal versus pelagic; see Ruddiman et al., 1989c; Tiedemann et al., 1989).

A more robust perspective of African climatic evolution during the Late Neogene is possible through analysis of very detailed (a resolution of 1–3 kyr) and continuous paleoclimatic records that extend from the Late Miocene. Long and continuous records are needed to examine African climatic variability both before and after the onset of Northern Hemispheric glaciation near 2.8 myr and the subsequent intensification of glacial-interglacial climatic cycles. Very detailed records are needed to examine orbital-band (10^3–10^5 kyr) variability in the record as a function of time. Much of our understanding of high-latitude climatic change during the Late Neogene has resulted from frequency-domain analyses of $\delta^{18}O$, SST, and ice-rafting records. Similar analyses of low-latitude records can be used to identify the dominant climatic mechanisms affecting African climate over this same interval. Unfortunately, very few records of sufficient length and detail have been available to address the problem fully.

In this chapter we present several long (4–7 myr) and very high-resolution (1–3 kyr sample interval) marine records of aeolian dust contributions from West Africa and East Africa / Arabia. Mineral dust and other aeolian indexes are used as proxy indicators of the climate of subtropical African dust-source areas. Climatic linkages between high- and low-latitudes are examined through analysis of orbital-band climatic variations through time. Numerical climate models are used to investigate the sensitivity of African climate to changes in high-latitude ice-sheet size and to North Atlantic sea-surface temperature distributions. These experiments identify the dynamical mechanisms through which African climate responds to elements of high-latitude climate variability. Finally, the data and model results are combined to develop a conceptual model of Plio-Pleistocene climatic change in subtropical Africa, with particular emphasis on the paleoenvironment of hominid evolution.

Regional Climatology

In contrast to high latitudes, where temperature is the dominant climatic variable, low latitudes exhibit extreme variations in precipitation associated with seasonal migrations of monsoonal systems. Monsoonal circulation results from the differing heat capacities of land and water: sensible heating warms land surfaces much more rapidly than it does the ocean mixed layer. The Asian winter monsoon results because the south Asian landmass cools relative to the Indian Ocean and a broad high-pressure cell develops over the Tibetan Plateau. Dry and variable northeast trade winds develop over the south Asian region from October to April (fig. 19.1a). Sensible heating during summer in the Northern Hemisphere initiates a strong low-pressure cell over the Tibetan Plateau, which enables regional cyclonic circulation to prevail over south Asia from May to September (Hastenrath and Lamb, 1979). Strong moisture-laden southwest winds parallel the Arabian and East African coasts, bringing the monsoon rains to southern Asia (fig. 19.1b). Cool, nutrient-rich waters upwell off Arabia and Oman owing to the eastward displacement of Arabian Sea surface waters by Ekman transport.

Atmospheric haze measurements, aerosol samples, satellite images, and sediment-trap studies have demonstrated that aeolian dust is entrained from Arabian and northeast African sources during the peak months of the summer monsoon: June, July, and August (Pye, 1987, p. 76; Goldberg and Griffin, 1970; Nair et al., 1989; Sirocko and Sarnthein, 1989; Kolla et al., 1976, 1981). Precipitation in northwest Arabia falls mainly in winter and early spring (following the Mediterranean pattern). Arabian dust generation results from early summer drying of immature soils, which are subsequently deflated by strong northwest (Shamal) summer winds (Petrov, 1976). Sediment-trap data from the Arabian Sea indicate that 80 percent of the annual terrigenous flux to the western Arabian Sea occurs during the summer months (Nair et al., 1989).

Over West Africa, summer heating over central North

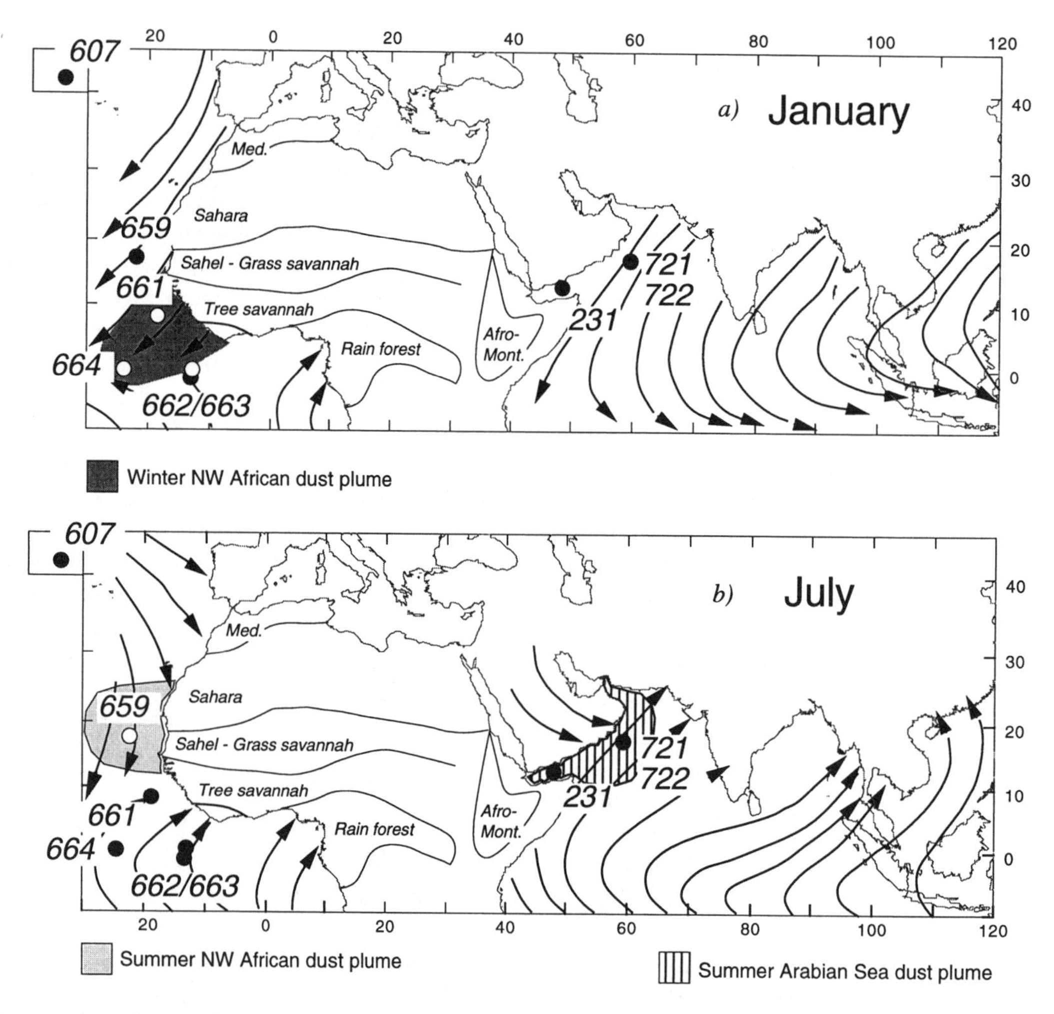

Fig. 19.1. West and East African DSDP/ODP site locations showing a) winter and b) summer surface-wind and aeolian dust trajectories.

Africa during boreal summer drives the inflow of moisture-laden air from the adjacent eastern equatorial Atlantic. Strong Southern Hemispheric trade winds cross the equator and penetrate into central Africa. The summer monsoon winds deliver sporadic but intense precipitation deep into central subtropical Africa. Atmospheric circulation reverses in boreal winter and northeast trade winds blow over Africa and the adjacent subtropical Atlantic. Boreal winter is the dry season in the Sahel and sub-Saharan regions.

The Sahel and Sahara regions are prolific sources of atmospheric dust. There are two dust plumes associated with the seasonality of precipitation. The summer plume is centered at 10°–25°N, and its dust is derived mainly from Saharan sources. The winter dust plume occurs between 10°N and 5°S and is associated with the winter northeast trade-wind trajectory that carries dust from dry Sahelian and sub-Saharan sources. Interannual variations in African dust export have been tied to occurrences of drought conditions in the Sahel and sub-Sahara regions (Prospero and Nees, 1977).

East African climate is considerably more complex owing to its great topographic variability and local microclimates. The main rains in the regions of East Africa and the Horn of Africa occur semiannually. March through May is the main rainy season (known as the "long rains"), whereas the more variable "short rains" occur from October through November. Runoff from the Ethiopian and Kenyan highlands feeds the numerous large lakes of the East African Rift Valley. Much of low-lying northeast Africa (below 1,000 m) remains semiarid owing to the combined effects of (1) the high East African topography, which blocks eastward penetration of Atlantic moisture; (2) the presence of cool, upwelled water near the Somali coast; and (3) the frictionally induced subsidence of the Somali jet, which parallels the East African coast (Flohn, 1965).

Site Locations, Methods, and Time Scale

The sites off West Africa (Ocean Drilling Program Sites 661, 662, 663) were drilled in an area presently within the winter season dust plume (fig. 19.1a). Regional terrigenous flux distributions (Ruddiman and Janecek, 1989) indicate that aeolian dust is the main supply of terrigenous material. At Sites 662 and 663 (see table 19.1 for site locations and water depths), terrigenous percentages were calculated as the residual fraction remaining after calcium carbonate analyses (by coulometry) and biogenic opal analyses (wet reduction-spectrophotomery technique; Mortlock and Froelich, 1989; Ruddiman and Janecek, 1989). An oxygen isotopic stratigraphy was used to establish the Site 663 aeolian time series to 0.9 myr (2.7 kyr sample resolution) (deMenocal et al., 1993). Abundances of opal phytoliths, wind-borne grass cuticles from Sahelian grasslands, were determined using a quantitative settling technique (Pokras and Mix, 1985; deMenocal et al., 1993), followed by optical counting of prepared microscope slides. Phytolith abundance in marine sediments reflects the relative proximity of the Sahelian grassland boundary to the site (fig. 19.1) and, perhaps, the strength of the transporting trade winds.

Table 19.1. Drill Site Locations

Site	Latitude	Longitude	Water depth (m)	Location
607	41° 00.0 N	32° 58.0 W	3,427	North Atlantic
661	9° 26.8 N	19° 23.2 W	4,012	West Africa
662	1° 23.4 S	11° 44.4 W	3,824	West Africa
663	1° 11.9 S	11° 52.7 W	3,708	West Africa
721	16° 40.6 N	59° 51.9 E	1,945	Arabian Sea
722	16° 37.3 N	59° 47.8 E	2,028	Arabian Sea
231	11° 53.4 N	48° 14.7 E	2,161	Gulf of Aden

A terrigenous percentage record extending to 4.5 myr, with 1.5 kyr resolution, was developed at Site 661 (table 19.1) using magnetic susceptibility as an aeolian proxy. Magnetic susceptibility is a measure of the concentration of magnetic grains in a sample; it can be measured continuously with a pass-through loop sensor on whole, unsplit core sections. Because aeolian dust carries with it a trace amount of magnetic particles, magnetic susceptibility can be used as a rapid, nonintrusive, and continuous measure of aeolian concentrations, provided that susceptibility is strongly correlated to terrigenous percentage. The measurements were also used to develop a complete, composite section at Site 661. Forty-eight random samples from Site 661 were subjected to traditional aeolian extraction techniques (isolation of the mineral fraction by sequential removal of carbonate, biogenic opal, and organic carbon, following a technique developed by Clemens et al., 1989). A strongly linear correlation between magnetic susceptibility and terrigenous percentage ($r = 0.96$; fig. 19.2a) demonstrates that susceptibility is an excellent aeolian proxy at this site (Bloemendal and deMenocal, 1989). The resulting regression equation was used to transform the susceptibility record into a continuous aeolian time series.

An identical technique was used to develop an ae-

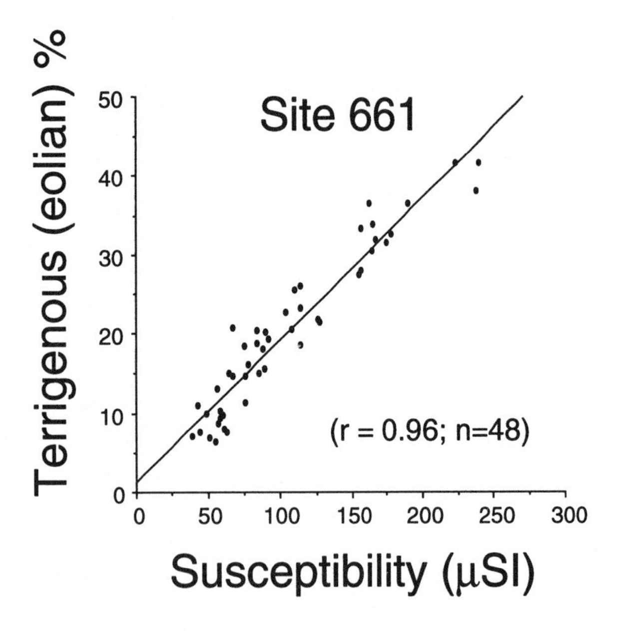

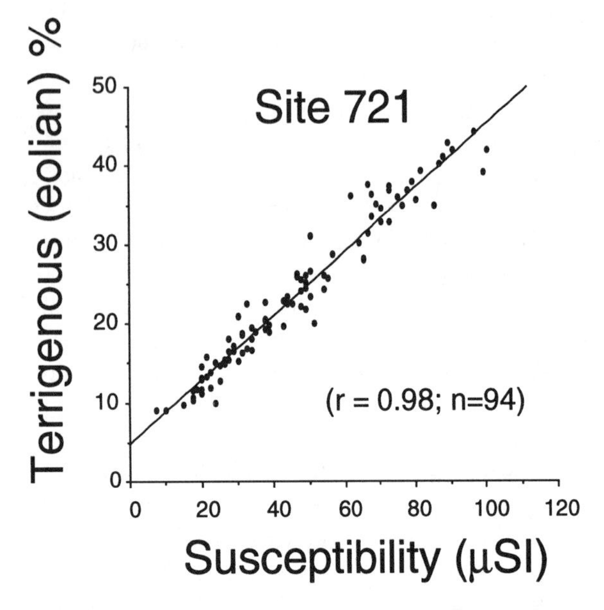

Fig. 19.2 Correlations between magnetic susceptibility and terrigenous percentage as measured by a sequential extraction technique: above, West African ODP Site 661; below, East African/Arabian Sea ODP Site 721.

olian time series from ca. 7.3 myr, with 1.5 kyr resolution, at Sites 721 and 722 on the Owen Ridge in the Arabian Sea (Bloemendal and deMenocal, 1989; deMenocal et al., 1991). The Owen Ridge is a southwest-northeast trending ridge that parallels the Arabian coast and rises to about 2,000 m. Susceptibility data were used to construct a composite section from the 5 holes drilled at Sites 721 and 722. The correlation between terrigenous percentage and susceptibility at this site was high (r = 0.98; fig. 19.2b; deMenocal et al., 1991), confirming that susceptibility is an excellent aeolian proxy at this site as well. The resulting regression equation was used to calculate a continuous aeolian time series back to 7.3 myr, with 1.5 kyr resolution. Reduction diagenesis (loss of fine-grained magnetite) was observed in the uppermost 7 m of the Site 721/722 record (see Bloemendal et al., 1993). Therefore core-based analytical terrigenous percentage data were patched in using data presented by Clemens et al. (1991). Oxygen isotopic data at Site 722 (Clemens et al., 1991) were used to establish the aeolian time series to 1.1 myr; magnetic polarity and biostratigraphic data constrained the time series to 7.3 myr.

An aeolian record is also developed for Deep Sea Drilling Project Site 231; this site is located south of the Sheba Ridge in the Gulf of Aden, approximately 70 km north of Somalia (fig. 19.1). These sediments contain several ash layers that have been correlated to tephra horizons at East African hominid localities in Ethiopia and Kenya (Sarna-Wojcicki et al., 1985; Brown et al., 1992). Proximity to the African continent, and the ash correlations to East African localities, make this site ideal for monitoring East African paleoclimate and for establishing correlations between the terrestrial fossil record and the marine paleoclimatic records.

All time-series records employ the time scale of Shackleton et al. (in press, a), who used an orbital tuning strategy at several sites from the central equatorial Pacific to develop a more precise (and, we hope, more accurate) chronology for the excellent paleomagnetic and biostratigraphic stratigraphies developed at these sites. This time scale is identical to the Cande and Kent (1992) magnetic polarity chronology to the base of the Matuyama chron (2.6 myr); reversal ages before 2.6 myr are generally older by 2–4 percent. Shackleton et al. (in press, a, b) discuss more fully the tuning strategy and the resulting paleomagnetic and biostratigraphic datum levels.

The Blackman-Tukey method of spectral analysis is used to identify dominant periods of variation within the various records (Blackman and Tukey, 1958; using linear detrend, no prewhitening, one-third lag for autocovariance, Hamming window smoothing). Unless otherwise noted, variance spectra are shown as scaled (normalized) variance.

Data and Results

West African Sites

Site 663. The aeolian and phytolith time series of Site 663 exhibit maximum values during glacial maxima over the last 0.9 myr (fig. 19.3; data from deMenocal et al., 1993). Covariance between the two indexes demonstrates that increased aeolian dust generation (aridity) and a more equatorward position of the Sahelian grassland boundary (fig. 19.1) characterize West African climate during glacial maxima. An aeolian flux record was calculated from the aeolian percentage, isotopic time-scale sedimentation rates, and dry-bulk density data, and the resulting flux record demonstrates that aeolian dust supply from West Africa to the adjacent Atlantic is much greater during glacial maxima (by a factor of 2 to 3; fig. 19.3). DeMenocal et al. (1993) presented several lines of evidence demonstrating that the observed terrigenous variations reflect real variations in terrigenous (aeolian) supply rather than effects of carbonate dissolution. Enhanced West African aridity during glacial maxima has been demonstrated by a number of marine and terrestrial studies (Hays and Perruzza, 1972; Sarnthein et al., 1981; Pokras and Mix, 1985; Gasse et al., 1990; Street-Perrott and Perrott, 1990; Lezine, 1991; Dupont and Hoogheimstra, 1989; Tiedemann et al., 1989).

Spectral analysis of the aeolian records demonstrates that they are dominated by 100 and 41 kyr periodicities, the same periods that dominate high-latitude paleoclimatic records for this interval (fig. 19.3). This is particularly true for the aeolian flux record in that glacial intervals have both high terrigenous percentages and higher sedimentation rates. Despite the strong 100 and 41 kyr variance, none of the aeolian records bears a strong resemblance to the oxygen isotopic record (fig. 19.3), which suggests that West African climate has responded to some other component of high-latitude climatic variability.

Meteorological observations and climate models have demonstrated that the moisture balance of the Sahel region is very sensitive to remote forcing by SST anomalies in the North and South Atlantic. Relatively warm South Atlantic and cold North Atlantic SSTs tend to promote relatively dry conditions in the Sahelian and sub-Saharan regions of West Africa (Lamb, 1978; Folland et al., 1986; Druyan, 1989). The mechanisms behind this relationship are discussed later.

Extending this relationship into the more distant past, the aeolian records of Site 663 are highly covariant with a record of North Atlantic (41°N) SST variability from Site 607 (fig. 19.3; data from Ruddiman et al., 1989b). Increased aeolian concentrations are coincident with cool SST anomalies. The aeolian and SST records exhibit the same long-term and orbital trends over the full 0.9 myr interval, and there are periods of sharp, rapid SST coolings between 0.65 and 0.90 myr that have correlative sharp aeolian increases. Cross-spectral analysis between Site 607 SST and Site 663 aeolian percentage demonstrates that both records exhibit coherent variance at the 100 and 41 kyr periodicities and that they are out of phase (i.e., maximum dust during minimum SST; fig. 19.4; see deMenocal et al., 1993).

Site 662. Site 662 is adjacent to Site 663 and is also within the modern winter season dust trajectory (fig. 19.1). The Site 662 aeolian record extends from roughly 1.7 to 4.0 myr (Ruddiman and Janecek, 1989; deMenocal et al., 1993). Although this record does not have a detailed oxygen isotopic stratigraphy, available biostratigraphic control permits construction of a preliminary time series (fig. 19.5). Examination of the interval prior to the expansion of high-latitude ice sheets (prior to 3 myr) reveals a distinctly different mode of variability than that which was observed during the Late Pleistocene at Site 663. The dominant periodicity of aeolian variation between 3.0 and 3.8 myr occurs at precessional periodicities (23–19 kyr) rather than the 100 kyr and 41 kyr periods that dominate the most recent 0.9 myr.

Site 661. A more complete picture of the nature and timing of West African climatic evolution during the Plio-Pleistocene is afforded by the long and continuous aeolian record at Site 661 (fig. 19.6). Site 661 is located northwest of Sites 662 and 663, toward the western margin of the winter-season dust trajectory (fig. 19.1). Using available paleomagnetic and biostratigraphic control, there is a marked change in the pattern of aeolian variability during the Late Pliocene. Specifically, the aeolian record exhibits strong precessional (23–19 kyr) variability until ca. 2.8 myr, after which the dominant period of variation shifts to the obliquity period (41 kyr). The timing of the shift at 2.8 myr coincides with the onset of Northern Hemispheric glaciation and the development of glacial-interglacial climate cycles (fig. 19.6). This shift is not accompanied by a dramatic change in aeolian dust

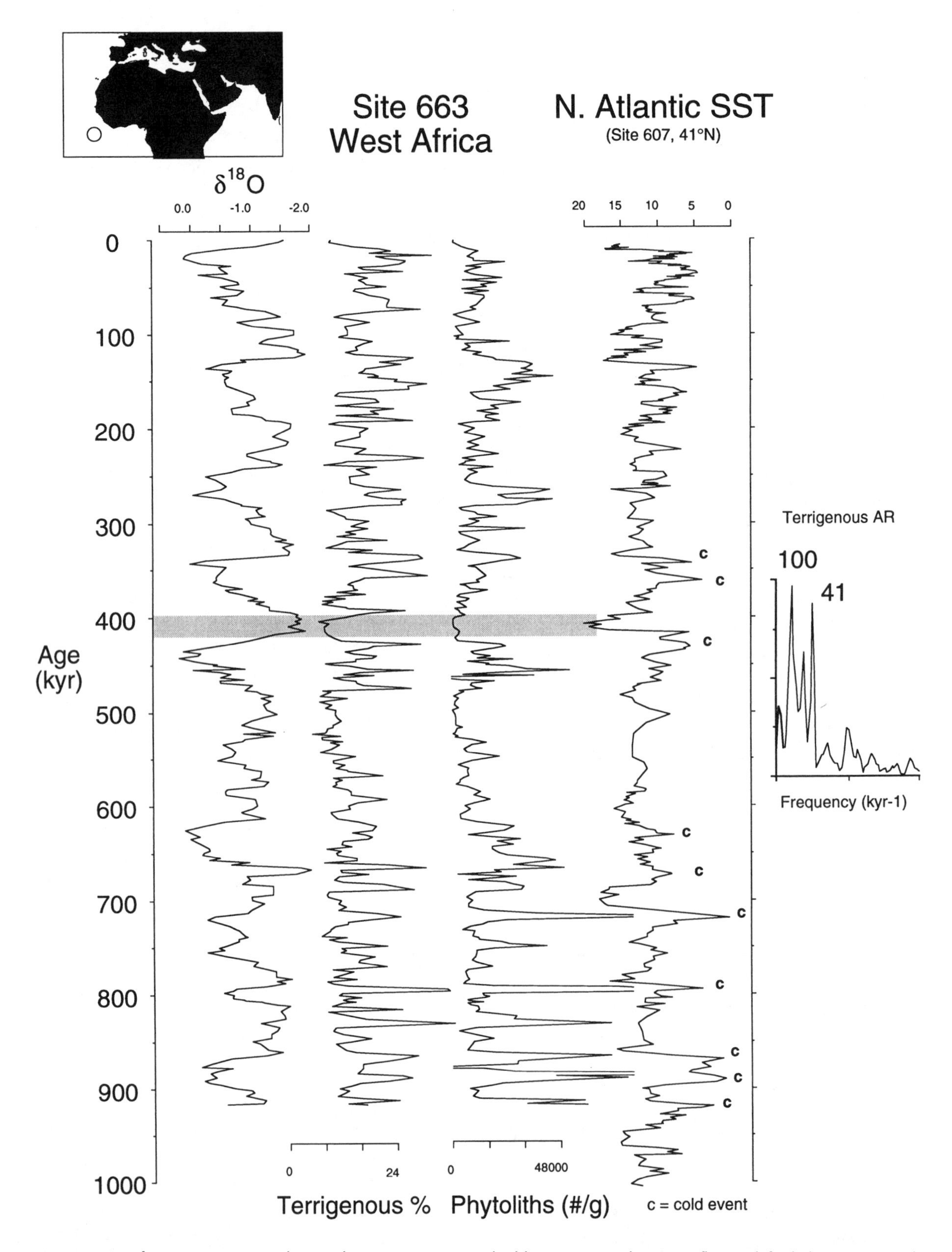

Fig. 19.3. West African Site 663 $\delta^{18}O$ (*G. ruber,* per mil), terrigenous percentage, phytolith concentration, and terrigenous flux records for the last 0.9 myr. Note the dominance of 100 kyr and 41 kyr variance, strong correspondence between the terrigenous and phytolith records, and their correlation with the record of North Atlantic sea-surface temperatures (SSTs) from DSDP Site 607 (41°N; Ruddiman et al., 1986). Shaded interval represents anomolously warm isotopic stage 11, which is reflected by low concentrations of dust and phytoliths from West Africa.

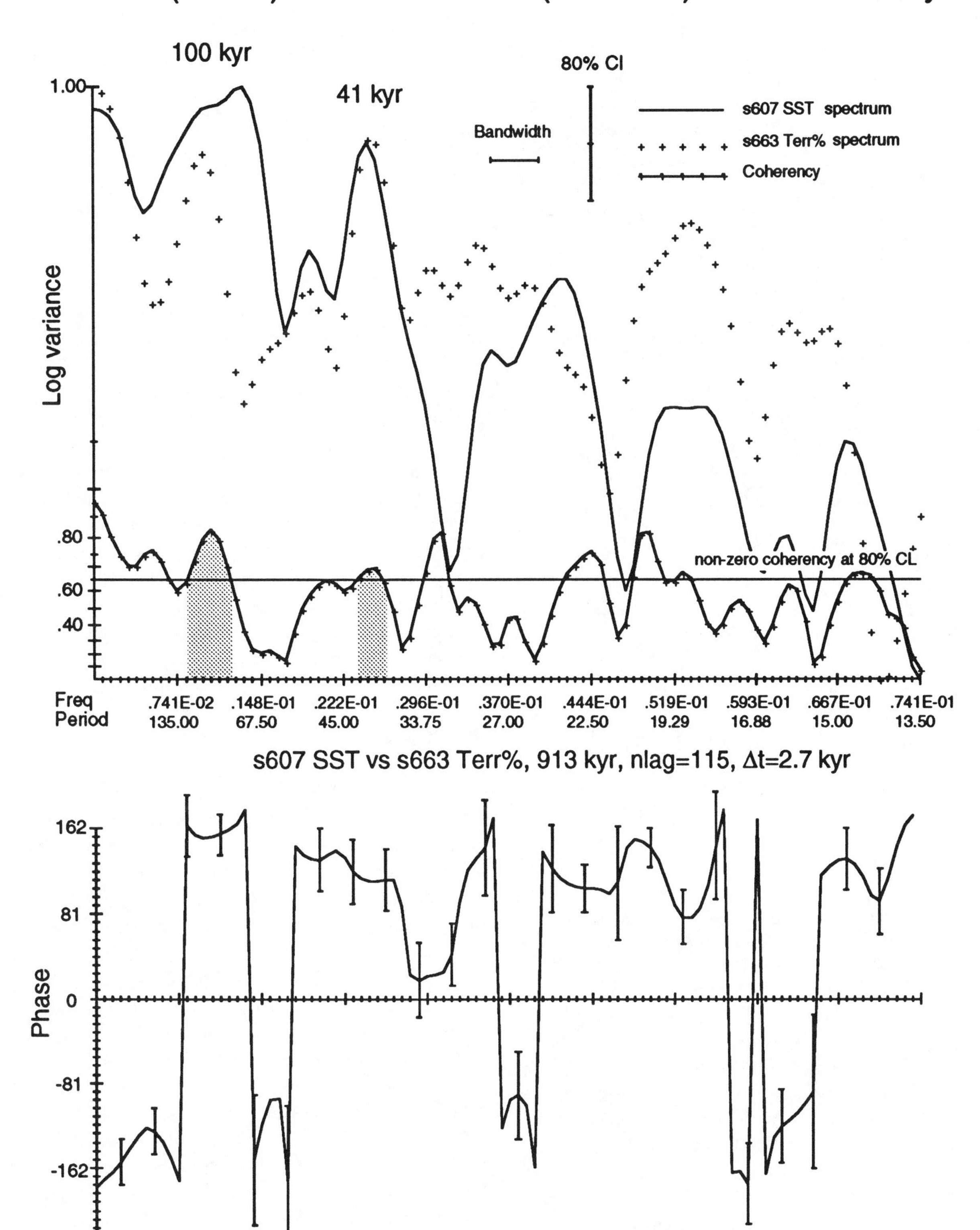

Fig. 19.4. Cross-spectral analysis of the West African Site 663 terrigenous percentage record and the North Atlantic Site 607 SST record (data from Ruddiman et al., 1986). The Site 607 SST record has been adjusted to the time scale of Shackleton et al. (1990). The two records are coherent and out of phase at the 100 kyr and 41 kyr orbital periodicities (cold SST = maximum dust) at the 80 percent confidence level. Solid line represents the Site 607 SST spectrum, crosses represent the Site 663 terrigenous percentage spectrum, and the line with crosses is the coherency spectrum. (CI = confidence interval, Δt = time-sampling interval, nlag = number of lags used for spectrum calculation)

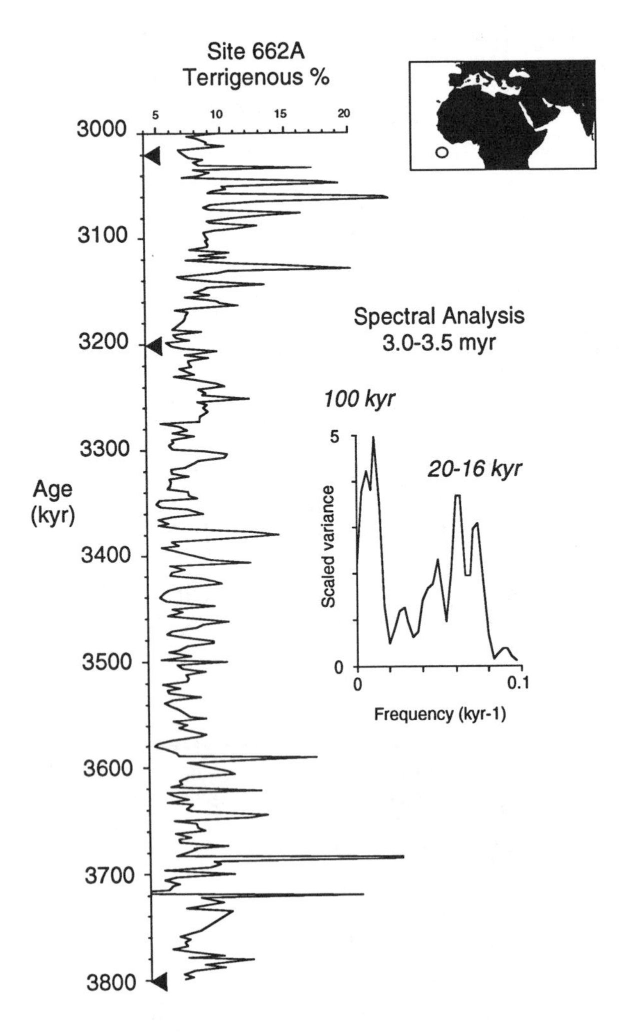

Fig. 19.5. West African Site 662A terrigenous percentage time series with power spectrum of the interval between 3.0 and 3.5 myr. Note the dominance of precessional-band variance (23–19 kyr) and its eccentricity modulation (100 kyr). Biostratigraphic age-control points are shown by solid triangles.

concentration, which argues against a wholesale increase in African aridity after 2.8 myr. After 0.9 myr, there is an increase in variance at the 100 kyr periodicity; the uppermost cores of this site were slightly disturbed during coring, however, so that about the uppermost 500 kyr may not be fully representative. This site is situated in relatively deep water (4,012 m), so dissolution may be an additional factor affecting the carbonate-terrigenous deposition through time. However, shifts in aeolian variability near 2.8 and 0.9 myr indicated at this site are also observed at Sites 662 and 663: A summary of the West African aeolian records for the last 4 myr is shown in figure 19.7. A detailed and well-dated aeolian record from Site 659 near the western Sahara extending to 5 myr exhibits similar aeolian variance-mode shifts at 2.8 and 0.9 myr (fig. 19.1; Tiedemann et al., 1994).

East African and Arabian Sites

Sites 721/722. A continuous, composite aeolian record extending to 7.3 myr, with 1.5-kyr resolution, was constructed from susceptibility measurements at Sites 721 and 722. This record provides a detailed and robust picture the variability of subtropical East African and Arabian climate since the latest Miocene. The time series is constrained by an oxygen isotopic stratigraphy to 1.1 myr (Clemens et al., 1991) and by paleomagnetic and biostratigraphic control to 7.3 myr. Clemens et al. (1991) demonstrated that aeolian flux at this site is coherent and in phase with ice volume over the past 1.1 myr and that aeolian supply is roughly four times greater during glacial stages.

Individual 800 kyr-long segments of the preliminary (untuned) aeolian time series of Site 721/722 are shown to illustrate the changes in subtropical East African climatic variability that have occurred during the Plio-Pleistocene (fig. 19.8). The 0–0.8 myr segment is dominated by 100 kyr variance; the 0.8–1.6 myr segment is dominated by 41 kyr variance, whereas an interval representing 5.0–5.6 myr is dominated by variance at the 23–19 kyr periodicities. The transition from dominant 23–19 kyr variability to dominant 41 kyr variability occurs between 3.2 and 2.3 myr in this untuned record (see earlier versions in Bloemendal and deMenocal, 1989, and deMenocal et al., 1991).

The Site 721/722 aeolian time series was further constrained by taking advantage of the strong precessional (23–19 kyr) character of the record prior to 2.6 myr and its correlation with ice volume ($\delta^{18}O$) after 2.6 myr. The preliminary aeolian time series was tuned to a precession-tilt orbital composite (Berger and Loutre, 1991) between 7.3 and 2.6 myr, and it was phase-locked to glacial ice volume after 2.6 myr (to Site 849 benthic $\delta^{18}O$ from Shackleton et al., in press, a). Tuning was accomplished using the CORPAC correlation program (Martinson et al., 1982). In no case did the tuning process violate an original age datum or did the resulting age-depth curve deviate markedly from the preliminary age model.

The aeolian record of Site 721/722 shares many characteristics with the aeolian records off West Africa. The Site 721/722 aeolian percentage and flux records

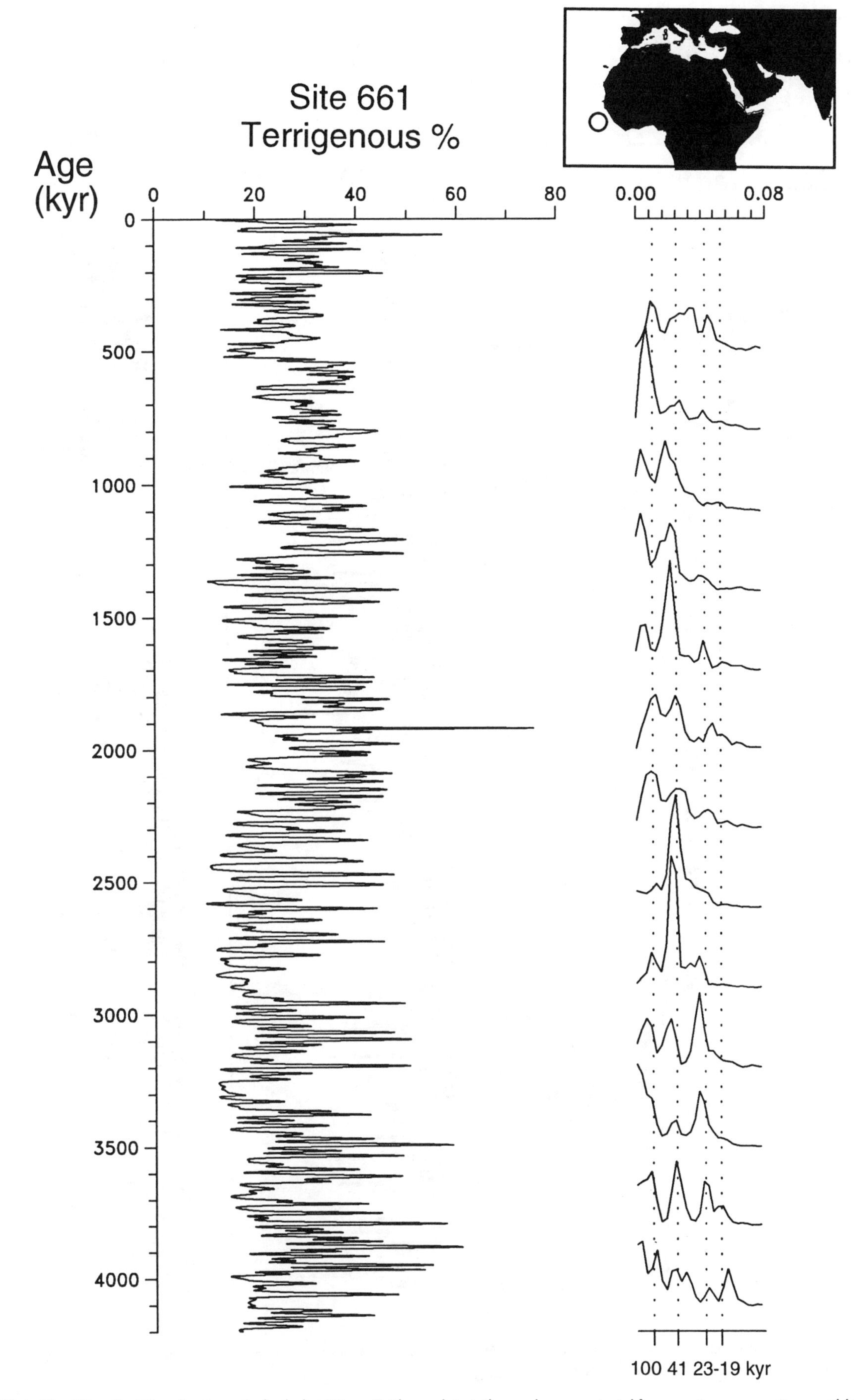

Fig. 19.6. West African Site 661 aeolian time series for the last 4.2 myr (1.5 kyr resolution). The record was reconstructed from a continuous, composite susceptibility depth-series (see text); age control was provided by biostratigraphic data only. Evolutive power spectra are shown on the right (500 kyr window, stepped 300 kyr). Note the shift in aeolian variance from 23–19 kyr to 41 kyr periodicities, which occurs near 2.8 myr.

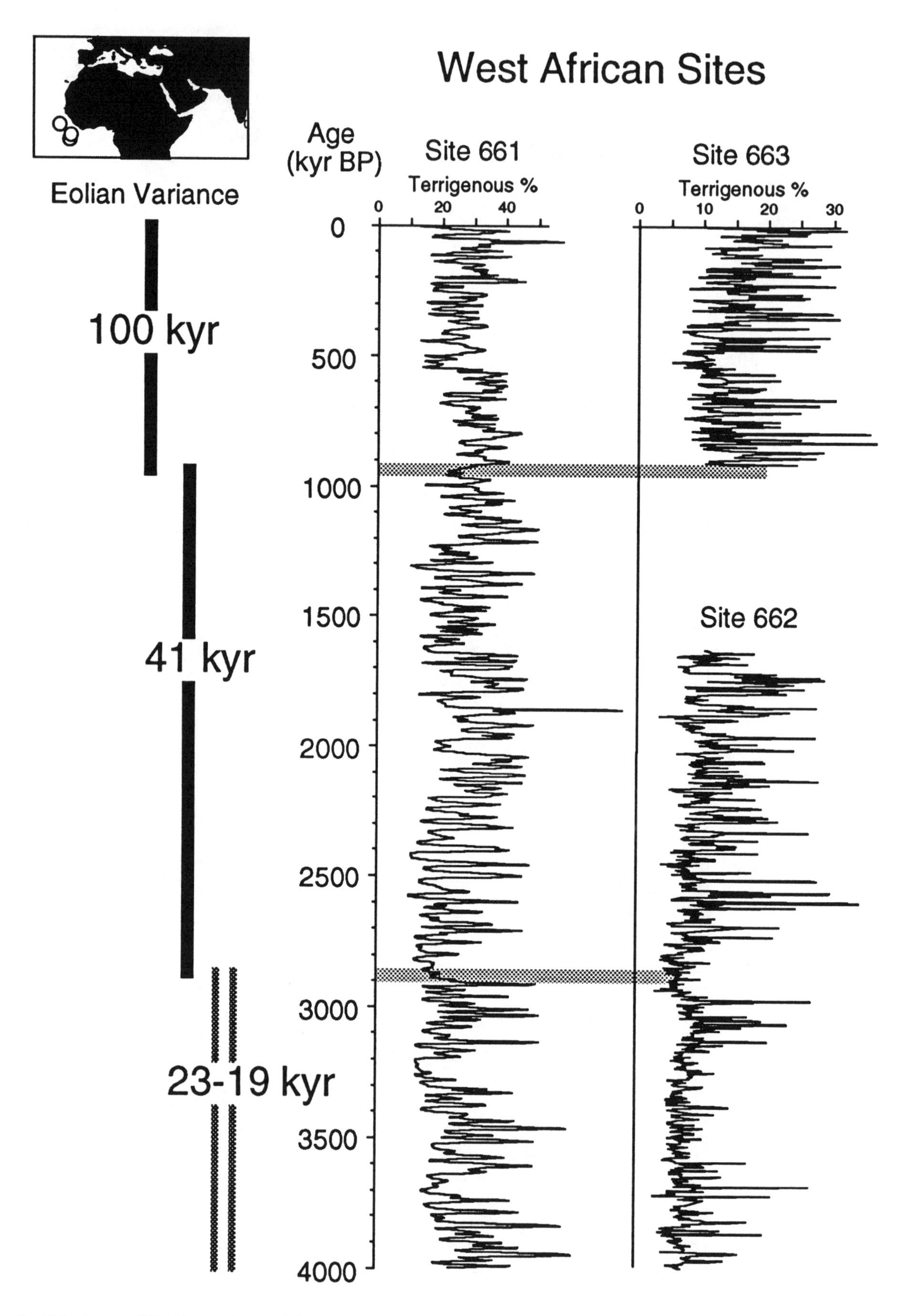

Fig. 19.7. Summary of West African aeolian records from Sites 661, 662, and 663. The records exhibit similar long-term as well as orbital-band trends in aeolian concentration. A summary of the orbital-band changes in aeolian variability is shown at center.

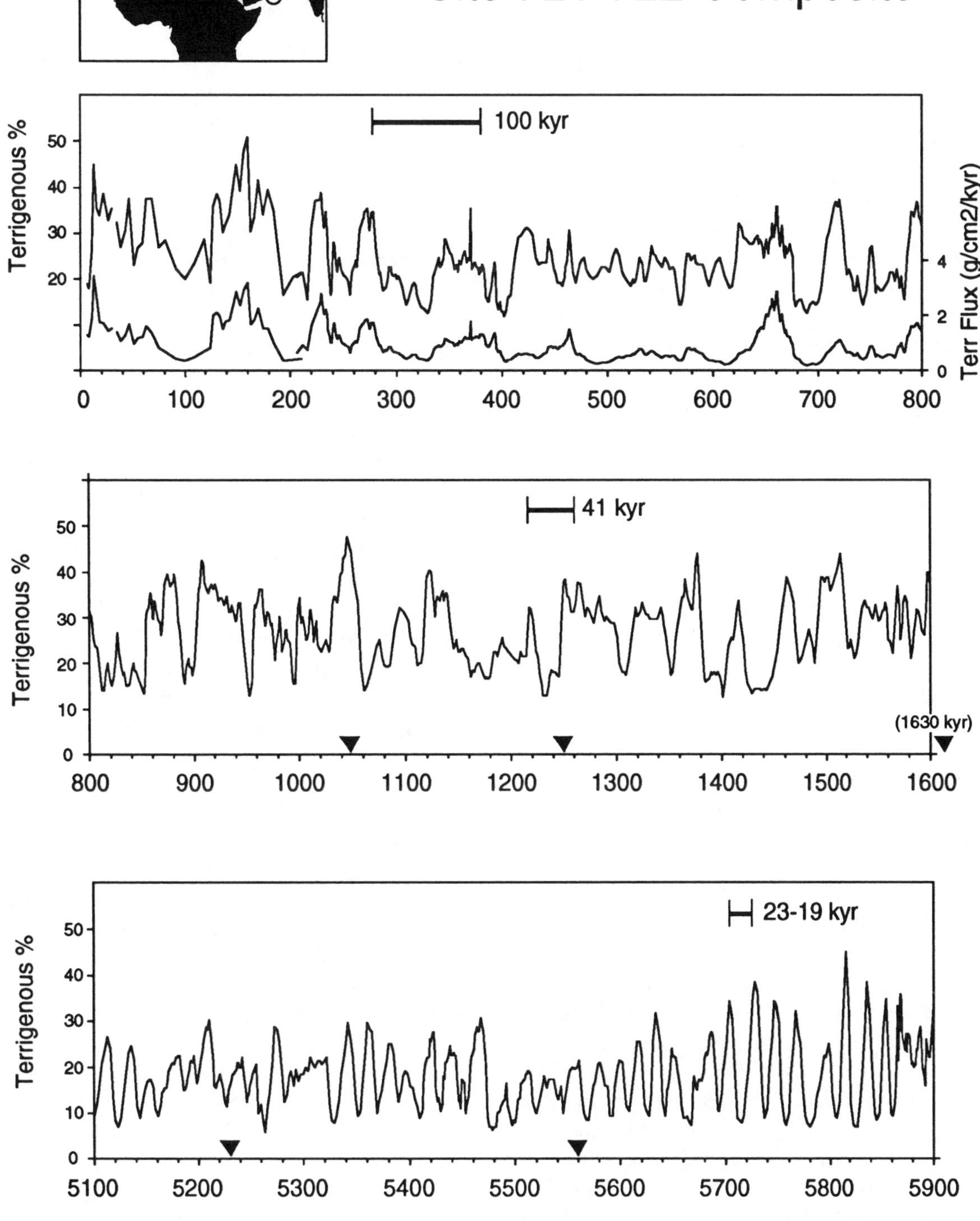

Fig. 19.8. Selected 800-kyr intervals of the ODP Site 721/722 aeolian time series from the Arabian Sea. The records are constrained by oxygen isotopic stratigraphy to 1.1 myr and by magnetic polarity and biostratigraphic data to 7.3 myr (solid triangles). Data representing the last 200 kyr were obtained from Clemens et al. (1991). Note the different modes of aeolian variance for the three intervals; 100 kyr variance dominates the last 800 kyr (particularly in the flux record), 41 kyr dominates the interval from 1.6–0.8 myr, and 23–19 kyr variance dominates the 5.1–5.9 myr interval. The transition from dominant 23–19 kyr to 41 kyr variance occurs between 3.0 and 2.6 myr.

are shown with their respective evolutive power spectra in fig. 19.9. Three principle results can be drawn from these data: 1) the shift from dominant 23–19 kyr variance to dominant 41 kyr variance occurs between 3.0 and 2.6 myr; 2) there is a marked increase in aeolian flux variability and a shift to 100-kyr variations after 0.9 myr; and 2) there is a broad but temporary increase in aeolian concentration near 1.7–1.8 myr. Additionally, there is a marked increase in aeolian variability from the base of the record to about 5 myr. Interestingly, there is no significant change in aeolian concentration or variance associated with the terminal Messinian flooding of the Mediterranean Basin, now dated near 5.35 myr (Hilgen and Langereis, 1993).

Origin of the Pre-2.8 myr Precessional Cycles

The paleoclimatic origin of the pre-2.8 myr precessional aeolian cycles at Site 721/722 can be tested by examining their relationship with indexes of Arabian Sea upwelling. Upwelling and productivity in the Arabian Sea are closely tied to the intensity of the summer monsoon owing to Ekman pumping by the southwest monsoon winds. Arabian Sea sediment-trap studies have demonstrated that both aeolian and biogenic (carbonate, biogenic opal, and organic carbon) fluxes reach their highest values during June, July, and August. Similarly, the export flux of planktonic foraminifera and the relative abundance of upwelling-sensitive species are highest during these summer months (Curry et al., 1992; Anderson and Prell, 1993).

A 400 kyr interval between approximately 5.5 and 5.9 myr was sampled at 15 cm (3 kyr) intervals; these samples were analyzed for biogenic opal (diatoms and radiolaria; Mortlock and Froelich, 1989 method) and organic carbon concentration (Verardo et al., 1990 method). Figure 19.10 shows the magnetic susceptibility and upwelling indexes of Site 721/722 plotted according to composite depth placed adjacent to the orbital composite and the tuned 721/722 aeolian time series. These data demonstrate that the aeolian maxima coincide with maximum values of upwelling indicators and therefore reflect relative increases in the strength of the Asian monsoon (fig. 19.10). This figure also demonstrates the strong precessional variability in these data; this mode of variability is clearly apparent in the depth-series plot and is not a tuning artifact. A section of this interval was also analyzed for relative abundances of upwelling-sensitive coccolithophorid species (*F. profunda;* Molfino and McIntyre, 1990), which confirmed that the aeolian, opal, and organic carbon maxima coincided with maximum values of this biotic upwelling index. Cross-spectral analyses demonstrate that upwelling indexes are coherent and in phase with each other and with the aeolian record, although amplitude differences between the individual records are apparent. These results identify changes in Asian monsoon intensity as the physical mechanism behind the Site 721/722 precessional aeolian variations prior to 2.8 myr.

Pliocene Aeolian Variability in the Gulf of Aden

Site 231 in the Gulf of Aden is of particular interest to this study because it is proximal to the African continent (70 km from the Somali coast; fig. 19.1) and because the sediments contain several Plio-Pleistocene ash layers that have been correlated to radiometrically dated tephra layers in the Turkana Basin of East Africa (Mac Dougall et al., 1992; Sarna-Wojcicki et al., 1985; Brown et al., 1992). Microprobe analyses of major elements were employed to "fingerprint" ash shards and geochemically correlate them to East African tephra horizons. These same tephra horizons have been used to date hominid fossil specimens recovered from several East African sites and to establish the chronology of hominid evolution (Brown et al., 1992; Mac Dougall, 1985; Feibel et al., 1989).

An interval representing 4.4 to 3.3 myr was selected for sampling at Site 231 because it contains several of the ash horizons that have been correlated to East African terrestrial sequences. Discrete samples were taken at 10 cm intervals from cores 19–23X of Site 231 (159–206 meters below sea floor (mbsf); samples were freeze-dried, weighed, and analyzed for calcium carbonate.

The resulting data demonstrate that the Site 231 terrigenous (aeolian) record exhibits the same mode of precessional (23–19 kyr) variability as observed at Sites 721/722 (fig. 19.11) for this interval. Absolute correlations between the records of Sites 231 and 721/722 were difficult to establish, although some degree of correlation is evident (fig. 19.11); the site was drilled over thirty-two years ago using rotary coring in unconsolidated sediments. That Site 231 has the same pattern of variability as Site 721/722 argues that this near-African site went through a similar pattern of climatic variability during the earliest Pliocene and, presumably, throughout the Pleistocene.

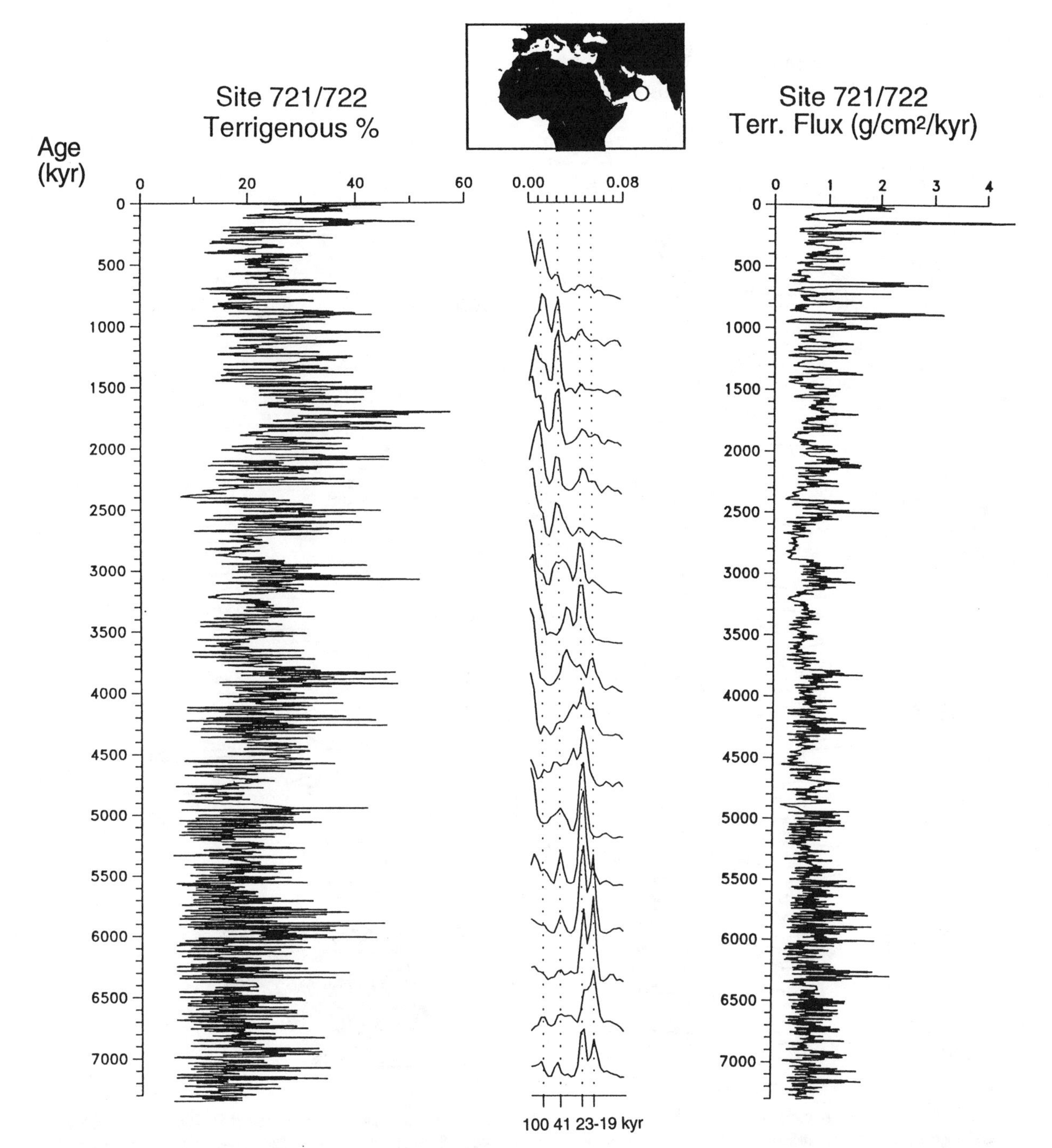

Fig. 19.9. The orbitally tuned Site 721/722 aeolian percentage (left) and flux (right) records with evolutive power spectra (aeolian percent record: 700 kyr window, stepped 400 kyr). Note the shifts in aeolian variance (and flux amplitude) that occur at ca. 2.8 myr and ca. 1.0 myr and the increased aeolian concentrations near 1.8–1.6 myr.

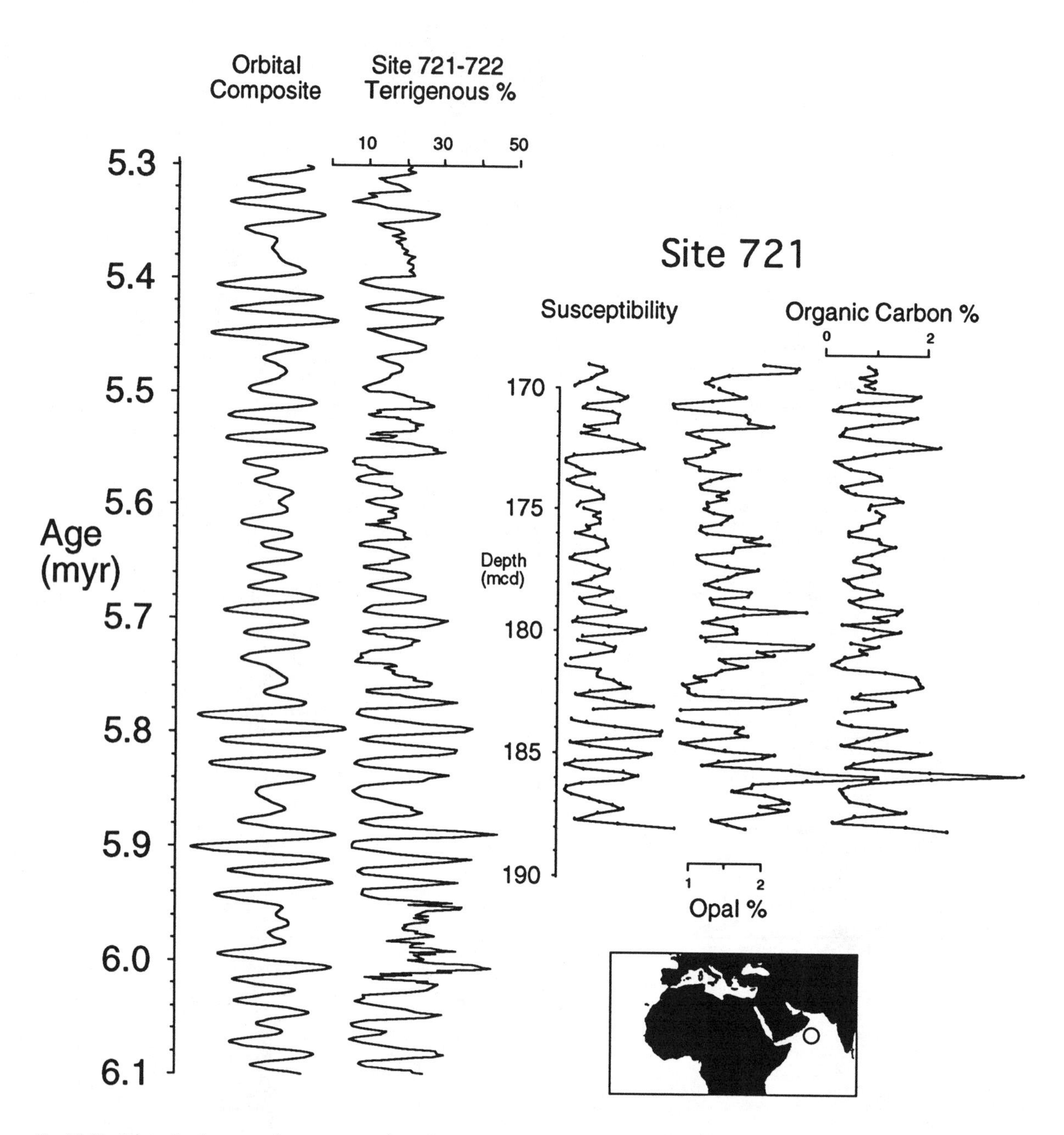

Fig. 19.10. Relationships between aeolian percentage and upwelling/productivity indexes for the interval between 5.5 and 5.9 myr at Site 721/722 (Arabian Sea). Increased biogenic opal (diatom and radiolarian tests) and organic carbon concentrations coincide with aeolian maxima. These results demonstrate that the precessional aeolian variability that dominates the record prior to 2.8 myr is related to variations in monsoon intensity.

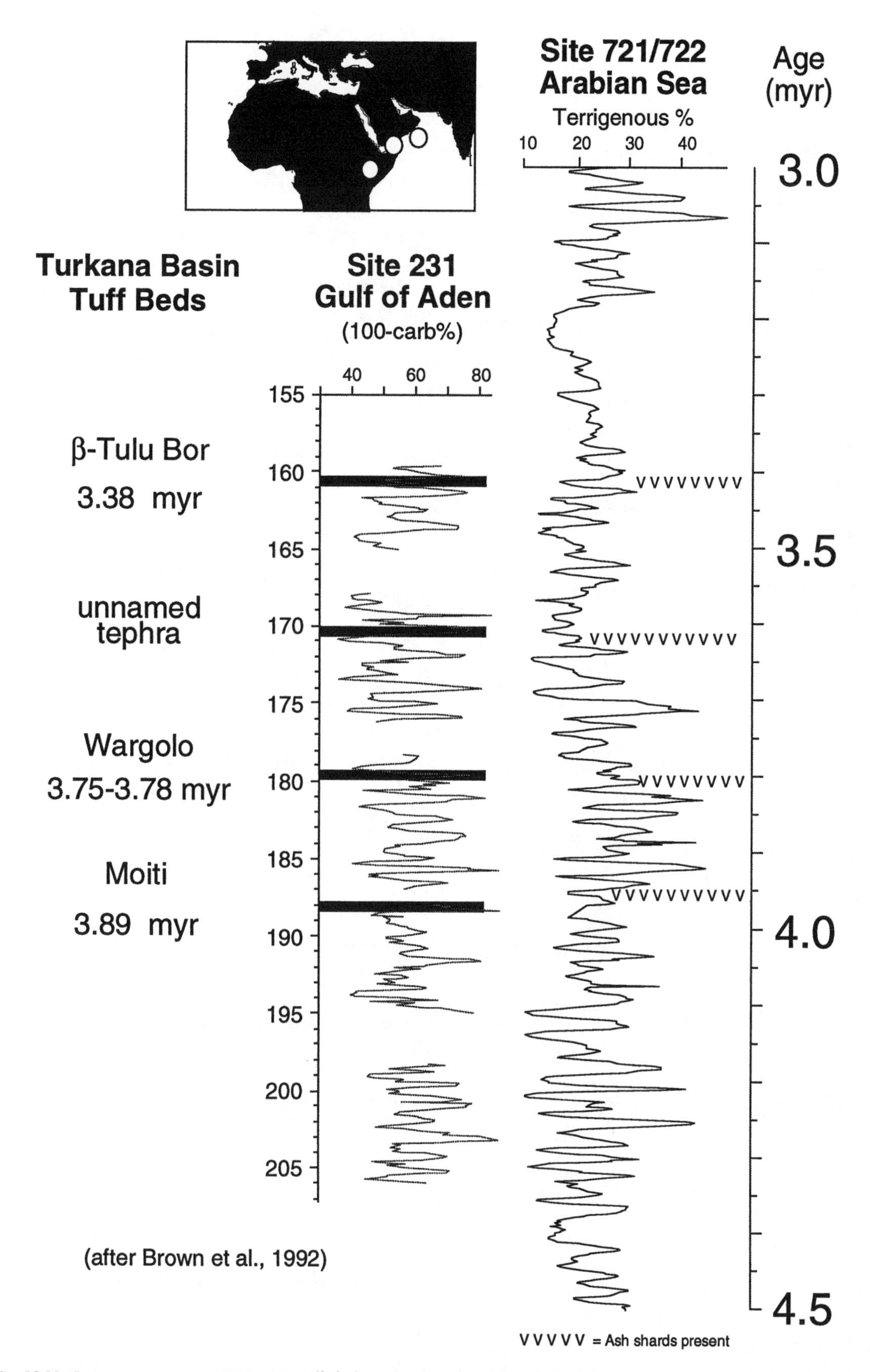

Fig. 19.11. Terrigenous variations at DSDP Site 231 (Gulf of Aden) with stratigraphic positions of ash layers that have been correlated geochemically to tephra horizons in East Africa (from Brown et al., 1992). Site 231 demonstrates a similar mode of (precessional) aeolian variability to that observed at Site 721/722, although precise intercorrelations between the two records are difficult to establish.

Pliocene Floral and Faunal Changes in East Africa

A broad variety of terrestrial paleoclimatic indexes suggest that East African climate changed from warmer, wetter conditions to a more seasonally contrasted, cooler, and drier climate during the Late Pliocene (between ca. 3.0 and 2.0 myr). Although none of these records is continuous, they collectively point to profound changes in the East African paleoenvironment near this time. East Africa experienced significant tectonic uplift (1 km) between 4 and 2 myr (Partridge et al., chap. 2, this vol.). Although this would certainly contribute to a gradual cooling and drying during the Plio-Pleistocene, many of the paleoclimatic and fossil faunal records indicate relatively rapid changes that would presumably exclude tectonic mechanisms.

Pollen spectra from a diatomite sequence in the Ethiopian highlands (Gadeb) indicate a shift to cooler and drier vegetation types (an increase in shrubs, heath, and grasses) between two radiometrically dated tuff layers at 2.51 and 2.35 myr (Bonnefille, 1983). A similar shift has also been described for the Turkana Basin region (Bonnefille, 1976) and Lake Chad (Coppens and Koeniguer, 1976). Stable isotopic analyses of pedogenic carbonates from the Turkana and Olduvai Basins indicate a gradual replacement of closed forest woodland by open savannah grasslands between 3 and 1 myr, with sharp increases in savannah vegetation near 1.8, 1.2, and 0.6 myr (ages rescaled using McDougall et al., 1992; Cerling, 1992; Cerling and Hay, 1988). An absence of carbonate nodule samples between 2.8 and 2.2 myr precludes interpretations over this important interval. Stable isotopic analyses of Turkana Basin gastropods indicate a shift toward reduced precipitation between two radiometrically dated tuffs at 3.2 and 1.9 myr (Abell, 1982). Both pollen and soil carbonate data indicate that parts of East Africa were still vegetated by closed-canopy rain forest during Middle and Late Miocene time (ca. 15–8 myr; Bonnefille, 1984; Yemane et al., 1985; Cerling, 1992).

The fauna of East African changed markedly during the Late Pliocene. A major change occurs between the Omo I and Omo II faunas (ca. 2.6 myr) of the Shungura Formation (southern Ethiopia; Coppens, 1975). A biochronology of African Bovidae indicates multiple first appearances of arid-adapted species near 2.7–2.5 myr (Vrba, 1988, and chap. 27, this vol.). Micromammal assemblages from the Shungura indicate a shift from high forest vegetation near 3.0 myr to arid and open conditions near 2.4 myr (Wesselman, 1985).

In summary, a variety of floral and faunal evidence supports a change in East African climate to cooler, drier, and more seasonal conditions after the late Pliocene (ca. 3–2 myr). The data are of insufficient temporal or spatial resolution to identify the precise timing or areal extent of climatic change, but they do point to a marked change in the East African paleoenvironment near this time.

Climate Model Sensitivity Experiments

The Model

The general circulation model (GCM) used in these experiments is the Goddard Institute for Space Studies (GISS) model II with 8° × 10° medium grid resolution (see Hansen et al., 1983). The model runs are used to investigate the sensitivity of subtropical African and Asian climate to prescribed changes in high-latitude climate. The strategy is to compare a given model run configured with a changed boundary condition (e.g., increased glacial ice cover) with a control version of the model (modern climate). The resulting climatic anomalies can then be identified and related to the given boundary-condition change. The model runs are not designed to represent a past climate; rather, they are designed to test the sensitivity of low-latitude climates to changes in high-latitude boundary conditions. The reader is directed to deMenocal and Rind (1993) for a more thorough discussion of experiment configurations, results, and interpretations.

The model solves the equations for conservation of mass, energy, momentum, and moisture for nine atmospheric layers. It calculates cloud cover, snow cover, soil moisture, and full radiative processes with a diurnal and seasonal cycle. The control (CONT) run has averaged topography and produces generally realistic temperature and precipitation fields when modern SSTs are prescribed (Hansen et al., 1983). In particular, topography over East Africa is greatly smoothed by the grid scheme. Precipitation is calculated for both subgrid-scale convection and also large-scale supersaturation, which occurs when relative humidity exceeds 100 percent; the moisture condensed in both cases is allowed to reevaporate in unsaturated atmospheric layers below. SST fields are prescribed and noninteractive in all runs.

Experiment Configurations

The effects of cold (full glacial) North Atlantic SSTs and full glacial ice-sheet extent on subtropical African climate are tested explicitly. The effects of cooler, full glacial (18 kyr B.P. radiocarbon years) North Atlantic SSTs on low-latitude climate are investigated by comparing the CONT model run against an identically configured run with cooler 18 kyr B.P. SSTs (CONT.SST) prescribed in the North Atlantic sector above 25°N. Sea ice was set at its modern latitudinal limit. To examine the effects of high-latitude glacial ice cover, a model configured with full 18 ka B.P. boundary conditions (18K.FULL glacial ice extent, SSTs, 120 m lower sea level, and orbital configurations) is compared with an identically configured run, except that the ice sheets were restored to their modern extent (18K.NOICE).

Sensitivity of Subtropical African Climate to North Atlantic SSTs

Winter Climate. Prescribing full glacial (18 kyr B.P.) SSTs above 25°N in the North Atlantic causes significant cooling, drying, and increased winter trade-wind circulation over northwest Africa during the December, January, and February (DJF) season. Northwest Africa and Arabia are cooler by 4°–2°C. Increased surface pressures develop anticyclonic circulation over the northeast Atlantic, which strengthens northeast trade winds over northwest Africa (fig. 19.12a). Modest precipitation reductions (of 0.5–2 mm per day) are observed over northwest Africa (fig. 19.12a). As noted by Rind et al. (1986), the major changes can be related directly to dynamical effects associated with cooler North Atlantic SSTs.

Summer Climate. Cooler North Atlantic SSTs reduce monsoonal inflow into North Africa (1–2 m/sec; fig. 19.12b) and produce decreases in surface air temperature (−2 to −4°C; fig. 19.12b) and precipitation (up to 1 mm per day; fig. 19.12b). Summer southwest African monsoon winds are reduced by the same high-pressure cell over the North Atlantic that strengthened the winter northeast trade-wind circulation over northwestern Africa. The precipitation decreases over Africa and Arabia (1–2 mm/day; fig. 19.12b) can be related to increased surface air outflow from the respective regions owing to the thermal effects of the cooler North Atlantic SSTs.

Comparison with Other Model Results. Rind et al. (1986) and Overpeck et al. (1989) considered the effects of cooler North Atlantic SSTs on European and Asian climate and concluded that subtropical West African climate is very responsive to cooler North Atlantic SST anomalies. GCM experiments have been used to examine historical occurrences of sub-Saharan drought. Meteorological and modeling studies have demonstrated that the surface climate of northwest Africa is very sensitive to SST anomalies in the adjacent North and South Atlantic. Using an interactive ocean GCM, Druyan (1987) demonstrated that arid conditions prevail when North Atlantic SSTs (30°–45°N) are relatively cool and South Atlantic SSTs (10°–30°S) are relatively warm. The drought conditions were attributed to a weakening of the land-sea pressure gradient from the eastern South Atlantic and a strengthening of the high-pressure cell over the North Atlantic, both of which act to decrease the advection of moisture from the adjacent Atlantic to the northwest African interior. Similar correlations between Sahelian aridity and Atlantic SST anomalies have been proposed based on meteorological data (Lamb, 1978; Folland et al., 1986).

Sensitivity of Subtropical African Climate to High-Latitude Glacial Ice Cover

Winter Climate. Increased glacial ice cover over North America, northern Europe, and Antarctica enhanced winter trade-wind circulation over southwest Asia (by 3–6 m/sec) and decreased precipitation over Arabia and East Africa; northwest Africa was relatively unaffected (fig. 19.13a). The stronger winds are attributable to an intensification of the high-pressure cell over the Himalayan region. The eastern Mediterranean, south Asian, and East African regions are drier by up to 4 mm/day. For the Arabian and northeast African regions, these decreases are equivalent to the entire winter-season precipitation in the control run. The largest cooling occurs over the same regions experiencing reduced precipitation: southern Asia and the Arabian Peninsula (fig. 19.13a). The winter low-latitude cooling and drying appears to result from downstream advection of cooler and drier air from the high-latitude ice sheets. Thin snow and ice cover (< 500 m thick) over the Himalayan-Tibetan region in the 18K.FULL simulation clearly affects the winter circulation and climate of south Asia.

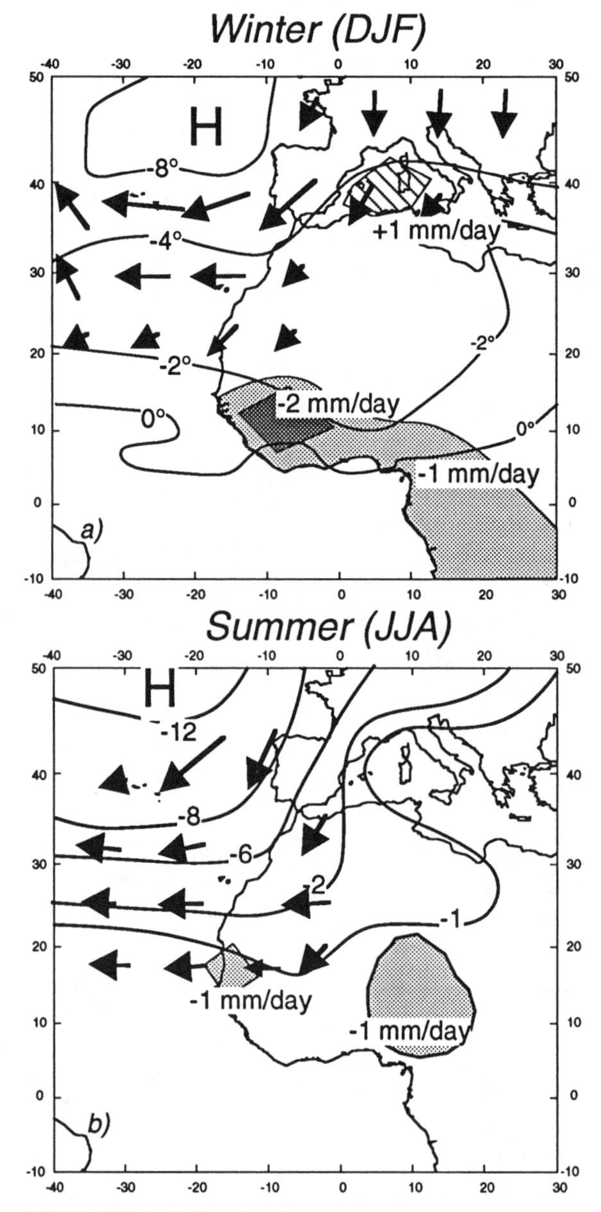

Fig. 19.12. Climatic anomaly summaries of the effects of cold CLIMAP North Atlantic SSTs on winter (December, January, February [DJF], above) and summer (June, July August [JJA], below) model climate. CLIMAP SSTs were imposed above 25°N in the North Atlantic only. This model run (CONST.SST) was compared to the control run (CONT) and climatic anomalies were computed (CONST.SST minus CONT). West African climate was most sensitive to cold North Atlantic SSTs. Cold North Atlantic SSTs promote stronger winter season trade-wind circulation over northwestern Africa, surface cooling, and drying. The African monsoon intensity is reduced during the summer season, and subtropical Africa is cooler and drier. See text for discussion; summarized from deMenocal and Rind (1993).

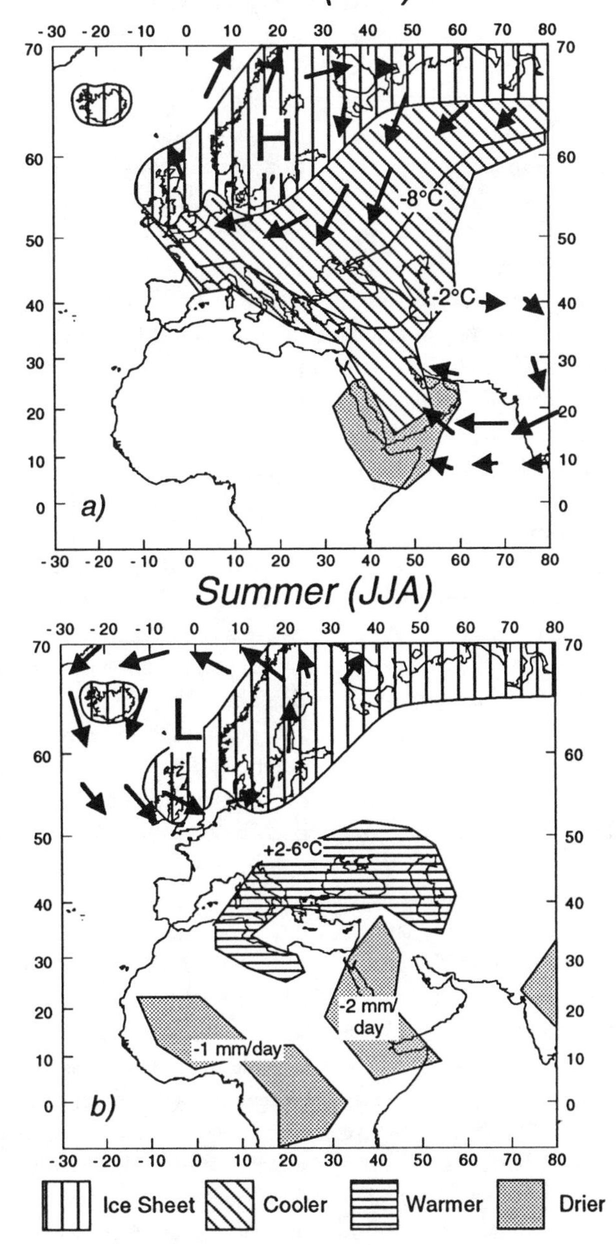

Fig. 19.13. Climatic anomaly summaries of the effects of 18 ka B.P. glacial ice cover on winter (DJF) and summer (JJA) climate. The full glacial model run (18K.FULL; last glacial ice cover, SSTs, sea level, orbital values) was reconfigured such that only ice-sheet size and areal extent were restored to their modern limits. This model run (18K.NOICE) was compared to the full glacial run (18K.FULL) and climatic anomalies were computed (18K.FULL minus 18K.NOICE). The resulting climatic anomalies are attributable to the effects of ice sheets alone. East African, Arabian, and eastern Mediterranean climates were most sensitive to the prescribed changes in ice-sheet size and extent. These regions became more seasonal, with dramatically cooler and drier winters. See text for discussion; summarized from deMenocal and Rind (1993).

Summer Climate. The inclusion of full 18 kyr B.P. ice cover and topography reduces the summer monsoon winds of southwest Asia, and there are new, "real" decreases in precipitation over a broad swath including West and East Africa, Arabia, and southwest Asia (fig. 19.13b). The differenced surface wind field indicates reduced Asian monsoonal circulation (lower by 1–5 m/sec; fig. 19.13b). Increased snow and ice cover over south Asia inhibits the northward penetration of the summer heat low over that area, and greatly reduced precipitation over East Africa and South Asia results (−2 to −6 mm/day; fig. 19.13b). Comparisons with the control wind field indicates that the monsoon is much weaker in absolute terms, although this comparison reflects the combined effect of changing all glacial boundary conditions, not just glacial ice cover.

Reduced Arabian and East African rainfall is attributable to downstream cooling and other dynamic effects (such as increased planetary wave amplitude) of the high-elevation Fennoscandian ice sheet. Additionally, thin snow and ice cover over south Asian topography inhibits sensible and latent heating, which drives the monsoon.

Comparison with Other Model Results. These results emphasize the sensitivity of East African, Arabian, and southwest Asian moisture balance to high-latitude ice cover. Many of these regions have Mediterranean-type (wet winter, dry summer) rainfall patterns, so the winter-season rainfall decreases are important. Specifically, the model results suggest that expanded high-latitude ice cover leads to enhanced aridification of the Asian monsoon dust-source areas. However, the response is less clearly developed than in the North Atlantic SST experiment, and more work is required before this response can be interpreted as definitive.

When full glacial boundary conditions are considered, many models show reduced Asian precipitation (Gates, 1976; Manabe and Hahn, 1977; Rind and Peteet, 1985; Rind, 1987; Prell and Kutzbach, 1987). Few of these studies have isolated the effects of increased ice cover alone, so it is difficult to assess a common response to glacial ice cover. Full glacial ice extent and elevation were responsible for precipitation decreases in south Asia in the GISS model (Rind, 1987). Prell and Kutzbach (1987, 1992) observed that the inclusion of glacial ice to their NCAR-CCM simulation of 18 kyr B.P. climate caused a 20 percent decrease in the south Asian precipitation as well as a reduction of 2 m/sec in the v-component of Arabian Sea winds.

Prell and Kutzbach (1987) explicitly considered the effects of full glacial boundary conditions (SSTs, land- and sea-ice distributions, sea level, trace-gas composition) on the climate of the summer Asian and African monsoons. Although their results indicate a weakened summer monsoon and a reduction in south Asian precipitation, northeast African precipitation increased owing to 1.5°C warmer CLIMAP SSTs in the western Indian Ocean. Low-latitude precipitation patterns are very sensitive to small changes in tropical SST distributions.

Sensitivity of Subtropical African Climate to Precessional Insolation Variations

Previous studies have demonstrated that the summer African and Asian monsoon circulation is highly responsive to 23–19 kyr summer insolation variations owing to orbital precession (Kutzbach, 1981; Kutzbach and Otto-Bliesner, 1982; Rind et al., 1986; Prell and Kutzbach, 1987; Kutzbach and Street-Perrott, 1985; Kutzbach and Guetter, 1984; deMenocal and Rind, 1993). The response is intense because the summer monsoon is effectively a heat engine driven directly (sensible heating) and indirectly (latent heating) by summer insolation.

Prell and Kutzbach (1987, 1992) conducted a series of GCM experiments in which summer insolation was varied from −6 percent to +13 percent of its modern value. They noted that southern Asian monsoonal precipitation increased by 38 percent in response to the 19 percent change in solar insolation; monsoon wind intensity increased by 75 percent. The precipitation and wind-speed responses were approximately linearly related to summer insolation intensity. The high gain factors of these responses (ca. 2 and ca. 4, respectively) emphasize the importance of insolation variations to monsoon paleointensity in the absence of other boundary condition changes.

Discussion

Synthesis of Data and Model Results

Prior to the onset of Northern Hemispheric glaciation near 2.8 myr, the West and East African aeolian records are dominated by 23–19 kyr cycles that are shown to reflect variations in monsoon intensity. Climate models indicate that subtropical African climate is extremely sensitive to precessional insolation forcing of monsoonal climate in the absence of changes in high-latitude glacial boundary conditions. Hence, the dominant 23–19 kyr aeolian variations prior to ca. 2.8 myr are interpreted to reflect variations in African and Asian monsoon intensity forced by low-latitude orbital insolation variations.

Both West and East African aeolian records indicate a shift to 41 kyr variations near 2.8 myr, coincident with the onset of Northern Hemispheric glaciation and the subsequent development of glacial-interglacial climatic cycles at the 41 kyr period. After ca. 0.9 myr the aeolian records exhibit markedly increased variance at the 100 kyr periodicity, again coincident with development of 100 kyr-period glacial-interglacial oscillations. The GCM sensitivity experiments suggest that both West and East African and Arabian dust-source areas are sensitive to changes in high-latitude climate. Specifically, the West African dust-source areas were most sensitive to changes in the North Atlantic SST field; glacial aridity there was related to a broad high-pressure cell over the cool North Atlantic, which advects cool, dry air from Europe over northwest Africa. Arabia and East Africa were most sensitive to changes in glacial ice cover; the development of regionally cooler and drier conditions there is linked to elevation and albedo effects of the high Fennoscandian ice sheets.

Conceptual Model of Plio-Pleistocene African Climatic Change

The data and modeling results can be combined to develop a conceptual model for the Plio-Pleistocene evolution of subtropical African climate. We propose that high- and low-latitude climatic regimes were independent when ice sheets were relatively small and invariant, but when ice sheets expanded such that large climatic oscillations were sustained (after ca. 2.8 myr), low-latitude climate became dependent upon the glacial-interglacial rhythm of high-latitude climatic variability.

Prior to ca. 2.8 myr, African climate primarily responded to variations in monsoon intensity related to summer-season insolation variations owing to orbital precession. Monsoon intensity, and its attendant precipitation, was greatest when summer insolation was highest (perihelion during boreal summer). Subtropical Africa would have experienced a continuum of wet-dry climatic cycles at the 23–19 kyr periodicities. Hence, African climatic variability was related to low-latitude orbital insolation variations and was effectively independent of high-latitude forcing prior to 2.8 myr.

African climate thus became dependent upon the onset of high-latitude glacial climatic cycles after 2.8 myr. The climatic effects of increased high-latitude ice cover propagated to low latitudes such that subtropical African climate was affected by the characteristic 41-kyr periods and, later, 100 kyr periods of high-latitude climatic change. The model results demonstrate that both West and East African regions are sensitive to high-latitude glacial conditions. West African climate was most sensitive to cool North Atlantic SSTs, whereas East African and Arabian climate was most sensitive to increases in the size and elevation of the Fennoscandian ice sheet. Hence, African climate after 2.8 myr was subjected to periodic cold-arid cycles corresponding to the succession of glacial stages that characterize the Plio-Pleistocene.

On the Paleoenvironmont of Hominid Evolution

Major evolutionary events punctuate the East African hominid fossil record over the last 4 myr. Several authors have previously proposed that these events were climatically mediated (Vrba, 1985 and chap. 3, this vol.; Hill, 1987), and numerous adaptational and behavioral theories have been advanced (Brain, 1981; Grine, 1986; Vrba et al., 1989; Stanley, 1992). Based on the results presented here, we can begin to place the record of hominid evolution within the context of an evolving subtropical African paleoclimate.

Although the chronology and structure of hominid phylogeny is broadly debated (see contributions in Grine, 1988; review by Wood, 1992; and the paleoanthropological contributions in this volume), certain aspects of evolutionary history are constrained by available ages and data. Using a cladogram developed by Wood (1992), at least two separate lineages emerge from a single ancestral lineage (*Australopithecus afarensis*) between 2 and 3 myr (fig. 19.14). These lineages include the "robust" australopithecines (larger boned, broad tooth crowns, sagittal crest, smaller cranial capacity), which first appear between 2.7 and 2.3 myr and our genus, *Homo* (more gracile, narrow tooth crowns, larger cranial capacity), which first appears near 1.9 myr. *Homo spp.* may have appeared as early as 2.5–2.4 myr based on the recent discovery in Malawi of an *H. rudolfensis* mandible (Schrenk et al., 1993) and an analysis of a single temporal

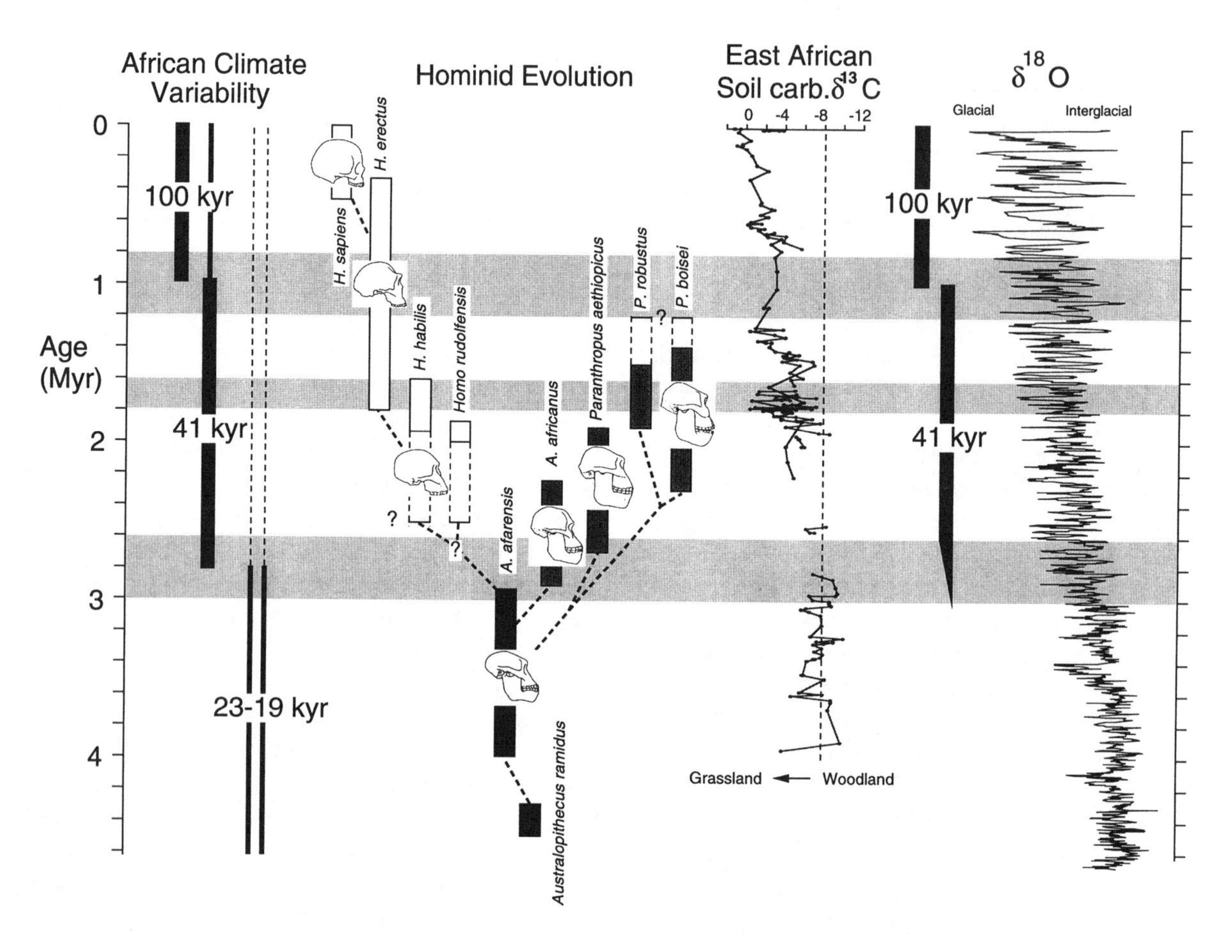

Fig. 19.14. Summary of the West and East African aeolian records compared to an interpretation of human phylogeny (after Wood, 1992). A composite (benthic) marine oxygen isotopic record is shown at right to summarize the temporal evolution of high-latitude climate (data from Shackleton et al., in press, a; Ruddiman et al., 1989). Shaded intervals near 2.8 and 1.0 myr represent mode shifts in aeolian variance that occur at both West and East African sites (see figs. 19.6, 19.8, 19.9); the 2.8 myr event represents the initiation of high-latitude glacial forcing of African climate. The shaded interval near 1.6–1.8 myr represents increased aeolian concentrations at Sites 721/722 (fig. 19.9). The question mark by the first appearances of early *Homo* reflects new fossil material discussed in Schrenk et al. (1993). The East African soil carbonate $\delta^{13}C$ data reflect variations in African vegetation (data from Cerling, 1992; Cerling and Hay, 1988). These data have been adjusted to the new time scale of Shackleton et al. (1990) and reflect the long-term shift from closed woodland to open savannah vegetation over the Plio-Pleistocene.

bone fragment from Chemeron, in Kenya (Hill et al., 1992). The earliest known stone tools have been radiometrically dated near 2.5 myr at two East African localities (e.g. Harris, 1983); these tools may represent both *Homo* and robust australopithecene industry.

The earliest major geographic expansion within the *Homo* lineage occurs near 1 myr, when *H. erectus* appears to have radiated from Africa into a variety of other regions and habitats. It was near this time that the entire robust australopithicine lineage became extinct and our direct ancestor, *H. erectus*, expanded and occupied sites in Europe and western Asia (Clark, 1980; Hamilton, 1982, p. 249; Vrba et al., 1989; Wood, 1992). Azzaroli (1983) describes the 1.0–0.9 myr interval as a "turning point in the history of Eurasia," when cold-adapted mammalian fossil assemblages become widely dispersed throughout Eurasia, involving a series of related extinctions and speciations.

If climate had a role in determining hominid evolution, the most parsimonious interpretation of the available data is that it was a change in mode of subtropical climatic *variability* rather than a wholesale, stepwise change in climate that prompted evolutionary responses. The aeolian records indicate that African climate has been a continuum of change over the entire Neogene, so apparently it was the onset and intensification of glacial cool and arid cycles after 2.8 myr that distinguished subtropical African climatic variability of the latest Plio-

Pleistocene (fig. 19.14). The onset and recurrence of extreme arid events after 2.8 myr, and their associated effects on African flora and fauna, would have been the primary climatic forcing agent behind any evolutionary responses. The development of glacial arid cycles near 2.8 myr, and their intensification after 0.9 myr, would have had profound influences on subtropical African ecology and created cyclic opportunities for species extinction and innovation.

Summary

Subtropical African climatic change during the Plio-Pleistocene (0–5 myr) is examined using deep-sea records of aeolian supply from West and East Africa. Both regions exhibit marked changes in climatic variability near 2.8 myr and again at 0.9 myr, coinciding with major changes in high-latitude climate. The East African records show a pronounced dry period between 1.8 and 1.6 myr. Aeolian variability prior to 2.8 myr occurs at the 23–19 kyr periodicities associated with orbital precession; these are shown to reflect variations in monsoon paleointensity. After 2.8 myr the records exhibit increased 41 kyr variance, with further increases in 100 kyr variance after 0.9 myr. Experiments with climate models demonstrate that both West and East Africa become seasonally cooler and drier owing to increased high-latitude glacial ice cover. In the absence of other factors the African climate model is very sensitive to direct insolation forcing. It is proposed that African climate was independent of high-latitude climate prior to 2.8 myr, when ice sheets were small and invariant, whereas African climate became dependent upon the rhythm of high-latitude climatic change after 2.8 myr, when ice sheets grew to sufficient size to sustain glacial-interglacial cycles. These changes coincide with several major steps in hominid evolution. If climatic change played a role in mediating hominid evolution, the onset and intensification of glacial arid cycles after 2.8 myr would have profoundly affected subtropical African ecology and created cyclic opportunities for faunal extinction, innovation, and radiation.

Acknowledgments

Special thanks are due to Elisabeth Vrba for inviting the first author to attend the Airlie conference. We thank Bill Ruddiman for providing the Site 663 and 662 terrigenous percentage records and Nick Shackleton for kindly providing his Site 847 isotope record. Thure Cerling generously supplied his East African soil carbonate stable isotopic data. Linda Baker and Pat Malone are thanked for their laboratory assistance; Ned Pokras performed the Site 663 phytolith counts. We are grateful to Bill Ruddiman, Andy McIntyre, Frank Brown and Paul Olsen for their helpful comments and criticisms; and to Chris Mato, Jerry Bode, and John Miller, from ODP, for fulfilling endless sample requests. This work was supported by funding from the National Science Foundation, Marine Geology and Geophysics Division. This is Lamont-Doherty Earth Observatory Contribution 5210.

References

Abell, P. I. 1982. Paleoclimates at Lake Turkana, Kenya, from oxygen isotope ratios of gastropod shells. *Nature* 297:321–323.

Anderson, D. M., and Prell, W. L. 1993. A 300 kyr record of upwelling off Oman during the Late Quaternary: Evidence of the Asian southwest monsoon. *Paleoceanography* 8:193–208.

Azzaroli, A. 1983. Quaternary mammals and the "End-Villefranchian" dispersal event: A turning point in the history of Asia. *Paleogeogr., Paleoclimatol., Paleoecol.* 44:117–139.

Berger, A., and Loutre, M. F. 1991. Insolation values for the climate of the last 10 million years. *Quat. Sci. Reviews* 10:297–317.

Blackman, R. B., and Tukey, J. W. 1958. *The measurement of power spectra from the point of communications engineering.* Dover, New York.

Bloemendal, J., and deMenocal, P. B. 1989. Evidence for a change in the periodicity of tropical climate cycles at 2.4 myr from whole-core magnetic susceptibility measurements. *Nature* 342:897–899.

Bloemendal, J., King, J. W., Hunt, A., deMenocal, P. B., and Hayashida, A. 1993. Origin of the sedimentary magnetic record at Ocean Drilling Program Sites on the Owen Ridge, western Arabian Sea. *J. Geophys. Res.* 98:4199–4219.

Bonnefille, R. 1976. Palynological evidence for an important change in the vegetation of the Omo Basin between 2.5 and 2.0 Ma. In *Earliest man and environments in the Lake Rudolf Basin*, pp. 421–432 (ed. Coppens, Y. et al.). University of Chicago Press, Chicago.

———. 1983. Evidence for a cooler and drier climate in the Ethiopian Uplands towards 2.5 Myr ago. *Nature* 303:487–491.

Bonnefille, R., Vincens, A., and Buchet, G. 1987. Palynology, stratigraphy, and paleoenvironment of a Pliocene hominid site (2.9–3.3 Ma) at Hadar, Ethiopia. *Paleogeogr., Paleoclimatol., Paleoecol.* 60:249–281.

Brain, C. K. 1981. The evolution of man in Africa: Was it the result of Cainozoic cooling? *Annex. Transv. Geol. Soc. S. Afr. J. Sci.* 84:1–19.

Brown, F. H., Sarna-Wojcicki, A. M., Meyer, C. E., and Haileab, B. 1992. Correlation of Pliocene and Pleistocene tephra layers between the Turkana Basin of East Africa and the Gulf of Aden. *Quat. Internat.* 13/14:55–67.

Cande, S. C., and Kent, D. V. 1992. A new geomagnetic polarity time scale for the Late Cretaceous and Cenozoic. *J. Geophys. Res.* 97:13917–13951.

Cerling, T. E. 1992. Development of grasslands and savannas in East Africa during the Neogene. *Paleogeog., Paleoclimatol., Paleoecol.* 97:241–247.

Cerling, T. E., and Hay, R. L. 1988. An isotopic study of paleosol carbonates from Olduvai Gorge. *Quat. Res.* 25:63–78.

Clark, J. D. 1980. Early human occupation of African savanna environments. In *Human ecology in savanna environments,* pp. 41–71 (ed. Harris, D. R.). Academic Press, London.

Clemens, S. C., and Prell, W. L. 1990. Late Pleistocene variability of Arabian Sea summer monsoon winds and continental aridity: Eolian records from the lithogenic component of deep-sea sediments. *Paleoceanography* 5:109–145.

———. 1991. One million year record of summer monsoon winds and continental aridity from the Owen Ridge (Site 722), Northwest Arabian Sea. In *Proc. Ocean Drill. Prog.,* vol. 17, pp. 365–388 (ed. Prell, W. L., and Niitsuma, N., et al.). Ocean Drill. Prog., College Station, Tex.

Clemens, S., Prell, W., Murray, D., Shimmield, G., and Weedon, G. 1991. Forcing mechanisms of the Indian Ocean monsoon. *Nature* 353:720–725.

Coppens, Y., and Koeniguer, J. C. 1976. Paléoflores ligneuses tertiaire et quaternaire du Tchad. *Paleoecol. Africa* 9:105–106.

Curry, W. B., Ostermann, D. R., Gupta, M. V. S., and Ittekot, V. 1992. Foraminiferal production and monsoonal upwelling in the Arabian Sea: Evidence from sediment traps. In *Upwelling systems: Evolution since the Early Miocene,* pp. 93–106 (ed. Summerhayes, C. P., Prell, W. L., and Emeis, K. C.). Geol. Soc. Spec. Publ.

deMenocal, P. B., Bloemendal, J., and King, J. W. 1991. A rock-magnetic record of monsoonal dust deposition to the Arabian Sea: Evidence for a shift in the mode of deposition at 2.4 Ma. In *Proc. Ocean Drill. Prog.,* vol. 117, pp. 389–407 (ed. Prell, W. L., Niitsuma, N., et al.). Ocean Drill. Prog., College Station, Tex.

deMenocal, P. B., and Rind, D. 1993. Sensitivity of Asian and African climate to variations in seasonal insolation, glacial ice cover, sea-surface temperature, and Asian orography. *J. Geophys. Res.* 98:7265–7287.

deMenocal, P. B., Ruddiman, W. F., and Pokras, E. M. 1993. Influences of high- and low-latitude processes on African climate: Pleistocene eolian records from equatorial Atlantic Ocean Drilling Program Site 663. *Paleoceanography* 8:209–242.

Denton, G. H., and Hughes, T. J. 1981. *The last great ice sheets.* Wiley-Interscience, New York.

Dupont, L., and Hoogheimstra, H. 1989. The Saharan-Sahelian boundary during the Brunhes chron. *Acta Bot. Nerrl.* 38:405–415.

Druyan, L. M. 1987. GCM studies of the African monsoon. *Clim. Dyn.* 2:117–126.

Druyan, L., and Koster, R. D. 1989. Sources of Sahel precipitation for simulated drought and rainy seasons. *J. of Climate* 3:1438–1446.

Feibel, C. S., Brown, F. H., and Mac Dougall, I. 1989. Stratigraphic context of fossil hominids from the Omo Group deposits: Northern Turkana Basin, Kenya and Ethiopia. *Amer. J. Phys. Anthropol.* 78:595–622.

Flohn, H. 1965. *Studies on the meteorology of tropical Africa.* Bonner Meteorologische Abhandlungen, vol. 5.

Folland, C., Palmer, T., and Parker, D. 1986. Sahel rainfall and worldwide sea temperatures. *Nature* 320:602–607.

Gasse, F., Lédée, V., Massault, M., and Fontes, J. C. 1989. Water level fluctuations of Lake Tanganyika in phase with oceanic changes during the last glaciation and deglaciation. *Nature* 342:57–59.

Gasse, F., Téhet, R., Durand, A., Gibert, E., and Fontes, J. C. 1990. The arid-humid transition in the Sahara and Sahel during the last deglaciation. *Nature* 346:141–146.

Gates, W. L. 1976. Modelling the ice-age climate. *Science* 191:1138–1144.

Grine, F. E. 1986. Ecological causality and the pattern of Plio-Pleistocene hominid evolution in Africa. *S. Afr. J. Sci.* 82:87–89.

———. 1988. *The evolutionary history of the robust australopithecines.* Aldine, New York.

Hamilton, A. C. 1982. *Environmental history of East Africa,* Academic Press, New York.

Hamilton, A. C., and Taylor, D. 1991. History of climate and forests in tropical Africa during the last 8 million years. *Climatic Change* 19:65–78.

Harris, J. W. K. 1983. Cultural beginnings: Plio-Pleistocene archaeological occurrences from the Afar, Ethiopia. In *African archaeological review,* pp. 3–31 (ed. David, N.). Cambridge University Press, Cambridge.

Hastenrath, S., and Lamb, P. 1978. Some aspects of circulation and climate over the eastern equatorial Atlantic. *Mon. Weath. Rev.* 105:1019–1023.

Hays, J. D., and Perruzza, A. 1972. The significance of calcium carbonate oscillations in eastern equatorial Atlantic deep-sea sediments for the end of the Holocene warm interval. *Quat. Res.* 2:355–362.

Hilgen, F. J., and Langereis, C. G. 1989. Periodicities of $CaCO_3$ cycles in the Pliocene of Sicily: Discrepancies with the quasi-periods of the earth's orbital cycles? *Terra Nova* 1:409–415.

———. 1993. A critical re-evaluation of the Miocene/Pliocene boundary as defined in the Mediterranean. *Earth and Planet. Sci. Lett.* 118:167–179.

Hill, A. 1987. Causes of perceived faunal change in the later Neogene of East Africa. *J. Hum. Evol.* 16:583–596.

Imbrie, J., Berger, A., and Shackleton, N. J. 1993. Role of orbital forcing: A two-million-year perspective. In *Global changes and the perspective of the past,* pp. 263–277 (ed. Eddy, J. A., and Oeschger, H.). J. Wiley, London.

Jansen, E., Bleil, V., Heinrich, R., Kringsrand, L., and Slettemark, B. 1988. Paleoenvironmental history of the Norwegian Sea and northeast Atlantic during the last 2.8 Ma: Deep-Sea Drilling Project/Ocean Drilling Program Sites 610, 642, 643, 644. *Paleoceanography* 3:563–581.

Keigwin, L. D. 1979. Late Cenozoic isotope stratigraphy and paleoceanography of DSDP sites from the east equatorial and central North Pacific Ocean. *Earth Planet. Sci. Lett.* 45:361–382.

Kolla, V., and Biscaye, P. E. 1977. Distribution and origin of quartz in the sediments of the Indian Ocean. *J. Sed. Petrol.* 47:642–649.

Kolla, V., Biscaye, P. E., and Hanley, A. F. 1979. Distribution of quartz in Late Quaternary Atlantic sediments in relation to climate. *Quat. Res.* 11:261–277.

Kutzbach, J. E. 1981. Monsoon climate of the Early Holocene:

Climatic experiment with the earth's orbital parameters for 9000 years ago. *Science* 214:59–61.

Kutzbach, J. E., and Guetter, P. J. 1984. Sensitivity of late-glacial and Holocene climates to the combined effects of orbital parameter changes and lower boundary condition changes: "Snapshot" simulations with a general circulation model for 18, 9, and 6 ka BP. *Ann. of Glaciology* 5:85–87.

Kutzbach, J. E., and Otto-Bliesner, B. L. 1982. The sensitivity of the African-Asian monsoonal climate to orbital parameter changes for 9000 years BP in a low-resolution general circulation model. *J. Atmos. Sci.* 39:1177–1188.

Lamb, P. J. 1978. Case studies of tropical Atlantic surface circulation patterns during recent sub-Saharan weather anomalies: 1967–1968. *Mon. Weath. Rev.* 106:482.

Lezine, A.-M. 1991. West African paleoclimates during the last climatic cycle inferred from an Atlantic deep-sea pollen record. *Quat. Res.* 35:456–463.

McDougall, I., Brown, F. H., Cerling, T. E., and Hillhouse, J. W. 1992. A reappraisal of the geomagnetic polarity timescale to 4 Ma using data from the Turkana Basin, East Africa. *Geophys. Res. Lett.* 19:2349–2352.

McIntyre, A., Ruddiman, W. F., Karlin, K., and Mix, A. C. 1989. Surface water response of the equatorial Atlantic Ocean to orbital forcing. *Paleoceanography* 4:19–55.

Manabe, S., and Hahn, D. G. 1977. Simulation of the tropical climate of an ice age. *J. Geophys. Res.* 82:3889–3911.

Martinson, D. G., Menke, W., and Stoffa, P. 1982. An inverse approach to signal correlation. *J. Geophys. Res.* 87:B2, 4807–4818.

Molfino, B., and McIntyre, A. 1990. Precessional forcing of nutricline dynamics in the equatorial Atlantic. *Science* 249:766–769.

Mortlock, R. A., and Froelich, P. N. 1989. A simple method for the rapid determination of biogenic opal in pelagic marine sediments. *Deep-Sea Research* 36:1415–1426.

Nair, R. R., Ittekot V., Manganini, S., Ramaswamy, V., Haake, B., Degens, E., Desai, B., and Honjo, S. 1989. Increased particle flux to the deep ocean related to monsoons. *Nature* 338:749–751.

Overpeck, J. T., Peterson, L. C., Kipp, N., Imbrie, J., and Rind, D. 1989. Climate change in the circum-North Atlantic region during the last deglaciation. *Nature* 338:553–557.

Petrov, M. P. 1976. *Deserts of the world.* Wiley and Sons, New York.

Pokras, E. M., and Mix, A. C. 1985. Eolian evidence for spatial variability of Quaternary climates in tropical Africa. *Quat. Res.* 24:137–149.

———. 1987. Earth's precession cycle and Quaternary climatic change in tropical Africa. *Nature* 326:486–487.

Prell, W. L., and Kutzbach, J. E. 1987. Monsoon variability over the past 150,000 years. *J. Geophys. Res.* 92:8411–8425.

———. 1992. Sensitivity of the Indian monsoon to forcing parameters and implications for its evolution. *Nature* 360:647–652.

Prospero, J. M., and Nees, R. T. 1977. Dust concentration in the atmosphere of the equatorial North Atlantic: Possible relationship to Sahelian drought. *Science* 196:1196–1198.

Raymo, M. E., Hodell, D., and Jansen, E. 1992. Response of deep ocean circulation to initiation of Northern Hemisphere glaciation (3–2 Ma). *Paleoceanography* 7:645–672.

Raymo, M. E., Ruddiman, W. F., Backman, J., Clement, B., and Martinson, D. G. 1989b. Late Pliocene variation in Northern Hemisphere ice sheets and North Atlantic deep water circulation. *Paleoceanography* 4:413–446.

Raymo, M. E., Ruddiman, W. F., and Froelich, P. N. 1988. Influence of Late Cenozoic mountain building on ocean geochemical cycles. *Geology* 16:649–653.

Rea, D. K., and Schrader, H. 1985. Late Pliocene onset of glaciation: Ice-rafting and diatom stratigraphy of North Pacific DSDP cores. *Paleogeogr., Paleoclimatol., Paleoecol.* 49:313–325.

Rind, D. 1987. Components of the ice age circulation. *J. Geophys. Res.* 92:4241–4281.

Rind, D., and Peteet, D. 1985. Terrestrial conditions at the last glacial maximum and CLIMAP sea-surface temperature estimates: Are they consistent? *Quat. Res.* 24:1.

Rossignol-Strick, M. 1983. African monsoons, an immediate climatic response to orbital insolation forcing. *Nature* 303:46–49.

Ruddiman, W. F., and Janecek, T. 1989. Pliocene-Pleistocene biogenic and terrigenous fluxes at equatorial Atlantic Sites 662, 663, and 664. In *Proc. Ocean Drill. Prog.*, vol. 108, pp. 211–240 (ed. Ruddiman, W. F., and Sarnthein, M., et al.). Ocean Drill. Prog., College Station, Tex.

Ruddiman, W. F., Prell, W. L., and Raymo, M. E. 1989a. Late Cenozoic uplift in southern Asia and the American West: Rationale for general circulation modeling experiments. *J. Geophys. Res.* 94:18,379–18,391.

Ruddiman, W. F., Raymo, M. E., Martinson, D. G., Clement, B. M., and Backman, J. 1989b. Pleistocene evolution: Northern Hemisphere ice sheets and North Atlantic Ocean. *Paleoceanography* 4:353–412.

Ruddiman, W. F., Sarnthein, M., Baldauf, J., Backman, J., Curry, W., Dupont, L. M., Janecek, T., Pokras, E. M., Raymo, M. E., Stabell, B., Stein, R., and Tiedemann, R. 1989c. Late Miocene to Pleistocene evolution of climate in Africa in Africa and the low-latitude Atlantic: Overview of Leg 108 results. In *Proc. Ocean Drill. Prog.*, vol. 108, pp. 463–484 (ed. Ruddiman, W., and Sarnthein, M., et al.). Ocean Drill. Prog., College Station, Tex.

Ruddiman, W. F., Shackleton, N. J., and McIntyre, A. 1986. North Atlantic sea-surface temperatures for the last 1.1 million years. In *North Atlantic paleoceanography,* vol. 21, pp. 155–173 (ed. Summerhayes, C. P., and Shackleton, N. J.). Geological Society of London, London.

Sarna-Wojcicki, A. M., Meyer, C. E., Roth, P. H., and Brown, F. H. 1985. Ages of tuff beds at East African early hominid sites and sediments in the Gulf of Aden. *Nature* 313:306–308.

Sarnthein, M., Tetzlaff, G., Koopman, B., Wolter, K., and Pflaumann, U. 1981. Glacial and interglacial wind regimes over the eastern subtropical Atlantic and northwest Africa. *Nature* 293:193–196.

Schrenk, F., Bromage, T. G., Betzler, C. G., Ring, U., and Juwayeyi, Y. M. 1993. Oldest *Homo . . . ?* and Pliocene biogeography of the Malawi Rift. *Nature* 365:833–836.

Shackleton, N. J., Backman, J., Zimmerman, H., Kent, D. V., Hall, M. A., Roberts, D. G., Schnitker, D., Baldauf, J., Despraires, A., Homrighausen, R., Huddlestun, P., Keene, J.,

Kaltenback, A. J., Krumsiek, K. A. O., Morton, A. C., Murray, J. W., and Westberg-Smith, J. 1984. Oxygen isotope calibration of the onset of ice-rafting and history of glaciation in the North Atlantic region. *Nature* 307:620–623.

Shackleton, N. J., Baldauf, J., Flores, J. A., Iwai, M., Moore, T. C., Raffi, I., and Vincent, E. (in press, b). Biostratigraphic summary: ODP Leg 138. In *Proc. Ocean Drill. Prog.*, vol. 138, (ed. Mayer, L., Pisias, N. J., and Janecek, T.). Ocean Drill. Prog., College Station, Tex.

Shackleton, N. J., Berger, A., and Peltier, W. R. 1990. An alternative astronomical calibration of the Lower Pleistocene timescale based on ODP Site 677. *Trans. Royal Soc. Edinburgh: Earth Sciences* 81:251–261.

Shackleton, N. J., Crowhurst, S., Hagelberg, T., Pisias, N. J., and Schneider, D. A. (in press, a). A new Late Neogene time scale: Application to Leg 138 Sites. In *Biostratigraphic summary, ODP Leg 138*, vol. 138, (ed. Mayer, L., Pisias, N. J., and Janecek, T.). Ocean Drill. Prog., College Station, Tex.

Sirocko, F., and Sarnthein, M. 1989. Wind-borne deposits in the northwestern Indian Ocean: Record of Holocene sediments versus modern satellite data. In *Paleoclimatology and paleometeorology: Modern and past pattern of global atmospheric transport*, pp. 401–433 (ed. M. Leinen, and M. Sarnthein). Kluwer, Berlin.

Sirocko, F., Sarnthein, M., Lange, H., and Erlenkeuser, H. 1991. Atmospheric summer circulation and coastal upwelling in the Arabian Sea during the Holocene and the Last Glaciation. *Quat. Res.* 36:72–93.

Stanley, S. M. 1992. An ecological theory for the origin of *Homo. Paleobiology* 18:237–257.

Stein, R. 1985. Late Neogene changes of paleoclimate and paleoproductivity off NW Africa (DSDP Site 397). *Paleogeogr., Paleoclimatol., Paleoecol.* 49:47–59.

Stein, R., and Sarnthein, M. 1986. Late Neogene events of atmospheric and oceanic circulation patterns offshore northwest Africa: High-resolution record from the deep-sea sediments. *Paleoecol. Africa* 16:9–36.

Street-Perrott, F. A., and Perrott, R. A. 1990. Abrupt climate fluctuations in the tropics: The influence of Atlantic Ocean circulation. *Nature* 343:607–612.

Street-Perrott, F. A., and Harrison, S. A. 1984. Temporal variations in lake levels since 30,000 yr BP: An index of the global hydrological cycle. In *Climate processes and climate sensitivity*, pp. 118–129 (ed. Hansen, J. E., and Takahashi, T.). American Geophysical Union, Washington, D.C.

Suc, J.-P. 1984. Origin and evolution of Mediterranean vegetation and climate in Europe. *Nature* 307:429–432.

Tiedemann, R., Sarnthein, M., and Shackleton, N. J. 1994. Astronomic timescale for the Pliocene Atlantic $\delta^{18}O$ and dust flux records of ODP Site 659. *Paleoceanography* 9:619–638.

Tiedemann, R., Sarnthein, M., and Stein, R. 1989. Climatic changes in the western Sahara: Aeolio-marine sediment record of the last 8 million years (Sites 657–661). In *Proc. Ocean Drill. Prog.*, vol. 108, pp. 241–261 (ed. Ruddiman, W. F., and Sarnthein, M., et al.). Ocean Drill. Prog., College Station, Tex.

van Zinderen Bakker, E. M., and Mercer, J. H. 1986. Major Late Cainozoic climatic events and paleoenvironmental changes in Africa viewed in a world wide context. *Paleogeogr., Paleoclimatol., Paleoecol.* 56:217–235.

Verardo, D., Froelich, P. N., and McIntyre, A. 1990. Determination of organic carbon and nitrogen in marine sediments using the Carlo Erba NA-1500 Analyzer. *Deep-Sea Res.* 37:157–165.

Vrba, E. S. 1985. Ecological and adaptive changes associated with early hominids. In *Ancestors: The hard evidence*, pp. 63–71 (ed. Delson, E.). Liss, New York.

Vrba, E. S., Denton, G. H., and Prentice, M. L. 1989. Climatic influences on early Hominid behavior. *Ossa* 14:127–156.

Wesselman, H. B. 1985. Fossil micromammals as indicators of climatic change about 2.4 myr ago in the Omo Valley, Ethiopia. *S. Afr. J. Sci.* 81:260–261.

Wood, B. 1992. Origin and evolution of the genus *Homo. Nature* 355:783–790.

Yemane, K., Bonnefille, R., and Faure, H. 1985. paleoclimatic and tectonic implications of Neogene microflora from the northwestern Ethiopian highlands. *Nature* 318:653–656.

Chapter 20

Steps toward Drier Climatic Conditions in Northwestern Africa during the Upper Pliocene

Lydie M. Dupont and Suzanne A. G. Leroy

This chapter deals with vegetational and climatic change in northwestern Africa during the Lower and the Upper Pliocene, from 3.7 to 1.7 million years (myr). Northwestern Africa is now characterized by aridity and covered by deserts and dry vegetation. A 200 m-long marine pollen record from the Ocean Drilling Program (ODP) Site 658, at 21°N and 19°W, reveals cyclic fluctuations and long-term variations in vegetation and continental climate. So far, it is the best Upper Pliocene pollen record for this part of Africa in terms of resolution and time coverage.

Marine records have an advantage over terrestrial ones, because they frequently cover long sedimentation periods, fit into a firm global stratigraphy, and have strict time control. Deep-sea pollen records contain grains that have been transported long distances. Consequently, they integrate the vegetation record of a large area and contain an impoverished pollen flora that represents only those plants that produce much pollen. These features make marine palynology a suitable tool for studying large vegetation zones over long periods. Additionally, terrestrial chronology relative to oceanic chronology can be established on the same material.

After introducing ODP Site 658 record, we focus on the long-term variations and on the comparison of two periods from the Late Pliocene, one between 3.5 and 3.2 myr and the other between 2.2 and 1.7 myr. Both periods show strong short-term fluctuations of climate but differ in the timing and extent of the response of the vegetation, as indicated by cross-correlation of pollen influx values with insolation values and oxygen isotopes. A full description of the palynology of ODP Site 658 is given in Dupont et al. (1989) and Leroy and Dupont (1994).

ISBN 0-300-06348-2.

Sedimentology

ODP Site 658 (leg 108) is situated northwest of Africa (fig. 20.1) at a water depth of 2,263 m on the continental slope 160 km west of Cape Blanc. It is located below an important near-shore upwelling cell induced by the trade winds. Its position at a terrace of the continental slope between two major canyon systems restricts to a minimum disturbance of the sediment record by lateral down-slope transport (Ruddiman et al., 1988). A hiatus spanning the Lower Pleistocene separates the upper 100 m of sediment covering the Brunhes chron from the lower 200 m covering the late Lower and Upper Pliocene (Sarnthein and Tiedemann, 1989). The high sedimentation rate, due to high organic production in the upwelling zone combined with high Saharan dust influx, provides a Plio-Pleistocene record of high quality in which bioturbation hardly obscures the fine-scale resolution up to 1,000 years (Tiedemann et al., 1989).

Chronology

The time scale of the sequence (figs. 20.2–20.5) is provided by biostratigraphy, paleomagnetism, and oxygen isotope stratigraphy (Ruddiman et al., 1988; Sarnthein and Tiedemann, 1989; Tiedemann, 1991). The ages of the isotope stages of ODP Site 658 are derived by comparison with ODP site 659 (18°N, 21°W; Tiedemann, pers. comm.). The orbitally tuned time scale of ODP 659 reaches back 5 myr (Tiedemann et al., 1994). The ages of the ODP Site 659 time scale are similar to the independent calibrations of Shackleton et al. (1990) on Pacific sediments (ODP Site 677) and of Hilgen (1991) on Mediterranean sapropels. Therefore, the ages of ODP Site 659, and thus of ODP Site 658, are consistently 130 kyr older than those obtained by Raymo et al. (1989) from the Deep Sea Drilling Project (DSDP) Site 607 in the North Atlantic. Correlation of ODP Site 658 with ODP Site 659 and DSDP Site 607 reveals a hiatus at 158 m, spanning the period from 2.25 to 2.46 myr, and several coring gaps (Tiedemann, 1991). The time resolution of the Pliocene pollen record of ODP Site 658 (using 401 samples) exceeds 1 sample per 5 thousand years (kyr), except for the gaps mentioned, and spans the period from 3.7 to 1.7 myr.

Wind Transport of Pollen into the Marine Sediments

Presently, at latitudes between 19°N and 21°N, the climatically sensitive vegetation of the Sahel gives way to the desert (fig. 20.1). In the adjacent East Atlantic, at 21°N, ODP Site 658 id located where northeasterly trade winds are overlaid by the mid-tropospheric African Easterly Jet (AEJ), namely the summer maximum of the Saharan Air Layer. Trade winds transport pollen from their source areas in the Mediterranean and the Sahara to the marine site (Hooghiemstra et al., 1986). Strong, heat-induced squall lines bring dust and pollen from the Sahel and the southern Sahara into altitudes of the AEJ (1,000–5,000 m). Then, the AEJ carries pollen from latitudes between 16°N and 20°N westward and northward over the Atlantic.

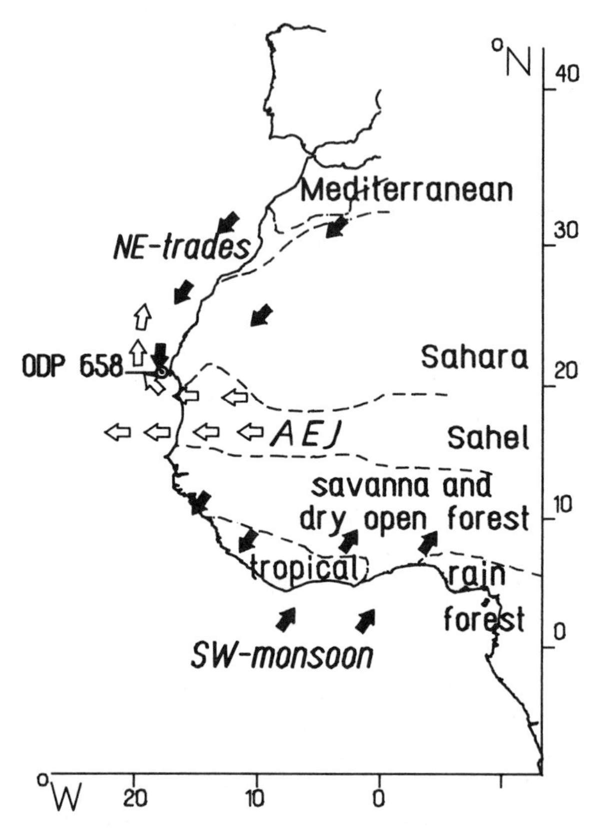

Fig. 20.1. Location of ODP Site 658, at 21°N 19°W; surface winds—northeasterly trades and southwesterly monsoon (black arrows), the mid-tropospheric African Easterly Jet (AEJ, open arrows); and the modern position of the main northwest-African vegetation zones—Mediterranean, Mediterranean Steppes, Sahara, Sahel, savanna and dry open forest, tropical rain forest.

Long-term Variation

We define *long-term variation* (fig. 20.2) as one that lasts over a period of several hundred thousand to a few million years and *short-term variation* as one that lasts up to one hundred thousand years.

River Discharge. For the Pliocene and the Pleistocene, the sedimentology of ODP Site 658 shows dust transport into the Atlantic by winds (trades and the African Easterly Jet) as well as clay transport by rivers (Tiedemann et al., 1989; Tiedemann, 1991). Quartz content and siliciclastics (eolian dust > 6 μm) indicate wind vigor, whereas the clay content illustrates the importance of river discharge to the formation of the sediment. The declining percentage maxima of Cyperaceae in the ODP Site 658 pollen record (fig. 20.4) confirms the conclusion drawn from sedimentary analysis (Tiedemann, 1991) of persistent river discharge before 3.4 myr followed by subsequent decline. Generally, river-borne pollen seems numerous until 2.97 myr (Leroy and Dupont, 1994). Thereafter, transport of pollen grains by river discharge is insignificant, which is comparable to the modern situation (Hooghiemstra, 1989; Dupont and Agwu, 1991).

Mangrove Swamps and Tropical Forests. A humid and probably warm climate prevailed before 3.5 myr. According to the percentage average of *Rhizophora* pollen that exceeds 4 percent (fig. 20.2), mangrove swamps were growing near Cape Blanc around 3.70 myr, probably accompanying a paleoriver. At present, 10 percent *Rhizophora* pollen grains are found at the mouth of the Senegal River, which is 5° to the south (Dupont and

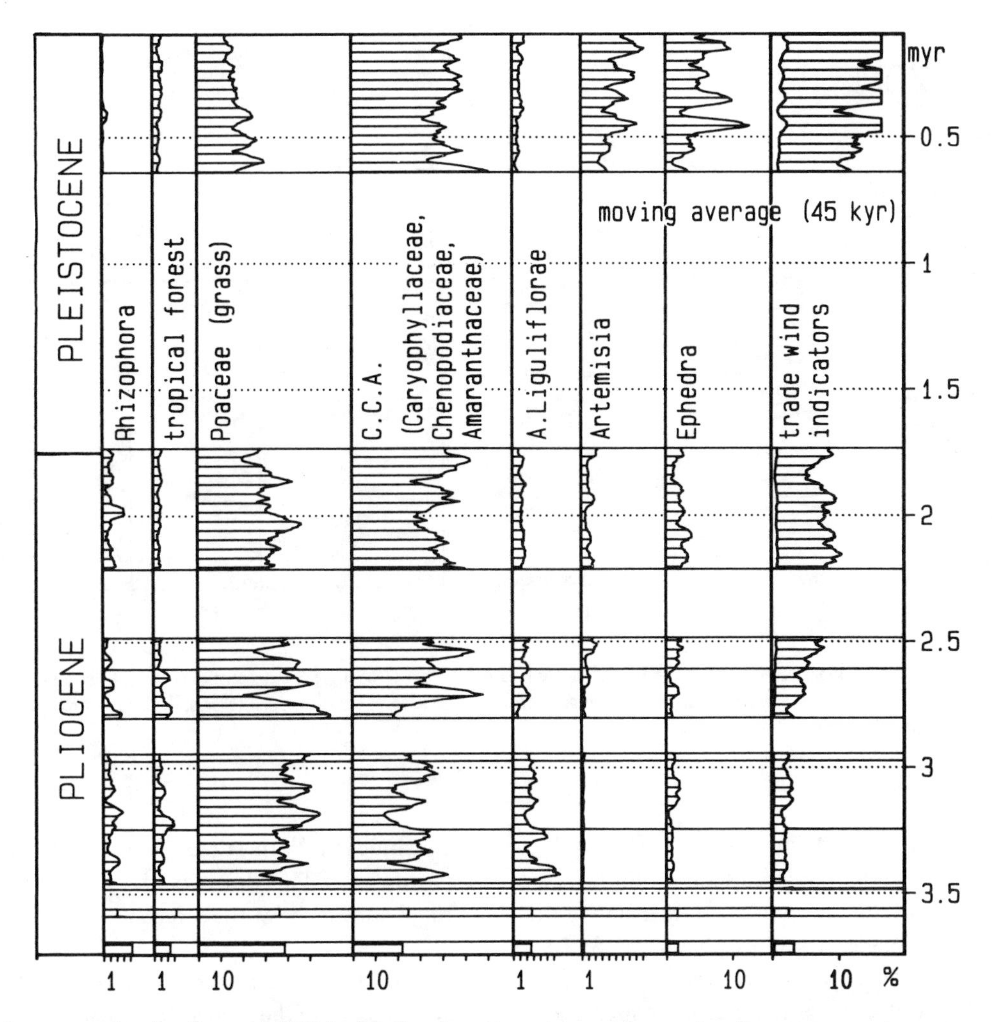

Fig. 20.2. Long-term trends in the pollen record of ODP Site 658. Plots of a moving average over nine successive pollen spectra expressed as percentages of the pollen total and interpolated at steps of 5 kyr: *Rhizophora* (mangrove tree), sum of tropical forest elements (dry open and wet lowland forest; Guinean and Sudanian vegetation zones), Poaceae (grasses), CCA (the sum of Caryophyllaceae and Amaranthaceae-Chenopodiaceae), Asteraceae Liguliflorae (A. Liguliflorae), *Artemisia, Ephedra,* and trade-wind indicators (sum of *Artemisia, Ephedra,* Mediterranean elements, and A. Liguliflorae). The horizontal axis of each curve starts at 0 and increases from left to right. Time scale in myr on the vertical axis.

Agwu, 1991). Percentages of the sum of pollen from Sudanian and Guinean vegetation, that is, wooded savanna, woodland, and tropical forest, repeatedly exceed 5 percent, indicating that forest and savanna had a distribution at least as far as 21°N during those periods. This latitude is 5° to 10° farther to the north than the modern situation (Dupont and Agwu, 1991). Before 3.5 myr, and between 3.25 and 2.6 myr, percentage averages of tropical forest element (>2 percent) indicate a northern extent of tropical forests that probably shifted southward after 2.6 myr.

The Development of a Desert in the Western Sahara. High pollen percentages of Poaceae (grasses) in the lower part of the sequence indicate that extensive savannas were already present in the Lower Pliocene (fig. 20.2). After 2.8 myr, a declining trend in percentage averages of Poaceae indicates a reduction of savanna vegetation (including wooded savanna and dry, open forest), probably as a result of the development of a desert in the western Sahara. Percentages of Asteraceae Liguliflorae reach high values before 3.2 myr but decline to low values afterward (fig. 20.3). Nowadays, many species of the A.

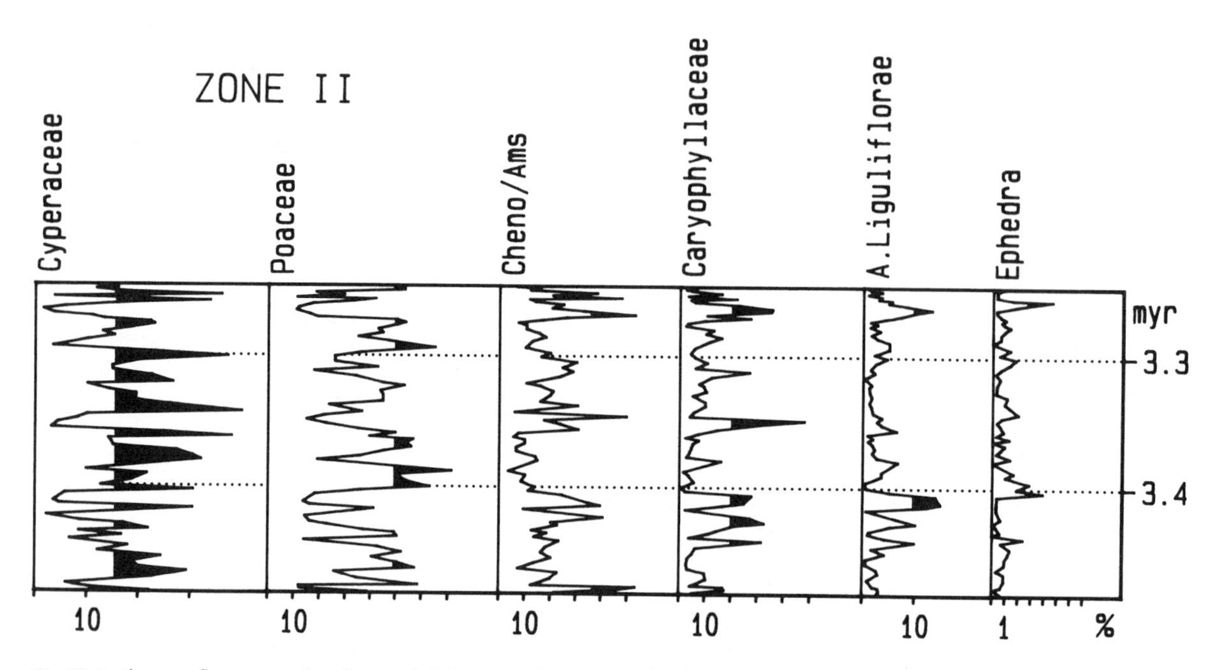

Fig. 20.3. Short-term fluctuations in the pollen record of ODP Site 658 between 3.5 and 3.2 myr (Zone II). Plots of six selected pollen taxa expressed as percentages of total pollen: Cyperaceae, Poaceae, Chenopodiaceae-Amaranthaceae (Cheno/Ams), Caryophyllaceae, A. Liguliflorae, and *Ephedra*. The horizontal axis of each curve starts at 0 and increases from left to right.

Liguliflorae grow in the Mediterranean area, but only a few of them grow in the savannas south of the Sahara. This may be a reflection of a westward extension of the desert driving a wedge in the distribution of A. Liguliflorae species. From 2.7 myr on, percentage averages of CCA (the sum of Caryophyllaceae, Chenopodiaceae, and Amaranthaceae) reach high values (around 40 percent), comparable to those of the Brunhes chron record indicating aridity in northwestern Africa. Mean percentages of *Ephedra* and *Artemisia* during the Pliocene are five times lower than those of the Upper Pleistocene, indicating that arid periods were even shorter and/or milder during the Pliocene.

The pollen record of the Brunhes chron registered, through the African Easterly Jet, latitudinal shifts of up to 10° for desert and wooded grassland. In particular, time-transgressive percentage maxima of Cyperaceae, Poaceae, CCA, *Artemisia,* and *Ephedra* (Dupont and Hooghiemstra, 1989; Dupont et al., 1989) reflect shifting of the desert-savanna boundary (Saharan-Sahelian boundary). We interpret each maximum as a southward extension of the desert and, therefore, as a reflection of drier climate. A few shifts were recorded before 3 myr, but they occur regularly from 2.6 myr onward (fig. 20.5). We conclude that by this time a desert established in the western Sahara. This desert built a permanent though shifting boundary with the savanna south of it.

Trade Winds. An estimation of trade-wind vigor is given by the sum of those pollen taxa that have their main source areas in the northern Sahara and North Africa: *Ephedra, Artemisia,* and *Pinus,* as well as A. Liguliflorae for periods after 2.9 myr. Generally, the strength of the trades was much lower during the Pliocene than during the Upper Pleistocene. Trade winds were very weak until 3.2 myr, resulting in low transport of pollen from North Africa. At 3.26 myr, however, trade-wind vigor probably increased for a short period. Then, at 3.2, 2.8, and 2.6 myr, the level of trade-wind strength increases stepwise (fig. 20.2). During the final Pliocene, trade winds are rather strong, with the exception of the period between 1.87 and 1.85 myr, when oxygen isotopes indicate reduced extension of ice sheets. The record of trade-wind indicators corroborates the estimates of wind strength by grain-size analysis of ODP Site 658, ODP Site 659, and DSDP Site 397, showing an increase of the trade winds from 3.2 to 2.6 myr (Tiedemann et al., 1989; Tiedemann, 1991).

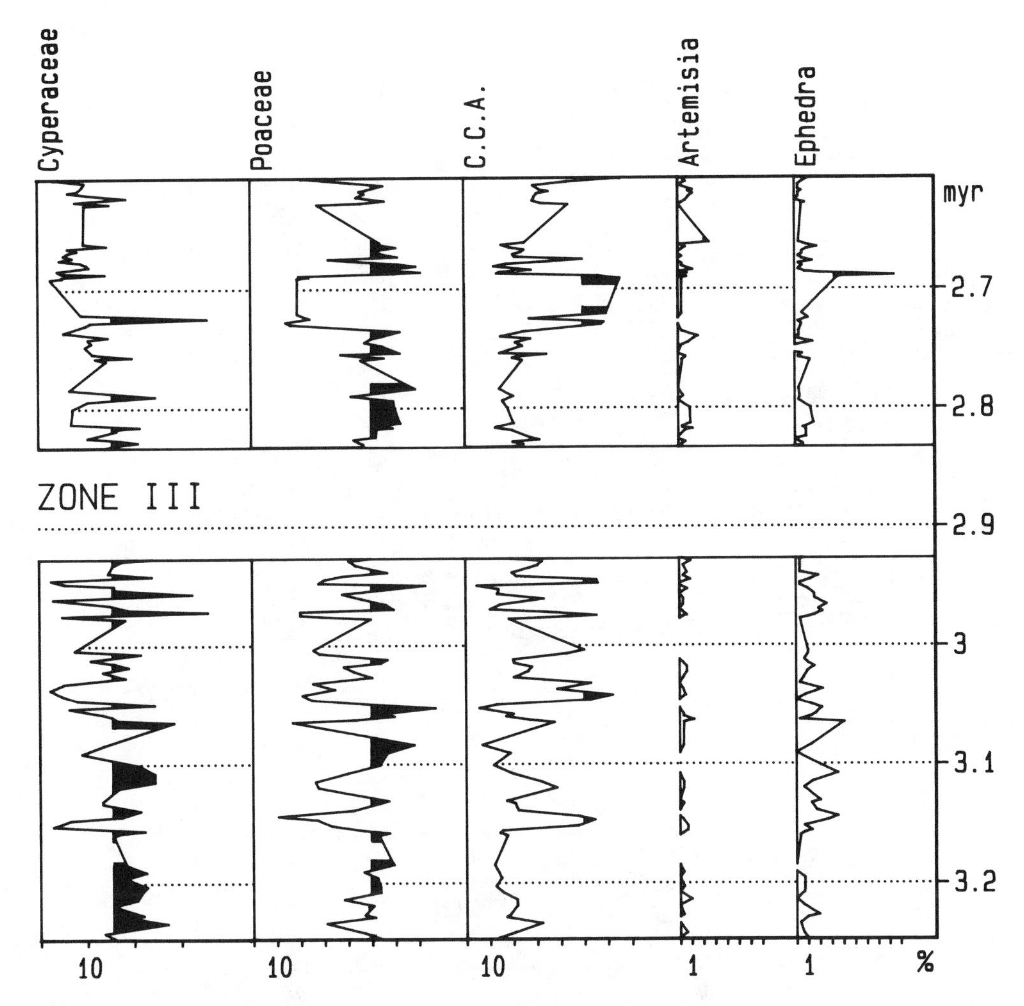

Fig. 20.4. Short-term fluctuations in the pollen record of ODP Site 658 between 3.2 and 2.6 myr (Zone III). Plots of five selected pollen taxa expressed as percentages of total pollen and arranged from left to right in the order of their indication value for climatic conditions at 21°N, from humid to arid: Cyperaceae, Poaceae, CCA (= Carophyllaceae, Chenopodiaceae, and Amaranthaceae), *Artemisia*, and *Ephedra*. The horizontal axis of each curve starts at 0 and increases from left to right. Coring gaps interrupt the record between 2.93 and 2.83 myr, between 2.72 and 2.69 myr, and between 2.66 and 2.63 myr.

Vegetation and Climatic Development before and after 2.5 Million Years

The establishment of desert and the onset of trade winds mark important changes in vegetation and climate of northwestern Africa that correlate with the strong development of ice sheets in the Northern Hemisphere during oxygen isotope stages 100 and 98, around 2.5 myr. In this section, we examine the difference in response of vegetation and climate to orbital forcing by means of spectral analyses of pollen taxa, oxygen isotopes, and insolation for two periods: 3.5–3.2 myr and 2.2–1.7 myr.

The Period between 3.5 and 3.2 myr (Zone II). Five dry phases characterize this period (fig. 20.3). Three of them, at 3.48, 3.35, and 3.26–3.27 myr, show high percentages of Amaranthaceae-Chenopodiaceae pollen (50 percent), indicating arid conditions. Two other dry phases, at 3.44–3.40 and 3.31 myr, are less prominent. Although they show high percentages of Caryophyllaceae, percentages of Amaranthaceae-Chenopodiaceae hardly exceed 40 percent. High percentage values (>10 percent) are found for A. Liguliflorae between 3.44 and 3.40 myr and at ca. 3.27 myr, presenting a unique situation without a modern analogue. The youngest and most

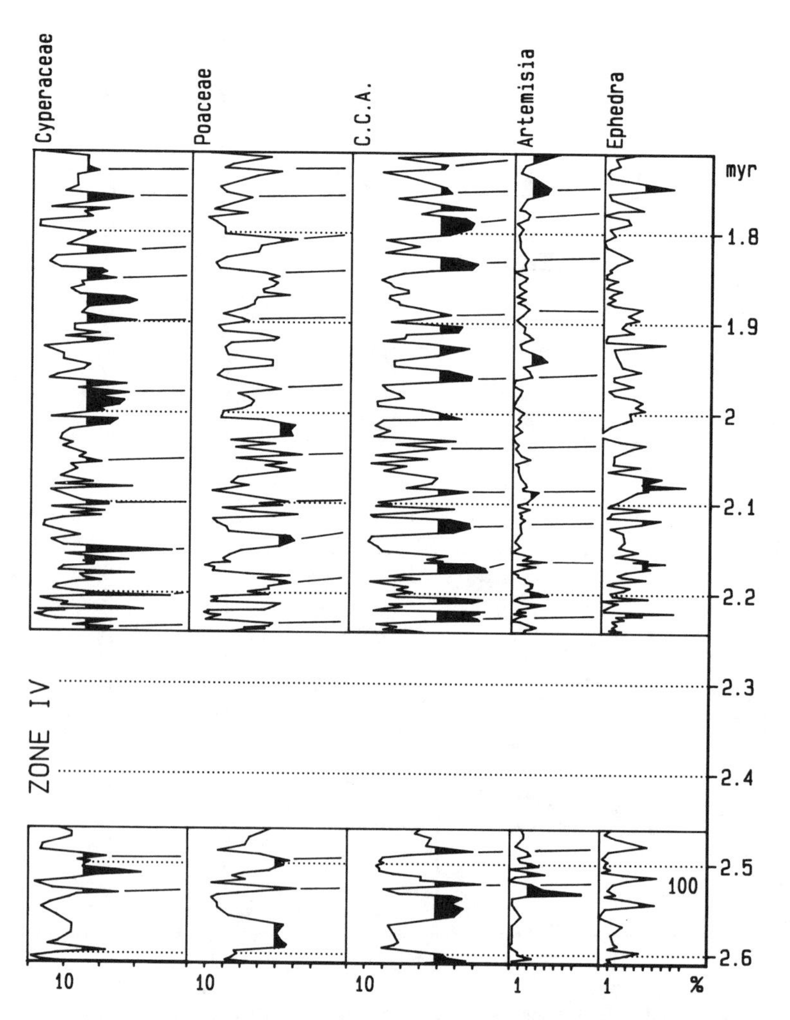

Fig. 20.5. Short-term fluctuations in the pollen record of ODP Site 658 between 2.6 and 1.7 myr (Zone IV). Plots of 5 selected pollen taxa expressed as percentages of total pollen and arranged from left to right in the order of their indication value for climatic conditions at 21°N, from humid to arid: Cyperaceae, Poaceae, CCA (Carophyllaceae, Chenopodiaceae, and Amaranthaceae) *Artemisia*, and *Ephedra*. Bars highlight transgressive maxima indicating a southward shift of the Saharan-Sahelian boundary (desert-savanna boundary). The horizontal axis of each curve starts at 0 and increases from left to right. A sedimentary hiatus interrupts the record from 2.46 to 2.25 myr. Oxygen isotope stage 100 is indicated at the right-hand side.

arid phase shows a percentage maximum of *Ephedra* of 4 percent at 3.26 myr. From the sedimentary record of ODP Site 658, Tiedemann (1991) concluded an increase of eolian activity, especially of the trade winds, around 3.26 myr.

Power spectra of all pollen influx curves for the period from 3.5 to 3.2 myr show very little variance at the 23 kyr precession band. Some variance concentrates around the 41 kyr obliquity band for the spectra of Chenopodiaceae-Amaranthaceae, Caryophyllaceae, *Ephedra*, A. Liguliflorae, and Asteraceae Tubuliflorae. The first four groups show significant coherency with the oxygen isotopes of benthic foraminifers (*Cibicidoides wuellerstorfi;* ODP Site 658; Tiedemann, 1991) recording global ice volume and deep-sea temperatures. Only the last group, A. Tubuliflorae, shows significant coherency with the insolation maxima (July, 65°N; Berger and Loutre, 1991).

Coherency between insolation and oxygen isotopes is lacking at the 41 kyr band. From 3.5 to 3.2 myr, there is probably no direct forcing between obliquity and deep-sea temperatures or ice volume as recorded by the oxygen isotopes of benthic foraminifers (Tiedemann et al., 1994). The signal of the A. Tubuliflorae leads the insolation signal. Despite coherency, a forcing relationship between obliquity and A. Tubuliflorae is, therefore, also unlikely. However, the positive correlation of minimum deep-sea temperatures (oxygen isotopes) with pollen influx curves recording drier conditions in northwestern Africa indicates a link between the low deep-sea temperatures and aridity existing well before 3 myr (fig. 20.6A).

The Period between 3.25 and 2.61 myr (Zone III, fig. 20.4). During this intermediate period, humid conditions were reestablished between 3.25 and 3.19 myr, as indicated by high percentages of tropical forest (>2 percent) and Cyperaceae (>15 percent). Afterward, the climate again became progressively drier, and percentages of Cyperaceae declined. During the next part (from 2.97 to 2.61 myr), high percentages of grass pollen (Poaceae up to 70 percent) were followed by percentage maxima of CCA (>50 percent) at 2.73 and 2.69 myr; of *Ephedra* (>5 percent) at 2.69 myr; and the first maximum of *Artemisia* (2 percent) at 2.66 myr. A slight increase in trade-wind strength occurred at 2.76 myr.

The Period between 2.61 and 1.74 myr (Zone IV, fig. 20.5). At 2.60 myr, 2.53 myr, and 2.49 myr (isotope stages 104, 100, and 98, respectively), severe dry periods are recorded by high percentages of *Ephedra* (ca. 3 percent), *Artemisia* (>2 percent), and CCA (>50 percent). They mark the start of a climatic regime in northwestern Africa resembling glacial to interglacial cycles that result in arid-cold and humid-warm phases. These phases show the above described sequence of time-transgressive percentage maxima of Cyperaceae, Poaceae, CCA, *Artemisia,* and *Ephedra* that characterizes the shifting Saharan-Sahelian boundary. Within the period between 2.6 and 1.7 myr, only two extended humid periods occurred, corresponding to the weakly developed isotope stages 76 and 68.

Power spectra of pollen influx values for the period between 2.2 and 1.7 myr show a concentration of variance in the 41 kyr obliquity band. Cross-correlation spectra of *Rhizophora* (mangrove tree), sum of tropical forest elements (Sudanian and Guinean elements), Poaceae (grasses), *Ephedra,* A. Tubuliflorae, Chenopodiaceae-Amaranthaceae, and Caryophyllaceae show coherent phase shifts with insolation as well as with benthic isotopes. The phase shifts of pollen influx versus isotopes are consistent to those of pollen influx relative to insolation (fig. 20.6B). Pollen influx of "wet" elements like *Rhizophora,* Poaceae, and tropical forest lag maximal insolation (65°N, July) by 5 to 7 kyr. Pollen influx of "dry" elements like *Ephedra,* A. Tubuliflorae, Chenopodiaceae-Amaranthaceae, and Caryophyllaceae lag minimal insolation with about the same amount of time. These results strongly suggest forcing by obliquity of northwestern African climate between 2.2 and 1.7 myr.

Some of the variance in the power spectra of Chenopodiaceae-Amaranthaceae, A. Tubuliflorae, tropical forest elements, and Poaceae also concentrates around the 100 kyr band (fig. 20.6C). The latter three groups have coherent phase shifts with the oxygen isotopes of *C. wuellerstorfi.* The lag between the influx values of the "dry" elements like A. Tubuliflorae with the isotope maximum (maximum deep-sea temperatures, minimum ice volume) is about 6 kyr, and the lag between the "wet" elements like Poaceae and tropical forest with the isotopes is about 10 kyr. Owing to large statistical errors, however, the difference is not significant.

Discussion

Without marking a prominent boundary, the beginning of a declining trend in the benthic oxygen isotopes of the eastern equatorial Pacific started around 3 myr, signaling colder deep-sea temperature minima and increased maxima of global ice volume (Shackleton et al., 1995; Shackleton, this vol.). The same result was found by Tiedemann (1991) for the eastern tropical Atlantic. However, because elements of a biocoenose respond only if certain thresholds are passed, parts will react sensitively to certain ranges while staying indifferent to others. These ranges are not necessarily congruent in all parts of the system, implying that one fixed starting point for global change may be an illusion and that a number of changes during a limited time range marked the onset of the Ice Age period.

The pollen record of ODP Site 658 shows comparably gradual change in sensitivity to possible forcing mechanisms. Long-term variation indicates a first step toward drier climate between 3.5 and 3.2 myr and a second, stronger one starting at about 2.6 myr. Before 3 myr, a correlation between "dry" pollen taxa and oxygen isotopes indicates a link between drought in northwestern Africa and low deep-sea temperatures. There is, however, no synchronization between insolation and the pollen record over that period.

This situation changes after 2.6 myr (de Menocal et al., 1993; deMenocal and Bloemendal, this vol.). The period of 2.2–1.7 myr records a lag of about 5 kyr be-

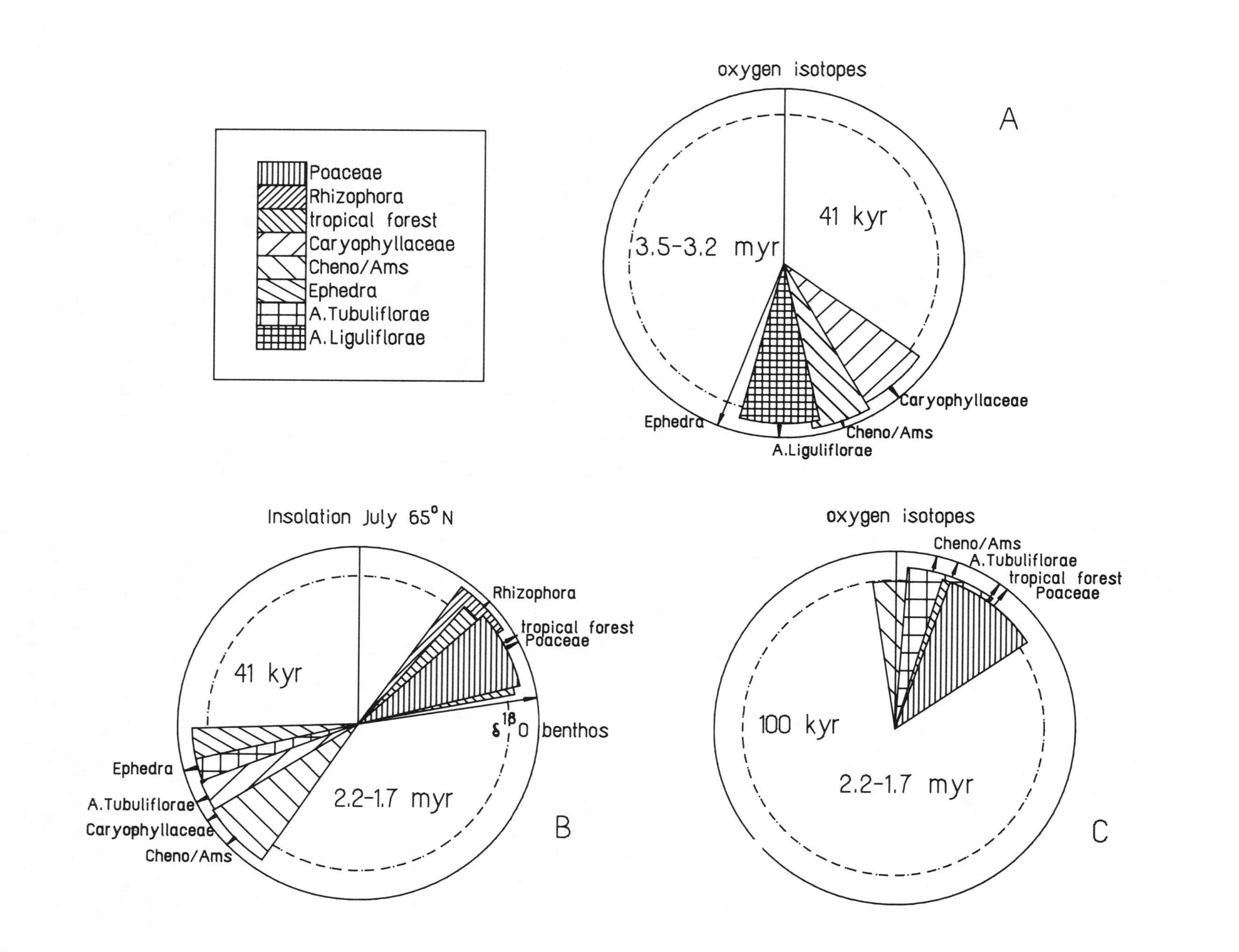
Poaceae
Rhizophora
tropical forest
Caryophyllaceae
Cheno/Ams
Ephedra
A.Tubuliflorae
A.Liguliflorae
oxygen isotopes
A
41 kyr
3.5-3.2 myr
Caryophyllaceae
Cheno/Ams
A.Liguliflorae
Ephedra
Insolation July 65°N
41 kyr
Rhizophora
tropical forest
Poaceae
δ 18 O benthos
Ephedra
A.Tubuliflorae
Caryophyllaceae
Cheno/Ams
2.2-1.7 myr
B
oxygen isotopes
Cheno/Ams
A.Tubuliflorae
tropical forest
Poaceae
100 kyr
2.2-1.7 myr
C

tween the maxima of the "wet" elements and the minima of the "dry" elements, with maximum insolation at 65°N. The Brunhes part of ODP Site 658 also shows correlation between pollen influx and ETP (the modeled curve of Eccentricity, Tilt, and Precession) and a lag of about 3 kyr between "wet" elements (Cyperaceae and Poaceae pollen) and insolation (Dupont et al., 1989 and unpublished data).

There are two possible nonexclusive explanations of forcing at the obliquity (41 kyr) band. One is a link between the climate of northwestern Africa and sea-surface temperatures of the North Atlantic advocated by de Menocal and Rind (1993) and deMenocal et al. (1993). The other is a correlation between the strength of the Hadley circulation, trade-wind vigor, and aridity in northwestern Africa. Model results reducing southern Asian orography show a reduction of the Hadley circulation through reduction of the high-pressure cell over the Himalayas. Trade winds, which are the surface component of the Hadley circulation, increase in periods with large ice volume, which is likewise attributable to an intensification of the high-pressure cell over the Himalayas (Kutzbach et al., 1989; Ruddiman et al., 1989; de Menocal and Rind, 1993). Trade winds are connected with sinking airflow and therefore contribute to aridity.

In conclusion, the short-term pulses of climatic cycles (less than 100 kyr) are already present well before 3 myr (and probably well before the Pliocene). Spectral analysis shows a positive correlation between low deep-sea temperatures and aridity in northwestern Africa. After 2.6 myr, when glaciations in the Northern Hemisphere increasingly mark global climate, our results strongly suggest forcing by obliquity of northwestern African climate. Trade winds in combination with North Atlantic sea surface temperatures then largely determine the climate of northwestern Africa.

Summary

Short-term fluctuations and long-term trends in the pollen record of ODP Site 658 reveal vegetation and climatic changes in northwestern Africa during the Upper Pliocene (3.7–1.7 myr). The pollen record correlates with oxygen isotopes indicating that aridity corresponds to low deep-sea temperatures and large global ice volume. Long-term variation involves decline in river discharge, southward retreat of mangrove swamps and tropical forest by at least 5°, development of desert vegetation in the western Sahara, and the onset of trade winds in three steps. Comparison of short-term fluctuations of periods before and after 2.5 myr shows that obliquity forcing of northwestern African climate started with the first large glaciations in the Northern Hemisphere.

Acknowledgments

We are indebted to R. Tiedemann for making available his time scale and partly unpublished stratigraphic data, to N. J. Shackleton for sending the manuscript of the leg 138 time scale and oxygen isotope stratigraphy, and to H.-J. Beug for his critical reading of the manuscript. L. Dupont thanks E. Vrba for organizing the highly stimulating Airlie Conference.
The research project was financed by a grant of the Bundesministerium für Forschung und Technologie (Bonn, 07KF018) to the Institute of Palynologie and Quaternary Sciences (Göttingen) and by additional financial support to S. Leroy by M. Sarnthein (Kiel). Postanalytical work of S. Leroy was made possible by the Impulse Program "Global Change," which is supported by the Prime Minister's Services of Belgium. Postanalytical work of L. Dupont was made possible by the Deutsche Forschungs Gemeinschaft (Du 221/1).

References

Berger, A., and Loutre, M. F. 1991. Insolation values for the climate of the last 10 million of years. *Quaternary Sciences Review* 10:297–317.

Fig. 20.6. A summary of phase relationships is given by phase wheels of the 41 kyr band for the periods 3.5–3.2 myr (A) and 2.2–1.7 myr (B) and of the 100 kyr band for the period 2.2–1.7 myr (C). Zero phase on wheels A and C is set for the oxygen isotope signal of *Cibicidoides wuellerstorfi* (minimum ice volume, maximum deep-sea temperature) and on wheel B for the maximum July insolation at 65°N. Vectors are used to represent the phase difference between pollen taxa and oxygen isotopes or between pollen taxa and insolation. Counterclockwise orientation represents a lead, clockwise represents a lag, and a phase of 180° represents a negative correlation. Phases are plotted if coherency is statistically significant at the 80-percent level (dashed circles). Hatched areas represent the phase error for each taxon. Power and coherency spectra of the influx values of nine different pollen taxa have been calculated after the Blackman-Tukey method (Jenkins and Watts, 1968). Pollen influx values have been interpolated to an equidistant step size of 3 kyr using a Gaussian filter of 9 kyr; the older sequence (3.5–3.2 myr) has an average sample distance of 3 kyr, and the younger one (2.2–1.7 myr) of 4 kyr. Gaps of more than two interpolated values between two measured ones are not closed but marked as missing values. All series are linearly detrended. The older series has a length of 100 and is analyzed using 60 lags, resulting in a bandwidth of 0.0074 kyr^{-1} and a level of nonzero coherency of 0.856. The younger series has a length of 176 and was analyzed using 100 lags resulting in a bandwidth of 0.0044 kyr^{-1} and a level of nonzero coherency of 0.835. The confidence interval is set at 80 percent. Poaceae = grasses, *Rhizophora* = mangrove tree, Cheno/Ams = Chenopodiaceae-Amaranthaceae, A. Tubuliflorae = Asteraceae Tubuliflorae, A. Liguliflorae = Asteraceae Liguliflorae, Asteraceae = Compositae.

deMenocal, P. B., and Rind, D. 1993. Sensitivity of Asian and African climate to variations in seasonal insolation, glacial ice cover, sea surface temperature, and Asian orography. *Journal of Geophysical Research* 98 D4:7265–7287.

deMenocal, P. B., Ruddiman, W. F., and Pokras, E. M. 1993. Influences of high- and low-latitude processes on African terrestrial climate: Pleistocene eolian records from equatorial Atlantic Ocean Drilling Program Site 663. *Paleoceanography* 8:209–242.

Dupont, L. M., and Agwu, C. O. C. 1991. Environmental control of pollen grain distribution patterns in the Gulf of Guinea and offshore NW-Africa. *Geologische Rundschau* 80:567–589.

Dupont, L. M., Beug, H.-J., Stalling, H., and Tiedemann, R. 1989. First palynological results from Site 658 at 21°N off northwest Africa: Pollen as climate indicators. In *Proceedings ODP, scientific results 108* (ed. W. Ruddiman et al.). College Station, Tex. (Ocean Drilling Program): 93-111.

Dupont, L. M., and Hooghiemstra, H. 1989. The Saharan-Sahelian boundary during the Brunhes chron. *Acta Botanica Neerlandica* 38: 405–415.

Hilgen, F. J. 1991. Astronomical calibration of Gauss to Matuyama sapropels in the Mediterranean and implication for the geomagnetic polarity time scale. *Earth and Planetary Science Letters* 104:226–244.

Hooghiemstra, H. 1989. Variations of the NW African trade wind regime during the last 140,000 years: Changes in pollen flux evidenced by marine sediment records. In *Paleoclimatology and paleometeorology: Modern and past patterns of global atmospheric transport* (ed. M. Leinen and M. Sarnthein). NATO ASI C 282, Kluwer, Dordrecht: 733–770.

Hooghiemstra, H., Agwu, C. O. C., and Beug, H. J. 1986. Pollen and spore distribution in recent marine sediments: A record of NW-African seasonal wind patterns and vegetation belts. *"Meteor"-Forschungs-Ergebnisse* C 40:87–135.

Jenkins, G. M., and Watts, D. G. 1968. *Spectral analysis and its applications.* Holden-Day, San Francisco.

Kutzbach, J. E., Guetter, P. J., Ruddiman, W. F., and Prell, W. L. 1989. Sensitivity of climate to Late Cenozoic uplift in southern Asia and American West: Numerical experiments. *Journal of Geophysical Research* 94 D15:18393–18407.

Leroy, S., and Dupont L. 1994. Development of vegetation and continental aridity in northwestern Africa during the Upper Pliocene: The pollen record of ODP 658. *Palaeogeography, Palaeoclimatology, Palaeoecology* 109:295–316.

Raymo, M., Ruddiman, W., Backman, J., Clement, B., and Martinson, D. 1989. Late Pliocene variation in Northern Hemisphere ice sheets and North Atlantic deep water circulation. *Paleoceanography* 4:413–446.

Ruddiman, W. F., Sarnthein, M., Backman, J., Baldauf, J. G., Curry, W., Dupont, L. M., Janecek, T., Pokras, E. M., Raymo, M. E., Stabell, B., Stein, R., and Tiedemann, R. 1989. Late Miocene to Pleistocene evolution of climate in Africa and the low-latitude Atlantic: Overview of leg 108 results. In *Proceedings ODP, scientific results 108* (ed. W. Ruddiman et al.). College Station, Tex. (Ocean Drilling Program): 463–484.

Ruddiman, W. F., Sarnthein, M., Baldauf, J., et al. 1988. *Proceedings ODP, initial reports 108(A).* College Station, Tex. (Ocean Drilling Program): 931–946.

Sarnthein, M., and Tiedemann, R. 1989. Towards a high-resolution stable isotope stratigraphy of the last 3.4 million years: Sites 658 and 659 off northwest Africa. In Ruddiman, W., Sarnthein, M. et al. *Proceedings ODP, scientific results 108* (ed. W. Ruddiman et al.). College Station, Tex. (Ocean Drilling Program): 167–185.

Shackleton, N. J., Berger, A., and Peltier, W. R. 1990. An alternative astronomical calibration of the lower Pleistocene timescale based on ODP Site 677. *Transactions Royal Society Edinburgh, Earth Sciences* 81:251–261.

Shackleton, N. J., Crowhurst, S., Hagelberg, T., Pisias, N. G., and Schneider, D. A. 1995. A new Late Neogene time scale: Application to leg 138 sites. In *Proceedings ODP, scientific results 138* (ed. N. Pisias, et al.). College Station, Tex. (Ocean Drilling Program).

Tiedemann, R. 1991. Acht Millionen Jahre Klimageschichte von Nordwest Afrika und Paläo-Ozeanographie des angrenzenden Atlantiks: Hochauflösende Zeitreihen von ODP-Sites 658–661. Ph.D. diss., University of Kiel.

Tiedemann, R., Sarnthein, M., and Shackleton, N. J. 1994. Astronomic time scale for the Pliocene Atlantic $\delta^{18}O$ and dust flux records of ODP Site 659. *Paleoceanography* 9:619–638.

Tiedemann, R., Sarnthein, M., and Stein, R. 1989. Climatic changes in the western Sahara: Aeolo-marine sediment record of the last 8 million years (sites 657–661). In Ruddiman, W., Sarnthein, M. et al. *Proceedings ODP, scientific results 108* (ed. W. Ruddiman et al.). College Station, Tex. (Ocean Drilling Program): 241–277.

Chapter 21

A Reassessment of the Plio-Pleistocene Pollen Record of East Africa

Raymonde Bonnefille

Palynological data from Plio-Pleistocene deposits in East Africa were first obtained in 1970–75. Although the record remained far too discontinuous to provide a complete reconstruction of paleovegetation through time, it did furnish information on the environmental setting of hominid sites (Bonnefille, 1985). Since then, no more fossil pollen has been gathered. Progress has centered on the pollen representation of various modern ecosystems and environmental settings, for which abundant data are now available, providing new insight into the earlier fossil pollen data.

Since publication of the last synthesis ten years ago, significant revisions of stratigraphical correlations have been made among the Shungura (Omo), Koobi Fora (East Turkana), Olduvai, Laetotil, Gadeb, and Hadar geological formations. Vegetational changes reported earlier, which appeared to be contemporaneous with critical steps in faunal and hominids evolution, are here placed in an up-to-date chronostratigraphy. They are also reexamined in the light of better understanding of the environmental and climatic forcing of past vegetation dynamics in the tropical regions.

The East African Plio-Pleistocene record includes 120 pollen assemblages (spectra), one of which has been excluded from this presentation because it has no precise stratigraphic attribution. The pollen spectra were obtained from sediments deposited into six separate basins, five of them in the East African Rift, and span the interval between 4 and 1.2 million years (myr). The assemblages from distinct stratigraphic levels can be placed fairly accurately against a continuous time scale, thanks to the exceptional effort to achieve greater precision in chronostratigraphical and geological studies (see Brown, this vol.). Samples for pollen studies were placed within local stratigraphy to allow for possible revisions. With the exception of one sample (U3-2) from the Usno Beds, the new intercorrelations among all the Plio-Pleistocene sites do not modify the relative positions of the pollen spectra from different basins through time, as was reported in Bonnefille (1985). But the new chronostratigraphy slightly modifies the ages assigned to the pollen assemblages and the time spans between them (fig. 21.1).

Revised Chronostratigraphy of Pollen Data

Omo/Turkana Basin. The stratigraphic correlation between the Shungura and Koobi Fora Formations of the Turkana Lake Basin are based on the most recent geological data (Feibel et al., 1989; Brown and Feibel, 1991) and on the accurate chronometric framework proposed for the Omo Group as summarized in Harris (1991). These correlations provide evidence for a hiatus in the Koobi Fora Formation between the lower and upper part of the Burgi Member, representing 500 thousand years (kyr).

Changes in the age assignment of fossil pollen spectra from the Shungura Formation are as follows: unit U-3-2 from the Usno Beds is now dated around 4.1 myr (Feibel et al., 1989) instead of 3.55 myr (de Heinzelin, 1983). This unit provides the oldest information for the Turkana Basin and all the other investigated Plio-Pleistocene

ISBN 0–300–06348–2.

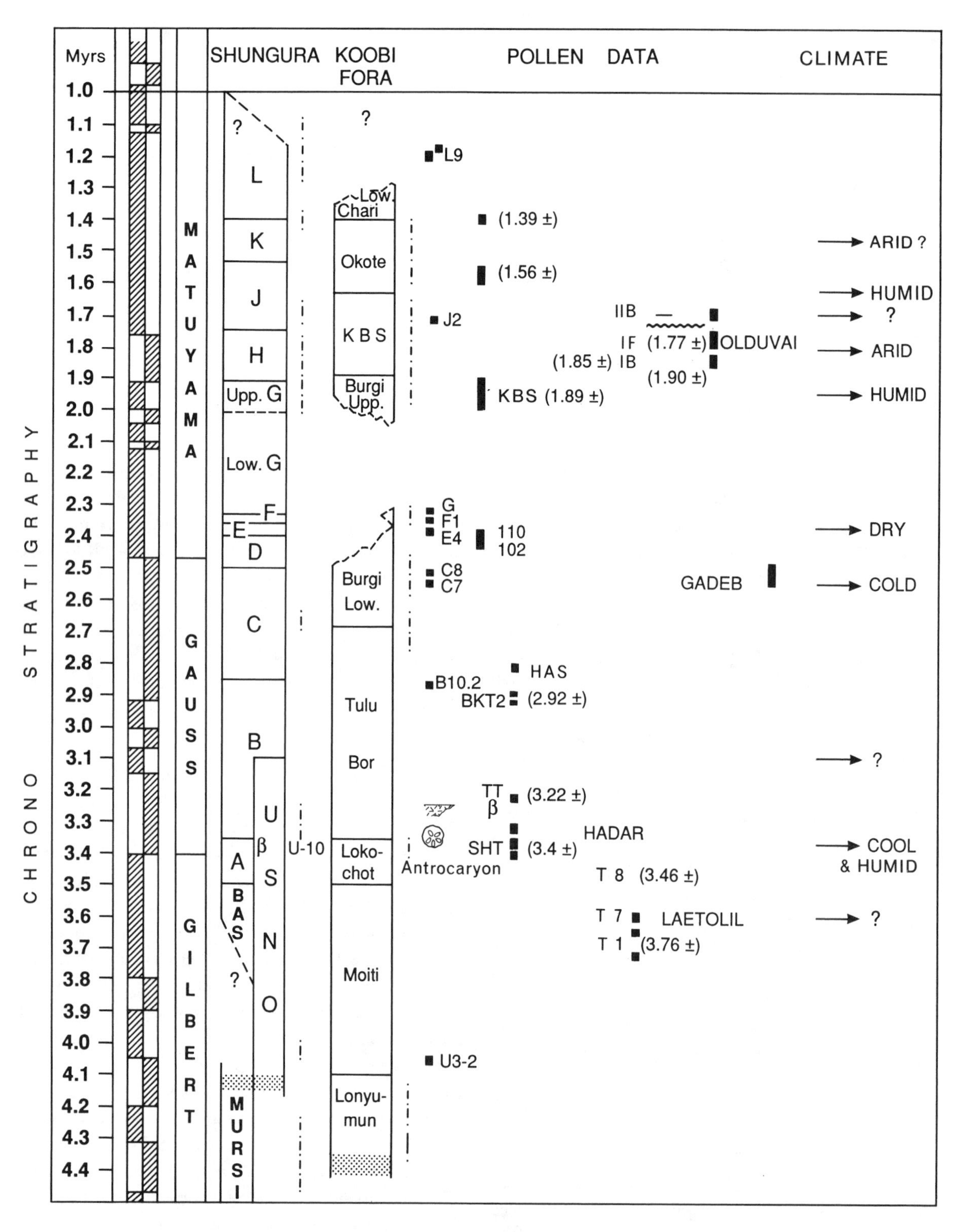

Fig. 21.1. Chronostratigraphic succession of fossil pollen data from the East African hominid sites. The chronostratigraphy has been revised after Brown and Feibel, 1991; Brown et al., 1992; Harris et al., 1991; Haileab and Brown, 1994; Walter and Aronson, 1993; Walter et al., 1992. The symbol ! indicates shallow lacustrine conditions. Stratigraphic units are designed by letters and numbers (C7, SHT, etc.); absolute date are indicated as follows: 3.4 ±, 1.39 ±, etc.

sites. One minor change concerns units C-7 and C-9, placed in the Gauss reversed chronozone, below Tuff D (= Lokalalei) and now dated 2.5 myr instead of 2.45 myr (Feibel et al., 1989). Using the new integrated chronometric framework implies that members D, E, and F span no more than about 200 kyr. Therefore the pollen spectra from horizons E-4, F-1, and G (just below Tuff G) are now included in the interval 2.4 to 2.3 myr, instead of being placed at 2 myr (Bonnefille, 1976a), but are in agreement with dates given later (Bonnefille, 1985).

In the Koobi Fora Formation, pollen samples from lacustrine deposits, below the KBS Tuff in areas 102 and 110 (Vincens, 1982), are now assigned to the Lower Burgi Formation (Brown and Feibel, 1986; Harris et al., 1991), and have been given an age of 2.4 myr (Brown and Feibel, 1991). This estimate is much older than the 1.8 myr age previously assigned to these samples (Bonnefille and Vincens, 1985). Initially, the hiatus was placed after deposition of the KBS Tuff, whereas the unconformity is now recognized between the upper and lower parts of the Burgi Members (Brown and Feibel, 1991). The new stratigraphic correlation places the lower Burgi Member of the Koobi Fora Formation at the base of the Matuyama chronozone, contemporaneous with members E and F of the Shungura Formation (Feibel et al., 1989).

As a result of these revisions, ten pollen samples now come from the time interval 2.5–2.3 myr and provide better information on past vegetation in Turkana near the Gauss/Matuyama boundary (fig. 21.1).

Pollen samples from between the Lorenyang and the KBS Tuffs in area 102 and below the KBS (= Tuff H2) in areas 110 and 116 (Vincens, 1982) are now included in the Upper Burgi Member of normal polarity. The age of 1.88 ± 0.02 myr for the KBS retained here places these pollen data very close in time to those recovered from the base of the Olduvai Lower Bed I.

Pollen samples from the Okote Member were collected in areas 118 and 131 in the type section and from sediment that includes the archaeological sites. They are stratigraphically close to the Okote Tuff Complex (McDougall et al., 1985) lying between 1.64 ± 0.03 myr and 1.52 myr, the age of Tuff J-7 (Brown and Feibel, 1985). They document past vegetational environments contemporaneous with the *Homo erectus* finds (Leakey and Leakey, 1978). The pollen sample from archaeological site FxJji 63, just below the Chari Tuff dated at 1.39 ± 0.02 myr (McDougall et al., 1985) and equivalent to Tuff L, predates the pollen sample of unit L-9 in the Shungura Formation, situated below the Jaramillo subchron.

In conclusion, a good set of ten pollen assemblages is situated around 1.9 myr; another set of eight pollen spectra is placed near 1.6 myr. Information in the upper part is dispersed across an interval of a few hundred thousand years.

In the Turkana Basin, pollen sampling was done regularly across the outcrops of the entire stratigraphic record, both for the Shungura and Koobi Fora Formations, but the results have been limited by pollen preservation. Incidently, it appears that fossil pollen seems to have been better preserved in shallow lacustrine deposits than in fluviatile sediment (fig. 21.1). The lacustrine beds of the Mursi Formation are the only strata that have not yet been investigated.

Olduvai and Laetoli Beds. The Laetoli site is located at about 4°S of the equator in the Eastern Rift. Palynological data were obtained from various types of samples in the Upper Laetolil Beds (Bonnefille and Riollet, 1987). Five pollen spectra come from samples collected below Tuff #1, which is 10 m above the base dated at 3.76 ± 0.03 myr (Hay, 1987). Four pollen spectra were extracted from chamber fill of termite-mound hives (Sands, 1987), which are found just above Tuff #1. The youngest pollen spectra is located between Tuff #6 and Footprint Tuff #7 (Leakey and Harris, 1987). The K-Ar dates of Tuff #8 and the xenolith-bearing horizon near the top of the upper unit are now dated to an average age of 3.46 ± 0.12 myr, whereas Tuff #7 is dated 3.56 ± 0.2 myr (Drake and Curtis, 1987). Given the average depositional rate of 1.7 m per 10 kyr (Hay, 1987) for the aeolian volcanic ash, the oldest pollen spectrum at Laetoli dates to about 3.7 myr, whereas the youngest one is older than 3.56 myr and can be estimated to be 3.6 myr old. This implies a time span of about one hundred thousand years (between 3.7 and 3.6 myr) for pollen data at Laetoli.

At Olduvai, palynological information is restricted to Bed I and lower Bed II (Bonnefille, 1976; Bonnefille and Riollet, 1980; Bonnefille et al., 1982). Recent single-crystal, laser-fusion $^{40}Ar/^{39}Ar$ dating of Bed I (Walter et al., 1991) provided an age of 1.9 myr for the lower Bed I, which fits the lower estimate given by Hay (1976). The oldest pollen spectrum at hominid site FLKNN comes from a level just below Tuff IB, a marker ignimbrite dated between 1.86 ± 0.01 myr and 1.80 ± 0.01 myr (Walter et al., 1992). It is situated above the base of the Olduvai subchron (Endale et al., in press). One pollen sample comes from below Tuff ID, dated 1.76 myr (Walter et al., 1991). The most abundant pollen sample was extracted from lacustrine deposits just below and above Tuff IF, which has been attributed an age of 1.75 myr (Walter et al., 1991). The results of recent dating indicate that the time interval between Tuffs IB and IF is about 50 kyr (Hay, 1992), that is, half the time previously reported in Bonnefille (1976; Bonnefille et al., 1982).

For lower Bed II, pollen data were obtained at site HWK East, from a level just below Tuff IIA, which marks the upper limit of the Olduvai subchron. The youngest pollen data at Olduvai come from marshy claystone deposits at site FC West, surrounding Tuff IIB. They are situated above the disconformity of the Lemuta member and the faunal break and were attributed an age of 1.6 myr after Hay (1976). This pollen information predates the first appearance datum (FAD) of the Acheulean prehistoric culture (Leakey, 1971).

In conclusion, nine pollen spectra from Olduvai document a period of about 200 kyr, from 1.9 to 1.7 myr, corresponding to the time span of the Olduvai subchron. They slightly postdate the KBS Tuff in the Koobi Fora Formation. Two other pollen spectra document past vegetation above the faunal break in Middle Bed II; these should be slightly younger than pollen spectrum from Shungura unit J-2.

Hadar. Correlation between the Hadar Formation and the Omo Group from Lake Turkana follows Brown (1982), Walter et al. (1984), and Sarna-Wojcicki et al. (1985). Pollen data were obtained from sediment samples of the Sidi Hakoma and Kada Hadar Members that yield the assemblages of fossil hominids (Kimbel et al., 1994). The oldest pollen spectrum from Hadar comes from a clay layer overlain by the Sidi Hakoma Tuff (SHT) at locality 398. Recent dating using the laser-probe technique for $^{40}Ar/^{39}Ar$ has provided a revised age of 3.4 ± 0.03 myr (Walter and Aronson, 1993), which fits cross-correlation with Tulu Bor Tuff of the Turkana Basin (Brown, 1982) and marine cores from the Indian Ocean (Brown et al., 1992).

A set of fourteen pollen assemblages was extracted from samples collected in a 20 m section of organic clay exposed at the Ourda section (Tiercelin, 1986), 30 m above the SHT and 80 m below the KM basalt. The dating of the KM basalt remains a subject of debate. However, a recent date of 3.28 ± 0.04 myr (Walter and Aronson, 1993) seems to support the new astronomically calibrated magnetostratigraphic data (Renne et al., 1993). The sediment that yielded the pollen have normal polarity and are situated below the lower limit of the Mammoth subchron. On the basis of this new revision (Renne et al., 1993), the assemblages have been attributed an estimated age of 3.36 myr. The time span included in these fourteen pollen assemblages (Bonnefille et al., 1987) is estimated at less than 20 kyr. On the basis of the new dating and calculation of the sedimentation rate (Walter, 1994), the four pollen spectra collected from around the gastropod layers close to hominid locality AL 266 have an interpolated age of 3.36 myr. Five samples statigraphically close to the top of the Triple Tuff (TT-4) can be assigned the TT-4 recent date of 3.22 ± 0.01 myr (Walter, 1994). In the revised stratigraphy, there is no pollen data for the Denen Dora Member situated between the TT-4 and the Kada Hadar Tuff (KHT), dated 3.18 myr ± 0.01 myr (Walter and Aronson, 1993). One pollen sample collected from a 2 m section of brown clay just below the Bouroukie Tuff (BKT-2) is attributed an age of 2.95 myr (Walter and Aronson, 1982). It is situated above the Kaena subchron and documents the environment for the youngest hominid discoveries (Kimbel et al., 1994). The stratigraphic position of the pollen spectrum from a clay sample 60 cm below the artifact horizon at Gona (Hadar archaeological site HAS; Harris, 1983) has not been changed. It is certainly younger than 2.9 myr but not yet firmly correlated with the Hadar-type section.

Gadeb. A continuous and well-documented pollen diagram (thirty-five pollen spectra) was extracted from Pliocene lacustrine diatomites of the Gadeb Formation exposed on the right bank near the head waters of the Webi Shebele River on the southern plateau of Ethiopia at an elevation of 2,350 m (Bonnefille, 1983). The sedimentary sequence was deposited in the middle of a Pliocene lake and was obtained from a 4.5 m core at the middle (from 18 to 22.5 m) of the type section at Melka Likimi 6B, but the outcrops are now covered by water as a result of damming the river. Basalt and volcanic ash underlying the diatomite were dated from 2.71 to 2.51 myr (Assefa et al., 1982), whereas the welded tuff, dated 2.35 myr (Williams et al., 1979; Eberg et al., 1988), caps the Gadeb Formation. Recent reinvestigation of volcanic ash at Meribo, a few kilometers to the south of Melka Likimi, gives evidence for cross-correlation of tephras with the Lokalalei Tuff (= Tuff D, 2.52 myr) of the Turkana Basin (Haileab and Brown, 1994). This agrees with the older dates and the fact that the diatomites have normal polarity. Assuming a steady sedimentation rate for the 50 m-thick diatomite deposits, the Gadeb pollen diagram represents a span of 10–20 kyr at most. It documents highland vegetation during a short time span at the bottom of the Gauss normal chronozone just below the Gauss/Matuyama boundary and contemporaneous to lower Member D (Shungura Formation; fig. 21.1).

Synthetic Presentation

The 120 pollen spectra recovered from Plio-Pleistocene sediments include total pollen counts of several hundred to 1,000 grains for each sample, distributed among more than 250 distinct taxa (fig. 21.2; table 21.1). Such an abundant data set is statistically significant, although still too fragmentary to provide a complete reconstruction of all the vegetation changes that may have occurred during the last 4 myr. I do not intend to reconstruct in detail the habitats and paleoenvironmental conditions at each site for each period. This task would require consideration of taphonomic preservation, sedimentary conditions, paleogeographical reconstructions, and information from other biological indicators, such as was done for the Turkana area (Harris, 1991) and Laetoli (Leakey and Harris, 1987). I shall discuss the significance of events illustrated in previous publications, enhanced by the accurate chronostratigraphy that provides a link with continuous marine and continental records (Dupont and Leroy, deMenocal and Bloemendal, Hooghiemstra, all in this vol.).

The synthetic pollen diagram presented here differs from previous ones by the calculation for the respective percentages of trees and shrubs, arboreal pollen (AD), and nonarboreal pollen (NAP, i.e., herbs) of a pollen sum that excluded pollen inferred to the aquatics (Cyperaceae and *Typha*). This exclusion of local and aquatic taxa allows a better representation of the regional plant cover more directly comparable from site to site.

Local Signal. The percentages of aquatic pollen are lower at Laetoli, a pattern that confirms that land surface was recovered by ash fall (Hay, 1987). At Hadar, high proportions of the aquatic Cyperaceae and *Typha* pollen have no equivalent in any of the other hominid sites. Their abundance can be attributed to the proximity of marshy, freshwater conditions and to periodic flooding (Bonnefille et al., 1987).

In most of the sites, we noted that the percentages of aquatic pollen fluctuated inversely with the grasses. This finding should indicate that Gramineae and Cyperaceae are issued from two distinct communities and have different responses to environmental changes. Therefore we consider that the Gramineae curves are predominantly a regional signal.

Among the Gramineae pollen, however, a certain amount comes from some semiaquatic plants occurring on the lake shore and riverbanks. Unfortunately it is not possible to distinguish between them.

The group Chenopodiaceae/Amaranthaceae (C/A) is also treated separately from the nonarboreal pollen component (NAP). Their abundance in modern sediments and in a Holocene core from Lake Turkana (Mohammed et al., in press) indicates that they are good pollen markers of aridity. Most of the C/A come from plants developed on saline soils and dry wadis and reflect a local negative balance between precipitation and evaporation. They have the same climatic significance as the CCA category (Dupont and Leroy, this vol.).

Regional Signal. The arboreal pollen signal (AP) of the Plio-Pleistocene deposits registers great variations in relative percentages and also in the qualitative composition. These taxonomic variations have already been discussed in the literature (Bonnefille, 1976a, 1976b; Bonnefille et al., 1982, 1987).

The arboreal pollen signal reflects the tree cover, but in the tropics it underestimates the density of trees, especially in dry woodlands and open savanna, because most tropical trees have a much lower pollen production than grasses and Cyperaceae (Mworia et al., 1988). However, modern pollen rain studies from southwestern Ethiopia, in the north of the Omo/Turkana Basin, show that woodland and forest can be distinguished from pollen analysis (Bonnefille et al., 1993). For this reason, quantitative fossil pollen data are best interpreted through comparison with modern pollen data rather than relative to modern plant ecological studies in the area (Carr, 1976). The distribution of key pollen taxa are now available for the dry subdesertic steppe (Bonnefille and Vincens, 1977; Vincens, 1984) and along altitudinal gradients from Ethiopian mountains (Bonnefille et al., 1987, fig. 7; Bonnefille et al., 1993; Bonnefille and Buchet, 1986). These surface samples provide critical information on pollen assemblages along gradients of decreasing temperature and increasing rainfall.

In the earlier interpretation, we distinguished between montane forest, riverine forest, and steppe or woodland. This subdivision relied on the assumption

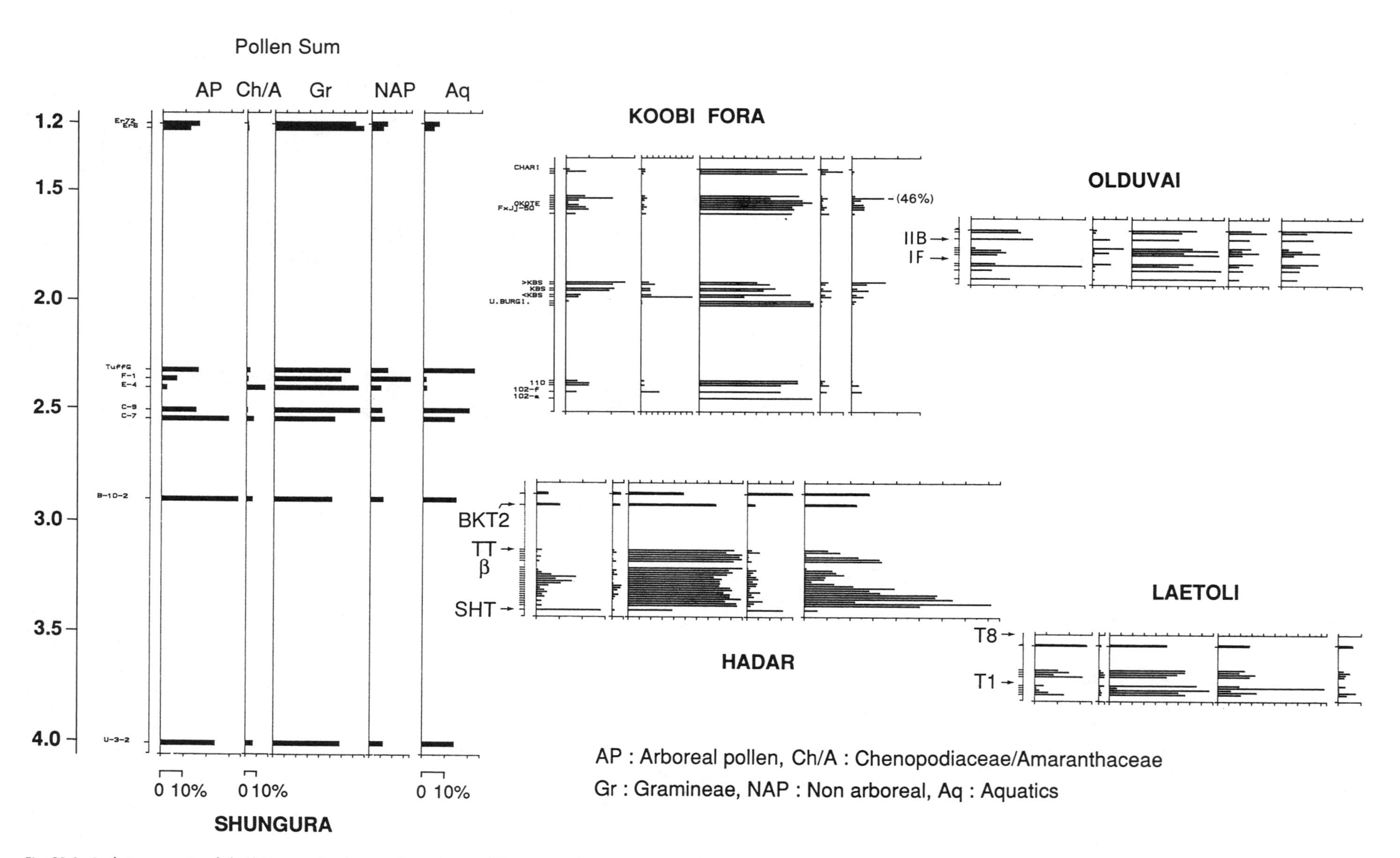

Fig. 21.2. Synthetic presentation of Plio-Pleistocene pollen data according to the revised chronostratigraphy (percentages are calculated versus a selected pollen sum, which includes arboreal pollen (AP), Chenopodiaceae/Amaranthaceae, and nonarboreal pollen (NAP). Pollen referred to aquatic plants (Cyperaceae and *Typha*) are excluded from this sum.

Table 21.1. Pollen Data Obtained from Plio-Pleistocene Samples Collected at the Hominid Sites, East Africa

Sites	No. of spectra	No. of taxa	Pollen count
Shungura	11	92	6,322
Laetoli	10	87	4,353
Hadar	27	120	17,289
Koobi Fora	25	111	13,480
Olduvai	11	109	6,812
Gadeb ML6B	35	85	11,544

that the plant communities were organized in the same ecosystems as they are now. Such an assumption can be a matter of debate. As an example, pollen results from the Shungura Formation show that AP percentages range from 5 to 30 percent (fig. 21.2). Among them 60 to 77 percent of the arboreal pollen (Bonnefille, 1976a) was attributed to montane forest element, recently described in modern pollen studies from southwestern Ethiopia (Bonnefille et al., 1993). This proportion is greater than percentages recorded either in surface soil samples (Bonnefille and Vincens, 1977) or lacustrine deposits (Vincens, 1984) under modern conditions and has been explained by river transport (Buchet, 1982), a conclusion that agrees with the existence of a perennial Omo River flowing from the Ethiopian highlands over the last 4 myr (Woldegabriel and Aronson, 1987). At Koobi Fora, the pollen assemblage KFFP1 (Bonnefille, 1986b) showed allochtonous pollen, especially *Juniperus*, a component of the dry mountains, which does not exist today in western Ethiopia (Friis, 1992). This pollen was apparently introduced by a large river that fits the more recent paleogeographical reconstructions for the area (Brown and Feibel, 1991).

When assigning pollen taxa to ecological categories, one should be aware that the lack of species identification leads to uncertainties and errors, some of which can perhaps be avoided if the total of arboreal pollen is used as an index of the tree cover in the whole basin, as in figure 21.2. It has been argued that arboreal pollen frequencies could simply reflect paleogeographic changes (Feibel et al., 1991) that caused the total amount of allochtonous input to vary from deltaic to lacustrine conditions (Vincens, 1984). But this factor alone would not explain the variations of paleoflora throughout time, as well illustrated in Harris (1991, table 6.4). Such changes are unlikely to be attributable to plant colonization of new areas, because they involve pollen taxa that belong to different phytogeographic kingdoms whose present-day distribution is conditioned by different climatic limits (Trochain, 1980).

In contrast to faunal list, there is no last appearance datum (LAD) in the floral list for pollen taxa. The record of *Alangium chinense* at Hadar 3.4 myr ago was mentioned as a relict tertiary taxon not existing in the modern flora from Ethiopia (Bonnefille et al., 1987). Yet it was recently discovered during additional plant collecting in the mountain rain forest of Ethiopia, at an elevation of 1,600–1,800 m (Friis, 1993).

Revised Interpretation

Around 4 myr. Before 4 myr, the paleobotanical reconstruction relies on fossil wood identification that indicates a deciduous woodland with components of the Sudano-Zambezian flora, among which Combretaceae and Caesalpiniaceae were abundant (Dechamps and Maes, 1985). This finding is consistent with a northern extension of deciduous Sudano-Zambezian woodland that was more recently documented in pollen flora from lignite deposits north of Lake Tana (Yemane et al., 1985).

Around 4 myr, a single pollen spectrum from the lower Usno Beds contains taxa from both montane forests and warmer woodland. This finding agrees with the fact that the Turkana Basin included highlands in the watershed of the Omo River topographic environment as established later on geological studies (WoldeGabriel and Aronson, 1987).

The "Golden Age," 3.7–3.2 myr. Later, around 3.4 myr, stratigraphically close to the ash marker bed (Tuff Bβ = Tulu Bor = SHT), paleobotanical evidence of fossil fruit *Antrocaryon* (Bonnefille and Letouzey, 1976), rain-forest prosobranch *Potadoma* (Williamson, 1985), and fossil microfauna (Wesselman, this vol.) indicate floristic affinities with present floristic composition of the rain forest. The fossil record documents eastward extension of plants now restricted much farther to the southwest (Bonnefille et al., 1987). The first African record of the primarily Eurasian bivalve *Corbicula* (Williamson, 1985) is contemporaneous with the occurrence at Hadar (SHT) of pollen from *Alangium chinense,* the only African genus of a mainly Asiatic family (Bonnefille et al., 1987). At the same time, the occurrence of many pollen grains of *Garcinia* at Hadar provides a first link with fossil wood remains from the Usno Formation in the Turkana Basin. The modern distribution of *Garcinia* includes twelve species from the Guineo-Congolian rain-forest massif, but *G. huillensis*, identified as fossil wood, is also common in savannah and riparian forest south of the equator. These humid indicators are not dominant, however, in the SHT pollen spectra from Hadar, which has more components of the undifferentiated mountain forest of Ethiopia (Friis, 1992) than of the equatorial rain forest proper, as it is known today in the Guineo-Congolian massif.

Coincident with the occurrence of humid indicators in the paleobotanical record of East Africa is a small but noticeable increase in the tropical-forest signal in deep-sea sediments off West Africa, also dated 3.4–3.2 myr (Dupont and Leroy, this vol.). In West Africa, this increase can be interpreted as indicating a noticeable expansion of the Guineo-Congolian rain-forest massif. An eastern expansion that reached the Turkana Basin was proposed earlier (Williamson, 1985). Today, the West African rain forest occupies areas with the wettest and warmest climates of the continent, with no strong seasonal or diurnal contrasts (Trochain, 1980). The establishment of a rain forest at Turkana would imply a 400-fold increase in rainfall, with an even distribution throughout the year or with a dry season of less than one month. A rain belt situated north of its present-day limit could have been the result of an atmospheric circulation pattern during a period characterized by unipolar glaciation (Flohn, 1983). However, the simultaneous occurrence of *Juniper, Euclea,* and *Ekebergia* at Hadar and of *Commiphora* and Capparidaceae at Usno from dry evergreen submontane bushland and subdesertic steppe implies seasonal contrast and a long dry season that precludes any occurrence of a true rain-forest belt from 4–3 myr ago at these sites. Therefore northern extension of the rain-forest belt cannot be postulated on the grounds of paleobotanical and paleontological data presently available. The only firm conclusion regarding this problem is that more humid and evergreen species existed in both lowlands and highlands around 3.4 myr than at present. Permanent water bodies could have maintained a few species with rain-forest affinities in local riverine habitats. Further investigations in older deposits should add information to this problem.

A Cold Spell during the Golden Age? At Hadar, several changes are observed in the pollen data. One is situated between the Sidi Hakoma Tuff (SHT) and the Ourda section after 3.4 myr, when trees decreased and Chenopodiaceae/Amaranthaceae increased, most probably indicating drier conditions (Bonnefille et al., 1987). But the best-documented change comes from the middle part of the homogenous clay section, starting 10 m above the SHT, and that contains an organic layer a few meters thick (fig. 21.2). Several pollen spectra from this layer show an increase in AP, associated with a floristic change. Pollen of *Myrica, Hagenia,* and *Juniper,* all high-altitude markers, replace the former association of *Juniper, Euclea* (Ebenaceae), and *Ekebergia* of lower-altitude woodland. It is most probable that tectonics associated with rifting strongly modified the topography since the deposition of the Hadar Formation (Adamson and Williams, 1987). The pollen data cannot be understood on the basis of the present paleogeograhic setting of the Hadar site in the lowlands today. Nevertheless, the introduction of cold-pollen indicators is unlikely to be explained by tectonic movements, because (1) such tectonism is expected to have occurred after 3 myr and (2) no strong changes are seen in the sedimentary deposition at that time. A more likely explanation is that a "cold" short-term event would have interrupted the Golden Age. The cold event at Hadar is synchronous with a noticeable increase in CCA arid-pollen markers of the marine record in the western Atlantic (Dupont and Leroy, this vol.) and with a shift in $d^{18}O$, well marked at that time in the new isotopic oceanique record of DSDP (Shackleton, this vol.). At Hadar, it is indicated by pollen spectra from around the 3.3–3.2 myr that show a much lower proportion of trees, which can belong to cool high-altitude forest rather than to lowland warm steppe or woodland (Bonnefille et al., 1987).

Regarding paleoenvironmental reconstruction at all the hominid sites, the above discussions should not hide the fact that Gramineae pollen always dominate the pollen data. Gramineae are proof of openings in forest and woodland or indicate wooded grassland and grasslands in the surroundings of the Omo, Turkana, Hadar and Laetoli Basins. A great diversity of plant ecosystems, habitats, and food resources should therefore have been available for various types of fauna from 4 to 1 myr. No clear sign of a trend toward increasing aridity can be assessed from the pollen data. Rather, we interpret vegetational changes as reversible shifts, as documented in continuous cores from South America (Hooghiemstra, this vol.). Only a few of these changes are found in the East African Plio-Pleistocene outcrops, either because they correspond to exceptional preservation of fossil pollen or because they represent major thresholds of the past climate.

Highland Cooling (2.50–2.35 myr). Cooling was documented at Gadeb in lacustrine diatomite deposited on the southern plateau of Ethiopia (Bonnefille, 1983). The evidence published ten years ago came from a pollen diagram that registered the history of the upper forest and the Afro-alpine vegetation, whereas the site is located in the lower montane forest zone at 2,350 m. This difference in altitude implies a 1,000 m descent of vegetation belts, relative to the present-day situation. The Pliocene pollen diagram shares similar features with pollen diagrams from sites in the East African mountains during the last glacial period. The climatic interpretation based on these data indicate a cooling of 5°–6° in mean annual temperature on highlands (Bonnefille et al., 1993). Recent palynological data (Bonnefille and Mohammed, 1994) from above the tree line in the same area reinforce the interpretation presented earlier. The timing of the Gadeb cooling was placed between 2.51 and 2.35 myr, dates that were obtained from an underlying basalt and an upper-volcanic ash (Williams et al., 1979), which is now correlated and identical to tuff Lokalalei (= Tuff D) in the Turkana Basin (Haileab and Brown, 1994). Therefore the colder conditions at Gadeb occur simultaneously with the strong decrease in tree cover documented in the contemporaneous deposits of the Shungura and Koobi Fora Formations, regardless of sedimentary conditions.

The development of glaciation in the Northern Hemisphere had a great effect on patterns of general atmospheric circulation (deMenocal and Bloemendal, this vol.). Simulations, using general circulation models, show that high-latitude glacial ice cover induces cooler winters and drier summers in East Africa (deMenocal and Rind, 1993). In tropical regions under a summer rainfall regime, these climatic conditions could have produced both a lowering and a shrinking of the montane forest belt 2.5–2.35 myr ago, such as observed at Gadeb. Stronger atmospheric circulation also would result from a more extensive northern ice cap (Flohn, 1983), and aridity in the lowlands would be reinforced by increased dry northeast trade winds. As a result of stronger atmospheric circulation and stronger upwelling, consecutive decreases in atmospheric CO_2 and H_2O content of the atmosphere could also directly lead to reduced tree growth at high altitude (such as Gadeb), though it would favor expansion of grasslands. This postulated mechanism is consistent with observed changes in the palynological record of East Africa both in the highlands and lowlands.

Arid/Humid Oscillations in the Lowlands. Following the new chronostratigraphy (fig. 21.1), Members D, E and F from the Shungura Formation now correlate with the Lower Burgi Member at Koobi Fora. Therefore changes in the vegetation cover first documented at Omo (Bonnefille, 1976) become synchronous with the grassland expansion that occurred prior to the deposition of the KBS (area 110, 102; Bonnefille, 1985). In the old chronology, the two distinct changes in vegetation were delayed by a few hundred thousand years (Bonnefille and Vincens, 1985).

The interpretation of pollen changes in terms of climatic change remains a matter of debate (Feibel et al., 1991). It is true that distortion in pollen composition from lacustrine or deltaic samples was illustrated by analysis of modern sediments from Lake Turkana (Vincens, 1982). Among the differences observed, the deltaic sediments showed a greater proportion of pollen from montane forest than did the lacustrine units. In the fossil samples collected above and below the KBS Tuff, in which it was possible to sample contemporaneous fluviatile and lacustrine deposition, we found no great difference in the values of the pollen groupings, as illustrated on figure 21.2 (Bonnefille and Vincens, 1985). On the other hand, differential deposition could not explain the shift in relative proportions from *Podocarpus*, *Ilex*, and *Myrica* components of the wettest broad-leaved forest to *Juniper* and *Olea*, which are associated with the driest type of forests in Ethiopia (Friis, 1992). Later, additional independent evidence for changes toward more arid conditions was documented, for example, by examining microfaunal assemblages in the same strata of the Shungura Formation or in strata very close to levels yielding the pollen data (Wesselman, this vol.). The bovids also clearly show similar changes (Vrba, chap. 27, this vol.).

Abundant fossil wood from lower G has provided a puzzling assemblage of forest components (Bonnefille and Dechamps, 1983) that is not yet fully understood. Pollen spectra from lacustrine deposits below that KBS (1.9 myr) yielded abundant Gramineae but much less C/A than in modern lacustrine sediments from the Turkana Lake (Vincens, 1982; Mohammed et al., in press). This is a clear indication that conditions during the period that was interpreted as an arid event remained more humid in comparison with the modern situation. This remark has gained in validity because of our increasing knowledge of pollen composition of submodern sediments. Pollen data dated 1.9 myr from and above the KBS show an increase in the arboreal component regardless of whether the depositional site was lacustrine, fluviatile, or archaeological (Bonnefille, 1983). These levels from Turkana are synchronous with Lower Bed I at Olduvai. Therefore, several shifts from arid to humid conditions seem to have occurred over close stratigraphical intervals spaced less than 100 kyr apart.

Environmental changes encountered in the Olduvai Gorge deposits were fairly well documented by several studies (Hay, 1990). Pollen from drier and warmer vegetation than is found today characterizes a trend toward increasing aridity in Upper Bed I at 1.77 myr (Bonnefille, 1976, 1979), in agreement with other studies (Kappelman, 1984); Cerling (1992) showed that the abundance of C_4 plants increased markedly at that time. After the aridity trend, the vegetation became progressively more wooded with *Acacia* in Tuff II A. Woodland with *Alchornea* and montane forest components was established after the faunal break, synchronous with the Acheulean FAD. In the Turkana Basin the pollen spectrum J-2 also indicates an important increase in *Podocarpus*, which appears synchronous with a change toward more humid conditions following the arid 1.7 myr episode at Olduvai.

The composition of pollen spectra around the Okote (1.56 myr) and Chari (1.4 myr) Tuffs in the Turkana Basin has great affinities with modern vegetation. The strongest difference lies in the proportion of C/A, which was ten times lower than in any modern fluviatile (Buchet, 1982), lacustrine (Vincens, 1984), or aerial samples (Bonnefille and Vincens, 1977). The lower proportion of C/A reflects a much lower deficit of evapotranspiration over precipitation. Although climatic conditions are interpreted as more arid than before the lower Pleistocene, they appear less arid than today (Mohammed, in press).

Conclusion

From 3 to 1 myr the paleobotanical samples of six hominid sites document a great variety of plant ecosystems, ranging from the subdesertic steppe to Afro-alpine heath moorland without evidence of any closed rain forest. There have been strong modifications and reorganization in the plant communities through time. The main pollen events documented include more humid conditions before 3.2 myr throughout East Africa, but mainly at Hadar, colder conditions in the highlands and drier conditions in the lowlands at 2.35 myr, and greatest aridity at 1.8 myr. Although paleogeographic conditions that differed at each site and changed through time played an important role in the reorganization of plant communities, the pollen signal in East Africa highlights the critical period that corresponds in time with changes in vegetation and climate in other tropical regions of West Africa and South America. To use pollen properly as a proxy for climate, however, requires a complete record from a long core that could be collected by drilling a big lake.

Acknowledgments

I wish to thank E. Vrba, C. Whitelock, P. Barthein, and T. Johnson for help with the English version of this chapter.

References

Adamson, D. A., and Williams, M. A. J. 1987. Geological setting of Pliocene rifting and deposition in the Afar depression of Ethiopia. *Journal of Human Evolution* 16:597–610.

Assefa, G., Clark, J. D., and Williams, M. A. J. 1982. Late Cenozoic history and archaeology of the upper Webi Shebele basin, east-central Ethiopia. *Sinet* 5:27–46.

Bonnefille, R. 1976a. Palynological evidence for an important change in the vegetation of the Omo Basin between 2.5 and 2 million years ago. In *Earliest man and environment in the Lake Rudolf Basin*, pp. 421–431 (ed. Y. Coppens et al.). University of Chicago Press, Chicago.

———. 1976b. Implications of pollen assemblage from the Koobi Fora Formation, East Rudolf, Kenya. *Nature* 264:403–407.

———. 1983. Evidence for a cooler and drier climate in the Ethiopian uplands towards 2.5 myr ago. *Nature* 303:487–491.

———. 1984. Palynological research at Olduvai Gorge. *National Geographic Society* 17:227–243.

———. 1985. Evolution of the continental vegetation: The palaeobotanical record from East Africa. *South African Journal of Science* 81:267–270.

Bonnefille, R., and Buchet, G. 1987. Contribution palynologique à l'histoire récente de la forêt de Wenchi (Ethiopie). In *Palynologie et milieux tropicaux* Ninth Conference of the Ass. Palynol. Langue Fr., Montpellier, October 1985, pp. 143–158. Ecole Pratique des Hautes Etudes, Montpellier.

Bonnefille, R., Buchet, G., Friis, I., Kelbessa, E., and Mohammed, M. U. 1993. Modern pollen rain on an altitudinal range of forests and woodlands in south west Ethiopia. *Opera Botanica* 121:71–84.

Bonnefille, R., and Dechamps, R. 1983. Data on fossil flora. In *The Omo group: Archives of the international Omo research expedition*, pp. 191–207 (ed. J. de Heinzelin). Musée Royal de l'Afrique Centrale, Tervuren, Belgium.

Bonnefille, R., and Letouzey, R. 1976. Fruits fossiles d'Antrocaryon dans la vallée de l'Omo (Ethiopie). *Adansonia* 16: 65–82.

Bonnefille, R., Lobreau, D., and Riollet, G. 1982. Pollen fossile de Ximenia (Olacaceae) dans le Pléistocène inférieur d'Olduvai en Tanzanie: Implications paléoécologiques. *Journal of Biogeography* 9:469–486.

Bonnefille, R., and Mohammed, U. 1994. Pollen inferred climatic fluctuations during the last 3000 years. *Palaeogeography, Palaeoclimatology, Palaeoecology* 109:331–343.

Bonnefille, R., and Riollet, G. 1980. Palynologie, végétation et climats de Bed 1 et Bed 2 à Olduvai, Tanzanie. In *Actes du Huitiéme Congr. Panafr. Préhist. et Quaternaire, Nairobi, September 1977,* pp. 123–127 (ed. R. E. Leakey and Ogot, B. A.). Tillmiap, Nairobi.

———. 1987. Palynological spectra from the Upper Laetolil Beds. In *The Pliocene site of Laetoli, northern Tanzania,* pp. 52–61 (ed. M. D. Leakey and J. M. Harris). Oxford University Press, Oxford.

Bonnefille, R., and Vincens, A. 1977. Représentation pollinique d'environnements arides à l'Est du lac Turkana (Kenya). In *Recherches françaises sur le Quaternaire hors de France,* suppl. bull. 50/1, pp. 235–247. Ass. Fr. Quater., Paris.

——— 1985. Apport de la palynologie à l'environnement des Hominidés d'Afrique orientale. In *L'environnement des Hominidés au Plio-Pléistocène,* pp. 237–278 (ed. Fondation Singer-Polignae). Masson, Paris.

Bonnefille, R., Vincens, A., and Buchet, G. 1987. Palynology, stratigraphy and palaeoenvironment of a Pliocene hominid site (2.9–3.3 M.Y.) at Hadar, Ethiopia. *Palaeogeography, Palaeoclimatology, Palaeoecology* 60:249–281.

Brown, F. H. 1982. Tulu Bor Tuff at Koobi Fora correlated with the Sidi Hakoma Tuff at Hadar. *Nature* 300:631–633.

Brown, F. H., and Feibel, C. S. 1985. Stratigraphical notes on the Okote tuff complex at Koobi Fora, Kenya. *Nature* 316: 794–797.

———. 1986. Revision of lithostratigraphic nomenclature in the Koobi Fora region, Kenya. *Journal of the Geological Society* 143:297–310.

———. 1991. Stratigraphy, depositional environments, and palaeogeography of the Koobi Fora formation. In *Koobi Fora research project.* Vol. 3, *The fossil ungulates: Geology, fossil artiodactyls, and palaeoenvironments,* pp. 1–30 (ed. J. M. Harris). Clarendon Press, Oxford.

Brown, F. H., Sarna-Wojcicky, A. M., Meyer, C. E., and Haileab, B. 1992. Correlation of Pliocene and Pleistocene tephra layers between the Turkana basin of East Africa and the Gulf of Aden. In *Tephrochronology: Stratigraphic applications of tephra,* pp. 55–67 (ed. J. A. Westgate, R. C. Walter, and N. Naeser). Pergamon, Oxford.

Buchet, G. 1982. *Transport des pollens dans les fleuves Omo et Awash (Ethiopie): Etude des vases actuelles.* Mémoire de l'Ecole Pratique des Hautes Etudes, Montpellier.

Carr, C. 1976. Plant ecological variation and pattern in the lower Omo Basin. In *Earliest man and environment in the Lake Rudolf Basin,* pp. 432–467 (eds. Y. Coppens et al.). University of Chicago Press, Chicago.

Cerling, T. 1992. Development of grass and savannas in East Africa during the Neogene. *Palaeogeography, Palaeoclimatology, Palaeoecology* 9713:241.

Dechamps, R., and Maes, F. 1985. Essai de reconstitution des climats et des végétations de la basse vallée de l'Omo au Plio-Pléistocène à l'aide de bois fossiles. In *L'environnement des Hominidés au Plio-Pléistocène*, pp. 175–232 (ed. Fondation Singer-Polignac. Masson, Paris.

de Heinzelin, J. 1983. *The Omo Group.* Musée Royal de l'Afrique Cenrale, Tervuren, Belgium.

deMenocal, P., and Rind, D. 1993. Sensitivity of Asian and African climates to variations in seasonal insolation, glacial ice cover, sea-surface temperature and Asian orography. *Journal of Geophysical Research* 98:7263–7287.

Drake, R., Curtis, G. H. 1987. K-Ar geochronology of the Laetoli fossil localities. In *Laetoli: A Pliocene site in northern Tanzania* (ed. M. D. Leakey and J. M. Harris). Clarendon Press, Oxford.

Eberz, G. W., Williams, F. M., and Williams, M. A. J. 1988. Plio-Pleistocene volcanism and sedimentary facies changes at Gadeb prehistoric site, Ethiopia. *Geologische Rundschau* 77: 513–527.

Endale, T., Thouveny, N., Taieb, M., and Opdyke, N. D. (in press). Revised magnetostratigraphy of the Plio-Pleistocene sedimentary sequence of the Olduvai Formation (Tanzania). *Palaeogeography, Palaeoclimatology, Palaeoecology.*

Feibel, C. S., Brown, F. H., and MacDougall, I. 1989. Stratigraphic context of fossil hominids from the Omo group deposits, northern Turkana Basin, Kenya and Ethiopia. *American Journal of Physical Anthropology,* n.s., 78:595–622.

Feibel, C. S., Harris, J. M., and Brown, F. H. 1991. Palaeoenvironmental context for the Late Neogene of the Turkana Basin. In *Koobi Fora research project.* Vol. 3, *The fossil ungulates: Geology, fossil artiodactyls, and palaeoenvironments,* pp. 321–346 (ed. J. M. Harris). Clarendon Press, Oxford.

Flohn, H. 1983. Climate evolution in the Southern Hemisphere and the equatorial region during the Late Cenozoic. In *Late Cainozoic palaeoclimates of the Southern Hemisphere,* pp. 5–20 (ed. J. C. Vogel). Balkema, Rotterdam.

Friis, I. 1992. *Forests and forest trees of northeast tropical Africa.* Her Majesty's Stationery Office, London.

Haileab, B., and Brown, F. H. 1994. Tephra correlations between the Gadeb prehistoric site and the Turkana Basin. *Journal of Human Evolution* 26:167–173.

Harris, J. M. (ed.). 1991. *Koobi Fora research project.* Vol. 3, *The fossil ungulates: Geology, fossil artiodactyls, and palaeoenvironments.* Clarendon Press, Oxford.

Harris, J. W. K. 1983. Cultural beginnings: Plio-Pleistocene archaeological occurrences from the Afar, Ethiopia. *African Archaeological Review* 1:3–31.

Hay, R. L. 1976. *Geology of the Olduvai Gorge.* University of California Press, Berkeley.

———. 1987. Geology of the Laetoli area. In *Laetoli. A Pliocene site in northern Tanzania,* pp. 23–47 (ed. M. D. Leakey and J. M. Harris). Clarendon Press, Oxford.

———. 1990. Olduvai Gorge: A case history in the interpretation of hominid paleoenvironments in East Africa. In *Establishment of a geologic framework for paleoanthropology,* pp. 23–37 (ed. L. F. Laporte). Geological Society of America, Boulder.

———. 1992. Potassium-argon dating of Bed 1, Olduvai Gorge, 1961–1972. In *Tephrochronology: Stratigraphic applications of tephra,* pp. 31–36 (ed. J. A. Westgate, R. C. Walter, and N. Naeser). Pergamon, Oxford.

Kappelman, J. 1984. Plio-Pleistocene environments of Bed I and

Tanzania. *Palaeogeography, Palaeoclimatology, Palaeoecology* 48:197–213.

Kibunjia, M., Roche, H., Brown, F. H., and Leakey, R. E. 1992. Pliocene and Pleistocene archaeological sites west of Lake Turkana, Kenya. *Journal of Human Evolution* 23:431–438.

Kimbel, W. H., Johanson, D. C., and Rak, Y. 1994. The first skull and other new discoveries of *Australopithecus afarensis* at Hadar, Ethiopia. *Nature* 368:449–451.

Leakey, M. D. 1971. *Olduvai Gorge.* Vol. 3, *Excavations in Bed I and Bed II, 1960–1963.* Cambridge University Press, Cambridge.

Leakey, M. D., and Harris, J. M. (eds.). *Laetoli: A Pliocene site in northern Tanzania.* Clarendon Press, Oxford.

Leakey, M. G., and Leakey, R. E. 1978. *Koobi Fora research project.* Vol. 1, *The fossil hominids and an introduction to their context, 1968–1974.* Clarendon Press, Oxford.

McDougall, I., Davies, T., Maier, R., and Rudowski, R. 1985. Age of the Okote tuff complex at Koobi Fora, Kenya. *Nature* 316:792–794.

Mohammed, M. U., Bonnefille, R., Johnson, T. (in press). Comparison of the pollen record and other proxies of past climatic change in Late Holocene sediments from Lake Turkana, Kenya. *Palaeogeography, Palaeoclimatology, Palaeoecology.*

Mworia, J., Dallmeijer, A., and Jacobs, B. 1988. Vegetation and modern pollen rain at Olorgesailie, Kenya. *Utafiti, Occasional Papers of the National Museums of Kenya* 1:1–22.

Renne, P., Walter, R., Verosub, K., Sweitzer, M., and Aronson, J. 1993. New data from Hadar (Ethiopia) support orbitally tuned time scale to 3.3 Ma. *Geophysical Research Letters* 20:1067–1070.

Sands, W. A. 1987. Ichnocoenoses of probable termite origin from Laetoli. In *Laetoli. A Pliocene site in northern Tanzania,* pp. 409–432 (ed. M. D. Leakey and J. M. Harris). Clarendon Press, Oxford.

Sarna-Wojcicki, A. M., Meyer, C. E., Roth, P. H., and Brown, F. H. 1985. Ages of tuff beds at East African early hominid sites and sediments in the Gulf of Aden. *Nature* 313:306–308.

Tiercelin, J. J. 1986. The Pliocene Hadar Formation, Afar depression of Ethiopia. In *Sedimentation in the African Rifts,* pp. 221–240 (ed. L. E. Frostick et al.). Geological Society, London.

Trochain, J. L. 1980. Ecologie végétale de la zone intertropicale non désertique. Université Paul-Sabatier, Toulouse.

Vincens, A. 1982. Palynologie, environnements actuels et plio-pléistocènes à l'Est du lac Turkana (Kenya). Ph.D. diss., Université de Aix-Marseille 2.

———. 1984. Environnement végétal et sédimentation pollinique lacustre actuelle dans le bassin du lac Turkana (Kenya). *Revue Paléobiologie,* special issue, 235–242.

Walter, R. C. 1994. The age of Lucy and the first family: Single crystal $^{40}Ar/^{39}Ar$ dating of the Denen Dora and lower Kada Hadar Members of the Hadar Formation. *Geology* 22:6–10.

Walter, R. C., and Aronson, J. L. 1993. Age and source of the Sidi Hakoma Tuff, Hadar Formation, Ethiopia. *Journal of Human Evolution* 25:229–240.

Walter, R. C., Manega, P. C., and Hay, R. L. 1992. Tephrochronology of Bed 1, Olduvai Gorge: An application of laser-fusion $^{40}Ar/^{39}Ar$ dating to calibrating biological and climatic change. In *Tephrochronology: Stratigraphic applications of tephra,* pp. 37–46 (ed. J. A. Westgate, R. C. Walter, and N. Naeser). Pergamon, Oxford.

Walter, R. C., Manega, P. C., Hay, R. L., Drake, R. E., and Curtis, G. H. 1991. Laser-fusion $^{40}Ar/^{39}Ar$ dating of Bed 1, Olduvai Gorge, Tanzania. *Nature* 354:145–149.

Walter, R. C., Westgate, J. A., Hart, W. K., and Aronson, J. L. 1984. Tephrostratigraphic correlation of the Sidi Hakoma and Tulu Bor Tuffs: Nd isotope and new trace elements. *Geological Society of America–Abstracts* 16:65.

Williams, M. A. J., Williams, F. H., Gasse, F., Curtis, G. H., and Adamson, D. A. 1979. Pleistocene environments at Gadeb prehistoric site, Ethiopia. *Nature* 282:29–33.

Williamson, P. G. 1985. Evidence for an early Plio-Pleistocene rainforest expansion in East Africa. *Nature* 315:487–489.

WoldeGabriel, G., and Aronson, J. L. 1987. Chew Bahir rift: A "failed" rift in southern Ethiopia. *Geology* 15:430–433.

WoldeGabriel, G., Walter, R. C., Aronson, J. L., and Hart, W. K. 1992. Geochronology and distribution of silicic volcanic rocks of Plio-Pleistocene age from the central sector of the main Ethiopian rift. In *Tephrochronology: Stratigraphic applications of tephra,* pp. 69–76 (ed. J. A. Westgate, R. C. Walter, and N. Naeser). Pergamon, Oxford.

Yemane, K., Bonnefille, R., and Faure H. 1985. Paleoclimatic and tectonic implications of Neogene microflora from the northwestern Ethiopian highlands. *Nature* 318:653–656.

Chapter 22

The "Elephant-*Equus*" and the "End-Villafranchian" Events in Eurasia

Augusto Azzaroli

The well-known sea-level fluctuations and climatic shifts of the Pliocene and Pleistocene, now calibrated to the paleomagnetic scale and correlated with the climatic record evidenced by stable isotopes, provide a basis for a detailed investigation of major vegetational and faunal revolutions in the Eurasian continent. Among the many fluctuations of oceans and seas, two major sea-level drops stand out, not only for their amplitude but also for the wide-ranging environmental changes that accompanied or preceded them in the continental area (see Azzaroli et al., 1988; Azzaroli, 1991). The older of these sea-level falls took place about halfway through the Pliocene and was called the "Acquatraversan erosional phase" (originally the "Fase Glaciale dell'Acquatraversa") in the Mediterranean. The name was derived from Fosso dell' Acquatraversa, the little valley of a creek in the outskirts of Rome. The same drop in sea level was observed in the Gulf of Mexico and was correlated with the Nebraskan glaciation (Beard et al., 1982; the curve of Pliocene and Pleistocene sea levels obtained by these authors in the Gulf of Mexico matches the curve obtained in the Tiber delta, near Rome). The younger, major sea retreat, the "Cassian erosional phase" in the Mediterranean, falls in the late Matuyama paleomagnetic epoch. Between these, a minor retreat, the "Aullan erosional phase" in the Mediterranean, corresponds with a sea-level fall that has been correlated with the Kansan glaciation of North America (Beard et al., 1982). This Aullan erosional phase had only minor impact on vegetation and faunas but coincides, with good approximation, with the arrival of the well-known "northern guests" in the Mediterranean and with the conventional Neogene-Quaternary boundary.

The strongly marked Mid-Pliocene (Aquatraversan) sea-level fall was accompanied by a general cooling of ocean waters and by massive extinctions in marine faunas. The crisis was in fact so dramatic that several students proposed to assume this event for a revised Neogene-Quaternary boundary. The crisis was also accompanied by the beginning of loess sedimentation in northern China (Kukla and An, 1989) and by wide-ranging changes in vegetation and faunas on the continents: the transition from the warm Reuverian to the cooler Pretiglian in western Europe (Zagwijn, 1974), the extinction of warm-forest faunal elements such as zygodont mastodont and tapir, and the arrival of elephants and monodactyl equids (the "Elephant-*Equus* event" of Lindsay et al., 1980).

The Cassian sea-level fall also coincided with a new cooling of ocean waters but did not affect marine faunas to any great extent. On the other hand, it deeply affected continental vegetation and mammalian faunas, the "end-Villafranchian dispersal event." According to available evidence, changes in continental vegetation and faunas appear to have been approximately, but not exactly, synchronous with major shifts in ocean-water temperatures, age differences being in excess of random errors. Accurate datings of these phenomena may provide a clue to understanding cause-and-effect relationships.

The second of the major events—or better, the set of events referred to here, the end-Villafranchian dispersal event and the Cassian sea-level fall—will be discussed first. This set of events offers ground for extensive analysis.

ISBN 0-300-06348-2.

The end-Villafranchian event marks the transition from the Late Villafranchian (Early Pleistocene) to the new mammalian assemblages of a time interval that corresponds with good approximation to the Middle Pleistocene.[1] Stages characterized by early post-Villafranchian faunal associations have received various names: Tiraspolian (Gromov, 1948) in Russia; Olyorian (Sher, 1971) in northeastern Siberia; Galerian (Ambrosetti et al. 1972) in western Europe (the Biharian, based on associations of smaller mammals, mainly rodents, straddles part of the Late Villafranchian and the Galerian).

The close of the Villafranchian marked the demise of a faunal complex that retained affinities with the Late Pliocene (Early and Middle Villafranchian) faunas and the incoming of a new fauna of modern type. Over twenty species of large mammals, and possibly a similar number of small rodents and insectivores, disappeared, but faunal diversity was practically unaffected: some species disappeared by evolutionary change; other lineages became extinct and were replaced by immigrants. The picture of western Europe gives the impression of a once-flourishing fauna suddenly subjected to stress by the arrival of competitors better adapted to new conditions created by a climatic change. Not a single Villafranchian species survives today, although several representatives of the present-day fauna made their first appearance in the Galerian. New, previously unknown patterns of body adaptations appeared: the gigantic cervids of the genera *Megaloceros, Megaceroides,* and *Cervalces*—the latter two having less specialized forerunners in the Villafranchian of Europe—and the heavy, massive-bodied bovids *Bison, Bos,* and *Praeovibos.*

The faunal turnover at the end of the Villafranchian in Europe was matched by the transition from the Early Pleistocene Gongwangling to the early Mid-Pleistocene Chenjavo faunal complexes in northern China (Xu, 1989).

Faunas with mixed assemblages of Late Villafranchian and Galerian elements have been reported from few localities, and their scantiness may suggest that the faunal turnover was rapid, possibly a matter of a few ten thousand years. One of these faunas, Solilhac, in the Massif Central of France, has been calibrated to the Jaramillo episode (Thouveny and Bonifay, 1984). In Lakhuti, Tadzhikistan, two typical Galerian immigrants were recorded, one from the beginning and one from the end of the Jaramillo episode (Azzaroli et al., 1988). In northeastern Siberia, on the other hand, the first elements of the Olyorian fauna, including lemmings, are recorded between approximately 1.4 and 1.2 million years (myr), with good approximation (Sher, 1987), suggesting that the faunal revolution began in the cold northern areas and expanded to the temperate latitudes in the course of few hundred thousand years.

The record of vegetation also points to a significant climatic crisis around 1.0–0.9 myr. In the Pannonian plain of central Europe vegetational changes were accurately recorded in the cores of deep wells. Vegetation indicating a cold, dry, fully glacial climate appeared at the top of the Jaramillo episode (Cooke, 1981). In the Netherlands the change was more gradual: Zagwijn and De Jong (1984) consider the beginning of the Menapian cold stage older than the Jaramillo episode; this was followed, in rapid sequence, by three warm-cold oscillations of short duration within the interval from 0.9 to 0.7 myr.

In northern Israel the pollen record shows three marked cold episodes, beginning at about 1.2 myr and ending around 0.9 myr, whereas south of the Dead Sea the oldest cold episode began later, shortly before 1.0 myr (Horowitz, 1989).

According to Suzuki and Manabe (in Azzaroli et al., 1988) a spectacular change took place in the vegetation of Japan at the beginning of the Jaramillo episode, with the appearance of an "alpine-type" flora. Itihara and Yoshikawa (1991 p. 147) held a somewhat different view: the decline of the warm "*Metasequoia* flora" took place in the course of an appreciable time interval, the beginning of which was not specified; the end of the extinction process is dated to the Late Matuyama epoch, "about 15 m. below the base of the Brunhes epoch," when the *Metasequoia* flora was completely superseded by the new alpine flora.

The picture evidenced by loess sedimentation is also significant. Loess deposition began in China around 2.5 myr. There is evidence of shifts toward colder and drier conditions at 2.4, 1.2, and 0.5 myr, and of minor shifts at 1.65, 0.8 and 0.2 myr (Kukla and An, 1989). The cool episode at 1.65 myr is practically synchronous with the conventional Neogene-Quaternary boundary: the major climatic shift at 1.2 myr has its counterpart in the advent of the Olyorian faunal complex in northeastern Siberia. In Tadzhikistan loess deposition became widespread during the Late Matuyama.

It appears that the continental record gives a more diversified picture than the relatively simple pattern of the marine record. Changes of great consequence took place in Eurasia. Faunal and vegetational turnover epi-

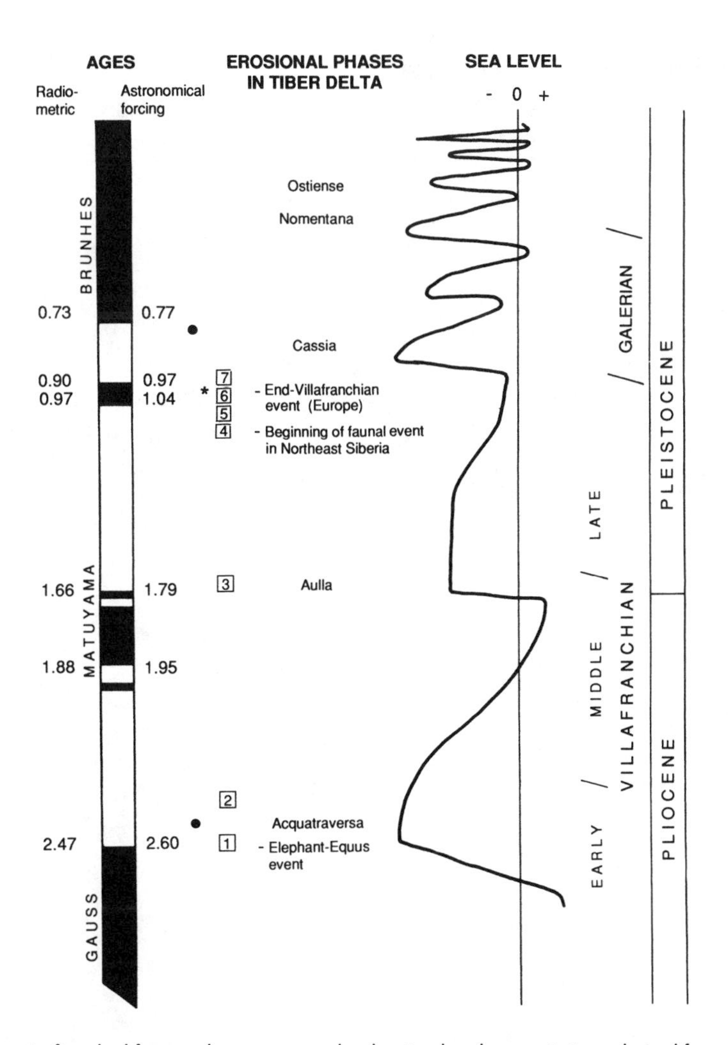

Fig. 22.1. Sea-level fluctuations, significant local faunas and turnover events plotted against the paleomagnetic time scale. Local faunas and faunal complexes. 1: Tatrot-Pinjor transition in Upper Siwaliks; 2: Montopoli, Lower Valdarno; 3: Matassino, Upper Valdarno; 4: first appearance of *Dicrostonyx* in northeastern Siberia; 5: Lakhuti 1, Tadzhikistan; 6: Solilhac, France; 7: Lakhuti 2, Tadzhikistan. The star close to 1 myr indicates the estimated time of climax in Early Pleistocene Himalayan uplift, as evidenced by the Upper Siwaliks Boulder Bed that was overridden by deposits from the Lower Siwaliks. The two solid circles, between faunas 1 and 2 and after .7 myr, indicate times of major shifts in ocean-water temperature.

sodes were apparently rapid events that have been recognized in areas scattered throughout the continent, and although detailed investigations do not yet cover the whole area, it is natural to assume that all Eurasia was affected. Single events were not strictly contemporary: they occurred at appreciably different times but each within an apparently short span, of the order of tens of thousand years; and all these regional events occurred between roughly 1.2 and 0.9 myr. The area of loess deposition also expanded considerably in the Late Matuyama. It is reasonable to assume that some major geological cause lay at the root of this revolution.

The mountain ranges of the Himalayan complex, with their eastern and western extensions, and the adjoining Tibetan Plateau deeply influenced the climate of southern Asia on one side and of central and northern Asia on the other. The history of this complex of mountains and high plateaus may provide a clue to the causes of climatic changes and their accompanying phenomena. This history, of course, was no simple process. Continental faunas to the south and to the north of the mountain belt are widely different. The Indian subcontinent is a separate bioprovince from inner Asia, and its history indicates that it already was so at the beginning of the

Table 22.1. Stratigraphic Distribution of Significant Mammalian Species Evidencing Faunal Turnover Events in Europe

	Villafranchian							
	early	middle		late			Galerian	
Faunal Units	Tr	// Mo	SV	/ Ol	Ta	Fa	///	Ga
Mammut borsoni	+							
Tapirus arvernensis	+							
Sus minor	+							
Pseudodama pardinensis	+							
Ursus minimus	+							
Prolagus savagei	+							
Mimomys stehlini	+							
Mimomys polonicus	+							
Glirulus pusillus	+							
Pseudodama lyra	+	?						
Dicerorhinus jeanvireti	+	+						
Leptobos stenometopon	+	?						
Anancus arvernensis	+	+	+					
Pachycrocuta perrieri	+	+	+					
Nyctereutes megamastoides	+	+	+	+	+			
Croizetoceros ramosus	+	+	+	+	+			
Lynx issiodorensis	+	+	+	+	+			
Acynonyx pardinensis	+	+	+	+	+			
Homotherium crenatidens	+	+	+	+	+	+	+	
Archidiskodon gromovi		+						
Equus livenzovensis		+						
Eucladoceros falconeri		+						
Gallogoral meneghinii		+	+					
Archidiskodon cf. *meridionalis* (primitive)			+					
Pseudodama rhenanus			+					
Eucladoceros tegulensis			+					
Leptobos merlai			+					
Ursus cf. *etruscus* (primitive)			+	+	+	+		
Dicerorhinus etruscus			+	+	+	+		
Equus stenonis			+	+	+	+	+	
Sus strozzii			+	+	+			
Leptobos etruscus			+	+	+	+		
Archidiskodon meridionalis meridionalis				+	+			
Ursus etruscus				+	+	+	+	
Canis etruscus				+	+	+		
Panthera toscana				+	+			
Mimomys pliocaenicus				+	+			
Hystrix etrusca				+	?			
Castor plicidens				+	+	+		
Pseudodama nestii				+	+			
Eucladoceros dicranios olivolanus				+				
Pachycrocuta brevirostris				+	+	+		
Eucladoceros ctenoides				+				
Eucladoceros dicranios dicranios				+	+			
Leptobos vallisarni				+	+			
Mimomys savini				+	+			
Allophaiomys pliocaenicus				+	+			
Canis arnensis				+	+			
Canis falconeri				+	?	+		
Hippopotamus antiquus				+	+	?		
Pseudodama farnetensis					+			
Eucladoceros tetraceros					+			
Equus bressanus					+			

Table 22.1. (*Continued*)

	Villafranchian							
	early	middle		late			Galerian	
Faunal Units	Tr	// Mo	SV	/ Ol	Ta	Fa	///	Ga
Archidiskodon meridionalis vestinus/cromerensis					+			
Eobison degiulii							+	
Elephas antiquus								+
Mammuthus armeniacus								+
Equus caballus								+
Equus süssenbornensis								+
Dicerorhinus hundsheimensis								+
Bison schoetensacki								+
Praeovibos priscus								+
Hippopotamus tiberinus								+
Sus scrofa								+
Cervus elaphus acoronatus								+
Capreolus capreolus								+
Megaceroides verticornis								+
Megaloceros savini								+
Cervalces latifrons								+
Canis lupus mosbachensis								+
Ursus deningeri								

Note: Mammal faunal units and events: Tr = Triversa; Mo = Montopli; SV = Saint Vallier; Ol = Olivola; Ta = Tasso; Fa = Farneta; Ga = Ponte Galeria; // = Elephant-*Equus* event; / = Pliocene-Pleistocene boundary; /// = end-Villafranchian event.

Late Miocene. The Nagri fauna is approximately the equivalent of the Vallesian of western Europe. There is a sharp contrast between the endemic faunas of the Indian subcontinent and the faunas north of the mountain belt, which range with only minor differences from China to western Europe. A physiographic barrier between the southern peninsulas and central Asia has existed for a long time, but the generally molassic, fine-grained sediments of the southern foothills of the Himalayan belt suggest a barrier that was not as high and rugged as the present one. The history of the more recent evolution of the Himalayan ranges may be read, to some extent at least, in the sedimentary record of the Upper Siwaliks of India and Pakistan.

In the Marwat and Bhittanni ranges of Trans-Indus Pakistan, the Marwat molassic formation contains a fairly rich fauna characteristic of the Pinjor stage (Late Pliocene to Early Pleistocene). The total thickness of the formation exceeds 2,000 m in the Bhittanni range. A boulder bed (the "Bain Boulder Bed" of Morris, 1938) is intercalated in the lower part of the sequence and was first interpreted as made of "englacial material of a floating ice tongue" (Morris, 1938, p. 403) but was later reinterpreted as debris flow triggered by a catastrophic flood (Sve-Djin and Taseer Hussain, 1981). The Marwat Formation is overlain by the Malagan Formation, a coarser detrital deposit than the former. The series was not calibrated to the paleomagnetic scale, but on faunal evidence West (1979) estimated that the Bain Boulder Bed dates approximately from the Olduvai episode, that is, it is near the conventional Neogene-Quaternary boundary. Johnson et al. (1982) on the other hand reported an age of about 2.0 myr for the boulder bed, based on an unpublished report by Reynolds.

Farther east, in the Jhelum valley, southeast of Rawalpindi, coarse conglomerates and boulder beds with metasediments and igneous rocks from the Lesser and Higher Himalayas were reported from three sections. Conglomeratic sedimentation begins at slightly different ages in each section: in the Rohtas anticline, at the level of the Olduvai episode; in the Mangla-Samwal section, slightly later than the Olduvai; in the Chambal area, in the late Matuyama (Opdyke et al., 1982).

In northern India, Siwalik sediments are exposed close to the foot of the Himalayan ranges. Ranga Rao et al. (1988) calibrated two sections in the Jammu Hills with the paleomagnetic scale. The Nagrota section, north of the town of Jammu, is molassic in its lower part, which contains a fauna typical of the Pinjor stage. Around 1.7 myr, during the Olduvai paleomagnetic episode, the section becomes coarsely conglomeratic; the thickness of conglomerates exceeds 400 m. Farther southeast, in the Parmandal-Uttarbeni section, the series is also molassic in its lower part; conglomeratic intercalations appear at the level of the Jaramillo episode, and at the beginning of the Brunhes epoch the series becomes entirely conglomeratic. A rich fauna typical of the Pinjor stage extends from the beginning of the Matuyama epoch to the Jaramillo episode. Still farther east, in the area of the town of Pinjor, near Chandigarh, the molasse of the Pinjor stage is richly fossiliferous in its lower part. Coarse conglomeratic intercalations appear at the level of the Olduvai episode, and higher up conglomerates become increasingly frequent. The series ends with massive conglomerates several hundred meters thick (the "Lower Boulder Conglomerate"), the base of which has been correlated with the beginning of the Jaramillo episode (Azzaroli and Napoleone, 1982). The series lies at the foot of the Lesser Himalayas and provides further evidence on physiographic evolution: the Lower Boulder Bed is overlapped by fine-grained molasse of the Lower Siwaliks and then by increasingly older rocks in a series of superimposed thrusts heading south or southwest.

The record of the Siwaliks thus provides evidence for strong uplift in the mountain belts to the north during the Early Pleistocene, which reached a climax, it seems, around the end of the Early Pleistocene, shortly before the temperature shift in the Late Matuyama that is recorded by stable isotopes in the oceans.

The Thirteenth INQUA Congress (Beijing, 1991) dedicated two symposia to the uplifting of the Tibetan Plateau and adjoining mountain ranges. There is evidence of a strong uplift in relatively recent times and of movements still under way, but accurate dating of the movements seems to be an especially difficult problem. Qian (1991) estimated that the plateau was lifted by 2,000 m during the last 4.9 myr. He recognized a strong uplifting phase in the Early Pleistocene, without dating it more specifically.

According to Kuhle (1987, 1991) an ice cap of 2.0 to 2.4 million km^2 covered Tibet and the surrounding mountains during the glacial periods of the Pleistocene; Kuhle sees in this huge inland ice cap, which exceeded the size of the present Greenland ice cap, the main cause of the Pleistocene glaciations.

To conclude, a major climatic crisis affected Eurasia during a relatively short time span, beginning approximately at 1.2 myr in the higher latitudes of the continent and around 1.0–0.9 myr in the middle latitudes. As a result of this crisis a new, more cold-adapted fauna of modern type superseded the Late Villafranchian faunas. The climatic and faunal revolutions are older than the major cooling of ocean waters of the Late Matuyama. Tectonic activity and ensuing glaciations seem to have been the main causes of the revolution on the continent. The cooling of the oceans in the Late Matuyama was, it seems, the effect, not the cause, of the climatic change.

The faunal changes of the Middle Pliocene, the Elephant-*Equus* event of Lindsay et al. (1980), has been recognized in Eurasia and in other continents. During this major crisis, the early Pliocene faunas, characterized by warm forest species such as zygodont mastodont, and tapir, were wiped out. The Reuverian flora, which also characterized a warm climate, gave way to the cooler Pretiglian flora. Regrettably, the significance of this crisis was understood relatively late, owing to inadequate taxonomical analysis of vertebrate faunas and to the lack of the valuable tool now represented by palynology. As a result, the currently used term "Villafranchian," proposed by Pareto in 1865 and interpreted by Gignoux (1916) as the time equivalent of the marine Late Pliocene (the Calabrian, now included in the Pleistocene), actually ranges from a date older than 3.0 myr to about 1.0 myr, straddling the major event of faunal turnover in the whole Pliocene.

The Elephant-*Equus* event was dated with accuracy in the Siwaliks of India and Pakistan. The boundary between the Tatrot and Pinjor stages, evidenced by a marked change in the fauna, and in some cases also by a clear-cut difference in lithology, has been correlated with the Gauss-Matuyama transition in several localities (Lindsay et al., 1980; Opdyke et al., 1982; Azzaroli and Napoloene, 1982; Azzaroli, 1985). The faunal change was marked but not as sharp as the approximately contemporary change in Europe (faunas of central Asia are not as well known). It can only be noted that in the Siwaliks monodactyl horses and *Elephas hysudricus* appeared at the Elephant-*Equus* event, but the more primitive *Elephas* (or *Protelephas;* Garutt, 1957) *planifrons* appeared in the Tatrot, well before this event (Sahni and Khan, 1988).

In western Europe the faunal revolution involved taxa different from those of the Indian subcontinent and seems to have been slightly older. The Montopoli local fauna of central Italy, a fairly rich assemblage including elephant and monodactyl horse, was calibrated by Lindsay et al. (1980) to the Gauss-Matuyama transition, but an elephant skeleton from Laiatico, not far, and downsection, from Montopoli, was collected in a brackish marine deposit that has not been dated exactly but that is older than the Montopoli local fauna. The Rincon 1 local fauna of eastern Spain, similar in composition to the Montopoli fauna, was dated at 2.6 myr by Leone (reported by Azzaroli et al., 1988).

Pollen analysis gives divergent results. In Israel Horowitz (1989) has found evidence of a shift toward a cooler climate at 2.6 myr, with good approximation, whereas Zagwijn and De Jong (1984) date the Reuverian-Pretiglian transition to the Early Matuyama, which is much later than the end of the Gauss epoch. In the northern Atlantic, morainic material rafted by ice is first recorded at 2.5 myr and is followed by a major pulse at 2.4 myr (Shackleton et al., 1984). A cooling of water is indicated at the same time by stable isotopes.

The Mid-Pliocene event has also been recognized in terrestrial mammalian faunas from Africa (Vrba, 1985, and chap. 27, this vol.) and in the vegetation of South America (Hoogiemstra, 1989, and this vol.) and appears to have been an event of worldwide scope, perhaps even more wide-ranging in is effects than the end-Villafranchia event; but its various aspects—turnover in terrestrial faunas and vegetation, marine climatic changes, glacial phenomena—have not yet been dated with the necessary time resolution to disentangle cause-and-effect relationships.

Summary

The continental Plio-Pleistocene of Eurasia was characterized by two major faunal revolutions, the Mid-Pliocene Elephant-*Equus* event and the end-Villafranchian event around the transition from Early to Middle Pleistocene. The latter event has been analyzed in detail: it appears that climatic, vegetational, and faunal changes on the continent preceded cooling of ocean waters by 100–200 thousand years. The climatic crisis is believed to have been triggered by a strong upheaval of the mountains and plateaus of central Asia and the ensuing land glaciation. The Mid-Pliocene Elephant-*Equus* event had comparably wide-ranging effects on the continents and also coincided approximately with a cooling of ocean waters. It has not yet been dated with sufficient time resolution, however, to distinguish causes from effects.

Notes

1. The boundary between Early and Middle Pleistocene was provisionally agreed upon as the Matuyama-Brunhes paleomagnetic transition at the Twelfth INQUA Congress (held in Ottawa in 1987), but it was discussed again at the Thirteenth INQUA Congress (Beijing, 1991) and has not yet been formally defined.

References

Ambrosetti, P., Azzaroli, A., Bonadonna F. P., and Follieri, M. 1972. A scheme of Pleistocene stratigraphy for the Tyrrhenian side of central Italy. *Bollettino della Società Geologica Italiana* 91:169–184.

Azzaroli, A. 1985. Provinciality and turnover events in Late Neogene and Early Quaternary vertebrate faunas of the Indian subcontinent. *Contributions to Himalayan Geology* 3:27–37.

———. 1991. Major events at the transition from Early to Middle Pleistocene. *Il Quaternario* 4:5–11.

Azzaroli, A., De Giuli, C., Ficcarelli, G., and Torre, D. 1988. Late Pliocene to early Mid-Pleistocene mammals in Eurasia: Faunal succession and dispersal events. *Palaeogeography, Palaeoclimatology, Palaeoecology* 66:77–100.

Azzaroli, A., and Napoleone, G. 1982. Magnetostratigraphic investigation in the Upper Sivaliks near Pinjor, India. *Rivista Italiana di Paleontologia e Stratigrafia* 87:739–762.

Beard, J. H., Sangree, J. B., and Smith, L. A. 1982. Quaternary chronology, paleoclimate, depositional sequence and eustatic cycles. *American Association of Petroleum Geologists Bulletin* 66: 158–169.

Cooke, H. B. S., 1981. Age control of Quaternary sedimentary-climatic record from deep boreholes in the Great Hungarian Plain. In *Quaternary paleoclimate,* pp. 1–12 (ed. W. C. Mahaney). Geological Abstracts Ltd., Norwich, Eng.

Garutt, V. E. 1957. Novye dannye o drevnejshikh slonakh. Rod *Protelephas* gen. nov. *Doklady Akademii Nauk SSSR 1957,* 144 (1):189–191.

Gignoux, M. 1916. L'étage calabrien (Pliocène supérieur marin) sur le versant NE de l'Apennin, entre Monte Gargano et Plaisance. *Bulletin de la Société Géologique de France,* ser. 4, 14:324–348.

Gromov, V. 1948. Paleontologicheskoe i arkheologicheskoe obosnovanie stratigrafii kontinental'nikh otlozhenij chetvertichnogo perioda na territorii SSSR. *Trudy Instituta Geologicheskikh Nauk Akademii Nauk SSSR,* no. 64. Geologicheskaja Serija, no. 17.

Horowitz, A. 1989. Continuous pollen diagrams for the last 3.5 m.y. from Israel: Vegetation, climate and correlation with the oxygen isotope record. *Palaeogeography, Palaeoclimatology, Palaeoecology* 72:63–78.

Itihara, M., and Yoshikawa, S. 1991. Major subdivision of Quaternary deposits in and around Osaka, Konki, Japan. *Thirteenth INQUA Congress, Beijing, 1991,* Abstracts: 147.

Johnson, G. D., Zeitler, P., Naeser, C. M., Johnson, N. M., Summers, D. M., Frost, C. D., Opdyke, N. D., and Tahirkeli, R. A. K. 1982. The occurrence of fission-track ages of Late Neogene and Quaternary volcanic sediments, Siwalik group, northern Pakistan. *Palaeogeography, Palaeoclimatology, Palaeoecology* 37:63–93.

Kuhle, M. 1987. Subtropical mountain- and highland-glaciation as Ice Age trigger and the waning of the glacial period in the Pleistocene. *GeoJournal* 14:393–421.
———. 1991. Observations supporting the Pleistocene inland glaciation of high Asia. *GeoJournal* 25:133–231.
Kukla, G., and An, J. 1989. Loess stratigraphy in central China. *Palaeogeography, Palaeoclimatology, Palaeoecology* 72:203–225.
Lindsay, E. H., Opdyke, N. D., and Johnson, N. M. 1980. Pliocene dispersal of horse *Equus* and late Cenozoic mammalian dispersal events. *Nature* 287:135–138.
Morris, T. O. 1938. The Bain Boulder Bed: A glacial episode in the Siwalik series of the Marwat Kundi Range and Shekh Budin, Northwest Frontier Province, India. *Quarterly Journal of the Geological Society of London* 94:385–421.
Opdyke, N. D., Johnson, N. M., Johnson, G. D., Lindsay, E. H., and Tahirkeli, R. A. K. 1982. Palaeomagnetism of the middle Siwalik formations of northern Pakistan and rotation of the Salt Range decollement. *Palaeogeography, Palaeoclimatology, Palaeoecology* 37:1–16.
Pareto, L. 1865. Sur les subdivisions que l'on pourrait établir dans les terrains tertiaires de l'Appenin septentrional. *Bulletin de la Société Géologique de France*, ser. 2, 22:210–277.
Qian, F. 1991. The age of *Hipparion guizhongensis* and the upheaval of the Qinghai-Xizang Plateau. *Thirteenth INQUA Congress, Beijing, 1991,* Abstracts: 292.
Ranga Rao, A., Agrawal, R. P., Sharma, U. N., Bhalla, M. B., and Nanda, A. C. 1988. Magnetic polarity stratigraphy and vertebrate palaeontology of the Upper Siwalik subgroup of Jammu Hills, India. *Journal of the Geological Society of India* 31:361–385.
Sahni, M. R., and Khan, E. 1988. *Pleistocene vertebrate fossils and prehistory of India.* Books and Books, New Delhi.
Shackleton, N. J., Backman, J., Zimmermann, H., Kent, D. V., Hall, M. A., Roberts, D. G., Schnitker, D., Baldauf, J. G., Desprairies, A., Homrighausen, R., Huddleston, P., Keene, J. B., Kaltenback, A. J., Krumsier, K. A. O., Monton, A. C., Murray, J. W., and Westberg Smith, J. 1984. Oxygen isotopic calibration of the onset of ice-rafting and history of glaciation in the North Atlantic region. *Nature* 307:620–623.
Sher, A. V. 1971. Mlekopitajushchie i stratigrafija Pleistozena krajnegø Severo-Vostoka SSSR is Severnoj Ameriki. Nauka, Moscow.
Sher, A. V. 1987. Olyorian land mammal age of northeastern Siberia. *Palaeontographia Italica* 74:97–112.
Sve-Dijn, N., and Taseer Hussain, S. 1981. Sedimentary studies on Neogene/Quaternary fluvial deposits on the Bhittanni Range, Pakistan. *Proceedings of the Neogene/Quaternary Boundary Field Conference, India, 1979,* pp. 177–183. Geological Survey of India, Calcutta.
Thouveny, N., and Bonifay, E. 1984. New chronological data on European Plio-Pleistocene faunas and hominid occupation sites. *Nature* 308:355–358.
Vrba, E. S. 1985. African Bovidae: Evolutionary events since the Miocene. *Suid-Afrikaanse Tydskrift vir Wetenskap* 81:263–266.
West, R. M. 1979. Plio-Pleistocene fossil vertebrates and biostratigraphy, Bhittanni and Marwat Ranges, north-east Pakistan. *Proceedings of the Neogene/Quaternary Boundary Field Conference, India, 1979,* pp. 211–215. Geological Survey of India, Calcutta.
Xu, Q. 1989. Late Cenozoic mammalian events in north China. *Proceedings of the International Symposium on Pacific Neogene Continental and Marine Events,* pp. 129–136. International Geological Correlation Program, no. 246.
Zagwijn, V. A. 1974. The Pliocene-Pleistocene boundary in western and southern Europe. *Boreas* 3:75–97.
Zagwijn, V. A., and De Jong, J. 1984. Die Interglaziale von Bavel und Leerdam und ihre stratigraphische Stellung im Niederländischen Früh-Pleistozän. *Mededelingen Rijks Geologische Dienst* 37:155–169.

Chapter 23

The Potential of the Turkana Basin for Paleoclimatic Reconstruction in East Africa

Francis H. Brown

A sedimentary record from latest Oligocene to Late Pleistocene exists in the Turkana Basin, most parts of which have been reasonably well calibrated chronologically. The sole exception is the site of Lothagam, which will also fall into line given the present investigations by I. McDougall and C. S. Feibel. Geological as well as faunal and floral clues to ancient climates are preserved in the oldest deposits (the Oligocene Eragaleit Beds; Boschetto et al., 1992), in Miocene strata at many sites (Lothidok, Locherangan, Buluk, Lothidok, Kajong), in the Pliocene and Pleistocene Formations of the Omo Group (the Mursi, Usno, Shungura, Koobi Fora, and Nachukui Formations), and in the Middle and Late Pleistocene strata of the Turkana Group (the Galana Boi and Kibish Formations). Still unstudied are sedimentary sequences in the temporal interval from about 10–7 million years (myr), but these too may yield paleoclimatic information.

The time span from which evidence is derived is sufficiently long that we should not neglect slow motions of the African plate when making paleoclimatic interpretations. Nor should we neglect building of volcanic complexes, possibly of high elevation, followed by their subsequent erosion. Furthermore, we must consider the general elevation of East Africa related to the formation of the Rift Valley; and changes in the water balance of the region caused by tectonic movements and volcanic construction both within the depositional area of the basin and also in the upper regions of the Omo River Basin. At present it is not possible to evaluate all the effects of these geological changes, but it is perhaps worthwhile to point out certain problems and to mention research that may lead to greater understanding of changes in the area.

ISBN 0-300-06348-2.

Stratigraphic Summary

Data relating to paleoclimates in the Turkana Basin or any other area must come from sedimentary sections preserved in the region. Strata from the latest Oligocene to the Late Pleistocene are preserved in the Turkana Basin, but the sampling of time is far less complete than this statement might imply. In a diagram of times represented by strata, particularly fossiliferous strata (fig. 23.1), it is evident that only about half the time from 27.5 myr to the present is represented by sediments. Without doubt, this is an overestimate of the actual fraction of time represented, for within each of the sequences there are many smaller temporal gaps. These hiatuses severely limit our ability to trace climatic changes through time in any particular section, except when a continuous section happens to coincide with a significant shift.

In contrast to later deposits in the region, Miocene strata have not been correlated in detail, despite reasonably good temporal control. Each of the Miocene sites has been treated geologically more or less as if no other strata of comparable age existed in the region. This approach has been adopted not only because the sites have been investigated by different workers but also because the stratigraphy of each site appears to be quite distinct from that of other sites. Below I give a brief summary of the stratigraphy and fossil content of these deposits, followed by an equally brief recapitulation of the principal features of the Pliocene and Pleistocene deposits in the region.

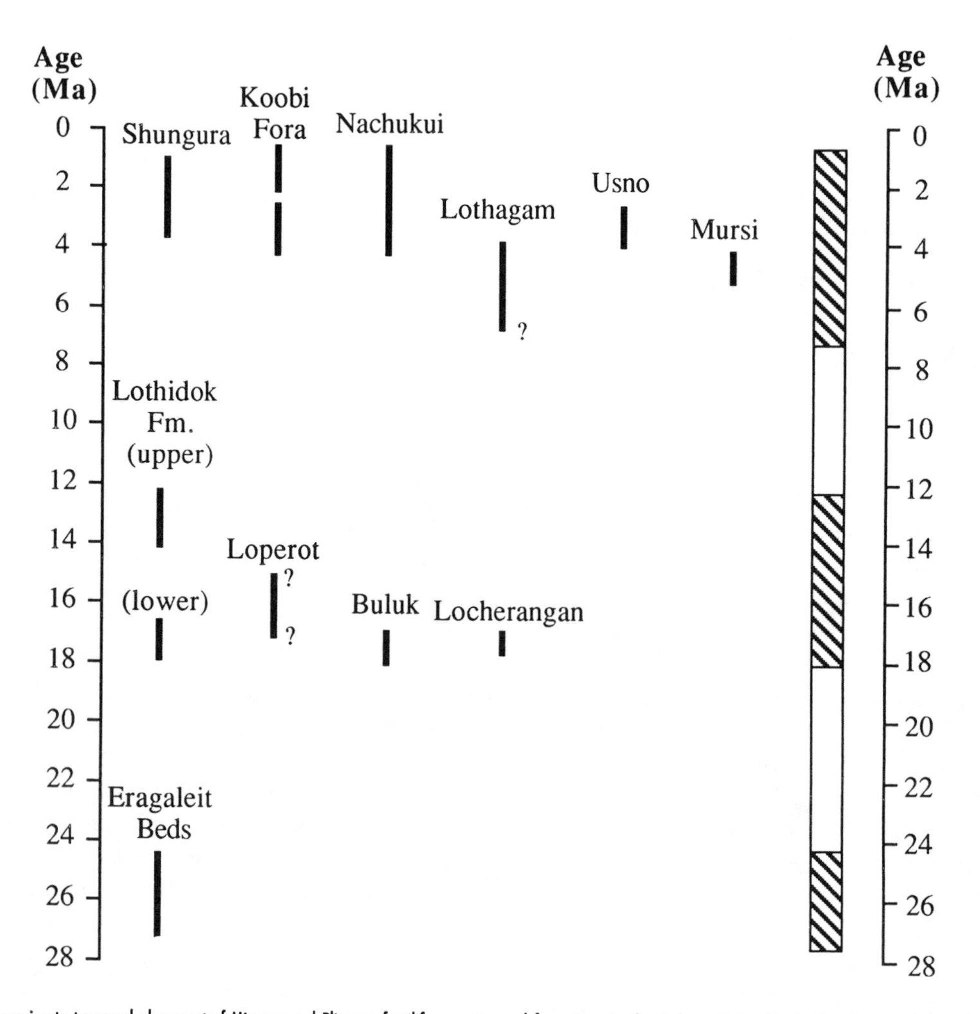

Fig. 23.1. Approximate temporal placement of Miocene and Pliocene fossiliferous sites and formations in the Turkana Basin. The hachured parts of the column on the right indicate the maximum amount of time sampled.

Lothidok. At least 1,540 m of lava flows, tuffs, and terrestrial clastic sediments are exposed in the Lothidok Range west of Lake Turkana. Near the base of the section is a fossiliferous interval up to 50 m thick designated the Eragaleit beds, dated between 24 and 27.5 myr by K/Ar determinations on overlying and underlying lavas. Reddish volcaniclastic conglomerates make up most of the section, with thinner sequences of siltstone and tuff. A reasonable fauna preserved within the upper part contains *Etheria* (freshwater oyster), seven nonprimate mammalian taxa, and primate fossils that have not yet been described. The Eragaleit beds are overlain by at least 600 m of lava flows, informally named the Kalokol basalts, with minor intercalated sedimentary lenses. Many of these lenses are fossiliferous, though no collections have been made from them.

Overlying the Kalokol basalts is the Lothidok Formation (18–12 myr), which is approximately 580 m thick and divided into five members. It consists principally of reddish volcaniclastic conglomerates, sandstones, and siltstones, together with lahars and airfall tuffs in its lower part. These are overlain by the Kalatum Phonolite lava flow, followed by sandstones that increase in quartz content upward. A sequence of lahars and volcaniclastic rocks forms the upper part of the section. A rich fauna recovered from the Lothidok Formation (Leakey and Leakey, 1986a, 1986b) include *Etheria,* gastropods, insects, reptiles, birds, a good sample of mammals, and the type specimens of the primates *Turkanapithecus, Afropithecus,* and *Simiolus enjiessi,* in addition to *Proconsul.* Most of the fauna is 16.8–17.5 myr old, but a smaller collection from the upper part of the formation is between 12.2 and ca. 13.5 myr in age. Fossil leaves preserved in the Lothidok Formation have not yet been studied in detail, but from their size and form they appear to be from plants that require more water than the area receives today.

Lothagam. Presently the lithostratigraphy and chronostratigraphy is under revision by a team under the direction of Dr. Meave Leakey, so no formational names are used in this synopsis. It is reasonably certain that the upper part of the Lothagam sequence is equivalent to the Lonyumun member of the Nachukui Formation and must thus be approximately 4 myr old. Powers (1980) described a thick lower sequence of coarse volcaniclastic rocks and lavas that is unfossiliferous, overlain by about 500 m of fossiliferous sedimentary section at the site. The lower part of the fossiliferous section is composed of markedly reddened volcaniclastic sandstones and siltstones, and the upper part is composed principally of coarse, pale orange arkosic sandstones overlain by laminated claystones. Powers (1980) interpreted the fossiliferous part of the section as the product of fluvial deposition; claystones of the upper part of the sequence were interpreted as being of lacustrine origin. A hominid (or hominoid) mandible is known from the upper part of the sequence at Lothagam in deposits that were termed Lothagam IC by Behrensmeyer (1976).

Other Miocene Localities. Several other fossiliferous Miocene localities are also known in the Turkana Basin and have been described by other workers. These include Kajong, where a 220 m-thick section with fluvial and lacustrine rocks is exposed (Savage and Williamson, 1978); Buluk, with a 50 m-thick section of sandstones, finer deposits, and tuffs (Harris and Watkins, 1974; McDougall and Watkins, 1985); Locherangan, where a dominantly lacustrine section about 60 m thick is exposed (Anyonge, 1991); Loperot, where a largely fluvial section 215 m thick crops out (Joubert, 1966); and Napedet, where Dodson (1971) describes fluvial and lacustrine sandstones and claystones about 30 m thick.

Pliocene and Pleistocene Deposits. Nearly 800 m of strata give the Turkana Basin the most complete lithostratigraphic record in East Africa for the period 4.3 to 1.2 (–0.6) myr. These strata, exposed discontinuously around the northern end of Lake Turkana in the form of an inverted U, have been described as five formations; the Nachukui Formation, west of the lake; the Koobi Fora Formation, east of the lake; and the Mursi, Shungura, and Usno Formations, north of the lake.

De Heinzelin and Haesaerts (1983c) describe the Mursi Formation as 110 m of principally fluvial sandstones and siltstones with minor lacustrine clays. The formation is capped by a basalt approximately 4 myr old, making this formation an important one, as it is the oldest Pliocene sedimentary sequence in the basin.

The stratigraphy of the Shungura Formation has been documented by de Heinzelin and Haesearts (1983a). The formation crops out in two major areas west of the Omo River in southwestern Ethiopia: a northern, or Type Area, and a southern, Kalam Area. In these two areas a section 765 m thick is exposed, but the highest levels are exposed only in the Kalam Area, and the lowest only in the Type Area. Most of the section records fluvial deposition, but there is an important lacustrine interval in the middle of the section (upper Member G), and thereafter fluvial and lacustrine conditions alternate. Through the work of the Omo Research Expedition, a very complete vertebrate fauna was amassed from this formation,and microvertebrates from various levels have led to paleoclimatic interpretations for particular times in the past (Wessleman, this vol.). The rich tephra record of the Shungura Formation has provided a key section to which other sections may be referred through tephrostratigraphic correlation over a large part of East Africa (e.g., Haileab, 1988; Haileab and Brown, 1994).

Exposures of the Usno Formation lie about 25 km northeast of the northern part of the Shungura Formation. The formation is 172 m thick (de Heinzelin and Haesaerts, 1983b), but most of its vertebrate fauna is derived from a very restricted interval of fluvial section in the upper part of the formation. This interval lies just above the Tuff U-10 (= Tulu Bor Tuff), between magnetozones attributed to the Mammoth and Kaena subchrons.

The Koobi Fora Formation is exposed east of Lake Turkana in northern Kenya over an area of about 1,200 km^2. The lithostratigraphic nomenclature of the region has been revised several times, the most current version having been given by Brown and Feibel (1986). A principal difficulty in constructing a stratigraphic column (565 m thick) for this region is that the exposures are discontinuous, isolated from one another by extensive tracts of alluvial cover. Unlike the Shungura or Nachukui Formations, there is a substantial hiatus (ca. 0.5 myr) *within* the sequence at Koobi Fora from about 2.5–2 myr ago. Now that the stratigraphy is understood, the abundant vertebrate fauna has been interpreted paleoclimatically (Feibel et al., 1991) and integrated with the pollen record so that a series of probable vegetational communities has been postulated for the area at different times in the past.

The Nachukui Formation (Harris et al., 1988a, 1988b) crops out over about 500 km^2 west of Lake Turkana, but as with the Koobi Fora Formation, exposures are separated by substantial areas of cover. Approximately 715 m of section are exposed, which correspond in large part to the Shungura Formation but which extend to both younger and older time intervals. Harris et al. (1988) and Feibel et al. (1991) discuss the paleoenvironmental relevance of the fossil record of the Nachukui Formation. Both lacustrine and fluvial environments are apparent from the lithostratigraphy of the formation, but without information from related formations, any climate model constructed from these data alone would be far off the mark.

Late Pleistocene and Holocene Deposits. Strata ranging in age from perhaps 100 thousand years (kyr) to approximately 3 kyr have been described as the Kibish Formation, in the lower Omo Valley (Butzer et al., 1969) and as the Galana Boi Formation, in the Koobi Fora region (Owen and Renault, 1986). The Kibish Formation has been divided into four members (I–IV) with the uppermost split into submembers. These members are separated by erosional surfaces and possibly record climatically induced fillings of ancient Lake Turkana. If these members correlate with astronomical precessional cycles, then the estimated age of the oldest member would be between 92 and 115 kyr, depending on whether four or five cycles are recorded. Three partial skulls of archaic *Homo sapiens* have been collected from the lowest member of this sequence.

In the southern part of the Omo Valley, and east and west of the lake, only the latest part of this sequence is well represented, although thicker deposits occur high on the north side of the Kokoi and in Area 103 at Koobi Fora. Several hominid skeletons are known from these deposits, and the existence of the deposits is proof enough that water supply to the area was considerably greater about 11 kyr ago than it is today.

Possible Paleoclimatic Implications of Plate Tectonic Motions

In many reconstructions of continental positions in past times, Africa is held geographically fixed, and other continents are moved with respect to it. Depending on the remoteness of the time under consideration, it is not necessarily appropriate to consider the continent as having remained immobile (contra Ruddiman et al., 1989). For the bulk of the Neogene, Africa has been moving northward at a rate of approximately 5 cm per year, that is, 50 km every million years. At this rate, what is presently the northern end of Lake Turkana was at a latitude very near the present southern end when sedimentation of the Omo Group began ca. 4 myr ago. Six myr ago Lothagam was located at the latitude of what is now Lake Bogoria, and 17 myr ago Lothidok was latitudinally positioned about 70 km south of Lake Manyara in what is presently northern Tanzania. Just what effect these changes in latitude would have on the local climate prevailing at sites of different ages cannot be determined without considerable effort in modeling. If, as a first approximation, one considered the precipitation patterns fixed by the equatorial Hadley cells, however, simple northward latitudinal motion of Africa would result in a long-term trend toward drier conditions from the Mid-Miocene until the Mid-Pliocene. The effects of these motions should certainly be explored by those interested in hominoid evolution, because it is likely that the central and West African forests were far more extensive in the Miocene than they are today; at that time the equator was through Conakry, Guinea, to Ras Hafun, Somalia, now about 10° farther north.

Topographic Changes and Volcanic Construction

Most of the present highland areas in eastern Kenya today are of very recent vintage. Mt. Kenya was probably built almost entirely during the Pliocene, and the Nyambeni Hills northeast of it are still younger. Mt. Marsabit, the Huri Hills, Esie, and Mt. Kulal all date from Late Pliocene to Pleistocene time. Each of these highlands creates local climatic and vegetational zones of its own, so that the flora and fauna at higher elevations reflect more mesic conditions than those pertaining in the surrounding lowlands. In the past, these highland areas may have acted as refugia for animals and plants during exceptionally dry intervals. Hence, even when considering Pliocene climatic conditions in the Turkana Basin, the effects of these local highlands should not be ignored.

At earlier times, there is also considerable evidence for local volcanic highlands. These highlands have been well documented for Miocene sites in Uganda and Kenya (Bishop, 1965) and also for some Miocene sites in the

Turkana Basin. Boschetto et al. (1992), for example, describe thick volcanic mudflows that require a considerable highland area for their origin. At present, however, almost no vestiges of these highlands remain aside from Mt. Moiti, east of Lake Turkana, and possibly the core of the Morua Rith Range, west of the lake. There is a significant gap in the stratigraphic record in the Turkana Basin between about 8 and 12 myr, during which these highlands may have been considerably reduced in elevation by erosion, though structural adjustments may also partly account for the lack of physical expression. As a result, it is nearly impossible to make statements about Miocene climates in the area, for the size, orientation, and elevation of these highlands, and thus their potential climatic effects, are presently unknown.

The timing of uplift of the present highlands of Ethiopia and Kenya is still not well understood. Vertical structural motions on the order of 1 km have occurred within the past 0.7 myr in the Turkana Basin, and in many other parts of the East African Rift system it is possible to demonstrate that downfaulting of the rift is reasonably young (less than 2 myr old). By contrast, most workers assume that the building of the highlands began with extensive volcanism on the order of 20–30 myr ago, depending on location. Until more is known about the time of formation of the present topography, it is dangerous to consider that Miocene and Pliocene topographic elements were similar to those of the present.

Basinal Models and Climatic Reconstructions

According to early interpretations of Pliocene and Pleistocene sedimentation in the Turkana Basin, deposition occurred in response to fluctuation in a large stable lake (Vondra and Bowen, 1978). The fluctuations in turn were believed to be caused by climatic changes and tectonic activity. More recent research has replaced this view with one in which climate still has a role but in which sedimentation is believed to be controlled dominantly by tectonic movements and volcanic construction, pointing up a general difficulty with paleoclimatic proxies. If the history of a region is incorrectly understood, then one may erroneously attribute sedimentological differences to climatic causes. Even isotopic records may respond to changes in local conditions rather than to regional changes, and just what is meant by climatic change is often left rather ambiguous. Our present understanding of the depositional history followed the establishment of tephrostratigraphic correlations between the Koobi Fora, Nachukui, and Shungura Formations.

During the 1970s many workers documented extensive lacustrine deposits in the Turkana Basin about 2 myr in age and suggested that because of climatic change a lake expanded, entering the lower Omo Valley and leaving its record in the Shungura Formation (Brown, 1981). Instead of expansion and contraction of a single lacustrine system, the record documents temporary closing, or partial closing, of the basin to produce different lakes at different times. It is now possible to reconstruct paleogeographic elements over the entire basin for many different time intervals, and it is seen that the history of the basin was one in which either a fluvial system occupied the entire basin, or the central part of the basin was occupied by a lake. Transition between the two states was rapid.

Need for Chronological Control in Paleoclimatic Studies

One of the principal contributions of research in the Turkana Basin has been to establish a well-controlled chronology from 4.3 to 0.6 myr ago. Brown (1994) has recently reviewed the development of the chronology, and it is quite obvious that the chief credit must go to Ian McDougall, who satisfactorily resolved existing problems and produced the stratigraphically consistent set of dates on which we presently rely (McDougall, 1981; McDougall et al., 1980, 1985; Brown et al., 1985; McDougall, 1985; McDougall et al., 1985; Feibel et al., 1989). This chronology has been extended from the Turkana Basin to other sites through tephrostratigraphic correlations, so that sediments in the Hadar area (Brown, 1982), the Middle Awash Valley (Haileab and Brown, 1992), the Baringo Basin (Namwamba, 1993), the Gadeb area (Haileab and Brown, 1994), the Albert Rift (Pickford et al., 1991), and the deep-sea record in the Gulf of Aden (Rahman and Roth, 1988; Brown et al., 1992) can all be referred, at least in part, to a single reference column (see fig. 23.4). More recently, it has become possible to import dates from the Hadar region and the Middle Awash areas to the Turkana Basin, which have helped refine the chronology of the Omo Group (White et al., 1993; Walter and Aronson, 1993). These additional results are incorporated in Figure 23.2. A significant effect of increased temporal resolution is that climatic signals recorded in the deep sea thus become integrated with those on land. In addition the correla-

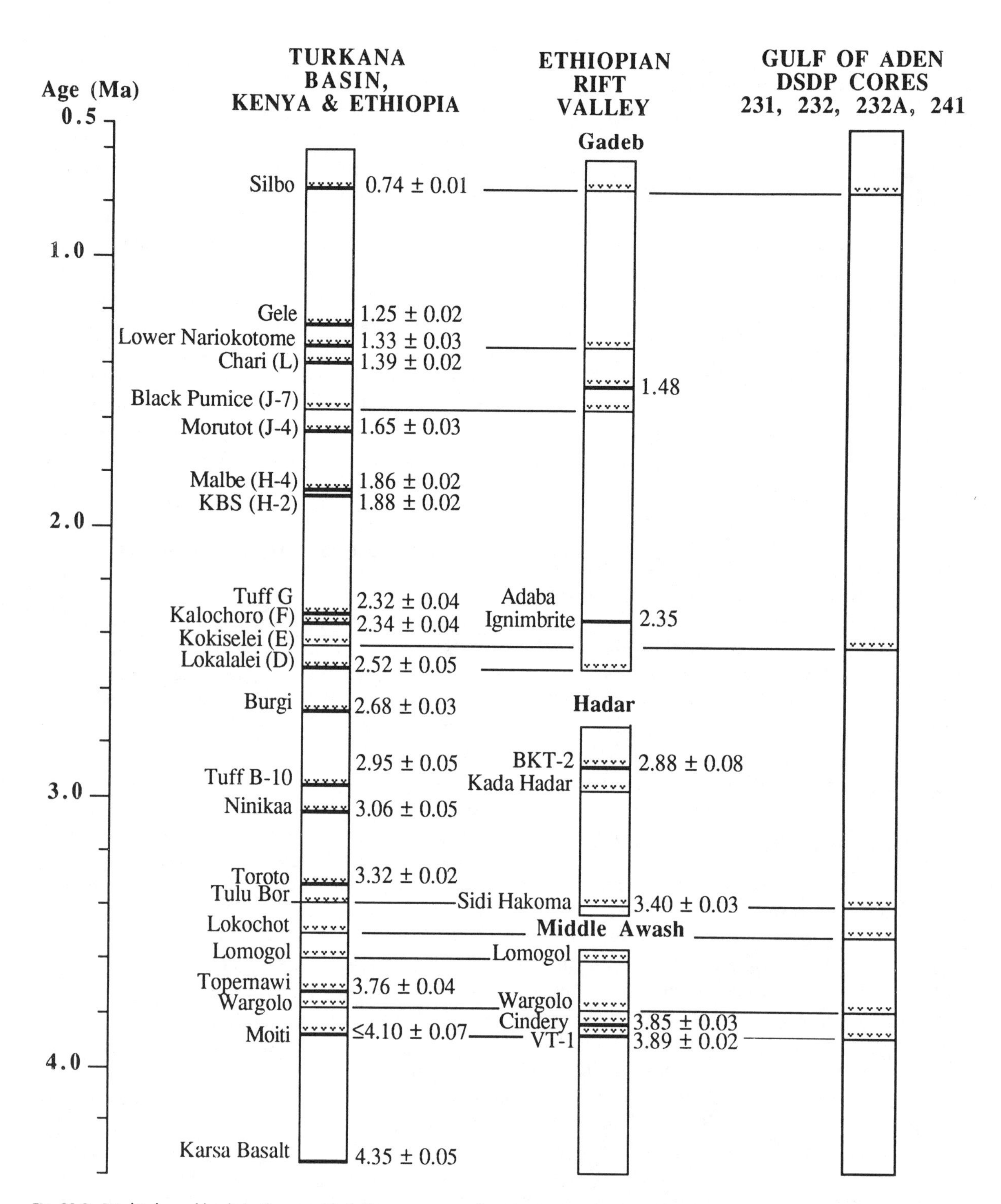

Fig. 23.2. Dated tephra and basalts in Pliocene and Early Pleistocene strata in the Turkana Basin and Ethiopian Rift Valley (Middle Awash and Hadar areas), with tephrostratigraphic correlations between them. Dated tephra layers are shown with a heavy line; undated defining or correlated tephra layers are shown with a thin line. Tephra correlated to deep-sea cores are also shown.

tions have the effect of freeing the fossil faunas from doing double duty, that is, their ecological and possible paleoclimatic significance can be considered independent of time.

An additional advance flowing from the chronology is that the times of paleomagnetic polarity change for the past 4 myr have been refined, so that the apparent discrepancy between the time scales for deposits in the Turkana Basin based on paleomagnetic stratigraphy and on K/Ar dating has now been removed (McDougall et al., 1992). Times of polarity transition computed from data in the Turkana Basin agree exceptionally well with those derived from astronomical considerations, as do data from the Hadar Formation (Renne et al., 1993).

Isotopic and Biotic Paleoclimatic Indicators

Situated at low elevation in an interior equatorial region, deriving the main part of its water from highlands at great distance, and principally controlled by tectonic and volcanic events, the Turkana Basin, through its geologic record, has given only broad indications of climatic change (Feibel et al., 1991). From calcareous soil nodules, it appears that for the past 4 myr the region has undergone a net deficit of evapotranspiration over precipitation. Vertisols in the sections provide evidence for cyclic wetting and drying, owing to either precipitation or flooding. Estimates of 40–80 cm of rain per year, with approximately half of the year dry, have been suggested previously on the basis of fauna (Brown, 1981). Similarly, the mean annual temperature likely lay between 25 and 30°C, and diurnal temperatures ranges were likely 20°–25°C. Cerling (1992) has shown that changes in isotopic composition of carbon and oxygen in soil carbonates relate to the fraction of C_4 plants in modern ecosystems and that the proportion of C_4 plants has increased at Olduvai, Laetoli, and in the Turkana Basin over the past 4 myr. He further showed that the abundance of C_4 plants increased markedly at both Olduvai and in the Turkana Basin about 1.7 myr and decreased thereafter before rising to its modern high. In the Turkana Basin a shift in oxygen isotope composition precedes the carbon isotopic shift by about 0.1 myr, which implies a change in the isotopic composition of soil water. At Olduvai, however, the two isotopic shifts appear to be synchronous. Positive excursions of $\delta^{18}O$ were also identified in the Turkana Basin near 3.4 and 3.1 myr in age, but without corresponding shifts in $\delta^{13}C$, which may be of climatic origin or may instead reflect changes in local conditions. For an isotopic change to be considered climatically significant, it should be shown to be synchronous, or nearly so, at widely separated localities. Tephrostratigraphic correlations between widely separated sites make such comparisons possible, but to date this has not been attempted.

Potential Links with Paleoclimatic Data in Deep-Sea Cores

Beginning with the tephrostratigraphic work of Sarna-Wojcicki et al. (1985), detailed correlations between the deep-sea record and the terrestrial record in the Turkana Basin and the Ethiopian Rift Valley became possible for the period from 0 to 4 myr. Those authors worked with a core from DSDP Site 231, in the Gulf of Aden, and the work was subsequently refined by Brown et al. (1992). Even in the initial publication, it was recognized that paleoclimatic information, so well recorded in the deep sea, might be applied to hominid sites on land.

More recently, deMenocal and Bloemendal (this vol.) produced a detailed record of the aeolian dust content of ODP cores 721 and 722 in the Arabian Sea that extends from the present to the Late Miocene and showed that African climate became dependent upon high-latitude glacial-interglacial climate after 2.8 myr. Further, they demonstrate correlations with an aeolian record at DSDP Site 231, which is some 1,000 km closer to the Turkana Basin and directly linked to it through tephrostratigraphy.

The pronounced cyclic nature of deposition in the Shungura Formation has been known since the first modern work began there (Brown, 1969) and was admirably documented by de Heinzelin (1983). Haesaerts et al. (1983) recognized five phases of deposition in the Shungura Formation. The two earliest of these are fluvial, and each sedimentary unit accumulates over periods of 30–60 kyr, seemingly too long to accommodate a cyclicity with a period of 20 kyr. Within the Matuyama chron, accumulation periods drop to 10–17 kyr and have greater potential to be fitted to some sort of scheme of cyclic deposition. At shorter intervals, where age control is good, it is possible and there are apparent good fits to deposition in response to a 20 kyr climatic cycle.

Within the basin, the interval between Tuff H-2 and Tuff J-4 in the Shungura Formation corresponds with the interval between the KBS Tuff and the Morutot Tuff at Koobi Fora. Dates can be assigned at 1.89 myr (KBS Tuff/Tuff H-2); 1.87 myr (Malbe Tuff/Tuff H-4); 1.78

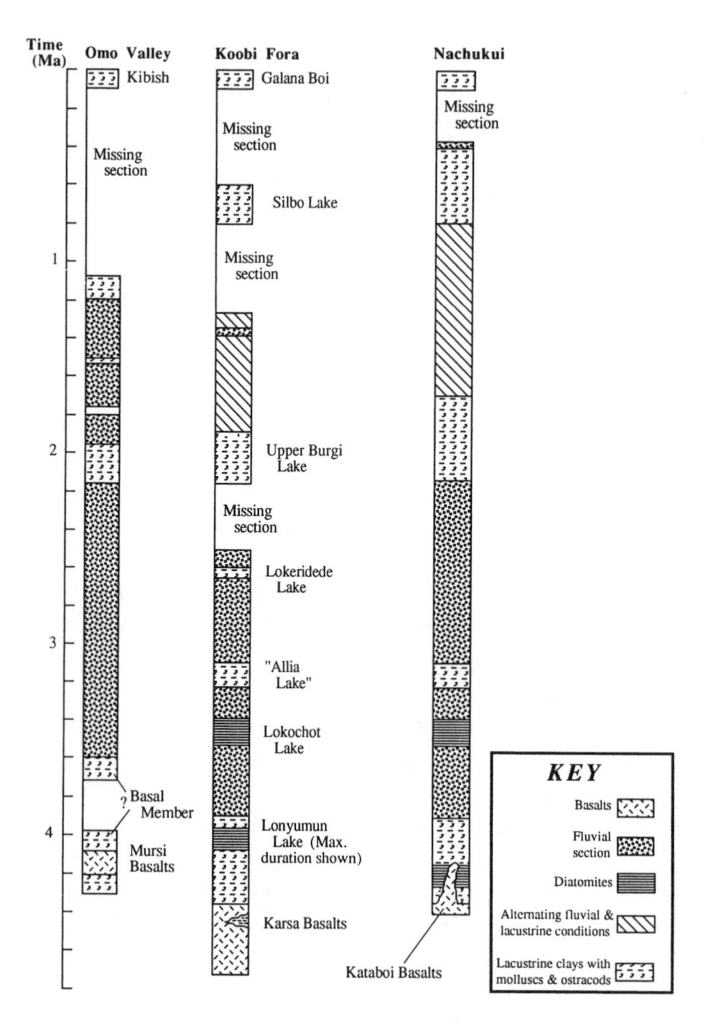

Fig. 23.3. Interpretive columns of Pliocene and Early Pleistocene strata in the Turkana Basin, showing general depositional conditions through time. Note that the record of lacustrine deposition is not synchronous in all sections.

myr (top of Olduvai event); and 1.65 myr (Tuff J-4; Morutot Tuff). Sedimentation proceeded at markedly different rates in the two areas; it was reasonably slow (147 m per myr) in the Shungura Formation, but much faster (295 m per myr) in the Koobi Fora Formation. Nonetheless, some geologic features probably correspond between the two areas, although they are differently developed. The molluscan sandstone C4 (Feibel, 1984), which is such a prominent marker along the western part of Koobi Fora Ridge, for example, almost certainly corresponds to the prominent molluscan level in submember H-5 of the Shungura Formation. Thus some sedimentologic events at widely separated locations in the basin appear to respond to the same causative factors.

Between about 2.1 and 1.6 myr ago a lake existed in the Turkana Basin (fig. 23.3), which appears to have been initiated by volcanic construction and / or tectonic movements near the southern end of the present lake. Claystones and siltstones of upper Member G of the Shungura Formation, of the upper Burgi Member of the Koobi Fora Formation, and of the upper Kalochoro and Kaitio Members of the Nachukui Formation were deposited in this lake.

Most previous studies focused on why lakes were present at some times but not at others. As noted above, lakes formed in the basin in response to tectonic, not climatic, events. There are, however, clear packages of strata within the lakes of longer duration (for example, the upper Burgi Lake), for which no general explanation has been proposed. Lake levels fluctuate in response to

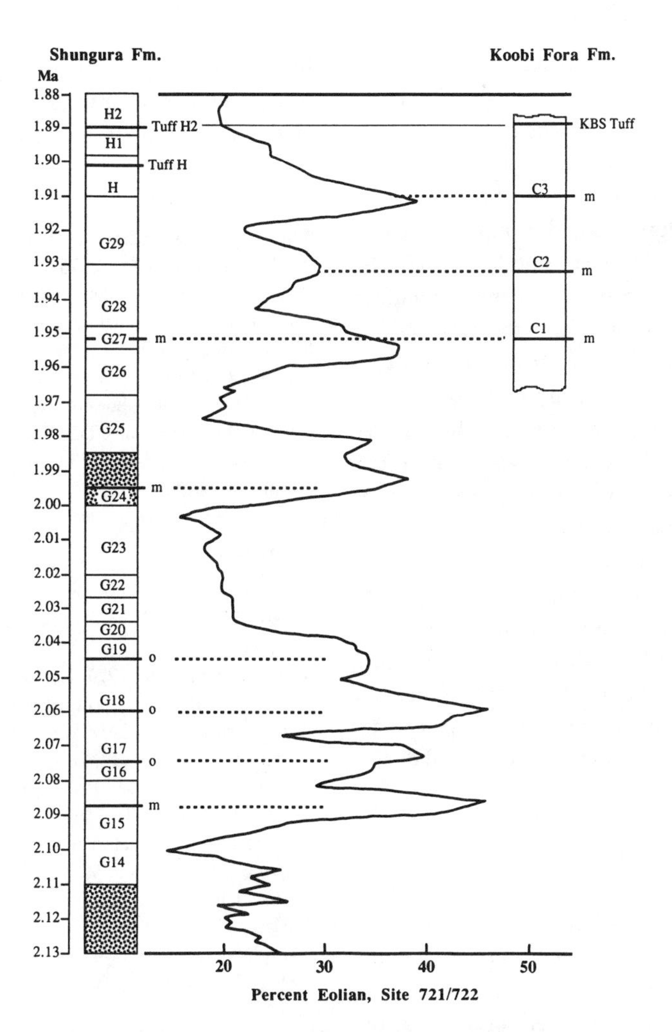

Fig. 23.4. Probable correlations of dust peaks in deep-sea cores 721 and 722, from the Arabian Sea (deMenocal and Bloemendal, this vol.) with stratigraphic indicators of low lake levels (molluscan [m] and ostracod [o] layers) in the Turkana Basin for the period 1.88–2.13 myr. Darkened areas represent deltaic sedimentation. C1–C3 are bioclastic marker beds at Koobi Fora (see Brown and Feibel, 1986).

changes in precipitation in their catchments, to changes in evaporation and, if externally drained, to changes in local base level. Some depositional areas are more sensitive to changes in lake level than others. In the deepest parts of a lake, which are farthest from sediment sources, changes in sedimentary character are likely to be less dramatic than changes along a shoreline. Because gradients are low, shorelines shift substantially in position with only small changes in lake level. Where there is little detrital input, a vast plain may be exposed and inundated repeatedly with little change in the character of the sediment that is deposited—Sanderson's Gulf is a good example of this. In the late 1960s this area lateral to the main Omo drainage was flooded every year, because the level of Lake Turkana was quite high, but for the past decade it has remained nearly dry, because the level of Lake Turkana has fallen. By contrast the Omo delta has migrated southward in the axial part of the basin over the same time. It must now carry coarse detritus farther southward than it did in the late 1960s. Such changes should be detectable in the stratigraphic record.

In fact, several stratigraphic units in the temporal interval from 2.11 to 1.89 myr most likely reflect low stands in the lake. Chronological control is sufficient in this interval to suggest correlation of these low stands with individual aeolian dust peaks in the Arabian Sea documented by deMenocal and Bloemendal (this vol.). In the Shungura Formation these possible correlations are submember G-15, which contains a mollusc bed, abundant fossil wood, and sparse mammals with a dust peak near 2.088 myr; the three ostracod beds of submembers G-17, G-18, and G-19, with dust peaks near 2.075, 2.061, and 2.045 myr, respectively; and submember G-24, containing a mollusc bed and mammalian fossils, with the dust peak at 1.994 myr. The latter may also correlate with the molluscan packstone marker horizon C1 of Feibel (1984) at Koobi Fora; the two ensuing marker horizons C2 and C3 may correlate with dust peaks at 1.931 and 1.913 myr (fig. 23.4). If these correlations are borne out, the implication is that the regional climate in East Africa is closely linked with that in Arabia. Cool, dry conditions in Arabia then imply reduced precipitation in the highlands of eastern Africa, which supply most of the water to Lake Turkana.

Several dust peaks in the interval from 1.89 to 1.65 myr have possible correlative events in the Turkana Basin. Chronological control is still too poor to confirm correlations in this interval, but it is expected that the chronology will become better through new paleomagnetic work and K/Ar dating of samples collected during the last field season.

Concluding Remarks

Additional work along several lines is still necessary in the Turkana Basin in order to realize the potential of the region for paleoclimatic information. Comparison of soil carbonate isotopes at widely separated localities of the same age is desirable but has not yet been done. Such work might allow us to distinguish between local effects and regional ones. Close stratigraphic investigation of the more extended lacustrine records, where chronological control is good, should result in additional correlations with the deep-sea record. Recent work by deMenocal on diatomite sections in the Turkana Basin is very encouraging in this regard. With each additional correlation, more information about the world climatic regime from deep-sea studies becomes applicable to hominid sites. Continued refinement of the chronology of all the strata in the Turkana Basin and elsewhere is, of course, also needed. Without exceptionally close control on the age of deposits that contain hominid fossils and artifacts, it is not possible to make use of the wonderful climatic record of the deep sea. Age control with errors of less than 0.02 myr is necessary, because this length of time encompasses an entire precessional cycle. Renewed work on the Galana Boi and Kibish Formations would seem to be opportune for determining phase relations and magnitudes of climatic effects as expressed in the record on land and in deep marine conditions.

Acknowledgments

This work was supported by National Science Foundation (NSF) grants BNS-884-06737, 86-05687, 88-05371, and 90-07662 and drew on data gathered under NSF support for 1967 onward. Additional support for various phases of the fieldwork was provided by the L. S. B. Leakey Foundation, the National Geographic Society, and the National Museums of Kenya. Parts of the laboratory work were supported by funds from the University of Utah. S. K. Williams and J. M. Harris kindly read the manuscript in draft form and provided many useful suggestions.

References

Anyonge, W. 1991. Fauna from a new lower Miocene locality west of Lake Turkana, Kenya. *Journal of Vertebrate Paleontology* 11:387–390.

Behrensmeyer, A. K. 1976. Lothagam, Kanapoi, Ekora: A general summary of stratigraphy and fauna. In *Earliest man and environments of the Lake Rudolf Basin,* pp. 163–172 (ed. Y. Coppens, F. C. Howell, G. L. Isaac, and R. E. F. Leakey). University of Chicago Press, Chicago.

Bishop, W. W. 1965. The later Tertiary in East Africa: Volcanics, sediments and faunal inventory. In *Background to evolution in Africa,* pp. 31–56 (ed. W. W. Bishop). University of Chicago Press, Chicago.

Boschetto, H. B., Brown, F. H., and McDougall, I. 1992. Stratigraphy of the Lothidok Range, northern Kenya, and K/Ar ages of its Miocene primates. *Journal of Human Evolution* 22: 47–71.

Brown, F. H. 1969. Observations on the stratigraphy and radiometric age of the Omo Beds, southern Ethiopia. *Quaternaria* 11:7–14.

———. 1981. Environments in the lower Omo Basin from one to four million years ago. In *Hominid sites: Their geologic settings,* pp. 149–163, AAAS Symposium 63 (ed. G. Rapp, Jr., and C. F. Vondra). Westview Press, Boulder, Colo.

———. 1982. Tulu Bor Tuff at Koobi Fora correlated with Sidi Hakoma Tuff at Hadar. *Nature* 300:631–635.

———. 1994. Development of Pliocene and Pleistocene chronology of the Turkana Basin, East Africa, and its relation to other sites. In *Integrative paths to the past,* pp. 285–312 (ed. R. S. Corrucini and R. L. Ciochon). Prentice-Hall, Englewood Cliffs, N.J.

Brown, F. H., and Feibel, C. S. 1985. Stratigraphical notes on the Okote Tuff Complex at Koobi Fora, Kenya. *Nature* 316:794–797.

———. 1986. Revision of lithostratigraphic nomenclature in the Koobi Fora region, Kenya. *Journal of the Geological Society of London* 143:297–310.

Brown, F. H., McDougall, I., Davies, T., and Maier, R. 1985. An integrated Plio-Pleistocene chronology of the Turkana Basin. In *Paleoanthropology: The hard evidence,* pp. 82–90. (ed. E. Delson) Alan R. Liss, New York.

Brown, F. H., Sarna-Wojcicki, A. M., Meyer, C. E., and Haileab, B. 1992. Correlation of Pliocene and Pleistocene tephra layers between the Turkana Basin of East Africa and the Gulf of Aden. *Quaternary International* 13/14:55–67.

Butzer, K. W., Brown, F. H., and Thurber, D. L. 1969. Horizontal sediments of the lower Omo Valley: The Kibish Formation. *Quaternaria* 11:15–29.

Cerling, T. E. 1992. Development of grasslands and savannas in East Africa during the Neogene. *Palaeogeography, Palaeoclimatology, Palaeoecology* 97:241–247.

de Heinzelin, J. 1983. *The Omo Group.* Musée Royal de l'Afrique Centrale, Tervuren, ser. 8, Sciences Géologiques, no. 85.

de Heinzelin, J., and Haesaerts, P. 1983a. The Shungura Formation. In *The Omo Group.* Musée Royal de l'Afrique Centrale, Tervuren, ser. 8, Sciences Géologiques, no. 85:25–128.

———. 1983b. The Usno Formation. In *The Omo Group.* Musée Royal de l'Afrique Centrale, Tervuren, ser. 8, Sciences Géologiques, no. 85:129–140.

———. 1983c. The Mursi Formation. In *The Omo Group.* Musée Royal de l'Afrique Centrale, Tervuren, ser. 8, Sciences Géologiques, no. 85:141–143.

Dodson, R. G. 1971. *Geology of the area south of Lodwar.* Report no. 87, Geological Survey of Kenya.

Feibel, C. S. 1983. Stratigraphy and paleoenvironments of the Koobi Fora Formation along the western Koobi Fora Ridge, East Turkana, Kenya. Master's thesis, Iowa State University.

———. 1988. Reconstruction of paleoenvironments from the Koobi Fora Formation, northern Kenya. Ph.D. diss., University of Utah.

Feibel, C. S., Brown, F. H., and McDougall, I. 1989. Stratigraphic context of fossil hominids from the Omo Group deposits, northern Turkana Basin, Kenya and Ethiopia. *American Journal of Physical Anthropology* 78:595–622.

Feibel, C. S., Harris, J. M., and Brown, F. H. 1991. Neogene paleoenvironments of the Turkana Basin. In *Koobi Fora Research Project.* Vol. 3, *Stratigraphy, artiodactyls and paleoenvironments,* pp. 321–346 (ed. J. M. Harris). Clarendon Press, Oxford.

Haesaerts, P., Stoops, G., and Van Vliet-Lanoë, B. 1983. Data on sediments and fossil soils, In *The Omo Group.* Musée Royal de l'Afrique Centrale, Tervuren, ser. 8, Sciences Géologiques, no. 85:149–186.

Haileab, B. 1988. Characterization of tephra from the Shungura Formation, southwestern Ethiopia. Master's thesis, University of Utah.

Haileab, B., and Brown, F. H. 1992. Turkana Basin–Awash Valley correlations and the age of the Sagantole and Hadar Formations. *Journal of Human Evolution* 22:453–468.

———. 1994. Tephra correlations between Gadeb prehistoric site, Ethiopia, and the Lake Turkana Basin. *Journal of Human Evolution* 26:167–173.

Harris, J. M., Brown, F. H., and Leakey, M. G. 1988a. *Stratigraphy and paleontology of Pliocene and Pleistocene localities west of Lake Turkana, Kenya.* Contributions in Science, Los Angeles County Museum of Natural History.

Harris, J. M., Brown, F. H., Leakey, M. G., Walker, A., and Leakey, R. E. 1988b. Plio-Pleistocene hominid-bearing sites from west of Lake Turkana, Kenya. *Science* 239:27–33.

Harris, J. M., and Watkins, R. 1974. New Early Miocene vertebrate locality near Lake Rudolf, Kenya. *Nature* 252:576–577.

Joubert, P. 1966. *Geology of the Loperot area.* Report no. 74, Geological Survey of Kenya.

Leakey, R. E., and Leakey, M. G. 1986a. A new Miocene hominoid from Kenya. *Nature* 324:143–146.

———. 1986b. A second new Miocene hominoid from Kenya. *Nature* 324:146–148.

McDougall, I. 1981. $^{40}Ar/^{39}Ar$ age spectra from the KBS Tuff, Koobi Fora Formation. *Nature* 294:120–124.

———. 1985. K-Ar and $^{40}Ar/^{39}Ar$ dating of the hominoid-bearing Pliocene-Pleistocene sequence at Koobi Fora, Lake Turkana, northern Kenya. *Geological Society of American Bulletin* 96:159–175.

McDougall, I., Brown, F. H., Cerling, T. E., and Hillhouse, J. W. 1992. A reappraisal of the Geomagnetic Time Scale to 4 Ma using data from the Turkana Basin, East Africa. *Geophysical Research Letters* 19:2349–2352.

McDougall, I., Davies, T., Maier, R., and Rudowski, R. 1985. Age of the Okote Tuff complex at Koobi Fora, Kenya. *Nature* 316:792–794.

McDougall, I., Maier, R., Sutherland-Hawkes, P., and Gleadow, A. J. W. 1980. K-Ar age estimate for the KBS Tuff, East Turkana, Kenya. *Nature* 284:230–234.

McDougall, I., and Watkins, R. T. 1985. Age of the hominoid-bearing sequence at Buluk, northern Kenya. *Nature* 318:175–178.

Namwamba, F. 1992. Tephrostratigraphy of the Baringo Basin, Kenya. Master's thesis, University of Utah.

Owen, R. B., and Renault, R. W. 1986. Sedimentology, stratigraphy, and palaeoenvironments of the Holocene Galana Boi Formation, NE Lake Turkana, Kenya. In *Sedimentation in the African Rifts,* pp. 311–322 (ed. L. E. Frostick, R. W. Renaut, I. Reid, and J. J. Tiercelin). Blackwell, Oxford.

Pickford, M., Senut, B., Poupeau, G., Brown, F. H., and Haileab, B. 1991. Correlation of tephra layers from the Turkana Basin to the western Rift Valley. *Comptes Rendus Académie Sciences* (Paris) 313:223–229.

Powers, D. 1980. Geology of the Mio-Pliocene sediments of the lower Kerio River Valley, Kenya. Ph.D. diss., Princeton University.

Rahman, A., and Roth, P. H. 1988. Late Neogene paleoceanography of the Gulf of Aden region based on calcareous nannofossils. *EOS* 69:1253.

Renne, P. R., Walter, R. C., Verosub, K. L., Sweitzer, M., and Aronson, J. L. 1993. New data from Hada (Ethiopia) support orbitally tuned time scale to 3.3 million years ago. *Geophysical Research Letters* 20:1067–1070.

Ruddiman, W. F., Sarnthein, M., Backman, J., Baldauf, J. G., Curry, W., Dupont, L. M., Janecek, T., Pokras, E. M., Raymo, M. E., Stabell, B., Stein, R., and Tiedemann, R. 1989. Late Miocene to Pleistocene evolution of climate in Africa and the low-latitude Atlantic: Overview of Leg 108 results. In *Proceedings of the Ocean Drilling Program, scientific results,* vol. 108, pp. 463–484 (ed. W. Ruddiman, M. Sarnthein et al.) U.S. Government Printing Office, Washington, D.C.

Savage, R. J. G., and Williamson, P. W. 1978. Early history of the Turkana depression. In *Geological background to fossil man,* pp. 375–394 (ed. W. W. Bishop). Scottish Academic Press, Edinburgh.

Sarna-Wojcicki, A. M., Meyer, C. E., Roth, P. H., and Brown, F. H. 1985. Ages of tuff beds at East African early hominid sites and sediments in the Gulf of Aden. *Nature* 313:306–308.

Vondra, C. F., and Bowen, B. E. 1978. Stratigraphy, sedimentary facies, and paleoenvironments. East Lake Turkana, Kenya. In *Geological background to fossil man,* pp. 395–414 (ed. W. W. Bishop). Scottish Academic Press, Edinburgh.

Walter, R. C., and Aronson, J. L. 1993. Age and source of the Sidi Hakoma Tuff, Hadar Formation, Ethiopia. *Journal of Human Evolution* 25:229–240.

White, T. D., Suwa, G., Hart, W. K., Walter, R. C., WoldeGabriel, G., de Heinzelin, J., Clark, J. D., Asfaw, F., and Vrba, E. 1993. New discoveries of *Australopithecus* at Maka, Ethiopia. *Nature* 366:261–265.

Chapter 24

The Influence of Global Climatic Change and Regional Uplift on Large-Mammalian Evolution in East and Southern Africa

Timothy C. Partridge, Bernard A. Wood, and Peter B. deMenocal

The notion that periods of rapid evolutionary change are triggered by changes in the physical environment, as embodied in the "turnover pulse" hypothesis, has stimulated vigorous debate on the mechanisms and timing of evolutionary events, particularly those that gave rise to the earliest human ancestors. A major difficulty in matching evolutionary pulses with climatic changes on a global scale lies in defining the modulating role of local or regional influences, which may either augment or overprint global signals. One must bear in mind, too, that some regional events, such as Cenozoic plateau uplift, are likely to have had an important influence on global patterns of climate (Partridge et al., chap. 2, this vol.); under such circumstances the hierarchical distinction between global and regional factors may become somewhat blurred, and it is often difficult to discriminate between cause and effect (Molnar and England, 1990).

Although our object is to explore the strength of any relationship between global and regional changes in climate and the pattern of mammalian evolution of the African Neogene, we will not, for reasons of space, dwell on the details of the paleontological evidence. Wood addresses these details elsewhere in the volume, as do several of the other contributors. We aim to review the evidence for specific regional climatic signals in the two African regions for which there is a relatively rich mammalian fossil record: East Africa and southern Africa. For each region we discuss the possible causes of changes in landforms that are on a scale likely to bring consequential changes in climate. We discuss how these changes may have resulted in the type of fragmentation of the landscape that is conducive to evolutionary change, expressed as either the generation or the extinction of species. It is important to understand that faunal turnover cannot be used both as evidence of the effect of changes in paleoclimate and to demonstrate that paleoclimatic change had taken place. Finally, we consider whether any mismatch between changes in the *global* paleoclimate and patterns of mammalian evolution may be due to the confounding effects of *regional* changes in climate.

ISBN 0-300-06348-2.

Potential Regional Influences

The principal regional influences that may modify global climatic changes are:

—lateral plate movements
—vertical tectonics, in the form of uplift and faulting
—volcanism, associated both with aerosol emission and with the construction of volcanic landforms
—erosional modification of the landscape.

All these have affected East and southern Africa to varying degrees. The slow northward drift of the African Plate has caused Africa to migrate steadily through some 14° since the Cretaceous (Smith and Briden, 1982); the Late Neogene component of this drift is far too small to have been associated with any significant changes in climate. The Rift Valley of East Africa separates the African Plate into Nubian (western) and Somalian (eastern) parts, the slow, progressive separation of which has been linked to major uplift and volcanism during the Neogene. Evolution of the rift system was manifested in the development of a large number of discrete fault basins along its eastern and western branches, the more or less concurrent rise of two major domal structures (the Afar Plateau and the East African Plateau), and the superimposition of additional relief upon these elevated areas through the rise of the rift shoulders and the development of volcanic massifs, chiefly along the eastern branch of the rift system. The overall effect of these events was the augmentation of relief adjoining the rifts by more than a kilometer in many areas, while the topography within the confines of the rifts themselves became fragmented. Although significant locally, lava and ash outpourings and the generation of volcanic aerosols (gases and dust) were small by global standards and could hardly have given rise to important climatic effects.

Erosional processes operating on the newly uplifted rift margins would, given sufficient time, have reduced their environmental significance, at least locally; landscape degradation is, however, a slow and progressive process and is thus unlikely to have been associated with rapid changes in local environments in comparison with those wrought by periodic faulting movements. The generation of new relief through tectonic and volcanic activity seems, on the whole, to have counterbalanced erosion. Progressive sedimentation within the rift basins would, however, have tended to reduce the topographic fragmentation generated during recurrent faulting.

In southern Africa, Neogene tectonism followed a different style. There, large-scale epeirogenic uplift and flexuring with the eastern coastal hinterland were associated with a lesser degree of topographic fragmentation than in East Africa, except in some local areas. Late Neogene uplifts were somewhat smaller than their East African counterparts, with correspondingly fewer impacts on regional climate.

The timing of these tectonic and volcanic events within the eastern hinterland of Africa is of special interest, because they ostensibly coincided with major evolutionary changes within its large mammal faunas, including the emergence of the genus *Homo.* Coppens (1982, 1988a, 1988b) has repeatedly drawn attention to the impact of rift tectonics on East Africa climates; and, as Adamson and Williams (1987, p. 597) have observed, "On the premise that major environmental changes can drive evolutionary events, the rough coincidence in time between major geological, environmental and evolutionary changes may not be trivial." The possible influences on evolution of such geological events, through their contribution to climatic changes, are explored in the remainder of this chapter.

An Example. The effects of diastrophism on mammalian evolution are well illustrated by Cenozoic events in the southern part of South America. During the Early Miocene, much of Argentina was subject to a major marine transgression; the withdrawal of this vast Paraná sea in the Middle Miocene marked the development of the terrestrial Araucanian Fauna in the period between ca. 11 and 3 million years (myr) (Pascual et al., 1985). During this interval the first development of the major river basins of the Negro and Paraná, which drained eastward from the Andean spine, took place; along the latter, uplift and exhumation had been occurring since the Mesozoic. Further elevation of the southern Andes during the Middle and Upper Miocene was marked by a change from marine to terrestrial sedimentation and tilting of earlier depositional sequences. The major phase of Andean uplift began at the end of the early Pliocene and involved all the major ranges in existence today. The timing of these movements is well bracketed both by changes in style of sedimentation (during a period when the confounding influences of global climate change were not prominent) and by the further deformation of dated terrestrial sequences. The rise of the eastern ranges created a barrier to moisture-laden winds from the Atlantic Ocean and resulted in progressive desiccation of the Chaco-Pampean plains through the interposition of rainshadow effects. According to Pascual et al. (1985), the twofold effects of topographic fragmentation, caused by orogenesis along multiple axes, and of leeward aridification resulted in the creation of much more varied environmental subdivisions than had existed previously; these are reflected in continental sediments of the Friasian Stage. Forest species hitherto widely represented disappeared and were replaced by a range of specialized herbivores, including huge edentates, such as glyptodonts and armadillos, sloths and caviomorph rodents. The rodents, in particular, became varied and locally adapted, giving rise to the largest species ever known within this group.

This major faunal turnover may well bear an imprint of the effects of global climatic events toward the end of the Miocene, but the evidence presented by Pascual et al. (1985) suggests that the incremental proliferation of specialized taxa within the Araucanian Fauna may be attributable, in large measure, to environmental fragmentation and deterioration related directly to major orogenic events of the Late Neogene. Parallels with the rise of the rift margins and the development of leeward rain-shadow zones in East Africa during the Late Neogene are noteworthy, as we indicate below.

East Africa

Uplift and Volcanism. The East African Rift System extends over some 3,200 km from the Afar triple junction, where the Red Sea meets the Gulf of Aden, to the Zambezi River in southern Africa. The uplifted Afar Plateau flanks the northern extremity of the rift (Fig. 24.1). To the south of the Turkana Depression the rift system bifurcates into eastern and western branches around the Nyanza Craton; the latter coincides in part with the uplifted East African Plateau (fig. 24.2). Along both the eastern and western branches of the rift significant additional relief is superimposed upon the broad domal topography of the East African Plateau. That associated with the eastern rift (in Kenya) is known as the Kenya Dome; about 140,000 km^3 of volcanics, which have erupted since the Miocene, form a significant part of this feature (Williams, 1978). Elsewhere the high rift-margin topography is largely tectonic in origin.

Recent models of rift evolution help to explain the genesis of the unique suite of geological and topographic features that make up the East African Rift System. Broad negative Bouguer gravity anomalies coincide with the topographic domes of the Afar and East African plateaus (figs. 24.2 and 24.6), and this suggested to Ebinger et al. (1989a) that these uplifts are, for the most part, isostatically compensated and are the result of elevated geotherms within the lithosphere beneath these regions. In terms of this model the plateau topography is supported by buoyancy forces generated by low-density material beneath the uplifted areas, which, in turn, produces the negative gravity features. The East African Plateau, however, appears overcompensated, and Ebinger et al. (1989a) consider that convection within the asthenosphere must be operating to maintain its elevation. Within such a convectively upwelling region, the extensional stresses that cause rifting are generated largely by heating and thinning within the subcrustal lithosphere.

Achauer et al. (1992) cite the results of gravimetric, explosion seismic, and teleseismic investigations in confirming that the Kenya Rift is underlain by low-density mantle material. They suggest that partially molten material may have ascended close to the base of the crust and even intruded it, leading to the penetration of the upper crust by dykes and magma chambers along the line of the rift. The resulting density distribution does not, in their view, necessarily satisfy mechanisms of isostatic compensation, and, as Ebinger et al. (1989a) also suggested, Achauer et al. conclude that a mushroom-shaped mantle plume exists beneath the apex of the Kenya Dome. However, the relatively small volumes of surface volcanics and the limited extent of intrusives predicted by the geophysics suggested to Achauer et al. that the amount of crystal extension above the deep anomalous mantle zone has been small—perhaps no more than a few tens of kilometers. By comparison, post-Miocene extension in the main Ethiopian Rift is estimated as less than 25 km (Ebinger et al., 1993).

A somewhat different interpretation of the geophysical data has been offered by Nyblade and Pollack (1990), who propose that the rift is located along the line of an ancient crustal suture and that the observed gravity anomaly is a complex of both "rift" and "suture" signatures. The latter reflects the existence of a deep crustal root beneath the Nyanza Craton to the west and the Pan-African Mozambique Belt to the east. If such a suture exists, buoyancy forces generated by it would cause crustal uplift along the line of fracture of the rift, and estimates of lithospheric thinning and crustal extension, based on a model of rifting under the sole influence of convecting, low-density asthenospheric material, may be too high.

The pattern and timing of vertical tectonic events associated with East African rift evolution are far from simple. In the 1000 km-wide Afar Plateau in the north, the eruption of flood basalts occurred in the Paleogene between 49 and 33 myr (Davidson and Rex, 1980; Woldegabriel et al., 1990; Ebinger et al., 1993); faults bounding the Ethiopian Rift began to develop in Late Oligocene–Early Miocene times (Davidson and Rex, 1980); and by the Mid-Miocene the rift was well established (Adamson and Williams, 1987). Some uplift preceded this early faulting, and an episode of alkali basalt and trachyte volcanism occurred between 18 and 11 myr (Ebinger et al., 1993). However, pollen recovered from lignites, sandwiched between Late Miocene (8 myr) basalt flows near Gondar, display spectra similar to those of contemporary lowland rain forest (Yemane et al., 1985), indicating that elevation of at least the northwestern part of the Afar Plateau to its present altitude of ca. 2,000 m has occurred since that time (fig. 24.3). Baker et al. (1972) refered the major component of this movement (1,000–1,500 m) to the Late Pliocene; Adamson and Williams (1987) cite more specific evidence, which links

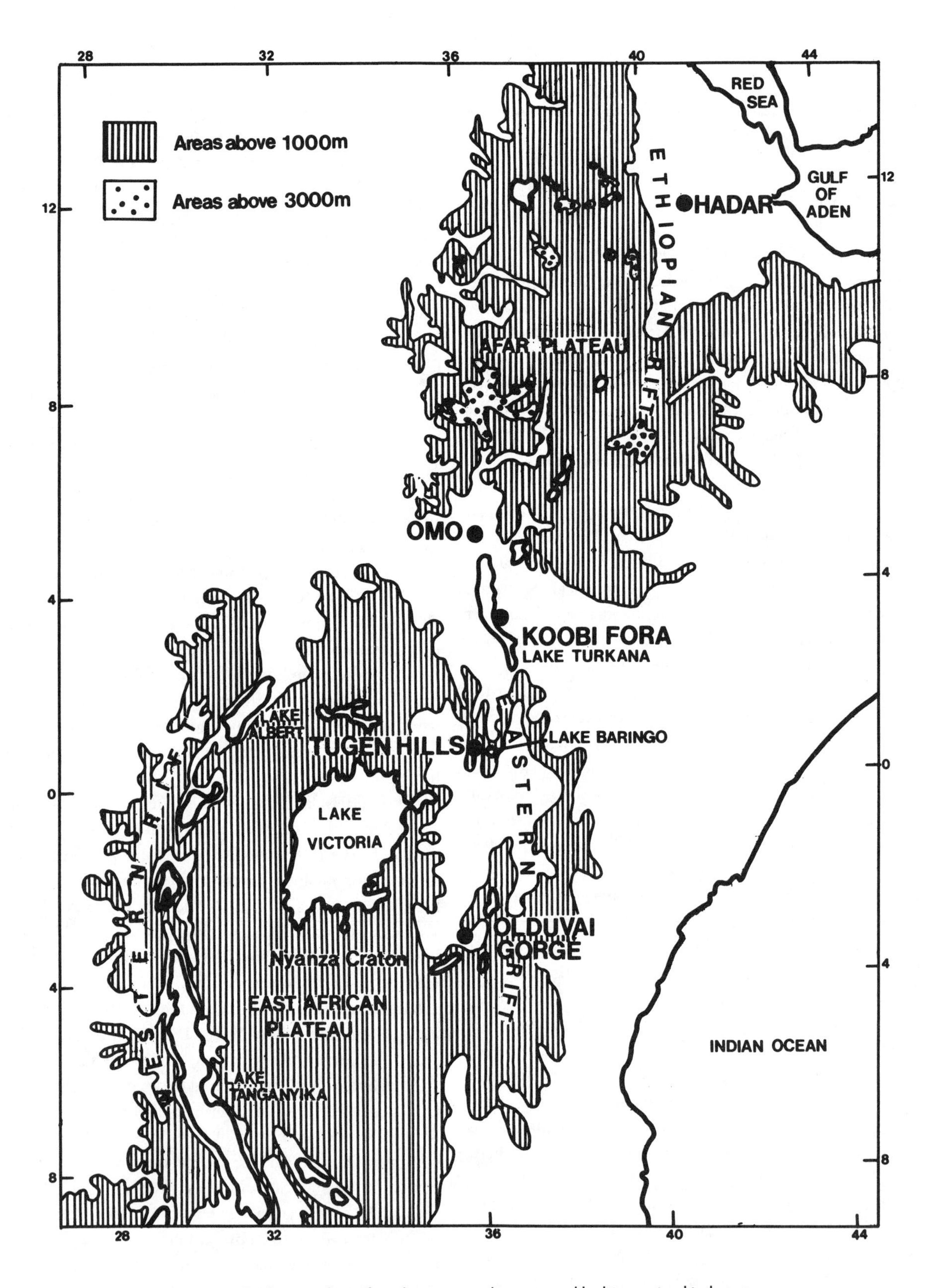

Fig. 24.1. The rift system of East Africa, showing topographic contours and localities mentioned in the text.

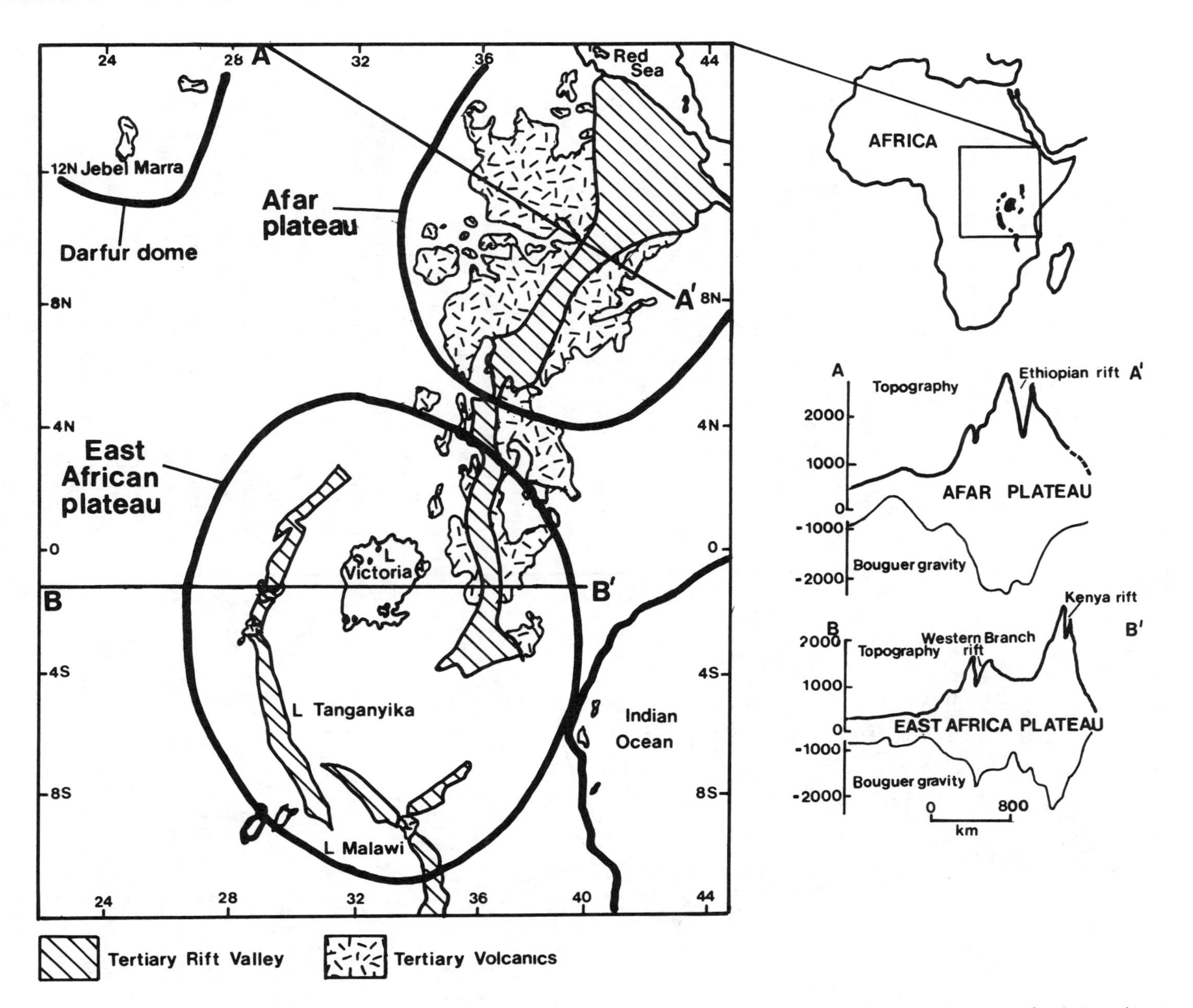

Fig. 24.2. East Africa, showing location of uplifted domes or plateaus and profiles of gravity and topography. The heavy solid lines enclose elevations greater than 800 m (after Ebinger et al., 1989a).

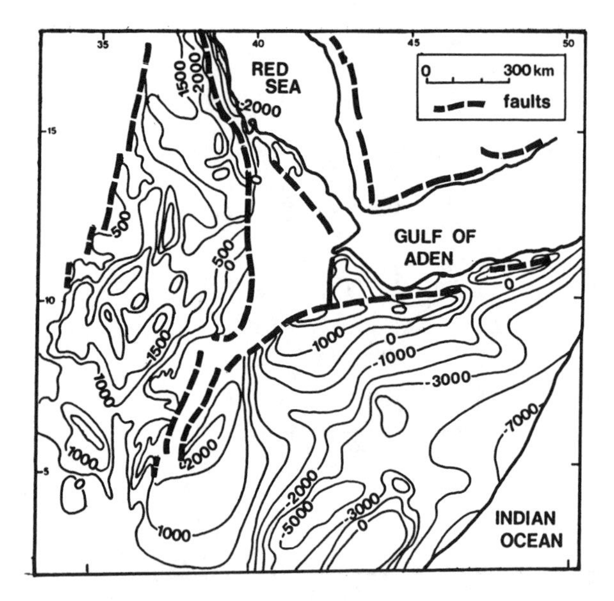

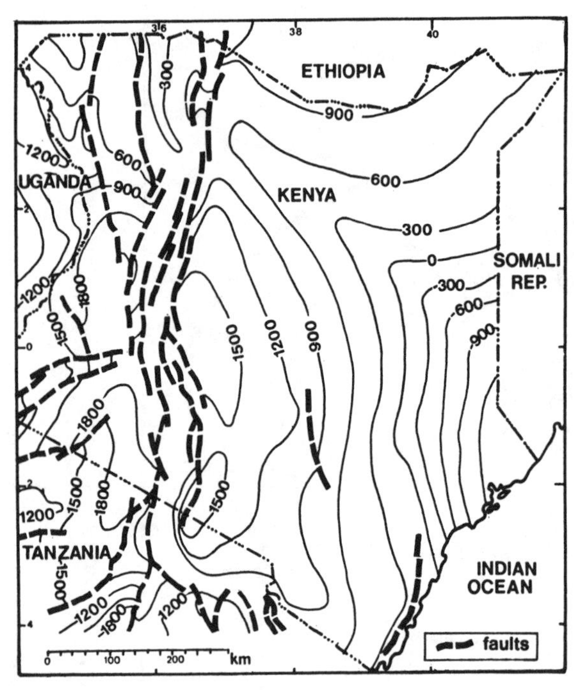

Fig. 24.3. (above) Isohypsals of the Precambrian basement in Ethiopia; (below) Isobases of the sub-Miocene erosion surface in Kenya (equivalent to the Post-African I surface in southern Africa). Elevations in meters (after Baker et al., 1972).

a change from lacustrine to fluviatile sedimentation in the Middle Awash Valley, dated to between 4.0 and 3.8 myr, to tectonic disruption of Early Pliocene lake systems following continued movements along the western fault scarp of the Ethiopian Rift. This faulting was associated with both the lowering of the Awash Graben and the continued rise of the Afar Plateau and served to accentuate climatic contrasts between lowland and upland through the medium of both altitudinal and rain-shadow effects. Denys et al. (1986) bracket a major phase of tectonic movement along the western scarp of the Ethiopian Rift between 2.9 and 2.4 myr, over which interval there was a massive influx of detrital sediments into the Hadar Basin. The geological events of this period were certainly of a magnitude sufficient to induce major environmental changes: Bonnefille et al. (1987) have noted that pollen spectra from the Hadar hominid site within the Ethiopian Rift indicate that the vegetation present between 3.3 and 2.9 myr has no analogues in the subdesertic steppe flora that characterize the area today and is most similar to modern vegetation communities occurring between elevations of 1,600 and 2,200 m and receiving two to three times the present rainfall. They favor downfaulting within the graben of ca. 1,000 m as the most likely cause of these changes. WoldeGabriel et al. (1990) believe that graben subsidence may, in fact, have been as great as 2 km on the evidence of a marker tuff (the Munesa Crystal Tuff dated to 3.5 myr), which is exposed on both rift margins and is present also in a geothermal well beneath the rift floor. It should be noted, however, that graben lowering on this scale could not have occurred without major concomitant uplift of the rift flanks as envisaged by Adamson and Williams (1987). The available evidence is thus strongly suggestive of at least 1,000 m of uplift within the Afar Plateau, with concurrent graben subsidence within the Ethiopian Rift, beginning after 3.5 myr (probably around 2.9 myr). The greater part of these tectonic adjustments can, on the available evidence, probably be referred to the period prior to 2.4 myr, when the large influx of sediments into the Hadar Basin ceased. The arguments advanced by Molnar and England (1990) against the use of sedimentological and paleobotanical data to infer uplift do not apply in this case in that evidence points to *local* changes, which are most compatible with graben development, rather than to a regional response.

The 1,300 km-wide East African Plateau is separated from the Afar Plateau by the Anza Graben, which impinges on the eastern short of Lake Turkana (fig. 24.1). Averaging about 1,200 m in elevation, the East African Plateau supports narrower belts of more elevated topography associated with the Kenya and Western rift systems. The rift-flank areas are some 100–200 km wide

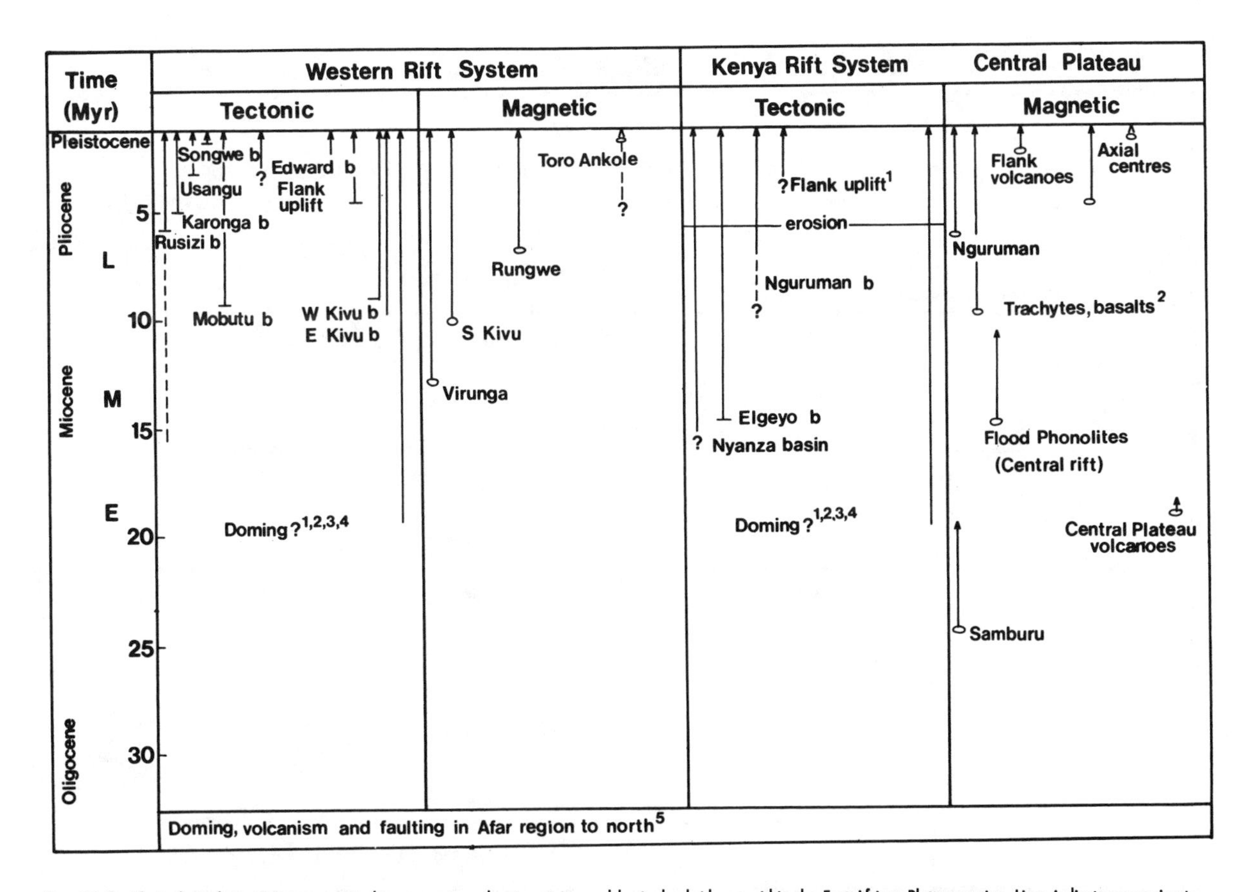

Fig. 24.4. Chronological constraints on vertical movements, volcanic activity, and basinal subsidence within the East African Plateau region. Lines indicate approximate time span of activity; dashes and question marks are used where dating is uncertain (after Ebinger, 1989). Sources: 1) Saggerson and Baker, 1965; 2) Baker, 1986; 3) Dixey, 1956; 4) Shackleton, 1978; 5) Mohr, 1987.

and have been uplifted above the surrounding plateau by ca. 1,000 m. Along the Kenya Rift, volcanism and regional uplift began in the Early Miocene (Saggerson and Baker, 1965; Williams, 1978; Baker, 1986). The doming brought about by these events probably dates to around 20 myr and amounted to no more than about 300–500 m on the basis of deformations of planation surfaces associated with deposits of well-established age (Saggerson and Baker, 1965; Baker et al., 1972). Rapid sedimentation within the Tana Basin resulted. Unlike the Western Rift, where the elevated rift flanks are almost exclusively the result of uplift, a considerable proportion of the topography was associated with the growth of major volcanic edifices, particularly toward the northern end of the rift. In total, some 140,000 km^3 of volcanic material have erupted in the vicinity of the Kenya Rift since the Early Miocene (Williams, 1978); and the tempo of volcanic activity increased with time: flood phonolites were erupted from about 15 myr and were followed by trachytes and basalts after about 10 myr (Baker, 1986; fig. 24.4). Important new volcanic centers were added around 5 myr. Major uplift, concentrated in the vicinity of the Kenya Dome, occurred above a deep anomalous mantle zone identified in teleseismic studies (Achauer et al., 1992). This phase of movement, documented by the deformation of well-dated planation surfaces, by the faulting and deformation of deposits of known age, and by renewed basin sedimentation, occurred after about 3 myr and reached a maximum during the Plio-Pleistocene interval (Saggerson and Baker, 1965; Baker et al., 1972; Baker, 1986; Fairhead, 1986; Ebinger, 1989; Ebinger et al., 1989b; Pickford et al., 1993; see fig. 24.3). In the course of these movements the rift shoulders were raised by 1,200–1,500 m in places. The rift itself was characterized by the formation of a number of discrete fault basins, some containing relief of as much as 1,800 m between rift floor and rift shoulder.

The evolution of the Western Rift differed from that

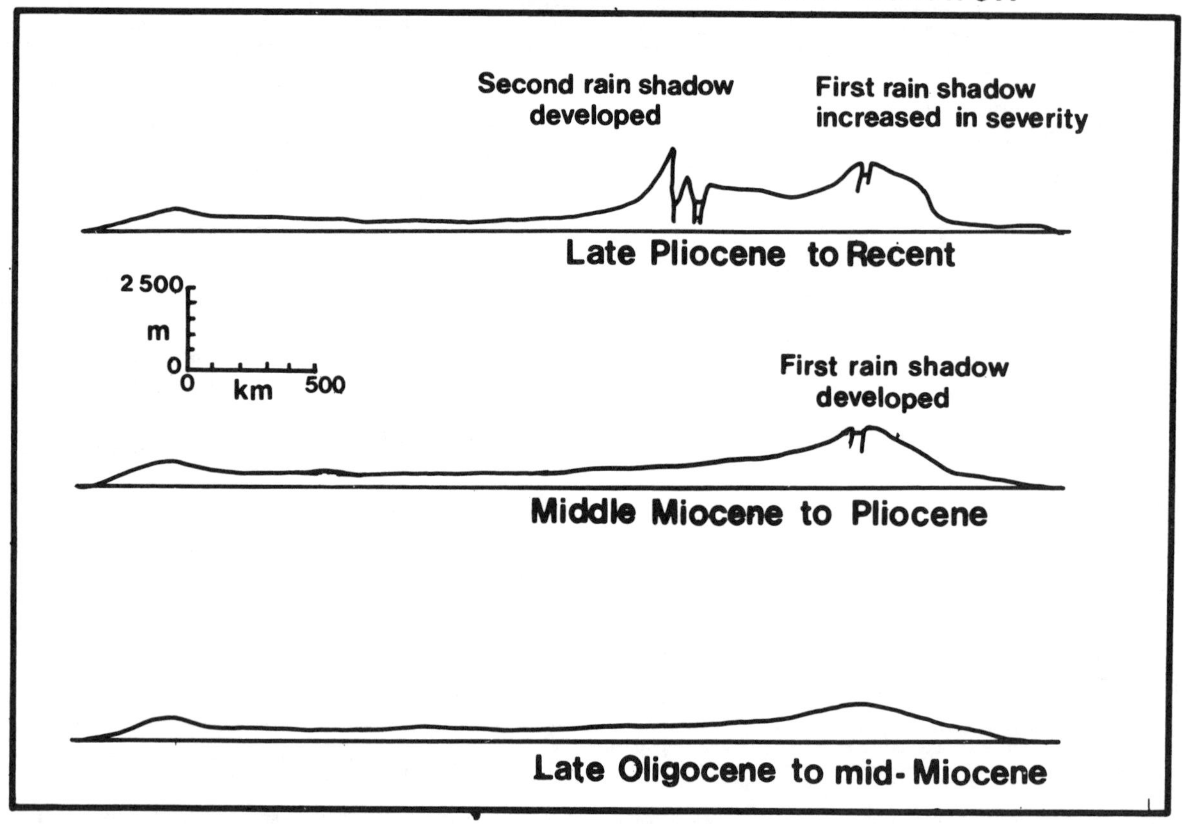

Fig. 24.5. Schematic cross-sections across the East African rifts at the equator, showing three stages in the evolution of the associated topography (after Pickford et al., 1993).

of the Kenya Rift in that Early Miocene doming around the incipient rift structure was of smaller amplitude and volcanic activity began later (12–10 myr) and was restricted mainly to fault-bounded basins (Ebinger et al., 1989b; Ebinger, 1989; Pickford et al., 1993). This early volcanism is thought to have signaled the time at which the East African mantle plume reached the base of the lithosphere (Westaway, 1993). Rifting began during the Middle Miocene and led to the first lacustrine sedimentation ca. 12 myr; major subsequent periods of faulting occurred at about 5.0, 2.8, and 2.4 myr (Pickford et al., 1993). Uplift of the rift shoulders averaged about 1,500 m but reached 4,300 m in the Ruwenzori Mountains; on the evidence of sedimentary and volcanic sequences the major part of these movements occurred in the period 3–2 myr. Figure 24.5 represents a reconstruction of the three major stages of topographic and rift evolution in East Africa (Pickford et al., 1993). By the time of maximum uplift the rift had become segmented into numerous separate basins, some of which had floors that extended below sea level. In large parts of both the Kenya and Western rifts this fragmentation persisted until about 2 myr, after which progressive sedimentation largely obliterated the effects of local rift structures; in any consideration of evolutionary events within East Africa one should keep in mind that both these local structures and the rifts themselves obstructed migration and may have led to the unusually high degree of endemism that is evident, particularly in micromammalian faunas, up to the end of the Pliocene (Denys et al., 1986). Uplift, particularly in the Ruwenzori area, continued well into the Pleistocene, and the ponding of Lake Victoria can, in fact, be linked to upwarping at the northern end of the Albertine Rift in the terminal Pleistocene. Figure 24.6 indicates the typical amplitude of Late Neogene uplifts in the eastern hinterland of Africa.

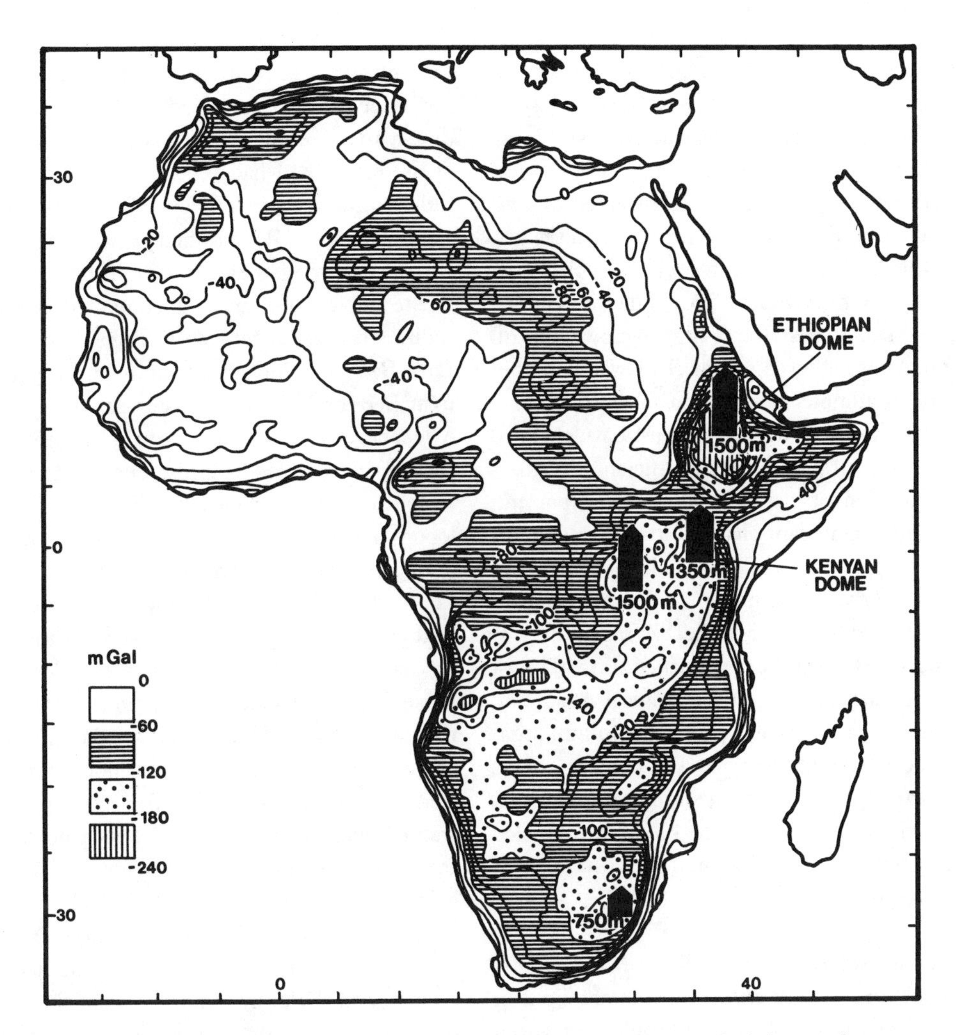

Fig. 24.6. Bouguer gravity map of Africa (after Sletlene et al., 1973), showing typical late Neogene uplifts associated with the principal negative Bouguer gravity anomalies that characterize the eastern hinterland of the continent.

In summary, an early phase of faulting and uplift occurred in the Afar Plateau area in Oligocene to Early Miocene times and in the vicinity of the East African Plateau during the Early Miocene. Vertical movements totaling 300–500 m were involved in both areas. Along the Kenya Rift the subsequent growth of large volcanoes, accompanied by further tectonic movements, created sufficient relief by 7 myr to produce important local climatic effects. In both the Afar and East African plateau areas the major period of subsidence of the rift floors, with concomitant uplift of the rift shoulders by some 1,000–1,500 m, took place after about 3 myr and reached a maximum in the Plio-Pleistocene interval. We now turn to the more important regional climatic influences of these latter movements.

Climatic Effects of Uplift. East African topography tremendously influences the climatology of the region. Surface temperatures cool markedly with increasing elevation owing to the influence of atmospheric lapse rate. Seasonal temperature variability also increases with elevation in that there is less atmospheric mass to regulate infrared radiative losses. Rainfall is highly variable over short distances owing to local rain-shadow effects that fragment regional precipitation patterns. The Kenyan and Ethiopian Highlands extract atmospheric moisture through forced adiabatic ascent of moist, tropical air masses that originate over the Atlantic and Indian Oceans.

Effects of the East African Highlands on atmospheric dynamics also contribute to regional aridity. Strong southwesterly winds (the "Findlater" jet) are focused along the East African–Arabian coasts during the summer months as a consequence of the summer Asian monsoon circulation. This low-level jet is most intense on its westernmost boundary, where it is constrained by East African topography. Boundary-layer friction causes vertical shear, which, in turn, promotes regional subsidence of warm, dry air over northeast Africa. The Findlater jet also develops strong coastal and Ekman upwelling off Somalia and Arabia; the cool, upwelled water further promotes arid conditions in some coastal areas.

Hence, the climate of this region must have changed markedly once topography reached sufficient elevations to trap moisture and become dynamically important. Perhaps the most significant changes attributable to uplift would have been the adiabatic cooling of highland temperatures and the fragmentation of the regional precipitation field. Additionally, high East African topography would have restricted the zonal penetration of moist maritime air masses into eastern subtropical Africa and orographically focused available precipitation into organized drainage and runoff systems. The climatic effects of East African uplift can be assessed using climate-model sensitivity tests; however, the model would need to be of sufficient grid resolution to take account of regional topographic variability.

The Late Neogene Paleoclimatic Record in East Africa: Regional and Global Influences. In the light of the tectonic history outlined above, it would be surprising if paleoclimatic records from East Africa did not bear a strong imprint of local environmental changes. At particular times and in specific localities, uplift, particularly of the rift shoulders, with graben subsidence in the intervening areas, occurred sufficiently rapidly to induce relatively sudden changes in climatic proxies in response to changes in altitude and to rain-shadow effects. As Hill (1987) has pointed out, the imprint of such local factors sometimes tends to reduce the resolution of correlation between apparent faunal changes and global climatic events and makes it difficult to establish a functional relationship between the two. In the light of this difficulty it is important to identify those tectonic or volcanic events that are likely to have been of major local significance, either in terms of their role as primary agents of environmental change or in their cumulative effect on causing environmentally important thresholds to be crossed.

Although records from Miocene sites are few and discontinuous, some useful conclusions can be drawn for the latter part of this period. Pickford et al. (1993) have drawn attention to the fact that, by ca. 7 myr, the flanks of the Kenya Rift had been elevated to a sufficient altitude to produce rain-shadow effects within the rift itself and between it and the Indian Ocean. Changes associated with the passing of this threshold may well have given rise to the faunal changes documented by Hill (1987) in sedimentary sequences of the Tugen Hills, to the west of Lake Baringo. The earliest units in this area, the Muruyur Beds and the Ngorora Formation, date to the Mid-Miocene. The first major faunal change is to be seen in the overlying Mpesida Beds, dated to about 6.4 myr, in which new families such as the Elephantidae and the first leporids make their appearance (Hill, this vol.). The overlying Lukeino Formation (ca. 6.3–6.0 myr) is better sampled and includes new pig and elephant species, as well as a number of new carnivores. Fauna in the succeeding Chemeron Formation (< 5.6 myr) are similar to those from the Lukeino Formation. Hill points out that the entire Tugen Hills sequence shows gradual faunal change but that the most significant change occurs at the transition to the Mpesida Beds around 6.4 myr. A substantial positive shift in the $\delta^{18}O$ of paleosol carbonates from East Africa Rift sites between 8.5 and 6.5 myr had been recorded by Cerling (1992) and is interpreted as reflecting an increase in the abundance of C4 (grassland) biomass over that interval. This picture would certainly accord with the growing influence of the rift shoulders during the latter part of the Miocene, as recognized by Pickford and his co-workers, rather than with a response to climatic signals of a global nature alone. In a recent study Cerling et al. (1993) argue that $\delta^{18}O$ decreased on a global basis between 7 and 5 myr, probably as the result of a decrease in atmospheric CO_2. Molnar et al. (1993) have presented persuasive evidence for major rapid uplift of the Tibetan Plateau around 8 myr, which they consider to be the cause of the intensification of the Asian monsoon, as revealed in a number of oceanic sequences. They believe that the rapid exhumation of silicate rocks, and the increased chemical weathering precipitated by these events, may have played a major role in this change in atmospheric chemistry. Although a significant decrease in global temperatures does not appear to have followed the reduction in CO_2, there is evidence for a marked increase in the extent of the west Antarctic ice sheet about this time (Burckle, chap. 1, this vol.).

Evidence of important climatic changes in East Africa coinciding with the Messinian salinity crisis is not forthcoming, at least at present, from the patchy terrestrial record, but data from cores off the west coast of North Africa provide important information on the onset of hyperarid conditions within that part of the continent. Sediment studies by Tiedemann et al. (1989) indicate a large increase in siliciclastic dust within these cores between 4.6 and 3.7 myr, reflecting substantial enlargement of the summer dust plume. The authors interpret these changes as resulting from major intensification of the African Easterly Jet–Saharan Air Layer circulation system over North Africa, the desiccating effects of which were further augmented by an increase in the northeast trade winds from about 3.2 myr onward. The effects of the gradual tectonic narrowing of the Mediterranean that culminated in the Messinian events between 5.6 and 4.8 myr are not manifested in these records. The closing of the Panamanian isthmus around 3 myr likewise appears to have had a negligible affect on African climates (Partridge et al., chap. 2, this vol.).

A number of influences of global significance may have contributed to North African aridification in the Late Miocene and Early Pliocene: uplift of the Tibetan Plateau, which apparently included a significant Late Miocene component (Molnar et al., 1993), would have created a flow of dry, northeasterly winds over the central and eastern parts of North Africa and accentuated atmospheric subsidence over its western regions (Partridge et al., chap. 2, this vol.). At the same time global CO_2 values appear to have decreased significantly between the Late Miocene and the Early Pliocene (Cerling, 1991; Cerling et al., 1993); this decrease, too, may have had as its primary forcing mechanism the increased weathering of silicate rocks on a worldwide basis as a result of Tibetan Plateau uplift (Ruddiman and Kutzbach, 1989; Molnar et al., 1993). These global influences may have been accentuated by a reduction in moisture, brought from the Indian Ocean by the southeast trade winds, as a result of the rise of the East African Plateau. The aeolian sands, which blanket areas currently occupied by equatorial rain forest (e.g., the Séries des Sables Ocres of Zaire), were probably first distributed over a Mid-Tertiary land surface during the Mio-Pliocene and may conceivably reflect a central African response to these influences. As better data became available from East African sites, the Early Pliocene may provide important evidence on changing vegetation distributions and patterns of mammalian adaptation during a period that was clearly one of major climatic variations elsewhere on the African continent.

A substantial body of evidence summarized in the previous section now points to the period between about 3 and 2 myr as one in which some of the most dramatic tectonic events of the Neogene occurred in East Africa. Because this interval spans a period during which major global climatic changes occurred in response to the growth of ice sheets in the Northern Hemisphere, it is not entirely surprising that some of the East African signals of major environmental change around this time are stronger than would be expected in the tropics, where the response to changes in the high latitudes is frequently ambiguous. As Pickford (1990) has pointed out, the most important factor in the evolution of East African climates and faunas was the uplift of the mountain ranges flanking the rifts during the Pliocene and Pleistocene, but these changes were, in turn, themselves affected by global-scale changes related to Neogene cooling, which culminated in the glaciations of the last 3 myr.

Some of the best evidence for the advent of a cooler and drier climate in East Africa around 2.5 myr comes from the work of Gasse (1980) and Bonnefille (1983) on the sediments of Lake Gadeb on the Afar Plateau. Here diatomaceous sediments between tuffs dated at 2.51 and 2.35 myr have yielded pollens that reflect the cool, montane conditions now found over 1,200 m above Lake Gadeb's 2,300 m elevation. The pollen spectra indicate a temperature decline of between 4° and 6°C around 2.5 myr (Bonnefille, 1983). At the hominid site of Hadar, within the Ethiopian Rift, pollen from deposits radiometrically bracketed between 3.3 and 2.9 myr is interpreted as representative of an "Afromontane" flora characteristic of present highland areas and differing markedly from the subdesertic steppe vegetation that characterizes this part of the rift today (Bonnefille et al., 1987). Possible altitudinal changes as a result of rift tectonics have been discussed previously. These effects apart, the evidence suggests the existence of a climate significantly moister than today's or that which followed the 2.5-myr cooling. Palynological data from the delta of the Omo River in Ethiopia indicate an expansion of grasslands at this time (Bonnefille, 1976), and macrobotanical studies by Bonnefille and Letouzey (1976) show that fossilized wood and fruits disappeared more or less simultaneously from the lower Omo area. The work of Wesselman (1985) and Vrba (1985, chap. 27, this vol.) on the Omo mammalian faunas also indicates a shift to greater aridity at ca. 2.5 myr: Vrba's work records a pronounced increase in open grassland species of bovids, whereas Wesselman's analyses of a number of micromammalian assemblages spanning this interval show the increasing dominance of species adapted to grassland environments by around 2.4 myr (recent recalibration of the Plio-Pleistocene paleomagnetic time scale by Hilgen [1991a, 1991b] and Shackleton et al. [1990, 1993] indicates that these dates for the Omo area should now be increased by about 5 percent). Further east, in the Turkana Basin, the replacement of *Hipparion* by *Equus* and other species, and notable turnovers in bovid taxa, together constitute the widespread pulse at ca. 2.7–2.5 myr associated with an increase in areas of open, dry grassland (Feibel et al., 1991). In gastropod assemblages and pedogenic carbonate horizons from deposits of Lake Turkana, changes in stable isotope ratios documented by Abell (1982) and Cerling et al. (1977) show a trend toward reduced precipitation between tuffs dated radiometrically at 3.2 and 1.9 myr. The most dramatic increase in both $\delta^{18}O$ and $\delta^{13}C$, indicating a major expansion of grasslands, occurred, however, in the Turkana Basin around 1.8 myr (Cerling, 1992); this event suggests that the largest spread of C_4 grasslands occurred later than suggested by some other lines of evidence and, in fact, postdated many of the important evolutionary events of the Late Pliocene—except, perhaps, the appearance of *Homo erectus.*

Further insights into the Plio-Pleistocene evolution of African climate can be gleaned from studies of long marine records of aeolian deposition. The Plio-Pleistocene interval, representing approximately the last 5.3 myr, is punctuated by several climatic mode shifts, including the onset of glaciation in the Northern Hemisphere near 2.8 myr. Records of African climate variability that span this interval can therefore monitor the sensitivity of African climate to these changes in high-latitude climate. Long and detailed records (with a resolution of 1–2 thousand years [kyr]) are now available in marine cores from subtropical West and East Africa (deMenocal and Bloemendal, this vol.). Fluxes of aeolian dust in these cores reflect changes in the intensity of seasonal wind systems carrying dust from sources in Arabia and East Africa. Both the West and East African dust records exhibit a marked change in variability at ca. 2.8 myr and again near 1 myr. Prior to 2.8 myr, and extending back to at least 7.2 myr, aeolian records are dominated by variations at precessional (23–19 kyr) periodicities, indicating changes in the intensities of the African and Asian monsoons. After ca. 2.8 myr both regions exhibit a marked increase in aeolian variance at the 41 kyr periodicity, whereas after ca. 1 myr the dominant period of variation changes to 100 kyr. These shifts in variance mode coincide with the onset of Northern Hemispheric glaciation near 2.8 myr and the subsequent intensification of glacial cycles at 100 kyr intervals after 1 myr (Ruddiman et al., 1989). In terms of relationships between climate and uplift, these data suggest that African climate must have responded to a long-term cooling-drying trend associated with regional uplift over the last 4 myr, as well as reflecting a dependence on the timing and pattern of high-latitude climate variability after 2.8 myr.

In a recent assessment of the fossil evidence for large-mammal evolution in the Late Pliocene of East Africa, Turner and Wood (1993a) concluded that although there was a strong signal of taxonomically widespread evolutionary change beginning ca. 2.6 myr ago, an equally strong, and for some taxa an even stronger, signal for evolutionary change occurred at 2.3–2.4 myr, and perhaps for as long as 500 kyr after the 2.6 myr datum. Dental evidence for this more prolonged period of evolutionary activity is given in Wood (this vol.). There is evidence from the early hominin[1] fossil record of at least two events associated with the period between ca. 2.6 and 2.3 myr or less. The first is the appearance of the earliest species that is currently included in our own genus, *Homo rudolfensis.* At present, the earliest sound morphological evidence for *H. rudolfensis* comes from Malawi and dates, albeit on the basis of biostratigraphy, from around 2.5 myr (Schrenk et al., 1993). More reliably dated, but morphologically more ambiguous, remains of approximately the same age come from the Baringo region of Kenya (Hill et al., 1992). The other significant hominid evolutionary event around this time is the appearance of the first specimens of *Paranthropus.* It is not until near the 2.3 myr datum, however, that there occurs probably the best evidence we have for a speciation event in early hominids. This is the earliest appearance of *Paranthropus boisei* (Suwa, 1988; Wood et al., 1994). The reliability that has been established for the timing of these events in East Africa is due to the painstaking research carried out by Brown and Feibel and their co-workers, who have developed a system for correlating tuff layers based on the characteristic isotopic signature of each volcanic eruption (Feibel et al., 1989; Brown, this vol.). Complementary paleontological evidence comes from the Albertine section of the Western Rift, where Pickford et al. (1993) have shown that the faunas of the Late Miocene Nkondo Formation and the Lower Pliocene Warwire and Kyeoro Formations (the latter dated to ca. 3 myr) were all adapted to a dense semideciduous forest environment; the presence of fossilized fruits supports these conclusions. The Hohwa-Kaiso Village Formation, on the other hand, spans the interval 2.6–2.3 myr and contains faunal and botanical remains that contrast strongly with those of the earlier units—a dominance of grasses is accompanied by a fauna rich in hypsodont elephants and suids, suggesting a shift to a wooded savannah vegetation (Pickford et al., 1993). Pickford and his co-workers believe that these changes occurred in response to both local tectonic and global climatic changes.

Southern Africa

Cenozoic Uplift. Viewed on a continental scale, eastern and southern Africa form part of what Nyblade and Robinson (1994) have named the African Superswell. This area, which is characterized by anomalously elevated topography that extends into the surrounding oceans to the edge of the Africa Plate (fig. 24.7), is almost an order of magnitude larger than the Tibetan Plateau. Over most of it, the positive anomalies, defined in relation to global-mean continental and ocean-floor elevations, exceed 500 m; in fact, as has already been indicated, within large parts of East Africa the anomaly is greater than 1,000 m; this is also the case for elevated areas of the eastern hinterland of southern Africa.

The broad, low-amplitude nature of the Superswell suggests that its origin should be sought in a buoyancy residing deep within the earth's mantle, a supposition confirmed by the recent studies on global tomography of Su et al. (1992) and Masters et al. (1992) and the seismic studies of Tanaka and Hamaguchi (1992). The smaller, high-amplitude anomalies superimposed upon the Superswell, such as the raised flanks of the East African Rift System, are linked to more localized mantle and lithospheric sources, as discussed previously. The deep-seated buoyancy reflected by the Superswell is thought by Nyblade and Robinson (1994) to be attributable to the movement of the African plate over numerous hotspots during the last 200 myr. The mushrooming of mantle-plume heads beneath the lithosphere can produce uplifts of about 1 km over diameters of 2,000 km, indicating that large regional effects can be generated by movement over a relatively small number of hotspots. The presence of multiple hotspot tracks across southern Africa, however, suggest to Nyblade and Robinson (1994) that the observed uplift may have been caused by thermal alteration of the lithosphere by plume tails rather than by plume heads. The presence of plume-derived material with temperatures elevated by no more than 10°–20°C could conceivably have provided the requisite density contrast. However, evidence for such pervasive thermal alteration of the lithosphere beneath both eastern and southern Africa is not unequivocally forthcoming from seismic studies, although measurements do suggest that, within the Kalahari Craton and the surrounding mobile belts, heat flow is raised sufficiently to account for regional uplifts of up to 500 m (Nyblade and Robinson, 1994). The extensive negative Bouguer gravity anomalies within much of eastern and southern Africa (fig. 24.6) indicate the probable presence of widespread areas of low-density mantle lithosphere (with density deficiencies of 50–100 kg/m^3), but these low-density areas are too large to be attributed to lithospheric heating alone; rather, they imply extensive thinning of the mantle lithosphere and its replacement by hot asthenospheric material, a circumstance that, again, is not firmly substantiated by the seismic evidence, although Bloch et al. (1969) and Clouser and Langston (1990) see evidence for a reduction in lithospheric thickness by some 50 km beneath parts of southern Africa, which may be a result of basal alteration or replacement. If sufficiently recent, these changes need not necessarily be associated with surface thermal anomalies.

The timing of regional uplift has also been a matter of debate. Nyblade and Robinson (1994) argue that the generation of the Great Escarpment of southern Africa implies that the Superswell must have existed prior to continental rifting or have been generated during the rifting process; they believe that large-scale uplift at a later stage would have been inhibited by the existence of extensive sedimentary sequences on the continental shelf and slope along the passive margins of the subcontinent. Partridge and Maud (1987) have, however, shown that the Great Escarpment was primarily a product of the high prerifting elevation of Africa, which occupied a central position within the Gondwanaland mosaic. In addition, there seems to be no good reason why shelf sediments, which off the east coast do not attain thicknesses much greater than 3,000 m (Dingle et al., 1983; Martin, 1987), should have such an inhibiting effect on uplifts generated by deep-seated mantle-lithospheric processes under the influence of the thermal anomalies for which evidence is now forthcoming (see below).

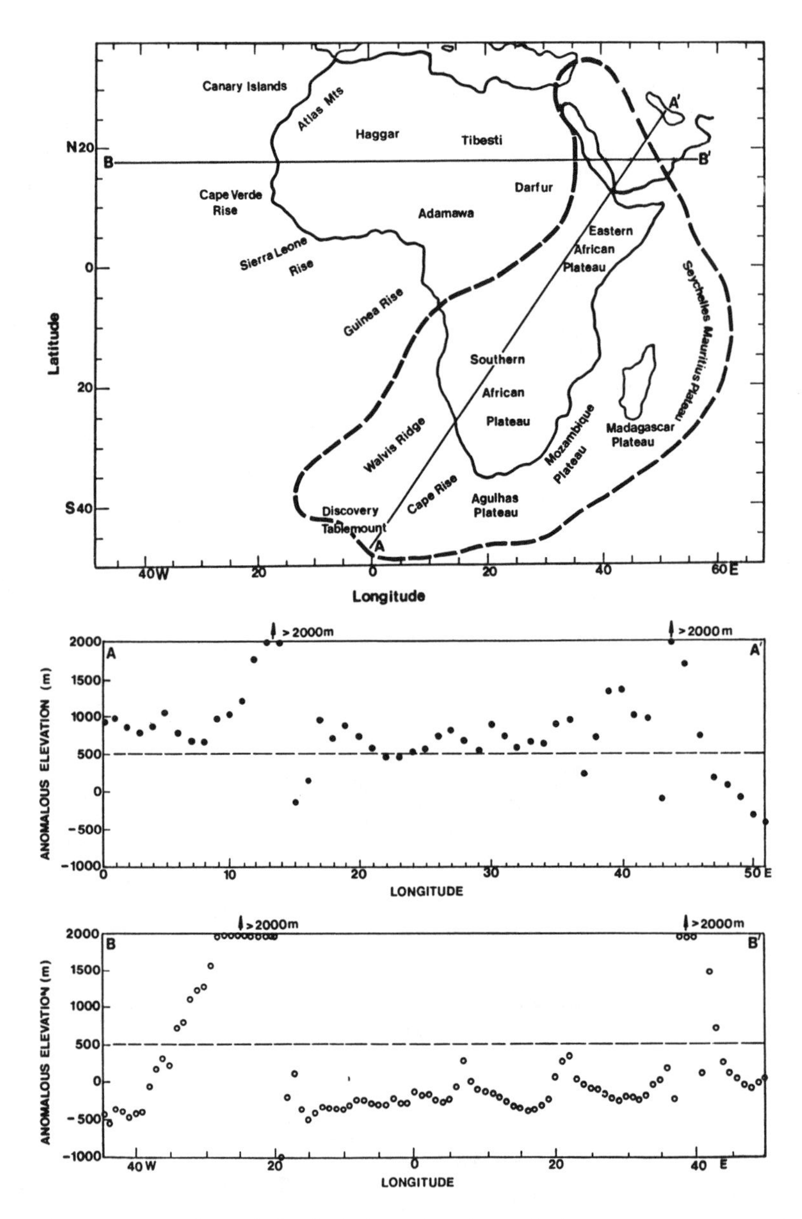

Fig. 24.7. The African Superswell (upper diagram, outlined by bold dashed line) coinciding with anomalously elevated topography on the African Plate. A–A′ and B–B′ (lower two diagrams) mark the cross-sections (after Nyblade and Robinson, 1994).

Erosion Cycles. Evidence for late uplift in southern Africa can be gleaned from a variety of sources. Among these are the tectonic interludes separating the cycles of erosion that have dominated the geomorphic evolution of the subcontinent since the Mesozoic (Partridge and Maud, 1987). The two cycles relevant to this analysis are the African, initiated by continental rifting near the beginning of the Cretaceous period, and the Post-African I, set in train by epeirogenic uplift in the eastern part of the subcontinent in the Lower Miocene. The prolonged period during which the African cycle was current (more than 100 myr) resulted in the formation of a vast plantation surface of low relief, above which a few erosional massifs such as the highlands of Lesotho, the Namaqua Highlands, and the Cape Fold Mountains remained as upland remnants (fig. 24.8). Several tectonic episodes occurred during this cycle, some associated with interludes of alkali volcanism (Partridge and Maud, 1989).

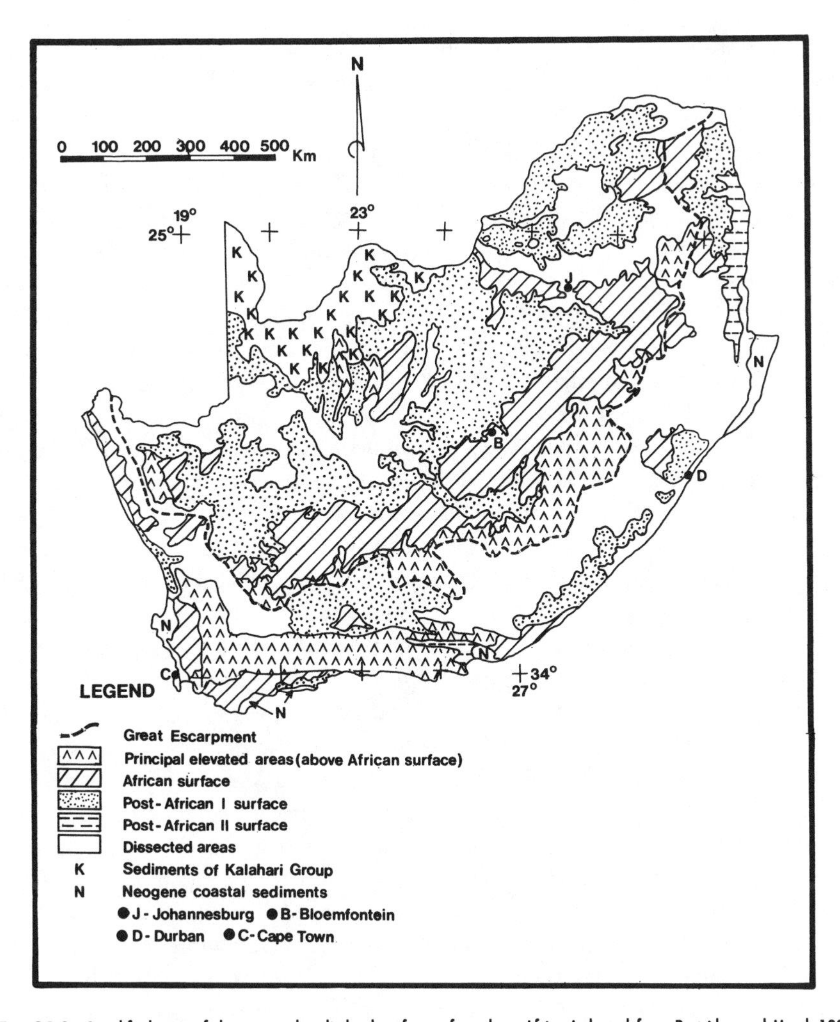

Fig. 24.8. Simplified map of the principal cyclic land surfaces of southern Africa (adapted from Partridge and Maud, 1987).

However, the localized subcycles that these generated are, on the subcontinental scale, indistinguishable from the broad sweep of the African surface. By the end of the Cretaceous the surface had become armored by duricrusts (laterite in the east and calcrete in the west); throughout its extent remnants of terrestrial and marginal marine deposits ranging from Upper Cretaceous to Eocene in age are preserved (Partridge and Maud, 1987, 1989). Uplift in the Early Miocene (ca. 18 myr) resulted in gentle monoclinal warping of the African surface in the southeastern hinterland of the subcontinent. These movements initiated the Post-African I erosion cycle, which resulted in the formation of an imperfectly planed surface below remnants of the African pediplain. Their timing was constrained both by the advent of renewed sedimentation on the continental shelf and by the earliest fossiliferous deposits, marine and terrestrial, preserved on the Post-African I surface (Partridge and Maud, 1987).

The Post-African I cycle was brought to an end by major tectonic uplift within the southeastern hinterland of southern Africa, which was concentrated along an axis extending from Port Elizabeth in the south to Swaziland in the north. The timing of this important event has long been a matter of debate, but, the evidence of cyclic land surfaces notwithstanding, there is an array of information pointing to its age and extent. These data are best considered in the context of the mechanisms through which such large-scale regional uplifts can be explained. The tectonic movements that have influenced southern Africa since the rifting of Gondwanaland are of two principal types: (1) movements related to the rifting process itself and (2) subsequent flexural movements, which were largely restricted to the coastal hinterland. The breakup of Gondwanaland was preceded, and partly accompanied, by a volcanic cycle that terminated the Karoo Sequence in southern Africa. This cycle was marked by a progression from Early Jurassic foci in the center of the subcontinent to Cretaceous centers along the eastern and western margins as volcanism spread into the zones of active rifting associated with the incipient development of the African continental margin (Cahen et al., 1984). As in East Africa, updoming of adjacent zones preceded rifting and continental separation; evidence for this early episode of uplift can be seen in the existence, immediately inland of both the eastern and western limbs of the Great Escarpment of southern Africa, of a cordon of high land. This raised area extends up to 350 km inland of the present coastline and stands above the level of the African surface (fig. 24.8).

Although these early movements would certainly have contributed to the amplitude of the African Superswell, they were superimposed upon a landscape that already stood at high elevation. The prerifting altitude of southern Africa can, in fact, be reconstructed with some confidence from the morphology of the remaining sections of numerous kimberlite pipes, whose vertical zonation is an accurate measure of depth below the land surface at the time of their eruption (Hawthorne, 1975; Partridge and Maud, 1987). Thus, near the crest of the erosional remnant of the Lesotho Highlands, only about 300 m has been eroded since the Mid-Cretaceous, whereas in the Kimberley area as much as 1,800 m has been removed. Other recent estimates of the extent of postrifting erosion range from 1,800 m, on the basis of data on offshore sediment volume (Rust and Summerfield, 1990), to upwards of 2,500 m, calculated from apatite fission-track analyses (Brown et al., 1990). Reconstructions based on these figures, and that take into account the present elevation of the land surface and likely isostatic adjustments during erosion, yield original elevations for the western part of the subcontinent ranging from 1,200 to 1,600 m (Gilchrist and Summerfield, 1990; de Wit, 1993). This finding compares well with the estimate of ca. 1,500 m of Partridge and Maud (1987), based on the kimberlite evidence and allowing for isostatic compensation of 250 m. As Partridge and Maud have pointed out, however, the prerifting elevation of the Lesotho area was some 2,350 m; allowing for 300 m of subsequent erosion, and taking into account the present elevation of the area of around 3,200 m, uplift of some 1,150 m must have occurred within the eastern hinterland of the subcontinent after the breakup of Gondwanaland. When did this uplift occur and through what mechanism?

Gilchrist and Summerfield (1990), in their analysis of the western margin of southern Africa, have proposed a model of isostatic response to contrasting denudation rates on either side of the rift shoulder formed during continental separation. In terms of this model a flexural bulge up to 600 m in amplitude would have migrated inland from the coast over a period of some 100 myr following rifting. In support of this prolonged period of adjustment, they cite recalculated sedimentation rates off the west coast of southern Africa, based on the work of Rust and Summerfield (1990). These revised figures suggest that sedimentation rates increased through the Cretaceous and peaked in the Paleocene and Eocene. This conclusion is controversial, however, in light of the fact that almost all offshore records show a decline in sedimentation rates throughout the Cretaceous, which became more pronounced following climatic deterioration at the end of the Cretaceous (Dingle et al., 1983; Partridge and Maud, 1987). It seems likely that the estimates of Rust and Summerfield are in error as a result of sparse borehole data and the imperfect calibration of seismic reflectors, upon which their calculations are based. In any case, the magnitude of postrifting uplifts within the western hinterland can be shown to have been too small (no more than 200–300 m) to match the sedimentary inputs that they claim (Partridge and Maud, 1987).

The amplitude of uplift of the southeastern coast is, in contrast, considerable larger. Ten Brink and Stern (1992) estimate that the effective elastic thickness of the lithosphere beneath the Great Escarpment of southern Africa is similar to that of the cratonic interior (80–120 km). This suggestion implies the presence of a thermally thick lithosphere within which residual heat from the time of rift-flank uplift prior to continental separation would have decayed progressively over more than 100 myr, leading to lower elevations than are, in fact, present in the vicinity of the escarpment. At the same time, this elastic thickness is too great to permit flexural uplift in response to continental denudation of 1.5–3 km (based on the fission-track analyses and geological evidence cited previously). Ten Brink and Stern (1992) therefore consider uplift to have occurred early after rifting in response to both thermal and isostatic influences. Their conclusion is not, however, supported by a considerable body of geomorphic and other evidence (see below). This evidence indicates that although a major component of the total uplift in the southeastern hinterland occurred during the Cenozoic, nowhere within the marine sequences offshore of this area have pulses of Cenozoic sedimentation, and consequent loading of the continental shelf, occurred on a scale that could conceivably have caused so large an isostatic response. Rather, a preponderantly thermal mechanism must be invoked. This mechanism is in line with the evidence of seismic tomography showing that much of the elevated eastern part of Africa is associated with the presence of anomalously hot-mantle material at depth (Partridge et al., chap. 2, this vol.).

Data on the extent of these Cenozoic movements come from several sources. Deformation of the remnants of well-mapped erosion surfaces provides one line of evidence: undisturbed gradients across such areas of advanced planation are usually less than 1 m/km and seldom rise to more than twice this value. In the coastal hinterland of Natal, however, gradients inland from the coast across accordant remnants of the African surface commonly attain 30 m/km and are only marginally less steep in profiles orthogonal to the coast reconstructed across surviving areas of the Post-African I surface. Within these profiles the zone of maximum uplift commonly occurs some 80 km inland of the present coastline (Partridge and Maud, 1987).

Some researchers have voiced the view that the recognition of such remnants of cyclic-planation surfaces is, in reality, subjective and does not provide a good basis for the quantification of tectonic movements. There is, however, nothing subjective about the criteria used for identifying and mapping key surfaces: in this area the African cycle is always associated with well-developed laterite duricrusts overlying kaolinized weathering profiles up to 50 m thick. Pedogenesis and weathering beneath the Post-African I surface are markedly less well developed (Partridge et al., in press); both surfaces clearly transgress geological structure and are characterized by excellently developed summit or shoulder accordances over large areas. Divergence between the African and Post-African I surfaces, across the crest of the uplift axis, indicates that maximum vertical movements that terminated the African cycle were of the order of 250 m. Subsequent deformations of the Post-African I surface ranged from 600 to 900 m (fig. 24.9).

Supporting evidence for these deformations is provided by the long profiles of major rivers on the east coast, (fig. 24.10), which are convex upward, clearly reflecting the influence of recent uplift and warping. These anomalous profiles can be linked to the most recent tectonic episode that deformed the Post-African I surface. Uplifts calculated from the river long profiles are somewhat smaller than those derived from the upwarping of this land surface, averaging 400–500 m. Taken together, all sources of data converge to provide robust evidence for uplifts in the southeastern hinterland of southern Africa totaling between 650 and 1,150 m since the termination of the African erosion cycle.

The timing of these movements can now be placed within a reasonably secure chronological framework. The earliest fossiliferous deposits which overlie the Post-African I surface in transgressive marine sequences close to the coast are of Early to Mid-Miocene age (Maud and Orr, 1975; Siesser and Miles, 1979; Partridge and Maud, 1987; le Roux, 1989). In terrestrial settings, fossiliferous fluviatile sediments associated with drainages on the Post-African I surface are of similar antiquity (Hendey, 1978; Dingle and Hendey, 1984; Partridge and Maud, 1987; de Wit, 1993). A small but widespread resurgence in shelf sedimentation dates to the Early Miocene (Dingle et al., 1993; Partridge and Maud, 1987) and represents a response to the modest uplift that terminated the African cycle. An Early Miocene age (ca. 18 myr) for the onset of this movement accords best with the various lines of evidence.

The far more significant uplift that raised the southeastern hinterland of the subcontinent, by perhaps as much as 900 m, is of more recent vintage. The clearest evidence for its timing is forthcoming from a major pulse of terrigenous sediment onto the continental shelf from the mouths of major east-coast rivers such as the Tugela and Limpopo. Martin (1987), in his detailed analysis of Cenozoic sedimentation rates off the east coast of southern Africa, has defined the base of this flux as coinciding with the regional acoustic seismic reflector "Jimmy," which has been dated to ca. 5 myr. Post-Jimmy sedimentation was about 50 percent more rapid than the average for the last 145 myr, which includes the extremely high rates of sediment accumulation immediately following continental rifting during the terminal Jurassic and Early Cretaceous. The total volume of offshore sediments in this area corresponds reasonably well with the estimate of 1,400 m of erosion from the hinterland since the Early Cretaceous (Martin, 1987). The significantly increased rate of deposition over the last 5 myr is therefore of some importance as an indicator of onshore events. Such ac-

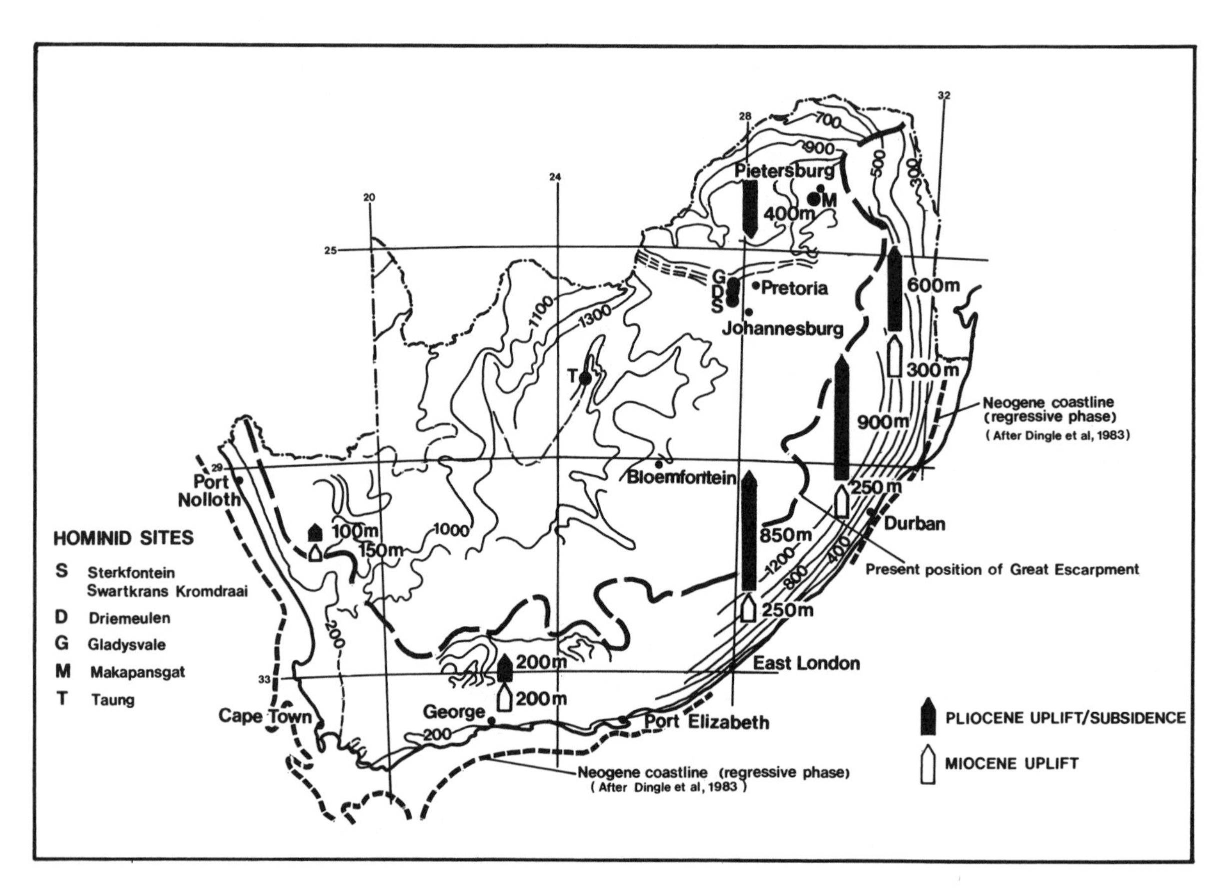

Fig. 24.9. Generalized contours on the Post-African I erosion surface. Open arrows indicate amplitude of Early Miocene uplift; solid arrows show Pliocene uplift or subsidence. The position of the present Great Escarpment is shown by a long broken line (after Partridge and Maud, 1987).

celerated sedimentation must be viewed in the context of the deep gorge cutting initiated in the coastal hinterland by this major episode of uplift, which is manifested in such features as the 500-m-deep dissection within the Valley of a Thousand Hills in Natal.

Two other sources of supporting evidence argue for a late age for this movement. First, Late Miocene and Early Pliocene marine deposits east of Port Elizabeth have been raised to elevations of up to 330 m along the southeastern flank of the axis of uplift (King, 1972; le Roux, 1989). These fossiliferous sediments are located some 35 km from the coast, that is, less than half of the distance to the axis of maximum uplift. Taking into account a maximum elevation of 80–90 m for the Early Pliocene marine transgression, which was the highest after the Mid-Miocene (Haq et al., 1987), this elevation of more than 300 m accords well with a maximum uplift of some 600 m within the last 5 myr. Secondly, along the coast of Natal the underformed 110 m marine terrace coincides with the outer edge of the Post-African II surface. The 170 m terrace, in contrast, shows clear evidence of seaward warping, in conformity with local gradients on the Post-African I surface (Partridge and Maud, 1987). On the basis of known rates of post-Pliocene isostatic uplift on the east coast of North America (Dowsett and Cronin, 1990) and on other passive margins similar to those of southern Africa, the 110 m terrace can be linked with reasonable confidence to a sea-level stand of 55–70 m, and the 170 m terrace to a stand between 80 and 110 m. In terms of the Neogene eustatic sea-level curve of Haq et al. (1987), this evidence would bracket the main period of uplift between 5 and 3 myr, in conformity with other independent lines of evidence.

Unlike the situation in East Africa, where uplift was accompanied by rift faulting, this uplift was a regional event that did not induce any marked degree of environmental fragmentation, except in areas marginal to the elevated zones. One such area is the Bankenveld region,

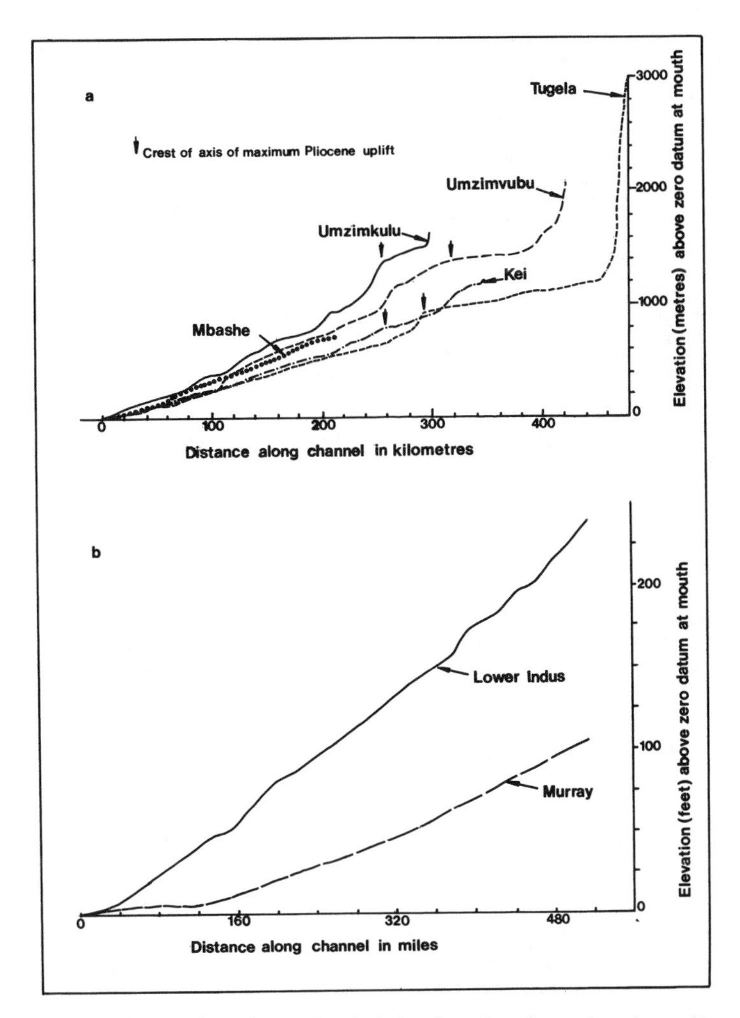

Fig. 24.10. (a) Long profiles of major rivers that cross the southeastern hinterland of southern Africa; (b) typical river long profiles in areas unaffected by recent uplift (after Leopold et al., 1963).

which is transitional from the Highveld of the Transvaal to the Bushveld Basin. Du Toit (1933) presented evidence for subsidence of this basin during the Pliocene, which may have amounted to as much as 400 m (Partridge and Maud, 1987). This movement may well have been more or less synchronous with the regional uplift of surrounding areas documented above. The importance of this basin-margin area for the study of hominid evolution in southern Africa stems from the fact that all of the Pliocene hominid-bearing cave deposits of the Transvaal are located within it (fig. 24.9). A result of the Pliocene movements was the incision of valleys within this area of strong lithological contrasts, leading to a high degree of environmental fragmentation. It is tempting to speculate that the major evolutionary events documented by its hominid and faunal sites may have been precipitated, at least in part, by this fragmentation.

In the absence of records of Neogene volcanism in southern Africa, the mechanism of an uplift of 600–900 m remains problematical. Of importance is the fact that, as in East Africa, the most strongly uplifted areas correspond with large negative Bouguer gravity anomalies. Hartnady (1990) has presented data indicating that the rift system of East Africa may be continuous with seismically active zones in the eastern part of southern Africa. In summing up the available evidence he concludes that "a variety of tectonothermal phenomena in this part of south-eastern Africa, including the seismicity in the Lesotho-Natal region, late Tertiary-Recent faulting, offshore submarine slumping, epeirogenic or 'cymatogenic'

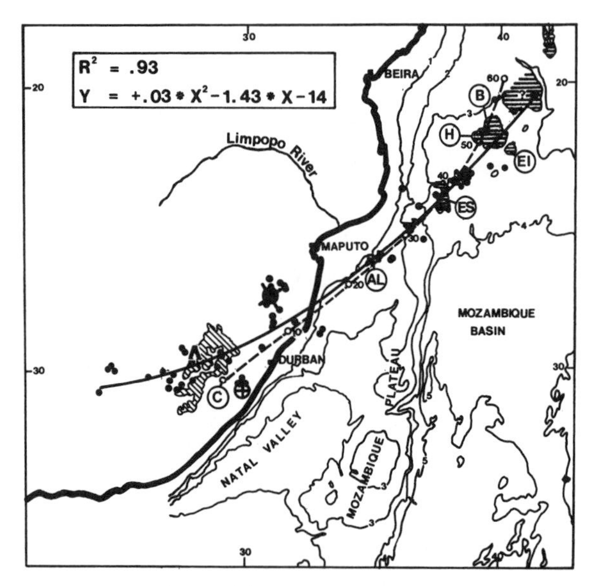

Fig. 24.11. Major features of the southeastern African continental margin and adjacent oceanic basins (bathymetric contours at 1 km intervals; after Fisher, 1982), illustrating the hypothetical Lesotho-Natal hotspot track (dashed line with open circles at 10 myr intervals; after Morgan, 1983). Annotated, horizontally hatched features along this track are as follows: C - Cedarville Flats fault-trough (0 myr location); AL - Almirante Leite Bank; ES - Eggert seamount (1,576 m); H - Hall Bank (117 m); B - Bassas da India (+ 2 m); EI - Europa Island (+ 12 m). The slanting diagonal hachure marks the area of the Drakensberg Mountains above the + 2 km topographic contour. The solid circle with cross marks the site of the hottest (40.7°C) thermal spring in Natal (Lurula Natal Spa; Kent, 1981). The open circle with cross marks the site of the Bongwan CO_2 gas exhalation (Gevers, 1941). The inverted V marks the site of the 1983 Lesotho eruption. The solid line is a seismicity axis defined by a second-order polynomial regression of epicentral location data weighted by earthquake magnitude. Framed insert gives correlation coefficient (R^2) and regression equation (after Hartnady, 1985, 1990).

uplift, thermal springs, and CO_2 gas exhalations, may be related to a single deep-seated cause, namely, the passage of the African Plate over an asthenospheric mantle hotspot now located near southern Lesotho" (p. 478). The reconstructed track of this hotspot, adduced from a variety of evidence, is shown in figure 24.11. Such motion relative to an asthenospheric heat source goes far toward explaining the axial, Late Neogene uplift that characterized this part of the subcontinent.

Support is given to this hypothesis by the occurrence, in 1983, of a small volcanic eruption within the Lesotho Mountains east of Maseru (Armstrong et al., in press). Previously dismissed as the result of the remelting of local Drakensberg basalt by a lightning strike, recent work has shown that the small volume of lava erupted has a deep-seated source and may be the first manifestation of volcanism associated with the hotspot track postulated by Hartnady. This event would provide a plausible explanation for the shallow heat-flow anomaly recorded beneath Lesotho by Jones (1992). It also lends to support to the notion that the African Superswell may have developed in response to the (late) influence of multiple or large-scale sublithospheric heat anomalies, as discussed previously.

Paleoclimatic Effects: Paleontological Evidence. Although extremely fragmentary, the evidence from Neogene deposits in southern Africa does tend to reflect global climatic trends. Early to Mid-Miocene vertebrate fossils recovered from Elizabeth Bay and Arris Drift along the Atlantic seaboard of Namibia suggest that a warm, mesic woodland environment had developed by this time (Stromer, 1926; Hopwood, 1929; Hendey, 1978). This interpretation is supported by the discovery of early Mid-Miocene fossils in alluvial deposits at Bosluispan in the now-arid interior of Bushmanland in the northwestern part of South Africa (Dingle and Hendey, 1984).

Farther to the south, the important sites of Langebaanweg and Noordhoek contain lagoonal stratigraphies that span significant segments of the Neogene. The Early Miocene levels of the Noordhoek sequence contain pollen spectra that Coetzee (1978) and Coetzee et al. (1983) regard as indicative of a warm, humid rain forest with abundant palms. The earliest deposits at Langebaanweg belong to the Upper Miocene Elandsfontein Formation, and the studies of Coetzee and Rogers (1982) show these to contain a significantly higher proportion of those shrubland elements that characterize the contemporary fynbos (macchia) vegetation of the area. These elements, together with grassland taxa, are still more strongly represented in peats of the succeeding Lower Pliocene Varswater Formation, indicating marked cooling, probably associated with a more strongly seasonal rainfall pattern (see also Scott, this vol). Hendey's (1983) analyses of the associated vertebrate fauna provide broad confirmation of these results and indicate the presence of a shrubland-grassland-forest mosaic with local intergrades during the Early Pliocene. This evidence for conspicuous climatic change across the Miocene-Pliocene boundary correlates well with offshore sedimentological data obtained by Siesser (1978) from DSDP Site 362 on the Walvis Ridge off the west coast of Namibia; Siesser's results indicate that although weak upwelling of cool waters, associated with the incipient development of the Benguela Current, began as early as the Middle or Late Oligocene, upwelling was greatly intensified during the Late Miocene. To the north of the Orange River, marked aridification followed, owing to the inception of a stable anticyclonic circulation pattern in response to the new oceanic temperature regime. The aeolian Tsondab Sandstone Formation of the Namib Desert, which was laid down upon a Mid-Miocene planation surface, bears testimony to this change. Maximum aridity was attained during the Pleistocene with the establishment of the present Namib sand sea, and it is likely that the marked aridification of the Kalahari and the northwestern Cape was a contemporary phenomenon, although no well-dated sequences are available to confirm this. It is of interest to note, however, that the evidence for a change to cool, dry conditions along the west coast of southern Africa near the end of the Miocene contrasts strongly with the results of studies of marine microfaunas off the southeastern coast of the continent: in that area warm-water conditions during the Miocene were succeeded by very warm conditions in the Pliocene (Dingle et al., 1983).

Progressive aridification of southern Africa during the Neogene is supported by the findings of de Wit (1993), whose evidence suggests that semiarid conditions prevailed throughout most of this interval. Two notable periods of increased fluvial activity are, however, apparent. The first spans the Early Miocene, when the fossiliferous sediments of Bosluispan were deposited by the now-defunct Koa River. The second dates to the latter part of the Pliocene, when major alluvial deposits accumulated in the Carnarvon Leegte and Sak paleodrainage systems, a few hundred kilometers to the east in the Northern Cape Province. Paleontological evidence, in the form of wood and mammalian remains, suggests that the latter period of increased rainfall coincided with maximum Pliocene warming (de Wit, 1993). No clear imprint of Neogene tectonic events is apparent in these alluvial records, although the beheading of the Koa drainage may be linked directly to the Miocene rise of the Griqualand-Transvaal axis of warping.

More profound changes, probably brought about by the combined effects of uplift and global cooling, can be discerned in the hominid-bearing cave deposits of the Transvaal, which fall within the area affected directly by Pliocene uplift of the southeastern hinterland. Several data sets provide independent lines of evidence pointing to increased aridification at this time. At the Makapansgat Limeworks and at Sterkfontein, the advent of cyclic sedimentation with a significant increase in the proportion of coarse debris occurs within the upper *Australopithecus africanus* levels—Member 4 in each case (Partridge, 1985, 1986). Aridification, with a more seasonal distribution of rainfall, is indicated. Vrba's (1974, 1975) analyses of the bovid faunas from Sterkfontein, Swartkrans, and Kromdraai show that this transition was also associated with a change from species indicative of a relatively greater bush cover to others reflecting a significantly higher proportion of grass. On the basis of paleontological comparisons she places this change, which corresponds with a major shift in the composition of the antelope fauna, sometime between 2.6 and 2.0 myr (Vrba, chap. 27, this vol.). Simultaneous changes in local hominid populations, from those present in Member 4 at Sterkfontein to the admixture of robust forms and early representatives of genus *Homo* evident in Sterkfontein Member 5, at Swartkrans and at Kromdraai, accompanied this faunal turnover.

Summary

It is important to bear in mind that the turnover pulse hypothesis goes further than the suggestion that evolutionary change is linked to environmental influences. It postulates that evolutionary change occurs in a series of synchronous pulses during which diverse groups of organisms show evidence of species generation or extinction. Synchronous turnover of species in two geographically separated regions of Africa in the Neogene was apparently powerful corroboration of the hypothesis. Evidence from the distribution of several monophyletic groups of the larger mammals of the African Plio-Pleistocene, however, suggests that there was "considerable contact" between the two regions (Turner and Wood, 1993b), and recent discoveries in Malawi, midway between the two regions, support that proposal. In the two regions, there is now strong prima facie evidence linking these faunal changes both to climatic events of a global nature, associated with growth of ice sheets in the Northern Hemisphere, and to large-scale tectonic uplift. The relative contribution of each is impossible to determine, but there can be no doubt that the critically important evolutionary events that occurred during this period of the Late Neogene must be attributed to the combined environmental impact of two separate forcing mechanisms within the eastern hinterland of Africa.

Notes

1. Wood (1992) has suggested that "hominin" should now be used in place of the term "hominid" when referring to the *Homo* clade.

References

Abell, P. I. 1982. Paleoclimates at Lake Turkana, Kenya, from oxygen isotope ratios of gastropod shells. *Nature,* 297:231–233.

Achauer, U., Maguire, P. K. H., Mechie, J., Green, W. V., and the KRISP Working Group. 1992. Some remarks on the structure and geodynamics of the Kenya Rift. *Tectonophysics,* 213:257–268.

Adamson, D. A., and Williams, M. A. J. 1987. Geological setting of Pliocene rifting and deposition in the Afar Depression of Ethiopia. *Journal of Human Evolution,* 16:597–610.

Armstrong, R. A., Maud, R. R., and Partridge, T. C. (in preparation). The 1983 volcanic eruption in Lesotho.

Baker, B. H. 1986. Tectonics and volcanism of the southern Kenya Rift Valley and its influence on rift sedimentation. In *Sedimentation in the African rifts* (ed. L. E. Frostick et al.). Special Publication of the Geological Society of London, 25: 45–57.

Baker, B. H., Mohr, P. A., and Williams, L. A. J. 1972. *The geology of the Eastern Rift System of Africa.* Geological Society of America, Special Paper, no. 136.

Bloch, S., Hales, A. L., and Landisman, M. 1969. Velocities in the crust and upper mantle of southern Africa from multimode surface wave dispersion. *Bulletin of Seismological Society of America,* 59:1599–1629.

Bonnefille, R. 1976. Palynological evidence for an important change in the vegetation of the Omo Basin between 2.5 and 2 million years ago. In *Earliest man and environments in the Lake Rudolf Basin,* pp. 421–431 (ed. Y. Coppens et al.). University of Chicago Press, Chicago.

———. 1983. Evidence for a cooler and drier climate in the Ethiopian uplands towards 2.5 myr ago. *Nature,* 303:487–491.

Bonnefille, R., and Letouzey, R. 1976. Fruits fossiles D'Antrocaryon dans la vallée de l'Omo (Ethiopie). *Adansonia,* 16: 65–82.

Bonnefille, R., Vincens, A., and Buchet, G. 1987. Palynology, stratigraphy and palaeoenvironment of a Pliocene hominid site (2.9–3.3 m.y.) at Hadar, Ethiopia. *Palaeogeography, Palaeoclimatology, Palaeoecology,* 60:249–281.

Brown, R. W., Rust, D. J., Summerfield, M. A., Gleadow, A. J. W., and de Wit, M. C. J. 1990. An early Cretaceous phase of accelerated erosion on the south-western margin of Africa: Evidence from apatite fission track analysis and the offshore sedimentary record. *International Journal of Radiation and Applied Instrumentation. Pt. D, Nuclear Tracks and Radiation Measurements,* 17:339–350.

Cahen, L., Snelling, N. J., Delhal, J., and Vail, J. R. 1984. The geochronology and evolution of Africa. Clarendon Press, Oxford.

Cerling, T. E. 1991. Carbon dioxide in the atmosphere: Evidence from Cenozoic and Mesozoic paleosols. *American Journal of Science,* 291:337–400.

———. 1992. Development of grasslands and savannas in East Africa during the Neogene. *Palaeogeography, Palaeoclimatology, Palaeoecology,* 97:241–247.

Cerling, T. E., Hay, R. L., and O'Neil, J. R. 1977. Isotopic evidence for dramatic climate changes in East Africa during the Pleistocene. *Nature,* 267:137–138.

Cerling, T. E., Wang, Y., and Quade, J. 1993. Expansion of C4 ecosystems as an indicator of global ecological change in the late Miocene. *Nature,* 361:344–345.

Clouser, R. H., and Langston, C. A. 1990. Upper mantle structure of southern Africa from Pnl waves. *Journal of Geophysical Research,* 95:17403–17415.

Coetzee, J. A. 1978. Climatic and biological changes in southwestern Africa during the Late Cainozoic. *Palaeoecology of Africa,* 10/11:13–29.

Coetzee, J. A., and Rogers, J. 1982. Palynological and lithological evidence for the Miocene palaeoenvironment in the Saldanha region (South Africa). *Palaeogeography, Palaeoclimatology, Palaeoecology,* 39:71–85.

Coetzee, J. A., Scholtz, A., and Deacon, H. J. 1983. Palynological studies and the vegetation history of the fynbos. In *Fynbos palaeoecology: A preliminary synthesis,* pp. 156–173 (ed. H. J. Deacon et al.). South African National Programmes Report, no. 75. Council for Scientific and Industrial Research, Pretoria.

Coppens, Y. 1982. Les plus anciens fossiles d'hominidés. *Pontifical Academy of Science, Scripta Varia,* 50:1–9.

———. 1988a. L'origine des hominidés et de l'homme. In *L'Évolution dans sa réalité et ses diverses modalités.* Fondation Singer-Polignac, Masson, Paris. 8:171–178.

———. 1988b. Les vicissitudes de l'évolution humaine. *Bulletin de l'Académie Nationale de Medicine* (Paris), 172:1289–1296.

Davidson, A., and Rex, D. C. 1980. Age of volcanism and rifting in southwestern Ethiopia. *Nature,* 283:657–658.

Denys, C., Chorowicz, J., and Tiercelin, J. J. 1986. Tectonic and environmental control on rodent diversity in the Plio-Pleistocene sediments of the African Rift System. In *Sedimentation in the African rifts* (ed. L. E. Frostick et al.). Special Publication of the Geological Society of London, 25:363–372.

de Wit, M. C. J. 1993. Cainozoic evolution of drainage systems in the north-western Cape. Ph.D. diss., University of Cape Town.

Dingle, R. V., and Hendey, Q. B. 1984. Late Mesozoic and Tertiary sediment supply to the eastern Cape Basin (SE Atlantic) and palaeo-drainage systems in southwestern Africa. *Marine Geology,* 56:13–26.

Dingle, R. V., Siesser, W. G., and Newton, A. R. 1983. Mesozoic and Tertiary geology of southern Africa. Balkema, Rotterdam.

Dixey, F. 1956. *The East African Rift System.* Supplementary Bulletin, Overseas Geological Mineral Resources, no. 1. H. M. Stationary Office, London.

Dowsett, H. J., and Cronin, T. M. 1990. High eustatic sea level during the Middle Pliocene: Evidence from the southeastern U.S. Atlantic Coastal Plain. *Geology,* 18:435–438.

du Toit, A. L. 1933. Crustal movements as a factor in the geographical evolution of South Africa. *South African Geographical Journal,* 16:3–20.

Ebinger, C. J. 1989. Tectonic development of the western branch of the East African Rift System. *Geological Society of American Bulletin*, 101:885–903.

Ebinger, C. J., Bechtel, T. D., Forsyth, D. W., and Bowin, C. O. 1989a. Effective elastic plate thickness beneath the East African and Afar Plateaus and dynamic compensation of the uplifts. *Journal of Geophysical Research*, 94:2883–2901.

Ebinger, C. J., Deino, A. L., Drake, R. E., and Tesha, A. L. 1989b. Chronology of volcanism and rift basin propagation: Rungwe volcanic province, East Africa. *Journal of Geophysical Research*, 94:15785–15803.

Ebinger, C. J., Yemano, T., WoldeGabriel, G., Aronson, J. L., and Walter, R. C. 1993. Late Eocene–Recent volcanism and faulting in the southern main Ethiopian rift. *Journal of Geological Society of London*, 150:99–108.

Fairhead, J. D. 1986. Geophysical controls on sedimentation within the African Rift Systems. In *Sedimentation in the African Rifts* (ed. L. E. Frostick et al.). Special Publication of the Geological Society of London, 25:19–27.

Feibel, C. S., Brown, F. H., and McDougall, I. 1989. Stratigraphic context of fossil hominids from the Omo group deposits: Northern Turkana Basin, Kenya and Ethiopia. *American Journal of Physical Anthropology*, 78:595–622.

Feibel, C. S., Harris, J. M., and Brown, F. H. 1991. Palaeoenvironmental context for the Late Neogene of the Turkana Basin. In *Koobi Fora Research Project*, 3:321–346 (ed. J. M. Harris). Clarendon Press, Oxford.

Fisher, R. L. 1982. General bathymetric charts of the oceans (GEBCO), sheet 5.09. Commission of Hydrological Survey, Ottawa.

Gasse, F. 1980. Les diatomées lacustres plio-pléistocènes du Gadeb (Ethiopie): Systématique, paléoécologie, biostratigraphie. *Revnue Algologique*, 3:1–249.

Gevers, T. W. 1941. Carbon dioxide springs and exhalations in North Pondoland and Alfred County, Natal. *Transactions of the Geological Society of South Africa*, 44:233–301.

Gilchrist, A. R., and Summerfield, M. A. 1990. Differential denudation and flexural isostasy in formation of rifted-margin upwarps. *Nature*, 346:739–742.

Haq, B. U., Hardenbol, J., and Vail, P. R. 1987. Chronology of fluctuating sea levels since the Triassic. *Science*, 235:1156–1167.

Hartnady, C. J. H. 1985. Uplift, faulting, seismicity, thermal spring and possible incipient volcanic activity in the Lesotho-Natal region, SE Africa: The Quathlamba hotspot hypothesis. *Tectonics*, 4:371–377.

———. 1990. Seismicity and plate boundary evolution in southeastern Africa. *South African Journal of Geology*, 93:473–484.

Hawthorne, J. B. 1975. Model of Kimberlite pipe. In *Physics and chemistry of the earth*, 9:1–15 (ed. L. H. Ahrens et al.). Pergamon, Oxford.

Hendey, Q. B. 1978. Preliminary report on the Miocene vertebrates from Arrisdrift, South West Africa. *Annals of South African Museum*, 76:1–41.

———. 1983. Palaeoenvironmental implications of the Late Tertiary vertebrate fauna of the fynbos region. In *Fynbos palaeoecology: A preliminary synthesis*, pp. 100–115 (ed. H. J. Deacon et al.). South African National Programmes Report, no. 75. Council for Scientific and Industrial Research, Pretoria.

Hilgen, F. J. 1991a. Astronomical calibration of Gauss to Matuyama sapropels in the Mediterranean and implication for the geomagnetic polarity time scale. *Earth and Planetary Science Letters*, 104:226–244.

———. 1991b. Extension of the astronomically calibrated (polarity) time scale to the Miocene-Pliocene boundary. *Earth and Planetary Science Letters*, 107:349–368.

Hill, A. 1987. Causes of perceived faunal change in the later Neogene of East Africa. *Journal of Human Evolution*, 16:583–596.

Hill, A., Ward, S., Deino, A., Curtis, G., and Drake, R. 1992. Earliest *Homo*. *Nature*, 355:719–722.

Hopwood, A. J. 1929. New and little known mammals from the Miocene of Africa. *America Museum Novitates*, 344:1–9.

Jones, M. Q. W. 1992. Heat flow anomaly in Lesotho: Implications for the southern boundary of the Kaapvaal craton. *Geophysical Research Letters*, 19:2031–2034.

Kent, L. E. 1981. The thermal springs of the southeastern Transvaal and northern Natal. *Annals of the Geological Survey of South Africa*, 15:51–67.

King, L. C. 1972. Pliocene marine fossils from the Alexandria Formation in the Paterson District, Eastern Cape Province, and their geomorphic significance. *Transactions of the Geological Society of South Africa*, 75:159–160.

Leopold, L. B., Wolman, M. G., and Miller, J. P. 1964. *Fluvial processes in geomorphology*. Freeman, San Francisco.

le Roux, F. G. 1989. The lithostratigraphy of Cenozoic deposits along the south-east Cape coast as related to sea-level changes. Master's thesis, University of Stellenbosch.

Martin, A. K. 1987. Comparison of sedimentation rates in the Natal Valley, south-western Indian Ocean, with modern sediment yields in east coast rivers of southern Africa. *South African Journal of Science*, 83:716–724.

Masters, G., Bolton, H., and Shearer, P. 1992. Large-scale 3-dimensional structure of the mantle. *EOS (American Geophysical Union Transactions)*, 73:201.

Maud, R. R., and Orr, W. M. 1975. Aspects of past-Karoo geology in the Richards Bay area. *Transactions of the Geological Society of South Africa*, 78:101–109.

Mohr, P. A. 1987. Structural style of continental rifting in Ethiopia: Reverse décollements. *EOS (American Geophysical Union Transactions)*, 68:721-729.

Molnar, P., and England, P. 1990. Late Cenozoic uplift of mountain ranges and global climate change: Chicken or egg? *Nature*, 346:29–34.

Molnar, P., England, P., and Martinod, J. 1993. Mantle dynamics, uplift of the Tibetan Plateau, and the Indian monsoon. *Reviews of Geophysics*, 31:357–396.

Morgan, W. J. 1983. Hotspot tracks and the early rifting of the Atlantic. *Tectonophysics*, 94:123–139.

Nyblade, A. A., and Pollack, H. N. 1992. A gravity model for the lithosphere in western Kenya and northeastern Tanzania. *Tectonophysics*, 212:257–267.

Nyblade, A. A., and Robinson, S. W. 1994. The African Superswell. *Geophysical Research Letters*, 21:765–768.

Partridge, T. C. 1985. The palaeoclimatic significance of Cainozoic terrestrial stratigraphic and tectonic evidence from Southern Africa: A review. *South African Journal of Science,* 81:245–247.

———. 1986. Palaeoecology of the Pliocene and Lower Pleistocene hominids of southern Africa: How good is the chronological and palaeoenvironmental evidence? *South African Journal of Science,* 82:80–83.

Partridge, T. C., de Villiers, J. M., Fitzpatrick, R. W., and Maud, R. R. (in press). Soil-landscape-climate relations in southern Africa. *Catena.*

Partridge, T. C., and Maud, R. R. 1987. Geomorphic evolution of southern Africa since the Mesozoic. *South African Journal of Geology,* 90:179–208.

———. 1989. The end-Cretaceous event. New evidence from the Southern Hemisphere. *South African Journal of Science,* 85:428–430.

Pascual, R., Vucetich, M. C., Scillato-Yane, G. J., and Bond, M. 1985. Main pathways of mammalian diversification in South America. In *The great American biotic interchange,* pp. 219–247 (ed. F. G. Stehli and S. D. Webb). Plenum Press, New York.

Pickford, M. 1990. Uplift of the roof of Africa and its bearing on the evolution of mankind. *Human Evolution,* 5:1–20.

Pickford, M., Senut, B., and Hadoto, D. 1993. *Geology and palaeobiology of the Albertine Rift Valley, Uganda-Zaire.* Vol. 1, *Geology,* pp. 190. International Centre for Training and Exchanges in the Geosciences (CIFEG), Occasional Paper 1993/24, Orléans, France.

Ruddiman, W. F., and Kutzbach, J. E. 1989. Forcing of Late Cenozoic Northern Hemisphere climate by plateau uplift in southern Asia and the America west. *Journal of Geophysical Research,* 94:18409–18427.

Ruddiman, W. F., Raymo, M. E., Martinson, D. G., Clement, B. M., and Backman, J. 1989. Pleistocene evolution: Northern Hemisphere ice sheets and North Atlantic Ocean. *Paleoceanography,* 4:353–412.

Rust, D. J., and Summerfield, M. A. 1990. Isopach and borehole data as indicators of rifted margin evolution in southwestern Africa. *Marine and Petroleum Geology,* 7:287–297.

Saggerson, E. P., and Baker, B. H. 1965. Post-Jurassic erosion surfaces in eastern Kenya and their deformation in relation to rift structure. *Quarterly Journal of the Geological Society of London,* 121:51–72.

Schrenk, F., Bromage, T. G., Betzler, C. G., Ring, U., and Juwayeyi, Y. M. 1993. Oldest *Homo* and Pliocene biogeography of the Malawi Rift. *Nature,* 365:833–836.

Shackleton, N. J. 1978. Structural development of the East African rift system. In *Geologic background to fossil man,* pp. 20–28. (ed. W. W. Bishop). Scottish Academic Press, Edinburgh.

Shackleton, N. J., Berger, A., and Peltier, W. R. 1990. An alternative astronomical calibration of the Lower Pleistocene timescale based on ODP Site 677. *Transactions of the Royal Society of Edinburgh: Earth Sciences,* 81:251–261.

Shackleton, N. J., Crowhurst, S., Hagelberg, T., Pisias, N. G., and Schneider, D. A. (in press). A new Late Neogene time scale: Application to Leg 138 sites. In *Proceedings ODP: Science results,* vol. 138 (ed. N. Pisias et al.). College Station, Tex. (Ocean Drilling Program).

Siesser, W. G. 1978. Aridification of the Namib Desert: Evidence from oceanic cores. In *Antarctic glacial history and world palaeoenvironments,* pp. 105–113 (ed. E. M. van Zinderen Bakker). Balkema, Rotterdam.

Siesser, W. G., and Miles, G. A. 1979. Calcareous nannofossils and planktic foraminifers in Tertiary limestones, Natal and Eastern Cape, South Africa. *Annals of South African Museum,* 79:139–158.

Sletlene, L., Wilcox, L. E., Blouse, R. S., and Sanders, J. R. 1973. A Bouguer gravity map of Africa. DMAAC Technical Paper 73-003, Defense Mapping Agency, St. Louis.

Smith, A. G., and Briden, J. C. 1982. *Mesozoic and Cenozoic palaeocontinental maps.* Cambridge University Press, Cambridge.

Stromer, E. 1926. Resteland-und Süsswasser-bewohnender Wirbeltiere aus dem Daimantenfelden Deutsch-Südwestafrika. In *Die Diamantewüste Südwestafrikas,* 2:107–153 (ed. E. Kaiser). Dietrich Reimer, Berlin.

Su, W. J., Woodward, R. L., and Dziemwonski, A. M. 1992. Joint inversion of travel time and waveform data for 3-D models of the Earth up to Degree 12. *EOS (American Geophysical Union Transactions),* 73:201.

Suwa, G. 1988. Evolution of the "robust" australopithecines in the Omo succession: Evidence from mandibular premolar morphology. In *Evolution history of the "robust" australopithecines,* pp. 199–222 (ed. F. E. Grine). Aldine de Gruyter, New York.

Tanaka, S., and Hamaguchi, H. 1992. Heterogeneity in the lower mantle beneath Africa as revealed from S and ScS phases. *Tectonophysics,* 209:213–222.

Ten Brink, U., and Stern, T. 1992. Rift flank uplifts and hinterland basins: Comparison of the Transantarctic Mountains with the Great Escarpment of southern Africa. *Journal of Geophysical Research,* 97:569–585.

Tiedemann, R., Sarnthein, M., and Stein, R. 1989. Climatic changes in the western Sahara: Aeolo-marine sediment record of the last 8 million years. In *Proceedings ODP: Scientific results,* 108:241–261 (ed. W. F. Ruddiman et al.). College Station, Tex. (Ocean Drilling Program).

Turner, A., and Wood, B. 1993a. Comparative palaeontological context for the evolution of the early hominid masticatory system. *Journal of Human Evolution,* 24:301–318.

———. 1993b. Taxonomic and geographic diversity in robust australopithecines and other African Plio-Pleistocene larger mammals. *Journal of Human Evolution,* 24:147–168.

Vrba, E. S. 1974. Chronological and ecological implications of the fossil Bovidae at the Sterkfontein australopithecine site. *Nature,* 250:19–23.

———. 1975. Some evidence of chronology and palaeoecology of Sterkfontein, Swartkrans and Kromdraai from the fossil Bovidae. *Nature,* 254:301–304.

———. 1985. African Bovidae: Evolutionary events since the Miocene. *South African Journal of Science,* 81:263–266.

Wesselman, H. B. 1985. Fossil micromammals as indicators of climatic change about 2.4 m.y. ago in the Omo Valley, Ethiopia. *South African Journal of Science,* 81:260–261.

Westaway, R. 1993. Forces associated with mantle plumes. *Earth and Planetary Science Letters,* 119:331–348.

Williams, L. A. J. 1978. The volcanological development of the

Kenya Rift. In *Petrology and geochemistry of continental rifts.* Vol. 1, *Proceedings of NATO Advanced Study Institute on Paleorift Systems,* p. 36 (ed. H. J. Neumann and I. B. Ramberg). NATO Advanced Study Institute. Reidel, Dordrecht.

WoldeGabriel, G., Aronson, J. L., and Walter, R. C. 1990. Geology, geochronology and rift basin development in the central sector of the Main Ethiopian Rift. *Geological Society of America Bulletin,* 102:439–458.

Wood, B. A. 1992. Origin and evolution of the genus *Homo. Nature,* 355:783–790.

Wood, B. A., Wood, C. G., and Konigsberg, L. W. 1994. *Paranthropus boisei:* An example of evolutionary stasis? *American Journal of Physical Anthropology,* 95:117–136.

Yemane, K., Bonnefille, R., and Faure, H. 1985. Palaeoclimatic and tectonic implications of Neogene microflora from the northwestern Ethiopian highlands. *Nature,* 318:653–656.

Chapter 25

Of Mice and Almost-Men: Regional Paleoecology and Human Evolution in the Turkana Basin

Henry B. Wesselman

The Omo River and its tributaries drain the southwestern region of the Ethiopian highlands, emerging from a gorge onto the floodplain of the lower Omo Valley, a northern extension within the Turkana Basin, which contains sedimentary formations accumulated during the Pliocene and Pleistocene epochs (fig. 25.1). The deposits are composed primarily of fluviatile sediments, with channels, floodplains and backswamps predominating, although deep-water lacustrine and deltaic situations also occur (de Heinzelin, 1983). Intercalated into the sands, silts, and clays are numerous volcanic tuffs, which, through magnetostratigraphic studies, have yielded radiometric age determinations (Brown et al., 1978, 1985; Feibel et al., 1989; McDougall et al., 1992). Stimulated by the need for evidence relevant to the origin and evolution of early Hominidae in Africa, an international team directed by F. Clark Howell and Yves Coppens carried out a remarkable number of interdisciplinary studies within the Omo Valley between 1967 and 1974. The wealth of fossils recovered from the Omo Beds currently exceeds forty thousand specimens, more than two hundred of which have been identified as fragmentary bones and teeth of early hominids (Howell and Coppens, 1974, 1976; Howell et al., 1987).

An attempt to evaluate the paleoecological character of the hominid-bearing fossil localities through the recovery and analysis of micromammals was carried out during the 1972, 1973, and 1974 field seasons. Twenty thousand kg of excavated sediments from thirty fossiliferous sites were screenwashed, producing roughly 2,200 kg of concentrated residues, which were then hand-sorted for microfauna. The fossils obtained in this manner were all extracted from excavated sediments, although some of the larger taxa such as the hystricids and thryonomyids were discovered as surface finds. The number of specimens recovered was sparse owing to the fluviatile nature of the depositional environments. The species diversity, however, was impressive. Ten localities within the Shungura Formation, spanning the period from approximately 3.0 to 1.9 million years (myr), yielded forty-seven species within thirty-four genera (table 25.1). Wesselman (1984) has presented their systematic analysis.

Many contemporary small mammals are strongly habitat-specific and live comparatively short lives within relatively small home ranges. Coe (1972), for example, has shown that the micromammals living in the Turkana Basin today are not dispersed across a whole spectrum of environments but are restricted to distinct areas of habitat preference. The use of small mammals as paleoenvironmental indicators rests on the principle of actualism, which implies that the fossil taxa had ecological requirements similar to those of the contemporary species they most closely resemble. This principle must be qualified, however, because the range of any contemporary taxon may be limited by factors that did not exist in the past. Similarly, some of the paleospecies may have been limited by circumstances that do not exist in the present.

The taphonomic analysis of the Omo micromammals suggests that the fossil assemblages were derived from raptor pellets incorporated into the edges of channel lags within close proximity to their original points of accumulation and were not subject to reworking (Wesselman, 1984). Korth (1979) has revealed that the osseous re-

ISBN 0–300–06348–2.

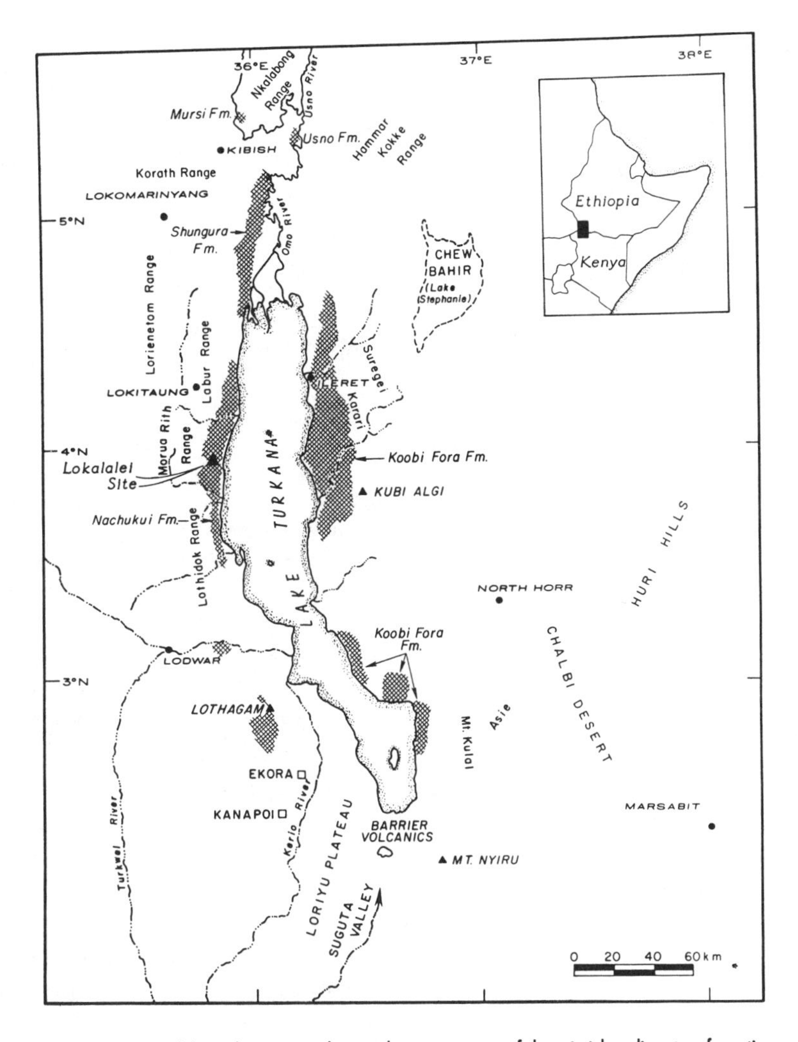

Fig. 25.1. Map of the Turkana Basin, showing the outcrop areas of the principle sedimentary formations.

mains of small mammals do not survive postmortem transport of any distance in fluviatile situations. The Omo assemblages thus represent point localities in space and time and preserve a record of more local paleoenvironmental conditions than do the macrovertebrate or macrobotanical remains, which have generally been collected from stratigraphic levels that are more broadly defined and of greater lateral extent. It is likely, given the range of most modern raptors, that the assemblages include representatives from the mosaic of paleocommunities that were present along the ancient river system. Systematic analysis has revealed that 86 percent of the taxa are closely related to extant species. This number rises to 90 percent if we consider only the forty identifiable species from the five most productive localities, making the paleoenvironmental reconstruction of the Omo Beds unusually strong.

Paleoenvironments of Hominid-Bearing Localities

Member B. A large assemblage of micromammals (26 species) dated as close to 3 myr has been recovered from 3 localities in Member B of the Shungura Formation (Localities 793 and Omo 229 in unit B-2; and Locality 1 in unit B-10). Remains of *Australopithecus afarensis* have been found at Locality 1 and at other sites of comparable age within the Shungura Formation and the Usno Formation, farther to the north (fig. 25.2).

The micromammals from Member B include a predominance of species characteristic of tropical forests

Table 25.1. Micromammal Taxa Recovered from the Shungura Formation

Taxon	Member						
	B	C	D	E	F	G	H
Insectivora							
Soricidae							
Crocidura aithiops	—	—	—	—	—		
Crocidura aff. *dolichura*	—						
Myosorex robinsoni	—						
Suncus haesaertsi	—						
Suncus aff. *lixus*	—						
Suncus shungurensis	—						
Chiroptera							
Pteropodidae							
Eidolon aff. *helvum*	—						
Emballonuridae							
Taphozous abitus	—	—	—	—	—		
Coleura muthokai					—		
Hipposideridae							
Hipposideros aff. *cyclops*	—						
Hipposideros aff. *camerunensis*	—						
Hipposideros kaumbului					—		
Primates							
Lorisidae							
Galago howelli	—						
Galago senegalensis						—	
Galgoides demidovii	—						
Carnivora							
Viverridae							
Helogale kitafe	—	—					
Helogale hirtula					—	—	
Hyracoidea							
Procaviidae							
Gigantohyrax maguirei	—	—					
Heterohyrax brucei	—						
Lagomorpha							
Leporidae							
Lepus capensis				—	—	—	
Rodentia							
Cricetidae							
Tatera aff. *minuscula*	—	—	—	—	—	—	
Gerbillus aff. *pusillus*					—		
Muridae							
Mastomys minor	—	—	—	—	—	—	
Golunda gurai	—	—					
Lemniscomys aff. *striatus*	—	—	—	—	—		
Pelomys sp. indet.					—		
Aethomys deheinzelini					—	—	
Thallomys jaegeri	—	—					
Thallomys quadrilobatus					—	—	
Thallomys sp. indet.	—						
Oenomys sp. indet.		—					
Acomys sp. indet.					—		
Arvicanthis sp. indet.	—	—	—	—	—	—	—
Mus aff. *minutoides*	—						
Muridae							
Saidomys sp. indet.							—
Murinae gen. et sp. indet. A						—	
Murinae gen. et sp. indet. B	—						
Dipodidae							
Jaculus orientalis					—	—	
Bathyergidae							
Heterocephalus atikoi					—	—	
Sciuridae							
Xerus erythropus	—	—					
Xerus sp. indet.					—	—	
Paraxerus ochraceus	—	—	—	—	—		
Thryonomyidae							
Thryonomys gregorianus	—	—	—	—	—	—	
Thryonomys swinderianus						—	—
Hystricidae							
Xenohystrix crassidens	—						
Hystrix makapanensis	—	—	—	—	—	—	—
Hystrix cristata	—	—	—	—	—	—	—

and mesic woodlands, with only a few reflecting more open savanna-woodland habitats (fig. 25.2). Many closely resemble contemporary taxa restricted in their distribution to the rain forests of central and West Africa. These include a soricid indistinguishable from the contemporary *Crocidura dolichura,* two species of *Hipposideros* bats, a paleospecies of *Taphozous* allied with the contemporary *T. peli, Galagoides demidovii,* and the pteropodid *Eidolon helvum.* Taxa that suggest forest biotopes more obliquely include *Galago howelli,* a paleospecies intermediate in size and morphology between *Galago alleni* of the central African forests and *Otolemur crassicaudatus,* which is found in moist riverine forests, woodlands, and thickets. Contemporary representatives of the murids *Mastomys* and *Mus,* the sciurid *Paraxerus ochraceus,* and the large soricid *Crocidura aithiops,* a paleospecies allied with *C. flavenscens,* are all highly successful and ubiquitous forms occurring in a wide variety of habitats, including forests. It is the inclusion of the true central and West African rain-forest species, however, that strongly suggests the presence of this ecosystem within the Turkana Basin during the Pliocene and provides evidence for a biogeographic connection with the equatorial forests of central Africa.

Support for this paleoscenario has been provided by the recovery of fossilized fruits of *Antrocaryon,* a West and central African rain-forest tree, from sediments within Member A and the Usno Formation, dated between 3.3 and 3.4 myr (Bonnefille and Letouzey, 1976).

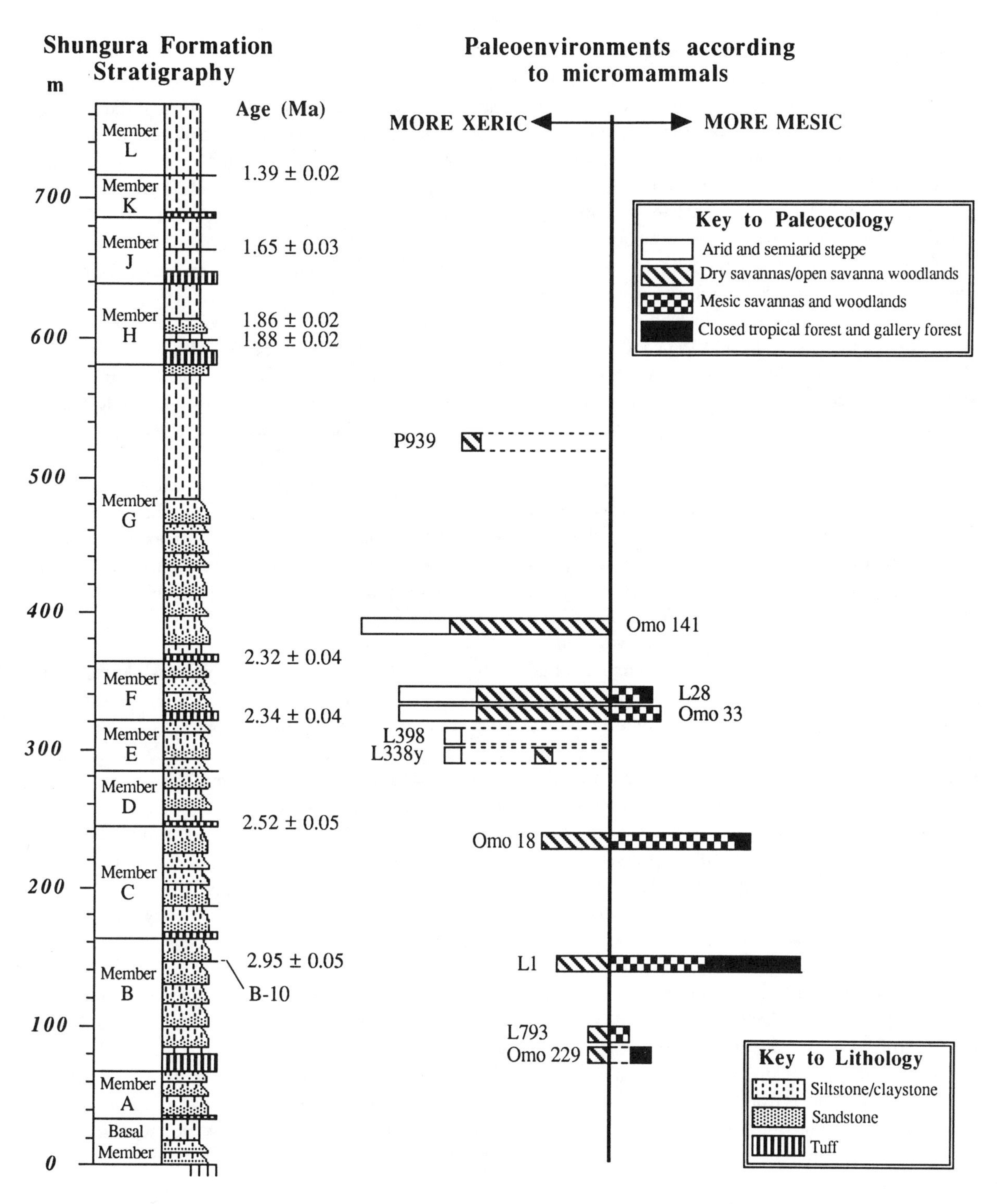

Fig. 25.2. Paleoecological analysis of the Omo micromammal assemblages.

In addition, Williamson (1985) has commented on the presence of *Potadoma*, a rain-forest prosobranch gastropod, in stratigraphic horizons of the same age in the Nachukui Formation west of Lake Turkana, noting that the joint distribution of *Antrocaryon* and *Potadoma* essentially delimits the extent of tropical high forest in central Africa today. Fossilized wood from both Member A and the Usno Formation, dated at about 2.9 myr, includes trees (*Fiscus, Diospyros,* and *Xylia*, another West African forest species) as well as woody-stemmed vines (such as *Landolphia*) all of which suggest mesic forests (Bonnefille and Dechamps, 1983; Dechamps and Maes, 1985). The presence of developed paleosols in Members A and B has also been considered as indicative of greater rainfall between 3.3 and 2.9 myr (de Heinzelin, 1983).

This evidence suggests the presence of a significant tropical rain-forest extension in eastern Africa during the Pliocene. Its temporal duration and paleogeographical extent are difficult to determine. It may have existed as a developed gallery forest bordering the river, or it may have occupied much or most of the floodplain. Brown and Feibel (1988) have suggested that there was no long-lived lake in the basin between 4.0 and 2.4 myr and that the paleoriver maintained its integrity as a watercourse, flowing south and east to drain directly into the Indian Ocean until the uplift of Mount Kulal created a barrier (and thus a lake) sometime between 2.5 and 2.0 myr. The high forests of central Africa may thus have extended from the Atlantic to the Indian Ocean during much of the Pliocene, a connection that could explain some puzzling contemporary faunal distributions. One such is the presence of *Cercocebus galeritus*, the crested mangabey, in the forests along the Tana River in northeastern Kenya. This population is separated from the rest of its species in central Zaire by nearly 1,500 km and many ecologic and geographic barriers. Another puzzle is the presence of the central African *Galagoides demidovii* in the forests of Mount Marsabit, 120 km east of Lake Turkana.

Eight micromammals from Member B suggest mesic woodlands, savanna-woodlands, and moist savannas. These include the shrews of the genus *Suncus;* a paleospecies of the dwarf mongoose *Helogale;* the murids *Lemniscomys, Arvicanthis* and *Golunda;* and *Thryonomys gregorianus.* Only three species (of twenty-six) represent the drier savanna-woodlands so typical of much of eastern Africa today. These include the gerbil *Tatera*, the *Acacia* rat *Thallomys*, and the ground squirrel *Xerus erythropus*, a northern savanna species whose range today extends across central and West Africa. The lower frequency of these three taxa (11 percent) suggests that xeric ecosystems were more distal. Much or most of the floodplain may have been densely covered by a succession of tropical forest (as suggested by 32 percent of the taxa), and mesic woodlands and thickets (31 percent).

The rest of this early assemblage is made up of ubiquitous species and a few extinct forms. Of interest is the presence of the large porcupine *Xenohystrix crassidens*, originally described by Greenwood (1955) from the Makapansgat Limeworks in southern Africa, where it is associated with the remains of *Australopithecus africanus. Xenohystrix* is also known from the older beds at Laetoli (Denys, 1987), dated at 3.5–3.7 myr (though it is absent in the younger Ndolanya Beds, dated at 2.6 myr). It has also been recovered at Hadar (Sabatier, 1982) from Locality AL 133, currently dated at between 3.20 and 3.22 myr (Walter, 1994). Its association with these sites confirms the antiquity of the gracile australopithecines in southern Africa and implies that the habitat preferences of this porcupine may have been more ubiquitous than has previously been suggested.

Member C. Locality Omo 18, in upper Member C (unit C-8), dated at slightly more than 2.5 myr, is notable in that it produced an edentulous hominid mandible, the type specimen of *Australopithecus aethiopicus.* Nine micromammal species recovered from this site by Jean-Jacques Jaeger suggest a decline of tropical high forest by this time. Only one, the murid *Oenomys*, could be considered as a forest form; in reality, it is more of a forest-edge taxon, occurring today in clearings or in the annual growth along road cuts that penetrate forested areas. *Helogale, Lemniscomys, Golunda, Mastomys, Paraxerus,* and *Thryonomys* reflect the continued presence of mesic woodlands and dense thickets reminiscent of Member B time. This finding has been confirmed by what has been observed in the sedimentology (de Heinzelin, 1983), in the pollen data, and in the macrofloral remains (Bonnefille and Letouzey, 1976; Dechamps and Maes, 1985). Only two species (*Thallomys* and *Xerus*) suggest more xeric *Acacia* savannas.

Member D. No micromammal localities have been found in Member D, nor have any pollen yet been recovered, but macrobotanical remains from low in the member suggest mesic plant communities, with large forest galleries persisting as well as some evidence for wooded savannas.

Member E. The discovery of the partial cranium of an immature robust australopithecine (now assigned to *Australopithecus aethiopicus* by Howell et al., 1987) at Locality 338y in Member E (unit E-3) led to a large excavation from which the remains of the ground squirrel *Xerus* and the hare *Lepus* have been recovered. This find marks the earliest appearance of *Lepus* in the Omo Beds, a taxon considered by some to be Palearctic in origin and part of a widespread dispersal event that occurred near the end of the Pliocene. Its presence is considered a reliable indicator of aridity. A pollen spectrum recovered by Bonnefille (1976, 1980) from unit E-4 reveals the predominance of grasses (75 percent of the pollen recovered) and the almost complete absence of riverine taxa (only 2.9 percent), reflecting a marked shift toward arid climatic conditions sometime between 2.5 and 2.4 myr.

Member F. This paleoclimatic shift is confirmed by the micromammals from Member F. Jaeger excavated eleven species from a silty tuffite of Tuff F′ at Locality Omo 33, 2.5 m above unit E-4. Locality 28 in unit F-1, immediately above Tuff F, dated at 2.34 myr, has yielded eighteen species. These taxa reflect a drastic reduction in riverine and forest taxa, a decrease in the species diversity of murid rodents, and an increase in the number and relative frequency of gerbilline rodent taxa. Only two species of the eleven recovered from Omo 33 (*Lemniscomys* and *Thryonomys*) and one of the eighteen from Locality 28 (*Pelomys*) reveal the presence of mesic savannas or riverine environments. Only one from Locality 28 (*Taphozous*) provides evidence for the continued presence of some gallery forest. Xeric biotopes are suggested by nine species (72 percent of the taxa) at Omo 33 (*Lepus, Helogale, Tatera, Acomys, Aethomys, Thallomys, Xerus, Heterocephalus,* and *Jaculus*) and by twelve species (67 percent) at Locality 28 (all of the above plus *Hipposideros, Coleura,* and *Gerbillus*). Several strongly arid-adapted forms represent 27 and 28 percent of the taxa at the two localities, respectively. *Jaculus orientalis* is a saltating dipodid, completely independent of water, found today in the open desert terrain of the northern Sahara. The bathyergid (a paleospecies of *Heterocephalus*) and *Helogale hirtula* (a dwarf mongoose species different from that of Members B and C) are associated today with the Somali arid zone. The *Gerbillus* at Locality 28 is allied with the contemporary *G. pusillus,* which favors open, sandy flats in arid, semidesertic conditions. The contemporary species of *Coleura* is found in dry savanna-woodlands and subdesertic steppe thicket. *Lepus* favors arid, waterless plains and *Acacia* scrub. The small amount of fossil wood found in Member F is also dominated by forms reflecting very dry conditions.

Member G. Nine micromammal species recovered from Locality Omo 141 in unit G-3 about 12 m above Tuff G, dated at 2.32 myr, include no taxa reflecting forested or mesic environments. Five species (*Galago senegalensis, Helogale hirtula, Tatera, Aethomys,* and *Thallomys*) suggest arid *Acacia* grasslands and scrub (reflected also in the fossil wood from Member G). With the exception of the prosimian, all the taxa from Omo 141 are also present in Member F (including the ubiquitous *Mastomys*). *Heterocephalus, Lepus,* and *Jaculus* continue to represent 30 percent of the specimens, revealing the depth and extent of the climatic shift. This shift is confirmed by a small assemblage of micromammals (N = 15) dated at about 1.6 myr from the Koobi Fora Formation at East Turkana, which includes *Aethomys, Thallomys, Tatera,* and *Jaculus,* suggesting that the aridity may have been of long duration (Black and Kristalka, 1986). Prentice and Denton (1988), Shackleton et al. (1948), Vrba (1985, 1988), and others (this vol.) have confirmed that this shift was not localized within the Turkana Basin but reflected worldwide trends that accompanied the onset of the Plio-Pleistocene ice-age climatic fluctuations.

Of interest is the high percentage of micromammals (86 percent) that are closely related to or virtually indistinguishable from modern forms despite their antiquity. This suggests that evolutionary stasis occurred within the majority of lineages and implies that the expansion and contraction of the sub-Saharan ecosystems in response to the Plio-Pleistocene climatic fluctuations must have been accompanied by the successful migration and survival of many of the small mammals that inhabited them. As has been discussed (Jaeger and Wesselman, 1976), the virtual absence of any archaic lineages from the earlier Tertiary is notable and suggests that the adaptive radiation that gave rise to the modern tropical African micromammal fauna occurred earlier in the Pliocene. In general, the Omo micromammal assemblages are taxonomically closer to those of East Turkana (Black and Kristalka, 1986), Olduvai (Jaeger, 1976), and Laetoli (Denys, 1987) than they are to those of Hadar, with its predominance of Asiatic genera (Sabatier, 1982) or to those of the southern African sites, with their high percentage of endemic species, suggesting that a fauna distinctive to the East African region developed during the Pliocene.

Also of note is the turnover among the micromammals in response to the climatic shift. Of the forty-seven species recovered from the Shungura Formation, twenty are limited to Members B and C, and seventeen are restricted to Members E, F, and G. Only 10 lineages are common to both the earlier and later localities (table 25.1), and these represent ubiquitous species or taxa characteristic of xeric biotopes. This finding resembles what Vrba (1985, 1988, and chap. 27, this vol.) has observed among the African bovids, thereby providing support for the turnover pulse hypothesis. Extinction of some earlier micromammal lineages is probable (*Golunda, Xenohystrix,* and *Gigantohyrax*). Migration is equally certain for some of the others (the arid-adapted forms). Evidence for punctuation may be present within the genus *Thallomys,* in which a smaller, less derived species in Members B and C is abruptly replaced by a larger, more derived form in Members F and G. Recovery of intermediates from Members D and E would clarify whether this represents migration or an evolutionary pulse within a single lineage. The same point could obviously be made with relation to the appearance of the robust australopithecines and the genus *Homo.*

The earliest hominid occupation sites in the Turkana Basin appear as in situ scatters of stone artifacts at seven localities in Members E and F of the Shungura Formation, dated between 2.34 and 2.4 myr (Chavaillon, 1976; Howell et al., 1987; Merrick et al., 1973; Merrick and Merrick, 1976), and at the Lokalalei site in the Kalochoro Member of the Nachukui Formation, west of Lake Turkana, dated at 2.35 myr (Kibunjia et al., 1992). The earliest fossil evidence for *Homo* is currently revealed by a series of isolated teeth from Members E and F of the Shungura Formation (Howell et al., 1987), by a temporal bone fragment dated at 2.4 myr from the Chemeron Formation, near Lake Baringo, to the south (Hill et al., 1992), and by a mandible provisionally aged at 2.3–2.5 myr from the Chiwondo Beds in the Malawi Rift (Schrenk et al., 1993). This evidence suggests that early *Homo* may have been responsible for these sites and that the paleoclimatic shift may have played a major role in the appearance of this genus.

Implications for Human Evolution

In considering the relationship between climatic change and evolution, many investigators tend to focus on shifts in mean ambient temperatures (cooler versus warmer) as the primary causative factor in evolutionary change. Within tropical ecosystems, however, it may be more useful to think in terms of climatic stability versus instability or, from a faunal point of view, "more predictable" versus "less predictable" climatic regimes. It is generally known that extensive tropical forest ecosystems are supported by greater (and more predictable) annual rainfall. A shift in climatic stability resulting in reduced precipitation patterns will produce more spatially heterogeneous ecosystems (mosaics of various forest-wooded savanna–open savanna types, etc.), which become more arid (and less predictable) as the dry seasons achieve greater depth and duration.

It is also generally known that climatic factors affect the primary production within contemporary ecosystems and that the stability of primary production is a major determinant of species diversity (Paine, 1966). The effects of climate are thus manifested first in the overall complexity of the vegetation, which, in turn, determines the numbers of animal species present. By implication, the effects of the Late Pliocene climatic shift should theoretically be visible not only within the faunal and floral composition of regional paleocommunities but also within the number of fossil species collected. In the tropics, greater rainfall before the climatic shift should be expressed in greater diversity, a direct result of greater (and more stable) primary production. Reduced precipitation and increased seasonality after the shift should result in lesser numbers of fossil species recovered.

This theoretical situation is confirmed by the species diversity of Omo micromammal assemblages. Member B discloses a mosaic of tropical forest, mesic woodlands, and more distal wooded savannas, whose greater primary production is reflected in the greater diversity of micromammal species recovered (twenty-six taxa). *Australopithecus afarensis* is associated with this more predictable paleoenvironment, a fact revealed also by the microfauna at Hadar (Sabatier, 1982).

Subsequent to the paleoclimatic shift, the mean annual rainfall must have decreased dramatically as the dry seasons achieved greater depth and duration. The lowered primary production, lessened climatic predictability, and increased seasonality are reflected in the types of micromammals present and result in an immediate decrease in the species diversity (eighteen at Locality 28, eleven at Omo 33, nine at Omo 141). It should be noted that the beginnings of the shift may be apparent by 2.52 myr at Omo 18, where the forest taxa of Member B are lacking and where only nine micromammal species were recovered in association with *Australopithecus aethiopicus.* It is also notable that the earliest artifacts and fossils assignable to the genus *Homo* do not appear until the full-fledged aridity of Member E time.

Several questions then arise: Was the divergence of both hominid lineages (*Australopithecus* and *Homo*) the result of an evolutionary pulse in response to the climatic shift, or was it the product of an earlier radiation? Were both lineages indigenous residents within the basin? Or was this true only of *Australopithecus,* and was *Homo* a migrant who appeared among the arid-adapted species, arriving from the Rift Valley to the south, from the Somali arid zone to the northeast, or from the arid biozones that stretch westward across Africa and northward along the drainage of the Nile?

Wood (1992) has suggested that three synchronous early species of *Homo (H. habilis, H. rudolfensis,* and *H. ergaster)* were present in the Turkana Basin by 1.9 myr (and possibly earlier, as revealed by the fossils from Members E, F, and G of the Shungura Formation). More questions arise: Does this diversity reflect allopatry within the ancestral stock(s) living in different drainage systems? Did the approximation of arid ecosystems close to the depositional environments along the Omo River encourage a migration of similar but morphologically distinct hominids into the basin from more than one source area? And once again, did this diversification occur before or after the climatic shift?

Brown and Feibel (1988) have observed that the Turkana Basin lies on a drainage divide between the Mediterranean Sea, to the north, and the Indian Ocean, to the east. It is not bounded by steep escarpments (except in the south), and if the basin was periodically filled with sediments (see Cerling, 1986), the water from the Omo River would have flowed into one ocean or the other. The evidence suggests that the paleoriver flowed into the Indian Ocean for much of the Pliocene, until volcanic damming by the uplift of Mount Kulal between 2.5 and 2.0 myr and upwarp along the eastern margin of the basin caused it to be inundated by a large, stable lake. The appearance of sudano-nilotic fish in Member G suggests that the paleoriver periodically drained westward into central Africa and possibly into the Nile, establishing biogeographic connections with the west and north and enabling faunal and floral elements from these regions to migrate into the basin.

Support for this view is offered by the arrival in Members F and G of *Jaculus orientalis,* a taxon found today in open desert terrain in the northern Sahara near the Mediterranean Sea, and by the appearance in Member G (Locality P 939 in unit G-20) of the murid *Saidomys,* an invader from Asia known from the Mid-Pliocene deposits at Wadi El Natrun, in Egypt (James and Slaughter, 1974), and from Hadar (Sabatier, 1982), where it is common. *Saidomys* is known also from the Upper Member of the Nawata Formation at Lothagam Hill, dated at about 5 myr, and from Tabarin, near Lake Baringo, further to the south (Winkler, 1992). Its absence from the earlier Shungura micromammal assemblages suggests that it became extinct in the Turkana Basin during the Pliocene and that its reappearance in Member G may be due to migration from the north. Also of note is the appearance in Member G (units G-15 to G-23) of fossilized wood of *Populus euphratica,* a tree that extends today from southern Spain across northern Africa through Asia Minor to Pakistan (Bonnefille and Dechamps, 1983). Taken together, the evidence suggests that the Turkana Basin served as a biogeographic crossroad joining eastern, central, and northern Africa for much of the later Pliocene and that latitudinal shifts in biome zonation associated with the paleoclimatic shift encouraged the equatorward dispersion of northern faunal and floral elements. Migration may thus have been responsible for the appearance of *Homo* in the Turkana Basin and for the early diversity seen within this genus.

With reference to the relationship of paleoclimates to migration and early hominid behavior, Klopfer and MacArthur (1960), who studied species diversity among nonpasserine and passerine birds, may offer some insight. Broadly speaking, nonpasserines express more stereotyped behavior and are better adapted to exploit the more consistent tropical ecosystems than are passerines, whose generally more plastic behavior allows them to inhabit less predictable habitats. Passerines migrate and are found to be more diverse in the temperate regions, whereas nonpasserines migrate less and are more diverse in the tropics. Klopfer (1959) and others have suggested that to exist and to exploit an environment successfully, a species must have behavioral flexibility that is roughly inversely proportional to the predictability of the environment in which it lives. It has also been suggested that regions with relatively more stable climates, high primary production, and relative constancy of resources allow for the evolution of finer specializations and adaptations than do areas with more erratic, and therefore less predictable, climates.

The hominid lineages present in the Turkana Basin

subsequent to the climatic shift express derived morphological features that distinguish them from each other and from their ancestor. Those manifested by the robust australopithecines involve the enlargement of the masticatory complex; a finer specialization reflective of dependency on preferred food items, which encouraged the inflation of the postcanine teeth; and the modification of the attendant facial and cranial architecture (Beynon and Wood, 1987; Dean and Wood, 1982; Suwa, 1990; Wood et al., 1988; and others). Whether continued involvement with a relatively more predictable habitat characterized by more stable primary production reminiscent of their ancestor's preferred environment (forests and woodlands) was part of this econiche remains to be determined, but it may have been so.

A different suite of morphological features distinguish the genus *Homo* (Wood, 1992). Included among these are the expansion and reorganization of the neocortex of the brain, a finer specialization presumably associated with a shift away from the relatively more stereotyped behavior of its ancestors, and close relations presumably living in what woodlands remained along river and lake margins after the climatic shift. It is tempting to speculate that this behavioral shift occurred in response to the decreasing predictability of climatic regimes and food sources in less productive xeric habitats and involved a successful transition into a new econiche that required more plastic, and possibly migratory, behavior. Many investigators have offered ideas about the nature of this transition, and most now concur that the new niche involved increased dependency upon meat, a highly concentrated but unpredictable food source found in seasonably variable densities in open, less predictable habitats. Many have theorized that increased carnivory involved opportunistic scavenging and hunting, which was facilitated by the invention of stone-tool technology, a product of the more plastic behavior encouraged by uncertain environments and enabled by an expanding brain (Binford, 1985; Blumenschine, 1989; Blumenschine and Cavallo, 1992; Bunn and Kroll, 1986; Howell et al., 1987; Isaac, 1984; Potts, 1988; Shipman, 1986, Stanley, 1992; Speth, 1989; and others).

Many of the investigators cited above have speculated on the importance of fat in the diet of these early hominids. During the increasingly prolonged dry seasons subsequent to the climatic shift, it is likely that the body fat of both predators and prey dropped to only a small percentage of total body weight, much like the present-day ratio. It has been suggested that members of the *Homo* lineage obtained dietary fat from seasonally impoverished carcasses by crushing bones to obtain the marrow, an activity facilitated by the use of stone tools, another finer specialization that appears coincident with the climatic shift.

How dietary fat was obtained by the surviving australopithecines remains to be determined. A food source that may have been available in tropical riverine and perilacustrine forests and woodlands and that may have been exploited by *Australopithecus afarensis* is the oil palm *Elaeis guineensis.* Originally a native of West and central Africa, the oil palm has a wide geographic distribution today and does well in both tropical and semideciduous forests. A very productive species, the oil palm bears an average of ten bunches of two hundred nuts per year. The fibrous pulp and kernels contain up to 70 percent fat (Hill, 1952), making them a staple for chimpanzees and baboons (as well as humans) in such places as the Gombe National Park and elsewhere (Boesch and Boesch, 1983). The existence of a rain-forest extension in eastern Africa during the Pliocene suggests that the oil palm was present, giving rise to the speculation that the distribution of the early australopithecines may have been determined by the distribution of oil palms, a predictable fat source in a productive environment that could be obtained simply through the ability to climb trees. The adult male *Australopithecus afarensis* skull recovered from Locality AL. 444 at Hadar (Kimbel et al., 1994) reveals the presence of a strong masticatory complex that may well have been up to the task of processing palm nuts. Considered from this perspective, the absence of fossil pongids from the Plio-Pleistocene sites within the Rift Valley could well be an example of competitive exclusion.

Increased spatial heterogeneity through deforestation and habitat fragmentation in response to aridification must have resulted in the progressive rarity and eventual loss of oil palms in much of eastern Africa, encouraging a dietary shift into the only other readily available fat source: bones. To obtain the fat reservoirs stored within them, most investigators agree that the evolving populations of early *Homo* responded with the use of stone tools, a strategy also used by wild chimpanzees to open palm nuts in West Africa (Boesch and Boesch, 1983). This functional shift in focus from the teeth to the hands may have resulted in the easement of the selective pressures responsible for maintaining a strong masticatory complex, which in turn led to the gracilization of the teeth and jaws in many of the evolving populations of *Homo.* This reduction may have been the result of what

Brace (1988) and others term the "probable mutation effect," a phenomenon observable in the increased rate of dental reduction in humans during the last 10,000 years, presumably a response to changes in diet and diet-related technologies (Brace et al., 1987).

The surviving australopithecines may have responded to the loss of oil palms by turning to the same resource. Presumably less plastic in their behavior than the larger-brained *Homo,* they may have continued using their jaws to accomplish the same end. In the process they may have generated inflated postcanine teeth, thickened mandibles, increased orthognathism, and an increasingly robust and elevated attachment of the jaw musculature on the cranium (Beynon and Wood, 1987; Dean and Wood, 1982; Suwa, 1990; Wood, this vol.; Wood et al., 1988; and others), producing a morphological complex for bone chewing much like that of *Crocuta.* The more robust nuts of the doum palm, *Hyphaene,* associated with arid environments, may have provided another, alternative fat source. Eaten by elephants, their husks are often glazed with sugar crystals (F. Brown, pers. comm.) and may have been consumed by the australopithecines before the fruits reached full maturity. Whatever the cause, the first signs of this hypermasticatory trend are revealed by hominid remains recovered from the Shungura and Nachukui Formations, dated at about 2.5 myr (Suwa, 1990; Wood, this vol.). The enlarged premolars and facial architecture of *Homo rudolfensis* reveal that at least one of the early *Homo* types converged on this pattern as well (Wood, 1992).

It is of interest that the Omo micromammals reveal xeric ecosystems to have been present in Members B and C time, though perhaps at a comparatively greater distance from the river. Clearly, the opportunity for a behaviorally less stereotyped australopithecine to become involved in these habitats always existed, even before the climatic shift. The dry, open country revealed in the older beds at Laetoli (Denys, 1987) suggests that such biotopes were accessed or traversed by *Australopithecus afarensis* as early as 3.7 myr.

This last point gives rise to some final observations about paleoclimatic and paleogeographic events with relation to human evolution in the Late Pliocene. MacArthur (1965) has observed that the total species diversity of a geographic area has a theoretical upper limit set by the abundance or scarcity of the taxa that live there. If primary production remains high and the number of species increases, the number of individuals (species biomass) cannot grow. So, as the number of species increases, each species will become rarer in terms of numbers of individuals present. Because of fewer climatic hazards, species can be rarer in tropical ecosystems without running great danger of extinction. Seen in this light, the relative scarcity of hominid remains in comparison with those of other faunal elements recovered from Pliocene fossiliferous sites may reveal their status as rare tropical species existing in specialized marginal niches. Binford (1968) has sketched an elegant model showing how the stability of a system might be maintained by encouraging the surplus population to expand into marginal areas where the habitat might be less predictable and where niches might require redefinition. Others, including MacArthur, have observed that only in areas characterized by high productivity can a marginal niche support a species.

It is also generally known that there is a limiting similarity of species that can coexist within a habitat. Species that are more similar than this limiting value must occupy different habitats. This limiting value has been shown to be less where productivity is high, where family size is low, and where seasons are relatively uniform. MacArthur (1965) has suggested that this value should be less for "pursuing hunters" than for species that search for "stationary prey."

The question then reappears: Was the Late Pliocene hominid species diversity in the Turkana Basin the result of an evolutionary pulse in response to the climatic shift, or did the radiation occur before that? The micromammals from Members B and C reveal greater rainfall, greater primary production, and greater species diversity around 3.0 myr, with a decline occurring between 3.0 and 2.5 myr. If the australopithecines were rare tropical species, diversification into specialized marginal niches would be more likely during this earlier time. The appearance in the basin of *Australopithecus aethiopicus,* a taxon morphologically distinct from *Australopithecus afarensis,* reveals that a speciation event was occurring within this lineage during Member C time (roughly 2.8–2.5 myr) and implies that the hominid diversification may indeed have begun before the paleoclimatic and paleogeographic events that transformed the basin between 2.5 and 2.0 myr.

The limiting similarity of these diverging hominid types would have served to isolate them spatially as well as behaviorally. As long as conditions of high productivity and more predictable climate continued, competition between the diverging types would have been slight, allowing them to coexist in small groups as rare, evolving tropical species. As their biomass increased in response to the high productivity and they came into more frequent contact, increased competition for resources and the lessened limiting value for pursuing hunters could very well have encouraged a trend toward increased carnivory (increased niche specialization) in one or more of these hominid types in the Turkana Basin and elsewhere. Such an adaptive shift would have lessened competition between similar sympatric subspecies and encouraged increased spatial and behavioral isolation. Further segregation may also have been achieved through migratory behavior in response to seasonal and regional shifts in prey-species biomass.

The Late Pliocene climatic shift created conditions of lowered primary production and lessened climatic stability, which, combined with increased seasonality and spatial heterogeneity (deforestation and habitat fragmentation), must have subjected the diversifying hominid types to considerable stress. The surviving australopithecines coped in their way, and the evolving populations of *Homo* responded in theirs.

It is of interest that the seven archaeological occurrences from Members E and F of the Shungura Formation, as well as the Lokalalei site from the Nachukui Formation, all represent small, temporary occupations near the river. Also of interest is that similar sites have not been found on the more distal deltaic or dryland situations (see Howell et al., 1987). A partial *Homo habilis* cranium from Locality 894 in Member G comes from a part of the Shungura succession (unit G-28, dated at about 1.9 myr) that preserves well-developed paleosols and paleoland surfaces, but no traces of associated archaeological occurrences have yet been detected. This paucity of evidence and artifacts may suggest that the earliest *Homo* groups were only an ephemeral (seasonal?) presence in the Turkana Basin and may have been open-country migrants who were simply passing through.

Acknowledgments

Gratitude is expressed to Elisabeth Vrba for inviting me to participate in an excellent and thought-provoking conference; to F. Clark Howell, who invited me to go into the field and who has offered continued support, encouragement, and friendship over the years; and to Raymonde Bonnefille, Frank Brown, Jean-Jacques Jaeger, Don Johanson, Bob Walter, and Tim White, who offered useful information and critique. Much of the success of this project was due to the efforts of Noel Boaz, Claude Guillemot, Ato Getachew Ayele, Ato Daniel Touafe, Muthoka Kivingo, John Kaumbulu, and Lokiriakwanga (Atiko). The early phases of this research were partially supported by grants from the National Science Foundation to F. Clark Howell and by National Defense Education Act Fellowships to the University of California. For all who did fieldwork with the Omo Research Expedition, there is a special affection. Frank Brown kindly redrafted the figures.

References

Beynon, A. D., and Wood, B. A. 1987. Patterns and rates of enamel growth in the molar teeth of early hominids. *Nature* 326:493–496.

Binford, L. R. 1968. Post-Pleistocene adaptations. In *New perspectives in archeology,* pp. 5–32 (ed. S. R. Binford and L. R. Binford). Aldine, New York.

———. 1985. Human ancestors: Changing views of their behavior. *Journal of Anthropological Archeology* 4:292–327.

Black, C. C., and Kristalka, L. 1986. Rodents, bats and insectivores from the Plio-Pleistocene sediments to the east of Lake Turkana, Kenya. *Natural History Museum of Los Angeles County, Contributions in Science* 372:1–15.

Blumenschine, R. J. 1989. A landscape taphonomic model of the scale of prehistoric scavenging opportunities. *Journal of Human Evolution* 18:345–371.

Blumenschine, R. J., and Cavallo, J. A. 1992. Scavenging and human evolution. *Scientific American* 266:90–96.

Boesch, C., and Boesch, H. 1983. Optimization of nut-cracking with natural hammers by wild chimpanzees. *Behavior* 83: 265–86.

Bonnefille, R. 1976. Palynological evidence for an important change in the vegetation of the Omo Basin between 2.5 and 2 million years. In *Earliest man and environments in the Lake Rudolf Basin,* pp. 421–431 (ed. Y. Coppens, F. C. Howell, G. Isaac, and R. E. F. Leakey). University of Chicago Press, Chicago.

———. 1980. Vegetation history of savanna in East Africa during the Plio-Pleistocene. *Proceedings of the Fourth International Palynological Conference, Lucknow (1976–77)* 3:75–89.

———. 1983. Evidence for a cooler and drier climate in the Ethiopian uplands towards 2.5 myr ago. *Nature* 303:487–491.

Bonnefille, R., and Dechamps, R. 1983. Data on fossil flora. In *The Omo Group: Archives of the International Omo Research Expedition,* pp. 191–207 (ed. J. de Heinzelin). Annales, ser. 8, Sciences Géologiques, no. 85. Musée Royal de l'Afrique Centrale, Tervuren.

Bonnefille, R., and Letouzey, R. 1976. Fruits fossilés d'*Antrocaryon* dans la vallée de l'Omo (Ethiopie). *Adansonia* 16: 65–82.

Brace, C. L. 1988. *The stages of human evolution: Human and cultural origins.* 3d ed. Prentice-Hall, Englewood Cliffs, N.J.

Brace, C. L., Rosenberg, K. R., and Hunt, K. D. 1987. Gradual change in human tooth size in the Late Pleistocene and post-Pleistocene. *Evolution* 41:705–720.

Brown, F. H., and Feibel, C. S. 1988. "Robust" hominids and Plio-Pleistocene paleogeograhy of the Turkana Basin, Kenya and Ethiopia. In *Evolutionary history of the "robust" australopithecines,* pp. 325–341 (ed. F. E. Grine). Aldine, New York.

Brown, F. H., McDougall, I., Davies, T., and Maier, R. 1985. An integrated Plio-Pleistoene chronology for the Turkana Basin.

In *Ancestors: The hard evidence,* pp. 82–90 (ed. E. Delson). Alan R. Liss, New York.

Brown, F. H., Shuey, R. T., and Croes, M. K. 1978. Magnetostratigraphy of the Shungura and Usno Formations, southwestern Ethiopia: New data and comprehensive reanalysis. *Geophysical Journal of the Royal Astronomical Society* 54:519–538.

Bunn, H., and Kroll, E. 1986. Systematic butchery by Plio-Pleistocene hominids at Olduvai Gorge, Tanzania. *Current Anthopology* 27:431–452.

Cerling, T. E. 1986. A mass-balance approach to basin sedimentation: Constraints on the recent history of the Turkana Basin. *Palaeogeography, Palaeoclimatology and Palaeoecology* 54:63–86.

Chavaillon, J. 1976. Evidence for the technical practices of Early Pleistocene hominids, Shungura Formation, lower Omo Valley, Ethiopia. In *Earliest man and environments in the Lake Rudolf Basin,* pp. 565–573 (ed. Y. Coppens, F. C. Howell, G. Isaac, and R. E. F. Leakey). University of Chicago Press, Chicago.

Coe, M. 1972. The south Turkana expedition: Ecological studies of small mammals of south Turkana. Scientific Papers, no. 9. *Journal of Geography* 138:316–338.

Dean, M. C., and Wood, B. A. 1982. Basicranial anatomy of Plio-Pleistocene hominids from East and South Africa. *American Journal of Physical Anthropology* 59:157–174.

Dechamps, R., and Maes, F. 1985. Essai de reconstitution des climats et des végétations de la basse vallée de l'Omo au Plio-Pléistocène à l'aid des bois fossilés. In *L'environnement des hominidés au Plio-Pléistocène* (ed. Y. Coppens). Foundation Singer-Polignac, Paris.

de Heinzelin, J., ed. 1983. The Omo Group: Stratigraphic and related earth sciences studies in the lower Omo Basin, southern Ethiopia. Annales, ser. 8, Sciences Géologiques, no. 85. Musée Royal de l'Afrique Centrale, Tervuren.

Denys, C. 1987. Fossil rodents (other than Pedetidae) from Laetoli. In *Laetoli: A Pliocene site in northern Tanzania,* pp. 118–170 (ed. M. D. Leakey and J. M. Harris). Clarendon Press, Oxford.

Feibel, C. S., Brown, F. H., and McDougall, I. 1989. Stratigraphic context of fossil hominids from the Omo group deposits: Turkana Basin, Kenya and Ethiopia. *American Journal of Physical Anthropology* 78:595–622.

Greenwood, M. 1955. Fossil Hystricoidea from the Makapan Valley, Transvaal. *Paleontologica Africana* 3:77–85.

Hill, A. F. 1952. *Economic botany.* McGraw-Hill, New York.

Hill, A., Ward, S., Deino, A., Curtiss, G., and Drake, R. 1992. Earliest *Homo. Nature* 355:719–722.

Howell, F. C., and Coppens, Y. 1974. Inventory of remains of Hominidae from Pliocene-Pleistocene formations of the lower Omo Basin, Ethiopia (1967–1972). *American Journal of Physical Anthropology* 40:1–16.

———. 1976. An overview of the Hominidae from the Omo succession, Ethiopia. In *Earliest man and environments in the Lake Rudolf Basin,* pp. 522–532 (ed. Y. Coppens, F. C. Howell, G. Isaac, and R. E. F. Leakey). University of Chicago Press, Chicago.

Howell, F. C., Haesaerts, P., and de Heinzelin, J. 1987. Depositional environments, archeological occurrences and hominids from Members E and F of the Shungura Formation (Omo Basin, Ethiopia). *Journal of Human Evolution* 16:665–700.

Isaac, G. L. 1984. The archeology of human origins. *Advances in World Archeology* 3:1–87.

Jaeger, J.-J. 1976. Les rongeurs (Mammalia, Rodentia) du Pléistocène inférieur d'Olduvai Bed I (Tanzanie). Pt. 1, Les Muridés. In *Fossil vertebrates of Africa* 4:58–120 (ed. R. S. G. Savage and S. C. Coryndon). Academic Press, New York.

Jaeger, J.-J., and Wesselman, H. B. 1976. Fossil remains of micromammals from the Omo group deposits. In *Earliest man and environments in the Lake Rudolf Basin,* pp. 351–360 (ed. Y. Coppens, F. C. Howell, G. Isaac, and R. E. F. Leakey). University of Chicago Press, Chicago.

James, G. T., and Slaughter, B. H. 1974. A primitive new Middle Pliocene murid from Wadi El Natrun, Egypt. *Annals of the Geological Survey of Egypt:* 4:333–362.

Kibunjia, M., Roche, H., Brown, F. H., and Leakey, R. E. 1992. Pliocene and Pleistocene archeological sites west of Lake Turkana, Kenya. *Journal of Human Evolution* 23:431–438.

Kimbel, W. H., Johanson, D. C., and Rak, Y. 1994. The first skull and other new discoveries of *Australopithecus afarensis* at Hadar, Ethiopia. *Nature* 368:449–451.

Klopfer, P. H. 1959. Environmental determinants in faunal diversity. *American Naturalist* 94:337–342.

Klopfer, P. H., and MacArthur, R. H. 1960. Niche size and faunal diversity. *American Naturalist* 95:293–300.

Korth, W. W. 1979. Taphonomy of microvertebrate fossil assemblages. *Annals of the Carnegie Museum* 46:235–285.

MacArthur, R. H. 1965. Patterns of species diversity. *Biological Review* 40:510–533.

McDougall, I., Brown, F. H., Cerling, T. E., and Hillhouse, J. W. 1992. A reappraisal of the geomagnetic polarity time scale to 4 ma. using data from the Turkana Basin, East Africa. *Geophysical Research Letters* 19:2349–2352.

Merrick, H. V., Haesaerts, P., de Heinzelin, J., and Howell, F. C. 1973. Archaeological occurrences of Early Pleistocene age from the Shungura Formation, lower Omo Valley, Ethiopia. *Nature* 242:572–575.

Merrick, H. V., and Merrick, J. P. S. 1976. Archeological occurrences of earlier Pleistocene age from the Shungura Formation. In *Earliest man and environments in the Lake Rudolf Basin,* pp. 574–584 (ed. Y. Coppens, F. C. Howell, G. Isaac, and R. E. F. Leakey). University of Chicago Press, Chicago.

Paine, R. T. 1966. Food web complexity and species diversity. *American Naturalist* 100:65–75.

Potts, R. 1988. Early hominid activities at Olduvai. Aldine, New York.

Prentice, M. L., and Denton, G. H., 1988. The deep-sea oxygen isotope record, the global ice sheet system and hominid evolution. In *Evolutionary history of the "robust" australopithecines,* pp. 383–403 (ed. F. E. Grine). Aldine, New York.

Sabatier, M. 1982. Les rongeurs du site Pliocène à Hominidés de Hadar (Ethiopie). *Palaeovertebrata* 12:1–56.

Schrenk, F., Bromage, T. G., Betzler, C. G., Ring, U., and Juwayeyi, Y. M. 1993. Oldest *Homo* and Pliocene biogeography of the Malawi Rift. *Nature* 365:833–836.

Shackleton, N. J., Backman, J., Zimmerman, H., Kent, D. V.,

Hall, M. A., Roberts, D. G., Schnitaker, D., Baldauf, J. G., Desprairies, A., Homrighausen, R., Huddleston, P., Kennett, J. B., Kaltenback, A. J., Krumsiek, K.A.O., Morton, A. C., Murray, J. W., and Westberg-Smith, J. 1984. Oxygen isotope calibration of the onset of ice rafting and history of glaciation in the North Atlantic region. *Nature* 307:620–623.

Shipman, P. 1986. Scavenging or hunting in early hominids. *American Anthropologist* 88:27–43.

Speth, J. D. 1989. Early hominid hunting and scavenging: The role of meat as energy source. *Journal of Human Evolution* 18: 329–343.

Stanley, S. M. 1992. An ecological theory for the origin of *Homo. Paleobiology* 18:237–257.

Suwa, G. 1990. A comparative analysis of hominid dental remains from the Shungura and Usno Formations, Omo Valley, Ethiopia. Ph.D. diss., University of California, Berkeley.

Vrba, E. S. 1985. Ecological and adaptive changes associated with early hominid evolution. In *Ancestors: The hard evidence,* pp. 63–71 (ed. E. Delson). Alan R. Liss, New York.

———. 1988. Late Pliocene climatic events and hominid evolution. In *Evolutionary history of the "robust" australopithecines,* pp. 405–426 (ed. F. E. Grine). Aldine, New York.

Walter, R. C. 1994. Age of Lucy and the first family: Single-crystal $^{40}Ar/^{39}Ar$ dating of the Denen Dora and lower Kada Hadar Members of the Hadar Formation, Ethiopia. *Geology* 22:6–10.

Wesselman, H. B. 1984. *The Omo micromammals: Systematics and paleoecology of early man sites from Ethiopia.* Contributions to Vertebrate Evolution, no. 7. S. Karger, Basel.

———. 1985. Fossil micomammals as indicators of climatic change about 2.4 myr ago in the Omo Valley, Ethiopia. *South African Journal of Science* 81:260–261.

Williamson, P. G. 1985. Evidence for an Early Plio-Pleistocene rainforest expansion in East Africa. *Nature* 315:487–489.

Winkler, A. J. 1992. Small mammals from the Late Miocene Lothagam locality, West Turkana, Kenya. *Journal of Vertebrae Paleontology,* abstracts, 12:60A.

———. 1992. Systematics and biogeography of Middle Miocene rodents from the Muruyur Beds, Baringo District, Kenya. *Journal of Vertebrate Paleontology* 12:236–249.

Wood, B. A. 1992. Origin and evolution of the genus *Homo. Nature* 355:373–390.

Wood, B. A. Abbot, S. A., and Uytterschaut, H. T. 1988. Analysis of the dental morphology of Plio-Pleistocene hominids. Pt. 4, Mandibular post-canine root morphology. *Journal of Anatomy* 156:107–139.

Chapter 26

African Omnivores: Global Climatic Change and Plio-Pleistocene Hominids and Suids

Tim D. White

The zoological families of pigs and people, Suidae and Hominidae, have been inextricably linked in human evolutionary studies since Henry Fairfield Osborn misidentified a Nebraskan fossil suid tooth as North America's earliest hominoid in 1922. The association between fossil suids and hominids became less embarrassing for paleontologists after Basil Cooke pioneered the use of pigs in the biochronology of African early hominid sites. In the 1970s, pig paleontology got even more respect when its practitioners identified and contributed to the resolution of one of the most important problems concerning the evolution of our own genus—the age of the KBS tuff and earliest *Homo* at Koobi Fora, Kenya. Since then, suids have continued to serve with distinction in both biochronological and ecological studies of early hominids.

In this chapter I examine the Pliocene fossil records of suids and hominids and provide raw data to test the hypothesis, best articulated by Vrba (p. 385, this vol.), that "nearly all speciation and extinction events require initiation by climatic change." According to this hypothesis, speciations and extinctions are expected to covary. They are predicted to be concentrated in sets of pulses across diverse groups of organisms. These pulses are predicted to be in synchrony with climatic changes.

I first consider a variety of problems that are encountered when linking speciation and extinction among terrestrial mammals with records of global climatic change. Species origins and extinctions, the raw data of the analysis, are then delineated for African suids and hominids. Finally, hypotheses linking key events in the human career to changes in global climate are tested with the available data.

ISBN 0–300–06348–2.

This review of hominid and suid fossil records draws on field and laboratory results gathered by many workers across Africa. The synthesis includes reference to unpublished results of two ongoing Ethiopian investigations, the Middle Awash project and the Paleoanthropological Inventory of Ethiopia project. I thus acknowledge in advance my use of the resources and results of many field and laboratory geologists, geochronologists, archaeologists, and paleontologists. This assessment must be seen as an interim report, because data now being collected in Ethiopia, Kenya, and Tanzania hold great promise of refining the results outlined here.

Background

As witnessed by the title of this volume, fossils representing hominid species are afforded an inordinate amount of scrutiny and speculation owing to their close relations with modern authors. However, the fossil record of hominids is a limited one. The relatively few hominid specimens that exist are scattered widely in time and space. A variety of taphonomic, behavioral, phylogenetic, and situational constraints make it difficult for legions of hominid paleontologists to reconstruct an argument-free phylogeny.

African suid fossils are far more abundant but far less studied than those of hominids. Hundreds of suid fossils are collected for every scrap of hominid that is found in the fossil fields of eastern Africa. Yet no suid skulls grace the covers of *Nature* or garner headlines like "new pig skull completely overturns all previous theories of pig evolution." Few investigators study suids, even though this family is far more diverse than even the most diverse estimates of hominid lineages for the period between 6 and 1 million years (myr) ago. In eastern Africa, the suid record is particularly good, with fossils usually recovered from chronometrically controlled contexts. Several suid lineages have proven very useful for biochronological work. Paleontological work on the African suids has been dominated by phylogenetic and taxonomic considerations, leaving vast realms of function, development, adaptation, and locomotion uninvestigated. The paleoanthropological community seems not to care about suids very much until the need to estimate the age of some hominid site arises. For this reason alone, it is advantageous that suids are ubiquitous in African Pliocene sediments. However, the diverse and abundant African pigs provide a larger data set than do the available hominids, better enabling investigators to test hypotheses about whether mammals in Africa speciated and vanished in response to global climatic change.

Problems in Linking Mammalian Evolution with Global Climatic Change

How does one test the hypothesis that speciation and extinction of mammals are driven by global climatic change? In a perfect world, this question would be a simple matter of comparing the data set for global climatic change with the data set for species origins and extinctions across the Mammalia. In the real world, data sets are biased and incomplete. As Vrba notes in this volume, the first and last appearance dates (FADs and LADs) of mammalian taxa are the primary raw data that comprise any test of link between climate and evolution.

Ideal tests of the hypothesis on a global scale would employ complete faunal lists for precisely dated localities. Artifacts of preservation, deposition, ecology, and interpretation can, however, make even the most statistically robust tests evolutionarily meaningless, ambiguous, or even misleading. Many factors can play roles in obscuring *real* linkages between climate and evolution. Furthermore, many *apparent* linkages may be mere artifacts of the records or of their interpretations. In other words, many factors affect the raw data of a species' first and last appearances in the fossil record. It is useful to consider the most important of these factors with examples from the African record of suid evolution. Appendix 26.1 contains notes on fossil suid systematics and chronology.

Problem 1: The Chronological Imperfection of the Terrestrial Fossil Record (Time Gets Lost: There Are Gaps in the Record). Darwin recognized that the imperfection of the geological record biases our view of the history of life. In the century and a half since the publication of his book about the origin of species we have come to appreciate that the fossil record of mammalian history rarely, if ever, matches the continuity of the various records of global climatic change that have accumulated in relatively undisturbed oceanic depositories. Marine microorganisms rain down to the sea floor to be entombed in gradually accumulating sediments in deep-water settings, whereas the terrestrial counterparts for vertebrate fossilization provide for episodic, discontinuous deposition and fossilization. Stratigraphic proximity in such terrestrial depositories is only sometimes a good indicator of chronological relationships. The consequent terrestrial record of vertebrate evolution is broken by innumerable gaps of various sizes. It is as if the tape recorders of global change in the marine record were left in the "on" position, while the recorders on land (and even inland lakes; Tiercelin et al., 1992) were "off" most of the time, recording the passing of life only rarely and episodically, when sedimentation was switched "on" by local tectonic and climatic conditions.

Well-documented gaps in the chronological and physical aspects of the mammalian fossil record have profound effects on the data bases for evolutionary studies. The accuracy of first and last appearance data depends on fairly continuous sampling through time. This condition is usually not met on land, creating difficulty in obtaining reliable raw paleontological data necessary to test different models of mode and tempo in vertebrate evolution. Testing various hypotheses about the correlation of global climatic change and the origins and extinctions of vertebrate species suffers from the same constraints.

An obvious example of this problem of gaps in the African fossil artiodactyl record concerns the simultaneous appearance and disappearance of the species *Cainochoerus africanus.* When described by Hendey in 1976, this suid was the sole member of the peccary family Tayassuidae in Africa. It is only distantly related to the other African suids. It is known from a single occurrence on the southern tip of the continent, in the Pelletal Phosphorite unit (4–4.5 myr) of the Langebaanweg site (Cooke and Hendey, 1992). This find constitutes both its first and last appearance to date. The record is effectively mute on its dispersal to this remote outpost, its existence elsewhere on the continent, and its eventual extinction. There are no modern peccaries native to Africa.

Problem 2. The Ecological Imperfection of the Terrestrial Fossil Record (Habitats Go Unsampled: Many Animals Die in the "Wrong Place"). The accuracy of first and last appearance dates depends upon the record being sampled fairly continuously across ecological space. Because terrestrial mammals are preserved differentially across ecological space, there is the potential for dramatic bias to arise in our samples of evolution, and, as a result, species' first and last appearance dates may be misleading. Terrestrial mammalian species may often be generated in marginal habitat situations where peripheral isolates form incipient species. The last members of species approaching extinction may often disappear from ecological refugia. The geographically biased terrestrial fossil record (White, 1988) does a poor job of sampling such potential habitats.

The suid fossil record includes several taxa that are clearly ecologically constrained. For example, the Pliocene species *Nyanzachoerus kanamensis* is ubiquitous, and representatives are usually abundant in fossil fields both older and younger than the Laetolil Beds (ca. 3.5 myr). However, not a single specimen is known from this large Tanzanian fossil assemblage. In this sense, *Nyanzachoerus kanamensis* is like other water-dependent forms, such as crocodiles and hippopotami—also absent without a trace from the dry upland habitat recorded in the Laetolil Beds collection.

An example from the African suid record of how FADs and LADs might be ecologically biased involves the extant African bushpig, *Potamochoerus porcus.* This suid is obviously a daughter species of *Kolpochoerus afarensis,* a geographically and ecologically widespread Pliocene form. The last known specimen of the mother species is from Mid-Pliocene sediments. The first appearance of the daughter species is Late Pleistocene, where it is known from archaeological faunal assemblages. Both mother and daughter species are highly conservative, there being little morphological change over 3 myr. This presumed lineage, with its two recognized chronospecies, appears to have "disappeared from the scene" during the Pleistocene by inhabiting forested habitats, where it is still encountered today. We do not have the ability to monitor it through time in these habitats, because the vertebrate fossil record of Africa is largely confined to more open ecological settings. Without the neontological and archaeological data, we would have pegged the last appearance date for this species at 2.5 myr.

Problem 3. The Anatomical Imperfection of the Terrestrial Fossil Record (Parts Get Lost: Not All Fossils Are Complete). The accuracy of FADs and LADs depends upon our ability to recognize biological species units. Recognizing species is difficult enough in the modern world. In vertebrate paleontology, the biases of the record leave less to analyze and more to argue about. The "species" of paleontologists are morphospecies or paleospecies—sets of fossils that are grouped on the basis of shared anatomical (usually dental and cranial) characters. The limitations of designating species among fossil remains are well known but worth reiterating, because the basic data of this analysis concern these very paleontological "species."

All systematics among the African fossil suids have been accomplished exclusively with cranial and dental remains. Even this statement exaggerates the nature of the enterprise, because assessment of the upper and lower third molars has often been of paramount importance in decisions about classification. How might our ideas about the units we have called species change if there were as many suid crania as isolated teeth? Monitoring the third molars is not necessarily the best gauge of African suid speciation, but for now it is all we have. Dental analysis may work better for biochronological assessment than for recognizing real species in the record.

Among the African suids, the transitions of several species are recognized as gradual, given the progressive increase of third-molar length coupled with decreased premolar row length. Thus, "speciation" has been interpreted as anagenetic rather than cladogenetic and therefore of little relevance to questions of global climatic change (see problem 4 below). If complete suid crania were available in appropriate sample sizes, these admittedly gradualistic concepts might require alteration. For example, if a variety of cranial characters proved discordant and rectangular in contrast to the third molar progression seen in the *Kolpochoerus limnetes* to *Kolpochoerus phacochoeroides* transition, a "real" rather than artificial speciation would be indicated. Vrba (pers. comm.) has found this mosaic evolution to be the case in some bovid clades.

Problem 4. The Nomenclatorial Imperfection of the Terrestrial Fossil Record (Names Are Labels, Not Species: Not All Speciation Was Cladogenetic). Names are static entities and evolution is a dynamic process. If virtually all speciation is cladogenetic and the pattern of evolution is rectangular, as many have argued (Gould and Eldredge, 1993), the issue of naming is not a great concern. If, on the other hand, anagenetic speciation is commonplace, the species problem will play a role in the validity of first and last appearance dates of species unless a strictly cladistic concept of species is employed.

The utility of species' FADs and LADs in assessments of climatic change requires that these dates describe real evolutionary phenomena, not arbitrary taxonomic boundaries. An example of this problem comes within the genus *Kolpochoerus.* Granting the caveat issues in point 3, this genus is characterized by two chronospecies, the ancestor *Kolpochoerus limnetes* and the descendant *Kolpochoerus phacochoeroides.* The relationship appears to be anagenetic. Cooke and Maglio (1971) diagnosed these species on the basis of their third molar morphology, suggesting that once the third molar's trigonid/talonid (the rear portion of the tooth crown) became longer than the trigon/trigonid (the front portion), the species boundary had been crossed. This analysis worked fairly well because the period around 1.5–1.7 myr was not well sampled by the Olduvai and Omo collections. In the mid-1970s, however, the Koobi Fora fossil field yielded abundant specimens from this previously poorly represented time period. To apply Cooke's species distinction, half of the sub-Okote collection would end up in *K. limnetes,* the other, contemporaneous half in *K. phacochoeroides.* The species distinction is arbitrary—third-molar evolution (our proxy for organismal evolution lacking other parts of the body in sufficient quantity) suggests anagenetic "speciation." However, the arbitrary positioning of the species "boundary" is reflective of neither *rapid phyletic* evolution nor *cladogenetic* speciation (phenomena predicted to correlate with climatic change).

Problem 5. Imperfections in Relative Abundance in the Fossil Record (Some Species Are Rare: Absence Means Different Things). Some species in the same habitat are characterized by high population density and high susceptibility to death in environments suitable for fossilization. Other species, like hominids, may be rare and smart. For a given fossil record, the first and last appearance dates for rarely represented species are much more likely to be artifacts of the record than first and last appearances of abundantly represented species.

A good example of how relative abundance can affect the value and reliability of first and last appearance data among African fossil suids comes with the species *Metridiochoerus hopwoodi* and *M. modestus.* These species are extremely rare in the Late Pliocene, represented only by isolated teeth. In contrast, the first appearance of their sister species *Metridiochoerus compactus,* can be assumed to be more reliable owing to the relative abundance of this form in similar contexts.

Problem 6. The Imperfections of Interpreting the Fossil Record (Not All Species Are "Real": Splitters Versus Lumpers in Inferring Species Boundaries). The delineation of prehistoric species is accomplished by modern people interpreting static evidence from the fossil record. The predispositions of the analyst may bias the results of this exercise. The "lumper" will be in danger of submerging "real" species, the units of analysis, into artificial, composite species whose first and last appearance dates are no longer accurate. The "splitter" will be in danger of creating "false" species and thereby doubling the number of potentially inaccurate first and last appearance dates.

Examples from the suid fossil record are legion, and rather than cite a single example, it is perhaps more instructive and appropriate to note that Harris and White (1979) recognized only sixteen of the seventy-seven species of African Plio-Pleistocene suids named by paleontologists since 1900.

Testing Evolutionary Models

The potential effects of each of the problems outlined above are cause for considerable concern in using the first and last appearance dates for fossil species to test evolutionary models. These problems are not isolated; rather, they have the potential to act synergistically, thereby further reducing the reliability and accuracy of the appearance and disappearance data. For us to have any confidence in our tests, we must seriously consider these challenges to the integrity of the data. It is obvious that a simple compilation of information gleaned from species lists generated for different collections by different investigators will not be adequate to the task. Species-by-species studies by multiple, independent specialists intimately familiar with the fossil record in all aspects are required. These studies must pay particular attention to the field abundance, completeness, continuity, and chronological precision of the record. Such focus is critical to the success and believability of any hypothesis testing based on the analysis of first and last appearances. Furthermore, it is necessary to set forth explicitly the rationale behind each decision about a first or last appearance datum. The procedure outlined below addresses many of the concerns outlined above. This method may allow other workers to judge the value of these results and to incorporate selected portions of these results into larger syntheses required for more comprehensive hypothesis testing. This study concentrates on the species level. Identifying the origin of generic or other higher-level groups is a far more subjective exercise, the results of which will ultimately be based on species-level data in any case.

Ranking First and Last Appearances

Accurate determination of first and last appearance data for vertebrate species in the fossil record is difficult for all the reasons outlined above. The accuracy, and hence reliability, of FAD and LAD information is expected to span a range whose extent is determined by the various imprecisions inherent in the materials from the fossil record and in the methods of analysis. Some FADs will be extremely accurate, with the fossil record managing to have captured the newly formed species shortly after its cladogenesis. Other FADs will be accurate only to the degree of precision allowed by weak geochronology or a broken fossil record. Still other FADs are clearly artifacts of the arbitrary division of a continuous record. Each assigned FAD or LAD represents a variably subjective judgment on the part of the analyst. Each FAD and LAD is always something of an artifact of the record, because we cannot hope to find the truly first or last individual organism representing any species. How, then, do we rank the reliability of any FAD or LAD designation?

In making FAD and LAD assignments for each fossil species, we must assess the continuity and ecological balance of the record, the abundance of fossils, the precision of geochronological control, and the validity of our nomenclatorial interpretations. This is no small undertaking among workers prone to exaggerating the strength of their data sets.

For most vertebrate groups in most time spans, rigorous pruning of data sets to eliminate false FADs and LADs will result in insufficient data. Faced with this dilemma, it might prove useful to attempt some ranking of quality (valance, value, or reliability) of FAD and LAD determinations. One set of appearance and disappearance data that should be viewed askance comprises species that are truly demarcated as time-successive, anagenetically derived chronospecies. These "speciations" or "disappearances" (actually artifacts of our nomenclature) I have disregarded in hypothesis testing. For other "origins" and "extinctions" of suid and hominid species I have asked: Is the record sufficiently continuous, ecologically balanced, sufficiently sampled, and chronologically controlled? In other words, given the record available, is there reason to expect that the FAD or LAD based on the known fossil record is reflective of reality or merely an artifact of the record?

This question is easily asked but not easily answered. In reality, there is a quality continuum in FAD and LAD numbers that reflects the interplay of the problems described above. In the analysis below, I asked whether there was good reason to expect a significantly earlier FAD or later LAD. In answering this question, I looked to the closest expected field occurrence where the species was not found (sometimes this is possible within the same local stratigraphy; sometimes the closest occurrence is available only thousands of miles away). Many first and last appearance data fall away as artifacts of nomenclature or artifacts of the record during such analysis, but others are judged to be more reliable.

Materials and Methods

The suid species units used are set forth in White and Harris (1977) and Harris and White (1979). These treatments were significantly different from previous phylogenies (e.g., Cooke and Maglio, 1971). Phylogenies subsequently published by Cooke (1978, 1985) are closer to ours. All cases where my current systematics differ from those of Harris and White (1979) are noted and explained in appendix 26.1. The hominid species used are conventional.

The first and last appearance dates of each hominid and suid species were evaluated (appendixes 26.1 and 26.2) and tabulated (tables 26.1 and 26.2). Appendixes 26.1 and 26.2 provide discussions of the rationale for each FAD and LAD decision, with appropriate references. During the exercise, first and last appearance data that were judged to be "artifacts" of nomenclature were scored as 0, whereas FADs and LADs that were judged to be "artifacts" of an obviously broken fossil record were identified and scored as 1. These "artifactual" records

Table 26.1. First and Last Appearance Data for Suid Species

Taxon	First Appearance (FAD) Location	Age (myr)	Quality	Last Appearance (LAD) Location	Age (myr)	Quality
Nyanzachoerus						
Nyanzachoerus tulotos (syrticus)	Lukeino	5.9	1	Sahabi	5.5	1
Nyanzachoerus waylandi	Nyaburogo Fm.	5–6	1	Nyaburogo Fm.	5–6	1
Nyanzachoerus kanamensis	Lothagam 1C	5–6	2	Omo Shungura	2.7	3
Nyanzachoerus jaegeri	Lothagam 1C	5–6	2	Bodo, M. Awash	3.75	0
Notochoerus						
Notochoerus euilus	Bodo, M. Awash	3.75	0	Omo Shungura Upp. G	2.0	3
Notochoerus capensis	Kubi Algi	2.7	0	Kubi Algi	2.7	0
Notochoerus scotti	Omo Shungura C	2.8	0	Koobi Fora KBS+	1.8	3
Kolpochoerus						
Kolpochoerus afarensis	Baringo Chemeron	4–5	1	Matabaietu, M. Aw.	2.5	3
Kolpochoerus limnetes	Omo Shungura B11	2.9	3	Koobi Fora OK–	1.7	0
Kolpochoerus phacochoeroides	Koobi Fora	1.7	0	Olduvai Bed IV	0.78	2
Kolpochoerus majus	Konso-Gardula	1.44	1	Bodo, M. Awash	0.6	1
Potamochoerus						
Potamochoerus porcus	Archaeological/MSA	U. Ple.	1	modern	—	—
Hylochoerus						
Hylochoerus meinertzhageni	Gamble's cave	0.01	1	modern	—	—
Metridiochoerus						
Metridiochoerus andrewsi	Omo Shungura B10	2.95	2	Koobi Fora OK+	1.7	3
Metridiochoerus compactus	Koobi Fora OK–	1.6	3	Olduvai Bed IV	0.78	2
Metridiochoerus hopwoodi	Koobi Fora KBS–	1.89	3	Olduvai Bed IV	0.78	2
Metridiochoerus modestus	Koobi Fora KBS–	1.89	3	Olduvai Bed IV	0.78	1
Phacochoerus						
Phacochoerus antiquus	Olduvai Bed IV	0.78	1	unknown		0
Phacochoerus africanus	Bodo, M. Awash	0.06	0	modern	—	—

Note: Quality scores: 0 = arbitrary split of lineage; 1 = FAD or LAD probably artifact of the record; 2 = date *possibly* actual FAD OR LAD; 3 = date *probably* actual FAD or LAD. The rational for each FAD and LAD and its relative quality is presented in appendix 26.1.

Table 26.2. First and Last Appearance Data for Hominid Species

Taxon	First Appearance (FAD) Location	Age (myr)	Quality	Last Appearance (LAD) Location	Age (myr)	Quality
Australopithecus						
Australopithecus afarensis	Laetoli	3.5	1	Hadar	3.0	2
Australopithecus africanus	Makapansgat Grey B.	2.8	2	Sterkfontein	2.6	1
Australopithecus robustus	Swartkrans 1	1.7	1	Swartkrans 3	1.7	1
Australopithecus aethiopicus	Omo Shungura C6	2.7	3	Omo Shungura G3	2.3	0
Australopithecus boisei	Omo Shungura G3	2.3	0	Omo Shungura K4	1.45	2
Homo						
Homo habilis	Koobi Fora KBS–	1.89	1	Olduvai mid-Bed II	1.7	2
Homo erectus	Koobi Fora OK–	1.75	3	Variable	—	—
Homo neanderthalensis	Atapuerca	0.3	1	St. Cesaire	0.036	3
Homo sapiens	Qafzeh	0.1	2	modern	—	—

Note: Quality scores: 0 = arbitrary split of lineage; 1 = FAD or LAD probably artifact of the record; 2 = date *possibly* actual FAD or LAD; 3 = date *probably* actual FAD or LAD. The rationale for each FAD and LAD and its relative quality is presented in appendix 26.2. Where ongoing debate over the phylogenetics is in progress, the lowest quality score was entered in an effort to provide the most conservative, reliable assessment.

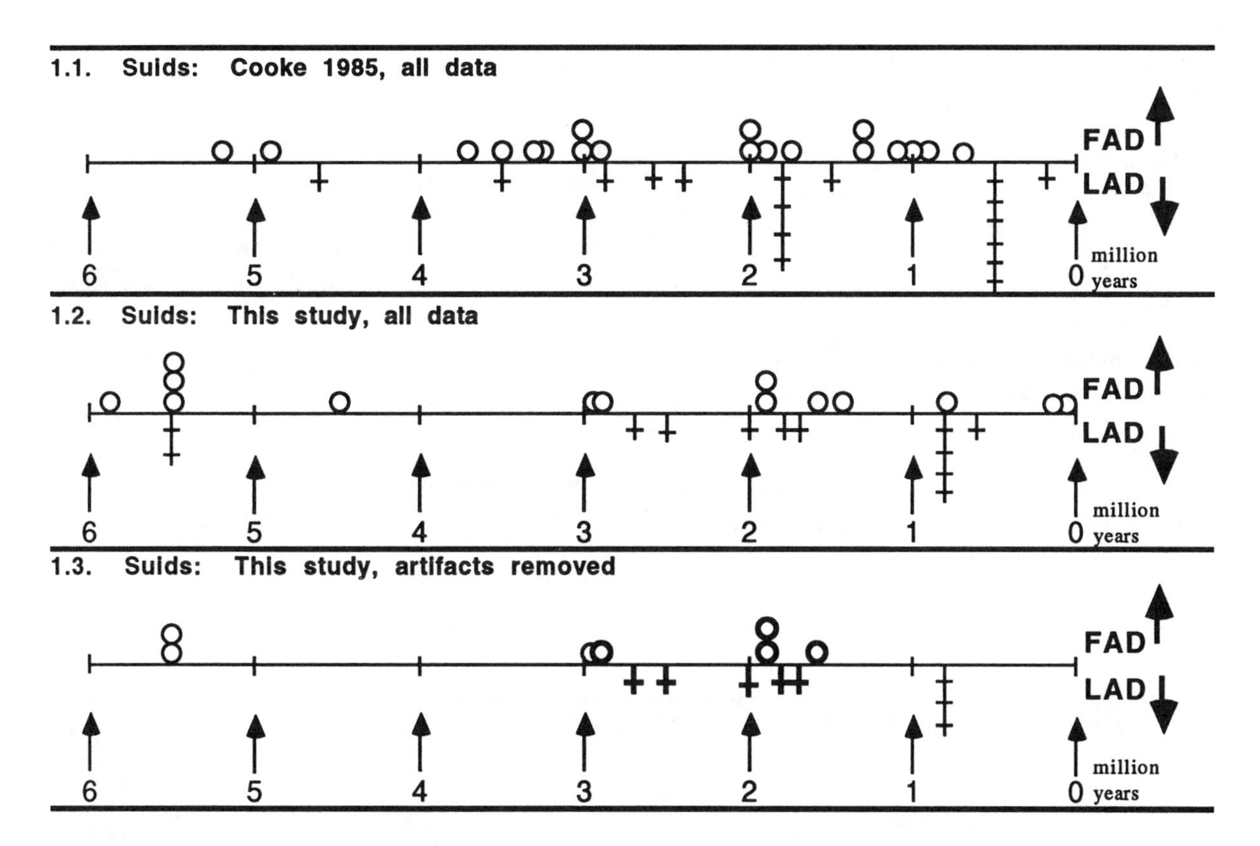

Fig. 26.1. First and last appearance data for fossil suids. See table 26.1 and appendix 26.1 for data. Circles indicate origination points (FADs); inverted crosses indicate extinction points (LADs). Bold symbols in histogram 1.3 indicate high-quality data (quality value = 3); nonbold symbols indicate data of slightly lower value (quality value = 2). Data points in histograms 1.1 and 1.2 include artifacts of sampling and nomenclature, whereas histogram 1.3 provides the most reliable data on biologically meaningful FADs and LADs.

should not figure in any rigorous testing of the correspondence between global climatic change and species origin and extinction. FADs and LADs that were judged to be possibly biologically significant were assigned quality scores of 2, whereas those judged to be probably biologically significant (probably real appearances or disappearances) earned scores of 3.

Suid Evolution: The Big Picture

Figure 26.1 illustrates the progressive "cleaning" of the suid FAD-LAD data set through elimination of nomenclatorial and stratigraphic artifacts. Histogram 1.1 plots first and last appearance data for species recognized by Cooke in his 1985 phylogeny. Cooke did not attempt to rid these data of arbitrary lineage divisions and/or artifacts of the fossil record, but they are included here for an alternative portrait of what another investigator of African suids might derive as an alternative to the results of the current study. Histograms 1.2 and 1.3 are plots of FADs and LADs based on the current study. Histogram 1.2 eliminates nomenclatorial artifacts of arbitrarily splitting continuous lineages (a quality score of 0), but includes all appearance and disappearance data, even those attributed to artifacts of the fossil record (a quality score of 1). Histogram 1.3 eliminates those nonbiological artifacts and represents the clearest view of real species origins and extinctions available, including only biologically possible or probable first and last appearances. Although the resultant data set is much reduced relative to the first two, the signals are much more meaningful.

Histogram 1.3, the plot of the most reliably timed speciation and extinction events, shows that the period before 3 myr was marked by little speciation and extinction relative to the period after 3 myr. This is also true of Cooke's 1985 data set (histogram 1.1). There are two suid species originations at ca. 2.8–3.0 myr and two extinctions 2.5 and 2.7 myr. Histogram 1.3 shows the most dramatic faunal turnover during the period between ca. 1.6 and 2.0 myr ago. Here, there are three suid species originations and three extinctions. The period after ca. 0.78 myr (Olduvai Bed IV) is also marked by three apparently simultaneous extinctions. These signals, perhaps reflecting turnover pulses, are both present in the less reliable data sets presented in histograms 1.1 and 1.2. Note that a pulse of species originations and extinctions at 2.5 myr is not evident in the suid record (see also Bishop, 1993). In fact, the fossil record for suid evolution seems to be relatively quiescent across this period, whether judged by an independent phylogenetic analysis (Cooke, 1985), pooled raw data (this study, histogram 1.2), or a highly cleaned data set (this study, histogram 1.3). This is true despite the fact that much suid evolution comprises taxa evolving parallel adaptations to more grass-dominated diets. These results stand in contrast to those derived for bovid taxa by Vrba (chap 27, this vol.).

Hominid Evolution: The Big Picture

As an experiment, the hominid fossil record was evaluated in the same manner as that of suids (fig. 26.2). Histogram 2.1 is based on an independent assessment of hominid first and last appearance data by Kimbel (this vol.). Histogram 2.2 shows origin and extinction data for hominid species as judged by the current study. Histogram 2.3 shows species origins and extinctions after eliminating probable anagenetic evolution and artifacts of the fossil record. The pattern of hominid species origin or extinction is not very different from the pattern elicited from analysis of the suid data. These data are not, however, interpreted as indicating that the bipedal and quadrupedal omnivores of Africa were entrained to the same determining environments, or that they precipitated one another's origins and extinctions.

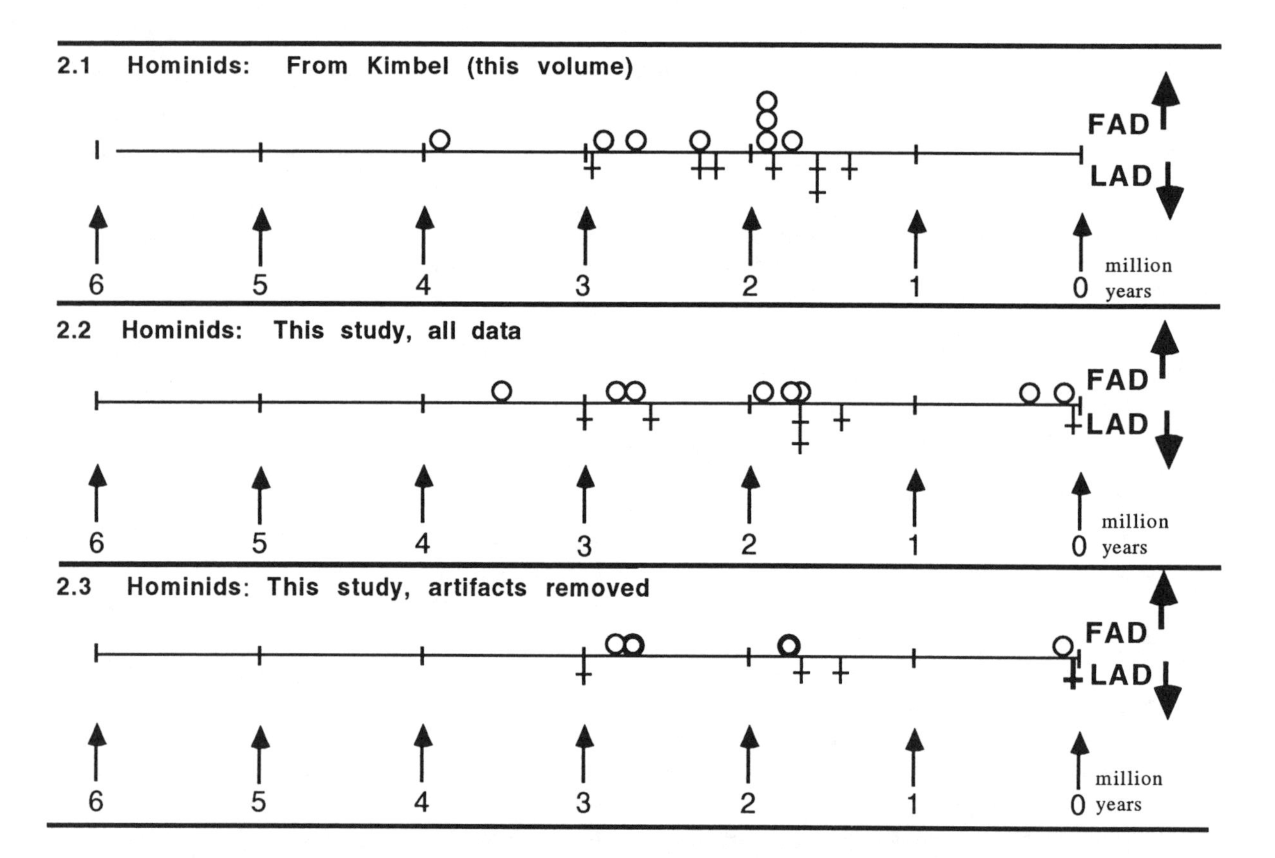

Fig. 26.2. First and last appearance data for fossil hominids. See table 26.2 and appendix 26.2 for data. Circles indicate origination points (FADS); inverted crosses indicate extinction points (LADS). Bold symbols (in histogram 2.3) indicate high-quality data (quality value = 3); nonbold symbols indicate data of slightly lower value (quality value = 2). Data points in histograms 2.1 and 2.2 include artifacts of sampling and nomenclature, whereas histogram 2.3 provides the most reliable data on biologically meaningful FADS and LADS.

Hypothesis Testing

The first step in hypothesis testing using FADs and LADs should be some form of "noise reduction"—eliminating bad and questionable FADs and LADs that result from the problems discussed above. Patterns of origins and extinctions then can be compared with patterns in various proxies of global climatic change in a search for correlations. It is necessary to consider cause only when correlations between valid, reliable species origin and/or extinction events and global climatic changes have been securely established. Causational linking of the climatic and evolutionary records is very difficult (Hill, 1987). Mere plausibility or reference to "authority" often constitutes the test of a causal-linkage hypothesis.

Before causality can be addressed, it is obvious that the correlational linkage of climatological proxies and paleontological first and last appearance dates must be secure. Linkage strength is a function of temporal calibration and density of events in the two records. When temporal calibration is poor, more correlations are permissible. In other words, the more uncertain the age of a speciation or an extinction event, the greater the potential for correlational linking of the event and some global climatic signal. The greater the number of climatic or evolutionary events available, the greater the chance of correlation. The correlation itself may be falsified if more precise dating of the evolutionary event shows noncontemporaneity. When simultaneity of global climatic and evolutionary events withstand critical analyses, hypotheses about causal linkage are generated and can best be tested by widening the temporal and taxonomic comparative nets.

The following causal-linkage hypotheses about hominids (and, by inference, suids) have been articulated since Brain's early work in this area (1981). They concern three of the major events in the human career:

Hypothesis 1. The phylogenetic and locomotor divergence of hominids from African apes has been hypothesized to be causally linked to global climatic change in the Late Miocene (Vrba, 1988).

Hypothesis 2. The technological and craniofacial divergence of early members of the genus *Homo* from the robust *Australopithecus* clade has been attributed to Middle Pliocene climatic change. Specifically, the period ca. 2.0–3.0 myr witnesses a turnover pulse in the evolution of many organisms on the planet that correlates with "global refrigeration" (Vrba, 1988, this vol.).

Hypothesis 3. Old World range expansion of *Homo erectus* and its Acheulean industry as well as the extinction of robust *Australopithecus* have been tentatively hypothesized as causally linked to global climatic changes at ca. 0.9 myr (Vrba, 1988).

Testing the correlational sides of these hypotheses calls for rigorous control of event timing. The results of several studies presented in this volume make it clear that the global climatic record is one characterized by a high frequency of variability. The good news is that it is easy to match something on land with some oceanic proxy of global climatic change. The bad news is that most such matches are unlikely to mean very much! Other studies in this volume consider how local physiographic effects may mimic global change in the terrestrial records (Brown, this vol.; Partridge et al., chap. 24, this vol.). Such cautions are well applied to the terrestrial settings that have yielded our bovids, pigs, and human ancestors. With these cautionary tales in mind, what evidence is there that hominid origins (hypothesis 1), diversification (hypothesis 2) and geographic expansion and extinction (hypothesis 3) are correlated with and caused by global climatic change?

Hypothesis 1: Hominid Origins

The hominid fossil record is not complete enough to test any hypothesis linking hominid origins to global climatic change (even if anthropologists could decide what they meant by the phrase origins of the family). Biochemically based diveregence estimates between living forms are poorly time-constrained. In short, we have no data of reliable precision to indicate when the hominids arose relative to global climatic change (app. 26.1; White, 1986; Hill and Ward, 1988; Kramer, 1986). Suids do show some species originations in the period between 5 and 6 myr ago, but the pre-6 myr record is so poor as to make interpretation of these data very ambiguous.

Hypothesis 2: Hominid Diversification

Both hominid and suid fossil records show a diversification of species and an increased extinction rate during the period between 2.0 and 3.0 myr, compared to the period between 3 and 5 myr (but with no clear pulse at ca.

2.5 myr). A significant amount of turnover among bovid and rodent species also occurs at this time (Vrba, chap. 27, this vol.; Wesselman, this vol.; but see Feibel et al., 1991). The diverse, specialized rodents and bovids are, of course, the best groups to examine for evidence of turnover pulses. The fossil bovid record was previously interpreted by Vrba to display evidence of much speciation and extinction at ca 2.5 myr. The challenge is to assess critically the patterning of these bovid and rodent data, first thought to correspond to a marked global climatic 2.5 myr event. Could they involve a combination of artifacts of the fossil or interpretive records, or are they strong and synchronous?

The strongest pulse of speciations and extinctions in both pigs and hominids, however, occurs between ca. 1.6 and 2.0 myr ago, during a period in which no dramatic global climatic oscillation signal is available. It seems likely that the end of this time window was marked by technological innovation involved with the production of the Acheulean industry (Asfaw et al., 1992; Manega, 1993).

Hypothesis 3: Geographic Expansion and Extinction

Recent work in the Main Ethiopian Rift by teams involved in the Paleoanthropological Inventory of Ethiopia project has refined the geochronology of this part of Africa. Discoveries in the Konso-Gardula area show that *Homo erectus* occupied this area by ca. 1.4+ myr, long before the pronounced changes in global ice budget between 0.9 and 0.7 myr (Asfaw et al., 1992). The evidence from Olduvai Gorge EF-HR, another early Acheulean locality, is consistent with this appraisal. Contemporary Turkana Basin sediments just to the west yield the youngest robust *Australopithecus* remains. Again, the density of the record is not adequate to test the hypothesis of linkage with global climatic change. Documenting the spread of *Homo erectus* populations out of Africa is made difficult by the paucity of well-dated sites, but evidence from Java, Israel, and Georgia suggest that 0.9 myr may be far too young for the original "out-of-Africa" expansion of the genus *Homo* (Curtis, 1981; Eisenmann et al., 1983; Geraads et al., 1986; Dzaparidze et al., 1989).

Summary

Understanding how climate affected mammalian evolution is a formidable undertaking, one that calls for bold hypothesis making and rigorous, critical hypothesis testing. The best hypotheses are those that can be tested by the most comprehensive data sets. The suid and hominid data sets presented here are incomplete and biased. When one accounts for these defects, the data available for these mammalian families stand ready to be employed in hypothesis testing. Such testing, as attempted above, focuses attention on a number of unresolved questions that remain unresolved because the basic data have not yet been recovered. Combining data presented here with data generated from similarly controlled analyses of other groups should, however, provide a more comprehensive understanding of evolution while more fieldwork is conducted.

Bob Brain and Elisabeth Vrba have challenged us with their ideas about the effect of global climatic change on mammalian evolution in Africa. As field research continues in an attempt to gather bigger and better data sets that will test these ideas and further illuminate the past, we owe thanks to those bold individuals who were willing to step forward with ideas that challenged us to be more precise, dared us to look at our data sets in novel ways, and prompted us to undertake ever more difficult basic data collection.

Acknowledgments

I am grateful to the conference organizers for inviting me to participate at the Lamont and Airlie meetings. In particular, I thank Elisabeth Vrba for her assistance, encouragement, and collegiality. I am grateful to Elisabeth and to John Harris and Laura Bishop for reviewing the manuscript and recommending needed changes. Field studies in Ethiopia have been made possible through the financial support of the National Science Foundation and the National Geographic Society. Conference travel was funded by the University of California. Thanks to all the field collectors who endured heat, cold, rain, wind, dust, mud, ticks, scorpions, mosquitos, snakes, disease, and other forms of assault to make available all the fossils considered here. Thanks also to the laboratory preparators responsible for cleaning, restoring, and curating the fossils and to the individuals who had the really hard job of granting me permission to study them.

References

Asfaw, B., Beyene, Y., Suwa, G., Walter, R. C., White, T. D., WoldeGabriel, G., and Yemane, T. 1992. The earliest Acheulean from Konso-Gardula. *Nature* 360:732–735.

Bishop, L. 1993. Hominids of the East African Rift Valley in a macroevolutionary context. Abst. in *American Journal of Physical Anthropology Suppl.* 16:57.

Boaz, N. T., et al. (eds.) 1987. Neogene Paleontology and Geology of Sahabi. Alan R. Liss, New York.

Brain, C. K. 1981. The evolution of man in Africa: Was it a consequence of Cainozoic cooling? *Annals of the Geological Society of South Africa* 84:1–19.

Clark, J. D., de Heinzelin, J., Schick, K., Hart, W. K., White, T. D., WoldeGabriel, G., Walter, R. C., Suwa, G., Asfaw, B., Vrba, E., and Haile Selassie, Y. 1994. African *Homo erectus:* Old radiometric ages and young Oldowan assemblages. *Science* 264:1907–1910.

Clarke, R. J. 1988. A new *Australopithecus* cranium from Sterkfontein and its bearing on the ancestry of *Paranthropus.* In *Evolutionary history of the "robust" australopithecines,* pp. 285–292 (ed. F. E. Grine). Aldine, New York.

Cooke, H. B. S. 1978a. Suid evolution and correlation of African hominid localities: An alternative taxonomy. *Science* 201:460–463.

———. 1978b. Plio-Pleistocene Suidae from Hadar, Ethiopia. *Kirtlandia* (Cleveland Museum of Natural History) 29:1–63.

———. 1982. *Phacochoerus modestus* from Bed I, Oluvai Gorge, Tanzania. *Zeitschrift für Geologie Wissenschaften* (Berlin) 10: 899–908.

———. 1985. Plio-Pleistocene Suidae in relation to African hominid deposits. In *L'environnement des hominidés au Plio-Pléistocène,* pp. 101–117 (ed. M. Beden et al.). Masson, Paris.

———. 1987. Fossil suidae from Sahabi, Libya. In *Neogene paleontology and geology of Sahabi,* pp. 255–266 (ed. N. Boaz). Alan R. Liss, New York.

Cooke, H. B. S., and Hendey, Q. B. 1992. *Nyanzachoerus* (Mammalia: Suidae: Tetraconodontinae) from Langebaanweg, South Africa. *Durban Museum Novitates* 17:1–20.

Cooke, H. B. S., and Maglio, V. J. 1971. Plio-Pleistocene stratigraphy in East Africa in relation to proboscidean and suid evolution. In *Calibration of hominoid evolution,* pp. 303–329 (ed. W. W. Bishop and J. D. Clark). Scottish Academic Press, Edinburgh.

Curtis, G. H. 1981. Man's immediate forerunners: Establishing a relevant time scale in anthropological and archaeological research. *Philosophical Transactions of the Royal Society of London,* ser. B., 292:7–20.

Delson, E. 1988. Chronology of South African australopith site units. In *Evolutionary history of the "robust" autralopithecines,* pp. 317–324 (ed. F. E. Grine). Aldine, New York.

Dzaparidze, V., Bosinski, G., et al. 1989. Der Altpaläolithische Fundplatz Dmanisi in Georgien (Kaukasus). *Jahrbuch des Römisch-Germanischen Zentralmuseums Mainz* 36:67–116.

Eisenmann, V., Ballesio, R., Beden, M., Faure, M., Geraads, D., Guérin, C., and Heintz, E. 1983. Nouvelle interprétation biochronologique des grandes mammifères d'Ubeidiya, Israël. *Geobios* 16:629–633.

Feibel, C. S., Brown, F. H., and McDougall, I. 1989. Stratigraphic context of fossil hominids from the Omo Group deposits: Northern Turkana Basin, Kenya and Ethiopia. *American Journal of Physical Anthropology* 78:595–622.

Feibel, C. S., Harris, J. M., and Brown, F. H. 1991. Palaeoenvironmental context for the Late Neogene of the Turkana Basin. In *Koobi Fora Research Project.* Vol. 3, *The fossil ungulates: Geology, fossil artiodactyls, and palaeoenvironments,* pp. 321–370 (ed. J. M. Harris). Clarendon Press, Oxford.

Geraads, D., Guérin, C., and Faure, M. 1986. Les suidés du Pléistocène ancien d'Oubeidiyeh (Israel). *Mémoires et Travaux du Centre de Recherche Française de Jérusalem* 5:93–105.

Gould, S. J., and Eldredge, N. 1993. Punctuated equilibrium comes of age. *Nature* 366:223–227.

Harris, J. M. (ed.) 1983. *Koobi Fora Resarch Project.* Vol. 2, *The fossil ungulates: Proboscidea, Perissodactyla, and Suidae.* Clarendon Press, New York.

Harris, J. M., and White, T. D. 1979. Evolution of the Plio-Pleistocene African Suidae. *Transactions of the American Philosophical Society* 69:1–128.

Hay, R. L. 1976. *Geology of the Olduvai Gorge.* University of California Press, Berkeley, California.

Hendey, Q. B. 1976. Fossil peccary from the Pliocene of South Africa. *Science* 192:787–789.

Hendey, Q. B., and Cooke, H. B. S. 1985. *Kolpochoerus paiceae* (Mammalia, Suidae) from Skurwerug, near Saldanha, South Africa, and its paleoenvironmental implications. *Annals of the South African Museum* 97:9–56.

Hill, A. 1987. Causes of perceived faunal change in the later Neogene of East Africa. *Journal of Human Evolution* 15:583–596.

Hill, A., and Ward, S. 1988. Origin of the Hominidae: The record of African large hominid evolution between 14 my and 4 my. *Yearbook of Physical Anthropology* 31:49–83.

Hill, A., Ward, S., and Brown, B. 1992. Anatomy and age of the Lothagam mandible. *Journal of Human Evolution* 22:439–451.

Hill, A., Ward, S., Deino, A., and Curtis, G. 1992. Earliest *Homo. Nature* 355:719–722.

Hooker, P. J., and Miller, J. A. 1979. K-Ar dating of the Pleistocene fossil hominid site at Chesowanja, north Kenya. *Nature* 282:710–712.

Kimbel, W. H., and White, T. D. 1988. Variation, sexual dimorphism and the taxonomy of *Austalopithecus.* In *Evolutionary history of the "robust" australopithecines,* pp. 175–192 (ed. F. E. Grine). Aldine, New York.

Kimbel, W. H., White, T. D., and Johanson, D. C. 1988. Implications of KNM-WT 17000 for the evolution of "robust" *Australopithecus.* In *Evolutionary history of the "robust" australopithecines,* pp. 259–268 (ed. F. E. Grine). Aldine, New York.

Kramer, A. 1986. Hominid-pongid distinctiveness in the Miocene-Pliocene fossil record: The Lothagam mandible. *American Journal of Physical Anthropology* 70:457–473.

Manega, P. C. 1993. Geochronology, geochemistry and isotopic study of the Plio-Pleistocene hominid sites and the Ngorongoro volcanic highland in northern Tanzania. Ph.D. diss., University of Colorado.

Pickford, M. 1989. New specimens of *Nyanzachoerus waylandi* (Mammalia, Suidae, Tetraconodontinae) from the type area, Nyaburogo (upper Miocene), Lake Albert Rift, Uganda. *Geobios* 22:641–651.

Pickford, M., Johanson, D., Lovejoy, O., White, T., and Aronson, J. 1983. A hominoid humeral fragment from the Pliocene of Kenya. *American Journal of Physical Anthropology* 60:337–346.

Renne, P., Walter, R. C., Verosub, K., Sweitzer, M., and Aronson, J. 1993. New data from Hadar (Ethiopia) support orbitally tuned time scale to 3.3 Ma. *Geophysical Research Letters* 20:1067–1070.

Stringer, C. B. 1993. Secrets of the pit of the bones. *Nature* 362: 501–502.

Stringer, C. B., and Gamble, C. 1993. *In search of the Neanderthals.* Thames and Hudson, London.

Suwa, G. 1988. Evolution of the "robust" australopithecines in the Omo succession: Evidence from mandibular premolar morphology. In *Evolutionary history of the "robust" australopithecines,* pp. 199–222 (ed. F. E. Grine). Aldine, New York.

———. 1990. A comparative analysis of hominid dental remains from the Shungura and Usno Formations, Omo Valley, Ethiopia. Ph.D. diss., University of California, Berkeley.

Tiercelin, J. J., Solreghan, M., Cohen, A. S., Lezzar, K. E., et al. 1992. Sedimentation in large rift lakes: Example from the Middle Pleistocene–Modern deposits of the Tanganyika Trough, East African Rift System. *Bulletin des Centres de Recherches Exploration-Production Elf-Aquitaine* 16:83–111.

Tobias, P. V. 1991. *Olduvai Gorge.* Vol. 4, *The skulls, endocasts, and teeth of* Homo habilis. Cambridge University Press, New York.

Vrba, E. S. 1988. Late Pliocene climatic events and hominid evolution. In *Evolutionary history of the "robust" australopithecines,* pp. 405–426 (ed. F. E. Grine). Aldine, New York.

Walker, A. C., and Leakey, R. E. 1988. The evolution of *Australopithecus boisei.* In *Evolutionary history of the "robust" australopithecines,* pp. 247–258 (ed. F. E. Grine). Aldine, New York.

Ward, S. C., and Hill, A. 1987. Pliocene hominid partial mandible from Tabarin, Baringo, Kenya. *American Journal of Physical Anthropology* 72:21–37.

White, T. D. 1986. *Australopithecus afarensis* and the Lothagam mandible. *Anthropos* (Brno) 23:79–90.

———. 1988. The comparative biology of "robust" *Australopithecus:* Clues from context. In *Evolutionary history of the "robust" australopithecines,* pp. 449–484 (ed. F. E. Grine). Aldine, New York.

White, T. D., and Harris, J. M. 1977. Suid evolution and correlation of African hominid localities. *Science* 198:13–21.

White, T. D., Suwa, G., Hart, W. K., Walter, R. C., WoldeGabriel, G., de Heinzelin, J., Clark, J. D., Asfaw, B., and Vrba, E. 1993. New discoveries of *Australopithecus* at Maka in Ethiopia. *Nature* 366:261–265.

Wood, B. A. 1988. Are "robust" australopithecines a monophyletic group? In *Evolutionary history of the "robust" australopithecines,* pp. 269–284 (ed. F. E. Grine). Aldine, New York.

———. 1991. *Koobi Fora Research Project.* Vol. 4, *Hominid cranial remains.* Oxford University Press, New York.

Appendix 26.1. Notes on Fossil Suid Systematics

1. *Nyanzachoerus tulotos (syrticus).* Cooke synonomized the eastern African *Ny. tulotos* with the North African *Ny. syrticus* in 1987. Cooke also recognized a smaller form from Sahabi, *Ny. devauxi,* but this taxon is not yet recognized here pending results of research by J. Harris and M. Leakey currently underway on new Lothagam *Nyanzachoerus* specimens (pers. comm.). This species complex forms the substrate for all later nyanzachoere / notochoere evolution, giving rise to *Ny. kanamensis* and *Ny. waylandi.*

The earliest record of *Ny. syrticus* is Lukeino, radiometrically dated between ca. 5.6 and 6.2 myr (Hill, this vol.). The latest record of *Ny. syrticus* is from Sahabi, biochronologically and geologically interpreted as immediately post-Mesinian, ca. 5.5 myr (Boaz et al., 1987). Both the FAD and LAD are probably artifacts of an incomplete fossil record, with very few sites having produced vertebrates in the 2 myr before Lukeino or the 1 myr after Lothagam.

2. *Nyanzachoerus waylandi.* Pickford (1989) established this most recently recognized species of the genus based on the single available sample from the Nyaburogo Formation, Uganda. The taxon is thus far restricted to those deposits, thought, on limited biostratigraphic grounds, to date to the upper Miocene, ca. 5–6 myr (Pickford, 1989). It is more derived than *Ny. syrticus* (and "*Ny. devauxi*"). The species may be a possible ancestor for *Ny kanamensis* or a small, forest-adapted form that went extinct.

The available Ugandan sample represents both the FAD adn LAD (or the OAD, "only appearance datum") at ca. 5–6 myr.

3. *Nyanzachoerus kanamensis.* The priority of the species name has now been recognized by Cooke (1987). The taxon is a long-lasting, ubiquitous one in Africa, and at least one subspecies has been proposed (Cooke, 1992). The species probably arose from *Ny. syrticus,* perhaps via a form such as *Nyanzachoerus waylandi.*

The earliest record of the species is Lothagam 1C (Hill et al., 1992), dated to 5–6 myr, largely with reference to the Baringo series. Reliability of this FAD is moderate, given the absence of the form from earlier units at Lothagam. The latest record of *Ny. kanamensis* is from Omo Shungura C-6, dated here at 2.7 myr based on radiometrics and stratigraphic scaling (Harris and White, 1979; Feibel et al., 1989). This LAD is judged to be reliable even though the species is increasingly rare in strata postdating ca. 3.2 myr in the Afar and in the Turkana Basin.

4. *Nyanzachoerus jaegeri.* This species is an obvious derivative of *Ny. kanamensis,* appearing at several locations between 4 and 5 myr. It is the obvious ancestor of *Not. euilus,* and the transition between the forms is seen best in the Middle Awash Pliocene section (White et al., 1993).

The first solid appearance of *Nyanzachoerus jaegeri* is in the Baringo succession, where it appears in the Chemeron Formation, dated between 5.03 and 4.02 myr (Hill, this vol.). Hill et al. (1992, p. 445) had previously suggested that "5.6 Ma, plus or minus a few hundred thousand, would seem to be a good estimate of the time of change from *N. syrticus* to *N. jaegeri,*" but this was before more precise dates for the lower Chemeron were available. A 5–6 myr FAD for the species is probably a reliable one, but only within the million-year error range. The LAD of *Nyanzachoerus jaegeri* is placed at 3.75 myr based on specimens between the Cindery Tuff and VT-3 (= Wargolo) of the Middle Awash succession (White et al., 1993). This is not a useful LAD, because it represents the arbitrary division of what appears to be a gradually evolving species lineage to the daughter chronospecies *Notochoerus euilus.*

5. *Notochoerus euilus.* This form is derived from *Ny. jaegeri* as part of the group's trend toward increased third molar height and length coupled with reduction in the premolar row. The earlier representatives of the species are more like the ancestral *Ny. jaegeri,* whereas the daughter species *Not. capensis / scotti* are highly derived.

The FAD of the taxon comes anagenetically (see 4 above) and is therefore not reliable for the purpose of this exercise. The LAD of the taxon comes in radiometrically dated upper Member G of the Omo Shungura Formation, at 2.0 myr (Harris and White, 1979; Feibel et al., 1989). This LAD is considered to be reliable because of the absence of this species in the well-sampled sediments between 1.5 and 2.0 myr in multiple eastern African locations.

6. *Notochoerus capensis.* This species is an artifact of nomenclature in the sense that it is morphologically and temporally intermediate between *Notochoerus euilus* and *Notochoerus scotti.* It has been retained as a separate species, because it types the genus and because this morph is seen in both southern and eastern Africa (Harris and White, 1979).

The FAD and LAD of this taxon are in the period centering around 2.7–2.8 myr in the Turkana Basin (Harris and White, 1979; Feibel et al., 1989). The species is the unstable state between the longer-lasting mother and daughter species. Thus, the species is interpreted here as a link in an anagenetically broken chain and of no value in this analysis.

7. *Notochoerus scotti.* This species is derived from *Not. capensis,* as described above. It changes little during its duration, going extinct ca. 1 myr after appearing.

The first specimens placed in *Not. scotti* are in the Omo Shungura Formation, lower Member C (Harris and White, 1979), dated radiometrically and by stratigraphic scaling to ca. 2.8 myr (Vrba, this vol.). This FAD is, of course, arbitrary and of little value in this analysis. The estimated extinction date for the species at ca. 1.8 myr (Koobi Fora, just above the KBS tuff) is, however, considered to be very reliable, because the record is well-sampled after this LAD in several very fossiliferous depositories in eastern Africa.

8. *Kolpochoerus afarensis.* White and Harris (1978) followed Cooke (1974) in suggesting that the genus name *Kolpochoerus* be suppressed in light of the widespread use of Cooke's name *Mesochoerus.* In his reply, however, Cooke (1978a) reversed himself, resurrecting *Kolpochoerus* (but curiously ignoring the priority species nomen "*phacochoeroides*"). The name *Kolpochoerus* has subsequently been widely used.

The species *afarensis* was assigned to the genus *Potamochoerus* by Harris and White (1979), largely on the basis of a shared primitive trait complex in the dentition. Subsequent description of cranial material from Hadar by Cooke (1978b) established *Kolpochoerus afarensis* as a valid species ancestral to later kolpochoeres and probably also ancestral to *Potamochoerus.*

The earliest representative of *Kolpochoerus afarensis* is from sediments of the Chemeron Formation (Pickford et al., 1983), dated between 5.03 and 4.02 myr (Hill, this vol.). This FAD is judged to be unreliable for the limited purposes of this analysis, as it is very likely that this genus is an Asian immigrant. The LAD is probably reliable; it is based on radiometrically dated occurrences in the Matabaietu section of the Middle Awash, at ca. 2.5 myr (White et al., in prep.).

9. *Kolpochoerus limnetes.* This taxon was extended to all segments of what appears to be a continuously evolving lineage by Harris and White (1979). Cooke (1978) has maintained the existence of two additional species, *K. paiceae* and *K. olduvaiensis,* both derived from *K. limnetes.* It is clear that the species *K. limnetes* evolved from *K. afarensis* and gave rise to a progressive lineage that featured increasingly long and hypsodont third molar teeth.

The first appearance of a morphology that can be distinguished from the earlier, more primitive *K. afarensis* is with molar teeth from Omo Shungura Member B-11, radiometrically and paleomagnetically dated to ca. 2.9 myr (Harris and White, 1979; Feibel et al., 1989). This FAD is judged to be reliable and reflects a temporally proximate speciation event resulting in the daughter species *K. limnetes.* This daughter species, however, then undergoes an evolutionary transformation that appears, on the basis of the available parts, to be gradual and progressive, culminating with the artificial truncation point for the morphospecies at ca. 1.7 myr, in the Turkana Basin successions (between the KBS and Okote tuff complexes; Feibel et al., 1989). Here, when the talon and trigon are shorter than the talonid and trigonid, the species boundary is considered crossed; hence this time is considered an arbitrary LAD for *K. limnetes.*

10. *Kolpochoerus phacochoeroides.* In the current analysis *K. paiceae* and *K. olduvaiensis* are synonomized and subsumed by *K. phacochoeroides.* The latter has obvious priority, whereas the former is not considered specifically distinct from the eastern African variant (contra Hendey and Cooke, 1985).

This taxon's FAD is arbitrary, as discussed above. The last species representatives are found in deposits dating to ca. 0.78 myr in Olduvai Bed IV (Manega, 1993). They are not found in younger Middle Pleistocene deposits, despite the fact that there are few such well-dated locations yielding suid fossils. As a result, this LAD is considered possibly valid.

11. *Kolpochoerus majus.* This species is clearly derived from *Kolpochoerus limnetes* sometime around 2.0 myr (*K. limnetes/phacochoeroides* lineage members after this time are too derived to be ancestral). It is poorly known relative to its nearest relatives.

The taxon is rare in some deposits but present as early as 1.44 myr at Konso-Gardula (Asfaw et al., 1992). This FAD is considered to be an artifact of the record. The species drops from the record in the Middle Pleistocene but is abundant at some sites such as Bodo, its youngest radiometrically dated occurrence (Clark et al., 1994). This LAD is judged to be an artifact of the record, and the extinction date for the species is an open question.

12. *Potamochoerus porcus.* The modern bushpig has a very poor fossil record. Cooke's (1978b) demonstration of *Kolpochoerus* affinities for the closely related material from the Pliocene at Hadar emphasized how highly conservative this suid species is. Once *Kolpochoerus afarensis* is accepted, the species *P. porcus* is left with an artificial, unreliable FAD in the Late Pleistocene archaeological record. This FAD comes little before the modern manifestation of this pig.

13. *Hylochoerus meinertzhageni.* The modern giant forest hog's record is no better than the bushpig's. This species was probably derived from *Kolpochoerus limnetes/majus* around 2 myr, but, like the bushpig, the modern species is first encounterd as food remains in the very recent archaeological record. The FAD is thus artificial and unreliable.

14. *Metridiochoerus andrewsi.* Cooke has resisted White and Harris's (1978) combining of all of the early southern and eastern African metridiochoeres into a single, evolving lineage. He maintains *Metridiochoerus jacksoni* as a species distinct from, and contemporary with, *M. andrewsi* in eastern Africa and maintains *Potamochoeroides shawi* as a distinct species in southern Africa. Little evidence has been marshaled to support this view, and here the Harris and White (1979) interpretation is followed.

The most primitive representative of this lineage is from Omo Shungura B10, dated radiometrically at ca. 2.95 myr (Harris and White, 1979; Feibel et al., 1989). This FAD is considered to be reliable in the sense that lower, dated, fossiliferous units have failed to yield earlier representatives of the species. However, because the form is thought to be a possible Asian immigrant, the FAD may be geographical instead of phyletic or cladogenetic (although it could be both an immigrant and a new species). The lineage of the mother species forms the evolutionary substrate for three other species of *Metridiochoerus*, with whom it continues to coexist before disappearing ca. 1.6 myr, just above the Okote tuff at Koobi Fora (Harris, 1983; Feibel et al., 1989). This LAD is considered to be reliable.

15. *Metridiochoerus compactus.* This species is placed in the genus *Stylochoerus* by Cooke (1978) on the basis of its specialized characters, but its phylogenetic placement is agreed upon by all authors. The first record of the taxon is just below the lower Okote tuff at Koobi Fora, ca. 1.62 myr (Harris, 1983; Feibel et al., 1989). This FAD is considered to be reliable. The last appearance of the species is in Olduvai Bed IV, dated to ca. 0.78 myr (Harris and White, 1979; Manega, 1993). *M. compactus* is not found in younger Middle Pleistocene deposits, despite the fact that there are few such well-dated locations yielding suid fossils until ca. 0.6 myr. As a result, this LAD is considered possibly valid.

16. *Metridiochoerus hopwoodi.* This species is roughly the equivalent of what Cooke (1978, 1985) calls *M. nyanzae.* It is obviously derived from *Metridiochoerus andrewsi.* The FAD for *M. hopwoodi* is two specimens just below the KBS tuff at Koobi Fora, radiometrically dated at 1.89 myr (Harris, 1983; Feibel et al., 1989). Despite the rarity of the taxon, this is a reasonable FAD. The species makes its last appearance in the fossil record at Olduvai Bed IV, dated to 0.78 myr (Harris and White, 1979; Manega, 1993). This disappearance, like that of *K. phacochoeroides* and *M. compactus,* is possibly a valid LAD.

17. *Metridiochoerus modestus.* This species, like *M. hopwoodi,* is rare, resulting in the differences in systematic treatment between Harris and Whie (1979) and Cooke (1982, 1985). All workers have noted the strong worthog resemblances of the species, and Cooke places the species in the warthog genus *Phacochoerus.* Cooke synonomizes this taxon with *P. antiquus,* whereas Harris and White recognize both taxa. With the exception of the species' first appearance just below the KBS tuff at Koobi Fora (radiometrically dated to 1.89 myr; Harris, 1983; Feibel et al., 1989) and the occurrence of a fine skull from Olduvai Bed I (Cooke, 1982), this taxon is very poorly known. The LAD is placed at 0.78 myr in Olduvai Bed IV (Harris and White, 1979; Manega, 1993), but unlike other suid taxa that go extinct here, the rarity of *Metridiochoerus modestus* make this LAD somewhat questionable.

18. *Phacochoerus antiquus.* This taxon is known from Kromdraai, Bolt's Farm, and Swartkrans, in southern Africa. All these occurrences are effectively undated. Additional specimens are known from Olduvai Bed IV, where the species co-occurs with *Metridiochoerus modestus.* Cooke (1982) does not separate the taxon from *M. modestus.* One problem with warthog systematics is discriminating fossilized teeth found on the surface (but derived from overlying Holocene or Late Pleistocene deposits) from fossilized teeth actually eroded from earlier sediments. Warthog and equid teeth seem to "fossilize" (mineralize, stain) rapidly, and the potential for contamination is great. This is a particularly acute problem for the species *M. modestus* and *P. antiquus*—species generally ancestral to modern warthogs and whose third molars are very similar to those of modern warthogs. The FAD of *P. antiquus* is set at ca. 0.78 myr at Olduvai Bed IV (Harris and White, 1979; Manega, 1993). The species, if valid, gives rise to the modern warthog, so the LAD (also Olduvai) is an artifact of nomenclature and not a valid LAD for the purpose of this study.

19. *Phacochoerus africanus.* This extant species appears to have arisen anagenetically, but with a poor fossil record, from either *Metridiochoerus modestus* or *Phacochoerus antiquus.* Hence, the first appearance date, in Middle Pleistocene deposits of the Middle Awash, is of little value relative to this study.

Appendix 26.2. Notes on Fossil Hominid Systematics

1. *Australopithecus afarensis.* After its naming in 1978 the systematics of this species were widely debated. Originally named for material at Hadar and Laetoli, the species is now recognized from the Omo Shungura, Koobi Fora, and the Middle Awash. Additional material has been collected at Hadar and Laetoli. The post-1978 discoveries have largely confirmed the existence of this species, although some maintain that the degree of variation seen in the hypodigm exceeds that expected in a single species.

Fossils from Lothagam, Tabarin, and Belohdelie have been tentatively attributed to *A. afarensis,* largely on the basis of primitive characters. It is possible that these fossils belong, but the Laetoli paratype series is taken here as the FAD for the taxon. These fossils are dated radiometrically at ca. 3.5 myr. This *A. afarensis* FAD is an artifact of the fossil record, because no locality earlier than Laetoli has yielded hominid remains for firm species attribution. The LAD for the species is radiometrically set at approximately 3.0 myr based on specimens from the upper part of the Kada Hadar Member (Renne et al., 1993; Kimbel, this vol.). This LAD is judged possibly reliable, within limits set by the lack of a hominid record in eastern Africa between ca. 2.7 and ca. 3.0 myr.

2. *Australopithecus africanus.* The systematics of this, the first *Australopithecus* species recognized, are among the most controversial within hominid paleontology. Much of the controversy comes from the fact that this species is exclusively South African in distribution, and hence temporal control for the holotype and much of the hypodigm is poor.

The FADs and LADs for all of the South African hominids come from complex but usually rapidly and episodically accumulating depositories that are best considered points in time (like Swartkrans) or merged sets of points (like Sterkfontein; White, 1988; Delson, 1988). The *A. africanus* FAD is from the Makapansgat gray breccia, set at ca. 2.8 myr by a good biochronological data set, based on two independent lineages of suids (Harris and White, 1979). The chronological placement is probably relatively accurate, but the FAD itself is an artifact of the record: there are few earlier localities in southern Africa, and none have hominids. The LAD of *A. africanus* is far more controversial, with the Taung specimen effectively undated because of its insecure provenance. The Sterkfontein occurrence would thus become the LAD for the species, but this deposit may span a great deal of time, and there are suggestions that the specimens from there usually attributed to *A. africanus* actually fall into two lineages (Kimbel and White, 1988), one of them ancestral to later *Homo* and the other ancestral to *A. robustus* (Clarke, 1988). Pending results of the ongoing Sterkfontein investigations, an estimated LAD of ca. 2.6 myr is applied to the species, with the recognition that this is certainly an artifact of the record and of little value to the current analysis.

3. *Australopithecus robustus.* This taxon is defined on material from South African depositories—Kromdraai and the three earliest Swartkrans units (Brain, this vol.). These are biochronologically indistinguishable (Harris and White, 1979; Delson, 1988), best matching eastern African faunas dated radiometrically to ca. 1.7 myr. Thus, the FAD and LAD become the same and are both artifacts of the record. If, however, Clarke's (1988) suggestion that the taxon began at Sterkfontein is accepted, or if Suwa's (1988)

suggestion that the robust lineage in eastern Africa speciated to *A. boisei* after 2.3 myr is confirmed, the FADs would be pushed earlier.

4. *Australopithecus aethiopicus.* This taxon is seen by some workers as the primitive form of *A. boisei* and by others as the chronospecies ancestral to *A. boisei.* The earliest occurrence of a definitively "robust" *Australopithecus* (derived relative to *A. afarensis*) comes in the form of L55s-33 from Shungura Member C6, radiometrically and stratigraphically dated at ca. 2.7 myr. This FAD is judged probably reliable to within 0.3 myr (2.7–3.0 myr), given the relatively good records without *A. aethiopicus* between 3 and 4 myr. All workers accept *A. aethiopicus* as the ancestor to *A. boisei,* but there is much disagreement on the question of the LAD of *A. aethiopicus.* Based on premolar morphology, Suwa (1988) suggests that the transition to a true *A. boisei* occurs at about 2.3 myr in the Shungura Formation. Kimbel (this vol.) recognizes 1.9 myr (the radiometrically dated KNM ER-1482 mandible) as the LAD for *A. aethiopicus,* implying, like Suwa, that the mother and daughter species coexisted for a time in eastern Africa. The available samples are not adequate to confirm these models. Therefore, the LAD for *A. aethiopicus* is either an arbitrary split point along a continuously evolving lineage at ca. 2.3 myr or a possibly reliable termination date at ca. 1.9 myr.

5. *Australopithecus boisei.* Better known than its mother taxon *Australopithecus aethiopicus, A. boisei* is seen by all workers as a species that went extinct in the lower Pleistocene. The first appearance of this species comes with specimens in lower Member G (radiometrically dated at ca. 2.3 myr). These specimens have premolars that are considered by Suwa (1988) to be "advanced" relative to earlier, contemporary, and later specimens, which he describes as "conservative." He does not apply species names to these morphs and does not comment on the issue of whether the "primitive" form should be considered *A. robustus* or *A. aethiopicus.* Indeed, Suwa is extremely cautious in his analysis, identifying a clear need for more character complexes to be studied in larger samples before species assignments are made. Therefore, the FAD for *A. boisei* depends very much on whether the southern and eastern African lineages were separate deep into the Pliocene (whether major parts of the hypermasticatory apparatus of *A. robustus* was derived from *A. africanus* independent of an *A. Aethiopicus* to *A. boisei* lineage). There is no resolution to this question in sight, so the FAD for *A. boisei* can be set at 2.3 myr as a cladogenesis or as an arbitrary anagenetic "speciation," or it can be set at >2.7 myr if *A. robustus* arose in parallel in southern Africa.

In contrast to the very dubious FAD for *A. boisei,* the LAD for this species is widely held to be around 1 myr. A close examination of the record, however, shows that the last occurrence is actually substantially earlier. The specimen most often cited as the youngest *A. boisei* is a deciduous molar from Olduvai BK, in upper Bed II. This specimen is from channels above Tuff IID. This unit is dated at 1.48 ± 0.05 myr, but this date is based on only five feldspars ranging as follows: 1.18, 1.27, 1.36, 1.46, 1.82 myr. As Manega (1983) notes, the date is not well constrained. Because the Chesowanja partial cranium is chronologically placed on an old basalt date (Hooker and Miller, 1979), a better LAD may be the F203-1 molar from Shungura Member K4, dated radiometrically to 1.45 myr (Feibel et al., 1989). This specimen is below the Chari tuff (= tuff L), dated to 1.36 myr, but there are very few sites in the interval between 1.4 and 1.0 myr, so the LAD is considered possible but will probably turn out to be too old as more collection is done.

6. *Homo habilis.* This species has been controversial since its recognition. At first, parts of the hypodigm were attributed to *A. africanus* and *H. erectus,* and many claimed that a separate species did not exist. More recently, subsequent to the recovery of a few more complete specimens in eastern and southern Africa, most workers now argue that two species comprise what has traditionally been called *H. habilis.* There is no agreement on what hypodigms these species have, although one of them has been given the name *H. rudolfensis* (see Wood, 1991). In the face of disagreement (and the inability of investigators to decide which new specimens belong to the same species as the original Olduvai holotype!), the now-traditional approach of Tobias (1991) will be taken here.

The first appearance of small-toothed forms that are diagnostically neither *A. afarensis* nor *A. aethiopicus* is in the Omo Shungura Member C (Suwa, 1990). There is a temporal bone from Baringo Chemeron dated to ca. 2.6 myr that has been attributed to *Homo* (Hill et al., 1992) and a new mandible fragment from Gamedah in the Middle Awash that probably belongs to this taxon (White et al., in prep.). The oldest firm FAD for *H. habilis,* however, is with the KNM ER-1470 specimen from radiometrically controlled, 1.9 myr sediments at Koobi Fora (Feibel et al., 1989), probably over 0.5 myr younger than the biological FAD.

The LAD for *H. habilis* is equally problematic. Olduvai Hominid 13 is widely but not universally accepted as different from *Homo erectus.* This specimen, from MNK, was found 2 m below tuff IIB (and the much larger *Homo* specimen OH 15 was found 1 m below IIB at the same site). Tuff IIB is unsuitable for radiometric dating. The radiometrically dated unit closest to OH 13 is within 1 m below the partial skull but separated from it by a recognizable, major disconformity. Tuff IIA, dated at 1.66 ± .1 myr (Hay, 1976; Manega, 1993), thus provides a maximum age for the specimen but not a strong LAD for the taxon. Thus, the LAD for *Homo habilis* is probably close to ca. 1.6–1.7 myr, and it remains possible that there is no actual temporal overlap between the youngest *Homo habilis* and the oldest *Homo erectus.* It is also possible that OH 13 is actually the female morph of early African *Homo erectus.* The significance of the uncertain *Homo habilis* LAD will become apparent in the discussion below.

7. *Homo erectus.* Did *Homo habilis* give rise to *Homo erectus* anagenetically, and rapidly, in eastern Africa? Detemining the FAD for *Homo erectus* necessitates a consideration of the Asian record. There are hints of very early (>1.7 myr) occurrences of this taxon at Dmanisi (Dzaparidze et al., 1989) and Modjokerto (Curtis, 1981). For the purposes of this discussion, only the eastern African record will be considered. Here, the FAD is set by KNM ER-3733, a cranium from Koobi Fora radiometrically dated at ca. 1.75 myr (the KNM-ER-2598 cf. *Homo erectus* specimen, dated at >1.9 myr, was a surface find in a lag deposit and may have weathered from stratigraphically younger deposits; Feibel et al., 1989; pers. obs.). It seems more likely that *Homo erectus* is an immigrant from Asia to the eastern African area than an anagenetic, in situ derivative from *Homo habilis.* Given the LAD for *Homo habilis,* the window for substantial morphological change is very narrow. The FAD for *Homo erectus* at 1.7 myr is therefore considered a reliable one, with one further nomenclatorial caveat: there is an ongoing debate about whether early eastern African fossils should be subsumed in *Homo erectus* or put in a different taxon, *Homo ergaster.* The outcome of this debate will, of course, affect the FAD for *H. erectus.*

Determining the LAD for *H. erectus* is extremely difficult. By the Middle Pleistocene this species had occupied most of the Old World, and the available fossils are widely scattered in time and space. Systematics of this diverse group are being done in a much finer-grained fashion than those practiced for earlier hominid fossils. The presence of specimens such as Bodo and Petralona (often given the name "archaic *Homo sapiens*"), which are intermediate between traditional *Homo erectus* and anatomically modern *Homo sapiens*, means that the LAD of *H. erectus* may be strictly an arbitrary matter of definition.

8. *Homo neanderthalensis.* It is apparent that European populations of later hominids were genetically, if not culturally, isolated from Asian and African populations of contemporary hominids. These European forms share a number of derived characters that set them apart. These characters appear at ca. 0.3 myr in the Atapuerca sample (Stringer, 1993; Stringer, this vol.). This FAD is, however, moderately reliable for this clade. The LAD for the neanderthals is set at ca. 0.036 by the St. Cesaire discovery (Stringer and Gamble, 1993).

9. *Homo sapiens.* The first known appearance of anatomically modern humans is in Israel and South Africa at between 0.08 and 0.1 myr, based on thermoluminescence and electron spin resonance dating (Bar-Yosef, this vol.). This species is modern.

Chapter 27

The Fossil Record of African Antelopes (Mammalia, Bovidae) in Relation to Human Evolution and Paleoclimate

Elisabeth S. Vrba

My general question is this: To what extent do climatic changes force the pace of biotic turnover, that is, of speciation, extinction, and migration? In asking that, I interpret "climatic change" broadly to include the climatic effects of structural changes in the earth's crust. One possible answer is that climate plays a subsidiary role: the initiating causes of turnover most often come from a myriad of idiosyncratic interactions among organisms and among species, occasionally accelerated by physical change, such that large turnover patterns approach randomness and constancy in time. Another possible answer is the turnover pulse hypothesis (Vrba, 1985a, 1993, and chap. 3, this vol.), which posits that nearly all speciation and extinction events require initiation by climatic change that systematically concentrates these turnover events in time. In particular, the problem of how biotic and physical factors combine during the origin of species, Darwin's (1859) question, has not yet been solved, in spite of the numerous models of speciation that have been explored. I suggest that there will be no definitive progress on this issue without analysis of the fossil record in its physical settings. This record is the only direct evidence of evolution, and any model of macroevolution must stand or fall against it. Every part of the fossil record can potentially make unique contributions in this respect. The Late Neogene evidence is particularly suitable for an interdisciplinary onslaught on the outstanding questions of evolution. It allows phylogenetic and ecological studies that combine abundant and well-dated data on fossil species and paleoclimate with information on the same or closely related living species and ecosystems.

My specific questions concern the relationship between mammalian turnover and major cooling trends during the Plio-Pleistocene. The Late Neogene records of the astronomical climatic cycles (see Shackleton; deMenocal and Bloemendal; and Dupont and Leroy, all in this vol.) indicate the following. The glacial-value envelope of the cycles of higher frequencies (the line approximately passing through the highest $\delta^{18}O$ values of the cycles with periods ca. 23–19 thousand years [kyr], 41 kyr, and 100 kyr) itself underwent oscillations of lower frequency or major warming-cooling trends over longer intervals. During the Late Neogene, such large-scale cooling trends appear to have reached an acme roughly every 0.7–0.9 million years (myr) (Shackleton, this vol.; I know of no proposal that there is a cycle of such periodicity) and descended to progressively more severe cooling minima. What is remarkable about one such cooling trend, the one after ca. 3.0 myr toward 2.5 myr ago, is not so much the rate at which its mean changed (climatologists differ on how sudden or gradual this cooling trend was) but that it involved fundamental and rare changes in the climatic system: it culminated in fully established Arctic glaciation—the onset of the "modern Ice Age"—by ca. 2.5 myr (Shackleton et al., 1984). DeMenocal and Bloemendal (this vol.) document in deep-sea records adjacent to eastern Africa that after ca. 2.8 myr ago there was a shift toward climatic dominance of the 41 kyr astronomical cycle, from previous dominant influence at 23–19 kyr variance (see also Dupont and Leroy, this vol., for western Africa); and that a second shift toward increasing dominance in 100 kyr variance occurred after 0.9 myr. The few available African cli-

ISBN 0-300-06348-2.

matic records for the Late Pliocene suggest that the cooling trend toward 2.5 myr strongly affected at least some African biomes. Dupont and Leroy (this vol.) document the stepwise onset of the extensive Sahara desert especially just after 2.8 myr (see also Bonnefille and others in this vol.). There is also evidence of uplift and rifting in eastern Africa during the Late Neogene (Partridge et al., chap. 2, this vol.). Quite recently, tephrostratigraphic cross-correlations between the deep-sea record of the Gulf of Aden and the East African fossil strata has enabled the integration of the marine climatic signals with the land record (Brown, this vol.).

These recent results make it worthwhile to take a fresh look at the old proposal that climatic change coincided with and was responsible for hominid and other faunal turnover during the Late Pliocene (proposed independently by Coppens, 1975, and Vrba, 1974, 1975): for South Africa, in the Transvaal cave successions, the kinds of bovid taxa that turned over concurrently with the replacement of *Australopithecus africanus* by robust australopithecines and *Homo* indicated a Late Pliocene change toward more open vegetation caused "by a decrease in rainfall . . . [or] a drop in temperature" (Vrba, 1974, p. 23). The fact that East African proliferation of open-adapted bovids seemed to coincide with the analogous South African development suggested (Vrba, 1975) that the changes were widespread and that "environmental change may have been responsible for the faunal change," including the hominid evolutionary changes toward robust australopithecines (p. 302) and toward *Homo* (p. 303, fig. 2; further discussed in Vrba, 1985b, 1988). For East Africa, Coppens (1975) proposed something similar based on mammalian turnover in the Shungura Formation between successive faunas that he called "Omo1–3" (p. 1696): "Quand apparait en [membre] E un changement [climatique] mineur, à l'intérieur de l'association faunique Omo 2, *Australopithecus* aff. *robustus* apparait. Quand un changement climatique majeur dans la base du membre G, l'association faunique Omo 2 passe à l'association Omo 3 et *Homo* aff. *habilis* arrive." At that time physical dating of the East African strata had barely begun, paleoclimatologists had not yet announced evidence for major global cooling at this time, and cladistic results had not yet provided a genealogical context for mammalian evolution. I shall now combine the greatly improved information of all three kinds to reexamine that old proposal. I investigate the record of African Bovidae in some depth and, using new methods, compare the results with major features of the hominid record as reported by Wood, Kimbel, and White (all in this vol.).

The Bovid Record

I surveyed the entire African bovid record, which covers a Miocene-to-Recent time span close to the global one for bovids. The fossils belong to assemblages from the sites and site subunits listed in appendix 27.1; Figure 27.1 shows the distribution of fossil assemblages over the past 6.5 million years (myr). Neogene African assemblages are restricted almost entirely to the northern, eastern, and southern parts of the continent. Bovid fossils totaling hundreds of thousands of specimens usually dominate strongly (typically 60–80 percent) among the large mammals. Following some rich Early and early Middle Miocene assemblages in East Africa, the interval between 14 and 4 myr is sparsely sampled (Hill, this vol.). Spanning the period from 4 myr to the Middle Pleistocene there are several excellent East African sequences that have been well dated and cross-correlated by radiometric dates from homologous tuffs (Brown, this vol.). There are also bovid-rich sequences for the past 3 myr from South Africa that are dated only by biochronology (Vrba, 1976, 1982). A proper understanding of the differences in fossil richness among the assemblages and through time requires counts of fossil specimens. Such differences are only partly represented in figure 27.2, at least for the Turkana Basin in Kenya and Ethiopia, by the number of mammalian species known from each assemblage. The decrease in numbers of species after 2.0 myr is peculiar to the Turkana Basin (there are many assemblages elsewhere in Africa that fit into this interval), but the decrease from 3 myr to earlier times correctly reflects the overall African pattern.

Appendix 27.2 lists the bovid species and the dates and sites of their first and last appearance data (FADs and LADs). I noted members and submembers and, where possible, even finer subdivisions of temporal placement

Fig. 27.1. Frequencies of African fossil assemblages over 6.5–0 myr (for acronyms of assemblages see app. 27.1, and for dates app. 27.2) from northern (N), eastern (E), and southern (S) Africa. Based on available earliest and latest dates for assemblages, some are plotted as representing successive 0.1 myr intervals. Note that this procedure may mask gaps in the record. For instance, although Shungura Member G (SH G) is shown from ca. 2.3 to 1.9 myr (Brown, this vol.), there may be one or more gap(s) that are ≥ 100 kyr within this member.

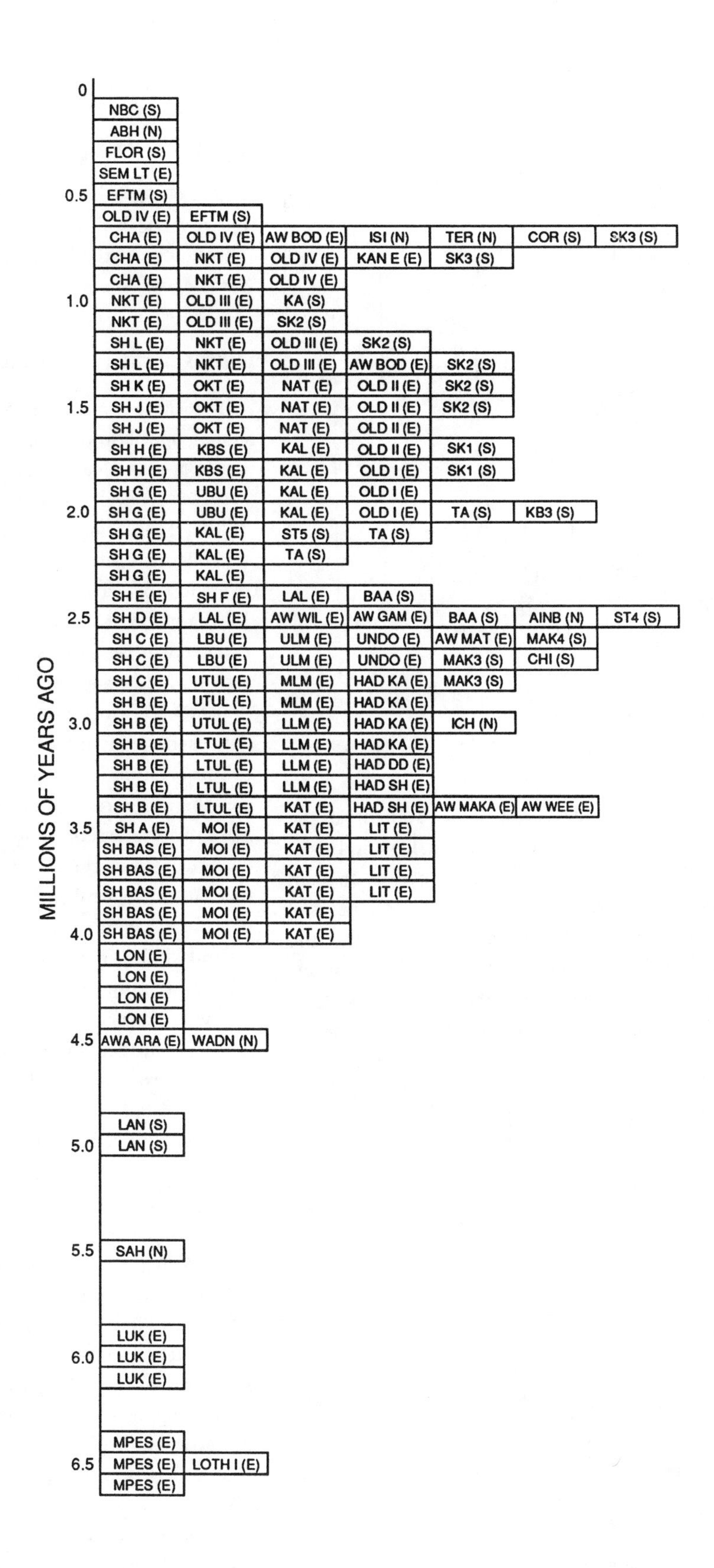
MILLIONS OF YEARS AGO
0
NBC (S)
ABH (N)
FLOR (S)
SEM LT (E)
0.5
EFTM (S)
OLD IV (E) EFTM (S)
CHA (E) OLD IV (E) AW BOD (E) ISI (N) TER (N) COR (S) SK3 (S)
CHA (E) NKT (E) OLD IV (E) KAN E (E) SK3 (S)
CHA (E) NKT (E) OLD IV (E)
1.0
NKT (E) OLD III (E) KA (S)
NKT (E) OLD III (E) SK2 (S)
SH L (E) NKT (E) OLD III (E) SK2 (S)
SH L (E) NKT (E) OLD III (E) AW BOD (E) SK2 (S)
SH K (E) OKT (E) NAT (E) OLD II (E) SK2 (S)
1.5
SH J (E) OKT (E) NAT (E) OLD II (E) SK2 (S)
SH J (E) OKT (E) NAT (E) OLD II (E)
SH H (E) KBS (E) KAL (E) OLD II (E) SK1 (S)
SH H (E) KBS (E) KAL (E) OLD I (E) SK1 (S)
SH G (E) UBU (E) KAL (E) OLD I (E)
2.0
SH G (E) UBU (E) KAL (E) OLD I (E) TA (S) KB3 (S)
SH G (E) KAL (E) ST5 (S) TA (S)
SH G (E) KAL (E) TA (S)
SH G (E) KAL (E)
SH E (E) SH F (E) LAL (E) BAA (S)
2.5
SH D (E) LAL (E) AW WIL (E) AW GAM (E) BAA (S) AINB (N) ST4 (S)
SH C (E) LBU (E) ULM (E) UNDO (E) AW MAT (E) MAK4 (S)
SH C (E) LBU (E) ULM (E) UNDO (E) MAK3 (S) CHI (S)
SH C (E) UTUL (E) MLM (E) HAD KA (E) MAK3 (S)
SH B (E) UTUL (E) MLM (E) HAD KA (E)
3.0
SH B (E) UTUL (E) LLM (E) HAD KA (E) ICH (N)
SH B (E) LTUL (E) LLM (E) HAD KA (E)
SH B (E) LTUL (E) LLM (E) HAD DD (E)
SH B (E) LTUL (E) LLM (E) HAD SH (E)
SH B (E) LTUL (E) KAT (E) HAD SH (E) AW MAKA (E) AW WEE (E)
3.5
SH A (E) MOI (E) KAT (E) LIT (E)
SH BAS (E) MOI (E) KAT (E) LIT (E)
SH BAS (E) MOI (E) KAT (E) LIT (E)
SH BAS (E) MOI (E) KAT (E) LIT (E)
SH BAS (E) MOI (E) KAT (E)
4.0
SH BAS (E) MOI (E) KAT (E)
LON (E)
LON (E)
LON (E)
LON (E)
4.5
AWA ARA (E) WADN (N)
LAN (S)
5.0
LAN (S)
5.5
SAH (N)
LUK (E)
6.0
LUK (E)
LUK (E)
MPES (E)
6.5
MPES (E) LOTH I (E)
MPES (E)

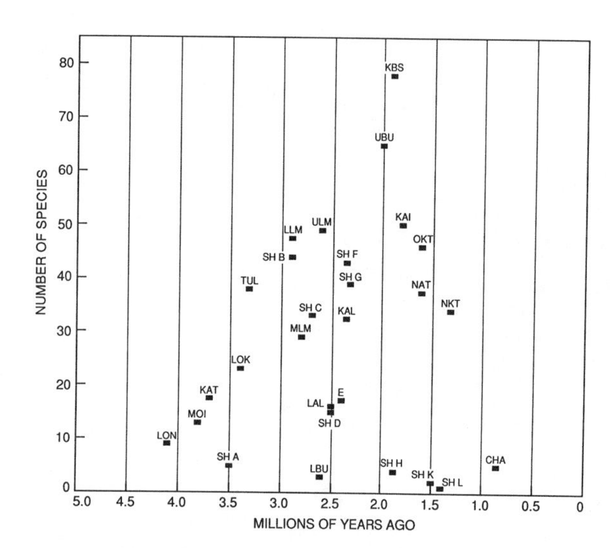

Fig. 27.2. Number of mammal species known from fossil assemblages in the Turkana Basin (from Feibel et al., 1989, table 6.3). For acronyms of assemblages see app. 27.1, and for dates app. 27.2.

of FADs and LADs. Such data, usually not available in the literature, have to be extracted from catalogs and used together with geological information, such as that on stratigraphic thicknesses. Appendix 27.2 gives sources of dates: about one-third are biochronological, and the others radiometric and paleomagnetic. All major systematic disagreements on species' FADs and LADs are noted in appendix 27.2. I analyzed the data using my own systematic evaluations as well as the entire set of evaluations that differ from mine. The summary of extinct and living bovid species in table 27.1 includes an estimate of the proportions of morphologically distinct and sibling species in the modern fauna. Figure 27.3 shows stratigraphic ranges of all African bovid species, and figure

Table 27.1. Numbers of Living and Extinct Species of Bovidae Known from Africa

Category of Species	Unknown as Fossils	Known as Fossils	Total
Living species			
A. Morphologically distinct	8		
—Recorded in app. 27.2		28	
—Late Quaternary records omitted from app. 27.2		14	50
B. Sibling species relative to taxa in category A	27		27
Extinct species		119	119
Total	35	161	196

27.4 a closer view of the past 7 myr. The survey of the entire African bovid record using new dating indicated chronological reevaluations of some South African assemblages (app. 27.3).

Biases in the Fossil Record

The term *turnover,* used without qualification, refers to actual speciation, extinction, and migration. Several kinds of biases can result in FADs and LADs that do not represent coeval turnover (Behrensmeyer and Hill, 1980; White, Hill, and Barry, all in this vol.). Reference to turnover as coeval with a FAD or LAD pulse means turnover close before the signature of that pulse in the record. In chapter 3, I discussed two kinds of biases that are expected to wax and wane with climatic changes from astronomical or tectonic sources: what might be termed "rarity bias" does not act to distort the relative abundances in the fossil record. It acts even if the probabilities of preservation for various organisms are the same: species that are rare because their habitats are rare will always be underrepresented in the record. If climatic change expands those habitats, such species may appear in a FAD pulse, while others in shrinking habitats are last recorded, even if no speciation or extinction occurred. A second kind of bias, by "abundance distortion," results in a deviation in the fossil record from the relative abundances of taxa in the living biota. Of particular importance is the general tendency of open, mesic-to-arid areas to preserve vertebrate fossils better than do the more forested, wetter ones (Hare, 1980). The greater abundance of African fossils after 4.0 myr than before (figs. 27.1, 27.2) is probably largely due to this bias. During widespread climatic changes, such abundance distortion can become intensified and act together with changes in rarity bias to result in artifactual turnover pulses and gaps in the record. For instance, even if no new species have been added, an increase in less densely vegetated areas during onset of a glacial period or uplift of a large area, could result in a FAD pulse of open-adapted forms that does not reflect coeval speciation or immigration. Both kinds of bias, by rarity and by abundance distortion, are here included under taphonomic biases.

Taphonomic biases are always operating. To the extent that their effects have remained averagely constant, any pulse perceived in the record must be accounted for by coeval species turnover. Thus, the danger for interpretation resides not in biases per se but in large-scale

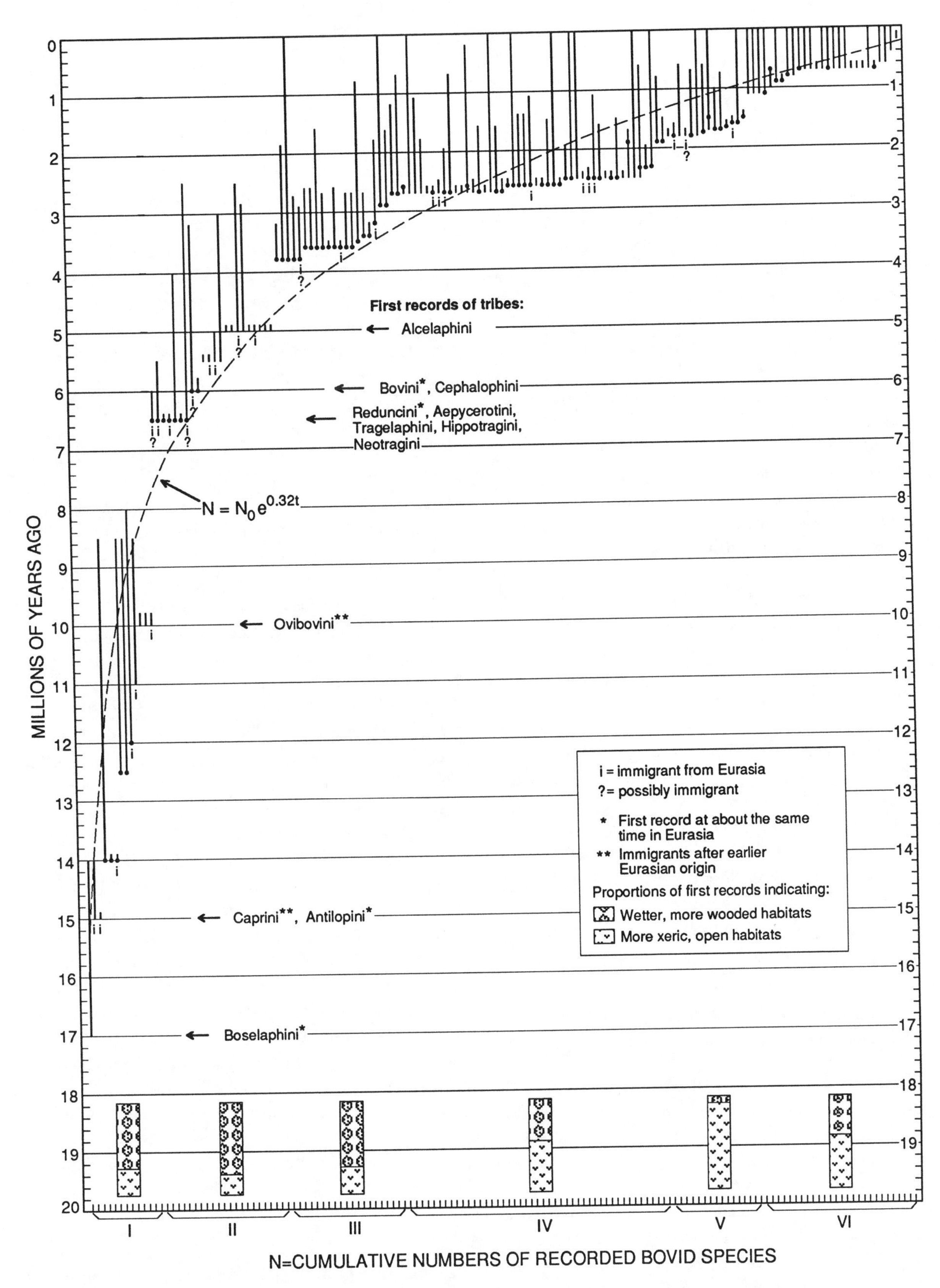

Fig. 27.3. Durations of African fossil bovid species. From left to right, in Groups I through VI, they correspond to taxon numbers in app. 27.2 as follows. Group I: 20, 60, 36, 21, 37, 62, 35, 22, 61, 63, 55, 38, 47. Group II: 10, 23, 33, 66, 109, 112, 124, 1, 25, 116, 9, 48, 125, 2, 12, 24, 27, 39, 58, 71, 72, 126. Group III: 17, 18, 110, 129, 143, 3, 34, 44, 73, 98, 113, 114, 123, 145, 43, 107, 128, 56, 130, 134. Group IV: 4, 8, 11, 26, 31, 40, 41, 49, 57, 59, 81, 99, 115, 117, 118, 135, 138, 141, 142, 144, 97, 16, 51, 54, 69, 87, 92, 100, 122, 146, 13, 14, 15, 64, 65, 67, 80, 86, 93, 96, 101, 102, 119, 50, 111, 137. Group V: 5, 83, 46, 70, 84, 85, 103, 147, 42, 90, 91, 104, 76, 68, 77, 79. Group VI: 28, 53, 88, 136, 105, 19, 127, 95, 140, 6, 30, 32, 45, 52, 74, 78, 82, 89, 94, 106, 108, 120, 132, 121, 139, 75, 133. See text for use of the equation of the dashed curve $N = N_0e^{0.32t}$. Radiometric or paleomagnetic dates for FADs are indicated by solid circles.

changes in biases. My particular focus is on the correct interpretation of a FAD pulse, that is, of a cluster in time of first appearances of distinct morphologies in the fossil record. (The basic arguments for FADs also apply, with some changes, to a pattern of LADs). The range of potential causes of a FAD pulse includes combinations of climatic change, biotic interaction, taphonomic bias, and turnover. In chapter 3 (especially in app. 3.4) I argue that if any significant pulses of FADs are found in the fossil record, then climate-driven widespread change in taphonomic conditions, or/and climate-driven change in turnover rates are the only candidate causes that we need to unravel. (Recall that climatic change may have more than one cause, including local or widespread tectonism. Thus, the concept of climatic forcing of taphonomic bias or turnover includes the possibility that tectonic changes forced the climatic changes.) In sum, a FAD and/or LAD pulse is only a pattern observed in the fossil record. It is not synonymous with a turnover pulse. Nor does it necessarily reflect a turnover pulse, although it may do so. The challenge is to distinguish the taphonomic and turnover contributions to that pulse. In keeping with this, I shall first address whether there are statistically significant FAD pulses in the Plio-Pleistocene African antelope record and then examine any such pulses to determine their causes. Unfortunately the African record for the Late Miocene to Early Pliocene is not yet of sufficient quality to permit such analyses (Figs. 27.1–27.3). Nevertheless, before I concentrate on the better Plio-Pleistocene record, I shall briefly comment on Late Miocene bovid turnover.

The Problem of Earliest Records of Major Monophyletic Groups during the Late Miocene

The latest Miocene marks the global advent of eight major distinct groups of antelopes, usually ranked as tribes, seven of which today dominate the African savanna biota (Greenacre and Vrba, 1984). Of the twelve tribes that are known in Africa over 17 myr, two certainly and a further four possibly originated in Eurasia (fig. 27.3). All remaining six probably arose in Africa, and appear 6.5–5.0 myr ago, that is, in less than 10 percent of the record. During this period there are no dramatic peaks at the level of species' originations (figs. 27.3, 27.4). But there is a high morphological diversity among those few species that do appear, as reflected by their membership of eight tribes.

It will remain difficult to analyze what this observed pattern means until the African Miocene record improves. Most of that record is sparsely sampled, and this bunching of FADs of high-ranking taxa between 6.5 and 5.0 myr ago could conceivably be nothing more than a taphonomic artifact. But it is worth noting that this pattern would also be consistent with several Late Miocene speciation and migration pulses, in relatively close succession, initiated by widespread climatic changes. There are three reasons why I regard this latter hypothesis as alive and worthy of future investigation: 1) Our cladistic results of mitochondrial DNA sequences (Gatesy et al., 1992), as well as other molecular analyses of bovids, suggest that most of the living tribes originated over a relatively short interval by repeated, rapidly successive lineage splitting (known as cladogenesis); and the numbers of molecular changes since basal tribal divergence, compared with rates of molecular evolution in other mammals, are compatible with Late Miocene divergences; 2) Similar paleontological patterns have been reported from other continents. For instance, the Miocene record of Eurasia, which is superior to the African one, shows marked turnover of mammals, especially 6.92–5.30 myr ago (Bernor, pers. comm; Bernor and Lipscomb, this vol.; see also in this vol. Webb, for North America, and Archer, for Australia); 3) There is evidence of major tectonic and regional and global climatic changes during the Late Miocene (Brain, 1981; Kennett, this vol.). The Late Miocene earth was not as cold as it became during later glacial cycles (Shackleton, this vol.). But what counts for organismal evolution is not the absolute magnitude of a climatic variable over a given time but its change relative to previous patterns. Were there major changes during the Late Miocene in dominance pattern of the astronomical cycles, such as deMenocal and Bloemendal (this vol.) report for ca. 2.8–2.7 and 0.9 myr ago? Were there climatic departures to new temperature minima that were severe relative to the previous patterns? And did these changes affect species turnover in many parts of the world? These questions remain open.

Plio-Pleistocene Antelopes: Analyses and Results

Accumulation Rate of Distinct Fossil Morphologies: An Exponential Null Model. The pattern of change in numbers of species per unit time has been measured as the difference between rates of speciation and extinction

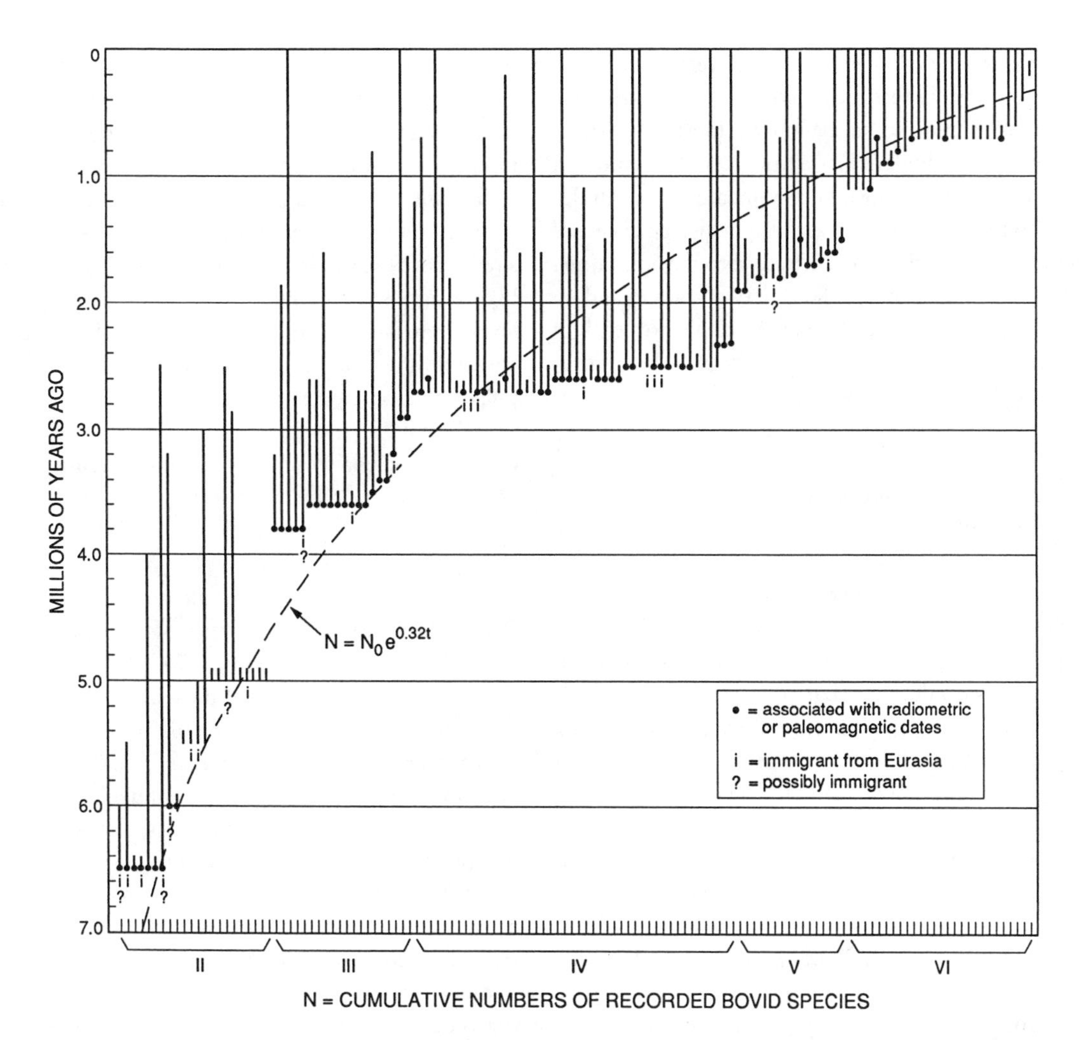

Fig. 27.4. Durations of African fossil bovid species over the past 7 myr. See fig. 27.3 for key to species in Groups II–VI, and text for use of the equation of the dashed curve $N = N_0e^{0.32t}$.

($R = S - E$; see Stanley, 1979). The analysis I present here does not require FADs to represent species, does not refer to extinctions, and acknowledges that FADs may or may not reflect coeval turnover. (I will not be able to avoid reference to "species" in what follows, because the taxa I discuss are generally recognized as separate species. But note that this model can be explored simply in relation to distinct taxa.) The analysis addresses the following question: How does the cumulative total of all FADs of distinct bovid morphologies recognized in the African fossil record change through time? This cumulative total is represented by the line formed when FADs are arranged in rank order, as in Figures 27.3 and 27.4. I modeled the null hypothesis, H_0, that the exponential rate of accumulation, A, is averagely constant through long time, using this equation:

$$N = N_0e^{At} \quad (1)$$

Over any interval, one can set the starting time at $t_0 = 0$, and t is the time elapsed since t_0.

In this sample, 16 myr ago at t_0, the number of distinct morphologies is $N_0 = 1$, and at time t the total is N. If all distinct taxa known as fossils (some are still be extant; see table 27.1) since 16 myr ago are counted, then $N = 161$ by time $t = 16$ in the Recent. Using equation (1), I estimated the accumulation rate A for this interval and plotted the cumulative curve under H_0 (figs. 27.3 and 27.4). The model in equation (1) provides a useful null hypothesis for the following reasons: Overall, the major cause of increase in recorded distinct morphologies in a group must surely be speciation by lineage splitting, which is expected under H_0 to accumulate the number of taxa at an averagely constant, exponential rate. Under H_0, taphonomic and other biases might also be expected to act at an averagely constant rate through long time, because they come from a wide variety of sources and, in this case, involve a taxon set and geographic area each of which is very extensive and heterogeneous. One can rewrite equation (1) as follows:

$$\ln (N/N_0) = At \qquad (2)$$

If one plots ln N against time, the slope $A = (\ln N_i - \ln N_0)/(t_i - t_0)$ of the line over any time interval, t_i to t_0 ($t = t_i - t_0$), is the FAD accumulation rate. This method allows quick visual assessment of any deviations from the straight line expected under H_0. Figure 27.5 depicts examples of such graphs; note the differences between the FAD pulses produced by **b** (a prior gap in the record), **c** (a subsequent gap) and **e** (coeval speciation and/or immigration). One can apply various significance tests to such a logarithmic plot of an observed FAD accumulation to determine which of the theoretical patterns in Figure 27.5, or additional ones, most closely resemble the fossil record.

Tests for Significant FAD Pulses and Gaps in the Plio-Pleistocene Bovid Record. I applied iterative χ^2 tests to the record of the past 3.6 myr (app. 27.4) to answer this question: Has the cumulative rate of appearance in the record of distinct morphologies remained averagely constant? If not, for which 0.1 myr intervals is H_0 rejected? These tests pointed out five 0.1 myr-long intervals of significantly higher numbers of FADs than expected under the null hypothesis. In decreasing order of importance, these are: ca. 2.7 myr, 2.5 myr, 0.7 myr, 2.6 myr, 3.6 myr, and 1.8 myr (app. 27.4). The period 2.7–2.5 myr stands out particularly.

For a second test, described in table 27.2, the bovid accumulation totals were analyzed as ln N relative to time (see fig. 27.5; and plot **a** in fig. 27.6, although the statistics in this figure do not refer to this test but to the next one). The short-term rate of FAD accumulation for each 0.3 myr-long interval is estimated by the slope a_i between the earliest and latest values of ln N in that interval. The mean α, of all the slopes over 0.3 myr-long segments, is taken to represent the background, mean, short-term rate of FAD accumulation. I obtained 99 percent confidence limits for α ($P[C_L \leq \alpha \leq C_U] = 0.99$). Table 27.2 gives all those 0.3 myr intervals over the past 3.6 myr having rates a_i that lie above or below those limits. These intervals are respectively taken to be characterized by significant pulses and gaps in bovid FAD accumulation.

The results agree with those of the previous test in singling out the FAD increases after 2.8 and 0.9 myr ago as highly significant pulses. At the 95 percent confidence level, there is also a pulse between 1.9 and 1.7 myr ago. I conclude that the bovid pattern does show significant pulses, the most prominent one occurring 2.7–2.5 myr ago. Without further analysis, however, the relative contributions of changes in taphonomic regimes and increased turnover remain unclear.

Are the Observed FAD Pulses Artifacts of Taphonomic Gaps, or Do They Reflect Increased Turnover, or Both? Because the dates for most of the FADs ca. 0.7 myr are in doubt (app. 27.2 and fig. 27.4), I shall not pursue further whether this is a real pulse or an artifact of the subsequent gap. The rate of the big pulse after 2.8 myr, with a slope of 2.18, falls more than three standard deviations above the upper 99 percent confidence limit in table 27.2. The pattern of slopes in figure 27.6 suggests that this pulse cannot be accounted for substantially by the gap that follows it, 2.3–2.0 myr ago. (Even if all gaps after 2.5 myr are excised and the other parts of the curve are spliced together, the rate from 2.8 to 2.5 myr remains significantly higher than the subsequent rate.) The possible effects on the 2.7–2.5 myr pulse of the preceding gaps (the 3.2–3.0 myr gap and especially the large 4.9–3.9 myr gap, table 27.2) need careful analysis. The method described in figure 27.6 is based on calculation of the 95 percent confidence limits (see line c) of the regression (line **b**) for the interval 6.5–5.5 myr before the big gap set in. The intersection of line **c** and the 2.7–2.5 myr FAD pulse (solid arrow) is the estimated separation between the proportion of FADs in the pulse that are artifacts of the previous gaps in the record and the proportion due to coeval speciation and immigration.

Lines **d** and **e** show regressions for the intervals before and after the 2.7–2.5 myr pulse. If one disregards periods earlier than 3.6 myr ago, one can note that these lines, **d** and **e**, do not differ significantly in slope, yet their standard error limits indicate a highly significant displacement across 2.7–2.5 myr. This suggests the pattern in figure 27.5, line **e**, of a pure speciation and/or immigration pulse close in time to the FAD pulse. Such a local analysis is unsatisfactory, however, because it misses

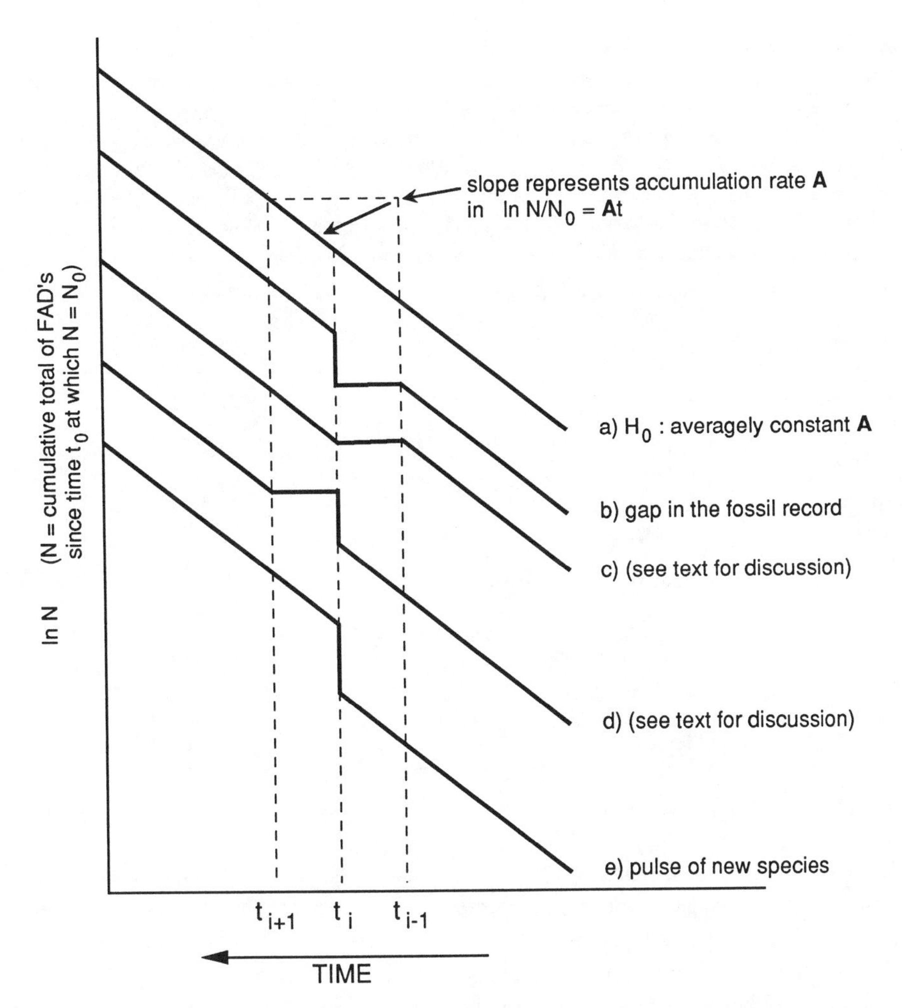

Fig. 27.5. Diagrammatic examples of the curve of the cumulative total of FADS (N, here converted to ln N) through time. The slope $A = (\ln N_i - \ln N_0)/(t_i - t_0)$ of the line over any time interval $t_i - t_0$ is the FAD accumulation rate. (Based on $N = N_0e^{At}$, where $t = t_i - t_0$, or time elapsed since the start of the interval, N_0 = number of taxa at the start). The interpretations of a to e are given as examples. One can think of other examples and combinations of the a to e versions. Turnover here refers to speciation and/or immigration.

a) H_0: no significant changes in background rate from either taphonomic or turnover sources.
b) A FAD pulse at time t_i that is an artifact of a gap in the fossil record between $t_{i-1} - t_i$, with recovery only from t_i onward of species that were added by turnover in the real world during $t_{i-1} - t_i$. An alternative, that turnover decreased only during t_{i-1} to t_i and showed an equal increase at t_i, is regarded as much less likely.
c) A decrease in rates of fossil preservation, a gap, without subsequent recovery of the taxa missing during the gap, and/or a decrease in turnover rates during t_{i-1} to t_i, with resumption of previous net rates from t_i onward. This can resemble a FAD pulse at time t_{i-1} in data that is not log-transformed.
d) An increase in rates of fossil preservation and/or turnover at t_i, followed by an equal decrease in one or both factors during t_i to t_{i+1}, with resumption of previous net rates from t_i onward.
e) A pulse of speciation and/or immigration at time t_i, with the accumulation pattern before and after that agreeing with H_0. An alternative, that only once in a long time did fossil preservation increase greatly, is regarded as much less likely.

the potential influence of the large gap in the record over 4.9–3.9 myr (figs. 27.1–27.3).

The results of the analysis using the regression over 6.5–5.0 myr (lines **b** and **c** in fig. 27.6) suggest that about twelve FADs, or one quarter of the pulse (below the solid arrow), do not represent coeval turnover but are "old species" that were not recovered in the previous record owing to gaps, whereas three quarters of the pulse does represent coeval turnover. That is, relative to the gap-pulse pattern characterizing the 6.5–5.0 myr interval, the subsequent gaps over 4.9–3.9 myr and 3.2–3.0 myr were compensated for approximately by the level of the FAD total indicated by the solid arrow in Figure 27.6; and relative to the 6.5–5.0 myr pattern there is a large com-

Table 27.2. FAD Pulses and Gaps in the African Bovid Record since 3.6 myr Ago
All 0.3 myr-long intervals are listed during which the short-term rate a_i of FAD accumulation (estimated by the slope between the earliest and latest values of ln N in that interval) departs significantly from a background mean (α) of rates per 0.3 myr. The mean rate α was calculated jointly over the periods 6.5–5.0 and 3.6–0 myr using two schemes of time subdivision in order to sample all possible 0.3 myr-long intervals within those periods. For each scheme, α, the 99 percent confidence limits for α ($P[C_L \leq \alpha \leq C_U] = 0.99$), and each 0.3 myr interval with an a_i outside those limits, were determined. For intervals not cited, there was no significant deviation from α, that is, $C_L \leq a_i \leq C_U$. The large gap between 4.9 and 3.9 myr (see fig. 27.4) is included to set the 3.6–0 myr period in context. It is cited in parentheses because it was not included in the analysis of α. Note that one additional interval showed a significant deviation at the 95 percent level: a pulse over 1.9–1.7 myr.

	The set of intervals (in myr): 6.5–6.3, 6.3–6.1, . . . 5.3–5.1, and 3.5–3.3, 3.3–3.1, . . . 0.3–0.1		The set of intervals (in myr): 6.4–6.2, 6.2–6.0, . . . 5.2–5.0, and 3.6–3.4, 3.4–3.2, . . . 0.2–0	
Number of intervals, n, each of which has a rate a_i	24		25	
Mean rate α	0.30		0.29	
Standard deviation	0.43		0.47	
99% confidence intervals:	$C_L = 0.06$, $C_U = 0.54$		$C_L = 0.03$, $C_U = 0.55$	
Resulting gaps and pulses:	*FAD Gap:*	*FAD Pulse:*	*FAD Gap:*	*FAD Pulse:*
	$a_i < C_L$	$a_i > C_U$	$a_i < C_L$	$a_i > C_U$
	(4.9–3.9)			
			3.2–3.0	
		2.9–2.7, 2.7–2.5		2.8–2.6, 2.6–2.4
	2.3–2.1		2.2–2.0	
	1.5–1.3		1.6–1.4, 1.4–1.2	
		0.9–0.7		0.8–0.6
	0.5–0.3		0.6–0.4, 0.4–0.2	

ponent of the 2.7–2.5 myr pulse that cannot be accounted for by the previous gaps. That still leaves questions. Could it be possible that this FAD pulse is, after all, due mostly to climate-driven taphonomic factors (recall again that I am using the term *climate* to include the effects of tectonism) and that there was not only recovery of species missed during the prior gaps—a recovery matching the highest preservational levels of the 6.5–5.0 myr interval—but also an additional unprecedented increase in preservation potential near this time? The regression analyses suggest that, for this to be true, nearly the entire ca. 13 myr-long bovid record before 2.7 myr and that after 2.5 myr would have to represent massive "gaps" (or at least times of very much lower preservation rates than during 2.7–2.5 myr). The 2.7–2.5 myr interval would be the only one during which such extraordinarily favorable taphonomic circumstances were ever attained in bovid evolution. I find this unparsimonious. But I shall nevertheless consider a way to falsify that hypothesis in a later section.

I conclude that a proportion of the 2.7–2.5 myr FAD pulse can be accounted for by the previous gaps, while a larger proportion (ca. 75 percent) cannot be explained in that way without recourse to a taphonomic event that was unique over the past 16 myr. This bovid pattern is a hybrid between those in figure 27.5, line **b**, of a gap and figure 27.5, line **e**, of a coeval pulse of new species. Three caveats should be considered: 1) Makapansgat Member 3, from which come several 2.7 myr FADs, may be older, 2.8 or even 2.9 myr (app. 27.3); 2) if all systematic evaluations that differ from mine are invoked together, the pulse is smaller by six FADs (app. 27.2); and 3) a more conservative confidence interval, of 99 instead of 95 percent, might be considered better for the test in figure 27.6. The earlier date for Makapansgat Member 3 would not diminish the pulse but only give it an extra "step." If both the second and third caveats are invoked together, eighteen FADs still exceed the upper 99 percent confidence limit, and there remains a significant turnover pulse in terms of the analyses in appendix 27.4, table 27.2, and figure 27.6.

Insofar as any pulse is produced by coeval turnover, the evolutionary phenomenon would have to have occurred in the 100 kyr (or 200 kyr?) interval preceding the onset of the observed pulse. I shall continue to discuss the observation in the record as the "2.7–2.5 myr FAD pulse." But any significant turnover component that un-

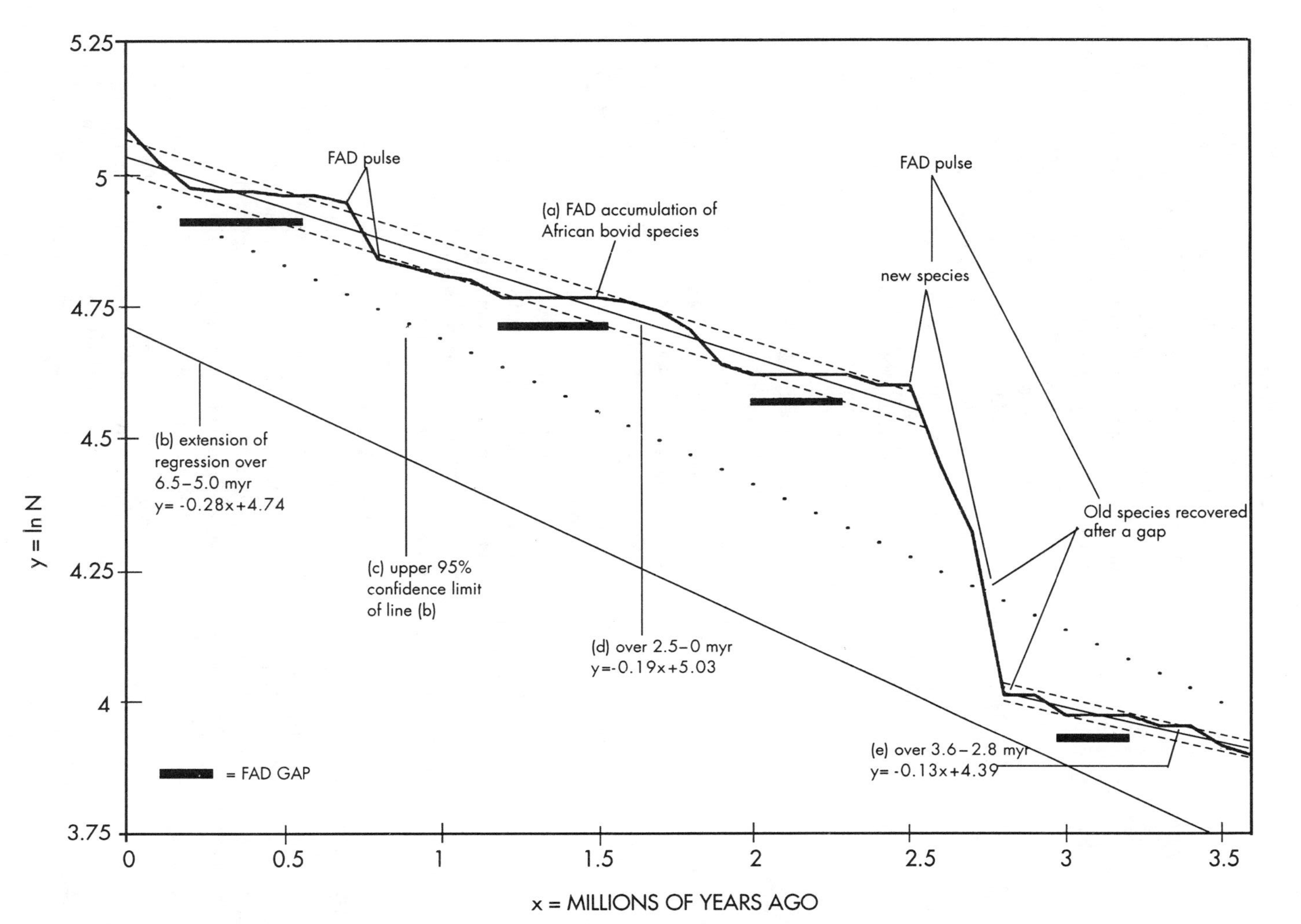

Fig. 27.6. The ln curve of the cumulative total (N) of FADS of African bovid species during 3.6–0 myr, a. The slope $A = (\ln N_i - \ln N_0)/(t_i - t_0)$ over any interval $t_i - t_0$ is the accumulation rate of FADS. (This slope is based on equation [2] in the text: $\ln (N/N_0) = At$, where $t = t_i - t_0$, the time elapsed since start of interval $N_0 = N$ at t_0.) The analysis in table 27.2 indicates the significant pulses over 2.7–2.5 myr and ca. 0.8–0.6 myr ago, as well as four gaps in the record as shown. A second analysis resulted in proportional separation of two components of the 2.7–2.5 myr pulse (shown by the intersection of lines a and c): the fraction of species below the line (12 species of 44, or 27 percent) is the estimated proportion accounted for by two previous gaps in the record (A, 4.9–3.9 myr; and B, 3.2–3.0 myr); while the fraction above (32/44, or 73 percent) is the proportion contributed by speciation and/or immigration during this interval. This separation is based as follows. Regression line b represents the background pulse-gap pattern over 6.0–5.0 myr, before gaps A and B set in. Its upper (line c) and lower 95 percent confidence limits represent the magnitude of the largest pulses and gaps over 6.5–5.0 myr. If the pattern after 3.0 myr returned to the 6.5–5.0 background pattern, with recovery of the excess of missing species owing to the gaps, the maximum value of a positive excursion after 3.0 myr should not exceed upper confidence limit c. Regressions d and e are discussed in the text. Statistics of regression lines are: b) $r = 0.87$, $SE(y) = 0.12$, $SE(\alpha) = 0.01$; c) $r = 0.97$, $SE(y) = 0.33$, $SE(\alpha) = 0.01$; d) $r = 0.98$, $SE(y) = 0.03$, $SE(\alpha) = 0.01$; e) $r = 0.95$, $SE(y) = 0.01$, $SE(\alpha) = 0.02$ (r = correlation coefficient, SE = standard error, SE(y) = SE of y estimates that are shown for d and e, $SE(\alpha)$ = SE of regression slope).

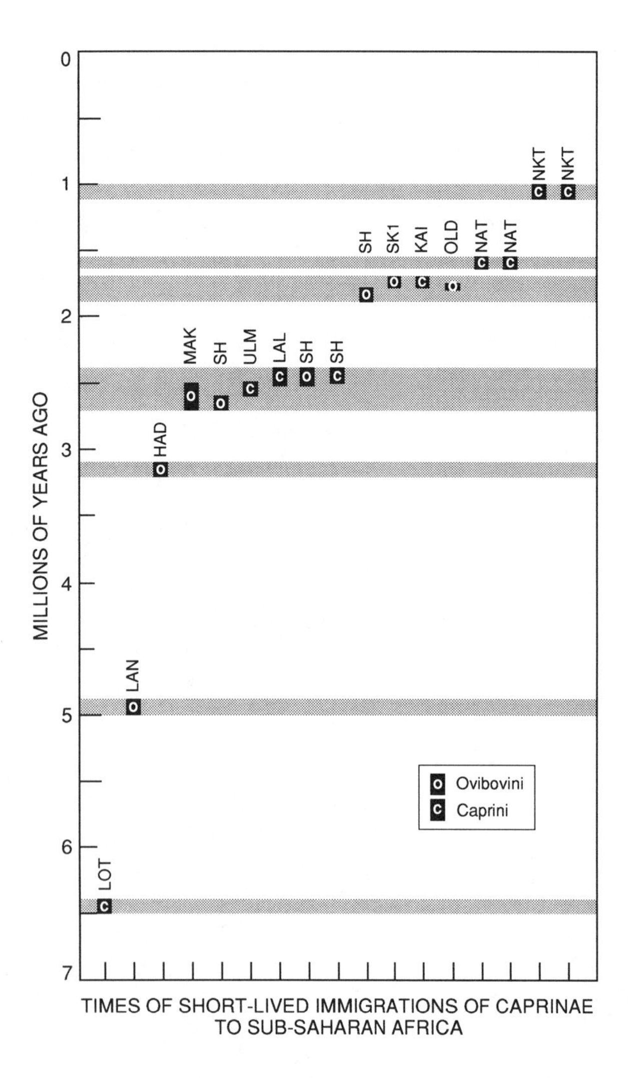

Fig. 27.7. Times of immigration to, and short-lived presence in, sub-Saharan Africa of species of Caprinae, tribes Ovibovini and Caprini. See app. 27.1 for acronyms of fossil sites, and app. 27.2 for taxon numbers of the seventeen records from left to right: 66, 58, 56, 57, 59, 69, 67, 59, 65, 59, 56, 59, 56, 68, 70, 69, 67. This figure illustrates the hypothesis that all these instances of Caprinae in African fossil assemblages since the Late Miocene represent separate immigration events from the north. Note that taxa 56, 59, 67, 69, and 70 in app. 27.2 all contain two or more records, which are here shown separately.

derlies it is more realistically assigned to the interval 2.8–2.5 myr, as in the "2.8–2.5 myr *turnover* pulse."

Eurasian-African Migrations can be inferred by vicariance biogeography, which indicates, for instance, that certain gazelle lineages, the genus *Antilope*, and the cold- and often high-altitude-adapted Caprinae originated and predominantly proliferated in Eurasia. Among these lineages, all of which have African FADs at certain times, the Caprinae are particularly suitable for analysis of immigration episodes, because their Eurasian origin is generally accepted, cladistic analysis of many species suggests their monophyly (Gatesy et al., 1992), and they include many fossil and living species. Thus, I scored incursions into Africa of species of Caprinae in the tribes Ovibovini and Caprini (fig. 27.7), and I argue that these data convey special information on global climatic changes. The features of this pattern include the following: most of the different morphologies involved during the successive episodes are distinct from one another and show no signs of close phylogenetic relationship with one another or with previous African lineages; each caprine taxon appears for only a short time in Africa; and the incursions after 3 myr ago each include more than one immigrant. Taken together, this suggests separate immigration events in response to onsets of global cooling and opening of landbridges, as I argued in chapter 3. As temperatures decreased, open habitats that were previously situated at higher latitudes and altitudes evidently moved toward the equator and into lowlands. The elevated cooler areas of eastern Africa are expected to have acted as conduits toward the African equator for such Eurasian incursions.

Of the seven episodes involved, the two earliest, ca. 6.6 and 5 myr, provide the least satisfactory evidence. They are dated by biochronology only and occur in sparsely sampled periods of the African record. The last caprine immigration is also unsatisfactory in that its date could belong anywhere between 1.39 and 0.8 myr (app. 27.2, note 26). The records ca. 3.2 myr, between 2.7 and 2.5 myr, ca. 1.8 myr, and ca. 1.6 myr are of most interest. The times 1.8 myr and 2.7–2.5 myr are marked both by high FAD totals (app. 27.4) and by caprine incursions. In the case of the 1.8 myr event, the caprine influx tips the balance toward a high FAD total at the 0.95 level of significance (table 27.2). But for the 2.7–2.5 myr interval, in which the immigrants total only seven alongside an estimated twenty-five new species by coeval speciation (out of forty-four FADs, twelve FADs are artifacts of gaps; see fig. 27.6), the two indicators give independent evidence of a climatic event.

There were two Pliocene emigration episodes. During the first, only one lineage, a *Hippotragus*, gave rise to an Eurasian immigrant, *H. brevicornis* (see tree in fig. 27.9), which appeared during the Tatrot stage in the Siwaliks ca. 3.05–2.6 myr ago (app. 27.2, note 33). As explained more fully in chapter 3, I suggest that this event occurred with the earliest onset of one of the

Pliocene warming phases, perhaps the beginning of the warming close to 3.0 myr ago (Dowsett et al., 1992; Poore, this vol.) just after the cooling spike ca. 3.1 myr ago (see Shackleton's data, this vol.). This time is expected to satisfy the requirement of a landbridge that was still available, as polar ice had not yet melted sufficiently to flood the landbridge, *and* the requirement that the mesic wooded savanna habitats inferred for *Hippotragus* would have already extended to higher latitudes owing to global warming.

The second Pliocene episode involved five African emigrants very close to or soon after 2.7 myr: A *Damalops* population left Africa, as evidenced by the records of *D. palaeindicus* at ca. 2.6 myr from Tadzhikistan (Dmitrieva, 1977) and from the Pinjor stage in the Siwaliks (Pilgrim, 1939). Also close to this time, emigrations by early *Oryx* and *Hippotragus* populations gave rise to *O. sivalensis* and *H. bohlini* (fig. 27.9), and the reduncine kob and lechwe lineages to *Vishnucobus patulicornis* and *Sivacobus palaeindicus*, respectively, all known from the Pinjor stage. I suggest that this excursion occurred as a result of the earliest simultaneous presence of two factors: 1) the appearance of a landbridge between Africa and eastern Asia across Arabia owing to sea-level lowering; and 2) the linking up of previously vicariated steppe habitats at high altitudes in East Africa with the southward-migrating, similar habitats from Eurasia.

An intriguing question is why more taxa migrate into than out of Africa (fig. 27.8). A possible explanation is the model I proposed in chapter 3, which invokes unequal conditions of habitat availability akin to a biased traffic light. Migrations require both a landbridge and suitable habitat across that landbridge. When the earth is cold, the landbridge is open (the traffic light is green) for a long time, because cool habitats extend to the equator and sea-level is low. Most migrants should traverse toward the equator during cooling, the vector dictated by

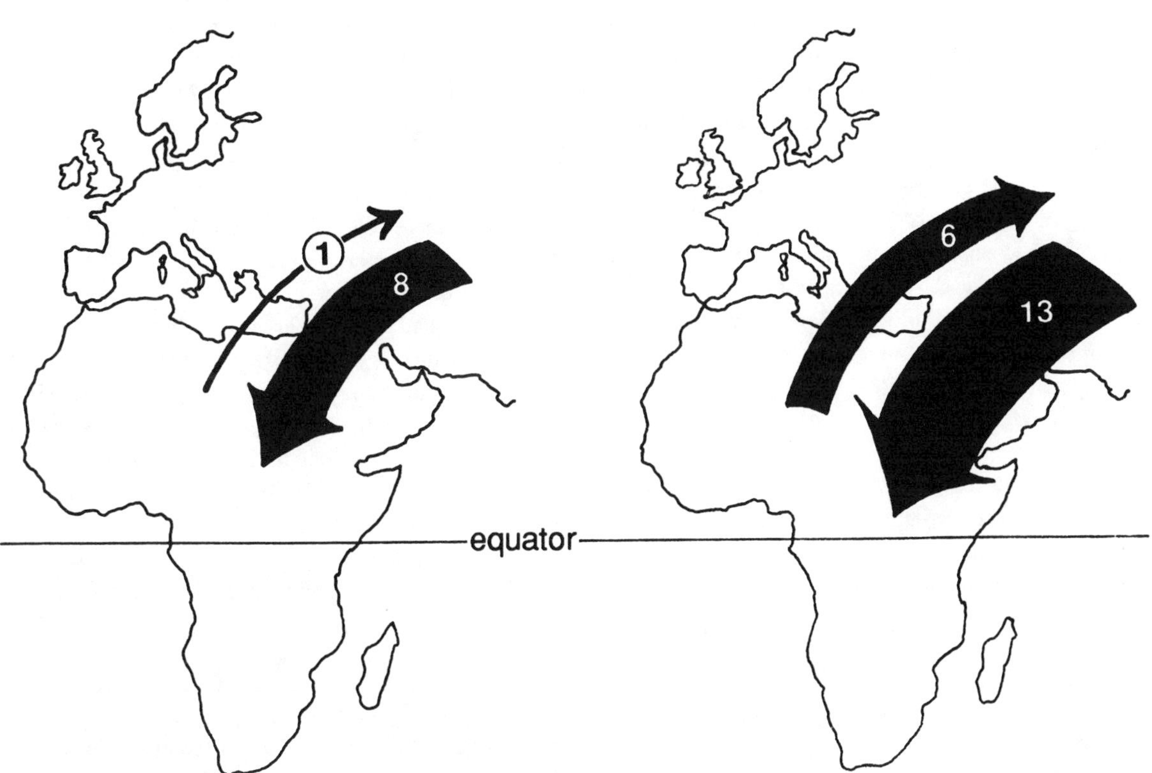

Fig. 27.8. Imbalance in transcontinental migration of Bovidae between Eurasia and Africa during the Miocene (Thomas, 1984) and the Plio-Pleistocene (fig. 27.4). The number of species that migrated between the continents is indicated on the arrows.

their moving habitats on a cooling earth. A few African species, such as *D. palaeindicus* and the four other African emigrants close to 2.7 myr that were confined to high altitudes during the previous warm phase, would also gain habitat passage to Asia to some extent as their habitats link up through lowlands with similar ones moving in from the north. In contrast, when the earth is warm, the landbridge is reduced or gone for most of the time. Only during the short lag between onset of global temperature change and ice-volume response can migration from the equator toward higher latitudes of warm-adapted forms occur. But such African emigrants are likely to be few, like the lone species of *Hippotragus* somewhere close to 3 myr ago, because this traffic light turns red again very quickly as sea-level rise reduces landbridges. Even if intergalacials and glacials were equivalent in length, the interplay of eustasy and temperature would ensure that more species should move toward lower latitudes than vice versa.

Cladistic Inference of Times of Speciation. According to a hypothesis of a stochastically constant accumulation rate of bovid FADs, the species in the 2.7–2.5 myr FAD pulse arose at various, randomly distributed, previous times such that the FAD pulse is merely a taphonomic artifact (see fig. 3.5, this vol., for a diagram illustrating this). A comparison of cladograms with fossil chronology offers the possibility to falsify this hypothesis. The cladistic tree in fig. 27.9, for extinct and living species of Hippotragini antelopes, provides an example of this approach (which is explored more fully in chap. 3). Of the thirty-seven endemic African taxa that appear between 2.7 and 2.5 myr ago (excluding the seven Eurasian immigrants of the total of forty-four FADs; fig. 27.4), four are hippotragine (fig. 27.9): *Oryx* sp. 1, *Hippotragus cookei, H. gigas*, and *H. equinus*. The FAD of each of these taxa happens to determine the minimum age stated on the nearest branching node below it. For instance, *Oryx* sp. 1 has the earliest securely dated FAD, 2.6 myr, in the *Oryx* clade from node 4 and thereby determines the minimum age of the clade. Note that *O.* (*Sivoryx*) *sivalensis* and *H.* (*Sivatragus*) *bohlini* are Asian taxa whose FADs in the Siwalik Pinjor zone at ca. 2.6 myr or later (app. 2, note 33). agree with those of their African relatives, while the FAD of the *Hippotragus* sp. from Sahabi, Libya, may agree as well but remains in doubt (app. 27.2, note 10).

One can inspect cladograms like the one in figure 27.9 for potential ancestors. Any taxon A that has no advanced characters of its own, but is wholly plesiomorphic relative to its sister-group B, allows the hypothesis that A is potentially the directly ancestral stem of B (see fig. 3.7, this vol., for a diagram illustrating this). For instance, *H. cookei*, with no advanced characters of its own, is a potential ancestor of the branching event at node 9 and has an appropriately timed LAD within 2.8–2.6 myr, which is also its FAD. Note that in this case one of the taxa with FADs during 2.7–2.5 myr is potentially ancestral to two others that also have FADs later during the same interval. For *H. cookei* itself there is no potential ancestor known. *H. brevicornis* is disqualified from ancestry by an autapomorphy, although its LAD dates to a time that would be appropriate. Thus, a hypothesis of potential ancestry requires that the relevant parts of the cladogram be correct and that the potential ancestor really be the direct ancestor that ceased to exist as it gave rise to the descendants. If these assumptions are valid, the branching at node 9 is limited to the interval 2.8–2.6 myr. The branching at node 12 is inferred to have occurred between 2.6 and 2.5 myr ago. For *Oryx* sp. 1, in contrast, there is no potential ancestor known, because the sister-group *Praedamalis* has a synapomorphy that disqualifies it from direct ancestry of oryxes.

Of the thirty-seven endemic African FADs between 2.7 and 2.5 myr ago, twenty-three have been cladistically analyzed, including the four Hippotragini (fig. 27.9). I applied this kind of analysis to the taxa in appendix 27.5, including the twenty-three endemic African FADs as well as Eurasian taxa that belong to the mainly African clades and emigrated. Figure 27.10 shows phylogenetic trees of the cases in which potential ancestors with FADs $\leq$ 2.9 myr were found. Note that these cases are not beset by the problem of gaps in the record insofar as the record

→

Fig. 27.9. Consensus tree for hippotragine and related fossil and Recent taxa, based on twenty-five skull characters with advanced (apomorphic) character states shown as solid black circles (from Vrba and Gatesy, in press, who discuss the numbered character changes). This is an example of how cladistic analysis can be used together with fossil chronology to estimate what proportion of taxa in a FAD pulse speciated closely before that FAD pulse. See app. 27.2 for minimum estimated dates of branching nodes. Note that *Oryx* participates in the 2.8–2.5 myr FAD pulse, yet no potential direct ancestor is known for this taxon. However, each of the two main sister lineages of the clade from node 9 within *Hippotragus* is represented in this FAD pulse by *H. gigas* and *H. equinus* FADs at 2.6 and 2.5 myr ago, and *H. cookei* is a potential direct ancestor based on both the character state distribution (it has no autapomorphies) and its temporal record (see tree for *Hippotragus* in fig. 27.10).

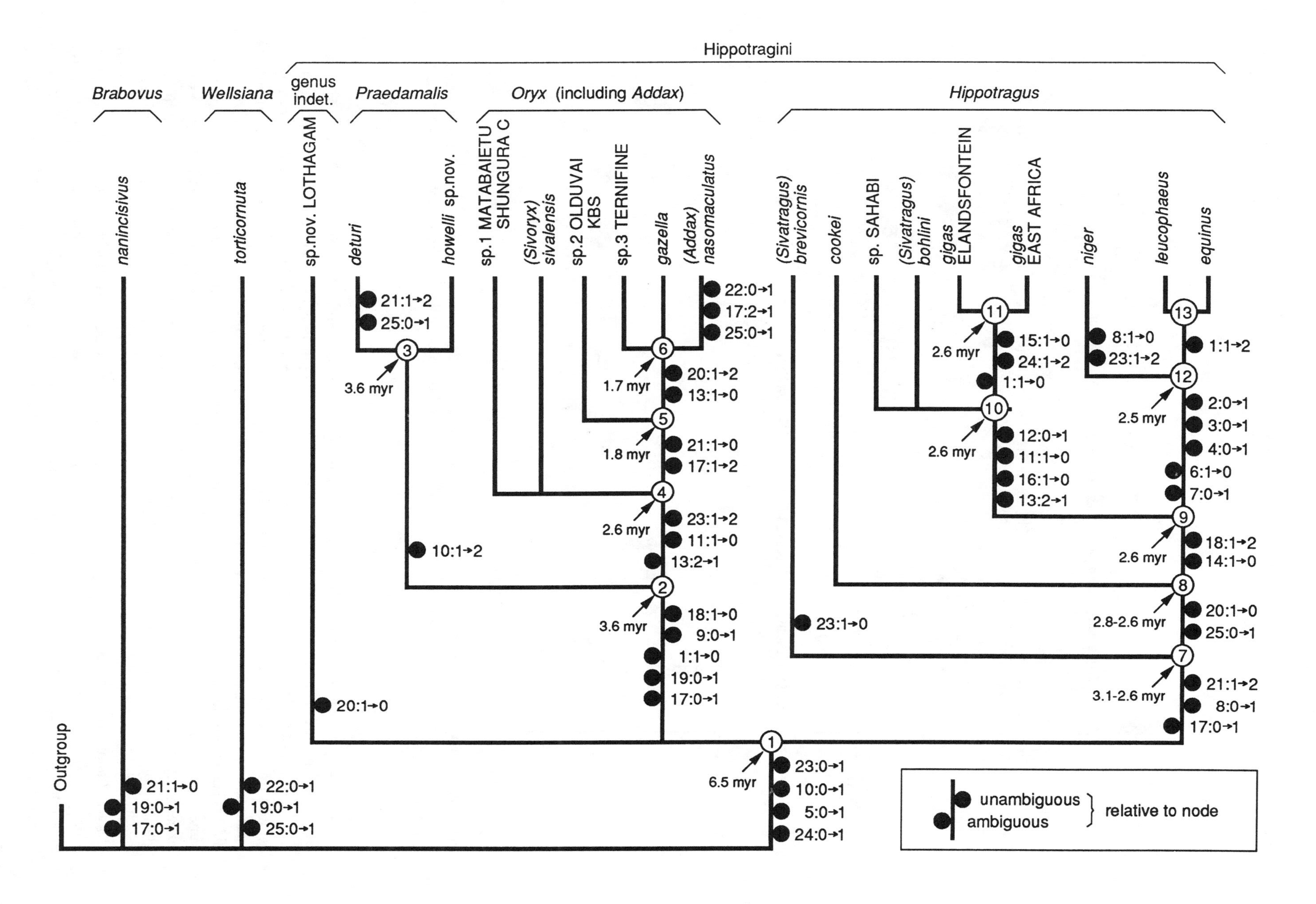

Hippotragini
Brabovus
Wellsiana
genus indet.
Praedamalis
Oryx (including Addax)
Hippotragus
nanincisivus
torticornuta
sp.nov. LOTHAGAM
deturi
howelli sp.nov.
sp.1 MATABAIETU SHUNGURA C
(Sivoryx) sivalensis
sp.2 OLDUVAI KBS
sp.3 TERNIFINE
gazella
(Addax) nasomaculatus
(Sivatragus) brevicornis
cookei
sp. SAHABI
(Sivatragus) bohlini
gigas ELANDSFONTEIN
gigas EAST AFRICA
niger
leucophaeus
equinus
21:1→2
25:0→1
3.6 myr
10:1→2
22:0→1
17:2→1
25:0→1
1.7 myr
20:1→2
13:1→0
1.8 myr
21:1→0
17:1→2
2.6 myr
23:1→2
11:1→0
13:2→1
3.6 myr
18:1→0
9:0→1
1:1→0
19:0→1
17:0→1
20:1→0
2.6 myr
15:1→0
24:1→2
1:1→0
2.6 myr
12:0→1
11:1→0
16:1→0
13:2→1
8:1→0
23:1→2
1:1→2
2.5 myr
2:0→1
3:0→1
4:0→1
6:1→0
7:0→1
2.6 myr
18:1→2
14:1→0
2.8-2.6 myr
20:1→0
25:0→1
23:1→0
3.1-2.6 myr
21:1→2
8:0→1
17:0→1
6.5 myr
23:0→1
10:0→1
5:0→1
24:0→1
Outgroup
21:1→0
19:0→1
17:0→1
22:0→1
19:0→1
25:0→1
unambiguous
ambiguous
relative to node

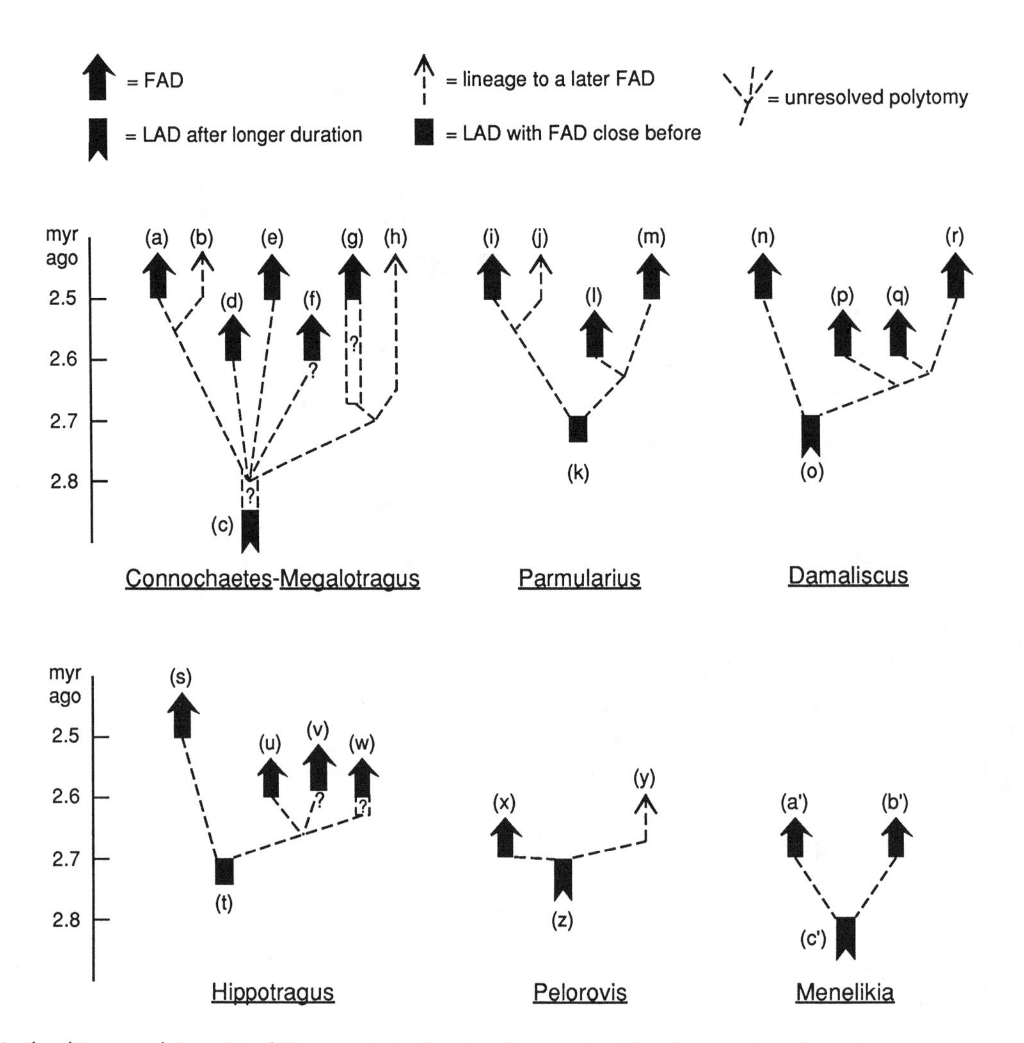

Fig. 27.10. The splitting events between 2.9 and 2.5 myr ago in six African bovid lineages inferred from cladistic analyses. The *Hippotragus* tree can be compared to the cladogram in fig. 27.9 as an example. See app. 27.5 for literature citations and for details of taxa (a) through (c′). Dating uncertainties are marked by a question mark (see app. 27.2).

indicates the interval for the splitting event relatively accurately in spite of any gaps.

Cladistics can effectively distinguish possible anagenesis from necessary cladogenesis by the distribution of character states along branches. The fossil record confirms cladogenesis in cases where sister taxa overlap in time. More than 90 percent of the taxa in appendix 27.5 arose by cladogenesis according to the character-state distributions. Second, for more than 70 percent (17) of the cladistically analyzed African taxa with FADs between 2.7 and 2.5 myr, there are potential ancestors with LADs close to the FADs of those 17 descendants (figure 27.10). The remaining taxa might also have had a splitting event $\leq$ 2.9 myr ago, but either the potential ancestor has not been found or its LAD occurs earlier (app. 27.5). If the same proportions were to be found in the remaining 14 African endemics that first appeared 2.7–2.5 myr ago, it would mean that for about 27 of the 37 nonimmigrants the evidence does indicate lineage splitting at this time, whereas for about 10 taxa it does not. This figure happens to be close to the estimated number (12) of species in the 2.7–2.5 myr FAD pulse that were present before that time yet were unrepresented in the record owing to gaps (see fig. 27.6). Cladistic analysis of all the taxa will allow a more rigorous estimate. However, there is already evidence that at least a substantial proportion of the African-endemic FADs of 2.7–2.5 myr ago participated in a coeval lineage-splitting pulse. The present cladistic analyses also indicate the intervention of taphonomic biases: after all, every direct ancestor that is absent from the record (and, for that matter, every piece of lineage that is absent) attests to factors that precluded its presence in the record. Thus, for about one-third of the African-endemic FADs 2.7–2.5 myr ago, obliteration of the record previous to 2.7 myr by taphonomic biases is suggested.

In sum, bovid phylogenetic trees exhibit a typical, or at least prevalent, Late Pliocene pattern (fig. 27.10): a potential direct ancestor is last recorded between 2.9 and 2.7 myr ago. Its descendants are often first recorded between 2.7 and 2.5 myr and sometimes later, like taxon (y), at 1.9 myr; taxon (h), at 1.8 myr; and taxon (j), at 1.7 myr. The majority of trees in figure 27.10 underwent multiple splitting events in a short interval between 2.9 and 2.5 myr. The symptom of rapid lineage splitting is often what cladists call "unresolved polytomy," as exemplified by two of the clades in figure 27.10.

Ecology and a Test of the Hypothesis of "Relay Speciation." Comparisons of ecological associations of living antelopes (Greenacre and Vrba, 1984) with cladograms show that broad habitat specificities for variables such as vegetation cover and temperature can be heritable and characteristic for entire clades through millions of years (Vrba, 1987c, 1992). Thus, changing species proportions among particular sets of clades should also be good climatic indicators for Plio-Pleistocene strata. I counted the relative proportions of bovid FADs in two major habitat categories: 1) the more cold, arid, highland- and grassland-adapted contingent in which Alcelaphini (gnus, hartebeests, and allies) and Antilopini (gazelles and allies) dominate overwhelmingly and which also includes the genera *Oryx* and *Pelea* and the tribes Caprini and Ovibovini; and 2) the more warmth-, moisture-, and wood-cover-loving remainder. Figure 27.3 shows in a crude way how the proportions of African FADs in these two habitat categories have changed since the early Middle Miocene. One can, of course, use finer subdivisions of habitat specificity. For the test I want to report here, however, this simple dichotomy is sufficient to make the point.

In chapter 3, I proposed a hypothesis that one can term the relay model. The model predicts that different categories of lineages speciate and become extinct during different phases of the climatic oscillations, rather like runners starting up and ending during different phases of a relay race. A central assumption of this model is that long-term vicariance needs to precede most speciations (vicariance is fragmentation by physical change of a species' geographic distribution into isolated populations). One prediction is that, during a long-term cooling trend such as occurred after 3.0 myr ago (e.g., Shackleton, this vol.), warm-adapted species should enter vicariance and speciate first, before the more cold- and open-adapted ones do. Table 27.3 shows that the African bovid pattern 2.7–2.5 myr ago strongly supports this prediction. Another prediction of this model is that there are relays among major successive cooling and warming episodes over much longer time periods (see figure 3.3, this vol.). A long-term warming emphasizes relatively FADs of the warmer-adapted group, and vice versa. The FAD totals for the two groups in table 27.3 are nearly equal, but relative to the proportions over 3.8–2.8 myr, speciations of the colder-adapted group did increase 2.7–2.5 myr ago, and even more so by 1.5 myr ago (see base of figure 27.3). I offer the relay model as a preliminary hypothesis to explain why lineages of pigs, monkeys, and giraffids tend to have first appearances before and up to ca. 3.0 myr ago, whereas in open-adapted forms, like most bovids and rodents, waves of first appearances occur later, between 2.7 and 2.4 myr ago: pigs show two FADs ca. 5.5 myr, then none for a long time, then two further FADs ca. 3.0 myr, followed by only three more after 2.0 myr (White, this vol.). White (1985) singled out 3.0 myr as a time of suid FADs, and for African monkeys, Delson (1985) showed a huge peak of originations at 3.5–3.0 myr. This interval was marked by substantial warmth

Table 27.3. Numbers of African Bovid FADs at 2.7, 2.6, and 2.5 myr Ago
A comparison of the more cool-, arid-, highland-, and grassland-adapted contingent (Alcelaphini, gnus, hartebeests, and allies; Antilopini, gazelles, and allies; *Oryx*, oryxes, *Pelea*, rhebok) with the more warmth-, moisture-, and wood-cover-loving remainder. Eurasian immigrants (fig. 27.4) are not included. Thus, these FADs are inferred to represent the species that originated in Africa. A χ^2 comparison with Yates's correction, of 2.7 myr with 2.6–2.5 myr, indicates that the 2.7 myr bovid FADs include a significantly higher proportion of warm-closed-adapted taxa, whereas the 2.6–2.5 myr interval has a significantly higher proportion of cool-open-adapted taxa ($\chi_1^2 = 7.92$; $\chi_{1;0.01}^2 = 6.64$).

	Bovid Taxa		
Age of FAD	Cooler- and open-adpated:	Warmer- and closed-adapted:	Total
2.7 myr	3	12	15
2.6–2.5 myr	16	6	22
Total	19	18	37

and forest extent (Dowsett et al., 1992; PRISM, this vol.). Pigs and monkeys therefore seem to show many FADs during the climatic regime opposite to that for the highest peak of bovid FADs. Similarly, the recent biostratigraphic data for giraffids from East Turkana (Harris, 1991) and for primates from West Turkana (Harris et al., 1988) show more FADs between 3.5 and 3.0 myr than between 3.0 and 2.0 myr. Under the relay model these two categories of taxa are expected to be "climatically out of phase" with each other: the generally warm-adapted pigs, monkeys, and giraffids had evolutionary heydays during the warm, wet, wooded intervals before and ca. 3.0 myr, while the cooler- and more open-adapted forms, like most bovids and rodents, diversified during the subsequent major cooling trend in the Late Pliocene.

Some Developmental and Morphological Correlates of Climate can be mentioned here briefly, although these phenomena have so far hardly been studied in bovids. Vrba (1994) reviewed aspects of Bergmann's Rule, which holds that large bodies, often attained by prolonged growth, are commonly associated with cold, and Allen's Rule, which holds that extremities of the body (horns, limbs, ears, muzzles) show a relative decrease as temperature declines. Allen phenotypes probably evolve by the kind of paedomorphosis that occurs by reduction in rates of shape growth relative to size growth. It results in descendants that appear juvenilized relative to the ancestor, because in the ancestor the juveniles had relatively shorter extremities than the adults (Vrba, 1994). A second form of evolutionary juvenilization of a structure involves body enlargement with relatively even greater enlargement of the structure (hyperpaedomorphosis; Vrba, 1994). Such structures were relatively large in ancestral juveniles by virtue of faster exponential growth rates earlier than later during ontogeny. Thus, they can be said to become juvenilized by becoming relatively larger. Examples are the hind feet of rodents and the brain volumes of primates especially but also of other taxa. If such an early growth phase of exponential rate is extended by general prolongation of growth during evolution, a substantial increase in hind-foot length in rodents, or in brain weight in primates, relative to body weight can result. Because evolution toward prolonged growth is often associated with cooling (part of Bergmann's Rule), so will hyperpaedomorphosis in characters with fast early growth rates be associated with cooling. This is discussed in detail in Vrba (1994.) Suffice it here to say that a high rate of appearance of taxa with large body sizes and juvenilized morphologies affords an independent indication of climatic cooling at that time, although one would still need to establish whether that climatic cause acted to promote speciation and/or simply altered the species abundances and preservational circumstances.

Most of the FADs during 2.7–2.5 myr are of species that are larger than their earlier relatives (or outgroups, in cladistic terms). For instance, this time marks the advent of *Pelorovis*, the lineage of giant buffaloes; *Megalotragus*, the giant hartebeests; the huge *Beatragus* sp. nov.; and *Hippotragus gigas* and *H. equinus*, the largest hippotragines. Note that these five are among the taxa for which potential direct ancestors are known that have FADs since 2.9 myr ago (fig. 27.10). If that inference is correct, one cannot resort to the hypothesis that these species were present previously and were seen in the record only by 2.7 myr. Many of these taxa or their close relatives are extant; and ontogenetic information on these survivors from Late Pliocene FADs indicates in several cases that larger size resulted from prolonged growth. Also, many juvenilized morphologies appear, including both Allen juvenilization, with reduction of extremities, and hyperpaedomorphosis, with enlargement of structures. For instance, among the taxa that first appear 2.7–2.5 myr ago, juvenilization of both kinds is evident in the antelopes *Kobus sigmoidalis* and *K. ellipsiprymnus* (Vrba et al., 1994), in *Oryx*, and in the numerous alcelaphine morphologies of decreased face-to-braincase angles and horns that are short relative to skull size. Several saltatory rodent taxa appear with hyperpaedomorphosed hind feet (see Wesselman, this vol.; Hafner and Hafner, 1988, for ontogenetic analyses). Many mammalian taxa appear to have become encephalized over this time by the evolution of longer growth to larger body sizes together with hyperpaedomorphic juvenilization (Vrba, 1994). Such taxa include many Alcelaphini, which perhaps account for more new species over this interval than any other large mammalian clade (fig. 27.3) and which are notably more encephalized than many other bovids (Oboussier, 1972).

Late Pliocene Bovid Extinction. The number of LADs in the interval between 2.8 and 2.3 myr is twenty-nine (fig. 27.4). This is high considering that the relatively densely sampled past 3.5 myr include eighty-six LADs, or an average of about twelve per 0.5 myr interval. The total of twenty-nine LADs includes seventeen very short-lived taxa that appeared and then disappeared during this interval. Of seventeen survivors past 3.0 myr, twelve became extinct by 2.5 myr. If the durations of bovid taxa in figure 27.3 are replotted in rank order of LADs, the most prominent extinction episode in the entire record—what appears to be the largest LAD pulse—starts ca. 2.8 myr. However, an analysis of the possible influence of gaps in the record, similar to the one done on the Late Pliocene FAD pattern, is needed to take this further. In particular, this extinction pattern may be affected by the gap in the record between 2.3 and 2.0 myr (table 27.2, fig. 27.6). For the moment, I conclude only that there is a LAD pulse during the Late Pliocene and that this pulse possibly reflects a substantial wave of coeval extinctions.

The Late Pliocene Bovid Evidence Taken Together is consistent with a turnover pulse that started 2.8 myr ago and is seen in the record by 2.7–2.5 myr and that was initiated by a major cooling trend. (See chap. 3 for a more detailed and expanded discussion of turnover pulses.) This pulse involved closely coeval speciation and immigration with attendant biological characteristics that are precisely as predicted for a major cooling trend. The nature and number of extinctions near this time is not yet clear. A turnover pulse should include events in diverse lineages. The bovid lineages turning over at this time are in fact very diverse (from the aquatic lechwes to desert oryxes and from giant buffaloes to minute mountain reedbuck). One might argue that this phenomenon, even if it were found to be confined to this group, could be termed a turnover pulse. Comparison with other groups is hampered by two factors: few others have a fossil record as rich in specimens and species or have been studied in as much detail. Nevertheless, there are already indications that some other lineages were similarly affected ca. 2.7–2.5 myr ago. Although the African rodent evidence reported by Wesselman (this vol.) comes from a restricted area, it does closely parallel the bovid evidence (fig. 27.11, col. d). Thus, the turnover pulse in Africa includes at least these two groups. That it was more widespread is suggested by evidence from Eurasia of the massive mammalian turnover at this time that Azzaroli (1983, this vol.) terms the elephant-*Equus* event (see also Lindsay et al., 1980).

Hominids in the Context of Climate and Fauna: Are There General Patterns?

Climatic change from astronomical causes (see Shackleton, for Africa; and Dupont and Leroy, both in this vol.) and tectonism (see Partridge et al., chap. 2, this vol.) are expected to have affected much of Africa during the Late Pliocene. The proposal that climate commonly acts through vicariance and selection to influence speciation (chap. 3) predicts general tendencies across phylogenies in relation to climate, with respect to both branching topology and phenotypic modification. There should be characteristically differing turnover responses among kinds of geographic areas, among biomes such as the African tropical forests and savannas (Vrba, 1992), and among contrasting ecological categories of organisms (see figs. 3.2, 3.3, this vol.). Yet these differences should be intelligible in terms of a single consistent body of underlying theory. Do hominids and other mammals that evolved together in the African savannas share features of phylogenetic branching topology? And do they share similar kinds of phenotypic evolution in response to common climatic causes? Definitive answers require more information on fossil and living morphology and genealogy in relation to heterochrony, chronology, and paleoclimate. I offer a preliminary comparison of hominids, bovids, and other mammals in support of a tentative yes in answer to both questions.

The Late Pliocene patterns of hominid and bovid genealogies and FAD frequencies in time can be broadly compared with references to the summary in figure 27.11 and the bovid trees in figure 27.10. In figure 27.11b–e one sees that the timing of trans-African bovid FADs closely matches changes in Ethiopian micromammals (Wesselman, this vol.) and that both conform to the major global cooling trend that culminated in the onset of Arctic glaciation (Shackleton, this vol.) and strongly affected at least some African ecosystems (e.g., Dupont and Leroy, this vol., document the onset of the Sahara desert 2.8–2.7 myr ago; see also Bonnefille, this vol.). The bovid turnover pulse starts together with the shift in dominant cyclic periodicity at ca. 2.8 myr, and its late phase coincides approximately with the onset of Arctic glaciation (deMenocal et al., this vol.). Figure 27.11b and c also show that a substantial peak of bovid FADs remains between 2.7 and 2.5 myr, even if all systematic interpretations for FADs that differ from mine are used, a peak that decisively survives the tests in figure 27.6, appendix 27.4, and table 27.2. Figure 27.11 suggests that the same would be true if all biochronologically dated FADs were removed.

Let us consider my first question: Do the hominids and other mammals share features of phylogenetic branching topology? Recall that the bovid phylogenetic trees

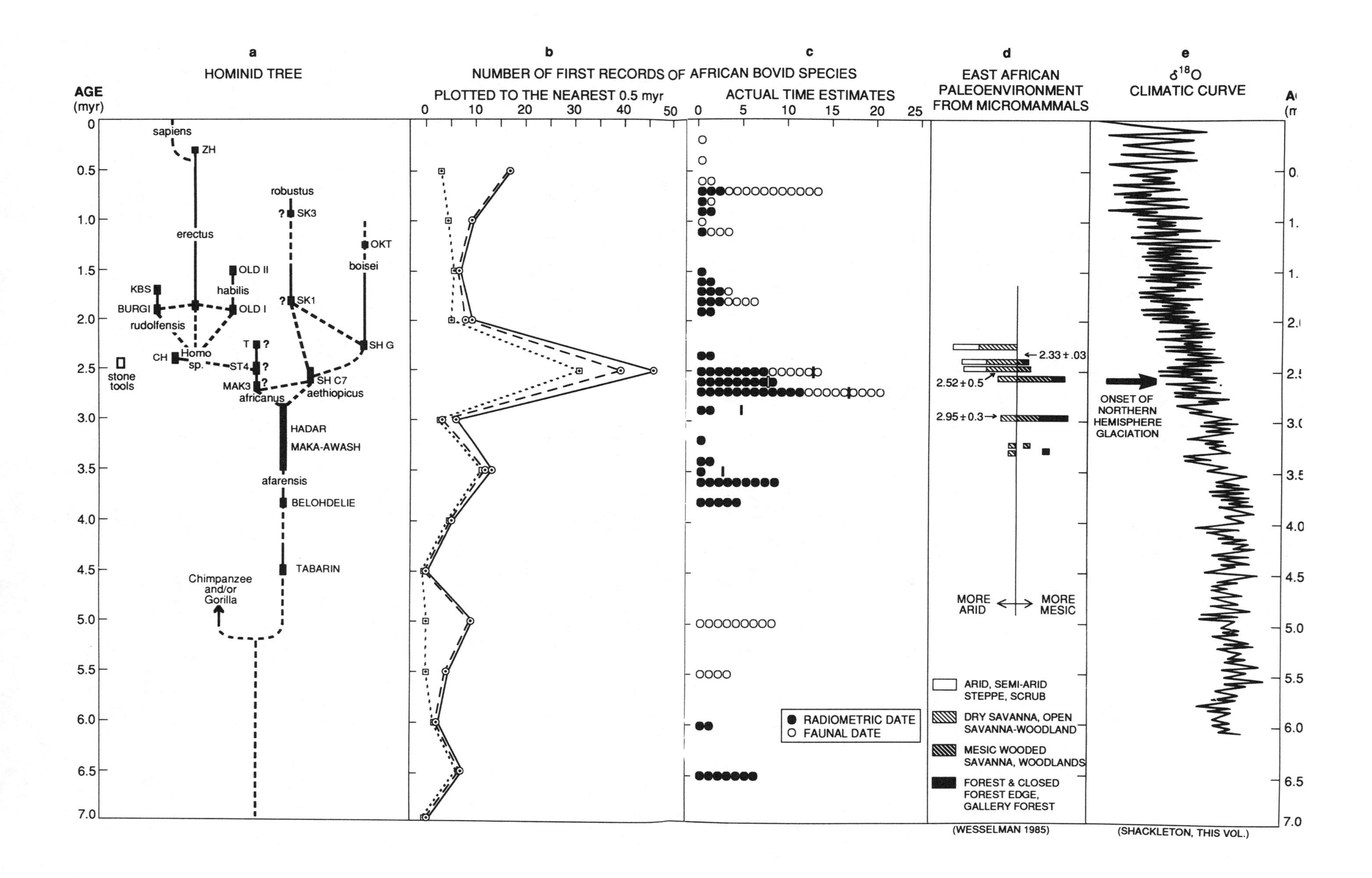

a
HOMINID TREE
AGE (myr)
sapiens
ZH
erectus
robustus
SK3
OKT
boisei
OLD II
habilis
KBS
BURGI
OLD I
SK1
rudolfensis
Homo sp.
CH
T
ST4
MAK3
SH G
SH C7
africanus
aethiopicus
stone tools
HADAR
MAKA-AWASH
afarensis
BELOHDELIE
TABARIN
Chimpanzee and/or Gorilla
b
NUMBER OF FIRST RECORDS OF AFRICAN BOVID SPECIES
PLOTTED TO THE NEAREST 0.5 myr
c
ACTUAL TIME ESTIMATES
RADIOMETRIC DATE
FAUNAL DATE
d
EAST AFRICAN PALEOENVIRONMENT FROM MICROMAMMALS
2.33 ± .03
2.52 ± 0.5
2.95 ± 0.3
MORE ARID
MORE MESIC
ARID, SEMI-ARID STEPPE, SCRUB
DRY SAVANNA, OPEN SAVANNA-WOODLAND
MESIC WOODED SAVANNA, WOODLANDS
FOREST & CLOSED FOREST EDGE, GALLERY FOREST
(WESSELMAN 1985)
e
δ18O
CLIMATIC CURVE
ONSET OF NORTHERN HEMISPHERE GLACIATION
(SHACKLETON, THIS VOL.)

exhibit a prevalent, Late Pliocene pattern (fig. 27.10): a potential direct ancestor is last recorded between 2.9 and 2.7 myr ago. Its descendants are often first recorded between 2.7 and 2.5 myr. After the first branching, multiple splitting events in a short interval are evident. Descendant FADs from repeated branching are distributed in the 2.7–1.8 myr interval. The similarities to the hominid pattern are striking (fig. 27.11a): the character states of *Australopithecus afarensis* have often been argued to qualify it as the direct ancestor of the branching to a *Homo* and a *Paranthropus* clade. *A. afarensis* is now recognized to persist in relative stasis until its LAD after 3.0 myr ago (Kimbel et al., 1994; Kimbel, this vol.). The exact sequence of branching in these basal stems, especially relative to *A. africanus*, is difficult to resolve and is often represented by polytomy (e.g., Wood, 1992), precisely as in some bovid trees over this interval (fig. 27.10) and precisely as expected to result from episodes of rapid lineage splitting. At least two distinct hominid taxa appear between ca. 2.7 and 2.5 myr ago, *A. africanus* (this FAD could be at 2.8 myr, see app. 27.3) and *P. aethiopicus*. If the Chemeron taxon is found to be *Homo* (Hill et al., 1992) and/or if the earliest stone tools are taken as evidence of a behavioral phenotype that is most parsimoniously attributable to *Homo*, then a third hominid taxon appears during this interval, and a fourth by 2.3 myr, *P. boisei*. I suggest that the answer to the first question is yes in comparison with bovids, and I predict that, as more chronologically constrained, phylogenetic trees of other mammals become available, some of these will also be found to conform to this pattern of showing their most dramatic periods of diversification between ca. 2.9 and 2.3 myr ago.

Do hominids and other mammals share similar kinds of phenotypic evolution in response to cooling? Wood (this vol.) concludes that they do from his analyses of the masticatory apparatus in hominid and other lineages. Two aspects of my agreement with this view are as follows.

1. Recall that larger sizes resulting from prolonged growth are generally associated with climates that are at least seasonally cold and that many bovids and other mammals during the Late Pliocene underwent this kind of evolution. No one disputes the evidence that human evolution has been marked by an increase in size and also by spectacular prolongation of growth.

2. As mentioned above, in many living cold-adapted taxa and in taxa that appeared during cold times such as 2.7–2.5 myr ago in the fossil record, evolutionary extension of growth phases resulted in simultaneous enlargement and juvenilization of characters with faster earlier than later exponential growth rates during ontogeny. Encephalization in *Homo* is such a character (Vrba, 1994). The size of the human brain in terms of heterochrony is the analogue of hind-foot size in cool- and desert-adapted rodents. The mammalian taxa that became encephalized during the Late Pliocene probably did so mostly by evolution of longer growth to larger body sizes, just as *Homo* apparently did (Vrba, 1994). Perhaps we have been so preoccupied with our own brain evolution since the Pliocene that we failed to notice how many other taxa became encephalized together with us! (See, e.g., Oboussier, 1972.) There are plausible and interesting explanations (Vrba, 1994) of why other cases of encephalization did not culminate in "super-brains," including why the robust australopithecines did not become more encephalized (fig. 27.12) in spite of a substantial increase in body size.

←

Fig. 27.11. Evolutionary events in hominids, antelopes, and micromammals in relation to climatic change.

a) Hominid genealogical tree (adapted from Wood, 1992, this vol., with minor modifications). Solid rectangles = dating estimates for FADs and LADs of hominid taxa; dating for East Africa after Brown and Feibel (1991); for Middle Awash, T. White (pers. comm.); for Tabarin, Hill et al. (1991); for South Africa, Vrba (1982, recent revisions). Solid rectangles with question marks indicate biochronological estimates. Dashed lines on lineages denote uncertain and alternative branch connections. In South Africa: SK1–3 = Swartkrans Members (M) 1–3; ST4 = Sterkfontein M4; MAK3 = Makapansgat M3; T = Taung. In Ethiopia: HADAR = Hadar Formation; BELOHDELIE, MAKA-AWASH = MAKA unit in Middle Awash. In Kenya: BURGI = Burgi Member at Koobi Fora; KBS = stratum below KBS tuff at Koobi Fora; OKT = Okote Member; CH = Chemeron; TABARIN. In Tanzania: OLD I, II = Oluvai Beds I and II, Level BK. In China: ZH = Zhoukoutien.

b) Earliest African records of antelope species (Bovidae) from app. 27.2. (i) Solid line: based on all records as interpreted in app. 27.2; (ii) dashed line: same data set as (i) except that all the alternative interpretations cited in app. 27.2, notes 4, 6, 9, 12, 14, 20, 23, 24, 25, 30, 38, 39, 40, are used; (iii) dotted line: same data set as (i) except that all biochronological records have been removed.

c) Earliest African records of antelope species (Bovidae). Solid bars show magnitude of histogram peaks when all the alternative interpretations are used, as in column b (ii).

d) Relative micromammal frequencies in different habitat categories (rectangles) indicate paleoenvironmental change in the Shungura Formation, Ethiopia (adapted from Wesselman, 1985: fig. 1, this vol.)

e) Deep-sea oxygen isotope data from a North Atlantic site (adapted from Shackleton, this vol.); ice-volume increase is indicated on the left.

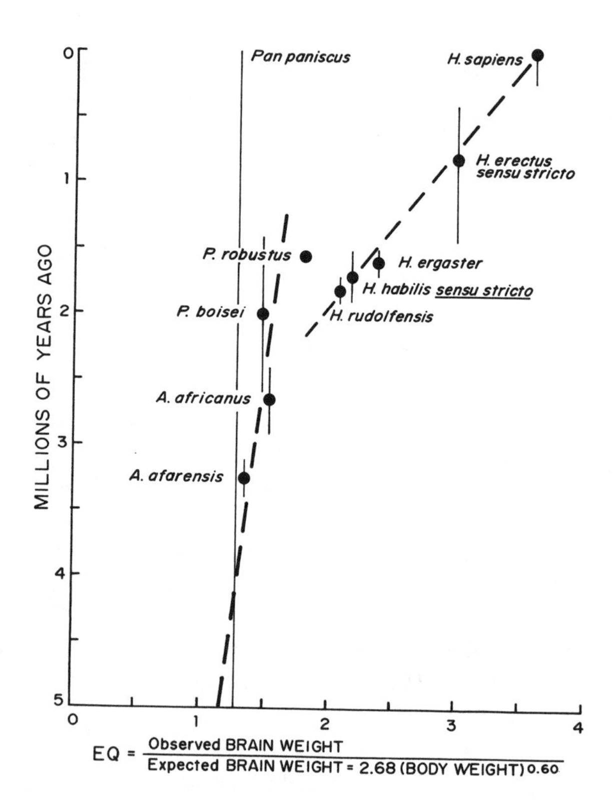

Fig. 27.12. Evolutionary trend in hominid encephalization quotients (EQs), using Martin's (1983) formula for Cercopithecoidea. The graph does not represent genealogy. Estimates of hominid body weights are from McHenry (1991, 1992); of cranial capacities, from Holloway (1970, 1972, 1978), Tobias (1971), Johanson and Edey (1981), and McHenry (1988); and of values for *Pan paniscus*, from Martin (1983). Wood (1992) and others he cites view the traditional *Homo habilis* as comprising at least two taxa, *H. rudolfensis* and *H. habilis*, the EQs for which were calculated based on individual specimens as classified in Wood (1992) and McHenry (1992). A similar procedure was used to subdivide *H. erectus* by separation of *H. ergaster* (Wood, 1992). Relative to the EQ axis, all solid dots are placed at mean EQ values. Relative to the time axis, solid dots for *A. africanus* and *P. robustus* are placed at age estimates after Vrba (1982, 1985a, dashed line = error estimate), and for other taxa at mean ages for crania within age ranges (solid lines), as reviewed in Wood (1992).

We have recent new evidence on the timing of the massive overall cooling trend since the Late Pliocene, on the hominid record and phylogeny, and on evolution of bovids and other mammals (fig. 27.11). To compare the net rate and timing of encephalization in hominids with this new evidence, I calculated hominid encephalization quotients (EQs) using recent estimates of early-hominid body weights and cranial capacities and recent estimates of chronology (fig. 27.12). There are several reasons for assessing the early hominid EQ data with caution. Disagreements remain on how to estimate hominid body weights (McHenry, 1991, 1992), encephalization quotients (McHenry, 1988; Martin, 1983), fossil taxonomic and sexual identities, taxon-branching sequence, and chronology (Howell, 1978; Wood, 1992). Also, there is a gap ca. 2.6–1.9 myr in the record of hominid crania. Figure 27.12 represents only two general trends, not genealogy (although the more dramatic trend, in *Homo*, is expressed among taxa that are nearly always considered to be a monophyletic group). In spite of these caveats, we can compare figure 27.12 with the cladistic tree in figure 27.11a and note certain implications: if *A. afarensis* is really the direct ancestor of the lineage split that led to *Homo* and robust australopithecines, as many have suggested, then the EQ trend in *Homo* is constrained to have commenced after 3.0 myr (see the new FAD for *A. afarensis*, Kimbel et al., 1994) and before 2.0 myr (fig. 27.12). The timing and pattern among the splitting events in the hominid tree (fig. 27.6a; and Wood, White, Kimbel, and others in this vol.) remain disputed. Nevertheless, the same interval of interest in the EQ trend, 3.0–2.0 myr, also certainly contains several of the hominid splitting events as well as the earliest known stone tools (fig. 27.11). The conclusion is inescapable that hominine encephalization in the latest Pliocene started a new trend, of higher evolutionary rates than before (fig. 27.12); and there is a reasonable basis in evidence for the hypothesis that this may be an extreme expression of a common mammalian evolutionary tendency in response to the onset and intensification of the modern Ice Age.

Conclusions and Summary

I have addressed the question of physical forcing of macroevolution by analysis of the first appearance data (FADs) of African bovids. I have also considered whether there is evidence that hominids speciated together with other faunal groups in response to common environmental causes. A FAD could appear in the record in an area X at time Y because it really appeared there at that time by a speciation or immigration event that was initiated by climatic change (including the effects of tectonism) or by biotic interactions. An alternative is the explanation that invokes taphonomic bias: the species was there previous to time Y but remained unrecorded because of a gap in the fossil record. I used a synthetic approach, which draws on different kinds of data and analyses, to distinguish these possibilities. The research programs concerned with speciation, ecological adaptation, embryological development, cladistic systematics, and patterns in the fossil record have traditionally been conducted more or less independently, and none of them has payed much attention to climatology. The Late Neogene offers an opportunity to integrate the theory and data from all these disciplines. Elements of such integration are evident in many chapters of this volume; for this chapter on the Late Pliocene they can be summarized as follows.

1. The Climatic Context. African climate was presumably influenced by uplift and rifting in eastern Africa over this time (Partridge et al., chap. 24, this vol.) and also by the major global cooling trend toward 2.5 myr ago that culminated in two remarkable and rare changes in the climatic system: deMenocal and Bloemendal (this vol.) document a shift ca. 2.8 myr ago toward climatic dominance of the 41 kyr cycle, from previous dominant influence at 23–19 kyr variance, and a second shift toward increases in 100 kyr variance after 0.9 myr. By ca. 2.5 myr extensive Arctic glaciation was in place (Shackleton et al., 1984; Shackleton, this vol.). The few African climatic records available do show that at least some African biomes were strongly affected. For instance, Dupont and Leroy (this vol.) document the onset of the Sahara desert after 2.8 myr ago.

2. The Model of Allopatric Speciation enjoys strong consensus among biologists. It implies that vicariance (the production of isolated populations) by physical change is of critical importance to speciation (chap. 3). I extended this model to the proposal that long-term vicariance is necessary for most speciations (the relay model; see chap. 3). One prediction is that during a long-term cooling trend such as occurred after 3.0 myr, warm-adapted species should be the first to become vulnerable to vicariance and should speciate before the more cold- and open-adapted ones do. The African bovids that have FADs between 2.7 and 2.5 myr ago strongly support this prediction. This result is significant; it cannot easily be explained either by an argument that competition was the initiating factor underlying those FADs or by an argument that those FADs are artifacts of a prior gap in the record.

3. Cladistic Systematics was used together with fossil chronology to estimate that about one-sixth of the forty-six bovid FADs of 2.7–2.5 myr ago represent Eurasian immigrants. For at least one-half there is evidence that cladogenesis occurred and that speciation dates to the time window 2.9–2.5 myr. I compared the branching topologies of the phylogenetic trees of bovid clades with those of the hominid tree. Several bovid trees share with the hominid tree a potential direct ancestor that is last recorded 2.9–2.7 myr ago, with descendant FADs from multiple, apparently rapidly successive splitting events (the cladistic polytomy is a symptom of this) that are distributed in the 2.7–1.8 myr interval, especially between 2.7 and 2.5 myr. These similar patterns suggest common external causes, specifically that climate acted through vicariance and selection to influence speciation.

4. Ecology, Morphology, and Embryological Development. Certain kinds of morphologies have long been taken to indicate evolutionary responses to (at least seasonal) cooling. Two examples are: 1) large body size often attained by prolongation of growth to maturity (Bergmann's Rule, 1847); and 2) juvenilized shapes in the form of relatively enlarged structures. The latter result by prolongation of all growth phases of a structure that has a faster exponential growth rate earlier rather than later in ontogeny (hyperpaedomorphosis; Vrba, 1994). Examples of such evolution are the encephalization that occurred in many mammals and enlargement of hind feet in rodents. Morphologies of kinds 1) and 2) are strongly represented among the bovids that first appear 2.7–2.5 myr ago (many "giants" and relatively encephalized forms appeared) and also among rodents that appear then (see Wesselman, this vol., for the fossil record; Hafner and Hafner, 1988, for ontogeny). One cannot invoke competition-induced speciation to make sense of this pattern. And the explanation that these morphologies were there before 3.0 myr ago, such that their FADs are artifacts of a taphonomic gap, is rejected for those sister groups with potential direct ancestors that are last recorded between 2.9 and 2.7 myr ago. The evolution of *Homo* also involved larger body size by growth prolongation and encephalization. Thus, not only do different bovid groups, as well as bovids and hominids, share aspects of how their phylogenetic trees relate to time, but the evolutionary innovations nested within those branching patterns show similarities in terms of the heterochrony and climatic selective forces they imply.

5. Macroevolutionary Patterns in the Fossil Record. My main analysis (quite independent of those mentioned above) is based on regression and other analyses of the FAD accumulation curve of African bovid species. The results indicate a number of significant gaps in the Plio-Pleistocene record as well as significant pulses of

FADs 0.9–0.7 and 2.7–2.5 myr ago. I tested whether the 2.7–2.5 myr FAD pulse could be an artifact of preceding and/or succeeding taphonomic gaps. Two different analyses of the bovid data give consistent estimates that about one-fourth of the 2.7–2.5 myr FAD pulse represents species that were present previously yet unseen owing to gaps in the record. About one-sixth are Eurasian immigrants, and the balance of just more than one-half of the pulse is estimated to comprise speciation events at this time. The immigrants indicate global cooling: these caprine and antilopine taxa, from ancestral Eurasian high-latitude and/or high-altitude habitats, "rode" an equatoward spreading "wave" of open habitats over an intercontinental landbridge that had appeared as sea levels became lower. The speciation processes that resulted in the radiometrically dated FADs at 2.7 myr presumably occurred in the preceding 100 kyr (or 200 kyr?) interval. Thus, one can refer to the "2.8–2.5 myr turnover pulse" that left the observation in the record of the "2.7–2.5 myr FAD pulse."

One of the most important results is this: the two rare shifts—toward new dominance of the 41 kyr astronomical cycle ca. 2.8 myr ago and dominance of the 100 kyr cycle 0.9 myr ago, which deMenocal and Bloemendal (this vol.) document in the deep-sea record off the African coast—fall almost exactly at the beginnings of the only two bovid FAD pulses since 3.6 myr ago that emerged as significant at the 99 percent level, 2.7–2.5 myr ago and 0.9–0.7 myr ago. DeMenocal and Bloemendal (this vol.) suggest that any strong climatic influence on African faunal and hominid evolution starting ca. 2.8 myr should be sought in a change in mode of subtropical variability rather than in a long-term trend in the cyclic mean. The evolutionary models support that view: if prolonged vicariance is important to speciation, then a shift toward dominance of a longer cyclic period in particular might be expected to promote speciation. Over most of the time since the start of this debate on possible climatic initiation of hominid and other biotic turnover, data that might have been used for tests were still very imprecise. Consequently, many of the arguments, including my own, were confined to citation of crude correlations, or non-correlations, between gross categories of "events" that were poorly constrained in time. Only recently has it become possible to advance beyond crude correlations, that is, to apply interdisciplinary tests and quantitative tests to turnover patterns, at least in the Middle Pliocene to Pleistocene record. I hope that my efforts in this regard illustrate the usefulness of simultaneous testing by predictions from differing disciplinary sources: bovids not only have FADs in different ecological categories in a temporal sequence expected (under the allopatric speciation model) if cooling promoted speciation through vicariance. They are also indicated, according to cladistic methods, to share similar topologies of their phylogenetic trees relative to time, with branching events that are limited to the time window 2.9–2.5 myr, during which a major mode shift in the astronomical cycles was followed by the onset of Arctic glaciation. In addition, their morphological innovations nested within those branching patterns show similarities in terms of heterochrony as predicted for cooling by ecology and by developmental theory. These consistent results in response to diverse predictions offer more valuable support for the hypothesis of climatic initiation of turnover than any single demonstrated correlation is likely to do.

Acknowledgments

I am grateful to Nick Shackleton for allowing me to use his data in my figure 27.11. I also thank Bernard Wood and Hank Wesselman for allowing me to reproduce versions of their data in my figures. Helpful comments on the manuscript were received from John Barry, Ray Bernor, Lloyd Burckle, George Denton, Andrew Hill, Tim White, and Bernard Wood. Susan Hochgraf prepared most of the figures.

References

Alberdi, M. T., and Bone, E. 1978. Macrovertèbres du gisement d'Arenas del Rey (Miocène supérieur du bassin de Grenade, Andalousie, Espagne). *Bull. Soc. Belge Géologie* (Bruxelles) 87: 199–214.

Allen, J. A. 1877. The influence of physical conditions in the genesis of species. *Radical Review* 1:108–140.

Arambourg, C. 1979. *Vertèbres villafranchiens d'Afrique du Nord.* Singer-Polignac, Paris.

Azzaroli, A. 1983. Quaternary mammals and the "End-Villafranchian" dispersal event: A turning point in the history of Eurasia. *Paleogeogr. Palaeoclimat., Palaeoecol.* 44:117–139.

Barry, J. C., Lindsay, E. H., and Jacobs, L. L. 1982. A biostratigraphic zonation of the Middle and Upper Siwaliks of the Potwar Plateau of northern Pakistan. *Palaeogeogr. Palaeoclimat. Palaeoecol.* 37:95–130.

Bergmann, C. 1847. Über die Verhältnisse der Wärmeokonomie der Thiere zu ihrer Grösse. *Göttinger Studien* 3. 1:596–708.

Bosscha Erdbrink, D. P. 1982. A fossil reduncine antelope from the locality K 2 east of Maragheh, N.W. Iran. *Mitteilungen der Bayerischen Staatssammlung für Palaeontologie und historische Geologie* 22:103–112.

Brain, C. K. 1981. The evolution of man in Africa: Was it a consequence of Cainozoic cooling? *Annex. Transv. Geol. Soc. S. Afr.* 84:1–19.

Brown, F. H., and Feibel, C. S. 1991. Stratigraphy, depositional environment, and paleogeography of the Koobi Fora Formation. In *Koobi Fora Research Project*. Vol. 3, *Stratigraphy, artiodactyls and paleoenvironments*, pp. 1–30 (ed. J. M. Harris). Clarendon Press, Oxford.

Brown, F. H., and Lajoie, K. R. 1971. Radiometric age determinations on Pliocene/Pleistocene formations in the lower Omo Basin, southern Ethiopia. *Nature* 229:483–485.

Brown, F. H., and Shuey, R. T. 1976. Magnetostratigraphy of the Shungura and Usno Formations, lower Omo Valley, Ethiopia. In *Earliest man and environments in the Lake Rudolf Basin*, pp. 64–78 (ed. Y. Coppens, F. C. Howell, G. L. l. Isaac, and R. E. F. Leakey). University of Chicago Press, Chicago.

Coppens, Y. 1975. Evolution des hominidés et de leur environnement au cours du Plio-Pléistocène dans la basse vallée de l'Omo en Ethiopie. *C.R. Acad. Sc.* (Paris), ser. D, 281:1693–1696.

Darwin, C. 1859. *On the origin of species by means of natural selection, or the preservation of favoured races in the struggle for life.* John Murray, London.

Delson, E. 1985. Neogene African catarrhine primates: Climatic influence on evolutionary patterns. *S. Afr. J. Sci.* 81:273–274.

Dmitrieva, E. L. 1977. Tajikistan's and India's fossil Alcelaphinae. *J. Palaeont. Soc. India* 20:97–101.

Dowsett, H. J., Cronin, T. M., Poore, R. Z., Thompson, R. S., Whatley, R. C., and Wood, A. M. 1992. Micropaleontological evidence for increased meridional heat transport in the North Atlantic Ocean during the Pliocene. *Science* 258:1133–1135.

Drake, R., and Curtis, G. H. 1987. Geochronology of the Laetoli fossil localities. In *Laetoli: A Pliocene site in northern Tanzania*, pp. 48–52 (ed. M. D. Leakey and J. M. Harris). Clarendon Press, Oxford.

Feibel, C. S., Brown, F. H., and McDougall, I. 1989. Stratigraphic context of fossil hominids from the Omo Group deposits: Northern Turkana Basin, Kenya and Ethiopia. *Am. J. Phys. Anthrop.* 78:595–622.

Gatesy, J. 1993. Cows, sheep, antelopes, and molecules. Ph.D. diss., Yale University.

Gatesy, J., Yelon, D., Desalle, R., and Vrba, E. S. 1992. Phylogeny of the Bovidae (Artiodactyla, Mammalia), based on mitochondrial ribosomal DNA sequences. *Mol. Biol. Evol.* 9: 433–466.

Gentry, A. W. 1970. The Bovidae of the Fort Ternan fossil fauna. In *Fossil Vertebrates of Africa*, pp. 243–323 (ed. L. S. B. Savage and R. J. G. Savage). Academic Press, London.

———. 1974. A new genus and species of Pliocene boselaphine (Bovidae, Mammalia) from South Africa. *Ann. S. Afr. Mus.* 65:145–188.

———. 1976. Bovidae of the Omo Group deposits. In *Earliest man and environments in the Lake Rudolf Basin*, pp. 275–292 (ed. Y. Coppens, F. C. Howell, G. L. l. Isaac and R. E. F. Leakey). University of Chicago Press, Chicago.

———. 1978. Bovidae. In *Evolution of African mammals*, pp. 540–572 (ed. V. J. Maglio and H. B. S. Cooke). Harvard University Press, Cambridge.

———. 1980. Fossil Bovidae from Langebaaweg, South Africa. *Ann. S. Afr. Mus.* (Cape Town) 79:213–337.

———. 1981. Notes on Bovidae from the Hadar Formation, Ethiopia. *Kirtlandia* (Cleveland) 33:1–30.

———. 1985. The Bovidae of the Omo Group deposits, Ethiopia. In *Les faunes plio-pléistocènes de la basse vallée de l'Omo (Ethiopie), Périssodactyles, Artiodactyles (Bovidae)*, vol. 1, pp. 119–191 (ed. Y. Coppens). Paris.

———. 1987. Pliocene Bovidae from Laetoli. In *The Pliocene site of Laetoli, northern Tanzania*, pp. 378–408 (ed. M. D. Leakey and J. M. Harris). Clarendon Press, Oxford.

———. 1990a. Evolution and dispersal of African Bovidae. In *Horns, pronghorns, and antlers: Evolution, morphology, physiology, and social significance*, pp. 195–233 (ed. G. A. Bubenik and A. B. Bubenik). Springer-Verlag, New York.

———. 1990b. The Semliki fossil bovids. *Virginia Mus. Nat. Hist. Memoir* 1:225–234.

Gentry, A. W., and Gentry, A. 1978. Fossil Bovidae of Olduvai Gorge, Tanzania. *Bull. Br. Mus. Nat. Hist.* (Geol.) 29:289–445.

Geraads, D. 1981. Bovidae et Giraffidae (Artiodactyla, Mammalia) du Pléistocène de Ternifine (Algérie). *Bull. Mus. Natn. Hist. Nat.* (Paris) 3:47–86.

———. 1987. Dating the Northern African cercopithecid fossil record. *Human Evolution* 2:19–27.

———. 1989. Phylogenetic analysis of the tribe Bovini (Mammalia: Artiodactyla). *Zool. J. Lin. Soc.* 104:193–207.

Gould, S. J. 1977. *Ontogeny and phylogeny.* Harvard University Press, Cambridge.

Greenacre, M. J., and Vrba, E. S. 1984. A correspondence analysis of biological census data. *Ecology* 65:984–997.

Hafner, J., and Hafner, M. 1988. Heterochrony in rodents. In *Heterochrony and evolution*, pp. 217–235 (ed. M. L. McKinney). Plenum Press, New York.

Haileab, B., and Brown, F. H. 1992. Turkana Basin–Middle Awash Valley correlations and the age of the Sagantole and Hadar Formations. *J. Hum. Evol.* 22:453–368.

Hamilton, W. R. 1973. The lower Miocene ruminants of Gebel Zelten, Libya. *Bull. Br. Mus. Nat. Hist. (Geol.)* 21:73–150.

Hare, P. E. 1980. Organic geochemistry of bone and its relation to the survival of bone in the natural environment. In *Fossils in the making*, pp. 208–219 (ed. A. P. Hill and A. K. Behrensmeyer). University of Chicago Press, Chicago.

Harris, J. M. 1991. Family Bovidae. In *Koobi Fora Research Project*. Vol. 3, *The fossil ungulates: Geology, fossil artiodactyls, and palaeoenvironments*, pp. 139–320 (ed. J. M. Harris). Clarendon Press, Oxford.

Harris, J. M., Brown, F. H., and Leakey, M. G. 1988. Stratigraphy and paleontology of Pliocene and Pleistocene localities west of Lake Turkana, Kenya. *Contributions in Science, Natural History Museum of Los Angeles County* 399:1–128.

Hay, R. L. 1976. *Geology of the Olduvai Gorge: A study of sedimentation in a semiarid basin.* University of California Press, Berkeley.

Hendey, Q. B. 1978. The age of the fossils from Baard's Quarry, Langebaanweg, South Africa. *Ann. S. Afr. Mus.* 69:215–247.

———. 1981. Geological succession of Langebaanweg, Cape Province, and global events of the Late Tertiary. *S. Afr. J. Sci.* 77:33–38.

Hilgen, F. J. 1991. Extension of the astronomically calibrated (polarity) time scale to the Miocene/Pliocene boundary. *Earth and Planetary Science Letters* 107:349–368.

Hill, A. 1985. Early hominid from Baringo, Kenya. *Nature* 315: 222–224.

Hill, A., Drake, R., Tauxe, L., Monaghan, M., Barry, J. C., Behrensmeyer, A. K., Curtis, G., Jacobs, B., Jacobs, L., Johnson, L., Johnson, N., and Pilbeam, D. 1985. Neogene paleontology and geochronology of the Baringo Basin, Kenya. *J. Hum. Evol.* 14:759–773.

Hill, A., Ward, S., Deino, A., Curtis, G., and Drake, R. 1992. Earliest *Homo*. *Nature* 355:719.

Holloway, R. L. 1970. New endocranial values for the australopithecines. *Nature* 227:199–200.

———. 1972. New australopithecine endocast SK 1585, from Swartkrans, South Africa. *Am. J. Phys. Anthrop.* 37:173–186.

———. 1978. Problems of brain endocast interpretation and African hominid evolution. In *Early hominids of Africa,* pp. 379–401 (ed. C. J. Jolly). Duckworth, London.

Hsu, K. M., Montadert, L., Bernoulli, D., Cita, M. S., Erickson, A., Garrison, R. E., Kidd, R. M., Melieres, F., Muller, C., and Wright, R. 1977. History of the Mediterranean Salinity Crisis. *Nature* 267:399–403.

Kimbel, W. H., Johanson, D. C., and Rak, Y. 1994. The first skull and other new discoveries of *Australopithecus afarensis* at Hadar, Ethiopia. *Nature* 368:449–451.

Johanson, D. C., and Edey, M. 1981. *Lucy: The beginnings of humankind.* Simon and Schuster, New York.

Kaufulu, Z., Vrba, E. S., and White, T. 1981. Age of the Chiwondo Beds, Northern Malawi. *Ann. Trans. Mus.* 33:1–8.

Klein, R. G. 1972. The Late Quaternary mammalian fauna of Nelson Bay Cave (Cape Province, South Africa): Its implications for megafaunal extinctions and environmental and cultural change. *Quatern. Res.* 2:135–142.

Klein, R. G., and Cruz-Uribe, K. 1991. The bovids from Elandsfontein, South Africa, and their implications for the age, palaeoenvironment, and origins of the site. *Afr. Archaeol. Rev.* 9:21–79.

Lehmann, U., and Thomas, H. 1987. Fossil Bovidae from the Mio-Pliocene of Sahabi (Libya). In *Neogene paleontology and geology of Sahabi,* pp. 323–335 (ed. N. T. Boaz, A. El-Arnauti, A. W. Gaziry, J. de Heinzelin, and D. D. Boaz). Liss, New York.

Lindsay, E. H., Opdyke, N. D., and Johnson, N. M. 1980. Pliocene dispersal of the horse *Equus* and Late Cenozoic mammalian dispersal events. *Nature* 287:135–138.

McHenry, H. M. 1988. New estimates of body weight in early hominids and their significance to encephalization and megadontia in "robust australopithecines." In *The evolutionary history of the robust australopithecines,* pp. 133–148 (ed. F. E. Grine). Aldine, New York.

———. 1991. Petite bodies of the "robust" australopithecines. *Am. J. Phys. Anthropol.* 86:445–454.

———. 1992. Body size and proportions in early hominids. *Am. J. Phys. Anthropol.* 87:407–431.

Manega, P. C. 1993. Geochronology, geochemistry and isotopic study of the Plio-Pleistocene hominid sites and the Ngorongoro volcanic highland in northern Tanzania. Ph.D. diss., submitted, University of Colorado.

Martin, R. D. 1983. *Human brain evolution in an ecological context.* 2d James Arthur Lecture on the Evolution of the Human Brain. American Museum of Natural History Publications, New York.

Oboussier, H. 1972. Morphologische und quantitative Neocortexuntersuchungen bei Boviden, ein Beitrag zur Phylogenie dieser Familie. III. Formen uber 75 kg Korpergewicht. *Mitt. Hamb. Zool. Mus. Inst.* 68:271–292.

Pilgrim, G. E. 1939. The fossil Bovidae of India. *Mem. Geol. Surv. India Paleont. Indica,* n.s., 26:1–356.

Rayner, R., Moon, B., and Masters, J. C. 1993. The Makapansgat australopithecine environment. *J. Hum. Evol.* 24:219–231.

Savage, D. E., and Russell, D. E. 1983. *Mammalian palaeofaunas of the world.* Addison-Wesley, London.

Shackleton, N. J., et al. 1984. Oxygen isotope calibration of the onset of ice-rafting and history of glaciation in the North Atlantic region. *Nature* 307:620–623.

Shackleton, N. J., Berger, A., and Peltier, W. R. 1990. An alternative astronomical calibration of the lower Pleistocene timescale based on ODP Site 677. *Transactions of the Royal Society of Edinburgh: Earth Sciences* 81:251–261.

Shipman, P., Walker, A., Van Couvering, J. A. V., Hooker, P. J., and Miller, J. A. 1981. The Fort Ternan hominoid site, Kenya: Geology, age, taphonomy and paleoecology. *J. Hum. Evol.* 10: 49–72.

Smart, C. L. 1976. The Lothagam 1 fauna: Its phylogenetic, ecological and biogeographic significance. In *Earliest man and environments in the Lake Rudolf Basin,* pp. 361–369 (ed. Y. Coppens, F. C. Howell, G. L. l. Isaac and R. E. F. Leakey). University of Chicago Press, Chicago.

Stanley, S. M. 1979. *Macroevolution: Pattern and process.* W. H. Freeman, San Francisco.

Thomas, H. 1979. *Miotragocerus cyrenaicus* sp. nov. (*Bovidae, Artiodactyla, Mammalia*) du Miocène supérieur de Sahabi (Libye) et ses rapports avec les autres *Miotragocerus. Geobios* 12:267–281.

———. 1980. Les bovidés du Miocène supérieur des couches de Mpesida et de la formation de Lukeino (district de Baringo, Kenya). In *Proc. Eighth Panafr. Congr. Prehistory and Quaternary Studies Nairobi,* pp. 82–91 (ed. R. E. F. Leakey and B. A. Ogot). TILLMIAP, Nairobi.

———. 1981. Les bovidés miocènes de la formation de Ngorora du bassin de Baringo (Kenya). *Proc. K. Ned. Akad.* (Amsterdam), ser. B, 84:335–409.

———. 1984a. Les Bovidae (Artiodactyla: Mammalia) du Miocène du sous-continent Indien, de la péninsule arabique et de l'Afrique: Biostratigraphie, biogéographie et écologie. *Palaeogeogr., Palaeoclimat., Palaeoecol.* 45:251–299.

———. 1984b. Les Giraffoidea et les Bovidae miocènes de la formation Nyakach (rift Nyanza, Kenya). *Palaeontographica* 183 A:64–89.

———. 1984c. Les origines africaines des Bovidae miocènes des lignites de Grosseto (Toscane, Italie). *Bull. Mus. Natn. Hist. Nat.* (Paris), ser. C, 4:81–101.

Thomas, H. R., Bernor, R., and Jaeger, J. J. 1982. Origine du peuplement mammalien en Afrique du Nord durant le Miocène terminal. *Geobios* 15:283–297.

Tobias, P. V. 1971. *The brain in hominid evolution.* Columbia University Press, New York.

Vrba, E. S. 1974. Chronological and ecological implications of the fossil Bovidae at the Sterkfontein australopithecine site. *Nature* 256:19–23.

———. 1975. Some evidence of chronology and palaeoecology of Sterkfontein, Swartkrans and Kromdraai from the fossil Bovidae. *Nature* 254:301–304.

———. 1976. The fossil Bovidae of Sterkfontein, Swartkrans and Kromdraai. *Transv. Mus. Mem.* (Pretoria) 21:1–166.

———. 1977a. New species of *Parmularius* Hopwood and *Damaliscus* Sclater and Thomas (Alcelaphini, Bovidae, Mammalia) from Makapansgat, and comments on faunal chronological correlation. *Palaeont. Afr.* 20:137–151.

———. 1977b. Problematical alcelaphine fossils from the Kromdraai faunal site (Mammalia: Bovidae). *Ann. Transv. Mus.* 31:21–39.

———. 1979. Phylogenetic analysis and classification of fossil and recent Alcelaphini. *Biol. J. Linn. Soc.* (London) 11:207–228.

———. 1980. Evolution, species and fossils: How did life evolve? *S. Afr. J. Sci.* 76:61–84.

———. 1982. Biostratigraphy and chronology, based particularly on Bovidae, of southern African hominid-associated assemblages. In *Proceedings of the Congrès International de Paléontologie Humaine* (ed. H. DeLumley and M. A. DeLumley). Union International des Sciences Préhistoriques et Protohistoriques, Nice.

———. 1985a. Environment and evolution: Alternative causes of the temporal distribution of evolutionary events. *S. Afr. J. Sci.* 81:229–236.

———. 1985b. Early hominids in southern Africa: Updated observations on chronological and ecological background. In *Hominid evolution: Past, present and future*, pp. 195–200 (ed. P. V. Tobias). Liss, New York.

———. 1987a. A revision of the Bovini (Bovidae) and a preliminary revised checklist of Bovidae from Makapansgat. *Palaeont. Afr.* 26:33–46.

———. 1987b. New species and a new genus of Hippotragini (Bovidae) from Makapansgat Limeworks. *Palaeont. Afr.* 26: 47–58.

———. 1987c. Ecology in relation to speciation rates: Some case histories of Miocene-Recent mammal clades. *Evolutionary Ecology* 1:283–300.

———. 1988. Late Pliocene climatic events and hominid evolution. In *The evolutionary history of the robust australopithecines*, pp. 405–426 (ed. F. E. Grine). Aldine, New York.

———. 1992. Mammals as a key to evolutionary theory. *J. Mamm.* 73:1–28.

———. 1993. Turnover-pulses, the Red Queen, and related topics. *Am. J. Sci.* 293-A:418–452.

———. 1994. An hypothesis of heterochrony in response to climatic cooling and its relevance to early hominid evolution. In *Integrative paths to the past: Paleoanthropological advances in honour of F. Clark Howell*, pp. 345–376 (ed. R. Corruccini and R. Ciochon). Prentice-Hall, Englewood Cliffs, N.J.

Vrba, E. S., and Gatesy, J. (in press). New hippotragine antelope fossils from the Middle Awash, Ethiopia, in the context of the phylogeny of Hippotragini (Bovidae, Mammalia). *Palaeont. Afr.*

Vrba, E. S., Vaisnys, J. R., Gatesy, J. E., DeSalle, R., and Wei, K.-Y. 1994. Tests of paedomorphosis using allometric characters: The example of extant Reduncini (Bovidae, Mammalia). *Systematic Biology* 43:92–116.

Walker, A., Leakey, R. E., Harris, J. N., and Brown, R. H. 1986. 2.5-myr *Australopithecus boisei* from west of Lake Turkana, Kenya. *Nature* 322:517–522.

Walter, R. C., Manega, P. C., Hay, R. L., Drake, R. E., and Curtis, G. H. 1991. Laser-fusion $^{40}Ar/^{39}Ar$ dating of Bed I, Olduvai Gorge, Tanzania. *Nature* 354:145–149.

Ward, S., and Hill, A. 1987. Pliocene hominid partial mandible from Tabarian, Baringo, Kenya. *Am. J. Phys. Anthropol.* 72: 21–27.

Wells, L. H., and Cooke, H. B. S. 1956. Fossil Bovidae from the Limeworks Quarry, Makapansgat, Potgietersrus. *Palaeont. Afr.* 4:1–55.

Wesselman, H. B. 1985. Fossil micromammals as indicators of climatic change about 2.4 Myr ago in the Omo Valley, Ethiopia. *S. Afr. J. Sci.* 81:260–261.

White, T. 1985. African suid evolution: The last six million years. *S. Afr. J. Sci.* 81:271.

White, T. W., Suwa, G., Hart, W. K., Walter, R. C., WoldeGabriel, G., de Heinzelin, J., Clark, J. D., Asfaw, B., and Vrba, E. S. 1994. New discoveries of *Australopithecus* at Maka, Ethiopia. *Nature* 366:261–265.

Wood, B. 1992. Origin and evolution of the genus *Homo. Nature* 355:1.

Appendix 27.1. A Key to the Fossil Sites and Localities Referred to in This Chapter

Locations and sites are listed alphabetically by acronym. Within each site, the stratigraphic subunits are arranged chronologically, from the latest (at the top) to the earliest (at the bottom). In appendix 27.2, additional subdivisions of areas within stratigraphic subunits are given by numbers.

ABH:	Abu Hugar, Sudan, unknown time in Late Pleistocene	
AINB:	Ain Boucherit, Algeria, Late Pliocene	
AINH:	Ain Hanech, Algeria, Early Pleistocene	
AW:	Middle Awash, Ethiopia, Plio-Pleistocene; includes the areas:	
	AW BOD:	Bodo
	AW GAM:	Gamedah
	AW WEE:	Wee-ee
	AW WIL:	Wilti Dora
	AW MAT:	Matabaietu
	AW MAKA:	Maka
	AW ARA:	Aramis
BAA:	Baards Quarry, lower level, Cape Province, South Africa, Late Pliocene	
BLED:	Bled Douarah, Tunisia, Middle to Late Miocene	
BOU:	Bou Hanifia (Oued el Hammam), Algeria, Middle to Late Miocene	

(*continued*)

Appendix 27.1. (*Continued*)

CHI: Chiwondo, Malawi, Late Pliocene, includes the area:
- CHI MWE: Mwenirondo

COR: Cornelia, Orange Free State, South Africa, early Middle Pleistocene
EFTM: Elandsfontein, Cape Province, South Africa, Middle Pleistocene
FLOR: Florisbad, Orange Free State, South Africa, Middle to Late Pleistocene
FTER: Fort Ternan, Kenya, Middle Miocene
HAD: Hadar Formation, Ethiopia, Middle Pliocene, includes members
- HAD KA: Kadar Hadar Member
- HAD DD: Denen Dora Member
- HAD SH: Sidi Hakoma Member

ICHK: Ichkeul, Tunisia, Middle to Late Pliocene
ISI: Isimila, southern Tanzania, Middle Pleistocene
KA: Kromdraai A cave breccia, Transvaal, South Africa, Middle Pleistocene
KB: Kromdraai B cave breccia, Transvaal, South Africa, lower Middle Pleistocene
KAN E: Kanam East, one of the Kavirondo Gulf sites in Kenya, Pleistocene
KF: Koobi Fora Formation, East of Lake Turkana, Kenya, Plio-Pleistocene, with members;
- KF CHA: Chari Member
- KF OKT: Okote Member
- KF KBS: KBS Member
- KF UBU: Upper Burgi Member
- KF LBU: Lower Burgi Member
- KF TUL: Tulu Bor Member
- KF UTUL: Upper Tulu Bor Member
- KF LTUL: Lower Tulu Bor Member
- KF LOK: Lokochot Member
- KF MOI: Moiti Member
- KF LON: Lonyumun Member

LAN: The Varswater Formation in E Quarry at Langebaanweg, Cape Province, South Africa, close to the Miocene-Pliocene boundary
LIT: Laetoli Beds, northern Tanzania, early Middle Pliocene
LOTH I: Lothagam I level, Turkana District, Kenya, Late Miocene, with stratigraphic unit:
- LOTH IB/C: stratigraphic units B and/or C

LUK: Lukeino formation, Baringo Basin, Kenya, Late Miocene
MAK: Makapansgat Formation, Transvaal, South Africa, Late Pliocene, with members:
- MAK4, MAK3: Makapansgat Members 4 and 3

MPES: Mpesida Beds, Baringo Basin, Kenya, Late Miocene.
MUR: Mursi Formation, lower Omo Basin, southern Ethiopia, Lower to Middle Pliocene
NA: Nachukui Formation, west of Lake Turkana, Kenya, Plio-Pleistocene, with members:
- NA NKT: Nariokotome Members
- NA NAT: Natoo Member
- NA KAI: Kaitio Member
- NA KAL: Kalochoro Member
- NA LAL: Lokalalei Member
- NA LM: Lomekwi Member
- NA ULM: Upper Lomekwi Member
- NA MLM: Middle Lomekwi Member
- NA LLM: Lower Lomekwi Member
- NA KAT: Kataboi Member
- NA LON: Lonyumun Member

NBC: Nelson Bay Cave, Cape Province, South Africa, Late Pleistocene
NGO A-E: Ngorora Formation, Baringo Basin, Kenya, Middle to Late Miocene, with stratigraphic units A to E
NYA: Nyakach Formation, Kenya, Middle Miocene
OLD: Olduvai Gorge, Tanzania, Pleistocene, with beds:
- OLD I–IV: Olduvai Beds I–IV

SAH: Sahabi, Lybia, latest Miocene
SEM: Semliki, Zaire, Pleistocene, with:
- SEM LT: Semliki Lower Terrace Complex

SH: Shungura Formation, Omo Group Deposits, Ethiopia, Plio-Pleistocene, with members:
- SH A-L: Shungura Members A to L
- SH BAS: Shungura Basal Member

SK: Swartkrans cave breccias, Transvaal, South Africa, Lower Pleistocene, with members:
- SK1–5: Swartkrans Members 1 to 5

ST: Sterkfontein cave breccias, Transvaal, South Africa, Late Pliocene, with members:
- ST4, ST5: Sterkfontein Members 4 and 5.

TA: Taung australopithecine breccia, Cape, South Africa, Late Pliocene
TER: Ternifine, Algeria, Middle Pleistocene
UNDO: Upper Ndolanya Beds, northern Tanzania, Late Pliocene.
WADN: Wadi Natrun, Egypt, earliest Pliocene
ZEL: Gebel Zelten, Lybia, Lower to Middle Miocene

Appendix 27.2. First and Last Appearance Data (FAD and LAD) for All Known African Taxa of Bovidae

For an explanation of acronyms, see appendix 27.1; for literature sources, see note 1; and for alternative interpretations of some of the FADs and LADs and other details, see additional numbered notes at the base of the table. The traditional genus names are not adjusted to conform with the cladistic classificatory practice that sister taxa should be assigned the same taxonomic rank. Sources of dates are specified as radiometric (rm), paleomagnetic (pm, corrected by astronomical calibration, Shackleton et al., 1990, this vol.; Hilgen, 1991.), stratigraphic (str, such as by sediment thickness), biochronological (bio), or a combination (such as rm/str). Radiometric FAD and LAD chronological limits are given in the following form: (mean date stratigraphically below FAD) ± (error) – (mean date above FAD) ± (error); the less accurate biochronological and stratigraphic estimates are preceded by ~ (approximate). The column on the right gives best estimates.

Taxon	Site and Locality FAD LAD	Date Estimate in myr FAD LAD	Source	Used in Figs. 1–4
Bovini	LUK	6.31 ± .20–5.65 ± .07 Recent	rm	6.0 0.0

Taxon	Site and Locality FAD LAD	Date Estimate in myr FAD LAD	Source	Used in Figs. 1–4
Ugandax				
1. cf. *gautieri*	LUK	6.31 ± .20–5.65 ± .07	rm	6.0[2]
	HAD DD-3	3.22 ± .01–3.18 ± .01	rm	3.2
Simatherium				
2. *demissum*	LAN	~ 5.0	bio	5.0
	LAN	~ 5.0	bio	5.0
3. *kohllarseni*	LIT	3.76 ± .03–3.46 ± .12	rm	3.6
	MAK4	~ 2.7 – ~ 2.5	bio	2.6
Pelorovis				
4. *oldowayensis*	SH C5	below 2.52 ± .05	rm/str	2.7[3,4]
	OLD III JK	1.33 ± 0.6–0.96	rm/pm	1.2[5]
5. *turkanensis*	KF UBU	~2.0–1.88 ± .02	rm/str	1.9[6]
	KF CHA	1.39 ± .01–0.67	rm/str	0.8[7]
6. *antiquus*	OLD IV GC	0.78–0.6	pm/str	0.7[8]
	NBC	0.012–0.009	rm	0.01
Syncerus				
7. *caffer*	NBC	0.018–0.011	rm	0.02
		Recent		0.0
8. *acoelotus*	CH C6	below 2.52 ± .05	rm/str	2.7[3,9]
	OLD IV GC	0.78–0.6	pm/str	0.7[8]
?*Leptobos*				
9. *syrticus*	SAH	~ 5.5	bio	5.5[10]
	SAH	~ 5.5	bio	5.5
Tragelaphini	MPES	(7.3 or 6.8)–6.3	rm	6.5
		Recent		0.0
Tragelaphus				
10. sp. aff.	MPES	(7.3 or 6.8)–6.3	rm	6.5
spekei	LUK	6.31 ± .20–5.65 ± .07	rm	6.0
11. *pricei*	MAK3	~ 2.8–2.6	bio	2.7
	SH C9	below 2.52 ± .05	rm/str	2.6[3]
12. sp. nov.	LAN	~ 5.0	bio	5.0
		Recent		0.0
13. *gaudryi*	AW GAM-1	2.5	rm/str	2.5[11,12]
	SH G13	2.33 ± .05–1.88 ± 02	rm/str	1.95
14. *strepsiceros*	NA LAL	2.52 ± .05–2.35 ± .05	rm	2.5
		Recent		0.0
15. cf. *angasi*	ST4	~ 2.6–2.4	bio	2.5
		Recent		0.0
16. cf. *buxtoni*	UNDO	below 2.41 ± .12	rm/str	2.6[13]
		Recent		0.0
17. *kyaloae*	KF MOI	4.10 ± .07–3.6	rm/str	3.8[14]
	NA LLM	above 3.36 ± .04	rm/str	3.2
18. *nakuae*	KF MOI	4.10 ± .07–3.6	rm/str	3.8[14]
	SH H3	1.88 ± .02–1.86 ± .02	rm	1.87
Taurotragus				
19. *arkelli-oryx* lineage	OLD IV LK-RK	0.96–0.6	pm/str	0.9[15]
		Recent		0.0
Boselaphini	ZEL	~ 17.0	bio	17.0
	LAN	~ 5.0	bio	5.0
Eotragus				
20. *Eotragus* sp.	ZEL	~ 17.0	bio	17.0
	FTER	14.0 ± .20–13.9 ± .30	rm	14.0

(*continued*)

Appendix 27.2. (*Continued*)

Taxon	Site and Locality FAD / LAD	Date Estimate in myr FAD / LAD	Source	Used in Figs. 1–4
Protragocerus				
21. *labidotus*	FTER	14.0 ± .20–13.9 ± .30	rm	14.0
	NGO D	8.5	rm/str	8.5
Sivoreas				
22. *eremita*	NGO A	< 13.15	rm/str	12.5
	NGO D	8.5	rm/str	8.5
Miotragocerus[16]				
23. *cyrenaicus*	LOTH IB/C	6.5	bio	6.5
	SAH	~ 5.5	bio	5.5
Mesembriportax				
24. *acrae*	LAN	~ 5.0	bio	5.0
	BAA	~ 2.5	bio	2.5
Cephalophini	LUK	6.31 ± .20–5.65 ± .07	rm	6.0[17]
		Recent		0.0
Cephalophus				
25. sp. nov.	LUK	6.31 ± .20–5.65 ± .07	rm	6.0
	LUK	6.31 ± .20–5.65 ± .07	rm	6.0
26. cf. *monticola*	MAK3	~ 2.8–2.6	bio	2.7
		Recent		0.0
Neotragini	MPES	(7.3 or 6.8)–6.3	rm	6.5[18]
		Recent		0.0
Raphicerus				
27. *paralios*	LAN	~ 5.0	bio	5.0
	MAK3	~ 2.8–2.6	bio	2.7
28. *campestris*	SK2	~ 1.1	bio	1.1[19]
		Recent		0.0
29. *melanotis*	EFTM	~ 0.6	bio	0.6
		Recent		0.0
Ourebia				
30. *ourebi*	SK3	~ 0.7	bio	0.7[19]
		Recent		0.0
Oreotragus				
31. *major*	MAK3	~ 2.8–2.6	bio	2.7
	SK2	~ 1.1	bio	1.1[19]
32. *oreotragus*	SK3	~ 0.7	bio	0.7[19]
		Recent		0.0
Madoqua				
33. sp. nov.	MPES	(7.3 or 6.8)–6.3	rm	6.5
		Recent		0.0
34. *avifluminis*	LIT	3.76 ± .03–3.46 ± .12	rm	3.6
	UNDO	below 2.41 ± .12	rm/str	2.6[13]
Antilopini	NYA	~ 15	bio	15.0
		Recent		0.0
Homoiodorcas[18]				
35. *tugenium*	NGO A	<13.15	rm/str	12.5
	NGO D	8.5	rm/str	8.5
Gazella				
36. ?*Gazella* sp.	NYA	~ 15	bio	15.0
	NYA	~ 15	bio	15.0
37. *Gazella* sp.	FTER	14.0 ± .20–13.9 ± .30	rm	14.0
	FTER	14.0 ± .20–13.9 ± .30	rm	14.0

Taxon	Site and Locality FAD / LAD	Date Estimate in myr FAD / LAD	Source	Used in Figs. 1–4
38. *praegaudryi*	BOU	~ 10	bio	10.0
	BOU	~ 10	bio	10.0
39. aff.	LAN	~ 5.0	bio	5.0
vanhoepeni	LAN	~ 5.0	bio	5.0
40. *vanhoepeni*	MAK3	~ 2.8–2.6	bio	2.7
	SK1	~ 1.8	bio	1.8
41. sp. nov. 1	MAK3	~ 2.8–2.6	bio	2.7
	MAK4	~ 2.7–~ 2.5	bio	2.6
42. sp. nov. 2	OLD I KK	1.80 ± .004–1.75 ± .01	rm	1.77
	EFTM	~ 0.6	bio	0.6
43. *praethomsoni*	KF LOK	3.58–3.36 ± .04	rm/pm	3.5[20]
	KF CHA	1.39 ± .01–0.67	rm/str	0.8
44. *janenschi*	LIT	3.76 ± .03–3.46 ± .12	rm	3.6
	NA NAT	1.65 ± .03–1.33 ± .03	rm	1.6
45. *dracula*	TER	~ 0.7	bio	0.7
	TER	~ 0.7	bio	0.7
46. *pomeli*	AINH	~ 1.8	bio	1.8
	AINH	~ 1.8	bio	1.8
Prostrepsiceros				
47. ?*P.* sp.	BOU	~ 10	bio	10
	BOU	~ 10	bio	10
48. *lybicus*	SAH	~ 5.5	bio	5.5[21]
	SAH	~ 5.5	bio	5.5
Antilope				
49. aff. *subtorta*	SH C5	below 2.52 ± .05	rm/str	2.7[3,22]
	SH C5	below 2.52 ± .05	rm/str	2.7
Antidorcas[23]				
50. *recki*	ST4	~ 2.6–2.4	bio	2.5
	SH F3	2.34 ± .04–2.33 ± .05	rm	2.34
	EFTM	~ 0.6	bio	0.6
51. sp. nov.	UNDO	below 2.41 ± .12	rm/str	2.6[13]
	OLD II BK	1.48 ± .05–1.33 ± .06	rm/pm	1.4
52. *marsupialis*	SK3	~ 0.7	bio	0.7[19]
		Recent		0.0
53. *bondi*	SK2	~ 1.1	bio	1.1[19,24]
		0.009–0.006		0.01
Gen. indet.				
54. sp. nov.	UNDO	below 2.41 ± .12	rm/str	2.6[13]
	OLD II BK	1.48 ± 0.5–1.33 ± .06	rm/pm	1.4
Ovibovini	BOU	~ 10	bio	10.0
	SK1	~ 1.8	bio	1.8
Damalavus				
55. *boroccoi*	BOU	~ 10	bio	10
	BOU	~ 10	bio	10
56. sp. cf. "*Bos*"	HAD DD?	3.22 ± .01–2.8	rm/str	3.1
makapaani	OLD I FLKN	1.80 ± .00–1.75 ± .01	rm	1.8
	SK1	~ 1.8	bio	1.8
Makapania				
57. *broomi*	MAK3	~ 2.8–2.6	bio	2.7
	ST4	~ 2.6–2.4	bio	2.5
Gen. indet.				
58. sp. nov.	LAN	~ 5.0	bio	5.0
	LAN	~ 5.0	bio	5.0

(*continued*)

Appendix 27.2. (*Continued*)

Taxon	Site and Locality FAD / LAD	Date Estimate in myr FAD / LAD	Source	Used in Figs. 1–4
59. SH Ovibovini	SH C6	below 2.52 ± .05	rm/str	2.7[3,25]
sp. indet.	SH G12–13	below 1.88 ± .02	rm/str	1.95
Caprini	NYA	~ 15	bio	15.0
		Recent		0.0
Caprotragoides (= *Pseudotragus* in part)				
60. *potwaricus*	NYA	~ 15	bio	15.0
	FTER	14.0 ± .20–13.9 ± .30	rm	14.0
61. *gentryi*	NGO B	~ 12	rm/str	12.0
	NGO E	~ 8	rm/str	8.0
Hypsodontus (= *Oioceros* in part)				
62. *tanyceras*	FTER	14.0 ± .20–13.9 ± .30	rm	14.0
	FTER	14.0 ± .20–13.9 ± .30	rm	14.0
Pachytragus				
63. *solignaci*	BLED	~ 10.5–8.5	bio	11.0
	NGO D	8.5	rm/str	8.5
Parantidorcas				
64. *latifrons*	AINB	~ 2.7–2.5	bio	2.5
	AINB	~ 2.7–2.5	bio	2.5
Gen. indet.				
?Caprini				
65. sp. 1	SH D	2.52 ± .05–2.34 ± .04	rm	2.5[6]
	SH F	2.34 ± .04–2.33 ± .05	rm	2.34
66. ?Caprini sp. 2	LOTH IB/C	6.5	bio	6.5
	LOTH IB	6.5	bio	6.5
67. sp. A	NA LAL LO7	2.52 ± .05–2.35 ± .05	rm	2.5
	NA NKT NC2	1.39 ± .02–0.8	rm	0.9[26]
68. sp. B	NA NAT KI2	1.65 ± .03–1.33 ± .03	rm	1.6
	NA NAT KI2	1.65 ± .03–1.33 ± .03	rm	1.6
69. sp. C	NA ULM LO9	below 2.52 ± .05	rm/str	2.6
	NA NKT NC1	above 1.39 ± .02	rm	0.9
70. sp. D	NA NAT KL1	1.65 ± .03–1.39 ± .02	rm	1.6
	NA NAT KL1	1.65 ± .03–1.39 ± .02	rm	1.6
Alcelaphini[27]	LAN	~ 5.0	bio	5.0
		Recent		0.0
Damalacra				
71. *acalla*	LAN	~ 5.0	bio	5.0
	WADN	~ 4.5	bio	4.5
72. *neanica*	LAN	~ 5.0	bio	5.0
	LAN	~ 5.0	bio	5.0
Damalops				
73. sp. aff. *palaeindicus*	LIT	3.76 ± .03–3.46 ± .12	rm	3.6[28]
	SH B11	above 2.95 ± .05	rm	2.9
	MAK3	~ 2.8–2.6	bio	2.7
Alcelaphus				
74. *buselaphus*	AW BOD-1	above 0.74 ± .03	rm/str	0.7
		Recent		0.0
Sigmoceros				
75. *lichtensteini*	SEM LT	~ 0.5–0.3	bio	0.4
		Recent		0.0
76. sp. nov.	OLD Tuff IIA	1.66 ± .10	rm	1.66
	OLD Tuff IIA	1.66 ± .10	rm	1.66

Taxon	Site and Locality FAD LAD	Date Estimate in myr FAD LAD	Source	Used in Figs. 1–4
Connochaetes				
77. *taurinus*	OLD II MNK	1.66 ± .10–1.48 ± .05	rm	1.6
		Recent		0.0
78. *gnou*	COR	~ 0.08–0.06 (42)	bio	0.7
		Recent		0.0
79. *africanus*	OLD II	1.75 ± .01–1.33 ± .06	rm	1.5
	OLD II	1.75 ± .01–1.33 ± .06	rm	1.5
80. *gentryi*	NA ULM KU2	close to 2.52 ± .05	rm	2.5
	OLD II HWKE	1.66 ± .10–1.48 ±.05	rm	1.6
Megalotragus				
81. *kattwinkeli*	CHI MW	~ 2.8–2.5	bio	2.7[29]
	ST4	~ 2.6–2.4	bio	2.5
	OLD IV GC	0.78–0.6	pm ± str	0.7[8]
82. *priscus*	COR	~ 0.08–0.06	bio	0.7
	NBC	0.012–0.009	rm	0.01
83. *isaaci*	KF UBU	~ 2.0–1.88 ± .02	rm/str	1.9
	KF OKT	1.6–1.39 ± .02	rm/str	1.5
Rabaticeras				
84. *porrocornutus-arambourgi*	SK1	~ 1.8	bio	1.8
	EFTM	~ 0.6	bio	0.6
Numidocapra				
85. *crassicornis*	AINH	~ 1.8	bio	1.8
	AINH	1.8	bio	1.8
Oreonagor				
86. *tournoueri*	AINB	~ 2.7–2.5	bio	2.5
	AINB	~ 2.7–2.5	bio	2.5
Parestigorgon				
87. *gadgingeri*	UNDO	below 2.41 ± .12	rm/str	2.6[13]
	UNDO	below 2.41 ± .12	rm/str	2.6[13]
Damaliscus				
88. *dorcas*	SK2	~ 1.1	bio	1.1[19]
		Recent		0.0
89. *lunatus*	SK3	~ 0.7	bio	0.7[19]
		Recent		0.0
90. *gentryi-niro* lineage	MAK5	~ 1.6–1.8	bio	1.7
	OLD II SHK	1.66 ± .10–1.48 ± .05	rm	1.5
	FLOR	~ 200–100 kyr	bio	0.2
91. *agelaius*	OLD II FLKW	1.75 ± .01–1.66 ± .10	rm	1.7
	OLD III JK2 GP8	1.33 ± .06–0.96	rm	1.0[5]
92. *asfawi*	AW GAM-1	2.5	rm/str	2.5[11]
	AW GAM-1	2.5	rm/str	2.5
93. sp. nov. "*hipkini*"	COR	~ 0.08–0.06	bio	0.7
	EFTM	~ 0.6	bio	0.6
Awashia				
94. *suwai*	AW MAT-3	below 2.518	rm/str	2.6
	AW MAT-3	below 2.518	rm/str	2.6
Beatragus				
95. *hunteri*	KF CHA	1.39 ± .01–0.67	rm/str	0.8[7]
		Recent		0.0
96. *antiquus*	AW GAM-1	2.5	rm/str	2.5[11]
	OLD II Kit K	> 1.48 ± .05	rm	1.5
97. *whitei*	AW MAT-4	below 2.518	rm/str	2.6
	AW MAT-4	below 2.518	rm/str	2.6

(continued)

Appendix 27.2. (*Continued*)

Taxon	Site and Locality FAD / LAD	Date Estimate in myr FAD / LAD	Source	Used in Figs. 1–4
Parmularius				
98. *pandatus*	LIT	3.76 ± .03–3.46 ± .12	rm	3.6
	LIT	3.76 ± .03–3.46 ± .12	rm	3.6
99. *braini*	MAK3	~ 2.8–2.6	bio	2.7[30]
	MAK3	~ 2.8–2.6	bio	2.7
100. *eppsi*	SH C7	below 2.52 ± .05	rm/str	2.6[3,30]
	KF OKT	1.6–1.39 ± .02	rm/str	1.5
101. *cuiculi*	AINB	~ 2.7–2.5	bio	2.5[30]
	AINB	~ 2.7–2.5	bio	2.5
102. *altidens*	AINB	~ 2.7–2.5	bio	2.5
	OLD I HWK	< 1.75 ± .007	rm	1.7
103. *angusticornis*	OLD I FLKN	1.80 ± .00–1.75 ± .01	rm	1.8
	ISI	~ 0.8–0.6 myr	bio	0.7
104. *rugosus*	OLD II HWK	1.75 ± .01–1.66 ± .10	rm	1.7
	OLD IV HWK	0.78–0.7	pm ± str	0.75[8]
105. *parvus*	KA	~ 1–0.7	bio	1.0
	OLD IV Upper	0.78–0.7	pm ± str	0.7[8]
106. *ambiguus*	TER	~ 0.7	bio	0.7
	TER	~ 0.7	bio	0.7
Gen. indet.				
107. sp. nov.	HAD SH-2	above 3.40 ± .03	rm	3.4
	MAK3	~ 2.8–2.6	bio	2.7
108. *helmoedi*	COR	~ 0.08–0.06	bio	0.7
	COR	~ 0.08–0.06	bio	0.7
Aepycerotini[31]	LOTH IB/C	6.5	bio	6.5
		Recent		0.0
Aepyceros				
109. sp. nov. 1	LOTH IB/C	6.5	bio	6.5
	MUR	> 3.99 ± .04	rm/str	4.0
110. *shungurae-melampus*	KF MOI	4.10 ± .07–3.6	rm/str	3.8[14,32]
		Recent		0.0
111. cf. *A.* sp. nov. 2	SH F	2.34 ± .04–2.33 ± .05	rm	2.34
	SH G13	2.33 ± .05–1.88 ± .02	rm/str	1.95
Hippotragini[33]	LOTH IB/C	6.5	bio	6.5
		Recent		0.0
112. Gen. nov.	LOTH IB/C	6.5	bio	6.5
	LOTH IB/C	6.5	bio	6.5
Praedamalis				
113. *deturi*	LIT	3.76 ± .03–3.46 ± .12	rm	3.6
	UNDO	below 2.41 ± .12	rm/str	2.6[13]
Brabovus				
114. *nanincisivus*	LIT	3.76 ± .03–3.46 ± .12	rm	3.6[34]
	LIT	3.76 ± .03–3.46 ± .12	rm	3.6
Wellsiana				
115. *torticornuta*	MAK3	~ 2.8–2.6	bio	2.7
	MAK3	~ 2.8–2.6	bio	2.7
Hippotragus				
116. sp. nov.	SAH	~ 5.5	bio	5.5[10]
	SAH	~ 5.5	bio	5.5
117. *gigas*	SH C8	below 2.52 ± .05	rm/str	2.6[3]
	AW GAM-1	2.5	rm/str	2.5[11]
	FLOR	~ 0.2–0.1	bio	0.2

Taxon	Site and Locality FAD / LAD	Date Estimate in myr FAD / LAD	Source	Used in Figs. 1–4
118. *cookei*	MAK3	~ 2.8–2.6	bio	2.7
	MAK3	~ 2.8–2.6	bio	2.7
119. *equinus*	ST4	~ 2.6–2.4	bio	2.5[24]
		Recent		0.0
120. *niger*	SK3	~ 0.7	bio	0.7[19]
		Recent		0.0
121. *leucophaeus*	EFTM	~ 0.6	bio	0.6
	Cape Province	1799 AD		0.0
Oryx[35]				
122. *Oryx* lineage several spp.	SH C9	Below 2.52 ± .05	rm/str	2.6[3]
		Recent		0.0
Gen. indet.				
123. sp. nov.	LIT	3.76 ± .03–3.46 ± .12	rm	3.6
	MAK3	~ 2.8–2.6	bio	2.7
Reduncini[36]	LOTH IB/C	6.5	bio	6.5
		Recent		0.0
Dorcadoxa				
124. *porrecticornis*	MPES	(7.3 or 6.8)–6.3	rm	6.5
	HAD SH-?	3.40 ± .03–3.22 ± .01	rm	3.4
	BAA	~ 2.5	bio	2.5
Kobus				
125. *subdolus*	SAH	~ 5.5	bio	5.5[37,10]
	LAN	~ 5.0	bio	5.0
	ICHK	~ 3.0	bio	3.0
126. sp. nov. 1	LAN	~ 5.0	bio	5.0
	LAN	~ 5.0	bio	5.0
127. *radiciformis*	OLD IV ?	0.96–0.7	pm ± str	0.9
	OLD IV ?	0.96–0.6	pm ± str	0.9
128. sp. nov. 2	HAD SH-3	3.40 ± .03–3.22 ± .01	rm	3.3
	HAD DD-3	3.22 ± .01–3.18 ± .01	rm	3.2
129. *oricornis*	KF MOI	4.10 ± .07–3.6	rm/str	3.8[14]
	NA LAL	2.52 ± .05–2.35 ± .05	rm	2.5
130. *kob*	SH B11	above 2.95 ± .05	rm	2.9
		Recent		0.0
131. sp. aff. *kob*	HAD SH-3	3.40 ± .03–3.22 ± .01	rm	3.3
	HAD SH-3	3.40 ± .03–3.22 ± .01	rm	3.3
132. sp. nov. 3	AW BOD-1	above 0.74 ± .03	rm/str	0.7
	AW BOD-1	above 0.74 ± .03	rm/str	0.7
133. sp. nov. 4	ABH	?Late Pleistocene	bio	0.2
	ABH	?Late Pleistocene	bio	0.2
134. *ancystrocera*	SH B11	above 2.95 ± .05	rm	2.9
	SH J6	1.65 ± .03– ~ 1.6	rm/str	1.65
135. *sigmoidalis*	SH C6	below 2.52 ± .05	rm/str	2.7[3,38]
	OLD II MNK	1.66 ± .10–1.48 ± .05	rm	1.6
136. *leche*	NA NKT NC3	1.39 ± .02–0.8	rm	0.9
		Recent		0.0
137. *ellipsiprymnus*	SH G1	just < 2.33 ± .05	rm	2.3
		Recent		0.0
Redunca				
138. *darti*	MAK3	~ 2.8–2.6	bio	2.7
	MAK3	~ 2.8–2.6	bio	2.7
139. *arundinum*	EFTM	~ 0.6	bio	0.6
		Recent		0.0

(*continued*)

Appendix 27.2. (*Continued*)

Taxon	Site and Locality FAD / LAD	Date Estimate in myr FAD / LAD	Source	Used in Figs. 1–4
140. *redunca*	KAN E	Middle Pleistocene ?	bio	0.8
		Recent		0.0
141. *fulvorufula* lineage	SH C5	below 2.52 ± .05	rm/str	2.7[3]
		Recent		0.0
Menelikia				
142. *lyrocera*	SH C6	below 2.52 ± .05	rm/str	2.7[3,39]
	SH J6	< 1.65 ± .03	rm/str	1.6
143. *leakeyi*	KF MOI	4.10 ± .07–3.6	rm/str	3.8[14]
	KF UTUL	> 2.68 ± .03	rm/str	2.8
144. *sp. nov*	SH C6	below 2.52 ± .05	rm/str	2.7[3,40]
	SH D1	above 2.52 ± .05	rm	2.5
Peleini	LIT	3.76 ± .03–3.46 ± .12	rm	3.6
		Recent		0.0
Pelea				
145. aff. *P.* sp. nov.	LIT	3.76 ± .03–3.46 ± .12	rm	3.6
	MAK3	~ 2.8–2.6	bio	2.7
146. ?*Pelea*	UNDO	below 2.41 ± .12	rm/str	2.6[13]
sp. nov.	UNDO	below 2.41 ± .12	rm/str	2.6[13]
147. *capreolus*	SK1	~ 1.8	bio	1.8
		Recent		0.0

[1]*Sources for bovid records* (Fossils or casts of all the taxa described by others were also studied by me): Ain Boucherit, Ain Hanech, Ichkeul, Wadi Natrun (Geraads, 1987, partly after Arambourg, 1979); Awash (Vrba, in press or prep.); Bou Hanifia (Gentry, 1990a); Chiwondo (Kaufulu et al., 1981); Elandsfontein (Klein and Cruz-Uribe, 1991); Fort Ternan (Gentry, 1970); Gebel Zelten (Hamilton, 1973); Hadar (Gentry, 1981); Koobi Fora Formation (Harris, 1991); Laetoli, Ndolanya (Gentry, 1987); Langebaanweg, Baards Quarry (Gentry, 1974, 1980); Lothagam (Smart, 1976); Lukeino and Mpesida (Thomas, 1980); Makapansgat (Wells and Cooke, 1956; Vrba, 1977a, 1987a, 1987b); Nachukui Formation (Harris et al., 1988); Nelson Bay Cave (Klein, 1972); Ngorora (Thomas, 1981); Nyakach (Thomas, 1984b); Olduvai, Abu Hugar, Cornelia, Florisbad, Kanam East (Gentry and Gentry, 1978); Sahabi (Thomas, 1979; Lehmann and Thomas, 1987); Semliki (Gentry, 1990b); Shungura and Mursi (Gentry, 1976, 1985); Sterkfontein, Swartkrans, Kromdrai (Vrba, 1975, 1976, 1978); Ternifine (Geraads, 1981). *Sources for chronology*: Abu Hugar, Bou Hanifia, Gebel Zelten, Kanam East, Nyakach (Gentry, 1990a; also Thomas et al., 1982, for Bou Hanifia); Ain Boucherit, Ain Hanech, Ichkeul, Wadi Natrun (Geraads, 1987, in part after Arambourg, 1979); Awash (White et al., 1994; White, pers. comm.); Chiwondo (Kaufulu et al., 1981); Cornelia, Elandsfontein, Florisbad, Kromdraai, Makapansgat, Sterkfontein, Swartkrans (Vrba, 1982, unpubl. revisions); Fort Ternan (Shipman et al., 1981); Gebel Zelten (Gentry, 1990a); Hadar (Kimbel, this vol.); Koobi Fora, Nachukui, Shungura, Mursi Formations (Brown and Lajoie, 1970; Brown and Shuey, 1976; Feibel et al., 1989; Brown and Feibel, 1991; Haileab and Brown, 1992; Brown, this vol.); Laetoli, Ndolanya (Hay, 1976; Drake and Curtis, 1987); Langebaanweg, Baards Quarry (Hendey, 1978, 1981, pers. obs.); Lothagam (pers. obs.); Lukeino, Mpesida, Ngorora (Hill et al., 1985; Hill, this vol.); Nelson Bay Cave (Klein, 1972); Nyakach (Thomas, 1984b); Olduvai (Manega, 1993; Walter et al., 1991); Sahabi (Thomas, 1979; Lehmann and Thomas, 1987); Semliki (Gentry, 1990b); Ternifine (Geraads, 1981).

[2]Because this monophyletic group may consist of a single unbranching lineage, it is treated here as an undifferentiated taxon.

[3]The age of Shungura Tuff C is 2.86 myr based on stratigraphic thicknesses between dated Tuffs B and D (Brown and Shuey, 1976; Brown and Feibel, 1991). Based on stratigraphic thicknesses of submembers C1–9 of Member C (C9 is just below Tuff D dated 2.52 ± .05 myr), I use the following estimates for submembers: C7–9 = 2.6 myr, C4–6 = 2.7 myr, C1–3 = 2.8 myr.

[4]Gentry (1976) thought that a left P_4, L369-2, from Shungura B12, ca. 3 myr, might belong to *Pelorovis*, with all other *Pelorovis* no earlier than C5 onward. I consider the tooth from B12 to be indistinguishable from the Makapansgat *Simatherium* (Vrba, 1987a); and I here use Gentry's later evaluation (1985) that the earliest secure identifications of *Pelorovis* start after Shungura B.

[5]I take sites of the Olduvai III JK group, JK1, and JK2 to be close above Tuff III-1 dated 1.33 ± .06 myr (Manega, 1993), and Olduvai III JK2 GP8 to be closer to the Bed III/IV junction, estimated on stratigraphic criteria to be about 0.96 myr old (Manega, 1993).

[6]L98-1 D and L52-114 F are horncore pieces that Gentry (1976) attributed to the caprine ?*Tossunnoria* sp. Gentry (1985) suggested that they may belong to *P. turkanensis*, which would change its FAD to 2.5 myr. I retain these fossils in ?Caprini.

[7]I assume that these fossils belong to the uppermost Chari Member (0.85–0.67 myr) above an unconformity that accounts for some 500 kyr of the total 690 kyr of this member, which contains most of the Chari fossils (Brown and Feibel, 1991).

[8]Based on Manega (1993) I estimate that in Olduvai Bed IV 1) the LK, RK, and HEB group sites date ca. 0.9 myr close below the Matuyama-Brunhes reversal (0.78 myr by astronomical calibration; Shackleton et al., 1991); 2) the GC, GTC, and WK group sites are ca. 0.7 myr late in Bed IV (Bed IV ends at 0.7–0.6 myr based on stratigraphic criteria); and 3) site PDK is between 1) and 2).

[9]An alternative FAD could be 2.9 myr: Gentry (1976) noted that two cranial fossils from Shungura B, L2-26 B and L1-238 B, might belong to *Syncerus* and that other *Syncerus* specimens start only from Shungura C6. I here agree with Gentry (1985, table 6) that the earliest secure *Syncerus* occurs later than Shungura B.

[10]If the date for Sahabi were correct and if *syrticus* is correctly attributed to *Leptobos,* this would be a surprisingly early African record of an otherwise Eurasian Late Pliocene genus. On morphological grounds I question Lehmann and Thomas's (1987) inclusion of this species in *Leptobos.* Geraads (1989) suggested that some of the the Sahabi fossils may be much later than the main Late Miocene assemblage. Although I have here followed the conventional chronology for all the Sahabi bovids, my own systematic assessments accord with Geraads's (1989) and suggest a Late Pliocene date for some of the taxa, such as the *Hippotragus* sp. 116.
[11]The Middle Awash assemblages WIL-2, WIL-3, and GAM-1 come from two sections that are not yet directly associated with physical dates. But one section, MAT, which on stratigraphic grounds represents the same part of the column as GAM and WIL, contains fossil assemblages MAT-1-MAT-6, which share many species with GAM-1, WIL-2, and WIL-3. The present estimate of 2.5 myr is based on the fact that these MAT assemblages closely overlie and underlie a tuff dated 2.52 myr (White, pers. comm.). It remains possible that the WIL and GAM records will turn out to be closer to 2.0 myr than to 2.5 myr.
[12]The isolated teeth in Shungura B listed as *T. gaudryi* by Gentry (1985) are excluded. The earliest secure record is from Shungura E4, dated 2.52–2.34 myr.
[13]The Olgol lava overlying the Upper Ndolanya Beds is dated as 2.41 ± .12 myr, and a calcrete layer representing perhaps 100 kyr lies between the lava and the beds (Hay, 1976; Drake and Curtis, 1987). Thus, the Upper Ndolanya fauna is probably only a little older than this lava namely, 2.6–2.7 myr (Drake and Curtis, 1987).
[14]This FAD could be earlier and nearer 4.1 myr. I use an alternative nearer to the top than to the base of the Moiti Member, because Harris (pers. comm.) regards the upper strata as the more probable source of the Moiti fossils.
[15]In include *T. arkelli* (Gentry and Gentry, 1978), *T. algericus* from Ternifine (Geraads, 1981), and the Elandsfontein record of the living species (Klein and Cruz-Uribe, 1991), because they appear to represent a single unbranching lineage (Gentry, 1990a).
[16]This genus persisted globally only until the latest Miocene or slightly later. I consider the *Miotragocerus* sp. at Lothagam IB/C as probably conspecific with, although less advanced than, *M. cyrenaicus* from Sahabi.
[17]Gentry (1978:550) thought that a maxillary fragment from Ngorora may represent a duiker, but I follow Thomas (1981) in regarding this as insecure.
[18]Thomas's (1981) tentative placement of *Homoiodorcas tugenium* in ?Neotrgini would make the neotragine FAD much earlier, ca. 12.5 myr. Thomas (1981) did note that this taxon may belong to Antilopini, an alternative that I use here.
[19]Using biochronology based on Bovidae, I tentatively place Swartkrans Member 2 ca. 1.1 myr and Member 3 ca. 0.7 myr.
[20]Harris's (1991) view is here followed. However, his extension of this FAD to an earlier time than the previous FAD in Shungura F (Gentry, 1985) depends on fragments of dentitions from Lokochot and Tulu Bor and three fragments of horncore from Lokochot, which he noted differ from later specimens. Because the dentitions of this species are hardly known and because there is doubt on the attribution of the Lokochot horncores, an alternative FAD is Shungura F, 2.34–2.33 myr.
[21]This is the largest and latest of all known *Prostrepsiceros.* Apart from these two North African records, this genus is known only from the southern part of eastern Europe and from southwest Asia (Lehmann and Thomas, 1987).
[22]*Antiolpe subtorta* was described from the Pinjor stage, Siwaliks (Pilgrim, 1939). *Antilope* is also recorded from Early and Middle Pleistocene China and Late Pleistocene eastern Asia (Savage and Russell, 1983).
[23]I agree with Gentry (1990a, p. 219) that "despite contrary claims by others and myself, I would not now accept as definite *Antidorcas* any remains prior to those horncores in the Upper Ndolanya Beds." That is, I exclude the dental and horncore fragments from above the Tulu Bor Tuff, later than 3.36 myr, that Harris (1991) assigned to this genus. I also exclude claimed records from Zelten and Ngorora (reviewed in Gentry, 1990a).
[24]A few fossils in the successive assemblages within each of Sterkfontein, Swartkrans, and Kromdraai may belong to adjacent strata (Vrba, 1976). For instance, alternative FADS for *Antidorcas bondi* and *Hippotragus equinus* could be later.
[25]Note that this record refers to a Shungura FAD of ovibovine species known only by dentitions, which could belong to *Makapania* and/or ovibovine species known from earlier African sites. I record this appearance here because it is significant to the argument illustrated by figure 27.4. Ovibovine dentitions are also known from Shungura Member D54, 2.52 ± 0.5–2.34 ± 0.4 myr.
[26]This LAD could be earlier in the cited time interval. WT 14941 is a frontlet from NC1 in the NKT Member and WT 16239 a horncore base from LO9 in the Lomekwi Member. LO9 is below the Emekwi Tuff in the Lomekwi Member. I use 0.9 myr because the bovids appear to belong to the later part of the quoted interval.
[27]Langebaanweg appears to be the earliest site with undisputed alcelaphine fossils. I regard the Lothagam taxon aff. *Damaliscus* (Smart, 1976) as more likely to be hippotragine. The Mpesida tooth identified by Gentry (1978) as possibly alcelaphine is considered by Thomas (1980) to be tragelaphine. Thomas (1984b) included *Maremmia* from the Turolian of Baccinello, Tuscany, ca. 6–6.5 myr old, in Alcelaphini. I here follow Gentry (1990a), who regards it as likely that *Maremmia* is a dentally precocious offshoot from the same ancestry as true Alcelaphini.
[28]Gentry (1980) suggested that the *Damalops* from Laetoli and Hadar and the Koobi Fora Formation (Gentry, pers. comm.) could be closely related to *D. palaeindicus* from the Pinjor (Pilgrim, 1939) and from Tadzhikistan (Dmitrieva, 1977), where it appears ca. 2.6 myr ago.
[29]The Chiwondo assemblages on which this date is based are the later, more northerly localities (Kaufulu et al., 1991), and this date is tentative. Early *Megalotragus* is also known from MAK3, ST4, and SH G (Gentry, 1985). The MAK3 record is based on sparse dental material that may belong to later MAK5. That would make the earliest secure record of *Megalotragus* 2.5 myr.
[30]I here follow Harris's (1991) separation of the species *eppsi* and Geraads's (1987) separation of *cuiculi* from *braini.* But note my alternative interpretation (Vrba, 1977a) that all three are conspecific.
[31]Gentry (1990a) considered that the Fort Ternan taxon previously described (Gentry, 1970) as *Gazella* sp. could possibly be related to *Aepyceros.*
[32]This appears to me be a single unbranching lineage that one could argue should bear a single species name. Gentry (1985) and Harris (1991) acknowledge the possibility of a single lineage but prefer to subdivide this lineage arbitrarily by recognizing the onset of the living species, *A. melampus,* either ca. 2.5 myr (Harris, 1991) or ca. 2.0 myr (Gentry, 1985).
[33]Hippotragines from the Siwalik formations discussed by Pilgrim (1939) and recent chronological estimates are as follows (from Barry, pers. comm., who suggests ≤ 2.5 myr for the earliest Pinjor, and the Gauss chron above the Kaena (3.05–2.60 myr based on astronomical calibration; Shackleton, this vol.) for most Tatrot fossils; and Azzaroli, this vol, whose estimate of 2.6 myr for the Tatrot-Pinjor boundary I use): *Sivatragus bohlini* from the Pinjor faunal zone, ca. 2.6–1.6 myr, *Sivatragus brevicornis* probably from the later part of the Tatrot zone (ca. 3.05–2.6 myr), and *Sivoryx sivalensis* of unknown provenience and *Sivoryx cautleyi* from the Pinjor zone, which probably represent the same species of *Oryx* (Gentry and Gentry, 1978).
[34]Gentry (1987) placed this species in Bovini. Vrba (1987b) suggested that it could belong to Hippotragini.

(*continued*)

Appendix 27.2. (*Continued*)

[35]Because the fossil record of this group remains scant and unstudied, *Oryx* is not subdivided here. This is the monophyletic lineage of the modern African oryxes and *Sivoryx sivalensis*. See note 33 above.
[36]The two small horncore fragments from Ngorora assigned to ?Hippotraginae ?Reducini by Thomas (1981) are doubtful identifications. Reduncini are known outside Africa. There are three Siwalik reduncine taxa (see revision of Pilgrim, 1939, in Gentry and Gentry, 1978), each of which had close relatives in Africa (for chronology of the Siwaliks, see Barry et al., 1982): *Dorcadoxa porrecticornis* from the Dhok Pathan and Tatrot zones, ca. 7.4–2.9 myr, *Sivacobus palaeindicus* from the Pinjor zone, ca. 2.6–1.6 myr, and *Vishnucobus patulicornis* from the Tatrot and Pinjor zones, ca. 3.05–1.6 myr old. Other claims of Reduncini outside Africa have not survived subsequent scrutiny: Bosscha Erdbrink (1982) erected *Redunca eremopolitana* based on horncores and dentition fragments from east of Maragheh; but he misinterpreted the orientation of the horncore, which could be from a caprine like *Pachytragus* (Gentry, pers. comm.). Thomas et al. (1982) discuss but do not support additional claims of Alberdi and Bone (1978) of Reducini from the Turolian (ca. 8–5 myr) of Arenas del Rey (based on an M_3) and from Concud, both in Spain.
[37]Following Gentry (1980, pers. comm.) and recent phylogenetic analyses of Reduncini (Vrba, in prep.), I here include *Redunca* aff. *darti* from Sahabi (Lehmann and Thomas, 1987); a Wadi Natrun horncore, ca. 4.5 myr; new material from Awash Maka-1; part of *Kobus* sp. A of Gentry (1981) from HAD DD-2; and *Redunca khroumirensis* (Arambourg, 1979) from Ichkeul.
[38]I examined the specimens from earlier strata in the Koobi Fora Formation that Harris (1991, p. 254) listed as *sigmoidalis*, and I concluded that they belong to other taxa.
[39]Two horncore fragments from Lower Tulu Bor, 3.36–3.0 myr, Koobi Fora Formation, that Harris (1991) assigned to *M.* cf. *M. lyrocera* are not included.
[40]Although Harris (1991) refers to this as the same species as *M. leakeyi* in one part of this paper (p. 184), he treats it as a separate species in this final summary (table 6.3).

Appendix 27.3. Stratigraphic Reevaluation of Some Late Pliocene Strata
Stratigraphic distributions of bovid and selected other taxa (*Xenohystrix* is a genus of porcupines, *Equus* of horses) used in reevaluation of the dating of Makapansgat Member 3 (MAK3), Sterkfontein Member 4 (ST4), and Baards Quarry (BAA). SH C, F, G = Shungura Members C, F, G; UNDO = Upper Ndolanya Beds; LKAL = Lower Kalochoro Member.

	Site, Locality, and Date (myr)							
Taxa	Strata ≥ 2.8	MAK3[1,2] 2.8 or 2.7	SH C4–9 2.7–2.6	UNDO 2.6	ST4[3] 2.5	BAA[3] 2.5	SH F-base G, LKAL 2.4–2.2	Strata < 2.0
Xenohystrix	x	x						
Simatherium	x	x						
Damalops	x	x?						
Tragelaphus pricei		x	x					
Parmularius braini/ eppsi		x	x					x
Makapania broomi and ?*M. broomi*		x	x		x			
Antidorcas				x	x	x	x	x
Equus					x	x	x	x
Gazella praethomsoni						x	x	x

[1]For MAK3, a previous estimate was 2.9–2.7 myr, with a best estimate ca. 2.8 myr (Vrba, 1982). I still regard MAK3 as the earliest of these assemblages that span 3.0–2.0 myr, because it includes LADS of several genera such as *Simatherium*. A close relationship with upper SH C is indicated by the shared earliest presence of taxa like *Tragelaphus pricei*, close relatives *Parmularius braini* and *P. eppsi*, and by the close resemblance of MAK3 dentitions of *Makapania broomi* to ovibovine teeth that first appear in SH C6. The fact that the earliest radiometric date for the Late Pliocene invasion of Caprinae across Africa is 2.7 myr (figure 27.4) suggests a date close to 2.7 myr for MAK3, with its large ovibovine sample. Thus, I used 2.8–2.6 myr, with a best estimate of ca. 2.7 myr in appendix 27.2. Nevertheless, I still consider my earlier estimate of 2.9–2.7 myr, centered on 2.8 myr, feasible, because MAK3 contains LADS of taxa like *Xenohystrix* and *Simatherium*.
[2]Several of the climatic records from within Africa, or from the marine margins of Africa (Dupont and Leroy, this vol.; deMenocal, this vol.) signify that by 2.8 myr the major long-term net cooling trend of the Late Pliocene had already begun. That is, both the earlier estimate of ca. 2.8 myr and the current one of ca. 2.7 myr place MAK3 within an averagely cool interval. This is consistent with some of the paleoecological inferences for MAK3 based on fauna (e.g., Vrba, 1987a). The pollen evidence for MAK3 remains equivocal: Rayner et al. (1993) claim palynological indications of a relatively warm, wet, and forested MAK3 environment. Scott (this vol.) rejects this; he points out, first, that the MAK3 pollen spectrum contains a high proportion of Chenopodiaceae and Amaranthaceae (indicative of some aridity) and, second, that it resembles the modern spectrum of the area and certainly partly, and maybe largely, reflects Recent contamination.
[3]For ST4, considered to be close after MAK3, a previous estimate was 2.8–2.4 myr (Vrba, 1982). ST4 and BAA each include both *Antidorcas* and *Equus* and both may be close to 2.5 myr.

Appendix 27.4. Statistical Tests of the H_0: The Cumulative Rate of Appearance in the Fossil Record of Morphologically Distinct Taxa Has Remained Averagely Constant.

1. The expected curve was calculated using equation (1) in the text.
2. The differences between the observed and expected numbers, $o_i - e_i$, of added FADS were calculated for each 0.1 myr interval.
3. To satisfy the χ^2 requirement that each $e_i > 5$, a χ^2 test was applied to 12 larger intervals (myr): $3.7 < t \leq 3.4$, $3.4 < t \leq 3.1$, $3.1 < t \leq 2.8$, $2.8 < t \leq 2.5$, $2.5 < t \leq 2.2$, $2.2 < t \leq 1.9$, $1.9 < t \leq 1.6$, $1.6 < t \leq 1.3$, $1.3 < t \leq 1.0$, $1.0 < t \leq 0.7$, $0.7 < t \leq 0.4$, $0.4 < t > 0$.
4. For that 0.1 myr interval with the largest $o_i - e_i$, e_i was substituted for o_i.

Steps 1 to 4 were repeated on the data adjusted by step 4 until H_0 was no longer rejected. This process pointed out several 0.1 myr-long intervals of significantly higher FADS.

Sequential rounds of tests	0.1 myr intervals ending at t_i with significantly higher FADS (o_i) than expected (e_i)	0.1 myr intervals ending at t_i in which e_i was substituted for o_i before repeating χ^2 test	χ^2_{11}; levels of significance: ** = 0.99, * = 0.95 ($\chi^2_{11;0.01}$ = 24.73)
1			265.94**
2	2.7	2.7	121.20**
3	2.5	2.7, 2.5	66.72**
4	0.7	2.7, 2.5, 0.7	62.59**
5	2.6	2.7, 2.5, 0.7, 2.6	47.25**
6	3.6	2.7, 2.5, 0.7, 2.6, 3.6	26.28**
7	1.8	2.7, 2.5, 0.7, 2.6, 3.6, 1.8	16.39*

Appendix 27.5. Sister-Taxa and Potential Ancestors of African Endemic Taxa with FADS 2.7–2.5 Myr (app. 27.2, figs. 27.3, 27.4)
A potential ancestor of a taxon is an unbranched, entirely plesiomorphic outgroup. Of 37 endemic taxa appearing during this interval (an additional 7 FADS are of Eurasian immigrants; fig. 27.4), 23 species were included below as they have been analyzed cladistically: Alcelaphini (Vrba, 1979; Gatesy, 1993; Vrba, in prep.), Hippotragini (see tree in fig. 27.9; Vrba, 1987a, in press), Reduncini (Vrba et al., 1994; Vrba, in prep.), Bovini (Gentry and Gentry, 1978; Vrba, 1987b), and many bovid species in different tribes based on mitochondrial DNA sequences (Gatesy et al., 1992). Phylogenetic trees, based on the cited analyses, for taxa listed by lowercase letters in parentheses are given in figure 27.10. Potential ancestors with LADS between 2.7 and 2.3 myr, ancestral to inferred descendants with FADS during this interval, are indicated by an ansterisk. Species numbers in parentheses refer to appendix 27.2.

(i) Taxon with FAD between 2.7 and 2.3 myr ago	(ii) Sister-taxon of (i) or taxa in the same polytomy as (i)	(iii) Potential ancestor of (i)-(ii)
Alcelaphini: *Connochaetes, Alcelaphus, Megalotragus,* and allies[1]		
(a) clade of *C. taurinus* and *C. gentryi* (77, 80)	(b) clase of *C. gnou* and *C. africanus* (78, 79)	(c) *Damalops* sp.*(73)[1]
(d) *Parestigorgon gadjingeri* (87)	polytomy with clades (a)–(b), (e), (f), and (g)–(h)	(c) *Damalops* sp.*(73)
(e) Oreonagor tournoueri (86)	polytomy with clades (a)–(b), (d), (f), and (g)–(h)	(c) *Damalops* sp.*(73)
(f) *Damalops palaeindicus*	polytomy with clades (a)–(b), (d), (e), and (g)–(h)	(c) *Damalops* sp.*(73)
(g) clade of *Megalotragus* spp. (81, 82, 83)	(h) clade of *Alcelaphus, Sigmoceros, Rabaticeras* spp. (74–76, 84)	(c) *Damalops* sp.*(73)
Alcelaphini: *Parmularius*[2]		
(i) clade of *P. altidens* and *P. angusticornis* (102, 103)	(j) clade of *P. rugosus* and *P. parvus* (104, 105)	(k) *P. braini** (99)
(k) *P. braini** (99)	clade (i)-(j)-(l)-(m)	*P. pandatus* (98)
(l) *P. eppsi* (100)	(m) *P. cuiculi* (101)	(k) *P. braini** (99)
(m) *P. cuiculi* (101)	(l) *P. eppsi* (100)	(k) *P. braini** (99)

(*continued*)

Appendix 27.5. (*Continued*)

(i) Taxon with FAD between 2.7 and 2.3 myr ago	(ii) Sister-taxon of (i) or taxa in the same polytomy as (i)	(iii) Potential ancestor of (i)-(ii)
Alcelaphini: *Damaliscus* and *Beatragus*[3]		
(n) clade of *D. niro* (90, 93)	clade (p)-(q)-(r)	(o) sp. nov.* (107)
(p) clade of *D. agelaius* (88, 89, 91, 92)	clade (q)–(r)	(o) sp. nov.* (107)
(q) *Beatragus* whitei (97)	clade (r)	none known[4]
(r) clade of *B. hunteri* and *B. antiquus* (95, 96)	(q) *Beatragus* sp. nov. (97)	none known[4]
Hippotragini: *Hippotragus*[5]		
(s) clade of *H. equinus*, *H. niger* and *H. leucophaeus* (119–121)	clade (u)-(v)-(w = *H.* sp. nov., 116)	(t) *H. cookei** (118)
(t) *H. cookei* (118)	clade (s)-(u)-(v)-(w)	none known
(u) *H. gigas* (117)	polytomy with (v) and (w)	(t) *H. cookei** (118)
(v) *H. bohlini*	polytomy with (u) and (w)	(t) *H. cookei** (118)
Hippotragini: *Oryx*[6]		
Oryx sp. nov. (part of 122)	polytomy with clade (122) of *O. gazella* and *O. sivalensis*	none known
Bovini		
(x) clade of *Pelorovis oldowayensis* (4, 6)	(y) *P. turkanensis* (5)	(z) *Simatherium kohllarseni** (3)
Syncerus acoelotus (8)	*S. caffer* (7)	*Ugandax* sp. (1)
Reduncini: *Menelikia*, *Redunca*, and *Kobus*		
(a′) *M. lyrocera* (142)	(b′) *M.* sp. nov. (144)[7]	(c) *M. leakeyi** (143)
(b′) *M.* sp. nov. (144)	(a′) *M. lyrocera* (142)	(c′) *M. leakeyi** (143)
R. fulvorufula lineage (141)	clade of *R. arundinum* and *R. redunca* (139, 140)	none known[8]
R. darti (138)	sp. nov. (128)	none known[8]
clade of *Kobus sigmoidalis* and *K. leche* (135, 136)	clade of *K. ellipsiprymnus* (137)	none known

[1]The cladistic results of Vrba (1979) showed the *Damalops* taxa from Laetoli and Hadar as a sistergroup, and this sistergroup as the outgroup of the clade (((*Alcelaphus*)(*Rabaticeras*)) (*Megalotragus*)) ((*Sigmoceros*) (*Connochaetes*))), with *Damalops palaeindicus* placed one branch lower than *Damalops* from Laetoli and Hadar. Preliminary results of recent revisions (Vrba, in prep.; see also Gatesy, 1993) suggest the following: The *Damalops* populations from Laetoli and Hadar belongs to a single species that may also be represented at Makapansgat. *Sigmoceros* is the sisterclade not of *Connochaetes* but of *Alcelaphus-Rabaticeras*. The five taxa *Damalops palaeindicus* from the Siwaliks, *Connochaetes*, *Parestigorgon*, *Oreonagor*, and the clade ((((*Alcelaphus*) (*Rabaticeras*)) (*Sigmoceros*)) (*Megalotragus*)) are part of an unresolved polytomy, of which *Damalops* sp. from Africa is the outgroup. The last record from Makapansgat of this African *Damalops* species is tentative. The next latest record is 2.9 myr from Shungura B11 (app. 27.2).

[2]Here I adhere to the conclusion from previous cladistic analyses that *P. braini* from Makapansgat is a potential ancestor of the clade containing *P. altidens*, *P. angusticornis*, *P. rugosus* (Vrba, 1979), and *P. parvus* (Vrba, 1977a and b). In these earlier publications, however, I considered that the material that has since been described as *P. eppsi* and *P. cuiculi* might belong to the same species as Makapansgat *P. braini*, whereas here I treat them as three separate species following Geraads (1987) and Harris (1991).

[3]Taxon 107 is an undescribed species from Hadar and Makapansgat. The present evaluation that it is a potential ancestor of the (n)-(p)-(q)-(r) clade is tentative, because this taxon was not included in the cladistic analysis of Vrba (1979).

[4]Although the ancestor of the (q)-(r) sistergroup is unknown, a potential ancestor of clade (p)-(q)-(r) has a LAD very close in time to the FADs of taxa (q) and (r).

[5]The cladogram of Vrba and Gatesy (in press) resulted in a trichotomy of new species 116 from Sahabi, *H.* (*Sivatragus*) *bohlini* from the Siwaliks in Asia, and *H. gigas*. Because I sympathize with Geraads's (1989) suggestion that part of the Sahabi fossil assemblages may be much later than the main Late Miocene assemblage, I consider that the earliest record of this monophyletic group is best determined by the other taxa and not by the Sahabi *Hippotragus* sp.

[6]This earliest *Oryx* species comes from Matabaietu and Shungura C. It forms part of the lineage 122 in app. 27.2. Vrba and Gatesy (in press) found an unresolved trichotomy uniting this species with *Oryx* (*Sivoryx*) *sivalensis* from the Siwaliks in Asia and with the clade containing the living oryxes.

[7]See note in app. 27.2 on the possibility that this is not separable from *M. lyrocera*. In that case this entry would refer not to a splitting but to an anagenetic event.

[8]Gentry and Gentry (1978) suggested that *R. darti* might be ancestral to the three living *Redunca* species.

Chapter 28

Hominid Speciation and Pliocene Climatic Change

William H. Kimbel

In this chapter I survey the African hominid record of the Early Pliocene through late Early Pleistocene (covering the period roughly between 4.0 and 1.0 million years ago) with an eye toward delimiting the first and last appearance data for hominid species. My principal goal is the evaluation of this record in terms of its reliability as a data base for assessing the proposed relationship between global climatic change and evolutionary events. As I shall show, the hominid record is far from ideal for the purpose of testing precise causal hypotheses about this relationship, but it is sufficiently informative to establish a broad picture of variation in taxonomic and adaptive diversity over time.

Much of the recent discussion about paleoclimate and human evolution during the Pliocene is due to Vrba's (1985, 1988, 1992, and chap. 3, this vol.) "strong environmentalist argument" for the origin, extinction, and dispersal of species: "Evolution is normally conservative and speciation does not occur unless forced by changes in the physical environment. Similarly, forcing by the physical environment is required to produce extinctions and most migration events. Thus, most lineage turnover in the history of life has occurred in pulses, nearly synchronous across diverse groups of organisms, and in predictable synchrony with changes in the physical environment" (1985, p. 232). Vrba has labeled this statement the "turnover pulse hypothesis" and has incorporated it into a "habitat theory" that challenges the role of biotic factors such as interspecific competition in explanations of clade radiations and extinctions (Vrba, 1992 and chap. 3, this vol.).

ISBN 0-300-06348-2.

Early attempts to test this idea against the African later Neogene fossil record have focused on ca. 2.5 million years (myr) ago, a time of increased cooling and aridification in the lower latitudes heralded by the onset of glaciation in the Northern Hemisphere (Shackleton et al., 1984; Vrba, 1985, 1988; Prentice and Denton, 1988). Now, with a more complete data set at hand, a consensus of paleoceanographers at the Airlie Conference doubted the sharp demarcation at 2.5 myr, favoring instead a broader, but nonetheless marked, increase in Northern Hemispheric glaciation and a resulting trend toward cooler, drier, and more seasonal climates during the Mid-Pliocene. The hominid record has, of course, been examined in this context, and correlations with both paleoclimatic and other evolutionary phenomena sought (e.g., Brain, 1981; Hill, 1987; Vrba, 1988 and chap. 27, this vol.; Stanley, 1992; Turner and Wood, 1993; Schrenk et al., 1993).

Critical for meaningful testing of the turnover pulse hypothesis against the record of African faunal evolution is a well-sampled, and well-calibrated, stratigraphic succession of mammalian faunas. To the extent that the data base falls short of this ideal, we are confronted with the difficult task of interpreting the first appearance data (FAD) and last appearance data (LAD) for taxa whose stratigraphic ranges may be misleading indicators of their true dispersion in time. As we shall see, this is indeed the case for much of the African hominid record over the time span deemed most critical for testing the hypothesis of global climatic influence on hominid speciation and extinction.

Another important ingredient of a successful test is a concept of the species category that recognizes cladogenesis—lineage splitting—as the sole source of taxonomic diversity. We must explicitly exclude the multiplication of species through phyletic (anagenic) speciation, which yields an increase in the number of species names without mirroring the increase in lineage diversity that the turnover pulse hypothesis attempts to explain. Although the reality of anagenic evolution is not debatable, its role in transspecific evolution lacks theoretical justification. This issue is relevant to the present topic because for hypotheses of ancestry and descent involving two time-sequential species whose stratigraphic ranges are not known to overlap, corroborating evidence for cladogenesis is lacking (this does not mean that anagenesis is thereby demonstrated, however; hypotheses of anagensis always rely on the absence of evidence of cladogenesis). We are helped out of this quandary by cladistic analysis of putative species taxa in that hypotheses of phylogenetic ancestry and descent based on an interpretation of the stratigraphic record must agree with cladistic estimates of relative splitting times between sister taxa. Cladistic analysis also reveals whether splitting times between sister taxa (putative speciation "events") correlate with the absolute chronology of large-scale climatic events invoked as a causal explanation of speciation.

Phylogeny and the Turnover Pulse Hypothesis

In the context of the turnover pulse hypothesis, lineage turnover encompasses speciation, extinction and migration of species. For present purposes, I focus on speciation, the generation of diversity via cladogenesis.

The turnover pulse hypothesis poses a causal chain linking climatic change to macroevolutionary change that may be depicted as follows:

Change in environment → vicariance → speciation → macroevolutionary change.

Environmental change, usually construed in this context as large-scale (e.g., tectonically or astronomically forced), triggers the fragmentation of species' ranges (vicariance) as ancestral habitats shift and break up. Vicariance, in turn, provides a necessary physical condition (allopatry) for the evolution and fixation by selection of suites of novel characters in newly formed small, geographically isolated daughter populations. Although speciation does not logically entail a specified amount of morphological change, it is nonetheless seen in this context as a necessary (but not sufficient) ingredient in the production of macroevolutionary change. (Here I follow Levinton's [1988, p. 2] definition of macroevolution as "the sum of those processes that explain the character-state transitions that diagnose evolutionary differences of major taxonomic rank." Used in this way, macroevolutionary change implies evolution by speciation.)

The efficacy of the causal chain depends on the prior acceptance of a phylogenetic hypothesis, the elements of which are monophyletic clades comprising two or more species hypothesized to have evolved from a common ancestor based on their common possession of evolutionary novelties (apomorphies). This is a critical component of any test of the turnover pulse hypothesis, because cladistic estimates of splitting times must be congruent with the timing of the paleoclimatic event(s) posited to play a causal role in cladogenesis. As an example, consider two cladistic arrangements of five species (fig. 28.1). In the phylogeny to the left, the timing of lineage turnover—extinction (species B and E) combined with replacement by speciation (species C and D)—correlates with the timing of a climatic event. However, an alternative cladistic arrangement of the same taxa (on the right) shifts the origins of species C and D to a time prior to the onset of the climatic event, and the temporal coincidence of lineage turnover and climatic change cannot be demonstrated.

Note that in the second cladistic hypothesis, the corresponding phylogeny is consistent with a missing fossil record for both species C and D. This conclusion would require thorough investigation, as the stratigraphically revealed FADs for the taxa must be misleading, underscoring the importance of distinguishing real from apparent first appearance data in the fossil record. In the present example, however, the persistence of species B and E in the same local section lessens the likelihood that geological or taphonomic factors operated to obliterate the records of species C and D in this section, while raising a subsidiary ecological hypothesis: species C and D shared a common ancestor in species A, but local ecological conditions prohibited the occupation of the geographical area by species C and D until the expansion of favorable habitats into the area. This scenario is close to the set of conditions promoting the paleontological pattern of punctuated equilibria, a set of empirical generalizations about the predominant pattern in the fossil record of phenotypic stasis interrupted by discontinuities generated by allopatric speciation, as described initially by Eldredge and Gould (1972). According to allopatric speciation, a species originates as a small peripheral isolate at or near the margin of the ancestral species range and expands only subsequently, with its

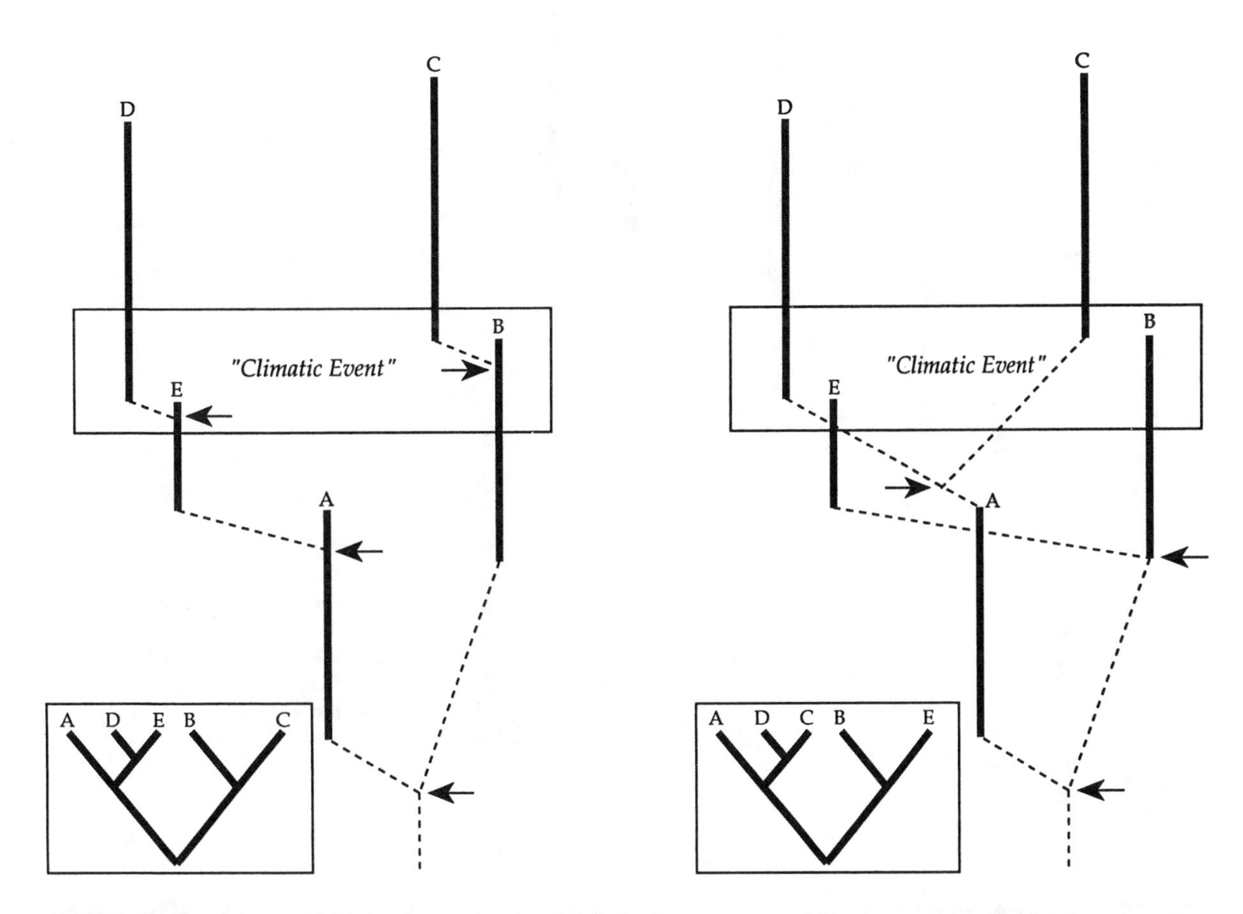

Fig. 28.1. The effect of alternative cladistic hypotheses on the temporal relationship between speciation and climatic events. The hypothesized cladistic ordering of taxa A–E (*left*) is consistent with a positive correlation between speciation and the climatic event. A different ordering of the taxa (*right*) results in no correlation and implies a missing fossil record for taxa C and D.

own novel set of adaptations, into the core of the ancestor's range as a new habitat shifts. In this case, the origin of species C and D is not necessarily correlated with the relevant climatic event, but the turnover pulse hypothesis remains a viable explanation of their appearance in a hypothetical local section because genuine lineage turnover results from the combination of extinction and replacement by immigration in response to climatic change and the shifting of habitats across space.

Species in the African Hominid Fossil Record

I have compiled (provisional) estimates of the FADs and LADs for eight African hominid species between 4.0 and 1.0 myr ago (fig. 28.2 and table 28.1). My approach is explicitly "taxic" in that I see the deployment of species in the stratigraphic record as reflected in the appearance and disappearance of novel morphologies, as discussed more fully elsewhere (Kimbel, 1991; Kimbel and Rak, 1993). Within the limits of resolution currently afforded by the fossil record, I believe that the appearance of a novel morphology records the result of speciation. Although in some paleontological cases we cannot eliminate the possibility that novel morphological characters evolved through long-term continuous change within an undivided lineage, my reading of the hominid data suggests that where a reasonably good record exists, there is evidence of morphological stability through time.

Australopithecus afarensis. The oldest remains reasonably attributable to *A. afarensis* are teeth and a mandible fragment from sediments about 3.95 myr old at Allia Bay, East Turkana, Kenya (Coffing et al., 1994). The Middle Awash (Ethiopia) hominid frontal bone from Belohdelie (Asfaw, 1987) is slightly younger than this, but its importance in the present context lies in the

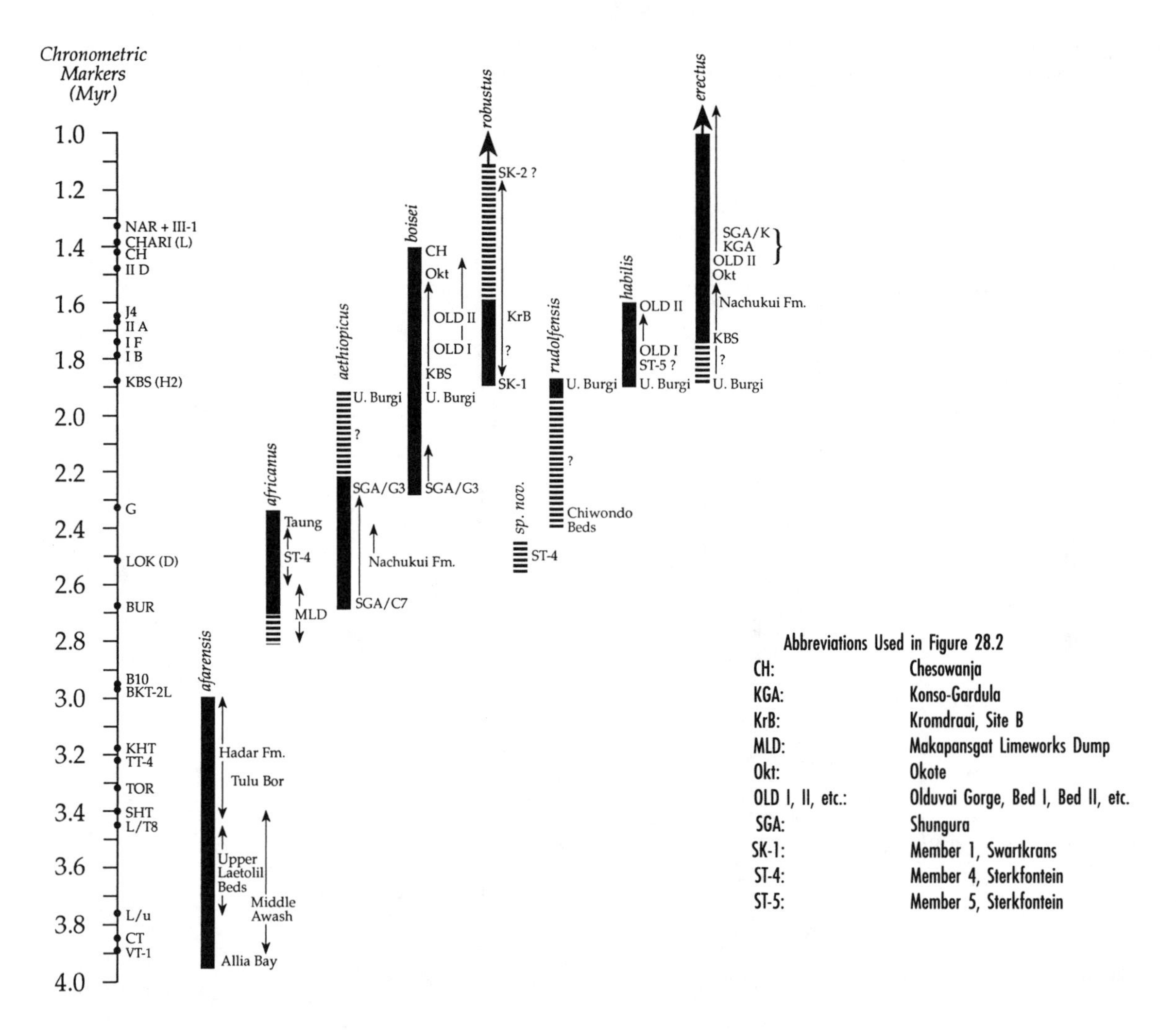

Fig. 28.2. Chronological distribution of hominid species between 4.0 and 1.0 myr, with key chronometric markers shown. Dashed extension of solid bar indicates uncertainty regarding geochronological and/or taxonomic information for that portion of the species' range.

Marker Unit		Provenience	Age (myr)	Reference
VT-1	VT-1 (= Moiti Tuff)	Middle Awash	3.89 ± 0.02	1
CT	Cindery Tuff	Middle Awash	3.85 ± 0.08	1
L/u	Tuff, base Upper unit	Laetolil Beds	3.76 ± 0.03	2
L/T8	Tuff 8	Laetolil Beds	3.46 ± 0.12	2
SHT	Sidi Hakoma Tuff	Hadar Fm.	3.40 ± 0.03	3
TOR	Toroto Tuff	Koobi Fora Fm.	3.32 ± 0.02	4
TT-4	Triple Tuff 4	Hadar Fm.	3.22 ± 0.01	5
KHT	Kada Hadar Tuff	Hadar Fm.	3.18 ± 0.01	5
BKT-2L	Bouroukie Tuff-2 (Lower)	Hadar Fm.	2.95 ± 0.02	9
B10	Tuff B-10	Shungura Fm.	2.95 ± 0.03	4
BUR	Burgi Tuff	Koobi Fora Fm.	2.68 ± 0.05	4
LOK (D)	Lokalalei Tuff (= Tuff D)	Koobi Fora Fm.	2.52 ± 0.05	4
G	Tuff G	Shungura Fm.	2.33 ± 0.03	4
KBS (H2)	KBS Tuff (= Tuff H2)	Koobi Fora Fm.	1.88 ± 0.02	4
IB	Tuff IB	Olduvai Gorge	1.798 ± 0.004	6
IF	Tuff IF	Olduvai Gorge	1.749 ± 0.007	6
IIA	Tuff IIA	Olduvai Gorge	1.66 ± 0.10	7
J4	Tuff J4	Shungura Fm.	1.65 ± 0.03	4
IID	Tuff IID	Olduvai Gorge	1.48 ± 0.05	7
CH	Chesowanja Basalt	Chesowanja	1.42 ± 0.07	8
CHARI (L)	Chari Tuff (= Tuff L)	Koobi Fora Fm.	1.39 ± 0.01	4
NAR	Nariokotome Tuff	Nachukui Fm.	1.33 ± 0.02	4
III-1	Tuff III-1	Olduvai Gorge	1.33 ± 0.06	7

References: 1. White et al., 1993; 2. Drake and Curtis, 1987; 3. Walter and Aronson, 1993; 4. Feibel et al., 1989; 5. Walter, 1994; 6. Walter et al., 1991; 7. Manega, 1993; 8. Hooker and Miller, 1979; 9. Walter, pers. comm.

Table 28.1. Chronostratigraphic Ranges of Middle Pliocene–Lower Pleistocene Hominid Species

Species / datum	Age / provenance
Australopithecus afarensis	ca. 3.95 (3.9)–3.0 myr
FAD: KNM-ER 20432, mandible	Lonyumun Member, Koobi Fora Formation, Kenya
(BEL-VP 1/1, frontal	Belohdelie, Middle Awash, Ethiopia)
LAD: A.L. 444-2, skull	Kada Hadar Member, Hadar Formation, Ethiopia
Australopithecus africanus	≤ 2.8–2.4 myr
FAD: MLD 37/38, cranium	Member 3, Makapansgat, South Africa
LAD: Taung, juvenile skull	Taung, South Africa
Australopithecus aethiopicus	ca. 2.7–1.9 myr
FAD: L. 55s-33, mandible	Unit C7, Shungura Formation, Ethiopia
LAD: KNM-ER 1482, mandible	Upper Burgi Member, Koobi Fora Formation, Kenya
Australopithecus boisei	ca. 2.3 (2.39)–1.4 myr
FAD: L. 74a-21, mandible	Units G-3 or G-5, Shungura Formation, Ethiopia
(L. 338y-6, juvenile calotte	Unit E-3, Shungura Formation, Ethiopia)
LAD: KNM-CH 1/302, cranium	Chesowanja Formation, Chesowanja, Kenya
Australopithecus robustus	ca. 1.8–?1.0 (1.6) myr
FAD: n/a	Member 1, Swartkrans cave, South Africa
LAD: n/a	Member 3, Swartkrans cave, South Africa
Homo rudolfensis	?2.4–1.9 myr
FAD: UR 501, mandible	Unit 3A, Chiwondo Beds, Malawi
LAD: KNM-ER 1470, cranium	Upper Burgi Member, Koobi Fora Formation, Kenya
Homo habilis	ca. 1.9–1.65 myr
FAD: KNM-ER 3735, frag. skeleton	Upper Burgi Member, Koobi Fora Formation, Kenya
LAD: OH 13, partial skull	Bed II, Olduvai Gorge, Tanzania
Homo erectus	≥ 1.75 (1.9) myr
FAD: KNM-ER 3733, cranium	KBS Member, Koobi Fora Formation, Kenya
(KNM-ER 2598, occipital	Upper Burgi Member, Koobi Fora Formation, Kenya)

Note: FAD = first appearance datum; LAD = last appearance datum. Specimens and ages in parentheses indicate alternative datum points based on less secure taxonomic or geochronological information. For *A. robustus,* no individual specimen can be selected as representing the earliest or latest appearance of the taxon. See notes to figure 28.2 for sources.

morphology it shares with the A.L. 444-2 skull from the Kada Hadar Member of the Hadar Formation, ca. 3.0 myr (Kimbel et al., 1994). While recognizing that it will likely turn out to be earlier, I show the *A. afarensis* FAD at ca. 4.0 myr. The LAD for *A. afarensis* at 3.0 myr must also be tentative, however; the upper part of the Kada Hadar Member is sparsely fossiliferous, and this taxon is not known from above the BKT-2 marker tephra (ca. 2.95 myr). Indeed, the record of hominid evolution between 3.0 and ca. 2.7 myr is miserably incomplete.

Australopithecus aethiopicus. The FAD at 2.7 myr for *A. aethiopicus* is based on Suwa's (1990) detailed analysis of mandibular premolar morphology, which demonstrates the presence of a relatively undifferentiated "robust" taxon spanning the stratigraphic interval between units C7 and G3 of the Shungura Formation, a period of some 400 kyr (see also Howell et al., 1987). The well-known KNM-WT 17000 cranium of *A. aethiopicus* falls at 2.5 myr (Leakey and Walker, 1988). I suggest that this taxon may also be present in the Upper Burgi Member at Koobi Fora, ca. 1.9 myr. The relevant specimen is mandible KNM-ER 1482, which is usually assigned to *A. boisei,* although from time to time it has been pointed out that in size and osseous morphology it departs from the typical *A. boisei* pattern. Suwa's (1990, p. 317) study of mandibular fourth premolar morphology also suggests that ER 1482 is more conservative than *A. boisei,* and he suggests that it represents either an odd variant of this species or a "persisting lineage of a conservative 'robust' australopithecine taxon." However, Wood (1991) finds evidence of relatively thin cheek-tooth enamel (primitive) and P_3 root simplification (derived) in ER 1482, characters that ally it with *Homo.* If his classification of the ER 1482 mandible as *Homo* sp. nov. (= *H. rudolfensis*) is correct, then it underscores the convergence of this species' craniodental morphology on that of "robust" *Australopithecus* taxa, already evinced by cranium KNM-ER 1470 from the Upper Burgi Member (see below). Thus the LAD of *A. aethiopicus* must be considered uncertain.

Australopithecus boisei. In the continuous stratigraphic sequence of the Shungura Formation, derived *A. boisei* mandibular premolar morphology first appears in unit G3, ca. 2.3 myr (Suwa, 1990). A slightly earlier appearance for this species is suggested by the juvenile calvaria L 338y-6 from Shungura Formation unit E3 (ca. 2.39 myr: Feibel et al., 1989), although Leakey and Walker (1988) have emphasized several primitive endocranial characters relative to those of *A. boisei,* and Howell et al. (1987) have actually reassigned it to *A. aethiopicus.* Although samples of *A. boisei* predating 1.9 myr are still fairly small, an age of 2.3–2.4 myr may tentatively be taken as the FAD for the taxon. Later *A. boisei* occurrences, based on the distribution of highly derived dental, facial, and cranial base morphology, span the interval between ca. 1.9 and ca. 1.4 myr. The absence of the previously abundant *A. boisei* in hominid-bearing units younger than 1.4 myr (such as Beds III and IV at Olduvai Gorge) probably signals a marked restriction, if not extinction, of the species soon after this time.

Homo rudolfensis. The FADs of two *Homo* species (*rudolfensis* and *habilis*) occur after 2.4 myr (fig. 28.2). These species historically have been lumped in a hypervariable "wastebasket" taxon containing specimens attributable to *Homo* but lacking the derived skull characters of later *H. erectus.* In recent years, however, there has been a tendency to separate them based on differences in facial architecture and dental morphology (see Wood, 1991, 1992). A recently announced mandible (UR 501) from the Chiwondo Beds of Malawi appears to fit well with previously diagnosed *H. rudolfensis* morphology (Schrenk et al., 1993), which emphasizes inter alia mandible corpus shape and mandibular premolar crown and root morphology (e.g., KNM-ER 1802: Wood, 1991). A best-fit, biochronologically determined age of ca. 2.4 myr has been suggested for this find, although most of the associated mammalian taxa range rather widely in time. I tentatively show the *H. rudolfensis* FAD at 2.4 myr, with the caveat that UR 501 may turn out to be a significantly younger than this. Wood's (1991) conclusion that *H. rudolfensis* is otherwise known only at Koobi Fora is correct, in my view, but, unlike him, I do not currently see evidence of this taxon above the KBS Tuff (i.e., KNM-ER 3891). Thus, in the Koobi Fora Formation, *H. rudolfensis* occurs in a very narrow temporal window centered on 1.9 myr. Either the absence of a later record of *H. rudolfensis* can be explained as a sampling artifact, which is a plausible explanation considering the relative rarity of remains of any early *Homo* species, or the Koobi Fora occurrence of *H. rudolfensis* samples the genuine LAD of the taxon, with its earlier record in the Koobi Fora Formation having been lost owing to the elimination of most of the time-stratigraphic interval between 2.0 and 2.4 myr through a depositional hiatus (Feibel et al., 1989). The second hypothesis would neatly explain why *H. rudolfensis* is not present at Olduvai Gorge, whose fossil-bearing sequence commences at 1.9 myr.

Homo habilis. In the Shungura Formation the FAD for *H. habilis* is the L. 894-1 fragmentary skull in unit G28 (ca. 1.9 myr). It is represented synchronically in the Upper Burgi Member at Koobi Fora (e.g., KNM-ER 1813, ER 3735) and at the taxon's type-site, Olduvai Gorge, in Bed I below Tuff IB (OH 24, $\geq$ 1.8 myr). At Olduvai, the record of *H. habilis* continues through Bed I and up into Bed II, where it occurs above Tuff IIA (OH 13, $\leq$ 1.67 myr). The Bed II occurrence makes it highly probable that *H. habilis* was synchronic with *H. erectus,* given the early age estimate for KNM-ER 3733 (ca. 1.75 myr: Feibel et al., 1989). If Wood's (1991) identification of the Upper Burgi Member occipital bone KNM-ER 2598 as "cf. *H.* aff. *erectus*" is borne out, then the origin of *H. erectus* may be extended even further back in time, which is highly probable if the new $^{40}Ar/^{39}Ar$ age of 1.8 myr for the Indonesian Modjokerto site (Swisher et al., 1994) actually "dates" the juvenile *H. erectus* calvaria found there in 1936.

South African Hominid Localities. Hominid localities in South Africa lack the continuous stratigraphic record of the sort encountered in the fluvio-lacustrine sequences of East African sites. We have only a rough idea of the ages for the cave deposits based on the correlation of faunas with radioisotopically calibrated East African stratigraphic records. Although the site samples can be arranged in approximate chronological succession (fig. 28.2), a precise ordering of the fossil hominid specimens against an absolute time scale within each of these deposits is probably impossible. Thus, FAD and LAD points for the hominid species cannot be interpreted as anything other than approximations, essentially average ages of the sediments in which they occur. Strictly speaking, they are not homologous with the East African data points, to the extent the latter are reliable.

Australopithecus africanus. The type-species of the genus *Australopithecus* is characterized by a variable mix of primitive and phylogenetically conflicting derived skull morphologies, which has made its systematics a vexing issue (see below). Of the three known *A. africanus* –bearing caves, Makapansgat is usually said to be the oldest, followed by Sterkfontein Member 4 and then Taung (Vrba, 1982, 1988, and chap. 27, this vol., on bovids; Delson, 1984, 1988, on cercopithecids). Based on these studies, the *A. africanus* FAD is unlikely to be older than ca. 2.8 myr. The age of the Taung hominid, type-specimen of the species, has always been problematic. Although Taung was once thought to be the oldest of the three sites, consensus shifted in the 1980s to the view that it was the youngest *A. africanus*–bearing site (see summary in McKee, 1993). An ongoing problem is the temporal relationship between the hominid locality, which no longer exists, and the nearby source of the temporally diagnostic fauna. Delson (1988) sees no geological reason to doubt that they were coeval, and based on cercopithecid fossils, he assigns an age of 2.3 myr to the Taung hominid. Renewed excavations at Taung have permitted McKee (1993) to assign an upper limit to the age of the Taung fauna of 2.4–2.6 myr, based on the number of species shared with other South African faunal localities. This would place Taung squarely in the range of estimated ages for Sterkfontein Member 4 (Vrba, chap. 27, this vol.). A temporal span of 2.8–2.4 myr for *A. africanus* is suggested by the known occurrences of this taxon, but there is no reason to believe that this represents all, or even most, of the actual duration of the taxon.

Australopithecus robustus. Neither the cercopithecid nor the bovid evidence has settled the issue of which site unit, Swartkrans Member 1 or Kromdraai B East, contains the earliest known *A. robustus,* whose apparent FAD can be constrained to lie somewhere within the interval of 1.9–1.7 myr (African Cercopithecid Zones 5 and 6 of Delson, 1984, 1988, pers. comm.). Based on recently completed field work, Brain (1993) has discerned three hominid-bearing Members (1–3) in the Swartkrans cave, each of which yields remains of *A. robustus.* The cercopithecid fauna is similarly composed in all three Members, favoring the view that they were all deposited within perhaps 200 kyr (Delson, 1993). However, the bovid fossils are said to support ages of 1.1 myr for SK-2 and 700 kyr for SK-3 (Vrba, chap. 27, this vol.). I tentatively depict the LAD for this taxon at ca. 1.0 myr. The same caveat applies here as to the *A. africanus*–bearing caves.

Other South African Hominid Taxa. The Sterkfontein hominid sample from Member 4 seems to me to contain, in addition to *A. africanus,* at least one other hominid species. The identification of a second hominid taxon in Member 4 stems from a restudy of the Sts. 19 cranial base (Kimbel and Rak, 1993), usually considered to represent *A. africanus.* This specimen departs from the relatively primitive cranial base morphology of *A. africanus* in the derived direction of *Homo* and otherwise exhibits none of the uniquely derived cranial base morphology of "robust" *Australopithecus* species. It is most similar to *H. habilis,* but it retains several primitive temporal bone characters relative to known examples of this taxon (including Stw. 53 from Member 5 at Sterkfontein). Kimbel and Rak (1993) concluded that it may represent a previously unknown species of *Homo* (in fig. 28.2 it is shown as *Homo* sp. nov.). Its temporal relationship to *A. africanus* is unknown; I depict a more or less arbitrary temporal datum of 2.5 myr for this taxon.

The hominid sample from Member 5 in the Sterkfontein deposits is small but includes one partial skull (Stw. 53) and other cranial fragments (e.g., Stw. 98) of *H. habilis.* The age of Member 5 is not well constrained, with the bovid data suggesting a loose fit within the broad range of estimates determined for the Member 1 deposits of Swartkrans, ca. 2.0–1.7 myr (Vrba, 1982, 1987).

Although it is most often compared to early *H. erectus* of eastern Africa (e.g., KNM-ER 3733), partial cranium SK 847 from Member 1 in the Swartkrans cave has recently been likened to *H. habilis* (i.e., KNM-ER 1813, Stw. 53) on the basis of relatively narrow nasal and postorbital regions and the relatively large mastoid process (Grine et al., 1993). On the other hand, the SK 847 cranial base lacks the apparent autapomorphies of the *H. habilis* glenoid region (Kimbel and Rak, 1993), and the anterior projection of the nasal bridge is strong (a synapomorphy of *H. erectus* and later *Homo*) compared to the condition in *H. habilis.* For the present, I opt for the conservative view that SK 847 documents *H. erectus,* which implies an estimated age of ca. 1.9–1.7 myr for this taxon in southern Africa (approximately coeval with early *H. erectus* in the Koobi Fora Formation).

Species Diversity through Time

In spite of the apparent deficiencies with regard to FADs and LADs in the hominid fossil record, periods of peak taxonomic diversity can be identified (fig. 28.3). Between ca. 3.0 and 4.0 myr ago one hominid species, *A. afarensis,* is documented. Its stratigraphic range extends back to at least 3.9 myr, and its geographic range was widespread in eastern Africa. The South African record before 2.7–2.8 myr is a blank, although this void says little, if anything, about the actual geographic distribution of hominids in southern Africa prior to this time.

It is clear that the hominid clade diversified, both taxonomically and adaptively, during the period between 3.0 and 2.5 myr, with at least two species appearing by ca. 2.7 myr, *A. aethiopicus,* in eastern Africa, and *A. africanus,* in southern Africa. A third species, identified (in Sterkfontein Member 4) as a basal part of the *Homo* clade, also appears to have originated during this period but thus far is poorly known. Adaptively, both *A. africanus* and *A. aethiopicus* yield the earliest documented glimpses of specialized masticatory apparatus (dentognathic) hypertrophy, while retaining a fairly large number of primitive cranial characters relative to subsequent hominid species.

Almost all of our knowledge of hominid evolution in the interval between 2.5 and 2.0 myr derives from eastern Africa. This fact is important to bear in mind, as scenarios of climatic influence on the pattern of hominid evolution are confronted by about a 500 kyr information void in the southern part of the African paleontological record. (Coincidentally, this span is approximately the same critical period missing from the Koobi Fora Formation in Kenya.) Nevertheless, comparisons of the macrofauna (including hominids) demonstrate that a dramatic biotic change had indeed occurred in the gap between the Sterkfontein/Makapansgat (ca. 2.8–2.4 myr) and Swartkrans (ca. 1.9–1.7 myr) accumulations (Vrba, 1988; Delson, 1988).

In eastern Africa the records of *A. boisei* (Shungura Formation, lower Member G) and possibly *H. rudolfensis* (if the proposed age of ca. 2.4 myr for the Chiwondo Beds mandible is borne out) first occur early in, and persist throughout, the 2.5–2.0 myr interval. Whereas the FADs of *H. habilis* and possibly *H. erectus* occur at the beginning of the succeeding interval, the probability of earlier actual origins for these two species is fairly high given the sizable Late Pliocene temporal gap (2.4–2.0 myr) in the Koobi Fora Formation, the spotty Shungura Formation record, and the great potential of the West Turkana localities to yield new fossils in this time window. I am inclined to see evidence of another species persisting through this period: *A. aethiopicus* is certainly present at 2.5 myr and may be present as late as 1.9 myr (but see above).

As many as six hominid species occur in the 2.0–1.5 myr interval; three FADs—of *A. robustus, H. habilis, H. erectus*—lie within it, although, as already noted, there are reasons to suspect actual origins in the previous interval. In eastern Africa, *H. habilis* and *H. erectus* persist, together with *A. boisei* alone among the earlier appearing taxa, throughout this period. South African *A. robustus* is first documented during this period; its actual time of origin is unknown. *Homo habilis* is also known in South Africa at this time; thus far it is the earliest hominid species known to have ranged in both southern and eastern Africa (cf. remarks above concerning SK 847), although recall that the previous 500 kyr interval in South

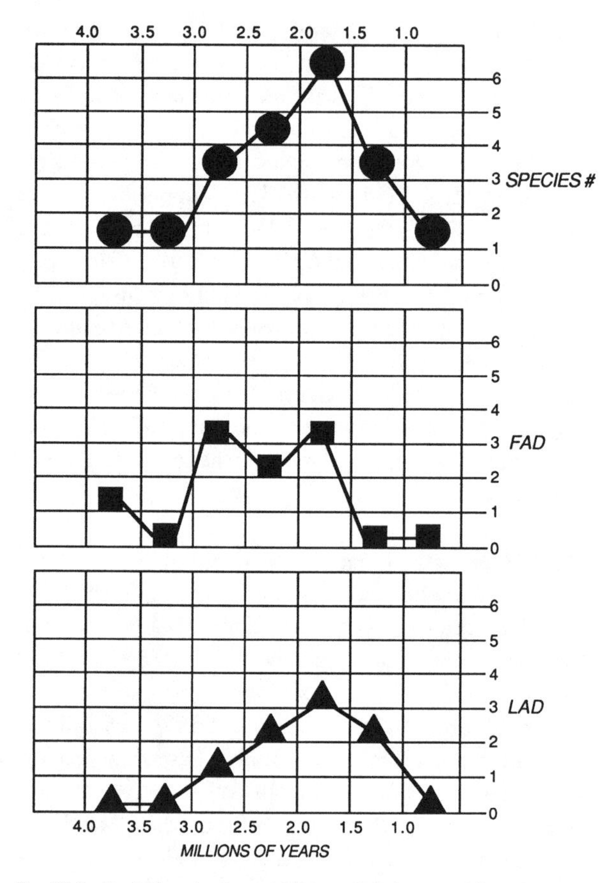

Fig. 28.3. Hominid species diversity (*top*), number of FADs (*middle*) and number of LADs (*bottom*) divided into 500 kyr intervals. The symbols for species diversity represent the number of species documented at some time during a given interval; that is, it includes FADs, LADs, and taxa with FADs in earlier intervals that persist through a succeeding interval. Strict synchronicity of all the taxa in a given interval is not necessarily implied by these data.

Africa is a void. The LADs of *H. rudolfensis* and *A. aethiopicus* (?) occur near the beginning of this interval, while the LAD of *H. habilis*, based on the approximate age of OH 13, is in the neighborhood of 1.65 myr or slightly younger.

The 1.5–1.0 myr interval seems largely to have been characterized by extinction, with *A. boisei* and *A. robustus* (whose LAD is uncertain) disappearing during this time. With the possible exception of *A. robustus* (see above), after 1.4 myr *H. erectus* is the only known hominid species on the African continent.

Hominid evolution between 3.0 and 2.0 myr ago was characterized chiefly by taxonomic and adaptive diversification (fig. 28.3). Owing partly to the ambiguity attached to most of the FAD points, the data do not suggest that ages of first appearance of hominid species over this time span were isochronous. Nevertheless, it is possible to conclude that peak taxonomic diversity was achieved by ca. 2.0 myr ago, with at least five, possibly six, broadly synchronic species documented, four of which are so far known to be endemic in either eastern (*A. aethiopicus, A. boisei, H. rudolfensis*) or southern (*A. robustus*) Africa. On the other hand, hominid evolution after 2.0 myr presents a starkly contrasting pattern of taxonomic stability, then depletion through extinction. Although this portrait is painted by necessity in fairly broad strokes, there is little question as to a general correlation between protracted climatic change through the Middle to Late Pliocene and a net increase in hominid taxonomic and adaptive diversity.

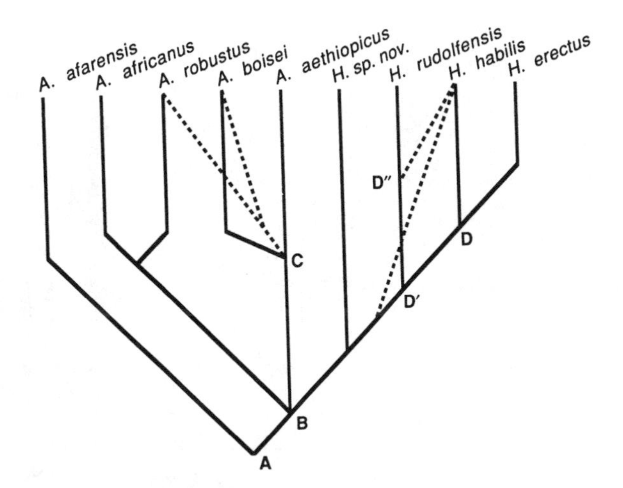

Fig. 28.4. Cladogram of hominid relationships, with alternative arrangements of taxa indicated by dashed lines.

Temporal Pattern of Speciation in Hominid Phylogeny

There remains a good deal of uncertainty concerning the hominid phylogenetic pattern in the Plio-Pleistocene. This uncertainty derives in part from conflicting sets of phylogenetic signals in the skull and dental morphology of some taxa, resulting in fairly high levels of homoplasy, no matter which phylogeny one prefers. Homoplasy due to parallel adaptive evolution across clades is theoretically explicable by hypotheses of climatic change (e.g., Turner and Wood, 1993), but inasmuch as the turnover pulse hypothesis attempts to delineate the causal factors underlying clade radiations (and extinctions), it is the pattern of speciation that is the focus here.

Figure 28.4 is a cladogram of Plio-Pleistocene hominid species relationships based on skull and dental characters (Kimbel et al., 1984, 1988; Kimbel and Rak, 1993), and figure 28.5 is a hypothesis of hominid phylogenetic relationships based on the cladogram and stratigraphically revealed FADs. Principal areas of uncertainty are shown (dashed lines in fig. 28.4, the question marks in fig. 28.5), although these are not necessarily the only ones. Morphologically as well as stratigraphically, the relatively plesiomorphic *A. afarensis* is the best candidate for the last common ancestor of subsequent hominids at node A in fig. 28.4 (Johanson and White, 1979; Kimbel et al., 1984, 1988).

The uncertain phylogenetic status of *A. africanus* has figured prominently in discussions about the cladistic relationships among the post–3.0 myr basal taxa of *Australopithecus* and *Homo* lineages (depicted here as an unresolved trichotomy at node B in fig. 28.4). As noted above, this uncertainty is partly due to high levels of homoplasy inherent in alternative phylogenetic hypotheses, but it may also relate to the 200–300 kyr gap in the hominid record after 3.0 myr (fig. 28.5). In this regard, the blank South African record prior to ca. 2.7 myr looms large. In light of the pronounced degree of provincialism among some African Pliocene mammal groups (e.g., hominids, papionins, bovids), it is not unreasonable to speculate about the existence of a deeper (pre–3.0 myr), endemic South African hominid lineage to which *A. africanus* was phylogenetically linked. If this speculation is supported by future discoveries, then the major division of the hominid clade would predate the *A. afarensis* LAD (ca. 3.0 myr).

The pattern of relationships among species of "robust" *Australopithecus* (including *Paranthropus* of some authors) is also an unresolved question (node C in fig. 28.4): Do the "robust" *Australopithecus* species constitute a monophyletic higher taxon, or are the shared craniofacial specializations of later South African (*A. robustus*) and East African (*A. boisei*) "robust" forms convergent? Either *A. africanus* or *A. aethiopicus* is viable as the relatively plesiomorphic sister taxon of *A. robustus* (Kimbel et al., 1988), but hypotheses of ancestry and descent are confronted by a 200–300 kyr temporal gap between *A. robustus* and either of its putative ancestors (this is true in the case of *A. aethiopicus* only if mandible ER 1482 does *not* belong to this taxon; see above).

Within the *Homo* subclade, *H. habilis* appears (on the basis of supraorbital structure, facial proportions, zygomatic morphology, and P_4 root number [Wood, 1991])

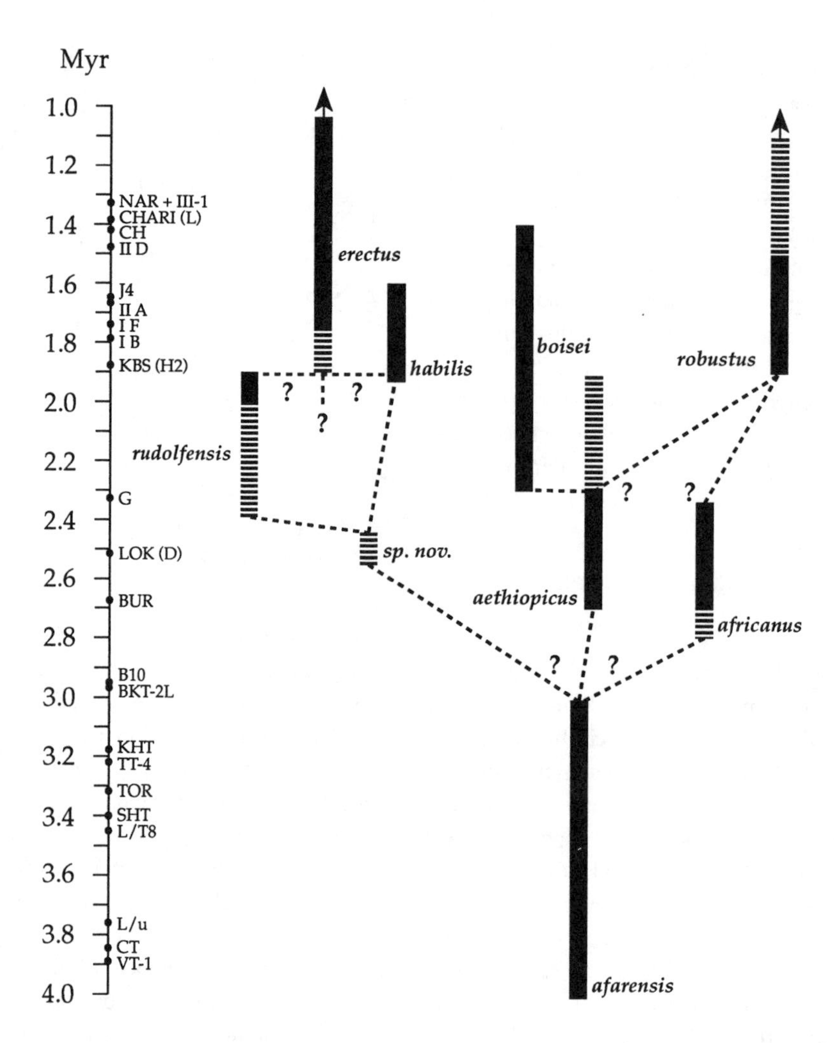

Fig. 28.5 Hominid phylogeny. Temporal scale, chronometric markers, and dashed bar extensions of species' ranges as in fig. 28.2. Thin dashed lines indicate potential ancestor-descendant relationships based on the cladogram in fig. 28.4.

to represent a better ancestral morphotype (at node D, fig. 28.4) for *H. erectus* than does *H. rudolfensis,* which exhibits "robust" *Australopithecus*-like character states in its facial structure and P_4 root number. Accepting this cladistic hypothesis, a cladogenic origin of *H. erectus* from *H. habilis* is corroborated by reference to the stratigraphic record (fig. 28.5). The alternative hypothesis linking *H. rudolfensis* and *H. erectus* as sisters is based chiefly on the inference of relatively large brain size (at node D′) in these taxa as compared to *H. habilis.* Even if this cladistic hypothesis is corroborated, a cladogenic origin of *H. erectus* from *H. rudolfensis* cannot be supported by reference to the stratigraphic record, as the *H. rudolfensis* LAD and the *H. erectus* FAD are time-successive. Finally, an exclusive sister relationship between *H. habilis* and *H. rudolfensis* is suggested by a few characters (at node D″), implying that these two taxa shared a more recent common ancestor than the one they shared with *H. erectus* and that the proximate ancestor of *H. erectus* is currently not documented.

In table 28.2 the stratigraphic FADs of hominid species are compared with the minimum-age FADs permitted by cladistic hypotheses of phylogenetic relationship (fig. 28.4). There is general agreement between stratigraphic and phylogenetic FADs, fitting the expectation that apomorphic species are geologically younger than their plesiomorphic relatives (ancestors). Disagreement arises only in two alternative cladistic solutions: "robust" *Australopithecus* monophyly and *H. habilis* as the sister of (*H. rudolfensis* + *H. erectus*). In these cases the stratigraphic

Table 28.2. Comparison of Stratigraphic and Cladistic FADs for Hominid Species

Species	Stratigraphic FAD/LAD	(1) Cladistic FAD	(2) Cladistic FAD
A. afarensis	4.0/3.0	4.0	4.0
A. africanus	2.8/2.4	2.8	2.8
A. aethiopicus	2.7/?1.9	2.7	2.7
A. robustus	1.9/?1.0	1.9	2.3
A. boisei	2.3/1.4	2.3	2.3
H. sp. nov.	?2.5/?	?2.5	?2.5
H. rudolfensis	?2.4/1.9	?2.4	?2.4
H. habilis	1.9/1.6	1.9	2.4
H. erectus	?1.8/ca. 0.4	?1.8	?1.8

Note: FAD/LAD points = myr. Cladistic FADs are minimum ages of origin permitted by alternative phylogenetic hypotheses discussed in the text and depicted in fig. 28.4.

FAD for the more plesiomorphic taxon (*A. robustus, H. habilis*) is geologically younger than that of the apomorphic relative (*A. boisei, H. rudolfensis*). The cladistic alternatives argue for minimum FADs of 2.3 myr for *A. robustus* and 2.4 myr for *H. habilis* (again, assuming the age estimate for the Chiwondo Beds mandible is accurate). As discussed above, the vagaries of the 2.0–2.4 myr time-stratigraphic interval in both southern and eastern Africa would easily tolerate such adjustments.

Regardless of which phylogenetic hypothesis is more accurate, it is clear that a pulse of speciation occurred in the hominid lineage between 3.0 and ca. 2.7 myr, producing at least three lineages. Identifying the proximate ancestor(s) of these lineages and delineating their cladistic relationships is prevented by the virtual absence of data from this period. The data are only slightly better between 2.6 and 2.3 myr, but the origins of as few as three and as many as five species are dispersed throughout, depending on the choice of cladistic hypothesis. Between one and three species originate around 1.8–1.9 myr, again depending on the cladistic hypothesis.

Conclusions

The following conclusions can be drawn from the foregoing survey of hominid species data:

1) In very few cases are stratigraphic records sufficiently complete to warrant high levels of confidence in species' FADs and LADs. Significant information gaps are pervasive, especially in the time-stratigraphic intervals of 4.0–3.6 myr, 3.0–2.7 myr, 2.4–2.0 myr, and 1.4–1.0 myr. *Of the three million years of hominid evolution discussed in this chapter, about one-half of that time remains virtually undocumented.* Given the coarse-grained portrait of hominid evolution now available, it would be inappropriate to place too much value on the perceived fit between apparent first and last appearances of African hominid species and the fine-grained record of paleoclimatic change revealed in later Neogene oceanic sediments.

2) Notwithstanding this restriction, the data presented here do reveal significant information on taxonomic diversity. In the African hominid record, the period between 3.0 and 2.0 myr is characterized by increasing diversity, with peak species numbers achieved in the terminal Pliocene (by roughly 2.0 myr). Five or six species were present at that time, four of which, on the basis of present knowledge, were endemic in either eastern or southern Africa. In contrast, the lower Pleistocene was a period of evolutionary persistence and extinction, with one or perhaps two species documented after 1.4 myr and definitely only one (*H. erectus*) after ca. 1.0 myr.

3) Inferences regarding the temporal pattern of hominid speciation are indeed sensitive to alternative cladistic hypotheses. Conflicts between stratigraphic and cladistic FADs for two species arise under certain phylogenetic hypotheses. Under the hypotheses of "robust" *Australopithecus* monophyly *and* of a *H. rudolfensis* + *H. erectus* sister-species relationship, seven species' origins are fairly evenly dispersed through the Middle Pliocene (2.8–2.3 myr). Under the hypotheses of independent South and East African "robust" clades *and* a *H. habilis* + *H. erectus* sister-species relationship, the data on speciation times appear somewhat more clustered: two species' origins occur at ca. 2.7–2.8 myr, three at ca. 2.3–2.5 myr, and three more at ca. 1.8–1.9 myr. However, cladistic (minimum-age) and stratigraphic FADs completely agree under the latter hypothesis, and consequently the main factor controlling origin times is the *absence* of data for several hundred thousand years prior to both 2.7 and 2.0 myr.

4) The apparent endemism of hominid species prior to ca. 2.0 myr may be an underappreciated problem for phylogenetic hypotheses, especially in light of the long time-stratigraphic gaps in the South African record (before 2.7 and between 2.4 and 2.0 myr). If hominid macroevolution was driven by allopatric speciation, then our knowledge of the geographic, as well as the temporal, ranges of early hominid species must be woefully incomplete.

5) The net increase in hominid taxonomic diversity

after 3.0 myr correlates with that documented by Vrba (chap. 27, this vol.) for the African Bovidae. Vrba's data show a strong pulse of bovid FADs ca. 2.7–2.5 myr ago (a combination of immigration and speciation events), with a predominance of cool, open-adapted (xeric) taxa appearing in the 2.6–2.5 myr interval. However, other African mammal records do not seem to document contemporaneous phases of diversification. The suids underwent a cladistic radiation by ca. 3.0 myr and were relatively stable (cladistically speaking) for the rest of the Pliocene (White, 1985), whereas a major turnover in the cercopithecid fauna occurred after 2.5 myr (Delson, 1985).

6) It is intuitively apparent that among the major mammalian clades there exist quite different sensitivity thresholds to the effects of climatic change (temperature, vegetation) and shift in habitat, which, to a large extent, reflect different breadths of resource utilization (e.g., specialists vs. generalists). Such differences will play a deterministic role with respect to the among-clade variation in the pattern of turnover pulses in the fossil record. Habitat generalists, may have a low rate of turnover compared to habitat specialists (Vrba, 1985), but it is worth considering whether they also exhibit a delayed response (i.e., higher threshold) to climatic change as compared to specialists. Perhaps the search for an exact temporal correspondence of speciation events in mammal groups as ecologically divergent as hominids, pigs, antelopes, and monkeys is, therefore, chimerical.

7) A general correlation can be made between the Late Pliocene onset of global climatic change and a net increase in hominid taxonomic and adaptive diversity over the period 3.0–2.0 myr. The hominid data are, however, insufficiently dense to support the hypothesis of discrete pulses of speciation within this time span. This paucity of data does not imply the absence of a relationship between Pliocene climatic change and hominid speciation, nor does it spell the failure of the turnover pulse hypothesis. It does mean that the hominid data base must improve considerably before we can move from proposing correlations to the meaningful testing of causally specific hypotheses.

Acknowledgments

I thank Elisabeth Vrba for the invitation to attend the Airlie Conference, for helpful discussion, and for seemingly unending patience during the preparation of the manuscript. In addition, several discussions with Bob Walter about the geochronology of East African hominid sites greatly improved the chapter.

References

Asfaw, B. 1987. The Belohdelie frontal: New evidence of early hominid cranial morphology from the Afar of Ethiopia. *Journal of Human Evolution* 16:611–624.

Brain, C. K. 1981. The evolution of man in Africa: Was it a consequence of Cainozoic cooling? *Transactions of the Geological Society of South Africa* 84:1–19.

———. 1993. Structure and stratigraphy of the Swartkrans cave in the light of the new excavations. In *Swartkrans: A cave's chronicle of early man*, pp. 23–34 (ed. C. K. Brain). Transvaal Museum Monograph, no. 8. Transvaal Museum, Pretoria.

Coffing, K., Feibel, C., Leakey, M., and Walker, A. 1994. Four-million-year-old hominids from East Lake Turkana, Kenya. *American Journal of Physical Anthropology* 93:55–66.

Delson, E. 1984. Cercopithecid biochronology of the African Plio-Pleistocene: correlation among eastern and southern hominid-bearing localities. *Courier Forschungsinstitut Senckenberg* 69:199–218.

———. 1985. Neogene African catarrhine primates: Climatic influence on evolutionary events. *South African Journal of Science* 81:273–274.

———. 1988. Chronology of South African australopith site units. In *The evolutionary history of the "robust" australopithecines*, pp. 317–324 (ed. F. E. Grine). Aldine de Gruyter, New York.

———. 1993. *Theropithecus* fossils from Africa and India and the taxonomy of the genus. In Theropithecus: *The rise and fall of a primate genus*, pp. 157–189 (ed. N. G. Jablonski). Cambridge University Press, Cambridge.

Drake, R., and Curtis, G. 1987. K-Ar geochronology of the Laetoli fossil localities. In *Laetoli: A Pliocene site in northern Tanzania*, pp. 48–52. (ed. M. D. Leakey and J. M. Harris). Clarendon Press, Oxford.

Eldredge, N., and Gould, S. 1972. Punctuated equilibria: An alternative to phyletic gradualism. In *Models in paleobiology*, pp. 82–115 (ed. T. Schopf). Freeman, San Francisco.

Feibel, C., Brown, F., and McDougall, I. 1989. Stratigraphic context of fossil hominids from the Omo Group deposits: Northern Turkana Basin, Kenya and Ethiopia. *American Journal of Physical Anthropology* 78:595–622.

Grine, F., Demes, B., Jungers, W. and Cole, T. 1993. Taxonomic affinity of the early *Homo* cranium from Swartkrans, South Africa. *American Journal of Physical Anthropology* 92:411–426.

Hill, A. 1987. Causes of perceived faunal change in the later Neogene of East Africa. *Journal of Human Evolution* 16:583–596.

Hooker, P., and Miller, J. 1979. K-Ar dating of the Pleistocene fossil hominid site at Chesowanja, north Kenya. *Nature* 282: 710–712.

Howell, F. C., Haesaerts, P., and de Heinzelin, J. 1987. Depositional environments, archeological occurrences and hominids from Members E and F of the Shungura Formation (Omo basin, Ethiopia). *Journal of Human Evolution* 16:665–700.

Johanson, D., and White, T. 1979. A systematic assessment of early African hominids. *Science* 203:321–329.

Kimbel, W. 1991. Species, species concepts and hominid evolution. *Journal of Human Evolution* 20:355–371.

Kimbel, W., White, T., and Johanson, D. 1984. Cranial morphology of *Australopithecus afarensis:* A comparative study based on a composite reconstruction of the adult skull. *American Journal of Physical Anthropology* 64:337–388.

———. 1988. Implications of KNM-WT 17000 for the evolution of "robust" *Australopithecus.* In *The evolutionary history of the "robust" australopithecines,* pp. 259–268 (ed. F. E. Grine). Aldine de Gruyter, New York.

Kimbel, W., and Rak, Y. 1993. The importance of species taxa in paleoanthropology and an argument for the phylogenetic concept of the species category. In *Species, species concepts and primate evolution,* pp. 461–484 (ed. W. Kimbel and L. Martin). Plenum, New York.

Kimbel, W., Johanson, D., and Rak, Y. 1994. The first skull and other new discoveries of *Australopithecus afarensis* at Hadar, Ethiopia. *Nature* 368:449–451.

Leakey, R., and Walker, A. 1988. New *Australopithecus boisei* specimens from east and west Lake Turkana, Kenya. *American Journal of Physical Anthropology* 76:1–24.

Levinton, J. 1988. *Genetics, paleontology and macroevolution.* Cambridge University Press, Cambridge.

McKee, J. 1993. Faunal dating of the Taung hominid fossil deposit. *Journal of Human Evolution* 25:363–376.

Manega, P. 1993. Geochronology, geochemistry and isotopic study of the Plio-Pleistsocene hominid sites and the Ngorongoro volcanic highland in northern Tanzania. Ph.D. diss., University of Colorado.

Prentice, M., and Denton, G. 1988. The deep-sea oxygen isotope record, the global ice sheet system and hominid evolution. In *The evolutionary history of the "robust" australopithecines,* pp. 383–403 (ed. F. E. Grine). Aldine de Gruyter, New York.

Schrenk, F., Bromage, T., Betzler, C., Ring, U., and Juwayeyi, Y. 1993. Oldest *Homo* and Pliocene biogeography of the Malawi Rift. *Nature* 365:833–836.

Shackleton, N., Backman, J. et al. 1984. Oxygen isotope calibration of the onset of ice-rafting and history of glaciation in the North Atlantic region. *Nature* 307:620–623.

Stanley, S. 1992. An ecological theory for the origin of *Homo. Paleobiology* 18:237–257.

Suwa, G. 1990. A comparative analysis of hominid dental remains from the Shungura and Usno Formations, Omo Valley, Ethiopia. Ph.D. diss. University of California, Berkeley.

Swisher, C., Curtis, G., Jacob, T., Getty, A., Suprijo, A., and Widiasmoro. 1994. Age of the earliest known hominids in Java, Indonesia. *Science* 263:1118–1121.

Turner, A., and Wood, B. 1993. Comparative palaeontological context for the evolution of the early hominid masticatory system. *Journal of Human Evolution* 24:301–318.

Vrba, E. 1982. Biostratigraphy and chronology, based particularly on Bovidae, of southern African hominid associated assemblages. Preprint, *Premier Congrès International de Paléontologie Humaine,* pp. 707–752. Union Internationale des Sciences Préhistoriques et Protohistoriques, Nice.

———. 1985. Environment and evolution: Alternative causes of temporal distribution of evolutionary events. *South African Journal of Science* 81:229–236.

———. 1987. A revision of the Bovini (Bovidae) and a preliminary revised checklist of Bovidae from Makapansgat. *Palaeontologia Africana* 26:33–46.

———. 1988. Late Pliocene climatic events and human evolution. In *The evolutionary history of the "robust" australopithecines,* pp. 405–426 (ed. F. E. Grine). Aldine de Gruyter, New York.

———. 1992. Mammals as a key to evolutionary theory. *Journal of Mammology* 73:1–28.

Walter, R. 1994. Age of Lucy and the first family: Single-crystal ^{40}Ar/^{39}Ar dating of the Denen Dora and lower Kada Hadar Members of the Hadar Formation, Ethiopia. *Geology* 22:6–10.

Walter, R., and Aronson, J. 1993. Age and source of the Sidi Hakoma Tuff, Hadar Formation, Ethiopia. *Journal of Human Evolution* 25:229–240.

Walter, R., Manega, P., Hay, R., Drake, R., and Curtis, G. 1991. Laser-fusion ^{40}Ar/^{39}Ar dating of Bed I, Olduvai Gorge, Tanzania. *Nature* 354:145–149.

White, T. 1985. African suid evolution: The last six million years. *South African Journal of Science* 81:271.

White, T., Suwa, G., Hart, W., Walter, R., WoldeGabriel, G., de Heinzelin, J., Clark, J., Asfaw, B., and Vrba, E. 1993. New Pliocene hominids from Maka, Ethiopia. *Nature* 366:261–265.

Wood, B. 1991. *Koobi Fora Research Project.* Vol. 4, *Hominid cranial remains.* Clarendon, Oxford.

———. 1992. Origin and evolution of the genus *Homo. Nature* 355:783–790.

Chapter 29

Evolution of the Early Hominin Masticatory System: Mechanisms, Events, and Triggers

Bernard A. Wood

The mammalian masticatory system consists of both hard and soft tissues. Its function is to reduce the particle size of the ingested food to facilitate the latter's passage into the foregut and its subsequent digestion. The shape of the teeth relates to the physical properties of the ingested food. Bladelike carnassials are effective at slicing meat but are ineffective for dealing with brittle foods. Likewise, teeth with a broad crushing platform can deal with hard, brittle foods but cope poorly with flesh. The masticatory system is thus a relatively sensitive indicator of an animal's diet. Diets are responsive, however, to changes in food availability, which is determined by the nature of a habitat. Habitat, in turn, is a function of location, topography, and climate, which together determine the paleoenvironment. The link between masticatory morphology and the paleoenvironment can therefore be used to explore the impact of external influences, such as climate, on evolutionary history as expressed in morphological changes to the masticatory system.

The soft-tissue components of the masticatory system—the muscles and their associated vessels and nerves—are seldom preserved for long after death; only in the most exceptional circumstances are they preserved in the fossil record. In contrast, the physical properties that make the hard-tissue components, the teeth and jaws, resistant to structural failure during life also confer relative protection from the taphonomic processes that normally cause the breakup and subsequent destruction of all but the hardest mammalian tissues. Thus, teeth and jaws usually form a disproportionately high percentage of any collection of vertebrate fossils. For example, more than half (52 percent) of the 250 or so hominid fossils collected at Koobi Fora, Kenya, either consist solely of dental or gnathic remains or include them as components. At other fossil sites, where high-energy depositional environments predominated, the proportion of dental and gnathic remains is even higher (Bishop, 1976).

Fossil evidence conveys two interrelated messages about evolutionary history: the first concerns phylogeny and the second the functional demands that faced the species group and its immediate antecedents during its existence. The morphological details of the masticatory system of early hominids convey information about both phylogeny and function, but what are the relative strengths of the two messages? Does the functional signal all but obliterate the phylogenetic one (Hartman, 1988, 1989)? Or can the phylogenetic signal be retrieved with the use of the appropriate analytical methods (Wood, 1993)?

In this chapter I shall explore one aspect of the many ways the fossil evidence of the masticatory system can be used to reconstruct the evolutionary history of the clade that comprises the African apes and modern humans. Whereas teeth, once their growth has been completed, undergo little or no modification other than abrasion, the bone of the jaws is distinguished by its plasticity. The masticatory system therefore offers a unique opportunity to investigate both phylogeny and function. I shall also suggest that because of the rapid accumulation of knowledge about the molecular basis of dental development, and thus about the proximate mechanisms for bringing about changes in dental morphology, the mas-

The term "hominin" is used in place of the traditional "hominid," which designates, as a separate family, the group comprising *Pan*, *Gorilla*, and *Homo*.

ISBN 0-300-06348-2.

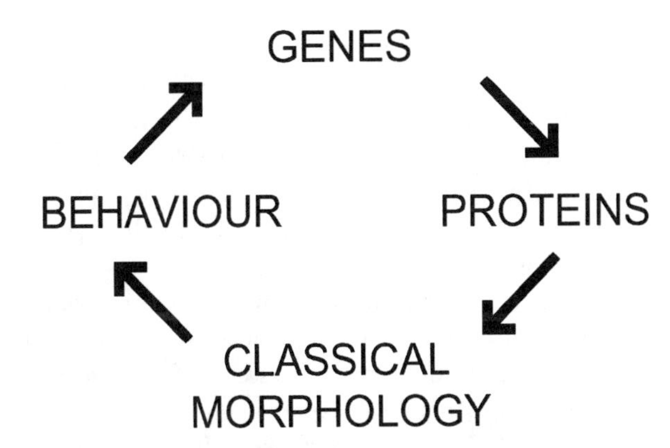

Fig. 29.1. Evolution occurs when behavior, which is ultimately determined by the structures that develop under the influence of genes, confers more, or less, reproductive success.

ticatory system may offer the first opportunity to track at the level of the genotype the complex interactions between behavior and the phenotype (fig. 29.1).

I address three events in the evolutionary history of the hominin clade whose living representatives are the African apes and modern humans. The first event, the emergence of the earliest evidence of the australopithecines, involves, among other changes, an alteration in the absolute and relative size of the dentition and an increase in the cross-sectional area of the mandibular corpus. The second, the development of what has been called a "hypermasticatory trend" in the "robust" australopithecines, mainly involves changes in the size and shape of the postcanine tooth crowns. The third involves a reduction in the absolute and relative size of the postcanine dentition and a reduction in the size of the mandibular corpus. Finally, I review evidence consistent with the hypothesis that the trigger for these events may have been changes in climate, either regional or global in scale.

Mechanisms

Primate teeth consist of a crown that projects into the oral cavity and a root (or roots) set in the jaw. The basis of a tooth's structure is a hollow skeleton of dentine. The dentine of the crown of the tooth is coated by enamel, and the dentine skeleton of the root is coated by cementum.

Development of the Crown and Root. The microanatomical changes that accompany dental development are well known (Scott and Symons, 1958). Two germ layers are involved, the ectoderm, which forms the epithelial layer lining the primitive oral cavity, and mesoderm, which underlies it. In both jaws, a horseshoe-shaped strip of thickened epithelium, the primary epithelial band, marks the beginning of the initiation phase of tooth development. The outer of the two processes to develop from the band marks the boundary between the lips and cheeks on one side and the gums on the other. The inner of the two processes, the dental lamina, gives rise to the teeth. By twelve to thirteen weeks of intrauterine life, swellings develop at intervals along the dental lamina; these are the enamel organs of the deciduous teeth (fig. 29.2). The enamel organs for the permanent teeth do not develop until much later, and it is not until after birth that the germs of the second and third permanent molars appear.

The enamel organs proceed through the so-called cap and bell stages; at the latter stage the developing tooth loses its connection with the dental lamina, and the cellular components become distinct. The inner of the two layers of the invaginated epithelium differentiates into the enamel-forming cells, the ameloblasts. The mesoderm that is "trapped" within the bell-shaped enamel

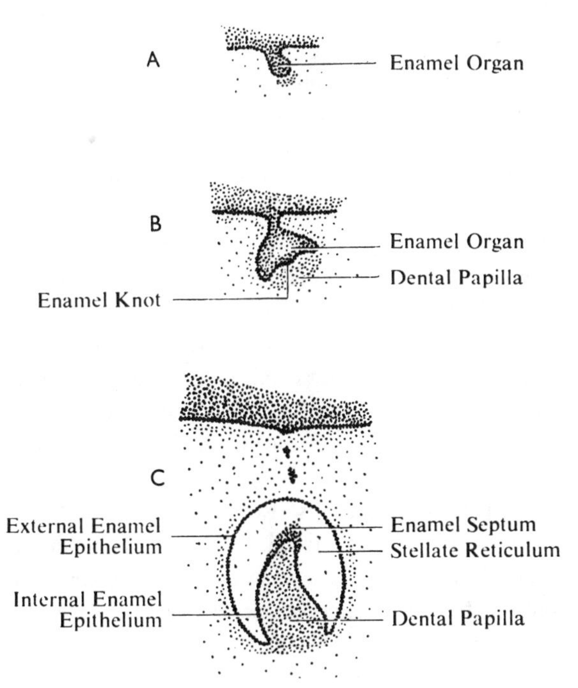

Fig. 29.2. Histological stages in the development of the enamel organ (after Scott and Symons, 1958): A) primary stage; B) "cap" stage; and C) "bell" stage. *Hox-8* is expressed by the internal enamel epithelium during the period between A and C.

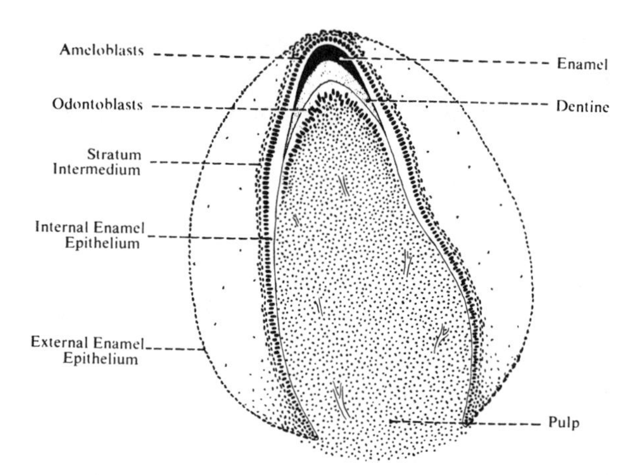

Fig. 29.3. Early stages of enamel and dentine formation. Note the attenuated lower edge of the bell-shaped enamel organ, which will elongate to form the sheath of Hertwig and eventually give rise to the root system (after Scott and Symons, 1958). *Hox-8* expression switches from the internal enamel epithelium to the odontoblasts, which differentiate from the underlying mesoderm within the dental papilla.

organ is called the dental papilla; from the cells of the papilla that abut the inner enamel epithelium the dentine-producing odontoblasts develop (fig. 29.3).

The number and size of the cusps of the tooth crown are determined by the subsequent behavior of the inner enamel epithelium: if the epithelium remains unfolded, the tooth will develop a crown with a single cusp; the greater the number of folds that develop, the greater the number of resulting cusps. It should be noted, however, that a recent study (Macho and Thackeray, 1993; Macho, in press) has confirmed that the form of the inner enamel epithelium, expressed as the shape of the dentino-enamel junction (DEJ), does not always correspond to the shape of the crown. The DEJ morphology may thus be the more reliable indicator of any phylogenetic signal.

The attenuated lower edge of the bell forms what is known as the sheath of Hertwig, from which the root system of the tooth forms. If the tooth has a single root, the sheath is a simple tubular structure; in teeth with multiple roots the lumen is subdivided by processes formed from proliferating cells that grow in from the sheath wall.

Molecular Mechanisms. The molecular basis of primate tooth development has yet to be explored, but investigations of dental development in mice suggests that the genes involved are so highly conserved that the systems controlling murine development are likely to be much the same as those determining the form of primate teeth. Indeed, so great is the extent of the evolutionary conservatism that similar regulatory genes also control development in insects; much of the terminology of molecular development is based on the effects these genes have on insect development.

The basis of the control system is a group of genes that contain conserved regions called homeoboxes. In insects there are three major classes of developmental control genes. Mutations of genes in one of these classes result in body parts that are recognizable but are inappropriate for that position. This phenomenon is termed "homeosis"—hence, the class of genes is called "homeotic." The general term "homeobox" (*Hox*) is used for these and the related regulatory genes that contain a highly conserved region.

Each component homeobox consists of approximately 180 base pairs of genomic DNA. Homeobox genes are classified according to the effects their mutations display in insects. For example, the gene that causes antennae to develop in place of legs is called *antennapedia* (*antp*), and the one that causes rudimentary wings, or halteres, to be replaced by functional wings is called *bithorax* (*bx*). Mutations of individual homeobox genes have spectacular effects, because the protein they encode binds to, and regulates, the expression of the so-called downstream genes. These are the parts of the genome that, by coding for structural rather than regulatory proteins, determine the details of development.

Molecular biological techniques can be used both to locate homeobox genes on the chromosomes of an animal and to detect whether the gene is active during development. Specific Hox genes can also be introduced into transgenic animals, such as mice, and their effects can be observed in their descendants (Kessel and Gruss, 1990).

Experiments using mouse embryos have suggested that a subclass of muscle-specific homeobox (msh) genes, namely, *Hox-7* and *Hox-8*, is implicated in dental development. These genes are expressed in several areas where epithelial-mesenchymal interactions are known to occur during the early development of the murine embryo (MacKenzie, Leeming et al., 1991; MacKenzie, Ferguson et al., 1991; Monaghan et al., 1991). *Hox-7* is expressed in the dental papilla throughout dental development, but it is the expression of *Hox-8* that has provoked most interest (MacKenzie et al., 1992). This homeobox is first seen in the epithelium at the dental placode stage, but its expression becomes more localized when the pattern of the cusps in the molar teeth begins to take shape. *Hox-8* expression is asymmetrical and is concentrated at the site of the first, buccal cusp inflection of the inner enamel epithelium; once this initial inflection is located the subsequent cusp pattern is determined by differential cell division (MacKenzie et al., 1992). This asymmetrical distribution of *Hox-8* activity in molars contrasts with its symmetrical expression in single-cusped incisor teeth, which lack inflections in the inner enamel epithelium. Once the inner enamel epithelium differentiates into ameloblasts, *Hox-8* expression ceases in the epithelium and switches to the mesenchymally derived odontoblasts. Although the histological changes that accompany root development and determine root form are well known (see Wood et al., 1988, for a summary), we remain ignorant about the molecular mechanisms controlling these events.

Information about the genetic control of the production of enamel, which determines the size of the crown and the thickness of the enamel, has come from studies on the molecular basis of a genetic disorder of enamel development called *amelogenesis imperfecta*. This defect has been linked to the amelogenin gene *AMEL* (Lagerstrom et al., 1989). In the mouse the only *AMEL* locus is on the X chromosome, but in humans there are two loci, *AMELX* and *AMELY*. Extracts of unfixed, deciduous teeth of male humans were found to include a protein that was not present in females and that was identified as an amelogenin, or amelogenin-like, polypeptide (Fincham et al., 1991). Amelogenins apparently regulate the form and size of hydroxyapatite crystallites during enamel biomineralization (Eastoe, 1979), and the products of this Y-linked gene may be responsible for the larger teeth of males. Furthermore, its differential expression may be responsible for differences in tooth size and enamel thickness among early hominin taxa (Beynon and Wood, 1986, 1987; Grine and Martin, 1988; Wood, 1991).

Events

Hominin Emergence. The first of the three events in which dental evolution is involved is the emergence of the earliest hominins, the australopithecines.

The earliest unambiguous evidence of a creature that departs, albeit not comprehensively, from the phenotype of the apes is that of *Ardipithecus ramidus* from 4.4 myr deposits at Aramis, in Ethiopia (White et al., 1995). The earliest hominid, however, for which we presently have more comprehensive information is *Australopithecus afarensis*, the earliest evidence of which may be the mandible from Tabarin (Hill, 1985; Ward and Hill, 1987). The taxon *A. afarensis* was established (Johanson et al., 1978) on the basis of hominid fossils from Pliocene sediments at the sites of Hadar, in Ethiopia, and Laetoli, in Tanzania. A mandibular corpus from Laetoli, LH4, was designated as the type specimen, and the paratypes include other specimens from the Upper Unit of the Laetoli Beds and from three Members (Sidi Hakoma, Denen Dora, and Kadar Hadar) of the Hadar Formation. The other substantial contribution to the hypodigm comes from Maka, also in Ethiopia (Clark et al., 1984; White et al., 1993).

The sensible way to compare tooth size across a range of taxa is to express it in relation to body mass. Data from samples of the living African apes and from two samples of modern humans with contrasting absolute tooth sizes reveal a remarkable consistency in relative sizes of teeth and mandibles (table 29.1). Yet when a representative sample of these data is compared with values of the same parameters for *A. afarensis*, there is, over all the variables, an average twofold increase in relative tooth and mandible size. Thus, the processes controlling dental development in the early hominins were adjusted toward an increase in the size of the postcanine tooth crowns. What is at issue, however, is whether this increase in tooth size was established in the common ancestor of hominins and *Pan*. There is precious little fossil evidence that is helpful in the task of reconstructing these Miocene creatures (Andrews, 1992), but Andrews and Martin (1991) specify that these ancestral forms differ from living *Pan* species in at least two ways. First, they believe that their diet placed more emphasis on hard fruits; and second, they suggest that their body mass was greater than that of the common chimpanzee. The implications for relative tooth size of the proposed dietary differences are difficult to predict, but they are unlikely to involve a reduction in postcanine tooth size. However, because of the negative allometry of postcanine tooth size with respect to body mass (Kay, 1975), which is evident in the differences between values for males and females in table 29.1, the effect of Andrews and Martin predicting a larger body weight for a Miocene common ancestor would be to exaggerate further the already substantially larger relative size of the postcanine teeth of the earliest hominins.

Table 29.1. Relative Tooth and Mandible Sizes Based on Computed Crown Areas and Mandibular Cross-Sectional Area at M_1: Living Hominids

Genus and Species		P_4		M_1		M_3		Corpus		Body Mass (kg)
		A	S	A	S	A	S	A	S	
Gorilla gorilla[1]	(M)	158	2.3	220	2.7	273	3.1	698	4.9	158[4]
	(F)	142	2.8	200	3.4	236	3.6	538	5.5	75
Pan troglodytes[1]	(M)	73	2.3	110	2.8	116	2.9	345	4.9	54[4]
	(F)	64	2.3	101	2.9	104	3.0	328	5.3	40
Pan paniscus[2]	(M)	55	2.0	87	2.6	76	2.4	—	—	48[4]
	(F)	53	2.3	86	2.9	78	2.8	—	—	33
Homo sapiens[1]	(M)	63	2.1	123	2.9	118	2.8	308	4.6	56[5]
(Bantu)	(F)	57	2.1	113	2.9	107	2.8	285	4.6	49
Homo sapiens[3]	(M)	70	2.0	143	2.9	132	2.8	—	—	71[5]
(Australian aboriginals)	(F)	65	2.0	132	2.9	127	2.9	—	—	62

Note: A = absolute mean values; S = mean values scaled by estimated body mass.
[1]Wood (1975)
[2]Johanson (1974)
[3]Barrett et al. (1963, 1964)
[4]McHenry (1992)
[5]Eveleth and Tanner (1976)

Hypermasticatory Trend. The second event, or, more accurately, series of events, is the emergence of the distinctive postcanine teeth of the "robust" australopithecines. The first signs of what Tobias (1991) has dubbed a "hypermasticatory trend" in hominin evolution comes from fossils belonging to *Paranthropus aethiopicus* recovered from the Omo-Shungura (Arambourg and Coppens, 1968) and the West Turkana (Walker et al., 1986; Leakey and Walker, 1988) sites, of which the earliest are dated at around 2.6 myr.

The initial expression of the trend is characterized by large-crowned premolars and molars with thick enamel and relatively massive mandibles. Species manifesting an exaggerated version of the trend appear later, namely, *Paranthropus boisei* in East Africa (2.3–1.1 myr) and *Paranthropus robustus / crassidens* (1.8–1.5 myr) in southern Africa. The modifications include, in *P. boisei* in particular, a further increase in the absolute size of the premolar and molar crowns, an increased emphasis on expansion of the distal (talonid) part of the tooth crowns, especially in the premolars, a thicker and more uniform covering of enamel, shallower fissures, a more complex mandibular premolar root system, and an increase in the size of the body of the mandible (Beynon and Wood, 1986; Suwa, 1988; Wood et al., 1988; Wood et al., 1994).

The excellent stratigraphic and chronological control for the hominids recovered from the Omo region (Feibel et al., 1989) has allowed the shift in morphology in the East African material to be dated relatively precisely to 2.3 myr (Wood et al., 1994). A further, but less dramatic, modification to the dentition occurs in *P. boisei* between 1.9 and 1.7 myr (Wood et al., 1994).

The morphological modifications are more marked in *P. boisei* than in *P. robustus*, as are the absolute increases in crown area, but when the latter are related to estimates of body mass (table 29.2), the discrepancy between the two taxa is a good deal less marked.

Emergence of Early African **Homo erectus** *or* **Homo ergaster** *(Hereinafter Referred To As* **H. ergaster**). The third event is the emergence, around 1.9 myr, of a hominin species that is either conspecific with or belongs to the same grade as *Homo erectus*, which is known from later assemblages in East Africa and from sites in the Far East (Wood, 1991). In addition to providing unequivocal evidence of upright posture and of being an obligatory biped (Brown et al., 1985; Walker and Leakey, 1993), this hominin is the first to show relationships in relative tooth and mandibular-corpus size that differ little from those of modern humans (table 29.2).

Table 29.2. Relative Tooth and Mandible Sizes Based on Computed Crown Areas and Mandibular Cross-Sectional Area at M_1: Fossil Hominins, with *P. troglodytes* and *H. sapiens* for Comparison

Genus and Species		P_4		M_1		M_3		Corpus		Body Mass (kg)
		A	S	A	S	A	S	A	S	
Pan troglodytes[1]	(M)	73	2.3	110	2.8	116	2.8	345	4.9	54[4]
	(F)	64	2.3	101	2.9	104	3.0	328	5.3	40[4]
Homo sapiens[1]	(M)	63	2.1	123	2.9	118	2.8	308	4.6	56[5]
(Bantu)	(F)	57	2.1	113	2.9	107	2.8	285	4.6	49
Australopithecus afarensis[2]		108	3.1	166	3.8	193	4.1	488	6.5	38
Australopithecus africanus[2]		121	3.4	179	4.1	218	4.5	568	7.3	35
Paranthropus robustus[2]		146	3.7	207	4.4	254	4.8	786	8.5	36[4]
Paranthropus boisei[2]		205	4.2	239	4.5	327	5.2	960	9.0	41[4]
Homo habilis[2]		104	3.2	166	4.1	201	4.5	421	6.5	31
Homo rudolfensis[2]		133	3.0	187	3.6	250	4.2	667	6.8	55[6]
Homo ergaster[3]		96	2.6	144	3.1	170	3.4	455	5.6	56

Note: A = absolute mean values; S = square root of mean values scaled by the cube root of estimated body mass.
[1]Wood (1975)
[2]Wood (1991)
[3]Brown et al. (1985)
[4]McHenry (1992)
[5]Eveleth and Tanner (1976)
[6]Aiello and Wood (1994)

Triggers

The proposal that environmental change is an important motor driving evolution (Vrba et al., 1985) has been formalized as the "turnover pulse hypothesis" (Vrba, 1985, 1988). This hypothesis suggests that lineage turnover across clades and regions should be approximately synchronous and coincident with geographically widespread shifts in climate. It is claimed that the broad scope of the hypothesis will effectively eliminate the danger that biases in the fossil record will mimic turnover, but some observers remain skeptical about its explanatory power (Hill, 1987; Foley, 1994). Recent research has also emphasized that regional changes in topography, uplift in particular, can have as profound an influence on local environments as does a global change in climate. This latter aspect does not weaken the link between evolution and climatic change but does suggest that to be influential climatic change may not always be a global phenomenon.

What evidence is there that the three episodes of morphological change referred to above are just one aspect of more widespread evolutionary change within the mammalian fossil record? And is there any evidence that any such widespread change is linked to climatic change on a global scale?

Although the Tabarin mandible apparently provided a *terminus post quem* for the appearance of *A. afarensis*, the recent discovery of remains attributed to *A. ramidus* ca. 4.5 myr means that there is uncertainty about the time range of *A. afarensis;* this uncertainty reduces the fruitfulness of any search for synchronous change in other clades. Some of the best evidence for mammalian faunal turnover around this period comes from research in the Tugen Hills west of Baringo, in Kenya (Hill et al., 1985; Hill, this vol.). This effort has produced a relatively good fossil record between approximately 15 and 4 myr. Evidence from molecular anthropology suggests that the divergence of the hominin and *Pan* clades took place between 9 and 5 myr (Caccone and Powell, 1989; Ruvolo et al., 1994), so that the Ngorora Formation, on the one hand, and the Mpesida Beds and Lukeino Formation, on the other, have the potential to provide information about any more widespread faunal change across the interval that is likely to have marked the emergence of

Table 29.3. Summary of the Evidence for Species Turnover and Anagenetic Change within African Large Non-Hominid Mammals between Approximately 2.5–2.0 myr

			2.5 myr				2.5–2.0 myr			
Family	Genus	Species	Species Turnover	Hypsodonty	Molars	Molar Complexity	Species Turnover	Hypsodonty	Molars	Molar Complexity
Elephantidae	*Loxodonta*	—	D[1]	•	•	•	•	•	•	•
	Elephas	*recki*	—	+	+	+	—	++	+	++
Suidae	*Notochoerus*	*euilus*	D	•	•	•	•	•	•	•
		scotti	—	+	—	—	—	+	+	+
	Kolpochoerus	*limnetes*	—	+	+	+	—	+	+	+
	Metridiochoerus	*andrewsi*	—	—	—	—	—	+	+	+
Bovidae	—	—	A[2]/D	—	+	—	A/D	—	+	—
Equidae	*Hipparion*	—	—	+	—	—	D	+	—	—
	Equus	—	A	—	—	—	—	—	—	—
Hippopotamidae	*Hexaprotodon*	—	—	—	—	—	—	—	—	+
	Hippopotamus	—	A	+	—	—	A	—	—	—
Cercopithecidae	*Theropithecus*	*brumpti*	—	—	—	—	D	—	—	—
		oswaldi	A	—	+	+	—	—	+	+
Rhinocerotidae	*Ceratotherium*[3]	—	—	—	—	—	—	?	—	—
	Diceros	—	—	—	—	—	—	?	—	—

Note: • = taxon extinct; +, ++ = degree of morphological change.
[1]D = Taxon/taxa disappear.
[2]A = Taxon/taxa appear.
[3]*Ceratotherium* shows cranial changes prior to and around 2.0 myr.
Source: Turner and Wood (1993).

hominins. Hill (this vol.) reports that while the fauna from the Ngorora Formation is typically Late Miocene in character, that from the Mpesida Beds and the Lukeino Formation shows evidence of the demise of some forms, the listriodont pigs, for example, and the appearance of new families such as the Elephantidae and the leporids. It may be tempting to relate these faunal changes with the Messinian Salinity Crisis at 6.5–5.3 myr, but elsewhere on the globe, in the Neogene Siwalik Formations, for example, evidence of major faunal turnover was occurring rather earlier, at around 7.5 myr (Barry, this vol.). A realistic assessment of the present evidence germane to the first of the three events is that the relatively imprecise chronological control presently rules out confirming or refuting a relationship between hominin emergence and more general faunal change. Nor is there presently any evidence to link hominin emergence to global, as opposed to regional, climatic change.

The more precise chronological control available for the East African evidence relevant to the second event, the emergence of a hypermasticatory trend within hominin evolution, bodes well for any attempt to test whether that event was part of a turnover pulse. Evidence of evolutionary change, expressed either as species turnover or as directional change in morphology within species, has recently been assessed across a range of terrestrial mammalian taxa (Turner and Wood, 1993), the results of which are summarized in table 29.3. These findings confirm that although species turnover around 2.5 myr is not confined to the well-known and widely cited evidence from bovid evolution (Vrba, 1988), in many of the mammalian groups that were investigated the evidence for it is not as marked as it is in the bovids. Even within the bovids, however, the evidence from the fauna of the Omo region points to a change closer to 2.3 myr; evidence from other animal groups is also consistent with evolutionary change persisting over a more prolonged period than just ca. 2.5 myr. Within hominins, although 2.3 myr is not the time of onset of the hypermasticatory trend, it is apparently the time when the specialized morphology of *P. boisei* makes its first appearance (Suwa, 1988; Wood et al., 1994).

What is the evidence for global climatic change around 2.5–2.3 myr? Various lines of evidence mark the period ca. 2.5 myr as a time of climatic change. They include the first major glacial event of the Plio-Pleistocene in the Northern Hemisphere (Shackleton et al., 1984), the advent of major ice rafting in the northern oceans (Shackleton et al., 1984; Jansen et al., 1988), the onset of loess deposition in Asia (Kukla, 1987), palynological evidence from South America (Hooghiemstra, 1986), and comparable indicators from the Netherlands and the Mediterranean (DeJong, 1988; Zagwijn and Suc, 1984). In addition there is evidence from offshore cores (Janecek and Ruddiman, 1987), from palynology (Bonnefille, 1983; Bonnefille and Vincens, 1985; Bonnefille et al., 1987), and from micromammals (Wesselman, 1984) of increased aridity. All this evidence, however, is mute about whether 2.3 myr was a time of especially rapid or substantial climatic change, either globally or locally in Africa.

The work of deMenocal and Bloemendal (this vol.) may shed some light on subtropical African climate after 2.5 myr. Whole-core measurements of magnetic susceptibility have been used as a surrogate for tracking the proportion of aeolian dust in cores from ODP sites 661 and 721. These results suggest that from 2.5 myr African subtropical climates have behaved with a 41 kyr periodicity that had not been evident before 2.5 myr. Although this might explain a generally higher level of evolutionary change subsequent to 2.5 myr, it does not necessarily explain the episodes of evolutionary change at 2.3 and 2.0–1.9 myr (see below).

The best-known evidence for *H. ergaster* (the earliest unambiguous *Homo* species), KNM-WT 15000 and KNM-ER 3733, dates from approximately 1.6 and 1.8 myr, respectively, but the earliest evidence is probably an occipital fragment, KNM-ER 2598, which may be as old as 1.88–1.9 myr (Feibel et al., 1987; Wood, 1991). Are there any indications from the paleontological or any other record for global or regional climatic change at that time?

At Olduvai Gorge there is evidence that by 1.67 myr (Cerling and Hay, 1986), and perhaps by as early as 1.8 myr (Hay, 1990), a significant shift toward a more arid habitat had occurred. This shift is reflected in the mammalian fauna, notably the bovids (Hay, 1973; Kappelman, 1984), the microfauna (Butler and Greenwood, 1965, 1976; Jaeger, 1976; Kappelman, 1984), the flora (Bonnefille and Riollet, 1980; Cerling and Hay, 1986), and the soil conditions (Cerling et al., 1977). In the Omo region there are indications of a gradual increase in aridity between 2.5 and 2.0 myr ago (e.g., Bonnefille, 1976; Jaeger and Wesselman, 1976 and see above), but there is also evidence to suggest an episode of evolutionary change at, or around, 2 myr ago. This evidence was used to demarcate the lower boundary of the *Metridiochoerus andrewsi* vertebrate biozones of Maglio (1972) and Harris (1976), and 2 myr was marked by rapid turnover in the bovid fauna, as recorded at Koobi Fora (Harris, 1991). It is now known that this period of heightened evolutionary activity approximately correlates with the formation and deposition of the KBS Tuff at Koobi Fora and with the laterally equivalent H-2 Tuff in the Shungura Formation (Brown and Feibel, 1991); the eruption that gave rise to both these tuff layers has been dated at 1.88 ± 0.02 myr (McDougall, 1985).

Paleontological, paleobotanical, and mineralogical evidence all supports the proposition that the changes referred to above are related to a change in climate. But was any such change part of a global modification to the climate, or was it predominantly a regional phenomenon? There is, in the event, no compelling evidence for any particular change in the global climate at, or around, 2 myr. A piston core from the western flank of the Rockall Bank in the North Atlantic does show one nannofossil extinction between 1.9 and 1.8 myr, but this has to be compared with the three extinctions that occur at, or around, 2.5 myr (Shackleton et al., 1984).

Conclusions

The emergence of hominins at some time between 9 and ca. 4 myr, the appearance of "robust" australopithecines around 2.6 myr, the first evidence of *P. boisei* around 2.3 myr, and the earliest evidence for early African *Homo erectus / Homo ergaster* are all events in hominin evolution that have been linked with changes in the local African or global climatic regime. In no case is the evidence for such a link with climate change conclusive, but the events within hominin evolution between 2.6 and 2.3 myr and the emergence of *H. ergaster* at around 2 myr are matched by a good deal of evolutionary activity within other groups of African mammals.

Instead of being concerned that there is not simultaneous turnover of species in all groups of large mammals, perhaps it would be more fruitful to investigate why some mammal groups are relatively unaffected by these events. It is also worth remembering that evolutionary change is not confined to speciation and extinction. Directional morphological change in the surface area and height of tooth crowns within species may also indicate a response to modifications in the habitat. Indeed, such changes, together with population migration, might be predicted to be the more likely response of the generalist.

The dental and mandibular changes that have been described, both at the onset of the hominin lineage and during its evolutionary history, are of a kind consistent with relatively modest modifications of the molecular control of dental development. The knowledge gained from dental development in the mouse is eminently transferable to a primate model. Thus, changes in the cusp pattern of hominin postcanine teeth could have been mediated by extending or modifying the sites of expression of *Hox-8,* and any increase or decrease in the size of the teeth, or any change in the thickness of the enamel, could be achieved by modifying the expression and/or the activity of control factors such as *AMELY.*

At present it is not conceivable that hominin evolution will ever be mimicked in the laboratory, but it is quite probable that we shall soon come to understand enough about the control of some aspects of development at the molecular level to enable future researchers to relate the morphological modifications that have occurred during the course of hominin evolution to specific modifications of the genome. These are, however, only the proximate mechanisms for evolutionary change. They may be able to provide the how, but not the why. We do not yet understand why some taxa respond to xeric conditions by hypsodonty, which increases the functional life of teeth without increasing the area available for food processing, whereas others increase the thickness of the enamel, or the size of the teeth, or both. These research questions involve topics as diverse as comparative energetics and the mechanics of food breakdown. The pursuit of solutions to such questions will usher in yet another new and exciting phase in the quest to understand the factors that have shaped the evolution of the hominin masticatory system.

Acknowledgments

I am particularly grateful to the organizers for their invitation and support. Special thanks are due Elisabeth Vrba for her ability to identify the important issues and for her persistence in making sure her colleagues did not evade them. Some aspects of the research touched upon in this chapter were carried out in collaboration with Alan Turner, whose contribution is gratefully acknowledged. Nearly all of my contributions were made possible by grants from The Leverhulme Trust and the Science-Based Archaeology Committee of the National Environment Research Council. The comments of Gabriele Macho, David Edgar, and Hugh Rees on the manuscript were much appreciated.

References

Aiello, L., and Wood, B. 1994. Cranial variables as predictors of hominine body mass. *American Journal of Physical Anthropology* 95:409–426.

Andrews, P. 1992. Evolution and environment in the Hominoidea. *Nature* 360:641–646.

Andrews, P., and Martin, L. 1991. Hominoid dietary evolution. *Philosophical Transactions of the Royal Society of London,* ser. B, 334:199–209.

Arambourg, C., and Coppens, Y. 1968. Dćouverte d'un Australopithécien nouveau dans le gisement de l'Omo (Ethiopie) en 1967. *South African Journal of Science* 64:58–59.

Barrett, M. J., Brown, T., Arato, G., and Ozols, I. V. 1964. Dental observations on Australian aboriginals: Buccolingual crown diameters of deciduous and permanent teeth. *Australian Dental Journal* 9:280–285.

Barrett, M. J., Brown, T., and Macdonald, M. R. 1963. Dental observations on Australian aborigines: Mesiodistal crown diameters of permanent teeth. *Australian Dental Journal* 8:150–155.

Beynon, A. D., and Wood, B. A. 1986. Variations in enamel thickness and structure in East African hominids. *American Journal of Physical Anthropology* 70:177–193.

———. 1987. Patterns and rates of enamel growth in the molar teeth of early hominids. *Nature* 326:493–496.

Bishop, W. W. 1976. Thoughts on the workshop: "Stratigraphy, paleoecology and evolution in the Lake Rudolf Basin." In *Earliest man and environments in the Lake Rudolf Basin,* pp. 585–589 (ed. Y. Coppens, F. C. Howell, G. L. Isaac, and R. E. F. Leakey). University of Chicago Press, Chicago.

Bonnefille, R. 1976. Palynological evidence for an important change in the vegetation of the Omo Basin between 2.5 and 2 million years. In *Earliest man and environments in the Lake Rudolf Basin,* pp. 421–431 (ed. Y. Coppens, F. C. Howell, G. L. Isaac, and R. E. F. Leakey). University of Chicago Press, Chicago.

———. 1983. Evidence for a cooler and drier climate in the Ethiopian uplands towards 2.5 myr ago. *Nature* 303:487–491.

Bonnefille, R., and Riollet, G. 1980. Palynologie, végétation et climats de Bed I et de Bed II à Olduvai, Tanzanie. In *Proc. Eighth Pan African Congress of Prehistory and Quaternary Studies,* pp. 123–127. TILLMIAP, Nairobi.

Bonnefille, R., and Vincens, A. 1985. Apport de la palynologie à l'environnement des hominidés d'Afrique orientale. In *L'environnement des hominidés au Plio-Pléistocène,* pp. 237–278. (ed. Y. Coppens). Masson, Paris.

Bonnefille, R., Vincens, A., and Buchet, G. 1987. Palynology, stratigraphy and palaeoenvironment of a Pliocene hominid site (2.9–3.3 M.Y.) at Hadar, Ethiopia. *Palaeogeogr., Palaeoclimatol., Palaeoecol.* 60:249–281.

Brown, F. H., and Feibel, C. S. 1991. Stratigraphy, depositional environments, and palaeogeography of the Koobi Fora Formation. In *Koobi Fora Research Project.* Vol. 3, *The fossil ungulates: Geology, artiodactyls and palaeoenvironments,* pp. 1–30 (ed. J. M. Harris). Clarendon Press, Oxford.

Brown, F. H., Harris, J., Leakey, R., and Walker, A. 1985. Early

Homo erectus skeleton from West Lake Turkana, Kenya. *Nature* 316:788–792.
Butler, P. M., and Greenwood, M. 1965. Order: Insectivora. In *Olduvai Gorge, 1951–1961.* Vol. 1., *A preliminary report on the geology and the fauna,* pp. 13–14 (ed. L. S. B. Leakey). Cambridge University Press, Cambridge.
———. 1976. Elephant-shrews (Macroscelididae) from Olduvai and Makapansgat. In *Fossil vertebrates of Africa,* Vol. 4, pp. 1–56 (ed. R. J. G. Savage and S. C. Coryndon). Academic Press, London.
Caccone, A., and Powell, J. R. 1989. DNA divergence among hominoids. *Evolution* 43:925–942.
Cerling, T. E., and Hay, R. L. 1986. An isotopic study of paleosol carbonates from Olduvai Gorge. *Quaternary Research* 25: 63–78.
Cerling, T. E., Hay, R. L., and O'Neil, J. R. 1977. Isotopic evidence for dramatic changes in East Africa during the Pleistocene. *Nature* 267:137–138.
Clark, J. D., Asfaw, B., Assefa, G., Harris, J. W. K., Kurashina, H., Walker, R. C., White, T. D., and Williams, M. A. J. 1984. Palaeoanthropological discoveries in the Middle Awash Valley, Ethiopia. *Nature* 307:423–428.
DeJong, J. 1988. Climatic variability during the past three million years, as indicated by vegetational evolution in northwest Europe and with emphasis on data from the Netherlands. *Philosophical Transactions of the Royal Society of London,* ser. B, 318:603–617.
Eastoe, J. E. 1979. Enamel protein chemistry—past, present and future. *Journal of Dental Research* 58:753–764.
Eveleth, P. B., and Tanner, J. M., eds. 1976. *International Biological Programme 8: Worldwide variation in human growth,* Cambridge University Press, Cambridge.
Feibel, C. S., Brown, F. H., and McDougall, I. 1989. Stratigraphic context of fossil hominids from the Omo Group deposits: North Turkana Basin, Kenya and Ethiopia. *American Journal of Physical Anthropology* 78:595–622.
Fincham, A. G., Bessem, C. C., Lau, E. C., Pavlova, Z., Shuler, C., Slavkin, H. C., and Snead, M. L. 1991. Human developing enamel proteins exhibit a sex-linked dimorphism. *Calcified Tissue International* 48:288–290.
Foley, R. A. 1994. Speciation, extinction and climatic change in hominid evolution. *Journal of Human Evolution* 26:275–289.
Grine, F. E., and Martin, L. B. 1988. Enamel thickness and development in *Australopithecus* and *Paranthropus.* In *Evolutionary history of the "robust" australopithecines,* pp. 3–43 (ed. F. E. Grine). Aldine de Gruyter, New York.
Harris, J. M. 1976. Rhinocerotidae from the East Rudolf succession. In *Earliest man and environments in the Lake Rudolf Basin,* pp. 222–224 (ed. Y. Coppens, F. C. Howell, G. L. Isaac, and R. E. F. Leakey) University of Chicago Press, Chicago.
———. 1991. Family Bovidae. In *Koobi Fora Research Project.* Vol. 3, *The Fossil ungulates: geology, artiodactyls and palaeoenvironments,* pp. 139–320 (ed. J. M. Harris). Clarendon Press, Oxford.
Hartman, S. E. 1988. A cladistic analysis of hominoid molars. *Journal of Human Evolution* 17:489–502.
———. 1989. Stereophotogrammetric analysis of occlusal morphology of extant hominoid molars: Phenetics and function. *American Journal of Physical Anthropology* 80:145–166.
Hay, R. L. 1973. Lithofacies and environments of Bed I, Olduvai Gorge, Tanzania. *Quaternary Research* 3:541–560.
———. 1990. Olduvai Gorge: A case history in the interpretation of hominid palaeoenvironments in East Africa. *Geological Society of America,* special paper 242:23–27.
Hill, A. 1985. Early hominid from Baringo, Kenya. *Nature* 315: 222–224.
———. 1987. Causes of perceived faunal change in the later Neogene of East Africa. *Journal of Human Evolution* 16:583–596.
Hill, A., Drake, R., Tauxe, L., Mongahan, M., Barry, J. C., Behrensmeyer, A. K., Curtis, G., Finejacobs, B., Jacobs, L., Johnson, N., and Pilbeam, D. 1985. Neogene palaeontology and geochronology of the Baringo Basin, Kenya. *Journal of Human Evolution* 14:759–773.
Hooghiemstra, H. 1986. A high-resolution palynological record of 3.5 million years of northern Andean climatic history: The correlation of 26 "glacial cycles" with terrestrial, marine and astronomical data. *Zentralblatt für Geologie und Paläontologie* 1:1363–1366.
Jaeger, J. J. 1976. Les rongeurs (Mammalia, Rodentia) du Pléistocène inférieur d'Olduvai Bed I (Tanzanie) Ière partie: Les muridés. In *Fossil vertebrates of Africa,* Vol. 4, pp. 57–120 (ed. R. J. G. Savage and S. C. Coryndon). Academic Press, London.
Jaeger, J. J., and Wesselman, H. B. 1976. Fossil remains of micromammals from the Omo Group deposits. In *Earliest man and environments in the Lake Rudolf Basin,* pp. 351–360 (ed. Y. Coppens, F. C. Howell, G. L. Isaac, and R. E. F. Leakey). University of Chicago Press, Chicago.
Janecek, T. R., and Ruddiman, W. F. 1987. Plio-Pleistocene sedimentation in the south equatorial Atlantic divergence. *Twelfth INQUA Congress, Ottawa.* Abstracts: 193.
Jansen, E., Bleil, V., Henrich, R., Kringsrad, L., and Slettemark, B. 1988. Paleoenvironmental changes in the Norwegian Sea and the Northeast Atlantic during the last 2.8 m. y.: Deep Sea Drilling Project/Ocean Drilling Program Sites 610, 642, 643 and 644. *Paleoceanography* 3:563–581.
Johanson, D. C. 1974. Some metric aspects of the permanent and deciduous dentition of the pygmy chimpanzee (*Pan paniscus*). *American Journal of Physical Anthropology* 41:39–48.
Johanson, D. C., White, T. D., and Coppens, Y. 1978. A new species of the genus *Australopithecus* (primates: *Hominidae*) from the Pliocene of East Africa. *Kirtlandia* 28:1–14.
Kappelman, J. 1984. Plio-Pleistocene environment of Bed I and Lower Bed II, Olduvai Gorge, Tanzania. *Palaeogeography, Palaeoclimatology, Palaeoecology* 48:171–196.
Kay, R. F. 1975. Allometry and early hominids. *Science* 189:63.
Kessel, M., and Gruss, P. 1990. Murine developmental control genes. *Science* 249:374–379.
Kukla, G. 1987. Loess stratigraphy in central China. *Quaternary Science Review* 6:191–219.
Lagerstrom, M., Dahl, N., Iselius, L., Backman, B., and Pettersson, U. 1989. Linkage analysis of x-linked amelogenesis imperfecta. *Cytogenetics and Cell Genetics* 51:1028.
Leakey, R. E. F., and Walker, A. 1988. New *Australopithecus*

boisei specimens from East and West Lake Turkana, Kenya. *American Journal of Physical Anthropology* 76:1–24.

McDougall, I. 1985. K-Ar and $^{40}Ar/^{39}Ar$ dating of the hominid-bearing Plio-Pleistocene sequence at Koobi Fora, Lake Turkana, northern Kenya. *Geological Society of America Bulletin* 96:159–175.

McHenry, H. M. 1992. Body size and proportions in early hominids. *American Journal of Physical Anthropology* 87:407–431.

Macho, G. A. in press. The significance of hominid enamel thickness for phylogenetic and life-history reconstruction. In *Structure, function and evolution of teeth* (ed. P. Luckett and J. Moggi Cecchi).

MacKenzie, A., Ferguson, M. W. J., and Sharpe, P. T. 1991. *Hox-7* expression during murine craniofacial development. *Development* 113:601–611.

———. 1992. Expression patterns of the homeobox gene, *Hox-8,* in the mouse embryo suggest a role in specifying tooth initiation and shape. *Development* 115:403–420.

MacKenzie, A., Leeming, G., Jowett, A. K., Ferguson, M. W. J., and Sharpe, P. T. 1991. The homeobox gene *Hox-7.1* has specific regional and temporal expression patterns during early murine craniofacial embryogenesis, especially tooth development *in vivo* and *in vitro. Development* 111:269–285.

Macho, G. A., and Thackeray, J. F. 1985. Computed tomography and intercuspal angulation of maxillary molars of Plio-Pleistocene hominids from Sterkfontein, Swartkrans and Kromdraai (South Africa): An exploratory study. *Zeitschrift für Morphologie und Anthropologie* 79(1):261–269.

Maglio, V. J. 1972. Vertebrate faunas and inferred chronology of East Lake Rudolf, Kenya. *Nature* 239:379–385.

Monaghan, A. P., Hill, R. E., Bhattacharya, S. S., Graham, E., and Davidson, D. R. 1991. *Msh*-like homeobox genes define domains in the developing vertebrate eye. *Development* 112:1053–1061.

Ruvolo, M., Pan, D., Zehr, S., Goldberg, T., Disotell, T. R., and Von Dornum, M. 1994. Gene trees and hominoid phylogeny. *Proceedings National Academy of Sciences* 91:8900–8904.

Scott, J. H., and Symons, N. B. B. 1958. *Introduction to dental anatomy.* 2d ed. Livingstone, Edinburgh.

Shackleton, N. J., Backman, J., Zimmerman, H., Kent, D. V., Hall, M. A., Roberts, D. G., Schnitker, D., Baldauf, J. G., Desprairies, A., Homrighausen, R., Huddlestun, P., Keene, J. B., Kaltenback, A. J., Krumsiek, K. A. O., Morton, A. C., Murray, J. W., and Westberg-Smith, J. 1984. Oxygen isotope calibration of the onset of ice-rifting and history of glaciation in the North Atlantic region. *Nature* 307:620–623.

Suwa, G. 1988. Evolution of the "robust" australopithecines in the Omo succession: Evidence from mandibular premolar morphology. In *Evolutionary history of the "robust" australopithecines,* pp. 199–222 (ed. F. E. Grine). Aldine de Gruyter, New York.

Tobias, P. V. 1991. *Olduvai Gorge.* Vol. 4, *The skull, endocasts and teeth of* Homo habilis. Cambridge University Press, Cambridge.

Turner, A., and Wood, B. A. 1993. Comparative palaeontological context for the evolution of the early hominid masticatory system. *Journal of Human Evolution* 24:301–318.

Vrba, E. S. 1985. Environment and evolution: Alternative causes of the temporal distribution of evolutionary events. *South African Journal of Science* 81:229–236.

———. 1988. Late Pliocene climatic events and human evolution. In *Evolutionary history of the "robust" australopithecines,* pp. 405–426 (ed. F. E. Grine). Aldine de Gruyter, New York.

Vrba, E. S., Burckle, L. H., Denton, G. H., and Partridge, T. C. 1985. Palaeoclimate and evolution pt. 1. *South African Journal of Science* 81:224–275.

Walker, A. C., Leakey, R. E. F., Harris, J. M., and Brown, F. H. 1986. 2.5 million year old *Australopithecus boisei* from west of Lake Turkana, Kenya. *Nature* 322:517–522.

Walker, A. C., and Leakey, R. E., eds. 1993. *The Nariokotome* Homo erectus *skeleton.* Harvard University Press, Cambridge.

Ward, S., and Hill, A. 1987. Pliocene hominid partial mandible from Tabarin, Baringo, Kenya. *American Journal of Physical Anthropology* 72:21–37.

Wesselman, H. B. 1984. The Omo micromammals: Systematics and paleoecology of early man sites from Ethiopia. *Contributions to Vertebrate Evolution* 7:1–219.

White, T. D., Suwa, G., and Asfaw, B. 1994. *Australopithecus ramidus,* a new species of early hominid from Aramis, Ethiopia. *Nature* 371:306–312.

———. 1995. *Australopithecus ramidus,* a new species of early hominid from Aramis, Ethiopia. Corrigendum. *Nature* 375:88.

Wood, B. A. 1975. An analysis of sexual dimorphism in primates. Ph.D. diss. University of London.

———. 1991. *Koobi Fora Research Project.* Vol. 4, *Hominid cranial remains.* Clarendon Press, Oxford.

———. 1993. Hominid paleobiology: Recent achievements and challenges. In *Integrative pathways of the past,* pp. 147–165 (ed. R. C. Corrucini and R. Ciochon). Prentice-Hall, N.J.

Wood, B. A., Abbott, S. A., and Uytterschaut, H. 1988. Analysis of the dental morphology of Plio-Pleistocene hominids IV. Mandibular postcanine root morphology. *Journal of Anatomy* 156:107–139.

Wood, B., Wood, C., and Konigsberg, L. 1994. *Paranthropus boisei:* An example of evolutionary stasis? *American Journal of Physical Anthropology* 95:117–136.

Zagwijn, W. H., and Suc, J. P. 1984. Palynostratigraphie du Plio-Pléistocène d'Europe et de la Méditérranée nord-occidentales: Corrélations chronostratigraphiques, histoire de la végétation et du climat. *Paleobiological Contributions* 14:475–483.

Part Four

The Pleistocene

Chapter 30

The Influence of Climatic Changes on the Completeness of the Early Hominid Record in Southern African Caves, with Particular Reference to Swartkrans

C. K. Brain

A long-term and detailed investigation of the Swartkrans cave and its relics (Brain, 1993a) has provided insights into the influence of changing climate on the completeness of the evolutionary record that has been preserved there. Conclusions drawn from Swartkrans will surely have relevance, to a greater or lesser extent, for the other southern African australopithecine cave sites as well. It now seems likely that each of the fossiliferous deposits that form components of the time sequence at Swartkrans represents a brief interglacial accumulation, separated from its successor by a much longer hiatus during which deposition was interrupted and while erosion within the cave deposit may have occurred (table 30.1). The result is that the sequential glimpses that one has of evolutionary developments in this part of southern Africa probably reflect only the warmer intervals of the climatic continuum.

An Overview of the Swartkrans Stratigraphic Sequence

Situated one kilometer northwest of Sterkfontein in the Transvaal, the Swartkrans cave was formed by solution of Precambrian dolomite of the Transvaal Supergroup. As is evident from the plan of the cave (fig. 30.1), it is somewhat irregular in shape and measures a maximum of about 45 m in both east-west and north-south dimensions. Also shown are the positions of the permanent metal excavation grid and the Hanging Remnant, an unusual feature that provides a key to the understanding of the remarkable sequence of depositional and erosional cycles that have molded the complex form of the cave deposit. Following an early stratigraphic study (Butzer, 1976), the Swartkrans Formation was formally named, along with two members, designated 1 and 2. Since then, additional excavation (Brain, 1993a) has provided further insights into the complexity of the filling and has identified five members, each separated from its older counterparts by an erosional discontinuity. There is no question that many other separate deposits have filled erosional spaces in the Swartkrans cave fill during the 1.7 million years (myr) of its history, but these have been lost through erosion or mining. The important point is that by virtue of its structure the Swartkrans cave has formed a sediment trap, with a series of openings to the surface above and below, leading to lower caverns. Following each period of deposition, a mass of sediment rested for a while on the shelflike floor, to be carried away, entirely or in part, during a different phase of a subsequent climatic cycle. It now appears that Members 1 and 2 occupied much of the Outer Cave, which represents the area south of the Hanging Remnant; the positions of Members 3, 4, and 5 are indicated in figure 30.2.

Repeated attempts to obtain absolute dates for the Swartkrans deposits and fossils have so far proved fruitless, and we still rely on faunal estimates, particularly those of Vrba, which resulted from her extensive study of fossil bovids from the australopithecine caves. She suggested (Vrba, 1975) that fossils from the Hanging Remnant of Member 1 could be referred to the Swartkrans Faunal Span, with an age indication of between 2 and 1 myr. She later refined this estimate (Vrba, 1982) to place the Hanging Remnant deposit at between 1.8 and 1.5 myr. At that time, the fossils referred to Member 2 con-

ISBN 0–300–06348–2.

Table 30.1. The Five Stratigraphic Members Currently Recognized at Swartkrans, with Information on Their Fossil and Cultural Content

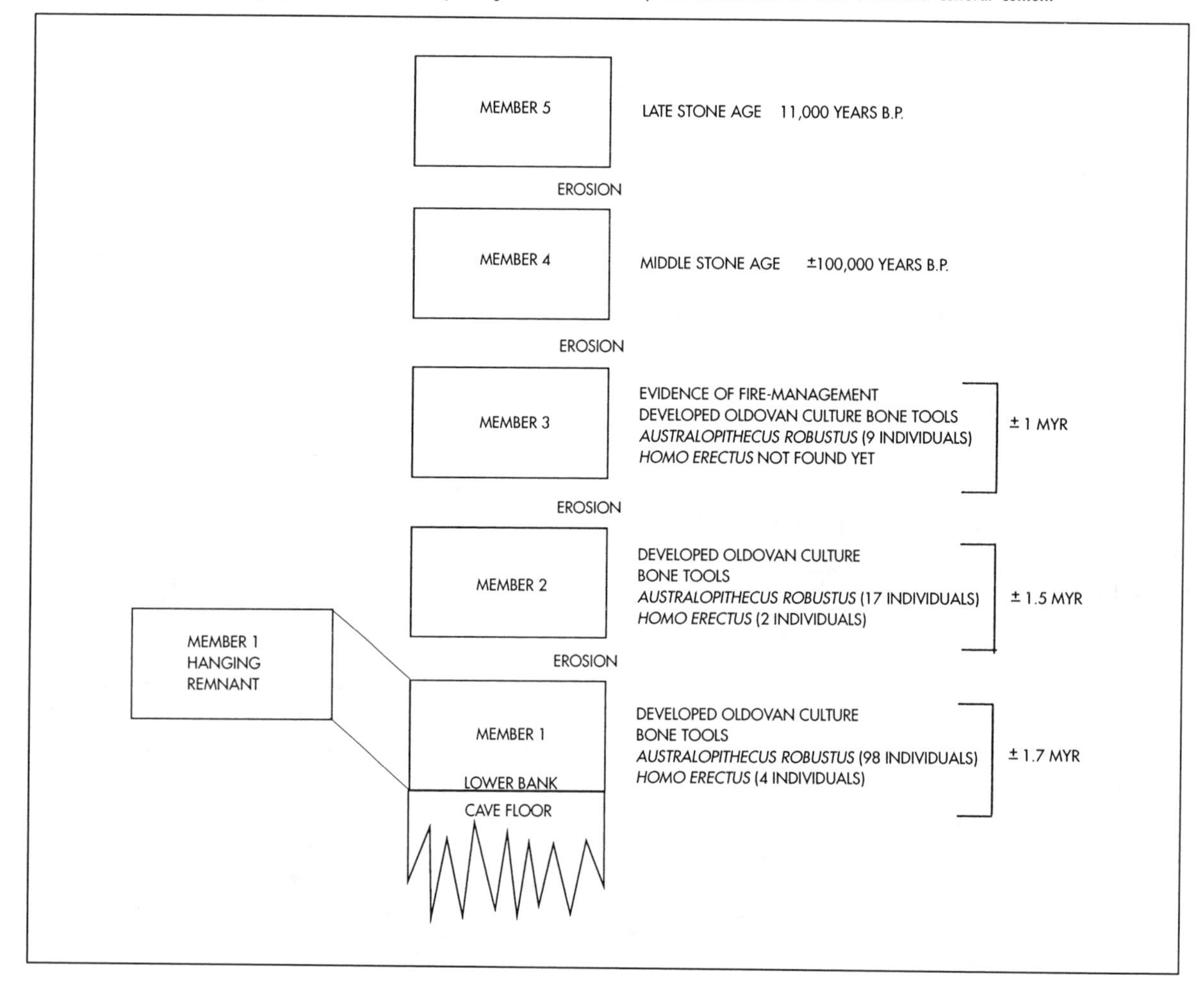

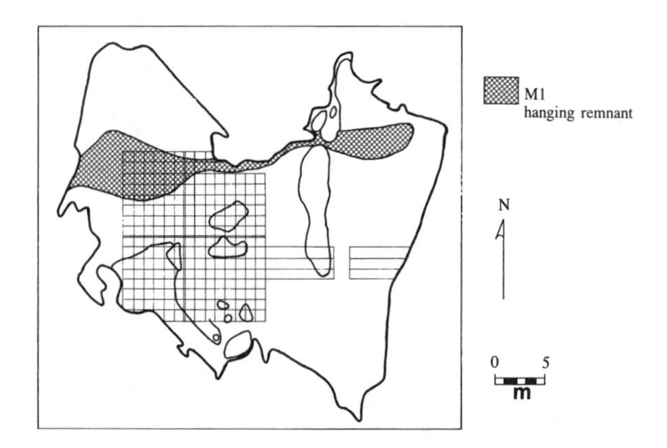

Fig. 30.1. A plan of the Swartkrans cave, showing the position of the Hanging Remnant and excavation grid.

stituted a very mixed assemblage from a variety of contexts and ages, with disturbance having been introduced by lime mining. As a result of more recent and carefully controlled excavations (Brain, 1993a), we now have additional faunal assemblages from the Lower Bank of Member 1, as well as from Members 2 and 3.

The two units of Member 1—the Hanging Remnant and the Lower Bank—have so far yielded the largest number of hominid fossils from Swartkrans, currently estimated to have come from ninety-eight individuals of *Australopithecus robustus* and 4 of *Homo erectus* (Grine, 1993). Stone artifacts attributed to the Developed Oldowan culture (Clark, 1993) are well represented, particularly in the Lower Bank, as are bone tools apparently used for digging (Brain and Shipman, 1993). An age of

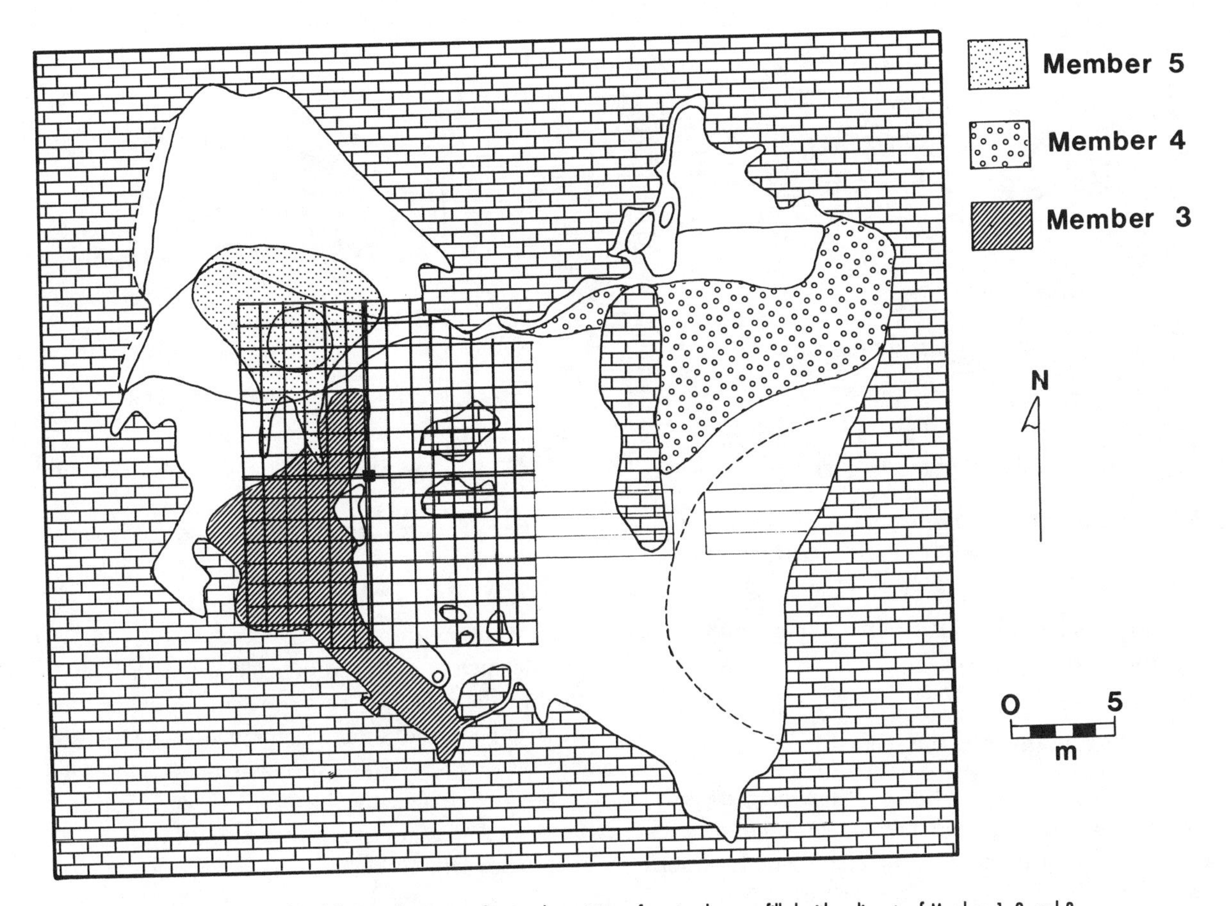

Fig. 30.2. A plan of the Swartkrans cave showing the positions of erosional spaces filled with sediments of Members 1, 2, and 3.

1.7 myr for these units is currently thought to be reasonable.

A sample of Member 2, excavated between 1979 and 1982, provided fossils from seventeen individuals of *Australopithecus robustus* and two of *Homo erectus,* together with Developed Oldowan stone artifacts and bone tools similar to those from the Lower Bank of Member 1. The overall faunal assemblage (Watson, 1993) is not strikingly different from that of Member 1, and an age of about 1.5 myr is suggested for Member 2.

The Member 3 deposit, excavated between 1982 and 1986, was found to occupy a deep gully eroded into older sediment along the west wall of the cave (see fig. 30.2). Diggings in Member 3 have produced remains of nine individuals of *Australopithecus robustus* but none, as yet, of *Homo* (Grine, 1993). Developed Oldowan stone artifacts and bone tools have been recovered, together with a fauna indistinguishable from that of Members 1 or 2 (Watson, 1993). Perhaps the most striking feature of the Member 3 fossil assemblage is the presence of about 270 pieces of burnt bone, the distribution of which through the deposit strongly suggests the repetitive management of fire by hominids within the cave (Brain and Sillen, 1988; Brain, 1993b). An age of between 1.0 and 1.5 myr is suggested for Member 3. The deposit designated Member 4 occupies the northeast corner of the Swartkrans cave and contains abundant Middle Stone Age artifacts. It has not yet yielded any fossils but is assumed to be about 100 thousand years (kyr) old. The youngest formally designated deposit in the cave is Member 5, occupying an erosional channel close to the north wall. It is rich in remains of the extinct springbuck, *Antidorcas bondi,* bones of which have been dated by ^{14}C to 11 kyr before the present (J. C. Vogel, pers. comm.).

Suggested Stages in the Formation of the Swartkrans Deposit

The Swartkrans stratigraphic sequence can best be visualized using a series of six reconstructed and somewhat

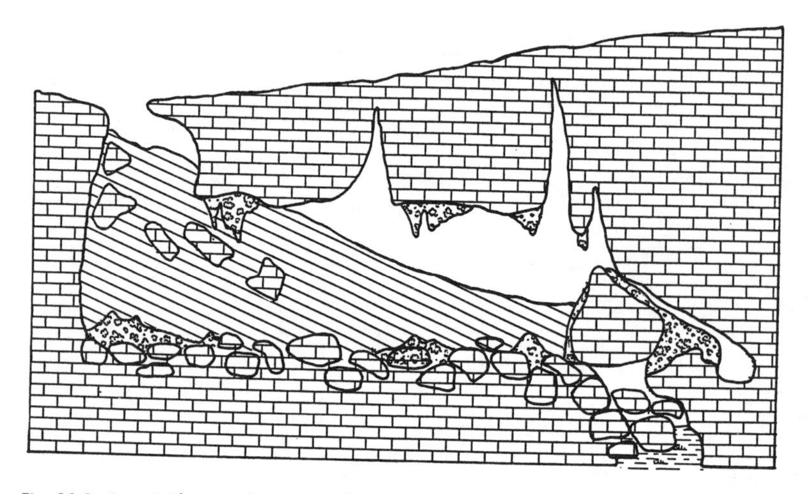

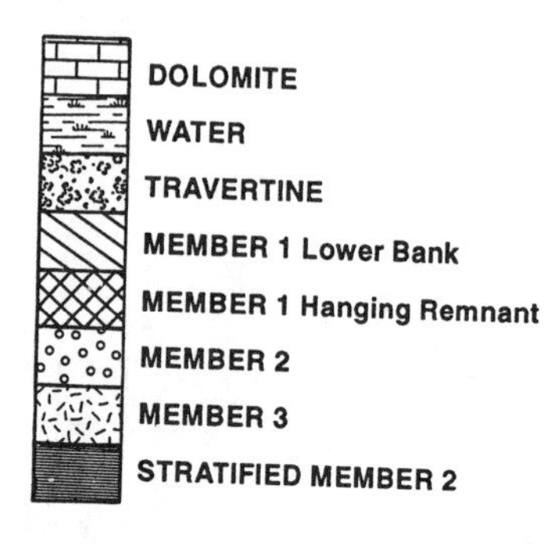

Fig. 30.3. Stage 1. The original entrance to the cave, above the southeast wall, has admitted a substantial talus cone, which, in calcified form, has become known as the Lower Bank of Member 1. This first infilling process may have taken about 20 kyr.

idealized sections (figs. 30.3–30.8) running southeast to northwest across the cave.

Climatic Implications of the Depositional Sequence

Perhaps the most striking feature of the Swartkrans depositional sequence is the repeated alternation between periods of sediment accumulation and periods of nondeposition, the latter generally accompanied by active erosion within the cave. The fact that deposition was interrupted from time to time, even though the entrance may not have been choked with sediment, suggests that outside factors such as changes in climate and the extent of ground-covering vegetation were involved.

In the course of his study of the lithostratigraphy of the Swartkrans Formation, Butzer (1976) inferred that prior to the first opening of the Swartkrans cave, a gently undulating land surface existed in the Swartkrans-Sterkfontein area at an altitude 1,500–1,510 m above sea level. Today, the top of the ridge above Swartkrans stands at 1,496 m, the top of the fossiliferous deposits at 1,489 m, and the valley floor below the cave at 1,454 m. It is estimated that during the last 1.7 myr, a bedrock incision of about 30 m has occurred along the valley axis and that 5–10 m of dolomite have been denuded from the hilltop. Butzer points out that during this time a smoothly undulating upland above the cave was reduced to a small hill with a relief of about 40 m and an average inclination of 16°. He suggests that at the time of the first cave opening, the Swartkrans hill had twice its present area but only half its relief. Today the top of the ridge is only 25 m beyond the northern wall of the cave, providing a catchment area of less than 600 m^2, with a longitudinal gradient of 25°. This situation allows for little more than pockets of lithosols and relict B-horizon soil among the rough, micro-karstic *lapies* topography. However, at the time of entry of the Member 1–Hanging Remnant sediment, Butzer estimates that the catchment area of the Swartkrans doline was twice as large, that the gradient was perhaps half that of today's, and that a subcontinuous mantle of soil covered most of the *lapies.* It is likely that the progressive denudation of the Swartkrans hill, which enclosed the cave, proceded cyclically in response to worldwide glacial-interglacial cycles that at the time of the deposition of Members 1–3 would presumably have had a 40 kyr periodicity. Butzer postulates that the very extended intervals of nonaccumulation within the cavern must have coincided with periods of soil deepening outside and with a vegetation mat that served to inhibit erosion. But questions remain: At what part of a cycle did deposition occur within the cave, and at what part was this followed by erosion? In answering these questions, insights provided by examination of the youngest of the Swartkrans fossiliferous deposits, Member 5, are instructive. Here a 4 m-thick deposit of earth accumulated in an irregular solution channel close to the north wall of the cave; and this took place, apparently rapidly, between 9 and 12 kyr before the present. The deposition reflects a time, at the start of the current interglacial, when storm water entering the cavern carried a good deal of soil eroded from the hillside, presumably as a result of an incomplete mat of

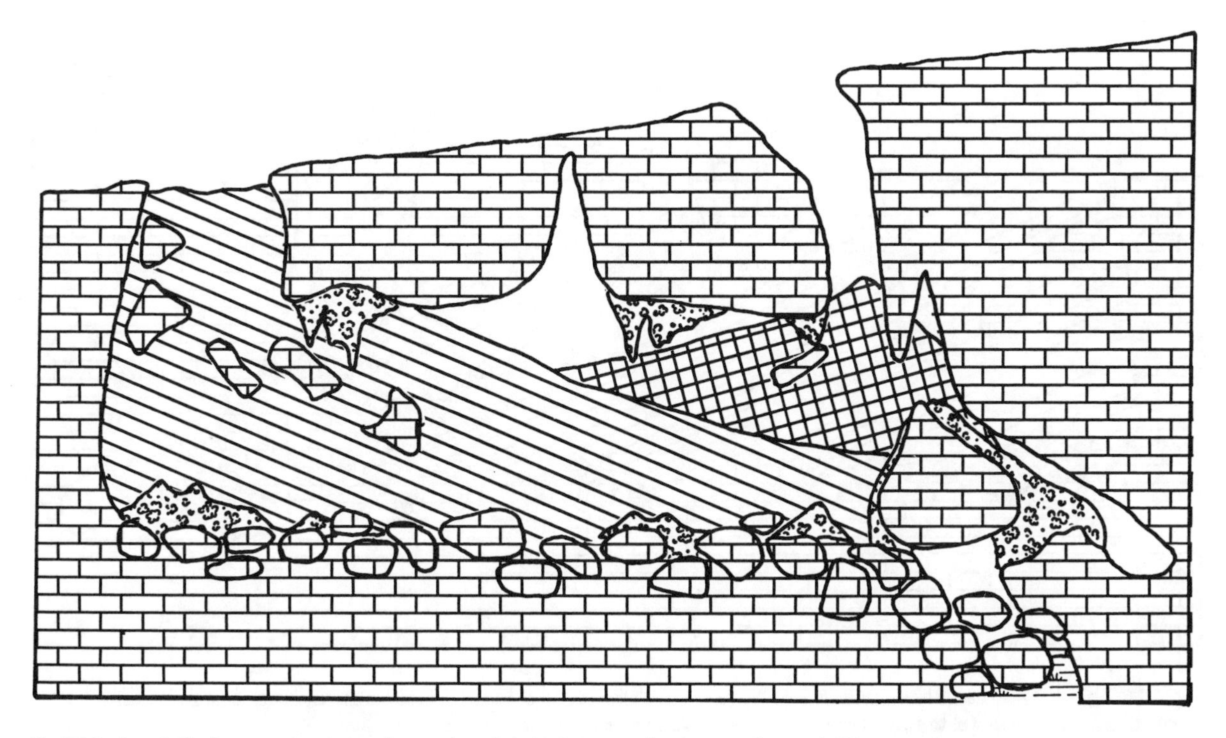

Fig. 30.4. Stage 2. The first entrance to the cave has now been choked with Lower Bank sediment, and a new shaftlike opening has formed above the north wall. This is admitting sediment destined to become the Hanging Remnant of Member 1; the richness of hominid remains in this deposit points to somewhat unusual taphonomic circumstances.

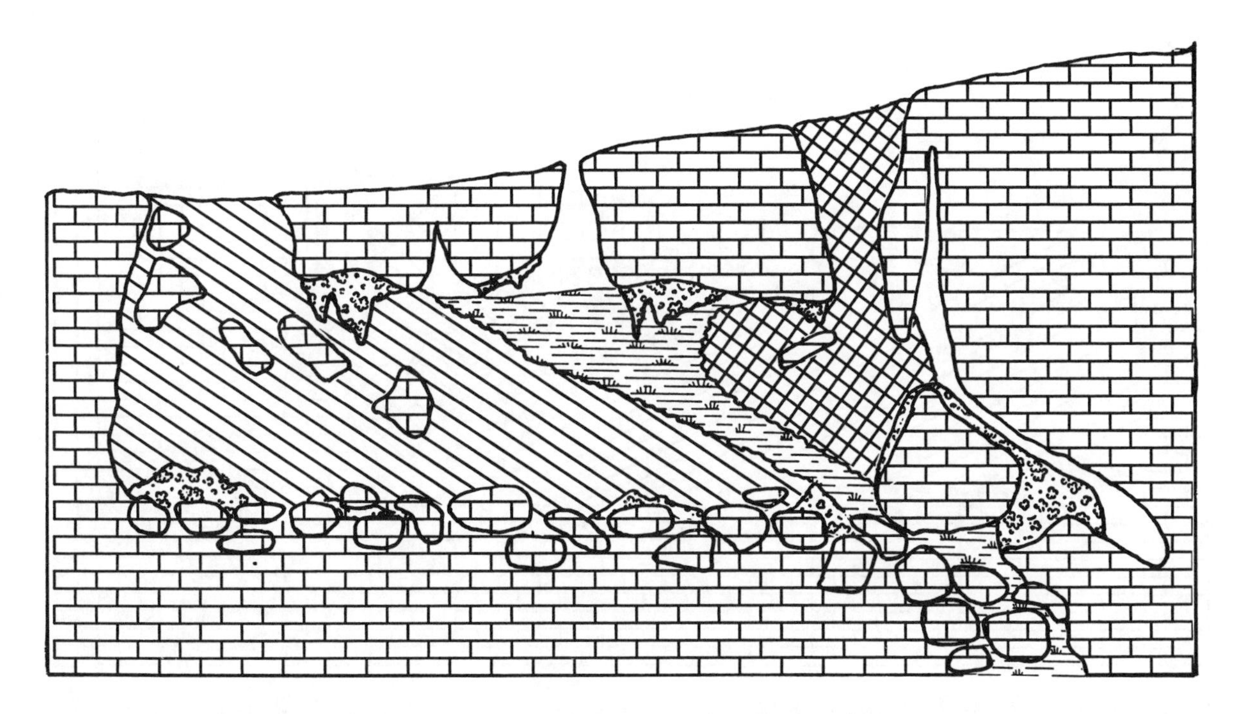

Fig. 30.5. Stage 3. A period of erosion within the cave occurs, with a good deal of water entering through a new shaft opening midway between the south and north walls. An irregular gap several meters wide is eroded between the top of the Lower Bank and the lower surface of the Hanging Remnant, as the sediment is trickled away to lower caverns in the dolomite.

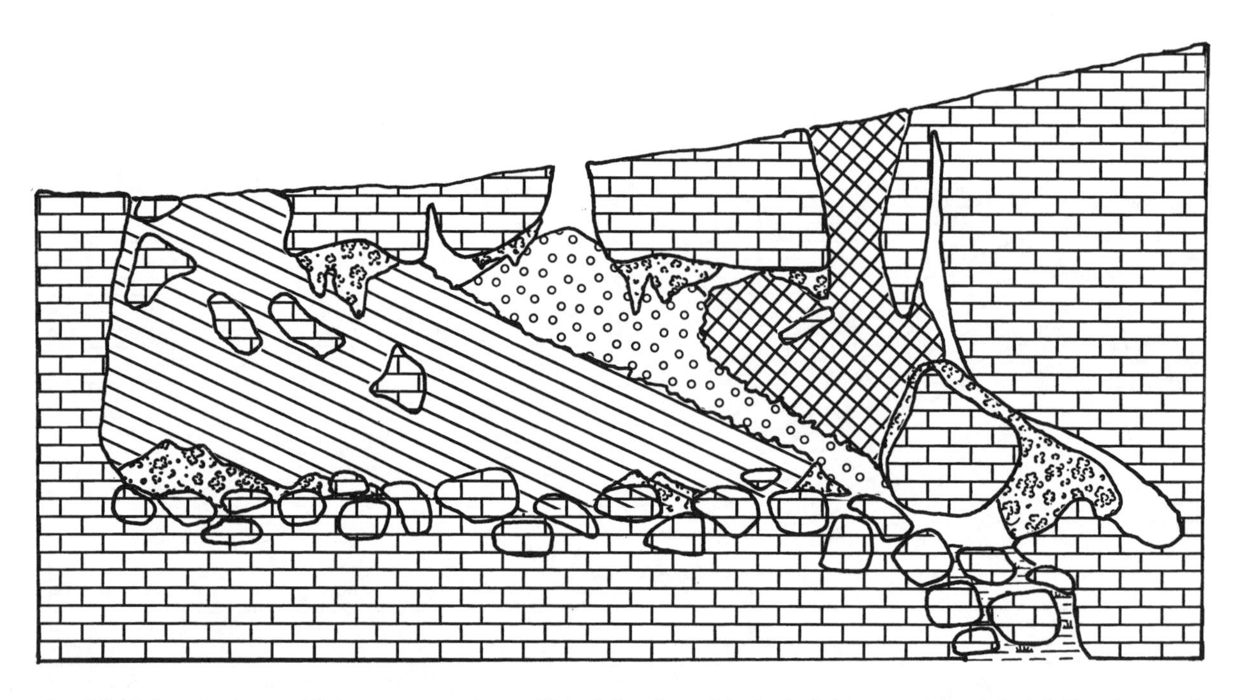

Fig. 30.6. Stage 4. The gap created by the erosion in Stage 3 is now filled with the sediment of Member 2, which has entered through the shaft. The sediment is rich in artifacts and bones as a result of hominid and other animal activity in the entrance area of the cave.

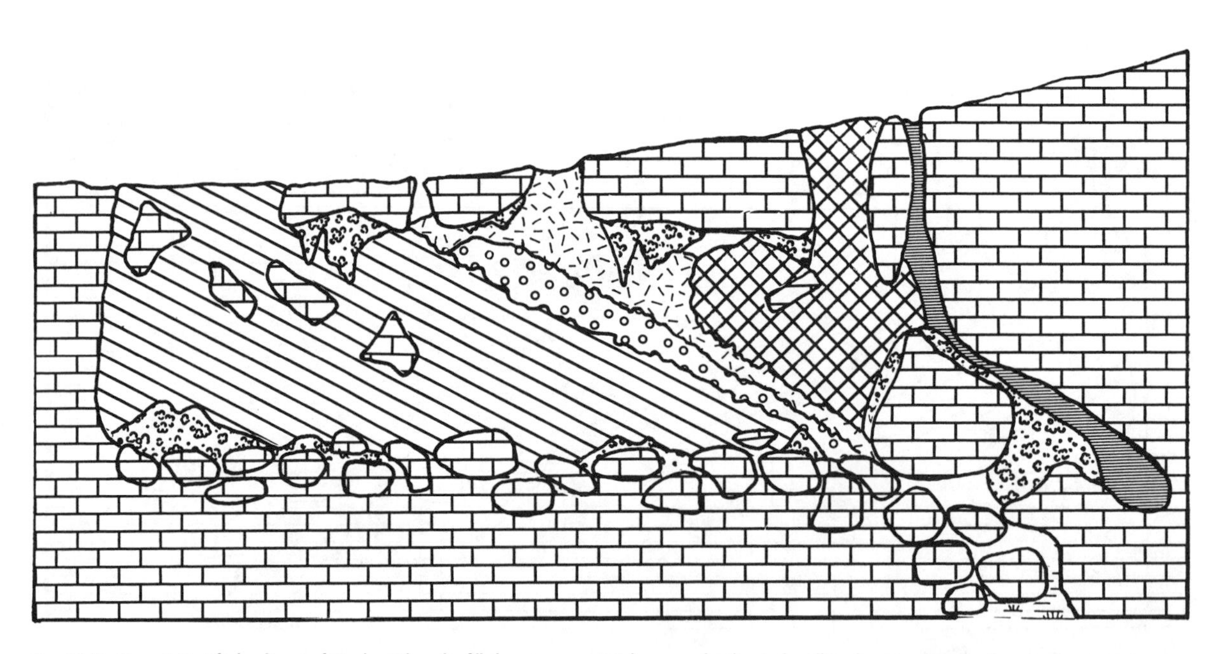

Fig. 30.7. Stage 5. Stratified sediment of Member 2 has also filled a narrow erosional space under the north wall, and a renewed cycle of erosion has created a gulley along the west wall of the cave, removing parts of the earlier deposits of Members 1 and 2. This has filled with sediment of Member 3, in which normal and burnt bones as well as artifacts are preserved. It appears that the entrance area to this new gulley was used by early hominids for shelter and that early experiments with the management of fire took place here.

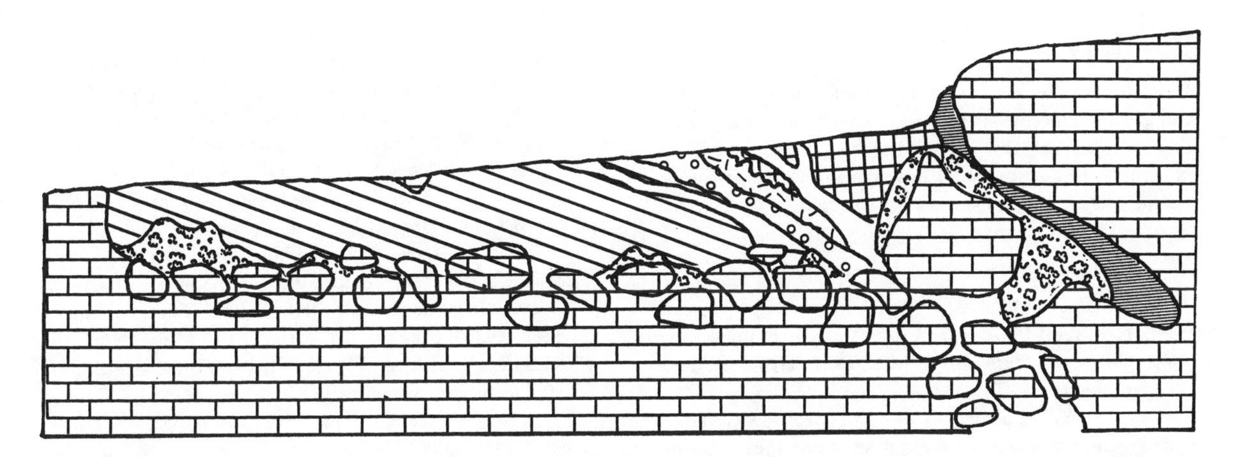

Fig. 30.8. Stage 6. This cross-section shows the contemporary situation before mining or excavation took place. Most of the dolomite roof of the cave has been removed by hillside erosion, and channels of various ages have ramified through the older deposits. Fillings of these channels include the sediments of Members 4 and 5.

vegetation in the catchment area. There are indications from other sites, such as the Wonderkrater peat site, in the central Transvaal, and the Wonderwerk cave, in the northern Cape, that the moisture index at that time was low (Thackeray and Lee-Thorp, 1992). The central interior of southern Africa was emerging then from the harsh conditions of the last glacial maximum, and as rainfall increased, either in quantity or intensity, hillside soil, unprotected by vegetation, was apparently prone to erosion.

The erosional space into which the Member 5 sediment was deposited, on the other hand, seems to have been formed under different conditions, when abundant water entering the cave was both devoid of a sediment load and acidic. Such conditions must have prevailed at some stage during the long, preceding "glacial" interval, when low temperatures led to reduced evaporation and to more water passing through a dense vegetation mat that was rich in humic acid.

If this was, in fact, the sequence of events of Swartkrans during the most recent glacial-interglacial cycle, then a similar pattern is likely to have been repeated many times during the last 1.7 myr, some results of which appear to be reflected in the earlier Swartkrans members.

Implications of the Swartkrans Conclusions for Other Australopithecine Caves in Southern Africa

The sensitivity of Swartkrans as an indicator of environmental change is a product of its structural configuration, where openings existed both above and below the cavern. Such a configuration is by no means unusual in dolomite caves and certainly exists at Sterkfontein (Partridge et al., 1991). It seems that some of Member 5 there may fill an erosional space in Member 4 (Clarke, 1985 and pers. comm.). Should this prove to be the case, I suspect that the infilling deposit was probably an interglacial accumulation, as were those at Swartkrans. Experience at Swartkrans has shown me that interfaces between adjacent breccias in the same cave are often indistinct and subtle and may not become apparent until careful excavation intersects and follows them. I suspect that a renewed search for such interfaces in other australopithecine cave deposits could be a rewarding one.

If the Transvaal cave deposits are, in fact, typically rather brief interglacial accumulations, the situation exists that long intervals of "glacial" time during the last 1.5 myr will be unrecorded in such breccias. The result is that significant evolutionary developments are bound to have occurred in unrecorded time, and an example of one of these appears to be the first control of fire by early hominids in southern Africa. Evidence for such control makes its sudden appearance in Swartkrans Member 3 (Brain and Sillen, 1988; Brain, 1993b) but is lacking completely in the preceding Member 2, despite the occurrence of hominid remains and culture in both members. If the unrecorded interval between Members 2 and 3 was, in fact, a "glacial" one, it is not unreasonable to think that the low temperatures of that time were an incentive to the management of fire for warmth and protection.

Acknowledgments

I wish to thank the organizers of the conference on paleoclimate and evolution for inviting me to participate and to congratulate them on what was achieved in the meetings. In addition, I am grateful to the Foundation for Research and Development for its assistance with travel expenses and its support of the research discussed here.

References

Brain, C. K. (ed.) 1993a. Swartkrans: A cave's chronicle of early man. *Transvaal Museum Monograph,* no. 8. Transvaal Museum, Pretoria.

———. 1993b. The occurrence of burnt bones at Swartkrans and their implications for the control of fire by early hominids. In *Swartkrans: A cave's chronicle of early man,* pp. 229–242 (ed. C. K. Brain). Transvaal Museum, Pretoria.

Brain, C. K., and Shipman, P. 1993. The Swartkrans bone tools. In *Swartkrans: A cave's chronicle of early man,* pp. 195–215 (ed. C. K. Brain). Transvaal Museum, Pretoria.

Brain, C. K., and Sillen, A. 1988. Evidence from the Swartkrans cave for the earliest use of fire. *Nature* 336:464–466.

Butzer, K. W. 1976. Lithostratigraphy of the Swartkrans Formation. *South African Journal of Science* 72:136–141.

Clark, J. D. 1993. Stone artefact assemblages from Members 1–3, Swartkrans cave. In *Swartkrans: A cave's chronicle of early man,* pp. 167–194 (ed. C. K. Brain). Transvaal Museum, Pretoria.

Clarke, R. J. 1985. Early Acheulean with *Homo habilis* at Sterkfontein. In *Hominid evolution, past, present and future,* pp. 287–298 (ed. P. V. Tobias). Alan R. Liss, New York.

Grine, F. E. 1993. Description and preliminary analysis of new hominid craniodental fossils from the Swartkrans Formation. In *Swartkrans: A cave's chronicle of early man,* pp. 75–116 (ed. C. K. Brain). Transvaal Museum, Pretoria.

Partridge, T. C., Tobias, P. V., and Hughes, A. R. 1991. Paléoécologie et affinités entre les Australopithecinés d'Afrique du Sud: Nouvelles données de Sterkfontein et Taung. *L'Anthropologie* 95:363–378.

Thackeray, J. F., and Lee-Thorp, J. A. 1992. Isotopic analysis of equid teeth from Wonderwerk Cave, northern Cape Province, South Africa. *Palaeogeography, Palaeoclimatology, Palaeoecology* 99:141–150.

Vrba, E. S. 1975. Some evidence of chronology and paleoecology of Sterkfontain, Swartkrans and Kromdraai from the fossil Bovidae. *Nature* 250:19–23.

———. 1982. Biostratigraphy and chronology, based particularly on Bovidae, of southern African hominid-associated assemblages. In *Congrès International de Paléontologie Humaine, 1er Congrès,* pp. 707–752. UNESCO, Nice.

Watson, V. 1993. Composition of the Swartkrans bone accumulations, in terms of skeletal parts and animals represented. In *Swartkrans: A cave's chronicle of early man,* pp. 35–73 (ed. C. K. Brain). Transvaal Museum, Pretoria.

Chapter 31

Southern Savannas and Pleistocene Hominid Adaptations: The Micromammalian Perspective

D. Margaret Avery

The theory that the history of the Hominidae is bound up with the development of savanna vegetation in Africa is gaining acceptance and has considerable importance for an improved understanding of hominid adaptations. As such, it deserves further consideration, and the limits of the Savanna Biome will repay closer inspection in this regard. The margins of any distribution pattern are likely to best illuminate factors governing that pattern, because organisms become increasingly ill adapted to their environment as the limits of their distribution are approached; the balance between surviving or not becomes correspondingly finer until even a slight change in one factor may be critical. It can thus be predicted that if hominids are endemic to savannas, there should be an initially strong correlation between their movements and fluctuations in the margins of savanna vegetation caused by climatic fluctuations at various times in the past. (It is implicit in this scenario that technological and sociological advancement would gradually reduce this correlation between environmental conditions and hominid distribution.)

Micromammalian evidence from the four Pleistocene sites of Swartkrans, Wonderwerk Cave, Border Cave, and Klasies River Mouth (fig. 31.1), supported by dating and other evidence such as site distribution, strongly suggests that various early hominid taxa moved into southern Africa at times when interglacial and interpleniglacial conditions obtained and when savanna vegetation would have attained its greatest distribution. Vegetational mosaics comprising dense riverine vegetation with relatively open bush or trees and grassland are indicated at all four sites. At Klasies River Mouth, the most southerly of these sites, the vegetation may not have been savanna per se during most of the occupation, but it was of similar structure; micromammalian samples for the very earliest occupation have not been seen, but it would appear that savanna may have reached this region along the eastern seaboard during the earliest part of the Last Interglacial. The current indication that these sites were deserted during full glacial periods would support the suggestion that the hominids concerned were originally adapted to a tropical or subtropical environment.

ISBN 0–300–06348–2.

Physical Setting

The area discussed lies south of 22°S, and altitudes range from approximately 1,650 m in the interior to near sea level along the south coast (fig. 31.1). Physiographically, the land comprises a high interior plateau (rising to maximum elevation in the west) and a low-altitude marginal zone, separated from the interior by an escarpment. Rainfall isohyets generally run from north to south, with mean annual rainfall decreasing from east to west. Average annual water deficiency ranges from under 100 mm in the east up to 1,000 mm in the western interior. Rainfall is largely a summer phenomenon, except in the extreme southwest, where most is received in winter, and in an intermediate zone with no well-defined rainy season. Temperatures are highest in southern Mozambique and the western interior, with mean annual range of temperature being greatest in the latter region. Lowest temperatures occur in the southeastern interior where there are high mountains. In the high interior the Savanna Biome is restricted to about 29°S; along the low-lying eastern coastal margin it reaches as far south as about 34°S.

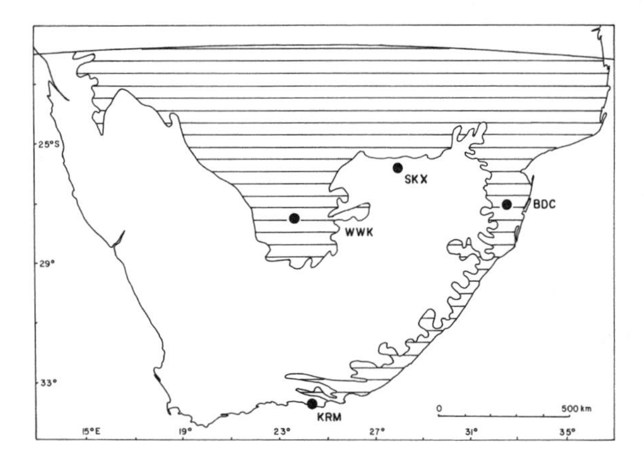

Fig. 31.1. Location of the four sites discussed and present-day southern extent of the Savanna Biome (shaded area). BDC = Border Cave; KRM = Klasies River Mouth; SKX = Swartkrans; WWK = Wonderwerk.

Savanna can exist in areas with relatively severe winter frosts provided that summer rainfall is low. The plants are mostly well adapted to withstand fire and drought, which means that they have remarkable persistence (Rutherford and Westfall, 1986). There is wide variation in the proportion of trees and bushes versus grasses in different areas.

The four sites to be discussed here range from low-altitude coastal to high-altitude continental locations, but all are located in a heterogeneous environment featuring hillsides and flats with a mix of closed and relatively open vegetation. The most northerly (and oldest) site is Swartkrans, which is located at an altitude of about 1,550 m. The cave is situated in the rocky hillsides of a shallow valley containing a small river. The area has been intensively farmed for a considerable time, but the natural vegetation of the area is grassland (veld type 61, Bankenveld of Acocks, 1988), with a wide variety of bushes and shrubs on the rocky hills and ridges; the nearby Krugersdorp Game Reserve comprises open grassland, bushveld, and woodland, which support many introduced game species (Stuart and Stuart, 1992). Climatic data for this and the other sites are given in table 31.1.

The second high-altitude inland site is Wonderwerk, which is a large cave cut into the foothills of the Kuruman Hills. It overlooks the Ghaap Plateau at an altitude of about 1,660 m. There is no permanent water in the immediate vicinity today, the nearest being a spring zone about 5 km to the south and a large sinkhole some 12 km to the southeast. The vegetation of the region is variably

Table 31.1. Modern Weather Data Relevant to Possible Past Conditions at Sites Discussed

Station	Co-ordinates	Altitude (m)	Rainfall[1]		Temperature[2]			
			ann.	sum.	max.	min.	range	< 0°C
Swartkrans	26°01′S; 27°44′E	1,550	—	—	—	—	—	—
Krugersdorp	26°06′S; 27°46′E	1,699	767	633	26.0	2.3	15.0	13.9
Zuurbekom	26°18′S; 27°48′E	1,578	664	554	26.6	−2.1	19.8	71.9
Brits	25°35′S; 27°49′E	1,158	621	522	30.3	1.3	20.2	23.6
Marico	25°30′S; 21°21′E	1,078	657	432	31.2	3.5	18.7	5.4
Wonderwerk	27°50′S; 23°33′E	1,660	± 420[3]	—	—	—	—	—
Kuruman	27°28′S; 23°26′E	1,312	455	357	31.5	1.1	18.7	25.9
Mancorp	28°18′S; 23°00′E	1,326	335	255	31.7	5.1	14.6	5.4
Sishen	27°47′S; 22°59′E	1,204	386	309	32.6	2.9	17.0	11.2
Koopmansfontein	28°12′S; 24°04′E	1,341	454	366	31.3	−0.1	19.4	52.0
Balkfontein	27°24′S; 26°30′E	1,280	564	454	30.2	0.7	19.7	40.4
Mafikeng	25°51′S; 25°39′E	1,278	553	462	30.4	3.0	17.5	12.3
Border Cave	27°51′S; 31°59′E	600	—	—	—	—	—	—
Siteki	26°27′S; 31°57′E	653	794	673	27.1	9.9	12.5	0
Pongola	27°23′S; 31°37′E	300	638	474	31.1	9.6	14.5	0.2
Klasies River Mouth	34°06′S; 24°24′E	8	—	—	—	—	—	—
Storms R. Mouth	34°02′S; 23°54′E	4	951	465	23.0	9.6	7.5	0
Cape St. Francis	34°12′S; 24°50′E	7	673	264	23.1	14.4	8.2	0

[1]ann. = mean annual rainfall in mm; sum. = mean annual rainfall for six summer months (September to March) in mm.

[2]max. = January mean daily maximum in °C; min. = mean July daily minimum in °C; range = maximum mean daily range in °C; < 0°C = mean annual number of days with minimum below 0°C.

[3]From Thackeray, 1984.

open or closed shrub savanna (veld type 16, Kalahari Thornveld of Acocks, 1988). It is to be noted that Wonderwerk lies within about 120 km southwest of Taung, the most southerly australopithecine site.

Border Cave is situated at a similar latitude to Wonderwerk but, unlike the latter, is approximately 75 km west of the Indian Ocean at an altitude of about 600 m. An important feature of the site is its location just below the crest of the Lebombo Mountains facing over a steep slope down to the Swaziland Lowveld at least 300 m below. This lower-lying area is readily accessible from the cave. Steep gradients result in varied vegetation that ranges from bush and forest (veld type 6, Zululand Thornveld of Acocks, 1988) to open *Acacia nigrescens–Sclerocarya–Themeda* Savanna (veld type 10, Lowveld of Acocks, 1988).

The southernmost site is Klasies River Mouth, which is situated some 6–8 m above sea level at the base of a cliff that rises steeply some 60–100 m to the coastal plain above. This plain, here about 10 km wide, is bounded on its landward side by a range of mountains averaging about 600 m in elevation and crossed by a number of rivers, including the eponymous Klasies River, which debouches into the sea about 1 km west of the site. Although the cave is currently within a few meters of the beach, it would have been farther from the shore at times when sea level was lower than it is today. Klasies River Mouth is situated in an area of botanical complexity, (Cowling, 1984), where the vegetation is notable for its mixed affinities and transitional nature. Significantly, there are several subtropical elements as well as tropical C_4 grasses (Moll et al., 1984) within the thicket, shrub, and grassland occurring in the area. In addition, the Afromontane Knysna Forest is located some 20 km to the west.

Periods Represented

The four sites represent discrete periods in hominid history within the last 2 million years (myr; table 31.2). At Swartkrans, deposits designed Members 1, 2, and 3 are thought to have been deposited between approximately 1.8 and 1.0 myr (Brain and Watson, 1992). There were possibly extended periods of erosion between each depositional phase, however, and the deposits may represent short periods relatively widely separated in time. The Middle Pleistocene deposits at Wonderwerk were initially thought to represent the period between about 450 kyr (kyr = thousand years) and 200 kyr (Beaumont,

Table 31.2. Approximate Periods Thought to Be Represented by Deposits at Sites Discussed in the Text

Site	First Occupation	Last Occupation	Source
Swartkrans	1.8 myr	1.0 myr	Brain and Watson, 1992
Wonderwerk	450 (or 800) kyr	200 (or 170) kyr	Beaumont, 1979 and pers. comm.
Border Cave	130 (or < 145) kyr	39 kyr	Grün et al., 1990a; Beaumont et al., 1992; Miller et al., 1992
Klasies River Mouth	120 kyr	70–40 kyr	Deacon and Geleijnse, 1988; Grün et al., 1990b; Thackeray, 1992

1979) but may span a period from approximately 800 to 170 kyr (P. B. Beaumont, pers. comm.). There are also Late Pleistocene and Holocene sequences (Beaumont, 1979), the latter of which has already been described in some detail (Avery, 1981; Van Zinderen Bakker, 1982; Humphreys and Thackeray, 1983) and will be referred to here for comparative purposes. Deposition at Border Cave probably began near the end of $\delta^{18}O$ stage 6 about 130 kyr (Grün et al., 1990a), although amino-acid dating has suggested that the earliest deposits may be up to 145 kyr old (Miller et al., 1992). Deposits continued to accumulate until about 21 kyr, although human occupation ceased around 39 kyr (Beaumont et al., 1992). Deposition at Klasies River Mouth is thought to have begun during $\delta^{18}O$ stage 5e (Thackeray, 1992), very generally at the same time as at Border Cave, and to have ended between about 70 kyr (Deacon and Geleijnse, 1988; Thackeray, 1992) and 40 kyr (Grün et al., 1990b) before the onset of the Last Glacial Maximum.

Material and Methods

The data base comprises minimum numbers of individual micromammals that are thought for a number of reasons detailed elsewhere (Avery, 1982) to have been accumulated by barn owls *Tyto alba*. Numbers are based on upper and lower jaws, in excavation units identified by the excavators. Where samples are small or greater generalization is required, scores from smaller units have been summed if this is justified by stratigraphic reality. Species or genera are characterized by the type of vegetation they inhabit and its place in the landscape. In addition,

Table 31.3. Data Base for Depositional Units of the Four Sites Studied

Units[1]	MNI[2]	Topography[3]			Vegetation[4]			Openness[5]		Dryness[6]		H[7]	Exotic Distributions[8]			
		F	H	W	G	S	T	C	O	D	M		A	B	C–E	F
Swartkrans																
Memb. 3-1	38	5.3	21.1	65.8	88.2	0.0	1.3	63.2	5.3	23.7	7.9	1.39	2.6	0.0	89.5	0.0
2	115	17.4	16.5	56.5	78.7	0.0	5.7	56.5	11.3	16.5	0.9	1.49	11.3	0.0	84.3	0.0
3	240	19.2	11.9	62.7	77.9	0.4	3.8	62.1	7.9	12.9	2.9	1.60	7.5	0.0	82.1	0.0
4	368	22.0	14.0	58.6	77.2	0.0	3.8	59.0	8.2	13.6	3.8	1.67	7.6	0.0	81.0	0.0
5	365	19.0	14.5	61.0	79.7	0.0	4.1	61.4	7.9	14.2	5.2	1.62	7.9	0.0	83.8	0.0
6	301	26.6	9.3	60.8	73.6	0.0	3.8	60.1	7.6	9.6	2.3	1.57	7.0	0.0	77.4	0.0
7	388	25.0	16.5	53.9	75.0	0.6	4.3	54.1	9.5	17.3	4.6	1.82	9.0	0.0	79.9	0.0
8	165	20.3	12.4	61.8	77.9	0.6	3.3	63.0	7.3	12.7	4.2	1.61	7.3	0.0	81.8	0.0
9	185	22.7	11.9	59.5	74.3	0.0	4.1	59.5	7.6	11.4	3.8	1.72	7.0	0.0	78.4	0.0
10	156	20.5	12.5	61.9	76.6	0.0	2.2	62.8	4.5	11.5	5.8	1.55	4.5	0.0	78.8	0.0
11	107	28.0	11.2	55.1	72.4	0.0	5.1	55.1	10.3	11.2	1.9	1.69	9.3	0.0	77.6	0.0
Memb. 2-1	78	23.1	10.3	57.7	71.8	0.6	4.5	55.1	9.0	12.8	2.6	1.86	7.7	0.0	76.9	0.0
2	176	31.8	11.6	50.9	67.0	0.0	4.5	50.0	10.2	12.5	2.3	1.76	9.1	0.0	71.6	0.0
3	226	34.1	15.5	46.5	69.0	0.4	7.5	46.0	15.5	16.4	3.5	1.85	15.0	0.0	77.0	0.0
4	87	31.0	12.1	53.4	71.8	0.0	6.3	51.7	14.9	13.8	2.3	1.76	12.6	0.0	78.2	0.0
5	202	22.8	12.1	59.7	75.7	0.0	3.5	59.9	6.9	11.9	1.5	1.50	6.9	0.0	79.2	0.0
6	201	26.1	10.7	58.2	74.6	0.5	5.0	58.2	11.4	11.4	2.0	1.58	10.4	0.5	80.1	0.0
7	20	30.0	15.0	55.0	75.0	0.0	0.0	55.0	0.0	15.0	0.0	1.11	0.0	0.0	75.0	0.0
8	97	26.8	12.4	52.6	71.1	0.0	4.1	53.6	8.2	11.3	5.2	1.86	8.2	0.0	75.3	0.0
9	43	30.2	7.0	60.5	70.9	0.0	3.5	60.5	7.0	7.0	2.3	1.50	7.0	0.0	74.4	0.0
Memb. 1-1	56	23.2	9.8	67.0	80.4	0.0	3.6	69.6	7.1	7.1	10.7	1.50	7.3	0.0	83.9	0.0
2	412	24.4	12.9	57.6	76.2	0.1	5.9	57.3	12.4	13.1	4.1	1.72	11.3	0.3	82.3	0.0
3	654	24.0	13.5	56.8	74.9	0.2	4.4	56.6	9.5	13.8	2.9	1.75	8.4	0.3	79.5	0.0
4	810	23.1	15.9	56.7	77.6	0.2	4.4	57.4	9.3	15.7	4.1	1.66	8.9	0.0	82.2	0.0
5	643	23.2	14.4	56.5	74.3	0.1	3.7	57.5	7.0	13.2	6.5	1.79	6.8	0.0	78.1	0.0
6	690	26.2	11.2	57.8	72.9	0.5	3.1	58.1	7.2	11.6	4.3	1.81	7.0	0.0	76.5	0.0
7	434	29.7	12.2	53.0	70.3	0.1	4.0	54.1	8.8	11.3	6.2	1.88	8.3	0.3	74.4	0.0
8	151	29.1	10.6	57.0	72.5	0.3	2.6	58.3	6.0	10.6	7.3	1.86	4.5	0.0	75.5	0.0
10	47	19.1	11.7	64.9	76.6	0.0	4.3	63.8	4.3	10.6	4.3	1.57	4.7	0.0	80.9	0.0
11	43	27.9	14.0	51.2	69.8	1.2	3.5	51.2	11.6	16.3	4.7	1.62	10.0	0.0	74.4	0.0
Wonderwerk																
6	87	77.0	10.3	9.2	30.5	4.6	7.5	12.6	33.3	36.8	8.0	2.50	0.0	0.0	42.5	0.0
7A	492	77.4	11.3	8.8	26.6	6.6	11.1	16.1	36.2	39.0	13.2	2.36	0.0	0.8	44.3	0.0
7B	403	74.9	11.9	10.2	26.8	6.1	12.3	17.9	35.5	35.0	15.4	2.36	0.0	0.0	45.2	0.0
7C	132	69.3	12.1	12.5	28.8	6.8	13.6	22.0	34.1	36.4	18.9	2.55	0.0	0.0	49.2	0.0
7D/E	95	62.1	9.5	21.1	32.1	2.6	17.9	37.9	18.9	25.3	31.6	2.31	0.0	0.0	52.6	0.0
8,9	76	68.4	11.8	14.5	26.3	5.3	14.5	27.6	26.3	30.3	18.4	2.46	0.0	0.0	46.1	0.0
10,11	78	62.2	9.6	16.7	26.9	6.4	14.1	29.5	25.6	28.2	20.5	2.53	0.0	0.0	47.4	0.0
12,13	267	44.4	6.9	23.2	25.8	2.4	12.5	41.6	11.6	12.0	21.0	2.27	0.0	0.0	40.8	0.0
14A	258	52.3	11.2	24.8	40.1	5.8	12.2	36.4	20.2	26.4	18.6	2.43	0.0	4.7	58.1	0.0
14B	458	63.8	9.0	12.7	29.5	5.3	8.4	17.9	32.3	34.1	9.4	2.51	0.0	0.7	43.2	0.0
14C	273	45.2	2.4	21.6	30.2	1.3	13.9	35.2	20.9	20.1	26.7	2.16	0.0	0.4	45.4	0.0
14DF	228	49.1	2.6	22.4	32.2	1.8	12.9	35.5	16.7	16.7	24.6	2.27	0.0	0.9	46.9	0.0
15,16	97	66.0	6.7	12.9	28.9	4.1	9.3	32.0	22.7	27.8	17.5	2.66	0.0	3.0	42.3	0.0
Border Cave																
1LRA	1849	71.5	16.2	10.1	27.9	5.3	14.0	35.7	11.4	10.6	29.2	2.39	1.2	6.4	47.1	0.0
1LRB	1070	75.0	13.0	10.3	26.6	1.7	11.6	37.9	3.8	3.5	33.4	2.14	0.1	7.2	39.9	0.0
1LRC	442	73.8	13.2	12.1	29.9	2.4	9.8	40.0	4.8	4.8	29.0	2.17	0.0	7.9	42.1	0.0
WA1	216	68.3	14.8	14.6	30.1	3.0	12.3	38.4	6.5	6.0	28.7	2.27	0.0	6.5	45.4	0.0
2UP	386	61.9	22.2	13.6	29.4	6.9	19.2	37.0	14.8	13.7	31.3	2.24	0.3	2.3	55.4	0.0
2LRB	365	66.6	17.3	14.0	27.4	4.2	21.8	46.3	8.8	8.5	43.0	2.26	0.0	4.9	53.4	0.0
WA2	242	54.1	32.9	11.8	27.1	3.5	15.3	51.7	7.4	7.0	49.2	2.06	0.0	2.5	45.9	0.0
BS3	149	53.4	20.1	23.8	33.2	3.4	21.8	57.7	6.7	6.7	51.7	2.20	0.0	6.0	58.4	0.0

Table 31.3. *(Continued)*

Units[1]	MNI[2]	Topography[3]			Vegetation[4]			Openness[5]		Dryness[6]		H[7]	Exotic Distributions[8]			
		F	H	W	G	S	T	C	O	D	M		A	B	C–E	F
WA3	172	62.8	15.7	17.4	32.0	0.0	18.6	55.8	1.2	0.0	54.7	2.29	0.0	12.2	50.6	0.0
BS4	237	74.1	8.0	15.8	34.8	0.0	17.1	46.4	2.1	0.0	43.9	2.48	0.0	11.0	51.9	0.0
WA4	1209	71.0	11.2	16.6	26.2	0.0	10.5	39.1	3.1	0.0	32.9	1.99	0.0	5.5	36.7	0.0
BS5	594	67.0	16.2	14.6	23.9	0.0	10.6	44.6	3.7	0.0	36.2	2.00	0.0	7.7	34.5	0.0
WA5	50	54.0	14.0	30.0	25.0	0.0	23.0	64.0	2.0	0.0	52.0	2.09	0.0	2.0	48.0	0.0
BS6	53	47.2	24.5	26.4	18.9	0.0	18.9	71.7	5.7	0.0	60.0	1.68	0.0	1.9	37.7	0.0
Klasies River Mouth																
E50	7328	36.1	44.2	19.7	77.0	0.1	8.1	45.0	0.0	19.2	44.8	1.62	0.0	0.1	85.2	0.5
H51	789	39.4	40.3	20.3	75.0	0.0	7.4	48.4	0.1	13.1	48.2	1.82	0.0	0.0	82.4	0.1
J51	980	37.0	44.3	18.6	75.8	0.0	7.4	44.7	0.0	19.0	44.2	1.63	0.0	0.0	83.2	0.5
K48	115	36.5	44.3	19.1	71.3	0.0	10.4	47.0	0.0	17.4	46.1	1.79	0.0	0.0	81.7	1.7
M50	455	34.6	42.6	22.7	72.2	0.2	10.4	56.3	0.0	10.8	56.0	1.71	0.0	0.0	82.9	1.3
T50	240	35.6	39.0	25.4	81.3	0.0	6.3	57.5	0.0	7.5	57.5	1.40	0.0	0.0	87.5	0.8
AA43	53	34.0	45.3	20.8	67.9	0.0	11.3	54.7	0.0	11.3	54.7	1.66	0.0	0.0	79.2	1.9

Note: Figures 31.3–31.8 are based on these data. The micromammalian taxa were characterized in terms of six ecological variables. The ecological scores for depositional units were then based on the relative abundances of individuals in ecological categories.

[1] Site units are depositional groupings identified by the excavators.

[2] MNI = minimum number of individuals per sample.

[3] Topography: F = flats; H = hillsides; W = waterside.

[4] Vegetation: G = grass; S = scrub; T = trees (and bushes).

[5] Openness: C = closed; O = open.

[6] Dryness: D = dry; M = moist.

[7] H = Shannon index of general diversity.

[8] Exotic distributions: A = xeric; B = relatively high temperature; C, D, and E = mesic; F = Mediterranean climate. See text and figure 31.3 for further details.

the general diversity of each sample was calculated, based on genera in this case to allow for correlation between sites, as a means of determining equability of climate. The proportion of the samples belonging to taxa no longer occurring near each site was also calculated, and the implications considered. The resulting data base is given in table 31.3.

Essentially six general distribution patterns can be discerned in modern micromammals (fig. 31.2). Climatic conditions associated with these patterns are given in tables 31.4 and 31.5. Three patterns are complementary in a general sense. The largest (A) of these can be designated the xeric pattern; the second (B) is the subtropical-mesic (savanna) pattern with a longer rainy season and mean annual rainfall above 500 mm; the smallest (C), in the southeast, is the temperate-mesic component. Of the three remaining patterns, one (D) is mesic, with both temperate and subtropical elements; another (E) is very similar except that it lacks the subtropical element of the previous pattern and is therefore restricted to rather cooler temperatures; the last (F) is equated with winter rainfall and fynbos vegetation. Currently, Swartkrans and Border Cave are located within distributions B, D, and E and near the boundary of C: Klasies River Mouth is situated within or near the boundary of all distributions, whereas Wonderwerk is associated only with distribution pattern A.

Results and Paleoenvironmental Interpretation

Results. Mystromys albicaudatus (the white-tailed rat) is the dominant species at Swartkrans and today occurs in particularly dense grass in the Transvaal (Rautenbach, 1982) but not necessarily elsewhere. This rat was treated as representative of the waterside habitat in all cases. At Swartkrans the waterside was consequently the dominant topographic element (fig. 31.3); at Wonderwerk and Border Cave the flats were best represented, whereas at Klasies River Mouth all three habitats were approximately equally represented. It is noticeable that there is rather more variation in geographic elements represented at Wonderwerk and Border Cave than at the other two sites, although proportions generally remain fairly

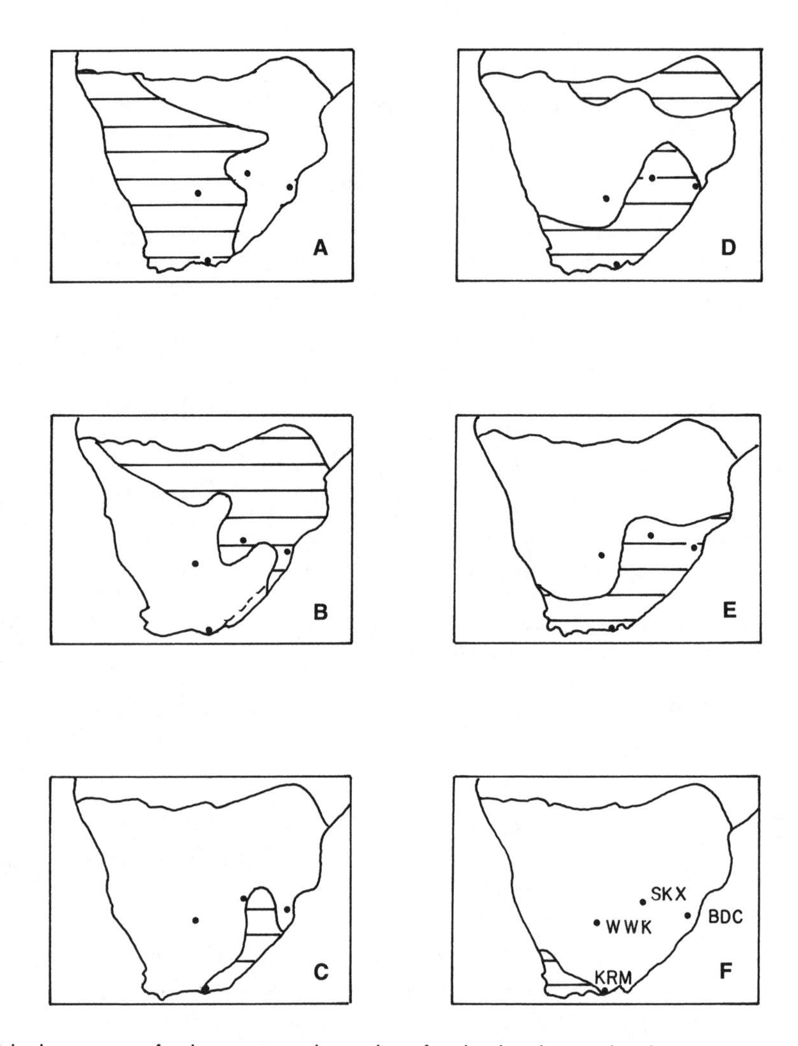

Fig. 31.2. Generalized distribution patterns of modern micromammals in southern Africa (based on Skinner and Smithers, 1990). See text and tables 31.2 and 31.3 for further details.

constant. In almost all cases grass is the dominant vegetational form, although the overall proportions vary from around 25 percent at Border Cave and Wonderwerk to around 75 percent at Klasies River Mouth and Swartkrans (fig. 31.4). Samples from Wonderwerk and Border Cave indicate some scrub and a higher proportions of trees and bush than do those from the other two sites. The remaining percentage, not shown on the charts, comprises those taxa whose preferences are not known, those taxa with catholic habits, and those (mainly burrowing) species that are constrained by ground type rather than by vegetation.

At Klasies River Mouth, Wonderwerk, and Border Cave the vegetation was more closed when the sites were first occupied (fig. 31.5) than it was by the time the sites were abandoned. This pattern was especially clear at Wonderwerk, where the open-habitat indicators are best represented in the latest samples. The pattern is less clear at Klasies River Mouth, where there is, in these general terms, virtually no representation of open vegetation. More detailed analysis indicated a period of relatively open vegetation while the site was occupied (Avery, 1987), but desertion of the site must have been due to other causes. At Swartkrans the pattern is complicated by the fact that the micromammalian samples do not represent the entire depositional sequence (C. K. Brain, pers. comm.), so no inferences can be made as to the situation at the beginning and end of deposition. There is, however, little variation between existing samples, so conditions may well have been relatively stable through-

Table 31.4. Moisture Characteristics Associated with Various Distribution Patterns of Micromammals in Southern Africa

Pattern[1]	Ann. Rain[2]	AAWD[3]	Rainy Season[4]	Jan. Precip[5]	Regions[6]
A	< 500	> 100	≤ 4 summer months (none > 125 mm)	< 100	3–6
B	± 500–1000	100–600	≥ 4 summer months (some > 125 mm)	> 100	mainly 3–5
C	gen. > 750	gen. < 400 (mainly < 100)	extended summer or year-round	± 100–200	gen. 1–3
D	gen. > 500	< 400 (surplus gen. > 100)	± year-round, winter, or extended summer	variable	1–4
E	gen. > 500	gen. < 400 (mainly < 200)	± year-round, winter, or extended summer	gen. < 150	gen. 1–4
F	125–750	gen. 200–600	≤ 5 winter months	< 50	gen. 3–5

[1]Generalized distribution patterns of various micromammalian species (based on Skinner and Smithers, 1990). See fig. 31.2.
[2]Mean annual rainfall in mm (based on Cooke, 1964).
[3]Average annual water deficit, in mm (based on Schulze and McGee, 1978).
[4]Number of months, or season, receiving ≥ 50 mm precipitation per month (based on Schulze and McGee, 1978).
[5]Mean January precipitation in mm (based on Schulze and McGee, 1978).
[6]Regions (based on Schulze and McGee, 1978): 1 and 2 = Humid; 3 and 4 = Subhumid; 5 = Semiarid; 6 = Arid.

out. The patterns of moist and dry representation (fig. 31.6) confirm the open-or-closed pattern. There is a tendency toward increasing aridity at the top of the sequence, although the pattern is less clear at Swartkrans. The total lack of arid indicator species in the lower third of the Border Cave sequence is particularly striking, and the trend is also clear at Klasies River Mouth.

If one accepts that high diversity indicates climatic equability, the Shannon index of general diversity indicates that Border Cave and Wonderwerk were situated in the most equable environments (fig. 31.7). Diversity was lower when Border Cave was first occupied than it was subsequently, possibly owing to a strongly seasonal climate with long dry winters. At Wonderwerk, on the other hand, the earliest sample indicates particularly equable conditions. Likewise, at Klasies River Mouth conditions appear to have been more equable at first than they were later. The lowness of values of the index at Swartkrans and Klasies River Mouth is largely due to the overwhelming dominance of one species at both sites.

At each site a proportion of the sample comprises species that would be considered exotic today because they no longer occur in the area (fig. 31.8). It is not, however, a simple case of one group replacing another even when, as at Swartkrans, both xeric and mesic elements have to be accommodated in any interpretation. In the case of Swartkrans it is most likely that different microhabitats are indicated, though with a climate that is intermediate between the various patterns. At Wonderwerk it is noticeable that the northern element is better represented in the earlier half of the sequence. Moreover, both it and the other two elements indicate a longer rainy season, with higher rainfall than is presently the case. At Border Cave there are two significant exotic elements. The lower part of the sequence is distinguished by the presence of species that presently occur well to the north of the site, whereas the later half includes an arid indicator species now occurring only to the west. The presence

Table 31.5. Temperature Characteristics Associated with Various Distribution Patterns of Micromammals in Southern Africa

Pattern[1]	Mean Temp.[2]	Mean Range[3]	Jan. Max.[4]	Regions[5]
A	mainly < 20	no correlation	no correlation	4–6
B	± 18–24	gen. < 14 (most < 12)	≤ 30	1–4
C	mainly < 18	gen. ≤ 12	≤ 30	mianly 4–6
D	mainly < 20	gen. ≤ 12	≤ 30	4–6 (mainly 5–6)
E	mainly < 18	gen. ≤ 12	≤ 30	mainly 4–6
F	mainly < 16	gen. ≤ 12	≤ 27.5	mainly 4–6

[1]Generalized sitrubition patterns of various micromammalian species (based on Skinner and Smithers, 1990). See fig. 31.2.
[2]Mean annual temperature in °C (based on Schulze and McGee, 1978).
[3]Mean annual range of temperature in °C (based on Schulze and McGee, 1978).
[4]January mean daily maximum temperature in °C (based on Schulze and McGee, 1978).
[5]Thermal regions (based on Schulze and McGee 1978); 1 = Tropical; 2 = Subtropical; 3 and 4 = Warmer temperate; 5 and 6 = Cooler temperate; 7 = Sub-alpine.

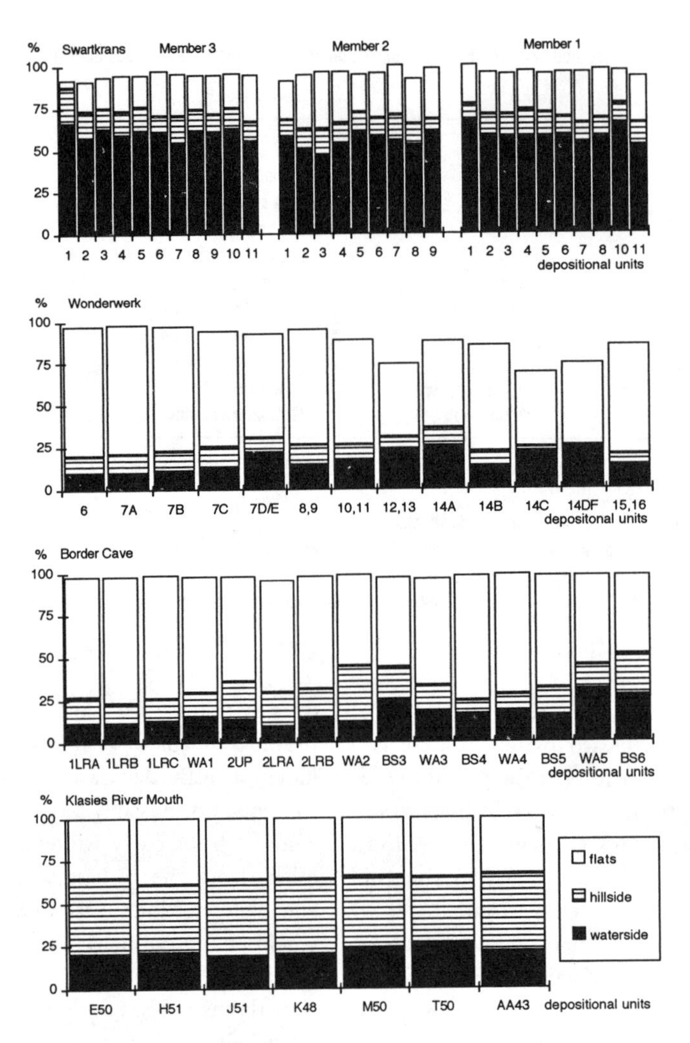

Fig. 31.3. Topographic distribution of vegetation, based on micromammalian evidence, with *Mystromys albicaudatus* representing the waterside element. Depositional units for each site are arranged from earliest, on the right, to latest, on the left.

of the northern element implies rather higher temperatures than those of today. There are no really significant exotic species at Klasies River Mouth, mainly because nearly all the distributional groupings fall close to the site; the proportion of species no longer shown to occur in the area is, in any case, low. Given, however, that both winter and summer rainfall regimes are represented, it would appear that, then as now, there was an extended rainfall season that was either bimodal or approximately year-round.

Swartkrans. The overall similarity of samples throughout the sequence is quite unlike the patterns found in later sites that cover glacial and interglacial conditions. Instead, it would appear that these data support the contention (Brain et al., 1988) that only interglacial conditions are represented. When the material from each member is taken as a unit, there is some indication that Member 2 represented slightly different conditions than the other two, but the difference is not significant.

In a general way, the micromammalian evidence supports that from other mammals, notably the bovids, in indicating that the vegetation of the area was relatively open throughout the period represented but that there was a water body in the area, as indicated by the presence of otters and hippopotamuses (Brain et al., 1988). The differences have to do with both the animals themselves and the predator species. If the emphasis of the bovids is

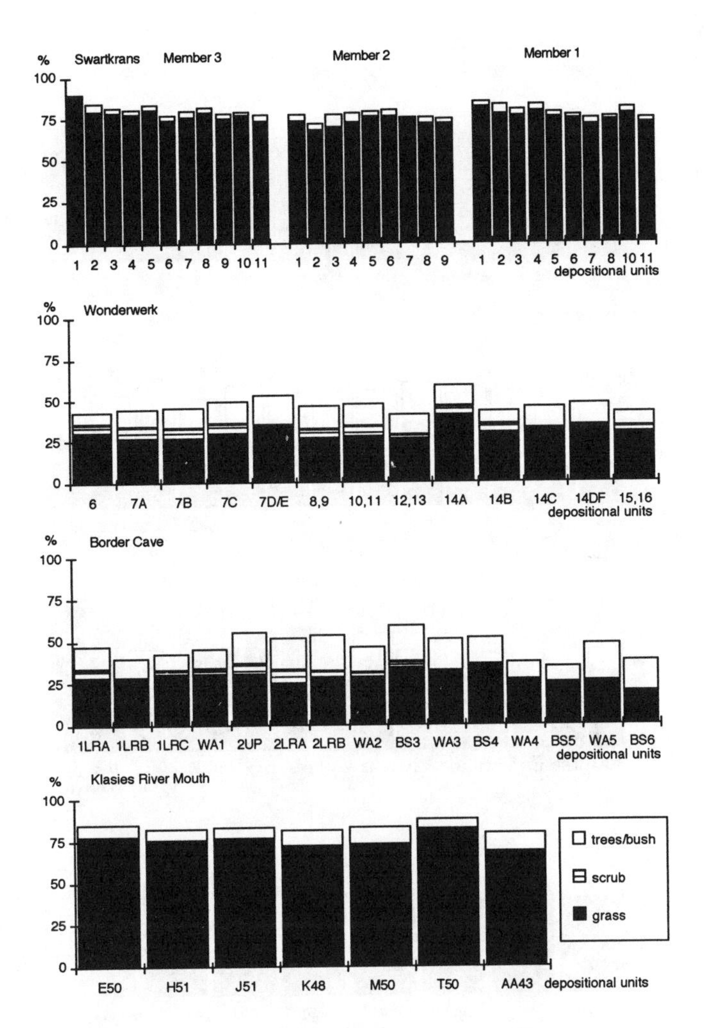

Fig. 31.4. Proportions of vegetation types, based on micromammalian evidence. Depositional units for each site are arranged from earliest, on the right, to latest, on the left.

on open savanna grassland, this is because the majority did not inhabit riverine vegetation but were probably preyed upon by large carnivores as they came to drink. Conversely, many of the micromammals in the samples would have inhabited the valley bottom. Moreover, the barn owl *Tyto alba* can be expected to overemphasize the waterside habitat for which it has a marked preference (Shawyer, 1987). Such a bias will have a particular bearing on interpretations concerning landscapes where there is a fairly sharp dichotomy between a restricted mesic waterside habitat and the general, relatively xeric environment. Thus, if the barn owl is the predator involved at Swartkrans, the riverine element, although important, was probably rather less extensive than the evidence indicates. Nevertheless, even after making allowance for bias, the floodplain must have been more prominent than it is at present (Brain and Watson, 1992). Semiaquatic vegetation close to the river would have given way to dense grass on marshy or damp ground that probably extended right across the valley floor. Beyond the valley there would have been open savanna woodland of a type categorized as tropical bush and savanna (bushveld) by Acocks (1988), such as presently occurs a minimum of 50 km north of Swartkrans and some 500 m lower in altitude. By analogy with present climate associated with the lower-altitude vegetation (table 31.1), rainfall may not have differed greatly from today's, falling during about a four-month summer rainy season. It is likely, however, to have been less effective because the mean daily maximum temperature for January was probably 4°–5°C higher than it is at present. In addition, the mean daily minimum temperature for July may have

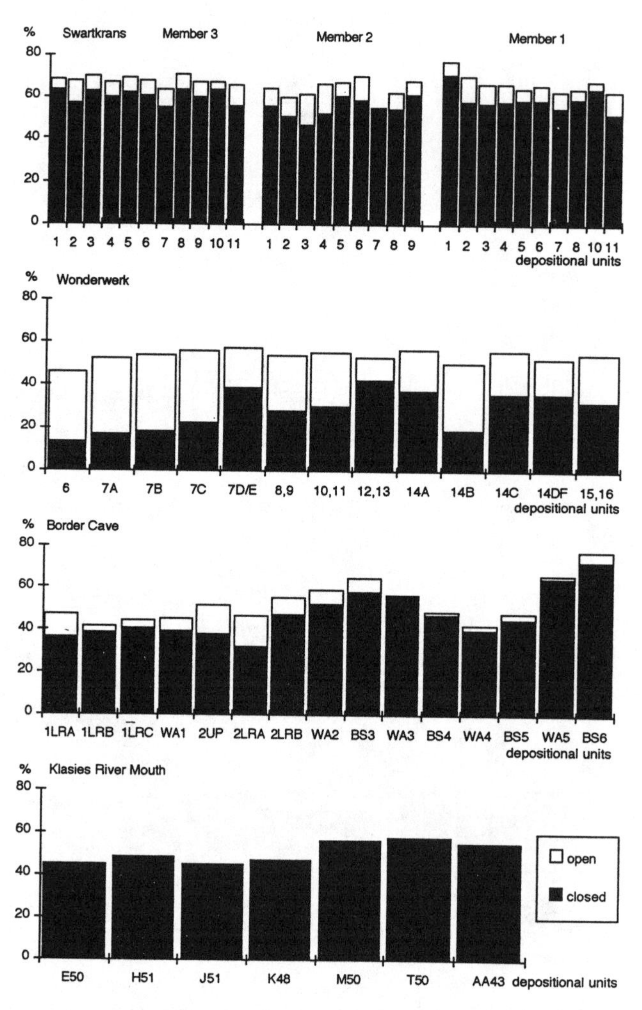

Fig. 31.5. Proportions of closed and open vegetation, based on micromammalian evidence. Depositional units for each site are arranged from earliest, on the right, to latest, on the left.

been as much as 3°C higher. Given an apparently fairly high summer rainfall, the latter factor would have been critical in allowing co-dominance of trees in the area, as was mentioned above.

Wonderwerk. The Middle Pleistocene vegetation was apparently similar to that of the Holocene and probably involved periodic shifts in variants of bush savanna (veld type 16, Kalahari Thornveld of Acocks, 1988) of the same general type postulated for Swartkrans. The vegetation probably varied between "an extremely open savanna of *Acacia eriobola* and *A. haematoxylon*, except along rivers and near ranges of hills and mountains" (Acocks, 1988), where vegetation is much denser, with bushes as well as trees and grass, and a generally dense *Tarchonanthus* bush savanna. If this suggestion is correct, Middle Pleistocene hominid occupation took place under interglacial conditions. Long-term decline in the waterside element, which indicates increasing desiccation, is suggested by an increase in open and dry vegetation and a concomitant decrease in the tree (and bush) element. It should be noted, however, that the overall proportions of physiognomic elements remain fairly constant. Conditions generally appear to have been at their most equable when the site was first occupied, and during the first third of the period of occupation the subtropical mesic element is most consistently represented. It seems likely that rainfall was around 500 mm and that there was a longer rainy season that is presently the case, with a consequent reduction in the average

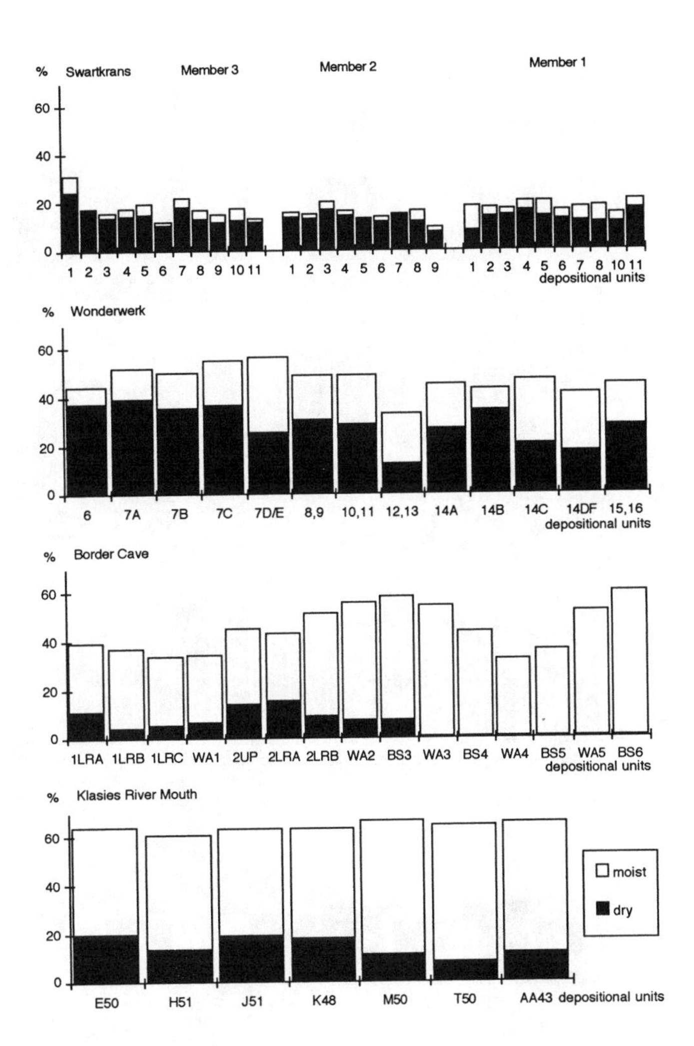

Fig. 31.6. Proportions of dry and moist vegetation, based on micromammalian evidence. Depositional units for each site are arranged from earliest, on the right, to latest, on the left.

annual water deficit. By analogy with modern weather data (table 31.1) it is probable that mean daily minima were higher than they are at present and that frost was less frequent, at least during the earlier part of the succession. Subsequent reduction in rainfall and lowering of temperature minima very probably caused this vegetation to be replaced by a form of *Tarchonanthus* bush savanna, as discussed above.

Border Cave. It is very clear that moist and closed vegetation types were at their optimum when the site was first occupied. Grass was at its lowest, and the waterside element was best represented in the lower half of the sequence. The micromammalian evidence indicates (Avery, 1992) that the vegetation around the site when it was first occupied was analogous to miombo (*Brachystegia*) woodland. Today, miombo is mainly a tropical vegetational form that does not occur within about 400 km of the site. At first there appears to have been relatively closed high-rainfall woodland with a sparse understory except along drainage lines. Thereafter, the vegetation became increasingly open, presumably owing to an overall decline in rainfall, until approximately 65 kyr ago, when a slight increase in rainfall took place. During the later part of its occupation, the site was apparently surrounded by an open *Acacia nigrescens–Sclerocarya* savanna. Vegetation in the lower-lying areas must have been more arid than that on the mountains, possibly analogous at first to the open *Acacia nigrescens–Sclerocarya* savanna and subsequently to mopane (*Colophospermum mopane*), which

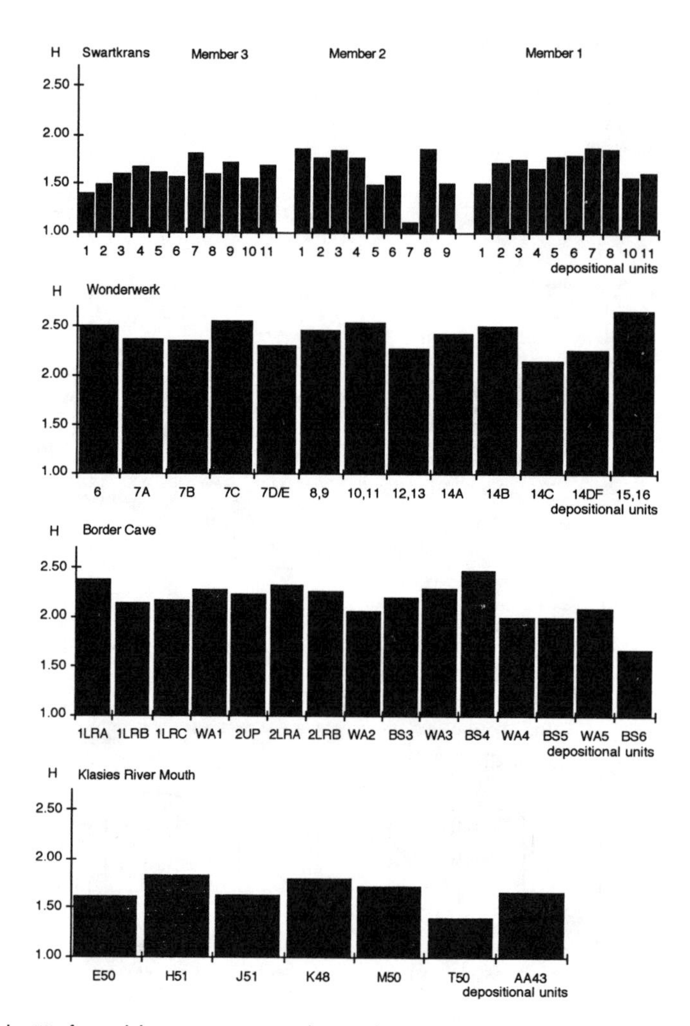

Fig. 31.7. Variation in Shannon index (H) of general diversity in micromammalian samples. Depositional units for each site are arranged from earliest, on the right, to latest, on the left.

currently occurs in the extreme north of South Africa and neighboring areas. It would appear that rainfall may initially have been as much as twice what it is today and more highly seasonal, with drier winters. Subsequently it fell to no more than about 60 percent of the present amount. Interestingly, the suggested changes in vegetation agree with those postulated by Cooke (1964) at 140–160 percent and 50–60 percent rainfall. Again, at this site as at the others, winter temperatures were probably higher, with a mean annual temperature range that was less at the beginning of the sequence than it was later.

Klasies River Mouth. The vegetation comprised a mosaic of changing proportions of grass, shrubland, and thicket that was physiognomically comparable to savanna vegetation occurring near the other sites. Indeed, savanna probably occurred in the area during the Late Pleistocene, when the site was first occupied; even today, the nearest savanna outlier is no more than 50 km away. Certainly the complexity of the present regional vegetation, which exhibits affinities with both subtropical and Afromontane types, provides evidence of past fluctuations and incursions from different areas (Cowling, 1984). Vegetation may have been a mid-dense grassy shrubland that gave way to a more open form at times when rainfall was lower and summers drier. The picture is almost certainly complicated by variation in the available land surface near the site, caused by changes in sea level, but grass apparently increased at the expense of trees and bushes later in the sequence. Even slight

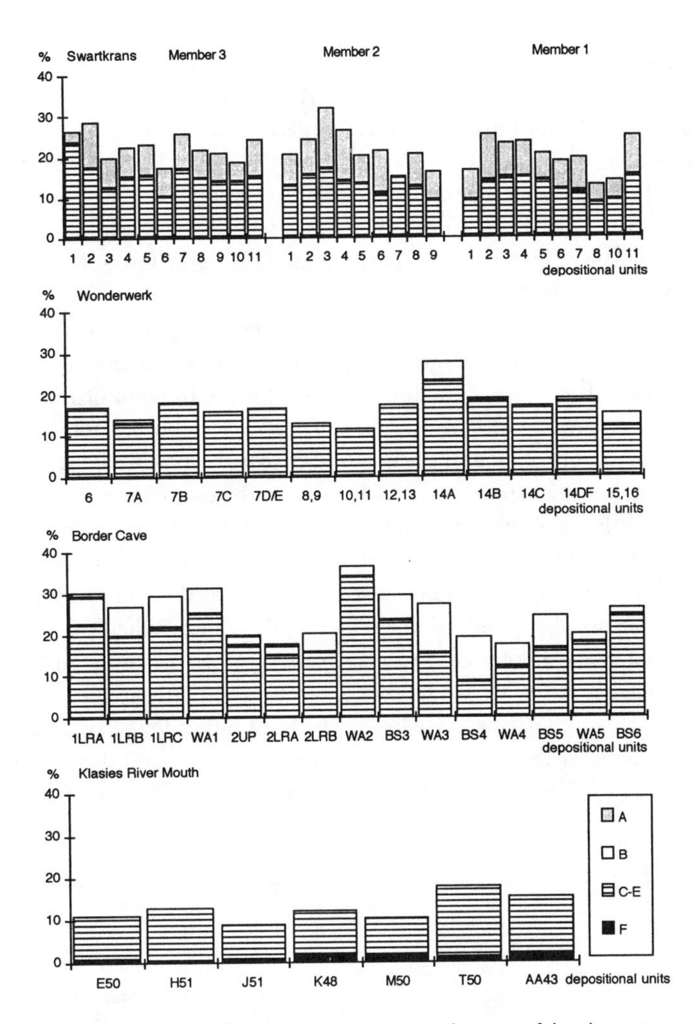

Fig. 31.8. Proportions of micromammalian samples comprised of species no longer occurring in the vicinity of the relevant site. A = xeric pattern; B = relatively high temperature pattern; C, D, and E = mesic patterns; F = Mediterranean climate pattern. See text, figure 31.2, and tables 31.2 and 31.3. for further details. Depositional units for each site are arranged from earliest, on the right, to latest, on the left.

changes in the season and amount of rainfall can change quite considerably both the structure of such mixed vegetation and the proportions of types represented. Rainfall probably remained between about 500 and 800 mm throughout the Late Pleistocene sequence at Klasies River Mouth; it was probably more seasonal than usual at various times, although it is unlikely that there was ever a clearly demarcated dry season. In general terms, the proportion of moist, closed vegetation showed a progressive decline.

Discussion

Timing. Dating of the four sites as presently understood can be read as supporting the micromammalian evidence that the sites were occupied during interglacial periods but not during major glacial periods. The suggested dates of ± 1.8 to 1.0 myr for the Swartkrans deposits coincide broadly with the period before the onset of strong glacial pulses (Burckle, chap. 16, this vol.). If accumulation episodes were of short duration (Brain, this vol.) they could well have occurred under pronounced interglacial conditions. If, on the other hand, they were originally of longer duration and deposits were subsequently reduced by erosion, the area may have been abandoned only under pronounced glacial conditions. At present this question does not seem resolvable. Dating of the latest Middle Pleistocene levels at Wonderwerk to about 170 kyr (P. B. Beaumont, pers. comm.) would imply abandonment of the site at the maximum of glacial

$\delta^{18}O$ stage 6, when a prolonged period of dry, cold conditions may well have made the region uninhabitable by humans. P. B. Beaumont (pers. comm.) maintains that the second highest Acheulean level is older than 350 kyr B.P. Unless there proves to be a major hiatus in the sequence, however, the micromammalian evidence tends to support a first-occupation date of ± 400 kyr, because there is no evidence of the glacial conditions that could be expected during $\delta^{18}O$ stages 12 and 16 (Burckle, this vol.). Clearly, the dating of the Middle Pleistocene levels at this site requires clarification. It is almost certainly no coincidence that the earliest occupation of both Border Cave and Klasies River Mouth occurred about the beginning of $\delta^{18}O$ stage 5, at a time when an apparently rapid increase in temperature in southeast Africa at the end of $\delta^{18}O$ stage 6 (Van Campo et al., 1990) must have been accompanied by a southward migration not only of vegetation types but of the animals that lived in them. Whereas the date of abandonment of Border Cave seems reasonably secure, that of Klasies River Mouth has not yet been firmly established, although it must have been earlier than the date for Border Cave in that deposits are beyond the range of ^{14}C dating. From an environmental point of view it is entirely possible that Klasies River Mouth was abandoned as early as 70 kyr, during the First Glacial Maximum, or $\delta^{18}O$ stage 4 (Avery, 1987; Deacon and Geleijnse, 1989), since it is to be expected that its effects would have been more severe at this southern location than at Border Cave, where people were able to remain until the Last Glacial Maximum, or $\delta^{18}O$ stage 2.

Distribution. All the australopithecine sites are located north of 28°S, the most southerly being Taung (see fig. 31.9 for the location of this and other sites mentioned below). Even allowing for the scarcity of known sites, the fact that the prolific Pliocene site of Langebaanweg in the southwestern Cape Province has yielded no hominids and virtually no primates (Grine and Hendey, 1981) must be significant. Lower Acheulean sites are likewise restricted to the more northerly part of the region (fig. 31.10). Upper Acheulean sites show a more extended pattern of distribution. Apart from the internal sector reaching the Ghaap Plateau near Wonderwerk and a high-altitude site in Namibia (Wendt, 1972), there is a series of sites round the coastal margin as far as central Namibia. Significantly, even today closed, large-leafed

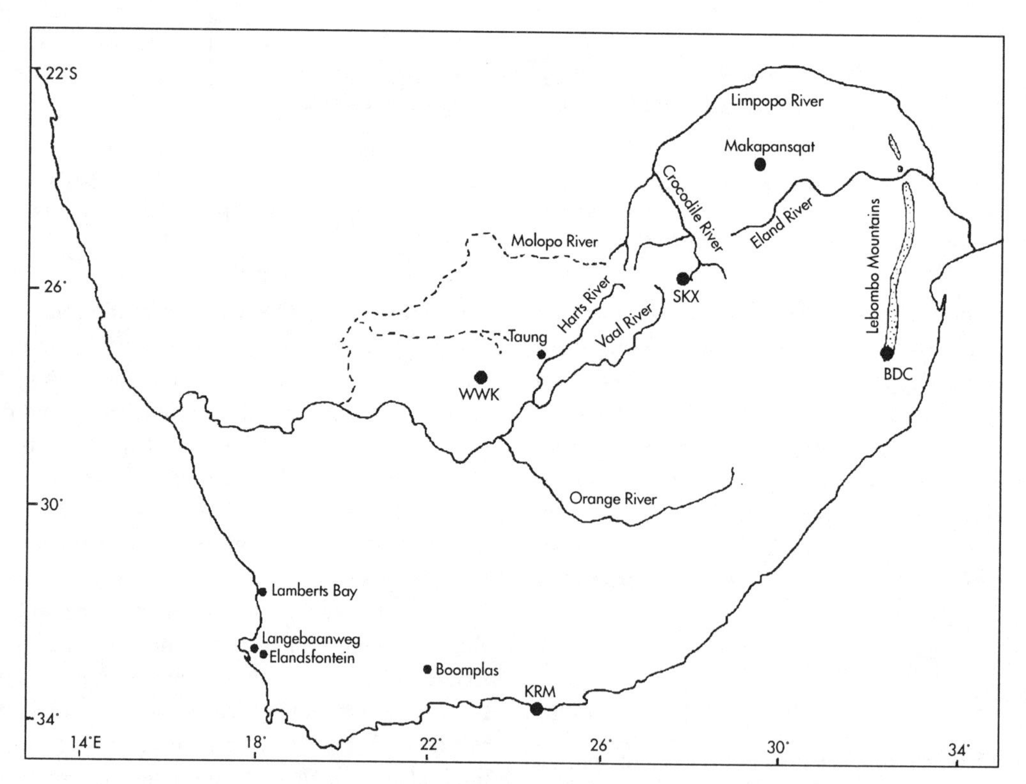

Fig. 31.9. Location of other sites and features mentioned. Dashed lines indicate generally dry riverbeds.

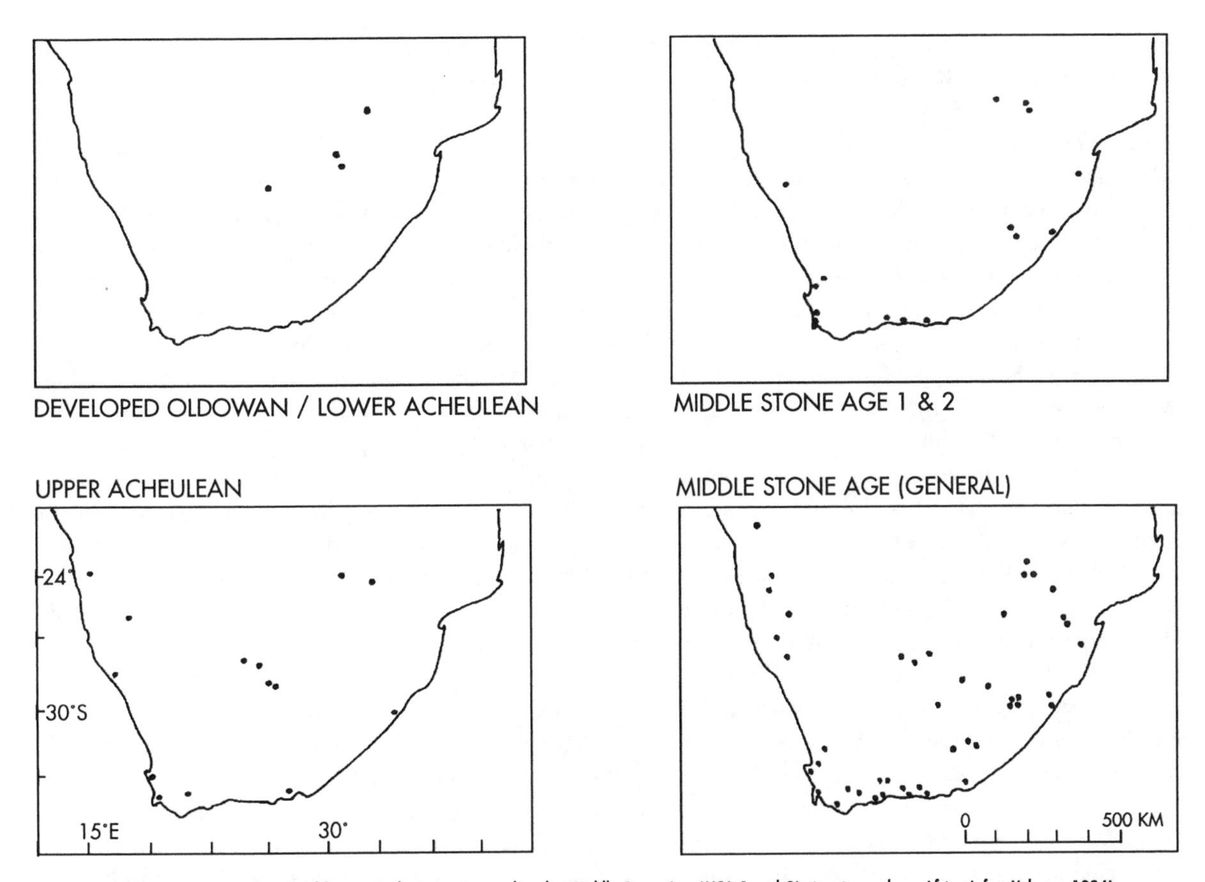

Fig. 31.10. Distribution of known Early Stone Age and earlier Middle Stone Age (MSA 1 and 2) sites in southern Africa (after Volman, 1984).

shrublands of essentially tropical and subtropical affinities extend from the southeast, near Klasies River Mouth, westward along the coast as far as Lamberts Bay (Cowling, 1984). These shrublands form thickets similar to others found throughout tropical and subtropical Africa. Elandsfontein, the most westerly of the South African Acheulean sites, lies within this area of distribution, and the vegetation there is thought to have been more grassy during the Middle Pleistocene than it is today (Klein, 1978), suggesting a savanna aspect to the vegetation. According to Volman (1984), the earliest Middle Stone Age (MSA1; Late Pleistocene) sites are also located in the southwestern Cape Province within this zone. MSA2 sites, which include Border Cave and Klasies River Mouth, show a similar distribution pattern to that of Acheulean sites, except that they have not been found on the Ghaap Plateau. Later and unspecified MSA sites are located in approximately the same area, with the addition of the higher-altitude eastern, presently grassland region.

The location of early archaeological sites generally indicates an avoidance of land under about 1,000 m in the more northerly interior. Present climate patterns suggest that this area would comprise the arid part of the subcontinent, notably in the western interior, where there are large annual water deficits. The consequent lack of surface water might well have rendered the region unsuitable for hominid occupation (Deacon, 1975). The existence of sites in currently arid western Namibia needs to be investigated from this point of view, but it could be significant that even today water deficiency is not as great as would be expected from the low annual rainfall. Sites occur mostly in areas that today have a moderate annual water deficit, as is shown particularly by the distribution of Late Pleistocene sites given by Parkington (1990). Moreover, large mammal faunas from Acheulean sites in southern Africa represent unusually moist conditions (Klein, 1988), which could support a contention that such sites were chosen preferentially.

So far there is no clear evidence to indicate where Early and Middle Pleistocene hominids may have been during times when they apparently left the areas discussed. During the apparently long periods at Swartkrans not represented by deposits, the entire area may have been abandoned by hominids. If this were the case, one would predict a contraction of savanna vegetation and its animal occupants to the lower-altitude valleys of the Eland or Crocodile Rivers, which flow into the Limpopo (fig. 31.9). At Wonderwerk, low artifact densities in the Acheulean deposits (Beaumont, 1979) may reflect periodic ephemeral occupation, perhaps during summer. The time scale of such occupations needs to be established, but such putatively short-term movements could have been to the valley of the Vaal and Harts Rivers. Longer-term resettlement may have been along the Molopo River, to the north, or the Orange River, to the south. Indeed, the distribution of sites near drainage lines (fig. 31.9) supports the hypothesis that, as an extension of the Rift Valley lake system (Schrenk et al., 1993) and as implied in Ethiopia, where early sites are concentrated along stream banks (Harris, 1983), these lines were of prime importance in the movements of Early and Middle Pleistocene hominids.

A Late Pleistocene example of a possible glacial refugium is Boomplaas Cave in the interior of the southern Cape Province. This site is situated in the foothills of the Swartberg Mountains, and its protected location would have provided a haven for humans and other animals under extreme glacial conditions (Avery, 1982). That this site was not occupied until about 80 kyr further supports the suggestion that its attractions were recognized only under the harsh conditions first experienced during the First Glacial Maximum. Although it is not without the bounds of possibility that the people from Klasies River Mouth retreated to Boomplaas, or a similar site, no evidence has yet been found to indicate where the occupants of Border Cave went during the Last Glacial Maximum. It could, however, be predicted that they withdrew northward along the Lebombo Mountains, possibly to the valley of the Limpopo River, just as movements into and out of this valley have been documented during the Iron Age (Huffman, 1993).

Hominid Paleoecology. According to Cerling (1992), the development of savanna grasslands in East Africa is a relatively recent phenomenon, with C_4 grasses reaching a peak about 1.7 myr; before that time the vegetation was probably more wooded than has previously been suggested, that is, grassy woodland or shrubland rather than wooded grassland. This hypothesis would imply that early hominids inhabited a wooded environment and only subsequent forms adapted to a more open savanna grassland. As this type of vegetation expanded, the woodland habitat of the hominids would have become reduced, presumably to gallery forest or riverine bush, as has been postulated in Ethiopia (Wesselman, 1984 and this vol.). The hominids would have had a choice, as Vrba (1985) has pointed out; they could pursue a generalist strategy that made use of both old and new habitats or they could specialize in the new habitat.

The apparent ability of two hominid genera to coexist at Swartkrans for some considerable time supports the idea (Klein, 1989) that the two species adapted in different ways to different microhabitats (Grine, 1985), with the *Homo* lineage developing a generalist habit. After *Australopithecus africanus,* which is thought to have inhabited forest or forest margin, as at Makapansgat (Cadman and Rayner, 1989), the *Homo* lineage probably gradually expanded from the riverine habitat to surrounding drier, more open vegetation, such as that eventually colonized by *Homo erectus* (Klein, 1989). The location of early archaeological sites in Ethiopia confirms that higher and drier open areas were increasingly occupied after about 1.5 myr (Harris, 1983). The robust australopithecines, on the other hand, apparently became specialist frugivores, focusing on more open vegetation at an early stage (Grine, 1985) and presumably coming down to the river only to drink, much as do baboons today. In either case, however, the presence of water nearby must have remained a prerequisite.

The location of all the sites in heterogeneous environments supports the suggestion that hominines adopted a generalist strategy. Even at Wonderwerk, where the divisions are less dramatic today, there is a clear difference between the biota of the Kuruman Hills and that of the surrounding plains. Although both Late Pleistocene sites were situated in closed, mesic vegetation when they were first occupied, a more open microhabitat nearby was probably also utilized by humans. Indeed, at Border Cave such utilization may have been preferential, because many of the large mammals likely to have been brought into the cave by humans would have been found on the plains below the site (Klein, 1977). The evidence for Klasies River Mouth needs further consideration, but there is some indication of hunting in more open vegetation (Klein, 1976).

Although the two lineages apparently developed different strategies, in a more general sense both were clearly adapted to the same conditions. There are, of course, many taphonomic difficulties to compound the problems of interpretation, but the present evidence strongly suggests that all four sites discussed were located in similar environments. The micromammalian evidence from both Swartkrans and Wonderwerk indicates that Early and Middle Pleistocene hominids inhabited areas with a moderate climate, that is, with winter minima around 3°C, summer maxima of about 30°C and a mean daily temperature range of 18°C; rainfall, which was probably around 500 mm per annum, would have been restricted mainly to a summer rainy season of about four months. The evidence from Border Cave and Klasies River Mouth suggests that in South Africa anatomically modern humans may initially have inhabited more-mesic habitats than did their predecessors, but this impression would be due to the smallness of the data base. In all cases this equable climate supported a larger or smaller riverside component in savanna, or its physiognomic equivalent, in an environment that would provide a diverse and predictable supply of food. Further, the consistent lack of evidence indicating glacial conditions would suggest that before the Late Pleistocene such conditions rendered Africa south of about 28°S unsuitable for hominid occupation. The corollary of this hypothesis is that hominids were savanna endemics who were able to move to higher latitudes or altitudes only at such times as savanna vegetation expanded to those areas. Even later evidence suggests that human populations in southern Africa were reduced during the Last Glacial Maximum (Deacon and Thackeray, 1984) and that parts of the interior were abandoned during the mid-Holocene altithermal (Deacon, 1974). Iron Age farmers were also obliged to desert certain areas of southern Africa when conditions became unfavorable (Huffman, 1993); even today certain areas are being abandoned for similar reasons. Many questions remain to be answered, but these preliminary investigations have produced patterns that are sufficiently suggestive to encourage further work in this direction.

Acknowledgments

I thank the organizers of the conference, and in particular Dr. E. S. Vrba, for their invitation to participate and for logistical support. The Foundation for Research Development covered travel expenses and supported the research on which this chapter is based. Permission to attend the conference was given by the Council of the South African Museum.

References

Acocks, J. P. H. 1988. Veld types of South Africa. 3d ed. *Memoirs of the Botanical Survey of South Africa* 57:1–146.

Avery, D. M. 1981. Holocene micromammalian faunas from the northern Cape Province, South Africa. *South African Journal of Science* 77:265–273.

———. 1982. Micromammals as palaeoenvironmental indicators and an interpretation of the Late Quaternary in the southern Cape Province, South Africa. *Annals of the South African Museum* 85:183–374.

———. 1987. Late Pleistocene coastal environment of the southern Cape Province of South Africa: Micromammals from Klasies River Mouth. *Journal of Archaeological Science* 14: 405–421.

———. 1992. The environment of early modern humans at Border Cave, South Africa: Micromammalian evidence. *Palaeogeography, Palaeoclimatology, Palaeoecology* 91:71–87.

Beaumont, P. B. 1979. A first account of recent excavations at Wonderwerk Cave. Unpubl. abstract, Biennial Conference of the Southern African Association of Archaeologists, Cape Town, September 1979.

Beaumont, P. B., Miller, G. H., and Vogel, J. C. 1992. Contemplating old clues to the impact of future greenhouse climates in South Africa. *South African Journal of Science* 88:490–498.

Brain, C. K., Churcher, C. S., Clark, J. D., Grine, F. E., Shipman, P., Susman, R. L., Turner, A., and Watson, V. 1988. New evidence of early hominids, their culture and environment from the Swartkrans cave, South Africa. *South African Journal of Science* 84:828–835.

Brain, C. K., and Watson, V. 1992. A guide to the Swartkrans early hominid cave site. *Annals of the Transvaal Museum* 35: 343–365.

Cadman, A., and Rayner, R. J. 1989. Climatic change and the appearance of *Australopithecus africanus* in the Makapansgat sediments. *Journal of Human Evolution* 18:107–113.

Cerling, T. E. 1992. Development of grasslands and savannas in East Africa during the Neogene. *Global and Planetary Change* 5:241–247.

Cooke, H. B. S. 1964. The Pleistocene environment in southern Africa. In *Ecological studies in southern Africa,* pp. 1–23 (ed. D. H. S. Davis). Junk, The Hague.

Cowling, R. M. 1984. A syntaxonomic and synecological study in the Humansdorp region of the Fynbos Biome. *Bothalia* 15: 175–227.

Deacon, H. J. 1975. Demography, subsistence, and culture during the Acheulian in southern Africa. In *After the australopithecines,* pp. 543–569 (ed. K. W. Butzer and G. L. Isaac). Mouton, Paris.

Deacon, H. J., and Geleijnse, V. B. 1988. The stratigraphy and sedimentology of the main site sequence, Klasies River, South Africa. *South African Archaeological Bulletin* 43:5–14.

Deacon, H. J., and Thackeray, J. F. 1984. Late Pleistocene environmental changes and implications for the archaeological record in southern Africa. In *Late Cainozoic palaeoclimates of the Southern Hemisphere*, pp. 375–390 (ed. J. C. Vogel). Balkema, Rotterdam.

Deacon, J. 1974. Patterning in radiocarbon dates for the Wilton/Smithfield complex in southern Africa. *South African Archaeological Bulletin* 29:3–18.

Grine, F. E. 1985. Was interspecific competition a motive force in early hominid evolution? In *Species and speciation* (ed. E. S. Vrba). *Transvaal Museum Monograph* 4:143–152.

Grine, F. E., and Hendey, Q. B. 1981. Earliest primate remains from South Africa. *South African Journal of Science* 77:374–376.

Grün, R., Beaumont, P. B., and Stringer, C. B. 1990a. ESR dating evidence for early modern humans at Border Cave in South Africa. *Nature* 344:537–539.

Grün, R., Shackleton, N. J., and Deacon, H. J. 1990b. Electron-spin-resonance dating of tooth enamel from Klasies River Mouth Cave. *Current Anthropology* 31:427–432.

Harris, J. W. K. 1983. Cultural beginnings: Plio-Pleistocene archaeological occurrences from the Afar, Ethiopia. *African Archaeological Review* 1:3–31.

Huffman, T. N. 1993. Climatic change during the Iron Age. Unpubl. abstract, Four Million Years of Hominid Evolution in Africa, an International Congress in Honour of Dr. Mary Douglas Leakey's Outstanding Contribution in Palaeoanthropology, Arusha, Tanzania.

Humphreys, A. J. H., and Thackeray, A. I. 1983. Ghaap and Gariep. *South African Archaeological Society Monograph Series* 2:1–328.

Klein, R. G. 1976. The mammalian fauna of the Klasies River Mouth sites, southern Cape Province, South Africa. *South African Archaeological Bulletin* 31:75–98.

———. 1977. The mammalian fauna from the Middle and later Stone Age (later Pleistocene) levels of Border Cave, Natal Province, South Africa. *South African Archaeological Bulletin* 32:14–27.

———. 1978. The fauna and overall interpretation of the "Cutting 10" Acheulean site at Elandsfontein (Hopefield), southwestern Cape Province, South Africa. *Quaternary Research* 10:69–83.

———. 1988. The archaeological significance of animal bones from Acheulean sites in southern Africa. *African Archaeological Review* 6:3–25.

———. 1989. *The human career.* University of Chicago Press, Chicago.

Meester, J., Rautenbach, I. L., Dippenaar, N. J., and Baker, C. M. 1986. Classification of southern African mammals. *Transvaal Museum Monograph* 5:1–359.

Miller, G. H., Beaumont, P. B., Jull, A. J. T., and Johnson, B. 1992. Pleistocene geochronology and palaeothermometry from protein diagenesis in ostrich eggshells: Implications for the evolution of modern humans. *Philosophical Transactions of the Royal Society of London*, ser. B, 337:194–157.

Moll, E. J., Campbell, B. M., Cowling, R. M., Bossi, L., Jarman, M. L., and Boucher, C. 1984. A description of major vegetation categories in and adjacent to the Fynbos Biome. *South African National Scientific Programmes Report* 83:1–29.

Parkington, J. 1990. A critique of the consensus view on the age of the Howieson's Poort assemblages in South Africa. In *The emergence of modern humans*, pp. 34–55 (ed. P. Mellars). Cornell University Press, Ithaca.

Pocock, T. N. 1987. Plio-Pleistocene mammalian microfauna in southern Africa: A preliminary report including description of two new fossil muroid genera (Mammalia: Rodentia). *Palaeontologica Africana* 26:69–91.

Rautenbach, I. L. 1982. *The mammals of the Transvaal.* Ecoplan, Pretoria.

Rutherford, M. C., and Westfall, R. H. 1986. Biomes of southern Africa: An objective categorization. *Memoirs of the Botanical Society of South Africa* 54:1–98.

Schrenk, F., Bromage, T. G., Betzler, C. G., Ring, U., and Yuwayeyi, Y. M. 1993. Oldest *Homo* and Pliocene biogeography of the Malawi Rift. *Nature* 365:833–836.

Schulze, B. R., and McGee, O. S. 1978. Climatic indices and classification in relation to the biogeography of southern Africa. In *Biogeography and ecology of southern Africa* (ed. M. J. A. Werger). *Monographiae Biologicae* 31:19–52.

Shawyer, C. R. 1987. *The Barn Owl in the British Isles.* The Hawk Trust, London.

Skinner, H. J., and Smithers, R. H. N. 1990. *The mammals of the southern African subregion.* 2d ed. University of Pretoria, Pretoria.

Stuart, C., and Stuart, T. 1992. *Guide to southern African game and nature reserves.* 2d ed. Struik, Cape Town.

Taylor, P. J., Meester, J., and Kearney, T. 1993. The taxonomic status of Saunders' vlei rat, *Otomys saundersiae* Roberts (Rodentia: Muridae: Otomyinae). *Journal of African Zoology* 107:571–596.

Thackeray, A. I. 1992. The Middle Stone Age south of the Limpopo River. *Journal of World Prehistory* 6:385–440.

Thackeray, J. F. 1984. Man, animals and extinctions: The analysis of Holocene faunal remains from Wonderwerk Cave, South Africa. Ph.D. diss., Yale University.

van Campo, E., Duplessy, J. C., Prell, W. L., Barratt, N., and Sabatier, R. 1990. Comparison of terrestrial and marine temperature estimates for the past 135 kyr off southeast Africa: A test for GCM simulations of palaeoclimate. *Nature* 348:209–212.

van Zinderen Bakker, E. M. 1982. Pollen analytical studies of the Wonderwerk Cave, South Africa. *Pollen et Spores* 24:235–250.

Volman, T. P. 1984. Early prehistory of southern Africa. In *Southern African prehistory and palaeoenvironments*, pp. 169–395 (ed. R. G. Klein). Balkema, Rotterdam.

Vrba, E. S. 1985. Ecological and adaptive changes associated with early hominid evolution. In: *Ancestors: The hard evidence*, pp. 63–71 (ed. E. Delson). Alan R. Liss, New York.

Wendt, W. E. 1972. Preliminary report on an archaeological research programme in South West Africa. *Cimbebasia*, ser. B, 2: 1–61.

Wesselman, H. B. 1984. The Omo micromammals: Systematics and paleoecology of early man sites in Ethiopia. *Contributions to Vertebrate Evolution* 7:1–219.

Appendix 31.1. Micromammalian Taxa Represented in the Four Sites Studied

Taxon[1]	Common Name	SKX3[2]	SKX2	SKX1	WWK	BDC	KRM
Insectivora							
Myosorex varius	forest shrew				x	x	x
Myosorex cafer	dark-footed forest shrew	cf.	cf.	cf.		x	
Crocidura flavescens	greater musk shrew				x	x	x
Crocidura hirta	lesser red musk shrew					x	
Crocidura cyanea	reddish-gray musk shrew				x	x	x
Crocidura fuscomurina	tiny musk shrew				x	x	
Suncus varilla	lesser dwarf shrew	x	x	x	x	x	
Suncus infinitesimus	least dwarf shrew					x	x
Chrysochloridae indet.	golden moles				x		
Chrysospalax villosus	rough-haired golden mole	cf.	cf.	cf.			
Chlorotalpa duthiae	Duthie's golden mole						x
Amblysomus gunningi	Gunning's golden mole	cf.	cf.	cf.			
Amblysomus hottentotus	Hottentot golden mole					x	x
Chiroptera							
Chiroptera indet.	bats				x		
Nycteris thebaica	Egyptian slit-faced bat					x	
Rhinolophus hildebrandtii	Hildebrandt's horseshoe bat					cf.	
Rhinolophus clivosus	Geoffroy's horseshoe bat					x	
Rhinolophus darlingi	Darling's horseshoe bat					?	
Rhinolophus capensis	Cape horseshoe bat	cf.	cf.	cf.	x		
Miniopterus schreibersi	Schreiber's long-fingered bat				x	x	
Myotis tricolor	Temmink's hairy bat			x		x	
Chalinolobus variegatus	butterfly bat					x	
Eptesicus hottentotus	long-tailed serotine bat		cf.	cf.		cf.	
Eptesicus capensis	Cape serotine bat				x	cf.	
Scotophilus dinganii	yellow house bat					x	
Tadarida sp.	free-tailed bat					x	
Rodentia							
Cryptomys hottentotus	common molerat	x	x	x	x	x	x
Georychus capensis	Cape molerat	x	x	x			x
Otomys laminatus	laminate vlei rat					x	x
Otomys angoniensis	Angoni vlei rat				x	x	
+ *Otomys karoensis*	Karoo vlei rat	cf.	cf.	cf.			x
Otomys irroratus	vlei rat						x
Desmodillus auricularis	short-tailed gerbil				x		
Gerbillurus paeba	hairy-footed gerbil				x		
Tatera sp.	gerbil	x	x	x	x		
Tatera leucogaster	bushveld gerbil					cf.	
Mystromys albicaudatus	white-tailed rat	cf.	cf.	cf.	x	x	x
* *Proodontomys cookei*		x	x	x			
Saccostomus campestris	pouched mouse					x	
Dendromus sp.	climbing mouse	x	x	x			
Dendromus melanotis	gray climbing mouse				x	x	x
Dendromus mesomelas	Brant's climbing mouse						x
Dendromus mystacalis	chestnut climbing mouse					x	
Malacothrix typica	large-eared mouse	cf.	cf.	cf.	x	x	
Steatomys sp.	fat mouse	x	x	x			
Steatomys pratensis	fat mouse					x	
Steatomys krebsii	Kreb's fat mouse				x		
Pelomys fallax	groove-toothed rat					x	
Acomys sp.	spiny mouse			x	x		
Acomys subspinosus	Cape spiny mouse						x
Lemniscomys rosalia	single-striped mouse					x	

(*continued*)

Appendix 31.1. (*Continued*)

Taxon[1]	Common Name	SKX3[2]	SKX2	SKX1	WWK	BDC	KRM
Rhabdomys pumilio	striped mouse	cf.	cf.	cf.			x
Dasymys incomtus	water rat	cf.	cf.	cf.	x	x	x
Grammomys dolichurus	woodland mouse					x	x
Mus sp.		x	x	x			
Mus triton	gray-bellied pygmy mouse					x	
Mus minutoides	pygmy mouse				x	x	x
Mastomys sp.		x	x	x			
Mastomys natalensis/coucha	multimammate mouse				x	x	
Myomyscus verreauxi	Verreaux's mouse						x
Thallomys sp.		x	x	x			
Thallomys paedulcus	tree rat					x	x
Aethomys sp.		x	x	x			
Aethomys chrysophilus	red veld rat					x	
Aethomys namaquensis	Namaqua rock mouse				x	x	
Graphiurus sp.	dormouse	x	x	x			
Graphiurus ocularis	spectacled dormouse						x
Graphiurus murinus	woodland dormouse					x	
Macroscelididae	elephant-shrews				x		
Macroscelides proboscideus	round-eared elephant-shrew	cf.	cf.	cf.			x
Elephantulus spp.		x	x	x			
Elephantulus myurus	rock elephant-shrew				x	cf.	

[1]Taxonomic list according to Meester et al. (1986), except for the extinct taxon (*), which is based on Pocock (1987), and the revised species (+), which is based on Taylor et al. (1993).

[2]See figure 31.1 for site abbreviations.

Chapter 32

Exploring Ungulate Diversity, Biomass, and Climate in Modern and Past Environments

J. Francis Thackeray

Evidence has previously been presented to show that changes in the distribution and abundance of ungulates in modern and paleocommunities are likely to be associated with changes in habitat resulting from environmental change (Greenacre and Vrba, 1984; Thackeray, 1985; Vrba, 1992). The nature of relationships between climatic and biotic variables deserves to be explored further. In this chapter, I analyze estimated ungulate biomass and species richness (the number of species within each sample area) for modern faunal communities in sub-Saharan Africa. Biomass is plotted in relation to mean annual temperature and rainfall for regions under consideration. These results are used in an attempt to understand changes in ungulate fauna in the Pliocene and Pleistocene in sub-Saharan Africa.

Method of Analysis

Sub-Saharan Africa was divided into equal areas of 50 ha, and modern distribution maps (Smithers, 1983) were used to record the presence or absence of each ungulate species *s* in every area *a*. Estimates of population density in each area were obtained for each species by using Damuth's (1981) general relationship between density (*D*) and mean adult body mass (*MAB*) for mammalian herbivores:

$$\log D = -0.75 \log MAB + 4.23 \qquad r = 0.86 \quad (1)$$

where D and MAB are expressed in units of $N\,\text{km}^{-1}$ and g, respectively; r is the correlation coefficient. Estimate of ungulate biomass B were then obtained for each area a by summing the products of population density and mean adult body mass:

$$B = \Sigma\, D \times MAB \quad (2)$$

Mean annual rainfall and mean annual temperature were calculated from meteorological data for the areas in which each of the ungulate taxa is known to have occurred, based on distribution maps prepared by Smithers (1983).

Results

Figure 32.1 shows the distribution of modern ungulate biomass in sub-Saharan Africa as estimated from the method described above. Figure 32.2 shows the distribution of ungulate biomass in relation to mean annual temperature and rainfall, based on this analysis of ungulates in sub-Saharan Africa.

A relationship between species richness, the number of species in each area (*NSP*), and ungulate biomass *B* in the same area is given by the following equation:

$$NSP = 0.272\, B\ \text{kg/ha} + 0.622 \qquad r^2 = 0.96 \quad (3)$$

Table 32.1 lists mean annual temperature and rainfall associated with these ungulate taxa in modern sub-Saharan environments.

Discussion and Conclusions

Areas of high ungulate biomass occur where mean annual temperature ranges between 19° and 22°C and where mean annual rainfall ranges between 750 and 1,000 mm (figs. 32.1 and 32.2). In sub-Saharan Africa, these areas are associated with mixed woodland savanna. Low ungu-

ISBN 0-300-06348-2.

Fig. 32.1. Distribution of ungulate biomass (kg/ha) estimated for regions in sub-Saharan Africa, based on modern distribution maps (Smithers, 1983) as well as on a general relationship between mean adult body mass and population density of ungulate species (Damuth, 1981).

late biomass is associated with areas where mean annual rainfall is low (less than 250 mm) and where mean annual temperature may range between 16° and 23°C. These areas are generally open grassland and desert. Low ungulate biomass is also associated, however, with forested areas where rainfall is high. Ungulate diversity tends to be high for areas associated with high mean annual temperature.

Can such observations be relevant to an understanding of relationships between ungulate species diversity and climate in paleoenvironmental contexts? When one is dealing with extinct taxa from paleoenvironments which may or may not have had a modern analogue, there is reason to be careful about applying relationships such as those given above, based on studies of modern sub-Saharan fauna in current environmental conditions. However, a few statements can be emphasised here since they are potentially relevant for future attempts to understand the impact of declining temperature on ungulates after 3.0 million years (myr) ago.

First, ungulates that are most tolerant of conditions that combine low temperature with low rainfall tend to have relatively small body mass and include species that graze only or that graze and browse. Second, low ungulate biomass is associated with desert areas with low rainfall as well as with forested areas with high rainfall. Third, the general relationship given by equation (3) suggests that ungulate species diversity is strongly corre-

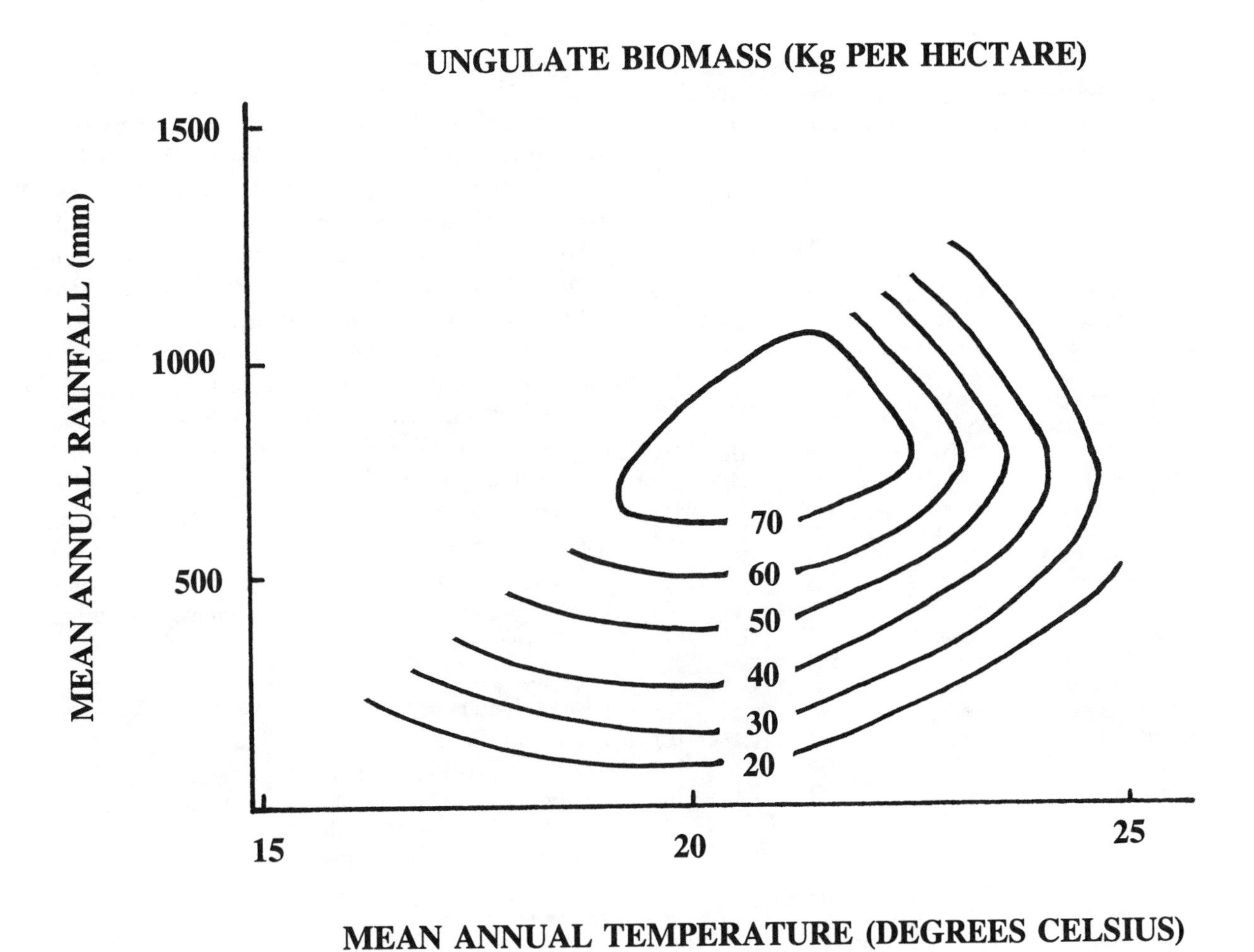

Fig. 32.2. Ungulate biomass in relation to mean annual rainfall and mean annual temperature in modern sub-Saharan Africa.

lated with ungulate biomass. Fourth, from figure 32.2 and equation (3) it is evident that biomass and diversity can be expected to be high in areas associated with certain ranges of mean annual temperature and mean annual rainfall; in modern sub-Saharan environments, these areas are represented by mixed woodland savanna habitats.

If we apply these observations to African paleoenvironments, we would expect changes in the distribution of ungulate biomass as a result of changes in both temperature and rainfall, which together affect habitat. At present it is not possible to generate equations that reliably estimate biomass from temperature and rainfall data, but one can estimate ungulate biomass from relationships such as equations 1 and 2. Preliminary analyses of bovid taxa represented in African paleoenvironments (Thackeray, in prep.) suggest that biomass as estimated from equations (1) and (2), using data from sites in South and East Africa, was generally around 60 kg/ha during the period between 3 and 2.5 myr, contrasting with higher biomass (approximately 70 kg/ha) after 2.5 myr. From our understanding of modern African fauna and environments, it would seem likely that this difference is a result of changes in habitat brought about by changes in climate.

A decline in global temperature around 2.5 myr is well documented (Shackleton, this vol.). By reference to figure 32.2, if temperatures in sub-Saharan environments declined from ca. 25° to 23°C, an increase in biomass could occur without a substantial increase in rainfall. However, a comparable increase in biomass can occur if mean annual rainfall increases from ca. 200 to 800 mm while mean annual temperature is relatively constant around 21°C. These possibilities illustrate the

Table 32.1. Mean Annual Temperature (MAT) and Mean Annual Rainfall (MAR) for Modern Sub-Saharan Environments in which Various Ungulate Taxa Are Known to Have Occurred; Ungulates Listed in Order of Mean Adult Body Mass (MAB)

Genus and Species	Common Name	MAB (kg)	Mean MAR (mm)	Mean MAT (°C)
Neotragus moschatus	Suni	5	643	20.8
Madoqua saltiana	Dik Dik	5	537	21.1
Raphicerus sharpei	Sharpe's grysbok	8	736	20.0
Raphicerus melanotis	Cape grysbok	10	210	14.3
Raphicerus campestris	Steenbok	11	440	18.5
Oreotragus oreotragus	Klipspringer	12	498	19.1
Cephalophus natalensis	Red Duiker	14	769	21.2
Ourebia ourebia	Oribi	14	823	21.9
Pelea capreolus	Rhebuck	20	356	15.6
Sylvicapra grimmia	Grimm's duiker	20	702	20.9
Redunca fulvorufula	Moutain reedbuck	30	459	19.1
Tragelaphus scriptus	Bushbuck	35	643	20.9
Antidorcas marsupialis	Springbok	39	186	18.3
Redunca arundinum	Reedbuck	45	852	20.7
Aepyceros melampus	Impala	48	687	20.5
Potamochoerus porcus	Bushpig	61	964	21.3
Damaliscus dorcas	Bontebok	62	284	14.6
Damaliscus dorcas	Blesbok	66	404	16.2
Phacochoerus aethiopicus	Warthog	68	669	22.0
Kobus vardoni	Puku	74	995	21.3

complexity of factors that potentially influence variability in the abundance of ungulate taxa in modern and paleocommunities.

It is recommended that future research be focused on attempts to quantify both paleotemperature and rainfall in order to understand better the nature and mechanisms of changes in ungulate communities resulting from habitat changes that are themselves influenced by climate. In addition, a detailed analysis of tolerance limits to temperature and to other environmental variables affecting modern ungulate distributions and abundance should be pursued.

Summary

Estimates of ungulate biomass in modern sub-Saharan African environments have been obtained from modern distribution data, taking advantage of Damuth's (1981) relationship between population density and mean adult body mass. A map of modern ungulate biomass in sub-Saharan Africa is presented, and the data are examined in relation to modern climatic variables and species richness. Highest ungulate biomass and species richness occur in areas where mean annual temperature ranges between 19° and 22°C and where mean annual rainfall ranges between 750 and 1,000 mm. Low ungulate biomass and low species richness are generally associated with areas of low mean annual rainfall (less than 250 mm) and where mean annual temperature ranges from 16° and 23°C. These observations are potentially useful for understanding changes in ungulate communities during the Pliocene and Pleistocene.

Acknowledgments

I wish to thank E. S. Vrba and co-organizers of the Airlie conference for the invitation to attend this enjoyable and stimulating meeting. The Foundation for Research Development (Republic of South Africa) and the National Science Foundation kindly contributed toward the costs of attending the meeting.

References

Damuth, J. 1981. Population density and body size in mammals. *Nature* 290:699–700.

Greenacre, M. J., and Vrba, E. S. 1984. A correspondence analysis of biological census data. *Ecology* 65:984–997.

Smithers, R. H. N. 1983. The mammals of the southern African subregion. University of Pretoria, Pretoria.

Thackeray, J. F. 1985. Ungulate biomass and species diversity in African environments. In *Extended abstracts of the Seventh Southern African Society for Quaternary Research Conference.* University of Stellenbosch, Stellenbosch.

———. (in preparation). Estimating ungulate biomass for paleoenvironments.

Vrba, E. S. 1992. Mammals as a key to evolutionary theory. *Journal of Mammalogy* 73:1–28.

Chapter 33

Diversity within the Genus *Homo*

G. Philip Rightmire

As described by Linnaeus in 1758, the genus *Homo* was made up only of living humans belonging to one species. The fossilized remains of earlier hominids were not uncovered until the middle of the next century. The first Neanderthal to be accorded scientific treatment was found in Germany in 1856 and referred to a new species in 1864. Relics of still more archaic humans were excavated in Java in the 1890s, and Eugène Dubois named *Pithecanthropus* (now *Homo*) *erectus* in 1894. Many subsequent discoveries have established that our genus has ancient roots in Africa, and recently Hill et al. (1992) have argued that a partial temporal bone from Kenya may document the presence of *Homo* in deposits that are 2.4 million years (myr) old. The morphological basis for this claim is sharply questioned by Tobias (1993), but there is now little doubt that representatives of *Homo* were living in eastern Africa prior to 2.0 myr ago. Whether the origin of this genus can be linked to a Late Pliocene cooling trend that resulted in increased aridity of the African landscape is an important question that has been addressed by several contributors to this volume. Another focus is the level of taxonomic diversity that is now being revealed within the human clade. Traditionally two extinct species have been recognized, in addition to living people. *Homo,* however, may contain as many as four or even five archaic species that antedate or in some cases overlap in time with *Homo sapiens.* In this chapter I shall comment on the evidence for this diversity and explore some of the current controversies surrounding the evolution of later Pliocene and Pleistocene humans.

ISBN 0–300–06348–2.

The Early African Radiation

Members of the genus *Homo* differ from *Australopithecus* (and from *Paranthropus* as recognized by Wood, this vol.) in a number of key features. The brain is variable in size but larger on average than that of other hominids. The cranium generally lacks a sagittal crest, and the area of nuchal muscle attachment is restricted. The face is small in proportion to the expanded vault. The facial profile is straight rather than protruding and there are differences in the architecture of the nose and cheek. The mandible is less massive than that of contemporary australopiths, and the dental arcade is rounded in contour. The cheek teeth tend to have smaller crowns, which, especially in earlier species, may be relatively narrow buccolingually. In the upper (and variably in the lower) rows, the last molar is reduced in size in comparison to the other molar teeth. The postcranial skeleton reflects adaptation to upright bipedal posture and use of the hand for manipulation. Members of the genus *Homo* may have been the first hominids to be dependent on culturally patterned behavior, including the preparation of stone tools.

One ancient species is *Homo habilis,* the taxon named by Leakey, Tobias, and Napier in 1964, principally on the strength of fossils from Beds I and II at Olduvai Gorge. Subsequently, there has been confusion over the composition of the hypodigm, although Tobias (1991) continues to defend his long-standing claim that this group is composed of all the specimens originally attributed to *Homo habilis* in addition to remains from Olduvai, the Turkana Basin, and perhaps other localities, including Sterkfontein, in South Africa. Such a species must have been quite strongly sexually dimorphic in body size, with females averaging perhaps 30 kg and males more than 50 kg (McHenry, 1992). There is much intragroup variation in the skull and teeth. Brain volume ranges from close to 500 cc to over 750 cc. Faces may be either delicate and relatively prognathic or massive, deep, and flattened, with resemblances to *Australopithecus* or *Paranthropus.* There are substantial differences in the size and proportions of the tooth crowns and structure of the roots.

Not all authorities accept this appraisal of *Homo habilis.* For some time it has been recognized that the smaller skulls differ from the larger ones in shape as well as size, and there may be more variation than would be expected between male and female conspecifics (Leakey and Walker, 1978). An alternative approach is to place the fossils in separate species. Wood (1992), who advocates this view, recognizes all of the Olduvai material as *Homo habilis sensu stricto.* To this group he adds certain finds from Koobi Fora, including the skulls KNM-ER 1805 and KNM-ER 1813, several partial mandibles, and the skeleton of KNM-ER 3735. *Homo habilis* treated in this strict sense displays less extreme dimorphism in body size, and postcranial elements are generally primitive, showing similarities to *Australopithecus.* Mean cranial capacity is close to 600 cc, and the faces, jaws, and teeth of most individuals are relatively small and lightly built. Other remains from Koobi Fora are attributed to a second species called *Homo rudolfensis.* To this assemblage, Wood (1992) assigns postcranial bones that are more modern in form, along with larger crania (such as KNM-ER 1470) and massive jaws (including KNM-ER 1802).

Other workers agree that separate taxa are present but find that it is appropriate to sort the Olduvai hominids into two groups, as has been done for the Turkana fossils. Here, the expanded parietals, lower jaw, and hand of OH 7 are lumped with the larger individuals from Koobi Fora, including KNM-ER 1470, KNM-ER 1590, and KNM-ER 1802. In this instance, premolar variation between specimens such as OH 7 and KNM-ER 1802, noted by Wood (1992), is considered to reflect polytypism (Stringer, 1986). Because OH 7 is the type for *Homo habilis,* this name must be applied to a hypodigm that is virtually the converse of *Homo habilis sensu stricto.* Such a taxon will include much of the material referred by Wood to *Homo rudolfensis.* Rightmire (1993) accordingly describes the cranium of *Homo habilis* as thin-walled and large relative to that of *Australopithecus,* with marked postorbital constriction. Neither sagittal nor compound temporal-nuchal crests are formed. The occiput is rounded, with a relatively long upper scale. The facial skeleton is massively constructed, with deep zygomatic bones, and the nasoalveolar clivus is broad and flattened. The lower jaw, as known for OH 7 and probably documented at Koobi Fora by KNM-ER 1802, has a thickened corpus and pronounced alveolar planum. Details of the mandibular dentition are as described by Leakey et al. (1964) and in part by Tobias (1991). The postcranial anatomy of *Homo habilis* may be close to that of later humans.

Given this reading of the evidence, a second species is represented by the partial cranium and jaw of OH 13, the braincase and teeth of OH 16, OH 24, KNM-ER 1813, and probably KNM-ER 1805. Still unnamed, this taxon has a smaller brain. Postorbital constriction is less marked, but there is more supratoral hollowing of the frontal. KNM-ER 1805, likely to be male, carries a bifid sagittal crest. This individual also exhibits lateral swelling of the mastoid process and its associated crest, whereas other crania of this species lack special inflation of the mastoid region. Occipital proportions resemble those of *Homo habilis,* but there are indications of a transverse torus. The face is relatively short, with low cheek bones and little transverse flattening below the nose. To the extent that OH 13 is a good guide, the mandible is more like that of *Homo erectus,* as are the teeth. Postcranial parts have been difficult to identify, but if OH 62 (a female weighing only about 25 kg) and KNM-ER 3735 are properly included in this species, they suggest a moderate level of sex dimorphism in body mass. These individuals also have chimpanzeelike forelimbs and short legs, relative to *Homo habilis* (Walker, 1993a).

Both of these taxa occur in deposits of the uppermost Burgi Member at Koobi Fora (Feibel et al., 1989), and probably both had evolved by 2.0 myr ago. A mandible recovered from Uraha in the Malawi Rift and described recently by Schrenk et al. (1993) may document the presence of one species as early as 2.3 or 2.4 myr ago, but this date, which is based on biostratigraphic correlation, remains to be confirmed. On the strength of similarities to KNM-ER 1802, the Malawi specimen is referred by its discoverers to *Homo rudolfensis,* but the two jaws probably match crania that are here called *Homo habilis.* Following this origin in eastern Africa, *Homo habilis* seems to remain in the record for a relatively brief interval. The last known representative is OH 7, from Bed I at Olduvai. The second species apparently survived into the Early Pleistocene. One specimen collected from the base of Bed II is dated to about 1.7 myr, and OH 13 from middle Bed II is 1.6 to 1.5 myr in age. This small-brained morph thus persists as a contemporary of *Homo erectus* (fig. 33.1).

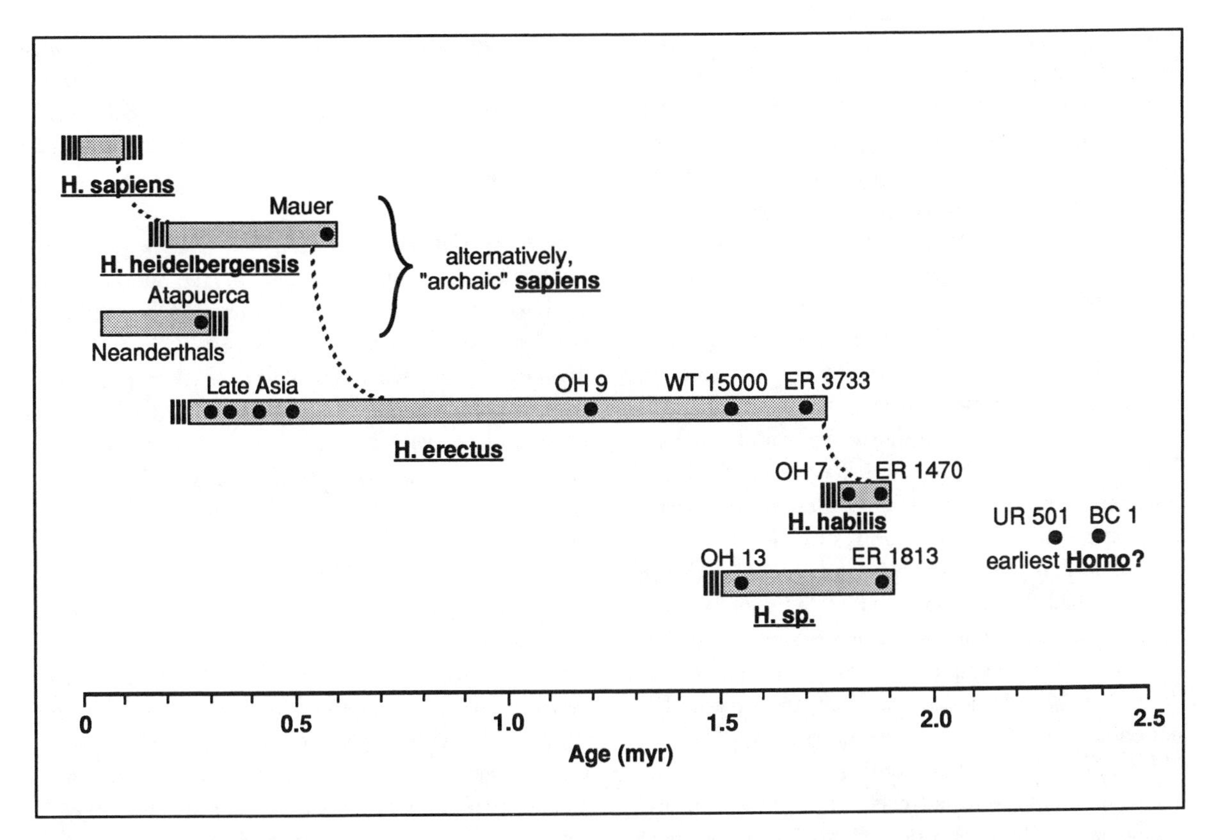

Fig. 33.1. Durations and possible evolutionary relationships of species belonging to the genus *Homo*. Two of the oldest fossils attributed to this genus are BC 1 (a temporal bone from the Chemeron Formation, Kenya) and UR 501 (a mandible from Uraha, Malawi). The latter shares some features with specimens that are here referred to *Homo habilis*. This relatively large-brained species seems only marginally more acceptable than contemporary *Homo sp.* as the ancestor to *Homo erectus*. As documented from sites in Africa and Asia, *Homo erectus* survived into the later Middle Pleistocene. This long-lasting lineage evolved toward anatomically advanced humans, grouped here as *Homo heidelbergensis* (Europe and Africa) and *Homo sapiens*. The Neanderthals now appear to have deep roots in western Europe, but their relationship to other Middle Pleistocene populations is uncertain.

The phylogenetic relationships of these most ancient members of the genus *Homo* are unclear. On various grounds, one can argue that the smaller species is unlikely to be ancestral to *Homo erectus*. Individuals such as OH 13, OH 24, and KNM-ER 1813 show some resemblances to this taxon in face and jaw form but differ from specimens referred to *Homo erectus* in brain volume and numerous details of cranial anatomy. Body size and postcranial proportions as documented by OH 62 and KNM-ER 3735 also set this group apart from other *Homo*. Although claims for or against ancestry based on stratigraphy are weak, the fact that the small-brained hominid is found at Olduvai at least 200 thousand years (kyr) after the first appearance of *Homo erectus* indicates that the former did not rapidly evolve more modern features.

Homo habilis appears also to differ from later humans. In cases where the face is preserved, as for KNM-ER 1470, there are resemblances to *Australopithecus* (or *Paranthropus*), as noted most recently by Wood (1992). Other characters such as the expanded vault, and particularly the morphology of the postcranial skeleton, suggest that this species is more like *Homo erectus*. However, an evolutionary link between the two taxa (as indicated in fig. 33.1) must be considered tentative. Our understanding of these early hominids is still very incomplete, and conclusions drawn from the present evidence will likely be modified in the light of new discoveries.

The Evolution of *Homo erectus*

Although its phylogenetic origin remains obscure, *Homo erectus* seems also to have evolved in Africa. The oldest fossils have been recovered from the Turkana Basin. Material from Koobi Fora, including the exceptionally well-preserved cranium of KNM-ER 3733, other less-complete crania, lower jaws, and the partial skeleton numbered KNM-ER 1808, are 1.8 to 1.7 myr old (Feibel et al., 1989), whereas the remains of a young male from Nariokotome are 1.5 myr in age (Brown and McDougall, 1993). Unfortunately, most of the limb fragments of KNM-ER 1808 are encrusted with deposits of coarse-woven bone; this pathology makes taking useful measurements difficult. More information concerning body size and proportions can be obtained from other specimens, including KNM-WT 15000. It is apparent that early *Homo erectus* was relatively tall and more modern in body form than *Australopithecus* and at least one antecedent species of *Homo*. These differences can be interpreted in relation to climate and thermoregulation, and an increase in size may help explain why *Homo erectus* was the first hominid to move out of Africa into other regions of the Old World.

The subadult male KNM-WT 15000 was about 160 cm tall at death and might have grown to a height of about 185 cm. This very complete skeleton from Nariokotome, along with five additional individuals from Koobi Fora and Olduvai, can be used to predict an average stature of 170 cm for early *Homo erectus* (Ruff and Walker, 1993). These hominids were tall even by modern standards. Insofar as can be determined from the reconstructed pelvis of KNM-WT 15000, they were also quite linear in body build. This finding coupled with studies of limb proportions suggests that African *Homo erectus* populations may have inhabited open, warm, and semiarid environments. In such a setting, where the body could be cooled efficiently by sweating, these people would have needed ready access to water and probably to salt as well. One source of salt might have been animal tissue, obtained either by scavenging or by active hunting (Walker, 1993b).

The skulls from Nariokotome and Koobi Fora also differ in important ways from those of *Australopithecus*, *Homo habilis*, and *Homo sp.* The face of KNM-ER 3733 is broad across the cheekbones and projects in its lower parts. The bone surrounding the nose is thin and platelike rather than thickened by pillars extending upward from the jaw, and the nasal bridge is low. Construction of the nose may support the suggestion that this hominid was adapted to life in an arid environment (Franciscus and Trinkaus, 1988). The brow is prominent but not as massive as that of OH 9 and some later *Homo erectus*. Brain size as assessed for KNM-ER 3733, KNM-ER 3883, and KNM-WT 15000 is greater than expected for either of the earlier species of *Homo*, and the occiput is strongly flexed. Koobi Fora adults, including KNM-ER 730, exhibit a moundlike transverse torus. Crests in the mastoid region are well developed, and the cranial base exhibits many of the features seen in specimens from Java and China.

Questions Concerning Taxonomy. Similarities to the Far Eastern assemblages indicate that the Turkana fossils are best lumped with other African and Eurasian material referred to *Homo erectus*. A species based on this hypodigm spans an interval of perhaps 1.5 myr, and some (Asian) populations may have survived even into latest Middle Pleistocene times (fig. 33.2). As documented by Weidenreich, von Koenigswald, Le Gros Clark, and others, members of this taxon share a suite of characters by which they can be distinguished from recent humans. Some of the principal differences relate to cranial capacity, keeling on the midline of the vault, parietal length, occipital proportions, the anatomy of the cranial base, form of the mandibular symphysis, tooth size, a relatively narrow pelvis, and length of the femoral neck. A large number of traits serve generally to describe *Homo erectus*, and to diagnose this species relative to living people.

For the past decade this view has been challenged by several workers, on diverse grounds. One question concerns the material from the Turkana Basin, and it has been claimed that the early African crania lack key features developed by the Asian populations. A midline keel on the vault, an angular torus at the posteroinferior corner of the parietal bone, certain characters of the base, and overall thickening of the braincase are said to be absent at Koobi Fora but well expressed in the remains from Trinil, Sangiran, and Zhoukoudian. These differences have prompted investigators, including Andrews (1984) and Tattersall (1986), to recognize two species and to suggest that *Homo erectus* as originally described by Dubois must be restricted geographically to the Far East. Wood (1991, 1994) concedes that Dubois's species may be present in Africa, but he argues that, on the basis of facial measurements, perhaps some aspects of temporal bone morphology, and dental differences, the Turkana hominids should indeed be set apart from later *Homo erectus*. Wood now refers the Koobi Fora specimens to *Homo ergaster* and advances a case for linking this species directly to *Homo sapiens*. In his opinion, *Homo erectus* as strictly defined is not likely to have played an important role in the evolution of later people.

Quite a different interpretation has been offered by Wolpoff et al. (1994), who now claim that the nomen

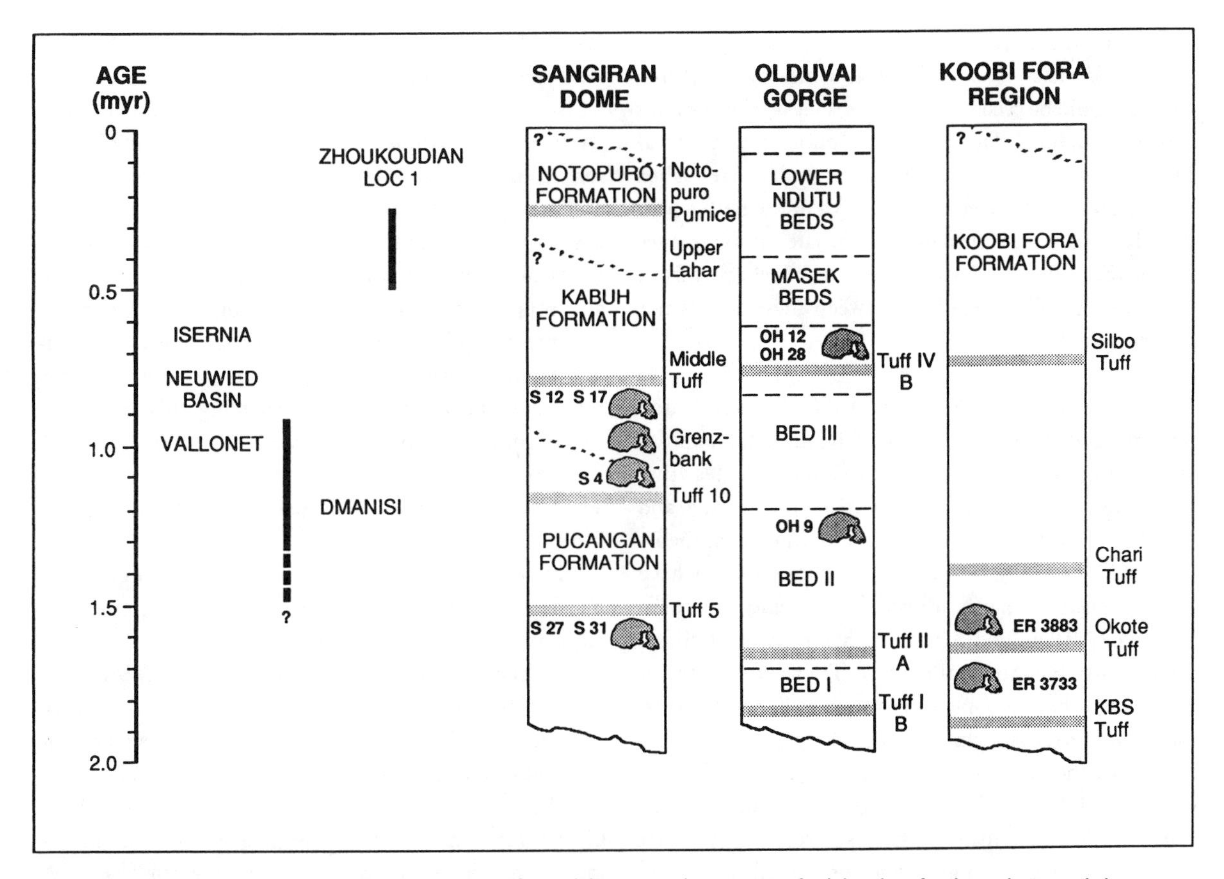

Fig. 33.2. Chronology for some localities yielding *Homo erectus* in Africa and the Far East. The most ancient fossils have been found at Koobi Fora, with the KNM-ER 3733 cranium dating to 1.8–1.7 myr ago. The large braincase from upper Bed II at Olduvai is at least 1.2 myr old. Much of the material from Sangiran is of uncertain stratigraphic provenience, and reliable dates have proved difficult to obtain. A few of the specimens from the Pucangan deposits may approach the age of the Koobi Fora hominids, but the partial cranium and upper jaw of S 4 are probably about 1.0 myr old. Many of the other fossils from Indonesia are younger, and dates for the important assemblage from Zhoukoudian (in China) are less than 0.5 myr. In addition, the mandible from Dmanisi (in Georgia) may document the presence of *Homo erectus* in western Asia by or before 1.0 myr ago; stone artifacts from several sites suggest that hominids were living in Europe at about this time.

Homo erectus is unnecessary and should be discarded altogether. Here the question is really whether there is continuity from earliest Pleistocene times to the present, that is, whether only one long lineage can be recognized, with no branches or extinctions. Such a lineage would include the ancient Turkana populations as well as those resident later in Africa and Eurasia. If it could be demonstrated that no splitting had occurred, one might argue that any division between taxa would have to be arbitrary. Wolpoff and his colleagues go a step further and say that there is simply no basis for keeping more than one species, which must then be *Homo sapiens.*

In my view, none of these scenarios can be supported fully, for reasons already adumbrated. Certainly there is geographic variation among the various *Homo erectus* assemblages, but the fossils from the Turkana Basin, Olduvai Gorge, and other sites in northwestern Africa exhibit essentially the same set of traits as do those from the Far East (Rightmire, 1990). Characters said to be unique to the Asian populations are variable in their expression, and in fact most of them can be identified in the earlier East African material (Bräuer and Mbua, 1992). The faces of KNM-ER 3733 and KNM-ER 3883 conform in many respects to the anatomy of *Homo erectus* as reconstructed from the Zhoukoudian specimens, and parietal and occipital proportions are remarkably similar to those of the better preserved Chinese crania. Vault thickness as measured near the junction of the frontal and parietal bones is about the same in the East African and Asian samples (Walker, 1993b). In addition, the teeth from both Koobi Fora and Nariokotome are close in size and shape to those from Zhoukoudian (Brown, 1993).

Apparently there are not many traits that can be used to diagnose *Homo ergaster*, and probably just one polytypic species should be recognized. Nevertheless, *Homo erectus* as broadly defined does possess a number of anatomical distinctions, extending not only to the skull and teeth but to the postcranial skeleton as well. All of the better preserved individuals, including even the late surviving ones from some of the Far Eastern sites, can be set apart from *Homo sapiens*. The boundary between these taxa is not arbitrarily defined.

The First Peopling of Eurasia. Although there is disagreement about how both the early and later populations of this species should be classified, it is clear that *Homo erectus* dispersed from Africa into Europe and Asia before 1.0 myr ago. These movements probably occurred over a long period, and the hominids may have made repeated sorties, introducing crude chopping tools, stone flakes, and Acheulean hand axes into different regions (Bar-Yosef, this vol.). They did not travel alone, and other large mammals such as lions, leopards, and spotted hyenas seem also to have reached temperate Eurasia by the onset of the Middle Pleistocene (Turner, 1984). Foley (1987) has suggested that the dispersal of *Homo erectus* in company with these predators was part of a general biogeographic event, prompted by dietary requirements and other factors acting on a variety of different species.

One attribute likely to have conferred an advantage on *Homo erectus* relative to earlier hominids is larger body size, as documented at Nariokotome and Koobi Fora. An increase in size might have allowed this species to range over greater distances in search of food. It is well established that among mammals, including primates, larger animals are freer of the constraints imposed by a patchy distribution of resources and can utilize a diverse array of food items. Larger bodies also require more energy, and it is probable that *Homo erectus* had an increased resting metabolic rate and expended more calories during daily activities than did other hominids (Leonard and Robertson, 1992). Although the extent to which earliest *Homo* scavenged or hunted is uncertain, it is likely that *Homo erectus* consumed at least some meat in addition to plant materials. Animal products are energetically dense, and hunting would have provided an efficient way of meeting elevated metabolic needs. Another correlate of large size is the decreased risk of running afoul of predators. Taller and probably stronger hominids would have less to fear of carnivores and might therefore venture farther from territories already established.

Such scenarios seem plausible but are difficult to verify, given the obvious problem of reconstructing behavior from the stone and bone scatters preserved in archaeological sites. Body size, energy requirements, and dietary preferences were probably not the only determinants of mobility among populations of early humans, and other factors almost certainly played a role in the expansion of *Homo erectus*. Intelligence linked to an enlarging brain, and increasing technical sophistication, must have been important in enabling this species to move from Africa into other regions of the Old World. The routes taken by the hominids are uncertain, but some of the migrants must have traveled through the Levant.

Evidence documenting the spread of ancient humans into Eurasia comes from ʿUbeidiya, in Israel. Remains of large mammals as well as rodents excavated from deposits accumulated in or near a lake at ʿUbeidiya suggest a relatively cool climate, at a date of perhaps 1.4–1.0 myr. In addition to the fauna, there are stone choppers, spheroids, picks, and bifaces, which Tchernov (1987) has compared to the lithic material from upper Bed II at Olduvai Gorge. How these sites in the Jordan Valley should be interpreted is unclear, but the tools were most probably utilized by groups of *Homo erectus*. Early traces of hominid activity are also found in Europe, at such localities as Le Vallonet, in France; Isernia, in central Italy; and the Neuwied Basin, in Germany (fig. 33.2). This archaeological evidence, consisting mainly of chopper assemblages lacking hand axes, demonstrates a human presence at least 1.0 myr ago (Villa, 1991). Unfortunately, none of the European sites has produced more than a few bits of human bone. An important exception is Dmanisi, in the Georgian Caucasus. Here a remarkably complete mandible was recovered in 1991, along with abundant animal bones and stone artifacts. Insofar as can be determined from the preliminary descriptions provided by its discoverers, this jaw with all of its teeth still in place displays a number of features characteristic of archaic *Homo* (Gabunia and Vekua, 1995). If the Dmanisi specimen is older than 1.0 myr, it will shed dramatic new light on the peopling of Europe and western Asia by *Homo erectus*.

Representatives of this species also reached the East Asian tropics before moving into more temperate regions. It has been assumed that movement into the Far

East began 1.0 myr ago or slightly earlier, but radiometric dates obtained recently from mineral samples suggest that the oldest Indonesian localities may be 1.8–1.6 myr in age (Swisher et al., 1994). Confirmation of these dates will indicate that *Homo erectus* spread quite rapidly across the Old World. Some of the larger and most informative collections of fossils are on record from Sangiran, in Java, and Zhoukoudian, in China. At Zhoukoudian and certain other localities, probably including Ngandong, it is likely that *Homo erectus* continued to flourish until relatively late in the Pleistocene, as noted above. These later populations resemble their more ancient relatives quite closely. Much has been said about the issue of (character) stasis versus gradual change in the evolution of *Homo erectus,* particularly in respect to brain size (Rightmire, 1985; Wolpoff, 1984). It now appears that, given the nature of the evidence available, questions concerning evolutionary rates cannot be answered very precisely. A tendency for cranial capacity to increase through time can be discerned, especially if the most recent Asian specimens are compared to those from the older sites in Java and eastern Africa. In many other respects, however, *Homo erectus* appears to change in only minor ways. There is fluctuation in vault dimensions and tooth size for skulls from different periods, but trends are difficult to document with certainty. Overall, the early and late specimens are remarkably alike, and the species seems to have remained quite stable morphologically.

Humans of the Later Pleistocene

Populations that are rather clearly different from *Homo erectus* evolved sometime in the Middle Pleistocene, probably in Africa or western Eurasia. Important fossils bearing on this period in human evolution have been discovered at a number of sites, including Broken Hill, in Zambia; Elandsfontein, in South Africa; Lake Ndutu, in Tanzania; and the Middle Awash localities of Ethiopia. Roughly comparable material has been recovered at Petralona, in Greece; Vértesszöllös, in Hungary; Arago Cave, in France; and Mauer and Bilzingsleben, in Germany. Dating is still very approximate, but even the Mauer mandible, arguably the oldest of the European hominids, may be less than 600 kyr in age (Zöller, 1991). Finds such as the Dali cranium suggest that groups more advanced than *Homo erectus* were also present in China later in the Middle Pleistocene.

The more complete skulls are decidedly archaic in some aspects of their anatomy, with massively constructed faces, prominent brows, and low foreheads. At the same time, cranial capacity is generally greater than expected for *Homo erectus,* and there is an indication that rates of brain evolution are increasing. Apart from size, there are differences of shape relative to earlier humans. The frontal displays less postorbital constriction, parietals are larger, and the occiput is more rounded, with an expanded upper scale. Certain features of the ear region and lower-jaw articulation also seem to be derived. Mandibles are not common at these African and Eurasian localities, but at least one of the specimens from Arago Cave has a bony chin. Postcranial bones, also scarce, are robust in their construction but provide little information concerning body proportions.

These hominids should be distinguished from *Homo erectus* at the species level. Many workers feel that the fossils from Broken Hill, Lake Ndutu, Petralona, and Arago Cave are best characterized as "archaic" *Homo sapiens.* This way of classifying them emphasizes broad similarities to all later populations, including the Neanderthals and living humans. One can argue, however, that the crania and jaws are in fact so different from modern ones that they must represent a separate species (Tattersall, 1986, 1992). If the mandible from Mauer is grouped with Petralona and the other remains, the entire assemblage can be referred to *Homo heidelbergensis,* as in figure 33.1. Although this species retains a number of archaic characters, it is clearly a close relative to *Homo sapiens* and may be regarded as the source from which recent humans are descended.

Stringer (this vol.) advocates instead an expanded use of the taxon *Homo neanderthalensis* to include Mauer, Petralona, and Arago as well as later hominids from Europe and southwestern Asia that display a full suite of Neanderthal characters. This view is strengthened by discoveries from the Sierra de Atapuerca in northern Spain, reported by Arsuaga et al (1993). The Atapuerca assemblage, in excess of 300 kyr in age, includes several crania, along with jaws and other body parts. There is variation in size and other aspects of morphology, but several individuals show the midfacial protrusion, arched supraorbital torus, and occipital proportions characteristic of Neanderthals. Although such special features are less obviously expressed, or absent, in the earlier remains from Europe, it can now be argued that the Neanderthal lineage has ancient roots, perhaps extending well back into the Middle Pleistocene. A problem with this interpretation concerns the role played by archaic African populations. Broken Hill, for example, is similar in many ways to Petralona but cannot easily be linked to the Neanderthals. It is not clear that Broken Hill and the other African fossils can be included with Late Pleistocene Europeans in a species presumed to have evolved separately from *Homo sapiens.*

Stringer (this vol.) explores these issues in greater depth, but it is apparent that systematic relationships among archaic groups of the later Middle Pleistocene are still only poorly understood. At least one major branch ending in the European Neanderthals probably died out. Whether extinctions also occurred in the Far East is debated, but there is some evidence for continuity of archaic and more recent people in Africa. To resolve these questions, it will be important to explore the patterns of evolutionary change that ultimately produced hominids of modern aspect, probably only in the last 100 kyr or so.

Summary

The most ancient representatives of *Homo* are known from eastern Africa. Traditionally, two extinct species have been recognized, along with living people, but probably the genus is substantially more diverse than this. *Homo habilis* was named in 1964, and workers including Tobias (1991) continue to argue that the fossils from Olduvai and Koobi Fora belong in one sexually dimorphic group. It now seems likely, however, that two taxa should be identified. Wood (1992) finds that the material from Olduvai Gorge is properly *Homo habilis* but refers the larger crania from Koobi Fora to *Homo rudolfensis.* By my reading of the evidence, OH 7 (the type) should be grouped instead with KNM-ER 1470 and other specimens as *Homo habilis,* whereas smaller individuals from both Olduvai and Koobi Fora represent a second species, presently unnamed.

If the fossils are sorted this way, then the cranium of *Homo habilis* may be described as thin-walled and large relative to *Australopithecus,* with marked postorbital constriction. The occiput is rounded, with a relatively long upper scale. The facial skeleton is massively constructed with deep zygomatic bones, and the lower jaw has a thickened corpus and pronounced alveolar planum. The postcranial skeleton may resemble that of later humans. The second species, including specimens such as OH 13, OH 24, and KNM-ER 1813, has a smaller brain. Postorbital constriction is less marked, and the face is relatively short. Mandibles are more like those of *Homo erectus,* as are the teeth. Limited postcranial evidence suggests smaller body size and long arms, relative to *Homo habilis.*

These taxa are present in deposits of the uppermost Burgi Member at Koobi Fora, and probably both evolved before 2.0 myr ago. Following this origin, *Homo habilis* seems to persist in the East African record for only a brief interval. The second species can be documented over a longer span and occurs as a contemporary of *Homo erectus* for a period of at least 200 kyr. The phylogenetic relationships of these most ancient members of the genus *Homo* are unclear. One can argue, however, that the smaller species is unlikely to be ancestral to later humans.

Homo erectus seems also to have evolved in Africa; the oldest fossils from the Turkana Basin are 1.8–1.7 myr old. Like other specimens from Olduvai, the skulls from Koobi Fora and Nariokotome differ from those of *Australopithecus, Homo habilis,* and *Homo sp.* The face is broad across the cheekbones, and the bone surrounding the nose is thin and platelike. There is a low nasal bridge. Brain size is greater than expected for either of the earlier species of *Homo,* and the occiput is strongly flexed. The cranial base exhibits many of the features seen in specimens from Java and China. The skeleton of KNM-WT 15000 demonstrates that early *Homo erectus* was tall and relatively slender, as would be expected for populations adapted to a warm and semiarid climate.

Following an African origin, populations dispersed into Europe and Asia well before 1.0 myr ago. Probably *Homo erectus* should be viewed as a widespread, polytypic species. Although there is regional variation, the African and Far Eastern assemblages share key characters and need not be set apart as distinct taxa. In fact, the species appears to change in only minor ways over an extended interval. Only later in the Middle Pleistocene do populations that are more advanced than *Homo erectus* evolve in Africa and Europe. These groups share with modern people such features as an expanded brain, a larger parietal, a more rounded occiput, and signs of a chin.

Many workers feel that the fossils from Broken Hill, Petralona, and other sites are best characterized as "archaic" *Homo sapiens,* but it can be argued that the crania and jaws are in fact so different from modern material that they must represent a separate species. A number of these hominids can be referred to *Homo heidelbergensis.*

Alternatively, the taxon *Homo neanderthalensis* may be expanded to include Petralona, Arago, and the Atapuerca remains, as well as later Neanderthals. This issue is presently unsettled, and there is uncertainty concerning the origin of the first people of fully modern aspect. It is likely that some later Middle Pleistocene populations of Africa or western Eurasia constitute the source from which recent humans are descended.

Acknowledgments

My studies of earlier *Homo* have been conducted with the assistance of many individuals and institutions, and I am grateful for this help. The governments of Ethiopia, Indonesia, Kenya, and Tanzania have granted me clearance to examine fossils in these countries. The National Science Foundation and the L. S. B. Leakey Foundation provided much of the funding for the research on which this chapter is based. For the invitation to participate in the Airlie conference on paleoclimate and evolution, I thank Elisabeth Vrba and her co-organizers.

References

Andrews, P. 1984. An alternative interpretation of characters used to define *Homo erectus. Courier Forschungsinstitut Senckenberg* 69:167–175.

Arsuaga, J.-L., Martinez, I., Gracia, A., Carretero, J.-M., and Carbonell, E. 1993. Three new human skulls from the Sima de los Huesos Middle Pleistocene site in Sierra de Atapuerca, Spain. *Nature* 362:534–537.

Bräuer, G., and Mbua, E. 1992. *Homo erectus* features used in cladistics and their variability in Asian and African hominids. *Journal of Human Evolution* 22:79–108.

Brown, B. 1993. Comparative dental anatomy of African *Homo erectus. Courier Forschungsinstitut Senckenberg* 171:175–184.

Brown, F. H., and McDougall, I. 1993. Geological setting and age. In *The Nariokotome* Homo erectus *skeleton,* pp. 9–20 (ed. A. Walker and R. Leakey). Harvard University Press, Cambridge.

Feibel, C. S., Brown, F. H., and McDougall, I. 1989. Stratigraphic context of fossil hominids from the Omo Group deposits: Northern Turkana Basin, Kenya and Ethiopia. *American Journal of Physical Anthropology* 78:595–622.

Foley, R. 1987. Another unique species: Patterns in human evolutionary ecology. Longman, Essex, Eng.

Franciscus, R. G., and Trinkaus, E. 1988. Nasal morphology and the emergence of *Homo erectus. American Journal of Physical Anthropology* 75:517–527.

Gabunia, L., and Vekua, A. 1995. A Plio-Pleistocene hominid from Dmanisi, East Georgia, Caucasus. *Nature* 373:509–512.

Hill, A., Ward, S., Deino, A., Curtis, G., and Drake, R. 1992. Earliest *Homo. Nature* 355:719–722.

Leakey, L. S. B., Tobias, P. V., and Napier, J. R. 1964. A new species of the genus *Homo* from Olduvai Gorge. *Nature* 202:7–9.

Leakey, R. E., and Walker, A. 1978. The hominids of East Turkana. *Scientific American* 239:54–66.

Leonard, W. R., and Robertson, M. L. 1992. Nutritional requirements and human evolution: A bioenergetics model. *American Journal of Human Biology* 4:179–195.

McHenry, H. M. 1992. How big were early hominids? *Evolutionary Anthropology* 1:15–20.

Rightmire, G. P. 1985. The tempo of change in the evolution of Mid-Pleistocene *Homo.* In *Ancestors: The hard evidence,* pp. 255–264 (ed. E. Delson). Liss, New York.

———. 1990. *The evolution of* Homo erectus: *Comparative anatomical studies of an extinct human species.* Cambridge University Press, Cambridge.

———. 1993. Variation among early *Homo* crania from Olduvai Gorge and the Koobi Fora region. *American Journal of Physical Anthropology* 90:1–33.

Ruff, C. B., and Walker, A. 1993. Body size and body shape. In *The Nariokotome* Homo erectus *skeleton,* pp. 234–265 (ed. A. Walker and R. Leakey). Harvard University Press, Cambridge.

Schrenk, F., Bromage, T. G., Betzler, C. G., Ring, U., and Juwayeyi, Y. M. 1993. Oldest *Homo* and Pliocene biogeography of the Malawi Rift. *Nature* 365:833–836.

Stringer, C. B. 1986. The credibility of *Homo habilis.* In *Major topics in primate and human evolution,* pp. 266–294 (ed. B. Wood, L. Martin, and P. Andrews). Cambridge University Press, Cambridge.

Swisher, C. C., Curtis, G. H., Jacob, T., Getty, A. G., Suprijo, A., and Widiasmoro. 1994. Age of the earliest known hominids in Java, Indonesia. *Science* 263:1118–1121.

Tattersall, I. 1986. Species recognition in human paleontology. *Journal of Human Evolution* 15:165–176.

———. 1992. Species concepts and species identification in human evolution. *Journal of Human Evolution* 22:341–349.

Techernov, E. 1987. The age of the ʿUbeidiya Formation, an Early Pleistocene hominid site in the Jordan Valley, Israel. *Israel Journal of Earth Science* 36:3–30.

Tobias, P. V. 1991. *Olduvai Gorge.* Vol. 4, *The skulls, endocasts and teeth of* Homo habilis. Cambridge University Press, Cambridge.

———. 1993. Earliest *Homo* not proven. *Nature* 361:307.

Turner, A. 1984. Hominids and fellow travellers: Human migration into high latitudes as part of a large mammal community. In *Hominid evolution and community ecology,* pp. 193–217 (ed. R. Foley). Academic Press, London.

Villa, P. 1991. Middle Pleistocene prehistory in southwestern Europe: The state of our knowledge and ignorance. *Journal of Anthropological Research* 47:193–217.

Walker, A. 1993a. The origin of the genus *Homo.* In *The origin and evolution of humans and humanness,* pp. 29–47 (ed. D. T. Rasmussen). Jones and Bartlett, Boston.

———. 1993b. Perspectives on the Nariokotome discovery. In *The Nariokotome* Homo erectus *skeleton,* pp. 411–430 (ed. A. Walker and R. Leakey). Harvard University Press, Cambridge.

Wolpoff, M. H. 1984. Evolution in *Homo erectus:* The question of stasis. *Paleobiology* 10:389–406.

Wolpoff, M. H., Thorne, A., Jelinek, J., and Zhang, Y. 1994. The case for sinking *Homo erectus:* 100 years of *Pithecanthropus* is enough! *Courier Forschungsinstitut Senckenberg* 171:341–361.

Wood, B. 1991. *Koobi Fora Research Project.* Vol. 4, *Hominid cranial remains.* Clarendon Press, Oxford.
———. 1992. Origin and evolution of the genus *Homo. Nature* 355:783–790.
———. 1994. Taxonomy and evolutionary relationships of *Homo erectus. Courier Forschungsinstitut Senckenberg* 171:159–165.
Zöller, L. 1991. The Palaeolithic site of Mauer: Approaches to the stratigraphy. Unpublished field guide.

Chapter 34

The Influence of Climate and Geography on the Biocultural Evolution of the Far Eastern Hominids

Geoffrey G. Pope

Recent analyses and advances in Asian paleoanthropology, paleolithic archaeology and mammalian paleontology, in combination with geological evidence, suggest that the early hominids that reached the Far East and Southeast Asia occupied regional habitats that did not undergo severe climatic fluctuation during the Pleistocene (01.65–0.001 million years [myr]). Furthermore, on the basis of present evidence, Premodern Asian hominids (early *Homo sapiens* from Asia, as defined by Pope, 1989) seem to have been largely restricted to stable and relatively equitable lowland environments, which ranged from tropical Southeast Asia to seasonally cool (but not glacial), temperate northern China. This conclusion is supported by both paleogeographic and biostratigraphic data. A number of subtle, but important, biogeographic and biostratigraphic contrasts characterize the Pleistocene hominid habitats of the Far East and support the interpretation that hominid anatomical and biobehavioral adaptations underwent relatively few changes throughout the Early (1.65–0.78 myr) and most of the Middle Pleistocene (0.78–0.125 myr). Hominid morphological diversity increased markedly, however, in the terminal Middle Pleistocene–early Late Pleistocene in China. In Java this same period is characterized by a marked increase in morphological homogeneity. Paleogeographic (but not paleoclimatic) diversity and a lack of dramatic climatic change are the most useful ways of characterizing the paleoenvironmental record of the Pleistocene Far East. Both the paleontological and paleolithic evidence from Asia indicates that Pleistocene hominid presence in the Far East began approximately 1 myr ago and can best be understood in terms of a process of adaption to environments that selected for the preservation of many of the initial adaptations to a variety of environments. In the terminal Pleistocene (ca. 0.2–0.12 myr) the appearance of novel morphological complexes and traits in East Asia may indicate increased gene flow from the West. This situation makes the scenario of cultural isolation and replacement (Stringer, 1990) highly unlikely.

ISBN 0-300-06348-2.

Tectonic Frameworks

The Tertiary and Quaternary evolution of the Far East (fig. 34.1) is most accurately understood in the context of two ongoing tectonic processes that continue to influence both the topography and climatic regimes of the that area. The first is the continuing collision of the Indian subcontinent with the Asian mainland. The uplift of the Himalayas and the Tibetan Highlands (Shan-Yunnan Massif) has resulted in the gradual increase in seasonality and desertification of western and northwestern China. The climatic effects of this tectonic activity have been linked to the deposition of loess-paleosol sequences (Liu et al., 1985; Liu 1991a, 1991b). In China, the uplift of the Shanxi block, the Qingling Shan Mountains (running from east to west) and the Taihang Shan ranges (running north to south) are also related to the collision of the Indian subcontinent with the Asian continent. The uplift of the Shanxi block not only continues to influence the course of the Huang He River but has also been directly linked to distinct erosional episodes represented by the Late Pliocene (ca. 3.0–1.65 myr) basin formation, gravel beds, and certain disconformities throughout northern China. In fact, all the great rivers of

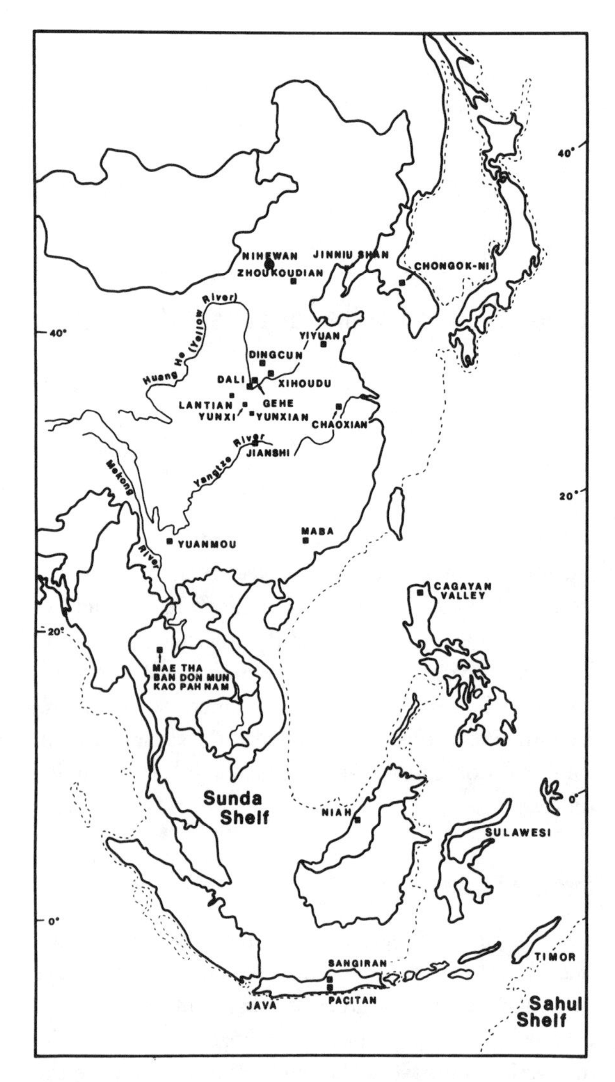

Fig. 34.1. Major paleoanthropological localities of the Far East. Squares indicate important archaeological and hominid sites that have yielded the principal evidence for Premodern hominid evolution in the region. The Nihewan Basin (shown by the dot) represents the primary biostratigraphic type section for the "Chinese Villafranchian" and the Quaternary in the temperate region of China.

the Far East have their genesis in the highlands resulting from the uplift of the Tibetan Plateau.

The Qingling Shan are the traditional dividing line between the subtropical *Stegodon-Ailuropoda* fauna of southern China and the Palearctic, temperate "Chinese Villafranchian" fauna or Nihewan fauna (fig. 34.2) of northern China (Aigner, 1981; Pope, 1982, 1983, 1984, 1985). The putatively distinct integrity of these faunal realms is blurred, however, in paleontological sites situated on the extensive East Chinese Coastal Plain. The dating of important early *Homo erectus* sites (fig. 34.3) such as Lantian (Gongwangling and Chenjiawo) and the Nihewan archaeological localities of Xiaochangliang and Donggutuo has depended on paleomagnetic studies (Li and Wang, 1982, 1991), whereas the dating of late *Homo erectus* sites such as Yunxian (Li, 1991; Li et al., 1991) and Hexian (Huang et al., 1982; Huang and Huang, 1985) has depended heavily on this regional tectonic-biostratigraphic framework and the use of uranium-series dates.

The second tectonic force in the Far East results from the continuing collision of the Australian plate (*sensu lato*) with mainland Southeast Asia. The Indonesian archipelago is the only area to yield fossil hominids; all of these finds come from Java, which is composed of a series of volcanic islands that coalesced throughout the Pleistocene and now comprise the central mountain spine of the island, running from east to west. The eruptions of extrusive, volcanic rock and ash that began in the Pliocene continue unabated. The low potassium content of the resulting marine-derived andesites has hindered attempts at $^{40}K/^{40}Ar$ dating. Volcanic activity has resulted in the deposition of high-energy flows from mud volcanoes that have intermingled fossils from different chronostratigraphic units (Pope, 1982, 1983). In addition to the confusion introduced by early attempts to understand the lithostratigraphy and biostratigraphy of the hominid-bearing Sangiran area (von Koenigswald, 1934, 1935), the activities of native collectors have also hampered efforts to establish the provenience of most of the Javanese hominid finds. Fluorine analysis has indicated that the earliest of these may date to as early as 1.3 myr (Matsu'ura, 1982, Pope, 1982, 1983), although the majority are Middle Pleistocene in age (Pope, 1984, 1985, 1988).

Paleoclimatic Frameworks

Java. Although virtually all workers are in agreement that the periodic rise and fall of eustatic sea level intermittently exposed the Sunda Shelf, there is ongoing controversy about the paleoenvironment represented by the now-inundated extension of the Southeast Asian mainland. The reconstruction of the Sunda Shelf as an open savanna is not supported by the data. The absence of any demonstrably open-dwelling forms precludes the recognition of any part of Pleistocene Southeast Asia as anything but a tropical forest environment (Pope, 1983,

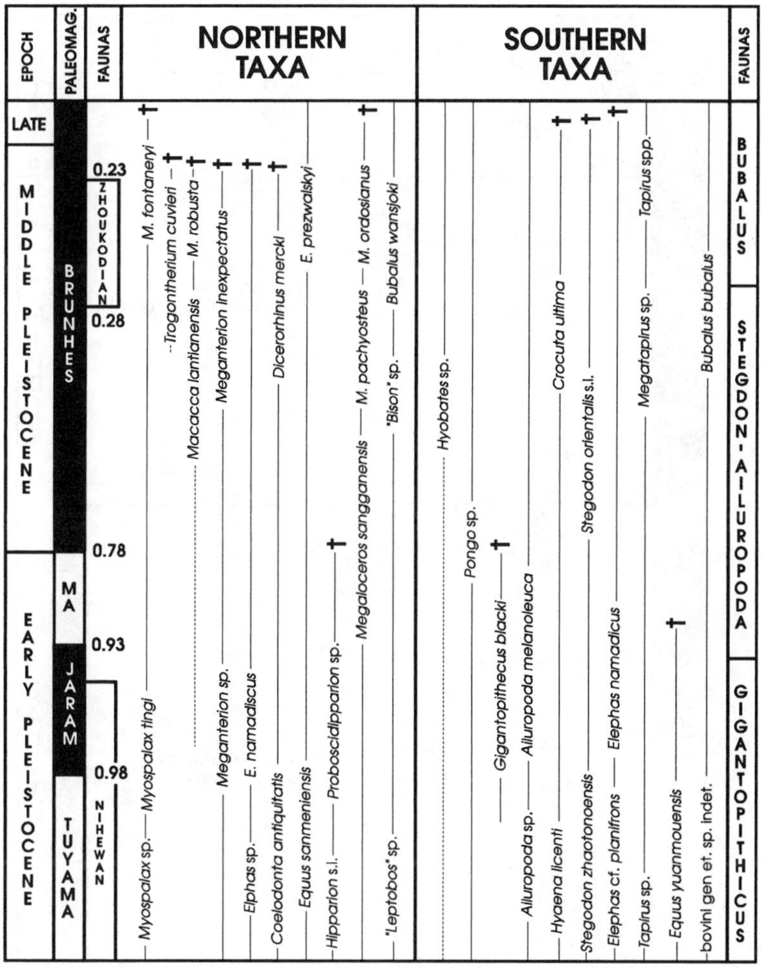

Fig. 34.2. Biostratigraphy of mammalian taxa commonly employed in the Far East for chronological estimates. No phylogenetic inferences should be drawn from the chart in that a single line includes taxa whose relationship ranges from congeneric to infraordinal. Also indicated from bottom right to top right are the southern regional faunas: The *Gigantopithecus* (GIGANTOPITHECUS.); the *Stegodon-Ailuropoda* (STEGODON-AILUROPODA.), and the *Bubalus* (BUBALUS.) In the temperate north the primary fossil faunas are the Nihewan and the Zhoukoudian. The dating of these faunas is based on the sites and techniques shown in fig. 34.3. Paleomag. = paleomagnetic polarity time scale; JARAM = Jaramillo; MA TUYAMA = Matuyama. Crosses indicate extinctions.

1985, 1988). Equids (*Hipparion, Equus*) camelids (*Paracamelus, Camelus*),and giraffoids (*Sivatherium*) present elsewhere in Eurasia are entirely absent in Southeast Asia. Conversely, demonstrably tropical forest-dwelling taxa such as orangutans, gibbons, macaques, tapirs, and many other forest forms are present in Far Eastern fossil faunas. Pollen analyses and modern phytogeographic evidence of the past and present existence of high multicanopied flora further support this conclusion (Pope, 1988). The Sunda Shelf and its present-day islands, regardless of the area of dry land, remained tropical because of its location between the warm, shallow South China Sea and the Indian Ocean. At the same time, the valley forests of mainland Southeast Asia, which run from north to south, provided dispersal routes for the vast majority of tropical animals that have inhabited both southern China and Southeast Asia since the Pleistocene.

China. The search for glacial periods that are synchronous and comparable in severity to those in Europe has a long history in Asia (Lee, 1939; Sun, 1991). Most of this research has been conducted in China, but notions of climatic and faunal oscillations in Java have also been predicated on this approach. Because most of the geological and paleontological work in China was conducted in

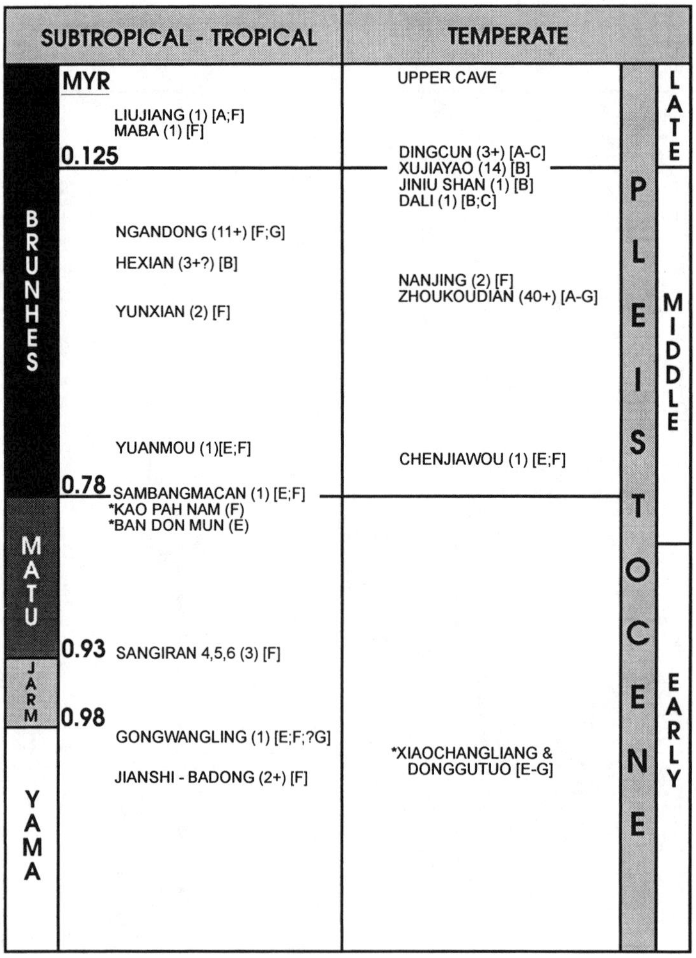

SUBTROPICAL - TROPICAL
TEMPERATE
MYR
UPPER CAVE
LIUJIANG (1) [A;F]
MABA (1) [F]
0.125
DINGCUN (3+) [A-C]
XUJIAYAO (14) [B]
JINIU SHAN (1) [B]
DALI (1) [B;C]
NGANDONG (11+) [F;G]
HEXIAN (3+?) [B]
NANJING (2) [F]
ZHOUKOUDIAN (40+) [A-G]
YUNXIAN (2) [F]
YUANMOU (1)[E;F]
CHENJIAWOU (1) [E;F]
0.78 SAMBANGMACAN (1) [E;F]
*KAO PAH NAM (F)
*BAN DON MUN (E)
0.93 SANGIRAN 4,5,6 (3) [F]
0.98
GONGWANGLING (1) [E;F;?G]
*XIAOCHANGLIANG & DONGGUTUO [E-G]
JIANSHI - BADONG (2+) [F]
BRUNHES
MATU
JARM
YAMA
PLEISTOCENE
LATE
MIDDLE
EARLY

Fig. 34.3. Hominid and archaeological finds indicating the presence of fossil hominids in the Far East. Hominids are placed according to the best estimate of their temporal position and the kind of fauna they are associated with. Note that the Nanjing hominids from southern China are not associated with a typical southern fauna. Numbers in parentheses indicate number of hominid individuals present. Letters in brackets indicate techniques used to estimate the age of the hominid finds: A = ^{14}C; B = uranium-series; C = thermoluminescence; D = electron spin resonance; E = paleomagnetic studies; F = biostratigraphic studies; G = lithostratigraphic studies. Indications of hominid presence based only on archaeological evidence are indicated by an asterisk. See fig. 34.1 for geographical locations and fig. 34.2 for abbreviations.

the north, there has been a concerted search for geomorphological and paleontological evidence that reified attempts to correlate glacial Europe with "glacial China." Although the idea of a northern Chinese Villafranchian has proved valid, the search for lowland geomorphological evidence of glacial periods in the tropics has all but proved the null hypothesis.

More recently, this approach has been abandoned in favor of a widely accepted framework linking the loess-paleosol sequence of the central loess plateau to glacial-integlacial oscillations, respectively. These connections have in turn been linked to oxygen isotope stages as established from deep-sea cores (Aigner, 1986, 1988). The central (and probably well-founded) assumption is that loess deposition took place during glacial periods and that paleosol formation occurred during relatively warmer and wetter periods, when seasonal, moisture-bearing paleomonsoons were active longer (An and Lu, 1984; An et al., 1987, 1991). One problem with this somewhat simplistic scheme is that loess deposition is actively occurring now (in interglacial times). Additionally, little consideration has been given in the literature to microhabitat variations as they have been discerned from stratigraphically equivalent deposits. In spite of this fact,the Luochuan section of the loess plateau now serves as the principal means of correlating climatic oscillations in northern China. This scheme has been much more difficult to apply in southern China, where authigenic alteration and the more limited distribution of loess make direct north-south correlations difficult. The problem of establishing a correspondence between the loess sequence and oxygen isotope record has been summarized elsewhere (Qi, 1989, 1990). For now, the scheme should be cautiously accepted as a means of providing broad and only provisional indications of both chronology and paleoclimatic contexts of hominid finds.

Paleontological indicators of Chinese paleoclimatic contexts of early hominid finds are somewhat more revealing. Generally accepted faunal indicators of glacial climates are few in China. Most conspicuous is the limited distribution of *Mammuthus* in China. Throughout the Pleistocene, the genus was strictly limited to the northeast and never penetrated farther south (Liu and Ding, 1984). Similarly, *Rangifer* is unknown from China, though *Marmota* has been reported as far south as Zhoukoudian. *Coelodonta* has been reported from a number of localities in both the north and south. This finding raises important questions about its usefulness as an indicator of glacial or periglacial habitat in China. On the basis of our current biogeographic knowledge, the genus should not be taken as a positive indication of periglacial climates in China. Pollen analysis (Luo et al., 1988) indicates a general shift in the north from warm, humid Neogene climates to drier Pleistocene climates. During the Pleistocene, however, both herbacious and arboreal pollen increased in diversity and abundance.

In southern China, reports of ostensibly cold-adapted forms in fossil aggregates that sample the subtropical habitats in Anhui, Jiangxi, and Hunan Provinces have always been only tentatively identified when found in association with subtropical forms. Conversely, previous reports of subtropical forms based on fragmentary dental finds of *Ailuropoda* and *Pongo* in northern China (cf. Aigner, 1981) have never been confirmed or replicated by modern research (Han and Xu, 1985, 1989). The Maba cranium from Guandong Province, for example, was originally reported as occurring in association with a cold-adapted taxa such as *Rhinoceros* sp. and *Ursus* sp. (Wu and Peng, 1959; Song and Zhang, 1988). The failure to identify unequivocal cold-adapted taxa beyond a generic level precludes the identification of specifically Palearctic animals. One recent description (Qiu, pers. comm.) of the fauna associated with the new Nanjing crania reports that the fauna is identical to that from the temperate Locality One of Zhoukoudian. By far the greatest factor in the distribution of extant forms may have been the impact of the depletion of once-widespread forms by modern humans. The overall picture that emerges is one in which both subtropical and temperate Pleistocene forms were distributed in a manner very similar to Recent forms.

Hominid Chronology

Java. The dating of hominid fossils (fig. 34.3) from the Far East has relied mostly on biostratigraphy, although attempts to use biostratigraphic divisions in Java have been hindered by the lack of secure provenience for the hominid finds and by lithostratigraphic inaccuracies introduced early on in Javanese paleontology. Despite claims to the contrary (De Vos, 1982), recent paleontological research has indicated that the biostratigraphic distinctions between the "Djetis" and "Trinil" faunas are subtle at best. Indeed, the most recent work has divided and reversed the antiquity of these units on the basis of unconvincing subspecific differences in taxa. The paleomagnetic studies (Sémah, 1982; Sémah et al., 1990) have not fared any better (Pope, 1988). The published paleomagnetic stratigraphy of Java is wholly at odds with the accepted geomagnetic polarity record.

The biostratigraphic record of Java continues to be the source of much controversy. Von Koenigswald's recognition (1934, 1935) of distinct litho- and biostratigraphic units, the presence of early "Villafranchian" faunal elements with "Siva-Malayan" and "Sino-Malayan" affinities has not held up to recent scrutiny (Itihara et al., 1991). Equally unlikely is the more recent conclusion that the "Djetis" fauna is actually older than the "Trinil" fauna. Although some modern workers have maintained this position (De Vos et al., 1982), in situ excavations have revealed that marked distinctions in the biostratigraphic succession in Java are extremely subtle, if not illusory. Most claims for faunal turnover have been based on arguable subspecific distinctions that supposedly resulted from the periodic invasion of extra-Javanese forms.

Southern China. In China, faunal successions have been divided into the Plio-Pleistocene *Gigantopithecus* fauna, the Middle Pleistocene *Stegodon-Ailuropoda* fauna, and the Late Pleistocene *Bubalus* fauna, all of which are difficult to define precisely. Paleomagnetic studies of early hominid localities have also been consistent in indicating that the earliest hominid finds are no older than the late Early Pleistocene (ca. 1.0 myr). The Yuanmou incisors, which were once accorded an age of ca. 1.7 myr (Hu, 1973), cannot be reliably tied into the Yuanmou sequence. The dental remains of this single individual are surface finds. Study of the extensive mammalian sequence has produced results that are in accord with other Eurasian sequences and indicate that the basin deposits span the Pliocene and Pleistocene. The presence of equids throughout the Yuanmou sequence is unique in southern China and is almost certainly indicative of a relict fauna that was preserved in the microhabitat of the relatively low and arid basin. However, no precise data on provenience has yet been reported.

Claims of 1.4 myr or more for the Lantian (Gongwangling) hominid are based on the combination of provenience in sediments that predate the Jaramillo event at > 0.97 myr and the extrapolation of loess sedimentation rates based on the study of the Luochuan loess sequence (An et al., 1987). In actuality, the slow rates used to calculate sedimentary accumulation are almost certainly too slow and neglect the fact that the Gongwangling site is very distant from the more northerly center of loess accumulation on the central loess plateau. A partial mandible from Wu Shan, Sichuan, has also been dated on paleomagnetic grounds as older than the Jaramillo (Huang and Fang, 1991), but personal observation suggests that this mandibular fragment is not that of a hominid.

Two other cave localities in Hubei (Badong and Jianshi) have been dated solely on the basis of biostratigraphy, which emphasizes the occurrence of isolated teeth of *Homo* found in association with *Gigantopithecus blacki* (Gao, 1975). This is the only verified co-occurrence of *Gigantopithecus* and *Homo,* although a much more questionable association has been reported from Vietnam (Hoang et al., 1979). Nonhominid primates in subtropical China and hominids in northern China may eventually prove to be some of the most useful biochronological indicators among fossil mammals. Although "morphological dating" is decried by human paleontologists, it is hard to avoid the conclusion that parameters of hominid morphology such as cranial thickness, supraorbital morphology (torus, tori, sulcus, postorbital constriction), height of maximum cranial breadth, temporomandibular joint (glenoid depth, tympanomastoid fissure), occipital region (roundness, inion-endion distance, torus development), and cranial capacity are useful in establishing the relative age of Pleistocene faunas in the Far East. The fact that such an approach disturbs paleoanthropologists cannot negate the value of using them as biostratigraphic indicators.

In southern China, there appears to be a general trend in which Middle Pleistocene forms, including ailuropodids, pongids, and tapirids, are larger than Early and Late Pleistocene forms (Pope, 1982). The same pattern for equids, cervids, and macaques may also be present in northern China. Of special interest in southern China is the putative observation that hylobatids and macaques of any sort cannot be shown to make a first appearance until the Middle Pleistocene. There is no good biogeographic (or phytogeographic) explanation of this phenomenon. *Pongo* sp., on the other hand, is certainly present in abundance in the Late Pliocene–Early Pleistocene "*Gigantopithecus* Faunas." Beginning in the Middle Pleistocene, few diagnostic faunal elements can be used to differentiate subdivisions of the Middle Pleistocene and Late Pleistocene faunas. This has caused considerable uncertainty about the ages of important hominid finds from Yunxian, Nanjing, and Maba. Again, among these faunas, it is the primates, especially the morphological "progressiveness" of the hominids, that currently serve as the best indicators of relative ages.

Northern China. The introduction of extraregional taxa during the Pleistocene is clearest in northern China. Thus, the overall biostratigraphic succession is most comparable to sequences in Europe (fig. 34.2). The Zhoukoudian fauna serves as the type for the late Middle Pleistocene and the Nihewan fauna serves as the comparative standard for the early Middle and Early Pleistocene. Equids, rhinocerotids, and camelids represent North American introductions that later reached Europe and Africa. Stegodons, cervids and canids represent indigenous Asian forms that also diffused into Europe and Africa. Bovines (? *Leptobos*) and elephantids represent introductions from Europe and Africa. Unfortunately, the first appearance datum (FAD) for all these forms considerably predates the FADs of hominids. The last appearance datum (LAD) for most of these taxa in Asia is sometime in the Late Pleistocene, after or just around the time of the FAD for early *Homo sapiens.* Most of Asian biostratigraphy has depended on the identification of morphological changes in lineages that were already present locally. The identification of reliable chronostratigraphic indicators has been made even more difficult by the fact that morphological change in Far Eastern lineages was slow. This fact has caused at least one worker to suggest that the pivotal Zhoukoudian locality spans only a single interglacial since the contrast between taxa from the lower and upper levels is minimal (Aigner, 1986, 1988). Aigner is correct in arguing that the hominid-bearing section of Locality One spans a shorter period than previously supposed, but a single, late interglacial period is almost certainly too short. In fact, numerous radiometric dates (cf. Wang, 1989) are in virtual agreement that the more than 40 m of hominid- and artifact-bearing deposits span at least 0.2–3 myr.

FADs at a generic level are highly problematical in northern China, though some LADs, such as those of *Proboscidipparion* (ca. 1 myr), machairodont cats (ca. 0.2 myr), and macaques (.02 myr), are useful for discriminating between sites that antedate or predate the upper levels at Zhoukoudian; but such forms are few. The lack of distinct Quaternary faunal breaks or turnovers either before or after the appearance of hominids has made the dating of terminal Middle Pleistocene occurrences of *Homo erectus* and *Homo sapiens* (Chen and Zhang, 1991) difficult. Dating has relied heavily on uranium-series dates (Chen and Yuan, 1988). The pattern of non-hominid mammalian evolution is even more interesting given the chronophenetic pattern of hominid evolution, which not only exhibits a pattern of increased diversity but also suggests accelerated gene flow both in and out of China during the terminal Middle and early Late Pleistocene (Pope, 1989, 1991, 1992).

Hominids

As mentioned previously, the patterns of hominid evolution in Java and China are dissimilar in that morphological diversity decreases in Java and increases in China. The insularized mammalian fauna of Java suggests that demic isolation was a major factor in the evolution of hominids there. Conversely, the great diversity of terminal Premodern Chinese *Homo sapiens* suggests that gene flow played a significant part in the evolution of continental Chinese populations. Both of these observations are extremely difficult to reconcile with the Replacement Model (Stringer and Andrews, 1988; Stringer, 1989, 1990; Wilson and Cann, 1992), which supposedly applies in every region of the Old World. The Asian evidence is much more in accord with a version of the Regional Continuity Model (Thorne and Wolpoff, 1981a, 1991b; Wolpoff et al., 1984).

Java. Although the plethora of taxonomic names that have been applied to the Javanese hominids results from both historical accidents and less than rigorous taxonomic approaches (cf. Pope, 1982), it is nonetheless possible that "*Meganthropus palaeojavanicus*" represents a taxon other than *Homo erectus.* No reliable evidence exists, however, for the possibility that robust australopithecines (Robinson, 1950, 1953; Sartono, 1982) or *Homo habilis* (Tobias and von Koenigswald, 1964) have been found in Java. Unfortunately, this notion was once used to support a similar (but now abandoned) interpretation for fragmentary Chinese finds. It is clear that the early Javanese hominids exhibit a range of robusticity that may be hard to reconcile with their inclusion in a single nonmodern taxon. This range of variation may be due to diachronic change and / or to the fact that workers have underestimated the range of variation even in isolated gene pools. These kinds of concerns are still at the heart of discussions about the Laetoli australopithecine and early *Homo* samples from Turkana. Whether or not "*Meganthropus palaeojavanicus*" warrants a separate taxonomic designation will have to await the recovery of much more complete specimens. For the time being, I consider it best to regard those samples informally assigned to this taxon as morphological variants of *Homo erectus* whose morphology diverges to some degree from continental representatives of the taxon. If we accept the Laetoli and Hadar australopithecines as a single taxon, and the Turkana specimens assigned to early *Homo* as another distinct taxa, then there should be little difficulty in including all the early Javanese hominids in a single species. Although there is considerable debate about the number of species of *Homo* present at Turkana (Wood, 1985; Tobias, 1991), it is obvious that this discussion depends on whether a particular worker is a "splitter" or a "lumper."

The later Premodern sample from Java (Ngandong and Ngawi) is easily recognizable as much more homogeneous than the early Javanese hominids. Even detailed study of the Ngandong hominids (Santa Luca, 1980) has been able to identify only a few debatably autapomorphic features that separate them from the Zhoukoudian sample. The homogeneity of the Ngandong and Zhoukoudian hominids (see below) may have resulted from intensified intraregional gene flow during the terminal Middle Pleistocene. Morphologically, both of these samples exhibit long crania, very thick cranial bones, a receding, flattened frontal, deep glenoid fossae, and a high frequency of angular tori (bilaterally variable in the same individual). The Hexian cranium also conforms to this general pattern. The Ngandong crania differ from most of the Chinese *Homo erectus* specimens in possessing slightly larger cranial capacities, large frontal sinuses, and a distinctively hook-shaped inion. With these observations in mind, it is still possible to observe that the Ngandong hominids preserve the essential "bauplan" (Santa Luca, 1980) of Chinese *Homo erectus,* as it is known from Zhoukoudian and Hexian.

China. Until recently, most of our understanding of Chinese *Homo erectus* was based largely on the Zhoukoudian sample and the work of Weidenreich (1939, 1943, 1945). Subsequent discoveries from Lantian, Hexian, Yunxian, and Nanjing have enlarged our perception of variation in Chinese *Homo erectus.* The recovery of specimens of early *Homo sapiens* from Xujiayao (Wu, 1986), Maba, Dali, and Jinniu Shan has added to the complexity of trying to piece together the relationship between *Homo erectus* and the extant anatomically modern populations of the Far East. Most East Asian specialists perceive distinct morphological commonalities between Chinese *Homo erectus* and early *Homo sapiens* (Wolpoff et al., 1984; Wu and Dong, 1985; Wu et al., 1989; Pope, 1992). These similarities are most apparent in the facial skeleton and include a short maxilla and low zygomatic root, horizontally oriented and flat cheekbones (Gongwangling, Dali), a distinct incisura malaris (Zhoukoudian, Dali), and an absence of lateral alveolar prognathism (Zhoukoudian and all Asian hominids). It is also apparent, however, that extraregional traits also occur for the first time in China in both late *Homo erectus* and early *Homo sapiens* (Pope, 1989). These include a larger middle face (Yunxian, Jinniu Shan), thinner cranial bones (Maba, Jinniu Shan), a taller maxilla, and a more arching inferior zygomaticomaxillary margin (Jinniu Shan). Dental features linking Chinese *Homo erectus* and Chinese *Homo sapiens* include the reduction or agenesis of M_3, a peg-shaped M^3, and shovel-shaped incisors.

The Far Eastern Paleolithic

Nowhere has the paleoanthropological evidence from the Far East been more misunderstood than in western interpretations of the Asian archaeological record. For decades, the non-Acheulian, non-Mousterian archaeological record has been interpreted as an indication that the Far East was an isolated cultural backwater (Movius, 1944) or, alternatively, that inadequate research had simply failed to locate "stage-mode" developments that characterized other regions of the Old World (Klein, 1992). Throughout the history of archaeological research in the Far East, there has also been a willingness to impose models and interpretations generated in the West on the Asian paleolithic evidence (Schick and Dong, 1993; Clarke, 1990). Further complicating an accurate understanding of the Asian evidence has been the concern of a few native Asian workers with the reification of such entities as the Acheulian and the Mousterian (Huang, 1987, 1990). Although the existence of elements of these traditions may be present in Siberia, after more than fifty years of concerted research, it is now obvious that no evidence for these broadly defined modes of technology has turned up in the Far East. "Levallois" artifacts and hand axes have occasionally been recovered at various localities in East Asia, but they continue to constitute insignificant fractions of any given assemblage. More important, Chinese paleolithic archaeologists have been much more occupied with the question of why some sites seem to preserve only large core-tool assemblages, whereas others yield only small flake-tool assemblages (Jia et al., 1972; cf. Jia and Huang, 1990; Jia and Wei, 1987). Although these assemblages were originally interpreted as indications of two distinct cultures, Chinese archeologists have recently begun to consider the possibility that they represent different activity facies (Jia and Huang, 1990).

My own work in the Far East convinces me that an emphasis on environmental variation of archaeological assemblages and nonlithic technology, in combination with variability in local-activity facies, is the best way of understanding the archaeological record of both early and late paleolithic assemblages (Pope et al., 1990; Pope, 1988). There is no justification for the notion that inadequate research is the source of the impression that the Far Eastern archaeological record is essentially the same as that for western Eurasia and Africa (contra Kline, 1992; Clarke, 1992). Cultural backwater models or contentions of inadequate research result from the paucity of western archaeologists who are able to read the primary Asian literature. I know of no Westerners or Chinese conversant with the data and literature who would support either contention. It is true that research techniques in Asia have been less sophisticated than in the

West, but more than half a century of even the most simply conceived research strategies have established beyond a doubt that the Far Eastern record contrasts markedly with that of Europe and Africa. The impression that Far Eastern assemblages are strictly informal, casual, or crude is belied by recent discoveries from Early Pleistocene sites in Southeast Asia and China indicating that early *Homo erectus* did in fact manufacture formalized artifacts (Pope et al, 1990; Pope, 1993). The argument that the lack of tractable raw material in Asia precluded the manufacture of standardized artifacts is also vitiated by the fact that suitable raw materials in the form of fine-grained cherts and (less frequently) basalts are present at or near sites that have yielded typical Far Eastern assemblages. Once again, geography and the availability of nonlithic resources are by far the most meaningful approaches to understanding the variability of Far Eastern assemblages in Southeast and East Asia.

Southeast Asia. An understanding of the paleolithic record in Southeast Asia has been hindered by geological circumstances that make the recovery of in situ assemblages difficult. The complexity of geological conditions in that area has often been confused and ignored in arguments holding that *Homo erectus* may not have manufactured stone artifacts at all (Hutterer, 1985; Bowdler, 1988, 1990). In Java, volcanic activity has hindered archaeological research as much as it has biostratigraphic research. Von Koenigswald's (1936a, 1936b) surface discovery of the Patjitanian has never been replicated by subsequent workers. In fact, hand axes have never been found in association with *Homo erectus* anywhere in the Far East, which has given rise to the suggestion that hand axes are associated with the arrival of anatomically modern *Homo sapiens* in Java (Bartstra, 1984; Bartstra et al., 1988). A few artifacts are known from Middle Pleistocene strata in Java; these are correlated with the Sambungmachan hominid find (Jacob et al., 1978).

On the mainland of Southeast Asia, the plethora of paleolithic cultures recognized by early workers is no longer tenable. It is now clear that many of these assemblages are either of natural origin or of much more recent than originally supposed. However, recent discoveries of radiometrically dated artifacts from northern Thailand (Kao Pah Nam, Ban Don Mun, Mae Tha) do document the presence of late Early Middle Pleistocene artifacts that display some formality and standardization (Pope et al., 1987, 1990). One characteristic of Southeast Asian (and southern Chinese) assemblages that continues to stand out is the small number of artifacts at any given site. This is true of the Early and Middle Pleistocene.

East Asia. The question of whether the Acheulian or Mousterian actually exists in China and adjacent countries should now be dismissed; instead, current research and strategies should recognize that although components of these assemblages do occur, they are rare or recent. The trihedral "picks" and "cleavers" from Lantian, Yunxian, Liang Shan, Chonggokni (Korea), Dingcun, and a few other areas in East Asia are always small portions of the total assemblages. Their very presence, however, indicates that early hominids did not lack the ability and "cultural capacity" (Hutterer, 1985; Binford and Ho, 1985; Binford and Stone, 1986) to manufacture such artifacts. It is also possible to see some chronological contrasts in temporally disparate assemblages (Zhang, 1990), even though they are extremely subtle.

Discussion

One admittedly sweeping conclusion supported by the Far Eastern paleoanthropological evidence is that climatic change and simple isolation of the entire region do not provide the appropriate framework for understanding hominid evolution in that part of the world. The paleoanthropological data do not fit well with evolutionary scenarios that have been developed for other parts of the Pleistocene Old World. Over one hundred years of research make it clear that cultural developments and biotic evolution were essentially conservative in this part of the world; future investigations are unlikely to alter this view. The lack of evidence for severe climatic oscillation, marked technological change, and diachronic morphological change has too often been equated with viewing the Far East as a "stagnant" peripheral arena of human evolution. A general Late Neogene trend toward progressive seasonality and aridification (especially in western China) continues today and is consistent with the interpretation that the past environments of Chinese fossil hominids were somewhat warmer and wetter than they are presently in this part of Eurasia. In tropical Java, the climate remained relatively constant and similar to present-day conditions. Insularization is postulated as a much more important factor, which resulted in the homogenization of the late Premodern fossil populations of the island.

A direct link between morphological and behavioral evolution in the Far East is unsupportable. Geographic variation (clinal?) seems to be the most parsimonious and useful means for understanding the morphology and archaeology of late *Homo erecus* and early *Homo sapiens.* The appearance of new morphologies more reminiscent of western Eurasia in the late Middle Pleistocene of China suggests genetic input from the West. At the same time, certain previous craniofacial morphologies continue on into modern East Asian populations (Pope, 1989, 1991, 1992). It is extremely interesting that few of the novel morphologies continue on in subsequent populations—for example, the large faces of specimens such as Yunxian and Nanjing. Taller maxillae do increase in frequency in comparison with earlier nonmodern samples; rather, it is a low maxilla with an inferiorly situated, horizontal maxillary root that characterizes fossil and extant East Asian populations. The continuity in dental features is well established (Turner, 1989, 1990).

No temporal overlap of Premodern *Homo sapiens* and anatomically modern *Homo sapiens* has yet been discovered in the Far East. The possibility that late *Homo erectus* overlaps with early *Homo sapiens* has been raised, however, by similar dates for late Zhoukoudian hominids and Dali (Chen and Zhang, 1991). At the same time it is also entirely possible that these two samples are separated by 0.05 myr or more. Neither punctuated equilibrium nor replacement need to be invoked as an explanation of accelerated morphological change in the terminal Middle Pleistocene. Instead, an increased selection pressure for less robust anatomical complexes in combination with increased extraregional gene flow present a very plausible interpretation of population dynamics in the Late Pleistocene.

General and Theoretical Conclusions. There is growing consideration of the proposition that climatic change in the form of the dessication of phytogeographic resources and increasing aridity selected for a shift in hominid biobehavioral adaptations. Specifically, a decrease in available plant foods facilitated an increasingly intense reliance on the utilization of animal resources obtained by scavenging or hunting. Such a scenario may account for both the hominidization and hominization process. In the Far East, however, climate appears to have remained relatively stable, and lithic technology cannot be said to have undergone any rapid or dramatic transformations. Technological and cultural revolutions of the kind that characterize parts of Upper Paleolithic Europe are not apparent in the Far East. The archaeological record in Australia contrasts markedly with that in the Far East (cf. Jones, 1989). Paleolithic rock art is restricted to Australia, but this perception may be expected to change as research continues. Already apparent on the basis of our current knowledge is that the ability to make standardized artifacts was present early on in the paleolithic record of the Far East. Early Pleistocene artifacts from Thailand and China show a simple, consistent, but sporadic standardization that indicates the ability to manufacture more formal tool categories. Blades, whether accidental or not, are present in both Early (Jia and Wei, 1987) and Late (Gai, 1991) Pleistocene Nihewan assemblages (Pope et al., 1990). Furthermore, a consistent mental template is evident in artifacts from Thailand (Pope et al., 1987). The early dating of the highly symmetrical and standardized Pacitanian remains difficult to demonstrate and may in fact be associated not with *Homo erectus* but with *Homo sapiens.* Unfortunately, various other putative industries that were originally recognized primarily on the basis of doubtful surface finds have tended to obscure the reality of Early Pleistocene lithic assemblages in Southeast Asia. The climatic history of the Far East makes it difficult to avoid the conclusion that nonlithic materials comprise a substantial resource for tool manufacture.

One traditional interpretation of both technology and morphology—posed largely in terms of increased efficiency in the technology of food processing and reduced selection pressure for masticatory efficiency (Brace et al., 1984; Brace and Hunt, 1990)—does not seem to fit in any obvious way with the Asian evidence. For instance, it is difficult to relate reduced dental and facial size to any even penecontemporaneous technological changes. At least partly for this reason, imperfectly comprehended selection pressures and simple gene flow seem the most useful models for understanding the current evidence. In combination with the relatively conservative nature of the archaeological evidence, this interpretation underscores the notion that gene flow can be quite independent of the introduction of technological change. A similar conclusion can be drawn from both Middle Eastern and European evidence of the Late Pleistocene.

Biobehavioral Thresholds of Humanness. Paleoanthropologists have long searched for rubicons of truly sapient human behavior. Language, art, foresight, and complexity and standardization of artifacts have been the most frequently discussed hallmarks of modern human behavior. Evidence for most of these behaviors has been suggested to derive from Africa and Europe. Evidence for language in the form of hominid endocasts and handedness is first known from Africa, but it predates the appearance of anatomically modern humans by more than 1 myr. There is no convincing anatomical or archaeological evidence to associate the appearance of language with the increase in complexity that characterizes the Upper Paleolithic in Europe or Africa. In fact "complexity" (at least as indicated by material culture) antedates the appearance of modern humans everywhere. Persistent arguments about the inadequacy of the vocal apparatus of hominids other than modern *Homo sapiens* are based on unconvincing reconstructions of soft tissue from their osseous substrates and are not compatible with the fully modern Neanderthal hyoid now known from Kebara (Frayer et al., 1993). Arguments that early *Homo sapiens* lacked the ability to reproduce the full range of modern human speech are especially inappropriate in large geographical areas of the Far East (i.e., China and Thailand) and in numerous North American languages (e.g., Amerind languages), where different tones of an identically produced syllable can carry as many as sixteen different meanings (pers. obs.).

On the bases of current evidence, the earliest art derives exclusively from Europe and possibly Australia. One conspicuous feature of the Far Eastern data is the seeming lack of portable and rock art. Evidence of body adornment in Asia may date back as far as 0.02 myr, but this date depends only on the much disputed age of the Upper Cave at Zhoukoudian. The geological context of Far Eastern finds that may have minimized the production of rock art. We know that it is present at an early age in Australia and that there are also provocative indications of early portable art from Siberia. In northern China, the source of most Chinese hominid discoveries, the relative paucity of rock outcrops and suitable caves may have had a substantial impact on both the execution and preservation of graphic depictions. The same is true in Java. It is important to point out that many areas of karstic Southeast Asia and southern China are highly suitable for the rendering of painted and etched rock art. The highly seasonable climate, however, may preclude long-term preservation of more than a few thousand years. To conclude that early developments of recognizably sapient and modern behavior took place first in Europe is premature. Once again, preservational context is more likely to account for this perception. One should keep in mind that although principal hominid-yielding regions of southern China and Southeast Asia are subtropical and tropical, they are subject to extreme seasonality, ranging from torrential monsoon rains to annual periods of no rain and high temperatures. Although these climates may have been less seasonal and more equitable during the Pleistocene, they are not as conducive to preservation as the rock shelters of Europe and North Africa. This contrast between the Far East and other regions of the Old World should always be considered when attempting to assess the data base used in deciphering hominid evolution in various parts of the world.

Claims for truly human complexity and foresight, as deduced from archaeological studies of lithic assemblages, have historically emphasized qualities perceived in the western paleolithic record. The emphasis on stone itself, though usually a practical necessity, almost certainly overlooks a fundamental difference between places such as Europe and the Far East. It is hard to imagine that early hominids in the Far East could have ignored resources as versatile as bamboo. In areas of the Far East where bamboo is absent, the standardization of lithic artifacts increases notably, even at Early Pleistocene localities. Perhaps an even deeper prejudice that Western scholars have brought to their studies of prehistoric Asian culture is a concern with aspects of food procurement and premasticatory food processing such as cutting edges, points, and the butchering of both large and small animals. The modern observation of hunters and gatherers in Southeast Asia suggests that lithic inventories must have been minuscule compared to nonlithic utilitarian items.

Conclusion

Severe climatic fluctuations and rapid faunal turnovers did not occur in Pleistocene East and Southeast Asia, as reflected at least partially by the obvious diachronic continuity of archaeological assemblages and hominid morphology. The earliest paleolithic samples, though indicative of sporadic formalization, are very similar to Late Pleistocene assemblages from the same regions. The picture of regional morphological changes is more complex. The hominid fossils in Java show marked diversity in the Early Pleistocene and marked homogeneity beginning in the terminal Middle Pleistocene. In China, the early hominids show general similarities that in terminal Middle Pleistocene times give way to a dramatic increase in morphological diversity. The reason for the contrasts between diachronic trends in Java and China can be related to the contrasting influences of relatively stable island versus relatively stable continental habitats, respectively. Specifically, it is suggested that Javanese hominids became increasingly isolated, while Chinese hominids became more morphologically diverse, at least in part owing to increased gene flow from the West. This hypothesis is in accord with independent mammalian biogeographic and biostratigraphic evidence from the Far East. A more general conclusion about biotic evolution in the Far East is that geographic variation and temporally stable clinal climatic contrasts were much more important in shaping human evolution in that area than was the diachronic climatic change that resulted in shifting environmental selection pressures. With these observations in mind, it is important to emphasize that the biocultural conservativeness of the Far East should not be equated with models of cultural retardation, stagnation, or phylogenetic dead ends, which can be used to support a Replacement Model for the origin of modern humans in Asia. It is further strongly argued that the Far Eastern record is sufficiently well studied to conclude that the impression of cultural and environmental contexts that contrast with Africa and Europe is not the result of inadequate research. The uniqueness of the Far Eastern record is much more accurately viewed as a set of new ecological strategies that arose as a response to different habitats that early and later species of *Homo* first encountered in the Southeast Asian tropics.

References

Aigner, J. S. 1981. *Archaeological remains in Pleistocene China.* C. H. Beck, Munich.

———. 1986. The age of Zhoukoudian locality 1: The newly proposed O[18] correspondence. In *Fossil man: New facts, new ideas* (ed. V. V. Novothy and A. Miserova). Papers in Honor of Jan Jelinek's Life Anniversary. *Anthropos. Brno.* 23:157–173.

———. 1988. Dating the earliest Chinese Pleistocene localities: The newly proposed O[18] correspondences. In *The palaeoenvironment of East Asia from the Mid-Tertiary,* vol. 2, pp. 1032–1061 (ed. P. Whyte, J. S. Aigner, N. G. Jablonski, G. Taylor, D. Walker, and P. Wang). Centre of Asian Studies of the University of Hong Kong, Hong Kong.

An, Z., Liu, T., Kan, X., Sun, J., Wang, J., Gao, W., Zhu, Y., and Wei, M. 1987. *Loess-paleosol sequence and chronology at Lantian Man localities: Aspects of loess research.* China Ocean Press, Beijing.

An, Z., and Lu, Y. 1984. A climatostratigrahic subdivision of late Pleistocene strata named by Malan Formation in North China. *Kexue Tongbao* 29:1240–1242.

An, Z., Wu, X., Wang, P., Wang S., Sun, X., and Lu, Y. 1991. An evolution model for paleomonsoon of China during the last 130,000 years. In *Quaternary geology and environment in China,* pp. 237–244 (ed. T. Liu). Science Press, Beijing.

Bartstra, G.-J. 1984. Dating the Pacitanian: Some thoughts. In *The early evolution of man with special emphasis on South East Asia and Africa* (ed. P. Andrews and J. L. Franzen). *Cour. Forsch. Senckenberg* 69:253–258.

Bartstra, G.-J., Soegondho, S., and van der Wijk, A. 1988. Ngandong man: Age and artifacts. *J. Hum. Evol.* 17:325–337.

Binford, L. R., and Ho, C. K. 1985. Taphonomy at a distance: Zhoukoudian, "The cave home of Beijing man"? *Curr. Anthropol.* 26:413–442.

Binford, L. R., and Stone, N. M. 1986. Zhoukoudian: A closer look. *Curr. Anthropol.* 27:453–475.

Bowdler, S. 1988. Early Southeast Asian prehistory: A view from Down Under (Abstract). Ass. Southeast Asian Archaeol. Western Europe, Sec. Internat. Conf., Paris, Musée Guimet, September, 1988.

Bowdler, S. 1990. The earliest Australian stone tools and implications for Southeast Asia. 14th Congress Indo-Pacific Prehistory Association, Yogyakarta, 25th Aug.–2nd Sept. 1990.

Bowdler, S. 1991. The evolution of modern humans, *Homo sapiens sapiens,* in East Asia: Implications of archeological evidence from Australia and Southeast Asia. Abstracts, p. 33. INQUA, Beijing.

Brace, C. L. and Hunt, K. D. 1990. A non-racial craniofacial perspective on human variation: A(ustralia) to Z(uni). *Am. J. Phys. Anthropol.* 83:341–360.

Brace, C. L., Shao, Z., and Zhang, Z. 1984. Prehistoric and modern tooth size in China. In *The origin of modern humans: A world survey of the fossil evidence,* pp. 485–516 (ed. F. H. Smith and F. Spencer). New York: Alan R. Liss, New York.

Chen, T., and Yuan, S. 1988. Uranium-series dating of bones and teeth from Chinese Palaeolithic sites. *Archaeometry* 30:59–76.

Chen, T., and Zhang, Y. 1991. Palaeolithic chronology and possible coexistence of *Homo erectus* and *Homo sapiens* in China. *World Arch.* 23:147–154.

Clark, R. J. 1990. The Ndutu cranium and the origin of *Homo sapiens. J. Hum. Evol.* 19:699–736.

De Vos, J., Sartono, S., Hardja-Sasmita, S., and Sondaar, P. 1982. The fauna from Trinil, type locality of *Homo erectus:* A reinterpretation. *Geol. Mijnbouw* 61:207–211.

Frayer, D., Wolpoff, M. H., Thorne, A. G., Smith, F. H., and Pope, G. G. 1993. Theories of modern human origins: The paleontological test. *Am. Anthropol.* 95:14–50.

Gai, P. 1991. Microblade tradition around the northern Pacific Rim: A Chinese perspective. In *Contributions to the Thirteenth INQUA, Beijing:* 21–31.

Gao, J. 1975. Australopithecine teeth associated with *Gigantopithecus. Vert. Palas.* 13:81–88.

Han, D., and Xu, C. 1985. Pleistocene mammalian faunas of China. In *Palaeoanthropology and palaeolithic archaeology in the People's Republic of China,* pp. 267–286 (ed. R. Wu and J. W. Olsen). Academic Press, Orlando.

———. 1989. Quaternary mammalian faunas and environment of fossil humans in South China. In *Early humankind in China* pp. 338–391 (ed. R. Wu, X. Wu, and Y. Zhang). Science Press, Beijing.

Hoang, X., Cuong, N., and Long, V. 1979. First discoveries of Pleistocene man, culture and fossilized fauna in Vietnam. In *Recent discoveries and new views on some archeological Problems in Vietnam,* pp. 14–20 (ed. Committee for Social Sciences of Vietnam). Institute of Archeology, Hanoi.

Hu, C. 1973. Ape-man teeth from Yunnan. *Acta Geol. Sin.* 1:65–71.

Huang, W. 1987. Bifaces in China. *Acta Anthropol. Sin.* 6:61–68.

———. 1990. Bifaces in China. *Hum. Evol.* 4:87–92.

Huang, W., Fang, D., and Ye, Y. 1982. Preliminary study of the fossil hominid and fauna from Hexian, Anhui. *Vert. PalAs.* 20:248–256.

Huang, W., and Fang. Q. 1991. *Wushan hominid site.* China Ocean Press, Beijing.

Huang, W., and Huang, C. 1985. Mammal fossils and sporo-pollen compositions at Hexian Man locality and their significance. In *(Selection) Symposium of National Conference on Quaternary Glacier and Periglacial,* pp. 180–183. Kexue Chubanshe, Beijing.

Hutterer, K. L. 1985. The Pleistocene archaeology of Southeast Asia in regional context. *Mod. Quatern. Res. Southeast Asia* 9:1–25.

Itihara, M., Arkhipov, S. A., Wang, P., and Kumai, H. 1991. Major subdivisions of the Quaternary of Asia, and their litho- and biostratigraphic significance. In *Special Proceedings Review Reports Thirteenth Inter. Cong. INQUA,* p. 75. Beijing.

Jia, L., Gai, P., and You, Y. 1972. A report on the excavation of Shanxi Shiyu paleolithic site. *Kaogu Xue Bao* 1:39–58.

Jia, L., and Huang, W. 1990. *The story of Peking Man.* Foreign Language Press and Oxford University Press, Hong Kong.

Jia, L., and Wei, Q. 1987. Stone artifacts from Lower Pleistocene at Donggutuo site near Nihewan (Nihowan), Hebei Province, China. *L'Anthropologie* 91:727–732.

Jones, R. 1989. East of Wallace's Line: Issues and problems in the colonization of the Australian Continent. In *The human revolution: Behavioural and biological perspectives on the origins of modern humans,* pp. 743–782 (ed. P. Mellars and C. B. Stringer). Edinburgh University Press, Edinburgh.

Klein, R. G. 1989. *The human career: Human biological and cultural origins.* University of Chicago Press, Chicago.

———. 1992. The archeology of modern human origins. *Evol. Anthropol.* 1:5–14.

Lee, J. S. 1939. *The geology of China.* Nordeman, New York.

Li, H., and Wang, J. 1982. Magnetostratigraphic study of several typical geologic sections in north China. In *Quaternary geology and environment of China.* Quaternary Research Association of China. China Ocean Press, Beijing.

———. 1991. The latest advance in Quaternary magneto-statigraphy of China. In *Quaternary geology in China,* pp. 158–167 (ed. Liu Tengsheng). Science Press, Beijing.

Li, J. 1991. The uplift of the Qinghai-Xizang Plateau and its effect on environment. In *Quaternary geology and environment in China,* pp. 265–272 (ed. T. Liu). Science Press, Beijing.

Li, T. 1991. Unearthing Chaoxian man's fossil skull. *China Cultural Report* 5:1.

Liu, T. 1991a. *Loess, environment and global change.* Science Press, Beijing.

———. 1991b. *Quaternary geology and environment.* Science Press, Beijing.

Liu, T., and Ding, M. 1984. Mammoths in China. In *Quaternary extinctions: A prehistoric revolution,* pp. 517–527 (ed. P. S. Martin and R. G. Klein). University of Arizona Press, Tucson.

Liu, T., Lu, Y., Zheng, H., Wu, Z., and Yuan, B. 1985. *Loess and the environment.* China Ocean Press, Beijing.

Luo, B., Wang, Y., Lin, Z., Chen, M., Lan, C., and Fu, M. 1988. In *Paleontological research in Nihewan Beds.* In *Study of Nihewan beds,* pp. 40–62 (ed. M. Chen) Ocean Press, Beijing.

Matsu'ra, S. 1982. A chronological framing for the Sangiran hominids: Fundamental study of the flourine dating method. *Bull. National Science Museum,* ser. D. (Anthropology), 8:1–53.

Movius, H. L. 1944. Early man and Pleistocene stratigraphy in southern and eastern Asia. *Papers of the Peabody Museum, Harvard University* 9:1–125.

Pope, G. G. 1977. Hominids from the Lower Pleistocene of South China. *Kroeber Anthropol. Soc. Pap.* 50:63–73.

———. 1982. Hominid evolution in East and Southeast Asia. Ph.D. diss., University of California, Berkeley.

———. 1983. Evidence of the age of the Asian hominidae. *Proc. Nat. Acad. Scien.* 80:4988–4992.

———. 1984. The antiquity and paleoenvironment of the Asian hominidae. In: *The evolution of the East Asian environment,* vol. 2, pp. 822–847 (ed. R. O. Whyte). Centre of Asian Studies of the University of Hong Kong, Hong Kong.

———. 1985. Taxonomy, dating, and paleoenvironment: The paleoecology of the early Far Eastern hominids. *Mod. Quat. Res. in Southeast Asia* 9:65–80.

———. 1988. Recent advances in Far Eastern paleoanthropology. *Ann. Rev. Anthropol.* 17:43–77.

———. 1989. Bamboo and human evolution. *Nat. Hist.* 98:48–57.

———. 1991. Evolution of the zygomaticomaxillary region in the genus *Homo* and its relevance to the origin of modern humans. *J. Hum. Evol.* 21:189–213.

———. 1992.The craniofacial evidence for the origin of modern humans in China. *Yearbook Phys. Anthropol.* 35:243–298.

———. 1993. Ancient Asia's cutting edge. *Nat. Hist.* 102:54–58.

Pope, G. G., An, Z., Keates, S., and Bakken, D. 1990. New discoveries in the Nihewan Basin, northern China. *East Asian Tertiary/Quaternary Newsletter* 11:68–73.

Pope, G. G., and Keates, S. G. 1994. The evolution of human cognition and cultural capacity: A view from the Far East. In *Integrative pathways to the past,* pp. 531–567 (ed. R. L. Ciohon and R. Courcinni) Prentice-Hall, Englewood Cliffs, N.J.

Pope, G. G., Nakabanlang, S., and Pitragool, S. 1987.Le paléolithique du nord de la Thaïlande: Découvertes et perspectives nouvelles. *L'Anthropologie* 91:749–754.

Qi, G. 1989. Quaternary mammalian faunas and environment of fossil humans in North China. In *Early humankind in China,* pp. 237–337 (ed. R. Wu, X. Wu, and S. Zhang). Science Press, Beijing.

———. 1990. The Pleistocene human environment of North China. *Acta Anthropol. Sin.* 9:340–349.

Robinson, J. T. 1950. The evolutionary significance of the australopithecines. *Am. J. of Phys. Anthropol.* 11:1–38.

Santa Luca, A. 1980. The Ngandong fossil hominids. *Yale Univers. Publ. Anthropol.* 78:1–175.

Sartono, S. 1982. Sagittal cresting in *Meganthropus palaeojavanicus* (v. Koenigswald). *Mod. Quat. Res. Southeast Asia* 7:201–210.

Schick, K., and Dong, Z. 1992. Early paleolithic of China and Eastern Asia. *Evol. Anthropol.* 3:22–70.

Sémah, F. 1982. Pliocene and Pleistocene geomagnetic reversals recorded in the Geomolong and Sangiran domes (Central Java). *Mod. Quatern. Res. Southeast Asia* 7:151–164.

Sémah, F., Sémah, A. M., and Djubiantono, T. 1990. *They discovered Java.* Pusat Penelitian Arkeologi, Jakarta, and Muséum National d'Histoire Naturelle, Paris.

Song, S., and Zhang, Z. 1988. The Maba fauna. In *Treatises in commemoration of the thirtieth anniversary of the discovery of Maba human cranium,* pp. 23–35. Guangdong Provincial Museum and the Museum of the Qujiang County, Beijing; Cultural Relics Publishing House, Beijing.

Stringer, C. B. 1988. The dates of Eden. *Nature* 331:565–566.

———. 1989. The origin of early modern humans: A comparison of the European and non-European evidence. In *The human revolution: Behavioural and biological perspectives on the origins of modern humans,* pp. 232–244 (ed. P. Mellars and C. Stringer). Edinburgh University Press, Edinburgh.

———. 1990. The Asian connection. *New Scientist* 178:33–37.

Stringer, C. B., and Andrews, P. 1988. Genetic and fossil evidence for the origin of modern humans. *Science* 239:1263–1268.

Sun, D. 1991. The hunting for the Quaternary glaciers in China. In *Quaternary geology and environment in China,* pp. 1–15 (ed. L. Teng). Science Press, Beijing.

Thorne, A. G., and Wolpoff, M. H. 1981a. Regional continuity in Australasian Pleistocene hominid evolution. *Am. J. Phys. Anthropol.* 55:337–349.

———. 1981b. The multiregional evolution of humans. *Sci. Amer.* 266:28–33.

Tobias, P. V. 1991. The skulls, endocasts and teeth of *Homo habilis.* Vol. 4. Cambridge University Press, Cambridge.

Tobias, P. V., and von Koenigswald, G. H. R. 1964. A comparison between the Olduvai hominines and those of Java and some implications for homini phylogeny. *Nature* 204:515–518.

Turner, C. G., II 1989. Teeth and prehistory in Asia. *Sci. Amer.* 260:88–96.

———. 1990. Major features of Sundadonty and Sinodonty, including suggestions about East Asian microevolution, population history, and Late Pleistocene relationships with Australian aboriginals. *Am. J. Phys. Anthropol.* 82:295–317.

von Koenigswald, G. H. R. 1934. Die Stratigraphie des javanischen Pleistocan. *Ingen. Ned. Indie* 1:85–201.

———. 1935. Die fossilen Saugertierfaunen Javas. *Proc. Kn. Akad. Wet. Amsterdam* 38:188–198.

———. 1936a. Early Palaeolithic stone implements from Java. *Bull. Raffles Museum,* ser. B, 1:52–60.

———. 1936b. Über altpalaeolithische Artefakte von Java. *T. K. Nederl. Aardrijkskd. Genoot.* 53:41–44.

Wang, L. 1989. New progress in Chinese paleoanthropology. In *Early humankind in China,* pp. 41–57 (ed. R. Wu, W. Wu, and S. Zhang). Beijing Science Press, Beijing.

Weidenreich, F. 1939. On the earliest representatives of modern mankind recovered on the soil of East Asia. *Bull. Nat. Hist. Soc.* (Peking) 13:161–174.

———. 1943. The skull of *Sinanthropus pekinensis:* A comparative study on a primitive hominid skull. *Pal. Sin.,* new ser. D, 10:1–484.

———. 1945. Giant early man from Java and south China. *Anthropol. Pap. Am. Mus. Nat.* 40:1–134.

Wilson, A. C., and Cann, R. L. 1992. The recent African genesis of humans. *Sci. Amer.* 266:66–73.

Wolpoff, M. H., Wu, X., and Thorne, A. G. 1984. Modern *Homo sapiens* origins: A general theory of hominid evolution involving the evidence from East Asia. In *The origin of modern humans: A world survey of the fossil evidence* pp. 411–483 (ed. F. Spencer and F. Smith). Alan R. Liss, New York.

Wood, B. 1985. Early *Homo* in Kenya and its systematic relationships. In *Ancestors: The hard evidence,* pp. 206–214 (ed. E. Delson). Alan R. Liss, New York.

Wu, M. 1986. Investigation of Xujiayao man fossils. *Acta Anthropol. Sin.* 5:220–226.

Wu, R., and Dong, X. 1985. *Homo erectus* in China. In *Paleoanthropology and paleolithic archaeology in the People's Republic of China,* pp. 79–88 (ed. R. Wu and J. W. Olsen). Academic Press, New York.

Wu, R., and Peng, R. 1959. Fossil human skull of early paleoanthropic stage found at Mapa, Shaokuan, Kwangtung Province. *Palaeovert. et Palaeoanthropol.* 4:159–164.

Wu, R., Wu, X., and Zhang, S. 1989. *Early humankind in China.* Science Press, Beijing.

Yuan, S., and Chen, T. 1991. Chinese Quaternary radioscopic chronology. In *Quaternary geology and environment in China,* pp. 179–184 (ed. T. Liu). Science Press, Beijing.

Zhang, S. 1990. Regional industrial gradual advance and cultural exchange of Paleolithic in north China. *Acta Anthropol. Sin.* 9:322–333.

Chapter 35

The Role of Climate in the Interpretation of Human Movements and Cultural Transformations in Western Asia

Ofer Bar-Yosef

Western Asia provides a unique opportunity to examine several crucial steps in human evolution that may have been triggered by climatic changes. The oldest movements of hominids resulted from major shifts in the distribution of food resources on the African landscape that occurred during and after the Olduvai subchron. A series of migrations out of Africa by *Homo erectus* is recognizable in the archaeological records of Eurasia, even when the fossils themselves are not found. Sites with either Oldowan-type core-choppers or Early Acheulean bifaces are interpreted as representing traditional ways of tool manufacture by hominids who may not have had the mental flexibility of modern humans.

It is tempting to see the dispersals of modern humans, who (according to nuclear and molecular studies) evolved in Africa sometime between ca. 500 and ca. 50 thousand years (kyr) (e.g., Stoneking, 1993), as resulting from environmental changes in their homeland. Without identifying the timing of these dispersal events, however, it is difficult to test their relationship to known climatic fluctuations during the Late Middle and Upper Pleistocene. In the desertic Saharo-Arabian belts, evidence indicates that humans did not survive in arid zones when glacial conditions prevailed in northern latitudes. Movements out of Africa are therefore expected to have taken place in the more humid intervals, either at the onset of glacial cycles (such as Isotope Stages 5d, ca. 115 kyr, and 5b, ca. 90 kyr) or perhaps later, during early Stage 3, ca. 60 or 50 kyr. Movement southward from northern latitudes was driven by the expansion of glaciated areas and the periglacial belt. Because there is no evidence that prehistoric technologies before the Upper Paleolithic enabled humans to survive in close proximity to the glaciers, they must have had to seek foraging territories in the Mediterranean Basin or in the lowlands around the Black and Caspian Seas. The Levant offered the most adequate refugium, with different vegetation associations that provided over one hundred species of edible seeds, fruits, and leaves, as well as a number of medium-sized mammals, reptiles, and birds that could have been hunted or trapped. Evidence for the presence of Neanderthal morphological features in Middle Paleolithic human fossils (ca. 80–55 kyr) in the Levant is therefore not surprising. Another example is the solid archaeological evidence for the Levantine coastal expansion of the Aurignacian culture, which originated in southeastern Europe (ca. 36–27 kyr).

The possible role of environmental changes in triggering the Middle-Upper Paleolithic revolution must also be tested. This revolution, ca. 47–45 kyr, represents a time of rapid technological and social change, as expressed in the European and Near Eastern archaeological record and by the incremental exploitation of the northern latitudes in Eurasia.

Finally, the establishment of farming communities ca. 10 kyr in the Levantine Corridor is interpreted to be the result of socioeconomic decisions made by sedentary hunter-gatherers facing the vagaries of the Younger Dryas (Bar-Yosef and Belfer-Cohen, 1992; Wright, 1993). Newly emerging lifeways are seen as the result of the "Neolithic Revolution" that in due course led to the establishment of urban centers and hierarchical, bureaucratic civilizations in the Near East. In the following pages I shall briefly explore each of these cultural revolutions in the light of current paleoclimatological, chrono-

ISBN 0-300-06348-2.

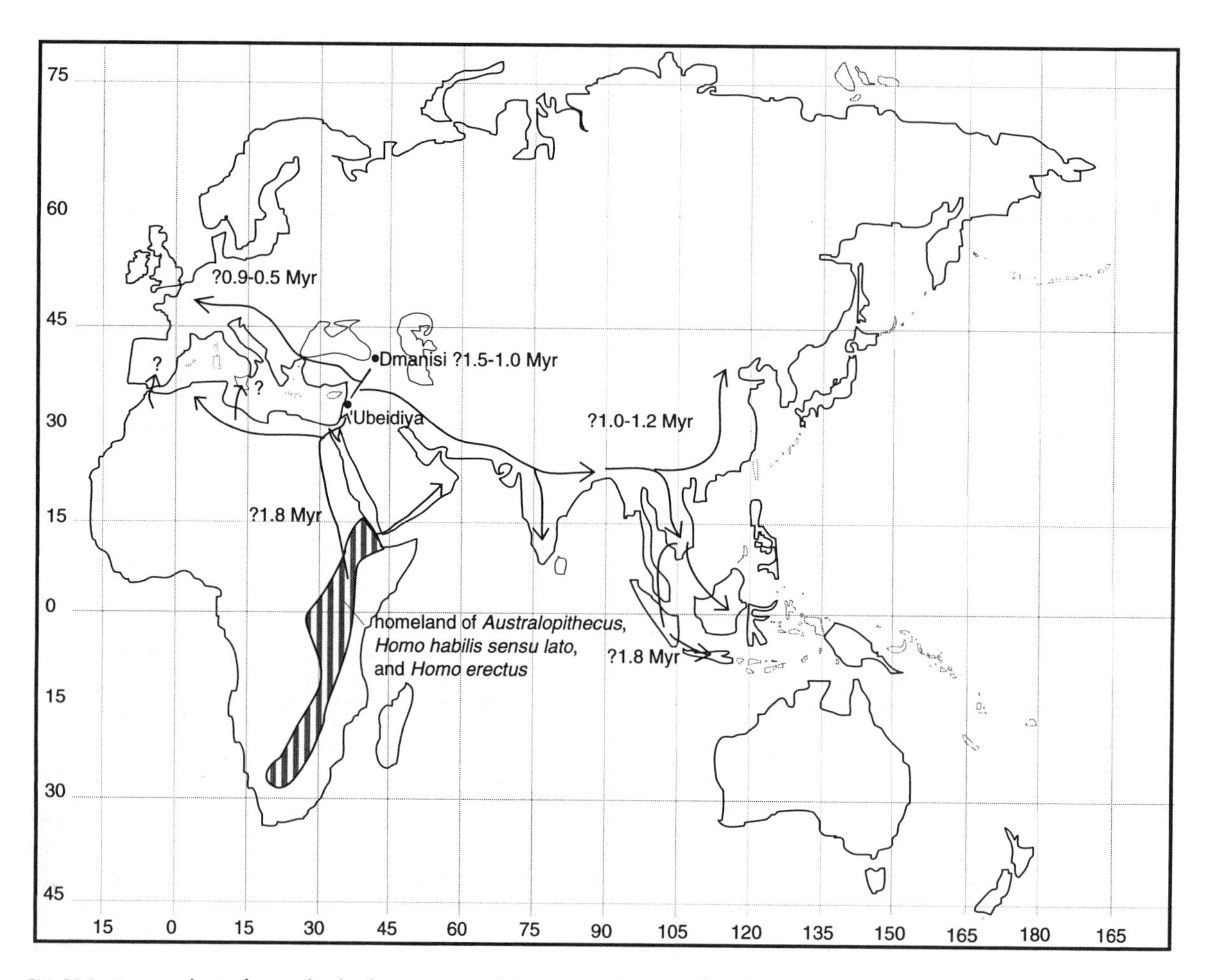

Fig. 35.1. A suggested map of potential early colonization routes of *Homo erectus,* taking into account the new dates for Java (Swisher et al., 1994) and the proposed dates for western Asia, Europe, and East Asia.

logical, and archaeological knowledge from western Asia with cursory reference to other regions of the Old World.

Identifying Populations in the Archaeological Record

Two general conclusions underlie the suggested interpretations for the Lower Paleolithic occurrences of human populations in western Asia. The first relates to the appearance of *Homo erectus* as an adaptation to the environmental settings during or immediately after the Olduvai subchron (1.95–ca. 1.84 myr). The Olduvai subchron seems to have resulted in considerable paleoecological changes. From about 3.7 to 1.0 myr ago, various australopithecines occupied the parkland-woodland region of at least East Africa, as did *Homo habilis sensu lato* from about 2.5 to 1.9 myr. Following major environmental shifts that resulted in the expansion of drier ecozones, groups of *Homo erectus* (a successful descendant of a certain *Homo habilis* population) must have had the necessary social and technical skills and biological capacities to colonize new regions (fig. 35.1). They moved into coastal North Africa, either across the Sahara, and/or along the Nile Valley, as well as into western and southeastern Asia. It is unlikely that all colonizations were successful; extinctions are expressed as gaps in the archaeological sequences. It was suggested (Bonifay and Vandermeersch, 1991) but later rejected (Roebroeks, 1994) that around 0.9 myr or a little earlier groups of late *Homo erectus* ventured into Europe. If such efforts were made, most were terminated abruptly by the glacial cycles. It was only later, with the emergence of Archaic *Homo sapiens* around 0.5 myr, that Europe became permanently inhabited (Roebroeks, 1994). At a later phase, this continuous occupation created the genetic background for the evolution of Neanderthals through periods of population isolation.

The second general conclusion is the assumption that prehistoric populations can be recognized, in many instances, on the basis of lithic studies. In the last three decades, in-depth investigations of lithic assemblages across the Old World and in Australia have demonstrated the feasibility of this contention. Lithic analysts and knappers have looked into how artifacts were made, used, and abandoned—a process better known as *chaîne opératoire,* or operational sequence (e.g., Geneste et al., 1990; Bar-Yosef, 1991; Bar-Yosef and Meignen, 1993; Toth and Schick, 1993). Others have studied the function of the artifacts and their "life histories" and have demonstrated that most of the recorded morphological variability in well-controlled samples of Lower, Middle, and Upper Paleolithic assemblages can be attributed to "style" rather than to function or to the limitations imposed by the accessibility of raw material. *Style* in this context means the guarding of long traditions of knapping techniques regardless of their efficiency (as measured by modern eyes). Quantitative studies based on type-lists, however, generally document the intensity of tool use (e.g., Dibble, 1991), site use, the presence of ad hoc–shaped stone artifacts versus curated tools, the amount of retooling, and other activities. From this kind of documentation, it would be presumptuous to assume that populations of *Homo erectus* and Archaic *Homo,* about whose social structure we know very little (Pilbeam, 1989), took the same innovative approaches with which we credit modern humans.

The general observation that stone-tool typology need not correspond to the state of cultural evolution can be illustrated by several examples. From a Eurocentric viewpoint we might expect that modern humans throughout the Old World and in Australia would have used the same knapping techniques or produced the same stone tools. Contrary to this expectation, people of Upper Paleolithic times in Tasmania fabricated a rather simplistic array of stone tools. These are essentially steep and flake scrapers of various forms and denticulates, cores and debitage (mainly of flakes). However, the bone points produced by these people testify to the fact that, despite ignorance of other tool types common to the Upper Paleolithic of western Europe, they were indeed *capable* of producing typically Upper Paleolithic forms (Bowdler, 1982; Jones, 1990; Cosgrove et al., 1990).

In small regions such as the Levant, where good-quality raw material is available and easily obtainable within a short distance of most sites (Jelinek, 1991), the forms of the shaped pieces reflect learned behaviors. This situation is especially apparent when comparing knapping techniques, raw materials, and final shapes of artifacts from the Upper Acheulean (ca. 500 kyr) through the Late Mousterian of Tabun cave (ca. 80–47 kyr). Similar conclusions can be reached concerning the Middle Paleolithic sites in the Taurus, in southwest Turkey (Yalçinkaya et al., 1993), in the Zagros (Baumler and Speth, 1993), and in southwest France (Geneste, 1988). Moreover, the near-total absence of Acheulean bifaces from central and southeastern Europe (to the Bosphoros straits), and from China and Southeast Asia, that is, beyond the "Movius line" (e.g., Clark, 1992; Schick and Zhuan, 1993), cannot be explained in most cases as a result of the lack of proper raw material. In Java, for example, large nodules of siliceous tuff would make exceptionally good raw material for stone knapping (R. Jones, pers. comm.). This is not to deny that within local traditions, when Lower and Middle Paleolithic sites were systematically occupied, one may observe signs of the diminution of good-quality raw material over time (e.g., Dibble, 1991).

By applying the concept of tool making as based on learned behavior, it is suggested that that once the knapping tradition and the basic shapes of the blanks were established within one population, they lasted for a very long time, regardless of environmental changes. As long as the shapes of the artifacts and their cutting edges were viewed by their manufacturers as satisfactory for cutting meat, scraping hides and wood, whittling, tattooing, and the like, no change occurred. This immutability is exemplified by the African Oldowan and the Karari industry (Leakey, 1971; Isaac, 1984, 1986), the presence of early core-chopper or non-handaxe assemblages such as the Chinese Lower Paleolithic (Schick and Zhuan, 1993), and the Clactonian in western Europe (which includes the Tayacian; Gamble 1986), all of which were made by different populations or groups of *Homo erectus* or Archaic *Homo sapiens.*

Taking into account the differences between tool assemblages and their geographic spread and time range (from the Late Pliocene through the Upper Pleistocene), it is concluded that the Oldowan-type industry, through the Clactonian and other flake industries, must have been made by different unrelated populations. Similarly, assemblages with bifaces (handaxes), generally defined as Early Acheulean or Developed Oldowan (Leakey, 1971;

Isaac, 1984), were not produced by every population across the world. The Early Acheulean appeared at least around 1.4 myr ago (e.g., Asfaw et al., 1992) but probably emerged earlier. In Eurasia, these Acheulean assemblages are associated with fossils defined as *H. erectus* and Archaic *Homo sapiens.* Many scholars have proposed that equating industries (identified by the use of particular stone-manufacturing techniques) with hominid populations can be employed as a means of enabling us, within a given geographic region and controlled temporal ranges, to trace the "history" of certain populations. This approach, developed in more detail elsewhere (Bar-Yosef, 1991), is exemplified below.

The Lower Paleolithic

The presence of both core-chopper and the Early Acheulean assemblages in western Asia is interpreted as evidence for early migrations out of Africa of different groups of *Homo erectus.* Most of the information concerning the Lower Pleistocene in this vast region is currently derived from the sites at Dmanisi (Georgia) and ʿUbeidiya (Israel), both of which are discussed below. Additional identifiable movement of African hominids into western Asia occurred during a later phase of the Acheulean sequence, ca. 0.8–0.7 myr, and will be presented briefly. Similar Mid-Pleistocene movements are probably discernible in other parts of Eurasia.

ʿUbeidiya. The excavations at ʿUbeidiya uncovered a series of stratified lithic and faunal assemblages. The bone assemblages contain the remains of over 120 species of mollusks, reptiles, birds, and mammals (Tchernov, 1986). The biogeographic origins of the mammalian species at the site were identified by Tchernov (1987, 1992a, 1992b) as a mixture from various regions but with a clear Palearctic stamp. Notably, only 6 species are considered of Ethiopian origin, 2 of Saharo-Arabian (*Oryx* sp. and *Gerbillus dasyurus*), and of North African (*Equus tabeti* and *Mus musculus*). The remaining 43 species (of which 16 are rodents and 3 insectivores) originated in Eurasia and the eastern Mediterranean. In addition, Tchernov and his associates identified a group of endemic species (mainly rodents and one hippopotamus), indicating that the ʿUbeidiya fauna had already been cut off from various sources for an undefined period (Tchernov, 1986). Although there may have been movements between North Africa and the Levant (such as the arrival of the cervids in the Maghreb), it seems that western Asia was isolated from the African world with the development of the arid Saharo-Arabian belt (Tchernov, 1992a, 1992b). This isolation probably resulted from the rapid uplift of the Tibetan Plateau around 2.5 myr (Zhongli et al., 1992), which established the Late Pliocene-Pleistocene pattern of atmospheric circulation.

The only African carnivore present at ʿUbeidiya is the *Crocuta crocuta,* considered a carcass destroyer (Turner, 1992). It is not known how early this species arrived in Asia, so it would be unfounded to suggest that *Homo erectus* moved into western Asia by following the trail of *Crocuta crocuta.*

The dating of ʿUbeidiya was the subject of considerable controversy. Horowitz (1989) suggested placing it around 0.8 myr, whereas Sanlaville (1988) proposed 1.4 to 0.8 myr. Repenning and Fejfar (1982) based their opinion on the older classifications of the ʿUbeidiya fauna, as published by Haas (1966, 1968), and suggested a range of 2.6 myr to 1.7 myr. Revised faunal studies by Tchernov and his associates concluded that the site should be dated to 1.4–1.0 myr (Tchernov, 1986, 1987), with higher probability of a date around 1.4 myr (Tchernov, 1992b). The chronological estimates for the rich paleontological and archaeological assemblages of this site are based on the following observations and/or age determinations of geologic formations below and above the site:

1. The major tectonic activities that formed the Jordan Rift Valley (or the Dead Sea Rift System) postdate the deposition of the Cover Basalt. This complex formation, around Lake Kinneret (Sea of Galilee), is currently dated to 3.11 ± 0.18 myr (Mor and Steinitz, 1982).
2. The lacustrine and fluvial sediments of the Erq el-Ahmar Formation were recently dated by paleomagnetic reversals (Verosub and Tchernov, 1989) to have lasted from the late Gilbert chron through the early part of the Matuyama chron. In the upper part of this sequence, which is considered to be slightly later than the Olduvai subchron, a few core-choppers and flakes were found.
3. The latter formation was dated to the Late Pliocene by the presence of *Hydrobia acuta* and *Dreissena chantrei* in its molluscan assemblage (Tchernov, 1975). In addition, it contains eight extinct species of mollusks not found in ʿUbeidiya or in later localities (Picard, 1943; Tchernov, 1975, 1986) and therefore indicates a hiatus between the existence of these two freshwater lake formations.

4. The ʿUbeidiya Formation (at least 150 m thick) was deposited following a tectonic movement that contorted the Erq el-Ahmar Formation. The deposition of the ʿUbeidiya Formation was halted by another tectonic movement that folded and faulted the Yarmuk Basalt (see below).
5. The Yarmuk Basalt, although it does not lie directly over the ʿUbeidiya Formation, is considered to postdate the latter. It was first K/Ar dated to 0.6 ± 0.05 and 0.64 ± 0.12 myr (Horowitz and Seidner, 1973). Later, nine samples were averaged to 0.79 ± 0.17 myr (Mor and Steinitz, 1985). Given the recalculated date for the Bruhnes-Matuyama boundary at 0.78 myr (Tauxe et al., 1992), it is not surprising that a normal polarity was reported for flows of the Yarmuk Basalt.

It should be stressed that the ʿUbeidiya Formation does not contain any tuffs, volcanic ashes, or other datable materials that are easily observable. However, no attempt have ever been made to search for microscopic volcanic ash that could be matched with known events of eruptions in the Near East, as was done in East Africa (Brown et al., 1992). The reversed paleomagnetic situation at ʿUbeidiya indicates merely an age within the Matuyama chron (Opdyke et al., 1985).

The date of the site relies on faunal correlation with European assemblages of known ages (Eisenman et al., 1983; Tchernov, 1986, 1988; Guérin and Faure, 1988). The presence of the following species with reference to the biozones, as defined by Guérin (1982), are currently considered to be the best indications of the age of ʿUbeidiya.

(1) The younger species (zone 19 and later; estimated age 1.5 myr and younger):
  Lagurodon arankae (zone 19, Final Villafranchian; Villanyian)
  Mammuthus meridionalis cf. *tamanensis* (zone 19 and early 20, Final Villafranchian and earliest Mid-Pleistocene)
  Praemegaceros verticornis (late Lower and Middle Pleistocene in Eurasia)
  Canis arnensis (zones 19–20)
  Pelovoris oldowayensis (present from Shungura Member C submember 5, dated close to 2.7 myr based on stratigraphic thicknesses below tuff D [2.5 ± .05 myr] through Oluvai III JK [1.33 ± .06–0.96]; see Vrba, this vol.)
  Apodemus (Sylvaticus) sylvaticus (reached Europe by Mid-Pleistocene from the Near East)
  Apodemus flavicollis (same as *A. sylvaticus*)

(2) The older species (zone 18 and younger or since 1.9 myr):
  Dicerorhinus etruscus (form of the latest evolutionary phase)
  Panthera gombaszoegensis (zones 18–20, Upper Villafranchian to Mid-Pleistocene)
  Kolpochoerus oldowayensis (in Shungura G, according to White [this vol.], from Koobi Fora, ca. 1.7 myr, through Olduvai IV, ca. 0.78 myr)
  Hippopotamus gorgops (present in the entire sequence of Olduvai)
  Hippopotamus behemoth (endemic species)

(3) The archaic species (zone 16 through zone 19 or later):
  Hypolagus brachygmathus (zones 16–20)
  Allocricetus bursae (in Eurasia from zone 17 through zone 21, seemingly survived later in the Near East)
  Cricetus cricetus (since zone 17, Middle Villafranchian)
  Gazellospira torticornis (through the entire Villafranchian)
  Sus strozzii (from zone 16 through zone 20)
  Ursus etruscus (through the entire Villafranchian)
  Pannonicitis ardea (through the entire Villafranchian into the Mid-Pleistocene)
  Megantereon cultridens (zones 16–19)
  Crocuta crocuta (since Shungura B)
  Herpestes sp. (since the Pliocene in Africa)

In sum, the fauna of ʿUbeidiya is essentially Late Villafranchian with a few Galerian elements. The lithic assemblages of the earliest layers (K/III-12, II-23,24) contain plenty of core-choppers, polyhedrons, and spheroids and are lacking bifaces, possibly indicating the presence of an early group of *Homo erectus*. In the rest of the sequence, bifaces occur in various frequencies and can be called 'Developed Oldowan' or Early Acheulean (Bar-Yosef and Goren-Inbar, 1993). In spite of the considerable similarity in the basic knapping techniques between the non-Acheulean and Acheulean assemblages, the presence of bifaces is taken to designate the additional incoming people.

Dmanisi. This site is situated on a basaltic block bordered by two rivers, tributaries of the larger Kura River. It was first excavated in the course of paleontological investigations (Vekua, 1987; Gabunia and Vekua, 1990). The stratified faunal assemblage, which immediately overlie a lava flow, contains a lithic industry consisting primarily of core-choppers and lacking bifaces (Dzaparidze et al., 1989; except perhaps for one piece in fig. 38). Among the reported flakes there are retouched pieces that can be classified as scrapers, as well as one burin. In addition, the excavators described a few worked bone objects.

Preliminary study of pollen preserved in coprolites indicate that the area was forested with the following trees: *Abies, Pinus, Fagus, Alnus Castanea, Tilia, Betula, Carpinus,* and rare *Ulmus* and *Salix.* Among the bushes represented were rhododendron, corylus, and myrtle; the herbaceous vegetation was dominated by Cyperaceae, Graminae, and Polygonaceae. The overall reconstructed environment consists of high mountains with Alpine associations and the well-watered woodland of an inland basin. This relatively wet environment with a few colder species is corroborated by the list of the fauna that is also used to support a Lower Pleistocene date (Dzaparaidze et al., 1989). The list includes the following species: *Struthio dmanisensis, Ursus etruscus, Canis etruscus, Pachycrocuta* sp., *Homotherium* sp., *Megantereon* cf. *megantereon, Archidiscodon meridionalis, Equus* cf. *stenonis, Equus* cf. *altidens, Dicerorhinus etruscus etruscus, Sus* sp., *Dama* cf. *nestii, Cervus* sp., *Dmanisibos georgicus, Caprini* gen., *Ovis* sp., Leporinae gen., *Cricetulus* sp., *Marmota* sp.

Originally the fauna from Dmanisi was attributed to the Upper Apscheronian or the Upper Villafranchian, as defined in the western Mediterranean Basin (Gabunia and Vekua, 1990). While reevaluating the assemblage following the discovery of the human mandible (Gabunia and Vekua, 1995), comparisons with faunas from Europe and ʿUbeidiya led the investigators (Dzaparidze et al., 1989) to suggest that the Dmanisi assemblage is contemporary with the Odessa fauna from southern Russia, which is considered to be slightly earlier than faunas of Senèze and Le Coupet, and thus earlier than ʿUbeidiya. In addition, the remains of the *Archidiskodon meridionalis* in Dmanisi are considered to be slightly more primitive than those described from ʿUbeidiya.

Therefore, Gabunia (in Dzaparidze et al., 1989) estimates that the site should be dated to the Olduvai subchron. According to the excavators, the latter attribution is supported by the normal polarity of the site, although the possible effects of demagnetization have not yet been taken into account. The K/Ar-dated lava flow under the site, with one reading of 1.8 ± 0.1 myr, is also cited to support the site's placement within the Olduvai subchron. Unfortunately, the lack of direct dating of the bone-bearing layers raises the possibility that they had accumulated over a long period. Estimating the age of the site within the range of 1.6–1.2 myr would be reasonable.

Gesher Benot Yaʿaqov. This site provided the only African-type assemblage within the Levantine Acheulean known from over 170 surface and in situ occurrences (Gilead, 1970; Hours, 1975; Bar-Yosef, 1975, 1987; Goren-Inbar et al., 1991). The site is located in the gorge of the upper Jordan Valley, and the available outcrops along the gorge form the type section for the Benot Yaʿaqov Formation (Horowitz, 1979). The nature of the deposits and the malacological assemblages, dominated by *Viviparus apameae,* indicate that the archaeological assemblages accumulated on the shores of an expanding lake that flooded the gorge.

The site was first excavated in the 1930s by Stekelis (1960) and is currently being excavated by Goren-Inbar et al. (1991, 1992a, 1992b). The complex sequence encompasses early layers with an African-type industry (Stekelis layers VI–V) dominated by the production of cleavers and bifaces from basalt (Stekelis, 1960). The cleavers were fabricated by the Kumbewa technique (Goren-Inbar et al., 1991). The upper layers in the Stekelis excavations (IV–II) contained bifaces made of flint, similar in form to other known Upper Acheulean assemblages in the Levant (Stekelis, 1960). The Gesher Benot Yaʿaqov sites lies on the eastern edge of a vast area covered by basalt (Gebel Druz and the Black Desert) within southern Syria and northern Jordan. Other parts of the Levant (such as in southern Jordan or the eastern Galilee in Israel) are also covered by more limited lava flows. It should be stressed, however, that there were no lava-made Acheulean assemblages noted in the surface surveys of these areas. On the contrary, in most cases, flint nodules derived from "island" outcrops, often of Eocene rocks, served as raw material for fabricating handaxes (e.g., Goren, 1979; Goren-Inbar, 1985; Ohel, 1991).

The archaeological horizons of Gesher Benot Yaʿaqov are embedded in a depositional sequence that accumulated above a lava flow with normal polarity. The lava flow, designated as the Yarda Basalt, was first K/Ar dated to 0.68 ± 0.12 myr (Horowitz et al., 1973) and later to 0.9 ± 0.15 myr (Goren-Inbar et al., 1992a). The fauna derived from the lower layers of the Stekelis excavations in the 1930s, as well as from the new excavations, included the following species: *Stegodon mediterraneus, Elephas trogontherii, Dicerorhinus merckii, Hippopotamus amphibius, Dama mesopotamica,* cf. *Bison priscus, Capra* sp., *Gazella gazella* (Hooijer, 1959, 1960; Goren-Inbar, 1992b). This assemblage falls within the general definition of the Galerian fauna that replaced the Late Villafranchian association around 0.9–0.7 myr (Azzaroli et al., 1988, this vol.). It should be noted that two broken leg bones, the exact proveniences of which within the site are unknown, were attributed to *Homo erectus* (Geraads and Tchernov, 1983).

The site of Gesher Benot Yaᶜaqov is interpreted as the residue left by hominids migrating from Africa. I suggest that this move was triggered by environmental change that occurred around the Jaramillo subchron or the Brunhes-Matuyama boundary. Paleoclimatic conditions in the Northern Hemisphere, as recorded by deep-sea cores and terrestrial fauna, indicate an increase in the intensity of the glacial cycles (e.g., Thunnell and Williams, 1983; Azzaroli et al., 1988; Forsten, 1988; deMenocal and Bloemendal, this vol.). Such cumulative change probably caused increased periods of aridity on the African continent. This drying trend might have resulted in an intense competition for resources that forced this group, and perhaps others, to look for alternative foraging grounds. It is as yet premature to propose the origin of these hominids within the African continent, although North Africa seems the likely candidate. After a period of undetermined length, the Gesher Benot Yaᶜaqov hominids either disappeared or intermingled with contemporary local groups who continued to produce the ordinary flint-made Levantine Acheulean industries such as those uncovered in Evron-Quarry (Ronen, 1991), Latamne (Clark, 1967, 1969), Umm Qatafa (Neuville, 1951), and Tabun F (Garrod and Bate, 1937).

In sum, it seems that the emergence of *Homo erectus* in sub-Saharan Africa (e.g., Klein, 1989; Rightmire, 1990) was triggered by the climatic changes that occurred around 1.9 myr ago. This appearance was followed by outward migrations into North Africa and Eurasia. The evidence from Lower Paleolithic sites in western Asia suggests that both ᶜUbeidiya and Dmanisi were among the early "stations" of *Homo erectus* in Eurasia. If the first groups of *Homo erectus* to leave their homeland were the bearers of a core-chopper industry, then the lowermost levels at ᶜUbeidiya and the assemblage of layer V at Dmanisi testify to their presence. The same would hold for the sequence of the Maghreb (Biberson, 1961). Bearers of this industry could have been among the first to colonize southeastern Asia (Schick and Zhuan, 1993; Swisher et al., 1994) and among those who much later ventured to colonize western Europe. The earliest dates for such trials are still debated (Klein, 1989; Bonifay and Vandermeersch, 1991; Roebroeks, 1994). Although the Middle Pleistocene inhabitants of central and eastern Europe made similar core-chopper assemblages, it should be stressed that this region is devoid of bifaces. Isolated occurrences of bifaces were reported from China (Schick and Zhuan, 1993), whereas western Europe and western Asia, from eastern Turkey to the Indian subcontinent, are strewn with Acheulean occurrences.

The various pathways used to cross the Mediterranean Sea may explain the differences in the lithic industries produced across Europe. The potential crossings were at the Strait of Gibraltar and through Sicily (Alimen, 1979; Freeman, 1975), especially during glacial periods, when sea levels were much lower than they are today. Nevertheless, extent to which early members of the *Homo erectus* lineage were able to traverse waterways has yet to be determined. Although this question has no satisfactory answer, a later, successful crossing is attributed to modern humans who colonize Australia by navigating a 100 km waterway. This event is currently estimated to have occurred as early as 55 kyr (Roberts et al., 1990; Davidson and Noble, 1992) and at the latest around 40 kyr (Bowdler, 1992). The Strait of Gibraltar is narrower than 100 km, and it has yet to be demonstrated that groups of *Homo erectus* were able to accomplish what is still viewed as an undertaking solely within the capacity of *Homo sapiens.*

Modern Humans. Although during less stressful climatic conditions, the Levant would have been an important two-way corridor for movement of humans between Africa and Eurasia, the number of African elements in the Near Eastern faunas decreased through the Pleistocene. Later arrivals were extremely rare and came mainly from North Africa (Tchernov, 1992a, 1992b). The region always enjoyed higher temperatures than did adjacent areas, as well as plant and animal food resources that were more predictable, stable, and reliable than those of most European environments. The Levant would therefore have been attractive to human groups living under conditions of diminishing resources and increasing social stress, for example, in such places as the Balkans, the Anatolian plateau, and the Taurus-Zagros ranges. Those who occupied the Caucasus area had their own refugium in the lowlands near the Black Sea and the Caspian Sea.

The Levant has provided an extensive collection of human skeletal remains that has been used to support the

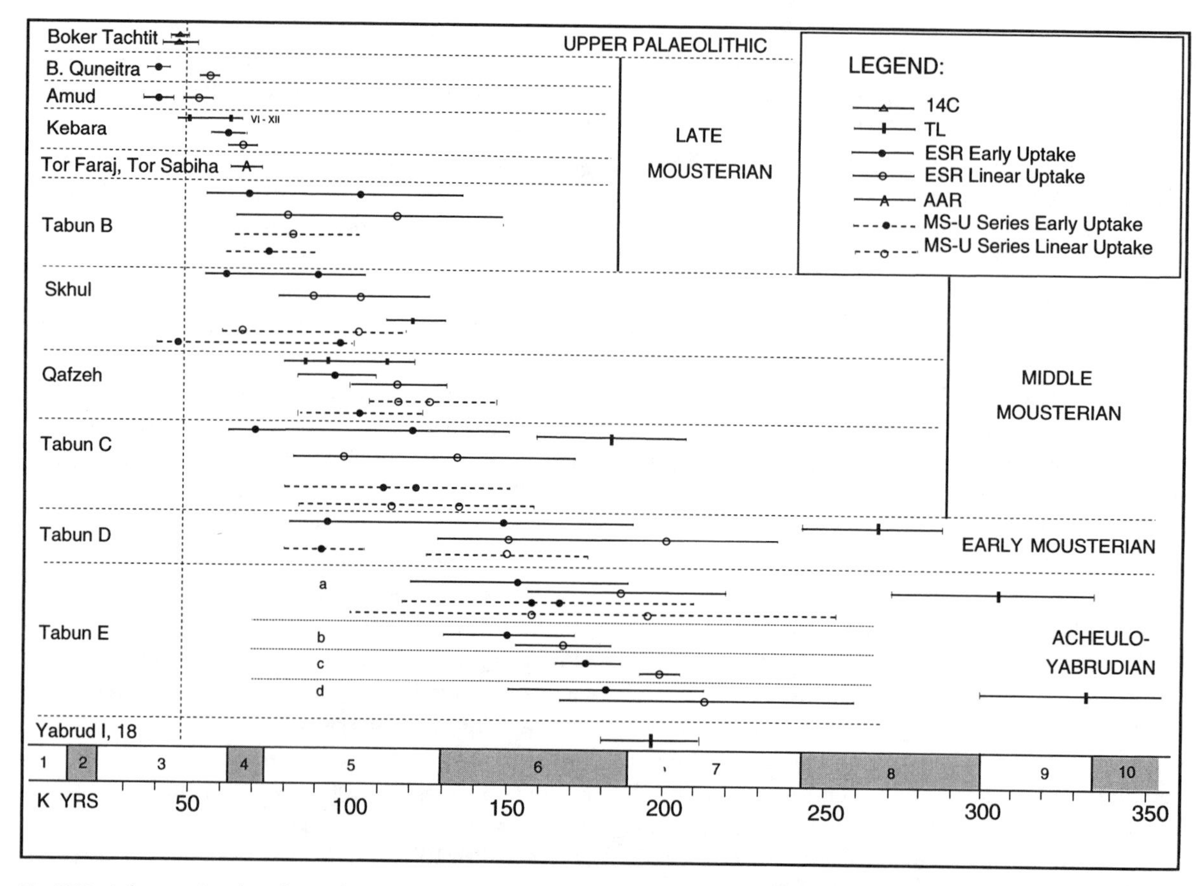

Fig. 35.2. Radiometric chronology of Late Acheulean through early Upper Paleolithic sites in the Levant, including dates obtained from thermoluminescence (TL), electron spin resonance (ESR), amino acid racemization (AAR), and mass spectrometer Uranium (MS-U) series. The sources of the dates appear in Grün and Stringer, 1991; Bar-Yosef, 1992; and Henry and Miller, 1992. Note that ESR early uptake and linear uptake as well as the MS-U series are given with maximum and minimum readings whenever available. For the Acheulo-Yabrudian of Tabun B, C, and D, only average dates are given. In the case of Qafzeh, the full range of TL dates is given. In Tabun E, a–d refer to stratigraphic subdivisions as defined by D. Garrod (Garrod and Bate, 1937).

two basic models for the emergence of modern humans (e.g., Stringer, 1992, this vol.; Wolpoff, 1989). Advocates of the recent African-origin model find the remains from Skhul and Qafzeh caves (from layers dated by thermoluminescence [TL] and electron spin resonance [ESR] to the time range of 120–85 kyr; see figs. 35.2, 35.3) compelling evidence for early modern humans in the Near East. Proponents of multiregional continuity regard the Mount Carmel sample (described by McCown and Keith [1939] and including the human relics from Tabun cave) as transitional fossils between Archaic *Homo* and forms closer to modern humans. Both parties agree that such forms as Qafzeh 9 or Skhul V are not fully modern; these forms occur only during the Upper Paleolithic (Howells, 1989). This means that additional morphological changes within this lineage took place in the ensuing period (85–45 kyr) and probably continued into the early part of the Upper Paleolithic. Both the two fragmentary robust Cro-Magnon skulls (as yet unpublished in full) found by Neuville and Stekelis in Qafzeh and the remains of early Cro-Magnons in Europe demonstrate that additional morphological changes occurred at least until the Neolithic Revolution.

The adult skeleton uncovered in the Mousterian deposits in Kebara cave is seen as a Mediterranean Neanderthal (e.g., Rak and Arensburg, 1987; Tillier, 1992; Vandermeersch, 1992: Bar-Yosef et al., 1992) or as another individual within the Middle Paleolithic Mount Carmel population (e.g., Frayer et al., 1993). The proposition that the Levantine Neanderthals reached the region only after one of the cold glacial spells, either Isotope Stage 4 or perhaps after 5b or 5d, finds support

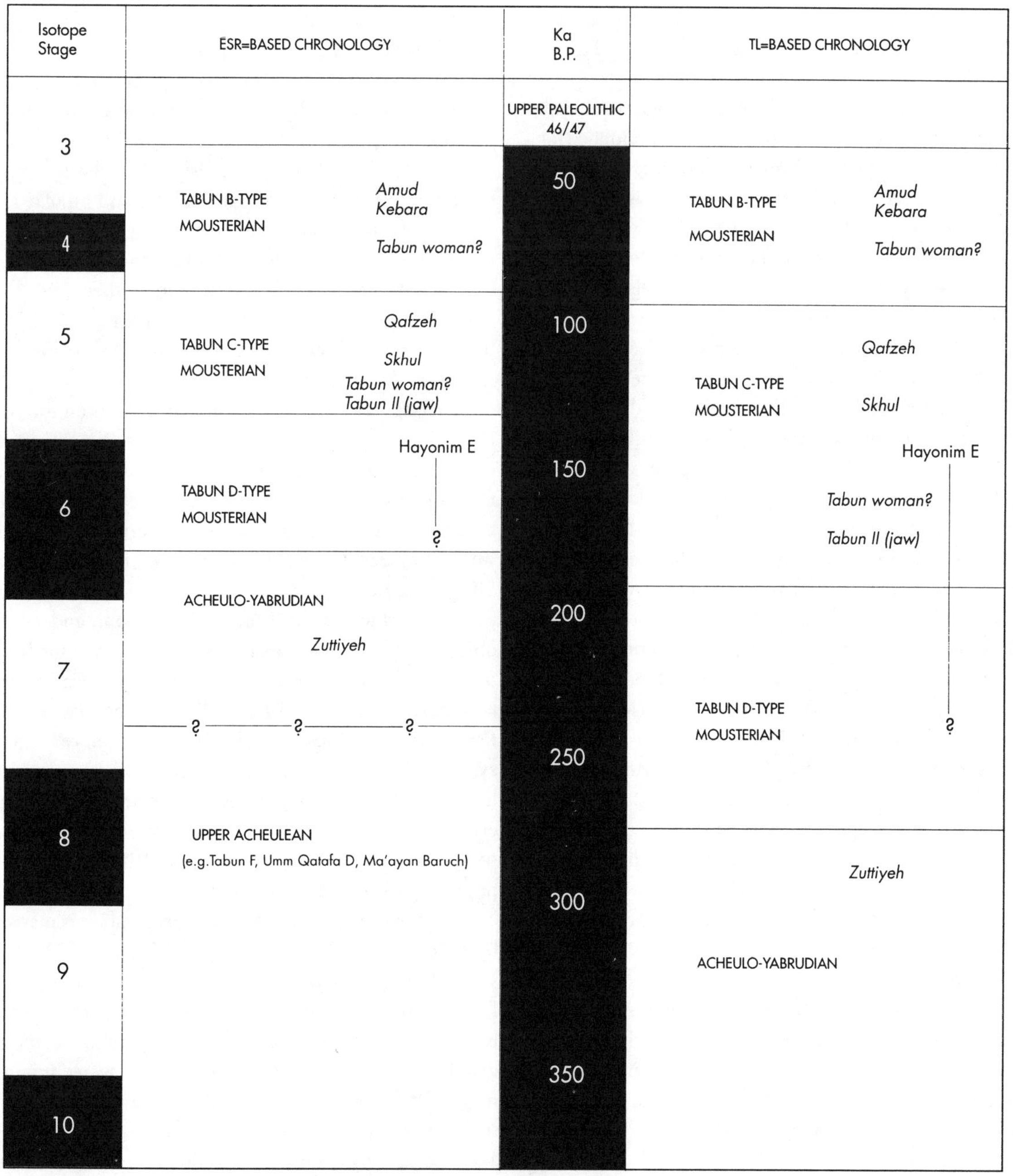

Isotope Stage

ESR=BASED CHRONOLOGY

Ka B.P.

TL=BASED CHRONOLOGY

UPPER PALEOLITHIC 46/47

3

4

5

6

7

8

9

10

50

100

150

200

250

300

350

TABUN B-TYPE MOUSTERIAN

Amud
Kebara

Tabun woman?

TABUN C-TYPE MOUSTERIAN

Qafzeh

Skhul

Tabun woman?
Tabun II (jaw)

Hayonim E

TABUN D-TYPE MOUSTERIAN

?

ACHEULO-YABRUDIAN

Zuttiyeh

? ? ?

UPPER ACHEULEAN
(e.g.Tabun F, Umm Qatafa D, Ma'ayan Baruch)

TABUN B-TYPE MOUSTERIAN

Amud
Kebara

Tabun woman?

Qafzeh

TABUN C-TYPE MOUSTERIAN

Skhul

Hayonim E

Tabun woman?

Tabun II (jaw)

TABUN D-TYPE MOUSTERIAN

?

Zuttiyeh

ACHEULO-YABRUDIAN

Fig. 35.3. Current differences between ESR- and TL-based chronologies for the Late, Middle, and Upper Pleistocene of the Mount Carmel–Galilee area. The main hominid specimens are marked in italics. Note the discrepancies between the two chronologies; according to the TL-based chronology, the Upper Acheulean would be positioned earlier than 400 kyr.

in the estimate of body size and shape as based on the biiliac breadth (Ruff, 1991, 1993). The Kebara KMH-2 individual and probably the Shanidar upper group, tentatively dated to 70–50 kyr (Trinkaus, 1983; Solecki and Solecki, 1993), are interpreted as the fossil evidence for a southward movement of humans from higher latitudes caused by the reduction of foraging grounds during the coldest periods of the Last Glaciation.

When one looks into the origin of modern humans, however, there is still not a basis for correlation with a climatic event. One may propose that the presence of modern-looking hominids (the Qafzeh-Skhul group) resulted from a move out of Africa during the Last Interglacial. Deteriorating environmental conditions in a hypothetical homeland would cause an influx into refugia farther away. Improved foraging potential in one region might also cause a population increase and ensuing outward migrations.

Unfortunately, we do not know which morphotypes were the manufacturers of the Tabun D–type industry, the earliest Mousterian in the Levant (Bar-Yosef, 1992a, 1992b). The particular stone knapping techniques of this industry make it an entity unto itself, and in Tabun cave it is currently dated to either 270 kyr by TL (Mercier and Valladas, 1994) or to 150–170 kyr by ESR (Grün and Stringer, 1991; see figs. 35.2, 35.3). No human relics have been found in sites and layers that contain this industry. Assuming that the producers of this entity came out of Africa, then the major cultural shift from the Acheulo-Yabrudian to the Early Mousterian is both clearly marked and significant. Finally, within the Middle Paleolithic sequence of the Near East, I view the presence of the so-called Mediterranean Neanderthals with the Late Mousterian (Tabun B–type assemblages) in the Levant as the best evidence for a climatically driven human movement. This migration was probably caused by the inability of Neanderthal populations to survive in the cold conditions of Isotope Stage 4 in the middle to northern latitudes of Eurasia.

The Middle-Upper Paleolithic Transition in the Levant

Contrary to previously held views, recent research has demonstrated that the anatomically early modern humans (so-called Proto-Cro-Magnons in Qafzeh and Skhul) that date to some 85–120 kyr had nothing to do with the major cultural change from the Middle to Upper Paleolithic that occurred around 47–45 kyr. The sequence of the Levantine Upper Paleolithic begins with what is called the Transitional Industry or the Emiran culture. The Transitional Industry is best known from the Ksar Akil rock shelter in Lebanon and from Boker Tachtit, an open-air site in the Negev highlands (Marks, 1983; Ohnuma and Bergman, 1990). It is characterized by the diminishing use of flat unipolar and bipolar Levallois core reduction, with increasing frequencies of prismatic cores. Blanks were modified into end scrapers, chamfered pieces (in Ksar Akil and other Lebanese sites), burins, and Emireh points. The Transitional Industry evolves directly into the full-blown, blade-rich Ahmarian tradition.

Archaeological evidence supports the contention that the technological transition to the Upper Paleolithic occurred within the Levant. It is still possible, however, that this transition could have been the result of rapid acculturation similar to the current interpretation of the west European Chatelperronian. The main behavioral differences between the Middle Paleolithic and the Upper Paleolithic are the production of bone and antler objects; use of worked marine shells (possibly for body decoration); development of rock art in western Europe (and perhaps in Australia as well) and the presence of a few art objects in western Asia; frequent use of red ochre; systematic presence of a few grinding tools; stone-circled hearths; and use of rocks for warmth banking. For western Asia the observable differences are interpreted as designating higher mobility of Upper Paleolithic bands, possibly within larger territories.

The Middle to Upper Paleolithic revolution is marked by innovation and invention (e.g., Mellars, 1992; Bar-Yosef, 1992a), including new or improved techniques for food acquisition (spear throwers and archery), perhaps basketry, and new tools for food preparation such as grinding stones. New trapping and storing techniques may have become available, although the evidence for this is not substantial. Any improved food-acquisition and -processing techniques would have led to better nourishment and the survival of newborns into adulthood. A slight increase in life expectancy may have secured the survival of older members of the group, thus extending the "living memory" of the group. The self-identification of certain populations was probably reflected in specialized lithic artifacts and body decorations. Long-range social alliances are expressed in mobile art objects (Gamble, 1986). The transportation of materials over long distances indicates higher mobility and knowledge of territories larger than the preceding

Mousterian "homelands." Socialization over vast areas reflects the ability to overcome seasonal or annual economic disasters.

The Middle to the Upper Paleolithic revolution can also be explained as resulting from changes in technological and social organization that took place within one region and later spread through other areas. Without locating the source of this revolution it will be difficult to test the hypothesis that it was triggered by environmental change. I have proposed to study this revolution by using the same approach employed to study the Neolithic Revolution (Bar-Yosef, 1992a, 1994). As with the Neolithic Revolution, there may have been dramatic changes within a single human population triggered by climatic changes in a certain region (see below). The emergence of the Upper Paleolithic cultures need not have required a sudden mutation resulting in the human capacity for modern language (e.g., Klein, 1989). However, Davidson and Noble (1992) proposed that the ability of modern humans some 55 kyr ago to cross the 100 km waterway to Australia was accomplished with boats whose construction required the use of modern languages. Brain scientists (Deacon, 1989) interpret the relationship between brain capacity and alignments as indicating the evolution of modern language at a much earlier time, that is, since the beginning of the Pleistocene. Archaeological markers of modern behavior are reflected in well-defined Mousterian *chaînes opératoires* of lithic techniques and the use of red ochre, intentional burials, spatial organization, and built-up hearths. This indicates evolving "planning depth" similar to that found in modern humans at least by the early part of the Upper Pleistocene. There may also be a morphological continuity from Middle Paleolithic modern-looking hominids to Upper Paleolithic humans (e.g., Howells, 1989; Arensburg et al., 1990; Stringer and Gamble, 1993). Consequently, there is no strong evidence for a biological threshold within the lineage of modern humans. Therefore, if we wish to locate the core area of the Middle-Upper Paleolithic revolution, we need to concentrate our efforts on the time span between 70 and 50–45 kyr.

Finally, within the Levantine Upper Paleolithic sequence there is evidence for a southward movement of humans from either the Balkans or Anatolia. The producers of the Levantine Aurignacian, some 32–27 kyr ago, lingered solely in the coastal ranges of the Levant, avoiding the semiarid and arid belt. Their assemblages are characterized in the early phase by the dominance of blades and bladelets with carinated and nosed scrapers, as well as El-Wad (Font Yves or Krems) points with a few bone and antler objects. The later phase is increasingly dominated by flake production, with nosed and carinated scrapers; El-Wad points; and a rich bone and antler industry, with a few split base points and deer-tooth pendants (reminiscent of the European types).

The Neolithic Revolution—Response to Climatic Change

Since the early years of this century, following the work of R. Pumpely in central Asia and the influential writings of V. Gordon Childe, the relationship of the Neolithic Revolution to climatic changes has been debated (Braidwood, 1972; Cauvin, 1978; Redman, 1978; Moore, 1985; Moore and Hillman, 1992; Henry, 1989; Bar-Yosef and Belfer-Cohen, 1989, 1991, 1992). The shift to intentional and systematic cultivation of cereals and legumes in the Levantine Corridor, essentially between Jericho and the Damascus Basin (van Zeist, 1986), occurred within a short time (ca. 10,300–9,000 B.P.). It was followed by the domestication of animals (goats, sheep, cattle, pigs) by Near Eastern agricultural communities at ca. 9,500–8,000 B.P.

A series of cultural changes led to the establishment of farming communities, all resulting from socioeconomic decisions triggered by abrupt climatic changes (fig. 35.4; Bar-Yosef and Belfer-Cohen, 1989, 1992, with references). The sequence actually begins with the Natufian sedentism that is interpreted to be a response to paleoecological shifts around 13,000–12,800 B.P. Evidence for climatic vagaries at the time of the "plateaus" in the radiocarbon calibration curve (Stuiver and Braziunas, 1993) is now reported from other regions (e.g., Lotter et al., 1992). Whether Natufian sedentism was a reaction to a short, dry, and cold interval at the onset of the Bölling period or was influenced by the attraction of the secure predictability of the plant resources in lush Mediterranean environments that resulted from rapidly increasing precipitation during the Bölling period (13,000–12,500 B.P.; Baruch and Bottema, 1991) is still unknown. Sedentism and/or the seasonal occupation of sites is demonstrable on the basis of bioarchaeological evidence. Nearly year-round habitation is indicated by the high frequencies of commensal rodents and birds (Auffrey et al., 1988; Tchernov, 1991). Seasonal information obtained from the study of cemen-

Uncalibrated radiocarbon years B.P.	Paleoclimate	Cultural Entities	Socio-economy
9,000	decreasing precipitation towards 8,000 B.P.	PPNB { Late, Middle, Early	establishment of large villages (12 ha) as part of a complex settlement pattern domestication of goat, sheep (9,500-8,500 BP), cattle, pig (8,000 BP) mobile hunter-gatherers in steppic and desertic areas
10,000	increase in precipitation espcieally in the northern and eastern parts of the Near East	PPNA { Mureybetian, Sultanian, Khiamian	change in settlement pattern establishment of cultivating communities along the Levantine Corridor; mobile hunter-gatherers in steppic and desertic areas
Younger Dryas 11,000	cold and dry	Harifian (in Negev and Sinai) Late Natufian	special semi-desertic adaptations return to mobile hunting and gathering except in well-watered areas
12,000 13,000	expansion of forests in the coastal ranges increasing precipitation and rising temperature	Early Natufian	mobile groups in the steppic and desertic belts Natufian sedentary base camps within the Mediterranean vegetation belt
14,000	cold and wet expansion of woodlands and parklands	Geometric Kebaran in the Mediterranean belt Mushabian and Hamran in semi-arid areas	semi-sedentism in coastal ranges and mobile small bands of hunter-gatheres exploitation of Sinai and Syro-Arabian desert
LGM	cold and dry Mediterranean forests limited to coastal ranges	Kebaran	mobile small bands of hunter-gatheres semi-sedentism in coastal ranges no evidence for exploitation of deserts

Fig. 35.4. Climatic states, cultural entities, and socioeconomic changes during the Terminal Pleistocene–Early Holocene of the Levant.

tum increments of gazelle teeth reflects the differences between soft and hard food consumed by these herbivores (Lieberman, 1993). Both kinds of evidence indicate that the Natufians had sedentary base camps and seasonal exploitation sites.

Two thousand radiocarbon years later, Natufian settlements faced the crisis brought about by the Younger Dryas (11,000–10,300 B.P. uncalibrated; fig. 35.4). It has been proposed that the main impact in the Levant was a decrease in the yields of wild cereal, a C_3 plant that was a staple food, along with diminishing precipitation (Bar-Yosef and Meadow, in press). Given the few options available to the already dispersed Late Natufians in the Levant, the shift to intentional cultivation involved only a change in the division of labor. The long-term impact of establishing farming communities in the Jordan Valley, where earlier there had only been Natufian seasonal camps, is expressed in the archaeological record by rapid population growth and by the budding of new communities. The crisis reflected in the 9,500 B.P. uncalibrated "plateau" in the radiocarbon curve had a minor influence, although it did occur contemporaneously with the earliest solid evidence for the domestication of the caprovines (Bar-Yosef and Meadow, in press). During the ensuing millennia, the expansion of Early Neolithic communities accompanied the diffusion of technology, domesticated plant species, and animal breeds northward into Anatolia and eastward into Iran and Pakistan, where climatic conditions had previously been unfavorable. The introduction of cereal cultivation to Anatolia, a region where such cereal species were not local and had no natural competitors, led to a human population explosion that drove demic-diffusion and acculturation farther westward (Ammerman and Cavalli-Sforza, 1984; Ammerman, 1989) through the Danube valley and, by coastal navigation, throughout the Mediterranean Basin.

Summary

Although a correlation between climatic and cultural changes does not necessarily imply causation in every case, it is suggested that the earliest migrations of *Homo erectus* out of Africa occurred sometime after 1.9–1.8 myr and were triggered by the environmental changes caused during the Olduvai subchron or its aftermath. It is also possible that the movements of hominids were accomplished by a series of isolated events interspersed with extinctions of certain populations. When isolated bursts of hominid migration are viewed from an overall Quaternary standpoint, they appear incremental over a long period. Therefore, bearers of various bifaces or core-choppers could have moved out of Africa at various times. The case of the Gesher Benot Yaʿaqov Acheulean assemblages is instructive, because it demonstrates that African-style Acheulean bifaces and cleavers indicating a definite preference for certain raw materials were present as a "foreign element" in the Near Eastern Acheulean sequence. It also means that the hominids who moved from regions such as the Maghreb or East Africa were accustomed to making regular bifaces through a specific operational sequence. Thus, noncontinuous archaeological sequences in well-researched areas may indicate temporary reductions in the number of people reflected in the drastic decreases in the number of sites or perhaps extinctions of regional groups of *Homo erectus.* Such extinctions may have occurred during the Lower and Middle Pleistocene and would help explain other archaeological phenomena related to the presence or absence of Archaic *Homo sapiens.* Regional archaeological studies have often ignored climatic changes in neighboring areas and their potential impact on the investigated region. Although the data on climatic fluctuations are not derived from the archaeological record, their effect, as transmitted through the "cultural filter," can be recognized.

Some archaeological paradigms of the past three decades have advocated the study of past human societies as regional continuities, proposing that cultural changes (often limited to lithics) should be interpreted as the results of local adaptations. In the current view, the ways the stone tools were shaped is seen as the outcome of learned behavior that was carried over many generations with only minor changes: when groups found themselves in different regions with other kinds of raw materials, they tried to continue making the same shapes they had fabricated in their homeland. In such a case we would be able to trace their origins. With this yardstick it was even possible to identify a non-Levantine Upper Pleistocene entity, the Aurignacian (34–27 kyr), that penetrated from southeastern Europe during the Upper Paleolithic.

Unfortunately, the paucity of dated African Middle Paleolithic or Middle Stone Age assemblages hampers the correlation of possible movements by modern humans with a known climatic event or events. It is, however, easier to demonstrate that the effects of the rapid environmental fluctuations during the terminal Pleistocene triggered the socioeconomic responses of the Neolithic Revolution. The shift to sedentism, the beginning of intentional cultivation, and the subsequent domestication of cereals and legumes are among the most crucial steps taken by Near Eastern human groups, or indeed by any human group at all.

Acknowledgments

I would like to thank Elisabeth Vrba for inviting me to participate in the conference and to express regret that I was unable to be there to enjoy the fruitful discussions. I am grateful to David Pilbeam, Richard Klein, Dan Leiberman, Greg Laden, and Martha Tappen for many useful comments on earlier drafts of this work. Needless to say, I am solely responsible for the interpretations put forth here. Without the constant assistance of Margot Fleischman, the research and writing of this chapter could never have been accomplished.

References

Alimen, M. H. 1979. Les "isthmes" hispano-marocain et siculo-tunisien aux temps acheuléens. *L'Anthropologie* 79:399–436.

Ammerman, A. J. 1989. On the Neolithic transition in Europe: A comment on Zveilbel and Zveilbel (1988). *Antiquity* 63:162–165.

Ammerman, A. J., and Cavalli-Sforza, L. L. 1984. *The Neolithic transition and the genetics of populations in Europe.* Princeton University Press, Princeton.

Arensburg, B., Schepartz, L. A., Tillier, A. M., Vandermeesch, B. and Rak, Y. 1990. A reappraisal of the anatomical basis for speech in Middle Palaeolithic hominids. *American Journal of Physical Anthropology* 83:137–146.

Asfaw, B., Beyene, Y., Suwa, G., Walter, R. C., White, T. D., WoldeGabriel, G., and Yemane, T. 1992. The earliest Acheulian from Konso-Gardula (Ethiopian archaeological region). *Nature* 360:732–35.

Aufray, J. C., Tchernov, E., and Nevo, E. 1988. Origine du commensalisme de la souris domestique (*Mus musculus domesticus*) vis-à-vis de l'Homme. *Comptes Rendus de l'Académie des Sciences* (Paris), ser. 3, 307:517–522.

Azzaroli, A., De Giuli, C., Ficcarelli, G., and Torre, D. 1988. Late Pliocene to Early Mid-Pleistocene mammals in Eurasia: Faunal succession and dispersal events. *Palaeogeography, Palaeoclimatology, Palaeoecology* 66:77–100.

Bar-Yosef, O. 1975. Archaeological occurrences in the Middle Pleistocene of Israel. In *After the australopithecines* (ed. K. W. Butzer and G. L. Isaac). Mouton, The Hague, pp. 571–604.

———. 1987. Pleistocene connexions between Africa and Southwest Asia: An archaeological perspective. *African Archaeological Review* 5:29–38.

———. 1991. Stone tools and social context in Levantine prehistory. In *Paradigmatic biases in Mediterranean hunter-gatherers research* (ed. G. A. Clark). University of Pennsylvania Press, Philadelphia, pp. 371–395.

———. 1992a. The role of western Asia in modern human origins. *Philosophical Transactions of the Royal Society,* ser. B, 337:193–200.

———. 1992b. Middle Paleolithic human adaptations in the Mediterranean Levant. In *The evolution and dispersal of modern humans in Asia* (ed. T. Akazawa, K. Aoki, and T. Kimura). Hokusen-sha, Tokyo, pp. 189–216.

———. 1994. The contributions of southwest Asia to the study of the origin of modern humans. In *Origins of anatomically modern humans* (ed. M. H. Nitecki and D. V. Nitecki). Plenum Press, New York, pp. 22–66.

Bar-Yosef, O., and Belfer-Cohen, A. 1989. The origins of sedentism and farming communities in the Levant. *Journal of World Prehistory* 3(4):447–498.

———. 1991. From sedentary hunter-gatherers to territorial farmers in the Levant. In *Between bands and states* (ed. S. A. Gregg). Center for Archaeological Investigations, Carbondale, pp. 181–202.

———. 1992. From foraging to farming in the Mediterranean Levant. In *Transitions to agriculture in prehistory* (ed. A. B. Gebauer and T. D. Price). Prehistory Press, Madison, Wis., pp. 21–48.

Bar-Yosef, O., and Goren-Inbar, N. 1993. *The lithic assemblages of ʿUbeidiya: A Lower Palaeolithic site in the Jordan Valley.* Hebrew University of Jerusalem, Jerusalem.

Bar-Yosef, O., and Meadow, R. H. (in press). The origins of agriculture in the Near East In *Last Hunters, First Farmers* (ed. T. D. Price and G. Gebauer). School of American Research, Santa Fe.

Bar-Yosef, O., and Meignen, L. 1992. Insights into Levantine Middle Paleolithic cultural variability. In *The Middle Paleolithic: Adaptation, behavior, and variability* (ed. H. L. Dibble and P. Mellars). The University Museum, University of Pennsylvania, Philadelphia, pp. 163–182.

Bar-Yosef, O., Vandermeersch, B., Arensburg, B., Belfer-Cohen, A., Goldberg, P., Laville, H., Meignen, L., Rak, Y., Speth, J. D., Tchernov, E., Tillier, A.-M., and Weiner, S. 1992. The excavations in Kebara Cave, Mt. Carmel. *Current Anthropology* 33(5):497–550.

Baruch, U., and Bottema, S. 1991. Palynological evidence for climatic changes in the Levant ca. 17,000–9,000 B.P. In *The Natufian culture in the Levant* (ed. O. Bar-Yosef and F. R. Valla). International Monographs in Prehistory, Ann Arbor, pp. 11–20.

Baumler, M. F., and Speth, J. D. 1993. A Middle Paleolithic assemblage from Kunji Cave, Iran. In *The Paleolithic prehistory of the Zagros-Taurus* (ed. D. I. Olszewski and H. L. Dibble). The University Museum, University of Pennsylvania, Philadelphia, pp. 1–74.

Biberson, P. 1961. *Le Paléolithique inférieur du Maroc atlantique.* Service des Antiquités du Maroc, Rabat.

Bonifay, E., and Vandermeersch, B. eds. 1991. *Les premiers européens.* Editions du C.T.H.S., Paris.

Bowdler, S. 1982. Prehistoric archaeology in Tasmania. In *Advances in world archaeology* (ed. F. Wendorf and A. Close). Academic Press, New York, pp. 1–49.

———. 1992. *Homo sapiens* in Southeast Asia and the Antipodes: Archaeological versus biological interpretations. In *The evolution and dispersal of modern humans in Asia* (ed. T.

Akazawa, K. Aoki, and T. Kimura). Hokusen-sha, Tokyo, pp. 559–590.

Braidwood, R. J. 1972. Prehistoric investigations in southwestern Asia. *Proceedings of the American Philosophical Society* 116(4):310–320.

Brown, F. H., Sarna-Wojcicki, A. M., Meyer, C. E., and Haileab, B. 1992. Correlation of Pliocene and Pleistocene tephra layers between the Turkana Basin of East Africa and the Gulf of Aden. *Quaternary International* 13/14:55–67.

Cauvin, J. S. 1978. *Les premiers villages de Syrie-Palestine* du neuvième au huitième millénaires av. Maison de l'Orient, Lyons.

Clark, J. D. 1967. The Middle Acheulian site at Latamne, northern Syria. *Quaternaria* 9:1–68.

———. 1969. The Middle Acheulian occupation site at Latamne, northern Syria. *Quaternaria* 10:1–68.

———. 1992. African and Asian perspectives on the origins of modern humans. *Philosophical Transactions of the Royal Society* 337(1280):201–215.

Cosgrove, R. J., and Marshall, B. 1990. Paleo-ecology and Pleistocene human occupation in south central Tasmania. *Antiquity* 64:59–78.

Davidson, I., and Noble, W. 1992. Why the first colonisation of the Australian region is the earliest evidence of modern human behaviour. *Archaeology of Oceania* 27:113–119.

Deacon, T. W. 1989. The neural circuitry underlying primate calls and human language. *Human Evolution* 4(5):367–401.

Dibble, H. 1991. Local raw material exploitation and its effects on Lower and Middle Paleolithic assemblage variability. In *Raw material economies among prehistoric hunter-gatherers* (ed. A. Montet-White and S. Holen). University of Kansas, Lawrence, pp. 33–48.

Dzaparidze, V., Bosinski, G., Bugianisvili, T., Gabunia, L., Justus, A., Klopotovskaja, N., Kvavadze, E., Lordkipanidze, D., Majsuradze, M., Mgeladze, N., Nioradze, M., Pavlenisvili, E., Schmincke, H.-U., and Sologasvili, D. 1989. Der altpaläolithische Fundplatz Dmanisi in Georgian (Kaukasus) *Jahrbuch des Römisch-Germanischen Zentralmuseums Mainz* 36:67–116.

Eisenmann, V., Bassesio, R., Beden, M., Faure, M., Geraads, D., Guerin, C., and Heints, E. 1983. Nouvelle interprétation biochronologique des grands mammifères d'Oubeidiyeh, Israel. *Geobois* 16:629–633.

Forsten, A. 1988. Middle Pleistocene replacement of stenonid horses by caballoid horses: Ecological implications. *Palaeogeograhy, Paleoclimatology, Palaeoecology* 65:23–33.

Frayer, D. W., Wolpoff, M. H., Thorne, A. G., Smith, F. D. and Pope, G. G. 1993. Theories of modern human origins: The paleontological test. *American Anthropologist* 95(1):14–50.

Freeman, L. G. 1975. Acheulian sites and stratigraphy in Iberia and the Maghreb. In *After the australopithecines* (ed. K. W. Butzer and G. L. Isaac). Mouton, The Hague, pp. 661–744.

Gabunia, L., and Vekua, A. 1990. L'évolution du paléoenvironnement au cours de l'anthropogène en Géorgie (Transcaucasie). *L'Anthropologie* 94(4):643–650.

———. 1995. A Plio-Pleistocene hominid from Dmanisi, east Georgia, Caucasus. *Nature* 373:509–512.

Gamble, C. 1986. *The Palaeolithic settlement of Europe.* Cambridge University Press, Cambridge.

Garrod, D. A. E., and Bate, D. M. 1937. *The Stone Age of Mount Carmel.* Clarendon Press, Oxford.

Geneste, J.-M. 1988. Economie des ressources lithiques dans le Moustérien du sud-ouest de la France. In *L'Homme de Néandertal.* Vol. 6, *La subsistance* (ed. M. Otte). ERAUL, Liège, pp. 75–97.

Geneste, J.-M., Boeda, E., and Meignen, L. 1990. Identification des chaînes opératoires lithiques du paléolithique ancien et moyen. *Paléo* 2:43–80.

Geraads, D., and Tchernov, E. 1983. Fémurs humains du Pléistocène moyen de Gesher Benot Yaʿacov (Israël). *L'Anthropologie* 87:1, 138–141.

Gilead, D. 1970. Handaxe industries in Israel and the Near East. *World Archaeology* 2(1):1–11.

Goren, N. 1979. An Upper Acheulian industry from the Golan Heights. *Quartär* 29(30):105–121.

Goren-Inbar, N. 1985. The lithic assemblage of the Berekhat Ram Acheulian site, Golan Heights. *Paléorient* 11(1):7–28.

Goren-Inbar, N., Belitzky, S., Goren, Y., Rabinovich, R., and Saragusti, I. 1992. Gesher Benot Yaʿaqov—the "Bar": An Acheulian assemblage. *Geoarchaeology* 7(1):27–40.

Goren-Inbar, N., Belitzky, S., Verosub, K., Werker, E., Kislev, M., Heimann, A., Carmi, I., and Rosenfeld, A. 1992. New discoveries at the Middle Pleistocene Acheulian site of Gesher Benot Yaʿaqov, Israel. *Quaternary Research* 38:117–128.

Goren-Inbar, N., Zohar, I., and Ben-Ami, D. 1991. A new look at old cleavers: Gesher Benot Yaʿaqov. *Mitekufat Haeven, Journal of the Israel Prehistoric Society* 24:7–33.

Grün, R., and Stringer, C. B. 1991. Electron spin resonance dating and the evolution of modern humans. *Archaeometry* 33(2):153–199.

Guérin, C. 1982. Première biozonation du Pléistocène européen, principal résultat biostratigraphique de l'étude des Rhinocerotidae (Mammalia, Perissodactyla) du Miocène terminal au Pléistocène supérieur d'Europe occidentale. *Geobios* 15(4):593–598.

Guérin, C., and M. Faure. 1988. Biostratigraphie comparée des grands mammifères du Pléistocène en Europe occidentale et au Moyen-Orient. *Paléorient* 14(2):50–56.

Haas, G. 1966. *On the vertebrate fauna of the Lower Pleistocene site of ʿUbeidiya.* Israel Academy of Sciences and Humanities, Jerusalem.

———. 1968. On the fauna of ʿUbeidiya. *Proceedings of the Israel Academy of Sciences and Humanities* 7:1–14.

Henry, D. O. 1989. *From foraging to agriculture: The Levant at the end of the Ice Age.* University of Pennsylvania Press, Philadelphia.

Henry, D. O., and Miller, G. H. 1992. The implications of amino acid racemization dates on Levantine Mousterian deposits in southern Jordan. *Paléorient* 18(2):45–52.

Hooijer, D. A. 1959. Fossil mammals from Jisr Banat Yaqub, south of Lake Hule, Israel. *Bulletin of the Research Council of Israel* 68:177–179.

———. 1960. A stegodon from Israel. *Bulletin of the Research Council of Israel* 68:104–107.

Horowitz, A. 1979. *The Quaternary of Israel.* Academic Press, New York.

———. 1989. Prehistoric cultures of Israel: Correlation with the oxygen isotope scale. In *Investigations in south Levantine prehistory* (ed. O. Bar-Yosef and B. Vandermeersch). British Archaeological Reports, International Series no. 497, Oxford, pp. 5–18.

Horowitz, A., Siedner, G., and Bar Yosef, O. 1973. Radiometric dating of the ʿUbeidiya Formation, Jordan Valley, Israel. *Nature* 242:186–187.

Hours, F. 1975.The Lower Paleolithic of Lebanon and Syria. In *Problems in prehistory: North African and the Levant* (ed. F. Wendorf and A. E. Marks). Southern Methodist University Press, Dallas, pp. 249–271.

Howells, W. W. 1989. *Skull shapes and the map: Craniometric analyses in the dispersion of modern* Homo. Peabody Museum, Harvard University, Cambridge.

Isaac, G. L. 1984. The archaeology of human origins: Studies of the Lower Pleistocene in East Africa, 1971–1981. In *Advances in world archaeology* (ed. F. Wendorf and A. E. Close). Academic Press, New York, pp. 1–87.

———. 1986. Foundation stones: Early artefacts as indicators of activities and abilities. In *Stone Age prehistory: Studies in memory of Charles McBurney* (ed. G. N. Bailey and P. Callow). Cambridge University Press, Cambridge, pp. 221–241.

Jelinek, A. J. 1991. Observations on reduction patterns and raw materials in some Middle Paleolithic industries in the Perigord. In *Raw material economies among prehistoric hunter-gatherers* (ed. A. Montet-White and S. Holen). University of Kansas, Lawrence, pp. 73–126.

Jones, R. 1990. From Kakadu to Kutikina: The southern continent at 18,000 years ago. In *The world at 18,000 BP* (ed. C. Gamble and O. Soffer). Unwin Hyman, London, pp. 264–295.

Klein, R. G. 1989.*The human career.* University of Chicago Press, Chicago.

Leakey, M. D. 1971. *Olduvai Gorge, excavations in Beds I and III, 1960–1963.* Cambridge University Press, Cambridge.

Lieberman, D. E. 1993. Life history variables preserved in dental cementum microstructure. *Science* 261:1162–1164.

Lotter, A. F., Ammann, B., Beer, J., Hajdas, I., and Sturm, M. 1992. A step towards an absolute time-scale for the Late-Glacial: Annually laminated sediments from Soppensee (Switzerland). In *The last deglaciation: Absolute and radiocarbon chronologies* (ed. E. Bard and W. Broecker). Springer-Verlag, Berlin.

McCown, T. D., and Keith, A. 1939. *The Stone Age of Mount Carmel II: The fossil human remains from the Levallioso-Mousterian.* Clarendon Press, Oxford.

Marks, A. 1983. The Middle to Upper Paleolithic transition in the Levant. In *Advances in world archaeology* (ed. F. Wendorf and A. E. Close). Academic Press, New York, pp. 51–98.

Mellars, P. A. 1992. Archaeology and the population-dispersal hypothesis of modern human origins in Europe. *Philosophical Transactions of the Royal Society* 337(1280):225–234.

Mercier, N., and Valladas, H. 1994. Thermoluminescence dates for the Paleolithic Levant. In *Late Quaternary chronology and paleoclimates of the eastern Mediterranean* (ed. O. Bar-Yosef and R. S. Kra). Radiocarbon and the American School of Prehistoric Research, Tucson and Cambridge, pp. 13–20.

Moore, A. 1985. The development of neolithic societies in the Near East. In *Advances in world archaeology* (ed. F. Wendorf and A. E. Close). Academic Press, New York, pp. 1–69.

Moore, A. M. T., and Hillman, G. C. 1992. The Pleistocene to Holocene transition and human economy in southwest Asia: The impact of the Younger Dryas. *American Antiquity* 57(3):482–494.

Mor, D., and Steinitz, G. 1982. K-Ar age of the Cover Basalt surrounding the Sea of Galilee, Interim report. ME/6/82, *Geological Survey of Israel.*

———. 1985. The history of the Yarmouk River based on K-Ar dating and its implication on the development of the Jordan Rift. GSI/40/85, *Geological Survey of Israel.*

Neuville, R. 1951. *Le Paléolithique et le Mésolithique du désert de Judée.* Masson, Paris.

Ohel, M. 1991. Prehistoric survey of the Baram Plateau. *Palestine Exploration Quarterly,* Jan.–June: 32–47.

Ohnuma, K., and Bergman, C. A. 1990. A technological analysis of the Upper Palaeolithic levels (XXV–VI) of Ksar Akil, Lebanon. In *The emergence of modern humans* (ed. P. Mellars). Edinburgh University Press, Edinburgh, pp. 91–138.

Opdyke, N. D., Lindsay, E., and Kukla, G. 1985. Evidence for earlier data of ʿUbeidiya, Israel hominid site. *Nature* 304:375.

Picard, L. 1943. Structure and evolution of Palestine. *Bulletin of the Geological Department* (Hebrew University) 4:1–134.

Pilbeam, D. 1989. Human fossil history and evolutionary paradigms. In *Evolutionary biology at the crossroads* (ed. M. K. Hecht). Queens College Press, Flushing, N.Y., pp. 117–138.

Rak, Y., and Arensburg, B. 1987. Kebara 2 Neanderthal pelvis: First look at a complete inlet. *American Journal of Physical Anthropology* 73:227–231.

Redman, C. 1978. *The rise of civilization.* Freeman, San Francisco.

Repenning, C. A., and O. Fejfar. 1982. Evidence for earlier data of ʿUbeidiya, Israel. *Nature* 299:344–347.

Rightmire, G. P. 1990. *The evolution of* Homo erectus. Cambridge University Press, Cambridge.

Roberts, R. G., Jones, R., and Smith, R. A. 1990. Thermoluminescence dating of a 50,000-year-old human occupation site in northern Australia. *Nature* 345:153–156.

Roebroeks, W. 1994. Updating the earliest occupation of Europe. *Current Anthropology* 35(3):301–305.

Ronen, A. 1991. The Lower Palaeolitic site Evron-Quarry in western Galilee, Israel. In *Festschrift Karl Brunnacker.* Geologisches Institut der Universität zu Köln, Köln, pp. 187–212.

Ruff, C. B. 1991. Climate, body size and body shape in hominid evolution. *Journal of Human Evolution* 21:81–105.

———. 1993. Climatic adaptation and hominid evolution: The thermoregulatory imperative. *Evolutionary Anthropology* 2(2):53–60.

Sanlaville, P. 1988. Synthèse sur le paléoenvironment. *Paléorient* 14(2):57–60.

Schick, K. D., and Zhuan, D. 1993. Early Paleolithic of China and eastern Asia. *Evolutionary Anthropology* 2:1.

Solecki, R. S., and Solecki, R. L. 1993. The pointed tools from the Mousterian occupations of Shanidar Cave, Northern Iraq. In *The Paleolithic prehistory of the Zagros-Taurus* (ed. D. I. Olszewski and H. L. Dibble). The University Museum, University of Pennsylvania, Philadelphia, pp. 119–146.

Stekelis, M. 1960. The Palaeolithic deposits of Jisr Banat Yaqub. *Bulletin of the Research Council of Israel* 9G(2–3):61–90.

Stoneking, M. 1993. DNA and recent human evolution. *Evolutionary Anthropology* 2(2):60–73.

Straus, L. G. 1989. Age of the modern Europeans. *Nature* 342:476–477.

Stringer, C., and Gamble, C. 1993. *In search of the Neanderthals.* Thames and Hudson, London.

Stringer, C. B. 1992. Reconstructing recent human evolution. *Philosophical Transactions of the Royal Society*, ser. B, 337:217–224.

Stuiver, M., and Braziunas, T. F. 1993. Modeling atmospheric ^{14}C influences and ^{14}C ages of marine Samples to 10,000m BC. *Radiocarbon* 35(1):137–189.

Swisher, C. C., Curtis, G. H., Jacob, T., Getty, A. G., Suprijo, A., and Widiasmoro. 1994. Age of the earliest known hominids in Java, Indonesia. *Science* 263:1118–1121.

Tauxe, L., Deino, A. D., Behrensmeyer, A. K., and Potts, R. 1992. Pinning down the Brunhes-Matuyama and upper Jaramillo boundaries. A reconciliation of orbital and isotopic time scales. *Earth and Planetary Science Letters* 109:561–572.

Tchernov, E. 1975. *The Early Pleistocene molluscs of Erq-el-Ahmar.* Israel Academy of Sciences and Humanities, Jerusalem.

———. 1986. *Les mammifères du Pléistocène inférieur de la vallée du Jourdain à Oubeidiyeh.* Association Paléorient, Paris.

———. 1987. The age of the ʿUbeidiya formation, an Early Pleistocene hominid site in the Jordan Valley, Israel. *Israel Journal of Earth Sciences* 36(1–2):3–30.

———. 1988. La biochronologie du site de ʿUbeidiya (vallée du Jourdain) et les plus anciens hominidés du Levant. *L'Anthropologie* 92(3):839–861.

———. 1991. Biological evidence for human sedentism in southwest Asia during the Natufian. In *The Natufian culture in the Levant* (ed. O. Bar-Yosef and F. R. Valla). International Monographs in Prehistory, Ann Arbor, pp. 315–340.

———. 1992a. The Afro-Arabian component in the Levantine mammalian fauna: A short biogeographical review. *Israel Journal of Zoology* 38:155–192.

———. 1992b. Eurasian-African biotic exchanges through the Levantine corridor during the Neogene and Quaternary. *Courier Forschungsinstitut Senckenberg* 153:103–123.

Thunell, R. C., and Williams, D. F. 1983. The stepwise development of Pliocene-Pleistocene paleoclimatic and paleoceanographic conditions in the Mediterranean: Oxygen isotopic studies of DSDP Sites 125 and 132. *Utrecht Micropaleontological Bulletin* 30:111–127.

Tillier, A.-M. 1992. The origins of modern humans in southwest Asia: Ontogenetic aspects. In *The evolution and dispersal of modern humans in Asia* (ed. T. Akazawa, K. Aoki, and T. Kimura). Hokusen-sha, Tokyo, pp. 15–28.

Toth, N., and Schick, K. 1993. Early stone industries and influences regarding language and cognition. In *Tools, language and cognition in human evolution* (ed. K. Gibson and T. Ingold). Cambridge University Press, Cambridge, pp. 346–362.

Trinkaus, E. 1983. *The Shanidar Neanderthals.* Academic Press, New York.

Turner, A. 1992. Large carnivores and earliest European hominids: Changing determinants of resource availability during the Lower and Middle Pleistocene. *Journal of Human Evolution* 22(2):109–126.

van Zeist, W. 1986. Some aspects of early Neolithic plant husbandry in the Near East. *Anatolica* 15:49–67.

Vandermeersch, B. 1992. The Near Eastern hominids and the origins of modern humans in Eurasia. In *The evolution and dispersal of modern humans in Asia* (ed. T. Akazawa, K. Aoki, and T. Kimura). Hokusen-sha, Tokyo, pp. 29–38.

Vekua, A. K. 1987. Lower Pleistocene mammalian fauna of Akhalkalaki (southern Georgia, USSR). *Palaeontographica Italica* 74:63–96.

Verosub, K. L., and E. Tchernov. 1989. Résultats préliminaires de l'étude magnétostratigraphique d'une séquence sédimentaire à industrie humaine en Israël. In *114ème Congrès National des Sociétés Savantes: Les premiers peuplements humains de l'Europe* (ed. E. Bonifay and B. Vandermeersch). Paris, pp. 237–242.

Wolpoff, M. H. 1989. Multiregional evolution: The fossil alternative to Eden. In *The human revolution: Behavioural and biological perspectives in the origins of modern humans* (ed. P. Mellars and C. Stringer). Edinburgh University Press, Edinburgh, pp. 62–108.

Wright, H. E., Jr. 1993. Environmental determinism in near eastern prehistory. *Current Anthropology* 34:458–469.

Yalçinkaya, I., Otte, M., Bar-Yosef, O., Kozlowski, J., Léotard, J.-M., and Taskiran, H. 1993. The excavations at Karain Cave, south-western Turkey: An interim report. In *The Paleolithic prehistory of the Zagros-Taurus* (ed. D. I. Olszewski and H. L. Dibble). The University Museum, University of Pennsylvania, Philadelphia, pp. 100–106.

Zhongli, D., Rutter, N., Jingtai, H., and Tungshen, L. 1992. A coupled environmental system formed at about 2.5 Ma in East Asia. *Palaeogeography, Palaeoclimatology, Palaeoecology* 94:223–242.

Chapter 36

The Evolution and Distribution of Later Pleistocene Human Populations

Christopher B. Stringer

As the record of human fossils increases in geographical range and local detail, new opportunities will arise to study the biogeography of modern human origins. Paleoclimatic data are increasingly available for regions inhabited in the Pleistocene; and with the development or wider application of new physical dating techniques, it should become more feasible to relate evolutionary events and dispersals to environmental changes. In this chapter, I discuss some of the outstanding problems and recent progress in looking at Neanderthal evolution from a biogeographic and adaptive perspective; I also examine data on modern human origins from this same perspective.

Plio-Pleistocene Hominids

Homo erectus is generally regarded as the first member of the genus *Homo* to have dispersed beyond the confines of the probable hominid homeland of Africa; we should therefore expect to see in this species the emergence of patterns of thermoregulatory adaptation paralleling those which we find in *Homo sapiens* today. A brief discussion of the origin and evolution of *Homo erectus* is perhaps appropriate before we consider the later Pleistocene record; for the purposes of this discussion I shall assume that specimens such as KNM-ER 3733 and WT 15000 represent *Homo erectus* (Rightmire, 1990) rather than *Homo ergaster* (Wood, 1992). It is usually assumed that *Homo erectus* originated from *Homo habilis* in the Late Pliocene of Africa. There are, however, factors that suggest caution concerning this assumption.

First, there are probably at least two species presently subsumed within the category of "*Homo habilis*," although there is no agreement as to how the individual specimens should be segregated (cf. Wood, 1992, with Stringer, 1986a, and Rightmire, 1993). Moreover, it is not at all clear which of the range of material presently allocated to *Homo habilis* might provide the most plausible ancestral form for early American *Homo erectus*. Second, it is possible from occipital KNM-ER 2598 that *Homo erectus* was present contemporaneously with Koobi Fora specimens assigned to *Homo habilis*, such as KNM-ER 1470, 1802, and 1813 (Wood, 1992). This in turn casts doubt on the assumed link, based on postcranial similarities, between large examples of *Homo habilis sensu lato* and *Homo erectus*, because *erectus*-like postcranial bones might actually derive from *Homo erectus* rather than from *Homo habilis sensu lato*. Third, there is possible evidence for a Late Pliocene presence of *Homo erectus* in Eurasia, depending on interpretations of specimens such as the Dmanisi mandible from Georgia (Gabunia and Vekua, 1995) and the revised dating of the earliest *Homo erectus* / "*Meganthropus*" fossils from Java (Swisher et al., 1994). This last point is linked with continuing claims for Eurasian artifacts of Late Pliocene age (Bonifay and Vandermeersch, 1991). From available fossil and geochronological evidence, it is still possible that *Homo erectus* originated outside of Africa, but clarification of this possibility will depend on further absolute dating of sites such as Dmanisi and Sangiran as well as on continuing reassessments of the African and non-African fossil evidence.

ISBN 0–300–06348–2.

The Middle Pleistocene

By the middle part of the Middle Pleistocene, ca. 400 thousand years (kyr), *Homo erectus* was apparently still present in China and Indonesia (with the Javanese Ngandong hominids, which I assign to *Homo erectus*, being perhaps even younger; Bartstra et al., 1988). But there is also evidence at this time for additional, more derived hominids in Africa (e.g., Elandsfontein, Ndutu, Broken Hill, and perhaps Bodo), Europe (Mauer, Petralona, Arago), and perhaps China (Yunxian) (fig. 36.1). Their precise taxonomic status is still uncertain, but they show departures from the typical *Homo erectus* morphology in such features as the reduced cortical bone in ectocranial buttresses and the tympanic, and the shape of cranial bones such as the temporal, parietal, and occipital. Many of the latter features appear to be related to an overall increase in endocranial volume, which seems to become accentuated during the course of the Middle Pleistocene. The face, when preserved, is intermediate in mean values measuring total projection between *Homo erectus* and those of later humans, and there is an increase in midfacial projection in several specimens, which is linked with greater nasal prominence. The postcranial skeleton is poorly known at present but retains features of *Homo erectus*, such as an external iliac buttress (Stringer, 1986b) and the relatively high levels of cortical thickening and femoral-shaft width found in archaic *Homo* generally (Ruff et al., 1993). There are, as yet, no data on body proportions to compare with the linear, heat-adapted physique inferred for the much earlier WT 15000 *Homo erectus* individual (Ruff, 1991).

In light of remarkable new finds from Atapuerca (Arsuaga et al., 1993; Stringer, 1993a), the status of the European Middle Pleistocene sample is currently the subject of some discussion. Extensive cranial and post-

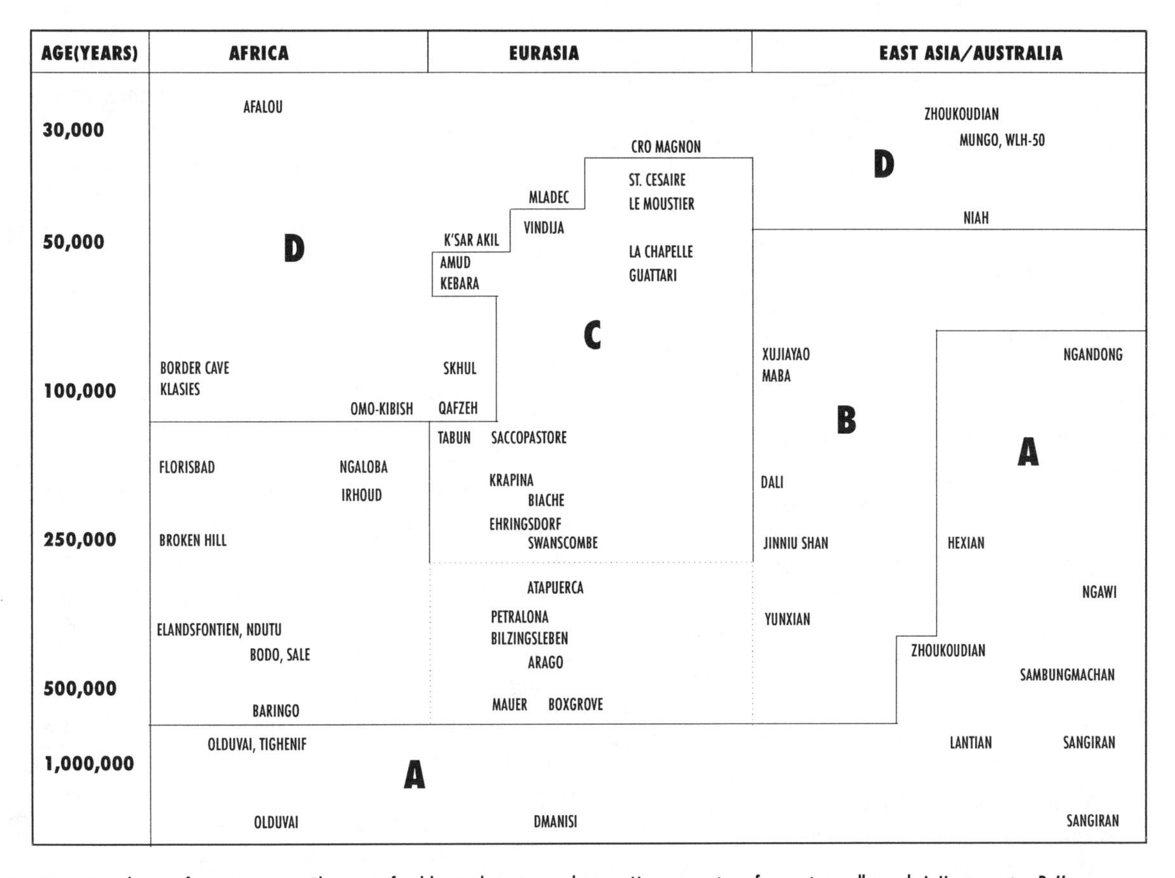

Fig. 36.1. Distribution of some important Pleistocene fossil hominids in time and space. My taxonomic preference is to call morph A *Homo erectus;* B *Homo heidelbergensis* or *Homo rhodesiensis;* C *Homo neanderthalensis,* perhaps extending back to Arago or even Mauer (thus affecting the status of *Homo heidelbergensis*); and D *Homo sapiens.*

cranial samples dated at ca. 200–300 kyr certainly show the suite of more derived, non-*erectus* features listed above, but they are combined with apparent Neanderthal characters of the face, occipital, mandible, and pubic ramus. The last item may turn out to be a plesiomorphy as more data become available, but the other features seem to represent real synapomorphies with the Neanderthals. If, as I advocate, the Neanderthal clade is given separate, specific status, then we can apparently date the origins of that species in the middle Pleistocene. By a reasonable extension of the variation found in the Atapuerca sample, it is possible that problematic European fossils such as Vértesszöllös, Steinheim, Arago, and Petralona will be drawn into the same general Middle Pleistocene lineage, and hence, for me, into the species *Homo neanderthalensis.* If the Mauer mandible were included, this would mean the end of the concept of a common Middle Pleistocene European-African species based on specimens such as Mauer, Petralona, Bilzingsleben, and Broken Hill, that is, *Homo heidelbergensis* (cf. Stringer, 1985; Rightmire, 1990). However, new discoveries from Atapuerca (Gran Dolina) and Boxgrove (England) will also be important to a resolution of the status of *Homo heidelbergensis.*

If *Homo neanderthalensis* is the sister species of *Homo sapiens,* then dating the origin of the *Homo neanderthalensis* clade would also date the origin of our own separate clade. But non-European specimens such as Broken Hill (Zambia) and Dali and Yunxian (China) provide little basis for recognizing synapomorphies with *Homo sapiens.* If the fossils are not regarded as representing *Homo heidelbergensis, Homo neanderthalensis,* or *Homo sapiens,* another taxon would need to be recognized here; based on the priorities of African names, *Homo rhodesiensis* or *Homo helmei* (for Florisbad) are available. When we move into the late Middle Pleistocene, there is evidence of possible modern synapomorphies in Chinese fossils such as Dali and Jinniu Shan and especially in African specimens such as Irhoud 2 and Omo-Kibish 2. The dating of these specimens is still uncertain, although it is likely that there was a temporal overlap with *Homo erectus* in China prior to 200 kyr (fig. 36.1; see also Chen and Zhang, 1991). Members of the African "late archaic" group such as Irhoud, Singa, and Ngaloba (Laetoli hominid 18) have been dated absolutely to between ca. 100–190 kyr (for Irhoud and Singa, see fig. 36.2).

The Neanderthals

Because of the introduction of the practice of burial by about 100 kyr, we are fortunate to have good data on the body form of both late Neanderthals and early *Homo sapiens* in western Eurasia. Neanderthal body form combined the general pattern of skeletal robusticity known from early *Homo* skeletal material with what, in modern terms, was certainly a cold-adapted body shape. Trinkaus's data on limb proportions (fig. 36.3), Holliday's data on limb / trunk proportions (Holliday and Trinkaus, 1991), and Ruff's data on pelvic width relative to stature (Ruff, 1991) show that Neanderthals resembled modern samples from high latitudes in body proportions, as characterized by a long and relatively wide trunk and short extremities. This body form would fit with their long-term adaptation to Eurasian climatic regimes that were predominantly cooler than the present interglacial. Neanderthals from both Europe and western Asia follow the pattern of modern high-latitude samples (fig. 36.3), in contrast to the WT 15000 *Homo erectus* skeleton, Broken Hill (inferred from the very long tibia), and the early modern samples. Neanderthal facial form may also have been related to cold adaptation. The configuration of the midface, particularly the voluminous and prominent nose, is unique. It may represent a combination of responses to high activity levels that generate internal heat, inhalation of relatively cold, dry air, and the functional demands of a face resistant to torsion centered around the anterior dentition (cf. Franciscus and Trinkaus, 1988, with Rak, 1986).

The extent to which the Neanderthal clade evolved in genetic isolation is difficult to establish owing to the lack of good comparative data from other regions, but in the Late Pleistocene, Neanderthals certainly existed as far east as Teshik Tash (Uzbekistan) and were present in the Levant (e.g., Amud, Tabun). The Levantine Neanderthals were arguably less extreme than their European counterparts in the brachial index and in certain cranial features, but I believe they show enough significant synapomorphies to warrant classification as *Homo neanderthalensis.* This similarity raises the question of their precise relationship to the European group. It has been argued (e.g., Bar-Yosef, this vol.) that the Asian Neanderthals moved into this region as a refuge from increasingly severe climatic conditions in Europe after about 70 kyr and that before this time the Levant may have been occupied by an evolving lineage of *Homo sapiens,* repre-

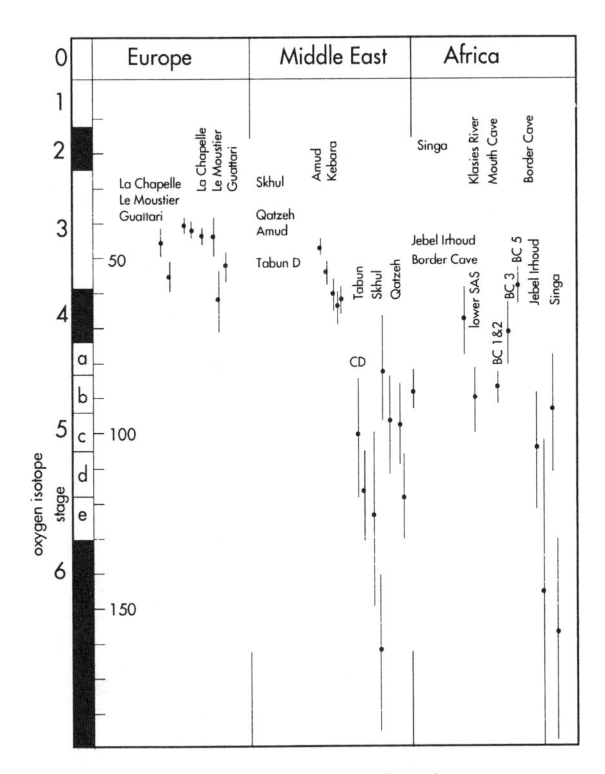

Fig. 36.2. Chronologies for some later Pleistocene hominid sites in western Eurasia and Africa (adapted from Grün and Stringer, 1991). New Uranium Series (U-S) and Electron Spin Resonance (ESR) dates on mammalian tooth enamel support the early uptake (EU) ESR age estimates for Tabun, Qafzeh, and Skhul but suggest that there could be an additional younger (oxygen isotope stage 3–4) fauna at Skhul (McDermott et al., 1993).

sented by the probable Middle Pleistocene Zuttiyeh craniofacial fragment and the succeeding Qafzeh and Skhul specimens. There are several problems with this argument. First, the Zuttiyeh fossil seems too fragmentary and plesiomorphous for definitive statements about its affinities. Second, Electron Spin Resonance (ESR) and new Uranium Series (U-S) dates for mammalian tooth enamel from layer C at Tabun (McDermott et al., 1993) suggest that the Neanderthal skeleton generally attributed to that layer dates from the early Late Pleistocene, making it penecontemporaneous with the Qafzeh specimens and significantly older than later Neanderthals from the sites of Kebara and Amud (figs. 36.1, 36.2). Third, it is not clear why European Neanderthal populations, having survived the rigors of oxygen isotope stage 6 in southern Europe, at least, would have been forced out by the less severe conditions of stage 4. Fourth, even if we disregard the third point and propose the need for a migration into Asia, it is not clear why Neanderthals would not previously have moved into the Levantine refuge during earlier cold stages, for instance, near the end of oxygen isotope stage 6.

Additionally, it seems likely that the most severe climatic conditions would have led to a marked restriction of migration routes between Europe and Asia. Although low sea levels would have opened up new migration routes, such as across the Bosphorus and Dardanelles, there would have been the additional obstacles of glaciers and lower or more persistent snow lines in regions such as Turkey and the Caucasus, as well as possible extensions of seas such as the Caspian and Aral. If such geographical barriers existed for significant portions of the later Pleistocene, they may explain the distinctive evolution of the Neanderthals in Europe, with the exchange of genes to and from Europe possible during briefer interglacial and interstadial stages rather than during the predominant cold stages, when severe selection could have acted on Neanderthal populations isolated in Europe, as Howell (1957) suggested many years ago.

Mode of Origin of Modern *Homo Sapiens:* Restricted or Multiregional?

I believe the evidence for a recent restricted origin of *Homo sapiens* is strong, both from considerations of present-day physical and genetic variation and from fossil data. Differences between modern crania are predominantly those of size rather than shape, whereas shape differences are much more marked between earlier Pleistocene samples. Thus, in my opinion, there is a relative decrease in morphological diversity immediately following the spread of *Homo sapiens.* As with the genetic data, the overall similarity of modern populations is most parsimoniously explained by recent common ancestry rather than by large, regular, global gene flow. The latter process seems to me to be inherently unlikely under Pleistocene conditions of low population density accompanied by constantly fluctuating climates, environments, and resources.

Dental morphological data have also been interpreted as favoring a single (although Asian) origin for modern humans. This view is based on phenetic similarities between geographically distant modern samples and on marked contrasts between those samples and supposedly ancestral archaic forms such as the Neanderthals and

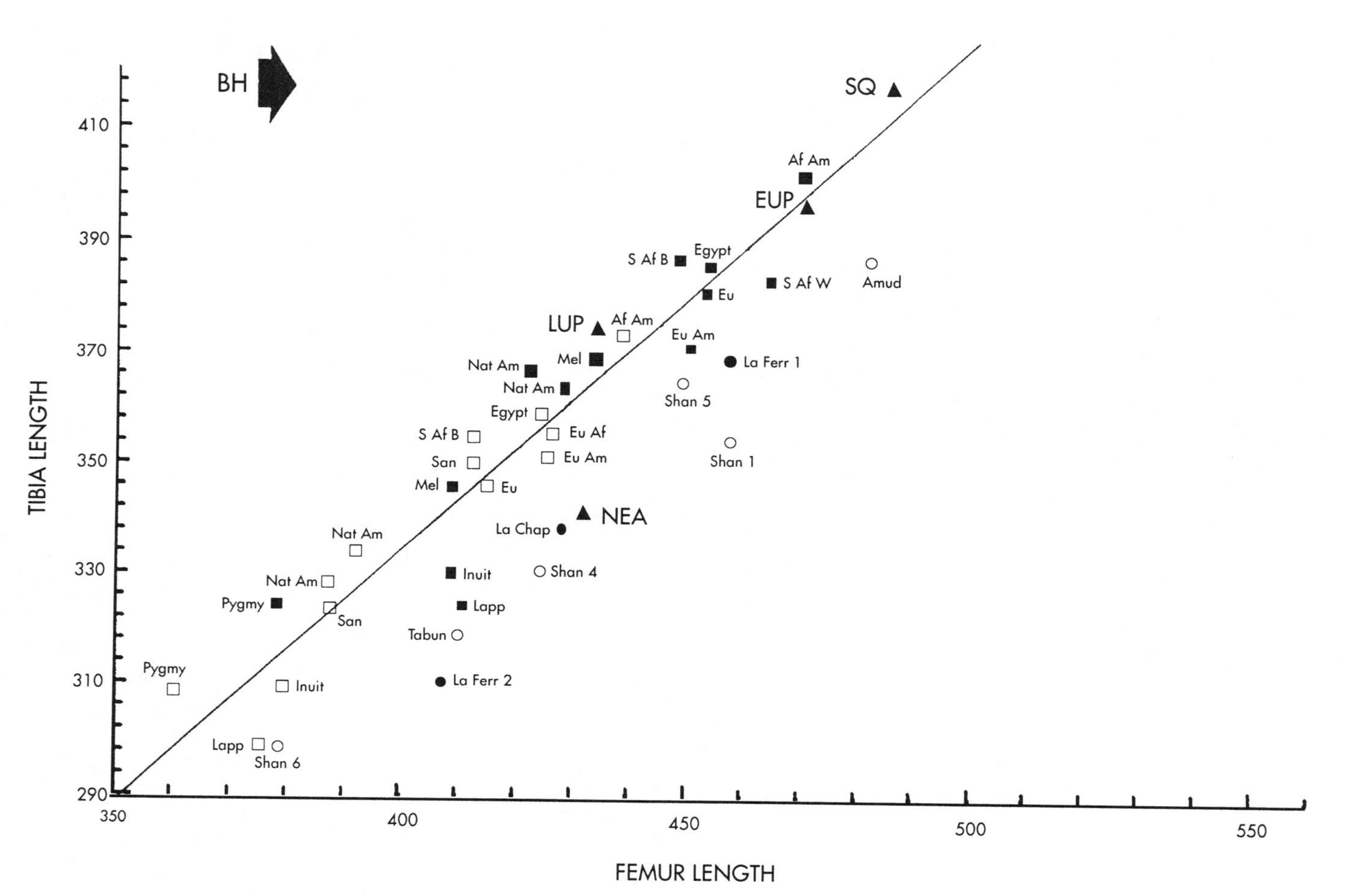

Fig. 36.3. Plot of tibia and femur lengths for recent and fossil hominids (modified from Trinkaus, 1981). Solid squares are recent male sample means; open squares are recent female sample means; solid circles are European Neanderthals; open circles are Asian Neanderthals; EUP and LUP are early and late European Upper Paleolithic means, respectively; SQ = Skhul-Qafzeh mean; NEA = overall Neanderthal mean. The relative position of the Broken Hill (BH) tibia value is also indicated. The value of WT 15000 early *Homo erectus* would be plotted near the LUP mean, with projected growth values beyond the SQ mean. The overall recent regression line is indicated (male and female lines are insignificantly different; cf. Ruff, 1993). Recent samples (from Trinkaus, 1981; Stringer, this vol.) are abbreviated as follows: Lapp = Lapps (Norway); Inuit = Eskimos (Alaska); Eu = Europeans (Yugoslavia); Eu Af = Euro-Africans; Eu Am = Euro-Americans; Nat Am = Amerindians (New Mexico); S Af B = South African Bantus; Nat Am = Amerindians (Arizona); San = San (Bushmen); Mel = Melanesians (New Caledonia); Pygmy = African Pygmies; Af Am = African Americans; Egypt = North Africans (Egypt).

Chinese *Homo erectus.* Southeast Asia is favored as the ancestral area based on the "average" dental morphology found there (Turner, 1992), but the remaining close phenetic similarity between African and Australian dentitions means that assumptions of constant evolutionary rates used by Turner must be wrong. Instead it seems more likely that the close African-Australian relationship is based on symplesiomorphies (Stringer, 1993b; Irish, 1994). Finally, and of great importance, early modern crania from Europe, the Levant, China, Java, and Australia predominantly resemble neither their archaic predecessors nor their recent successors. Instead they share characteristics with one another, which I would interpret as plesiomorphies retained from a recent common ancestry. It is also notable that western Eurasian early moderns display warm-adapted body shapes that contrast markedly with those of their Neanderthal counterparts (fig. 36.3). Modern patterns of regional differentiation were apparently poorly developed in early modern humans, thereby implying a recent development of such differentiation rather than a long Pleistocene gestation (Stringer, 1992a, 1992b).

Place of Origin of *Homo Sapiens*

Given the data on body shape mentioned above (and see fig. 36.3), we can infer a tropical or subtropical place of origin for the early Upper Paleolithic peoples of Europe and for the more primitive Levantine Skhul-Qafzeh samples. For the Upper Paleolithic populations, dated at 25–35 kyr, the region of origin could have been the Indian subcontinent, Southeast Asia, or Africa. I favor Africa because of marked similarities between the Late Pleistocene samples of Europe and North Africa (such as Nazlet Khater, Afalou, Taforalt) and the lack of comparable data from East Asia. However, the presence of early modern samples penecontemporaneous with those of the European early Upper Paleolithic in regions such as Sri Lanka, Zhoukoudian Upper Cave, and the Willandra Lakes of Australia should instill caution about assigning place of origin and remind us that Europe was by no means a center for modern human origins.

When we consider the Skhul-Qafzeh samples dated at ca. 80–120 kyr (although U-S and ESR data indicate that some Skhul fauna may be younger [McDermott et al., 1993]), there are associated faunal data from the sites that do suggest Afro-Arabian connections not so far apparent for Levantine Neanderthal sites (Tchernov, 1992). Together with the presence of comparably early modern or near-modern fossils (e.g., Omo-Kibish 1; Guomde KNM-ER 999 and 3884; Border Cave 3, 5 and, less certainly, 1, 2; Klasies 16424, 16425, 41815) and a feasible ancestral morphology (represented by fossils such as Irhoud 1-3, Omo-Kibish 2, Ngaloba, Eliye Springs, and Florisbad), both scattered across Africa, this region is certainly the most plausible one for the origin for *Homo sapiens* (Stringer, 1992a, 1992b). However, some caution should be injected because of the lack of good data from other localities. As I have already mentioned, the incomplete preservation of the Zuttiyeh craniofacial fragment makes its status uncertain, and my comparative analyses of cranial shape show that the probable late Middle Pleistocene cranium from Dali is not far behind the African late archaics in presenting a plausible ancestral morphology and shape for that of *Homo sapiens.* However, detailed study of this specimen is not possible at present, and its facial form would certainly look a lot less "modern" if it could be restored to its original proportions. Similarly, the Jinniu Shan partial skeleton is an extremely important specimen for future study, but it has already been the subject of two quite distinct cranial reconstructions, and its status remains unclear.

Some Questions for Future Research

I have concentrated on western Eurasia in the above discussion of the later Pleistocene hominid record, because this area has the best fossil, chronological and, arguably, paleoenvironmental data for this period; it is also the area with which I am most familiar. I would like to highlight, however, some spheres that remain largely unexplained or unexplored, where future research would be most valuable.

1. Roebroeks et al. (1992) have recently argued that northern European Pleistocene evidence negates claims for major differences in the ecology of human settlement between archaic and modern humans, for example, in the ability to adapt to peak interglacial fully forested conditions and to cold loess steppes. I would argue that Upper Paleolithic and comparable populations sustained and developed their occupations of these environments to a much greater extent than did archaic humans. I would also question the evidence for early adaptations to cold steppe in Europe. But how does the Asian record compare?

2. Chen and Zhang (1991) have argued that there is good evidence for the coexistence of *Homo erectus* and "archaic *Homo sapiens*" in the Middle Pleistocene of

China. I remain slightly cautious, while awaiting further dating evidence, but have previously argued for such contemporaneity in my comparisons of the eastern and western Old World records. If we accept this, and the validity of *Homo erectus* as a distinct species, there must have been cladogenesis in the hominid lineage by this time. However, could such distinct populations have coexisted in the same regions of Eurasia for any length of time? This certainly happened in the case of many closely related species of mammals in Africa during the Middle and later Pleistocene (Vrba, pers. comm.).

3. Linked with the last point, a late Middle to early Late Pleistocene date for the Ngandong (Solo) hominids seems likely, and they probably represent a late form of *Homo erectus.* To what extent was Java, as part of Pleistocene Sundaland, isolated from mainland Asia during the later Pleistocene? Van Andel (1989) depicts a wide Sundaland, one traversed by major rivers and subject to a more seasonal and drier climate than today's, which might have opened up the dense rain forests. This condition should have facilitated gene flow, but what was the predominant climatic and paleogeographic situation in this region during the later Pleistocene?

4. Comparing the "Out of Africa" and multiregional models: one model postulates the existence of largely separate hominid evolutionary lineages in different parts of the Old World, whereas the other requires regular and probably large-scale gene flow between all the inhabited regions. What have been the major factors encouraging or discouraging interregional gene flow during the Pleistocene? Can evidence from modern biogeography of faunal distributions and provinces provide useful comparative data here? There certainly appear to be useful analogues from other wide-ranging mammalian species (Vrba, pers. comm.; Groves, 1992).

5. If the "Out of Africa" model is largely accurate, what special factors could have led to the evolution of *Homo sapiens* in Africa, and what climatic or other factors could have led to Late Pleistocene dispersal events from there? In particular, can we reliably reconstruct North African paleoenvironments for oxygen isotope stages 4–6 by extrapolating from Late Pleistocene and Holocene data?

6. What behavioral or climatic factors could lie behind the early Late Pleistocene human population growth inferred from modern mtDNA diversity patterns (Harpending et al., 1993)? Genetic data can also be used to suggest a more complex multistaged dispersal pattern from Africa (cf. Cavalli-Sforza, 1991, with Nei and Roychoudhury, 1993). Is there evidence for a primary Asian-Australian dispersal through tropical and subtropical regions, followed by a separate European and/or Asian diversification?

7. Given the apparent arrival of humans in Australia through open-sea crossings by 50 kyr, could the Mediterranean have remained a significant biogeographic barrier between hominid populations in Europe and North Africa throughout the Late Pleistocene? Evidence of a late surviving Neanderthal population in southern Spain at Zaffaraya suggests that the (then-narrower) Straits of Gibraltar had not been crossed even by that time.

Summary

We have evidence that once *Homo erectus* dispersed from its assumed African homeland, regional evolution occurred under the constraints of local climatic conditions. The Neanderthals, in particular, are the first hominids known to show clear morphological evidence of cold adaptation. Modern humans appear to have had a single origin from a tropical/subtropical environment. Based on fossil and chronometric data, the region of origin probably lies in Africa or the Levant, although genetic analyses continue to favor Africa. We have much to learn, however, about the processes that produced *Homo sapiens* and shaped its dispersal and regional diversification.

Acknowledgments

I wish to thank the organizers and sponsors of the Airlie Conference for an excellent meeting in beautiful surroundings. For their assistance with this chapter I thank Rainer Grün, of the Australian National University, for the use of figure 36.2; Robert Kruszynski and Barbara West, for preparing figures 36.1 and 36.3; and Mrs Irene Baxter, for preparing the manuscript.

References

Arsuaga, J.-L., Martinez, I., Gracia, A., Carretero, J.-M., and Carbonell, E. 1993. Three new human skulls from the Sima de los Huesos Middle Pleistocene site in Sierra de Atapuerca, Spain. *Nature* 362:534–537.

Barstra, G.-J., Soegondho, S., and Van der Wijk, A. 1988. Ngandong Man: Age and artifacts. *Journal of Human Evolution* 17:325–337.

Bonifay, E., and Vandermeersch, B. (eds.). 1991. Les premiers Européens. Editions du Comité des Travaux Historiques et Scientifiques, Paris.

Cavalli-Sforza, L. L. 1991. Genes, peoples and languages. *Scientific American* 265(5):71–78.

Chen, T., and Zhang, Y. 1991. Palaeolithic chronology and possible co-existence of *Homo erectus* and *Homo sapiens* in China. *World Archaeology* 23:147–154.

Franciscus, R. G., and Trinkaus, E. 1988. Nasal morphology and the emergence of *Homo erectus. American Journal of Physical Anthropology* 75:517–527.

Gabunia, L., and Vekua, A. 1995. A Plio-Pleistocene hominid from Dmanisi, East Georgia, Caucasus. *Nature* 373:509–512.

Groves, C. P. 1992. How old are subspecies? A tiger's eye-view of human evolution. *Archaeology in Oceania* 27:153–160.

Grün, R., and Stringer, C. 1991. Electron spin resonance dating and the evolution of modern humans. *Archaeometry* 33:153–199.

Harpending, H. C., Sherry, S. T., Rogers, A. R., and Stoneking, M. 1993. The genetic structure of ancient human populations. *Current Anthropology* 34:483–496.

Holliday, T., and Trinkaus, E. 1991. Limb/trunk proportions in Neandertals and early anatomically modern humans. *American Journal of Physical Anthropology,* suppl., 12:93–94.

Howell, F. C. 1957. The evolutionary significance of variation and varieties of "Neanderthal" Man. *Quarterly Review of Biology* 32:330–347.

Irish, J. D. 1994. The African dental complex: Diagnostic morphological variants of modern sub-Saharan populations. *American Journal of Physical Anthropology,* suppl., 18:112.

McDermott, F., Grün, R., Stringer, C. B., and Hawkesworth, C. J. 1993. Mass-spectrometric U-series dates for Israeli Neanderthal/early modern hominid sites. *Nature* 363:252–255.

Nei, M., and Roychoudhury, A. K. 1993. Evolutionary relationships of human populations on a global scale. *Molecular Biology and Evolution* 10:927–943.

Rak, Y. 1986. The Neanderthal: A new look at an old face. *Journal of Human Evolution* 15:131–164.

Rightmire, G. P. 1990. *The evolution of* Homo erectus. Cambridge University Press, Cambridge.

———. 1993. Variation among early *Homo* crania from Olduvai Gorge and the Koobi Fora region. *American Journal of Physical Anthropology* 90:1–33.

Roebroeks, W., Conard, N. J., and van Kolfschoten, T. 1992. Dense forests, cold steppes, and the Palaeolithic settlement of northern Europe. *Current Anthropology* 33:551-586.

Ruff, C. 1991. Climate, body and body shape in hominid evolution. *Journal of Human Evolution* 21:81–105.

———. 1993. Climatic adaptation and hominid evolution: The thermoregulatory imperative. *Evolutionary Anthropology* 2:53–59.

Ruff, C., Trinkaus, E., Walker, A., and Larsen, C. S. 1993. Postcranial robusticity in *Homo* 1: Temporal trends and mechanical interpretation. *American Journal of Physical Anthropology* 91:21–53.

Stringer, C. B. 1985. Middle Pleistocene hominid variability and the origin of late Pleistocene humans. In *Ancestors: The hard evidence,* pp. 289–295 (ed. E. Delson). Liss, New York.

———. 1986a. An archaic character in the Broken Hill innominate E. 719. *American Journal of Physical Anthropology* 71:115–120.

———. 1986b. The credibility of *Homo habilis.* In *Major topics in primate and human evolution,* pp. 266–294 (ed. B. Wood, L. Martin, and P. Andrews). Cambridge University Press, Cambridge.

———. 1992a. Replacement, continuity and the origin of *Homo sapiens.* In *Continuity or replacement: Controversies in* Homo sapiens *evolution,* pp. 9–24 (ed. G. Bräuer and F. H. Smith). Balkema, Rotterdam.

———. 1992b. Reconstructing recent human evolution. *Phil. Trans. R. Soc. Lond.* B 337:217–224.

———. 1993a. Secrets of the Pit of the Bones. *Nature* 362:501–502.

———. 1993b. Understanding the fossil human record: Past, present and future. *Rivista di Antropologia* 71:91–100.

Swisher, C., Curtis, G., Jacob, T., Getty, A., Suprijo, A., and Widiasmoro. 1994. Age of the earliest known hominids in Java, Indonesia. *Science* 263:1118–1121.

Tchernov, E. 1992. Biochronology, paleoecology and dispersal events of hominids in the southern Levant. In *The evolution and dispersal of modern humans in Asia.* pp. 149–188 (ed. T. Akazawa, K. Aoki, and T. Kimura). Hokusen-Sha, Tokyo.

Trinkaus, E. 1981. Neanderthal limb proportions and cold adaptation. In *Aspects of human evolution,* pp. 187–224 (ed. C. B. Stringer). Taylor and Francis, London.

Turner, C. G. 1992. The dental bridge between Australia and Asia: Following Macintosh into the East Asian hearth of humanity. *Archaeology in Oceania* 27:143–152.

Van Andel, T. H. 1989. Late Quaternary sea-level changes and archaeology. *Antiquity* 63:733–745.

Wood, B. 1992. Origin and evolution of the genus *Homo. Nature* 355:783–790.

Contributors

Michael Archer
School of Biological Science
University of New South Wales
P.O. Box 1
Kensington, New South Wales, 2033
Australia

D. Margaret Avery
South African Museum
P.O. Box 61
Cape Town 8000
South Africa

Augusto Azzaroli
Museo di Geologia e Paleontologia
Via G. La Pira, 4
50121 Florence
Italy

John Barron
U.S. Geological Survey
MS 521 National Center
Reston, Virginia 22092

John C. Barry
Department of Anthropology
Harvard University
Cambridge, Massachusetts 02138

Ofer Bar-Yosef
Department of Anthropology
Harvard University
Cambridge, Massachusetts 02138

Raymond L. Bernor
College of Medicine
Department of Anatomy
Howard University
Washington, D.C. 20059

Jan Bloemendal
Geography Department
University of Liverpool
Liverpool
England

Gerard C. Bond
Lamont-Doherty Earth Observatory
Columbia University
Palisades, New York 10964

Raymonde Bonnefille
Laboratoire de Géologie du Quaternaire du CNRS
CEREGE
BP 80
13545 Aix-en-Provence Cedex 04
France

C. K. Brain
Transvaal Museum
Paul Kruger Street
P.O. Box 413
Pretoria 0001
South Africa

Francis H. Brown
Department of Geology and Geophysics
University of Utah
Salt Lake City, Utah 84112

Lloyd H. Burckle
Lamont-Doherty Earth Observatory
Columbia University
Palisades, New York 10964

Thomas Cronin
U.S. Geological Survey
MS 521 National Center
Reston, Virginia 22092

Peter B. deMenocal
Lamont-Doherty Earth Observatory
Columbia University
Palisades, New York 10964

George H. Denton
Department of Geological Sciences
University of Maine
Orono, Maine 04469

Harry Dowsett
U.S. Geological Survey
MS 521 National Center
Reston, Virginia 22092

Lydie M. Dupont
Institute for Palynology and Quarternary Sciences
University of Göttingen
Wilhelm-Weber-Strasse 2
D-37073 Göttingen
Germany

Véra Eisenmann
Institute of Paleontology
UA 12 du Centre National de la Recherche Scientifique
8 Rue Buffon
75005 Paris
France

Farley Fleming
U.S. Geological Survey
MS 521 National Center
Reston, Virginia 22092

Henk Godthelp
School of Biological Science
University of New South Wales
P.O. Box 1
Kensington, New South Wales, 2033
Australia

Suzanne J. Hand
School of Biological Science
University of New South Wales
P.O. Box 1
Kensington, New South Wales, 2033
Australia

Christopher J. H. Hartnady
Department of Geological Sciences
University of Cape Town
Cape Town 8000
South Africa

Andrew Hill
Department of Anthropology
Yale University
Box 208277, Yale Station
New Haven, Connecticut 06520-8277

Thomas Holtz, Jr.
U.S. Geological Survey
MS 521 National Center
Reston, Virginia 22092

Henry Hooghiemstra
Hugo DeVries Laboratory
Department of Palynology and Paleoactuoecology
University of Amsterdam
1098 Amsterdam
Netherlands

Richard C. Hulbert, Jr.
Department of Geology
University of Florida
Gainesville, Florida 32611

Scott Ishman
U.S. Geological Survey
MS 521 National Center
Reston, Virginia 22092

James P. Kennett
Marine Science Institute
University of California, Santa Barbara
Santa Barbara, California 93106

William H. Kimbel
Institute of Human Origins
1288 Ninth Street
Berkeley, California 94710

W. David Lambert
Department of Geology
University of Florida
Gainesville, Florida 32611

Suzanne A. G. Leroy
Institute for Palynology and Quarternary Sciences
University of Göttingen
Wilhelm-Weber-Strasse 2
D-37073 Göttingen
Germany

Diana Lipscomb
Department of Biological Sciences
George Washington University
Washington, D.C. 20052

Neil D. Opdyke
Department of Geology
University of Florida
Gainesville, Florida 32611

Timothy C. Partridge
Transvaal Museum
Paul Kruger Street
P.O. Box 413
Pretoria 0001
South Africa

Richard Poore
U.S. Geological Survey
MS 521 National Center
Reston, Virginia 22092

Geoffrey G. Pope
Department of Anthropology
The William Paterson College of New Jersey
Wayne, New Jersey 07470

Jelle W. F. Reumer
Natuurmuseum Rotterdam
Westzeedijk 345
P.O. Box 23452
NL-3001 KL Rotterdam
Netherlands

G. Philip Rightmire
Anthropology Department
State University of New York
Binghampton, New York 13901

William F. Ruddiman
Department of Environmental Science
Clark Hall
University of Virginia
Charlottesville, Virginia 22903

Louis Scott
Department of Botany and Genetics
University of the Orange Free State
P.O. Box 339
Bloemfontein 9300
South Africa

Nicholas J. Shackleton
Godwin Laboratory
University of Cambridge
Free School Lane
Cambridge, CB2 3RS
England

Christopher B. Stringer
Department of Paleontology
British Museum of Natural History
Cromwell Road
London SW7 5BD
England

J. Francis Thackeray
Transvaal Museum
Paul Kruger Street
P.O. Box 413
Pretoria 0001
South Africa

Robert Thompson
U.S. Geological Survey
MS 521 National Center
Reston, Virginia 22092

Elisabeth S. Vrba
Department of Geology and Geophysics
Yale University
P.O. Box 6666
New Haven, Connecticut 06521

S. David Webb
Department of Geology
University of Florida
Gainesville, Florida 32611

Henry B. Wesselman
Department of Anthropology
American River College
Sacramento, California 95841

Tim D. White
Department of Anthropology
University of California, Berkeley
Berkeley, California 94720

Debra Willard
U.S. Geological Survey
MS 521 National Center
Reston, Virginia 22092

Bernard A. Wood
Hominid Palaeontology Research Group
Department of Human Anatomy and Cell Biology
University of Liverpool
P.O. Box 147
Liverpool L69 3BX
England

Index